Regulated Chemicals Directory 1995

REGULATED CHEMICALS DIRECTORY

1995

Compiled by ChemADVISOR®, Inc.
Pittsburgh, Pennsylvania

VNR SPRINGER SCIENCE+BUSINESS MEDIA, LLC

For more information contact:

Van Nostrand Reinhold
115 Fifth Avenue
New York, NY 10003

International Thomson Publishing GmbH
Königswinterer Str. 418
53227 Bonn
Germany

International Thomson Publishing Europe
Berkshire House,168-173
High Holborn, London WC1V 7AA
England

International Thomson Publishing Asia
221 Henderson Road #05-10
Henderson Building.
Singapore 0315

Thomas Nelson Australia
102 Dodds Street
South Melbourne 3205
Victoria, Australia

International Thomson Publishing Japan
Hirakawacho Kyowa Building, 3F
2-2-1 Hirakawacho
Chiyoda-ku, 102 Tokyo
Japan

Nelson Canada
1120 Birchmount Road
Scarborough, Ontario
Canada M1K 5G4

International Thomson Editores
Campos Eliseos 385, Piso 7
Col. Polanco
11560 Mexico D.F. Mexico

This edition of *Regulated Chemicals Directory* ™ was prepared in collaboration with ChemADVISOR®, Inc., Regulatory Compliance Products & Services, Pittsburgh, PA, from the ChemADVISOR LOLI™ database, using software developed by Automated Publishing/ Pre-press Services, Phoenix, AZ.

1 2 3 4 5 6 7 8 9 10 BBR 01 00 99 98 97 96 95

ISSN 1058 1707
ISBN 978-0-442-02124-5   ISBN 978-94-011-4910-5 (eBook)
DOI 10.1007/978-94-011-4910-5

Contents

Acknowledgments

The *Regulated Chemicals Directory*™ is prepared from a database of chemical regulatory information maintained by ChemADVISOR®, Inc. Maintenance of the database is currently under the supervision of Scott A. Amoroso. The electronic database is updated on almost a daily basis to keep pace with regulatory changes and additions. The technical content and structure of the database are under the supervision of Patricia Dsida.

The *Regulated Chemicals Directory*™ is the product of a close collaboration between Van Nostrand Reinhold and Chapman and Hall scientific publishers and ChemADVISOR®, Inc. We wish to thank Dr. Barbara Goldman, Editor at Chapman & Hall, who originally approached Patricia Dsida with the idea of developing a reference book based on the ChemADVISOR® database, and who worked closely with ChemADVISOR® to create the *RCD*™.

Values from the ACGIH publication, "Threshold Limit Values and Biological Exposure Indices for 1994-95," were used for the ChemADVISOR® database with the kind permission of the American Conference of Governmental Industrial Hygienists.

Preface

The *Regulated Chemicals Directory*™ is meant to be a convenient source of information for everyone who needs to keep up-to-date regarding the regulations and recommendations that pertain to chemical substances. The *RCD*™ is designed to be the first reference book to consult when beginning compliance efforts. Every regulatory or advisory list used in the *RCD*™ is keyed to its source, to help readers who need more detailed information on regulations, recommendations, or guidelines readily locate source documents.

Some organizations now center their compliance efforts on computerized information stored in cross-referenced databases. A unique feature of the *RCD*™ is the availability of an electronic version suitable for use on IBM-compatible personal computers, download onto mainframes and CD-ROM players. Both the print and electronic versions are updated with the same timeliness. For more information on the electronic versions of the *Regulated Chemicals Directory*™, contact ChemADVISOR®, Inc. directly (750 William Pitt Way, Pittsburgh, PA 15238, phone 1-800-466-3750).

Many companies working on product development need information on what may be regulated in the future. The *RCD*™ provides selected information on pending regulations and in-progress testing lists, which can provide a starting place for tracking future regulatory considerations.

Information for the *RCD*™ is continually gathered and updated. Suggestions from readers for information that should be added to the *RCD*™ or for other ways to improve the book are welcomed by Van Nostrand Reinhold.

— Patricia L. Dsida, Pres.
ChemADVISOR®, Inc.

Clean Air Act (1990) - List of Hazardous Air Contaminants
[present]

CAA - HON Rule - SOCMI Chemicals
compliance by Oct. 24, 1994

CAA - HON Rule - Organic HAPs
[present]

EPA - Master Testing List
[present]

List of Pesticide Product Inert Ingredients
[present]

TSCA - Code of Federal Regulations Citations
40 CFR 712.30(d);   40 CFR 716.120(a);   40 CFR 799.2200

TSCA - PAIR - Reporting List
Reporting Date: November 19, 1982

TSCA - Health and Safety Reporting List
Effective Date: October 4, 1984

TSCA - Chemical Test Rules
Testing required by: manufacturers; processors (40 CFR 799.2200)

TSCA - Section 12(b) - Export Notification
export notification required - Section 4

INTERNATIONAL LISTS

Australian Exposure Standards - Time Weighted Averages
2 mg/m3 TWA

Canada - WHMIS: Ingredient Disclosure
1% item 852 (983)

Canada - NPRI (National Pollutant Release Inventory)
[present]

Canada - Alberta - 8 Hour Occupational Exposure Limit
2 mg/m3 TWA

Canada - Alberta - 15 Minute Occupational Exposure Limit
4 mg/m3 STEL

Canada - British Columbia - 8 Hour Exposure Limits
2 mg/m3 TWA

Canada - British Columbia - 15 Minute Exposure Limits
4 mg/m3 STEL

Canada - Ontario - OHSA - TWAEVs
2 mg/m3 TWAEV

Canada - Quebec - Time-Weighted Average Exposure Values
2 mg/m3 TWAEV

United Kingdom - Occupational Exposure Standards - TWAs
2 mg/m3 TWA

United Kingdom - Occupational Exposure Standards - STELs
4 mg/m3 STEL

German (DFG) - Peak Limitations
2 x normal MAK (5 min momentary value);   don't exceed 8 times during shift

German (DFG) - Carcinogens
animal evidence of carcinogenicity

German (DFG) - Pregnancy
classification not yet possible

Israel - Time Weighted Averages
2 mg/m3 TWA

Israel - Action Levels
1 mg/m3 AL

Mexico - Instruction No. 10 - TWAs
2 mg/m3 TWA

STATE LISTS

California - Air Bill 2588 Appendix A-I
6/91

California - Exposure Limits - PELs
2 mg/m3 PEL

California - Directors List of Hazardous Substances (8 CCR 339)
[present]

Florida Hazardous Substance List
[present]

Massachusetts Right To Know List
extraordinarily hazardous

Minnesota Hazardous Substance List
[present]

NJ Right to Know List (Total)
sn 1019

NJ Special Hazardous Substances
(mutagen)

Pennsylvania Right to Know List
environmental hazard

PROPOSED REGULATIONS

CERCLA/SARA - Proposed Hazardous Substance Additions
proposed RQ = 1 pound (.454 kg)

CERCLA/SARA - 1989 Proposed RQ Adjustments
proposed RQ = 100 pounds (45.4 kg)

Canada - Ontario - Proposed Occupational TWAEVs
0.5 mg/m3 TWAEV

Canada - Ontario - Proposed Occupational STEVs
1.5 mg/m3 STEV

M-HYDROXYBENZOIC ACID   99-06-9

HEALTH AND SAFETY LISTS

NFPA - Flash Points
flash point = 435 degrees F (224 degrees C)

NFPA - Hazard Identification Ratings
flammability-1;   reactivity-0

P-HYDROXYBENZOIC ACID   99-96-7

HEALTH AND SAFETY LISTS

NIOSH - Selected LD50s and LC50s
Oral, rat: LD50 = 3600 mg/kg

ENVIRONMENTAL LISTS

List of Pesticide Product Inert Ingredients
[present]

INTERNATIONAL LISTS

Canada - WHMIS: Ingredient Disclosure
1% item 853 (842)

1,4-BIS(3-HYDROXY-4-BENZOYLPHENOXY)BUTANE   RR-01691-1

HEALTH AND SAFETY LISTS

NIOSH - Selected LD50s and LC50s
Oral, rat: LD50 = 1950 mg/kg

ENVIRONMENTAL LISTS

List of Pesticide Product Inert Ingredients
[present]

INTERNATIONAL LISTS

Canada - WHMIS: Ingredient Disclosure
1% item 854 (94)

2-(2-HYDROXY-3-TERT-BUTYL-5-METHYL-   RR-01013-9
BENZYL)-4-METHYL-6-TERT-BUTYLPHENYL
METHACRYLATE

HEALTH AND SAFETY LISTS

NIOSH - Selected LD50s and LC50s
Oral, mouse: LD50 = 1500 mg/kg

2-HYDROXYDECALIN   825-51-4

ENVIRONMENTAL LISTS

CAA - HON Rule - SOCMI Chemicals
compliance by Oct. 23, 1995

HYDROXYDECANOIC ACID, GAMMA-LACTONE   706-14-9

ENVIRONMENTAL LISTS

TSCA - Chemicals with Significant New Use Rules
PMN number: P-89-507

HYDROXYDIMETHYLBENZENES   RR-01289-5

HEALTH AND SAFETY LISTS

IARC - Group 3 (not classifiable)
[present]

1-HYDROXYETHANE-1,1-DIPHOSPHONIC ACID   2809-21-4

HEALTH AND SAFETY LISTS

NIOSH - Selected LD50s and LC50s
Oral, mouse: LD50 = 2 gm/kg

3-(2-HYDROXYETHOXY)-4-PYRROLIDIN-1-YLBEN-   RR-01359-2
ZENEDAZONIUM ZINC CHLORIDE

HEALTH AND SAFETY LISTS

NIOSH - Selected LD50s and LC50s
Oral, mouse: LD50 = 2200 mg/kg

BETA-HYDROXYETHYLCARBAMATE   5395-01-7

ENVIRONMENTAL LISTS

TSCA - Chemicals with Significant New Use Rules
PMN number: P-93-483

HYDROXYETHYL CELLULOSE   9004-62-0

ENVIRONMENTAL LISTS

TSCA - Chemicals with Significant New Use Rules
PMN number: P-87-147

N-(HYDROXYETHYL)DIETHYLENETRIAMINE   1965-29-3

INTERNATIONAL LISTS

Canada - WHMIS: Ingredient Disclosure
1% item 855 (988)

2-HYDROXYETHYL DIMETHYL 3-OCTADECANAMI-   2764-13-8
DOPROPYL AMMONIUM NITRATE

INTERNATIONAL LISTS

Canada - WHMIS: Ingredient Disclosure
1% item 856 (1064)

N-HYDROXYETHYLETHYLENEDIAMINETRIACETIC   139-89-9
ACID TRISODIUM SALT

HEALTH AND SAFETY LISTS

U.S. DOT - Appendix B - Marine Pollutants
DOT regulated marine pollutant

1-(HYDROXYETHYL)-2-(HEPTADECYL)   95-19-2
IMIDAZOLINE

ENVIRONMENTAL LISTS

List of Pesticide Product Inert Ingredients
[present]

N-(2-HYDROXYETHYL)HEXADECANAMIDE   544-31-0

HEALTH AND SAFETY LISTS

U.S. DOT - Substances From 49 CFR 172.101
regulated by DOT (UN3035)

U.S. DOT - Hazard Classes
DOT hazard class = 4.1

2-HYDROXYETHYL METHACRYLATE   868-77-9

INTERNATIONAL LISTS

Canada - WHMIS: Ingredient Disclosure
1% item 858 (383)

N-(2-HYDROXYETHYL)OCTADECANAMIDE   111-57-9

ENVIRONMENTAL LISTS

List of Pesticide Product Inert Ingredients
[present]

N-(2-HYDROXYETHYL) PROPYLENE DIAMINE   10138-74-6

HEALTH AND SAFETY LISTS

NIOSH - Selected LD50s and LC50s
Oral, rat: LD50 = 4630 mg/kg Skin, rabbit: LD50 = 1041 mg/kg

N-(2-HYDROXYETHYL)TETRADECANAMIDE   142-58-5

ENVIRONMENTAL LISTS

List of Pesticide Product Inert Ingredients
[present]

4-HYDROXYHEPTANOIC ACID, GAMMA-LACTONE   105-21-5

ENVIRONMENTAL LISTS

List of Pesticide Product Inert Ingredients
[present]

HYDROXYLAMINE   7803-49-8

HEALTH AND SAFETY LISTS

NIOSH - Selected LD50s and LC50s
Oral, rat: LD50 = 3800 mg/kg

ENVIRONMENTAL LISTS

List of Pesticide Product Inert Ingredients
[present]

HYDROXYLAMINE, HYDROCHLORIDE   5470-11-1

ENVIRONMENTAL LISTS

List of Pesticide Product Inert Ingredients
[present]

HYDROXYL AMINE IODIDE   59917-23-6

HEALTH AND SAFETY LISTS

NIOSH - Selected LD50s and LC50s
Oral, mouse: LD50 = 5888 mg/kg

ENVIRONMENTAL LISTS

TSCA - Code of Federal Regulations Citations
40 CFR 712.30(d)

TSCA - PAIR - Reporting List
Reporting Date: November 19, 1982

INTERNATIONAL LISTS

Canada - WHMIS: Ingredient Disclosure
1% item 859 (1094)

HYDROXYLAMINE SULFATE (1:1)   10046-00-1

ENVIRONMENTAL LISTS

List of Pesticide Product Inert Ingredients
[present]

HYDROXYLAMINE SULFATE (2:1)   10039-54-0

HEALTH AND SAFETY LISTS

NFPA - Flash Points
flash point = 260 degrees F (127 degrees C)

NFPA - Hazard Identification Ratings
health-2; flammability-1; reactivity-0

STATE LISTS

Florida Hazardous Substance List
[present]

Massachusetts Right To Know List
[present]

Pennsylvania Right to Know List
[present]

2-HYDROXY-4-METHOXYBENZOPHENONE   131-57-7

ENVIRONMENTAL LISTS

List of Pesticide Product Inert Ingredients
[present]

2-HYDROXY-4-METHOXYBENZOPHENONE-5-SUL-FONIC ACID   4065-45-6

INTERNATIONAL LISTS

Canada - WHMIS: Ingredient Disclosure
1% item 860 (1065)

2-[(HYDROXYMETHYL)AMINO]-2-METHYL PROPANOL   52299-20-4

HEALTH AND SAFETY LISTS

NFPA - Flash Points
explodes at 265 degrees F (129 degrees C)

NFPA - Hazard Identification Ratings
health-2; flammability-0; reactivity-3

OSHA - List of Highly Hazardous Chemicals
threshhold quantity = 2500 pounds

ENVIRONMENTAL LISTS

TSCA - Code of Federal Regulations Citations
40 CFR 704.225

TSCA - CAIR - Reporting List
reporting required by: manufacturer, distributor, importer, processor

INTERNATIONAL LISTS

Canada - WHMIS: Ingredient Disclosure
1% item 861 (1003)

STATE LISTS

Florida Hazardous Substance List
[present]

Massachusetts Right To Know List
[present]

Pennsylvania Right to Know List
[present]

5-HYDROXY-1,4-NAPHTHAQUINONE   481-39-0

HEALTH AND SAFETY LISTS

NIOSH - Selected LD50s and LC50s
Oral, mouse: LD50 = 408 mg/kg

ENVIRONMENTAL LISTS

TSCA - Code of Federal Regulations Citations
40 CFR 704.225

TSCA - CAIR - Reporting List
reporting required by: manufacturer, distributor, importer, processor

3-HYDROXY-2-NAPHTHOIC ACID   92-70-6

HEALTH AND SAFETY LISTS

U.S. DOT - Hazard Classes
Forbidden from transport by the DOT

4-HYDROXY-3-NITROANILINE   119-34-6

ENVIRONMENTAL LISTS

TSCA - Code of Federal Regulations Citations
40 CFR 704.225

TSCA - CAIR - Reporting List
reporting required by: manufacturer, distributor, importer, processor

9-HYDROXYOCTADECANOIC ACID, BUTYL ESTER, HYDROGEN SULFATE, SODIUM SALT   42808-36-6

HEALTH AND SAFETY LISTS

U.S. DOT - Substances From 49 CFR 172.101
regulated by DOT (UN2865)

U.S. DOT - Hazard Classes
DOT hazard class = 8

ENVIRONMENTAL LISTS

List of Pesticide Product Inert Ingredients
[present]

TSCA - Code of Federal Regulations Citations
40 CFR 704.225

TSCA - CAIR - Reporting List
reporting required by: manufacturer, distributor, importer, processor

INTERNATIONAL LISTS

Canada - WHMIS: Ingredient Disclosure
1% item 862 (1523)

STATE LISTS

NJ Right to Know List (Total)
sn 1020

NJ Special Hazardous Substances
(corrosive)

4-HYDROXYOCTANOIC ACID LACTONE   104-50-7

HEALTH AND SAFETY LISTS

NIOSH - Selected LD50s and LC50s
Oral, rat: LD50 = 7400 mg/kg

NTP Chemical Status Reports - Testing Status and NTIS Number
Technical reports printed (PB93-126498)

ENVIRONMENTAL LISTS

List of Pesticide Product Inert Ingredients
[present]

2-HYDROXY-4-N-OCTOXYBENZOPHENONE   1843-05-6

ENVIRONMENTAL LISTS

List of Pesticide Product Inert Ingredients
[present]

2-(2-HYDROXY-5-TERT-OCTYLPHENYL) BENZOTRIAZOLE   3147-75-9

ENVIRONMENTAL LISTS

List of Pesticide Product Inert Ingredients
[present]

17A-HYDROXYPROGESTERONE CAPROATE   630-56-8

HEALTH AND SAFETY LISTS

NIOSH - Selected LD50s and LC50s
Oral, rat: LD50 = 112 mg/kg

2-HYDROXYPROPYL ACRYLATE   999-61-1

ENVIRONMENTAL LISTS

EPA - Master Testing List
[present]

TSCA - Code of Federal Regulations Citations
40 CFR 712.30(x); 40 CFR 716.120(d)

TSCA - PAIR - Reporting List
Reporting Date: November 27, 1991

TSCA - Health and Safety Reporting List
Effective Date: September 30, 1991

2-HYDROXYPROPYLAMINE NITRITE   7373-11-7

HEALTH AND SAFETY LISTS

IARC - Group 3 (not classifiable)
[present]

NTP Chemical Status Reports - Testing Status and NTIS Number
Technical reports printed (PB286189/AS)

NTP Chemical Status Reports - Evidence of Carcinogenicity
male rat-positive; female rat-equivocal; male mice-negative; female mice-negative

INTERNATIONAL LISTS

Canada - WHMIS: Ingredient Disclosure
1% item 863 (1004)

German (DFG) - Skin/Sensitizers
danger of cutaneous absorption

German (DFG) - Carcinogens
suspected carcinogen

STATE LISTS

Massachusetts Right To Know List
carcinogen; extraordinarily hazardous

N-(2-HYDROXYPROPYL)DECANAMIDE   23054-61-7

ENVIRONMENTAL LISTS

List of Pesticide Product Inert Ingredients
[present]

HYDROXYPROPYL ETHER OF CELLULOSE   9004-64-2

HEALTH AND SAFETY LISTS

NIOSH - Selected LD50s and LC50s
Oral, rat: LD50 = 4400 mg/kg

INTERNATIONAL LISTS

Canada - WHMIS: Ingredient Disclosure
1% item 864 (1066)

HYDROXYPROPYL GUAR GUM   39421-75-5

ENVIRONMENTAL LISTS

List of Pesticide Product Inert Ingredients
[present]

N-(2-HYDROXYPROPYL)OCTANAMIDE   23054-60-6

ENVIRONMENTAL LISTS

List of Pesticide Product Inert Ingredients
[present]

3-HYDROXYPYRIDINE   109-00-2

HEALTH AND SAFETY LISTS

IARC - Group Unspecified
[present] (Listed under 'Progestins')

8-HYDROXYQUINOLINE   148-24-3

HEALTH AND SAFETY LISTS

ACGIH 1995 - Time Weighted Averages
0.5 ppm TWA; 2.8 mg/m3 TWA

ACGIH 1995 - Skin Designations
skin - potential for cutaneous absorption

NFPA - Flash Points
flash point = 207 degrees F (97 degrees C)

NFPA - Hazard Identification Ratings
health-3; flammability-1; reactivity-2

NIOSH - Selected LD50s and LC50s
Oral, rat: LD50 = 250 mg/kg

OSHA - Vacated PELs - Time Weighted Averages
0.5 ppm TWA; 3 mg/m3 TWA

OSHA - Vacated PELs - Skin Designation
Prevent or reduce skin absorption

INTERNATIONAL LISTS

Australian Exposure Standards - Time Weighted Averages
0.5 ppm TWA; 2.8 mg/m3 TWA

Australian Exposure Standards - Skin Effects
skin absorption

Canada - WHMIS: Ingredient Disclosure
1% item 865 (160)

Canada - Alberta - 8 Hour Occupational Exposure Limit
0.5 ppm TWA; 3 mg/m3 TWA

Canada - Alberta - 15 Minute Occupational Exposure Limit
1.5 ppm STEL; 9 mg/m3 STEL

Canada - Alberta - Skin Designation
can be absorbed through the intact skin

Canada - Ontario - OHSA - TWAEVs
0.5 ppm TWAEV; 2.7 mg/m3 TWAEV

Canada - Ontario - OHSA - Skin Notations
absorption through skin, eyes, or mucous membranes

Canada - Quebec - Time-Weighted Average Exposure Values
0.5 ppm TWAEV; 2.8 mg/m3 TWAEV

Canada - Quebec - Skin Designations
absorbed through the skin

United Kingdom - Occupational Exposure Standards - TWAs
0.5 ppm TWA; 3 mg/m3 TWA

United Kingdom - Occupational Exposure Standards - Notes
can be absorbed through skin

Israel - Time Weighted Averages
0.5 ppm TWA; 2.8 mg/m3 TWA

Israel - Action Levels
0.25 ppm AL; 1.4 mg/m3 AL

Mexico - Instruction No. 10 - TWAs
0.5 ppm TWA; 3 mg/m3 TWA

Mexico - Instruction No. 10 - Skin designation
skin - potential for cutaneous absorption

STATE LISTS

California - Exposure Limits - PELs
0.5 ppm PEL; 3 mg/m3 PEL

California - Exposure Limits - Skin Notation
material may be absorbed through the skin, eyes or mucous membrane

California - Directors List of Hazardous Substances (8 CCR 339)
[present]

Florida Hazardous Substance List
[present]

Massachusetts Right To Know List
[present]

Minnesota Hazardous Substance List
skin

Pennsylvania Right to Know List
[present]

8-HYDROXYQUINOLINE SULFATE   134-31-6

ENVIRONMENTAL LISTS

List of Pesticide Product Inert Ingredients
[present]

HYDROXYSENKIRKINE   26782-43-4

ENVIRONMENTAL LISTS

List of Pesticide Product Inert Ingredients
[present]

HYDROXYUREA   127-07-1

HEALTH AND SAFETY LISTS

NIOSH - Selected LD50s and LC50s
Oral, rat: LD50 = 10200 mg/kg

ENVIRONMENTAL LISTS

List of Pesticide Product Inert Ingredients
[present]

HYPOCHLORITE ION   14380-61-1

ENVIRONMENTAL LISTS

List of Pesticide Product Inert Ingredients
[present]

HYPOCHLORITE SALTS   RR-01547-4

ENVIRONMENTAL LISTS

List of Pesticide Product Inert Ingredients
[present]

HYPOCHLORITE SOLUTIONS   RR-00337-2

HEALTH AND SAFETY LISTS

NIOSH - Selected LD50s and LC50s
Oral, bird: LD50 = 750 mg/kg

HYPONITROUS ACID   14448-38-5

HEALTH AND SAFETY LISTS

IARC - Group 3 (not classifiable)
[present]

NIOSH - Selected LD50s and LC50s
Oral, rat: LD50 = 1200 mg/kg

NTP Chemical Status Reports - Testing Status and NTIS Number
Technical reports printed (PB85213361/AS)

NTP Chemical Status Reports - Evidence of Carcinogenicity
male rat-no evidence; female rat-no evidence; male mice-no evidence; female mice-no evidence

INTERNATIONAL LISTS

Canada - WHMIS: Ingredient Disclosure
1% item 866 (1006)

HYPOPHOSPHORUS ACID   6303-21-5

ENVIRONMENTAL LISTS

List of Pesticide Product Inert Ingredients
[present]

HYTHERM BLUE E   6737-68-4

HEALTH AND SAFETY LISTS

IARC - Group 3 (not classifiable)
[present]

ICRF-159   21416-87-5

HEALTH AND SAFETY LISTS

NTP Chemical Status Reports - Testing Status and NTIS Number
Chronic studies exist for which technical reports were not prepared

STATE LISTS

NJ Right to Know List (Total)
sn 1021

NJ Special Hazardous Substances
(mutagen; teratogen)

IMAZALIL [1-[2-(2,4-DICHLOROPHENYL)-2-(2-PROPENYLOXY)ETHYL]-1H-IMIDAZOLE]   35554-44-0

PROPOSED REGULATIONS

Safe Drinking Water Act - Priority list
[present]

1-IMIDAZOLE   288-32-4

HEALTH AND SAFETY LISTS

IARC - Group 3 (not classifiable)
[present]

1H-IMIDAZOLE-1-ETHANOL, 2-(8-HEPTADACENYL)-4,5-DIHYDRO-   95-38-5

HEALTH AND SAFETY LISTS

U.S. DOT - Substances From 49 CFR 172.101
regulated by DOT (UN1791)

U.S. DOT - Hazard Classes
DOT hazard class = 8

STATE LISTS

NJ Right to Know List (Total)
sn 2477

1H-IMIDAZOLE-1-PROPANOIC ACID, 3-[2-(2-CARBOXYETHOXY)ETHYL]-2-HEPTYL-2,3-DIHYDRO-, DISODIUM SALT   68630-92-2

HEALTH AND SAFETY LISTS

U.S. DOT - Hazard Classes
Forbidden from transport by the DOT

2,4-IMIDAZOLIDINEDIONE, 5,5-DIMETHYL-3-[2-(OXIRANYLMETHOXY)PROPYL]-1-(OXIRANYLMETHYL)-   32568-89-1

ENVIRONMENTAL LISTS

List of Pesticide Product Inert Ingredients
[present]

2,4-IMIDAZOLIDINEDIONE, 3,3'-[2-(OXIRANYLMETHOXY)-1,3-PROPANEDIYL]BIS[5,5-DIMETHYL-1-(OXIRANYLMETHYL)-   38304-52-8

ENVIRONMENTAL LISTS

List of Pesticide Product Inert Ingredients
[present]

2-IMIDAZOLIDINONE, 4,5-DIHYDROXY-1,3-BIS   1854-26-8

HEALTH AND SAFETY LISTS

NTP Chemical Status Reports - Testing Status and NTIS Number
Technical reports printed (PB285853/AS)

NTP Chemical Status Reports - Evidence of Carcinogenicity
male rat-negative; female rat-positive; male mice-negative; female mice-positive

STATE LISTS

Massachusetts Right To Know List
carcinogen; extraordinarily hazardous

IMIDAZOLINIDYL UREA   39236-46-9

ENVIRONMENTAL LISTS

CERCLA/SARA - Section 313 - Emission Reporting
form R reporting required

2-IMIDAZOLINIUM, 1-(CARBOXYMETHYL)-1-(2-HYDROXYETHYL)-2-COCOYLNOR, CHLORIDE, MONOSODIUM SALT   RR-01087-7

HEALTH AND SAFETY LISTS

NIOSH - Selected LD50s and LC50s
Oral, mouse: LD50 = 880 mg/kg

2-IMIDAZOLINIUM, 1-(CARBOXYMETHYL)-1-(2-HYDROXYETHYL)-2-COCOYLNOR, HYDROXIDE, MONOSODIUM SALT   68390-66-9

HEALTH AND SAFETY LISTS

NIOSH - Selected LD50s and LC50s
Oral, rat: LD50 = 3130 mg/kg

ENVIRONMENTAL LISTS

List of Pesticide Product Inert Ingredients
[present]

1H-IMIDAZOLIUM, 1,3-BIS(CARBOXYMETHYL)-4,   68527-99-1
5-DIHYDRO-1-(2-HYDROXYETHYL)-2-UNDECYL-,
DIHYDROXIDE, DISODIUM SALT

ENVIRONMENTAL LISTS

List of Pesticide Product Inert Ingredients
[present]

1H-IMIDAZOLIUM, 1,1-BIS(CARBOXYMETHYL)-4,   61702-73-6
5-DIHYDRO-2-UNDECYL-, HYDROXIDE, DISODIUM
SALT
SEE ALSO:
GLYCIDOL (OXIRANEMETHANOL) AND ITS DERIVATIVES

ENVIRONMENTAL LISTS

EPA - Master Testing List
[present]

TSCA - Code of Federal Regulations Citations
40 CFR 716.120(c)

PROPOSED REGULATIONS

TSCA - Proposed Testing Rule for Glycidyl Ethers
member of Glycidyl subcategory V-A

1H-IMIDAZOLIUM, 1-(2-CARBOXYETHYL)-4,5-DIHY-   68630-96-6
DRO-3-(2-HYDROXYETHYL)-2-ISOHEPTADECYL-,
HYDROXIDE, INNER SALT
SEE ALSO:
GLYCIDOL (OXIRANEMETHANOL) AND ITS DERIVATIVES

ENVIRONMENTAL LISTS

EPA - Master Testing List
[present]

TSCA - Code of Federal Regulations Citations
40 CFR 716.120

PROPOSED REGULATIONS

TSCA - Proposed Testing Rule for Glycidyl Ethers
member of Glycidyl subcategory V-A

IMIDAZOLIUM COMPOUNDS, 2-(C16-18 ALKYL)-1-   RR-01088-8
(2-(C16-18 AMIDO)ETHYL)-4,5-DIHYDRO-1-METHYL,
METHYL SULFATES

ENVIRONMENTAL LISTS

EPA - Master Testing List
[present]

IMIDAZOLIUM COMPOUNDS, 1-[2-(CAR-   68650-39-5
BOXYMETHOXY)ETHYL]-1-CARBOXYMETHYL-4,
5-DIHYDRO-2-NORCOCO ALKYL, HYDROXIDES,
SODIUM SALTS

ENVIRONMENTAL LISTS

List of Pesticide Product Inert Ingredients
[present]

IMIDAZOLIUM COMPOUNDS, 4,5-DIHYDRO-1-   68122-86-1
METHYL-2-NORTALLOWALKYL-1-(2-TALLOW
AMIDOETHYL), METHYL SULFATES

ENVIRONMENTAL LISTS

List of Pesticide Product Inert Ingredients
[present]

IMIDAZOLIUM COMPOUNDS, 2-HEPTADECYL-4,5-   RR-01089-9
DIHYDRO-1-METHYL-1-(2-TALLOW AMIDOETHYL),
METHYL SULFATES

ENVIRONMENTAL LISTS

List of Pesticide Product Inert Ingredients
[present]

3,3'-IMINOBISPROPYLAMINE   56-18-8

ENVIRONMENTAL LISTS

List of Pesticide Product Inert Ingredients
[present]

IMINODIPROPIONITRILE   111-94-4

ENVIRONMENTAL LISTS

List of Pesticide Product Inert Ingredients
[present]

2-IMINO-1,3-THIAZIN-4-ONE-5,6-   RR-01210-2
DIHYDROMONOHYDROCHLORIDE

ENVIRONMENTAL LISTS

List of Pesticide Product Inert Ingredients
[present]

INDAN   496-11-7

ENVIRONMENTAL LISTS

List of Pesticide Product Inert Ingredients
[present]

INDENE   95-13-6

ENVIRONMENTAL LISTS

List of Pesticide Product Inert Ingredients
[present]

INDENO(1,2,3-CD)PYRENE (2,3-O-   193-39-5
PHENYLENEPYRENE)

ENVIRONMENTAL LISTS

List of Pesticide Product Inert Ingredients
[present]

TSCA - Code of Federal Regulations Citations
40 CFR 712.30(w); 40 CFR 716.120(a)

TSCA - PAIR - Reporting List
Reporting Date: August 18, 1988

TSCA - Health and Safety Reporting List
Effective Date: June 20, 1988

INDIUM   7440-74-6

ENVIRONMENTAL LISTS

List of Pesticide Product Inert Ingredients
[present]

INDIUM 109   14833-35-3

HEALTH AND SAFETY LISTS

U.S. DOT - Substances From 49 CFR 172.101
regulated by DOT (UN2269)

U.S. DOT - Hazard Classes
DOT hazard class = 8

NIOSH - Selected LD50s and LC50s
Oral, rat: LD50 = 738 mg/kg Skin, rabbit: LD50 = 110 mg/kg

INTERNATIONAL LISTS

Canada - WHMIS: Ingredient Disclosure
1% item 868 (1014)

STATE LISTS

NJ Right to Know List (Total)
sn 0230

NJ Special Hazardous Substances
(corrosive)

INDIUM 110   14133-75-6

HEALTH AND SAFETY LISTS
NIOSH - Selected LD50s and LC50s
Oral, rat: LD50 = 2700 mg/kg Skin, rabbit: LD50 = 2520 mg/kg

INDIUM 111   15750-15-9

ENVIRONMENTAL LISTS
TSCA - Chemicals with Significant New Use Rules
PMN number: P-91-101

INDIUM 112   14391-66-3

INTERNATIONAL LISTS
Canada - WHMIS: Ingredient Disclosure
1% item 869 (1015)

INDIUM 113M   RR-00443-3

HEALTH AND SAFETY LISTS
ACGIH 1995 - Time Weighted Averages
10 ppm TWA; 48 mg/m3 TWA

AIHA - Odor Threshold Values
no geometric mean air odor threshold

OSHA - Vacated PELs - Time Weighted Averages
10 ppm TWA; 45 mg/m3 TWA

ENVIRONMENTAL LISTS
TSCA - Health and Safety Reporting List
Effective Date: March 11, 1994; Sunset Date: March 11, 2004

INTERNATIONAL LISTS
Australian Exposure Standards - Time Weighted Averages
10 ppm TWA; 48 mg/m3 TWA

Canada - WHMIS: Ingredient Disclosure
1% item 870 (1016)

Canada - Alberta - 8 Hour Occupational Exposure Limit
10 ppm TWA; 47 mg/m3 TWA

Canada - Alberta - 15 Minute Occupational Exposure Limit
15 ppm STEL; 71 mg/m3 STEL

Canada - British Columbia - 8 Hour Exposure Limits
10 ppm TWA; 45 mg/m3 TWA

Canada - British Columbia - 15 Minute Exposure Limits
15 ppm STEL; 70 mg/m3 STEL

Canada - Ontario - OHSA - TWAEVs
10 ppm TWAEV; 47 mg/m3 TWAEV

Canada - Quebec - Time-Weighted Average Exposure Values
10 ppm TWAEV; 48 mg/m3 TWAEV

United Kingdom - Occupational Exposure Standards - TWAs
10 ppm TWA; 45 mg/m3 TWA

United Kingdom - Occupational Exposure Standards - STELs
15 ppm STEL; 70 mg/m3 STEL

Israel - Time Weighted Averages
10 ppm TWA; 48 mg/m3 TWA

Israel - Action Levels
5 ppm AL; 24 mg/m3 AL

Mexico - Instruction No. 10 - TWAs
10 ppm TWA; 45 mg/m3 TWA

Mexico - Instruction No. 10 - STELs
15 ppm STEL; 70 mg/m3 STEL

STATE LISTS
California - Exposure Limits - PELs
10 ppm PEL; 48 mg/m3 PEL

California - Directors List of Hazardous Substances (8 CCR 339)
[present]

Florida Hazardous Substance List
[present]

Massachusetts Right To Know List
[present]

Minnesota Hazardous Substance List
[present]

Pennsylvania Right to Know List
[present]

PROPOSED REGULATIONS
TSCA - ITC 32nd Report Priority Testing List
designated for dermal absorption testing

INDIUM 114M   RR-00442-2

HEALTH AND SAFETY LISTS
U.S. DOT - Appendix A Table 1 - Hazardous Substances
final RQ = 100 pounds (45.4 kg)

IARC - Group 2B (sufficient animal data)
[present] (Overall evaluation based only on evidence of carcinogenicity in monograph (32, 1983) or in Supplement 4)

NTP Seventh Report - Suspect Carcinogens
suspect carcinogen (Listed under 'Polycyclic aromatic hydrocarbons')

OSHA - Possible Select Carcinogens
[present]

ENVIRONMENTAL LISTS
ATSDR Priority List
Rank (of 275): 159

CERCLA/SARA - Section 313 - Emission Reporting
form R reporting required; (Listed under 'Polycyclic aromatic compounds')

CERCLA/SARA - Hazardous Substances and their Reportable Quantities
final RQ = 100 pounds (45.4 kg)

Clean Water Act - Priority Pollutants
[present]

EPA - Carcinogen Hazard Ranking for RQ Adjustment
Hazard ranking = Low

RCRA - U Series Wastes
waste number U137

RCRA - Hazardous Constituents-Appendix VIII
waste number U137

RCRA - Basis for Listing - Appendix VII
Included in waste streams: F039, K001, K035, K141, K142, K144, K147, K148

RCRA - Substances Banned From Land Disposal
[present]

RCRA - TSD Facilities Ground Water Monitoring
TM 8100 = 200 ug/L PQL; TM 8270 = 10 ug/L PQL

RCRA - Universal Treatment Standards (LDR)
WW: 0.0055 mg/l; NWW: 3.4 mg/kg

INTERNATIONAL LISTS
Canada - WHMIS: Ingredient Disclosure
0.1% item 871 (1017)

German (DFG) - Carcinogens
animal evidence of carcinogenicity

Mexico - Wastewater - Organic Toxic Pollutants and Heavy Metals
Listed under [Aromatic Hydrocarbons]

STATE LISTS
California - Air Bill 2588 Appendix A-I
known or potential carcinogen

California - Prop. 65 - Cancer list
carcinogen - initial date 1/1/88

California - Directors List of Hazardous Substances (8 CCR 339)
[present]

Florida Hazardous Substance List
[present]

Massachusetts Right To Know List
carcinogen; extraordinarily hazardous

Minnesota Hazardous Substance List
carcinogen

Pennsylvania Right to Know List
environmental hazard; special hazardous substance

Pennsylvania RTK - Special Hazardous Substances
[present]

INDIUM 115   14191-71-0

HEALTH AND SAFETY LISTS

ACGIH 1995 - Time Weighted Averages
as In: 0.1 mg/m3 TWA

OSHA - Vacated PELs - Time Weighted Averages
as In: 0.1 mg/m3 TWA

INTERNATIONAL LISTS

Australian Exposure Standards - Time Weighted Averages
as In: 0.1 mg/m3 TWA

Canada - WHMIS: Ingredient Disclosure
1% item 873 (1019)

Canada - Alberta - 8 Hour Occupational Exposure Limit
as In: 0.1 mg/m3 TWA

Canada - Alberta - 15 Minute Occupational Exposure Limit
as In: 0.3 mg/m3 STEL

Canada - British Columbia - 8 Hour Exposure Limits
as In: 0.1 mg/m3 TWA

Canada - British Columbia - 15 Minute Exposure Limits
as In: 0.3 mg/m3 STEL

Canada - Ontario - OHSA - TWAEVs
as In: 0.1 mg/m3 TWAEV

Canada - Quebec - Time-Weighted Average Exposure Values
0.1 mg/m3 TWAEV

Israel - Time Weighted Averages
as In: 0.1 mg/m3 TWA

Israel - Action Levels
as In: 0.05 mg/m3 AL

Mexico - Instruction No. 10 - TWAs
0.1 mg/m3 TWA

STATE LISTS

California - Exposure Limits - PELs
0.1 mg/m3 PEL

California - Directors List of Hazardous Substances (8 CCR 339)
[present]

Florida Hazardous Substance List
[present]

Massachusetts Right To Know List
[present]

Minnesota Hazardous Substance List
[present]

Pennsylvania Right to Know List
[present]

INDIUM 115M   RR-00441-1

HEALTH AND SAFETY LISTS

U.S. DOT - Appendix A Table 2 - Radionuclides
final RQ = 100 curies (3.7E 12 Bq)

ENVIRONMENTAL LISTS

CERCLA/SARA List of Radionuclides (Appendix B) and Their Reportable Quantities
final RQ = 100 curies (3.7E 12 Bq)

INDIUM 116M   RR-00439-7

HEALTH AND SAFETY LISTS

U.S. DOT - Appendix A Table 2 - Radionuclides
4.9 hour half-life: Final RQ = 10 curies (3.7E 11 Bq); 69.1 minute half-life: Final RQ = 100 curies (3.7E 12 Bq)

ENVIRONMENTAL LISTS

CERCLA/SARA List of Radionuclides (Appendix B) and Their Reportable Quantities
4.9 hour half-life: Final RQ = 10 curies (3.7E 11 Bq); 69.1 minute half-life: Final RQ = 100 curies (3.7E 12 Bq)

INDIUM 117   14914-66-0

HEALTH AND SAFETY LISTS

U.S. DOT - Appendix A Table 2 - Radionuclides
final RQ = 100 curies (3.7E 12 Bq)

ENVIRONMENTAL LISTS

CERCLA/SARA List of Radionuclides (Appendix B) and Their Reportable Quantities
final RQ = 100 curies (3.7E 12 Bq)

INDIUM 117M   RR-00438-6

HEALTH AND SAFETY LISTS

U.S. DOT - Appendix A Table 2 - Radionuclides
final RQ = 1000 curies (3.7E 13 Bq)

ENVIRONMENTAL LISTS

CERCLA/SARA List of Radionuclides (Appendix B) and Their Reportable Quantities
final RQ = 1000 curies (3.7E 13 Bq)

INDIUM 119M   RR-00437-5

HEALTH AND SAFETY LISTS

U.S. DOT - Appendix A Table 2 - Radionuclides
final RQ = 1000 curies (3.7E 13 Bq)

ENVIRONMENTAL LISTS

CERCLA/SARA List of Radionuclides (Appendix B) and Their Reportable Quantities
final RQ = 1000 curies (3.7E 13 Bq)

INDIUM COMPOUNDS, N.O.S.   RR-00600-8

HEALTH AND SAFETY LISTS

U.S. DOT - Appendix A Table 2 - Radionuclides
final RQ = 10 curies (3.7E 11 Bq)

ENVIRONMENTAL LISTS

CERCLA/SARA List of Radionuclides (Appendix B) and Their Reportable Quantities
final RQ = 10 curies (3.7E 11 Bq)

INDIUM NITRATE   13770-61-1

HEALTH AND SAFETY LISTS

U.S. DOT - Appendix A Table 2 - Radionuclides
final RQ = 0.1 curies (3.7E 9 Bq)

ENVIRONMENTAL LISTS

CERCLA/SARA List of Radionuclides (Appendix B) and Their Reportable Quantities
final RQ = 0.1 curies (3.7E 9 Bq)

INDIUM PHOSPHIDE   22398-80-7

HEALTH AND SAFETY LISTS

U.S. DOT - Appendix A Table 2 - Radionuclides
final RQ = 100 curies (3.7E 12 Bq)

ENVIRONMENTAL LISTS
   CERCLA/SARA List of Radionuclides (Appendix B) and Their Reportable Quantities
      final RQ = 100 curies (3.7E 12 Bq)

INDOLE   120-72-9
HEALTH AND SAFETY LISTS
   U.S. DOT - Appendix A Table 2 - Radionuclides
      final RQ = 100 curies (3.7E 12 Bq)

ENVIRONMENTAL LISTS
   CERCLA/SARA List of Radionuclides (Appendix B) and Their Reportable Quantities
      final RQ = 100 curies (3.7E 12 Bq)

3-INDOLEBUTYRIC ACID   133-32-4
HEALTH AND SAFETY LISTS
   U.S. DOT - Appendix A Table 2 - Radionuclides
      final RQ = 1000 curies (3.7E 13 Bq)

ENVIRONMENTAL LISTS
   CERCLA/SARA List of Radionuclides (Appendix B) and Their Reportable Quantities
      final RQ = 1000 curies (3.7E 13 Bq)

1H-INDOLE-5-SULFONIC ACID,2-(1,3-DIHYDRO-3-   860-22-0
OXO-5-SULFO-2H-
HEALTH AND SAFETY LISTS
   U.S. DOT - Appendix A Table 2 - Radionuclides
      final RQ = 100 curies (3.7E 12 Bq)

ENVIRONMENTAL LISTS
   CERCLA/SARA List of Radionuclides (Appendix B) and Their Reportable Quantities
      final RQ = 100 curies (3.7E 12 Bq)

3H-INDOL-3-ONE, 2-(1,3-DIHYDRO-3-OXO-2H-   482-89-3
HEALTH AND SAFETY LISTS
   U.S. DOT - Appendix A Table 2 - Radionuclides
      final RQ = 1000 curies (3.7E 13 Bq)

ENVIRONMENTAL LISTS
   CERCLA/SARA List of Radionuclides (Appendix B) and Their Reportable Quantities
      final RQ = 1000 curies (3.7E 13 Bq)

INDOMETHACIN   53-86-1
INTERNATIONAL LISTS
   Canada - WHMIS: Ingredient Disclosure
      1% item 872 (1018)
   Canada - Ontario - OHSA - TWAEVs
      as In: 0.1 mg/m3 TWAEV
   United Kingdom - Occupational Exposure Standards - TWAs
      as In: 0.1 mg/m3 TWA
   United Kingdom - Occupational Exposure Standards - STELs
      as In: 0.3 mg/m3 STEL
   Mexico - Instruction No. 10 - STELs
      0.3 mg/m3 STEL

STATE LISTS
   California - Exposure Limits - PELs
      0.1 mg/m3 PEL
   California - Directors List of Hazardous Substances (8 CCR 339)
      [present]

INKS   RR-00341-8
INTERNATIONAL LISTS
   Canada - WHMIS: Ingredient Disclosure
      1% item 874 (1206)

INORGANIC ARSENIC   RR-00065-7
HEALTH AND SAFETY LISTS
   NTP Chemical Status Reports - Testing Status and NTIS Number
      Assigned to laboratory for toxicology/carcinogenesis study

INORGANIC LEAD   RR-00538-9
HEALTH AND SAFETY LISTS
   NIOSH - Selected LD50s and LC50s
      Oral, rat: LD50 = 1000 mg/kg Skin, rabbit: LD50 = 790 mg/kg

ENVIRONMENTAL LISTS
   List of Pesticide Product Inert Ingredients
      [present]

INOSINE-5-MONOPHOSPHORIC ACID   4691-65-0
HEALTH AND SAFETY LISTS
   NIOSH - Selected LD50s and LC50s
      Oral, mouse: LD50 = 100 mg/kg

INOSITOL HEXANITRATE   RR-01429-9
HEALTH AND SAFETY LISTS
   NIOSH - Selected LD50s and LC50s
      Oral, rat: LD50 = 2 gm/kg

ENVIRONMENTAL LISTS
   List of Pesticide Product Inert Ingredients
      [present]

INSECTICIDE, DRY, N.O.S.   RR-00342-9
ENVIRONMENTAL LISTS
   EPA - Master Testing List
      [present]

INSECTICIDE GASES, N.O.S.   RR-00343-0
HEALTH AND SAFETY LISTS
   NIOSH - Selected LD50s and LC50s
      Oral, rat: LD50 = 2420 ug/kg

STATE LISTS
   Massachusetts Right To Know List
      [present]
   NJ Right to Know List (Total)
      sn 2480

INSECTICIDE, LIQUID, N.O.S.   RR-00345-2
HEALTH AND SAFETY LISTS
   U.S. DOT - Substances From 49 CFR 172.101
      regulated by DOT (UN1210)
   U.S. DOT - Hazard Classes
      DOT hazard class = 3

STATE LISTS
   NJ Right to Know List (Total)
      sn 2484; printer's: sn 2483

INTERFERON A (AIDS INITIATIVE)   76543-88-9
HEALTH AND SAFETY LISTS
   NIOSH - Health Standards - Exposure Limits
      C (15 min) 2 ug As/m3
   NIOSH - Health Standards - Health Effects and Precautions
      Lung and lymphatic cancer, dermatitis (Periodic chest X-ray required)

NIOSH - Health Standards - Carcinogenic Chemicals
potential human carcinogen

OSHA - 29 CFR 1910 Specifically Regulated Chemicals
*as As: 5 ug/m3 TWA action level;   10 ug/m3 TWA;   Cancer hazard
(see 29 CFR 1910.1018)*

OSHA - Select Carcinogens
[present]

STATE LISTS

California - Air Bill 2588 Appendix A-I
known or potential carcinogen

California - Prop. 65 - Cancer list
carcinogen - initial date 2/27/87

NJ Right to Know List (Total)
sn 2867

Pennsylvania Right to Know List
environmental hazard

INULIN TRINITRATE   135991-41-2

HEALTH AND SAFETY LISTS

IARC - Group 2B (sufficient animal data)
[present] (Listed under 'Lead and lead compounds')

NIOSH - Health Standards - Exposure Limits
*< 100 ug Pb/m3 TWA;   air level kept so that worker blood lead
remains < 60 ug/100 g of whole blood*

NIOSH - Health Standards - Health Effects and Precautions
Kidney, blood, and nervous system effects (Blood monitoring required)

OSHA - Possible Select Carcinogens
[present]

INTERNATIONAL LISTS

Canada - WHMIS: Ingredient Disclosure
1% item 940 (1434)

Canada - Alberta - 8 Hour Occupational Exposure Limit
as Pb: 0.05 mg/m3 TWA

Canada - British Columbia - 8 Hour Exposure Limits
as Pb: 0.05 mg/m3 TWA

Canada - British Columbia - 15 Minute Exposure Limits
as Pb: 0.45 mg/m3 STEL

Canada - British Columbia - Carcinogens
carcinogen - 0.05 mg/m3 TWA

Canada - Quebec - Time-Weighted Average Exposure Values
as Pb: 0.15 mg/m3 TWAEV

STATE LISTS

California - Air Bill 2588 Appendix A-I
known or potential carcinogen

NJ Right to Know List (Total)
sn 2873

PROPOSED REGULATIONS

ACGIH 1995 - Notice of Intended Changes
as Pb: 0.05 mg/m3 TWA;   A3-animal carcinogen

INULIN TRINITRATE   RR-01430-2

HEALTH AND SAFETY LISTS

NIOSH - Selected LD50s and LC50s
Oral, rat: LD50 = 15900 mg/kg

INVERT SUGAR   8013-17-0

HEALTH AND SAFETY LISTS

U.S. DOT - Hazard Classes
Forbidden from transport by the DOT

INVESTIGATIONAL ORAL CONTRACEPTIVES   RR-00013-5

STATE LISTS

NJ Right to Know List (Total)
sn 2485

IODACETIC ACID   64-69-7

HEALTH AND SAFETY LISTS

U.S. DOT - Substances From 49 CFR 172.101
regulated by DOT (UN1967)

U.S. DOT - Hazard Classes
*flammable: DOT hazard class = 2.1;   toxic: DOT hazard class = 2.3;
nonflammable and non-toxic: DOT hazard class = 2.2*

U.S. DOT - Substances Which Are Poisonous by Inhalation
gaseous hazardous material poisonous by inhalation (UN1967)

STATE LISTS

NJ Right to Know List (Total)
sn 2486;   sn 2487

IODINATED GLYCEROL   5634-39-9

STATE LISTS

NJ Right to Know List (Total)
sn 2488;   sn 2489

IODINE   7553-56-2

HEALTH AND SAFETY LISTS

NTP Chemical Status Reports - Testing Status and NTIS Number
Two year studies: pathology quality assessment in progress

IODINE 120   15480-34-9

HEALTH AND SAFETY LISTS

U.S. DOT - Hazard Classes
Forbidden from transport by the DOT

IODINE 120M   RR-00436-4

STATE LISTS

NJ Right to Know List (Total)
sn 2411

IODINE 121   15755-17-6

ENVIRONMENTAL LISTS

List of Pesticide Product Inert Ingredients
[present]

IODINE 123   15715-08-9

HEALTH AND SAFETY LISTS

IARC - Group Unspecified
[present] (Listed under 'Combined oral contraceptives')

IODINE 124   14158-30-6

HEALTH AND SAFETY LISTS

NIOSH - Selected LD50s and LC50s
Oral, mouse: LD50 = 83 mg/kg

IODINE 125   14158-31-7

HEALTH AND SAFETY LISTS

NTP Chemical Status Reports - Testing Status and NTIS Number
Technical reports printed (PB90259102)

NTP Chemical Status Reports - Evidence of Carcinogenicity
*male rat-some evidence;   female rat-no evidence;   male mice-no
evidence;   female mice-some evidence*

IODINE 126   14158-32-8

HEALTH AND SAFETY LISTS

ACGIH 1995 - Ceiling Limits
C 0.1 ppm; C 1.0 mg/m3

NIOSH - Selected LD50s and LC50s
Oral, rat: LD50 = 14 gm/kg

NIOSH 1990 - Pocket Guide - RELs
C 0.1 ppm; C 1 mg/m3

NIOSH 1990 - Pocket Guide - IDLHs
10 ppm IDLH

NIOSH 1990 - Pocket Guide - Target organs
respiratory system, eyes, skin, CNS, CVS

OSHA - Vacated PELs - Ceiling Limits
C 0.1 ppm; C 1 mg/m3

OSHA - Final PELs - Ceiling Limits
C 0.1 ppm; C 1 mg/m3

INTERNATIONAL LISTS

Australian Exposure Standards - Time Weighted Averages
Peak Limitation: 0.1 ppm; 1 mg/m3

Canada - WHMIS: Ingredient Disclosure
1% item 875 (1020)

Canada - Alberta - Ceiling Occupational Exposure Limit
C 0.1 ppm; C 1 mg/m3

Canada - British Columbia - Ceiling Exposure Limits
C 0.1 ppm; C 1 mg/m3

Canada - Ontario - OHSA - CEVs
0.1 ppm CEV; 1 mg/m3 CEV

Canada - Quebec - Ceiling Limits
P 0.1 ppm; P 1 mg/m3

United Kingdom - Occupational Exposure Standards - STELs
0.1 ppm STEL; 1 mg/m3 STEL

German (DFG) - MAK Values
0.1 ppm MAK; 1 mg/m3 MAK

German (DFG) - Peak Limitations
2 x normal MAK (5 min momentary value); don't exceed 8 times during shift

Israel - Ceiling Exposure Limits
C 0.1 ppm; C 1.0 mg/m3

Mexico - Instruction No. 10 - TWAs
0.1 ppm TWA; 1 mg/m3 TWA

STATE LISTS

California - Exposure Limits - Ceilings
C 0.1 ppm; C 1 mg/m3

California - Directors List of Hazardous Substances (8 CCR 339)
[present]

Florida Hazardous Substance List
[present]

Massachusetts Right To Know List
[present]

Minnesota Hazardous Substance List
[present]

NJ Right to Know List (Total)
sn 1026

Pennsylvania Right to Know List
[present]

IODINE 128   14391-72-1

HEALTH AND SAFETY LISTS

U.S. DOT - Appendix A Table 2 - Radionuclides
final RQ = 10 curies (3.7E 11 Bq)

ENVIRONMENTAL LISTS

CERCLA/SARA List of Radionuclides (Appendix B) and Their Reportable Quantities
final RQ = 10 curies (3.7E 11 Bq)

IODINE 129   15046-84-1

HEALTH AND SAFETY LISTS

U.S. DOT - Appendix A Table 2 - Radionuclides
final RQ = 100 curies (3.7E 12 Bq)

ENVIRONMENTAL LISTS

CERCLA/SARA List of Radionuclides (Appendix B) and Their Reportable Quantities
final RQ = 100 curies (3.7E 12 Bq)

IODINE 130   14914-02-4

HEALTH AND SAFETY LISTS

U.S. DOT - Appendix A Table 2 - Radionuclides
final RQ = 100 curies (3.7E 12 Bq)

ENVIRONMENTAL LISTS

CERCLA/SARA List of Radionuclides (Appendix B) and Their Reportable Quantities
final RQ = 100 curies (3.7E 12 Bq)

IODINE 131   10043-66-0

HEALTH AND SAFETY LISTS

U.S. DOT - Appendix A Table 2 - Radionuclides
final RQ = 10 curies (3.7E 11 Bq)

ENVIRONMENTAL LISTS

CERCLA/SARA List of Radionuclides (Appendix B) and Their Reportable Quantities
final RQ = 10 curies (3.7E 11 Bq)

IODINE-131   24267-56-9

HEALTH AND SAFETY LISTS

U.S. DOT - Appendix A Table 2 - Radionuclides
final RQ = 0.1 curies (3.7E 9 Bq)

ENVIRONMENTAL LISTS

CERCLA/SARA List of Radionuclides (Appendix B) and Their Reportable Quantities
final RQ = 0.1 curies (3.7E 9 Bq)

IODINE 132   14683-16-0

HEALTH AND SAFETY LISTS

U.S. DOT - Appendix A Table 2 - Radionuclides
final RQ = 0.01 curies (3.7E 8 Bq)

ENVIRONMENTAL LISTS

CERCLA/SARA List of Radionuclides (Appendix B) and Their Reportable Quantities
final RQ = 0.01 curies (3.7E 8 Bq)

IODINE 132M   RR-00435-3

HEALTH AND SAFETY LISTS

U.S. DOT - Appendix A Table 2 - Radionuclides
final RQ = 0.01 curies (3.7E 8 Bq)

ENVIRONMENTAL LISTS

CERCLA/SARA List of Radionuclides (Appendix B) and Their Reportable Quantities
final RQ = 0.01 curies (3.7E 8 Bq)

IODINE 133   14834-67-4

HEALTH AND SAFETY LISTS

U.S. DOT - Appendix A Table 2 - Radionuclides
final RQ = 1000 curies (3.7E 13 Bq)

ENVIRONMENTAL LISTS

CERCLA/SARA List of Radionuclides (Appendix B) and Their Reportable Quantities
final RQ = 1000 curies (3.7E 13 Bq)

IODINE 134   14914-27-3

HEALTH AND SAFETY LISTS

U.S. DOT - Appendix A Table 2 - Radionuclides
final RQ = 0.001 curies (3.7E 7 Bq)

ENVIRONMENTAL LISTS

CERCLA/SARA List of Radionuclides (Appendix B) and Their Reportable Quantities
final RQ = 0.001 curies (3.7E 7 Bq)

IODINE 135   14834-68-5

HEALTH AND SAFETY LISTS

U.S. DOT - Appendix A Table 2 - Radionuclides
final RQ = 1 curie (3.7E 10 Bq)

ENVIRONMENTAL LISTS

CERCLA/SARA List of Radionuclides (Appendix B) and Their Reportable Quantities
final RQ = 1 curie (3.7E 10 Bq)

IODINE AZIDE   14696-82-3

HEALTH AND SAFETY LISTS

U.S. DOT - Appendix A Table 2 - Radionuclides
final RQ = 0.01 curies (3.7E 8 Bq)

ENVIRONMENTAL LISTS

CERCLA/SARA List of Radionuclides (Appendix B) and Their Reportable Quantities
final RQ = 0.01 curies (3.7E 8 Bq)

IODINE MONOCHLORIDE   7790-99-0

STATE LISTS

California - Air Bill 2588 Appendix A-I
known or potential carcinogen: 9/89

California - Prop. 65 - Developmental Toxicity
developmental toxicity - initial date 1/1/89

IODINE PENTAFLUORIDE   7783-66-6

HEALTH AND SAFETY LISTS

U.S. DOT - Appendix A Table 2 - Radionuclides
final RQ = 10 curies (3.7E 11 Bq)

ENVIRONMENTAL LISTS

CERCLA/SARA List of Radionuclides (Appendix B) and Their Reportable Quantities
final RQ = 10 curies (3.7E 11 Bq)

2-IODOACETAMIDE   144-48-9

HEALTH AND SAFETY LISTS

U.S. DOT - Appendix A Table 2 - Radionuclides
final RQ = 10 curies (3.7E 11 Bq)

ENVIRONMENTAL LISTS

CERCLA/SARA List of Radionuclides (Appendix B) and Their Reportable Quantities
final RQ = 10 curies (3.7E 11 Bq)

IODOBENZENE   591-50-4

HEALTH AND SAFETY LISTS

U.S. DOT - Appendix A Table 2 - Radionuclides
final RQ = 0.1 curies (3.7E 9 Bq)

ENVIRONMENTAL LISTS

CERCLA/SARA List of Radionuclides (Appendix B) and Their Reportable Quantities
final RQ = 0.1 curies (3.7E 9 Bq)

1-IODOBUTANE   542-69-8

HEALTH AND SAFETY LISTS

U.S. DOT - Appendix A Table 2 - Radionuclides
final RQ = 100 curies (3.7E 12 Bq)

ENVIRONMENTAL LISTS

CERCLA/SARA List of Radionuclides (Appendix B) and Their Reportable Quantities
final RQ = 100 curies (3.7E 12 Bq)

IODOFORM   75-47-8

HEALTH AND SAFETY LISTS

U.S. DOT - Appendix A Table 2 - Radionuclides
final RQ = 10 curies (3.7E 11 Bq)

ENVIRONMENTAL LISTS

CERCLA/SARA List of Radionuclides (Appendix B) and Their Reportable Quantities
final RQ = 10 curies (3.7E 11 Bq)

IODO METHYLPROPANE   513-38-2

HEALTH AND SAFETY LISTS

U.S. DOT - Hazard Classes
Forbidden from transport by the DOT

2-IODO-2-METHYLPROPANE   558-17-8

HEALTH AND SAFETY LISTS

U.S. DOT - Substances From 49 CFR 172.101
regulated by DOT (UN1792)

U.S. DOT - Hazard Classes
DOT hazard class = 8

INTERNATIONAL LISTS

Canada - WHMIS: Ingredient Disclosure
1% item 876 (1169)

STATE LISTS

NJ Right to Know List (Total)
sn 1027

NJ Special Hazardous Substances
(corrosive)

IODOMETHYLPROPANES   RR-00146-7

HEALTH AND SAFETY LISTS

U.S. DOT - Substances From 49 CFR 172.101
regulated by DOT (UN2495)

U.S. DOT - Hazard Classes
DOT hazard class = 5.1

STATE LISTS

NJ Right to Know List (Total)
sn 1028

2-IODOPROPANE   75-30-9

INTERNATIONAL LISTS

Canada - WHMIS: Ingredient Disclosure
1% item 877 (1021)

1-IODOPROPANE   107-08-4

HEALTH AND SAFETY LISTS

NIOSH - Selected LD50s and LC50s
Inhalation, rat: LC50 = 16320 mg/m3 (8 hr) Oral, rat: LD50 = 1799 mg/kg

IODO PROPANE   26914-02-3

HEALTH AND SAFETY LISTS

NIOSH - Selected LD50s and LC50s
Inhalation, rat: LC50 = 6100 mg/m3 4 hr

3-IODO-2-PROPYNYL BUTYLCARBAMATE   55406-53-6

HEALTH AND SAFETY LISTS

ACGIH 1995 - Time Weighted Averages
0.6 ppm TWA; 10 mg/m3 TWA

AIHA - Odor Threshold Values
no geometric mean air odor threshold

NIOSH - Selected LD50s and LC50s
Inhalation, rat: LC50 = 165 ppm 7 hr Oral, rat: LD50 = 355 mg/kg Skin, rat: LD50 = 1184 mg/kg

NTP Chemical Status Reports - Testing Status and NTIS Number
Technical reports printed (PB286344/AS)

NTP Chemical Status Reports - Evidence of Carcinogenicity
male rat-negative; female rat-negative; male mice-negative; female mice-negative

OSHA - Vacated PELs - Time Weighted Averages
0.6 ppm TWA; 10 mg/m3 TWA

INTERNATIONAL LISTS

Australian Exposure Standards - Time Weighted Averages
0.6 ppm TWA; 10 mg/m3 TWA

Canada - Alberta - 8 Hour Occupational Exposure Limit
0.6 ppm TWA; 9.6 mg/m3 TWA

Canada - Alberta - 15 Minute Occupational Exposure Limit
1 ppm STEL; 16 mg/m3 STEL

Canada - British Columbia - 8 Hour Exposure Limits
0.6 ppm TWA; 10 mg/m3 TWA

Canada - British Columbia - 15 Minute Exposure Limits
1 ppm STEL; 20 mg/m3 STEL

Canada - Ontario - OHSA - TWAEVs
0.6 ppm TWAEV; 10 mg/m3 TWAEV

Canada - Quebec - Time-Weighted Average Exposure Values
0.6 ppm TWAEV; 10 mg/m3 TWAEV

United Kingdom - Occupational Exposure Standards - TWAs
0.6 ppm TWA; 10 mg/m3 TWA

United Kingdom - Occupational Exposure Standards - STELs
1 ppm STEL; 20 mg/m3 STEL

Israel - Time Weighted Averages
0.6 ppm TWA; 10 mg/m3 TWA

Israel - Action Levels
0.3 ppm AL; 5 mg/m3 AL

Mexico - Instruction No. 10 - TWAs
0.6 ppm TWA; 10 mg/m3 TWA

Mexico - Instruction No. 10 - STELs
1 ppm STEL; 20 mg/m3 STEL

STATE LISTS

California - Exposure Limits - PELs
0.6 ppm PEL; 10 mg/m3 PEL

California - Directors List of Hazardous Substances (8 CCR 339)
[present]

Florida Hazardous Substance List
[present]

Massachusetts Right To Know List
[present]

Minnesota Hazardous Substance List
[present]

NJ Right to Know List (Total)
sn 1030

Pennsylvania Right to Know List
[present]

PROPOSED REGULATIONS

Canada - Ontario - Proposed Occupational TWAEVs
0.2 ppm TWAEV; 3 mg/m3 TWAEV

IODOXY COMPOUNDS   RR-01431-3

HEALTH AND SAFETY LISTS

NIOSH - Selected LD50s and LC50s
Inhalation, rat: LC50 = 6700 mg/m3 4 hr

STATE LISTS

NJ Right to Know List (Total)
sn 1031

IONIZING RADIATION   RR-00158-1

STATE LISTS

NJ Right to Know List (Total)
sn 1032

IONONE, ALPHA (ALPHA-IONONE)   127-41-3

HEALTH AND SAFETY LISTS

U.S. DOT - Substances From 49 CFR 172.101
regulated by DOT (UN2391)

U.S. DOT - Hazard Classes
DOT hazard class = 3

IONONE   8013-90-9

HEALTH AND SAFETY LISTS

NIOSH - Selected LD50s and LC50s
Inhalation, rat: LC50 = 320000 mg/m3 (30 mn)

IONONE, BETA (BETA-IONONE)   14901-07-6

HEALTH AND SAFETY LISTS

NIOSH - Selected LD50s and LC50s
Inhalation, rat: LC50 = 73000 mg/m3 (30 mn)

IOXYNIL   618-76-8

HEALTH AND SAFETY LISTS

U.S. DOT - Substances From 49 CFR 172.101
regulated by DOT (UN2392)

U.S. DOT - Hazard Classes
DOT hazard class = 3

STATE LISTS

NJ Right to Know List (Total)
sn 1033

IPD   3458-22-8

ENVIRONMENTAL LISTS

CERCLA/SARA - Section 313 - Emission Reporting
form R reporting required

List of Pesticide Product Inert Ingredients
[present]

IQ   76180-96-6

HEALTH AND SAFETY LISTS

U.S. DOT - Hazard Classes
Forbidden from transport by the DOT

IRIDIUM 182   29054-62-4

INTERNATIONAL LISTS

Israel - Ceiling Exposure Limits
C 5 rem/year

IRIDIUM 184   27742-26-3

HEALTH AND SAFETY LISTS

NFPA - Flash Points
flash point > 212 degrees F (100 degrees C)

NFPA - Hazard Identification Ratings
flammability-1; reactivity-0

IRIDIUM 185   29054-43-1

HEALTH AND SAFETY LISTS

NIOSH - Selected LD50s and LC50s
Oral, rat: LD50 = 4590 mg/kg

IRIDIUM 186   24447-13-0

HEALTH AND SAFETY LISTS

NFPA - Flash Points
flash point > 212 degrees F (100 degrees C)

NFPA - Hazard Identification Ratings
flammability-1; reactivity-0

IRIDIUM 187   14834-71-0

HEALTH AND SAFETY LISTS

U.S. DOT - Appendix B - Marine Pollutants
DOT regulated marine pollutant

IRIDIUM 188   15752-22-4

HEALTH AND SAFETY LISTS

NTP Chemical Status Reports - Testing Status and NTIS Number
Technical reports printed (PB277455/AS)

NTP Chemical Status Reports - Evidence of Carcinogenicity
male rat-equivocal; female rat-equivocal; male mice-equivocal; female mice-equivocal

STATE LISTS

Massachusetts Right To Know List
carcinogen; extraordinarily hazardous

IRIDIUM 189   14265-84-0

HEALTH AND SAFETY LISTS

IARC - Group 2A (limited human data)
[present] (Other relevant data taken into account in making the overall evaluation)

OSHA - Possible Select Carcinogens
[present]

STATE LISTS

California - Air Bill 2588 Appendix A-II
known or potential carcinogen

California - Prop. 65 - Cancer list
carcinogen - initial date 4/1/90

California - Prop. 65 - No Significant Risk Levels
no significant risk level = 0.5 ug/day

California - Directors List of Hazardous Substances (8 CCR 339)
[present]

Massachusetts Right To Know List
carcinogen; extraordinarily hazardous

Minnesota Hazardous Substance List
carcinogen

IRIDIUM 190   14981-91-0

HEALTH AND SAFETY LISTS

U.S. DOT - Appendix A Table 2 - Radionuclides
final RQ = 1000 curies (3.7E 13 Bq)

ENVIRONMENTAL LISTS

CERCLA/SARA List of Radionuclides (Appendix B) and Their Reportable Quantities
final RQ = 1000 curies (3.7E 13 Bq)

IRIDIUM 190M   RR-00432-0

HEALTH AND SAFETY LISTS

U.S. DOT - Appendix A Table 2 - Radionuclides
final RQ = 100 curies (3.7E 12 Bq)

ENVIRONMENTAL LISTS

CERCLA/SARA List of Radionuclides (Appendix B) and Their Reportable Quantities
final RQ = 100 curies (3.7E 12 Bq)

IRIDIUM 192   14694-69-0

HEALTH AND SAFETY LISTS

U.S. DOT - Appendix A Table 2 - Radionuclides
final RQ = 100 curies (3.7E 12 Bq)

ENVIRONMENTAL LISTS

CERCLA/SARA List of Radionuclides (Appendix B) and Their Reportable Quantities
final RQ = 100 curies (3.7E 12 Bq)

IRIDIUM 192M   RR-00431-9

HEALTH AND SAFETY LISTS

U.S. DOT - Appendix A Table 2 - Radionuclides
final RQ = 10 curies (3.7E 11 Bq)

ENVIRONMENTAL LISTS

CERCLA/SARA List of Radionuclides (Appendix B) and Their Reportable Quantities
final RQ = 10 curies (3.7E 11 Bq)

IRIDIUM 194   14158-35-1

HEALTH AND SAFETY LISTS

U.S. DOT - Appendix A Table 2 - Radionuclides
final RQ = 100 curies (3.7E 12 Bq)

ENVIRONMENTAL LISTS

CERCLA/SARA List of Radionuclides (Appendix B) and Their Reportable Quantities
final RQ = 100 curies (3.7E 12 Bq)

IRIDIUM 194M   RR-00430-8

HEALTH AND SAFETY LISTS

U.S. DOT - Appendix A Table 2 - Radionuclides
final RQ = 10 curies (3.7E 11 Bq)

ENVIRONMENTAL LISTS

CERCLA/SARA List of Radionuclides (Appendix B) and Their Reportable Quantities
final RQ = 10 curies (3.7E 11 Bq)

IRIDIUM 195   15816-99-6

HEALTH AND SAFETY LISTS

U.S. DOT - Appendix A Table 2 - Radionuclides
final RQ = 100 curies (3.7E 12 Bq)

ENVIRONMENTAL LISTS

CERCLA/SARA List of Radionuclides (Appendix B) and Their Reportable Quantities
final RQ = 100 curies (3.7E 12 Bq)

IRIDIUM 195M   RR-00427-3

HEALTH AND SAFETY LISTS

U.S. DOT - Appendix A Table 2 - Radionuclides
final RQ = 10 curies (3.7E 11 Bq)

ENVIRONMENTAL LISTS

CERCLA/SARA List of Radionuclides (Appendix B) and Their Reportable Quantities
final RQ = 10 curies (3.7E 11 Bq)

IRIDIUM NITRATOPENTAMINE IRIDIUM NITRATE RR-01432-4

HEALTH AND SAFETY LISTS

U.S. DOT - Appendix A Table 2 - Radionuclides
final RQ = 1000 curies (3.7E 13 Bq)

ENVIRONMENTAL LISTS

CERCLA/SARA List of Radionuclides (Appendix B) and Their Reportable Quantities
final RQ = 1000 curies (3.7E 13 Bq)

IRIDIUM TETRACHLORIDE 10025-97-5

HEALTH AND SAFETY LISTS

U.S. DOT - Appendix A Table 2 - Radionuclides
final RQ = 10 curies (3.7E 11 Bq)

ENVIRONMENTAL LISTS

CERCLA/SARA List of Radionuclides (Appendix B) and Their Reportable Quantities
final RQ = 10 curies (3.7E 11 Bq)

IRON 7439-89-6

HEALTH AND SAFETY LISTS

U.S. DOT - Appendix A Table 2 - Radionuclides
final RQ = 100 curies (3.7E 12 Bq)

ENVIRONMENTAL LISTS

CERCLA/SARA List of Radionuclides (Appendix B) and Their Reportable Quantities
final RQ = 100 curies (3.7E 12 Bq)

IRON 52 14093-04-0

HEALTH AND SAFETY LISTS

U.S. DOT - Appendix A Table 2 - Radionuclides
final RQ = 100 curies (3.7E 12 Bq)

ENVIRONMENTAL LISTS

CERCLA/SARA List of Radionuclides (Appendix B) and Their Reportable Quantities
final RQ = 100 curies (3.7E 12 Bq)

IRON 55 14681-59-5

HEALTH AND SAFETY LISTS

U.S. DOT - Appendix A Table 2 - Radionuclides
final RQ = 10 curies (3.7E 11 Bq)

ENVIRONMENTAL LISTS

CERCLA/SARA List of Radionuclides (Appendix B) and Their Reportable Quantities
final RQ = 10 curies (3.7E 11 Bq)

IRON 59 14596-12-4

HEALTH AND SAFETY LISTS

U.S. DOT - Appendix A Table 2 - Radionuclides
final RQ = 1000 curies (3.7E 13 Bq)

ENVIRONMENTAL LISTS

CERCLA/SARA List of Radionuclides (Appendix B) and Their Reportable Quantities
final RQ = 1000 curies (3.7E 13 Bq)

IRON 60 32020-21-6

HEALTH AND SAFETY LISTS

U.S. DOT - Appendix A Table 2 - Radionuclides
final RQ = 100 curies (3.7E 12 Bq)

ENVIRONMENTAL LISTS

CERCLA/SARA List of Radionuclides (Appendix B) and Their Reportable Quantities
final RQ = 100 curies (3.7E 12 Bq)

IRON AND STEEL FOUNDING RR-00537-8

HEALTH AND SAFETY LISTS

U.S. DOT - Hazard Classes
Forbidden from transport by the DOT

IRON, C3-13-CARBOXYLATE NAPHTHENATE COMPLEXES 85763-69-5

HEALTH AND SAFETY LISTS

NIOSH - Selected LD50s and LC50s
Oral, rat: LD50 = 1560 mg/kg

INTERNATIONAL LISTS

Canada - WHMIS: Ingredient Disclosure
1% item 878 (1580)

STATE LISTS

Massachusetts Right To Know List
[present]

IRON DEXTRAN COMPLEX 9004-66-4

ENVIRONMENTAL LISTS

Safe Drinking Water Act - SMCLs
SMCL = 0.3 mg/L

List of Pesticide Product Inert Ingredients
[present]

INTERNATIONAL LISTS

Canada - Drinking Water Quality - AOs
<= 0.3 mg/L AO

Canada - Ontario - OHSA - TWAEVs
welding fume or particulate, as Fe: 5 mg/m3 TWAEV (listed as agent of variable composition)

Mexico - Wastewater - Organic Toxic Pollutants and Heavy Metals
Listed under [Heavy Metals]

Mexico - Drinking Water - Ecological Criteria
0.3 mg/l

STATE LISTS

California - Directors List of Hazardous Substances (8 CCR 339)
[present]

IRON DEXTRIN COMPLEX 8050-93-9

HEALTH AND SAFETY LISTS

U.S. DOT - Appendix A Table 2 - Radionuclides
final RQ = 100 curies (3.7E 12 Bq)

ENVIRONMENTAL LISTS

CERCLA/SARA List of Radionuclides (Appendix B) and Their Reportable Quantities
final RQ = 100 curies (3.7E 12 Bq)

IRON (III) DICHROMATE 10294-53-8

HEALTH AND SAFETY LISTS

U.S. DOT - Appendix A Table 2 - Radionuclides
final RQ = 100 curies (3.7E 12 Bq)

ENVIRONMENTAL LISTS

CERCLA/SARA List of Radionuclides (Appendix B) and Their Reportable Quantities
final RQ = 100 curies (3.7E 12 Bq)

IRON (II) GLUCONATE   22830-45-1

HEALTH AND SAFETY LISTS

U.S. DOT - Appendix A Table 2 - Radionuclides
final RQ = 10 curies (3.7E 11 Bq)

ENVIRONMENTAL LISTS

CERCLA/SARA List of Radionuclides (Appendix B) and Their Reportable Quantities
final RQ = 10 curies (3.7E 11 Bq)

IRON MANGANESE ZINC OXIDE   12645-49-7

HEALTH AND SAFETY LISTS

U.S. DOT - Appendix A Table 2 - Radionuclides
final RQ = 0.1 curies (3.7E 9 Bq)

ENVIRONMENTAL LISTS

CERCLA/SARA List of Radionuclides (Appendix B) and Their Reportable Quantities
final RQ = 0.1 curies (3.7E 9 Bq)

IRON NICKEL ZINC OXIDE   12645-50-0

HEALTH AND SAFETY LISTS

IARC - Group 1 (carcinogenic to humans)
[present]

OSHA - Select Carcinogens
[present]

STATE LISTS

Pennsylvania Right to Know List
special hazardous substance

Pennsylvania RTK - Special Hazardous Substances
[present]

IRON OXIDE   1309-37-1

ENVIRONMENTAL LISTS

List of Pesticide Product Inert Ingredients
[present]

IRON OXIDE   1332-37-2

HEALTH AND SAFETY LISTS

IARC - Group 2B (sufficient animal data)
[present]

NIOSH - Selected LD50s and LC50s
Oral, mouse: LD50 = 1 gm/kg

NTP Seventh Report - Suspect Carcinogens
suspect carcinogen

OSHA - Possible Select Carcinogens
[present]

ENVIRONMENTAL LISTS

RCRA - U Series Wastes
waste number U139

RCRA - Hazardous Constituents-Appendix VIII
waste number U139

STATE LISTS

California - Air Bill 2588 Appendix A-II
known or potential carcinogen

California - Prop. 65 - Cancer list
carcinogen - initial date 1/1/88

California - Directors List of Hazardous Substances (8 CCR 339)
[present]

Florida Hazardous Substance List
[present]

Massachusetts Right To Know List
carcinogen; extraordinarily hazardous

Minnesota Hazardous Substance List
carcinogen

NJ Special Hazardous Substances
(carcinogen)

Pennsylvania Right to Know List
environmental hazard; special hazardous substance

Pennsylvania RTK - Special Hazardous Substances
[present]

IRON OXIDE (FE2O3), HYDRATE   12259-21-1

HEALTH AND SAFETY LISTS

IARC - Group 3 (not classifiable)
[present]

STATE LISTS

California - Directors List of Hazardous Substances (8 CCR 339)
[present]

IRON OXIDE YELLOW   51274-00-1

STATE LISTS

Massachusetts Right To Know List
[present]

IRON PENTACARBONYL   13463-40-6

HEALTH AND SAFETY LISTS

NIOSH - Selected LD50s and LC50s
Oral, rat: LD50 = 4500 mg/kg

IRON SALTS (SOLUBLE)   RR-00521-0

INTERNATIONAL LISTS

Canada - WHMIS: Ingredient Disclosure
0.1% item 879 (1311)

IRON SORBITOL-CITRIC ACID COMPLEX   RR-01534-9

INTERNATIONAL LISTS

Canada - WHMIS: Ingredient Disclosure
0.1% item 880 (1312)

ISANO OIL   8001-86-3

SEE ALSO:
IRON

HEALTH AND SAFETY LISTS

ACGIH 1995 - Time Weighted Averages
as Fe: 5 mg/m3 TWA (welding fumes, total particulate (N.O.C.))

IARC - Group 3 (not classifiable)
[present]

NIOSH 1990 - Pocket Guide - RELs
as Fe: 5 mg/m3 TWA

NIOSH 1990 - Pocket Guide - Target organs
respiratory system

OSHA - Vacated PELs - Time Weighted Averages
fume: 10 mg/m3 TWA

OSHA - Final PELs - Time Weighted Averages
10 mg/m3 TWA

ENVIRONMENTAL LISTS

List of Pesticide Product Inert Ingredients
[present]

INTERNATIONAL LISTS

Australian Exposure Standards - Time Weighted Averages
fume, as Fe: 5 mg/m3 TWA

Canada - WHMIS: Ingredient Disclosure
1% item 762 (1327)

Canada - Alberta - 8 Hour Occupational Exposure Limit
5 mg/m3 TWA

Canada - Alberta - 15 Minute Occupational Exposure Limit
10 mg/m3 STEL

Canada - British Columbia - 8 Hour Exposure Limits
as Fe2O3: 5 mg/m3 TWA

Canada - British Columbia - 15 Minute Exposure Limits
as Fe2O3: 10 mg/m3 STEL

Canada - Ontario - OHSA - TWAEVs
total dust: 10 mg/m3 TWAEV (listed as a nuisance particulate)

Canada - Quebec - Time-Weighted Average Exposure Values
5 mg/m3 TWAEV

United Kingdom - Occupational Exposure Standards - TWAs
fume, as Fe: 5 mg/m3 TWA

United Kingdom - Occupational Exposure Standards - STELs
fume, as Fe: 10 mg/m3 STEL

German (DFG) - MAK Values
fine dust: 6 mg/m3 MAK

Israel - Time Weighted Averages
as Fe: 5 mg/m3 TWA (Welding fumes, total particulate (N.O.C.))

Israel - Action Levels
as Fe: 2.5 mg/m3 AL

Mexico - Instruction No. 10 - TWAs
5 mg/m3 TWA

Mexico - Instruction No. 10 - STELs
10 mg/m3 STEL

STATE LISTS

California - Exposure Limits - PELs
fume: 5 mg/m3 PEL

California - Directors List of Hazardous Substances (8 CCR 339)
[present]

Florida Hazardous Substance List
[present]

Massachusetts Right To Know List
[present]

Minnesota Hazardous Substance List
[present] as Fe

NJ Right to Know List (Total)
sn 1036

Pennsylvania Right to Know List
[present]

ISATIDINE   15503-86-3
SEE ALSO:
IRON

HEALTH AND SAFETY LISTS

U.S. DOT - Substances From 49 CFR 172.101
regulated by DOT (UN1376)

U.S. DOT - Hazard Classes
DOT hazard class = 4.2

ISETHIONIC ACID, OLEATE, SODIUM SALT   142-15-4

ENVIRONMENTAL LISTS

List of Pesticide Product Inert Ingredients
[present]

ISOAMYL ACETATE   123-92-2
SEE ALSO:
IRON

ENVIRONMENTAL LISTS

List of Pesticide Product Inert Ingredients
[present]

ISOAMYL ALCOHOL   123-51-3
SEE ALSO:
IRON

HEALTH AND SAFETY LISTS

ACGIH 1995 - Time Weighted Averages
as Fe: 0.1 ppm TWA; 0.23 mg/m3 TWA

ACGIH 1995 - Short Term Exposure Limits
as Fe: 0.2 ppm STEL; 0.45 mg/m3 STEL

U.S. DOT - Substances From 49 CFR 172.101
regulated by DOT (UN1994)

U.S. DOT - Hazard Classes
DOT hazard class = 6.1

U.S. DOT - Substances Which Are Poisonous by Inhalation
liquid hazardous material poisonous by inhalation (UN1994)

NFPA - Flash Points
flash point = 5 degrees F (-15 degrees C)

NFPA - Hazard Identification Ratings
health-2; flammability-3; reactivity-1 (avoid use of water)

NIOSH - Selected LD50s and LC50s
Inhalation, rat: LC50 = 44 mg/m3 8 hr Oral, rat: LD50 = 40 mg/kg
Skin, rabbit: LD50 = 240 mg/kg

OSHA - Vacated PELs - Time Weighted Averages
as Fe: 0.1 ppm TWA; 0.8 mg/m3 TWA

OSHA - Vacated PELs - Short Term Exposure Limits
as Fe: 0.2 ppm STEL; 1.6 mg/m3 STEL

OSHA - List of Highly Hazardous Chemicals
threshhold quantity = 250 pounds

ENVIRONMENTAL LISTS

CERCLA/SARA - Section 302 Extremely Hazardous Substances and TPQs
TPQ = 100 pounds

CERCLA/SARA - Section 313 - Emission Reporting
form R reporting required

CAA -Toxic Substances for Accidental Release Prevention
threshold quantity = 2,500 lbs

INTERNATIONAL LISTS

Australian Exposure Standards - Time Weighted Averages
as Fe: 0.1 ppm TWA; 0.23 mg/m3 TWA

Australian Exposure Standards - Short Term Exposure Limits
as Fe: 0.2 ppm STEL; 0.45 mg/m3 STEL

Canada - WHMIS: Ingredient Disclosure
1% item 881 (879)

Canada - Alberta - 8 Hour Occupational Exposure Limit
as Fe: 0.1 ppm TWA; 0.23 mg/m3 TWA

Canada - Alberta - 15 Minute Occupational Exposure Limit
as Fe: 0.2 ppm STEL; 0.46 mg/m3 STEL

Canada - British Columbia - 8 Hour Exposure Limits
0.01 ppm TWA; 0.08 mg/m3 TWA

Canada - Ontario - OHSA - TWAEVs
as Fe: 0.1 ppm TWAEV; 0.8 mg/m3 TWAEV

Canada - Ontario - OHSA - STEVs
0.2 ppm STEV; 1.6 mg/m3 STEV

Canada - Quebec - Time-Weighted Average Exposure Values
0.1 ppm TWAEV; 0.23 mg/m3 TWAEV

Canada - Quebec - Short-term Exposure Values
as Fe: 0.2 ppm STEV; 0.45 mg/m3 STEV

United Kingdom - Occupational Exposure Standards - TWAs
as Fe: 0.01 ppm TWA; 0.08 mg/m3 TWA

German (DFG) - MAK Values
0.1 ppm MAK; 0.8 mg/m3 MAK

German (DFG) - Peak Limitations
2 x normal MAK (30 min. average value); don't exceed 4 times during shift

Israel - Time Weighted Averages
as Fe: 0.1 ppm TWA; 0.23 mg/m3 TWA

Israel - Short Term Exposure Limits
as Fe: 0.2 ppm STEL; 0.45 mg/m3 STEL

Israel - Action Levels
as Fe: 0.05 ppm AL; 0.115 mg/m3 AL

Mexico - Instruction No. 10 - TWAs
0.01 ppm TWA; 0.8 mg/m3 TWA

Mexico - Instruction No. 10 - STELs
0.2 ppm STEL; 1.6 mg/m3 STEL

STATE LISTS

California - Exposure Limits - PELs
as Fe: 0.1 ppm PEL; 0.8 mg/m3 PEL

California - Exposure Limits - STELs
as Fe: 0.2 ppm STEL; 1.6 mg/m3 STEL

California - Directors List of Hazardous Substances (8 CCR 339)
[present]

Florida Hazardous Substance List
[present]

Massachusetts Right To Know List
extraordinarily hazardous

Minnesota Hazardous Substance List
[present] as Fe

NJ Right to Know List (Total)
sn 1037

NJ Special Hazardous Substances
(flammable - third degree)

Pennsylvania Right to Know List
environmental hazard

PROPOSED REGULATIONS

CERCLA/SARA - Proposed Hazardous Substance Additions
proposed·RQ = 1 pound (.454 kg)

CERCLA/SARA - 1989 Proposed RQ Adjustments
proposed RQ = 10 pounds (4.54 kg)

Canada - Ontario - Proposed Occupational TWAEVs
as Fe: 0.01 ppm TWAEV; 0.08 mg/m3 TWAEV

ISOAMYL ALCOHOL (SECONDARY)   528-75-6

HEALTH AND SAFETY LISTS

ACGIH 1995 - Time Weighted Averages
as Fe: 1 mg/m3 TWA

OSHA - Vacated PELs - Time Weighted Averages
as Fe: 1 mg/m3 TWA

INTERNATIONAL LISTS

Australian Exposure Standards - Time Weighted Averages
as Fe: 1 mg/m3 TWA

Canada - WHMIS: Ingredient Disclosure
1% item 882 (883)

Canada - Alberta - 8 Hour Occupational Exposure Limit
as Fe: 1 mg/m3 TWA

Canada - Alberta - 15 Minute Occupational Exposure Limit
as Fe: 2 mg/m3 STEL

Canada - British Columbia - 8 Hour Exposure Limits
as Fe: 1 mg/m3 TWA

Canada - British Columbia - 15 Minute Exposure Limits
as Fe: 2 mg/m3 STEL

Canada - Ontario - OHSA - TWAEVs
as Fe: 1 mg/m3 TWAEV

Canada - Quebec - Time-Weighted Average Exposure Values
as Fe: 1 mg/m3 TWAEV

United Kingdom - Occupational Exposure Standards - TWAs
as Fe: 1 mg/m3 TWA

United Kingdom - Occupational Exposure Standards - STELs
as Fe: 2 mg/m3 STEL

Israel - Time Weighted Averages
as Fe: 1 mg/m3 TWA

Israel - Action Levels
as Fe: 0.5 mg/m3 AL

Mexico - Instruction No. 10 - TWAs
1 mg/m3 TWA

Mexico - Instruction No. 10 - STELs
2 mg/m3 STEL

STATE LISTS

California - Exposure Limits - PELs
as Fe: 1 mg/m3 PEL

California - Directors List of Hazardous Substances (8 CCR 339)
[present] (refers only to water-soluble salts not mixed in food or animal feed)

Minnesota Hazardous Substance List
[present] as Fe

Pennsylvania Right to Know List
environmental hazard

ISOAMYL BUTYRATE   106-27-4

HEALTH AND SAFETY LISTS

IARC - Group 3 (not classifiable)
[present]

ISOAMYL CHLORIDE   107-84-6

HEALTH AND SAFETY LISTS

NFPA - Flash Points
exothermic reaction above 502 degrees F (261 degrees C) may become explosive

NFPA - Hazard Identification Ratings
flammability-1; reactivity-3 (may explode above 502 degrees F (261 degrees C))

STATE LISTS

Florida Hazardous Substance List
[present]

Massachusetts Right To Know List
[present]

Pennsylvania Right to Know List
[present]

ISOAMYL MERCAPTAN   541-31-1

HEALTH AND SAFETY LISTS

IARC - Group 3 (not classifiable)
[present]

STATE LISTS

California - Directors List of Hazardous Substances (8 CCR 339)
[present]

ISOBENZAN   297-78-9

ENVIRONMENTAL LISTS

List of Pesticide Product Inert Ingredients
[present]

1,3-ISOBENZOFURANDIONE, 4,5,6,7-TETRABROMO-   632-79-1

HEALTH AND SAFETY LISTS

ACGIH 1995 - Time Weighted Averages
100 ppm TWA; 532 mg/m3 TWA

AIHA - Odor Threshold Values
geometric mean air odor threshold = 0.22 ppm (detectable)

U.S. DOT - Appendix A Table 1 - Hazardous Substances
final RQ = 5000 pounds (2270 kg) (Listed under 'Amyl acetate')

NFPA - Flash Points
flash point = 77 degrees F (25 degrees C)
NFPA - Hazard Identification Ratings
health-1; flammability-3; reactivity-0
NIOSH - Selected LD50s and LC50s
Oral, rat: LD50 = 16600 mg/kg
NIOSH 1990 - Pocket Guide - RELs
100 ppm TWA; 525 mg/m3 TWA
NIOSH 1990 - Pocket Guide - IDLHs
3000 ppm IDLH
NIOSH 1990 - Pocket Guide - Target organs
eyes, skin, respiratory system
OSHA - Vacated PELs - Time Weighted Averages
100 ppm TWA; 525 mg/m3 TWA
OSHA - Final PELs - Time Weighted Averages
100 ppm TWA; 525 mg/m3 TWA

ENVIRONMENTAL LISTS

CERCLA/SARA - Hazardous Substances and their Reportable Quantities
final RQ = 5000 pounds (2270 kg) (Listed under 'Amyl acetate')
Clean Water Act - Hazardous Substances
[present] (Listed under 'Amyl acetate')
TSCA - PAIR - Reporting List
Effective Date: January 26, 1994; Reporting Date: March 28, 1994
TSCA - Health and Safety Reporting List
Effective Date: January 26, 1994; Sunset Date: January 26, 2004

INTERNATIONAL LISTS

Australian Exposure Standards - Time Weighted Averages
100 ppm TWA; 532 mg/m3 TWA
Canada - WHMIS: Ingredient Disclosure
1% item 883 (20)
Canada - Alberta - 8 Hour Occupational Exposure Limit
100 ppm TWA; 533 mg/m3 TWA
Canada - Alberta - 15 Minute Occupational Exposure Limit
125 ppm STEL; 665 mg/m3 STEL
Canada - British Columbia - 8 Hour Exposure Limits
100 ppm TWA; 525 mg/m3 TWA
Canada - British Columbia - 15 Minute Exposure Limits
125 ppm STEL; 655 mg/m3 STEL
Canada - Ontario - OHSA - TWAEVs
100 ppm TWAEV; 530 mg/m3 TWAEV
Canada - Quebec - Time-Weighted Average Exposure Values
100 ppm TWAEV; 532 mg/m3 TWAEV
United Kingdom - Occupational Exposure Standards - TWAs
100 ppm TWA; 525 mg/m3 TWA
United Kingdom - Occupational Exposure Standards - STELs
125 ppm STEL; 655 mg/m3 STEL
Israel - Time Weighted Averages
100 ppm TWA; 532 mg/m3 TWA
Israel - Action Levels
50 ppm AL; 266 mg/m3 AL
Mexico - Instruction No. 10 - TWAs
100 ppm TWA; 525 mg/m3 TWA
Mexico - Instruction No. 10 - STELs
125 ppm STEL; 655 mg/m3 STEL

STATE LISTS

California - Exposure Limits - PELs
100 ppm PEL; 532 mg/m3 PEL
California - Directors List of Hazardous Substances (8 CCR 339)
[present] (Listed under 'Amyl acetate, all isomers')
Florida Hazardous Substance List
[present]

Massachusetts Right To Know List
[present]
Minnesota Hazardous Substance List
[present]
NJ Special Hazardous Substances
(flammable - third degree)
Pennsylvania Right to Know List
environmental hazard

PROPOSED REGULATIONS

TSCA - ITC 31st Report Priority Testing List
designated to be tested

ISOBORNYL ACETATE   125-12-2

HEALTH AND SAFETY LISTS

ACGIH 1995 - Time Weighted Averages
100 ppm TWA; 361 mg/m3 TWA
ACGIH 1995 - Short Term Exposure Limits
125 ppm STEL; 452 mg/m3 STEL
AIHA - Odor Threshold Values
no geometric mean air odor threshold
NFPA - Flash Points
flash point = 109 degrees F (43 degrees C)
NFPA - Hazard Identification Ratings
health-1; flammability-2; reactivity-0
NIOSH - Selected LD50s and LC50s
Oral, rat: LD50 = 1300 mg/kg Skin, rabbit: LD50 = 3212 mg/kg
NIOSH 1990 - Pocket Guide - RELs
100 ppm TWA; 360 mg/m3 TWA; 125 ppm STEL; 450 mg/m3 STEL (Listed under 'Isoamyl alcohol (primary and secondary)')
NIOSH 1990 - Pocket Guide - IDLHs
10,000 ppm IDLH (Listed under 'Isoamyl alcohol (primary and secondary)')
NIOSH 1990 - Pocket Guide - Target organs
eyes, skin, respiratory system (Listed under 'Isoamyl alcohol (primary and secondary)')
OSHA - Vacated PELs - Time Weighted Averages
100 ppm TWA; 360 mg/m3 TWA
OSHA - Vacated PELs - Short Term Exposure Limits
125 ppm STEL; 450 mg/m3 STEL
OSHA - Final PELs - Time Weighted Averages
100 ppm TWA; 360 mg/m3 TWA

INTERNATIONAL LISTS

Australian Exposure Standards - Time Weighted Averages
100 ppm TWA; 361 mg/m3 TWA
Australian Exposure Standards - Short Term Exposure Limits
125 ppm STEL; 452 mg/m3 STEL
Canada - WHMIS: Ingredient Disclosure
1% item 884 (178)
Canada - Alberta - 8 Hour Occupational Exposure Limit
100 ppm TWA; 360 mg/m3 TWA
Canada - Alberta - 15 Minute Occupational Exposure Limit
125 ppm STEL; 450 mg/m3 STEL
Canada - British Columbia - 8 Hour Exposure Limits
100 ppm TWA; 360 mg/m3 TWA
Canada - British Columbia - 15 Minute Exposure Limits
125 ppm STEL; 450 mg/m3 STEL
Canada - Ontario - OHSA - TWAEVs
100 ppm TWAEV; 360 mg/m3 TWAEV
Canada - Ontario - OHSA - STEVs
125 ppm STEV; 450 mg/m3 STEV
Canada - Quebec - Time-Weighted Average Exposure Values
100 ppm TWAEV; 361 mg/m3 TWAEV
Canada - Quebec - Short-term Exposure Values
125 ppm STEV; 452 mg/m3 STEV

United Kingdom - Occupational Exposure Standards - TWAs
100 ppm TWA; 360 mg/m3 TWA
United Kingdom - Occupational Exposure Standards - STELs
125 ppm STEL; 450 mg/m3 STEL
German (DFG) - MAK Values
100 ppm MAK; 360 mg/m3 MAK
German (DFG) - Peak Limitations
2 x normal MAK (30 min. average value); don't exceed 4 times during shift
German (DFG) - Pregnancy
no risk to embryo/fetus if exposure limits adhered to
Israel - Time Weighted Averages
100 ppm TWA; 361 mg/m3 TWA
Israel - Short Term Exposure Limits
125 ppm STEL; 452 mg/m3 STEL
Israel - Action Levels
50 ppm AL; 180.5 mg/m3 AL
Mexico - Instruction No. 10 - TWAs
100 ppm TWA; 360 mg/m3 TWA
Mexico - Instruction No. 10 - STELs
125 ppm STEL; 450 mg/m3 STEL

STATE LISTS

California - Exposure Limits - PELs
100 ppm PEL; 360 mg/m3 PEL
California - Exposure Limits - STELs
125 ppm STEL; 450 mg/m3 STEL
California - Directors List of Hazardous Substances (8 CCR 339)
[present]
Massachusetts Right To Know List
[present]
Minnesota Hazardous Substance List
[present]
NJ Right to Know List (Total)
sn 1039
Pennsylvania Right to Know List
[present]

ISOBORNYL THIOCYANOACETATE   115-31-1

HEALTH AND SAFETY LISTS

NIOSH 1990 - Pocket Guide - RELs
100 ppm TWA; 360 mg/m3 TWA; 125 ppm STEL; 450 mg/m3 STEL (Listed under 'Isoamyl alcohol (primary and secondary)')
NIOSH 1990 - Pocket Guide - IDLHs
10,000 ppm IDLH (Listed under 'Isoamyl alcohol (primary and secondary)')
NIOSH 1990 - Pocket Guide - Target organs
eyes, skin, respiratory system (Listed under 'Isoamyl alcohol (primary and secondary)')

ISOBUTANE   75-28-5

HEALTH AND SAFETY LISTS

NFPA - Flash Points
flash point = 138 degrees F (59 degrees C)
NFPA - Hazard Identification Ratings
flammability-2

INTERNATIONAL LISTS

Canada - WHMIS: Ingredient Disclosure
1% item 885 (374)

ISOBUTYL ACETATE   110-19-0

HEALTH AND SAFETY LISTS

NFPA - Flash Points
flash point < 70 degrees F (21 degrees C)

NFPA - Hazard Identification Ratings
flammability-3

STATE LISTS

Florida Hazardous Substance List
[present]
Massachusetts Right To Know List
[present]
Pennsylvania Right to Know List
[present]

ISOBUTYL ACRYLATE   106-63-8

HEALTH AND SAFETY LISTS

U.S. DOT - Appendix B - Marine Pollutants
DOT regulated marine pollutant

ISOBUTYL ALCOHOL   78-83-1

HEALTH AND SAFETY LISTS

NIOSH - Selected LD50s and LC50s
Oral, rat: LD50 = 4800 ug/kg Skin, rat: LD50 = 5 mg/kg

ENVIRONMENTAL LISTS

CERCLA/SARA - Section 302 Extremely Hazardous Substances and TPQs
TPQ = 100/10,000 pounds

STATE LISTS

Florida Hazardous Substance List
effective March 13, 1992
Massachusetts Right To Know List
extraordinarily hazardous
NJ Right to Know List (Total)
sn 2494
Pennsylvania Right to Know List
environmental hazard

PROPOSED REGULATIONS

CERCLA/SARA - Proposed Hazardous Substance Additions
proposed RQ = 1 pound (.454 kg)
CERCLA/SARA - 1989 Proposed RQ Adjustments
proposed RQ = 100 pounds (45.4 kg)

ISOBUTYLAMINE   78-81-9

ENVIRONMENTAL LISTS

CAA - HON Rule - SOCMI Chemicals
compliance by April 24, 1995
TSCA - Code of Federal Regulations Citations
40 CFR 712.30(w); 40 CFR 716.120(a)
TSCA - PAIR - Reporting List
Reporting Dates: November 19, 1982; March 12, 1990
TSCA - Health and Safety Reporting List
Effective Date: January 11, 1990

ISOBUTYLATED, STYRENEATED CRESOLS   68457-75-0

HEALTH AND SAFETY LISTS

NFPA - Flash Points
flash point = 190 degrees F (88 degrees C)
NFPA - Hazard Identification Ratings
health-1; flammability-2; reactivity-0

ENVIRONMENTAL LISTS

List of Pesticide Product Inert Ingredients
[present]

ISOBUTYLBENZENE   538-93-2

INTERNATIONAL LISTS

Canada - WHMIS: Ingredient Disclosure
1% item 886 (1610)

ISOBUTYL BUTYRATE   539-90-2

HEALTH AND SAFETY LISTS

U.S. DOT - Substances From 49 CFR 172.101
regulated by DOT (UN1969)

U.S. DOT - Hazard Classes
DOT hazard class = 2.1

NFPA - Flash Points
gas (no flash point given)

NFPA - Hazard Identification Ratings
health-1; flammability-4; reactivity-0

ENVIRONMENTAL LISTS

CAA - Flammable Substances for Accidental Release Prevention
threshold quantity = 10,000 lbs

List of Pesticide Product Inert Ingredients
[present]

INTERNATIONAL LISTS

German (DFG) - MAK Values
1000 ppm MAK; 2350 mg/m3 MAK (Listed under 'Butane')

German (DFG) - Peak Limitations
2 x normal MAK (1 hour momentary value); don't exceed 3 times per shift (Listed under 'Butane')

STATE LISTS

Massachusetts Right To Know List
[present]

NJ Right to Know List (Total)
sn 1040

NJ Special Hazardous Substances
(flammable - fourth degree)

Pennsylvania Right to Know List
[present]

ISOBUTYL CHLORIDE   513-36-0

HEALTH AND SAFETY LISTS

ACGIH 1995 - Time Weighted Averages
150 ppm TWA; 713 mg/m3 TWA

AIHA - Odor Threshold Values
geometric mean air odor threshold = 1.1 ppm (detectable); 1.9 ppm (recognizable)

U.S. DOT - Substances From 49 CFR 172.101
regulated by DOT (UN1213)

U.S. DOT - Hazard Classes
DOT hazard class = 3

U.S. DOT - Appendix A Table 1 - Hazardous Substances
final RQ = 5000 pounds (2270 kg) (Listed under 'Butyl acetate')

NFPA - Flash Points
flash point = 64 degrees F (18 degrees C)

NFPA - Hazard Identification Ratings
health-1; flammability-3; reactivity-0

NIOSH - Selected LD50s and LC50s
Oral, rat: LD50 = 13400 mg/kg

NIOSH 1990 - Pocket Guide - RELs
150 ppm TWA; 700 mg/m3 TWA

NIOSH 1990 - Pocket Guide - IDLHs
7500 ppm IDLH

NIOSH 1990 - Pocket Guide - Target organs
eyes, skin, respiratory system

OSHA - Vacated PELs - Time Weighted Averages
150 ppm TWA; 700 mg/m3 TWA

OSHA - Final PELs - Time Weighted Averages
150 ppm TWA; 700 mg/m3 TWA

ENVIRONMENTAL LISTS

CERCLA/SARA - Hazardous Substances and their Reportable Quantities
final RQ = 5000 pounds (2270 kg) (Listed under 'Butyl acetate')

Clean Water Act - Hazardous Substances
[present] (Listed under 'Butyl acetate')

List of Pesticide Product Inert Ingredients
[present]

INTERNATIONAL LISTS

Australian Exposure Standards - Time Weighted Averages
150 ppm TWA; 713 mg/m3 TWA

Canada - WHMIS: Ingredient Disclosure
1% item 888 (21)

Canada - Alberta - 8 Hour Occupational Exposure Limit
150 ppm TWA; 713 mg/m3 TWA

Canada - Alberta - 15 Minute Occupational Exposure Limit
187 ppm STEL; 889 mg/m3 STEL

Canada - British Columbia - 8 Hour Exposure Limits
150 ppm TWA; 700 mg/m3 TWA

Canada - British Columbia - 15 Minute Exposure Limits
187 ppm STEL; 875 mg/m3 STEL

Canada - Ontario - OHSA - TWAEVs
150 ppm TWAEV; 710 mg/m3 TWAEV

Canada - Ontario - OHSA - STEVs
187 ppm STEV; 887 mg/m3 STEV

Canada - Quebec - Time-Weighted Average Exposure Values
150 ppm TWAEV; 713 mg/m3 TWAEV

United Kingdom - Occupational Exposure Standards - TWAs
150 ppm TWA; 700 mg/m3 TWA

United Kingdom - Occupational Exposure Standards - STELs
187 ppm STEL; 875 mg/m3 STEL

German (DFG) - MAK Values
200 ppm MAK; 950 mg/m3 MAK (Listed under 'Butyl acetate')

German (DFG) - Peak Limitations
2 x normal MAK (5 min momentary value); don't exceed 8 times during shift (Listed under 'Butyl acetate')

Israel - Time Weighted Averages
150 ppm TWA; 713 mg/m3 TWA

Israel - Action Levels
75 ppm AL; 356.5 mg/m3 AL

Mexico - Instruction No. 10 - TWAs
150 ppm TWA; 700 mg/m3 TWA

Mexico - Instruction No. 10 - STELs
187 ppm STEL; 875 mg/m3 STEL

STATE LISTS

California - Exposure Limits - PELs
150 ppm PEL; 700 mg/m3 PEL

California - Directors List of Hazardous Substances (8 CCR 339)
[present]

Florida Hazardous Substance List
[present]

Massachusetts Right To Know List
[present]

Minnesota Hazardous Substance List
[present]

NJ Right to Know List (Total)
sn 1041

NJ Special Hazardous Substances
(flammable - third degree)

Pennsylvania Right to Know List
environmental hazard

ISOBUTYL CHLOROFORMATE   543-27-1

HEALTH AND SAFETY LISTS

U.S. DOT - Substances From 49 CFR 172.101
regulated by DOT (UN2527)

U.S. DOT - Hazard Classes
DOT hazard class = 3

NFPA - Flash Points
flash point = 86 degrees F (30 degrees C)

NFPA - Hazard Identification Ratings
health-1; flammability-3; reactivity-1

NIOSH - Selected LD50s and LC50s
Oral, rat: LD50 = 7070 mg/kg Skin, rabbit: LD50 = 890 mg/kg

ENVIRONMENTAL LISTS

CAA - HON Rule - SOCMI Chemicals
compliance by Oct. 23, 1995

TSCA - Code of Federal Regulations Citations
40 CFR 712.30(d),(x)

TSCA - PAIR - Reporting List
Reporting Dates: November 19, 1982, November 27, 1991

TSCA - Health and Safety Reporting List
Effective Date: September 30, 1991

STATE LISTS

Florida Hazardous Substance List
[present]

Massachusetts Right To Know List
[present]

NJ Right to Know List (Total)
sn 1042

NJ Special Hazardous Substances
(flammable - third degree)

Pennsylvania Right to Know List
[present]

ISOBUTYLCYCLOHEXANE   RR-00863-9

SEE ALSO:
F005-HAZARDOUS WASTES
F039-HAZARDOUS WASTES

HEALTH AND SAFETY LISTS

ACGIH 1995 - Time Weighted Averages
50 ppm TWA; 152 mg/m3 TWA

AIHA - Odor Threshold Values
geometric mean air odor threshold = 3.6 ppm (detectable); 9.8 ppm (recognizable)

U.S. DOT - Substances From 49 CFR 172.101
regulated by DOT (UN1212)

U.S. DOT - Hazard Classes
DOT hazard class = 3

U.S. DOT - Appendix A Table 1 - Hazardous Substances
final RQ = 5000 pounds (2270 kg)

NFPA - Flash Points
flash point = 82 degrees F (28 degrees C)

NFPA - Hazard Identification Ratings
health-1; flammability-3; reactivity-0

NIOSH - Selected LD50s and LC50s
Oral, rat: LD50 = 2460 mg/kg Skin, rabbit: LD50 = 3400 mg/kg

NIOSH 1990 - Pocket Guide - RELs
50 ppm TWA; 150 mg/m3 TWA

NIOSH 1990 - Pocket Guide - IDLHs
8000 ppm IDLH

NIOSH 1990 - Pocket Guide - Target organs
eyes, skin, respiratory system

OSHA - Vacated PELs - Time Weighted Averages
50 ppm TWA; 150 mg/m3 TWA

OSHA - Final PELs - Time Weighted Averages
100 ppm TWA; 300 mg/m3 TWA

ENVIRONMENTAL LISTS

CERCLA/SARA - Hazardous Substances and their Reportable Quantities
final RQ = 5000 pounds (2270 kg)

EPA - Master Testing List
[present]

List of Pesticide Product Inert Ingredients
[present]

RCRA - U Series Wastes
waste number U140 (Ignitable waste; Toxic waste)

RCRA - Hazardous Constituents-Appendix VIII
waste number U140

RCRA - Basis for Listing - Appendix VII
Included in waste streams: F005, F039

RCRA - Substances Banned From Land Disposal
[present]

RCRA - TSD Facilities Ground Water Monitoring
TM 8015 = 50 ug/L PQL

RCRA - Universal Treatment Standards (LDR)
WW: 5.6 mg/l; NWW: 170 mg/kg

TSCA - Code of Federal Regulations Citations
40 CFR 716.120(a)

TSCA - Health and Safety Reporting List
Effective Date: March 7, 1986

TSCA - Multichemical Test Rules - Neurotoxicity
administrative stay for neurotoxicity tests effective June 27, 1994

INTERNATIONAL LISTS

Australian Exposure Standards - Time Weighted Averages
50 ppm TWA; 152 mg/m3 TWA

Canada - WHMIS: Ingredient Disclosure
1% item 887 (1032)

Canada - Alberta - 8 Hour Occupational Exposure Limit
50 ppm TWA; 152 mg/m3 TWA

Canada - Alberta - 15 Minute Occupational Exposure Limit
75 ppm STEL; 227 mg/m3 STEL

Canada - British Columbia - 8 Hour Exposure Limits
50 ppm TWA; 150 mg/m3 TWA

Canada - British Columbia - 15 Minute Exposure Limits
75 ppm STEL; 225 mg/m3 STEL

Canada - Ontario - OHSA - TWAEVs
50 ppm TWAEV; 150 mg/m3 TWAEV

Canada - Quebec - Time-Weighted Average Exposure Values
50 ppm TWAEV; 152 mg/m3 TWAEV

United Kingdom - Occupational Exposure Standards - TWAs
50 ppm TWA; 150 mg/m3 TWA

United Kingdom - Occupational Exposure Standards - STELs
75 ppm STEL; 225 mg/m3 STEL

German (DFG) - MAK Values
100 ppm MAK; 300 mg/m3 MAK (Listed under 'Butyl alcohol')

German (DFG) - Pregnancy
no risk to embryo/fetus if exposure limits adhered to

Israel - Time Weighted Averages
50 ppm TWA; 152 mg/m3 TWA

Israel - Action Levels
25 ppm AL; 76 mg/m3 AL

Mexico - Instruction No. 10 - TWAs
50 ppm TWA; 150 mg/m3 TWA

Mexico - Instruction No. 10 - STELs
75 ppm STEL; 225 mg/m3 STEL

STATE LISTS

California - Exposure Limits - PELs
50 ppm PEL; 150 mg/m3 PEL

California - Directors List of Hazardous Substances (8 CCR 339)
[present]

Florida Hazardous Substance List
[present]

Massachusetts Right To Know List
[present]

Minnesota Hazardous Substance List
[present]

NJ Right to Know List (Total)
sn 1043

NJ Special Hazardous Substances
(flammable - third degree)

Pennsylvania Right to Know List
environmental hazard

ISOBUTYLENE   115-11-7

HEALTH AND SAFETY LISTS

U.S. DOT - Substances From 49 CFR 172.101
regulated by DOT (UN1214)

U.S. DOT - Hazard Classes
DOT hazard class = 3

U.S. DOT - Appendix A Table 1 - Hazardous Substances
final RQ = 1000 pounds (454 kg) (Listed under 'Butylamine')

NFPA - Flash Points
flash point = 15 degrees F (-9 degrees C)

NFPA - Hazard Identification Ratings
health-2; flammability-3; reactivity-0

NIOSH - Selected LD50s and LC50s
Oral, rat: LD50 = 228 mg/kg

ENVIRONMENTAL LISTS

CERCLA/SARA - Hazardous Substances and their Reportable Quantities
final RQ = 1000 pounds (454 kg) (Listed under 'Butylamine')

Clean Water Act - Hazardous Substances
[present] (Listed under 'Butylamine')

INTERNATIONAL LISTS

German (DFG) - MAK Values
5 ppm MAK; 15 mg/m3 MAK (Listed under 'Butylamine')

German (DFG) - Peak Limitations
5 x normal MAK (30 min. average value); don't exceed 2 times during shift (Listed under 'Butylamine')

German (DFG) - Skin/Sensitizers
danger of cutaneous absorption (Listed under 'Butylamine')

STATE LISTS

California - Directors List of Hazardous Substances (8 CCR 339)
[present] (Listed under 'Butylamine, all isomers')

Florida Hazardous Substance List
[present]

Massachusetts Right To Know List
[present]

NJ Right to Know List (Total)
sn 1044

NJ Special Hazardous Substances
(flammable - third degree)

Pennsylvania Right to Know List
environmental hazard

ISOBUTYL FORMATE   542-55-2

ENVIRONMENTAL LISTS

List of Pesticide Product Inert Ingredients
[present]

ISOBUTYL HEPTYL KETONE   123-18-2

HEALTH AND SAFETY LISTS

NFPA - Flash Points
flash point = 131 degrees F (55 degrees C)

NFPA - Hazard Identification Ratings
health-2; flammability-2; reactivity-0

STATE LISTS

Florida Hazardous Substance List
[present]

Massachusetts Right To Know List
[present]

Pennsylvania Right to Know List
[present]

ISOBUTYL ISOBUTYRATE   97-85-8

HEALTH AND SAFETY LISTS

NFPA - Flash Points
flash point = 122 degrees F (50 degrees C)

NFPA - Hazard Identification Ratings
health-0; flammability-2

ISOBUTYL ISOCYANATE   1873-29-6

HEALTH AND SAFETY LISTS

NFPA - Flash Points
flash point < 70 degrees F (21 degrees C)

NFPA - Hazard Identification Ratings
health-2; flammability-3; reactivity-0

STATE LISTS

Florida Hazardous Substance List
[present]

Massachusetts Right To Know List
[present]

Pennsylvania Right to Know List
[present]

ISOBUTYL METHACRYLATE   97-86-9

HEALTH AND SAFETY LISTS

U.S. DOT - Substances From 49 CFR 172.101
regulated by DOT (NA2742)

U.S. DOT - Hazard Classes
DOT hazard class = 6.1

U.S. DOT - Substances Which Are Poisonous by Inhalation
liquid hazardous material poisonous by inhalation (NA2742)

ISOBUTYLNAPTHALENESULFONIC ACID, SODIUM SALT   28348-65-4

HEALTH AND SAFETY LISTS

NFPA - Hazard Identification Ratings
health-0; reactivity-0

ISOBUTYL NITRITE   542-56-3

HEALTH AND SAFETY LISTS

U.S. DOT - Substances From 49 CFR 172.101
regulated by DOT (UN1055)

U.S. DOT - Hazard Classes
DOT hazard class = 2.1

NFPA - Flash Points
gas (no flash point given)

NFPA - Hazard Identification Ratings
health-1; flammability-4; reactivity-0
NIOSH - Selected LD50s and LC50s
Inhalation, rat: LC50 = 620 gm/m3 4 hr
NTP Chemical Status Reports - Testing Status and NTIS Number
Prechronic studies completed; chemicals in review for further evaluation
ENVIRONMENTAL LISTS
CAA - Flammable Substances for Accidental Release Prevention
threshold quantity = 10,000 lbs
CAA - HON Rule - SOCMI Chemicals
compliance by Oct. 23, 1995
EPA - Master Testing List
[present]
STATE LISTS
Florida Hazardous Substance List
[present]
Massachusetts Right To Know List
[present]
NJ Right to Know List (Total)
sn 1045
NJ Special Hazardous Substances
(flammable - fourth degree)
Pennsylvania Right to Know List
[present]

ISOBUTYL PHENYLACETATE   102-13-6

HEALTH AND SAFETY LISTS
U.S. DOT - Substances From 49 CFR 172.101
regulated by DOT (UN2393)
U.S. DOT - Hazard Classes
DOT hazard class = 3
NFPA - Flash Points
flash point < 70 degrees F (21 degrees C)
NFPA - Hazard Identification Ratings
flammability-3
NIOSH - Selected LD50s and LC50s
Oral, rabbit: LD50 = 3064 mg/kg
STATE LISTS
Florida Hazardous Substance List
[present]
Massachusetts Right To Know List
[present]
NJ Right to Know List (Total)
sn 1046
NJ Special Hazardous Substances
(flammable - third degree)
Pennsylvania Right to Know List
[present]

ISOBUTYL PHOSPHATE   126-71-6

HEALTH AND SAFETY LISTS
NFPA - Flash Points
flash point = 195 degrees F (91 degrees C)
NFPA - Hazard Identification Ratings
health-2; flammability-2; reactivity-0
STATE LISTS
Florida Hazardous Substance List
[present]
Massachusetts Right To Know List
[present]
Pennsylvania Right to Know List
[present]

ISOBUTYL PROPIONATE   540-42-1

HEALTH AND SAFETY LISTS
U.S. DOT - Substances From 49 CFR 172.101
regulated by DOT (UN2528)
U.S. DOT - Hazard Classes
DOT hazard class = 3
NFPA - Flash Points
flash point = 101 degrees F (38 degrees C)
NFPA - Hazard Identification Ratings
health-0; flammability-2; reactivity-0
NIOSH - Selected LD50s and LC50s
Inhalation, rat: LC50 = 5000 ppm 6 hr Oral, rat: LD50 = 12800 mg/kg
STATE LISTS
NJ Right to Know List (Total)
sn 1047

ISOBUTYRALDEHYDE   78-84-2

HEALTH AND SAFETY LISTS
U.S. DOT - Substances From 49 CFR 172.101
regulated by DOT (UN2486)
U.S. DOT - Hazard Classes
DOT hazard class = 3
U.S. DOT - Substances Which Are Poisonous by Inhalation
liquid hazardous material poisonous by inhalation (UN2486)
INTERNATIONAL LISTS
Canada - WHMIS: Ingredient Disclosure
0.1% item 889 (1041)
STATE LISTS
NJ Right to Know List (Total)
sn 1048

ISOBUTYRIC ACID   79-31-2

HEALTH AND SAFETY LISTS
U.S. DOT - Substances From 49 CFR 172.101
regulated by DOT (UN2283)
U.S. DOT - Hazard Classes
DOT hazard class = 3
NIOSH - Selected LD50s and LC50s
Oral, mouse: LD50 = 11990 mg/kg
ENVIRONMENTAL LISTS
TSCA - Code of Federal Regulations Citations
40 CFR 712.30(d)
TSCA - PAIR - Reporting List
Reporting Date: November 19, 1982
STATE LISTS
NJ Right to Know List (Total)
sn 1049

ISOBUTYRIC ANHYDRIDE   97-72-3

ENVIRONMENTAL LISTS
List of Pesticide Product Inert Ingredients
[present]

ISOBUTYRONITRILE   78-82-0

HEALTH AND SAFETY LISTS
NTP Chemical Status Reports - Testing Status and NTIS Number
Two year studies scheduled for peer review

ISOBUTYRYL CHLORIDE   79-30-1

HEALTH AND SAFETY LISTS

NFPA - Flash Points
flash point > 212 degrees F (100 degrees C)
NFPA - Hazard Identification Ratings
health-0; flammability-1; reactivity-0

ISOCYANATES, ALL   RR-00294-8

HEALTH AND SAFETY LISTS

NFPA - Flash Points
flash point = 275 degrees F (135 degrees C)
NFPA - Hazard Identification Ratings
flammability-1

ENVIRONMENTAL LISTS

TSCA - Code of Federal Regulations Citations
40 CFR 712.30(x); 40 CFR 716.120(d)
TSCA - PAIR - Reporting List
Reporting Date: December 27, 1990
TSCA - Health and Safety Reporting List
Effective Date: October 29, 1990; Sunset Date: November 9, 1993

ISOCYANATE TERMINATED POLYOLS   RR-00955-2

HEALTH AND SAFETY LISTS

U.S. DOT - Substances From 49 CFR 172.101
regulated by DOT (UN2394)
U.S. DOT - Hazard Classes
DOT hazard class = 3
NIOSH - Selected LD50s and LC50s
Oral, rabbit: LD50 = 5599 mg/kg

STATE LISTS

NJ Right to Know List (Total)
sn 1050

ISOCYANATOBENZOTRIFLUORIDE   71121-36-3

HEALTH AND SAFETY LISTS

U.S. DOT - Substances From 49 CFR 172.101
regulated by DOT (UN2045)
U.S. DOT - Hazard Classes
DOT hazard class = 3
NFPA - Flash Points
flash point = -1 degrees F (-18 degrees C)
NFPA - Hazard Identification Ratings
health-2; flammability-3; reactivity-1
NIOSH - Selected LD50s and LC50s
Inhalation, mouse: LC50 = 39500 mg/m3 (2 hr) Oral, rat: LD50 = 2810 mg/kg Skin, rabbit: LD50 = 7130 mg/kg
NTP Chemical Status Reports - Testing Status and NTIS Number
Prechronic studies for which technical toxicity reports were not prepared; two year studies: pathology quality assessment in progress

ENVIRONMENTAL LISTS

CERCLA/SARA - Section 313 - Emission Reporting
form R reporting required for 1.0% de minimus concentration
EPA - Master Testing List
[present]
TSCA - Code of Federal Regulations Citations
40 CFR 712.30(x); 40 CFR 716.120(d)
TSCA - PAIR - Reporting List
Reporting Date: November 27, 1991
TSCA - Health and Safety Reporting List
Effective Date: September 30, 1991

INTERNATIONAL LISTS

Canada - WHMIS: Ingredient Disclosure
1% item 890 (1033)
Canada - NPRI (National Pollutant Release Inventory)
[present]

STATE LISTS

California - Air Bill 2588 Appendix A-II
6/91
Florida Hazardous Substance List
[present]
Massachusetts Right To Know List
[present]
NJ Right to Know List (Total)
sn 1051
Pennsylvania Right to Know List
environmental hazard

4-ISOCYANATO-N,N-BIS(4-ISOCYANATOPHENYL)-2,   106790-31-2
5-DIMETHOXYBENZENEAMINE

HEALTH AND SAFETY LISTS

U.S. DOT - Substances From 49 CFR 172.101
regulated by DOT (UN2529)
U.S. DOT - Hazard Classes
DOT hazard class = 3
U.S. DOT - Appendix A Table 1 - Hazardous Substances
final RQ = 5000 pounds (2270 kg) (Listed under 'Butyric acid')
NFPA - Flash Points
flash point = 132 degrees F (56 degrees C)
NFPA - Hazard Identification Ratings
health-1; flammability-2; reactivity-0
NIOSH - Selected LD50s and LC50s
Oral, rat: LD50 = 280 mg/kg Skin, rabbit: LD50 = 500 mg/kg

ENVIRONMENTAL LISTS

CERCLA/SARA - Hazardous Substances and their Reportable Quantities
final RQ = 5000 pounds (2270 kg) (Listed under 'Butyric acid')
Clean Water Act - Hazardous Substances
[present] (Listed under 'Butyric acid')
EPA - Master Testing List
[present]

INTERNATIONAL LISTS

Canada - WHMIS: Ingredient Disclosure
1% item 891 (95)

STATE LISTS

Massachusetts Right To Know List
[present]
NJ Right to Know List (Total)
sn 1052
NJ Special Hazardous Substances
(corrosive)
Pennsylvania Right to Know List
environmental hazard

PROPOSED REGULATIONS

TSCA - Proposed Substances for Developmental/Reproductive Testing
proposed testing for: Developmental Toxicity - oral

ISOCYANIC ACID, 3,4-DICHLOROPHENYL ESTER   102-36-3

HEALTH AND SAFETY LISTS

U.S. DOT - Substances From 49 CFR 172.101
regulated by DOT (UN2530)
U.S. DOT - Hazard Classes
DOT hazard class = 3
NFPA - Flash Points
flash point = 139 degrees F (59 degrees C)
NFPA - Hazard Identification Ratings
health-1; flammability-2; reactivity-1 (avoid use of water)

INTERNATIONAL LISTS

Canada - WHMIS: Ingredient Disclosure
1% item 892 (230)

STATE LISTS

Massachusetts Right To Know List
[present]

NJ Right to Know List (Total)
sn 1053

NJ Special Hazardous Substances
(corrosive)

ISOCYANIC ACID, TRIMETHYLCYCLOHEXYL ESTER   32052-51-0

HEALTH AND SAFETY LISTS

U.S. DOT - Substances From 49 CFR 172.101
regulated by DOT (UN2284)

U.S. DOT - Hazard Classes
DOT hazard class = 3

NFPA - Flash Points
flash point = 47 degrees F (8 degrees C)

NFPA - Hazard Identification Ratings
health-3; flammability-3; reactivity-0

NIOSH - Selected LD50s and LC50s
Oral, rat: LD50 = 102 mg/kg Skin, rabbit: LD50 = 310 mg/kg

NIOSH - Health Standards - Exposure Limits
8 ppm TWA; 22 mg/m3 TWA (Listed under 'Nitriles')

NIOSH - Health Standards - Health Effects and Precautions
Hepatic, renal, respiratory, cardiovascular, gastrointestinal, and nervous system effects (Periodic chest X-ray and pulmonary function testing required; prevent skin and eye contact; make first-aid kits and personnel available during use) (Listed under 'Nitriles')

ENVIRONMENTAL LISTS

CERCLA/SARA - Section 302 Extremely Hazardous Substances and TPQs
TPQ = 1000 pounds

CAA -Toxic Substances for Accidental Release Prevention
threshold quantity = 20,000 lbs

STATE LISTS

Florida Hazardous Substance List
[present]

Massachusetts Right To Know List
extraordinarily hazardous

Minnesota Hazardous Substance List
[present]

NJ Right to Know List (Total)
sn 1054

NJ Special Hazardous Substances
(flammable - third degree)

Pennsylvania Right to Know List
environmental hazard

PROPOSED REGULATIONS

CERCLA/SARA - Proposed Hazardous Substance Additions
proposed RQ = 1 pound (.454 kg)

CERCLA/SARA - 1989 Proposed RQ Adjustments
proposed RQ = 1000 pounds (454 kg)

ISOCYANURIC ACID   108-80-5

HEALTH AND SAFETY LISTS

U.S. DOT - Substances From 49 CFR 172.101
regulated by DOT (UN2395)

U.S. DOT - Hazard Classes
DOT hazard class = 3

INTERNATIONAL LISTS

Canada - WHMIS: Ingredient Disclosure
1% item 893 (503)

STATE LISTS

NJ Right to Know List (Total)
sn 1055

ISODECALDEHYDE   RR-00864-0

HEALTH AND SAFETY LISTS

U.S. DOT - Substances From 49 CFR 172.101
regulated by DOT (UN2206, UN2207, UN2478, UN3080)

U.S. DOT - Hazard Classes
flashpoint less than 23 degrees C: DOT hazard class = 3; flashpoint greater than or equal to 23 degrees C: DOT hazard class = 6.1

U.S. DOT - Substances Which Are Poisonous by Inhalation
liquid hazardous material poisonous by inhalation (alone or in solution) (UN2478)

INTERNATIONAL LISTS

Australian Exposure Standards - Time Weighted Averages
as NCO: 0.02 mg/m3 TWA

Australian Exposure Standards - Short Term Exposure Limits
as NCO: 0.07 mg/m3 STEL

Australian Exposure Standards - Skin Effects
as NCO: sensitiser

United Kingdom - Maximum Exposure Limits - TWAs
as NCO: 0.02 mg/m3 TWA

United Kingdom - Maximum Exposure Limits - STELs
as NCO: 0.07 mg/m3 STEL

United Kingdom - Maximum Exposure Limits - Notes
capable of causing respiratory sensitisation

STATE LISTS

California - Air Bill 2588 Appendix A-I
[present]

NJ Right to Know List (Total)
sn 2495; sn 2496; sn 2497

PROPOSED REGULATIONS

TSCA - ITC 33rd Report Priority Testing List
recommended with intent-to-designate

TSCA - ITC 34th Report Priority Testing List
recommended with intent-to-designate

ISODECANE   RR-00854-8

ENVIRONMENTAL LISTS

TSCA - Chemicals with Significant New Use Rules
PMN numbers: P-90-404; P-90-405; P-90-406

ISODECANOIC ACID   26403-17-8

HEALTH AND SAFETY LISTS

U.S. DOT - Substances From 49 CFR 172.101
regulated by DOT (UN2285)

U.S. DOT - Hazard Classes
DOT hazard class = 6.1

U.S. DOT - Substances Which Are Poisonous by Inhalation
liquid hazardous material poisonous by inhalation (UN2285)

INTERNATIONAL LISTS

Canada - WHMIS: Ingredient Disclosure
1% item 894 (916)

STATE LISTS

NJ Right to Know List (Total)
sn 1056

ISODECANOL   25339-17-7

ENVIRONMENTAL LISTS

TSCA - Section 12(b) - Export Notification
P-92-168; export notification required - Section 5

ISODECYL NEOPENTANOATE   60209-82-7
SEE ALSO:
DICHLOROPHENYL ISOCYANATE, ALL ISOMERS

ENVIRONMENTAL LISTS

CERCLA/SARA - Section 302 Extremely Hazardous Substances and TPQs
TPQ = 500/10,000 pounds
TSCA - Code of Federal Regulations Citations
40 CFR 712.30(x); 40 CFR 716.120(d)
TSCA - PAIR - Reporting List
Reporting Date: December 27, 1990
TSCA - Health and Safety Reporting List
Effective Date: October 29, 1990

STATE LISTS

Florida Hazardous Substance List
effective March 13, 1992
Massachusetts Right To Know List
extraordinarily hazardous
NJ Right to Know List (Total)
sn 0658
Pennsylvania Right to Know List
environmental hazard

PROPOSED REGULATIONS

CERCLA/SARA - Proposed Hazardous Substance Additions
proposed RQ = 1 pound (.454 kg)
CERCLA/SARA - 1989 Proposed RQ Adjustments
proposed RQ = 100 pounds (45.4 kg)

ISODODECANE   31807-55-3

ENVIRONMENTAL LISTS

TSCA - Code of Federal Regulations Citations
40 CFR 716.120(a)
TSCA - Health and Safety Reporting List
Effective Date: June 1, 1987; Sunset Date: November 9, 1993

ISODRIN   465-73-6

HEALTH AND SAFETY LISTS

AIHA - WEEL - Time Weighted Averages
total: 1.9 ppm TWA; 10 mg/m3 TWA; respirable: 0.95 ppm TWA; 5 mg/m3 TWA
NIOSH - Selected LD50s and LC50s
Oral, rat: LD50 = 7700 mg/kg

ENVIRONMENTAL LISTS

List of Pesticide Product Inert Ingredients
[present]

INTERNATIONAL LISTS

Canada - WHMIS: Ingredient Disclosure
1% item 895 (96)

STATE LISTS

Minnesota Hazardous Substance List
[present]

ISOEUGENOL   97-54-1

HEALTH AND SAFETY LISTS

NFPA - Flash Points
flash point = 185 degrees F (85 degrees C)

NFPA - Hazard Identification Ratings
health-0; flammability-2; reactivity-0

ISOFENPHOS   RR-01290-8

HEALTH AND SAFETY LISTS

NFPA - Hazard Identification Ratings
health-0; flammability-2; reactivity-0

ISOFENPHOS [2-[[ETHOXYL[(1-METHYLETHYL)   25311-71-1
AMINO]PHOSPHINOTHIOYL]OXY]BENZOIC ACID 1-METHYLETHYL ESTER]

HEALTH AND SAFETY LISTS

NFPA - Flash Points
flash point = 300 degrees F (149 degrees C)
NFPA - Hazard Identification Ratings
health-0; flammability-1; reactivity-0

ISOFLUORPHATE   55-91-4

HEALTH AND SAFETY LISTS

NFPA - Flash Points
flash point = 220 degrees F (104 degrees C)
NFPA - Hazard Identification Ratings
health-0; flammability-1; reactivity-0
NIOSH - Selected LD50s and LC50s
Oral, rat: LD50 = 6400 mg/kg Skin, rabbit: LD50 = 3150 mg/kg

ENVIRONMENTAL LISTS

List of Pesticide Product Inert Ingredients
[present]

ISOFLURANE   26675-46-7

ENVIRONMENTAL LISTS

List of Pesticide Product Inert Ingredients
[present]

ISOHEPTANE   591-76-4

STATE LISTS

NJ Right to Know List (Total)
sn 2498

ISOHEPTANE   31394-54-4
SEE ALSO:
F039-HAZARDOUS WASTES

HEALTH AND SAFETY LISTS

U.S. DOT - Appendix A Table 1 - Hazardous Substances
final RQ = 1 pound (0.454 kg)
NIOSH - Selected LD50s and LC50s
Oral, rat: LD50 = 7 mg/kg Skin, rat: LD50 = 23 mg/kg

ENVIRONMENTAL LISTS

CERCLA/SARA - Section 302 Extremely Hazardous Substances and TPQs
TPQ = 100/10,000 pounds
CERCLA/SARA - Section 313 - Emission Reporting
form R reporting required
CERCLA/SARA - Hazardous Substances and their Reportable Quantities
final RQ = 1 pound (0.454 kg)
RCRA - P Series Wastes
waste number P060
RCRA - Hazardous Constituents-Appendix VIII
waste number P060
RCRA - Basis for Listing - Appendix VII
Included in waste stream: F039
RCRA - Substances Banned From Land Disposal
[present]

RCRA - TSD Facilities Ground Water Monitoring
TM 8270 = 10 ug/L PQL

RCRA - Universal Treatment Standards (LDR)
WW: 0.021 mg/l; NWW: 0.066 mg/kg

TSCA - Code of Federal Regulations Citations
40 CFR 704.102

STATE LISTS

Florida Hazardous Substance List
effective March 13, 1992

Massachusetts Right To Know List
extraordinarily hazardous

NJ Right to Know List (Total)
sn 2499

Pennsylvania Right to Know List
environmental hazard

ISOHEPTENE   68975-47-3

HEALTH AND SAFETY LISTS

NFPA - Flash Points
flash point > 212 degrees F (100 degrees C)

NFPA - Hazard Identification Ratings
health-0; flammability-1; reactivity-0

NTP Chemical Status Reports - Testing Status and NTIS Number
Selected for general toxicology study by the NTP

ISOHEXADECANOL   36311-34-9

HEALTH AND SAFETY LISTS

U.S. DOT - Appendix B - Marine Pollutants
DOT regulated marine pollutant

ISOHEXANE   107-83-5

ENVIRONMENTAL LISTS

CERCLA/SARA - Section 313 - Emission Reporting
form R reporting required

ISOHEXENE   27236-46-0

HEALTH AND SAFETY LISTS

U.S. DOT - Appendix A Table 1 - Hazardous Substances
final RQ = 100 pounds (45.4 kg)

NIOSH - Selected LD50s and LC50s
Inhalation, rat: LC50 = 360 mg/m3 10 mn Oral, rat: LD50 = 5 mg/kg Skin, mouse: LD50 = 72 mg/kg

ENVIRONMENTAL LISTS

CERCLA/SARA - Section 302 Extremely Hazardous Substances and TPQs
TPQ = 100 pounds

CERCLA/SARA - Hazardous Substances and their Reportable Quantities
final RQ = 100 pounds (45.4 kg)

RCRA - P Series Wastes
waste number P043

RCRA - Hazardous Constituents-Appendix VIII
waste number P043

RCRA - Substances Banned From Land Disposal
[present]

STATE LISTS

Florida Hazardous Substance List
effective March 13, 1992

Massachusetts Right To Know List
extraordinarily hazardous

NJ Right to Know List (Total)
sn 2500

Pennsylvania Right to Know List
environmental hazard

TERT-ISOHEXYL ALCOHOL   77-74-7

STATE LISTS

California - Exposure Limits - PELs
2 ppm PEL; 15 mg/m3 PEL

1H-ISOINDOLE-1,3(2H)-DIONE, 2-(OXIRANYL-   5455-98-1
METHYL)-

HEALTH AND SAFETY LISTS

NFPA - Flash Points
flash point < 0 degrees F (-18 degrees C)

NFPA - Hazard Identification Ratings
health-0; flammability-3; reactivity-0

STATE LISTS

Florida Hazardous Substance List
[present]

Massachusetts Right To Know List
[present]

Pennsylvania Right to Know List
[present]

ISOINDOLINE, 1,3-DIIMINO   3468-11-9

HEALTH AND SAFETY LISTS

NFPA - Flash Points
Isoheptane and mixed monomers: flash point < 0 degrees F (-18 degrees C)

NFPA - Hazard Identification Ratings
health-0; flammability-3; reactivity-0; mixed isomers: health-1; flammability-3; reactivity-0

STATE LISTS

Pennsylvania Right to Know List
[present]

1H-ISOINDOL-1-ONE   5590-18-1

STATE LISTS

NJ Right to Know List (Total)
sn 1058

ISONICOTINIC ACID   55-22-1

ENVIRONMENTAL LISTS

List of Pesticide Product Inert Ingredients
[present]

ISONICOTINIC ACID HYDRAZINE   54-85-3

HEALTH AND SAFETY LISTS

NFPA - Flash Points
flash point < 20 degrees F (-7 degrees C); mixture of hexane isomers: flash point < -20 degrees F (-29 degrees C)

NFPA - Hazard Identification Ratings
health-1; flammability-3; reactivity-0

INTERNATIONAL LISTS

Canada - WHMIS: Ingredient Disclosure
1% item 1066 (1148)

STATE LISTS

Florida Hazardous Substance List
[present]

Massachusetts Right To Know List
[present]

Minnesota Hazardous Substance List
[present]

Pennsylvania Right to Know List
[present]

ISONONANOYL PEROXIDE   58449-37-9

HEALTH AND SAFETY LISTS
U.S. DOT - Substances From 49 CFR 172.101
regulated by DOT (UN2288)
U.S. DOT - Hazard Classes
DOT hazard class = 3

STATE LISTS
NJ Right to Know List (Total)
sn 1059

ISOOCTADECANOIC ACID   30399-84-9

HEALTH AND SAFETY LISTS
NFPA - Flash Points
flash point = 115 degrees F (46 degrees C)
NFPA - Hazard Identification Ratings
flammability-2; reactivity-0

ISOOCTADECANOL   27458-93-1

ENVIRONMENTAL LISTS
TSCA - Code of Federal Regulations Citations
40 CFR 712.30(d)
TSCA - PAIR - Reporting List
Reporting Date: November 19, 1982

ISOOCTANE   540-84-1

HEALTH AND SAFETY LISTS
NIOSH - Selected LD50s and LC50s
Oral, rat: LD50 = 273 mg/kg

ISOOCTANE (VAN)   26635-64-3

ENVIRONMENTAL LISTS
List of Pesticide Product Inert Ingredients
[present]

ISOOCTANOIC ACID   25103-52-0

HEALTH AND SAFETY LISTS
NIOSH - Selected LD50s and LC50s
Oral, rat: LD50 = 5000 mg/kg

ISOOCTENE   11071-47-9

HEALTH AND SAFETY LISTS
IARC - Group 3 (not classifiable)
[present]
NIOSH - Selected LD50s and LC50s
Oral, mouse: LD50 = 160 mg/kg

STATE LISTS
California - Directors List of Hazardous Substances (8 CCR 339)
[present]

ISOOCTYL ALCOHOL   26952-21-6

STATE LISTS
NJ Right to Know List (Total)
sn 1060

ISOOCTYL DIPHENYL PHOSPHITE   26401-27-4

ENVIRONMENTAL LISTS
List of Pesticide Product Inert Ingredients
[present]

ISOOCTYL NITRATE   RR-00865-1

ENVIRONMENTAL LISTS
List of Pesticide Product Inert Ingredients
[present]

ISOPARAFFINIC PETROLEUM HYDROCARBONS   64365-06-6

HEALTH AND SAFETY LISTS
NFPA - Flash Points
*flash point = 40 degrees F (4.5 degrees C);   *See list description*
NFPA - Hazard Identification Ratings
health-0; flammability-3; reactivity-0

ENVIRONMENTAL LISTS
CERCLA/SARA - Hazardous Substances and their Reportable Quantities
final RQ = 1 pound (.454 kg)
Clean Air Act (1990) - List of Hazardous Air Contaminants
[present]
CAA - HON Rule - SOCMI Chemicals
compliance by Oct. 23, 1995
CAA - HON Rule - Organic HAPs
[present]
TSCA - Code of Federal Regulations Citations
40 CFR 716.120(a)
TSCA - Health and Safety Reporting List
Effective Date: June 1, 1987

STATE LISTS
California - Air Bill 2588 Appendix A-I
6/91
Florida Hazardous Substance List
[present]
Massachusetts Right To Know List
[present]
NJ Right to Know List (Total)
sn 1061
NJ Special Hazardous Substances
(flammable - third degree)
Pennsylvania Right to Know List
[present]

ISOPENTALDEHYDE   590-86-3

STATE LISTS
Pennsylvania Right to Know List
[present]

ISOPENTANE   78-78-4

HEALTH AND SAFETY LISTS
NFPA - Flash Points
flash point = 270 degrees F (132 degrees C)
NFPA - Hazard Identification Ratings
health-0; flammability-1; reactivity-0

ISOPENTANOIC ACID   503-74-2

HEALTH AND SAFETY LISTS
U.S. DOT - Substances From 49 CFR 172.101
regulated by DOT (UN1216)
U.S. DOT - Hazard Classes
DOT hazard class = 3
NFPA - Flash Points
flash point < 20 degrees F (-7 degrees C)
NFPA - Hazard Identification Ratings
health-0; flammability-3; reactivity-0

STATE LISTS
NJ Right to Know List (Total)
sn 1062
NJ Special Hazardous Substances
(flammable - third degree)
Pennsylvania Right to Know List
[present]

ISOPENTENES   RR-00349-6

HEALTH AND SAFETY LISTS

ACGIH 1995 - Time Weighted Averages
50 ppm TWA; 266 mg/m3 TWA

ACGIH 1995 - Skin Designations
skin - potential for cutaneous absorption

NFPA - Flash Points
flash point = 180 degrees F (82 degrees C)

NFPA - Hazard Identification Ratings
health-0; flammability-2; reactivity-0

NIOSH - Selected LD50s and LC50s
Oral, rat: LD50 = 1480 mg/kg Skin, rabbit: LD50 = 2520 mg/kg

OSHA - Vacated PELs - Time Weighted Averages
50 ppm TWA; 270 mg/m3 TWA

OSHA - Vacated PELs - Skin Designation
Prevent or reduce skin absorption

INTERNATIONAL LISTS

Australian Exposure Standards - Time Weighted Averages
50 ppm TWA; 266 mg/m3 TWA

Australian Exposure Standards - Skin Effects
skin absorption

Canada - WHMIS: Ingredient Disclosure
1% item 896 (179)

Canada - Alberta - 8 Hour Occupational Exposure Limit
50 ppm TWA; 266 mg/m3 TWA

Canada - Alberta - 15 Minute Occupational Exposure Limit
75 ppm STEL; 400 mg/m3 STEL

Canada - Ontario - OHSA - TWAEVs
270 mg/m3 TWAEV (listed as an agent of variable composition)

Canada - Ontario - OHSA - Skin Notations
absorption through skin, eyes, or mucous membranes (listed under 'Agents of Variable Composition')

Canada - Quebec - Time-Weighted Average Exposure Values
50 ppm TWAEV; 266 mg/m3 TWAEV

Canada - Quebec - Skin Designations
absorbed through the skin

United Kingdom - Occupational Exposure Standards - TWAs
50 ppm TWA; 270 mg/m3 TWA

Israel - Time Weighted Averages
50 ppm TWA; 266 mg/m3 TWA

Israel - Action Levels
25 ppm AL; 133 mg/m3 AL

STATE LISTS

California - Exposure Limits - PELs
50 ppm PEL; 270 mg/m3 PEL

California - Exposure Limits - Skin Notation
material may be absorbed through the skin, eyes or mucous membrane

California - Directors List of Hazardous Substances (8 CCR 339)
[present]

Florida Hazardous Substance List
[present]

Massachusetts Right To Know List
[present]

Minnesota Hazardous Substance List
skin

NJ Right to Know List (Total)
sn 1063

Pennsylvania Right to Know List
[present]

ISOPHORONE   78-59-1

ENVIRONMENTAL LISTS

List of Pesticide Product Inert Ingredients
[present]

ISOPHORONE DIAMINE   2855-13-2

HEALTH AND SAFETY LISTS

U.S. DOT - Appendix B - Marine Pollutants
DOT regulated marine pollutant

NFPA - Flash Points
flash point = 205 degrees F (96 degrees C)

NFPA - Hazard Identification Ratings
flammability-1

ISOPHORONE DIISOCYANATE   4098-71-9

ENVIRONMENTAL LISTS

List of Pesticide Product Inert Ingredients
[present]

ISOPHOSPHAMIDE   3778-73-2

HEALTH AND SAFETY LISTS

NFPA - Flash Points
flash point = 48 degrees F (9 degrees C)

NFPA - Hazard Identification Ratings
health-2; flammability-3; reactivity-0

NIOSH - Selected LD50s and LC50s
Oral, rat: LD50 = 8910 mg/kg Skin, rabbit: LD50 = 3180 mg/kg

ENVIRONMENTAL LISTS

EPA - Master Testing List
[present]

TSCA - Code of Federal Regulations Citations
40 CFR 712.30(x); 40 CFR 716.120(d)

TSCA - PAIR - Reporting List
Reporting Date: November 27, 1991

TSCA - Health and Safety Reporting List
Effective Date: September 30, 1991

STATE LISTS

Florida Hazardous Substance List
[present]

Massachusetts Right To Know List
[present]

Pennsylvania Right to Know List
[present]

ISOPHTHALIC ACID   121-91-5

HEALTH AND SAFETY LISTS

NFPA - Flash Points
flash point < -60 degrees F (-51 degrees C)

NFPA - Hazard Identification Ratings
health-1; flammability-4; reactivity-0

ENVIRONMENTAL LISTS

CAA - Flammable Substances for Accidental Release Prevention
threshold quantity = 10,000 lbs

List of Pesticide Product Inert Ingredients
[present]

INTERNATIONAL LISTS

German (DFG) - MAK Values
1000 ppm MAK; 2950 mg/m3 MAK (Listed under 'Pentane')

German (DFG) - Peak Limitations
2 x normal MAK (1 hour momentary value); don't exceed 3 times per shift (Listed under 'Pentane')

STATE LISTS

Florida Hazardous Substance List
[present]

Massachusetts Right To Know List
[present]

NJ Right to Know List (Total)
sn 1064

NJ Special Hazardous Substances
(flammable - fourth degree)

Pennsylvania Right to Know List
[present]

ISOPHTHALIC ANHYDRIDE   4891-67-2

HEALTH AND SAFETY LISTS

NFPA - Hazard Identification Ratings
health-1; reactivity-0

NIOSH - Selected LD50s and LC50s
Oral, rat: LD50 = 2000 mg/kg Skin, rabbit: LD50 = 310 mg/kg

INTERNATIONAL LISTS

Canada - WHMIS: Ingredient Disclosure
1% item 897 (97)

STATE LISTS

Massachusetts Right To Know List
[present]

NJ Right to Know List (Total)
sn 1065

NJ Special Hazardous Substances
(corrosive)

ISOPHTHALOYL CHLORIDE   99-63-8

HEALTH AND SAFETY LISTS

U.S. DOT - Substances From 49 CFR 172.101
regulated by DOT (UN2371)

U.S. DOT - Hazard Classes
DOT hazard class = 3

STATE LISTS

NJ Right to Know List (Total)
sn 2501

ISOPRENE   78-79-5

HEALTH AND SAFETY LISTS

ACGIH 1995 - Ceiling Limits
(C 5) ppm; (C 28) mg/m3

AIHA - Odor Threshold Values
geometric mean air odor threshold = 0.19 ppm (detectable); 0.53 ppm (recognizable)

U.S. DOT - Appendix A Table 1 - Hazardous Substances
final RQ = 5000 pounds (2270 kg)

NFPA - Flash Points
flash point = 184 degrees F (84 degrees C)

NFPA - Hazard Identification Ratings
health-2; flammability-2; reactivity-0

NIOSH - Selected LD50s and LC50s
Oral, rat: LD50 = 2330 mg/kg Skin, rabbit: LD50 = 1500 mg/kg

NIOSH 1990 - Pocket Guide - RELs
4 ppm TWA; 23 mg/m3 TWA

NIOSH 1990 - Pocket Guide - IDLHs
800 ppm IDLH

NIOSH 1990 - Pocket Guide - Target organs
skin, respiratory system

NIOSH - Health Standards - Exposure Limits
4 ppm TWA; 23 mg/m3 TWA (Listed under 'Ketones')

NIOSH - Health Standards - Health Effects and Precautions
Irritation; liver, kidney, and nervous system effects (Urinalysis required) (Listed under 'Ketones')

NTP Chemical Status Reports - Testing Status and NTIS Number
Technical reports printed (PB86181823/AS)

NTP Chemical Status Reports - Evidence of Carcinogenicity
male rat-some evidence; female rat-no evidence; male mice-equivocal evidence; female mice-no evidence

OSHA - Vacated PELs - Time Weighted Averages
4 ppm TWA; 23 mg/m3 TWA

OSHA - Final PELs - Time Weighted Averages
25 ppm TWA; 140 mg/m3 TWA

ENVIRONMENTAL LISTS

CERCLA/SARA - Hazardous Substances and their Reportable Quantities
final RQ = 5000 pounds (2270 kg)

Clean Air Act (1990) - List of Hazardous Air Contaminants
[present]

CAA - HON Rule - SOCMI Chemicals
compliance by July 24, 1995

CAA - HON Rule - Organic HAPs
[present]

Clean Water Act - Priority Pollutants
[present]

Clean Water Act - Toxic Pollutants
[present]

List of Pesticide Product Inert Ingredients
[present]

RCRA - TSD Facilities Ground Water Monitoring
TM 8090 = 60 ug/L PQL; TM 8270 = 10 ug/L PQL

TSCA - Code of Federal Regulations Citations
40 CFR 712.30(d); 40 CFR 716.120(a)

TSCA - PAIR - Reporting List
Reporting Date: November 19, 1982

TSCA - Health and Safety Reporting List
Effective Date: October 4, 1982

INTERNATIONAL LISTS

Australian Exposure Standards - Time Weighted Averages
Peak Limitation: 5 ppm; 28 mg/m3

Canada - WHMIS: Ingredient Disclosure
1% item 898 (1047)

Canada - Alberta - Ceiling Occupational Exposure Limit
C 5 ppm; C 28 mg/m3

Canada - British Columbia - Ceiling Exposure Limits
C 5 ppm; C 25 mg/m3

Canada - Ontario - OHSA - CEVs
5 ppm CEV; 28 mg/m3 CEV

Canada - Quebec - Ceiling Limits
P 5 ppm; P 28 mg/m3

United Kingdom - Occupational Exposure Standards - STELs
5 ppm STEL; 25 mg/m3 STEL

German (DFG) - MAK Values
5 ppm MAK; 28 mg/m3 MAK

Israel - Ceiling Exposure Limits
C 5 ppm; C 28 mg/m3

Mexico - Instruction No. 10 - TWAs
5 ppm TWA; 12 mg/m3 TWA

Mexico - Instruction No. 10 - STELs
10 ppm STEL; 14 mg/m3 STEL

Mexico - Wastewater - Organic Toxic Pollutants and Heavy Metals
Listed under [Organic Toxic Pollutants]

Mexico - Drinking Water - Ecological Criteria
5.2 mg/l

STATE LISTS

California - Air Bill 2588 Appendix A-I
6/91

California - Exposure Limits - PELs
4 ppm PEL; 23 mg/m3 PEL

California - Directors List of Hazardous Substances (8 CCR 339)
[present]

Florida Hazardous Substance List
[present]

Massachusetts Right To Know List
[present]

Minnesota Hazardous Substance List
[present]

NJ Right to Know List (Total)
sn 1066

Pennsylvania Right to Know List
environmental hazard

PROPOSED REGULATIONS

ACGIH 1995 - Notice of Intended Changes
C 5 ppm; C 28 mg/m3; A3-animal carcinogen

Safe Drinking Water Act - Priority list
[present]

ISOPROCARB   2631-40-5

HEALTH AND SAFETY LISTS

U.S. DOT - Substances From 49 CFR 172.101
regulated by DOT (UN2289)

U.S. DOT - Hazard Classes
DOT hazard class = 8

INTERNATIONAL LISTS

Canada - WHMIS: Ingredient Disclosure
1% item 899 (1048)

STATE LISTS

NJ Right to Know List (Total)
sn 1067

NJ Special Hazardous Substances
(corrosive)

ISOPROPANOLAMINE   78-96-6

HEALTH AND SAFETY LISTS

ACGIH 1995 - Time Weighted Averages
0.005 ppm TWA; 0.045 mg/m3 TWA

U.S. DOT - Substances From 49 CFR 172.101
regulated by DOT (UN2290)

U.S. DOT - Hazard Classes
DOT hazard class = 6.1

NIOSH - Selected LD50s and LC50s
Inhalation, rat: LC50 = 123 mg/m3 4 hr Skin, rat: LD50 = 1060 mg/kg

NIOSH - Health Standards - Exposure Limits
5 ppb TWA; 45 ug/m3 TWA; C (10 min) 20 ppb; C (10 min) 180 ug/m3 (Listed under 'Diisocyanates')

NIOSH - Health Standards - Health Effects and Precautions
Respiratory effects and sensitization, pulmonary irritation (Periodic chest X-ray and pulmonary function testing required) (Listed under 'Diisocyanates')

OSHA - Vacated PELs - Time Weighted Averages
0.005 ppm TWA

OSHA - Vacated PELs - Short Term Exposure Limits
0.02 ppm STEL

OSHA - Vacated PELs - Skin Designation
Prevent or reduce skin absorption

ENVIRONMENTAL LISTS

CERCLA/SARA - Section 302 Extremely Hazardous Substances and TPQs
TPQ = 1000 pounds (This material is a reacitve solid. The TPQ does not default to 10,000 pounds for non-powder, non-molten, non-solution form)

CERCLA/SARA - Section 313 - Emission Reporting
form R reporting required; (Listed under 'Diisocyanates')

TSCA - Code of Federal Regulations Citations
40 CFR 712.30(x); 40 CFR 716.120(a)

TSCA - PAIR - Reporting List
Reporting Date: December 27, 1990

TSCA - Health and Safety Reporting List
Effective Date: June 1, 1987

INTERNATIONAL LISTS

Canada - WHMIS: Ingredient Disclosure
0.1% item 900 (718)

Canada - Alberta - 8 Hour Occupational Exposure Limit
0.01 ppm TWA; 0.09 mg/m3 TWA

Canada - Alberta - 15 Minute Occupational Exposure Limit
0.03 ppm STEL; 0.27 mg/m3 STEL

Canada - Alberta - Skin Designation
can be absorbed through the intact skin

Canada - British Columbia - 8 Hour Exposure Limits
0.01 ppm TWA; 0.09 mg/m3 TWA

Canada - British Columbia - Skin Notations
skin - potential for skin absorption

Canada - Ontario - OHSA - TWAEVs
0.005 ppm TWAEV; 0.2 micromoles/m3 TWAEV (designated substance regulation)

Canada - Ontario - OHSA - CEVs
0.02 ppm CEV; 0.8 micromoles/m3 CEV (designated substance regulation)

Canada - Ontario - OHSA - Designated Substances
0.005 ppm TWAEV; 0.2 micromoles/m3 TWAEV; See Ontario Reg. 842 for full information.

Canada - Quebec - Time-Weighted Average Exposure Values
0.005 ppm TWAEV; 0.045 mg/m3 TWAEV

German (DFG) - MAK Values
0.01 ppm MAK; 0.09 mg/m3 MAK

German (DFG) - Peak Limitations
2 x normal MAK (5 min momentary value); don't exceed 8 times during shift

German (DFG) - Skin/Sensitizers
danger of sensitization (skin or respiratory)

Israel - Time Weighted Averages
0.005 ppm TWA; 0.045 mg/m3 TWA

Israel - Action Levels
0.0025 ppm AL; 0.0225 mg/m3 AL

STATE LISTS

California - Exposure Limits - PELs
0.05 ppm PEL; 0.045 mg/m3 PEL

California - Exposure Limits - STELs
0.02 ppm STEL

California - Exposure Limits - Skin Notation
material may be absorbed through the skin, eyes or mucous membrane

California - Directors List of Hazardous Substances (8 CCR 339)
[present]

Florida Hazardous Substance List
[present]

Massachusetts Right To Know List
extraordinarily hazardous

Minnesota Hazardous Substance List
skin

NJ Right to Know List (Total)
sn 1068

Pennsylvania Right to Know List
environmental hazard

PROPOSED REGULATIONS

CERCLA/SARA - Proposed Hazardous Substance Additions
proposed RQ = 1 pound (.454 kg)

CERCLA/SARA - 1989 Proposed RQ Adjustments
proposed RQ = 100 pounds (45.4 kg)

ISOPROPANOLAMINE DODECYLBENZENE SULFONATE   42504-46-1

HEALTH AND SAFETY LISTS

IARC - Group 3 (not classifiable)
[present]

NTP Chemical Status Reports - Testing Status and NTIS Number
Technical reports printed (PB275677/AS)

NTP Chemical Status Reports - Evidence of Carcinogenicity
male rat-negative; female rat-positive; male mice-negative; female mice-positive

STATE LISTS

California - Air Bill 2588 Appendix A-II
9/90

California - Prop. 65 - Developmental Toxicity
developmental toxicity - initial date 7/1/90

California - Directors List of Hazardous Substances (8 CCR 339)
[present]

Massachusetts Right To Know List
carcinogen; extraordinarily hazardous

ISOPROPANOLAMINE DODECYLBENZENESUL-FONATE (1:1)   54590-52-2

HEALTH AND SAFETY LISTS

AIHA - WEEL - Time Weighted Averages
respirable dust: 5 mg/m3 TWA; total dust: 10 mg/m3 TWA

NIOSH - Selected LD50s and LC50s
Oral, rat: LD50 = 10400 mg/kg

ENVIRONMENTAL LISTS

CAA - HON Rule - SOCMI Chemicals
compliance by April 24, 1995

List of Pesticide Product Inert Ingredients
[present]

INTERNATIONAL LISTS

Canada - WHMIS: Ingredient Disclosure
1% item 901 (98)

ISOPROPENYL ACETATE   108-22-5

ENVIRONMENTAL LISTS

List of Pesticide Product Inert Ingredients
[present]

ISOPROPENYL ACETYLENE   78-80-8

HEALTH AND SAFETY LISTS

NFPA - Flash Points
flash point = 356 degrees F (180 degrees C)

NFPA - Hazard Identification Ratings
flammability-1; reactivity-0

ENVIRONMENTAL LISTS

TSCA - Code of Federal Regulations Citations
40 CFR 712.30(x); 40 CFR 716.120(d)

TSCA - PAIR - Reporting List
Reporting Date: November 27, 1991

TSCA - Health and Safety Reporting List
Effective Date: September 30, 1991

INTERNATIONAL LISTS

Canada - WHMIS: Ingredient Disclosure
1% item 902 (504)

2-ISOPROPENYL-2-OXAZOLINE   10471-78-0

HEALTH AND SAFETY LISTS

AIHA - WEEL - Time Weighted Averages
50 ppm TWA; 139 mg/m3 TWA

U.S. DOT - Substances From 49 CFR 172.101
regulated by DOT (UN1218)

U.S. DOT - Hazard Classes
DOT hazard class = 3

U.S. DOT - Appendix A Table 1 - Hazardous Substances
final RQ = 100 pounds (45.4 kg)

NFPA - Flash Points
flash point = -65 degrees F (-54 degrees C)

NFPA - Hazard Identification Ratings
health-1; flammability-4; reactivity-2

NIOSH - Selected LD50s and LC50s
Inhalation, rat: LC50 = 180 gm/m3 4 hr

NTP Chemical Status Reports - Testing Status and NTIS Number
Two year studies in progress; Post peer review technical reports in progress

ENVIRONMENTAL LISTS

CERCLA/SARA - Hazardous Substances and their Reportable Quantities
final RQ = 100 pounds (45.4 kg)

CAA - Flammable Substances for Accidental Release Prevention
threshold quantity = 10,000 lbs

Clean Water Act - Hazardous Substances
[present]

INTERNATIONAL LISTS

Canada - WHMIS: Ingredient Disclosure
1% item 903 (1049)

STATE LISTS

California - Directors List of Hazardous Substances (8 CCR 339)
[present]

Florida Hazardous Substance List
[present]

Massachusetts Right To Know List
[present]

NJ Right to Know List (Total)
sn 1069

NJ Special Hazardous Substances
(flammable - fourth degree; reactive - second degree)

Pennsylvania Right to Know List
environmental hazard

ISOPROPOXYETHANOL   109-59-1

HEALTH AND SAFETY LISTS

U.S. DOT - Appendix B - Marine Pollutants
DOT regulated marine pollutant

3-ISOPROPOXYPROPIONITRILE   110-47-4

HEALTH AND SAFETY LISTS

NFPA - Flash Points
flash point = 171 degrees F (77 degrees C)

NFPA - Hazard Identification Ratings
health-2; flammability-2; reactivity-0

NIOSH - Selected LD50s and LC50s
Oral, rat: LD50 = 4260 mg/kg; Skin, rabbit: LD50 = 1640 mg/kg

ENVIRONMENTAL LISTS

List of Pesticide Product Inert Ingredients
[present]

INTERNATIONAL LISTS

Canada - WHMIS: Ingredient Disclosure
1% item 905 (1051)

STATE LISTS

Florida Hazardous Substance List
[present]

Massachusetts Right To Know List
[present]

Pennsylvania Right to Know List
[present]

ISOPROPYL ACETATE   108-21-4

HEALTH AND SAFETY LISTS

U.S. DOT - Appendix A Table 1 - Hazardous Substances
final RQ = 1000 pounds (454 kg)

ENVIRONMENTAL LISTS

CERCLA/SARA - Hazardous Substances and their Reportable Quantities
final RQ = 1000 pounds (454 kg)

Clean Water Act - Hazardous Substances
[present]

STATE LISTS

California - Directors List of Hazardous Substances (8 CCR 339)
[present] (exempt when in solution)

Massachusetts Right To Know List
[present]

Pennsylvania Right to Know List
environmental hazard

ISOPROPYL ALCOHOL   67-63-0

STATE LISTS

NJ Right to Know List (Total)
sn 1070

Pennsylvania Right to Know List
environmental hazard

ISOPROPYL ALCOHOL MANUFACTURE (STRONG-   RR-00068-0
ACID PROCESS)

HEALTH AND SAFETY LISTS

U.S. DOT - Substances From 49 CFR 172.101
regulated by DOT (UN2403)

U.S. DOT - Hazard Classes
DOT hazard class = 3

NFPA - Flash Points
flash point = 60 degrees F (16 degrees C)

NFPA - Hazard Identification Ratings
health-2; flammability-3; reactivity-0

NIOSH - Selected LD50s and LC50s
Oral, rat: LD50 = 3000 mg/kg

STATE LISTS

Florida Hazardous Substance List
[present]

Massachusetts Right To Know List
[present]

NJ Right to Know List (Total)
sn 1071

NJ Special Hazardous Substances
(flammable - third degree)

Pennsylvania Right to Know List
[present]

ISOPROPYLAMINE   75-31-0

HEALTH AND SAFETY LISTS

NFPA - Flash Points
flash point < 19 degrees F (-7 degrees C)

NFPA - Hazard Identification Ratings
health-2; flammability-4; reactivity-2

STATE LISTS

Florida Hazardous Substance List
[present]

Massachusetts Right To Know List
[present]

Pennsylvania Right to Know List
[present]

ISOPROPYLAMINE, DISTILLATION RESIDUES   79771-08-7

HEALTH AND SAFETY LISTS

U.S. DOT - Substances Which Are Poisonous by Inhalation
liquid hazardous material poisonous by inhalation

ISOPROPYLAMINE   12068-04-1
METHYLNAPHTHALENESULFONATE
SEE ALSO:
GLYCOL ETHERS

HEALTH AND SAFETY LISTS

ACGIH 1995 - Time Weighted Averages
25 ppm TWA; 106 mg/m3 TWA

ACGIH 1995 - Skin Designations
skin - potential for cutaneous absorption

NFPA - Flash Points
flash point = 92 degrees F (33 degrees C)

NFPA - Hazard Identification Ratings
health-1; flammability-3; reactivity-0

NIOSH - Selected LD50s and LC50s
Inhalation, mouse: LC50 = 1930 ppm 7 hr Oral, rat: LD50 = 5660 mg/kg Skin, rabbit: LD50 = 1600 mg/kg

OSHA - Vacated PELs - Time Weighted Averages
25 ppm TWA; 105 mg/m3 TWA

INTERNATIONAL LISTS

Australian Exposure Standards - Time Weighted Averages
25 ppm TWA; 106 mg/m3 TWA

Canada - WHMIS: Ingredient Disclosure
1% item 719 (822)

Canada - Alberta - 8 Hour Occupational Exposure Limit
25 ppm TWA; 105 mg/m3 TWA

Canada - Alberta - 15 Minute Occupational Exposure Limit
75 ppm STEL; 320 mg/m3 STEL

Canada - Ontario - OHSA - TWAEVs
25 ppm TWAEV; 105 mg/m3 TWAEV

Canada - Quebec - Time-Weighted Average Exposure Values
25 ppm TWAEV; 106 mg/m3 TWAEV

Canada - Quebec - Skin Designations
absorbed through the skin

German (DFG) - MAK Values
5 ppm MAK; 22 mg/m3 MAK

German (DFG) - Peak Limitations
2 x normal MAK (30 min. average value); don't exceed 4 times during shift

German (DFG) - Skin/Sensitizers
danger of cutaneous absorption

German (DFG) - Pregnancy
no risk to embryo/fetus if exposure limits are adhered to

Israel - Time Weighted Averages
25 ppm TWA; 106 mg/m3 TWA

Israel - Action Levels
12.5 ppm AL; 53 mg/m3 AL

STATE LISTS

California - Exposure Limits - PELs
25 ppm PEL; 105 mg/m3 PEL

California - Directors List of Hazardous Substances (8 CCR 339)
[present]

Florida Hazardous Substance List
[present]

Massachusetts Right To Know List
[present]

Minnesota Hazardous Substance List
[present]

Pennsylvania Right to Know List
[present]

PROPOSED REGULATIONS

Canada - Ontario - Proposed Occupational TWAEVs
10 ppm TWAEV; 44 mg/m3 TWAEV

ISOPROPYLAMINE SALT OF OLEOYLISO-PROPANOLAMIDE DERIVATIVE OFSULFOSUC-CINIC ACID   RR-01090-2

HEALTH AND SAFETY LISTS

NFPA - Flash Points
flash point = 155 degrees F (68 degrees C)

NFPA - Hazard Identification Ratings
health-1; flammability-2; reactivity-1

ISOPROPYLAMINE SALT OF STEARYLISO-PROPANOLAMIDE DERIVATIVE OF SULFOSUC-CINIC ACID   RR-01091-3

HEALTH AND SAFETY LISTS

ACGIH 1995 - Time Weighted Averages
250 ppm TWA; 1040 mg/m3 TWA

ACGIH 1995 - Short Term Exposure Limits
310 ppm STEL; 1290 mg/m3 STEL

AIHA - Odor Threshold Values
geometric mean air odor threshold = 4.1 ppm (detectable); 6.1 ppm (recognizable)

U.S. DOT - Substances From 49 CFR 172.101
regulated by DOT (UN1220)

U.S. DOT - Hazard Classes
DOT hazard class = 3

NFPA - Flash Points
flash point = 35 degrees F (2 degrees C)

NFPA - Hazard Identification Ratings
health-1; flammability-3; reactivity-0

NIOSH - Selected LD50s and LC50s
Oral, rat: LD50 = 3000 mg/kg

NIOSH 1990 - Pocket Guide - RELs
See Appendix D

NIOSH 1990 - Pocket Guide - IDLHs
16,000 ppm IDLH

NIOSH 1990 - Pocket Guide - Target organs
eyes, skin, respiratory system

OSHA - Vacated PELs - Time Weighted Averages
250 ppm TWA; 950 mg/m3 TWA

OSHA - Vacated PELs - Short Term Exposure Limits
310 ppm STEL; 1185 mg/m3 STEL

OSHA - Final PELs - Time Weighted Averages
250 ppm TWA; 950 mg/m3 TWA

ENVIRONMENTAL LISTS

List of Pesticide Product Inert Ingredients
[present]

INTERNATIONAL LISTS

Australian Exposure Standards - Time Weighted Averages
250 ppm TWA; 1040 mg/m3 TWA

Australian Exposure Standards - Short Term Exposure Limits
310 ppm STEL; 1290 mg/m3 STEL

Canada - WHMIS: Ingredient Disclosure
1% item 906 (22)

Canada - Alberta - 8 Hour Occupational Exposure Limit
250 ppm TWA; 1044 mg/m3 TWA

Canada - Alberta - 15 Minute Occupational Exposure Limit
310 ppm STEL; 1295 mg/m3 STEL

Canada - British Columbia - 8 Hour Exposure Limits
250 ppm TWA; 950 mg/m3 TWA

Canada - British Columbia - 15 Minute Exposure Limits
310 ppm STEL; 1185 mg/m3 STEL

Canada - Ontario - OHSA - TWAEVs
250 ppm TWAEV; 1040 mg/m3 TWAEV

Canada - Ontario - OHSA - STEVs
310 ppm STEV; 1295 mg/m3 STEV

Canada - Quebec - Time-Weighted Average Exposure Values
250 ppm TWAEV; 1040 mg/m3 TWAEV

Canada - Quebec - Short-term Exposure Values
310 ppm STEV; 1290 mg/m3 STEV

United Kingdom - Occupational Exposure Standards - STELs
200 ppm STEL; 840 mg/m3 STEL

German (DFG) - MAK Values
200 ppm MAK; 840 mg/m3 MAK (Listed under 'Propyl acetate')

German (DFG) - Peak Limitations
2 x normal MAK (5 min momentary value); don't exceed 8 times during shift (Listed under 'Propyl acetate')

Israel - Time Weighted Averages
250 ppm TWA; 1040 mg/m3 TWA

Israel - Short Term Exposure Limits
310 ppm STEL; 1290 mg/m3 STEL

Israel - Action Levels
125 ppm AL; 520 mg/m3 AL

Mexico - Instruction No. 10 - TWAs
250 ppm TWA; 950 mg/m3 TWA

Mexico - Instruction No. 10 - STELs
310 ppm STEL; 1185 mg/m3 STEL

STATE LISTS

California - Exposure Limits - PELs
250 ppm PEL; 950 mg/m3 PEL

California - Exposure Limits - STELs
310 ppm STEL; 1185 mg/m3 STEL

California - Directors List of Hazardous Substances (8 CCR 339)
[present]

Florida Hazardous Substance List
[present]

Massachusetts Right To Know List
[present]

Minnesota Hazardous Substance List
[present]

NJ Right to Know List (Total)
sn 1074

NJ Special Hazardous Substances
(corrosive; flammable - third degree)

Pennsylvania Right to Know List
[present]

ISOPROPYLAMINE SULFONATE   26118-67-2

HEALTH AND SAFETY LISTS

ACGIH 1995 - Time Weighted Averages
400 ppm TWA; 983 mg/m3 TWA

ACGIH 1995 - Short Term Exposure Limits
500 ppm STEL; 1230 mg/m3 STEL

AIHA - Odor Threshold Values
geometric mean air odor threshold = 43 ppm (detectable); 19 ppm (recognizable)

U.S. DOT - Substances From 49 CFR 172.101
regulated by DOT (UN1219)

U.S. DOT - Hazard Classes
DOT hazard class = 3

IARC - Group 3 (not classifiable)
[present]

NFPA - Flash Points
flash point = 53 degrees F (12 degrees C); 87.9% iso: flash point = 57 degrees F (14 degrees C)

NFPA - Hazard Identification Ratings
health-1; flammability-3; reactivity-0

NIOSH - Selected LD50s and LC50s
Oral, rat: LD50 = 5045 mg/kg Skin, rabbit: LD50 = 12800 mg/kg

NIOSH 1990 - Pocket Guide - RELs
400 ppm TWA; 980 mg/m3 TWA; 500 ppm STEL; 1225 mg/m3 STEL

NIOSH 1990 - Pocket Guide - IDLHs
12,000 ppm IDLH

NIOSH 1990 - Pocket Guide - Target organs
eyes, skin, respiratory system

NIOSH - Health Standards - Exposure Limits
400 ppm TWA; 984 mg/m3 TWA; C (15 min) 800 ppm; C (15 min) 1968 mg/m3

NIOSH - Health Standards - Health Effects and Precautions
Mucous membrane irritation; possible cancer threat in manufacturing process (Stringent work practices and medical monitoring required for manufacturing workers)

OSHA - Vacated PELs - Time Weighted Averages
400 ppm TWA; 980 mg/m3 TWA

OSHA - Vacated PELs - Short Term Exposure Limits
500 ppm STEL; 1225 mg/m3 STEL

OSHA - Final PELs - Time Weighted Averages
400 ppm TWA; 980 mg/m3 TWA

ENVIRONMENTAL LISTS

ATSDR Priority List
Rank (of 275): 254

CERCLA/SARA - Section 313 - Emission Reporting
form R reporting required for 0.1% de minimus concentration (only if manufactured by the strong acid process)

EPA - Master Testing List
[present]

List of Pesticide Product Inert Ingredients
[present]

TSCA - Code of Federal Regulations Citations
40 CFR 712.30(t); 40 CFR 716.120(a); 799.2325

TSCA - PAIR - Reporting List
Reporting Date: February 12, 1987

TSCA - Health and Safety Reporting List
Effective Date: December 15, 1986

TSCA - Chemical Test Rules
Testing required by: manufacturers; importers; processors (40 CFR 799.2325)

TSCA - Section 12(b) - Export Notification
export notification required - Section 4

INTERNATIONAL LISTS

Australian Exposure Standards - Time Weighted Averages
400 ppm TWA; 983 mg/m3 TWA

Australian Exposure Standards - Short Term Exposure Limits
500 ppm STEL; 1230 mg/m3 STEL

Australian Exposure Standards - Under Review
exposure limits under review

Canada - WHMIS: Ingredient Disclosure
1% item 904 (1050)

Canada - NPRI (National Pollutant Release Inventory)
[present]

Canada - Alberta - 8 Hour Occupational Exposure Limit
400 ppm TWA; 983 mg/m3 TWA

Canada - Alberta - 15 Minute Occupational Exposure Limit
500 ppm STEL; 1228 mg/m3 STEL

Canada - Alberta - Skin Designation
can be absorbed through the intact skin

Canada - British Columbia - 8 Hour Exposure Limits
400 ppm TWA; 980 mg/m3 TWA

Canada - British Columbia - 15 Minute Exposure Limits
500 ppm STEL; 1225 mg/m3 STEL

Canada - British Columbia - Skin Notations
skin - potential for skin absorption

Canada - Ontario - OHSA - TWAEVs
400 ppm TWAEV; 980 mg/m3 TWAEV

Canada - Ontario - OHSA - STEVs
500 ppm STEV; 1225 mg/m3 STEV

Canada - Quebec - Time-Weighted Average Exposure Values
400 ppm TWAEV; 985 mg/m3 TWAEV

Canada - Quebec - Short-term Exposure Values
500 ppm STEV; 1230 mg/m3 STEV

United Kingdom - Occupational Exposure Standards - TWAs
400 ppm TWA; 980 mg/m3 TWA

United Kingdom - Occupational Exposure Standards - STELs
500 ppm STEL; 1225 mg/m3 STEL

United Kingdom - Occupational Exposure Standards - Notes
can be absorbed through skin

German (DFG) - MAK Values
400 ppm MAK; 980 mg/m3 MAK

German (DFG) - Peak Limitations
2 x normal MAK (30 min. average value); don't exceed 4 times during shift

German (DFG) - Pregnancy
classification not yet possible

Israel - Time Weighted Averages
400 ppm TWA; 983 mg/m3 TWA

Israel - Short Term Exposure Limits
500 ppm STEL; 1,230 mg/m3 STEL

Israel - Action Levels
200 ppm AL; 491.5 mg/m3 AL

Mexico - Instruction No. 10 - TWAs
400 ppm TWA; 980 mg/m3 TWA

Mexico - Instruction No. 10 - STELs
500 ppm STEL; 1225 mg/m3 STEL

Mexico - Instruction No. 10 - Skin designation
skin - potential for cutaneous absorption

STATE LISTS

California - Air Bill 2588 Appendix A-I
6/91

California - Exposure Limits - PELs
400 ppm PEL; 980 mg/m3 PEL

California - Exposure Limits - STELs
500 ppm STEL; 1225 mg/m3 STEL

California - Directors List of Hazardous Substances (8 CCR 339)
[present]

Florida Hazardous Substance List
[present]

Massachusetts Right To Know List
[present]

Minnesota Hazardous Substance List
[present]

NJ Right to Know List (Total)
sn 1076

NJ Special Hazardous Substances
(flammable - third degree)

Pennsylvania Right to Know List
environmental hazard

N-ISOPROPYLANILINE   643-28-7

HEALTH AND SAFETY LISTS

IARC - Group 1 (carcinogenic to humans)
[present]

OSHA - Select Carcinogens
[present]

STATE LISTS

Pennsylvania Right to Know List
special hazardous substance

Pennsylvania RTK - Special Hazardous Substances
[present]

N-ISOPROPYLANILINE   768-52-5

HEALTH AND SAFETY LISTS

ACGIH 1995 - Time Weighted Averages
5 ppm TWA; 12 mg/m3 TWA

ACGIH 1995 - Short Term Exposure Limits
10 ppm STEL; 24 mg/m3 STEL

AIHA - Odor Threshold Values
geometric mean air odor threshold = 0.21 ppm (detectable); 0.70 ppm (recognizable)

U.S. DOT - Substances From 49 CFR 172.101
regulated by DOT (UN1221)

U.S. DOT - Hazard Classes
DOT hazard class = 3

NFPA - Flash Points
flash point = -35 degrees F (-37 degrees C)

NFPA - Hazard Identification Ratings
health-3; flammability-4; reactivity-0

NIOSH - Selected LD50s and LC50s
Inhalation, rat: LC50 = 4000 ppm 4 hr Oral, rat: LD50 = 820 mg/kg Skin, rabbit: LD50 = 380 mg/kg

NIOSH 1990 - Pocket Guide - RELs
See Appendix D

NIOSH 1990 - Pocket Guide - IDLHs
4000 ppm IDLH

NIOSH 1990 - Pocket Guide - Target organs
eyes, skin, respiratory system

OSHA - Vacated PELs - Time Weighted Averages
5 ppm TWA; 12 mg/m3 TWA

OSHA - Vacated PELs - Short Term Exposure Limits
10 ppm STEL; 24 mg/m3 STEL

OSHA - Final PELs - Time Weighted Averages
5 ppm TWA; 12 mg/m3 TWA

OSHA - List of Highly Hazardous Chemicals
threshhold quantity = 5000 pounds

ENVIRONMENTAL LISTS

CAA - Flammable Substances for Accidental Release Prevention
threshold quantity = 10,000 lbs

List of Pesticide Product Inert Ingredients
[present]

INTERNATIONAL LISTS

Australian Exposure Standards - Time Weighted Averages
5 ppm TWA; 12 mg/m3 TWA

Australian Exposure Standards - Short Term Exposure Limits
10 ppm STEL; 24 mg/m3 STEL

Canada - WHMIS: Ingredient Disclosure
1% item 907 (1052)

Canada - Alberta - 8 Hour Occupational Exposure Limit
5 ppm TWA; 12 mg/m3 TWA

Canada - Alberta - 15 Minute Occupational Exposure Limit
10 ppm STEL; 24 mg/m3 STEL

Canada - British Columbia - 8 Hour Exposure Limits
5 ppm TWA; 12 mg/m3 TWA

Canada - British Columbia - 15 Minute Exposure Limits
10 ppm STEL; 24 mg/m3 STEL

Canada - Ontario - OHSA - TWAEVs
5 ppm TWAEV; 12 mg/m3 TWAEV

Canada - Ontario - OHSA - STEVs
10 ppm STEV; 24 mg/m3 STEV

Canada - Quebec - Time-Weighted Average Exposure Values
5 ppm TWAEV; 12 mg/m3 TWAEV

Canada - Quebec - Short-term Exposure Values
10 ppm STEV; 24 mg/m3 STEV

German (DFG) - MAK Values
5 ppm MAK; 12 mg/m3 MAK

German (DFG) - Peak Limitations
2 x normal MAK (30 min. average value); don't exceed 4 times during shift

Israel - Time Weighted Averages
5 ppm TWA; 12 mg/m3 TWA

Israel - Short Term Exposure Limits
10 ppm STEL; 24 mg/m3 STEL

Israel - Action Levels
2.5 ppm AL; 6 mg/m3 AL

Mexico - Instruction No. 10 - TWAs
5 ppm TWA; 12 mg/m3 TWA

Mexico - Instruction No. 10 - STELs
10 ppm STEL; 24 mg/m3 STEL

STATE LISTS

California - Exposure Limits - PELs
5 ppm PEL; 12 mg/m3 PEL

California - Exposure Limits - STELs
10 ppm STEL; 24 mg/m3 STEL

California - Directors List of Hazardous Substances (8 CCR 339)
[present]

Florida Hazardous Substance List
[present]

Massachusetts Right To Know List
[present]

Minnesota Hazardous Substance List
[present]

NJ Right to Know List (Total)
sn 1077

NJ Special Hazardous Substances
(flammable - fourth degree)

Pennsylvania Right to Know List
[present]

ISOPROPYLATED CRESOL   68987-86-0

ENVIRONMENTAL LISTS

TSCA - Code of Federal Regulations Citations
40 CFR 721.1250

TSCA - Chemicals with Significant New Use Rules
[present]

TSCA - Section 12(b) - Export Notification
P-80-289; export notification required - Section 5

ISOPROPYL BENZOATE   939-48-0
ENVIRONMENTAL LISTS
List of Pesticide Product Inert Ingredients
[present]

ISOPROPYL BICYCLOHEXYL   RR-00866-2
ENVIRONMENTAL LISTS
List of Pesticide Product Inert Ingredients
[present]

ISOPROPYL BIPHENYL   25640-78-2
ENVIRONMENTAL LISTS
List of Pesticide Product Inert Ingredients
[present]

ISOPROPYL BUTYRATE   638-11-9
ENVIRONMENTAL LISTS
List of Pesticide Product Inert Ingredients
[present]

ISOPROPYL CHLOROACETATE   105-48-6
HEALTH AND SAFETY LISTS
NIOSH - Selected LD50s and LC50s
Oral, rat: LD50 = 1180 mg/kg
INTERNATIONAL LISTS
Canada - WHMIS: Ingredient Disclosure
1% item 908 (1053)
Canada - Alberta - 8 Hour Occupational Exposure Limit
2 ppm TWA;   11 mg/m3 TWA
Canada - Alberta - Skin Designation
can be absorbed through the intact skin
STATE LISTS
California - Exposure Limits - PELs
2 ppm PEL;   10 mg/m3 PEL
California - Exposure Limits - Skin Notation
material may be absorbed through the skin, eyes or mucous membrane
California - Directors List of Hazardous Substances (8 CCR 339)
[present]
Florida Hazardous Substance List
[present]
Massachusetts Right To Know List
carcinogen;   extraordinarily hazardous

ISOPROPYL CHLOROFORMATE   108-23-6
HEALTH AND SAFETY LISTS
ACGIH 1995 - Time Weighted Averages
2 ppm TWA;   11 mg/m3 TWA
ACGIH 1995 - Skin Designations
skin - potential for cutaneous absorption
NFPA - Flash Points
flash point < 212 degrees F (100 degrees C)
NFPA - Hazard Identification Ratings
health-0;   flammability-1
OSHA - Vacated PELs - Time Weighted Averages
2 ppm TWA;   10 mg/m3 TWA
OSHA - Vacated PELs - Skin Designation
Prevent or reduce skin absorption
ENVIRONMENTAL LISTS
TSCA - Health and Safety Reporting List
Effective Date: March 11, 1994;   Sunset Date: March 11, 2004

INTERNATIONAL LISTS
Australian Exposure Standards - Time Weighted Averages
2 ppm TWA;   11 mg/m3 TWA
Australian Exposure Standards - Skin Effects
skin absorption
Canada - Alberta - 15 Minute Occupational Exposure Limit
5 ppm STEL;   22 mg/m3 STEL
Canada - Ontario - OHSA - TWAEVs
2 ppm TWAEV;   11 mg/m3 TWAEV
Canada - Ontario - OHSA - Skin Notations
absorption through skin, eyes, or mucous membranes
Canada - Quebec - Time-Weighted Average Exposure Values
2 ppm TWAEV;   11 mg/m3 TWAEV
Canada - Quebec - Skin Designations
absorbed through the skin
Israel - Time Weighted Averages
2 ppm TWA;   11 mg/m3 TWA
Israel - Action Levels
1 ppm AL;   5.5 mg/m3 AL
STATE LISTS
Massachusetts Right To Know List
[present]
Minnesota Hazardous Substance List
skin
Pennsylvania Right to Know List
[present]
PROPOSED REGULATIONS
TSCA - ITC 32nd Report Priority Testing List
designated for dermal absorption testing

ISOPROPYL-2-CHLOROPROPIONATE   40058-87-5
ENVIRONMENTAL LISTS
List of Pesticide Product Inert Ingredients
[present]

ISOPROPYLCUMYL HYDROPEROXIDE   RR-01433-5
HEALTH AND SAFETY LISTS
NFPA - Flash Points
flash point = 210 degrees F (99 degrees C)
NFPA - Hazard Identification Ratings
health-1;   flammability-1

ISOPROPYLCYCLOHEXANE   696-29-7
HEALTH AND SAFETY LISTS
NFPA - Flash Points
flash point = 255 degrees F (124 degrees C)
NFPA - Hazard Identification Ratings
health-0;   flammability-1;   reactivity-0

ISOPROPYL CYCLOHEXYLAMINE   1195-42-2
HEALTH AND SAFETY LISTS
NFPA - Flash Points
flash point = 285 degrees F (141 degrees C)
NFPA - Hazard Identification Ratings
health-0;   flammability-1;   reactivity-0
ENVIRONMENTAL LISTS
TSCA - Code of Federal Regulations Citations
40 CFR 712.30(k);   40 CFR 716.120(a)
TSCA - PAIR - Reporting List
Reporting Date: August 27, 1984
TSCA - Health and Safety Reporting List
Effective Date: June 28, 1984;   Sunset Date: November 9, 1993

ISOPROPYL ETHANOLAMINE   109-56-8

HEALTH AND SAFETY LISTS

U.S. DOT - Substances From 49 CFR 172.101
regulated by DOT (UN2405)

U.S. DOT - Hazard Classes
DOT hazard class = 3

STATE LISTS

NJ Right to Know List (Total)
sn 1078

ISOPROPYL ETHER   108-20-3

HEALTH AND SAFETY LISTS

U.S. DOT - Substances From 49 CFR 172.101
regulated by DOT (UN2947)

U.S. DOT - Hazard Classes
DOT hazard class = 3

STATE LISTS

NJ Right to Know List (Total)
sn 1079

ISOPROPYL FORMATE   625-55-8

HEALTH AND SAFETY LISTS

U.S. DOT - Substances From 49 CFR 172.101
regulated by DOT (UN2407)

U.S. DOT - Hazard Classes
DOT hazard class = 3

U.S. DOT - Substances Which Are Poisonous by Inhalation
liquid hazardous material poisonous by inhalation (UN2407)

NIOSH - Selected LD50s and LC50s
*Inhalation, mouse: LD50 = 299 ppm 1 hr Oral, rat: LD50 = 1070
mg/kg Skin, mouse: LD50 = 12 mg/kg*

ENVIRONMENTAL LISTS

CERCLA/SARA - Section 302 Extremely Hazardous Substances and
TPQs
TPQ = 1000 pounds

CAA -Toxic Substances for Accidental Release Prevention
threshold quantity = 15,000 lbs

INTERNATIONAL LISTS

Canada - WHMIS: Ingredient Disclosure
1% item 910 (438)

United Kingdom - Occupational Exposure Standards - TWAs
1 ppm TWA; 5 mg/m3 TWA

STATE LISTS

Florida Hazardous Substance List
effective March 13, 1992

Massachusetts Right To Know List
extraordinarily hazardous

NJ Right to Know List (Total)
sn 1080

NJ Special Hazardous Substances
(corrosive)

Pennsylvania Right to Know List
environmental hazard

PROPOSED REGULATIONS

CERCLA/SARA - Proposed Hazardous Substance Additions
proposed RQ = 1 pound (.454 kg)

CERCLA/SARA - 1989 Proposed RQ Adjustments
proposed RQ = 1000 pounds (454 kg)

ISOPROPYL GLYCIDYL ETHER (IGE)   4016-14-2

HEALTH AND SAFETY LISTS

U.S. DOT - Substances From 49 CFR 172.101
regulated by DOT (UN2934)

U.S. DOT - Hazard Classes
DOT hazard class = 3

STATE LISTS

NJ Right to Know List (Total)
sn 1081

4-ISOPROPYLHEPTANE   52896-87-4

HEALTH AND SAFETY LISTS

U.S. DOT - Hazard Classes
Forbidden from transport by the DOT

U.S. DOT - Organic Peroxides Table
Organic peroxide UN3109

ISOPROPYLIDENE, BIS(1,1-DIMETHYLPROPYL)   RR-01228-2
DERIVATIVE

HEALTH AND SAFETY LISTS

NFPA - Hazard Identification Ratings
health-1; reactivity-0

4,4'-ISOPROPYLIDENE DIPHENOL ALKYL (C12-15)   RR-01092-4
PHOSPHITES

HEALTH AND SAFETY LISTS

NFPA - Flash Points
flash point = 93 degrees F (34 degrees C)

NFPA - Hazard Identification Ratings
health-3; flammability-3; reactivity-0

STATE LISTS

Florida Hazardous Substance List
[present]

Massachusetts Right To Know List
[present]

Pennsylvania Right to Know List
[present]

ISOPROPYL ISOBUTYRATE   617-50-5

HEALTH AND SAFETY LISTS

NIOSH - Selected LD50s and LC50s
Oral, mouse: LD50 = 1250 mg/kg

ISOPROPYL ISOCYANATE   1795-48-8

HEALTH AND SAFETY LISTS

ACGIH 1995 - Time Weighted Averages
250 ppm TWA; 1040 mg/m3 TWA

ACGIH 1995 - Short Term Exposure Limits
310 ppm STEL; 1300 mg/m3 STEL

AIHA - Odor Threshold Values
*geometric mean air odor threshold = 0.017 ppm (detectable); 0.053
ppm (recognizable)*

U.S. DOT - Substances From 49 CFR 172.101
regulated by DOT (UN1159)

U.S. DOT - Hazard Classes
DOT hazard class = 3

NFPA - Flash Points
flash point = -18 degrees F (-28 degrees C)

NFPA - Hazard Identification Ratings
health-1; flammability-3; reactivity-1

NIOSH - Selected LD50s and LC50s
*Inhalation, rat: LC50 = 162 gm/m3 8 hr Oral, rat: LD50 = 8470
mg/kg Skin, rabbit: LD50 = 20 gm/kg*

NIOSH 1990 - Pocket Guide - RELs
500 ppm TWA; 2100 mg/m3 TWA

NIOSH 1990 - Pocket Guide - IDLHs
10,000 ppm IDLH

NIOSH 1990 - Pocket Guide - Target organs
skin, respiratory system

OSHA - Vacated PELs - Time Weighted Averages
500 ppm TWA; 2100 mg/m3 TWA

OSHA - Final PELs - Time Weighted Averages
500 ppm TWA; 2100 mg/m3 TWA

INTERNATIONAL LISTS

Australian Exposure Standards - Time Weighted Averages
250 ppm TWA; 1040 mg/m3 TWA

Australian Exposure Standards - Short Term Exposure Limits
310 ppm STEL; 1300 mg/m3 STEL

Canada - WHMIS: Ingredient Disclosure
1% item 911 (844)

Canada - Alberta - 8 Hour Occupational Exposure Limit
250 ppm TWA; 1044 mg/m3 TWA

Canada - Alberta - 15 Minute Occupational Exposure Limit
310 ppm STEL; 1316 mg/m3 STEL

Canada - British Columbia - 8 Hour Exposure Limits
250 ppm TWA; 1050 mg/m3 TWA

Canada - British Columbia - 15 Minute Exposure Limits
310 ppm STEL; 1320 mg/m3 STEL

Canada - Ontario - OHSA - TWAEVs
250 ppm TWAEV; 1045 mg/m3 TWAEV

Canada - Ontario - OHSA - STEVs
310 ppm STEV; 1295 mg/m3 STEV

Canada - Quebec - Time-Weighted Average Exposure Values
250 ppm TWAEV; 1040 mg/m3 TWAEV

Canada - Quebec - Short-term Exposure Values
310 ppm STEV; 1300 mg/m3 STEV

United Kingdom - Occupational Exposure Standards - TWAs
250 ppm TWA; 1050 mg/m3 TWA

United Kingdom - Occupational Exposure Standards - STELs
310 ppm STEL; 1320 mg/m3 STEL

German (DFG) - MAK Values
500 ppm MAK; 2100 mg/m3 MAK

Israel - Time Weighted Averages
250 ppm TWA; 1040 mg/m3 TWA

Israel - Short Term Exposure Limits
310 ppm STEL; 1300 mg/m3 STEL

Israel - Action Levels
125 ppm AL; 520 mg/m3 AL

Mexico - Instruction No. 10 - TWAs
250 ppm TWA; 1050 mg/m3 TWA

Mexico - Instruction No. 10 - STELs
310 ppm STEL; 1320 mg/m3 STEL

STATE LISTS

California - Exposure Limits - PELs
250 ppm PEL; 1050 mg/m3 PEL

California - Directors List of Hazardous Substances (8 CCR 339)
[present]

Florida Hazardous Substance List
[present]

Massachusetts Right To Know List
[present]

Minnesota Hazardous Substance List
[present]

NJ Right to Know List (Total)
sn 0730

NJ Special Hazardous Substances
(flammable - third degree)

Pennsylvania Right to Know List
[present]

ISOPROPYL LACTATE   617-51-6

HEALTH AND SAFETY LISTS

NFPA - Flash Points
flash point = 22 degrees F (-6 degrees C)

NFPA - Hazard Identification Ratings
health-2; flammability-3; reactivity-0

NIOSH - Selected LD50s and LC50s
Oral, guinea pig: LD50 = 1400 ug/kg

STATE LISTS

Florida Hazardous Substance List
[present]

Massachusetts Right To Know List
extraordinarily hazardous

NJ Right to Know List (Total)
sn 2504

Pennsylvania Right to Know List
environmental hazard

PROPOSED REGULATIONS

CERCLA/SARA - Proposed Hazardous Substance Additions
proposed RQ = 1 pound (.454 kg)

CERCLA/SARA - 1989 Proposed RQ Adjustments
proposed RQ = 100 pounds (45.4 kg)

ISOPROPYL LANOLIN   63393-93-1

SEE ALSO:
GLYCIDOL (OXIRANEMETHANOL) AND ITS DERIVATIVES

HEALTH AND SAFETY LISTS

ACGIH 1995 - Time Weighted Averages
50 ppm TWA; 238 mg/m3 TWA

ACGIH 1995 - Short Term Exposure Limits
75 ppm STEL; 356 mg/m3 STEL

NIOSH - Selected LD50s and LC50s
Inhalation, rat: LC50 = 1100 ppm 8 hr Oral, rat: LD50 = 4200 mg/kg Skin, rabbit: LD50 = 9650 mg/kg

NIOSH 1990 - Pocket Guide - RELs
C 50 ppm (15 min); C 240 mg/m3 (15 min)

NIOSH 1990 - Pocket Guide - IDLHs
1000 ppm IDLH

NIOSH 1990 - Pocket Guide - Target organs
eyes, skin, respiratory system

NIOSH - Health Standards - Exposure Limits
C (15 min) 50 ppm; C (15 min) 240 mg/m3 (Listed under 'Glycidyl ethers')

NIOSH - Health Standards - Health Effects and Precautions
Skin and mucous membrane effects; sensitization potential; possible hematopoietic and reproductive system effects (Medical monitoring required) (Listed under 'Glycidyl ethers')

OSHA - Vacated PELs - Time Weighted Averages
50 ppm TWA; 240 mg/m3 TWA

OSHA - Vacated PELs - Short Term Exposure Limits
75 ppm STEL; 360 mg/m3 STEL

OSHA - Final PELs - Time Weighted Averages
50 ppm TWA; 240 mg/m3 TWA

ENVIRONMENTAL LISTS

EPA - Master Testing List
[present]

TSCA - Code of Federal Regulations Citations
40 CFR 712.30(d); 40 CFR 716.120(c)

TSCA - PAIR - Reporting List
Reporting Date: November 19, 1982

INTERNATIONAL LISTS

Australian Exposure Standards - Time Weighted Averages
50 ppm TWA; 238 mg/m3 TWA

Australian Exposure Standards - Short Term Exposure Limits
75 ppm STEL; 356 mg/m3 STEL

Canada - WHMIS: Ingredient Disclosure
1% item 912 (826)

Canada - Alberta - 8 Hour Occupational Exposure Limit
50 ppm TWA; 237 mg/m3 TWA

Canada - Alberta - 15 Minute Occupational Exposure Limit
75 ppm STEL; 356 mg/m3 STEL

Canada - British Columbia - 8 Hour Exposure Limits
50 ppm TWA; 240 mg/m3 TWA

Canada - British Columbia - 15 Minute Exposure Limits
75 ppm STEL; 360 mg/m3 STEL

Canada - Ontario - OHSA - TWAEVs
50 ppm TWAEV; 237 mg/m3 TWAEV

Canada - Ontario - OHSA - STEVs
75 ppm STEV; 356 mg/m3 STEV

Canada - Quebec - Time-Weighted Average Exposure Values
50 ppm TWAEV; 238 mg/m3 TWAEV

Canada - Quebec - Short-term Exposure Values
75 ppm STEV; 356 mg/m3 STEV

United Kingdom - Occupational Exposure Standards - TWAs
50 ppm TWA; 240 mg/m3 TWA

United Kingdom - Occupational Exposure Standards - STELs
75 ppm STEL; 360 mg/m3 STEL

German (DFG) - Carcinogens
animal evidence of carcinogenicity

Israel - Time Weighted Averages
50 ppm TWA; 238 mg/m3 TWA

Israel - Short Term Exposure Limits
75 ppm STEL; 356 mg/m3 STEL

Israel - Action Levels
25 ppm AL; 119 mg/m3 AL

Mexico - Instruction No. 10 - TWAs
50 ppm TWA; 240 mg/m3 TWA

Mexico - Instruction No. 10 - STELs
75 ppm STEL; 360 mg/m3 STEL

STATE LISTS

California - Exposure Limits - PELs
50 ppm PEL; 240 mg/m3 PEL

California - Exposure Limits - STELs
75 ppm STEL; 360 mg/m3 STEL

California - Directors List of Hazardous Substances (8 CCR 339)
[present] (exempt when part of a cured epoxy or rubber)

Florida Hazardous Substance List
[present]

Massachusetts Right To Know List
[present]

Minnesota Hazardous Substance List
[present]

Pennsylvania Right to Know List
[present]

PROPOSED REGULATIONS

TSCA - Proposed Testing Rule for Glycidyl Ethers
member of Glycidyl subcategory I-A

ISOPROPYL MERCAPTAN   75-33-2

HEALTH AND SAFETY LISTS

NFPA - Hazard Identification Ratings
health-0; flammability-2; reactivity-0

ISOPROPYL METHANE SULPHONATE   926-06-7

ENVIRONMENTAL LISTS

TSCA - Chemicals with Significant New Use Rules
PMN number: P-85-648

ISOPROPYLMETHYLPYRAZOLYL DIMETHYLCARBAMATE   119-38-0

ENVIRONMENTAL LISTS

List of Pesticide Product Inert Ingredients
[present]

ISOPROPYL MYRISTATE   110-27-0

HEALTH AND SAFETY LISTS

U.S. DOT - Substances From 49 CFR 172.101
regulated by DOT (UN2406)

U.S. DOT - Hazard Classes
DOT hazard class = 3

STATE LISTS

NJ Right to Know List (Total)
sn 1083

ISOPROPYLNAPTHALENESULFONIC ACID, SODIUM SALT   28348-64-3

HEALTH AND SAFETY LISTS

U.S. DOT - Substances From 49 CFR 172.101
regulated by DOT (UN2483)

U.S. DOT - Hazard Classes
DOT hazard class = 3

U.S. DOT - Substances Which Are Poisonous by Inhalation
liquid hazardous material poisonous by inhalation (UN2483)

INTERNATIONAL LISTS

Canada - WHMIS: Ingredient Disclosure
0.1% item 913 (1042)

STATE LISTS

NJ Right to Know List (Total)
sn 1084

ISOPROPYL NITRATE   1712-64-7

HEALTH AND SAFETY LISTS

NFPA - Flash Points
flash point = 130 degrees F (54 degrees C)

NFPA - Hazard Identification Ratings
health-2; flammability-2; reactivity-0

STATE LISTS

Florida Hazardous Substance List
[present]

Massachusetts Right To Know List
[present]

Pennsylvania Right to Know List
[present]

ISOPROPYL OIL   RR-00813-9

ENVIRONMENTAL LISTS

List of Pesticide Product Inert Ingredients
[present]

ISOPROPYL PALMITATE   142-91-6

STATE LISTS

Massachusetts Right To Know List
[present]

NJ Right to Know List (Total)
sn 1085

ISOPROPYL PEROXYDICARBONATE   105-64-6

INTERNATIONAL LISTS

Canada - WHMIS: Ingredient Disclosure
0.1% item 914 (1102)

ISOPROPYL PHENOL   25168-06-3

HEALTH AND SAFETY LISTS

NIOSH - Selected LD50s and LC50s
Oral, rat: LD50 = 10800 ug/kg Skin, rat: LD50 = 5600 ug/kg

ENVIRONMENTAL LISTS

CERCLA/SARA - Section 302 Extremely Hazardous Substances and TPQs
TPQ = 500 pounds

STATE LISTS

Florida Hazardous Substance List
effective March 13, 1992

Massachusetts Right To Know List
extraordinarily hazardous

NJ Right to Know List (Total)
sn 2505

Pennsylvania Right to Know List
environmental hazard

PROPOSED REGULATIONS

CERCLA/SARA - Proposed Hazardous Substance Additions
proposed RQ = 1 pound (.454 kg)

CERCLA/SARA - 1989 Proposed RQ Adjustments
proposed RQ = 100 pounds (45.4 kg)

N-ISOPROPYL-N'-PHENYL-P-PHENYLENE-DIAMINE   101-72-4

ENVIRONMENTAL LISTS

EPA - Master Testing List
[present]

List of Pesticide Product Inert Ingredients
[present]

ISOPROPYL PHOSPHORIC ACID   1623-24-1

ENVIRONMENTAL LISTS

List of Pesticide Product Inert Ingredients
[present]

ISOPROPYL PROPIONATE   637-78-5

HEALTH AND SAFETY LISTS

U.S. DOT - Substances From 49 CFR 172.101
regulated by DOT (UN1222)

U.S. DOT - Hazard Classes
DOT hazard class = 3

NIOSH - Selected LD50s and LC50s
Inhalation, mouse: LC50 = 65 gm/m3 2 hr

STATE LISTS

NJ Right to Know List (Total)
sn 1086

ISOPROPYL STEARATE   112-10-7

HEALTH AND SAFETY LISTS

IARC - Group 3 (not classifiable)
[present]

INTERNATIONAL LISTS

German (DFG) - Carcinogens
suspected carcinogen (residue of isopropyl alcohol production)

ISOQUINOLINE   119-65-3

ENVIRONMENTAL LISTS

List of Pesticide Product Inert Ingredients
[present]

ISOSAFROLE   120-58-1

HEALTH AND SAFETY LISTS

U.S. DOT - Organic Peroxides Table
Organic peroxide UN3112;   UN3115

NFPA - Flash Points
rapid decomposition at 53 degrees F (12 degrees C)

NFPA - Hazard Identification Ratings
health-0; flammability-4; reactivity-4 (oxidizing properties)

NIOSH - Selected LD50s and LC50s
Oral, rat: LD50 = 2140 mg/kg Skin, rabbit: LD50 = 2025 mg/kg

OSHA - List of Highly Hazardous Chemicals
threshhold quantity = 7500 pounds

INTERNATIONAL LISTS

Canada - WHMIS: Ingredient Disclosure
1% item 915 (1368)

STATE LISTS

Florida Hazardous Substance List
[present]

Massachusetts Right To Know List
[present]

NJ Right to Know List (Total)
sn 0731

NJ Special Hazardous Substances
(flammable - fourth degree; reactive - fourth degree)

Pennsylvania Right to Know List
[present]

ISOSORBIDE DINITRATE MIXTURE   87-33-2

ENVIRONMENTAL LISTS

CAA - HON Rule - SOCMI Chemicals
compliance by April 24, 1995

TSCA - Code of Federal Regulations Citations
40 CFR 712.30(w);   40 CFR 716.120(a)

TSCA - PAIR - Reporting List
Reporting Date: June 13, 1989

TSCA - Health and Safety Reporting List
Effective Date: April 13, 1989;   Sunset Date: November 9, 1993

ISOTERTINATE   4759-48-2

HEALTH AND SAFETY LISTS

NIOSH - Selected LD50s and LC50s
Oral, rat: LD50 = 555 mg/kg

ENVIRONMENTAL LISTS

EPA - Master Testing List
[present]

INTERNATIONAL LISTS

Canada - WHMIS: Ingredient Disclosure
1% item 916 (1055)

3(2H)-ISOTHIAZOLONE, 2-OCTYL-   26530-20-1

HEALTH AND SAFETY LISTS

U.S. DOT - Substances From 49 CFR 172.101
regulated by DOT (UN1793)

U.S. DOT - Hazard Classes
DOT hazard class = 8

ENVIRONMENTAL LISTS

TSCA - Code of Federal Regulations Citations
40 CFR 712.30(x); 716.120(d)

TSCA - PAIR - Reporting List
Reporting Date: December 27, 1990

TSCA - Health and Safety Reporting List
Effective Date: October 29, 1990; Sunset Date: November 9, 1993

INTERNATIONAL LISTS

Canada - WHMIS: Ingredient Disclosure
1% item 917 (99)

STATE LISTS

NJ Right to Know List (Total)
sn 1075

NJ Special Hazardous Substances
(corrosive)

ISOTHIOCYANIC ACID   3129-90-6

HEALTH AND SAFETY LISTS

U.S. DOT - Substances From 49 CFR 172.101
regulated by DOT (UN2409)

U.S. DOT - Hazard Classes
DOT hazard class = 3

STATE LISTS

NJ Right to Know List (Total)
sn 1087

ISOTRIDECYL ALCOHOL   27458-92-0

ENVIRONMENTAL LISTS

List of Pesticide Product Inert Ingredients
[present]

ISOXATHION   RR-01291-9

HEALTH AND SAFETY LISTS

NIOSH - Selected LD50s and LC50s
Oral, rat: LD50 = 360 mg/kg Skin, rabbit: LD50 = 590 mg/kg

INTERNATIONAL LISTS

Canada - WHMIS: Ingredient Disclosure
1% item 918 (1056)

JACOBINE   6870-67-3

SEE ALSO:
F039-HAZARDOUS WASTES

HEALTH AND SAFETY LISTS

U.S. DOT - Appendix A Table 1 - Hazardous Substances
final RQ = 100 pounds (45.4 kg)

FDA - Controlled Substances Act - Precursor chemicals
Threshold by base weight = 4 kilograms

IARC - Group 3 (not classifiable)
[present]

NIOSH - Selected LD50s and LC50s
Oral, rat: LD50 = 1340 mg/kg

ENVIRONMENTAL LISTS

CERCLA/SARA - Section 313 - Emission Reporting
form R reporting required for 1.0% de minimus concentration

CERCLA/SARA - Hazardous Substances and their Reportable Quantities
final RQ = 100 pounds (45.4 kg)

EPA - Carcinogen Hazard Ranking for RQ Adjustment
Hazard ranking = Low

RCRA - U Series Wastes
waste number U141

RCRA - Hazardous Constituents-Appendix VIII
waste number U141

RCRA - Basis for Listing - Appendix VII
Included in waste stream: F039

RCRA - Substances Banned From Land Disposal
[present]

RCRA - TSD Facilities Ground Water Monitoring
TM 8270 = 10 ug/L PQL

RCRA - Universal Treatment Standards (LDR)
WW: 0.081 mg/l; NWW: 2.6 mg/kg

INTERNATIONAL LISTS

Canada - NPRI (National Pollutant Release Inventory)
[present]

STATE LISTS

California - Air Bill 2588 Appendix A-II
known or potential carcinogen: 9/90

California - Prop. 65 - Cancer list
carcinogen - initial date 10/1/89

California - Directors List of Hazardous Substances (8 CCR 339)
[present]

Florida Hazardous Substance List
[present]

Massachusetts Right To Know List
carcinogen; extraordinarily hazardous

NJ Special Hazardous Substances
(carcinogen)

Pennsylvania Right to Know List
environmental hazard; special hazardous substance

Pennsylvania RTK - Special Hazardous Substances
[present]

JAJOBA BEAN OIL   61789-91-1

STATE LISTS

NJ Right to Know List (Total)
sn 2508

JASMOLIN I   4466-14-2

STATE LISTS

California - Air Bill 2588 Appendix A-II
[present]

California - Prop. 65 - Developmental Toxicity
developmental toxicity - initial date 7/1/87

Massachusetts Right To Know List
teratogen

JASMOLIN II   1172-63-0

HEALTH AND SAFETY LISTS

NIOSH - Selected LD50s and LC50s
Oral, rat: LD50 = 550 mg/kg Skin, rabbit: LD50 = 690 mg/kg

ENVIRONMENTAL LISTS

TSCA - Code of Federal Regulations Citations
40 CFR 716.120(a)

TSCA - Health and Safety Reporting List
Effective Date: June 1, 1987

JASMONE   **488-10-8**

HEALTH AND SAFETY LISTS

   U.S. DOT - Hazard Classes
Forbidden from transport by the DOT

JET FUEL   **RR-01549-6**

HEALTH AND SAFETY LISTS

   NIOSH - Selected LD50s and LC50s
Oral, rat: LD50 = 17 gm/kg

ENVIRONMENTAL LISTS

   List of Pesticide Product Inert Ingredients
[present]

JET FUELS JET A AND JET A-1   **RR-00130-9**

HEALTH AND SAFETY LISTS

   U.S. DOT - Appendix B - Marine Pollutants
DOT regulated severe marine pollutant

JET FUELS JET B   **RR-00788-5**

HEALTH AND SAFETY LISTS

   IARC - Group 3 (not classifiable)
[present]

JET FUELS JP-4   **RR-00789-6**

ENVIRONMENTAL LISTS

   List of Pesticide Product Inert Ingredients
[present]

JET FUELS JP-6   **RR-00790-9**

STATE LISTS

   Massachusetts Right To Know List
[present]

K001-HAZARDOUS WASTES   **RR-00656-4**

STATE LISTS

   Massachusetts Right To Know List
[present]

K002-HAZARDOUS WASTES   **RR-00657-5**

INTERNATIONAL LISTS

   Canada - WHMIS: Ingredient Disclosure
1% item 919 (1058)

K003-HAZARDOUS WASTES   **RR-00658-6**

HEALTH AND SAFETY LISTS

   IARC - Group 3 (not classifiable)
[present]

K004-HAZARDOUS WASTES   **RR-00659-7**

HEALTH AND SAFETY LISTS

   NFPA - Flash Points
flash point = 110 to 150 degrees F (43 to 66 degrees C)

   NFPA - Hazard Identification Ratings
health-0; flammability-2; reactivity-0

K005-HAZARDOUS WASTES   **RR-00660-0**

HEALTH AND SAFETY LISTS

   NFPA - Flash Points
flash point = -10 to 30 degrees F (-23 to -1 degrees C)

   NFPA - Hazard Identification Ratings
health-1; flammability-3; reactivity-0

STATE LISTS

   Pennsylvania Right to Know List
[present]

K006-HAZARDOUS WASTES   **RR-00661-1**

HEALTH AND SAFETY LISTS

   NFPA - Flash Points
flash point = -10 to 30 degrees F (-23 to -1 degrees C)

   NFPA - Hazard Identification Ratings
health-1; flammability-3; reactivity-0

STATE LISTS

   Pennsylvania Right to Know List
[present]

K007-HAZARDOUS WASTES   **RR-00662-2**

HEALTH AND SAFETY LISTS

   NFPA - Flash Points
flash point = 100 degrees F (38 degrees C)

STATE LISTS

   Pennsylvania Right to Know List
[present]

K008-HAZARDOUS WASTES   **RR-00663-3**

HEALTH AND SAFETY LISTS

   U.S. DOT - Appendix A Table 1 - Hazardous Substances
final RQ = 1 pound (0.454 kg)

ENVIRONMENTAL LISTS

   CERCLA/SARA - Hazardous Substances and their Reportable Quantities
final RQ = 1 pound (0.454 kg)

   RCRA - K Series Wastes
Toxic waste

   RCRA - Substances Banned From Land Disposal
[present]

K009-HAZARDOUS WASTES   **RR-00664-4**

HEALTH AND SAFETY LISTS

   U.S. DOT - Appendix A Table 1 - Hazardous Substances
final RQ = 10 pounds (4.54 kg)

ENVIRONMENTAL LISTS

   CERCLA/SARA - Hazardous Substances and their Reportable Quantities
final RQ = 10 pounds (4.54 kg)

   RCRA - K Series Wastes
Toxic waste

   RCRA - Substances Banned From Land Disposal
[present]

K010-HAZARDOUS WASTES   **RR-00665-5**

HEALTH AND SAFETY LISTS

   U.S. DOT - Appendix A Table 1 - Hazardous Substances
final RQ = 10 pounds (4.54 kg)

ENVIRONMENTAL LISTS

   CERCLA/SARA - Hazardous Substances and their Reportable Quantities
final RQ = 10 pounds (4.54 kg)

   RCRA - K Series Wastes
Toxic waste

   RCRA - Substances Banned From Land Disposal
[present]

K011-HAZARDOUS WASTES   RR-00666-6

HEALTH AND SAFETY LISTS

U.S. DOT - Appendix A Table 1 - Hazardous Substances
final RQ = 10 pounds (4.54 kg)

ENVIRONMENTAL LISTS

CERCLA/SARA - Hazardous Substances and their Reportable Quantities
final RQ = 10 pounds (4.54 kg)

RCRA - K Series Wastes
Toxic waste

RCRA - Substances Banned From Land Disposal
[present]

K013-HAZARDOUS WASTES   RR-00667-7

HEALTH AND SAFETY LISTS

U.S. DOT - Appendix A Table 1 - Hazardous Substances
final RQ = 10 pounds (4.54 kg)

ENVIRONMENTAL LISTS

CERCLA/SARA - Hazardous Substances and their Reportable Quantities
final RQ = 10 pounds (4.54 kg)

RCRA - K Series Wastes
Toxic waste

RCRA - Substances Banned From Land Disposal
[present]

K014-HAZARDOUS WASTES   RR-00668-8

HEALTH AND SAFETY LISTS

U.S. DOT - Appendix A Table 1 - Hazardous Substances
final RQ = 10 pounds (4.54 kg)

ENVIRONMENTAL LISTS

CERCLA/SARA - Hazardous Substances and their Reportable Quantities
final RQ = 10 pounds (4.54 kg)

RCRA - K Series Wastes
Toxic waste

RCRA - Substances Banned From Land Disposal
[present]

K015-HAZARDOUS WASTES   RR-00669-9

HEALTH AND SAFETY LISTS

U.S. DOT - Appendix A Table 1 - Hazardous Substances
final RQ = 10 pounds (4.54 kg)

ENVIRONMENTAL LISTS

CERCLA/SARA - Hazardous Substances and their Reportable Quantities
final RQ = 10 pounds (4.54 kg)

RCRA - K Series Wastes
Toxic waste

RCRA - Substances Banned From Land Disposal
[present]

K016-HAZARDOUS WASTES   RR-00670-2

HEALTH AND SAFETY LISTS

U.S. DOT - Appendix A Table 1 - Hazardous Substances
final RQ = 10 pounds (4.54 kg)

ENVIRONMENTAL LISTS

CERCLA/SARA - Hazardous Substances and their Reportable Quantities
final RQ = 10 pounds (4.54 kg)

RCRA - K Series Wastes
Toxic waste

RCRA - Substances Banned From Land Disposal
[present]

K017-HAZARDOUS WASTES   RR-00671-3

HEALTH AND SAFETY LISTS

U.S. DOT - Appendix A Table 1 - Hazardous Substances
final RQ = 10 pounds (4.54 kg)

ENVIRONMENTAL LISTS

CERCLA/SARA - Hazardous Substances and their Reportable Quantities
final RQ = 10 pounds (4.54 kg)

RCRA - K Series Wastes
Toxic waste

RCRA - Substances Banned From Land Disposal
[present] (nonwastewaters)

K018-HAZARDOUS WASTES   RR-00672-4

HEALTH AND SAFETY LISTS

U.S. DOT - Appendix A Table 1 - Hazardous Substances
final RQ = 10 pounds (4.54 kg)

ENVIRONMENTAL LISTS

CERCLA/SARA - Hazardous Substances and their Reportable Quantities
final RQ = 10 pounds (4.54 kg)

RCRA - K Series Wastes
Toxic waste

RCRA - Substances Banned From Land Disposal
[present]

K019-HAZARDOUS WASTES   RR-00673-5

HEALTH AND SAFETY LISTS

U.S. DOT - Appendix A Table 1 - Hazardous Substances
final RQ = 10 pounds (4.54 kg)

ENVIRONMENTAL LISTS

CERCLA/SARA - Hazardous Substances and their Reportable Quantities
final RQ = 10 pounds (4.54 kg)

RCRA - K Series Wastes
Reactive waste; Toxic waste

RCRA - Substances Banned From Land Disposal
[present] (wastewaters)

K020-HAZARDOUS WASTES   RR-00674-6

HEALTH AND SAFETY LISTS

U.S. DOT - Appendix A Table 1 - Hazardous Substances
final RQ = 10 pounds (4.54 kg)

ENVIRONMENTAL LISTS

CERCLA/SARA - Hazardous Substances and their Reportable Quantities
final RQ = 10 pounds (4.54 kg)

RCRA - K Series Wastes
Reactive waste; Toxic waste

RCRA - Substances Banned From Land Disposal
[present] (wastewaters)

K021-HAZARDOUS WASTES   RR-00675-7

HEALTH AND SAFETY LISTS

U.S. DOT - Appendix A Table 1 - Hazardous Substances
final RQ = 5000 pounds (2270 kg)

ENVIRONMENTAL LISTS

CERCLA/SARA - Hazardous Substances and their Reportable Quantities
final RQ = 5000 pounds (2270 kg)

RCRA - K Series Wastes
Toxic waste

RCRA - Substances Banned From Land Disposal
[present] (wastewaters)

K022-HAZARDOUS WASTES   RR-00676-8

HEALTH AND SAFETY LISTS

U.S. DOT - Appendix A Table 1 - Hazardous Substances
final RQ = 10 pounds (4.54 kg)

ENVIRONMENTAL LISTS

CERCLA/SARA - Hazardous Substances and their Reportable Quantities
final RQ = 10 pounds (4.54 kg)

RCRA - K Series Wastes
Toxic waste

K023-HAZARDOUS WASTES   RR-00677-9

HEALTH AND SAFETY LISTS

U.S. DOT - Appendix A Table 1 - Hazardous Substances
final RQ = 1 pound (0.454 kg)

ENVIRONMENTAL LISTS

CERCLA/SARA - Hazardous Substances and their Reportable Quantities
final RQ = 1 pound (0.454 kg)

RCRA - K Series Wastes
Toxic waste

RCRA - Substances Banned From Land Disposal
[present]

K024-HAZARDOUS WASTES   RR-00678-0

HEALTH AND SAFETY LISTS

U.S. DOT - Appendix A Table 1 - Hazardous Substances
final RQ = 10 pounds (4.54 kg)

ENVIRONMENTAL LISTS

CERCLA/SARA - Hazardous Substances and their Reportable Quantities
final RQ = 10 pounds (4.54 kg)

RCRA - K Series Wastes
Toxic waste

RCRA - Substances Banned From Land Disposal
[present]

K025-HAZARDOUS WASTES   RR-00679-1

HEALTH AND SAFETY LISTS

U.S. DOT - Appendix A Table 1 - Hazardous Substances
final RQ = 10 pounds (4.54 kg)

ENVIRONMENTAL LISTS

CERCLA/SARA - Hazardous Substances and their Reportable Quantities
final RQ = 10 pounds (4.54 kg)

RCRA - K Series Wastes
Toxic waste

RCRA - Substances Banned From Land Disposal
[present]

K026-HAZARDOUS WASTES   RR-00680-4

HEALTH AND SAFETY LISTS

U.S. DOT - Appendix A Table 1 - Hazardous Substances
final RQ = 1 pound (0.454 kg)

ENVIRONMENTAL LISTS

CERCLA/SARA - Hazardous Substances and their Reportable Quantities
final RQ = 1 pound (0.454 kg)

RCRA - K Series Wastes
Toxic waste

RCRA - Substances Banned From Land Disposal
[present]

K027-HAZARDOUS WASTES   RR-00681-5

HEALTH AND SAFETY LISTS

U.S. DOT - Appendix A Table 1 - Hazardous Substances
final RQ = 1 pound (0.454 kg)

ENVIRONMENTAL LISTS

CERCLA/SARA - Hazardous Substances and their Reportable Quantities
final RQ = 1 pound (0.454 kg)

RCRA - K Series Wastes
Toxic waste

RCRA - Substances Banned From Land Disposal
[present]

K028-HAZARDOUS WASTES   RR-00682-6

HEALTH AND SAFETY LISTS

U.S. DOT - Appendix A Table 1 - Hazardous Substances
final RQ = 10 pounds (4.54 kg)

ENVIRONMENTAL LISTS

CERCLA/SARA - Hazardous Substances and their Reportable Quantities
final RQ = 10 pounds (4.54 kg)

RCRA - K Series Wastes
Toxic waste

RCRA - Substances Banned From Land Disposal
[present]

K029-HAZARDOUS WASTES   RR-00683-7

HEALTH AND SAFETY LISTS

U.S. DOT - Appendix A Table 1 - Hazardous Substances
final RQ = 1 pound (0.454 kg)

ENVIRONMENTAL LISTS

CERCLA/SARA - Hazardous Substances and their Reportable Quantities
final RQ = 1 pound (0.454 kg)

RCRA - K Series Wastes
Toxic waste

RCRA - Substances Banned From Land Disposal
[present]

K030-HAZARDOUS WASTES   RR-00684-8

HEALTH AND SAFETY LISTS

U.S. DOT - Appendix A Table 1 - Hazardous Substances
final RQ = 5000 pounds (2270 kg)

ENVIRONMENTAL LISTS

CERCLA/SARA - Hazardous Substances and their Reportable Quantities
final RQ = 5000 pounds (2270 kg)

RCRA - K Series Wastes
Toxic waste

RCRA - Substances Banned From Land Disposal
[present]

K031-HAZARDOUS WASTES   RR-00685-9

HEALTH AND SAFETY LISTS

U.S. DOT - Appendix A Table 1 - Hazardous Substances
final RQ = 5000 pounds (2270 kg)

ENVIRONMENTAL LISTS
  CERCLA/SARA - Hazardous Substances and their Reportable Quantities
  final RQ = 5000 pounds (2270 kg)
  RCRA - K Series Wastes
  Toxic waste
  RCRA - Substances Banned From Land Disposal
  [present]

K032-HAZARDOUS WASTES RR-00686-0

HEALTH AND SAFETY LISTS
  U.S. DOT - Appendix A Table 1 - Hazardous Substances
  final RQ = 10 pounds (4.54 kg)

ENVIRONMENTAL LISTS
  CERCLA/SARA - Hazardous Substances and their Reportable Quantities
  final RQ = 10 pounds (4.54 kg)
  RCRA - K Series Wastes
  Toxic waste
  RCRA - Substances Banned From Land Disposal
  [present]

K033-HAZARDOUS WASTES RR-00687-1

HEALTH AND SAFETY LISTS
  U.S. DOT - Appendix A Table 1 - Hazardous Substances
  final RQ = 1000 pounds (454 kg)

ENVIRONMENTAL LISTS
  CERCLA/SARA - Hazardous Substances and their Reportable Quantities
  final RQ = 1000 pounds (454 kg)
  RCRA - K Series Wastes
  Toxic waste
  RCRA - Substances Banned From Land Disposal
  [present]

K034-HAZARDOUS WASTES RR-00688-2

HEALTH AND SAFETY LISTS
  U.S. DOT - Appendix A Table 1 - Hazardous Substances
  final RQ = 10 pounds (4.54 kg)

ENVIRONMENTAL LISTS
  CERCLA/SARA - Hazardous Substances and their Reportable Quantities
  final RQ = 10 pounds (4.54 kg)
  RCRA - K Series Wastes
  Reactive waste; Toxic waste
  RCRA - Substances Banned From Land Disposal
  [present]

K035-HAZARDOUS WASTES RR-00689-3

HEALTH AND SAFETY LISTS
  U.S. DOT - Appendix A Table 1 - Hazardous Substances
  final RQ = 1 pound (0.454 kg)

ENVIRONMENTAL LISTS
  CERCLA/SARA - Hazardous Substances and their Reportable Quantities
  final RQ = 1 pound (0.454 kg)
  RCRA - K Series Wastes
  Toxic waste
  RCRA - Substances Banned From Land Disposal
  [present]

K036-HAZARDOUS WASTES RR-00690-6

HEALTH AND SAFETY LISTS
  U.S. DOT - Appendix A Table 1 - Hazardous Substances
  final RQ = 1 pound (0.454 kg)

ENVIRONMENTAL LISTS
  CERCLA/SARA - Hazardous Substances and their Reportable Quantities
  final RQ = 1 pound (0.454 kg)
  RCRA - K Series Wastes
  Toxic waste
  RCRA - Substances Banned From Land Disposal
  [present]

K037-HAZARDOUS WASTES RR-00691-7

HEALTH AND SAFETY LISTS
  U.S. DOT - Appendix A Table 1 - Hazardous Substances
  final RQ = 1 pound (0.454 kg)

ENVIRONMENTAL LISTS
  CERCLA/SARA - Hazardous Substances and their Reportable Quantities
  final RQ = 1 pound (0.454 kg)
  RCRA - K Series Wastes
  Toxic waste
  RCRA - Substances Banned From Land Disposal
  [present]

K038-HAZARDOUS WASTES RR-00692-8

HEALTH AND SAFETY LISTS
  U.S. DOT - Appendix A Table 1 - Hazardous Substances
  final RQ = 1 pound (0.454 kg)

ENVIRONMENTAL LISTS
  CERCLA/SARA - Hazardous Substances and their Reportable Quantities
  final RQ = 1 pound (0.454 kg)
  RCRA - K Series Wastes
  Toxic waste
  RCRA - Substances Banned From Land Disposal
  [present]

K039-HAZARDOUS WASTES RR-00693-9

HEALTH AND SAFETY LISTS
  U.S. DOT - Appendix A Table 1 - Hazardous Substances
  final RQ = 1 pound (0.454 kg)

ENVIRONMENTAL LISTS
  CERCLA/SARA - Hazardous Substances and their Reportable Quantities
  final RQ = 1 pound (0.454 kg)
  RCRA - K Series Wastes
  Toxic waste
  RCRA - Substances Banned From Land Disposal
  [present]

K040-HAZARDOUS WASTES RR-00694-0

HEALTH AND SAFETY LISTS
  U.S. DOT - Appendix A Table 1 - Hazardous Substances
  final RQ = 10 pounds (4.54 kg)

ENVIRONMENTAL LISTS
  CERCLA/SARA - Hazardous Substances and their Reportable Quantities
  final RQ = 10 pounds (4.54 kg)
  RCRA - K Series Wastes
  Toxic waste

RCRA - Substances Banned From Land Disposal
[present]

K041-HAZARDOUS WASTES   RR-00695-1

HEALTH AND SAFETY LISTS

U.S. DOT - Appendix A Table 1 - Hazardous Substances
final RQ = 10 pounds (4.54 kg)

ENVIRONMENTAL LISTS

CERCLA/SARA - Hazardous Substances and their Reportable Quantities
final RQ = 10 pounds (4.54 kg)

RCRA - K Series Wastes
Toxic waste

RCRA - Substances Banned From Land Disposal
[present]

K042-HAZARDOUS WASTES   RR-00696-2

HEALTH AND SAFETY LISTS

U.S. DOT - Appendix A Table 1 - Hazardous Substances
final RQ = 1 pound (0.454 kg)

ENVIRONMENTAL LISTS

CERCLA/SARA - Hazardous Substances and their Reportable Quantities
final RQ = 1 pound (0.454 kg)

RCRA - K Series Wastes
Toxic waste

RCRA - Substances Banned From Land Disposal
[present]

K043-HAZARDOUS WASTES   RR-00697-3

HEALTH AND SAFETY LISTS

U.S. DOT - Appendix A Table 1 - Hazardous Substances
final RQ = 1 pound (0.454 kg)

ENVIRONMENTAL LISTS

CERCLA/SARA - Hazardous Substances and their Reportable Quantities
final RQ = 1 pound (0.454 kg)

RCRA - K Series Wastes
Toxic waste

RCRA - Substances Banned From Land Disposal
[present]

K044-HAZARDOUS WASTES   RR-00698-4

HEALTH AND SAFETY LISTS

U.S. DOT - Appendix A Table 1 - Hazardous Substances
final RQ = 1 pound (0.454 kg)

ENVIRONMENTAL LISTS

CERCLA/SARA - Hazardous Substances and their Reportable Quantities
final RQ = 1 pound (0.454 kg)

RCRA - K Series Wastes
Toxic waste

RCRA - Substances Banned From Land Disposal
[present]

K045-HAZARDOUS WASTES   RR-00699-5

HEALTH AND SAFETY LISTS

U.S. DOT - Appendix A Table 1 - Hazardous Substances
final RQ = 10 pounds (4.54 kg)

ENVIRONMENTAL LISTS

CERCLA/SARA - Hazardous Substances and their Reportable Quantities
final RQ = 10 pounds (4.54 kg)

RCRA - K Series Wastes
Toxic waste

RCRA - Substances Banned From Land Disposal
[present]

K046-HAZARDOUS WASTES   RR-00700-1

HEALTH AND SAFETY LISTS

U.S. DOT - Appendix A Table 1 - Hazardous Substances
final RQ = 10 pounds (4.54 kg)

ENVIRONMENTAL LISTS

CERCLA/SARA - Hazardous Substances and their Reportable Quantities
final RQ = 10 pounds (4.54 kg)

RCRA - K Series Wastes
Toxic waste

RCRA - Substances Banned From Land Disposal
[present]

K047-HAZARDOUS WASTES   RR-00701-2

HEALTH AND SAFETY LISTS

U.S. DOT - Appendix A Table 1 - Hazardous Substances
final RQ = 10 pounds (4.54 kg)

ENVIRONMENTAL LISTS

CERCLA/SARA - Hazardous Substances and their Reportable Quantities
final RQ = 10 pounds (4.54 kg)

RCRA - K Series Wastes
Toxic waste

RCRA - Substances Banned From Land Disposal
[present]

K048-HAZARDOUS WASTES   RR-00702-3

HEALTH AND SAFETY LISTS

U.S. DOT - Appendix A Table 1 - Hazardous Substances
final RQ = 1 pound (0.454 kg)

ENVIRONMENTAL LISTS

CERCLA/SARA - Hazardous Substances and their Reportable Quantities
final RQ = 1 pound (0.454 kg)

RCRA - K Series Wastes
Toxic waste

RCRA - Substances Banned From Land Disposal
[present]

K049-HAZARDOUS WASTES   RR-00703-4

HEALTH AND SAFETY LISTS

U.S. DOT - Appendix A Table 1 - Hazardous Substances
final RQ = 10 pounds (4.54 kg)

ENVIRONMENTAL LISTS

CERCLA/SARA - Hazardous Substances and their Reportable Quantities
final RQ = 10 pounds (4.54 kg)

RCRA - K Series Wastes
Toxic waste

RCRA - Substances Banned From Land Disposal
[present]

K050-HAZARDOUS WASTES   RR-00704-5

HEALTH AND SAFETY LISTS

U.S. DOT - Appendix A Table 1 - Hazardous Substances
final RQ = 10 pounds (4.54 kg)

ENVIRONMENTAL LISTS

CERCLA/SARA - Hazardous Substances and their Reportable Quantities
  final RQ = 10 pounds (4.54 kg)
RCRA - K Series Wastes
  Toxic waste
RCRA - Substances Banned From Land Disposal
  [present]

K051-HAZARDOUS WASTES   RR-00705-6

HEALTH AND SAFETY LISTS

U.S. DOT - Appendix A Table 1 - Hazardous Substances
  final RQ = 10 pounds (4.54 kg)

ENVIRONMENTAL LISTS

CERCLA/SARA - Hazardous Substances and their Reportable Quantities
  final RQ = 10 pounds (4.54 kg)
RCRA - K Series Wastes
  Reactive waste
RCRA - Substances Banned From Land Disposal
  [present]

K052-HAZARDOUS WASTES   RR-00706-7

HEALTH AND SAFETY LISTS

U.S. DOT - Appendix A Table 1 - Hazardous Substances
  final RQ = 10 pounds (4.54 kg)

ENVIRONMENTAL LISTS

CERCLA/SARA - Hazardous Substances and their Reportable Quantities
  final RQ = 10 pounds (4.54 kg)
RCRA - K Series Wastes
  Reactive waste
RCRA - Substances Banned From Land Disposal
  [present]

K060-HAZARDOUS WASTES   RR-00707-8

HEALTH AND SAFETY LISTS

U.S. DOT - Appendix A Table 1 - Hazardous Substances
  final RQ = 10 pounds (4.54 kg)

ENVIRONMENTAL LISTS

CERCLA/SARA - Hazardous Substances and their Reportable Quantities
  final RQ = 10 pounds (4.54 kg)
RCRA - K Series Wastes
  Toxic waste
RCRA - Substances Banned From Land Disposal
  [present]

K061-HAZARDOUS WASTES   RR-00708-9

HEALTH AND SAFETY LISTS

U.S. DOT - Appendix A Table 1 - Hazardous Substances
  final RQ = 10 pounds (4.54 kg)

ENVIRONMENTAL LISTS

CERCLA/SARA - Hazardous Substances and their Reportable Quantities
  final RQ = 10 pounds (4.54 kg)
RCRA - K Series Wastes
  Reactive waste
RCRA - Substances Banned From Land Disposal
  [present]

K062-HAZARDOUS WASTES   RR-00709-0

HEALTH AND SAFETY LISTS

U.S. DOT - Appendix A Table 1 - Hazardous Substances
  final RQ = 10 pounds (4.54 kg)

ENVIRONMENTAL LISTS

CERCLA/SARA - Hazardous Substances and their Reportable Quantities
  final RQ = 10 pounds (4.54 kg)
RCRA - K Series Wastes
  Toxic waste
RCRA - Substances Banned From Land Disposal
  [present]

K064-HAZARDOUS WASTES   RR-00710-3

HEALTH AND SAFETY LISTS

U.S. DOT - Appendix A Table 1 - Hazardous Substances
  final RQ = 10 pounds (4.54 kg)

ENVIRONMENTAL LISTS

CERCLA/SARA - Hazardous Substances and their Reportable Quantities
  final RQ = 10 pounds (4.54 kg)
RCRA - K Series Wastes
  Toxic waste
RCRA - Substances Banned From Land Disposal
  [present]

K065-HAZARDOUS WASTES   RR-00711-4

HEALTH AND SAFETY LISTS

U.S. DOT - Appendix A Table 1 - Hazardous Substances
  final RQ = 10 pounds (4.54 kg)

ENVIRONMENTAL LISTS

CERCLA/SARA - Hazardous Substances and their Reportable Quantities
  final RQ = 10 pounds (4.54 kg)
RCRA - K Series Wastes
  Toxic waste
RCRA - Substances Banned From Land Disposal
  [present]

K066-HAZARDOUS WASTES   RR-00712-5

HEALTH AND SAFETY LISTS

U.S. DOT - Appendix A Table 1 - Hazardous Substances
  final RQ = 10 pounds (4.54 kg)

ENVIRONMENTAL LISTS

CERCLA/SARA - Hazardous Substances and their Reportable Quantities
  final RQ = 10 pounds (4.54 kg)
RCRA - K Series Wastes
  Toxic waste
RCRA - Substances Banned From Land Disposal
  [present]

K069-HAZARDOUS WASTES   RR-00713-6

HEALTH AND SAFETY LISTS

U.S. DOT - Appendix A Table 1 - Hazardous Substances
  final RQ = 10 pounds (4.54 kg)

ENVIRONMENTAL LISTS

CERCLA/SARA - Hazardous Substances and their Reportable Quantities
  final RQ = 10 pounds (4.54 kg)
RCRA - K Series Wastes
  Toxic waste

RCRA - Substances Banned From Land Disposal
[present]

K071-HAZARDOUS WASTES   RR-00714-7

HEALTH AND SAFETY LISTS

U.S. DOT - Appendix A Table 1 - Hazardous Substances
final RQ = 1 pound (0.454 kg)

ENVIRONMENTAL LISTS

CERCLA/SARA - Hazardous Substances and their Reportable Quantities
final RQ = 1 pound (0.454 kg)

RCRA - K Series Wastes
Toxic waste

RCRA - Substances Banned From Land Disposal
[present]

K073-HAZARDOUS WASTES   RR-00715-8

HEALTH AND SAFETY LISTS

U.S. DOT - Appendix A Table 1 - Hazardous Substances
final RQ = 10 pounds (4.54 kg)

ENVIRONMENTAL LISTS

CERCLA/SARA - Hazardous Substances and their Reportable Quantities
final RQ = 10 pounds (4.54 kg)

RCRA - K Series Wastes
Toxic waste

RCRA - Substances Banned From Land Disposal
[present]

K083-HAZARDOUS WASTES   RR-00716-9

HEALTH AND SAFETY LISTS

U.S. DOT - Appendix A Table 1 - Hazardous Substances
final RQ = 10 pounds (4.54 kg)

ENVIRONMENTAL LISTS

CERCLA/SARA - Hazardous Substances and their Reportable Quantities
final RQ = 10 pounds (4.54 kg)

RCRA - K Series Wastes
Corrosive waste; Toxic waste

RCRA - Substances Banned From Land Disposal
[present] (non CaSO4 form)

K084-HAZARDOUS WASTES   RR-00717-0

HEALTH AND SAFETY LISTS

U.S. DOT - Appendix A Table 1 - Hazardous Substances
final RQ = 10 pounds (4.54 kg)

ENVIRONMENTAL LISTS

CERCLA/SARA - Hazardous Substances and their Reportable Quantities
final RQ = 10 pounds (4.54 kg)

RCRA - K Series Wastes
Toxic waste

K085-HAZARDOUS WASTES   RR-00718-1

HEALTH AND SAFETY LISTS

U.S. DOT - Appendix A Table 1 - Hazardous Substances
final RQ = 10 pounds (4.54 kg)

ENVIRONMENTAL LISTS

CERCLA/SARA - Hazardous Substances and their Reportable Quantities
final RQ = 10 pounds (4.54 kg)

RCRA - K Series Wastes
Toxic waste

K086-HAZARDOUS WASTES   RR-00719-2

HEALTH AND SAFETY LISTS

U.S. DOT - Appendix A Table 1 - Hazardous Substances
final RQ = 10 pounds (4.54 kg)

ENVIRONMENTAL LISTS

CERCLA/SARA - Hazardous Substances and their Reportable Quantities
final RQ = 10 pounds (4.54 kg)

RCRA - K Series Wastes
Toxic waste

K087-HAZARDOUS WASTES   RR-00720-5

HEALTH AND SAFETY LISTS

U.S. DOT - Appendix A Table 1 - Hazardous Substances
final RQ = 10 pounds (4.54 kg)

ENVIRONMENTAL LISTS

CERCLA/SARA - Hazardous Substances and their Reportable Quantities
final RQ = 10 pounds (4.54 kg)

RCRA - K Series Wastes
Toxic waste

RCRA - Substances Banned From Land Disposal
[present]

K088-HAZARDOUS WASTES   RR-00721-6

HEALTH AND SAFETY LISTS

U.S. DOT - Appendix A Table 1 - Hazardous Substances
final RQ = 1 pound (0.454 kg)

ENVIRONMENTAL LISTS

CERCLA/SARA - Hazardous Substances and their Reportable Quantities
final RQ = 1 pound (0.454 kg)

RCRA - K Series Wastes
Toxic waste

RCRA - Substances Banned From Land Disposal
[present]

K090-HAZARDOUS WASTES   RR-00722-7

HEALTH AND SAFETY LISTS

U.S. DOT - Appendix A Table 1 - Hazardous Substances
final RQ = 10 pounds (4.54 kg)

ENVIRONMENTAL LISTS

CERCLA/SARA - Hazardous Substances and their Reportable Quantities
final RQ = 10 pounds (4.54 kg)

RCRA - K Series Wastes
Toxic waste

RCRA - Substances Banned From Land Disposal
[present]

K091-HAZARDOUS WASTES   RR-00723-8

HEALTH AND SAFETY LISTS

U.S. DOT - Appendix A Table 1 - Hazardous Substances
final RQ = 100 pounds (45.4 kg)

ENVIRONMENTAL LISTS

CERCLA/SARA - Hazardous Substances and their Reportable Quantities
final RQ = 100 pounds (45.4 kg)

RCRA - K Series Wastes
Toxic waste

RCRA - Substances Banned From Land Disposal
[present] (wastewaters)

K093-HAZARDOUS WASTES   RR-00724-9

HEALTH AND SAFETY LISTS

U.S. DOT - Appendix A Table 1 - Hazardous Substances
final RQ = 1 pound (0.454 kg)

ENVIRONMENTAL LISTS

CERCLA/SARA - Hazardous Substances and their Reportable Quantities
final RQ = 1 pound (0.454 kg)

RCRA - K Series Wastes
Toxic waste

RCRA - Substances Banned From Land Disposal
[present]

K094-HAZARDOUS WASTES   RR-00725-0

HEALTH AND SAFETY LISTS

U.S. DOT - Appendix A Table 1 - Hazardous Substances
final RQ = 10 pounds (4.54 kg)

ENVIRONMENTAL LISTS

CERCLA/SARA - Hazardous Substances and their Reportable Quantities
final RQ = 10 pounds (4.54 kg)

RCRA - K Series Wastes
Toxic waste

RCRA - Substances Banned From Land Disposal
[present] (wastewaters)

K095-HAZARDOUS WASTES   RR-00726-1

HEALTH AND SAFETY LISTS

U.S. DOT - Appendix A Table 1 - Hazardous Substances
final RQ = 10 pounds (4.54 kg)

ENVIRONMENTAL LISTS

CERCLA/SARA - Hazardous Substances and their Reportable Quantities
final RQ = 10 pounds (4.54 kg)

RCRA - K Series Wastes
Toxic waste

RCRA - Substances Banned From Land Disposal
[present] (solvent washes)

K096-HAZARDOUS WASTES   RR-00727-2

HEALTH AND SAFETY LISTS

U.S. DOT - Appendix A Table 1 - Hazardous Substances
final RQ = 100 pounds (45.4 kg)

ENVIRONMENTAL LISTS

CERCLA/SARA - Hazardous Substances and their Reportable Quantities
final RQ = 100 pounds (45.4 kg)

RCRA - K Series Wastes
Toxic waste

RCRA - Substances Banned From Land Disposal
[present]

K097-HAZARDOUS WASTES   RR-00728-3

HEALTH AND SAFETY LISTS

U.S. DOT - Appendix A Table 1 - Hazardous Substances
Statuatory RQ = 1 pound (the RQ is subject to change when the assessment of potential carcinogenicity is completed)

ENVIRONMENTAL LISTS

CERCLA/SARA - Hazardous Substances and their Reportable Quantities
Statuatory RQ = 1 pound (the RQ is subject to change when the assessment of potential carcinogenicity is completed)

RCRA - K Series Wastes
Toxic waste

K098-HAZARDOUS WASTES   RR-00729-4

HEALTH AND SAFETY LISTS

U.S. DOT - Appendix A Table 1 - Hazardous Substances
Statuatory RQ = 1 pound (the RQ is subject to change when the assessment of potential carcinogenicity is completed)

ENVIRONMENTAL LISTS

CERCLA/SARA - Hazardous Substances and their Reportable Quantities
Statuatory RQ = 1 pound (the RQ is subject to change when the assessment of potential carcinogenicity is completed)

RCRA - K Series Wastes
Toxic waste

K099-HAZARDOUS WASTES   RR-00730-7

HEALTH AND SAFETY LISTS

U.S. DOT - Appendix A Table 1 - Hazardous Substances
Statuatory RQ = 1 pound (the RQ is subject ot change when the assessment of potential carcinogenicity is completed)

ENVIRONMENTAL LISTS

CERCLA/SARA - Hazardous Substances and their Reportable Quantities
Statuatory RQ = 1 pound (the RQ is subject ot change when the assessment of potential carcinogenicity is completed)

RCRA - K Series Wastes
Toxic waste

K100-HAZARDOUS WASTES   RR-00731-8

HEALTH AND SAFETY LISTS

U.S. DOT - Appendix A Table 1 - Hazardous Substances
final RQ = 5000 pounds (2270 kg)

ENVIRONMENTAL LISTS

CERCLA/SARA - Hazardous Substances and their Reportable Quantities
final RQ = 5000 pounds (2270 kg)

RCRA - K Series Wastes
Toxic waste

RCRA - Substances Banned From Land Disposal
[present]

K101-HAZARDOUS WASTES   RR-00732-9

HEALTH AND SAFETY LISTS

U.S. DOT - Appendix A Table 1 - Hazardous Substances
final RQ = 5000 pounds (2270 kg)

ENVIRONMENTAL LISTS

CERCLA/SARA - Hazardous Substances and their Reportable Quantities
final RQ = 5000 pounds (2270 kg)

RCRA - K Series Wastes
Toxic waste

RCRA - Substances Banned From Land Disposal
[present]

K102-HAZARDOUS WASTES   RR-00733-0

HEALTH AND SAFETY LISTS

U.S. DOT - Appendix A Table 1 - Hazardous Substances
final RQ = 100 pounds (45.4 kg)

ENVIRONMENTAL LISTS

CERCLA/SARA - Hazardous Substances and their Reportable Quantities
final RQ = 100 pounds (45.4 kg)

RCRA - K Series Wastes
Toxic waste
RCRA - Substances Banned From Land Disposal
[present]

K103-HAZARDOUS WASTES   RR-00734-1

HEALTH AND SAFETY LISTS

U.S. DOT - Appendix A Table 1 - Hazardous Substances
final RQ = 100 pounds (45.4 kg)

ENVIRONMENTAL LISTS

CERCLA/SARA - Hazardous Substances and their Reportable Quantities
final RQ = 100 pounds (45.4 kg)
RCRA - K Series Wastes
Toxic waste
RCRA - Substances Banned From Land Disposal
[present]

K104-HAZARDOUS WASTES   RR-00735-2

HEALTH AND SAFETY LISTS

U.S. DOT - Appendix A Table 1 - Hazardous Substances
final RQ = 1 pound (0.454 kg)

ENVIRONMENTAL LISTS

CERCLA/SARA - Hazardous Substances and their Reportable Quantities
final RQ = 1 pound (0.454 kg)
RCRA - K Series Wastes
Toxic waste
RCRA - Substances Banned From Land Disposal
[present]

K105-HAZARDOUS WASTES   RR-00736-3

HEALTH AND SAFETY LISTS

U.S. DOT - Appendix A Table 1 - Hazardous Substances
final RQ = 1 pound (0.454 kg)

ENVIRONMENTAL LISTS

CERCLA/SARA - Hazardous Substances and their Reportable Quantities
final RQ = 1 pound (0.454 kg)
RCRA - K Series Wastes
Toxic waste
RCRA - Substances Banned From Land Disposal
[present]

K106-HAZARDOUS WASTES   RR-00737-4

HEALTH AND SAFETY LISTS

U.S. DOT - Appendix A Table 1 - Hazardous Substances
final RQ = 10 pounds (4.54 kg)

ENVIRONMENTAL LISTS

CERCLA/SARA - Hazardous Substances and their Reportable Quantities
final RQ = 10 pounds (4.54 kg)
RCRA - K Series Wastes
Toxic waste
RCRA - Substances Banned From Land Disposal
[present]

K107-HAZARDOUS WASTES   RR-00808-2

HEALTH AND SAFETY LISTS

U.S. DOT - Appendix A Table 1 - Hazardous Substances
final RQ = 10 pounds (4.54 kg)

ENVIRONMENTAL LISTS

CERCLA/SARA - Hazardous Substances and their Reportable Quantities
final RQ = 10 pounds (4.54 kg)
RCRA - K Series Wastes
Toxic waste
RCRA - Substances Banned From Land Disposal
[present]

K108-HAZARDOUS WASTES   RR-00810-6

HEALTH AND SAFETY LISTS

U.S. DOT - Appendix A Table 1 - Hazardous Substances
final RQ = 1 pound (0.454 kg)

ENVIRONMENTAL LISTS

CERCLA/SARA - Hazardous Substances and their Reportable Quantities
final RQ = 1 pound (0.454 kg)
RCRA - K Series Wastes
Toxic waste
RCRA - Substances Banned From Land Disposal
[present]

K109-HAZARDOUS WASTES   RR-00811-7

HEALTH AND SAFETY LISTS

U.S. DOT - Appendix A Table 1 - Hazardous Substances
final RQ = 1 pound (0.454 kg)

ENVIRONMENTAL LISTS

CERCLA/SARA - Hazardous Substances and their Reportable Quantities
final RQ = 1 pound (0.454 kg)
RCRA - K Series Wastes
Toxic waste
RCRA - Substances Banned From Land Disposal
[present]

K110-HAZARDOUS WASTES   RR-00804-8

HEALTH AND SAFETY LISTS

U.S. DOT - Appendix A Table 1 - Hazardous Substances
final RQ = 100 pounds (45.4 kg)

ENVIRONMENTAL LISTS

CERCLA/SARA - Hazardous Substances and their Reportable Quantities
final RQ = 100 pounds (45.4 kg)
RCRA - K Series Wastes
Toxic waste
RCRA - Substances Banned From Land Disposal
[present] (except in injection wells)

K111-HAZARDOUS WASTES   RR-00738-5

HEALTH AND SAFETY LISTS

U.S. DOT - Appendix A Table 1 - Hazardous Substances
final RQ = 10 pounds (4.54 kg)

ENVIRONMENTAL LISTS

CERCLA/SARA - Hazardous Substances and their Reportable Quantities
final RQ = 10 pounds (4.54 kg)
RCRA - K Series Wastes
Toxic waste
RCRA - Substances Banned From Land Disposal
[present] (except in injection wells)

K112-HAZARDOUS WASTES   RR-00739-6

HEALTH AND SAFETY LISTS

U.S. DOT - Appendix A Table 1 - Hazardous Substances
final RQ = 10 pounds (4.54 kg)

ENVIRONMENTAL LISTS

CERCLA/SARA - Hazardous Substances and their Reportable Quantities
final RQ = 10 pounds (4.54 kg)

RCRA - K Series Wastes
Toxic waste

RCRA - Substances Banned From Land Disposal
[present]

K113-HAZARDOUS WASTES   RR-00740-9

HEALTH AND SAFETY LISTS

U.S. DOT - Appendix A Table 1 - Hazardous Substances
final RQ = 1 pound (0.454 kg)

ENVIRONMENTAL LISTS

CERCLA/SARA - Hazardous Substances and their Reportable Quantities
final RQ = 1 pound (0.454 kg)

RCRA - K Series Wastes
Toxic waste

RCRA - Substances Banned From Land Disposal
[present]

K114-HAZARDOUS WASTES   RR-00741-0

ENVIRONMENTAL LISTS

RCRA - K Series Wastes
Corrosive waste; Toxic waste

RCRA - Substances Banned From Land Disposal
[present]

K115-HAZARDOUS WASTES   RR-00742-1

ENVIRONMENTAL LISTS

RCRA - K Series Wastes
Ignitable waste; Toxic waste

RCRA - Substances Banned From Land Disposal
[present]

K116-HAZARDOUS WASTES   RR-00743-2

ENVIRONMENTAL LISTS

RCRA - K Series Wastes
Toxic waste

RCRA - Substances Banned From Land Disposal
[present]

K117-HAZARDOUS WASTES   RR-00744-3

ENVIRONMENTAL LISTS

RCRA - K Series Wastes
Toxic waste

RCRA - Substances Banned From Land Disposal
[present]

K118-HAZARDOUS WASTES   RR-00745-4

HEALTH AND SAFETY LISTS

U.S. DOT - Appendix A Table 1 - Hazardous Substances
final RQ = 10 pounds (4.54 kg)

ENVIRONMENTAL LISTS

CERCLA/SARA - Hazardous Substances and their Reportable Quantities
final RQ = 10 pounds (4.54 kg)

RCRA - K Series Wastes
Corrosive waste; Toxic waste

RCRA - Substances Banned From Land Disposal
[present]

K123-HAZARDOUS WASTES   RR-00754-5

HEALTH AND SAFETY LISTS

U.S. DOT - Appendix A Table 1 - Hazardous Substances
final RQ = 10 pounds (4.54 kg)

ENVIRONMENTAL LISTS

CERCLA/SARA - Hazardous Substances and their Reportable Quantities
final RQ = 10 pounds (4.54 kg)

RCRA - K Series Wastes
Toxic waste

RCRA - Substances Banned From Land Disposal
[present]

K124-HAZARDOUS WASTES   RR-00751-2

HEALTH AND SAFETY LISTS

U.S. DOT - Appendix A Table 1 - Hazardous Substances
final RQ = 10 pounds (4.54 kg)

ENVIRONMENTAL LISTS

CERCLA/SARA - Hazardous Substances and their Reportable Quantities
final RQ = 10 pounds (4.54 kg)

RCRA - K Series Wastes
Toxic waste

RCRA - Substances Banned From Land Disposal
[present]

K125-HAZARDOUS WASTES   RR-00619-9

HEALTH AND SAFETY LISTS

U.S. DOT - Appendix A Table 1 - Hazardous Substances
final RQ = 10 pounds (4.54 kg)

ENVIRONMENTAL LISTS

CERCLA/SARA - Hazardous Substances and their Reportable Quantities
final RQ = 10 pounds (4.54 kg)

RCRA - K Series Wastes
Toxic waste

RCRA - Substances Banned From Land Disposal
[present]

K126-HAZARDOUS WASTES   RR-00597-0

HEALTH AND SAFETY LISTS

U.S. DOT - Appendix A Table 1 - Hazardous Substances
final RQ = 10 pounds (4.54 kg)

ENVIRONMENTAL LISTS

CERCLA/SARA - Hazardous Substances and their Reportable Quantities
final RQ = 10 pounds (4.54 kg)

RCRA - K Series Wastes
Toxic waste

RCRA - Substances Banned From Land Disposal
[present]

K131-HAZARDOUS WASTES   RR-00567-4

HEALTH AND SAFETY LISTS

U.S. DOT - Appendix A Table 1 - Hazardous Substances
final RQ = 10 pounds (4.54 kg)

ENVIRONMENTAL LISTS
  CERCLA/SARA - Hazardous Substances and their Reportable Quantities
    final RQ = 10 pounds (4.54 kg)
  RCRA - K Series Wastes
    Toxic waste
  RCRA - Substances Banned From Land Disposal
    [present]

K132-HAZARDOUS WASTES   RR-00596-9

HEALTH AND SAFETY LISTS
  U.S. DOT - Appendix A Table 1 - Hazardous Substances
    final RQ = 1 pound (0.454 kg)

ENVIRONMENTAL LISTS
  CERCLA/SARA - Hazardous Substances and their Reportable Quantities
    final RQ = 1 pound (0.454 kg)
  RCRA - K Series Wastes
    Toxic waste
  RCRA - Substances Banned From Land Disposal
    [present]

K136-HAZARDOUS WASTES   RR-00746-5

HEALTH AND SAFETY LISTS
  U.S. DOT - Appendix A Table 1 - Hazardous Substances
    final RQ = 1 pound (0.454 kg)

ENVIRONMENTAL LISTS
  CERCLA/SARA - Hazardous Substances and their Reportable Quantities
    final RQ = 1 pound (0.454 kg)
  RCRA - K Series Wastes
    Toxic waste
  RCRA - Substances Banned From Land Disposal
    [present]

K141-HAZARDOUS WASTES   RR-00888-8

HEALTH AND SAFETY LISTS
  U.S. DOT - Appendix A Table 1 - Hazardous Substances
    final RQ = 10 pounds (4.54 kg)

ENVIRONMENTAL LISTS
  CERCLA/SARA - Hazardous Substances and their Reportable Quantities
    final RQ = 10 pounds (4.54 kg)
  RCRA - K Series Wastes
    Toxic waste
  RCRA - Substances Banned From Land Disposal
    [present]

K142-HAZARDOUS WASTES   RR-00889-9

HEALTH AND SAFETY LISTS
  U.S. DOT - Appendix A Table 1 - Hazardous Substances
    final RQ = 10 pounds (4.54 kg)

ENVIRONMENTAL LISTS
  CERCLA/SARA - Hazardous Substances and their Reportable Quantities
    final RQ = 10 pounds (4.54 kg)
  RCRA - K Series Wastes
    Corrosive waste; Toxic waste
  RCRA - Substances Banned From Land Disposal
    [present]

K143-HAZARDOUS WASTES   RR-00890-2

HEALTH AND SAFETY LISTS
  U.S. DOT - Appendix A Table 1 - Hazardous Substances
    final RQ = 10 pounds (4.54 kg)

ENVIRONMENTAL LISTS
  CERCLA/SARA - Hazardous Substances and their Reportable Quantities
    final RQ = 10 pounds (4.54 kg)
  RCRA - K Series Wastes
    Toxic waste
  RCRA - Substances Banned From Land Disposal
    [present]

K144-HAZARDOUS WASTES   RR-00891-3

HEALTH AND SAFETY LISTS
  U.S. DOT - Appendix A Table 1 - Hazardous Substances
    final RQ = 10 pounds (4.54 kg)

ENVIRONMENTAL LISTS
  CERCLA/SARA - Hazardous Substances and their Reportable Quantities
    final RQ = 10 pounds (4.54 kg)
  RCRA - K Series Wastes
    Toxic waste
  RCRA - Substances Banned From Land Disposal
    [present]

K145-HAZARDOUS WASTES   RR-00892-4

HEALTH AND SAFETY LISTS
  U.S. DOT - Appendix A Table 1 - Hazardous Substances
    final RQ = 100 pounds (45.4 kg)

ENVIRONMENTAL LISTS
  CERCLA/SARA - Hazardous Substances and their Reportable Quantities
    final RQ = 100 pounds (45.4 kg)
  RCRA - K Series Wastes
    Corrosive waste; Toxic waste
  RCRA - Substances Banned From Land Disposal
    [present]

K147-HAZARDOUS WASTES   RR-00893-5

HEALTH AND SAFETY LISTS
  U.S. DOT - Appendix A Table 1 - Hazardous Substances
    final RQ = 1000 pounds (454 kg)

ENVIRONMENTAL LISTS
  CERCLA/SARA - Hazardous Substances and their Reportable Quantities
    final RQ = 1000 pounds (454 kg)
  RCRA - K Series Wastes
    Toxic waste
  RCRA - Substances Banned From Land Disposal
    [present]

K148-HAZARDOUS WASTES   RR-00894-6

HEALTH AND SAFETY LISTS
  U.S. DOT - Appendix A Table 1 - Hazardous Substances
    final RQ = 1 pound (0.454 kg)

ENVIRONMENTAL LISTS
  CERCLA/SARA - Hazardous Substances and their Reportable Quantities
    final RQ = 1 pound (0.454 kg)
  RCRA - K Series Wastes
    Toxic waste

RCRA - Substances Banned From Land Disposal
[present]

K149-HAZARDOUS WASTES   RR-00895-7

HEALTH AND SAFETY LISTS

U.S. DOT - Appendix A Table 1 - Hazardous Substances
Statuatory RQ = 1 pound (0.454 kg)

ENVIRONMENTAL LISTS

CERCLA/SARA - Hazardous Substances and their Reportable Quantities
statuatory RQ = 1 pound (0.454 kg)

RCRA - K Series Wastes
Toxic waste

K150-HAZARDOUS WASTES   RR-00896-8

HEALTH AND SAFETY LISTS

U.S. DOT - Appendix A Table 1 - Hazardous Substances
Statuatory RQ = 1 pound (0.454 kg)

ENVIRONMENTAL LISTS

CERCLA/SARA - Hazardous Substances and their Reportable Quantities
statuatory RQ = 1 pound (0.454 kg)

RCRA - K Series Wastes
Toxic waste

K151-HAZARDOUS WASTES   RR-00897-9

HEALTH AND SAFETY LISTS

U.S. DOT - Appendix A Table 1 - Hazardous Substances
statuatory RQ = 1 pound (0.454 kg)

ENVIRONMENTAL LISTS

CERCLA/SARA - Hazardous Substances and their Reportable Quantities
statuatory RQ = 1 pound (0.454 kg)

RCRA - K Series Wastes
Toxic waste

KAEMPFEROL   520-18-3

HEALTH AND SAFETY LISTS

U.S. DOT - Appendix A Table 1 - Hazardous Substances
statuatory RQ = 1 pound (0.454 kg)

ENVIRONMENTAL LISTS

CERCLA/SARA - Hazardous Substances and their Reportable Quantities
statuatory RQ = 1 pound (0.454 kg)

RCRA - K Series Wastes
Toxic waste

KANAMYCIN   59-01-8

HEALTH AND SAFETY LISTS

U.S. DOT - Appendix A Table 1 - Hazardous Substances
statuatory RQ = 1 pound (0.454 kg)

ENVIRONMENTAL LISTS

CERCLA/SARA - Hazardous Substances and their Reportable Quantities
statuatory RQ = 1 pound (0.454 kg)

RCRA - K Series Wastes
Toxic waste

KANECHLOR 500   25429-29-2

HEALTH AND SAFETY LISTS

U.S. DOT - Appendix A Table 1 - Hazardous Substances
statuatory RQ = 1 pound (0.454 kg)

ENVIRONMENTAL LISTS

CERCLA/SARA - Hazardous Substances and their Reportable Quantities
statuatory RQ = 1 pound (0.454 kg)

RCRA - K Series Wastes
Toxic waste

KANECHLOR 500   61788-33-8

HEALTH AND SAFETY LISTS

U.S. DOT - Appendix A Table 1 - Hazardous Substances
statuatory RQ = 1 pound (0.454 kg)

ENVIRONMENTAL LISTS

CERCLA/SARA - Hazardous Substances and their Reportable Quantities
statuatory RQ = 1 pound (0.454 kg)

RCRA - K Series Wastes
Toxic waste

KAOLIN   1332-58-7

HEALTH AND SAFETY LISTS

U.S. DOT - Appendix A Table 1 - Hazardous Substances
final RQ = 10 pounds (4.54 kg)

ENVIRONMENTAL LISTS

CERCLA/SARA - Hazardous Substances and their Reportable Quantities
final RQ = 10 pounds (4.54 kg)

RCRA - K Series Wastes
Toxic waste

KARAYA   9000-36-6

HEALTH AND SAFETY LISTS

U.S. DOT - Appendix A Table 1 - Hazardous Substances
final RQ = 10 pounds (4.54 kg)

ENVIRONMENTAL LISTS

CERCLA/SARA - Hazardous Substances and their Reportable Quantities
final RQ = 10 pounds (4.54 kg)

RCRA - K Series Wastes
Toxic waste

KEPONE   143-50-0

HEALTH AND SAFETY LISTS

U.S. DOT - Appendix A Table 1 - Hazardous Substances
final RQ = 10 pounds (4.54 kg)

ENVIRONMENTAL LISTS

CERCLA/SARA - Hazardous Substances and their Reportable Quantities
final RQ = 10 pounds (4.54 kg)

RCRA - K Series Wastes
Toxic waste

KEROSENE   8008-20-6

HEALTH AND SAFETY LISTS

IARC - Group 3 (not classifiable)
[present]

KETENE   463-51-4

HEALTH AND SAFETY LISTS

NIOSH - Selected LD50s and LC50s
Oral, mouse: LD50 = 20500 mg/kg

STATE LISTS

Massachusetts Right To Know List
teratogen

NJ Right to Know List (Total)
sn 1088

KETONES, LIQUID, N.O.S.   RR-00350-9

HEALTH AND SAFETY LISTS

NTP Seventh Report - Suspect Carcinogens
suspect carcinogen (Listed under 'Polychlorinated biphenyls')

KRYPTON   7439-90-9

ENVIRONMENTAL LISTS

TSCA - Code of Federal Regulations Citations
40 CFR 704.45; 40 CFR 716.120(a)
TSCA - Health and Safety Reporting List
Effective Date: October 4, 1982

KRYPTON 74   28522-15-8

HEALTH AND SAFETY LISTS

ACGIH 1995 - Time Weighted Averages
respirable dust: 2 mg/m3 TWA (The value is for total dust containing no asbestos and < 1% crystalline silica)
OSHA - Vacated PELs - Time Weighted Averages
total dust: 10 mg/m3 TWA; respirable fraction: 5 mg/m3 TWA
OSHA - Final PELs - Time Weighted Averages
total dust: 15 mg/m3 TWA; respirable fraction: 5 mg/m3 TWA

ENVIRONMENTAL LISTS

List of Pesticide Product Inert Ingredients
[present]

INTERNATIONAL LISTS

Australian Exposure Standards - Time Weighted Averages
10 mg/m3 TWA
Canada - Alberta - 8 Hour Occupational Exposure Limit
respirable mass: 5 mg/m3 TWA; total mass: 10 mg/m3 TWA
Canada - British Columbia - 8 Hour Exposure Limits
nuisance dust: 10 mg/m3 TWA
Canada - British Columbia - 15 Minute Exposure Limits
20 mg/m3 STEL
Canada - Ontario - OHSA - TWAEVs
total dust: 10 mg/m3 TWAEV (listed as nuisance particulate)
Canada - Quebec - Time-Weighted Average Exposure Values
total dust: 10 mg/m3 TWAEV; respirable dust: 5 mg/m3 TWAEV
Israel - Time Weighted Averages
10 mg/m3 TWA (The value is for total dust containing no asbestos and < 1% crystalline silica)
Israel - Action Levels
5 mg/m3 AL
Mexico - Instruction No. 10 - TWAs
10 mg/m3 TWA; (nuisance particulate)

STATE LISTS

Minnesota Hazardous Substance List
[present] (includes inert or nuisance dust)
Pennsylvania Right to Know List
[present]

KRYPTON 76   28522-17-0

HEALTH AND SAFETY LISTS

NIOSH - Selected LD50s and LC50s
Oral, rat: LD50 = 9100 mg/kg

KRYPTON 77   14983-72-3

SEE ALSO:
F039-HAZARDOUS WASTES

HEALTH AND SAFETY LISTS

U.S. DOT - Appendix A Table 1 - Hazardous Substances
final RQ = 1 pound (0.454 kg)

IARC - Group 2B (sufficient animal data)
[present] (Overall evaluation based only on evidence of carcinogenicity in monograph (20, 1979) or in Supplement 4)
NIOSH - Selected LD50s and LC50s
Oral, rat: LD50 = 95 mg/kg Skin, rabbit: LD50 = 345 mg/kg
NIOSH - Health Standards - Exposure Limits
1 ug/m3 TWA
NIOSH - Health Standards - Health Effects and Precautions
Liver cancer; nervous system effects (Liver function testing required)
NIOSH - Health Standards - Carcinogenic Chemicals
potential human carcinogen
NTP Chemical Status Reports - Testing Status and NTIS Number
Technical reports printed (PB264041)
NTP Seventh Report - Suspect Carcinogens
suspect carcinogen
OSHA - Possible Select Carcinogens
[present]

ENVIRONMENTAL LISTS

ATSDR Priority List
Rank (of 275): 170
CERCLA/SARA - Hazardous Substances and their Reportable Quantities
final RQ = 1 pound (0.454 kg)
Clean Water Act - Hazardous Substances
[present]
EPA - Carcinogen Hazard Ranking for RQ Adjustment
Hazard ranking = Medium
RCRA - U Series Wastes
waste number U142
RCRA - Hazardous Constituents-Appendix VIII
waste number U142
RCRA - Basis for Listing - Appendix VII
Included in waste stream: F039
RCRA - Substances Banned From Land Disposal
[present]
RCRA - TSD Facilities Ground Water Monitoring
TM 8270 = 10 ug/L PQL
RCRA - Universal Treatment Standards (LDR)
WW: 0.0011 mg/l; NWW: 0.13 mg/kg

INTERNATIONAL LISTS

German (DFG) - Carcinogens
suspected carcinogen

STATE LISTS

California - Air Bill 2588 Appendix A-II
known or potential carcinogen
California - Prop. 65 - Cancer list
carcinogen - initial date 1/1/88
California - Prop. 65 - Developmental Toxicity
developmental toxicity - initial date 1/1/89
California - Prop. 65 - No Significant Risk Levels
no significant risk level = 0.04 ug/day
California - Directors List of Hazardous Substances (8 CCR 339)
[present]
Florida Hazardous Substance List
[present]
Massachusetts Right To Know List
carcinogen; extraordinarily hazardous
Minnesota Hazardous Substance List
carcinogen
NJ Right to Know List (Total)
sn 1090
NJ Special Hazardous Substances
(carcinogen)

Pennsylvania Right to Know List
environmental hazard; special hazardous substance

Pennsylvania RTK - Special Hazardous Substances
[present]

KRYPTON 79   15478-11-2

HEALTH AND SAFETY LISTS

U.S. DOT - Substances From 49 CFR 172.101
regulated by DOT (UN1223)

U.S. DOT - Hazard Classes
DOT hazard class = 3

NFPA - Flash Points
*flash point = 100 to 162 degrees F (38 to 72 degrees C);   *See list description*

NFPA - Hazard Identification Ratings
health-0; flammability-2; reactivity-0

NIOSH - Selected LD50s and LC50s
Oral, rat: LD50 = 26 gm/kg

NIOSH - Health Standards - Exposure Limits
100 mg/m3 TWA (Listed under 'Refined petroleum solvents')

NIOSH - Health Standards - Health Effects and Precautions
Eye, nose, and throat irritation; dermatitis; nervous sytem effects (Blood and urine monitoring required; prevent skin contact) (Listed under 'Refined petroleum solvents')

NTP Chemical Status Reports - Testing Status and NTIS Number
Technical reports printed (PB87131678/AS)

NTP Chemical Status Reports - Evidence of Carcinogenicity
male mice-no evidence; female mice-no evidence

ENVIRONMENTAL LISTS

List of Pesticide Product Inert Ingredients
[present]

STATE LISTS

Florida Hazardous Substance List
[present]

Massachusetts Right To Know List
[present]

NJ Right to Know List (Total)
sn 1091

Pennsylvania Right to Know List
[present]

KRYPTON 81   15678-91-8

HEALTH AND SAFETY LISTS

ACGIH 1995 - Time Weighted Averages
0.5 ppm TWA; 0.86 mg/m3 TWA

ACGIH 1995 - Short Term Exposure Limits
1.5 ppm STEL; 2.6 mg/m3 STEL

NIOSH - Selected LD50s and LC50s
Oral, rat: LD50 = 1300 mg/kg

NIOSH 1990 - Pocket Guide - RELs
0.5 ppm TWA; 0.9 mg/m3 TWA; 1.5 ppm STEL; 3 mg/m3 STEL

NIOSH 1990 - Pocket Guide - Target organs
eyes, skin, respiratory system

OSHA - Vacated PELs - Time Weighted Averages
0.5 ppm TWA; 0.9 mg/m3 TWA

OSHA - Vacated PELs - Short Term Exposure Limits
1.5 ppm STEL; 3 mg/m3 STEL

OSHA - Final PELs - Time Weighted Averages
0.5 ppm TWA; 0.9 mg/m3 TWA

OSHA - List of Highly Hazardous Chemicals
threshhold quantity = 100 pounds

INTERNATIONAL LISTS

Australian Exposure Standards - Time Weighted Averages
0.5 ppm TWA; 0.86 mg/m3 TWA

Australian Exposure Standards - Short Term Exposure Limits
1.5 ppm STEL; 2.6 mg/m3 STEL

Canada - WHMIS: Ingredient Disclosure
1% item 920 (397)

Canada - Alberta - 8 Hour Occupational Exposure Limit
0.5 ppm TWA; 0.86 mg/m3 TWA

Canada - Alberta - 15 Minute Occupational Exposure Limit
1.5 ppm STEL; 2.6 mg/m3 STEL

Canada - British Columbia - 8 Hour Exposure Limits
0.5 ppm TWA; 0.9 mg/m3 TWA

Canada - British Columbia - 15 Minute Exposure Limits
1.5 ppm STEL; 3 mg/m3 STEL

Canada - Ontario - OHSA - TWAEVs
0.5 ppm TWAEV; 0.9 mg/m3 TWAEV

Canada - Ontario - OHSA - STEVs
1.5 ppm STEV; 2.6 mg/m3 STEV

Canada - Quebec - Time-Weighted Average Exposure Values
0.5 ppm TWAEV; 0.86 mg/m3 TWAEV

Canada - Quebec - Short-term Exposure Values
1.5 ppm STEV; 2.6 mg/m3 STEV

United Kingdom - Occupational Exposure Standards - TWAs
0.5 ppm TWA; 0.9 mg/m3 TWA

United Kingdom - Occupational Exposure Standards - STELs
1.5 ppm STEL; 3 mg/m3 STEL

German (DFG) - MAK Values
0.5 ppm MAK; 0.9 mg/m3 MAK

German (DFG) - Peak Limitations
2 x normal MAK (5 min momentary value); don't exceed 8 times during shift

Israel - Time Weighted Averages
0.5 ppm TWA; 0.86 mg/m3 TWA

Israel - Short Term Exposure Limits
1.5 ppm STEL; 2.6 mg/m3 STEL

Israel - Action Levels
0.25 ppm AL; 0.43 mg/m3 AL

Mexico - Instruction No. 10 - TWAs
0.5 ppm TWA; 0.9 mg/m3 TWA

Mexico - Instruction No. 10 - STELs
1.5 ppm STEL; 3 mg/m3 STEL

STATE LISTS

California - Exposure Limits - PELs
0.5 ppm PEL; 0.9 mg/m3 PEL

California - Exposure Limits - STELs
1.5 ppm STEL; 3 mg/m3 STEL

California - Directors List of Hazardous Substances (8 CCR 339)
[present]

Florida Hazardous Substance List
[present]

Massachusetts Right To Know List
[present]

Minnesota Hazardous Substance List
[present]

Pennsylvania Right to Know List
[present]

KRYPTON 83M   RR-00426-2

HEALTH AND SAFETY LISTS

U.S. DOT - Substances From 49 CFR 172.101
regulated by DOT (UN1224)

U.S. DOT - Hazard Classes
DOT hazard class = 3

STATE LISTS
NJ Right to Know List (Total)
sn 2510

KRYPTON 85   13983-27-2

HEALTH AND SAFETY LISTS
U.S. DOT - Substances From 49 CFR 172.101
regulated by DOT (UN1970, UN1056)
U.S. DOT - Hazard Classes
DOT hazard class = 2.2

STATE LISTS
NJ Right to Know List (Total)
sn 1093

KRYPTON 85M   RR-00425-1

HEALTH AND SAFETY LISTS
U.S. DOT - Appendix A Table 2 - Radionuclides
final RQ = 10 curies (3.7E 11 Bq)

ENVIRONMENTAL LISTS
CERCLA/SARA List of Radionuclides (Appendix B) and Their Reportable Quantities
final RQ = 10 curies (3.7E 11 Bq)

KRYPTON 87   14809-68-8

HEALTH AND SAFETY LISTS
U.S. DOT - Appendix A Table 2 - Radionuclides
final RQ = 10 curies (3.7E 11 Bq)

ENVIRONMENTAL LISTS
CERCLA/SARA List of Radionuclides (Appendix B) and Their Reportable Quantities
final RQ = 10 curies (3.7E 11 Bq)

KRYPTON 88   14995-61-0

HEALTH AND SAFETY LISTS
U.S. DOT - Appendix A Table 2 - Radionuclides
final RQ = 10 curies (3.7E 11 Bq)

ENVIRONMENTAL LISTS
CERCLA/SARA List of Radionuclides (Appendix B) and Their Reportable Quantities
final RQ = 10 curies (3.7E 11 Bq)

LACQUER   RR-01500-9

HEALTH AND SAFETY LISTS
U.S. DOT - Appendix A Table 2 - Radionuclides
final RQ = 100 curies (3.7E 12 Bq)

ENVIRONMENTAL LISTS
CERCLA/SARA List of Radionuclides (Appendix B) and Their Reportable Quantities
final RQ = 100 curies (3.7E 12 Bq)

LACTIC ACID   50-21-5

HEALTH AND SAFETY LISTS
U.S. DOT - Appendix A Table 2 - Radionuclides
final RQ = 1000 curies (3.7E 13 Bq)

ENVIRONMENTAL LISTS
CERCLA/SARA List of Radionuclides (Appendix B) and Their Reportable Quantities
final RQ = 1000 curies (3.7E 13 Bq)

LACTOFEN   77501-63-4

HEALTH AND SAFETY LISTS
U.S. DOT - Appendix A Table 2 - Radionuclides
final RQ = 1000 curies (3.7E 13 Bq)

ENVIRONMENTAL LISTS
CERCLA/SARA List of Radionuclides (Appendix B) and Their Reportable Quantities
final RQ = 1000 curies (3.7E 13 Bq)

LACTONITRILE   78-97-7

HEALTH AND SAFETY LISTS
U.S. DOT - Appendix A Table 2 - Radionuclides
final RQ = 1000 curies (3.7E 13 Bq)

ENVIRONMENTAL LISTS
CERCLA/SARA List of Radionuclides (Appendix B) and Their Reportable Quantities
final RQ = 1000 curies (3.7E 13 Bq)

BETA-D-LACTOSE   63-42-3

HEALTH AND SAFETY LISTS
U.S. DOT - Appendix A Table 2 - Radionuclides
final RQ = 100 curies (3.7E 12 Bq)

ENVIRONMENTAL LISTS
CERCLA/SARA List of Radionuclides (Appendix B) and Their Reportable Quantities
final RQ = 100 curies (3.7E 12 Bq)

LANOLIN   8006-54-0

HEALTH AND SAFETY LISTS
U.S. DOT - Appendix A Table 2 - Radionuclides
final RQ = 10 curies (3.7E 11 Bq)

ENVIRONMENTAL LISTS
CERCLA/SARA List of Radionuclides (Appendix B) and Their Reportable Quantities
final RQ = 10 curies (3.7E 11 Bq)

LANOLIN, ETHOXYLATED   61790-81-6

HEALTH AND SAFETY LISTS
U.S. DOT - Appendix A Table 2 - Radionuclides
final RQ = 10 curies (3.7E 11 Bq)

ENVIRONMENTAL LISTS
CERCLA/SARA List of Radionuclides (Appendix B) and Their Reportable Quantities
final RQ = 10 curies (3.7E 11 Bq)

LANOLIN, ETHOXYLATED PROPOXYLATED   68458-88-8

STATE LISTS
NJ Right to Know List (Total)
sn 2628; sn 2511; sn 2512

LANTHANUM 131   15715-04-5

HEALTH AND SAFETY LISTS
NIOSH - Selected LD50s and LC50s
Oral, rat: LD50 = 3730 mg/kg

ENVIRONMENTAL LISTS
List of Pesticide Product Inert Ingredients
[present]

INTERNATIONAL LISTS
Canada - WHMIS: Ingredient Disclosure
1% item 921 (100)

LANTHANUM 132   15066-93-0

ENVIRONMENTAL LISTS
CERCLA/SARA - Section 313 - Emission Reporting
form R reporting required

STATE LISTS

California - Air Bill 2588 Appendix A-II
known or potential carcinogen: 9/89

California - Prop. 65 - Cancer list
carcinogen - initial date 1/1/89

PROPOSED REGULATIONS

Safe Drinking Water Act - Priority list
[present]

LANTHANUM 135   15816-85-0

HEALTH AND SAFETY LISTS

NFPA - Flash Points
flash point = 171 degrees F (77 degrees C)

NFPA - Hazard Identification Ratings
health-4; flammability-2; reactivity-1

NIOSH - Selected LD50s and LC50s
Oral, rat: LD50 = 87 mg/kg Skin, rabbit: LD50 = 20 mg/kg

ENVIRONMENTAL LISTS

CERCLA/SARA - Section 302 Extremely Hazardous Substances and TPQs
TPQ = 1000 pounds

EPA - Master Testing List
[present]

TSCA - Code of Federal Regulations Citations
40 CFR 716.120(a)

TSCA - Health and Safety Reporting List
Effective Date: June 1, 1987

STATE LISTS

Florida Hazardous Substance List
[present]

Massachusetts Right To Know List
extraordinarily hazardous

Pennsylvania Right to Know List
environmental hazard

PROPOSED REGULATIONS

CERCLA/SARA - Proposed Hazardous Substance Additions
proposed RQ = 1 pound (.454 kg)

CERCLA/SARA - 1989 Proposed RQ Adjustments
proposed RQ = 1000 pounds (454 kg)

LANTHANUM 137   14834-69-6

ENVIRONMENTAL LISTS

List of Pesticide Product Inert Ingredients
[present]

LANTHANUM 138   15816-87-2

HEALTH AND SAFETY LISTS

NFPA - Flash Points
flash point = 460 degrees F (238 degrees C)

NFPA - Hazard Identification Ratings
health-0; flammability-1; reactivity-0

ENVIRONMENTAL LISTS

List of Pesticide Product Inert Ingredients
[present]

STATE LISTS

Pennsylvania Right to Know List
[present]

LANTHANUM 140   13981-28-7

ENVIRONMENTAL LISTS

List of Pesticide Product Inert Ingredients
[present]

LANTHANUM 141   15816-88-3

ENVIRONMENTAL LISTS

List of Pesticide Product Inert Ingredients
[present]

LANTHANUM 142   15816-89-4

HEALTH AND SAFETY LISTS

U.S. DOT - Appendix A Table 2 - Radionuclides
final RQ = 1000 curies (3.7E 13 Bq)

ENVIRONMENTAL LISTS

CERCLA/SARA List of Radionuclides (Appendix B) and Their Reportable Quantities
final RQ = 1000 curies (3.7E 13 Bq)

LANTHANUM 143   16729-61-6

HEALTH AND SAFETY LISTS

U.S. DOT - Appendix A Table 2 - Radionuclides
final RQ = 100 curies (3.7E 12 Bq)

ENVIRONMENTAL LISTS

CERCLA/SARA List of Radionuclides (Appendix B) and Their Reportable Quantities
final RQ = 100 curies (3.7E 12 Bq)

LANTHANUM CHLORIDE   10099-58-8

HEALTH AND SAFETY LISTS

U.S. DOT - Appendix A Table 2 - Radionuclides
final RQ = 1000 curies (3.7E 13 Bq)

ENVIRONMENTAL LISTS

CERCLA/SARA List of Radionuclides (Appendix B) and Their Reportable Quantities
final RQ = 1000 curies (3.7E 13 Bq)

LANTHANUM NITRATE   10099-59-9

HEALTH AND SAFETY LISTS

U.S. DOT - Appendix A Table 2 - Radionuclides
final RQ = 10 curies (3.7E 11 Bq)

ENVIRONMENTAL LISTS

CERCLA/SARA List of Radionuclides (Appendix B) and Their Reportable Quantities
final RQ = 10 curies (3.7E 11 Bq)

LARD   61789-99-9

HEALTH AND SAFETY LISTS

U.S. DOT - Appendix A Table 2 - Radionuclides
final RQ = 1 curie (3.7E 10 Bq)

ENVIRONMENTAL LISTS

CERCLA/SARA List of Radionuclides (Appendix B) and Their Reportable Quantities
final RQ = 1 curie (3.7E 10 Bq)

LASIOCARPINE   303-34-4

HEALTH AND SAFETY LISTS

U.S. DOT - Appendix A Table 2 - Radionuclides
final RQ = 10 curies (3.7E 11 Bq)

ENVIRONMENTAL LISTS

CERCLA/SARA List of Radionuclides (Appendix B) and Their Reportable Quantities
final RQ = 10 curies (3.7E 11 Bq)

STATE LISTS

California - Air Bill 2588 Appendix A-II
known or potential carcinogen: 9/89

California - Prop. 65 - Cancer list
carcinogen - initial date 1/1/89

PROPOSED REGULATIONS

Safe Drinking Water Act - Priority list
[present]

LANTHANUM 135   15816-85-0

HEALTH AND SAFETY LISTS

NFPA - Flash Points
flash point = 171 degrees F (77 degrees C)

NFPA - Hazard Identification Ratings
health-4; flammability-2; reactivity-1

NIOSH - Selected LD50s and LC50s
Oral, rat: LD50 = 87 mg/kg Skin, rabbit: LD50 = 20 mg/kg

ENVIRONMENTAL LISTS

CERCLA/SARA - Section 302 Extremely Hazardous Substances and
TPQs
TPQ = 1000 pounds

EPA - Master Testing List
[present]

TSCA - Code of Federal Regulations Citations
40 CFR 716.120(a)

TSCA - Health and Safety Reporting List
Effective Date: June 1, 1987

STATE LISTS

Florida Hazardous Substance List
[present]

Massachusetts Right To Know List
extraordinarily hazardous

Pennsylvania Right to Know List
environmental hazard

PROPOSED REGULATIONS

CERCLA/SARA - Proposed Hazardous Substance Additions
proposed RQ = 1 pound (.454 kg)

CERCLA/SARA - 1989 Proposed RQ Adjustments
proposed RQ = 1000 pounds (454 kg)

LANTHANUM 137   14834-69-6

ENVIRONMENTAL LISTS

List of Pesticide Product Inert Ingredients
[present]

LANTHANUM 138   15816-87-2

HEALTH AND SAFETY LISTS

NFPA - Flash Points
flash point = 460 degrees F (238 degrees C)

NFPA - Hazard Identification Ratings
health-0; flammability-1; reactivity-0

ENVIRONMENTAL LISTS

List of Pesticide Product Inert Ingredients
[present]

STATE LISTS

Pennsylvania Right to Know List
[present]

LANTHANUM 140   13981-28-7

ENVIRONMENTAL LISTS

List of Pesticide Product Inert Ingredients
[present]

LANTHANUM 141   15816-88-3

ENVIRONMENTAL LISTS

List of Pesticide Product Inert Ingredients
[present]

LANTHANUM 142   15816-89-4

HEALTH AND SAFETY LISTS

U.S. DOT - Appendix A Table 2 - Radionuclides
final RQ = 1000 curies (3.7E 13 Bq)

ENVIRONMENTAL LISTS

CERCLA/SARA List of Radionuclides (Appendix B) and Their Reportable Quantities
final RQ = 1000 curies (3.7E 13 Bq)

LANTHANUM 143   16729-61-6

HEALTH AND SAFETY LISTS

U.S. DOT - Appendix A Table 2 - Radionuclides
final RQ = 100 curies (3.7E 12 Bq)

ENVIRONMENTAL LISTS

CERCLA/SARA List of Radionuclides (Appendix B) and Their Reportable Quantities
final RQ = 100 curies (3.7E 12 Bq)

LANTHANUM CHLORIDE   10099-58-8

HEALTH AND SAFETY LISTS

U.S. DOT - Appendix A Table 2 - Radionuclides
final RQ = 1000 curies (3.7E 13 Bq)

ENVIRONMENTAL LISTS

CERCLA/SARA List of Radionuclides (Appendix B) and Their Reportable Quantities
final RQ = 1000 curies (3.7E 13 Bq)

LANTHANUM NITRATE   10099-59-9

HEALTH AND SAFETY LISTS

U.S. DOT - Appendix A Table 2 - Radionuclides
final RQ = 10 curies (3.7E 11 Bq)

ENVIRONMENTAL LISTS

CERCLA/SARA List of Radionuclides (Appendix B) and Their Reportable Quantities
final RQ = 10 curies (3.7E 11 Bq)

LARD   61789-99-9

HEALTH AND SAFETY LISTS

U.S. DOT - Appendix A Table 2 - Radionuclides
final RQ = 1 curie (3.7E 10 Bq)

ENVIRONMENTAL LISTS

CERCLA/SARA List of Radionuclides (Appendix B) and Their Reportable Quantities
final RQ = 1 curie (3.7E 10 Bq)

LASIOCARPINE   303-34-4

HEALTH AND SAFETY LISTS

U.S. DOT - Appendix A Table 2 - Radionuclides
final RQ = 10 curies (3.7E 11 Bq)

ENVIRONMENTAL LISTS

CERCLA/SARA List of Radionuclides (Appendix B) and Their Reportable Quantities
final RQ = 10 curies (3.7E 11 Bq)

LAURYLDIMETHYLAMINE OXIDE   1643-20-5

ENVIRONMENTAL LISTS

List of Pesticide Product Inert Ingredients
[present]

LAURYL METHACRYLATE   142-90-5

ENVIRONMENTAL LISTS

List of Pesticide Product Inert Ingredients
[present]

LAURYL POLYETHYLENE GLYCOL ETHER   9002-92-0

HEALTH AND SAFETY LISTS

U.S. DOT - Organic Peroxides Table
Organic peroxide UN3106

IARC - Group 3 (not classifiable)
[present]

OSHA - List of Highly Hazardous Chemicals
threshhold quantity = 7500 pounds

INTERNATIONAL LISTS

Canada - WHMIS: Ingredient Disclosure
1% item 923 (1366)

German (DFG) - Skin/Sensitizers
negligible effects on skin

STATE LISTS

Florida Hazardous Substance List
[present]

Massachusetts Right To Know List
[present]

NJ Right to Know List (Total)
sn 1095; maximum 42% in water: sn 2340

Pennsylvania Right to Know List
[present]

LAURYL SULFATE   151-41-7

HEALTH AND SAFETY LISTS

NFPA - Flash Points
flash point = 260 degrees F (127 degrees C)

NFPA - Hazard Identification Ratings
health-0; flammability-1; reactivity-0

NIOSH - Selected LD50s and LC50s
Oral, rat: LD50 = 12800 mg/kg

ENVIRONMENTAL LISTS

EPA - Master Testing List
[present]

List of Pesticide Product Inert Ingredients
[present]

INTERNATIONAL LISTS

Canada - WHMIS: Ingredient Disclosure
1% item 924 (180)

LEAD   7439-92-1

HEALTH AND SAFETY LISTS

NIOSH - Selected LD50s and LC50s
Oral, rat: LD50 = 23 gm/kg

ENVIRONMENTAL LISTS

TSCA - Code of Federal Regulations Citations
40 CFR 712.30(x); 40 CFR 716.120(d)

TSCA - PAIR - Reporting List
Reporting Date: November 27, 1991

TSCA - Health and Safety Reporting List
Effective Date: September 30, 1991

INTERNATIONAL LISTS

Canada - WHMIS: Ingredient Disclosure
1% item 925 (189)

LEAD 195M   RR-00421-7

HEALTH AND SAFETY LISTS

NFPA - Flash Points
flash point = 291 degrees F (144 degrees C)

NFPA - Hazard Identification Ratings
health-1; flammability-1; reactivity-0

LEAD 198   16646-00-7

HEALTH AND SAFETY LISTS

NIOSH - Selected LD50s and LC50s
Oral, rat: LD50 = 740 mg/kg

ENVIRONMENTAL LISTS

EPA - Master Testing List
[present]

INTERNATIONAL LISTS

Canada - WHMIS: Ingredient Disclosure
1% item 926 (1067)

LEAD 199   27486-00-6

ENVIRONMENTAL LISTS

List of Pesticide Product Inert Ingredients
[present]

INTERNATIONAL LISTS

Canada - WHMIS: Ingredient Disclosure
1% item 927 (1313)

LEAD 200   16645-99-1

ENVIRONMENTAL LISTS

TSCA - Code of Federal Regulations Citations
40 CFR 712.30(d)

TSCA - PAIR - Reporting List
Reporting Date: November 19, 1982

INTERNATIONAL LISTS

Canada - WHMIS: Ingredient Disclosure
1% item 928 (1095)

LEAD 201   17239-87-1

HEALTH AND SAFETY LISTS

NFPA - Flash Points
flash point > 200 degrees F (93 degrees C)

NFPA - Hazard Identification Ratings
health-0; flammability-1; reactivity-0

NIOSH - Selected LD50s and LC50s
Oral, rat: LD50 = 8600 mg/kg

ENVIRONMENTAL LISTS

List of Pesticide Product Inert Ingredients
[present]

INTERNATIONAL LISTS

Canada - WHMIS: Ingredient Disclosure
1% item 688 (181)

LEAD 202   15752-86-0

HEALTH AND SAFETY LISTS

NIOSH - Selected LD50s and LC50s
Oral, rat: LD50 = 1300 mg/kg

LEAD 202M   **RR-00420-6**
SEE ALSO:
  K052-HAZARDOUS WASTES
  K061-HAZARDOUS WASTES
  K062-HAZARDOUS WASTES
  K051-HAZARDOUS WASTES
  K046-HAZARDOUS WASTES
  K048-HAZARDOUS WASTES
  K049-HAZARDOUS WASTES
  K086-HAZARDOUS WASTES
  K100-HAZARDOUS WASTES
  LEAD COMPOUNDS
  K069-HAZARDOUS WASTES
  K064-HAZARDOUS WASTES
  K065-HAZARDOUS WASTES
  K066-HAZARDOUS WASTES
  K003-HAZARDOUS WASTES
  K005-HAZARDOUS WASTES
  F039-HAZARDOUS WASTES
  K002-HAZARDOUS WASTES

HEALTH AND SAFETY LISTS

ACGIH 1995 - Time Weighted Averages
  inorg. dusts and fumes, as Pb: (0.15) mg/m3 TWA

ACGIH 1995 - Biological Exposure Indices
  Lead in blood: (50 ug/100 ml), (B); Lead in urine: 150 ug/g creati-
  nine (B); Zinc protoporphyrin in blood: 250 ug/100 ml erythrocytes
  or 100 ug/100 ml blood, after 1 month exposure (B)

U.S. DOT - Appendix A Table 1 - Hazardous Substances
  final RQ = 10 pounds (4.54 kg) (No reporting of releases of this
  hazardous substance is required if the diameter of the pieces of the
  solid metal released is equal to or exceeds 0.004 inches)

NIOSH 1990 - Pocket Guide - RELs
  as Pb: 0.100 mg/m3 TWA (Air concentration to be maintained so that
  worker blood lead remains below 0.060 mg/100 g of whole blood)

NIOSH 1990 - Pocket Guide - IDLHs
  as Pb: 700 mg/m3 IDLH

NIOSH 1990 - Pocket Guide - Target organs
  GI tract, CNS, kidneys, blood, gingival tissue

NTP Chemical Status Reports - Testing Status and NTIS Number
  mixed with soil: Prechronic studies completed: chemicals in review for
  further evaluation

OSHA - 29 CFR 1910 Specifically Regulated Chemicals
  as Pb: 50 ug/m3 TWA PEL; 30 ug/m3 action level; Poison (see 29
  CFR 1910.1025)

ENVIRONMENTAL LISTS

ATSDR Priority List
  Rank (of 275): 001

CERCLA/SARA - Section 313 - Emission Reporting
  form R reporting required for 0.1% de minimus concentration

CERCLA/SARA - Hazardous Substances and their Reportable Quantities
  final RQ = 10 pounds (4.54 kg)

Clean Water Act - Priority Pollutants
  [present]

Clean Water Act - Toxic Pollutants
  [present] (Listed under 'Lead and compounds')

Safe Drinking Water Act - MCLs
  MCL = 0.05 mg/L

Safe Drinking Water Act - MCLGs
  MCLG = Zero

RCRA - D Series - Maximum Concentration of Contaminants
  waste number D008; regulatory level = 5.0 mg/L

RCRA - D Series - Chronic Toxicity Reference Levels
  chronic toxicity reference level = 0.05 mg/L

RCRA - Hazardous Constituents-Appendix VIII
  hazardous constituent - no waste number

RCRA - Basis for Listing - Appendix VII
  Included in waste streams: F039, K002, K003, K005, K046, K048,
  K049, K052, K061, K062, K064, K065, K066, K069, K086, K100

RCRA - Substances Banned From Land Disposal
  [present] (except for lead materials stored before secondary smelting
  by owners involved in the reclamation of lead-bearing hazardous
  materials)

RCRA - TSD Facilities Ground Water Monitoring
  TM 6010 = 40 ug/L PQL; TM 7420 = 1000 ug/L PQL; Tm 7421 =
  10 ug/L PQL (all species in the ground water that contain this element
  are included)

RCRA - Universal Treatment Standards (LDR)
  WW: 0.69 mg/l; NWW: 0.37 mg/l TCLP

INTERNATIONAL LISTS

Australian Exposure Standards - Time Weighted Averages
  dusts and fumes, as Pb: 0.15 mg/m3 TWA

Australian Exposure Standards - Under Review
  exposure limits under review

Canada - WHMIS: Ingredient Disclosure
  0.1% item 937 (1435)

Canada - NPRI (National Pollutant Release Inventory)
  [present]

Canada - CEPA Schedule I - Toxic Substances
  limited atmospheric releases from secondary lead smelters

Canada - CEPA Schedule III Part II - Restricted Substances (Ocean
Dumping)
  [present]

Canada - Drinking Water Quality - MACs
  0.01 mg/L MAC

Canada - British Columbia - Skin Notations
  skin - potential for skin absorption

Canada - Ontario - OHSA - TWAEVs
  0.15 mg/m3 TWAEV (designated substance regulation)

Canada - Ontario - OHSA - STEVs
  0.30 mg/m3 STEV (designated substance regulation)

Canada - Ontario - OHSA - Designated Substances
  0.15 mg/m3 TWAEV; See Ontario Reg. 843 for full information.

Canada - Quebec - Time-Weighted Average Exposure Values
  as Pb: 0.15 mg/m3 TWAEV

German (DFG) - MAK Values
  total dust: 0.1 mg/m3 MAK

German (DFG) - Peak Limitations
  10 x normal MAK (30 min average value); don't exceed during shift

German (DFG) - Pregnancy
  risk to embryo/fetus probable

Israel - Time Weighted Averages
  inorganic dust and fumes, men and women over 45 years: 0.10 mg/m3
  TWA; organic: 0.05 mg/m3

Israel - Ceiling Exposure Limits
  inorganic dusts and fumes, for women under 45 years: C 0.05 mg/m3

Israel - Action Levels
  inorganic dusts and fumes: women over 45 and men: 0.05 mg/m3
  AL; women under 45: 0.025 mg/m3 AL

Mexico - Instruction No. 10 - TWAs
  0.15 mg/m3 TWA

Mexico - Instruction No. 10 - STELs
  0.45 mg/m3 STEL

Mexico - Wastewater - Organic Toxic Pollutants and Heavy Metals
  Listed under [Heavy Metals]

Mexico - Drinking Water - Ecological Criteria
  0.05 mg/l

STATE LISTS

California - Air Bill 2588 Appendix A-I
  known or potential carcinogen

California - Prop. 65 - Cancer list
carcinogen - initial date 10/1/92
California - Prop. 65 - Developmental Toxicity
developmental toxicity - initial date 2/27/87
California - Prop. 65 - Reproductive - Female
female reproductive toxicity - initial date 2/27/87
California - Prop. 65 - Reproductive - Male
male reproductive toxicity - initial date 2/27/87
California - Prop. 65 - No Significant Risk Levels
NOEL = 0.5 ug/day
California - Exposure Limits - PELs
metallic and inorganic compounds, dust and fume, as Pb: 0.05 mg/m3 PEL; poison- avoid eating or smoking; see also section 5216
California - Directors List of Hazardous Substances (8 CCR 339)
[present]
Florida Hazardous Substance List
[present]
Massachusetts Right To Know List
teratogen
Minnesota Hazardous Substance List
[present] as Pb
NJ Right to Know List (Total)
sn 1095
NJ Special Hazardous Substances
(teratogen)
Pennsylvania Right to Know List
environmental hazard (any compound of this substance is also an environmental hazard)
PROPOSED REGULATIONS
ACGIH 1995 - Notice of Intended Changes
as Pb: 0.05 mg/m3 TWA; A3-animal carcinogen
ACGIH 1995 - Proposed Biological Exposure Indices
Lead in blood: 30 ug/100ml creatinine, not critical (B)
Canada - Ontario - Proposed Occupational TWAEVs
0.05 mg/m3 TWAEV

LEAD 203   14687-25-3

HEALTH AND SAFETY LISTS
U.S. DOT - Appendix A Table 2 - Radionuclides
final RQ = 1000 curies (3.7E 13 Bq)

ENVIRONMENTAL LISTS
CERCLA/SARA List of Radionuclides (Appendix B) and Their Reportable Quantities
final RQ = 1000 curies (3.7E 13 Bq)

LEAD 205   14119-28-9

HEALTH AND SAFETY LISTS
U.S. DOT - Appendix A Table 2 - Radionuclides
final RQ = 100 curies (3.7E 12 Bq)

ENVIRONMENTAL LISTS
CERCLA/SARA List of Radionuclides (Appendix B) and Their Reportable Quantities
final RQ = 100 curies (3.7E 12 Bq)

LEAD 209   14119-30-3

HEALTH AND SAFETY LISTS
U.S. DOT - Appendix A Table 2 - Radionuclides
final RQ = 100 curies (3.7E 12 Bq)

ENVIRONMENTAL LISTS
CERCLA/SARA List of Radionuclides (Appendix B) and Their Reportable Quantities
final RQ = 100 curies (3.7E 12 Bq)

LEAD 210   14255-04-0

HEALTH AND SAFETY LISTS
U.S. DOT - Appendix A Table 2 - Radionuclides
final RQ = 100 curies (3.7E 12 Bq)

ENVIRONMENTAL LISTS
CERCLA/SARA List of Radionuclides (Appendix B) and Their Reportable Quantities
final RQ = 100 curies (3.7E 12 Bq)

LEAD 211   15816-77-0

HEALTH AND SAFETY LISTS
U.S. DOT - Appendix A Table 2 - Radionuclides
final RQ = 100 curies (3.7E 12 Bq)

ENVIRONMENTAL LISTS
CERCLA/SARA List of Radionuclides (Appendix B) and Their Reportable Quantities
final RQ = 100 curies (3.7E 12 Bq)

LEAD 212   15092-94-1

HEALTH AND SAFETY LISTS
U.S. DOT - Appendix A Table 2 - Radionuclides
final RQ = 1 curie (3.7E 10 Bq)

ENVIRONMENTAL LISTS
CERCLA/SARA List of Radionuclides (Appendix B) and Their Reportable Quantities
final RQ = 1 curie (3.7E 10 Bq)

LEAD 214   15067-28-4

HEALTH AND SAFETY LISTS
U.S. DOT - Appendix A Table 2 - Radionuclides
final RQ = 10 curies (3.7E 11 Bq)

ENVIRONMENTAL LISTS
CERCLA/SARA List of Radionuclides (Appendix B) and Their Reportable Quantities
final RQ = 10 curies (3.7E 11 Bq)

LEAD ACETATE   301-04-2

HEALTH AND SAFETY LISTS
U.S. DOT - Appendix A Table 2 - Radionuclides
final RQ = 100 curies (3.7E 12 Bq)

ENVIRONMENTAL LISTS
CERCLA/SARA List of Radionuclides (Appendix B) and Their Reportable Quantities
final RQ = 100 curies (3.7E 12 Bq)

LEAD (IV) ACETATE   546-67-8

HEALTH AND SAFETY LISTS
U.S. DOT - Appendix A Table 2 - Radionuclides
final RQ = 100 curies (3.7E 12 Bq)

ENVIRONMENTAL LISTS
CERCLA/SARA List of Radionuclides (Appendix B) and Their Reportable Quantities
final RQ = 100 curies (3.7E 12 Bq)

LEAD ANTIMONATE   13510-89-9

HEALTH AND SAFETY LISTS
U.S. DOT - Appendix A Table 2 - Radionuclides
final RQ = 1000 curies (3.7E 13 Bq)

ENVIRONMENTAL LISTS
CERCLA/SARA List of Radionuclides (Appendix B) and Their Reportable Quantities
final RQ = 1000 curies (3.7E 13 Bq)

LEAD ANTIMONIDE   12266-38-5

HEALTH AND SAFETY LISTS

U.S. DOT - Appendix A Table 2 - Radionuclides
final RQ = 0.01 curies (3.7E 8 Bq)

ENVIRONMENTAL LISTS

ATSDR Priority List
Rank (of 275): 169

CERCLA/SARA List of Radionuclides (Appendix B) and Their Reportable Quantities
final RQ = 0.01 curies (3.7E 8 Bq)

LEAD ARSENATE   3687-31-8

HEALTH AND SAFETY LISTS

U.S. DOT - Appendix A Table 2 - Radionuclides
final RQ = 100 curies (3.7E 12 Bq)

ENVIRONMENTAL LISTS

CERCLA/SARA List of Radionuclides (Appendix B) and Their Reportable Quantities
final RQ = 100 curies (3.7E 12 Bq)

LEAD ARSENATE   7784-40-9

HEALTH AND SAFETY LISTS

U.S. DOT - Appendix A Table 2 - Radionuclides
final RQ = 10 curies (3.7E 11 Bq)

ENVIRONMENTAL LISTS

CERCLA/SARA List of Radionuclides (Appendix B) and Their Reportable Quantities
final RQ = 10 curies (3.7E 11 Bq)

LEAD ARSENATE (PB3(ASS04)2)   10102-48-4

HEALTH AND SAFETY LISTS

U.S. DOT - Appendix A Table 2 - Radionuclides
final RQ = 100 curies (3.7E 12 Bq)

ENVIRONMENTAL LISTS

CERCLA/SARA List of Radionuclides (Appendix B) and Their Reportable Quantities
final RQ = 100 curies (3.7E 12 Bq)

LEAD ARSENATES   RR-00132-1

SEE ALSO:
LEAD
LEAD
LEAD COMPOUNDS

HEALTH AND SAFETY LISTS

U.S. DOT - Substances From 49 CFR 172.101
regulated by DOT (UN1616)

U.S. DOT - Hazard Classes
DOT hazard class = 6.1

U.S. DOT - Appendix B - Marine Pollutants
DOT regulated marine pollutant

U.S. DOT - Appendix A Table 1 - Hazardous Substances
final RQ = 10 pounds (4.54 kg)

NTP Chemical Status Reports - Testing Status and NTIS Number
Prechronic studies for which toxicity technical reports were not prepared

NTP Seventh Report - Suspect Carcinogens
suspect carcinogen

OSHA - Possible Select Carcinogens
[present]

ENVIRONMENTAL LISTS

CERCLA/SARA - Hazardous Substances and their Reportable Quantities
final RQ = 10 pounds (4.54 kg)

Clean Water Act - Hazardous Substances
[present]

RCRA - U Series Wastes
waste number U144

RCRA - Hazardous Constituents-Appendix VIII
waste number U144

RCRA - Substances Banned From Land Disposal
[present]

INTERNATIONAL LISTS

Canada - WHMIS: Ingredient Disclosure
0.1% item 929 (30)

STATE LISTS

California - Air Bill 2588 Appendix A-I
known or potential carcinogen

California - Prop. 65 - Cancer list
carcinogen - initial date 1/1/88

California - Prop. 65 - No Significant Risk Levels
no significant risk level = 3 ug/day

Florida Hazardous Substance List
[present]

Massachusetts Right To Know List
carcinogen; extraordinarily hazardous

Minnesota Hazardous Substance List
carcinogen

NJ Right to Know List (Total)
sn 1097

NJ Special Hazardous Substances
(carcinogen; teratogen)

Pennsylvania Right to Know List
environmental hazard; special hazardous substance

Pennsylvania RTK - Special Hazardous Substances
[present]

LEAD ARSENATE, UNSPECIFIED   7645-25-2

SEE ALSO:
LEAD

STATE LISTS

Massachusetts Right To Know List
[present]

LEAD ARSENITE   10031-13-7

STATE LISTS

Massachusetts Right To Know List
[present]

LEAD ARSENITES   RR-01292-0

STATE LISTS

Massachusetts Right To Know List
[present]

LEAD AZIDE   13424-46-9

SEE ALSO:
LEAD
ARSENIC

HEALTH AND SAFETY LISTS

NIOSH 1990 - Pocket Guide - Carcinogens
occupational carcinogen

INTERNATIONAL LISTS

Australian Exposure Standards - Time Weighted Averages
as Pb3(AsO4): 0.15 mg/m3 TWA

Canada - Alberta - 8 Hour Occupational Exposure Limit
as Pb: 0.15 mg/m3 TWA

Canada - Alberta - 15 Minute Occupational Exposure Limit
as Pb: 0.45 mg/m3 STEL

Canada - Alberta - Designated Substances
designated substance - requires code of practice
Canada - British Columbia - 8 Hour Exposure Limits
as Pb: 0.15 mg/m3 TWA
Canada - British Columbia - 15 Minute Exposure Limits
as Pb: 0.45 mg/m3 STEL
Canada - Quebec - Time-Weighted Average Exposure Values
as Pb: 0.15 mg/m3 TWAEV
German (DFG) - Carcinogens
proven carcinogen (Listed under 'Arsenic trioxide')
Mexico - Instruction No. 10 - TWAs
0.15 mg/m3 TWA
Mexico - Instruction No. 10 - STELs
0.45 mg/m3 STEL

LEAD BORATE   14720-53-7
SEE ALSO:
LEAD
LEAD ARSENATE, UNSPECIFIED
ARSENIC

HEALTH AND SAFETY LISTS
ACGIH 1995 - Time Weighted Averages
as Pb3(AsO4)2: 0.15 mg/m3 TWA
U.S. DOT - Appendix A Table 1 - Hazardous Substances
final RQ = 1 pound (0.454 kg)
NTP Seventh Report - Known Carcinogens
known carcinogen (Listed under 'Arsenic and certain arsenic compounds')
OSHA - Select Carcinogens
[present]

ENVIRONMENTAL LISTS
CERCLA/SARA - Hazardous Substances and their Reportable Quantities
final RQ = 1 pound (0.454 kg)
Clean Water Act - Hazardous Substances
[present]
EPA - Carcinogen Hazard Ranking for RQ Adjustment
Hazard ranking = High

INTERNATIONAL LISTS
Canada - WHMIS: Ingredient Disclosure
1% item 931 (262)
Israel - Time Weighted Averages
as PbHAsO4: 0.15 mg/m3 TWA
Israel - Action Levels
as PbHAsO4: 0.075 mg/m3 AL

STATE LISTS
Massachusetts Right To Know List
[present]
NJ Right to Know List (Total)
sn 1098
NJ Special Hazardous Substances
(carcinogen)
Pennsylvania Right to Know List
environmental hazard

LEAD CARBONATE   598-63-0
SEE ALSO:
LEAD ARSENATE, UNSPECIFIED

HEALTH AND SAFETY LISTS
U.S. DOT - Appendix A Table 1 - Hazardous Substances
final RQ = 1 pound (0.454 kg) (Listed under 'Lead arsenate')

ENVIRONMENTAL LISTS
CERCLA/SARA - Hazardous Substances and their Reportable Quantities
final RQ = 1 pound (0.454 kg) (Listed under 'Lead arsenate')
Clean Water Act - Hazardous Substances
[present] (Listed under 'Lead arsenate')

STATE LISTS
Florida Hazardous Substance List
[present]
Massachusetts Right To Know List
[present]
Minnesota Hazardous Substance List
[present] as Pb3 (AsO4)2
Pennsylvania Right to Know List
environmental hazard

LEAD CARBONATE HYDROXIDE   1319-46-6

HEALTH AND SAFETY LISTS
U.S. DOT - Substances From 49 CFR 172.101
regulated by DOT (UN1617)
U.S. DOT - Hazard Classes
DOT hazard class = 6.1
U.S. DOT - Appendix B - Marine Pollutants
DOT regulated marine pollutant

LEAD CHLORIDE   7758-95-4
SEE ALSO:
ARSENIC
LEAD

HEALTH AND SAFETY LISTS
U.S. DOT - Appendix A Table 1 - Hazardous Substances
final RQ = 1 pound (0.454 kg) (Listed under 'Lead arsenate')
NIOSH - Selected LD50s and LC50s
Oral, rat: LD50 = 100 mg/kg

ENVIRONMENTAL LISTS
CERCLA/SARA - Hazardous Substances and their Reportable Quantities
final RQ = 1 pound (0.454 kg) (Listed under 'Lead arsenate')
Clean Water Act - Hazardous Substances
[present] (Listed under 'Lead arsenate')

INTERNATIONAL LISTS
Canada - CEPA Schedule II Part II - Toxic Substances (Export)
[present] (with the molecular formula PbHASO4 or its basic form that has the molecular formula Pb4(PbOH)(AsO4)3)

STATE LISTS
Massachusetts Right To Know List
[present]
Pennsylvania Right to Know List
environmental hazard

LEAD CHLORIDE SILICATE   39390-00-6
SEE ALSO:
LEAD
ARSENIC

STATE LISTS
NJ Right to Know List (Total)
sn 1099

LEAD CHLORIDE (VAN)   12612-47-4

HEALTH AND SAFETY LISTS
U.S. DOT - Substances From 49 CFR 172.101
regulated by DOT (UN1618)
U.S. DOT - Hazard Classes
DOT hazard class = 6.1

U.S. DOT - Appendix B - Marine Pollutants
DOT regulated marine pollutant

LEAD CHROMATE   7758-97-6
SEE ALSO:
LEAD

HEALTH AND SAFETY LISTS
U.S. DOT - Substances From 49 CFR 172.101
regulated by DOT (UN0129)
U.S. DOT - Hazard Classes
Forbidden from transport by the DOT

STATE LISTS
Massachusetts Right To Know List
[present]
NJ Right to Know List (Total)
sn 1100
NJ Special Hazardous Substances
(teratogen)

LEAD CHROMATE OXIDE   18454-12-1
STATE LISTS
Massachusetts Right To Know List
[present]

LEAD CHROMATE SILICATE   11113-70-5
SEE ALSO:
LEAD
LEAD
LEAD COMPOUNDS

INTERNATIONAL LISTS
Canada - WHMIS: Ingredient Disclosure
0.1% item 932 (389)

STATE LISTS
Massachusetts Right To Know List
[present]

LEAD COMPOUNDS   RR-00630-4
STATE LISTS
Massachusetts Right To Know List
[present]

LEAD CYANAMIDE   20837-86-9
SEE ALSO:
LEAD
INORGANIC LEAD

HEALTH AND SAFETY LISTS
U.S. DOT - Appendix A Table 1 - Hazardous Substances
final RQ = 10 pounds (4.54 kg)

ENVIRONMENTAL LISTS
CERCLA/SARA - Hazardous Substances and their Reportable Quantities
final RQ = 10 pounds (4.54 kg)
Clean Water Act - Hazardous Substances
[present]

INTERNATIONAL LISTS
Canada - WHMIS: Ingredient Disclosure
0.1% item 933 (521)

STATE LISTS
Massachusetts Right To Know List
[present]
NJ Right to Know List (Total)
sn 1101

NJ Special Hazardous Substances
(teratogen)
Pennsylvania Right to Know List
environmental hazard

LEAD CYANIDE   592-05-2
STATE LISTS
Massachusetts Right To Know List
[present]

LEAD DIMETHYLDITHIOCARBAMATE   19010-66-3
STATE LISTS
Pennsylvania Right to Know List
environmental hazard

LEAD 2-ETHYLHEXOATE   16996-40-0
SEE ALSO:
LEAD
LEAD
CHROMIUM (VI) COMPOUNDS
CHROMIUM COMPOUNDS
INORGANIC LEAD
LEAD COMPOUNDS

HEALTH AND SAFETY LISTS
ACGIH 1995 - Time Weighted Averages
as Pb: 0.05 mg/m3 TWA;   as Cr: 0.012 mg/m3 TWA
ACGIH 1995 - Carcinogens
as Pb and as Cr: A2-suspected human carcinogen
NIOSH - Selected LD50s and LC50s
Oral, mouse: LD50 = 12 gm/kg
NTP Seventh Report - Known Carcinogens
known carcinogen (Listed under 'Chromium and certain chromium compounds')
OSHA - Select Carcinogens
[present]

INTERNATIONAL LISTS
Australian Exposure Standards - Time Weighted Averages
as Cr: 0.05 mg/m3 TWA
Australian Exposure Standards - Carcinogens
as Cr: probable carcinogen
Canada - WHMIS: Ingredient Disclosure
0.1% item 934 (550)
Canada - Alberta - 8 Hour Occupational Exposure Limit
as Cr: 0.05 mg/m3 TWA
Canada - Alberta - 15 Minute Occupational Exposure Limit
as Cr: 0.15 mg/m3 STEL
Canada - Alberta - Designated Substances
designated substance - requires code of practice
Canada - British Columbia - 8 Hour Exposure Limits
as Cr: 0.05 mg/m3 TWA
Canada - British Columbia - Carcinogens
carcinogen - as Cr: 0.05 mg/m3 TWA
Canada - Quebec - Time-Weighted Average Exposure Values
as Cr: 0.012 mg/m3 TWAEV (substance of which the recirculation is prohibited)
Canada - Quebec - Carcinogens
as Pb: C2 carcinogen: effect suspected in humans
German (DFG) - Carcinogens
suspected carcinogen
Israel - Time Weighted Averages
as Cr: (0.05) mg/m3 TWA
Israel - Action Levels
as Cr: 0.025 mg/m3 AL

STATE LISTS

California - Air Bill 2588 Appendix A-I
known or potential carcinogen: 6/91

California - Exposure Limits - PELs
as Pb: 0.05 mg/m3 PEL; poison-avoid eating or smoking; see also section 5216

Massachusetts Right To Know List
carcinogen; extraordinarily hazardous

Minnesota Hazardous Substance List
carcinogen

NJ Right to Know List (Total)
sn 1102

NJ Special Hazardous Substances
(carcinogen)

Pennsylvania Right to Know List
environmental hazard; special hazardous substance

Pennsylvania RTK - Special Hazardous Substances
[present]

LEAD FLUOBORATE   13814-96-5
SEE ALSO:
LEAD

HEALTH AND SAFETY LISTS

OSHA - Select Carcinogens
[present]

OSHA - Possible Select Carcinogens
[present]

INTERNATIONAL LISTS

Canada - WHMIS: Ingredient Disclosure
0.1% item 935 (556)

German (DFG) - Carcinogens
suspected carcinogen (Listed under 'Lead chromate')

STATE LISTS

Florida Hazardous Substance List
[present]

Massachusetts Right To Know List
carcinogen; extraordinarily hazardous

Pennsylvania Right to Know List
environmental hazard; special hazardous substance

Pennsylvania RTK - Special Hazardous Substances
[present]

LEAD FLUORIDE   7783-46-2
SEE ALSO:
LEAD
CHROMIUM COMPOUNDS
CHROMIUM (VI) COMPOUNDS

STATE LISTS

Massachusetts Right To Know List
[present]

LEAD FLUOROSILICATE   25808-74-6

HEALTH AND SAFETY LISTS

U.S. DOT - Substances From 49 CFR 172.101
regulated by DOT (UN2291)

U.S. DOT - Hazard Classes
DOT hazard class = 6.1

U.S. DOT - Appendix B - Marine Pollutants
DOT regulated marine pollutant

IARC - Group 3 (not classifiable)
[present]

ENVIRONMENTAL LISTS

CERCLA/SARA - Section 313 - Emission Reporting
form R reporting reqired for 0.1% (inorganic), 1.0% (organic) de minimus concentration

Clean Air Act (1990) - List of Hazardous Air Contaminants
[present] (includes any unique chemical substance that contains Lead as part of that chemical's infrastructure)

Clean Water Act - Toxic Pollutants
[present] (Listed under 'Lead and compounds')

RCRA - Hazardous Constituents-Appendix VIII
hazardous constituent - no waste number

INTERNATIONAL LISTS

Canada - NPRI (National Pollutant Release Inventory)
[present]

Canada - CEPA Schedule III Part II - Restricted Substances (Ocean Dumping)
[present]

United Kingdom - Maximum Exposure Limits - TWAs
as Pb: 0.15 mg/m3 TWA (does not include Tetraethyl lead)

STATE LISTS

California - Air Bill 2588 Appendix A-I
other than inorganic: 6/91

California - Prop. 65 - Cancer list
carcinogen - initial date 10/1/92

California - Directors List of Hazardous Substances (8 CCR 339)
[present]

NJ Right to Know List (Total)
sn 2515; sn 3139; sn 2873

LEAD FORMATE   811-54-1
SEE ALSO:
LEAD COMPOUNDS
LEAD

STATE LISTS

Massachusetts Right To Know List
[present]

LEAD HYDROXIDE   19783-14-3
SEE ALSO:
LEAD
CYANIDE ANION
LEAD COMPOUNDS

HEALTH AND SAFETY LISTS

U.S. DOT - Substances From 49 CFR 172.101
regulated by DOT (UN1620)

U.S. DOT - Hazard Classes
DOT hazard class = 6.1

U.S. DOT - Appendix B - Marine Pollutants
DOT regulated marine pollutant

STATE LISTS

NJ Right to Know List (Total)
sn 1103

NJ Special Hazardous Substances
(teratogen)

LEAD HYDROXYSALICYLATE   87903-39-7
SEE ALSO:
LEAD

HEALTH AND SAFETY LISTS

NTP Chemical Status Reports - Testing Status and NTIS Number
Technical reports printed (PB298512/AS)

NTP Chemical Status Reports - Evidence of Carcinogenicity
male rat-negative; female rat-negative; male mice-negative; female mice-negative

INTERNATIONAL LISTS

Canada - WHMIS: Ingredient Disclosure
1% item 936 (740)

STATE LISTS

Massachusetts Right To Know List
[present]

LEAD IODATE   25659-31-8

STATE LISTS

Massachusetts Right To Know List
[present]

LEAD IODIDE   10101-63-0
SEE ALSO:
LEAD

HEALTH AND SAFETY LISTS

U.S. DOT - Appendix A Table 1 - Hazardous Substances
final RQ = 10 pounds (4.54 kg)

ENVIRONMENTAL LISTS

CERCLA/SARA - Hazardous Substances and their Reportable Quantities
final RQ = 10 pounds (4.54 kg)

Clean Water Act - Hazardous Substances
[present]

INTERNATIONAL LISTS

Canada - WHMIS: Ingredient Disclosure
1% item 938 (889)

STATE LISTS

Massachusetts Right To Know List
[present]

NJ Right to Know List (Total)
sn 1105

Pennsylvania Right to Know List
environmental hazard

LEAD LINOLEATE   16996-51-3
SEE ALSO:
LEAD

HEALTH AND SAFETY LISTS

U.S. DOT - Appendix A Table 1 - Hazardous Substances
final RQ = 10 pounds (4.54 kg)

ENVIRONMENTAL LISTS

CERCLA/SARA - Hazardous Substances and their Reportable Quantities
final RQ = 10 pounds (4.54 kg)

Clean Water Act - Hazardous Substances
[present]

STATE LISTS

Massachusetts Right To Know List
[present]

NJ Right to Know List (Total)
sn 1106

NJ Special Hazardous Substances
(teratogen)

Pennsylvania Right to Know List
environmental hazard

LEAD METHACRYLATE   1068-61-7
SEE ALSO:
LEAD

INTERNATIONAL LISTS

Canada - WHMIS: Ingredient Disclosure
1% item 939 (898)

STATE LISTS

Massachusetts Right To Know List
[present]

LEAD (II) METHYLTHIOLATE   35029-96-0

STATE LISTS

Massachusetts Right To Know List
[present]

LEAD MOLYBDATE   10190-55-3
SEE ALSO:
LEAD

STATE LISTS

Massachusetts Right To Know List
[present]

LEAD MONOXIDE   1317-36-8

STATE LISTS

Massachusetts Right To Know List
[present]

LEAD MYRISTATE   20403-41-2

STATE LISTS

Massachusetts Right To Know List
[present]

LEAD NAPHTHALATE   50825-29-1
SEE ALSO:
LEAD

HEALTH AND SAFETY LISTS

U.S. DOT - Appendix A Table 1 - Hazardous Substances
final RQ = 10 pounds (4.54 kg)

ENVIRONMENTAL LISTS

CERCLA/SARA - Hazardous Substances and their Reportable Quantities
final RQ = 10 pounds (4.54 kg)

Clean Water Act - Hazardous Substances
[present]

STATE LISTS

Massachusetts Right To Know List
[present]

NJ Right to Know List (Total)
sn 1107

NJ Special Hazardous Substances
(teratogen)

Pennsylvania Right to Know List
environmental hazard

LEAD NAPHTHENATE   61790-14-5

STATE LISTS

Massachusetts Right To Know List
[present]

LEAD NEOBATE   12034-88-7

STATE LISTS

Massachusetts Right To Know List
[present]

LEAD NEODECANOATE   27253-28-7

STATE LISTS

Massachusetts Right To Know List
[present]

LEAD NITRATE   10099-74-8
SEE ALSO:
  LEAD
STATE LISTS
  Massachusetts Right To Know List
  [present]

LEAD NITRORESORCINATE   51317-24-9
SEE ALSO:
  LEAD COMPOUNDS
  INORGANIC LEAD
  LEAD
  LEAD
HEALTH AND SAFETY LISTS
  NTP Chemical Status Reports - Testing Status and NTIS Number
  Prechronic studies for which toxicity technical reports were not prepared
INTERNATIONAL LISTS
  Canada - WHMIS: Ingredient Disclosure
  1% item 941 (1175)
STATE LISTS
  Massachusetts Right To Know List
  [present]

LEAD OCTOATE   15696-43-2
STATE LISTS
  Massachusetts Right To Know List
  [present]

LEAD OLEATE   1120-46-3
SEE ALSO:
  LEAD
STATE LISTS
  Massachusetts Right To Know List
  [present]

LEAD ORES   RR-01021-9
SEE ALSO:
  LEAD
  LEAD COMPOUNDS
  LEAD
ENVIRONMENTAL LISTS
  TSCA - Code of Federal Regulations Citations
  40 CFR 712.30(h); 40 CFR 716.120(a)
  TSCA - PAIR - Reporting List
  Reporting Date: September 20, 1983
  TSCA - Health and Safety Reporting List
  Effective Date: July 1, 1983
INTERNATIONAL LISTS
  Canada - WHMIS: Ingredient Disclosure
  1% item 942 (1183)

LEAD OXALATE   814-93-7
STATE LISTS
  Massachusetts Right To Know List
  [present]

LEAD OXIDE PHOSPHONATE, HEMIHYDRATE   1344-40-7
STATE LISTS
  Massachusetts Right To Know List
  [present]

LEAD PERCHLORATE   13637-76-8
SEE ALSO:
  LEAD
  LEAD
  INORGANIC LEAD
  LEAD COMPOUNDS
HEALTH AND SAFETY LISTS
  U.S. DOT - Substances From 49 CFR 172.101
  regulated by DOT (UN1469)
  U.S. DOT - Hazard Classes
  DOT hazard class = 5.1
  U.S. DOT - Appendix B - Marine Pollutants
  DOT regulated marine pollutant
  U.S. DOT - Appendix A Table 1 - Hazardous Substances
  final RQ = 10 pounds (4.54 kg)
ENVIRONMENTAL LISTS
  CERCLA/SARA - Hazardous Substances and their Reportable Quantities
  final RQ = 10 pounds (4.54 kg)
  Clean Water Act - Hazardous Substances
  [present]
INTERNATIONAL LISTS
  Canada - WHMIS: Ingredient Disclosure
  0.1% item 943 (1209)
STATE LISTS
  Florida Hazardous Substance List
  [present]
  Massachusetts Right To Know List
  [present]
  NJ Right to Know List (Total)
  sn 1108
  NJ Special Hazardous Substances
  (teratogen)
  Pennsylvania Right to Know List
  environmental hazard

LEAD PEROXIDE   1309-60-0
HEALTH AND SAFETY LISTS
  U.S. DOT - Hazard Classes
  Forbidden from transport by the DOT

LEAD PHOSPHATE   7446-27-7
SEE ALSO:
  LEAD
STATE LISTS
  Massachusetts Right To Know List
  [present]

LEAD PHTHALATE   6838-85-3
STATE LISTS
  Massachusetts Right To Know List
  [present]

LEAD PICRATE   25721-38-4
HEALTH AND SAFETY LISTS
  NTP Chemical Status Reports - Testing Status and NTIS Number
  Prechronic studies for which toxicity technical reports were not prepared

LEAD PYROPHOSPHATE   13453-66-2
STATE LISTS
  Massachusetts Right To Know List
  [present]

LEAD B-RESORCYLATE   41453-50-3
SEE ALSO:
 LEAD

HEALTH AND SAFETY LISTS
 U.S. DOT - Substances From 49 CFR 172.101
 regulated by DOT (UN2989)
 U.S. DOT - Hazard Classes
 DOT hazard class = 4.1

LEAD SEBACATE   29473-77-6
SEE ALSO:
 LEAD

HEALTH AND SAFETY LISTS
 U.S. DOT - Substances From 49 CFR 172.101
 regulated by DOT (UN1470)
 U.S. DOT - Hazard Classes
 DOT hazard class = 5.1
 U.S. DOT - Appendix B - Marine Pollutants
 solid or solution: DOT regulated marine pollutant

STATE LISTS
 Massachusetts Right To Know List
 [present]
 NJ Right to Know List (Total)
 sn 1109
 NJ Special Hazardous Substances
 (teratogen)

LEAD SELENATE   7446-15-3
SEE ALSO:
 LEAD

HEALTH AND SAFETY LISTS
 U.S. DOT - Substances From 49 CFR 172.101
 regulated by DOT (UN1872)
 U.S. DOT - Hazard Classes
 DOT hazard class = 5.1

STATE LISTS
 Massachusetts Right To Know List
 [present]
 NJ Right to Know List (Total)
 sn 1104

LEAD SELENIDE   12069-00-0
SEE ALSO:
 LEAD

HEALTH AND SAFETY LISTS
 U.S. DOT - Appendix A Table 1 - Hazardous Substances
 final RQ = 10 pounds (4.54 kg)
 NTP Seventh Report - Suspect Carcinogens
 suspect carcinogen
 OSHA - Possible Select Carcinogens
 [present]

ENVIRONMENTAL LISTS
 CERCLA/SARA - Hazardous Substances and their Reportable Quantities
 final RQ = 10 pounds (4.54 kg)
 RCRA - U Series Wastes
 waste number U145
 RCRA - Hazardous Constituents-Appendix VIII
 waste number U145
 RCRA - Substances Banned From Land Disposal
 [present]

INTERNATIONAL LISTS
 Canada - WHMIS: Ingredient Disclosure
 0.1% item 944 (1394)

STATE LISTS
 California - Air Bill 2588 Appendix A-I
 known or potential carcinogen
 California - Prop. 65 - Cancer list
 carcinogen - initial date 4/1/88
 California - Directors List of Hazardous Substances (8 CCR 339)
 [present]
 Florida Hazardous Substance List
 [present]
 Massachusetts Right To Know List
 carcinogen; extraordinarily hazardous
 Minnesota Hazardous Substance List
 carcinogen
 NJ Right to Know List (Total)
 sn 1110
 NJ Special Hazardous Substances
 (carcinogen)
 Pennsylvania Right to Know List
 environmental hazard; special hazardous substance
 Pennsylvania RTK - Special Hazardous Substances
 [present]

LEAD SELENITE   7488-51-9
STATE LISTS
 Massachusetts Right To Know List
 [present]

LEAD SILICATE   11120-22-2
HEALTH AND SAFETY LISTS
 U.S. DOT - Hazard Classes
 Forbidden from transport by the DOT

LEAD SILICATE SULFATE   67711-86-8
STATE LISTS
 Massachusetts Right To Know List
 [present]

LEAD STEARATE   7428-48-0
STATE LISTS
 Massachusetts Right To Know List
 [present]

LEAD STEARATE DIBASIC   52652-59-2
STATE LISTS
 Massachusetts Right To Know List
 [present]

LEAD STYPHNATE   63918-97-8
STATE LISTS
 Massachusetts Right To Know List
 [present]

LEAD SUBACETATE   1335-32-6
STATE LISTS
 Massachusetts Right To Know List
 [present]

LEAD SULFATE   7446-14-2
STATE LISTS
 Massachusetts Right To Know List
 [present]

LEAD SULFATE 15739-80-7
SEE ALSO:
  LEAD COMPOUNDS
  LEAD
  LEAD
STATE LISTS
  Massachusetts Right To Know List
  [present]

LEAD SULFIDE (PBS) 1314-87-0
STATE LISTS
  Massachusetts Right To Know List
  [present]

LEAD TANTALATE 12065-68-8
SEE ALSO:
  LEAD
  LEAD COMPOUNDS
HEALTH AND SAFETY LISTS
  U.S. DOT - Appendix A Table 1 - Hazardous Substances
  final RQ = 10 pounds (4.54 kg)
ENVIRONMENTAL LISTS
  CERCLA/SARA - Hazardous Substances and their Reportable Quantities
  final RQ = 10 pounds (4.54 kg)
  Clean Water Act - Hazardous Substances
  [present]
STATE LISTS
  Massachusetts Right To Know List
  [present]
  NJ Right to Know List (Total)
  sn 1111
  Pennsylvania Right to Know List
  environmental hazard

LEAD TELLURIDE 1314-91-6
SEE ALSO:
  LEAD STEARATE
HEALTH AND SAFETY LISTS
  U.S. DOT - Appendix A Table 1 - Hazardous Substances
  final RQ = 5000 pounds (2270 kg) (the RQ is subject to change when the assessment of potential carcinogenicity is completed)
ENVIRONMENTAL LISTS
  CERCLA/SARA - Hazardous Substances and their Reportable Quantities
  final RQ = 5000 pounds (2270 kg) (the RQ is subject to change when the assessment of potential carcinogenicity is completed)
  Clean Water Act - Hazardous Substances
  [present] (Listed under 'Lead stearate')
STATE LISTS
  Massachusetts Right To Know List
  [present]
  Pennsylvania Right to Know List
  environmental hazard

LEAD TELLURITE 13845-35-7
HEALTH AND SAFETY LISTS
  U.S. DOT - Substances From 49 CFR 172.101
  regulated by DOT (UN0130)
  U.S. DOT - Hazard Classes
  DOT hazard class = 1.1A
STATE LISTS
  NJ Right to Know List (Total)
  sn 1112

LEAD TETRAOXIDE 1314-41-6
SEE ALSO:
  LEAD
HEALTH AND SAFETY LISTS
  U.S. DOT - Appendix A Table 1 - Hazardous Substances
  final RQ = 10 pounds (4.54 kg)
ENVIRONMENTAL LISTS
  CERCLA/SARA - Hazardous Substances and their Reportable Quantities
  final RQ = 10 pounds (4.54 kg)
  RCRA - U Series Wastes
  waste number U146
  RCRA - Hazardous Constituents-Appendix VIII
  waste number U146
  RCRA - Substances Banned From Land Disposal
  [present]
INTERNATIONAL LISTS
  Canada - WHMIS: Ingredient Disclosure
  0.1% item 930 (31)
STATE LISTS
  California - Air Bill 2588 Appendix A-I
  known or potential carcinogen: 9/90
  California - Prop. 65 - Cancer list
  carcinogen - initial date 10/1/89
  California - Prop. 65 - No Significant Risk Levels
  no significant risk level = 20 ug/day
  California - Directors List of Hazardous Substances (8 CCR 339)
  [present]
  Florida Hazardous Substance List
  [present]
  Massachusetts Right To Know List
  carcinogen; extraordinarily hazardous
  Pennsylvania Right to Know List
  environmental hazard; special hazardous substance
  Pennsylvania RTK - Special Hazardous Substances
  [present]

LEAD THIOCYANATE 592-87-0
SEE ALSO:
  LEAD
  INORGANIC LEAD
  LEAD
  LEAD COMPOUNDS
HEALTH AND SAFETY LISTS
  U.S. DOT - Substances From 49 CFR 172.101
  regulated by DOT (UN1794, NA2291)
  U.S. DOT - Hazard Classes
  DOT hazard class = 8
  U.S. DOT - Appendix A Table 1 - Hazardous Substances
  final RQ = 10 pounds (4.54 kg)
ENVIRONMENTAL LISTS
  CERCLA/SARA - Hazardous Substances and their Reportable Quantities
  final RQ = 10 pounds (4.54 kg)
  Clean Water Act - Hazardous Substances
  [present]
INTERNATIONAL LISTS
  Canada - WHMIS: Ingredient Disclosure
  1% item 945 (1528)
STATE LISTS
  Massachusetts Right To Know List
  [present]

NJ Right to Know List (Total)
sn 1114

NJ Special Hazardous Substances
(corrosive, teratogen)

Pennsylvania Right to Know List
environmental hazard

LEAD THIOSULFATE   13478-50-7
SEE ALSO:
LEAD

STATE LISTS

Massachusetts Right To Know List
[present]

NJ Right to Know List (Total)
sn 3032

Pennsylvania Right to Know List
environmental hazard

LEAD TITANIUM OXIDE (PB.TI.O3)   12060-00-3
SEE ALSO:
LEAD

HEALTH AND SAFETY LISTS

U.S. DOT - Appendix A Table 1 - Hazardous Substances
final RQ = 10 pounds (4.54 kg)

NTP Chemical Status Reports - Testing Status and NTIS Number
Prechronic studies for which technical reports were not prepared

ENVIRONMENTAL LISTS

CERCLA/SARA - Hazardous Substances and their Reportable Quantities
final RQ = 10 pounds (4.54 kg)

Clean Water Act - Hazardous Substances
[present]

STATE LISTS

Massachusetts Right To Know List
[present]

NJ Right to Know List (Total)
sn 1113

Pennsylvania Right to Know List
environmental hazard

LEAD TRIOXIDE   1314-27-8
STATE LISTS

Massachusetts Right To Know List
[present]

LEAD TUNGSTEN OXIDE   7759-01-5
STATE LISTS

Massachusetts Right To Know List
[present]

LEAD VANADATE   10099-79-3
STATE LISTS

Massachusetts Right To Know List
[present]

LEAD ZIRCONATE   12060-01-4
SEE ALSO:
LEAD
LEAD
LEAD COMPOUNDS

STATE LISTS

Massachusetts Right To Know List
[present]

LECITHINS, PHOSPHOLIPASE A2-HYDROLYZED   RR-01722-1
SEE ALSO:
LEAD

HEALTH AND SAFETY LISTS

U.S. DOT - Appendix A Table 1 - Hazardous Substances
final RQ = 10 pounds (4.54 kg)

ENVIRONMENTAL LISTS

CERCLA/SARA - Hazardous Substances and their Reportable Quantities
final RQ = 10 pounds (4.54 kg)

Clean Water Act - Hazardous Substances
[present]

STATE LISTS

Florida Hazardous Substance List
[present]

Massachusetts Right To Know List
[present]

NJ Right to Know List (Total)
sn 1115

NJ Special Hazardous Substances
(teratogen)

Pennsylvania Right to Know List
environmental hazard

LEMONGRASS OIL   8007-02-1
STATE LISTS

Massachusetts Right To Know List
[present]

LEMON OIL   8008-56-8
SEE ALSO:
LEAD

STATE LISTS

Massachusetts Right To Know List
[present]

LENACIL   2164-08-1
STATE LISTS

Massachusetts Right To Know List
[present]

LEPTOPHOS   21609-90-5
STATE LISTS

Massachusetts Right To Know List
[present]

LETHANE   112-56-1
STATE LISTS

Massachusetts Right To Know List
[present]

LEVULINIC ACID   123-76-2
STATE LISTS

Massachusetts Right To Know List
[present]

LEWISITE   541-25-3
ENVIRONMENTAL LISTS

TSCA - Chemicals with Significant New Use Rules
PMN number: P-93-333

LIGHT AROMATIC DISTILLATE (PETROLEUM)   67891-80-9
ENVIRONMENTAL LISTS

List of Pesticide Product Inert Ingredients
  [present]
INTERNATIONAL LISTS
  Canada - WHMIS: Ingredient Disclosure
    1% item 946 (974)

LIGHT DISTILLATE SOLVENT EXTRACT   64742-03-6
(PETROLEUM)
  HEALTH AND SAFETY LISTS
    NIOSH - Selected LD50s and LC50s
      Oral, rat: LD50 = 2840 mg/kg

  ENVIRONMENTAL LISTS
    List of Pesticide Product Inert Ingredients
      [present]

LIGHT GREEN SF   5141-20-8
  HEALTH AND SAFETY LISTS
    NIOSH - Selected LD50s and LC50s
      Oral, rat: LD50 = 11000 mg/kg

LIGHT NAPHTHENIC OIL   64741-52-2
  HEALTH AND SAFETY LISTS
    NIOSH - Selected LD50s and LC50s
      Oral, rat: LD50 = 19 mg/kg Skin, rat: LD50 = 44 mg/kg
  ENVIRONMENTAL LISTS
    CERCLA/SARA - Section 302 Extremely Hazardous Substances and
    TPQs
      TPQ = 500/10,000 pounds
  INTERNATIONAL LISTS
    Canada - CEPA Schedule II Part I - Prohibited Substances (Export)
      [present]
  STATE LISTS
    Florida Hazardous Substance List
      effective March 13, 1992
    Massachusetts Right To Know List
      extraordinarily hazardous
    NJ Right to Know List (Total)
      sn 2516
    Pennsylvania Right to Know List
      environmental hazard
  PROPOSED REGULATIONS
    CERCLA/SARA - Proposed Hazardous Substance Additions
      proposed RQ = 1 pound (.454 kg)
    CERCLA/SARA - 1989 Proposed RQ Adjustments
      proposed RQ = 100 pounds (45.4 kg)

LIGHT PARAFFINIC DISTILLATE SOLVENT EX-   64742-05-8
TRACT (PETROLEUM)
  HEALTH AND SAFETY LISTS
    NIOSH - Selected LD50s and LC50s
      Oral, rat: LD50 = 90 mg/kg Skin, rat: LD50 = 250 mg/kg
  STATE LISTS
    NJ Right to Know List (Total)
      sn 1116

LIGHT VACUUM GAS OIL (PETROLEUM)   64741-58-8
  HEALTH AND SAFETY LISTS
    NIOSH - Selected LD50s and LC50s
      Oral, rat: LD50 = 1850 mg/kg
  INTERNATIONAL LISTS
    Canada - WHMIS: Ingredient Disclosure
      1% item 947 (101)

LIGNIN, ALKALI, REACTION PRODUCTS WITH   105859-97-0
DISODIUM SULFITE ANDFORMALDEHYDE
  SEE ALSO:
    ARSENIC
  HEALTH AND SAFETY LISTS
    NIOSH - Selected LD50s and LC50s
      Inhalation, rat: LC50 = 580 mg/m3 1 hr Skin, rat: LD50 = 15 mg/kg
  ENVIRONMENTAL LISTS
    CERCLA/SARA - Section 302 Extremely Hazardous Substances and
    TPQs
      TPQ = 10 pounds
  STATE LISTS
    Florida Hazardous Substance List
      effective March 13, 1992
    Massachusetts Right To Know List
      extraordinarily hazardous
    NJ Right to Know List (Total)
      sn 2517
    Pennsylvania Right to Know List
      environmental hazard
  PROPOSED REGULATIONS
    CERCLA/SARA - Proposed Hazardous Substance Additions
      proposed RQ = 1 pound (.454 kg)
    CERCLA/SARA - 1989 Proposed RQ Adjustments
      proposed RQ = 10 pounds (4.54 kg)

LIGNIN SULPHATE   8068-05-1
  ENVIRONMENTAL LISTS
    List of Pesticide Product Inert Ingredients
      [present]

LIGNOSULFONATE   8062-15-5
  STATE LISTS
    Massachusetts Right To Know List
      carcinogen; extraordinarily hazardous

LIGNOSULFONIC ACID, AMMONIUM SALT   8061-53-8
  HEALTH AND SAFETY LISTS
    IARC - Group 3 (not classifiable)
      [present]
  STATE LISTS
    California - Directors List of Hazardous Substances (8 CCR 339)
      [present]
    NJ Right to Know List (Total)
      sn 3140

LIGNOSULFONIC ACID, ETHOXYLATED, SODIUM   68611-14-3
SALTS
  ENVIRONMENTAL LISTS
    List of Pesticide Product Inert Ingredients
      [present]
  STATE LISTS
    Massachusetts Right To Know List
      carcinogen; extraordinarily hazardous

LIGNOSULFONIC ACID, MAGNESIUM SALT   8061-54-9
  STATE LISTS
    Massachusetts Right To Know List
      carcinogen; extraordinarily hazardous

LIGROINE   8032-32-4

HEALTH AND SAFETY LISTS

IARC - Group Unspecified
[present] (Listed under 'Occupational exposures in petroleum refining')

LIME GREEN PIGMENT   RR-01093-5

ENVIRONMENTAL LISTS

List of Pesticide Product Inert Ingredients
[present]

D-LIMONENE   5989-27-5

ENVIRONMENTAL LISTS

List of Pesticide Product Inert Ingredients
[present]

L-LIMONENE   5989-54-8

ENVIRONMENTAL LISTS

List of Pesticide Product Inert Ingredients
[present]

LINALOOL (EX BOIS DE ROSE; SYNTHETIC)   78-70-6

ENVIRONMENTAL LISTS

List of Pesticide Product Inert Ingredients
[present]

LINDANE   58-89-9

ENVIRONMENTAL LISTS

List of Pesticide Product Inert Ingredients
[present]

LINEAR ALKYLBENZENE (C-20 TO C-48)   68555-09-9

ENVIRONMENTAL LISTS

List of Pesticide Product Inert Ingredients
[present]

LINOLEAMIDE   3999-01-7

HEALTH AND SAFETY LISTS

ACGIH 1995 - Time Weighted Averages
300 ppm TWA; 1370 mg/m3 TWA

NFPA - Flash Points
flash point = 104 degrees F (40 degrees C)

NFPA - Hazard Identification Ratings
health-0; flammability-2; reactivity-0

NIOSH - Selected LD50s and LC50s
Inhalation, rat: LC50 = 3400 ppm 4 hr

OSHA - Vacated PELs - Time Weighted Averages
300 ppm TWA; 1350 mg/m3 TWA

OSHA - Vacated PELs - Short Term Exposure Limits
400 ppm STEL; 1800 mg/m3 STEL

ENVIRONMENTAL LISTS

List of Pesticide Product Inert Ingredients
[present]

INTERNATIONAL LISTS

Canada - WHMIS: Ingredient Disclosure
1% item 1693 (1180)

Canada - Alberta - 8 Hour Occupational Exposure Limit
300 ppm TWA; 1350 mg/m3 TWA

Canada - Alberta - 15 Minute Occupational Exposure Limit
400 ppm STEL; 1800 mg/m3 STEL

Israel - Time Weighted Averages
300 ppm TWA; 1370 mg/m3 TWA

Israel - Action Levels
150 ppm AL; 685 mg/m3 AL

Mexico - Instruction No. 10 - TWAs
300 ppm TWA; 1350 mg/m3 TWA

Mexico - Instruction No. 10 - STELs
400 ppm STEL; 1800 mg/m3 STEL

STATE LISTS

NJ Right to Know List (Total)
sn 3131; sn 0206

Pennsylvania Right to Know List
[present]

LINOLEIC ACID   60-33-3

ENVIRONMENTAL LISTS

List of Pesticide Product Inert Ingredients
[present]

LINOLEIC ACID DIMER-DIETHYLENETRIAMINE   37189-83-6
POLYMER

HEALTH AND SAFETY LISTS

AIHA - WEEL - Time Weighted Averages
30 ppm TWA; 165.6 mg/m3 TWA

IARC - Group 3 (not classifiable)
[present]

NIOSH - Selected LD50s and LC50s
Oral, rat: LD50 = 4400 mg/kg

NTP Chemical Status Reports - Testing Status and NTIS Number
Technical reports printed (PB90231416/AS)

NTP Chemical Status Reports - Evidence of Carcinogenicity
male rat-clear evidence; female rat-no evidence; male mice-no evidence; female mice-no evidence

ENVIRONMENTAL LISTS

List of Pesticide Product Inert Ingredients
[present]

INTERNATIONAL LISTS

Canada - WHMIS: Ingredient Disclosure
1% item 948 (1068)

LINSEED OIL   8001-26-1

INTERNATIONAL LISTS

Canada - WHMIS: Ingredient Disclosure
1% item 949 (1069)

LINSEED OIL, TUNG OIL, 4-(T-BUTYL)PHENOL,   68153-88-8
BISPHENOL A, FORMALDEHYDE POLYMER

HEALTH AND SAFETY LISTS

NFPA - Flash Points
flash point = 160 degrees F (71 degrees C)

NFPA - Hazard Identification Ratings
flammability-2; reactivity-0

LINURON   330-55-2

SEE ALSO:

F039-HAZARDOUS WASTES
HEXACHLOROCYCLOHEXANE (MIXED ISOMERS)
TERT-BUTYLDECALIN

HEALTH AND SAFETY LISTS

ACGIH 1995 - Time Weighted Averages
0.5 mg/m3 TWA

ACGIH 1995 - Skin Designations
skin - potential for cutaneous absorption

U.S. DOT - Appendix B - Marine Pollutants
DOT regulated severe marine pollutant

U.S. DOT - Appendix A Table 1 - Hazardous Substances
final RQ = 1 pound (0.454 kg)

IARC - Group Unspecified
[present] (Listed under 'Hexachlorocylcohexanes (HCH)')
NIOSH - Selected LD50s and LC50s
Oral, rat: LD50 = 76 mg/kg; Skin, rat: LD50 = 500 mg/kg
NIOSH 1990 - Pocket Guide - RELs
0.5 mg/m3 TWA
NIOSH 1990 - Pocket Guide - IDLHs
1000 mg/m3 IDLH
NIOSH 1990 - Pocket Guide - Target organs
eyes, CNS, blood, liver, kidneys, skin
NIOSH 1990 - Pocket Guide - Skin list
Potential for dermal absorption
NTP Chemical Status Reports - Testing Status and NTIS Number
Technical reports printed (PB273480/AS)
NTP Chemical Status Reports - Evidence of Carcinogenicity
male rat-negative; female rat-negative; male mice-negative; female mice-negative
NTP Seventh Report - Suspect Carcinogens
suspect carcinogen (Listed under 'Lindane and other hexachlorocyclohexane isomers')
OSHA - Vacated PELs - Time Weighted Averages
0.5 mg/m3 TWA
OSHA - Vacated PELs - Skin Designation
Prevent or reduce skin absorption
OSHA - Final PELs - Time Weighted Averages
0.5 mg/m3 TWA
OSHA - Final PELs - Skin Notations
prevent or reduce skin absorption
OSHA - Possible Select Carcinogens
[present]

ENVIRONMENTAL LISTS
ATSDR Priority List
Rank (of 275): 022
CERCLA/SARA - Section 302 Extremely Hazardous Substances and TPQs
TPQ = 1000/10,000 pounds
CERCLA/SARA - Section 313 - Emission Reporting
form R reporting required for 0.1% de minimus concentration
CERCLA/SARA - Hazardous Substances and their Reportable Quantities
final RQ = 1 pound (0.454 kg)
Clean Air Act (1990) - List of Hazardous Air Contaminants
[present]
Clean Water Act - Hazardous Substances
[present]
Clean Water Act - Priority Pollutants
[present]
Safe Drinking Water Act - MCLs
MCL = 0.0002 mg/L
Safe Drinking Water Act - MCLGs
MCLG = 0.0002 mg/L
EPA - Carcinogen Hazard Ranking for RQ Adjustment
Hazard ranking = Medium
RCRA - D Series - Maximum Concentration of Contaminants
waste number D013; regulatory level = 0.4 mg/L
RCRA - D Series - Chronic Toxicity Reference Levels
chronic toxicity reference level = 0.004 mg/L
RCRA - U Series Wastes
waste number U129
RCRA - Hazardous Constituents-Appendix VIII
waste number U129
RCRA - Basis for Listing - Appendix VII
Included in waste stream: F039

RCRA - Substances Banned From Land Disposal
[present]
RCRA - TSD Facilities Ground Water Monitoring
TM 8080 = 0.05 ug/L PQL; TM 8250 = 10 ug/L PQL
RCRA - Universal Treatment Standards (LDR)
WW: 0.0017 mg/l; NWW: 0.066 mg/kg

INTERNATIONAL LISTS
Australian Exposure Standards - Time Weighted Averages
0.5 mg/m3 TWA
Australian Exposure Standards - Skin Effects
skin absorption
Canada - CEPA Schedule II Part II - Toxic Substances (Export)
[present]
Canada - Drinking Water Quality - MACs
0.004 mg/L MAC
Canada - Alberta - 8 Hour Occupational Exposure Limit
0.5 mg/m3 TWA
Canada - Alberta - 15 Minute Occupational Exposure Limit
1.5 mg/m3 STEL
Canada - Alberta - Skin Designation
can be absorbed through the intact skin
Canada - British Columbia - 8 Hour Exposure Limits
0.5 mg/m3 TWA
Canada - British Columbia - 15 Minute Exposure Limits
1.5 mg/m3 STEL
Canada - British Columbia - Skin Notations
skin - potential for skin absorption
Canada - Ontario - OHSA - TWAEVs
0.5 mg/m3 TWAEV
Canada - Ontario - OHSA - Skin Notations
absorption through skin, eyes, or mucous membranes
Canada - Quebec - Time-Weighted Average Exposure Values
0.5 mg/m3 TWAEV
Canada - Quebec - Skin Designations
absorbed through the skin
United Kingdom - Occupational Exposure Standards - TWAs
0.5 mg/m3 TWA
United Kingdom - Occupational Exposure Standards - STELs
1.5 mg/m3 STEL
United Kingdom - Occupational Exposure Standards - Notes
can be absorbed through skin
German (DFG) - MAK Values
total dust: 0.5 mg/m3 MAK
German (DFG) - Peak Limitations
10 x normal MAK (30 min average value); don't exceed during shift
German (DFG) - Skin/Sensitizers
danger of cutaneous absorption
Israel - Time Weighted Averages
0.5 mg/m3 TWA
Israel - Action Levels
0.25 mg/m3 AL
Mexico - Instruction No. 10 - TWAs
0.5 mg/m3 TWA
Mexico - Instruction No. 10 - STELs
1.5 mg/m3 STEL
Mexico - Instruction No. 10 - Skin designation
skin - potential for cutaneous absorption
Mexico - Wastewater - Organic Toxic Pollutants and Heavy Metals
Listed under [Hexachlorocyclohexane]
Mexico - Drinking Water - Ecological Criteria
0.003 mg/l This level has been extrapolated by using a mathematic model

STATE LISTS
  California - Air Bill 2588 Appendix A-I
  known or potential carcinogen: 9/90
  California - Prop. 65 - Cancer list
  carcinogen - initial date 10/1/87
  California - Prop. 65 - No Significant Risk Levels
  no significant risk level = 0.6 ug/day
  California - Exposure Limits - PELs
  0.5 mg/m3 PEL
  California - Exposure Limits - Skin Notation
  material may be absorbed through the skin, eyes or mucous membrane
  California - Directors List of Hazardous Substances (8 CCR 339)
  [present]
  Florida Hazardous Substance List
  [present]
  Massachusetts Right To Know List
  carcinogen; extraordinarily hazardous
  Minnesota Hazardous Substance List
  carcinogen; skin (includes other hexachlorocyclohexane isomers)
  NJ Right to Know List (Total)
  sn 1117
  NJ Special Hazardous Substances
  (carcinogen)
  Pennsylvania Right to Know List
  environmental hazard; special hazardous substance
  Pennsylvania RTK - Special Hazardous Substances
  [present]

LIQUIFIED NATURAL GAS   RR-00375-8
  ENVIRONMENTAL LISTS
    CAA - HON Rule - SOCMI Chemicals
    compliance by Oct. 24, 1994

LITHIUM   7439-93-2
  ENVIRONMENTAL LISTS
    List of Pesticide Product Inert Ingredients
    [present]

LITHIUM ACETYLIDE ETHYLENEDIAMINE   50475-76-8
  ENVIRONMENTAL LISTS
    List of Pesticide Product Inert Ingredients
    [present]
  INTERNATIONAL LISTS
    Canada - WHMIS: Ingredient Disclosure
    1% item 950 (102)

LITHIUM ALKYLS   RR-01350-3
  ENVIRONMENTAL LISTS
    List of Pesticide Product Inert Ingredients
    [present]

LITHIUM ALUMINUM HYDRIDE   16853-85-3
  HEALTH AND SAFETY LISTS
    NFPA - Flash Points
    flash point = 432 degrees F (222 degrees C); boiled: flash point = 403 degrees F (206 degrees C)
    NFPA - Hazard Identification Ratings
    health-0; flammability-1; reactivity-0
  ENVIRONMENTAL LISTS
    List of Pesticide Product Inert Ingredients
    [present]

INTERNATIONAL LISTS
  Canada - WHMIS: Ingredient Disclosure
  1% item 951 (975)
STATE LISTS
  Pennsylvania Right to Know List
  [present]

LITHIUM ALUMINUM TRI-TERT-BUTOXY-   17476-04-9
HYDRIDE
  ENVIRONMENTAL LISTS
    List of Pesticide Product Inert Ingredients
    [present]

LITHIUM AMIDE   7782-89-0
  ENVIRONMENTAL LISTS
    CERCLA/SARA - Section 313 - Emission Reporting
    form R reporting required
  STATE LISTS
    California - Directors List of Hazardous Substances (8 CCR 339)
    [present]

LITHIUM BOROHYDRIDE   16949-15-8
  STATE LISTS
    Pennsylvania Right to Know List
    [present]

LITHIUM CARBONATE   554-13-2
  HEALTH AND SAFETY LISTS
    U.S. DOT - Substances From 49 CFR 172.101
    regulated by DOT (UN1415)
    U.S. DOT - Hazard Classes
    DOT hazard class = 4.3
  STATE LISTS
    Florida Hazardous Substance List
    [present]
    Massachusetts Right To Know List
    [present]
    NJ Right to Know List (Total)
    sn 1119
    Pennsylvania Right to Know List
    [present]

LITHIUM CHLORATE   13453-71-9
  STATE LISTS
    NJ Right to Know List (Total)
    sn 1120

LITHIUM CHLORIDE   7447-41-8
  HEALTH AND SAFETY LISTS
    U.S. DOT - Substances From 49 CFR 172.101
    regulated by DOT (UN2445)
    U.S. DOT - Hazard Classes
    DOT hazard class = 4.2

LITHIUM CHROMATE   14307-35-8
  SEE ALSO:
    LITHIUM
  HEALTH AND SAFETY LISTS
    U.S. DOT - Substances From 49 CFR 172.101
    regulated by DOT (UN1410, UN1411)
    U.S. DOT - Hazard Classes
    DOT hazard class = 4.3

INTERNATIONAL LISTS
  Canada - WHMIS: Ingredient Disclosure
    1% item 952 (1009)

STATE LISTS
  Florida Hazardous Substance List
    [present]
  Massachusetts Right To Know List
    [present]
  NJ Right to Know List (Total)
    sn 1121
  Pennsylvania Right to Know List
    [present]

LITHIUM CITRATE   919-16-4
  SEE ALSO:
    LITHIUM
  INTERNATIONAL LISTS
    Canada - WHMIS: Ingredient Disclosure
      1% item 953 (1638)

LITHIUM DICHROMATE   13843-81-7
  SEE ALSO:
    LITHIUM
  STATE LISTS
    NJ Right to Know List (Total)
      sn 1122

LITHIUM FERROSILICON   64082-35-5
  SEE ALSO:
    LITHIUM
  HEALTH AND SAFETY LISTS
    U.S. DOT - Substances From 49 CFR 172.101
      regulated by DOT (UN1413)
    U.S. DOT - Hazard Classes
      DOT hazard class = 4.3
  STATE LISTS
    NJ Right to Know List (Total)
      sn 1123

LITHIUM FLUORIDE   7789-24-4
  SEE ALSO:
    LITHIUM
  HEALTH AND SAFETY LISTS
    NIOSH - Selected LD50s and LC50s
      Oral, rat: LD50 = 525 mg/kg
  ENVIRONMENTAL LISTS
    CERCLA/SARA - Section 313 - Emission Reporting
      form R reporting required
    List of Pesticide Product Inert Ingredients
      [present]
  STATE LISTS
    California - Air Bill 2588 Appendix A-II
      6/91
    California - Prop. 65 - Developmental Toxicity
      developmental toxicity - initial date 1/1/91
    Massachusetts Right To Know List
      teratogen
    NJ Special Hazardous Substances
      (teratogen)

LITHIUM HYDRIDE   7580-67-8
  ENVIRONMENTAL LISTS
    List of Pesticide Product Inert Ingredients
      [present]

LITHIUM HYDROXIDE   1310-65-2
  SEE ALSO:
    LITHIUM
  HEALTH AND SAFETY LISTS
    NIOSH - Selected LD50s and LC50s
      Oral, rat: LD50 = 526 mg/kg
  ENVIRONMENTAL LISTS
    List of Pesticide Product Inert Ingredients
      [present]

LITHIUM HYDROXIDE MONOHYDRATE   1310-66-3
  SEE ALSO:
    CHROMIUM
  HEALTH AND SAFETY LISTS
    U.S. DOT - Appendix A Table 1 - Hazardous Substances
      final RQ = 10 pounds (4.54 kg)
  ENVIRONMENTAL LISTS
    CERCLA/SARA - Hazardous Substances and their Reportable Quantities
      final RQ = 10 pounds (4.54 kg)
    Clean Water Act - Hazardous Substances
      [present]
    EPA - Carcinogen Hazard Ranking for RQ Adjustment
      Hazard ranking = High
  STATE LISTS
    Massachusetts Right To Know List
      [present]
    NJ Right to Know List (Total)
      sn 1125
    Pennsylvania Right to Know List
      environmental hazard

LITHIUM HYPOCHLORITE   13840-33-0
  STATE LISTS
    California - Air Bill 2588 Appendix A-II
      6/91
    California - Prop. 65 - Developmental Toxicity
      developmental toxicity - initial date 1/1/91

LITHIUM NEODECANOATE   27253-30-1
  STATE LISTS
    Florida Hazardous Substance List
      [present]
    Massachusetts Right To Know List
      [present]
    Pennsylvania Right to Know List
      [present]

LITHIUM NITRATE   7790-69-4
  HEALTH AND SAFETY LISTS
    U.S. DOT - Substances From 49 CFR 172.101
      regulated by DOT (UN2830)
    U.S. DOT - Hazard Classes
      DOT hazard class = 4.3
  STATE LISTS
    NJ Right to Know List (Total)
      sn 1126

LITHIUM NITRIDE   26134-62-3
SEE ALSO:
LITHIUM

INTERNATIONAL LISTS

Canada - WHMIS: Ingredient Disclosure
1% item 954 (907)

LITHIUM OXIDE   12057-24-8
SEE ALSO:
LITHIUM

HEALTH AND SAFETY LISTS

ACGIH 1995 - Time Weighted Averages
0.025 mg/m3 TWA

U.S. DOT - Substances From 49 CFR 172.101
regulated by DOT (UN2805, UN1414)

U.S. DOT - Hazard Classes
DOT hazard class = 4.3

NIOSH 1990 - Pocket Guide - RELs
0.025 mg/m3 TWA

NIOSH 1990 - Pocket Guide - IDLHs
55 mg/m3 IDLH

NIOSH 1990 - Pocket Guide - Target organs
skin, eyes, respiratory system

OSHA - Vacated PELs - Time Weighted Averages
0.025 mg/m3 TWA

OSHA - Final PELs - Time Weighted Averages
0.025 mg/m3 TWA

ENVIRONMENTAL LISTS

CERCLA/SARA - Section 302 Extremely Hazardous Substances and TPQs
TPQ = 100 pounds (This material is a reactive solid. The TPQ does not default to 10,000 pounds for non-powder, non-molten, non-solution form)

INTERNATIONAL LISTS

Australian Exposure Standards - Time Weighted Averages
0.025 mg/m3 TWA

Canada - WHMIS: Ingredient Disclosure
1% item 955 (1010)

Canada - Alberta - 8 Hour Occupational Exposure Limit
0.025 mg/m3 TWA

Canada - Alberta - 15 Minute Occupational Exposure Limit
0.075 mg/m3 STEL

Canada - British Columbia - 8 Hour Exposure Limits
0.025 mg/m3 TWA

Canada - Ontario - OHSA - TWAEVs
0.025 mg/m3 TWAEV

Canada - Quebec - Time-Weighted Average Exposure Values
0.025 mg/m3 TWAEV

United Kingdom - Occupational Exposure Standards - TWAs
0.025 mg/m3 TWA

Israel - Time Weighted Averages
0.025 mg/m3 TWA

Israel - Action Levels
0.0125 mg/m3 AL

Mexico - Instruction No. 10 - TWAs
0.025 mg/m3 TWA

STATE LISTS

California - Exposure Limits - PELs
0.025 mg/m3 PEL

California - Directors List of Hazardous Substances (8 CCR 339)
[present]

Florida Hazardous Substance List
[present]

Massachusetts Right To Know List
extraordinarily hazardous

Minnesota Hazardous Substance List
[present]

NJ Right to Know List (Total)
sn 1127

NJ Special Hazardous Substances
(flammable - fourth degree; reactive - second degree)

Pennsylvania Right to Know List
environmental hazard

PROPOSED REGULATIONS

CERCLA/SARA - Proposed Hazardous Substance Additions
proposed RQ = 1 pound (.454 kg)

CERCLA/SARA - 1989 Proposed RQ Adjustments
proposed RQ = 10 pounds (4.54 kg)

LITHIUM PEROXIDE   12031-80-0
SEE ALSO:
LITHIUM

HEALTH AND SAFETY LISTS

AIHA - WEEL - Ceilings or Short Term Time Weighted Averages
1 mg/m3 (one minute)

ENVIRONMENTAL LISTS

List of Pesticide Product Inert Ingredients
[present]

INTERNATIONAL LISTS

Canada - WHMIS: Ingredient Disclosure
1% item 956 (993)

United Kingdom - Occupational Exposure Standards - STELs
1 mg/m3 STEL

STATE LISTS

Minnesota Hazardous Substance List
[present] (includes monohydrate)

LITHIUM SILICON   68848-64-6
SEE ALSO:
LITHIUM

HEALTH AND SAFETY LISTS

U.S. DOT - Substances From 49 CFR 172.101
regulated by DOT (UN2680, UN2679)

U.S. DOT - Hazard Classes
DOT hazard class = 8

STATE LISTS

Minnesota Hazardous Substance List
[present]

NJ Right to Know List (Total)
sn 1128

NJ Special Hazardous Substances
(corrosive)

LITHIUM STEARATE   4485-12-5
SEE ALSO:
LITHIUM

HEALTH AND SAFETY LISTS

U.S. DOT - Substances From 49 CFR 172.101
regulated by DOT (UN1471)

U.S. DOT - Hazard Classes
DOT hazard class = 5.1

STATE LISTS

NJ Right to Know List (Total)
sn 1129

LITHIUM SULFATE   10377-48-7

ENVIRONMENTAL LISTS

List of Pesticide Product Inert Ingredients
[present]

LITHOCHOLIC ACID   434-13-9
SEE ALSO:
LITHIUM

HEALTH AND SAFETY LISTS

U.S. DOT - Substances From 49 CFR 172.101
regulated by DOT (UN2722)

U.S. DOT - Hazard Classes
DOT hazard class = 5.1

STATE LISTS

NJ Right to Know List (Total)
sn 1130

LITHOL RUBINE   5281-04-9
SEE ALSO:
LITHIUM

HEALTH AND SAFETY LISTS

U.S. DOT - Substances From 49 CFR 172.101
regulated by DOT (UN2806)

U.S. DOT - Hazard Classes
DOT hazard class = 4.3

STATE LISTS

NJ Right to Know List (Total)
sn 1131

LOCUST BEAN GUM   9000-40-2
SEE ALSO:
LITHIUM

HEALTH AND SAFETY LISTS

AIHA - WEEL - Ceilings or Short Term Time Weighted Averages
1 mg/m3 (one minute) STEL

STATE LISTS

Minnesota Hazardous Substance List
[present]

LONDON PURPLE   8012-74-6
SEE ALSO:
LITHIUM

HEALTH AND SAFETY LISTS

U.S. DOT - Substances From 49 CFR 172.101
regulated by DOT (UN1472)

U.S. DOT - Hazard Classes
DOT hazard class = 5.1

STATE LISTS

NJ Right to Know List (Total)
sn 1132

LORAZEPAM   846-49-1

STATE LISTS

NJ Right to Know List (Total)
sn 1133

LOVASTATIN   75330-75-5
SEE ALSO:
LITHIUM

HEALTH AND SAFETY LISTS

NIOSH - Selected LD50s and LC50s
Oral, rat: LD50 = 15 gm/kg

STATE LISTS

California - Exposure Limits - PELs
10 mg/m3 PEL

L.P.G. (LIQUIFIED PETROLEUM GAS)   68476-85-7
SEE ALSO:
LITHIUM

HEALTH AND SAFETY LISTS

NIOSH - Selected LD50s and LC50s
Oral, mouse: LD50 = 1190 mg/kg

LUBRICATING OILS   RR-00384-9

HEALTH AND SAFETY LISTS

NTP Chemical Status Reports - Testing Status and NTIS Number
Technical reports printed (PB288476/AS)

NTP Chemical Status Reports - Evidence of Carcinogenicity
male rat-negative; female rat-negative; male mice-negative; female mice-negative

LUTEOSKYRIN   21884-44-6

ENVIRONMENTAL LISTS

EPA - Master Testing List
[present]

LUTETIUM 169   15715-05-6

HEALTH AND SAFETY LISTS

NTP Chemical Status Reports - Testing Status and NTIS Number
Technical reports printed (PB82163320)

NTP Chemical Status Reports - Evidence of Carcinogenicity
male rat-negative; female rat-negative; male mice-negative; female mice-negative

ENVIRONMENTAL LISTS

List of Pesticide Product Inert Ingredients
[present]

LUTETIUM 170   15741-32-9

HEALTH AND SAFETY LISTS

U.S. DOT - Substances From 49 CFR 172.101
regulated by DOT (UN1621)

U.S. DOT - Hazard Classes
DOT hazard class = 6.1

U.S. DOT - Appendix B - Marine Pollutants
DOT regulated marine pollutant

STATE LISTS

NJ Right to Know List (Total)
sn 1134

LUTETIUM 171   15752-27-9

HEALTH AND SAFETY LISTS

NIOSH - Selected LD50s and LC50s
Oral, rat: LD50 = 4500 mg/kg

STATE LISTS

California - Air Bill 2588 Appendix A-II
9/90

California - Prop. 65 - Developmental Toxicity
developmental toxicity - initial date 7/1/90

NJ Right to Know List (Total)
sn 1135

NJ Special Hazardous Substances
(teratogen)

LUTETIUM 172   14093-12-0

STATE LISTS

California - Prop. 65 - Developmental Toxicity
developmental toxicity - initial date 10/1/92

LUTETIUM 173   14391-24-3

HEALTH AND SAFETY LISTS

ACGIH 1995 - Time Weighted Averages
1000 ppm TWA; 1800 mg/m3 TWA

NIOSH 1990 - Pocket Guide - RELs
1000 ppm TWA; 1800 mg/m3 TWA

NIOSH 1990 - Pocket Guide - IDLHs
19,000 ppm IDLH (lower explosive level)

NIOSH 1990 - Pocket Guide - Target organs
CNS, respiratory system

OSHA - Vacated PELs - Time Weighted Averages
1000 ppm TWA; 1800 mg/m3 TWA

OSHA - Final PELs - Time Weighted Averages
1000 ppm TWA; 1800 mg/m3 TWA

INTERNATIONAL LISTS

Australian Exposure Standards - Time Weighted Averages
1000 ppm TWA; 1800 mg/m3 TWA

Canada - Alberta - 8 Hour Occupational Exposure Limit
1000 ppm TWA; 1800 mg/m3 TWA

Canada - Alberta - 15 Minute Occupational Exposure Limit
1250 ppm STEL; 2250 mg/m3 STEL

Canada - British Columbia - 8 Hour Exposure Limits
1000 ppm TWA; 1800 mg/m3 TWA

Canada - British Columbia - 15 Minute Exposure Limits
1250 ppm STEL; 2250 mg/m3 STEL

Canada - Ontario - OHSA - TWAEVs
1800 mg/m3 TWAEV (listed as an agent of variable composition)

Canada - Quebec - Time-Weighted Average Exposure Values
1000 ppm TWAEV; 1800 mg/m3 TWAEV

United Kingdom - Occupational Exposure Standards - TWAs
1000 ppm TWA; 1800 mg/m3 TWA

United Kingdom - Occupational Exposure Standards - STELs
1250 ppm STEL; 2250 mg/m3 STEL

Israel - Time Weighted Averages
1000 ppm TWA; 1800 mg/m3 TWA

Israel - Action Levels
500 ppm AL; 900 mg/m3 AL

Mexico - Instruction No. 10 - TWAs
1000 ppm TWA; 1800 mg/m3 TWA

Mexico - Instruction No. 10 - STELs
1250 ppm STEL; 2250 mg/m3 STEL

STATE LISTS

California - Exposure Limits - PELs
1000 ppm PEL; 1800 mg/m3 PEL

Massachusetts Right To Know List
[present]

Minnesota Hazardous Substance List
[present]

NJ Right to Know List (Total)
sn 1118

Pennsylvania Right to Know List
[present]

LUTETIUM 174   14914-12-6

HEALTH AND SAFETY LISTS

NFPA - Flash Points
spindle: flash point = 169 degrees F (76 degrees C); turbine: flash point = 400 degrees F (204 degrees C)

NFPA - Hazard Identification Ratings
spindle: health-0; flammability-2; reactivity-0; turbine: health-0; flammability-1; reactivity-0

STATE LISTS

California - Air Bill 2588 Appendix A-II
known or potential carcinogen: 9/89

LUTETIUM 174M   RR-00419-3

HEALTH AND SAFETY LISTS

IARC - Group 3 (not classifiable)
[present]

STATE LISTS

California - Directors List of Hazardous Substances (8 CCR 339)
[present]

LUTETIUM 176   14452-47-2

HEALTH AND SAFETY LISTS

U.S. DOT - Appendix A Table 2 - Radionuclides
final RQ = 10 curies (3.7E 11 Bq)

ENVIRONMENTAL LISTS

CERCLA/SARA List of Radionuclides (Appendix B) and Their Reportable Quantities
final RQ = 10 curies (3.7E 11 Bq)

LUTETIUM 176M   RR-00415-9

HEALTH AND SAFETY LISTS

U.S. DOT - Appendix A Table 2 - Radionuclides
final RQ = 10 curies (3.7E 11 Bq)

ENVIRONMENTAL LISTS

CERCLA/SARA List of Radionuclides (Appendix B) and Their Reportable Quantities
final RQ = 10 curies (3.7E 11 Bq)

LUTETIUM 177   14265-75-9

HEALTH AND SAFETY LISTS

U.S. DOT - Appendix A Table 2 - Radionuclides
final RQ = 10 curies (3.7E 11 Bq)

ENVIRONMENTAL LISTS

CERCLA/SARA List of Radionuclides (Appendix B) and Their Reportable Quantities
final RQ = 10 curies (3.7E 11 Bq)

LUTETIUM 177M   RR-00410-4

HEALTH AND SAFETY LISTS

U.S. DOT - Appendix A Table 2 - Radionuclides
final RQ = 10 curies (3.7E 11 Bq)

ENVIRONMENTAL LISTS

CERCLA/SARA List of Radionuclides (Appendix B) and Their Reportable Quantities
final RQ = 10 curies (3.7E 11 Bq)

LUTETIUM 178   14683-30-8

HEALTH AND SAFETY LISTS

U.S. DOT - Appendix A Table 2 - Radionuclides
final RQ = 100 curies (3.7E 12 Bq)

ENVIRONMENTAL LISTS

CERCLA/SARA List of Radionuclides (Appendix B) and Their Reportable Quantities
final RQ = 100 curies (3.7E 12 Bq)

LUTETIUM 178M   RR-00409-1

HEALTH AND SAFETY LISTS

U.S. DOT - Appendix A Table 2 - Radionuclides
final RQ = 10 curies (3.7E 11 Bq)

ENVIRONMENTAL LISTS
CERCLA/SARA List of Radionuclides (Appendix B) and Their Reportable Quantities
final RQ = 10 curies (3.7E 11 Bq)

LUTETIUM 179   15755-89-2
HEALTH AND SAFETY LISTS
U.S. DOT - Appendix A Table 2 - Radionuclides
final RQ = 10 curies (3.7E 11 Bq)

ENVIRONMENTAL LISTS
CERCLA/SARA List of Radionuclides (Appendix B) and Their Reportable Quantities
final RQ = 10 curies (3.7E 11 Bq)

LYNALYL ACETATE (EX BOIS DE ROSE; SYNTHETIC)   115-95-7
HEALTH AND SAFETY LISTS
U.S. DOT - Appendix A Table 2 - Radionuclides
final RQ = 1 curie (3.7E 10 Bq)

ENVIRONMENTAL LISTS
CERCLA/SARA List of Radionuclides (Appendix B) and Their Reportable Quantities
final RQ = 1 curie (3.7E 10 Bq)

LYNOESTRENOL   52-76-6
HEALTH AND SAFETY LISTS
U.S. DOT - Appendix A Table 2 - Radionuclides
final RQ = 1000 curies (3.7E 13 Bq)

ENVIRONMENTAL LISTS
CERCLA/SARA List of Radionuclides (Appendix B) and Their Reportable Quantities
final RQ = 1000 curies (3.7E 13 Bq)

LYNOESTRENOL AND OESTROGENS   RR-00098-6
HEALTH AND SAFETY LISTS
U.S. DOT - Appendix A Table 2 - Radionuclides
final RQ = 100 curies (3.7E 12 Bq)

ENVIRONMENTAL LISTS
CERCLA/SARA List of Radionuclides (Appendix B) and Their Reportable Quantities
final RQ = 100 curies (3.7E 12 Bq)

D-LYSERGIC ACID   82-58-6
HEALTH AND SAFETY LISTS
U.S. DOT - Appendix A Table 2 - Radionuclides
final RQ = 10 curies (3.7E 11 Bq)

ENVIRONMENTAL LISTS
CERCLA/SARA List of Radionuclides (Appendix B) and Their Reportable Quantities
final RQ = 10 curies (3.7E 11 Bq)

D-LYSERGIC ACID SALTS, OPTICAL ISOMERS, AND SALTS OF OPTICALISOMERS   RR-01784-5
HEALTH AND SAFETY LISTS
U.S. DOT - Appendix A Table 2 - Radionuclides
final RQ = 1000 curies (3.7E 13 Bq)

ENVIRONMENTAL LISTS
CERCLA/SARA List of Radionuclides (Appendix B) and Their Reportable Quantities
final RQ = 1000 curies (3.7E 13 Bq)

L-LYSINE HCL   657-27-2
HEALTH AND SAFETY LISTS
U.S. DOT - Appendix A Table 2 - Radionuclides
final RQ = 1000 curies (3.7E 13 Bq)

ENVIRONMENTAL LISTS
CERCLA/SARA List of Radionuclides (Appendix B) and Their Reportable Quantities
final RQ = 1000 curies (3.7E 13 Bq)

MAGENTA   632-99-5
HEALTH AND SAFETY LISTS
U.S. DOT - Appendix A Table 2 - Radionuclides
final RQ = 1000 curies (3.7E 13 Bq)

ENVIRONMENTAL LISTS
CERCLA/SARA List of Radionuclides (Appendix B) and Their Reportable Quantities
final RQ = 1000 curies (3.7E 13 Bq)

MAGENTA II   26261-57-4
HEALTH AND SAFETY LISTS
NFPA - Flash Points
flash point = 185 degrees F (85 degrees C)
NFPA - Hazard Identification Ratings
flammability-2; reactivity-0

MAGENTA III   3248-91-7
HEALTH AND SAFETY LISTS
IARC - Group Unspecified
[present] (Listed under 'Progestins')

MAGENTA-MANUFACTURE OF   RR-00540-3
HEALTH AND SAFETY LISTS
IARC - Group Unspecified
[present] (Listed under 'Combined oral contraceptives')

MAGNESITE (MG(CO3))   13717-00-5
HEALTH AND SAFETY LISTS
FDA - Controlled Substances Act - Precursor chemicals
Threshold by base weight = 10 grams

MAGNESIUM   7439-95-4
HEALTH AND SAFETY LISTS
FDA - Controlled Substances Act - Precursor chemicals
Threshold by base weight = 10 grams

MAGNESIUM 28   15092-71-4
HEALTH AND SAFETY LISTS
NIOSH - Selected LD50s and LC50s
Oral, rat: LD50 = 10 gm/kg

MAGNESIUM ALKYLS   RR-00568 5
HEALTH AND SAFETY LISTS
IARC - Group 2B (sufficient animal data)
[present]
IARC - Group Unspecified
[present] (Listed under 'Magenta (containing CI Basic Red 9')
OSHA - Possible Select Carcinogens
[present]

MAGNESIUM ALUMINUM SILICATE   1327-43-1
HEALTH AND SAFETY LISTS
IARC - Group Unspecified
[present] (Listed under 'Magenta (containing CI Basic Red 9')

MAGNESIUM ARSENATE   10103-50-1

HEALTH AND SAFETY LISTS

IARC - Group Unspecified
[present] (Listed under 'Magenta (containing CI Basic Red 9')

MAGNESIUM BROMATE   7789-36-8

HEALTH AND SAFETY LISTS

IARC - Group 1 (carcinogenic to humans)
[present]

OSHA - Select Carcinogens
[present]

STATE LISTS

Pennsylvania Right to Know List
special hazardous substance

Pennsylvania RTK - Special Hazardous Substances
[present]

MAGNESIUM CARBONATE   546-93-0

STATE LISTS

Pennsylvania Right to Know List
[present]

MAGNESIUM CHLORATE   10326-21-3

HEALTH AND SAFETY LISTS

U.S. DOT - Substances From 49 CFR 172.101
regulated by DOT (UN1869, UN1416, UN2950)

U.S. DOT - Hazard Classes
metal or alloys: DOT hazard class = 4.1;   granules or powder: DOT hazard class = 4.3

INTERNATIONAL LISTS

Mexico - Wastewater - Organic Toxic Pollutants and Heavy Metals
Listed under [Heavy Metals]

STATE LISTS

California - Directors List of Hazardous Substances (8 CCR 339)
[present]

Florida Hazardous Substance List
[present]

Massachusetts Right To Know List
[present]

NJ Right to Know List (Total)
sn 1136

NJ Special Hazardous Substances
(reactive - second degree)

Pennsylvania Right to Know List
[present]

MAGNESIUM CHLORIDE   7786-30-3

HEALTH AND SAFETY LISTS

U.S. DOT - Appendix A Table 2 - Radionuclides
final RQ = 10 curies (3.7E 11 Bq)

ENVIRONMENTAL LISTS

CERCLA/SARA List of Radionuclides (Appendix B) and Their Reportable Quantities
final RQ = 10 curies (3.7E 11 Bq)

MAGNESIUM CHLORIDE.6H2O   7791-19-7

HEALTH AND SAFETY LISTS

U.S. DOT - Substances From 49 CFR 172.101
regulated by DOT (UN3053)

U.S. DOT - Hazard Classes
DOT hazard class = 4.2

MAGNESIUM CHLORIDE, HEXAHYDRATE   7791-18-6

ENVIRONMENTAL LISTS

List of Pesticide Product Inert Ingredients
[present]

MAGNESIUM DIAMIDE   7803-54-5

SEE ALSO:
ARSENIC

HEALTH AND SAFETY LISTS

U.S. DOT - Substances From 49 CFR 172.101
regulated by DOT (UN1622)

U.S. DOT - Hazard Classes
DOT hazard class = 6.1

U.S. DOT - Appendix B - Marine Pollutants
DOT regulated marine pollutant

NIOSH - Selected LD50s and LC50s
Oral, mouse: LD50 = 315 mg/kg

INTERNATIONAL LISTS

Canada - WHMIS: Ingredient Disclosure
1% item 957 (261)

STATE LISTS

NJ Right to Know List (Total)
sn 1137

MAGNESIUM DICHROMATE   14104-85-9

HEALTH AND SAFETY LISTS

U.S. DOT - Substances From 49 CFR 172.101
regulated by DOT (UN1473)

U.S. DOT - Hazard Classes
DOT hazard class = 5.1

NIOSH - Selected LD50s and LC50s
Oral, rat: LD50 = 5250 mg/kg

STATE LISTS

NJ Right to Know List (Total)
sn 1138

MAGNESIUM DIPHENYL   555-54-4

HEALTH AND SAFETY LISTS

ACGIH 1995 - Time Weighted Averages
10 mg/m3 TWA (The value is for total dust containing no asbestos and < 1% crystalline silica)

OSHA - Vacated PELs - Time Weighted Averages
total dust: 15 mg/m3 TWA; respirable fraction: 5 mg/m3 TWA

OSHA - Final PELs - Time Weighted Averages
total dust: 15 mg/m3 TWA; respirable fraction: 5 mg/m3 TWA

ENVIRONMENTAL LISTS

List of Pesticide Product Inert Ingredients
[present]

INTERNATIONAL LISTS

Australian Exposure Standards - Time Weighted Averages
10 mg/m3 TWA

Canada - Alberta - 8 Hour Occupational Exposure Limit
respirable mass: 5 mg/m3 TWA; total mass: 10 mg/m3 TWA

Canada - British Columbia - 8 Hour Exposure Limits
nuisance dust, mists, and fumes: 10 mg/m3 TWA

Canada - British Columbia - 15 Minute Exposure Limits
20 mg/m3 STEL

Canada - Ontario - OHSA - TWAEVs
total dust: 10 mg/m3 TWAEV (listed as a nuisance particulate)

Canada - Quebec - Time-Weighted Average Exposure Values
total dust: 10 mg/m3 TWAEV; respirable dust: 5 mg/m3 TWAEV

United Kingdom - Occupational Exposure Standards - TWAs
total inhalable dust: 10 mg/m3 TWA; respirable dust: 5 mg/m3 TWA

Israel - Time Weighted Averages
10 mg/m3 TWA (The value is for total dust containing no asbestos and < 1% crystalline silica)

Israel - Action Levels
5 mg/m3 AL

Mexico - Instruction No. 10 - TWAs
10 mg/m3 TWA; (nuisance particulate)

Mexico - Instruction No. 10 - STELs
20 mg/m3 STEL

STATE LISTS

Minnesota Hazardous Substance List
[present] (includes inert or nuisance dust)

MAGNESIUM FLUORIDE   7783-40-6

HEALTH AND SAFETY LISTS

U.S. DOT - Substances From 49 CFR 172.101
regulated by DOT (UN2723)

U.S. DOT - Hazard Classes
DOT hazard class = 5.1

NIOSH - Selected LD50s and LC50s
Oral, rat: LD50 = 6348 mg/kg

STATE LISTS

NJ Right to Know List (Total)
sn 1139

MAGNESIUM HYDRIDE   60616-74-2

HEALTH AND SAFETY LISTS

NIOSH - Selected LD50s and LC50s
Oral, rat: LD50 = 2800 mg/kg

ENVIRONMENTAL LISTS

List of Pesticide Product Inert Ingredients
[present]

MAGNESIUM HYDROXIDE   1309-42-8

HEALTH AND SAFETY LISTS

NIOSH - Selected LD50s and LC50s
Oral, rat: LD50 = 5250 mg/kg

MAGNESIUM LACTATE   18917-93-6

HEALTH AND SAFETY LISTS

NIOSH - Selected LD50s and LC50s
Oral, rat: LD50 = 8100 mg/kg

MAGNESIUM NITRATE   10377-60-3

HEALTH AND SAFETY LISTS

U.S. DOT - Substances From 49 CFR 172.101
regulated by DOT (UN2004)

U.S. DOT - Hazard Classes
DOT hazard class = 4.2

STATE LISTS

NJ Right to Know List (Total)
sn 1140

MAGNESIUM OXIDE   1309-48-4

STATE LISTS

Massachusetts Right To Know List
[present]

MAGNESIUM OXIDE SULFATE   12286-12-3

HEALTH AND SAFETY LISTS

U.S. DOT - Substances From 49 CFR 172.101
regulated by DOT (UN2005)

U.S. DOT - Hazard Classes
DOT hazard class = 4.2

STATE LISTS

NJ Right to Know List (Total)
sn 1141

MAGNESIUM PERCHLORATE   10034-81-8

INTERNATIONAL LISTS

Canada - WHMIS: Ingredient Disclosure
1% item 958 (908)

MAGNESIUM PEROXIDE   14452-57-4

HEALTH AND SAFETY LISTS

U.S. DOT - Substances From 49 CFR 172.101
regulated by DOT (UN2010)

U.S. DOT - Hazard Classes
DOT hazard class = 4.3

STATE LISTS

NJ Right to Know List (Total)
sn 1142

MAGNESIUM PHOSPHIDE   12057-74-8

ENVIRONMENTAL LISTS

List of Pesticide Product Inert Ingredients
[present]

MAGNESIUM SILICATE   1343-88-0

ENVIRONMENTAL LISTS

List of Pesticide Product Inert Ingredients
[present]

MAGNESIUM SILICATE, HYDRATE   1343-90-4

HEALTH AND SAFETY LISTS

U.S. DOT - Substances From 49 CFR 172.101
regulated by DOT (UN1474, UN2724)

U.S. DOT - Hazard Classes
DOT hazard class = 5.1

ENVIRONMENTAL LISTS

List of Pesticide Product Inert Ingredients
[present]

STATE LISTS

Florida Hazardous Substance List
[present]

Massachusetts Right To Know List
[present]

NJ Right to Know List (Total)
sn 1143

Pennsylvania Right to Know List
[present]

MAGNESIUM SILICIDE   39404-03-0

HEALTH AND SAFETY LISTS

ACGIH 1995 - Time Weighted Averages
10 mg/m3 TWA

NIOSH 1990 - Pocket Guide - RELs
See Appendix D

NIOSH 1990 - Pocket Guide - Target organs
eyes, respiratory system

OSHA - Vacated PELs - Time Weighted Averages
total particulate: 10 mg/m3 TWA
OSHA - Final PELs - Time Weighted Averages
total particulate: 15 mg/m3 TWA

ENVIRONMENTAL LISTS
List of Pesticide Product Inert Ingredients
[present]

INTERNATIONAL LISTS
Australian Exposure Standards - Time Weighted Averages
fume: 10 mg/m3 TWA
Canada - WHMIS: Ingredient Disclosure
1% item 959 (1314)
Canada - Alberta - 8 Hour Occupational Exposure Limit
as Mg: 10 mg/m3 TWA
Canada - Alberta - 15 Minute Occupational Exposure Limit
20 mg/m3 STEL
Canada - British Columbia - 8 Hour Exposure Limits
as Mg: 10 mg/m3 TWA
Canada - Ontario - OHSA - TWAEVs
10 mg/m3 TWAEV
Canada - Quebec - Time-Weighted Average Exposure Values
as Mg: 10 mg/m3 TWAEV
United Kingdom - Occupational Exposure Standards - TWAs
*fume and respirable dust, as Mg: 5 mg/m3 TWA;   total inhalable dust,
as Mg: 10 mg/m3 TWA*
United Kingdom - Occupational Exposure Standards - STELs
fume and respirable dust, as Mg: 10 mg/m3 STEL
German (DFG) - MAK Values
fine dust: 6 mg/m3 MAK (includes magnesium oxide fume)
German (DFG) - Peak Limitations
*2 x normal MAK (30 min. average value);   don't exceed 4 times during
shift Glycidyl subcategory II-A)*
Israel - Time Weighted Averages
10 mg/m3 TWA
Israel - Action Levels
5 mg/m3 AL
Mexico - Instruction No. 10 - TWAs
10 mg/m3 TWA

STATE LISTS
California - Exposure Limits - PELs
fume, as Mg: 10 mg/m3 PEL
California - Directors List of Hazardous Substances (8 CCR 339)
[present]
Florida Hazardous Substance List
[present]
Massachusetts Right To Know List
[present]
Minnesota Hazardous Substance List
[present]
Pennsylvania Right to Know List
[present]

PROPOSED REGULATIONS
Canada - Ontario - Proposed Occupational TWAEVs
5 mg/m3 TWAEV
Canada - Ontario - Proposed Occupational STEVs
10 mg/m3 STEV

MAGNESIUM SILICOFLUORIDE   **18972-56-0**
INTERNATIONAL LISTS
German (DFG) - Carcinogens
as fibrous dust: suspected carcinogen

MAGNESIUM STEARATE   **557-04-0**
HEALTH AND SAFETY LISTS
U.S. DOT - Substances From 49 CFR 172.101
regulated by DOT (UN1475)
U.S. DOT - Hazard Classes
DOT hazard class = 5.1

STATE LISTS
Florida Hazardous Substance List
[present]
Massachusetts Right To Know List
[present]
NJ Right to Know List (Total)
sn 1145
Pennsylvania Right to Know List
[present]

MAGNESIUM SULFATE   **7487-88-9**
HEALTH AND SAFETY LISTS
U.S. DOT - Substances From 49 CFR 172.101
regulated by DOT (UN1476)
U.S. DOT - Hazard Classes
DOT hazard class = 5.1

STATE LISTS
NJ Right to Know List (Total)
sn 1146

MAGNESIUM SULFATE HEPTAHYDRATE   **10034-99-8**
HEALTH AND SAFETY LISTS
U.S. DOT - Substances From 49 CFR 172.101
regulated by DOT (UN2011)
U.S. DOT - Hazard Classes
DOT hazard class = 4.3

INTERNATIONAL LISTS
Canada - WHMIS: Ingredient Disclosure
1% item 960 (1411)

STATE LISTS
Massachusetts Right To Know List
[present]
NJ Right to Know List (Total)
sn 1147

MALACHITE GREEN   **569-64-2**
ENVIRONMENTAL LISTS
List of Pesticide Product Inert Ingredients
[present]

MALACHITE GREEN OXALATE   **2437-29-8**
ENVIRONMENTAL LISTS
List of Pesticide Product Inert Ingredients
[present]

MALAOXON   **1634-78-2**
HEALTH AND SAFETY LISTS
U.S. DOT - Substances From 49 CFR 172.101
regulated by DOT (UN2624)
U.S. DOT - Hazard Classes
DOT hazard class = 4.3

STATE LISTS
NJ Right to Know List (Total)
sn 1148

MALATHION   121-75-5

HEALTH AND SAFETY LISTS

U.S. DOT - Substances From 49 CFR 172.101
regulated by DOT (UN2853)

U.S. DOT - Hazard Classes
DOT hazard class = 6.1

STATE LISTS

NJ Right to Know List (Total)
sn 1149

MALEIC ACID   110-16-7

ENVIRONMENTAL LISTS

List of Pesticide Product Inert Ingredients
[present]

STATE LISTS

California - Exposure Limits - PELs
10 mg/m3 PEL

MALEIC ACID, DIBUTYL ESTER   105-76-0

ENVIRONMENTAL LISTS

List of Pesticide Product Inert Ingredients
[present]

MALEIC ACID N-ETHYLIMIDE   128-53-0

ENVIRONMENTAL LISTS

List of Pesticide Product Inert Ingredients
[present]

MALEIC ANHYDRIDE   108-31-6

HEALTH AND SAFETY LISTS

NIOSH - Selected LD50s and LC50s
Oral, mouse: LD50 = 80 mg/kg

NTP Chemical Status Reports - Testing Status and NTIS Number
Project leader assigned/study in design

ENVIRONMENTAL LISTS

CERCLA/SARA - Section 313 - Emission Reporting
form R reporting required for 1.0% de minimus concentration

List of Pesticide Product Inert Ingredients
[present]

INTERNATIONAL LISTS

Canada - WHMIS: Ingredient Disclosure
1% item 961 (1715)

Canada - NPRI (National Pollutant Release Inventory)
[present]

STATE LISTS

California - Air Bill 2588 Appendix A-II
6/91

Massachusetts Right To Know List
[present]

NJ Right to Know List (Total)
sn 0448

Pennsylvania Right to Know List
environmental hazard

MALEIC ANHYDRIDE, COMPOUND WITH TALL-   68139-89-9
OIL FATTY ACIDS

HEALTH AND SAFETY LISTS

NTP Chemical Status Reports - Testing Status and NTIS Number
Project leader assigned/study in design

MALEIC ANHYDRIDE-DIISOBUTYLENE POLYMER,   37199-81-8
SODIUM SALT

HEALTH AND SAFETY LISTS

NTP Chemical Status Reports - Testing Status and NTIS Number
Technical reports printed (PB299858/AS)

NTP Chemical Status Reports - Evidence of Carcinogenicity
male rat-negative; female rat-negative; male mice-negative; female mice-negative

MALEIC ANHYDRIDE - METHYLVINYL ETHER   9011-16-9
COPOLYMER

HEALTH AND SAFETY LISTS

ACGIH 1995 - Time Weighted Averages
10 mg/m3 TWA

ACGIH 1995 - Skin Designations
skin - potential for cutaneous absorption

U.S. DOT - Appendix B - Marine Pollutants
DOT regulated marine pollutant

U.S. DOT - Appendix A Table 1 - Hazardous Substances
final RQ = 100 pounds (45.4 kg)

IARC - Group 3 (not classifiable)
[present]

NIOSH - Selected LD50s and LC50s
Oral, rat: LD50 = 370 mg/kg Skin, rat: LD50 = 4444 mg/kg

NIOSH 1990 - Pocket Guide - RELs
10 mg/m3 TWA

NIOSH 1990 - Pocket Guide - IDLHs
5000 mg/m3 IDLH

NIOSH 1990 - Pocket Guide - Target organs
respiratory system, liver, blood cholinesterase, CNS, CVS, GI tract

NIOSH 1990 - Pocket Guide - Skin list
Potential for dermal absorption

NIOSH - Health Standards - Exposure Limits
15 mg/m3 TWA

NIOSH - Health Standards - Health Effects and Precautions
Nervous system effects (Prevent skin contact; blood monitoring required)

NTP Chemical Status Reports - Testing Status and NTIS Number
Technical reports printed (PB278527/AS) (PB8300301/AS)

NTP Chemical Status Reports - Evidence of Carcinogenicity
PB83165761: male rat-negative; female rat-negative; PB278527/AS: male rat-negative; female rat-negative; male mice-negative; female mice-negative

OSHA - Vacated PELs - Time Weighted Averages
total dust: 10 mg/m3 TWA

OSHA - Vacated PELs - Skin Designation
Prevent or reduce skin absorption

OSHA - Final PELs - Time Weighted Averages
total dust: 15 mg/m3 TWA

OSHA - Final PELs - Skin Notations
prevent or reduce skin absorption

ENVIRONMENTAL LISTS

ATSDR Priority List
Rank (of 275): 239

CERCLA/SARA - Section 313 - Emission Reporting
form R reporting required

CERCLA/SARA - Hazardous Substances and their Reportable Quantities
final RQ = 100 pounds (45.4 kg)

Clean Water Act - Hazardous Substances
[present]

INTERNATIONAL LISTS

Australian Exposure Standards - Time Weighted Averages
10 mg/m3 TWA

Australian Exposure Standards - Skin Effects
skin absorption

Canada - Drinking Water Quality - MACs
0.19 mg/L MAC

Canada - Alberta - 8 Hour Occupational Exposure Limit
10 mg/m3 TWA

Canada - Alberta - 15 Minute Occupational Exposure Limit
20 mg/m3 STEL

Canada - Alberta - Skin Designation
can be absorbed through the intact skin

Canada - British Columbia - 8 Hour Exposure Limits
10 mg/m3 TWA

Canada - British Columbia - Skin Notations
skin - potential for skin absorption

Canada - Ontario - OHSA - TWAEVs
10 mg/m3 TWAEV

Canada - Ontario - OHSA - Skin Notations
absorption through skin, eyes, or mucous membranes

Canada - Quebec - Time-Weighted Average Exposure Values
10 mg/m3 TWAEV

Canada - Quebec - Skin Designations
absorbed through the skin

United Kingdom - Occupational Exposure Standards - TWAs
10 mg/m3 TWA

United Kingdom - Occupational Exposure Standards - Notes
can be absorbed through skin

German (DFG) - Pregnancy
classification not yet possible

Israel - Time Weighted Averages
10 mg/m3 TWA

Israel - Action Levels
5 mg/m3 AL

Mexico - Instruction No. 10 - TWAs
10 mg/m3 TWA

Mexico - Instruction No. 10 - Skin designation
skin - potential for cutaneous absorption

STATE LISTS

California - Exposure Limits - PELs
10 mg/m3 PEL

California - Exposure Limits - Skin Notation
material may be asbsorbed through the skin, eyes or mucuous membrane

California - Directors List of Hazardous Substances (8 CCR 339)
[present]

Florida Hazardous Substance List
[present]

Massachusetts Right To Know List
neurotoxin

Minnesota Hazardous Substance List
skin

NJ Right to Know List (Total)
sn 1150

NJ Special Hazardous Substances
(mutagen)

Pennsylvania Right to Know List
environmental hazard

MALEIC ANHYDRIDE, POLYMER WITH ETHYL ACRYLATE AND VINYL ACETATE, HYDROLYZED   113221-69-5

HEALTH AND SAFETY LISTS

U.S. DOT - Appendix A Table 1 - Hazardous Substances
final RQ = 5000 pounds (2270 kg)

NIOSH - Selected LD50s and LC50s
Oral, rat: LD50 = 708 mg/kg Skin, rabbit: LD50 = 1560 mg/kg

ENVIRONMENTAL LISTS

CERCLA/SARA - Hazardous Substances and their Reportable Quantities
final RQ = 5000 pounds (2270 kg)

Clean Water Act - Hazardous Substances
[present]

List of Pesticide Product Inert Ingredients
[present]

INTERNATIONAL LISTS

Canada - WHMIS: Ingredient Disclosure
1% item 962 (103)

STATE LISTS

California - Directors List of Hazardous Substances (8 CCR 339)
[present]

Massachusetts Right To Know List
[present]

NJ Right to Know List (Total)
sn 1151

Pennsylvania Right to Know List
environmental hazard

MALEIC ANHYDRIDE-STYRENE POLYMER   9011-13-6

HEALTH AND SAFETY LISTS

NFPA - Flash Points
flash point = 285 degrees F (141 degrees C)

NFPA - Hazard Identification Ratings
health-1; flammability-1; reactivity-0

NIOSH - Selected LD50s and LC50s
Oral, rat: LD50 = 3730 mg/kg Skin, rabbit: LD50 = 10 gm/kg

ENVIRONMENTAL LISTS

EPA - Master Testing List
[present]

MALEIC HYDRAZIDE SODIUM SALT   28330-26-9

INTERNATIONAL LISTS

Canada - WHMIS: Ingredient Disclosure
1% item 963 (869)

MALIC ACID   6915-15-7
SEE ALSO:
K093-HAZARDOUS WASTES
K023-HAZARDOUS WASTES

HEALTH AND SAFETY LISTS

ACGIH 1995 - Time Weighted Averages
0.25 ppm TWA; 1.0 mg/m3 TWA

AIHA - Odor Threshold Values
no geometric mean air odor threshold

U.S. DOT - Substances From 49 CFR 172.101
regulated by DOT (UN2215)

U.S. DOT - Hazard Classes
DOT hazard class = 8

U.S. DOT - Appendix A Table 1 - Hazardous Substances
final RQ = 5000 pounds (2270 kg)

NFPA - Flash Points
flash point = 215 degrees F (102 degrees C)

NFPA - Hazard Identification Ratings
health-3; flammability-1; reactivity-1

NIOSH - Selected LD50s and LC50s
Oral, rat: LD50 = 400 mg/kg Skin, rabbit: LD50 = 2620 mg/kg

NIOSH 1990 - Pocket Guide - RELs
1 mg/m3 TWA; 0.25 ppm TWA

NIOSH 1990 - Pocket Guide - Target organs
eyes, skin, respiratory system

OSHA - Vacated PELs - Time Weighted Averages
0.25 ppm TWA; 1 mg/m3 TWA

OSHA - Final PELs - Time Weighted Averages
0.25 ppm TWA; 1 mg/m3 TWA

ENVIRONMENTAL LISTS

CERCLA/SARA - Section 313 - Emission Reporting
form R reporting required for 1.0% de minimus concentration

CERCLA/SARA - Hazardous Substances and their Reportable Quantities
final RQ = 5000 pounds (2270 kg)

Clean Air Act (1990) - List of Hazardous Air Contaminants
[present]

CAA - HON Rule - SOCMI Chemicals
compliance by Oct. 24, 1994

CAA - HON Rule - Organic HAPs
[present]

Clean Water Act - Hazardous Substances
[present]

List of Pesticide Product Inert Ingredients
[present]

RCRA - U Series Wastes
waste number U147

RCRA - Hazardous Constituents-Appendix VIII
waste number U147

RCRA - Basis for Listing - Appendix VII
Included in waste streams: K023, K093

RCRA - Substances Banned From Land Disposal
[present]

TSCA - Code of Federal Regulations Citations
40 CFR 712.30(d); 40 CFR 716.120(a)

TSCA - PAIR - Reporting List
Reporting Date: November 19, 1982

TSCA - Health and Safety Reporting List
Effective Date: September 10, 1984

INTERNATIONAL LISTS

Australian Exposure Standards - Time Weighted Averages
0.25 ppm TWA; 1 mg/m3 TWA

Australian Exposure Standards - Skin Effects
sensitiser

Canada - WHMIS: Ingredient Disclosure
0.1% item 964 (231)

Canada - NPRI (National Pollutant Release Inventory)
[present]

Canada - Alberta - 8 Hour Occupational Exposure Limit
0.25 ppm TWA; 1 mg/m3 TWA

Canada - Alberta - 15 Minute Occupational Exposure Limit
0.75 ppm STEL; 3 mg/m3 STEL

Canada - British Columbia - 8 Hour Exposure Limits
0.25 ppm TWA; 1 mg/m3 TWA

Canada - Ontario - OHSA - TWAEVs
0.25 ppm TWAEV; 1 mg/m3 TWAEV

Canada - Quebec - Time-Weighted Average Exposure Values
0.25 ppm TWAEV; 1 mg/m3 TWAEV

German (DFG) - MAK Values
0.1 ppm MAK; 0.4 mg/m3 MAK

German (DFG) - Peak Limitations
2 x normal MAK (5 min momentary value); don't exceed 8 times during shift

German (DFG) - Skin/Sensitizers
danger of sensitization (skin or respiratory)

Israel - Time Weighted Averages
0.25 ppm TWA; 1.0 mg/m3 TWA

Israel - Action Levels
0.125 ppm AL; 0.5 mg/m3 AL

Mexico - Instruction No. 10 - TWAs
0.25 ppm TWA; 1 mg/m3 TWA

STATE LISTS

California - Air Bill 2588 Appendix A-I
[present]

California - Exposure Limits - PELs
0.25 ppm PEL; 1 mg/m3 PEL

California - Directors List of Hazardous Substances (8 CCR 339)
[present]

Florida Hazardous Substance List
[present]

Massachusetts Right To Know List
[present]

Minnesota Hazardous Substance List
[present]

NJ Right to Know List (Total)
sn 1152

Pennsylvania Right to Know List
environmental hazard

MALONALDEHYDE   **542-78-9**

ENVIRONMENTAL LISTS

List of Pesticide Product Inert Ingredients
[present]

MALONALDEHYDE, SODIUM SALT   **24382-04-5**

ENVIRONMENTAL LISTS

List of Pesticide Product Inert Ingredients
[present]

MALONIC ACID   **141-82-2**

ENVIRONMENTAL LISTS

List of Pesticide Product Inert Ingredients
[present]

MALONONITRILE   **109-77-3**

ENVIRONMENTAL LISTS

List of Pesticide Product Inert Ingredients
[present]

MALTOSE   **69-79-4**

HEALTH AND SAFETY LISTS

NIOSH - Selected LD50s and LC50s
Oral, rat: LD50 = 21 gm/kg

ENVIRONMENTAL LISTS

List of Pesticide Product Inert Ingredients
[present]

MANCOZEB   **8018-01-7**

HEALTH AND SAFETY LISTS

NIOSH - Selected LD50s and LC50s
Oral, rat: LD50 = 6950 mg/kg

MANDELONITRILE   **532-28-5**

HEALTH AND SAFETY LISTS

U.S. DOT - Substances From 49 CFR 172.101
regulated by DOT (NA2215)

U.S. DOT - Hazard Classes
DOT hazard class = 8

NIOSH - Selected LD50s and LC50s
Oral, rat: LD50 = 4730 mg/kg

ENVIRONMENTAL LISTS

CAA - HON Rule - SOCMI Chemicals
compliance by Oct. 24, 1994

List of Pesticide Product Inert Ingredients
[present]

INTERNATIONAL LISTS

Canada - WHMIS: Ingredient Disclosure
1% item 965 (104)

MANEB   12427-38-2

HEALTH AND SAFETY LISTS

IARC - Group 3 (not classifiable)
[present]

INTERNATIONAL LISTS

Canada - WHMIS: Ingredient Disclosure
1% item 966 (1071)

MANGANESE   7439-96-5

HEALTH AND SAFETY LISTS

NTP Chemical Status Reports - Testing Status and NTIS Number
Technical reports printed (PB89204010/AS)

NTP Chemical Status Reports - Evidence of Carcinogenicity
male rat-clear evidence; female rat-clear evidence; male mice-no evidence; female mice-no evidence

MANGANESE 51   14392-03-1

HEALTH AND SAFETY LISTS

NIOSH - Selected LD50s and LC50s
Oral, rat: LD50 = 1310 mg/kg

MANGANESE 52   14092-99-0

HEALTH AND SAFETY LISTS

U.S. DOT - Substances From 49 CFR 172.101
regulated by DOT (UN2647)

U.S. DOT - Hazard Classes
DOT hazard class = 6.1

U.S. DOT - Appendix A Table 1 - Hazardous Substances
final RQ = 1000 pounds (454 kg)

NIOSH - Selected LD50s and LC50s
Oral, rat: LD50 = 60800 ug/kg

NIOSH - Health Standards - Exposure Limits
3 ppm TWA; 8 mg/m3 TWA (Listed under 'Nitriles')

NIOSH - Health Standards - Health Effects and Precautions
Hepatic, renal, respiratory, cardiovascular, gastrointestinal, and nervous system effects (Periodic chest X-ray and pulmonary function testing required; prevent skin and eye contact; make first-aid kits and personnel available during use) (Listed under 'Nitriles')

ENVIRONMENTAL LISTS

CERCLA/SARA - Section 302 Extremely Hazardous Substances and TPQs
TPQ = 500/10,000 pounds

CERCLA/SARA - Section 313 - Emission Reporting
form R reporting required

CERCLA/SARA - Hazardous Substances and their Reportable Quantities
final RQ = 1000 pounds (454 kg)

RCRA - U Series Wastes
waste number U149

RCRA - Hazardous Constituents-Appendix VIII
waste number U149

RCRA - Substances Banned From Land Disposal
[present]

TSCA - Code of Federal Regulations Citations
40 CFR 716.120; 40 CFR 799.5055(c),(d)(1),(e)(1)

TSCA - Health and Safety Reporting List
Effective Date: March 7, 1986

TSCA - Multichemical Test Rules - Waste Constituents
soil adsorption testing for chemical fate; subchronic toxicity testing for Health Effects

TSCA - Section 12(b) - Export Notification
export notification required - Section 4

INTERNATIONAL LISTS

Canada - WHMIS: Ingredient Disclosure
1% item 967 (1073)

STATE LISTS

Florida Hazardous Substance List
effective March 13, 1992

Massachusetts Right To Know List
extraordinarily hazardous

Minnesota Hazardous Substance List
[present]

NJ Right to Know List (Total)
sn 1153

Pennsylvania Right to Know List
environmental hazard

MANGANESE 52M   RR-00408-0

HEALTH AND SAFETY LISTS

NIOSH - Selected LD50s and LC50s
Oral, rat: LD50 = 34800 mg/kg

MANGANESE 53   14999-33-8

HEALTH AND SAFETY LISTS

AIHA - WEEL - Time Weighted Averages
1 mg/m3 TWA

STATE LISTS

California - Air Bill 2588 Appendix A-II
known or potential carcinogen: 9/90

California - Prop. 65 - Cancer list
carcinogen - initial date 1/1/90

MANGANESE 54   13966-31-9

INTERNATIONAL LISTS

Canada - WHMIS: Ingredient Disclosure
1% item 968 (1074)

Canada - Quebec - Time-Weighted Average Exposure Values
0.05 ppm TWAEV; 0.32 mg/m3 TWAEV

MANGANESE 56   14681-52-8
SEE ALSO:
MANGANESE

HEALTH AND SAFETY LISTS

U.S. DOT - Substances From 49 CFR 172.101
regulated by DOT (UN2210, UN2968)

U.S. DOT - Hazard Classes
DOT hazard class = 4.3

U.S. DOT - Appendix B - Marine Pollutants
DOT regulated marine pollutant

IARC - Group 3 (not classifiable)
[present]

NIOSH - Selected LD50s and LC50s
Oral, rat: LD50 = 3 gm/kg

ENVIRONMENTAL LISTS

CERCLA/SARA - Section 313 - Emission Reporting
form R reporting required for 1.0% de minimus concentration

STATE LISTS

California - Air Bill 2588 Appendix A-II
known or potential carcinogen: 9/90

California - Prop. 65 - Cancer list
carcinogen - initial date 1/1/90

Massachusetts Right To Know List
[present]

NJ Right to Know List (Total)
sn 1154

NJ Special Hazardous Substances
(carcinogen)

Pennsylvania Right to Know List
environmental hazard

MANGANESE(II) ACETATE   638-38-0
SEE ALSO:
MANGANESE COMPOUNDS, N.O.S.

HEALTH AND SAFETY LISTS

ACGIH 1995 - Time Weighted Averages
dust and compounds, as Mn: (5) mg/m3 TWA; fume, as Mn: (1) mg/m3 TWA

ACGIH 1995 - Short Term Exposure Limits
fume, as Mn: (3) mg/m3 STEL

NIOSH - Selected LD50s and LC50s
Oral, rat: LD50 = 9 gm/kg

NIOSH 1990 - Pocket Guide - RELs
as Mn: 1 mg/m3 TWA; 3 mg/m3 STEL

NIOSH 1990 - Pocket Guide - Target organs
respiratory system, CNS, blood, kidneys

OSHA - Vacated PELs - Time Weighted Averages
fume, as Mn: 1 mg/m3 TWA

OSHA - Vacated PELs - Ceiling Limits
compounds as Mn: C 5 mg/m3

OSHA - Final PELs - Time Weighted Averages
compounds as Mn: C 5 mg/m3; fume, as Mn: C 5 mg/m3

OSHA - Final PELs - Ceiling Limits
compounds as Mn: C 5 mg/m3; fume, as Mn: C 5 mg/m3

ENVIRONMENTAL LISTS

ATSDR Priority List
Rank (of 275): 050

CERCLA/SARA - Section 313 - Emission Reporting
form R reporting required for 1.0% de minimus concentration

Safe Drinking Water Act - SMCLs
SMCL = 0.05 mg/L

INTERNATIONAL LISTS

Australian Exposure Standards - Time Weighted Averages
dust and compounds, as Mn: 5 mg/m3 TWA; fume, as Mn: 1 mg/m3 TWA

Australian Exposure Standards - Short Term Exposure Limits
fume: 3 mg/m3 STEL

Canada - WHMIS: Ingredient Disclosure
1% item 974 (1077)

Canada - NPRI (National Pollutant Release Inventory)
[present]

Canada - Drinking Water Quality - AOs
<= 0.05 mg/L AO

Canada - Alberta - Ceiling Occupational Exposure Limit
as Mn: C 5 mg/m3

Canada - British Columbia - Ceiling Exposure Limits
as Mn: C 5 mg/m3

Canada - Ontario - OHSA - TWAEVs
dust, as Mn: 5 mg/m3 TWAEV; fume, as Mn: 1 mg/m3 TWAEV

Canada - Ontario - OHSA - STEVs
fume: 3 mg/m3 STEV

Canada - Quebec - Time-Weighted Average Exposure Values
Dust and compounds as Mn: 5 mg/m3 TWAEV

United Kingdom - Occupational Exposure Standards - TWAs
fume, as Mn: 1 mg/m3 TWA

United Kingdom - Occupational Exposure Standards - STELs
fume, as Mn: 3 mg/m3 STEL

German (DFG) - MAK Values
total dust: 0.5 mg/m3 MAK

German (DFG) - Peak Limitations
10 x normal MAK (30 min average value);   don't exceed during shift

German (DFG) - Pregnancy
no risk to embryo/fetus if exposure limits adhered to

Israel - Time Weighted Averages
dust and compounds, as Mn: 5 mg/m3 TWA; fume, as Mn: 1 mg/m3 TWA

Israel - Short Term Exposure Limits
fume, as Mn: 3 mg/m3 STEL

Israel - Action Levels
dust & compounds as Mn: 2.5 mg/m3 AL; fume, as Mn: 0.5 mg/m3 AL

Mexico - Instruction No. 10 - Ceiling Limits
as dust: P 5 mg/m3

Mexico - Wastewater - Organic Toxic Pollutants and Heavy Metals
Listed under [Heavy Metals]

Mexico - Drinking Water - Ecological Criteria
0.1 mg/l

STATE LISTS

California - Air Bill 2588 Appendix A-I
[present]

California - Exposure Limits - PELs
fume, as Mn: 1 mg/m3 PEL

California - Exposure Limits - STELs
fume, as Mn: 3 mg/m3 STEL

California - Exposure Limits - Ceilings
C 5 mg/m3

California - Directors List of Hazardous Substances (8 CCR 339)
[present]

Florida Hazardous Substance List
[present]

Massachusetts Right To Know List
[present]

Minnesota Hazardous Substance List
[present] as Mn

NJ Right to Know List (Total)
sn 1155

Pennsylvania Right to Know List
environmental hazard (any compound of this substance is also an environmental hazard)

PROPOSED REGULATIONS

ACGIH 1995 - Notice of Intended Changes
as Mn: 0.2 mg/m3 TWA

Safe Drinking Water Act - Priority list
[present]

MANGANESE BORON NEODECANOATE   68442-99-9

HEALTH AND SAFETY LISTS

U.S. DOT - Appendix A Table 2 - Radionuclides
final RQ = 1000 curies (3.7E 13 Bq)

ENVIRONMENTAL LISTS

CERCLA/SARA List of Radionuclides (Appendix B) and Their Reportable Quantities
final RQ = 1000 curies (3.7E 13 Bq)

MANGANESE CARBONATE   598-62-9

HEALTH AND SAFETY LISTS

U.S. DOT - Appendix A Table 2 - Radionuclides
final RQ = 10 curies (3.7E 11 Bq)

ENVIRONMENTAL LISTS

CERCLA/SARA List of Radionuclides (Appendix B) and Their Reportable Quantities
final RQ = 10 curies (3.7E 11 Bq)

MANGANESE(II) CHLORIDE   7773-01-5

HEALTH AND SAFETY LISTS

U.S. DOT - Appendix A Table 2 - Radionuclides
final RQ = 1000 curies (3.7E 13 Bq)

ENVIRONMENTAL LISTS

CERCLA/SARA List of Radionuclides (Appendix B) and Their Reportable Quantities
final RQ = 1000 curies (3.7E 13 Bq)

MANGANESE COMPOUNDS, N.O.S.   RR-00602-0

HEALTH AND SAFETY LISTS

U.S. DOT - Appendix A Table 2 - Radionuclides
final RQ = 1000 curies (3.7E 13 Bq)

ENVIRONMENTAL LISTS

CERCLA/SARA List of Radionuclides (Appendix B) and Their Reportable Quantities
final RQ = 1000 curies (3.7E 13 Bq)

MANGANESE CYCLOPENTADIENYL   12079-65-1
TRICARBONYL

HEALTH AND SAFETY LISTS

U.S. DOT - Appendix A Table 2 - Radionuclides
final RQ = 10 curies (3.7E 11 Bq)

ENVIRONMENTAL LISTS

CERCLA/SARA List of Radionuclides (Appendix B) and Their Reportable Quantities
final RQ = 10 curies (3.7E 11 Bq)

MANGANESE DIOXIDE   1313-13-9

HEALTH AND SAFETY LISTS

U.S. DOT - Appendix A Table 2 - Radionuclides
final RQ = 100 curies (3.7E 12 Bq)

ENVIRONMENTAL LISTS

CERCLA/SARA List of Radionuclides (Appendix B) and Their Reportable Quantities
final RQ = 100 curies (3.7E 12 Bq)

MANGANESE, 2-ETHYLHEXANOATE NAPHTHEN-   68609-86-9
ATE COMPLEXES
SEE ALSO:
MANGANESE

HEALTH AND SAFETY LISTS

NIOSH - Selected LD50s and LC50s
Oral, rat: LD50 = 2940 mg/kg

INTERNATIONAL LISTS

Canada - WHMIS: Ingredient Disclosure
1% item 969 (23)

MANGANESE ISOOCTANOATE   37449-19-7

ENVIRONMENTAL LISTS

List of Pesticide Product Inert Ingredients
[present]

MANGANESE METHYLCYCLOPENTADIENYL   RR-01699-9
TRICARBONYL
SEE ALSO:
MANGANESE
MANGANESE
MANGANESE COMPOUNDS, N.O.S.

INTERNATIONAL LISTS

Canada - WHMIS: Ingredient Disclosure
0.1% item 970 (387)

MANGANESE MONOXIDE   1344-43-0
SEE ALSO:
MANGANESE

HEALTH AND SAFETY LISTS

NIOSH - Selected LD50s and LC50s
Oral, mouse: LD50 = 1715 mg/kg

ENVIRONMENTAL LISTS

List of Pesticide Product Inert Ingredients
[present]

INTERNATIONAL LISTS

Canada - WHMIS: Ingredient Disclosure
0.1% item 971 (505)

MANGANESE NAPHTHENATE   1336-93-2

ENVIRONMENTAL LISTS

CERCLA/SARA - Section 313 - Emission Reporting
form R reporting required for 1.0% de minimus concentration

Clean Air Act (1990) - List of Hazardous Air Contaminants
[present] (includes any unique chemical substance that contains Manganese as part of that chemical's infrastructure)

INTERNATIONAL LISTS

Canada - WHMIS: Ingredient Disclosure
1% item 972 (1075)

Canada - NPRI (National Pollutant Release Inventory)
[present]

Canada - Ontario - OHSA - TWAEVs
as Mn: 5 mg/m3 TWAEV

United Kingdom - Occupational Exposure Standards - TWAs
as Mn: 5 mg/m3 TWA

Mexico - Instruction No. 10 - Ceiling Limits
as Mn: P 5 mg/m3

STATE LISTS

California - Air Bill 2588 Appendix A-I
9/89

California - Exposure Limits - Ceilings
as Mn: C 5 mg/m3

California - Directors List of Hazardous Substances (8 CCR 339)
[present]

NJ Right to Know List (Total)
sn 3141

PROPOSED REGULATIONS

ACGIH 1995 - Notice of Intended Changes
as Mn: 0.2 mg/m3 TWA

MANGANESE NEONONOATE   93918-16-2
SEE ALSO:
MANGANESE

HEALTH AND SAFETY LISTS

ACGIH 1995 - Time Weighted Averages
as Mn: 0.1 mg/m3 TWA

ACGIH 1995 - Skin Designations
as Mn: skin - potential for cutaneous absorption

NIOSH - Selected LD50s and LC50s
Oral, mouse: LD50 = 150 mg/kg

OSHA - Vacated PELs - Time Weighted Averages
as Mn: 0.1 mg/m3 TWA

OSHA - Vacated PELs - Skin Designation
Prevent or reduce skin absorption

INTERNATIONAL LISTS

Australian Exposure Standards - Time Weighted Averages
as Mn: 0.1 mg/m3 TWA

Australian Exposure Standards - Skin Effects
as Mn: skin absorption

Canada - WHMIS: Ingredient Disclosure
1% item 973 (1076)

Canada - Alberta - 8 Hour Occupational Exposure Limit
as Mn: 0.1 mg/m3 TWA

Canada - Alberta - 15 Minute Occupational Exposure Limit
0.3 mg/m3 STEL

Canada - Alberta - Skin Designation
can be absorbed through the intact skin

Canada - British Columbia - 8 Hour Exposure Limits
as Mn: 0.1 mg/m3 TWA

Canada - British Columbia - 15 Minute Exposure Limits
as Mn: 0.3 mg/m3 STEL

Canada - British Columbia - Skin Notations
as Mn: skin - potential for skin absorption

Canada - Ontario - OHSA - TWAEVs
0.1 mg/m3 TWAEV

Canada - Ontario - OHSA - Skin Notations
as Mn: absorption through skin, eyes, or mucous membranes

Canada - Quebec - Time-Weighted Average Exposure Values
as Mn: 0.1 mg/m3 TWAEV

Canada - Quebec - Skin Designations
as Mn: absorbed through the skin

United Kingdom - Occupational Exposure Standards - TWAs
as Mn: 0.1 mg/m3 TWA

United Kingdom - Occupational Exposure Standards - STELs
as Mn: 0.3 mg/m3 STEL

United Kingdom - Occupational Exposure Standards - Notes
can be absorbed through skin

Israel - Time Weighted Averages
as Mn: 0.1 mg/m3 TWA

Israel - Action Levels
as Mn: 0.05 mg/m3 AL

Mexico - Instruction No. 10 - TWAs
0.2 mg/m3 TWA

Mexico - Instruction No. 10 - STELs
0.6 mg/m3 STEL

STATE LISTS

California - Exposure Limits - PELs
as Mn: 0.1 mg/m3 PEL

California - Exposure Limits - Skin Notation
material may be absorbed through the skin, eyes or mucous membrane

California - Directors List of Hazardous Substances (8 CCR 339)
[present]

Florida Hazardous Substance List
[present]

Massachusetts Right To Know List
[present]

Minnesota Hazardous Substance List
as Mn: skin

Pennsylvania Right to Know List
[present]

MANGANESE NITRATE 10377-66-9

SEE ALSO:
MANGANESE COMPOUNDS, N.O.S.
MANGANESE
MANGANESE

ENVIRONMENTAL LISTS

TSCA - Code of Federal Regulations Citations
40 CFR 712.30(w)

TSCA - PAIR - Reporting List
Reporting Date: July 13, 1988

STATE LISTS

NJ Right to Know List (Total)
sn 1157

MANGANESE PROPIONATE 21129-18-0

ENVIRONMENTAL LISTS

List of Pesticide Product Inert Ingredients
[present]

MANGANESE RESINATE 9008-34-8

SEE ALSO:
MANGANESE COMPOUNDS, N.O.S.
MANGANESE

ENVIRONMENTAL LISTS

List of Pesticide Product Inert Ingredients
[present]

MANGANESE(II) SULFATE 7785-87-7

INTERNATIONAL LISTS

Canada - Quebec - Skin Designations
as Mn: absorbed through the skin

MANGANESE SULFATE MONOHYDRATE 10034-96-5

SEE ALSO:
MANGANESE
MANGANESE COMPOUNDS, N.O.S.
MANGANESE

HEALTH AND SAFETY LISTS

OSHA - Vacated PELs - Short Term Exposure Limits
as Mn: 3 mg/m3 STEL

ENVIRONMENTAL LISTS

List of Pesticide Product Inert Ingredients
[present]

INTERNATIONAL LISTS

Canada - Alberta - 8 Hour Occupational Exposure Limit
1 mg/m3 TWA

Canada - Alberta - 15 Minute Occupational Exposure Limit
3 mg/m3 STEL

Canada - Quebec - Time-Weighted Average Exposure Values
1 mg/m3 TWAEV

Canada - Quebec - Short-term Exposure Values
3 mg/m3 STEV

Mexico - Instruction No. 10 - TWAs
1 mg/m3 TWA

Mexico - Instruction No. 10 - STELs
3 mg/m3 STEL

MANGANESE TETROXIDE 1317-35-7

SEE ALSO:
MANGANESE

ENVIRONMENTAL LISTS

List of Pesticide Product Inert Ingredients
[present]

MANGANOUS CHLORIDE   13446-34-9

ENVIRONMENTAL LISTS

List of Pesticide Product Inert Ingredients
[present]

MANNITAN, COCONUT OIL ESTER   91031-95-7
SEE ALSO:
MANGANESE
MANGANESE COMPOUNDS, N.O.S.
MANGANESE

HEALTH AND SAFETY LISTS

U.S. DOT - Substances From 49 CFR 172.101
regulated by DOT (UN2724)

U.S. DOT - Hazard Classes
DOT hazard class = 5.1

STATE LISTS

NJ Right to Know List (Total)
sn 1158

MANNITAN TETRANITRATE   RR-01436-8

ENVIRONMENTAL LISTS

List of Pesticide Product Inert Ingredients
[present]

D-MANNITOL   69-65-8

HEALTH AND SAFETY LISTS

U.S. DOT - Substances From 49 CFR 172.101
regulated by DOT (UN1330)

U.S. DOT - Hazard Classes
DOT hazard class = 4.1

STATE LISTS

NJ Right to Know List (Total)
sn 2526

MANNITOL HEXANITRATE   15825-70-4
SEE ALSO:
MANGANESE
MANGANESE
MANGANESE COMPOUNDS, N.O.S.

ENVIRONMENTAL LISTS

List of Pesticide Product Inert Ingredients
[present]

TSCA - Code of Federal Regulations Citations
40 CFR 712.30(w)

TSCA - PAIR - Reporting List
Reporting Date: July 13, 1988

INTERNATIONAL LISTS

Canada - WHMIS: Ingredient Disclosure
1% item 975 (1524)

MANNICH-BASED ADDUCT   RR-01656-8

HEALTH AND SAFETY LISTS

NTP Chemical Status Reports - Testing Status and NTIS Number
Prechronic studies for which toxicity technical reports were not prepared; Toxicity reports printed

MANNOMUSTINE   551-74-6
SEE ALSO:
MANGANESE
MANGANESE COMPOUNDS, N.O.S.
MANGANESE

HEALTH AND SAFETY LISTS

OSHA - Vacated PELs - Time Weighted Averages
as Mn: 1 mg/m3 TWA

INTERNATIONAL LISTS
Canada - Alberta - 8 Hour Occupational Exposure Limit
1 mg/m3 TWA
Canada - Alberta - 15 Minute Occupational Exposure Limit
2 mg/m3 STEL
Canada - British Columbia - 8 Hour Exposure Limits
1 mg/m3 TWA
Canada - Ontario - OHSA - TWAEVs
as Mn: 1 mg/m3 TWAEV
Canada - Ontario - OHSA - Skin Notations
as Mn: absorption through skin, eyes, or mucous membranes
Canada - Quebec - Time-Weighted Average Exposure Values
1 mg/m3 TWAEV
United Kingdom - Occupational Exposure Standards - TWAs
1 mg/m3 TWA
German (DFG) - MAK Values
total dust: 1 mg/m3 MAK
German (DFG) - Peak Limitations
10 x normal MAK (30 min average value); don't exceed during shift

STATE LISTS

California - Exposure Limits - PELs
1 mg/m3 PEL
Florida Hazardous Substance List
[present]
Massachusetts Right To Know List
[present]
Minnesota Hazardous Substance List
[present]
Pennsylvania Right to Know List
[present]

MANURE   RR-01094-6
SEE ALSO:
MANGANESE

HEALTH AND SAFETY LISTS

NIOSH - Selected LD50s and LC50s
Oral, rat: LD50 = 1484 mg/kg

MAPP   59355-75-8

ENVIRONMENTAL LISTS

List of Pesticide Product Inert Ingredients
[present]

MARBLE   RR-01193-8

HEALTH AND SAFETY LISTS

U.S. DOT - Hazard Classes
Forbidden from transport by the DOT

MATE   68916-96-1

HEALTH AND SAFETY LISTS

NIOSH - Selected LD50s and LC50s
Oral, rat: LD50 = 13500 mg/kg

NTP Chemical Status Reports - Testing Status and NTIS Number
Technical reports printed (PB83129080)

NTP Chemical Status Reports - Evidence of Carcinogenicity
male rat-negative; female rat-negative; male mice-negative; female mice-negative

ENVIRONMENTAL LISTS

List of Pesticide Product Inert Ingredients
[present]

MEAT MEAL   68131-12-4

HEALTH AND SAFETY LISTS

U.S. DOT - Substances From 49 CFR 172.101
regulated by DOT (NA0133, UN0133)

U.S. DOT - Hazard Classes
dry: Forbidden from transport by the DOT; wetted: DOT hazard class = 1.1A;

STATE LISTS

NJ Right to Know List (Total)
sn 1160

MEDPHALAN   13045-94-8

ENVIRONMENTAL LISTS

TSCA - Chemicals with Significant New Use Rules
PMN number: P-93-66

MEDROXYPROGESTERONE ACETATE   71-58-9

HEALTH AND SAFETY LISTS

IARC - Group 3 (not classifiable)
[present]

STATE LISTS

California - Directors List of Hazardous Substances (8 CCR 339)
[present]

MEGESTROL ACETATE   595-33-5

ENVIRONMENTAL LISTS

List of Pesticide Product Inert Ingredients
[present]

MEGESTROL ACETATE AND OESTROGENS   RR-00096-4

HEALTH AND SAFETY LISTS

ACGIH 1995 - Time Weighted Averages
1000 ppm TWA; 1640 mg/m3 TWA

ACGIH 1995 - Short Term Exposure Limits
1250 ppm STEL; 2050 mg/m3 STEL

U.S. DOT - Substances From 49 CFR 172.101
regulated by DOT (UN1060)

U.S. DOT - Hazard Classes
DOT hazard class = 2.1

NIOSH 1990 - Pocket Guide - RELs
1000 ppm TWA; 1800 mg/m3 TWA; 1250 ppm STEL; 2250 mg/m3 STEL

NIOSH 1990 - Pocket Guide - IDLHs
15,000 ppm IDLH (lower explosive level)

NIOSH 1990 - Pocket Guide - Target organs
CNS, skin, eyes

OSHA - Vacated PELs - Time Weighted Averages
1000 ppm TWA; 1800 mg/m3 TWA

OSHA - Vacated PELs - Short Term Exposure Limits
1250 ppm STEL; 2250 mg/m3 STEL

OSHA - Final PELs - Time Weighted Averages
1000 ppm TWA; 1800 mg/m3 TWA

INTERNATIONAL LISTS

Australian Exposure Standards - Time Weighted Averages
1000 ppm TWA; 1640 mg/m3 TWA

Australian Exposure Standards - Short Term Exposure Limits
1250 ppm STEL; 2250 mg/m3 STEL

Canada - Alberta - 8 Hour Occupational Exposure Limit
1000 ppm TWA; 1640 mg/m3 TWA

Canada - Alberta - 15 Minute Occupational Exposure Limit
1250 ppm STEL; 2050 mg/m3 STEL

Canada - British Columbia - 8 Hour Exposure Limits
1000 ppm TWA; 1800 mg/m3 TWA

Canada - British Columbia - 15 Minute Exposure Limits
1250 ppm STEL; 2250 mg/m3 STEL

Canada - Ontario - OHSA - TWAEVs
1800 mg/m3 TWAEV (listed as an agent of variable composition)

Canada - Ontario - OHSA - STEVs
2250 mg/m3 STEV (listed as 'Agent of variable composition')

Canada - Quebec - Time-Weighted Average Exposure Values
1000 ppm TWAEV; 1640 mg/m3 TWAEV

Canada - Quebec - Short-term Exposure Values
1250 ppm STEV; 2050 mg/m3 STEV

Israel - Time Weighted Averages
1000 ppm TWA; 1640 mg/m3 TWA

Israel - Short Term Exposure Limits
1250 ppm STEL; 2050 mg/m3 STEL

Israel - Action Levels
500 ppm AL; 820 mg/m3 AL

Mexico - Instruction No. 10 - TWAs
1000 ppm TWA; 1800 mg/m3 TWA

Mexico - Instruction No. 10 - STELs
1250 ppm STEL; 2250 mg/m3 STEL

STATE LISTS

California - Exposure Limits - PELs
1000 ppm PEL; 1800 mg/m3 PEL

California - Exposure Limits - STELs
1250 ppm STEL; 2250 mg/m3 STEL

California - Directors List of Hazardous Substances (8 CCR 339)
[present]

Florida Hazardous Substance List
[present]

Massachusetts Right To Know List
[present]

Minnesota Hazardous Substance List
[present]

Pennsylvania Right to Know List
[present]

MEIQ (2-AMINO-3,4-DIMETHYLIMIDAZO[4,5-F]   77094-11-2
QUINOLINE)

INTERNATIONAL LISTS

Canada - Alberta - 8 Hour Occupational Exposure Limit
respirable mass: 5 mg/m3 TWA; total mass: 10 mg/m3 TWA

Canada - British Columbia - 8 Hour Exposure Limits
10 mg/m3 TWA

Canada - British Columbia - 15 Minute Exposure Limits
20 mg/m3 STEL

United Kingdom - Occupational Exposure Standards - TWAs
total inhalable dust: 10 mg/m3 TWA; respirable dust: 5 mg/m3 TWA

Mexico - Instruction No. 10 - TWAs
10 mg/m3 TWA; (nuisance particulate)

Mexico - Instruction No. 10 - STELs
20 mg/m3 STEL

MEIQX (2-AMINO-3,8-DIMETHYLIMIDAZO[4,5-F]   77500-04-0
QUINOXALINE)

HEALTH AND SAFETY LISTS

IARC - Group 2A (limited human data)
[present] (when drinking hot mate)

IARC - Group 3 (not classifiable)
[present]

OSHA - Possible Select Carcinogens
[present]

MELAMINE   **108-78-1**

ENVIRONMENTAL LISTS

List of Pesticide Product Inert Ingredients
[present]

MELPHALAN   **148-82-3**

HEALTH AND SAFETY LISTS

IARC - Group 3 (not classifiable)
[present]

MEMTETRAHYDROPHTHALIC ANHYDRIDE   **25134-21-8**

HEALTH AND SAFETY LISTS

IARC - Group 2B (sufficient animal data)
[present] (Listed under 'Progestins')

OSHA - Possible Select Carcinogens
[present]

STATE LISTS

California - Air Bill 2588 Appendix A-II
known or potential carcinogen

California - Prop. 65 - Cancer list
carcinogen - initial date 1/1/90

California - Prop. 65 - Developmental Toxicity
developmental toxicity - initial date 4/1/90

California - Directors List of Hazardous Substances (8 CCR 339)
[present]

Minnesota Hazardous Substance List
carcinogen

MENDELEVIUM 257   **15752-34-8**

HEALTH AND SAFETY LISTS

IARC - Group Unspecified
[present] (Listed under 'Progestins')

STATE LISTS

California - Air Bill 2588 Appendix A-II
6/91

California - Prop. 65 - Developmental Toxicity
developmental toxicity - initial date 1/1/91

MENDELEVIUM 258   **29665-18-7**

HEALTH AND SAFETY LISTS

IARC - Group Unspecified
[present] (Listed under 'Combined oral contraceptives')

MENHADEN OIL   **8002-50-4**

HEALTH AND SAFETY LISTS

IARC - Group 2B (sufficient animal data)
[present]

IARC - Group 3 (not classifiable)
[present]

OSHA - Possible Select Carcinogens
[present]

STATE LISTS

California - Prop. 65 - Cancer list
carcinogen - initial date 10/01/94

MENOTROPINS   **9002-68-0**

HEALTH AND SAFETY LISTS

IARC - Group 2B (sufficient animal data)
[present]

IARC - Group 3 (not classifiable)
[present]

OSHA - Possible Select Carcinogens
[present]

STATE LISTS

California - Prop. 65 - Cancer list
carcinogen - initial date 10/01/94

P-MENTHANE-1,8-DIAMINE   **80-52-4**

HEALTH AND SAFETY LISTS

IARC - Group 3 (not classifiable)
[present]

NIOSH - Selected LD50s and LC50s
Oral, rat: LD50 = 3161 mg/kg

NTP Chemical Status Reports - Testing Status and NTIS Number
Technical reports printed (PB83202630)

NTP Chemical Status Reports - Evidence of Carcinogenicity
male rat-positive; female rat-negative; male mice-negative; female mice-negative

ENVIRONMENTAL LISTS

EPA - Master Testing List
[present]

INTERNATIONAL LISTS

Canada - WHMIS: Ingredient Disclosure
1% item 976 (1081)

STATE LISTS

Massachusetts Right To Know List
[present]

Pennsylvania Right to Know List
environmental hazard

P-MENTHANE HYDROPEROXIDE   **80-47-7**

HEALTH AND SAFETY LISTS

U.S. DOT - Appendix A Table 1 - Hazardous Substances
final RQ = 1 pound (0.454 kg)

IARC - Group 1 (carcinogenic to humans)
[present]

NIOSH - Selected LD50s and LC50s
Oral, rat: LD50 = 11200 ug/kg

NTP Chemical Status Reports - Testing Status and NTIS Number
Chronic studies exist for which technical reports were not prepared

NTP Seventh Report - Known Carcinogens
known carcinogen

OSHA - Select Carcinogens
[present]

ENVIRONMENTAL LISTS

CERCLA/SARA - Hazardous Substances and their Reportable Quantities
final RQ = 1 pound (0.454 kg)

EPA - Carcinogen Hazard Ranking for RQ Adjustment
Hazard ranking = High

RCRA - U Series Wastes
waste number U150

RCRA - Hazardous Constituents-Appendix VIII
waste number U150

RCRA - Substances Banned From Land Disposal
[present]

STATE LISTS

California - Air Bill 2588 Appendix A-II
known or potential carcinogen

California - Prop. 65 - Cancer list
carcinogen - initial date 2/27/87

California - Prop. 65 - Developmental Toxicity
developmental toxicity - initial date 7/1/90

California - Prop. 65 - No Significant Risk Levels
no significant risk level = 0.005 ug/day

California - Directors List of Hazardous Substances (8 CCR 339)
[present]

Florida Hazardous Substance List
[present]

Massachusetts Right To Know List
carcinogen; extraordinarily hazardous

Minnesota Hazardous Substance List
carcinogen

NJ Right to Know List (Total)
sn 1162

NJ Special Hazardous Substances
(carcinogen; mutagen)

Pennsylvania Right to Know List
environmental hazard; special hazardous substance

Pennsylvania RTK - Special Hazardous Substances
[present]

MENTHOL   1490-04-6

HEALTH AND SAFETY LISTS

NIOSH - Selected LD50s and LC50s
Oral, rat: LD50 = 914 mg/kg

ENVIRONMENTAL LISTS

TSCA - Code of Federal Regulations Citations
40 CFR 712.30(d)

TSCA - PAIR - Reporting List
Reporting Date: November 19, 1982

STATE LISTS

NJ Right to Know List (Total)
sn 2534

DL-MENTHOL   15356-70-4

HEALTH AND SAFETY LISTS

U.S. DOT - Appendix A Table 2 - Radionuclides
final RQ = 100 curies (3.7E 12 Bq)

ENVIRONMENTAL LISTS

CERCLA/SARA List of Radionuclides (Appendix B) and Their Reportable Quantities
final RQ = 100 curies (3.7E 12 Bq)

MEPHOSFOLAN   950-10-7

HEALTH AND SAFETY LISTS

U.S. DOT - Appendix A Table 2 - Radionuclides
final RQ = 1 curie (3.7E 10 Bq)

ENVIRONMENTAL LISTS

CERCLA/SARA List of Radionuclides (Appendix B) and Their Reportable Quantities
final RQ = 1 curie (3.7E 10 Bq)

MEPROBAMATE   57-53-4

HEALTH AND SAFETY LISTS

NFPA - Flash Points
flash point = 435 degrees F (224 degrees C)

NFPA - Hazard Identification Ratings
health-0; flammability-1; reactivity-0

ENVIRONMENTAL LISTS

List of Pesticide Product Inert Ingredients
[present]

MERCAPTAMINE HYDROCHLORIDE   156-57-0

STATE LISTS

California - Air Bill 2588 Appendix A-II
9/90

California - Prop. 65 - Developmental Toxicity
developmental toxicity - initial date 4/1/90

California - Directors List of Hazardous Substances (8 CCR 339)
[present]

MERCAPTANS, LIQUID, N.O.S.   RR-00360-1

HEALTH AND SAFETY LISTS

NIOSH - Selected LD50s and LC50s
Oral, rat: LD50 = 704 mg/kg Skin, rabbit: LD50 = 292 mg/kg

2-MERCAPTOBENZIMIDAZOLE   583-39-1

STATE LISTS

NJ Right to Know List (Total)
sn 1163

2-MERCAPTOBENZOTHIAZOLE   149-30-4

ENVIRONMENTAL LISTS

List of Pesticide Product Inert Ingredients
[present]

2-MERCAPTOETHANOL   60-24-2

HEALTH AND SAFETY LISTS

NIOSH - Selected LD50s and LC50s
Oral, rat: LD50 = 2900 mg/kg

NTP Chemical Status Reports - Testing Status and NTIS Number
Technical reports printed (PB288761/AS)

NTP Chemical Status Reports - Evidence of Carcinogenicity
male rat-negative; female rat-negative; male mice-negative; female mice-negative

B-MERCAPTOETHYLAMINE HCL   60-23-1

HEALTH AND SAFETY LISTS

U.S. DOT - Appendix B - Marine Pollutants
DOT regulated marine pollutant

NIOSH - Selected LD50s and LC50s
Oral, rat: LD50 = 9 mg/kg Skin, rabbit: LD50 = 28700 ug/kg

ENVIRONMENTAL LISTS

CERCLA/SARA - Section 302 Extremely Hazardous Substances and TPQs
TPQ = 500 pounds

STATE LISTS

Florida Hazardous Substance List
effective March 13, 1992

Massachusetts Right To Know List
extraordinarily hazardous

NJ Right to Know List (Total)
sn 2535

Pennsylvania Right to Know List
environmental hazard

PROPOSED REGULATIONS

CERCLA/SARA - Proposed Hazardous Substance Additions
proposed RQ = 1 pound (.454 kg)

CERCLA/SARA - 1989 Proposed RQ Adjustments
proposed RQ = 100 pounds (45.4 kg)

3-MERCAPTOPROPYLTRIMETHOXYSILANE   4420-74-0

HEALTH AND SAFETY LISTS

NIOSH - Selected LD50s and LC50s
Oral, rat: LD50 = 1000 mg/kg

STATE LISTS

California - Prop. 65 - Developmental Toxicity
developmental toxicity - initial date 1/1/92

NJ Right to Know List (Total)
sn 1164

NJ Special Hazardous Substances
(teratogen)

6-MERCAPTOPURINE   50-44-2

HEALTH AND SAFETY LISTS

NIOSH - Selected LD50s and LC50s
Oral, mouse: LD50 = 1352 mg/kg

MERCAPTOPURINE   6112-76-1

HEALTH AND SAFETY LISTS

U.S. DOT - Substances From 49 CFR 172.101
regulated by DOT (UN1228, UN3071)

U.S. DOT - Hazard Classes
flash point less than 23 degrees C: DOT hazard class = 3;   flash point 23 degrees C or more: DOT hazard class = 6.1

STATE LISTS

NJ Right to Know List (Total)
sn 2536

MERCAPTOSUCCINIC ACID   70-49-5

HEALTH AND SAFETY LISTS

NIOSH - Selected LD50s and LC50s
Oral, rat: LD50 = 476 mg/kg

NTP Chemical Status Reports - Testing Status and NTIS Number
Prechronic studies for which toxicity technical reports were not prepared

5-MERCAPTOTETRAZOL-1-ACETIC ACID   RR-01332-1

HEALTH AND SAFETY LISTS

NIOSH - Selected LD50s and LC50s
Oral, rat: LD50 = 1680 mg/kg

NTP Chemical Status Reports - Testing Status and NTIS Number
Technical reports printed (PB88245154/AS)

NTP Chemical Status Reports - Evidence of Carcinogenicity
male rat-some evidence;   female rat-some evidence;   male mice-no evidence;   female mice-equivocal evidence

ENVIRONMENTAL LISTS

CERCLA/SARA - Section 313 - Emission Reporting
form R reporting required

List of Pesticide Product Inert Ingredients
[present]

TSCA - Code of Federal Regulations Citations
40 CFR 712.30(m);   40 CFR 716.120(a);   40 CFR 799.2475

TSCA - PAIR - Reporting List
Reporting Date: February 26, 1985

TSCA - Health and Safety Reporting List
Effective Date: December 28, 1984

TSCA - Chemical Test Rules
Testing required by: manufacturers;   importers;   processers (40 CFR 799.2475)

TSCA - Section 12(b) - Export Notification
export notification required - Section 4

MERCARBAM   RR-01293-1

HEALTH AND SAFETY LISTS

AIHA - WEEL - Time Weighted Averages
0.2 ppm TWA;   0.6 mg/m3 TWA

AIHA - WEEL - Skin Absorption Designations
skin absorber

U.S. DOT - Substances From 49 CFR 172.101
regulated by DOT (UN2966)

U.S. DOT - Hazard Classes
DOT hazard class = 6.1

NFPA - Flash Points
flash point = 165 degrees F (74 degrees C)

NFPA - Hazard Identification Ratings
health-2;   flammability-2

NIOSH - Selected LD50s and LC50s
Inhalation, mouse: LC50 = 13200 mg/m3 (15 mn) Oral, rat: LD50 = 244 mg/kg Skin, guinea pig: LD50 = 300 mg/kg

INTERNATIONAL LISTS

Canada - WHMIS: Ingredient Disclosure
1% item 1558 (1616)

STATE LISTS

Florida Hazardous Substance List
[present]

Massachusetts Right To Know List
[present]

NJ Right to Know List (Total)
sn 2821

Pennsylvania Right to Know List
[present]

MERCURIC ACETATE   1600-27-7

HEALTH AND SAFETY LISTS

NIOSH - Selected LD50s and LC50s
Oral, mouse: LD50 = 625 mg/kg

MERCURIC ARSENATE   7784-37-4

HEALTH AND SAFETY LISTS

NIOSH - Selected LD50s and LC50s
Oral, rat: LD50 = 2940 mg/kg Skin, rabbit: LD50 = 5880 mg/kg

MERCURIC BENZOATE   583-15-3

HEALTH AND SAFETY LISTS

IARC - Group 3 (not classifiable)
[present]

NTP Chemical Status Reports - Testing Status and NTIS Number
Chronic studies exist for which technical reports were not prepared

MERCURIC BROMIDE   7789-47-1

STATE LISTS

California - Air Bill 2588 Appendix A-II
9/90

California - Prop. 65 - Developmental Toxicity
developmental toxicity - initial date 7/1/90

MERCURIC CHLORIDE   7487-94-7

HEALTH AND SAFETY LISTS

NIOSH - Selected LD50s and LC50s
Oral, rat: LD50 = 800 mg/kg

MERCURIC CYANIDE   592-04-1

HEALTH AND SAFETY LISTS

U.S. DOT - Substances From 49 CFR 172.101
regulated by DOT (UN0448)

U.S. DOT - Hazard Classes
DOT hazard class = 1.4C

MERCURIC GLUCONATE   RR-01295-3

HEALTH AND SAFETY LISTS

U.S. DOT - Appendix B - Marine Pollutants
DOT regulated marine pollutant

MERCURIC IODIDE   7774-29-0
  SEE ALSO:
    MERCURY

  HEALTH AND SAFETY LISTS
    U.S. DOT - Substances From 49 CFR 172.101
     regulated by DOT (UN1629)
    U.S. DOT - Hazard Classes
     DOT hazard class = 6.1
    U.S. DOT - Appendix B - Marine Pollutants
     DOT regulated severe marine pollutant
    NIOSH - Selected LD50s and LC50s
     Oral, rat: LD50 = 40900 ug/kg Skin, rat: LD50 = 570 mg/kg

  ENVIRONMENTAL LISTS
    CERCLA/SARA - Section 302 Extremely Hazardous Substances and TPQs
     TPQ = 500/10,000 pounds

  INTERNATIONAL LISTS
    Canada - WHMIS: Ingredient Disclosure
     0.1% item 977 (36)

  STATE LISTS
    Florida Hazardous Substance List
     effective March 13, 1992
    Massachusetts Right To Know List
     extraordinarily hazardous
    NJ Right to Know List (Total)
     sn 1166
    Pennsylvania Right to Know List
     environmental hazard

  PROPOSED REGULATIONS
    CERCLA/SARA - Proposed Hazardous Substance Additions
     proposed RQ = 1 pound (.454 kg)
    CERCLA/SARA - 1989 Proposed RQ Adjustments
     proposed RQ = 100 pounds (45.4 kg)

MERCURIC NITRATE   10045-94-0
  SEE ALSO:
    ARSENIC
    MERCURY

  HEALTH AND SAFETY LISTS
    U.S. DOT - Substances From 49 CFR 172.101
     regulated by DOT (UN1623)
    U.S. DOT - Hazard Classes
     DOT hazard class = 6.1
    U.S. DOT - Appendix B - Marine Pollutants
     DOT regulated severe marine pollutant

  STATE LISTS
    NJ Right to Know List (Total)
     sn 1167

MERCURIC OXIDE   21908-53-2
  SEE ALSO:
    MERCURY

  HEALTH AND SAFETY LISTS
    U.S. DOT - Substances From 49 CFR 172.101
     regulated by DOT (UN1631)
    U.S. DOT - Hazard Classes
     DOT hazard class = 6.1
    U.S. DOT - Appendix B - Marine Pollutants
     DOT regulated severe marine pollutant

  STATE LISTS
    NJ Right to Know List (Total)
     sn 1168

MERCURIC OXYCYANIDE   1335-31-5
  SEE ALSO:
    MERCURY

  HEALTH AND SAFETY LISTS
    U.S. DOT - Appendix B - Marine Pollutants
     DOT regulated severe marine pollutant
    NIOSH - Selected LD50s and LC50s
     Oral, rat: LD50 = 40 mg/kg Skin, rat: LD50 = 100 mg/kg

  INTERNATIONAL LISTS
    Canada - WHMIS: Ingredient Disclosure
     1% item 979 (345)

  STATE LISTS
    NJ Right to Know List (Total)
     sn 1169

MERCURIC POTASSIUM CYANIDE   591-89-9
  SEE ALSO:
    MERCURY

  HEALTH AND SAFETY LISTS
    U.S. DOT - Substances From 49 CFR 172.101
     regulated by DOT (UN1624)
    U.S. DOT - Hazard Classes
     DOT hazard class = 6.1
    U.S. DOT - Appendix B - Marine Pollutants
     DOT regulated severe marine pollutant
    NIOSH - Selected LD50s and LC50s
     Oral, rat: LD50 = 1 mg/kg Skin, rat: LD50 = 41 mg/kg
    NTP Chemical Status Reports - Testing Status and NTIS Number
     Technical reports printed (no NTIS number given)
    NTP Chemical Status Reports - Evidence of Carcinogenicity
     male rat: some evidence; female rat: equivocal evidence; male mice: equivocal evidence; female mice: no evidence

  ENVIRONMENTAL LISTS
    CERCLA/SARA - Section 302 Extremely Hazardous Substances and TPQs
     TPQ = 500/10,000 pounds

  INTERNATIONAL LISTS
    Canada - WHMIS: Ingredient Disclosure
     0.1% item 980 (544)
    Canada - CEPA Schedule II Part II - Toxic Substances (Export)
     [present]

  STATE LISTS
    California - Air Bill 2588 Appendix A-I
     [present]
    Florida Hazardous Substance List
     effective March 13, 1992
    Massachusetts Right To Know List
     extraordinarily hazardous
    NJ Right to Know List (Total)
     sn 1170
    Pennsylvania Right to Know List
     environmental hazard

  PROPOSED REGULATIONS
    CERCLA/SARA - Proposed Hazardous Substance Additions
     proposed RQ = 1 pound (.454 kg)
    CERCLA/SARA - 1989 Proposed RQ Adjustments
     proposed RQ = 100 pounds (45.4 kg)

MERCURIC SUBSULFATE   **1312-03-4**
SEE ALSO:
  MERCURY
  CYANIDE ANION

HEALTH AND SAFETY LISTS
  U.S. DOT - Substances From 49 CFR 172.101
  regulated by DOT (UN1636)
  U.S. DOT - Hazard Classes
  DOT hazard class = 6.1
  U.S. DOT - Appendix B - Marine Pollutants
  DOT regulated severe marine pollutant
  U.S. DOT - Appendix A Table 1 - Hazardous Substances
  final RQ = 1 pound (0.454 kg)
  NIOSH - Selected LD50s and LC50s
  Oral, mouse: LD50 = 33 mg/kg

ENVIRONMENTAL LISTS
  CERCLA/SARA - Hazardous Substances and their Reportable Quantities
  final RQ = 1 pound (0.454 kg)
  Clean Water Act - Hazardous Substances
  [present]

INTERNATIONAL LISTS
  Canada - WHMIS: Ingredient Disclosure
  1% item 981 (598)

STATE LISTS
  Florida Hazardous Substance List
  [present]
  Massachusetts Right To Know List
  [present]
  NJ Right to Know List (Total)
  sn 1171
  Pennsylvania Right to Know List
  environmental hazard

MERCURIC SULFATE   **7783-35-9**
HEALTH AND SAFETY LISTS
  U.S. DOT - Appendix B - Marine Pollutants
  DOT regulated severe marine pollutant

MERCURIC THIOCYANATE   **592-85-8**
SEE ALSO:
  MERCURY

HEALTH AND SAFETY LISTS
  U.S. DOT - Appendix B - Marine Pollutants
  DOT regulated marine pollutant
  NIOSH - Selected LD50s and LC50s
  Oral, rat: LD50 = 18 mg/kg Skin, rat: LD50 = 75 mg/kg

INTERNATIONAL LISTS
  Canada - WHMIS: Ingredient Disclosure
  0.1% item 982 (1031)

STATE LISTS
  NJ Right to Know List (Total)
  sn 1172

MERCUROL   **12002-19-6**
SEE ALSO:
  MERCURY

HEALTH AND SAFETY LISTS
  U.S. DOT - Substances From 49 CFR 172.101
  regulated by DOT (UN1625)
  U.S. DOT - Hazard Classes
  DOT hazard class = 6.1

U.S. DOT - Appendix B - Marine Pollutants
  DOT regulated severe marine pollutant
U.S. DOT - Appendix A Table 1 - Hazardous Substances
  final RQ = 10 pounds (4.54 kg)
NIOSH - Selected LD50s and LC50s
  Oral, rat: LD50 = 26 mg/kg Skin, rat: LD50 = 75 mg/kg

ENVIRONMENTAL LISTS
  CERCLA/SARA - Hazardous Substances and their Reportable Quantities
  final RQ = 10 pounds (4.54 kg)
  Clean Water Act - Hazardous Substances
  [present]

INTERNATIONAL LISTS
  Canada - WHMIS: Ingredient Disclosure
  1% item 983 (1216)

STATE LISTS
  Massachusetts Right To Know List
  [present]
  NJ Right to Know List (Total)
  sn 1173
  Pennsylvania Right to Know List
  environmental hazard

MERCUROUS ACETATE   **631-60-7**
SEE ALSO:
  MERCURY

HEALTH AND SAFETY LISTS
  U.S. DOT - Appendix B - Marine Pollutants
  DOT regulated severe marine pollutant
  NIOSH - Selected LD50s and LC50s
  Oral, rat: LD50 = 18 mg/kg Skin, rat: LD50 = 315 mg/kg

ENVIRONMENTAL LISTS
  CERCLA/SARA - Section 302 Extremely Hazardous Substances and TPQs
  TPQ = 500/10,000 pounds

INTERNATIONAL LISTS
  Canada - WHMIS: Ingredient Disclosure
  0.1% item 984 (1328)

STATE LISTS
  Florida Hazardous Substance List
  effective March 13, 1992
  Massachusetts Right To Know List
  extraordinarily hazardous
  NJ Right to Know List (Total)
  sn 2537
  Pennsylvania Right to Know List
  environmental hazard

PROPOSED REGULATIONS
  CERCLA/SARA - Proposed Hazardous Substance Additions
  proposed RQ = 1 pound (.454 kg)
  CERCLA/SARA - 1989 Proposed RQ Adjustments
  proposed RQ = 100 pounds (45.4 kg)

MERCUROUS AZIDE   **38232-63-2**
SEE ALSO:
  MERCURY
  CYANIDE ANION

HEALTH AND SAFETY LISTS
  U.S. DOT - Substances From 49 CFR 172.101
  regulated by DOT (UN1642)
  U.S. DOT - Hazard Classes
  regular: Forbidden from transport by the DOT: desensitized: DOT hazard class = 6.1

U.S. DOT - Appendix B - Marine Pollutants
desensitized form: DOT regulated severe marine pollutant

STATE LISTS
NJ Right to Know List (Total)
sn 1174

MERCUROUS BISULPHATE   **RR-01296-4**
SEE ALSO:
CYANIDE ANION
MERCURY

HEALTH AND SAFETY LISTS
U.S. DOT - Substances From 49 CFR 172.101
regulated by DOT (UN1626)
U.S. DOT - Hazard Classes
DOT hazard class = 6.1
U.S. DOT - Appendix B - Marine Pollutants
DOT regulated severe marine pollutant

STATE LISTS
NJ Right to Know List (Total)
sn 1175

MERCUROUS CHLORIDE   **7546-30-7**
SEE ALSO:
MERCURY

STATE LISTS
NJ Right to Know List (Total)
sn 1176

MERCUROUS IODIDE   **7783-30-4**
SEE ALSO:
MERCURY

HEALTH AND SAFETY LISTS
U.S. DOT - Appendix B - Marine Pollutants
DOT regulated severe marine pollutant
U.S. DOT - Appendix A Table 1 - Hazardous Substances
final RQ = 10 pounds (4.54 kg)
NIOSH - Selected LD50s and LC50s
Oral, rat: LD50 = 57 mg/kg Skin, rat: LD50 = 625 mg/kg

ENVIRONMENTAL LISTS
CERCLA/SARA - Hazardous Substances and their Reportable Quantities
final RQ = 10 pounds (4.54 kg)
Clean Water Act - Hazardous Substances
[present]

INTERNATIONAL LISTS
Canada - WHMIS: Ingredient Disclosure
1% item 985 (1537)

STATE LISTS
Massachusetts Right To Know List
[present]
NJ Right to Know List (Total)
sn 1177
Pennsylvania Right to Know List
environmental hazard

MERCUROUS NITRATE   **778-26-7**
SEE ALSO:
MERCURY

HEALTH AND SAFETY LISTS
U.S. DOT - Appendix B - Marine Pollutants
DOT regulated severe marine pollutant
U.S. DOT - Appendix A Table 1 - Hazardous Substances
final RQ = 10 pounds (4.54 kg)

ENVIRONMENTAL LISTS
CERCLA/SARA - Hazardous Substances and their Reportable Quantities
final RQ = 10 pounds (4.54 kg)
Clean Water Act - Hazardous Substances
[present]

INTERNATIONAL LISTS
Canada - WHMIS: Ingredient Disclosure
1% item 986 (1609)

STATE LISTS
Massachusetts Right To Know List
[present]
NJ Right to Know List (Total)
sn 1194
Pennsylvania Right to Know List
environmental hazard

MERCUROUS NITRATE   **7782-86-7**

HEALTH AND SAFETY LISTS
U.S. DOT - Substances From 49 CFR 172.101
regulated by DOT (UN1639)
U.S. DOT - Hazard Classes
DOT hazard class = 6.1
U.S. DOT - Appendix B - Marine Pollutants
DOT regulated severe marine pollutant

STATE LISTS
NJ Right to Know List (Total)
sn 1178

MERCUROUS NITRATE   **10415-75-5**

HEALTH AND SAFETY LISTS
U.S. DOT - Appendix B - Marine Pollutants
DOT regulated severe marine pollutant

MERCUROUS OXIDE   **15829-53-5**

HEALTH AND SAFETY LISTS
U.S. DOT - Hazard Classes
Forbidden from transport by the DOT

MERCUROUS SULFATE   **7783-36-0**

HEALTH AND SAFETY LISTS
U.S. DOT - Appendix B - Marine Pollutants
DOT regulated severe marine pollutant

MERCURY   **7439-97-6**
SEE ALSO:
MERCURY

HEALTH AND SAFETY LISTS
U.S. DOT - Appendix B - Marine Pollutants
DOT regulated severe marine pollutant
NIOSH - Selected LD50s and LC50s
Oral, rat: LD50 = 166 mg/kg Skin, rat: LD50 = 1500 mg/kg

INTERNATIONAL LISTS
Canada - WHMIS: Ingredient Disclosure
1% item 987 (543)
Canada - CEPA Schedule II Part II - Toxic Substances (Export)
[present]

STATE LISTS
NJ Right to Know List (Total)
sn 1179

MERCURY 193   **15116-82-2**

STATE LISTS

NJ Right to Know List (Total)
sn 1189

MERCURY 193M   RR-00407-9

HEALTH AND SAFETY LISTS

U.S. DOT - Appendix B - Marine Pollutants
DOT regulated severe marine pollutant

MERCURY 194   15064-97-8

HEALTH AND SAFETY LISTS

U.S. DOT - Appendix A Table 1 - Hazardous Substances
final RQ = 10 pounds (4.54 kg)

ENVIRONMENTAL LISTS

CERCLA/SARA - Hazardous Substances and their Reportable Quantities
final RQ = 10 pounds (4.54 kg)

Clean Water Act - Hazardous Substances
[present]

STATE LISTS

Massachusetts Right To Know List
[present]

Pennsylvania Right to Know List
environmental hazard

MERCURY 195   15756-15-7

SEE ALSO:
MERCURY

HEALTH AND SAFETY LISTS

U.S. DOT - Substances From 49 CFR 172.101
regulated by DOT (UN1627)

U.S. DOT - Hazard Classes
DOT hazard class = 6.1

U.S. DOT - Appendix A Table 1 - Hazardous Substances
final RQ = 10 pounds (4.54 kg)

NIOSH - Selected LD50s and LC50s
Oral, rat: LD50 = 170 mg/kg Skin, rat: LD50 = 2330 mg/kg

ENVIRONMENTAL LISTS

CERCLA/SARA - Hazardous Substances and their Reportable Quantities
final RQ = 10 pounds (4.54 kg)

Clean Water Act - Hazardous Substances
[present]

INTERNATIONAL LISTS

Canada - WHMIS: Ingredient Disclosure
1% item 988 (1215)

STATE LISTS

Massachusetts Right To Know List
[present]

NJ Right to Know List (Total)
sn 1181

Pennsylvania Right to Know List
environmental hazard

MERCURY 195M   RR-00406-8

STATE LISTS

NJ Right to Know List (Total)
sn 1191

MERCURY 197   13981-51-6

SEE ALSO:
MERCURY

HEALTH AND SAFETY LISTS

U.S. DOT - Appendix B - Marine Pollutants
DOT regulated severe marine pollutant

NIOSH - Selected LD50s and LC50s
Oral, rat: LD50 = 205 mg/kg skin, rat: LD50 = 1175 mg/kg

STATE LISTS

NJ Right to Know List (Total)
sn 1182

MERCURY 197M   RR-00405-7

SEE ALSO:
K071-HAZARDOUS WASTES
K106-HAZARDOUS WASTES
MERCURY COMPOUNDS
F039-HAZARDOUS WASTES

HEALTH AND SAFETY LISTS

ACGIH 1995 - Time Weighted Averages
Alkyl compounds as Hg: 0.01 mg/m3 TWA; Aryl compounds, as Hg: 0.1 mg/m3 TWA; Inorganic forms including metallic mercury 0.025 mg/m3 TWA

ACGIH 1995 - Short Term Exposure Limits
alkyl compounds as Hg: 0.03 mg/m3 STEL

ACGIH 1995 - Skin Designations
as Hg: skin - potential for cutaneous absorption

ACGIH 1995 - Carcinogens
inorganic forms including metallic mercury: A4-not classifiable as a human carcinogen

ACGIH 1995 - Biological Exposure Indices
Total inorganic mercury in urine: 35 ug/g creatinine, preshift (B); Total inorganic mercury in blood: 15 ug/L, end of shift at end of workweek (B)

U.S. DOT - Substances From 49 CFR 172.101
regulated by DOT (UN2809)

U.S. DOT - Hazard Classes
DOT hazard class = 8

U.S. DOT - Appendix A Table 1 - Hazardous Substances
final RQ = 1 pound (0.454 kg)

NIOSH 1990 - Pocket Guide - RELs
0.05 mg/m3 TWA

NIOSH 1990 - Pocket Guide - IDLHs
28 mg/m3 IDLH

NIOSH 1990 - Pocket Guide - Target organs
skin, respiratory system, CNS, kidneys, eyes

NIOSH 1990 - Pocket Guide - Skin list
Potential for dermal absorption

OSHA - Vacated PELs - Time Weighted Averages
(organo) alkyl compounds as Hg: 0.01 mg/m3 TWA (Enforcement indefinitely stayed); vapor, as Hg: 0.05 mg/m3 TWA

OSHA - Vacated PELs - Short Term Exposure Limits
(organo) alkyl compounds as Hg: 0.03 mg/m3 STEL

OSHA - Vacated PELs - Ceiling Limits
aryl and inorganic, as Hg: C 0.1 mg/m3

OSHA - Vacated PELs - Skin Designation
Prevent or reduce skin absorption (all forms)

OSHA - Final PELs - Ceiling Limits
Aryl and inorganic compounds, vapor: C 1 mg/10m3; organo (alkyl) mercury compounds: 0.01 mg/m3 TWA; C 0.04 mg/m3

ENVIRONMENTAL LISTS

ATSDR Priority List
Rank (of 275): 003

CERCLA/SARA - Section 313 - Emission Reporting
form R reporting required for 1.0% de minimus concentration

CERCLA/SARA - Hazardous Substances and their Reportable Quantities
final RQ = 1 pound (0.454 kg)
Clean Water Act - Priority Pollutants
[present]
Clean Water Act - Toxic Pollutants
[present] (Listed under 'Mercury and compounds')
Safe Drinking Water Act - MCLs
MCL = 0.002 mg/L
Safe Drinking Water Act - MCLGs
MCLG = 0.002 mg/L
RCRA - D Series - Maximum Concentration of Contaminants
waste number D009; regulatory level = 0.2 mg/L
RCRA - D Series - Chronic Toxicity Reference Levels
chronic toxicity reference level = 0.002 mg/L
RCRA - U Series Wastes
waste number U151
RCRA - Hazardous Constituents-Appendix VIII
waste number U151
RCRA - Basis for Listing - Appendix VII
Included in waste streams: F039, K071, K106
RCRA - Substances Banned From Land Disposal
[present]
RCRA - TSD Facilities Ground Water Monitoring
TM 7470 = 2 ug/L PQL (all species in the groundwater that contain this element are included)
RCRA - Universal Treatment Standards (LDR)
NWW from retort: 0.20 mg/l TCLP; All others: WW:0.15 mg/l; NWW:0.025 mg/lTCLP

INTERNATIONAL LISTS

Australian Exposure Standards - Time Weighted Averages
elemental, as Hg: 0.05 mg/m3 TWA; alkyl compounds, as Hg: 0.01 mg/m3 TWA; aryl and inorganic compounds, as Hg: 0.1 mg/m3 TWA
Australian Exposure Standards - Short Term Exposure Limits
alkyl compounds, as Hg: 0.03 mg/m3 STEL
Australian Exposure Standards - Skin Effects
alkyl compounds, as Hg: skin absorption; aryl and inorganic compounds, as Hg: skin absorption
Australian Exposure Standards - Under Review
arly and inorganic compounds, as Hg: exposure limits under review
Canada - WHMIS: Ingredient Disclosure
0.1% item 990 (1080)
Canada - NPRI (National Pollutant Release Inventory)
[present]
Canada - CEPA Schedule I - Toxic Substances
limited atmospheric releases from chlor-alkali mercury plants
Canada - CEPA Schedule III Part I - Prohibited Substances (Ocean Dumping)
[present]
Canada - Drinking Water Quality - MACs
0.001 mg/L MAC
Canada - Alberta - 8 Hour Occupational Exposure Limit
0.05 mg/m3 TWA
Canada - Alberta - 15 Minute Occupational Exposure Limit
0.15 mg/m3 STEL
Canada - Ontario - OHSA - TWAEVs
all forms except alkyl, as Hg: 0.05 mg/m3 TWAEV; alkyl compounds of, as Hg: 0.01 mg/m3 TWAEV (designated substance regulation) These values apply to work placed in which the designated substance regulation does not apply.
Canada - Ontario - OHSA - STEVs
all forms except alkyl, as Hg: 0.15 mg/m3 STEV; alkyl compounds of, as Hg: 0.03 mg/m3 STEV (designated substance regulation)

Canada - Ontario - OHSA - CEVs
all forms except alkyl: 0.15 mg/m3 CEV; alkyl compounds: 0.03 mg/m3 TWAEV (designated substance regulation)
Canada - Ontario - OHSA - Designated Substances
all forms execpt alkyl, as Hg: 0.05 mg/m3 TWAEV; alkyl compounds of, as Hg: 0.01 mg/m3 TWAEV; See Ontario Reg. 844 for full information.
Canada - Quebec - Time-Weighted Average Exposure Values
0.05 mg/m3 TWAEV
German (DFG) - MAK Values
0.01 ppm MAK; organic compounds-total dust, as Hg: 0.01 mg/m3 MAK
German (DFG) - Peak Limitations
10 x normal MAK (30 min average value); don't exceed during shift
German (DFG) - Skin/Sensitizers
organic compounds: danger of cutaneous absorption; danger of sensitization (skin or respiratory)
Israel - Time Weighted Averages
Alkyl compounds, as Hg: 0.01 mg/m3 TWA; all forms except alkyl vapor, as Hg: 0.05 mg/m3 TWA; Aryl and inorganic compounds, as Hg: 0.1 mg/m3 TWA
Israel - Short Term Exposure Limits
alkyl compounds, as Hg: 0.03 mg/m3 STEL
Israel - Action Levels
alkyl compounds, as Hg: 0.005 mg/m3 AL; all forms except alkyl vapor, as Hg: 0.025 mg/m3 AL; aryl and inorganic compounds, as HG: 0.05 mg/m3 AL
Mexico - Instruction No. 10 - TWAs
0.05 mg/m3 TWA
Mexico - Wastewater - Organic Toxic Pollutants and Heavy Metals
Listed under [Heavy Metals]
Mexico - Drinking Water - Ecological Criteria
0.001 mg/l Substance presents persistence, bioaccumulations or risk of cancer, reduce human exposure to a minimum

STATE LISTS

California - Air Bill 2588 Appendix A-I
[present]
California - Exposure Limits - PELs
vapor, as Hg: 0.05 mg/m3 PEL; alkyls, as Hg: 0.01 mg/m3 PEL
California - Exposure Limits - STELs
alkyls, as Hg: 0.03 mg/m3 STEL
California - Exposure Limits - Ceilings
vapor, as Hg: C 0.1 mg/m3; alkyls, as Hg: C 0.04 mg/m3; aryl and inorganic compounds, as Hg: C 0.1 mg/m3
California - Exposure Limits - Skin Notation
material may be absorbed through the skin, eyes or mucous membrane (includes all compounds)
California - Directors List of Hazardous Substances (8 CCR 339)
[present]
Florida Hazardous Substance List
[present]
Massachusetts Right To Know List
[present]
Minnesota Hazardous Substance List
as Hg: skin
NJ Right to Know List (Total)
sn 1183
NJ Special Hazardous Substances
(corrosive)
Pennsylvania Right to Know List
environmental hazard (any compound of this substance is also an environmental hazard)

MERCURY 199M   RR-00404-6

HEALTH AND SAFETY LISTS
U.S. DOT - Appendix A Table 2 - Radionuclides
final RQ = 100 curies (3.7E 12 Bq)

ENVIRONMENTAL LISTS
> CERCLA/SARA List of Radionuclides (Appendix B) and Their Reportable Quantities
> *final RQ = 100 curies (3.7E 12 Bq)*

MERCURY 203   13982-78-0

HEALTH AND SAFETY LISTS
> U.S. DOT - Appendix A Table 2 - Radionuclides
> *final RQ = 10 curies (3.7E 11 Bq)*

ENVIRONMENTAL LISTS
> CERCLA/SARA List of Radionuclides (Appendix B) and Their Reportable Quantities
> *final RQ = 10 curies (3.7E 11 Bq)*

MERCURY ACETATES   RR-01297-5

HEALTH AND SAFETY LISTS
> U.S. DOT - Appendix A Table 2 - Radionuclides
> *final RQ = 0.1 curies (3.7E 9 Bq)*

ENVIRONMENTAL LISTS
> CERCLA/SARA List of Radionuclides (Appendix B) and Their Reportable Quantities
> *final RQ = 0.1 curies (3.7E 9 Bq)*

MERCURY ACETYLIDE   68833-55-6

HEALTH AND SAFETY LISTS
> U.S. DOT - Appendix A Table 2 - Radionuclides
> *final RQ = 100 curies (3.7E 12 Bq)*

ENVIRONMENTAL LISTS
> CERCLA/SARA List of Radionuclides (Appendix B) and Their Reportable Quantities
> *final RQ = 100 curies (3.7E 12 Bq)*

MERCURY, ALKYL COMPOUNDS   RR-00004-4

HEALTH AND SAFETY LISTS
> U.S. DOT - Appendix A Table 2 - Radionuclides
> *final RQ = 100 curies (3.7E 12 Bq)*

ENVIRONMENTAL LISTS
> CERCLA/SARA List of Radionuclides (Appendix B) and Their Reportable Quantities
> *final RQ = 100 curies (3.7E 12 Bq)*

MERCURY AMMONIUM CHLORIDE   10124-48-8

HEALTH AND SAFETY LISTS
> U.S. DOT - Appendix A Table 2 - Radionuclides
> *final RQ = 1000 curies (3.7E 13 Bq)*

ENVIRONMENTAL LISTS
> CERCLA/SARA List of Radionuclides (Appendix B) and Their Reportable Quantities
> *final RQ = 1000 curies (3.7E 13 Bq)*

MERCURY, ARYL AND INORGANIC COMPOUNDS   RR-00005-5

HEALTH AND SAFETY LISTS
> U.S. DOT - Appendix A Table 2 - Radionuclides
> *final RQ = 1000 curies (3.7E 13 Bq)*

ENVIRONMENTAL LISTS
> CERCLA/SARA List of Radionuclides (Appendix B) and Their Reportable Quantities
> *final RQ = 1000 curies (3.7E 13 Bq)*

MERCURY BASED PESTICIDES, N.O.S.   RR-00361-2

HEALTH AND SAFETY LISTS
> U.S. DOT - Appendix A Table 2 - Radionuclides
> *final RQ = 1000 curies (3.7E 13 Bq)*

ENVIRONMENTAL LISTS
> CERCLA/SARA List of Radionuclides (Appendix B) and Their Reportable Quantities
> *final RQ = 1000 curies (3.7E 13 Bq)*

MERCURY BISULPHATES   RR-01294-2

HEALTH AND SAFETY LISTS
> U.S. DOT - Appendix A Table 2 - Radionuclides
> *final RQ = 10 curies (3.7E 11 Bq)*

ENVIRONMENTAL LISTS
> CERCLA/SARA List of Radionuclides (Appendix B) and Their Reportable Quantities
> *final RQ = 10 curies (3.7E 11 Bq)*

MERCURY BROMIDE   15385-58-7

HEALTH AND SAFETY LISTS
> U.S. DOT - Appendix B - Marine Pollutants
> *DOT regulated severe marine pollutant*

MERCURY BROMIDE (HGBR)   10031-18-2

HEALTH AND SAFETY LISTS
> U.S. DOT - Hazard Classes
> *Forbidden from transport by the DOT*

MERCURY BROMIDES   RR-01342-3

HEALTH AND SAFETY LISTS
> NIOSH 1990 - Pocket Guide - RELs
> *as Hg: 0.01 mg/m3 TWA; 0.03 mg/m3 STEL*
> NIOSH 1990 - Pocket Guide - IDLHs
> *as Hg: 10 mg/m3 IDLH*
> NIOSH 1990 - Pocket Guide - Target organs
> *CNS, kidneys, eyes, skin*
> NIOSH 1990 - Pocket Guide - Skin list
> *as Hg: potential for dermal absorption*

INTERNATIONAL LISTS
> Canada - WHMIS: Ingredient Disclosure
> *1% item 989 (1079)*
> Canada - Alberta - 8 Hour Occupational Exposure Limit
> *0.01 mg/m3 TWA*
> Canada - Alberta - 15 Minute Occupational Exposure Limit
> *as Hg: 0.03 mg/m3 STEL*
> Canada - Alberta - Skin Designation
> *as Hg: can be absorbed through the intact skin*
> Canada - British Columbia - 8 Hour Exposure Limits
> *as Hg: 0.001 ppm TWA; 0.01 mg/m3 TWA*
> Canada - British Columbia - 15 Minute Exposure Limits
> *as Hg: 0.003 ppm STEL; 0.03 mg/m3 STEL*
> Canada - British Columbia - Skin Notations
> *skin - potential for skin absorption*
> Canada - Quebec - Time-Weighted Average Exposure Values
> *as Hg: 0.01 mg/m3 TWAEV*
> Canada - Quebec - Short-term Exposure Values
> *0.03 mg/m3 STEV*
> Canada - Quebec - Skin Designations
> *as Hg: absorbed through the skin*
> United Kingdom - Occupational Exposure Standards - TWAs
> *as Hg: 0.01 mg/m3 TWA*
> United Kingdom - Occupational Exposure Standards - STELs
> *as Hg: 0.03 mg/m3 STEL*
> United Kingdom - Occupational Exposure Standards - Notes
> *can be absorbed through skin*
> Mexico - Instruction No. 10 - TWAs
> *0.01 mg/m3 TWA*

Mexico - Instruction No. 10 - STELs
0.03 mg/m3 STEL

Mexico - Instruction No. 10 - Skin designation
skin - potential for cutaneous absorption

STATE LISTS

Massachusetts Right To Know List
[present]

Minnesota Hazardous Substance List
[present]

MERCURY CHLORIDE   10112-91-1

SEE ALSO:
MERCURY

HEALTH AND SAFETY LISTS

U.S. DOT - Substances From 49 CFR 172.101
regulated by DOT (UN1630)

U.S. DOT - Hazard Classes
DOT hazard class = 6.1

U.S. DOT - Appendix B - Marine Pollutants
DOT regulated severe marine pollutant

NIOSH - Selected LD50s and LC50s
Oral, rat; LD50 = 86 mg/kg Skin, rat; LD50 = 1325 mg/kg

INTERNATIONAL LISTS

Canada - WHMIS: Ingredient Disclosure
1% item 978 (506)

STATE LISTS

NJ Right to Know List (Total)
sn 1184

MERCURY (II) CHROMATE   13444-75-2

INTERNATIONAL LISTS

Canada - WHMIS: Ingredient Disclosure
1% item 989 (1079)

Canada - Alberta - 8 Hour Occupational Exposure Limit
0.1 mg/m3 TWA

Canada - Alberta - 15 Minute Occupational Exposure Limit
0.3 mg/m3 STEL

Mexico - Instruction No. 10 - TWAs
0.05 mg/m3 TWA

MERCURY (I) CHROMATE   13465-34-4

HEALTH AND SAFETY LISTS

U.S. DOT - Substances From 49 CFR 172.101
regulated by DOT (UN2777, UN2778, UN3011, UN3012)

U.S. DOT - Hazard Classes
toxic or toxic, flammable: DOT hazard class = 6.1; flammable, toxic: DOT hazard class = 3

U.S. DOT - Appendix B - Marine Pollutants
DOT regulated severe marine pollutant

STATE LISTS

NJ Right to Know List (Total)
sn 2538; sn 2539; sn 2540; sn 2541

MERCURY COMPOUNDS   RR-00138-7

HEALTH AND SAFETY LISTS

U.S. DOT - Appendix B - Marine Pollutants
DOT regulated severe marine pollutant

MERCURY FULMINATE   628-86-4

SEE ALSO:
MERCURY

HEALTH AND SAFETY LISTS

U.S. DOT - Appendix B - Marine Pollutants
DOT regulated severe marine pollutant

STATE LISTS

NJ Right to Know List (Total)
sn 1186

MERCURY GLUCONATE   63937-14-4

SEE ALSO:
MERCURY

HEALTH AND SAFETY LISTS

U.S. DOT - Appendix B - Marine Pollutants
DOT regulated severe marine pollutant

STATE LISTS

NJ Right to Know List (Total)
sn 1185

MERCURY, INORGANIC COMPOUNDS   RR-00569-6

HEALTH AND SAFETY LISTS

U.S. DOT - Substances From 49 CFR 172.101
regulated by DOT (UN1634)

U.S. DOT - Hazard Classes
DOT hazard class = 6.1

MERCURY IODIDE AQUABASIC AMMONOBASIC   RR-01437-9
(IODIDE OF MILTON'S BASE)

STATE LISTS

NJ Right to Know List (Total)
sn 1180

MERCURY IODIDE (VAN)   37320-91-5

STATE LISTS

Massachusetts Right To Know List
[present]

MERCURYMETHYLCHLORIDE   115-09-3

STATE LISTS

Massachusetts Right To Know List
[present]

MERCURY NITRIDE   12136-15-1

HEALTH AND SAFETY LISTS

U.S. DOT - Substances From 49 CFR 172.101
regulated by DOT (UN2024, UN2025)

U.S. DOT - Hazard Classes
DOT hazard class = 6.1

U.S. DOT - Appendix B - Marine Pollutants
DOT regulated severe marine pollutant

ENVIRONMENTAL LISTS

CERCLA/SARA - Section 313 - Emission Reporting
form R reporting required for 1.0% de minimus concentration

Clean Air Act (1990) - List of Hazardous Air Contaminants
[present] (includes any unique chemical substance that contains Mercury as part of that chemical's infrastructure)

Clean Water Act - Toxic Pollutants
[present] (Listed under 'Mercury and compounds')

RCRA - Hazardous Constituents-Appendix VIII
hazardous constituent - no waste number

INTERNATIONAL LISTS

Canada - NPRI (National Pollutant Release Inventory)
[present]

Canada - CEPA Schedule III Part I - Prohibited Substances (Ocean Dumping)
[present]

Canada - British Columbia - 8 Hour Exposure Limits
as Hg: 0.05 mg/m3 TWA

Canada - British Columbia - 15 Minute Exposure Limits
as Hg: 0.15 mg/m3 STEL

Canada - Quebec - Time-Weighted Average Exposure Values
as Hg: 0.1 mg/m3 TWAEV

Canada - Quebec - Skin Designations
absorbed through the skin

United Kingdom - Occupational Exposure Standards - TWAs
as Hg: 0.05 mg/m3 TWA (does not include mercury alkyls)

United Kingdom - Occupational Exposure Standards - STELs
as Hg: 0.15 mg/m3 STEL (does not include mercury alkyls)

STATE LISTS

California - Air Bill 2588 Appendix A-I
9/89

California - Prop. 65 - Developmental Toxicity
developmental toxicity - initial date 7/1/90

California - Directors List of Hazardous Substances (8 CCR 339)
[present]

NJ Right to Know List (Total)
sn 2542; sn 2543

MERCURY OLEATE   **1191-80-6**
SEE ALSO:
  MERCURY

HEALTH AND SAFETY LISTS

U.S. DOT - Substances From 49 CFR 172.101
regulated by DOT (UN0135)

U.S. DOT - Hazard Classes
wetted with not less than 20% alcohol and/or water: DOT hazard class = 1.1A; dry: Forbidden from transport by the DOT

U.S. DOT - Appendix A Table 1 - Hazardous Substances
final RQ = 10 pounds (4.54 kg)

ENVIRONMENTAL LISTS

CERCLA/SARA - Hazardous Substances and their Reportable Quantities
final RQ = 10 pounds (4.54 kg)

RCRA - P Series Wastes
waste number P065 (Reactive waste; Toxic waste)

RCRA - Hazardous Constituents-Appendix VIII
waste number P065

RCRA - Substances Banned From Land Disposal
[present]

STATE LISTS

Massachusetts Right To Know List
[present]

NJ Right to Know List (Total)
sn 1187

Pennsylvania Right to Know List
environmental hazard

MERCURY OXIDE (VAN)   **12653-71-3**
SEE ALSO:
  MERCURY

HEALTH AND SAFETY LISTS

U.S. DOT - Substances From 49 CFR 172.101
regulated by DOT (UN1637)

U.S. DOT - Hazard Classes
DOT hazard class = 6.1

U.S. DOT - Appendix B - Marine Pollutants
DOT regulated severe marine pollutant

STATE LISTS

NJ Right to Know List (Total)
sn 1188

MERCURY POTASSIUM CYANIDE   **RR-01298-6**

HEALTH AND SAFETY LISTS

NIOSH - Health Standards - Exposure Limits
0.05 mg Hg/m3 TWA (8 hr)

NIOSH - Health Standards - Health Effects and Precautions
Central nervous system and mental effects (Emphasize work practices, sanitation, and enviromental and medical monitoring)

STATE LISTS

NJ Right to Know List (Total)
sn 2874

MERCURY POTASSIUM IODIDE   **7783-33-7**

HEALTH AND SAFETY LISTS

U.S. DOT - Hazard Classes
Forbidden from transport by the DOT

MERCURY SALICYLATE   **5970-32-1**

HEALTH AND SAFETY LISTS

U.S. DOT - Substances From 49 CFR 172.101
regulated by DOT (UN1638)

U.S. DOT - Hazard Classes
DOT hazard class = 6.1

MERCURY SULFATES   **RR-01299-7**

STATE LISTS

Florida Hazardous Substance List
[present]

MERCURY SULFIDE   **1344-48-5**

HEALTH AND SAFETY LISTS

U.S. DOT - Hazard Classes
Forbidden from transport by the DOT

MERCURY THIOCYANATE (VAN)   **53408-91-6**
SEE ALSO:
  MERCURY

HEALTH AND SAFETY LISTS

U.S. DOT - Substances From 49 CFR 172.101
regulated by DOT (UN1640)

U.S. DOT - Hazard Classes
DOT hazard class = 6.1

U.S. DOT - Appendix B - Marine Pollutants
DOT regulated severe marine pollutant

STATE LISTS

NJ Right to Know List (Total)
sn 1190

MERPHALAN   **531-76-0**

HEALTH AND SAFETY LISTS

U.S. DOT - Substances From 49 CFR 172.101
regulated by DOT (UN1641)

U.S. DOT - Hazard Classes
DOT hazard class = 6.1

U.S. DOT - Appendix B - Marine Pollutants
DOT regulated severe marine pollutant

MESITYL OXIDE   **141-79-7**

HEALTH AND SAFETY LISTS

U.S. DOT - Appendix B - Marine Pollutants
DOT regulated severe marine pollutant

MESTRANOL   **72-33-3**
  SEE ALSO:
    MERCURY

  HEALTH AND SAFETY LISTS
    U.S. DOT - Substances From 49 CFR 172.101
      regulated by DOT (UN1643)
    U.S. DOT - Hazard Classes
      DOT hazard class = 6.1
    U.S. DOT - Appendix B - Marine Pollutants
      DOT regulated severe marine pollutant

  STATE LISTS
    NJ Right to Know List (Total)
      sn 1192

METAARSENIC ACID   **10102-53-1**
  SEE ALSO:
    MERCURY

  HEALTH AND SAFETY LISTS
    U.S. DOT - Substances From 49 CFR 172.101
      regulated by DOT (UN1644)
    U.S. DOT - Hazard Classes
      DOT hazard class = 6.1
    U.S. DOT - Appendix B - Marine Pollutants
      DOT regulated severe marine pollutant

  STATE LISTS
    NJ Right to Know List (Total)
      sn 1193

METABISULFITES   **RR-01551-0**

  HEALTH AND SAFETY LISTS
    U.S. DOT - Substances From 49 CFR 172.101
      regulated by DOT (UN1645)
    U.S. DOT - Hazard Classes
      DOT hazard class = 6.1
    U.S. DOT - Appendix B - Marine Pollutants
      DOT regulated severe marine pollutant

METAL ALKYL HALIDES, N.O.S.   **RR-00367-8**
  SEE ALSO:
    MERCURY

  HEALTH AND SAFETY LISTS
    U.S. DOT - Appendix B - Marine Pollutants
      DOT regulated marine pollutant

METAL ALKYL HYDRIDES, N.O.S.   **RR-00368-9**

  HEALTH AND SAFETY LISTS
    U.S. DOT - Substances From 49 CFR 172.101
      regulated by DOT (UN1646)
    U.S. DOT - Hazard Classes
      DOT hazard class = 6.1

METAL ALKYLS, N.O.S.   **RR-00369-0**

  HEALTH AND SAFETY LISTS
    IARC - Group 2B (sufficient animal data)
      [present] (Overall evaluation based only on evidence of carcinogenicity in monograph (9, 1975) or in Supplement 4)
    NIOSH - Selected LD50s and LC50s
      Oral, rat: LD50 = 105 mg/kg
    NTP Chemical Status Reports - Testing Status and NTIS Number
      Chronic studies exist for which technical reports were not prepared
    OSHA - Possible Select Carcinogens
      [present]

STATE LISTS
  California - Air Bill 2588 Appendix A-II
    known or potential carcinogen: 9/89
  California - Prop. 65 - Cancer list
    carcinogen - initial date 4/1/88
  California - Directors List of Hazardous Substances (8 CCR 339)
    [present]
  Florida Hazardous Substance List
    [present]
  Massachusetts Right To Know List
    carcinogen; extraordinarily hazardous
  Minnesota Hazardous Substance List
    carcinogen
  Pennsylvania Right to Know List
    special hazardous substance
  Pennsylvania RTK - Special Hazardous Substances
    [present]

METAL ALKYL SOLUTION, N.O.S.   **RR-00370-3**

  HEALTH AND SAFETY LISTS
    ACGIH 1995 - Time Weighted Averages
      15 ppm TWA; 60 mg/m3 TWA
    ACGIH 1995 - Short Term Exposure Limits
      25 ppm STEL; 100 mg/m3 STEL
    AIHA - Odor Threshold Values
      geometric mean air odor threshold = 0.017 ppm (detectable); 0.050 ppm (recognizable)
    U.S. DOT - Substances From 49 CFR 172.101
      regulated by DOT (UN1229)
    U.S. DOT - Hazard Classes
      DOT hazard class = 3
    NFPA - Flash Points
      flash point = 87 degrees F (31 degrees C)
    NFPA - Hazard Identification Ratings
      health-2; flammability-3; reactivity-1
    NIOSH - Selected LD50s and LC50s
      Inhalation, rat: LC50 = 9 gm/m3 4 hr Oral, rat: LD50 = 1120 mg/kg Skin, rabbit: LD50 = 5150 mg/kg
    NIOSH 1990 - Pocket Guide - RELs
      10 ppm TWA; 40 mg/m3 TWA
    NIOSH 1990 - Pocket Guide - IDLHs
      5000 ppm IDLH
    NIOSH 1990 - Pocket Guide - Target organs
      eyes, skin, CNS, respiratory system
    NIOSH - Health Standards - Exposure Limits
      10 ppm TWA; 40 mg/m3 TWA (Listed under 'Ketones')
    NIOSH - Health Standards - Health Effects and Precautions
      Irritation; liver, kidney, and nervous system effects (Urinalysis required) (Listed under 'Ketones')
    OSHA - Vacated PELs - Time Weighted Averages
      15 ppm TWA; 60 mg/m3 TWA
    OSHA - Vacated PELs - Short Term Exposure Limits
      25 ppm STEL; 100 mg/m3 STEL
    OSHA - Final PELs - Time Weighted Averages
      25 ppm TWA; 100 mg/m3 TWA

  ENVIRONMENTAL LISTS
    EPA - Master Testing List
      [present]
    List of Pesticide Product Inert Ingredients
      [present]
    TSCA - Code of Federal Regulations Citations
      40 CFR 712.30(i); 40 CFR 716.120(a); 40 CFR 799.2500; 40 CFR 799.5000
    TSCA - PAIR - Reporting List
      Reporting Date: October 8, 1984

TSCA - Health and Safety Reporting List
Effective Date: October 4, 1982

TSCA - Substances Subject to Testing Consent Orders
Test for: Health Effects

TSCA - Chemical Test Rules
Testing required by: manufacturers; processors (40 CFR 799.2500)

TSCA - Section 12(b) - Export Notification
export notification required - Section 4

INTERNATIONAL LISTS

Australian Exposure Standards - Time Weighted Averages
15 ppm TWA; 60 mg/m3 TWA

Australian Exposure Standards - Short Term Exposure Limits
25 ppm STEL; 100 mg/m3 STEL

Canada - WHMIS: Ingredient Disclosure
1% item 991 (1315)

Canada - Alberta - 8 Hour Occupational Exposure Limit
15 ppm TWA; 60 mg/m3 TWA

Canada - Alberta - 15 Minute Occupational Exposure Limit
25 ppm STEL; 100 mg/m3 STEL

Canada - British Columbia - 8 Hour Exposure Limits
25 ppm TWA; 100 mg/m3 TWA

Canada - Ontario - OHSA - TWAEVs
15 ppm TWAEV; 60 mg/m3 TWAEV

Canada - Ontario - OHSA - STEVs
25 ppm STEV; 100 mg/m3 STEV

Canada - Quebec - Time-Weighted Average Exposure Values
10 ppm TWAEV; 40 mg/m3 TWAEV

United Kingdom - Occupational Exposure Standards - TWAs
15 ppm TWA; 60 mg/m3 TWA

United Kingdom - Occupational Exposure Standards - STELs
25 ppm STEL; 100 mg/m3 STEL

German (DFG) - MAK Values
25 ppm MAK; 100 mg/m3 MAK

Israel - Time Weighted Averages
15 ppm TWA; 60 mg/m3 TWA

Israel - Short Term Exposure Limits
25 ppm STEL; 100 mg/m3 STEL

Israel - Action Levels
7.5 ppm AL; 30 mg/m3 AL

STATE LISTS

California - Exposure Limits - PELs
15 ppm PEL; 60 mg/m3 PEL

California - Exposure Limits - STELs
25 ppm STEL; 100 mg/m3 STEL

California - Directors List of Hazardous Substances (8 CCR 339)
[present]

Florida Hazardous Substance List
[present]

Massachusetts Right To Know List
[present]

Minnesota Hazardous Substance List
[present]

NJ Right to Know List (Total)
sn 1928

Pennsylvania Right to Know List
[present]

PROPOSED REGULATIONS

Canada - Ontario - Proposed Occupational TWAEVs
10 ppm TWAEV; 40 mg/m3 TWAEV

METALATED ALKYLPHENOL COPOLYMER   RR-00931-4

HEALTH AND SAFETY LISTS

IARC - Group Unspecified
[present] (Listed under 'Steroidal oestrogens')

NTP Seventh Report - Suspect Carcinogens
suspect carcinogen (Listed under 'Estrogens (not conjugated)')

OSHA - Possible Select Carcinogens
[present]

STATE LISTS

California - Air Bill 2588 Appendix A-II
known or potential carcinogen

California - Prop. 65 - Cancer list
carcinogen - initial date 4/1/88

California - Prop. 65 - Developmental Toxicity
developmental toxicity (when mixed with Norethisterone) - initial date 4/1/90

California - Directors List of Hazardous Substances (8 CCR 339)
[present]

Florida Hazardous Substance List
[present]

Massachusetts Right To Know List
carcinogen; extraordinarily hazardous

Minnesota Hazardous Substance List
[present]

NJ Right to Know List (Total)
sn 1196

NJ Special Hazardous Substances
(carcinogen; mutagen)

Pennsylvania Right to Know List
special hazardous substance

Pennsylvania RTK - Special Hazardous Substances
[present]

METALAXYL   57837-19-1

HEALTH AND SAFETY LISTS

U.S. DOT - Appendix B - Marine Pollutants
DOT regulated marine pollutant

METALDEHYDE   108-62-3

HEALTH AND SAFETY LISTS

IARC - Group 3 (not classifiable)
[present]

METALDEHYDE (VAN)   37273-91-9

HEALTH AND SAFETY LISTS

U.S. DOT - Substances From 49 CFR 172.101
regulated by DOT (UN3049)

U.S. DOT - Hazard Classes
DOT hazard class = 4.2

METAL SALT OF A COMPLEX INORGANIC OXYACID   RR-00917-6

HEALTH AND SAFETY LISTS

U.S. DOT - Substances From 49 CFR 172.101
regulated by DOT (UN3050)

U.S. DOT - Hazard Classes
DOT hazard class = 4.2

METAL SALTS OF METHYL NITRAMINE   RR-01438-0

HEALTH AND SAFETY LISTS

U.S. DOT - Substances From 49 CFR 172.101
regulated by DOT (UN2003)

U.S. DOT - Hazard Classes
DOT hazard class = 4.2

INTERNATIONAL LISTS

Canada - Quebec - Time-Weighted Average Exposure Values
10 mg/m3 TWAEV

STATE LISTS

NJ Right to Know List (Total)
sn 2547

METAL WORKING FLUIDS   RR-00814-0

HEALTH AND SAFETY LISTS

U.S. DOT - Substances From 49 CFR 172.101
regulated by DOT (NA9195)
U.S. DOT - Hazard Classes
DOT hazard class = 3

STATE LISTS

NJ Right to Know List (Total)
sn 2548

METAM-SODIUM   137-42-8

ENVIRONMENTAL LISTS

TSCA - Chemicals with Significant New Use Rules
PMN number: P-87-723

METANILIC ACID   121-47-1

PROPOSED REGULATIONS

Safe Drinking Water Act - Priority list
[present]

METANIL YELLOW   587-98-4

HEALTH AND SAFETY LISTS

NFPA - Flash Points
flash point = 97 degrees F (36 degrees C)
NFPA - Hazard Identification Ratings
health-1; flammability-3; reactivity-1

STATE LISTS

Florida Hazardous Substance List
[present]
Massachusetts Right To Know List
[present]
NJ Right to Know List (Total)
sn 1197
NJ Special Hazardous Substances
(flammable - third degree)
Pennsylvania Right to Know List
[present]

METAPHOSPHORIC ACID, TRISODIUM SALT   7785-84-4

HEALTH AND SAFETY LISTS

U.S. DOT - Substances From 49 CFR 172.101
regulated by DOT (UN1332)
U.S. DOT - Hazard Classes
DOT hazard class = 4.1

STATE LISTS

Pennsylvania Right to Know List
[present]

METHACHOLINE CHLORIDE   62-51-1

ENVIRONMENTAL LISTS

TSCA - Chemicals with Significant New Use Rules
PMN numbers: P-89-576; P-89-577

METHACROLEIN DIACETATE   10476-95-6

HEALTH AND SAFETY LISTS

U.S. DOT - Hazard Classes
Forbidden from transport by the DOT

METHACRYLALDEHYDE   78-85-3

INTERNATIONAL LISTS

German (DFG) - Carcinogens
suspected carcinogen

METHACRYLIC ACID   79-41-4

HEALTH AND SAFETY LISTS

U.S. DOT - Appendix B - Marine Pollutants
DOT regulated marine pollutant

METHACRYLIC ACID, BUTYL ESTER, POLYMER   9003-63-8

ENVIRONMENTAL LISTS

CAA - HON Rule - SOCMI Chemicals
compliance by Oct. 24, 1994
TSCA - Code of Federal Regulations Citations
40 CFR 716.120(a)
TSCA - Health and Safety Reporting List
Effective Date: June 1, 1987

INTERNATIONAL LISTS

Canada - WHMIS: Ingredient Disclosure
1% item 992 (105)

METHACRYLIC ACID, 3-(TRIMETHOXYSILYL)   2530-85-0
PROPYL ESTER

ENVIRONMENTAL LISTS

List of Pesticide Product Inert Ingredients
[present]

METHACRYLIC ANHYDRIDE   760-93-0

HEALTH AND SAFETY LISTS

U.S. DOT - Appendix A Table 1 - Hazardous Substances
final RQ = 5000 pounds (2270 kg) (Listed under 'Sodium phosphate, tribasic')
ENVIRONMENTAL LISTS

CERCLA/SARA - Hazardous Substances and their Reportable Quantities
final RQ = 5000 pounds (2270 kg) (Listed under 'Sodium phosphate, tribasic')
Clean Water Act - Hazardous Substances
[present] (Listed under 'Sodium phosphate, tribasic')

STATE LISTS

California - Directors List of Hazardous Substances (8 CCR 339)
[present] (Listed under 'Sodium phosphate, tribasic')
Massachusetts Right To Know List
[present]
Pennsylvania Right to Know List
environmental hazard

METHACRYLIC ESTER   RR-01011-7

HEALTH AND SAFETY LISTS

NIOSH - Selected LD50s and LC50s
Oral, rat: LD50 = 750 mg/kg

METHACRYLOYL CHLORIDE   920-46-7

HEALTH AND SAFETY LISTS

NIOSH - Selected LD50s and LC50s
Oral, rat: LD50 = 440 mg/kg Skin, rabbit: LD50 = 44 mg/kg

ENVIRONMENTAL LISTS

CERCLA/SARA - Section 302 Extremely Hazardous Substances and TPQs
TPQ = 1000 pounds

INTERNATIONAL LISTS

Canada - WHMIS: Ingredient Disclosure
1% item 1072 (619)

STATE LISTS

Florida Hazardous Substance List
effective March 13, 1992

Massachusetts Right To Know List
extraordinarily hazardous

Pennsylvania Right to Know List
environmental hazard

PROPOSED REGULATIONS

CERCLA/SARA - Proposed Hazardous Substance Additions
proposed RQ = 1 pound (.454 kg)

CERCLA/SARA - 1989 Proposed RQ Adjustments
proposed RQ = 1000 pounds (454 kg)

METHACRYLOYLOXYETHYL ISOCYANATE   30674-80-7

HEALTH AND SAFETY LISTS

U.S. DOT - Substances From 49 CFR 172.101
regulated by DOT (UN2396)

U.S. DOT - Hazard Classes
DOT hazard class = 3

NFPA - Flash Points
flash point = 35 degrees F (2 degrees C)

NFPA - Hazard Identification Ratings
health-3; flammability-3; reactivity-2

NIOSH - Selected LD50s and LC50s
Oral, rat: LD50 = 111 mg/kg Skin, rabbit: LD50 = 364 mg/kg

OSHA - List of Highly Hazardous Chemicals
threshhold quantity = 1000 pounds

ENVIRONMENTAL LISTS

TSCA - Code of Federal Regulations Citations
40 CFR 712.30(x); 40 CFR 716.120(d)

TSCA - PAIR - Reporting List
Reporting Date: November 27, 1991

TSCA - Health and Safety Reporting List
Effective Date: September 30, 1991

INTERNATIONAL LISTS

Canada - WHMIS: Ingredient Disclosure
1% item 993 (1087)

STATE LISTS

Florida Hazardous Substance List
[present]

Massachusetts Right To Know List
[present]

Pennsylvania Right to Know List
[present]

METHACYCLINE HYDROCHLORIDE   3963-95-9

HEALTH AND SAFETY LISTS

ACGIH 1995 - Time Weighted Averages
20 ppm TWA; 70 mg/m3 TWA

U.S. DOT - Substances From 49 CFR 172.101
regulated by DOT (UN2531)

U.S. DOT - Hazard Classes
DOT hazard class = 8

NFPA - Flash Points
flash point = 171 degrees F (77 degrees C)

NFPA - Hazard Identification Ratings
health-3; flammability-2; reactivity-2

NIOSH - Selected LD50s and LC50s
Skin, rabbit: LD50 = 500 mg/kg

OSHA - Vacated PELs - Time Weighted Averages
20 ppm TWA; 70 mg/m3 TWA

OSHA - Vacated PELs - Skin Designation
Prevent or reduce skin absorption

ENVIRONMENTAL LISTS

CAA - HON Rule - SOCMI Chemicals
compliance by Oct. 23, 1995

List of Pesticide Product Inert Ingredients
[present]

INTERNATIONAL LISTS

Australian Exposure Standards - Time Weighted Averages
20 ppm TWA; 70 mg/m3 TWA

Canada - WHMIS: Ingredient Disclosure
1% item 994 (106)

Canada - Alberta - 8 Hour Occupational Exposure Limit
20 ppm TWA; 70 mg/m3 TWA

Canada - Alberta - 15 Minute Occupational Exposure Limit
30 ppm STEL; 105 mg/m3 STEL

Canada - Ontario - OHSA - TWAEVs
20 ppm TWAEV; 70 mg/m3 TWAEV

Canada - Quebec - Time-Weighted Average Exposure Values
20 ppm TWAEV; 70 mg/m3 TWAEV

United Kingdom - Occupational Exposure Standards - TWAs
20 ppm TWA; 70 mg/m3 TWA

United Kingdom - Occupational Exposure Standards - STELs
40 ppm STEL; 140 mg/m3 STEL

Israel - Time Weighted Averages
20 ppm TWA; 70 mg/m3 TWA

Israel - Action Levels
10 ppm AL; 35 mg/m3 AL

STATE LISTS

California - Exposure Limits - PELs
20 ppm PEL; 70 mg/m3 PEL

California - Exposure Limits - Skin Notation
material may be absorbed through the skin, eyes or mucous membrane

California - Directors List of Hazardous Substances (8 CCR 339)
[present]

Florida Hazardous Substance List
[present]

Massachusetts Right To Know List
[present]

Minnesota Hazardous Substance List
[present]

NJ Right to Know List (Total)
sn 1199

NJ Special Hazardous Substances
(corrosive; reactive - second degree)

Pennsylvania Right to Know List
[present]

METHALLYL ALCOHOL   513-42-8

ENVIRONMENTAL LISTS

List of Pesticide Product Inert Ingredients
[present]

METHAMIDOPHOS   10265-92-6

ENVIRONMENTAL LISTS

TSCA - Code of Federal Regulations Citations
40 CFR 712.30(d)

TSCA - PAIR - Reporting List
Reporting Date: November 19, 1982

INTERNATIONAL LISTS

Canada - WHMIS: Ingredient Disclosure
1% item 996 (1100)

METHANE   74-82-8

HEALTH AND SAFETY LISTS

NIOSH - Selected LD50s and LC50s
Inhalation, mouse: LC50 = 450 mg/m3 (2 hr)

ENVIRONMENTAL LISTS

CERCLA/SARA - Section 302 Extremely Hazardous Substances and
TPQs
TPQ = 500 pounds

STATE LISTS

Florida Hazardous Substance List
effective March 13, 1992

Massachusetts Right To Know List
extraordinarily hazardous

NJ Right to Know List (Total)
sn 2550

Pennsylvania Right to Know List
environmental hazard

PROPOSED REGULATIONS

CERCLA/SARA - Proposed Hazardous Substance Additions
proposed RQ = 1 pound (.454 kg)

CERCLA/SARA - 1989 Proposed RQ Adjustments
proposed RQ = 100 pounds (45.4 kg)

METHANE, BROMODIFLUORO-   1511-62-2

ENVIRONMENTAL LISTS

TSCA - Chemicals with Significant New Use Rules
PMN number: P-86-650

P-METHANE HYDROPEROXIDE   52061-60-6

HEALTH AND SAFETY LISTS

NIOSH - Selected LD50s and LC50s
Inhalation, rat: LC50 = 60 mg/m3 4 hr

OSHA - List of Highly Hazardous Chemicals
threshhold quantity = 150 pounds

ENVIRONMENTAL LISTS

CERCLA/SARA - Section 302 Extremely Hazardous Substances and
TPQs
TPQ = 100 pounds

STATE LISTS

Florida Hazardous Substance List
effective March 13, 1992

Massachusetts Right To Know List
extraordinarily hazardous

Pennsylvania Right to Know List
environmental hazard

PROPOSED REGULATIONS

CERCLA/SARA - Proposed Hazardous Substance Additions
proposed RQ = 1 pound (.454 kg)

CERCLA/SARA - 1989 Proposed RQ Adjustments
proposed RQ = 100 pounds (45.4 kg)

METHANESULFONIC ACID, AMINOIMINO-   1758-73-2

HEALTH AND SAFETY LISTS

NIOSH - Selected LD50s and LC50s
Inhalation, rat: LC50 = 4 ppm 6 hr Oral, rat: LD50 = 670 mg/kg

OSHA - List of Highly Hazardous Chemicals
threshhold quantity = 100 pounds

ENVIRONMENTAL LISTS

CERCLA/SARA - Section 302 Extremely Hazardous Substances and
TPQs
TPQ = 100 pounds

TSCA - Code of Federal Regulations Citations
40 CFR 712.30(x); 40 CFR 716.120(d)

TSCA - PAIR - Reporting List
Reporting Date: December 27, 1990

TSCA - Health and Safety Reporting List
Effective Date: October 29, 1990

STATE LISTS

Florida Hazardous Substance List
effective March 13, 1992

Massachusetts Right To Know List
extraordinarily hazardous

Pennsylvania Right to Know List
environmental hazard

PROPOSED REGULATIONS

CERCLA/SARA - Proposed Hazardous Substance Additions
proposed RQ = 1 pound (.454 kg)

CERCLA/SARA - 1989 Proposed RQ Adjustments
proposed RQ = 100 pounds (45.4 kg)

METHANE SULFONYL CHLORIDE   124-63-0

STATE LISTS

California - Air Bill 2588 Appendix A-II
6/91

California - Prop. 65 - Developmental Toxicity
developmental toxicity - initial date 1/1/91

METHANESULFONYL FLUORIDE   558-25-8

HEALTH AND SAFETY LISTS

U.S. DOT - Substances From 49 CFR 172.101
regulated by DOT (UN2614)

U.S. DOT - Hazard Classes
DOT hazard class = 3

NFPA - Flash Points
flash point = 92 degrees F (33 degrees C)

NFPA - Hazard Identification Ratings
health-2; flammability-3; reactivity-0

STATE LISTS

Florida Hazardous Substance List
[present]

Massachusetts Right To Know List
[present]

NJ Right to Know List (Total)
sn 1200

NJ Special Hazardous Substances
(flammable - third degree)

Pennsylvania Right to Know List
[present]

2,4-METHANO-2H-INDENO [1,2-B:5,6-B']   81-21-0
BISOXIRENE, OCTAHYDRO-

HEALTH AND SAFETY LISTS

U.S. DOT - Appendix B - Marine Pollutants
DOT regulated marine pollutant

NIOSH - Selected LD50s and LC50s
Inhalation, rat: LD50 = 9 mg/kg 8 hr Oral, rat: LD50 = 7500 ug/kg
Skin, rat: LD50 = 50 mg/kg

ENVIRONMENTAL LISTS

CERCLA/SARA - Section 302 Extremely Hazardous Substances and TPQs
TPQ = 100/10,000 pounds

STATE LISTS

California - Directors List of Hazardous Substances (8 CCR 339)
[present]

Florida Hazardous Substance List
effective March 13, 1992

Massachusetts Right To Know List
extraordinarily hazardous; neurotoxin

NJ Right to Know List (Total)
sn 1201

Pennsylvania Right to Know List
environmental hazard

PROPOSED REGULATIONS

CERCLA/SARA - Proposed Hazardous Substance Additions
proposed RQ = 1 pound (.454 kg)

CERCLA/SARA - 1989 Proposed RQ Adjustments
proposed RQ = 100 pounds (45.4 kg)

4,7-METHANOISOBENZOFURAN-1,3-DIONE, 4,5,6,7,   115-27-5
8,8-HEXACHLORO-3A,4,7,7A-TETRAHYDRO-

HEALTH AND SAFETY LISTS

ACGIH 1995 - Time Weighted Averages
simple asphyxiant

U.S. DOT - Substances From 49 CFR 172.101
regulated by DOT (UN1971, UN1972)

U.S. DOT - Hazard Classes
DOT hazard class = 2.1

NFPA - Flash Points
gas (no flash point given)

NFPA - Hazard Identification Ratings
health-1; flammability-4; reactivity-0

ENVIRONMENTAL LISTS

ATSDR Priority List
Rank (of 275): 056

CAA - Flammable Substances for Accidental Release Prevention
threshold quantity = 10,000 lbs

List of Pesticide Product Inert Ingredients
[present]

INTERNATIONAL LISTS

Australian Exposure Standards - Time Weighted Averages
Asphyxiant at < 18% oxygen by volume; explosion hazard

Canada - British Columbia - 8 Hour Exposure Limits
asphyxiant substance

Canada - Ontario - OHSA - TWAEVs
simple asphyxiant

Canada - Quebec - Time-Weighted Average Exposure Values
simple asphixiant

Israel - Time Weighted Averages
Asphyxiant

Mexico - Instruction No. 10 - TWAs
simple asphyxiant

STATE LISTS

California - Exposure Limits - PELs
asphyxiant (limit depends on level of oxygen)

Massachusetts Right To Know List
[present]

Minnesota Hazardous Substance List
[present]

NJ Right to Know List (Total)
sn 1202

NJ Special Hazardous Substances
(flammable - fourth degree)

Pennsylvania Right to Know List
[present]

1-METHANOL   2216-51-5

ENVIRONMENTAL LISTS

Class 1 Ozone Depletors
ozone depletion potential = 0.74

TSCA - Code of Federal Regulations Citations
40 CFR 721.1296

TSCA - Chemicals with Significant New Use Rules
PMN number: P-89-1093

TSCA - Section 12(b) - Export Notification
P-89-1093; export notification required - Section 5

METHANOL, TRICHLORO-, CARBONATE (2:1)   32315-10-9

STATE LISTS

NJ Right to Know List (Total)
sn 1456

METHAPYRIDINE HYDROCHLORIDE   135-23-9

ENVIRONMENTAL LISTS

EPA - Master Testing List
[present]

METHAPYRILENE   91-80-5

HEALTH AND SAFETY LISTS

U.S. DOT - Substances Which Are Poisonous by Inhalation
liquid hazardous material poisonous by inhalation (UN3246)

METHAZOIC ACID   RR-01439-1

HEALTH AND SAFETY LISTS

NIOSH - Selected LD50s and LC50s
Oral, rat: LD50 = 2 mg/kg

ENVIRONMENTAL LISTS

CERCLA/SARA - Section 302 Extremely Hazardous Substances and TPQs
TPQ = 1000 pounds

STATE LISTS

Florida Hazardous Substance List
effective March 13, 1992

Massachusetts Right To Know List
extraordinarily hazardous

Pennsylvania Right to Know List
environmental hazard

PROPOSED REGULATIONS

CERCLA/SARA - Proposed Hazardous Substance Additions
proposed RQ = 1 pound (.454 kg)

CERCLA/SARA - 1989 Proposed RQ Adjustments
proposed RQ = 100 pounds (45.4 kg)

METHAZOLE [2-(3,4-DICHLOROPHENYL)-4-   20354-26-1
METHYL-1,2,4-OXADIAZOLIDINE-3,5-DIONE]

HEALTH AND SAFETY LISTS

NIOSH - Selected LD50s and LC50s
Oral, rat: LD50 = 210 mg/kg Skin, rabbit: LD50 = 8000 mg/kg

ENVIRONMENTAL LISTS

TSCA - Code of Federal Regulations Citations
40 CFR 712.30(d)

TSCA - PAIR - Reporting List
Reporting Date: November 19, 1982

METHDILAZINE   1982-37-2

ENVIRONMENTAL LISTS

TSCA - Code of Federal Regulations Citations
40 CFR 712.30(d)

TSCA - PAIR - Reporting List
Reporting Date: November 19, 1982

METHENAMINE   100-97-0

HEALTH AND SAFETY LISTS

NIOSH - Selected LD50s and LC50s
Oral, rat: LD50 = 3300 mg/kg

METHIDATHION   950-37-8

ENVIRONMENTAL LISTS

TSCA - Code of Federal Regulations Citations
40 CFR 721.1298

TSCA - Chemicals with Significant New Use Rules
PMN number: P-90-1535

TSCA - Section 12(b) - Export Notification
P-90-1535; export notification required - Section 5

METHIMAZOLE   60-56-0

HEALTH AND SAFETY LISTS

NTP Chemical Status Reports - Testing Status and NTIS Number
Chronic studies exist for which technical reports were not prepared; Prechronic studies completed: in review for further evaluation; prechronic studies for which toxicity technical reports were not prepared

METHIOCARB   2032-65-7
SEE ALSO:
F039-HAZARDOUS WASTES

HEALTH AND SAFETY LISTS

U.S. DOT - Appendix A Table 1 - Hazardous Substances
final RQ = 5000 pounds (2270 kg)

NIOSH - Selected LD50s and LC50s
Oral, mouse: LD50 = 182 mg/kg

ENVIRONMENTAL LISTS

CERCLA/SARA - Hazardous Substances and their Reportable Quantities
final RQ = 5000 pounds (2270 kg)

RCRA - U Series Wastes
waste number U155

RCRA - Hazardous Constituents-Appendix VIII
waste number U155

RCRA - Basis for Listing - Appendix VII
Included in waste stream: F039

RCRA - Substances Banned From Land Disposal
[present]

RCRA - TSD Facilities Ground Water Monitoring
TM 8270 = 10 ug/L PQL

RCRA - Universal Treatment Standards (LDR)
WW: 0.081 mg/l; NWW: 1.5 mg/kg

STATE LISTS

Florida Hazardous Substance List
[present]

Massachusetts Right To Know List
carcinogen; extraordinarily hazardous

NJ Special Hazardous Substances
(carcinogen)

Pennsylvania Right to Know List
environmental hazard

METHIONINE   63-68-3

HEALTH AND SAFETY LISTS

U.S. DOT - Hazard Classes
Forbidden from transport by the DOT

METHOMYL   16752-77-5

ENVIRONMENTAL LISTS

CERCLA/SARA - Section 313 - Emission Reporting
form R reporting required

METHOTREXATE   59-05-2

HEALTH AND SAFETY LISTS

NTP Chemical Status Reports - Testing Status and NTIS Number
Prechronic studies for which toxicity technical reports were not prepared

METHOTREXATE SODIUM   15475-56-6

HEALTH AND SAFETY LISTS

U.S. DOT - Substances From 49 CFR 172.101
regulated by DOT (UN1328)

U.S. DOT - Hazard Classes
DOT hazard class = 4.1

ENVIRONMENTAL LISTS

CAA - HON Rule - SOCMI Chemicals
compliance by Oct. 24, 1994

List of Pesticide Product Inert Ingredients
[present]

INTERNATIONAL LISTS

Canada - WHMIS: Ingredient Disclosure
0.1% item 997 (1104)

STATE LISTS

NJ Right to Know List (Total)
sn 0996

METHOXONE-SODIUM SALT ((4-CHLORO-2-   3653-48-3
METHYLPHENOXY) ACETATE SODIUM SALT)

HEALTH AND SAFETY LISTS

U.S. DOT - Appendix B - Marine Pollutants
DOT regulated marine pollutant

NIOSH - Selected LD50s and LC50s
*Inhalation, rat: LC50 = 50 mg/m3 4 hr Oral, rat: LD50 = 20 mg/kg
Skin, rat: LD50 = 25 mg/kg*

ENVIRONMENTAL LISTS

CERCLA/SARA - Section 302 Extremely Hazardous Substances and TPQs
TPQ = 500/10,000 pounds

STATE LISTS

California - Directors List of Hazardous Substances (8 CCR 339)
[present]

Florida Hazardous Substance List
effective March 13, 1992

Massachusetts Right To Know List
extraordinarily hazardous; neurotoxin

NJ Right to Know List (Total)
sn 1206

Pennsylvania Right to Know List
environmental hazard

PROPOSED REGULATIONS

CERCLA/SARA - Proposed Hazardous Substance Additions
proposed RQ = 1 pound (.454 kg)

CERCLA/SARA - 1989 Proposed RQ Adjustments
proposed RQ = 100 pounds (45.4 kg)

2-METHOXY-1,4-BENZENEDIAMINE 5307-02-8

HEALTH AND SAFETY LISTS

NIOSH - Selected LD50s and LC50s
Oral, rat: LD50 = 2250 mg/kg

STATE LISTS

California - Air Bill 2588 Appendix A-II
9/90

California - Prop. 65 - Developmental Toxicity
developmental toxicity - initial date 10/1/90

6-METHOXY-2-BENZOTHIAZOLAMINE 1747-60-0

HEALTH AND SAFETY LISTS

U.S. DOT - Appendix B - Marine Pollutants
DOT regulated marine pollutant

U.S. DOT - Appendix A Table 1 - Hazardous Substances
final RQ = 10 pounds (4.54 kg)

NIOSH - Selected LD50s and LC50s
Oral, rat: LD50 = 15 mg/kg Skin, rat: LD50 = 350 mg/kg

ENVIRONMENTAL LISTS

CERCLA/SARA - Section 302 Extremely Hazardous Substances and TPQs
TPQ = 500/10,000 pounds

CERCLA/SARA - Section 313 - Emission Reporting
form R reporting required

CERCLA/SARA - Hazardous Substances and their Reportable Quantities
final RQ = 10 pounds (4.54 kg)

Clean Water Act - Hazardous Substances
[present]

STATE LISTS

California - Directors List of Hazardous Substances (8 CCR 339)
[present]

Florida Hazardous Substance List
effective March 13, 1992

Massachusetts Right To Know List
extraordinarily hazardous; neurotoxin

NJ Right to Know List (Total)
sn 1165

Pennsylvania Right to Know List
environmental hazard

4-METHOXYBENZYL ALCOHOL 105-13-5

ENVIRONMENTAL LISTS

CAA - HON Rule - SOCMI Chemicals
compliance by Oct. 24, 1994

3-METHOXYBUTANOL 2517-43-3

HEALTH AND SAFETY LISTS

ACGIH 1995 - Time Weighted Averages
2.5 mg/m3 TWA

U.S. DOT - Appendix B - Marine Pollutants
DOT regulated marine pollutant

U.S. DOT - Appendix A Table 1 - Hazardous Substances
final RQ = 100 pounds (45.4 kg)

NIOSH - Selected LD50s and LC50s
Inhalation, rat: LC50 = 77 ppm 8 hr Oral, rat: LD50 = 17 mg/kg Skin, rabbit: LD50 = 5880 mg/kg

OSHA - Vacated PELs - Time Weighted Averages
2.5 mg/m3 TWA

ENVIRONMENTAL LISTS

CERCLA/SARA - Section 302 Extremely Hazardous Substances and TPQs
TPQ = 500/10,000 pounds

CERCLA/SARA - Hazardous Substances and their Reportable Quantities
final RQ = 100 pounds (45.4 kg)

RCRA - P Series Wastes
waste number P066

RCRA - Hazardous Constituents-Appendix VIII
waste number P066

RCRA - Substances Banned From Land Disposal
[present]

INTERNATIONAL LISTS

Australian Exposure Standards - Time Weighted Averages
2.5 mg/m3 TWA

Canada - Alberta - 8 Hour Occupational Exposure Limit
2.5 mg/m3 TWA

Canada - Alberta - 15 Minute Occupational Exposure Limit
5 mg/m3 STEL

Canada - Alberta - Skin Designation
can be absorbed through the intact skin

Canada - British Columbia - 8 Hour Exposure Limits
2.5 mg/m3 TWA

Canada - British Columbia - Skin Notations
skin - potential for skin absorption

Canada - Ontario - OHSA - TWAEVs
2.5 mg/m3 TWAEV

Canada - Quebec - Time-Weighted Average Exposure Values
2.5 mg/m3 TWAEV

United Kingdom - Occupational Exposure Standards - TWAs
2.5 mg/m3 TWA

United Kingdom - Occupational Exposure Standards - Notes
can be absorbed through skin

Israel - Time Weighted Averages
2.5 mg/m3 TWA

Israel - Action Levels
1.25 mg/m3 AL

Mexico - Instruction No. 10 - TWAs
2.5 mg/m3 TWA

Mexico - Instruction No. 10 - Skin designation
skin - potential for cutaneous absorption

STATE LISTS

California - Exposure Limits - PELs
2.5 mg/m3 PEL

California - Exposure Limits - Skin Notation
material may be absorbed through the skin, eyes or mucous membrane

California - Directors List of Hazardous Substances (8 CCR 339)
[present]

Florida Hazardous Substance List
[present]

Massachusetts Right To Know List
extraordinarily hazardous; neurotoxin

Minnesota Hazardous Substance List
skin

NJ Right to Know List (Total)
sn 1208

Pennsylvania Right to Know List
environmental hazard

PROPOSED REGULATIONS

Safe Drinking Water Act - Priority list
[present]

3-METHOXYBUTYRALDEHYDE 5281-76-5

HEALTH AND SAFETY LISTS

IARC - Group 3 (not classifiable)
[present]

NIOSH - Selected LD50s and LC50s
Oral, rat: LD50 = 135 mg/kg
NTP Chemical Status Reports - Testing Status and NTIS Number
Chronic studies exist for which technical reports were not prepared

STATE LISTS

California - Air Bill 2588 Appendix A-II
9/89

California - Prop. 65 - Developmental Toxicity
developmental toxicity - initial date 1/1/89

Massachusetts Right To Know List
teratogen

NJ Right to Know List (Total)
sn 1209

NJ Special Hazardous Substances
(mutagen; teratogen)

METHOXYCHLOR   72-43-5

STATE LISTS

California - Air Bill 2588 Appendix A-II
9/90

California - Prop. 65 - Developmental Toxicity
developmental toxicity - initial date 4/1/90

3-METHOXY-2,4-DIHYDROXYPENTANE   921-20-0

ENVIRONMENTAL LISTS

CERCLA/SARA - Section 313 - Emission Reporting
form R reporting required

2-METHOXYETHANOL   109-86-4
SEE ALSO:
2-METHOXY-1,4-BENZENEDIAMINE

ENVIRONMENTAL LISTS

TSCA - Code of Federal Regulations Citations
40 CFR 712.30(f); 40 CFR 716.120

TSCA - PAIR - Reporting List
Reporting Date: August 17, 1983

INTERNATIONAL LISTS

Canada - WHMIS: Ingredient Disclosure
1% item 998 (1105)

2-METHOXYETHYL ACETATE   110-49-6

HEALTH AND SAFETY LISTS

NTP Chemical Status Reports - Testing Status and NTIS Number
Prechronic studies for which toxicity technical reports were not prepared

2-METHOXYETHYL ACRYLATE   3121-61-7

HEALTH AND SAFETY LISTS

NIOSH - Selected LD50s and LC50s
Oral, rat: LD50 = 1200 mg/kg

METHOXYETHYL HYDROGEN MALEATE   10232-93-6

HEALTH AND SAFETY LISTS

NFPA - Flash Points
flash point = 165 degrees F (74 degrees C)
NFPA - Hazard Identification Ratings
health-1; flammability-2; reactivity-0

METHOXYETHYLMERCURIC ACETATE   151-38-2

HEALTH AND SAFETY LISTS

NFPA - Flash Points
flash point = 140 degrees F (60 degrees C)
NFPA - Hazard Identification Ratings
health-0; flammability-2; reactivity-0

METHOXY ETHYL PHTHALATE   16501-01-2
SEE ALSO:
TRIAMYLBENZENE
F039-HAZARDOUS WASTES

HEALTH AND SAFETY LISTS

ACGIH 1995 - Time Weighted Averages
10 mg/m3 TWA

U.S. DOT - Appendix A Table 1 - Hazardous Substances
final RQ = 1 pound (0.454 kg)

IARC - Group 3 (not classifiable)
[present]

NIOSH - Selected LD50s and LC50s
Oral, rat: LD50 = 5000 mg/kg Skin, rat: LD50 = 7600 mg/kg

NIOSH 1990 - Pocket Guide - Carcinogens
occupational carcinogen

NIOSH 1990 - Pocket Guide - Target organs
none known

NTP Chemical Status Reports - Testing Status and NTIS Number
Technical reports printed (PB278271/AS)

NTP Chemical Status Reports - Evidence of Carcinogenicity
male rat-negative; female rat-negative; male mice-negative; female mice-negative

OSHA - Vacated PELs - Time Weighted Averages
total dust: 10 mg/m3 TWA

OSHA - Final PELs - Time Weighted Averages
total dust: 15 mg/m3 TWA

ENVIRONMENTAL LISTS

ATSDR Priority List
Rank (of 275): 055

CERCLA/SARA - Section 313 - Emission Reporting
form R reporting required for 1.0% de minimus concentration

CERCLA/SARA - Hazardous Substances and their Reportable Quantities
final RQ = 1 pound (0.454 kg)

Clean Air Act (1990) - List of Hazardous Air Contaminants
[present]

Clean Water Act - Hazardous Substances
[present]

Safe Drinking Water Act - MCLs
MCL = 0.04 mg/L

Safe Drinking Water Act - MCLGs
MCLG = 0.04 mg/L

RCRA - D Series - Maximum Concentration of Contaminants
waste number D014; regulatory level = 10.0 mg/L

RCRA - D Series - Chronic Toxicity Reference Levels
chronic toxicity reference level = 0.1 mg/L

RCRA - U Series Wastes
waste number U247

RCRA - Hazardous Constituents-Appendix VIII
waste number U247

RCRA - Basis for Listing - Appendix VII
Included in waste stream: F039

RCRA - Substances Banned From Land Disposal
[present]

RCRA - TSD Facilities Ground Water Monitoring
TM 8080 = 2 ug/L PQL; TM 8270 = 10 ug/L PQL

RCRA - Universal Treatment Standards (LDR)
WW: 0.25 mg/l; NWW: 0.18 mg/kg

INTERNATIONAL LISTS

Australian Exposure Standards - Time Weighted Averages
10 mg/m3 TWA

Canada - Drinking Water Quality - MACs
0.9 mg/L MAC

Canada - Alberta - 8 Hour Occupational Exposure Limit
10 mg/m3 TWA

Canada - Alberta - 15 Minute Occupational Exposure Limit
20 mg/m3 STEL

Canada - British Columbia - 8 Hour Exposure Limits
10 mg/m3 TWA

Canada - Ontario - OHSA - TWAEVs
10 mg/m3 TWAEV

Canada - Quebec - Time-Weighted Average Exposure Values
10 mg/m3 TWAEV

United Kingdom - Occupational Exposure Standards - TWAs
10 mg/m3 TWA

German (DFG) - MAK Values
total dust: 15 mg/m3 MAK

German (DFG) - Peak Limitations
10 x normal MAK (30 min average value); don't exceed during shift

German (DFG) - Pregnancy
classification not yet possible

Israel - Time Weighted Averages
10 mg/m3 TWA

Israel - Action Levels
5 mg/m3 AL

Mexico - Instruction No. 10 - TWAs
10 mg/m3 TWA

STATE LISTS

California - Air Bill 2588 Appendix A-I
6/91

California - Exposure Limits - PELs
10 mg/m3 PEL

California - Directors List of Hazardous Substances (8 CCR 339)
[present]

Florida Hazardous Substance List
[present]

Massachusetts Right To Know List
[present]

Minnesota Hazardous Substance List
[present]

NJ Right to Know List (Total)
sn 1210

Pennsylvania Right to Know List
environmental hazard

METHOXYFLURANE   76-38-0

ENVIRONMENTAL LISTS

List of Pesticide Product Inert Ingredients
[present]

METHOXYMETHANOL   4461-52-3
SEE ALSO:
GLYCOL ETHERS

HEALTH AND SAFETY LISTS

ACGIH 1995 - Time Weighted Averages
5 ppm TWA; 16 mg/m3 TWA

ACGIH 1995 - Skin Designations
skin - potential for cutaneous absorption

AIHA - Odor Threshold Values
geometric mean air odor threshold = 2.4 ppm (detectable); 4.4 ppm (recognizable)

U.S. DOT - Substances From 49 CFR 172.101
regulated by DOT (UN1171)

U.S. DOT - Hazard Classes
DOT hazard class = 3

NFPA - Flash Points
flash point = 102 degrees F (39 degrees C)

NFPA - Hazard Identification Ratings
health-2; flammability-2; reactivity-0

NIOSH - Selected LD50s and LC50s
Inhalation, rat: LC50 = 1500 ppm 7 hr Oral, rat: LD50 = 2460 mg/kg Skin, rabbit: LD50 = 1280 mg/kg

NIOSH 1990 - Pocket Guide - RELs
Reduce exposure to lowest feasible concentration

NIOSH 1990 - Pocket Guide - IDLHs
2000 ppm IDLH (not applicable because of the NIOSH REL)

NIOSH 1990 - Pocket Guide - Target organs
CNS, blood, skin, eyes, kidneys

NIOSH - Health Standards - Exposure Limits
reduce exposure to lowest feasible concentration (Listed under 'Glycol ethers')

NIOSH - Health Standards - Health Effects and Precautions
Male and female reproductive effects; teratogenicity (Prevent skin contact) (Listed under 'Glycol ethers')

NTP Chemical Status Reports - Testing Status and NTIS Number
Technical reports printed (PB94-118106)

OSHA - Vacated PELs - Time Weighted Averages
25 ppm TWA; 80 mg/m3 TWA

OSHA - Vacated PELs - Skin Designation
Prevent or reduce skin absorption

OSHA - Final PELs - Time Weighted Averages
25 ppm TWA; 80 mg/m3 TWA

OSHA - Final PELs - Skin Notations
prevent or reduce skin absorption

ENVIRONMENTAL LISTS

CERCLA/SARA - Section 313 - Emission Reporting
form R reporting required for 1.0% de minimus concentration

CAA - HON Rule - SOCMI Chemicals
compliance by Oct. 24, 1994

INTERNATIONAL LISTS

Australian Exposure Standards - Time Weighted Averages
5 ppm TWA; 16 mg/m3 TWA

Australian Exposure Standards - Skin Effects
skin absorption

Canada - WHMIS: Ingredient Disclosure
0.1% item 999 (1106)

Canada - NPRI (National Pollutant Release Inventory)
[present]

Canada - Alberta - 8 Hour Occupational Exposure Limit
5 ppm TWA; 16 mg/m3 TWA

Canada - Alberta - 15 Minute Occupational Exposure Limit
10 ppm STEL; 32 mg/m3 STEL

Canada - Alberta - Skin Designation
can be absorbed through the intact skin

Canada - British Columbia - 8 Hour Exposure Limits
25 ppm TWA; 80 mg/m3 TWA

Canada - British Columbia - 15 Minute Exposure Limits
35 ppm STEL; 120 mg/m3 STEL

Canada - British Columbia - Skin Notations
skin - potential for skin absorption

Canada - Ontario - OHSA - TWAEVs
5 ppm TWAEV; 16 mg/m3 TWAEV

Canada - Ontario - OHSA - Skin Notations
absorption through skin, eyes, or mucous membranes

Canada - Quebec - Time-Weighted Average Exposure Values
5 ppm TWAEV; 16 mg/m3 TWAEV

Canada - Quebec - Skin Designations
absorbed through the skin

United Kingdom - Maximum Exposure Limits - TWAs
5 ppm TWA; 16 mg/m3 TWA

United Kingdom - Maximum Exposure Limits - Notes
can be absorbed through skin

German (DFG) - MAK Values
5 ppm MAK; 15 mg/m3 MAK

German (DFG) - Peak Limitations
2 x normal MAK (30 min. average value); don't exceed 4 times during shift

German (DFG) - Skin/Sensitizers
danger of cutaneous absorption

German (DFG) - Pregnancy
risk to embryo/fetus probable

Israel - Time Weighted Averages
5 ppm TWA; 16 mg/m3 TWA

Israel - Action Levels
2.5 ppm AL; 8 mg/m3 AL

Mexico - Instruction No. 10 - TWAs
25 ppm TWA; 80 mg/m3 TWA

Mexico - Instruction No. 10 - STELs
35 ppm STEL; 120 mg/m3 STEL

Mexico - Instruction No. 10 - Skin designation
skin - potential for cutaneous absorption

STATE LISTS

California - Air Bill 2588 Appendix A-I
9/89

California - Prop. 65 - Developmental Toxicity
developmental toxicity initial date 1/1/89

California - Prop. 65 - Reproductive - Male
male reproductive toxicity - initial date 1/1/89

California - Exposure Limits - PELs
5 ppm PEL; 16 mg/m3 PEL

California - Exposure Limits - Skin Notation
material may be absorbed through the skin, eyes or mucous membrane

California - Directors List of Hazardous Substances (8 CCR 339)
[present]

Florida Hazardous Substance List
[present]

Massachusetts Right To Know List
teratogen

Minnesota Hazardous Substance List
skin

NJ Right to Know List (Total)
sn 1211

Pennsylvania Right to Know List
environmental hazard

PROPOSED REGULATIONS

Proposed Amendments to OSHA Regulated Chemicals
proposed PEL: 0.1 ppm TWA; 0.5 ppm Excursion Limit; 0.05 ppm TWA Action Level

METHOXYMETHYL ISOCYANATE   6427-21-0
SEE ALSO:
  GLYCOL ETHERS

HEALTH AND SAFETY LISTS

ACGIH 1995 - Time Weighted Averages
5 ppm TWA; 24 mg/m3 TWA

ACGIH 1995 - Skin Designations
skin - potential for cutaneous absorption

AIHA - Odor Threshold Values
geometric mean air odor threshold = 0.33 ppm (detectable); 0.64 ppm (recognizable)

U.S. DOT - Substances From 49 CFR 172.101
regulated by DOT (UN1189)

U.S. DOT - Hazard Classes
DOT hazard class = 3

NFPA - Flash Points
*flash point = 120 degrees F (49 degrees C); *See list description*

NFPA - Hazard Identification Ratings
health-1; flammability-2

NIOSH - Selected LD50s and LC50s
Oral, rat: LD50 = 3390 mg/kg Skin, rabbit: LD50 = 5250 mg/kg

NIOSH 1990 - Pocket Guide - RELs
Reduce exposure to lowest feasible concentration

NIOSH 1990 - Pocket Guide - IDLHs
4000 ppm IDLH (not applicable becasue of the NIOSH REL)

NIOSH 1990 - Pocket Guide - Target organs
kidneys, brain, CNS, PNS

OSHA - Vacated PELs - Time Weighted Averages
25 ppm TWA; 120 mg/m3 TWA

OSHA - Vacated PELs - Skin Designation
Prevent or reduce skin absorption

OSHA - Final PELs - Time Weighted Averages
25 ppm TWA; 120 mg/m3 TWA

OSHA - Final PELs - Skin Notations
prevent or reduce skin absorption

INTERNATIONAL LISTS

Australian Exposure Standards - Time Weighted Averages
5 ppm TWA; 24 mg/m3 TWA

Australian Exposure Standards - Skin Effects
skin absorption

Canada - WHMIS: Ingredient Disclosure
0.1% item 1000 (24)

Canada - NPRI (National Pollutant Release Inventory)
[present]

Canada - Alberta - 8 Hour Occupational Exposure Limit
5 ppm TWA; 24 mg/m3 TWA

Canada - Alberta - 15 Minute Occupational Exposure Limit
10 ppm STEL; 48 mg/m3 STEL

Canada - Alberta - Skin Designation
can be absorbed through the intact skin

Canada - British Columbia - 8 Hour Exposure Limits
25 ppm TWA; 120 mg/m3 TWA

Canada - British Columbia - 15 Minute Exposure Limits
35 ppm STEL; 170 mg/m3 STEL

Canada - British Columbia - Skin Notations
skin - potential for skin absorption

Canada - Ontario - OHSA - TWAEVs
5 ppm TWAEV; 24 mg/m3 TWAEV

Canada - Ontario - OHSA - Skin Notations
absorption through skin, eyes, or mucous membranes

Canada - Quebec - Time-Weighted Average Exposure Values
5 ppm TWAEV; 24 mg/m3 TWAEV

Canada - Quebec - Skin Designations
absorbed through the skin

United Kingdom - Maximum Exposure Limits - TWAs
5 ppm TWA; 24 mg/m3 TWA

United Kingdom - Maximum Exposure Limits - Notes
can be absorbed through skin

German (DFG) - MAK Values
5 ppm MAK; 25 mg/m3 MAK

German (DFG) - Peak Limitations
2 x normal MAK (30 min. average value); don't exceed 4 times during shift

German (DFG) - Skin/Sensitizers
danger of cutaneous absorption

German (DFG) - Pregnancy
risk to embryo/fetus probable

Israel - Time Weighted Averages
5 ppm TWA; 24 mg/m3 TWA

Israel - Action Levels
2.5 ppm AL;   12 mg/m3 AL

Mexico - Instruction No. 10 - TWAs
20 ppm TWA;   120 mg/m3 TWA

Mexico - Instruction No. 10 - STELs
35 ppm STEL;   170 mg/m3 STEL

Mexico - Instruction No. 10 - Skin designation
skin - potential for cutaneous absorption

STATE LISTS

California - Air Bill 2588 Appendix A-I
9/90

California - Prop. 65 - Developmental Toxicity
developmental toxicity - initial date 1/1/93

California - Prop. 65 - Reproductive - Male
male reproductive toxicity - initial date 1/1/93

California - Exposure Limits - PELs
5 ppm PEL;   24 mg/m3 PEL

California - Directors List of Hazardous Substances (8 CCR 339)
[present]

Florida Hazardous Substance List
[present]

Massachusetts Right To Know List
[present]

Minnesota Hazardous Substance List
skin

NJ Right to Know List (Total)
sn 1212

Pennsylvania Right to Know List
[present]

PROPOSED REGULATIONS

Proposed Amendments to OSHA Regulated Chemicals
proposed PEL: 0.1 ppm TWA;   0.5 ppm Excursion Limit;   0.05 ppm TWA Action Level

4-METHOXY-4-METHYLPENTAN-2-ONE   107-70-0

HEALTH AND SAFETY LISTS

NFPA - Flash Points
flash point = 180 degrees F (82 degrees C)

NFPA - Hazard Identification Ratings
health-0;   flammability-2;   reactivity-0

NIOSH - Selected LD50s and LC50s
Inhalation, rat: LC50 = 500 ppm 4 hr Oral, rat: LD50 = 810 mg/kg Skin, rabbit: LD50 = 250 mg/kg

ENVIRONMENTAL LISTS

TSCA - PAIR - Reporting List
Effective Date: January 26, 1994;   Reporting Date: March 28, 1994

TSCA - Health and Safety Reporting List
Effective Date: January 26, 1994;   Sunset Date: January 26, 2004

PROPOSED REGULATIONS

TSCA - ITC 31st Report Priority Testing List
recommended for testing

4-METHOXYPHENOL   150-76-5

HEALTH AND SAFETY LISTS

NIOSH - Selected LD50s and LC50s
Oral, rat: LD50 = 3340 mg/kg Skin, rabbit: LD50 = 1940 mg/kg

2-METHOXY-1-PROPANOL   1589-47-5

SEE ALSO:
MERCURY

HEALTH AND SAFETY LISTS

NIOSH - Selected LD50s and LC50s
Oral, rat: LD50 = 25 mg/kg

ENVIRONMENTAL LISTS

CERCLA/SARA - Section 302 Extremely Hazardous Substances and TPQs
TPQ = 500/10,000 pounds

STATE LISTS

Florida Hazardous Substance List
effective March 13, 1992

Massachusetts Right To Know List
extraordinarily hazardous

NJ Right to Know List (Total)
sn 2554

Pennsylvania Right to Know List
environmental hazard

PROPOSED REGULATIONS

CERCLA/SARA - Proposed Hazardous Substance Additions
proposed RQ = 1 pound (.454 kg)

CERCLA/SARA - 1989 Proposed RQ Adjustments
proposed RQ = 100 pounds (45.4 kg)

3-METHOXYPROPANOL   1589-49-7

HEALTH AND SAFETY LISTS

NFPA - Flash Points
flash point = 275 degrees F (135 degrees C)

NFPA - Hazard Identification Ratings
health-0;   flammability-1;   reactivity-0

2-[2-(2-METHOXYPROPOXY)PROPOXY]-1-   10213-77-1
PROPANOL

HEALTH AND SAFETY LISTS

NIOSH - Selected LD50s and LC50s
Inhalation, rat: LC50 = 123 gm/m3 4 hr Oral, mammal: LD50 = 3600 mg/kg

NIOSH - Health Standards - Exposure Limits
C (1 hr) 2 ppm (Listed under 'Waste anasthetic gases and vapors')

NIOSH - Health Standards - Health Effects and Precautions
Reproductive system effects and audio-visual performance decrements (Advise workers of potential effects) (Listed under 'Waste anasthetic gases')

STATE LISTS

California - Exposure Limits - PELs
2 ppm PEL;   13 mg/m3 PEL

2-METHOXYPROPYL-1-ACETATE   70657-70-4

ENVIRONMENTAL LISTS

EPA - Master Testing List
[present]

3-METHOXYPROPYLAMINE   5332-73-0

HEALTH AND SAFETY LISTS

U.S. DOT - Substances From 49 CFR 172.101
regulated by DOT (UN2605)

U.S. DOT - Hazard Classes
DOT hazard class = 3

U.S. DOT - Substances Which Are Poisonous by Inhalation
liquid hazardous material poisonous by inhalation (UN2605)

INTERNATIONAL LISTS

Canada - WHMIS: Ingredient Disclosure
0.1% item 1001 (1043)

STATE LISTS

NJ Right to Know List (Total)
sn 1214

METHOXYPROPYLAMINE DODECYLBENZENESULFONATE   RR-01159-6

HEALTH AND SAFETY LISTS

U.S. DOT - Substances From 49 CFR 172.101
regulated by DOT (UN2293)

U.S. DOT - Hazard Classes
DOT hazard class = 3

NIOSH - Selected LD50s and LC50s
Oral, mouse: LD50 = 2050 mg/kg

ENVIRONMENTAL LISTS

List of Pesticide Product Inert Ingredients
[present]

STATE LISTS

NJ Right to Know List (Total)
sn 1215

1-METHOXY-4-PROPYLBENZENE   104-45-0

HEALTH AND SAFETY LISTS

ACGIH 1995 - Time Weighted Averages
5 mg/m3 TWA

NFPA - Flash Points
flash point = 270 degrees F (132 degrees C)

NFPA - Hazard Identification Ratings
flammability-1;   reactivity-0

NIOSH - Selected LD50s and LC50s
Oral, rat: LD50 = 1600 mg/kg

OSHA - Vacated PELs - Time Weighted Averages
5 mg/m3 TWA

ENVIRONMENTAL LISTS

List of Pesticide Product Inert Ingredients
[present]

TSCA - Health and Safety Reporting List
Effective Date: March 11, 1994;   Sunset Date: March 11, 2004

INTERNATIONAL LISTS

Australian Exposure Standards - Time Weighted Averages
5 mg/m3 TWA

Canada - WHMIS: Ingredient Disclosure
1% item 1002 (1107)

Canada - Alberta - 8 Hour Occupational Exposure Limit
5 mg/m3 TWA

Canada - Alberta - 15 Minute Occupational Exposure Limit
10 mg/m3 STEL

Canada - Ontario - OHSA - TWAEVs
5 mg/m3 TWAEV

Canada - Quebec - Time-Weighted Average Exposure Values
5 mg/m3 TWAEV

United Kingdom - Occupational Exposure Standards - TWAs
5 mg/m3 TWA

Israel - Time Weighted Averages
5 mg/m3 TWA

Israel - Action Levels
2.5 mg/m3 AL

STATE LISTS

California - Exposure Limits - PELs
5 mg/m3 PEL

California - Directors List of Hazardous Substances (8 CCR 339)
[present]

Florida Hazardous Substance List
[present]

Massachusetts Right To Know List
[present]

Minnesota Hazardous Substance List
[present]

NJ Right to Know List (Total)
sn 1216

Pennsylvania Right to Know List
[present]

PROPOSED REGULATIONS

TSCA - ITC 32nd Report Priority Testing List
designated for dermal absorption testing

8-METHOXYPSORALEN   298-81-7

INTERNATIONAL LISTS

German (DFG) - MAK Values
20 ppm MAK;   75 mg/m3 MAK

German (DFG) - Peak Limitations
2 x normal MAK (30 min. average value);   don't exceed 4 times during shift

German (DFG) - Pregnancy
risk to embryo/fetus probable

5-METHOXYPSORALEN   484-20-8

HEALTH AND SAFETY LISTS

NIOSH - Selected LD50s and LC50s
Oral, rat: LD50 = 5710 mg/kg Skin, rabbit: LD50 = 5660 mg/kg

8-METHOXYPSORALEN PLUS ULTRAVIOLET RADIATION   RR-00056-6

HEALTH AND SAFETY LISTS

NIOSH - Selected LD50s and LC50s
Oral, rat: LD50 = 3300 mg/kg

ENVIRONMENTAL LISTS

TSCA - Code of Federal Regulations Citations
40 CFR 712.30;   40 CFR 716.120(c)

METHOXYTRIGLYCOL ACETATE   3610-27-3

INTERNATIONAL LISTS

German (DFG) - MAK Values
20 ppm MAK;   110 mg/m3 MAK

German (DFG) - Peak Limitations
2 x normal MAK (30 min. average value);   don't exceed 4 times during shift

German (DFG) - Pregnancy
risk to embryo/fetus probable

METHYL ABIETATE   127-25-3

HEALTH AND SAFETY LISTS

AIHA - WEEL - Time Weighted Averages
10 ppm TWA

AIHA - WEEL - Ceilings or Short Term Time Weighted Averages
20 ppm STEL

NFPA - Flash Points
flash point = 90 degrees F (32 degrees C)

NFPA - Hazard Identification Ratings
health-2;   flammability-3;   reactivity-0

STATE LISTS

Florida Hazardous Substance List
[present]

Massachusetts Right To Know List
[present]

Minnesota Hazardous Substance List
[present]

Pennsylvania Right to Know List
[present]

N-METHYLACETAMIDE   79-16-3
ENVIRONMENTAL LISTS
List of Pesticide Product Inert Ingredients
[present]

METHYL ACETATE   79-20-9
ENVIRONMENTAL LISTS
List of Pesticide Product Inert Ingredients
[present]

METHYLACETOACETATE   105-45-3
HEALTH AND SAFETY LISTS
NTP Chemical Status Reports - Testing Status and NTIS Number
Technical reports printed (PB90110164/AS)
NTP Chemical Status Reports - Evidence of Carcinogenicity
male rat-clear evidence; female rat-no evidence
STATE LISTS
California - Prop. 65 - Cancer list
carcinogen - initial date 2/27/87
California - Directors List of Hazardous Substances (8 CCR 339)
[present]

P-METHYLACETOPHENONE   122-00-9
HEALTH AND SAFETY LISTS
IARC - Group 2A (limited human data)
[present] (Other relevant data, as given in Supplement 7, influenced the making of the overall evaluation)
OSHA - Possible Select Carcinogens
[present]
STATE LISTS
California - Air Bill 2588 Appendix A-II
known or potential carcinogen
California - Prop. 65 - Cancer list
carcinogen - initial date 10/1/88
California - Directors List of Hazardous Substances (8 CCR 339)
[present]
Massachusetts Right To Know List
carcinogen; extraordinarily hazardous

METHYL ACETYLENE   74-99-7
HEALTH AND SAFETY LISTS
IARC - Group 1 (carcinogenic to humans)
[present]
NTP Seventh Report - Known Carcinogens
known carcinogen
OSHA - Select Carcinogens
[present]
STATE LISTS
Minnesota Hazardous Substance List
carcinogen
Pennsylvania Right to Know List
special hazardous substance
Pennsylvania RTK - Special Hazardous Substances
[present]

2-METHYLACRYLAMIDE   79-39-0
HEALTH AND SAFETY LISTS
NFPA - Flash Points
flash point = 260 degrees F (127 degrees C)
NFPA - Hazard Identification Ratings
health-0; flammability-1; reactivity-0

N-METHYLACRYLAMIDE   1187-59-3
HEALTH AND SAFETY LISTS
NFPA - Flash Points
flash point = 356 degrees F (180 degrees C)
NFPA - Hazard Identification Ratings
health-0; flammability-1; reactivity-0
INTERNATIONAL LISTS
Canada - WHMIS: Ingredient Disclosure
1% item 1003 (1)

METHYL ACRYLATE   96-33-3
HEALTH AND SAFETY LISTS
NIOSH - Selected LD50s and LC50s
Oral, rat: LD50 = 5000 mg/kg

METHYLACRYLONITRILE   126-98-7
HEALTH AND SAFETY LISTS
ACGIH 1995 - Time Weighted Averages
200 ppm TWA; 606 mg/m3 TWA
ACGIH 1995 - Short Term Exposure Limits
250 ppm STEL; 757 mg/m3 STEL
AIHA - Odor Threshold Values
geometric mean air odor threshold = 180 ppm (detectable); 300 ppm (recognizable)
U.S. DOT - Substances From 49 CFR 172.101
regulated by DOT (UN1231)
U.S. DOT - Hazard Classes
DOT hazard class = 3
NFPA - Flash Points
flash point = 14 degrees F (-10 degrees C)
NFPA - Hazard Identification Ratings
health-1; flammability-3; reactivity-0
NIOSH - Selected LD50s and LC50s
Oral, rat: LD50 = 5450 mg/kg
NIOSH 1990 - Pocket Guide - RELs
200 ppm TWA; 610 mg/m3 TWA; 250 ppm STEL; 760 mg/m3 STEL
NIOSH 1990 - Pocket Guide - IDLHs
10,000 ppm IDLH
NIOSH 1990 - Pocket Guide - Target organs
skin, eyes, respiratory system
OSHA - Vacated PELs - Time Weighted Averages
200 ppm TWA; 610 mg/m3 TWA
OSHA - Vacated PELs - Short Term Exposure Limits
250 ppm STEL; 760 mg/m3 STEL
OSHA - Final PELs - Time Weighted Averages
200 ppm TWA; 610 mg/m3 TWA
ENVIRONMENTAL LISTS
CAA - HON Rule - SOCMI Chemicals
compliance by July 24, 1995
TSCA - PAIR - Reporting List
Effective Date: January 26, 1994; Reporting Date: March 28, 1994
TSCA - Health and Safety Reporting List
Effective Date: January 26, 1994; Sunset Date: January 26, 2004
INTERNATIONAL LISTS
Australian Exposure Standards - Time Weighted Averages
200 ppm TWA; 606 mg/m3 TWA
Australian Exposure Standards - Short Term Exposure Limits
250 ppm STEL; 757 mg/m3 STEL
Canada - WHMIS: Ingredient Disclosure
1% item 1004 (26)
Canada - Alberta - 8 Hour Occupational Exposure Limit
200 ppm TWA; 605 mg/m3 TWA

Canada - Alberta - 15 Minute Occupational Exposure Limit
250 ppm STEL;   760 mg/m3 STEL
Canada - British Columbia - 8 Hour Exposure Limits
200 ppm TWA;   610 mg/m3 TWA
Canada - British Columbia - 15 Minute Exposure Limits
250 ppm STEL;   760 mg/m3 STEL
Canada - British Columbia - Skin Notations
skin - potential for skin absorption
Canada - Ontario - OHSA - TWAEVs
200 ppm TWAEV;   605 mg/m3 TWAEV
Canada - Ontario - OHSA - STEVs
250 ppm STEV;   755 mg/m3 STEV
Canada - Quebec - Time-Weighted Average Exposure Values
200 ppm TWAEV;   606 mg/m3 TWAEV
Canada - Quebec - Short-term Exposure Values
250 ppm STEV;   760 mg/m3 STEV
United Kingdom - Occupational Exposure Standards - TWAs
200 ppm TWA;   610 mg/m3 TWA
United Kingdom - Occupational Exposure Standards - STELs
250 ppm STEL;   760 mg/m3 STEL
German (DFG) - MAK Values
200 ppm MAK;   610 mg/m3 MAK
German (DFG) - Peak Limitations
*2 x normal MAK (5 min momentary value);   don't exceed 8 times
during shift*
German (DFG) - Pregnancy
classification not yet possible
Israel - Time Weighted Averages
200 ppm TWA;   606 mg/m3 TWA
Israel - Short Term Exposure Limits
250 ppm STEL;   757 mg/m3 STEL
Israel - Action Levels
100 ppm AL;   303 mg/m3 AL
Mexico - Instruction No. 10 - TWAs
200 ppm TWA;   610 mg/m3 TWA
Mexico - Instruction No. 10 - STELs
250 ppm STEL;   760 mg/m3 STEL

STATE LISTS
California - Exposure Limits - PELs
200 ppm PEL;   610 mg/m3 PEL
California - Exposure Limits - STELs
250 ppm STEL;   760 mg/m3 STEL
California - Directors List of Hazardous Substances (8 CCR 339)
[present]
Florida Hazardous Substance List
[present]
Massachusetts Right To Know List
[present]
Minnesota Hazardous Substance List
[present]
NJ Right to Know List (Total)
sn 1217
NJ Special Hazardous Substances
(flammable - third degree)
Pennsylvania Right to Know List
[present]

PROPOSED REGULATIONS
TSCA - ITC 31st Report Priority Testing List
designated to be tested

METHYLAL   109-87-5
HEALTH AND SAFETY LISTS
NFPA - Flash Points
flash point = 170 degrees F (77 degrees C)
NFPA - Hazard Identification Ratings
health-2; flammability-2; reactivity-0
INTERNATIONAL LISTS
Canada - WHMIS: Ingredient Disclosure
1% item 1005 (39)
STATE LISTS
Florida Hazardous Substance List
[present]
Massachusetts Right To Know List
[present]
Pennsylvania Right to Know List
[present]

METHYL ALCOHOL   67-56-1
HEALTH AND SAFETY LISTS
NFPA - Flash Points
flash point = 205 degrees F (96 degrees C)
NFPA - Hazard Identification Ratings
health-0; flammability-1; reactivity-0
NIOSH - Selected LD50s and LC50s
Oral, rat: LD50 = 1400 mg/kg

METHYLALUMINUM SESQUIBROMIDE   12263-85-3
HEALTH AND SAFETY LISTS
ACGIH 1995 - Time Weighted Averages
1000 ppm TWA;   1640 mg/m3 TWA
NFPA - Flash Points
gas (no flash point given)
NFPA - Hazard Identification Ratings
health-2; flammability-4; reactivity-2
NIOSH 1990 - Pocket Guide - RELs
1000 ppm TWA;   1650 mg/m3 TWA
NIOSH 1990 - Pocket Guide - IDLHs
15,000 ppm IDLH (lower explosive level)
NIOSH 1990 - Pocket Guide - Target organs
CNS
OSHA - Vacated PELs - Time Weighted Averages
1000 ppm TWA;   1650 mg/m3 TWA
OSHA - Final PELs - Time Weighted Averages
1000 ppm TWA;   1650 mg/m3 TWA

ENVIRONMENTAL LISTS
CAA - Flammable Substances for Accidental Release Prevention
threshold quantity = 10,000 lbs

INTERNATIONAL LISTS
Australian Exposure Standards - Time Weighted Averages
1000 ppm TWA;   1640 mg/m3 TWA
Canada - WHMIS: Ingredient Disclosure
1% item 1006 (1108)
Canada - Alberta - 8 Hour Occupational Exposure Limit
1000 ppm TWA;   1638 mg/m3 TWA
Canada - Alberta - 15 Minute Occupational Exposure Limit
1250 ppm STEL;   2048 mg/m3 STEL
Canada - British Columbia - 8 Hour Exposure Limits
1000 ppm TWA;   1650 mg/m3 TWA
Canada - British Columbia - 15 Minute Exposure Limits
1250 ppm STEL;   2040 mg/m3 STEL
Canada - Ontario - OHSA - TWAEVs
1000 ppm TWAEV;   1635 mg/m3 TWAEV

Canada - Ontario - OHSA - STEVs
1250 ppm STEV; 2045 mg/m3 STEV

Canada - Quebec - Time-Weighted Average Exposure Values
1000 ppm TWAEV; 1640 mg/m3 TWAEV

German (DFG) - MAK Values
1000 ppm MAK; 1650 mg/m3 MAK

German (DFG) - Peak Limitations
2 x normal MAK (1 hour momentary value); don't exceed 3 times per shift

Israel - Time Weighted Averages
1000 ppm TWA; 1640 mg/m3 TWA

Israel - Action Levels
500 ppm AL; 820 mg/m3 AL

Mexico - Instruction No. 10 - STELs
1250 ppm STEL; 2040 mg/m3 STEL

STATE LISTS

California - Exposure Limits - PELs
1000 ppm PEL; 1650 mg/m3 PEL

California - Directors List of Hazardous Substances (8 CCR 339)
[present]

Florida Hazardous Substance List
[present]

Massachusetts Right To Know List
[present]

Minnesota Hazardous Substance List
[present]

NJ Special Hazardous Substances
(flammable - fourth degree; reactive - second degree)

Pennsylvania Right to Know List
[present]

METHYL ALUMINUM SESQUICHLORIDE   12542-85-7

HEALTH AND SAFETY LISTS

NIOSH - Selected LD50s and LC50s
Oral, rat: LD50 = 459 mg/kg

INTERNATIONAL LISTS

Canada - WHMIS: Ingredient Disclosure
1% item 1007 (1109)

METHYLAMINE   74-89-5

ENVIRONMENTAL LISTS

List of Pesticide Product Inert Ingredients
[present]

INTERNATIONAL LISTS

Canada - WHMIS: Ingredient Disclosure
0.1% item 1008 (1110)

METHYLAMINE DINITRAMINE AND SALTS   RR-01443-7

HEALTH AND SAFETY LISTS

ACGIH 1995 - Time Weighted Averages
(10) ppm TWA; (35) mg/m3 TWA

ACGIH 1995 - Skin Designations
skin - potential for cutaneous absorption

AIHA - Odor Threshold Values
no geometric mean air odor threshold

U.S. DOT - Substances From 49 CFR 172.101
regulated by DOT (UN1919)

U.S. DOT - Hazard Classes
DOT hazard class = 3

IARC - Group 3 (not classifiable)
[present]

NFPA - Flash Points
flash point = 27 degrees F (-3 degrees C)

NFPA - Hazard Identification Ratings
health-3; flammability-3; reactivity-2

NIOSH - Selected LD50s and LC50s
Inhalation, rat: LC50 = 1350 ppm 4 hr Oral, rat: LD50 = 277 mg/kg Skin, rabbit: LD50 = 1243 mg/kg

NIOSH 1990 - Pocket Guide - RELs
10 ppm TWA; 35 mg/m3 TWA

NIOSH 1990 - Pocket Guide - IDLHs
1000 ppm IDLH

NIOSH 1990 - Pocket Guide - Target organs
skin, eyes, respiratory system

NIOSH 1990 - Pocket Guide - Skin list
Potential for dermal absorption

OSHA - Vacated PELs - Time Weighted Averages
10 ppm TWA; 35 mg/m3 TWA

OSHA - Vacated PELs - Skin Designation
Prevent or reduce skin absorption

OSHA - Final PELs - Time Weighted Averages
10 ppm TWA; 35 mg/m3 TWA

OSHA - Final PELs - Skin Notations
prevent or reduce skin absorption

ENVIRONMENTAL LISTS

CERCLA/SARA - Section 313 - Emission Reporting
form R reporting required for 1.0% de minimus concentration

CAA - HON Rule - SOCMI Chemicals
compliance by Oct. 23, 1995

TSCA - Code of Federal Regulations Citations
40 CFR 712.30(d)

TSCA - PAIR - Reporting List
Reporting Date: November 19, 1982

INTERNATIONAL LISTS

Australian Exposure Standards - Time Weighted Averages
10 ppm TWA; 35 mg/m3 TWA

Australian Exposure Standards - Skin Effects
skin absorption; sensitiser

Canada - WHMIS: Ingredient Disclosure
1% item 1009 (161)

Canada - NPRI (National Pollutant Release Inventory)
[present]

Canada - Alberta - 8 Hour Occupational Exposure Limit
10 ppm TWA; 35 mg/m3 TWA

Canada - Alberta - 15 Minute Occupational Exposure Limit
20 ppm STEL; 70 mg/m3 STEL

Canada - Alberta - Skin Designation
can be absorbed through the intact skin

Canada - British Columbia - 8 Hour Exposure Limits
10 ppm TWA; 35 mg/m3 TWA

Canada - British Columbia - Skin Notations
skin - potential for skin absorption

Canada - Ontario - OHSA - TWAEVs
10 ppm TWAEV; 35 mg/m3 TWAEV

Canada - Ontario - OHSA - Skin Notations
absorption through skin, eyes, or mucous membranes

Canada - Quebec - Time-Weighted Average Exposure Values
10 ppm TWAEV; 35 mg/m3 TWAEV

Canada - Quebec - Skin Designations
absorbed through the skin

United Kingdom - Occupational Exposure Standards - TWAs
10 ppm TWA; 35 mg/m3 TWA

German (DFG) - MAK Values
5 ppm MAK; 18 mg/m3 MAK

German (DFG) - Peak Limitations
2 x normal MAK (5 min momentary value); don't exceed 8 times during shift

German (DFG) - Skin/Sensitizers
danger of sensitization (skin or respiratory)

Israel - Time Weighted Averages
10 ppm TWA; 35 mg/m3 TWA

Israel - Action Levels
5 ppm AL; 17.5 mg/m3 AL

Mexico - Instruction No. 10 - TWAs
10 ppm TWA; 35 mg/m3 TWA

Mexico - Instruction No. 10 - Skin designation
skin - potential for cutaneous absorption

STATE LISTS

California - Air Bill 2588 Appendix A-II
6/91

California - Exposure Limits - PELs
10 ppm PEL; 35 mg/m3 PEL

California - Exposure Limits - Skin Notation
material may be absorbed through the skin, eyes or mucous membrane

California - Directors List of Hazardous Substances (8 CCR 339)
[present]

Florida Hazardous Substance List
[present]

Massachusetts Right To Know List
[present]

Minnesota Hazardous Substance List
skin

NJ Right to Know List (Total)
sn 1219

NJ Special Hazardous Substances
(flammable - third degree; reactive - second degree)

Pennsylvania Right to Know List
environmental hazard

PROPOSED REGULATIONS

ACGIH 1995 - Notice of Intended Changes
(skin) 2 ppm TWA; 7 mg/m3 TWA; A4-not classifiable as a human carcinogen

METHYLAMINE NITROFORM 14147-71-8
SEE ALSO:
F039-HAZARDOUS WASTES

HEALTH AND SAFETY LISTS

ACGIH 1995 - Time Weighted Averages
1 ppm TWA; 2.7 mg/m3 TWA

ACGIH 1995 - Skin Designations
skin - potential for cutaneous absorption

AIHA - Odor Threshold Values
geometric mean air odor threshold = 6.9 ppm (detectable)

U.S. DOT - Substances From 49 CFR 172.101
regulated by DOT (UN3079)

U.S. DOT - Hazard Classes
DOT hazard class = 6.1

U.S. DOT - Substances Which Are Poisonous by Inhalation
liquid hazardous material poisonous by inhalation (inhibited form) (UN3079)

U.S. DOT - Appendix A Table 1 - Hazardous Substances
final RQ = 1000 pounds (454 kg)

NFPA - Flash Points
flash point = 34 degrees F (1.1 degrees C)

NFPA - Hazard Identification Ratings
health-2; flammability-3; reactivity-2

NIOSH - Selected LD50s and LC50s
Inhalation, rat: LC50 = 328 ppm 4 hr Oral, rat: LD50 = 120 mg/kg Skin, rat: LD50 = 2080 mg/kg

NTP Chemical Status Reports - Testing Status and NTIS Number
Prechronic studies completed: in review for further evaluation; approved for toxicology/carcinogenesis study

OSHA - Vacated PELs - Time Weighted Averages
1 ppm TWA; 3 mg/m3 TWA

OSHA - Vacated PELs - Skin Designation
Prevent or reduce skin absorption

OSHA - List of Highly Hazardous Chemicals
threshhold quantity = 250 pounds

ENVIRONMENTAL LISTS

CERCLA/SARA - Section 302 Extremely Hazardous Substances and TPQs
TPQ = 500 pounds

CERCLA/SARA - Section 313 - Emission Reporting
form R reporting required

CERCLA/SARA - Hazardous Substances and their Reportable Quantities
final RQ = 1000 pounds (454 kg)

CAA -Toxic Substances for Accidental Release Prevention
threshold quantity = 10,000 lbs

RCRA - U Series Wastes
waste number U152 (Ignitable waste; Toxic waste)

RCRA - Hazardous Constituents-Appendix VIII
waste number U152

RCRA - Basis for Listing - Appendix VII
Included in waste stream: F039

RCRA - Substances Banned From Land Disposal
[present]

RCRA - TSD Facilities Ground Water Monitoring
TM 8015 = 5 ug/L PQL; TM 8240 = 5 ug/L PQL

RCRA - Universal Treatment Standards (LDR)
WW: 0.24 mg/l; NWW: 84 mg/kg

INTERNATIONAL LISTS

Australian Exposure Standards - Time Weighted Averages
1 ppm TWA; 2.7 mg/m3 TWA

Australian Exposure Standards - Skin Effects
skin absorption

Canada - WHMIS: Ingredient Disclosure
1% item 1010 (1099)

Canada - Alberta - 8 Hour Occupational Exposure Limit
1 ppm TWA; 2.7 mg/m3 TWA

Canada - Alberta - 15 Minute Occupational Exposure Limit
2 ppm STEL; 5.5 mg/m3 STEL

Canada - Alberta - Skin Designation
can be absorbed through the intact skin

Canada - British Columbia - 8 Hour Exposure Limits
1 ppm TWA; 3 mg/m3 TWA

Canada - British Columbia - 15 Minute Exposure Limits
2 ppm STEL; 6 mg/m3 STEL

Canada - British Columbia - Skin Notations
skin - potential for skin absorption

Canada - Ontario - OHSA - TWAEVs
1 ppm TWAEV; 2.7 mg/m3 TWAEV

Canada - Ontario - OHSA - Skin Notations
absorption through skin, eyes, or mucous membranes

Canada - Quebec - Time-Weighted Average Exposure Values
1 ppm TWAEV; 2.7 mg/m3 TWAEV

Canada - Quebec - Skin Designations
absorbed through the skin

United Kingdom - Occupational Exposure Standards - TWAs
1 ppm TWA; 3 mg/m3 TWA

United Kingdom - Occupational Exposure Standards - Notes
can be absorbed through skin
Israel - Time Weighted Averages
1 ppm TWA; 2.7 mg/m3 TWA
Israel - Action Levels
0.5 ppm AL; 1.35 mg/m3 AL
Mexico - Instruction No. 10 - TWAs
1 ppm TWA; 3 mg/m3 TWA
Mexico - Instruction No. 10 - STELs
2 ppm STEL; 6 mg/m3 STEL
Mexico - Instruction No. 10 - Skin designation
skin - potential for cutaneous absorption

STATE LISTS
California - Exposure Limits - PELs
1 ppm PEL; 3 mg/m3 PEL
California - Exposure Limits - Skin Notation
material may be absorbed through the skin, eyes or mucous membrane
California - Directors List of Hazardous Substances (8 CCR 339)
[present]
Florida Hazardous Substance List
[present]
Massachusetts Right To Know List
extraordinarily hazardous
Minnesota Hazardous Substance List
skin
Pennsylvania Right to Know List
environmental hazard

METHYLAMINE PERCHLORATE   RR-01444-8

HEALTH AND SAFETY LISTS
ACGIH 1995 - Time Weighted Averages
1000 ppm TWA; 3110 mg/m3 TWA
U.S. DOT - Substances From 49 CFR 172.101
regulated by DOT (UN1234)
U.S. DOT - Hazard Classes
DOT hazard class = 3
NFPA - Flash Points
flash point = -26 degrees F (-32 degrees C)
NFPA - Hazard Identification Ratings
health-2; flammability-3; reactivity-2
NIOSH - Selected LD50s and LC50s
Inhalation, rat: LC50 = 15000 ppm 8 hr Oral, rabbit: LD50 = 5708 mg/kg
NIOSH 1990 - Pocket Guide - RELs
1000 ppm TWA; 3100 mg/m3 TWA
NIOSH 1990 - Pocket Guide - IDLHs
15,000 ppm IDLH (lower explosive level)
NIOSH 1990 - Pocket Guide - Target organs
CNS, skin, respiratory system
OSHA - Vacated PELs - Time Weighted Averages
1000 ppm TWA; 3100 mg/m3 TWA
OSHA - Final PELs - Time Weighted Averages
1000 ppm TWA; 3100 mg/m3 TWA

ENVIRONMENTAL LISTS
TSCA - Code of Federal Regulations Citations
40 CFR 716.120(a)
TSCA - Health and Safety Reporting List
Effective Date: June 1, 1987

INTERNATIONAL LISTS
Australian Exposure Standards - Time Weighted Averages
1000 ppm TWA; 3110 mg/m3 TWA
Canada - WHMIS: Ingredient Disclosure
1% item 1011 (1111)

Canada - Alberta - 8 Hour Occupational Exposure Limit
1000 ppm TWA; 3112 mg/m3 TWA
Canada - Alberta - 15 Minute Occupational Exposure Limit
1250 ppm STEL; 3891 mg/m3 STEL
Canada - British Columbia - 8 Hour Exposure Limits
1000 ppm TWA; 3100 mg/m3 TWA
Canada - Ontario - OHSA - TWAEVs
1000 ppm TWAEV; 3110 mg/m3 TWAEV
Canada - Quebec - Time-Weighted Average Exposure Values
1000 ppm TWAEV; 3110 mg/m3 TWAEV
United Kingdom - Occupational Exposure Standards - TWAs
1000 ppm TWA; 3100 mg/m3 TWA
United Kingdom - Occupational Exposure Standards - STELs
1250 ppm STEL; 3880 mg/m3 STEL
German (DFG) - MAK Values
1000 ppm MAK; 3100 mg/m3 MAK
Israel - Time Weighted Averages
1000 ppm TWA; 3110 mg/m3 TWA
Israel - Action Levels
500 ppm AL; 1555 mg/m3 AL
Mexico - Instruction No. 10 - TWAs
1000 ppm TWA; 3100 mg/m3 TWA
Mexico - Instruction No. 10 - STELs
1250 ppm STEL; 3875 mg/m3 STEL

STATE LISTS
California - Exposure Limits - PELs
1000 ppm PEL; 3100 mg/m3 PEL
California - Directors List of Hazardous Substances (8 CCR 339)
[present]
Florida Hazardous Substance List
[present]
Massachusetts Right To Know List
[present]
Minnesota Hazardous Substance List
[present]
NJ Right to Know List (Total)
sn 1221
NJ Special Hazardous Substances
(flammable - third degree; reactive - second degree)
Pennsylvania Right to Know List
[present]

METHYLAMINE SALTS   RR-01782-3
SEE ALSO:
F003-HAZARDOUS WASTES

HEALTH AND SAFETY LISTS
ACGIH 1995 - Time Weighted Averages
200 ppm TWA; 262 mg/m3 TWA
ACGIH 1995 - Short Term Exposure Limits
250 ppm STEL; 328 mg/m3 STEL
ACGIH 1995 - Skin Designations
skin - potential for cutaneous absorption
ACGIH 1995 - Biological Exposure Indices
Methanol in urine: 15 mg/L, end of shift (B, Ns); (Formic acid in urine): (80 mg/g creatinine), (prior to the last shift of workweek) (B, Ns)
AIHA - Odor Threshold Values
geometric mean air odor threshold = 160 ppm (detectable); 690 ppm (recognizable)
U.S. DOT - Substances From 49 CFR 172.101
regulated by DOT (UN1230)
U.S. DOT - Hazard Classes
DOT hazard class = 3

U.S. DOT - Appendix A Table 1 - Hazardous Substances
final RQ = 5000 pounds (2270 kg)

NFPA - Flash Points
flash point = 52 degrees F (11 degrees C)

NFPA - Hazard Identification Ratings
health-1; flammability-3; reactivity-0

NIOSH - Selected LD50s and LC50s
Inhalation, rat: LC50 = 64000 ppm 4 hr Oral, rat: LD50 = 5628 mg/kg Skin, rabbit: LD50 = 15800 mg/kg

NIOSH 1990 - Pocket Guide - RELs
200 ppm TWA; 260 mg/m3 TWA; 250 ppm STEL; 325 mg/m3 STEL

NIOSH 1990 - Pocket Guide - IDLHs
25,000 ppm IDLH

NIOSH 1990 - Pocket Guide - Target organs
skin, eyes, CNS, GI tract

NIOSH 1990 - Pocket Guide - Skin list
Potential for dermal absorption

NIOSH - Health Standards - Exposure Limits
200 ppm TWA; 262 mg/m3 TWA; C (15 min) 800 ppm; C (15 min) 1048 mg/m3

NIOSH - Health Standards - Health Effects and Precautions
Blindness, metabolic acidosis

NTP Chemical Status Reports - Testing Status and NTIS Number
Project leader assigned/study in design

OSHA - Vacated PELs - Time Weighted Averages
200 ppm TWA; 260 mg/m3 TWA

OSHA - Vacated PELs - Short Term Exposure Limits
250 ppm STEL; 325 mg/m3 STEL

OSHA - Vacated PELs - Skin Designation
Prevent or reduce skin absorption

OSHA - Final PELs - Time Weighted Averages
200 ppm TWA; 260 mg/m3 TWA

ENVIRONMENTAL LISTS

CERCLA/SARA - Section 313 - Emission Reporting
form R reporting required for 1.0% de minimus concentration

CERCLA/SARA - Hazardous Substances and their Reportable Quantities
final RQ = 5000 pounds (2270 kg)

Clean Air Act (1990) - List of Hazardous Air Contaminants
[present]

CAA - HON Rule - SOCMI Chemicals
compliance by July 24, 1995

CAA - HON Rule - Organic HAPs
[present]

List of Pesticide Product Inert Ingredients
[present]

RCRA - U Series Wastes
waste number U154 (Ignitable waste)

RCRA - Hazardous Constituents-Appendix VIII
waste number U154 (Ignitable waste)

RCRA - Substances Banned From Land Disposal
[present]

RCRA - Universal Treatment Standards (LDR)
WW: 5.6 mg/l; NWW: 0.75 mg/l TCLP

INTERNATIONAL LISTS

Australian Exposure Standards - Time Weighted Averages
200 ppm TWA; 262 mg/m3 TWA

Australian Exposure Standards - Short Term Exposure Limits
250 ppm STEL; 328 mg/m3 STEL

Australian Exposure Standards - Skin Effects
skin absorption

Canada - WHMIS: Ingredient Disclosure
1% item 1012 (183)

Canada - NPRI (National Pollutant Release Inventory)
[present]

Canada - Alberta - 8 Hour Occupational Exposure Limit
200 ppm TWA; 262 mg/m3 TWA

Canada - Alberta - 15 Minute Occupational Exposure Limit
250 ppm STEL; 328 mg/m3 STEL

Canada - Alberta - Skin Designation
can be absorbed through the intact skin

Canada - British Columbia - 8 Hour Exposure Limits
200 ppm TWA; 260 mg/m3 TWA

Canada - British Columbia - 15 Minute Exposure Limits
250 ppm STEL; 310 mg/m3 STEL

Canada - British Columbia - Skin Notations
skin - potential for skin absorption

Canada - Ontario - OHSA - TWAEVs
200 ppm TWAEV; 260 mg/m3 TWAEV

Canada - Ontario - OHSA - STEVs
250 ppm STEV; 325 mg/m3 STEV

Canada - Ontario - OHSA - Skin Notations
absorption through skin, eyes, or mucous membranes

Canada - Quebec - Time-Weighted Average Exposure Values
200 ppm TWAEV; 262 mg/m3 TWAEV

Canada - Quebec - Short-term Exposure Values
250 ppm STEV; 328 mg/m3 STEV

Canada - Quebec - Skin Designations
absorbed through the skin

United Kingdom - Occupational Exposure Standards - TWAs
200 ppm TWA; 260 mg/m3 TWA

United Kingdom - Occupational Exposure Standards - STELs
250 ppm STEL; 310 mg/m3 STEL

United Kingdom - Occupational Exposure Standards - Notes
can be absorbed through skin

German (DFG) - MAK Values
200 ppm MAK; 260 mg/m3 MAK

German (DFG) - Peak Limitations
2 x normal MAK (30 min. average value); don't exceed 4 times during shift

German (DFG) - Skin/Sensitizers
danger of cutaneous absorption

German (DFG) - Pregnancy
classification not yet possible

Israel - Time Weighted Averages
200 ppm TWA; 262 mg/m3 TWA

Israel - Short Term Exposure Limits
250 ppm STEL; 328 mg/m3 STEL

Israel - Action Levels
100 ppm AL; 131 mg/m3 AL

Mexico - Instruction No. 10 - TWAs
200 ppm TWA; 260 mg/m3 TWA

Mexico - Instruction No. 10 - STELs
250 ppm STEL; 310 mg/m3 STEL

Mexico - Instruction No. 10 - Skin designation
skin - potential for cutaneous absorption

STATE LISTS

California - Air Bill 2588 Appendix A-I
[present]

California - Exposure Limits - PELs
200 ppm PEL; 260 mg/m3 PEL

California - Exposure Limits - STELs
250 ppm STEL; 325 mg/m3 STEL

California - Exposure Limits - Ceilings
C 1000 ppm

California - Exposure Limits - Skin Notation
material may be absorbed through the skin, eyes or mucous membrane

California - Directors List of Hazardous Substances (8 CCR 339)
[present] (refers to solutions greater than or equal to 3%)

Florida Hazardous Substance List
[present]

Massachusetts Right To Know List
[present]

Minnesota Hazardous Substance List
skin

NJ Right to Know List (Total)
sn 1222

NJ Special Hazardous Substances
(flammable - third degree)

Pennsylvania Right to Know List
environmental hazard

PROPOSED REGULATIONS

ACGIH 1995 - Proposed Biological Exposure Indices
Methanol in urine: 15 mg/L, end of shift (B, Ns)

2-(METHYLAMINO)ETHANESULFONIC ACID   107-68-6

HEALTH AND SAFETY LISTS

NFPA - Flash Points
ignites spontaneously in air

NFPA - Hazard Identification Ratings
flammability-3;   reactivity-3 (do not use water, foam or halogenated extinguishing agents)

STATE LISTS

Florida Hazardous Substance List
[present]

Massachusetts Right To Know List
[present]

NJ Right to Know List (Total)
sn 2555

Pennsylvania Right to Know List
[present]

1-METHYLAMINO-4-   2475-46-9
ETHANOLAMINOANTHRAQUINONE

HEALTH AND SAFETY LISTS

NFPA - Flash Points
ignites spontaneously in air

NFPA - Hazard Identification Ratings
flammability-3;   reactivity-3 (do not use water, foam or halogenated extinguishing agents)

STATE LISTS

Florida Hazardous Substance List
[present]

Massachusetts Right To Know List
[present]

NJ Right to Know List (Total)
sn 1224

NJ Special Hazardous Substances
(flammable - third degree;   reactive - third degree)

Pennsylvania Right to Know List
[present]

METHYLAMMONIUM N-   21160-95-2
METHYLDITHIOCARBAMATE

HEALTH AND SAFETY LISTS

ACGIH 1995 - Time Weighted Averages
5 ppm TWA;   6.4 mg/m3 TWA

ACGIH 1995 - Short Term Exposure Limits
15 ppm STEL;   19 mg/m3 STEL

AIHA - Odor Threshold Values
geometric mean air odor threshold = 4.7 ppm (detectable)

U.S. DOT - Substances From 49 CFR 172.101
regulated by DOT (UN1235, UN1061)

U.S. DOT - Hazard Classes
DOT hazard class = 2.3

U.S. DOT - Substances Which Are Poisonous by Inhalation
gaseous hazardous material poisonous by inhalation (anhydrous form) (UN1061)

U.S. DOT - Appendix A Table 1 - Hazardous Substances
final RQ = 100 pounds (45.4 kg)

FDA - Controlled Substances Act - Precursor chemicals
Threshold by base weight = 1 kilogram

NFPA - Flash Points
gas (no flash point given)

NFPA - Hazard Identification Ratings
health-3;   flammability-4;   reactivity-0

NIOSH - Selected LD50s and LC50s
Inhalation, mouse: LC50 = 2400 mg/m3 (2 hr)

NIOSH 1990 - Pocket Guide - RELs
10 ppm TWA;   12 mg/m3 TWA

NIOSH 1990 - Pocket Guide - IDLHs
100 ppm IDLH

NIOSH 1990 - Pocket Guide - Target organs
skin, eyes, respiratory system

OSHA - Vacated PELs - Time Weighted Averages
10 ppm TWA;   12 mg/m3 TWA

OSHA - Final PELs - Time Weighted Averages
10 ppm TWA;   12 mg/m3 TWA

OSHA - List of Highly Hazardous Chemicals
anhydrous: threshhold quantity = 1000 pounds

ENVIRONMENTAL LISTS

CERCLA/SARA - Hazardous Substances and their Reportable Quantities
final RQ = 100 pounds (45.4 kg)

CAA - Flammable Substances for Accidental Release Prevention
threshold quantity = 10,000 lbs

CAA - HON Rule - SOCMI Chemicals
compliance by July 24, 1995

Clean Water Act - Hazardous Substances
[present]

INTERNATIONAL LISTS

Australian Exposure Standards - Time Weighted Averages
10 ppm TWA;   13 mg/m3 TWA

Canada - WHMIS: Ingredient Disclosure
0.1% item 1013 (1112)

Canada - Alberta - 8 Hour Occupational Exposure Limit
10 ppm TWA;   12 mg/m3 TWA

Canada - Alberta - 15 Minute Occupational Exposure Limit
20 ppm STEL;   25 mg/m3 STEL

Canada - British Columbia - 8 Hour Exposure Limits
10 ppm TWA;   12 mg/m3 TWA

Canada - Ontario - OHSA - TWAEVs
10 ppm TWAEV;   13 mg/m3 TWAEV

Canada - Quebec - Time-Weighted Average Exposure Values
10 ppm TWAEV;   13 mg/m3 TWAEV

United Kingdom - Occupational Exposure Standards - TWAs
10 ppm TWA;   12 mg/m3 TWA

German (DFG) - MAK Values
10 ppm MAK;   12 mg/m3 MAK

German (DFG) - Peak Limitations
2 x normal MAK (10 min momentary value);   don't exceed 4 times per shift

Israel - Time Weighted Averages
10 ppm TWA;   13 mg/m3 TWA

Israel - Action Levels
5 ppm AL; 6.5 mg/m3 AL
Mexico - Instruction No. 10 - TWAs
10 ppm TWA; 12 mg/m3 TWA
STATE LISTS
California - Exposure Limits - PELs
10 ppm PEL; 12 mg/m3 PEL
California - Directors List of Hazardous Substances (8 CCR 339)
[present]
Florida Hazardous Substance List
[present]
Massachusetts Right To Know List
[present]
Minnesota Hazardous Substance List
[present]
NJ Right to Know List (Total)
sn 1225
NJ Special Hazardous Substances
(flammable - fourth degree)
Pennsylvania Right to Know List
environmental hazard

METHYLAMYL ACETATE   7789-99-3
HEALTH AND SAFETY LISTS
U.S. DOT - Hazard Classes
Forbidden from transport by the DOT

METHYL N-AMYL KETONE   110-43-0
HEALTH AND SAFETY LISTS
U.S. DOT - Hazard Classes
Forbidden from transport by the DOT

5-METHYLANGELICIN   73459-03-7
HEALTH AND SAFETY LISTS
U.S. DOT - Hazard Classes
Forbidden from transport by the DOT

5-METHYLANGELICIN PLUS ULTRAVIOLET A   RR-01535-0
RADIATION
HEALTH AND SAFETY LISTS
FDA - Controlled Substances Act - Precursor chemicals
Threshold by base weight = 1 kilogram

N-METHYL ANILINE   100-61-8
ENVIRONMENTAL LISTS
List of Pesticide Product Inert Ingredients
[present]

METHYL ANTHRANILATE   134-20-3
HEALTH AND SAFETY LISTS
NIOSH - Selected LD50s and LC50s
Oral, rat: LD50 = 3000 mg/kg
ENVIRONMENTAL LISTS
List of Pesticide Product Inert Ingredients
[present]
INTERNATIONAL LISTS
Canada - WHMIS: Ingredient Disclosure
1% item 1014 (1113)

METHYLAZOXYMETHANOL   590-96-5
ENVIRONMENTAL LISTS
TSCA - Section 12(b) - Export Notification
P-84-1042; export notification required - Section 5

METHYLAZOXYMETHYL ACETATE   592-62-1
HEALTH AND SAFETY LISTS
NIOSH - Selected LD50s and LC50s
Oral, rat: LD50 = 7400 mg/kg

METHYLBENZOATE   93-58-3
HEALTH AND SAFETY LISTS
ACGIH 1995 - Time Weighted Averages
50 ppm TWA; 233 mg/m3 TWA
AIHA - Odor Threshold Values
no geometric mean air odor threshold
U.S. DOT - Substances From 49 CFR 172.101
regulated by DOT (UN1110)
U.S. DOT - Hazard Classes
DOT hazard class = 3
NFPA - Flash Points
flash point = 102 degrees F (39 degrees C)
NFPA - Hazard Identification Ratings
health-1; flammability-2; reactivity-0
NIOSH - Selected LD50s and LC50s
Oral, rat: LD50 = 1670 mg/kg Skin, rabbit: LD50 = 12600 mg/kg
NIOSH 1990 - Pocket Guide - RELs
100 ppm TWA; 465 mg/m3 TWA
NIOSH 1990 - Pocket Guide - IDLHs
4000 ppm IDLH
NIOSH 1990 - Pocket Guide - Target organs
skin, eyes, CNS, PNS, respiratory system
NIOSH - Health Standards - Exposure Limits
100 ppm TWA; 465 mg/m3 TWA (Listed under "Ketones")
NIOSH - Health Standards - Health Effects and Precautions
Irritation; liver, kidney, and nervous system effects (Urinalysis required) (Listed under 'Ketones')
OSHA - Vacated PELs - Time Weighted Averages
100 ppm TWA; 465 mg/m3 TWA
OSHA - Final PELs - Time Weighted Averages
100 ppm TWA; 465 mg/m3 TWA
ENVIRONMENTAL LISTS
List of Pesticide Product Inert Ingredients
[present]
INTERNATIONAL LISTS
Australian Exposure Standards - Time Weighted Averages
50 ppm TWA; 233 mg/m3 TWA
Canada - WHMIS: Ingredient Disclosure
1% item 1016 (1115)
Canada - Alberta - 8 Hour Occupational Exposure Limit
5 ppm TWA; 235 mg/m3 TWA
Canada - Alberta - 15 Minute Occupational Exposure Limit
100 ppm STEL; 465 mg/m3 STEL
Canada - British Columbia - 8 Hour Exposure Limits
100 ppm TWA; 465 mg/m3 TWA
Canada - British Columbia - 15 Minute Exposure Limits
150 ppm STEL; 710 mg/m3 STEL
Canada - Ontario - OHSA - TWAEVs
25 ppm TWAEV; 115 mg/m3 TWAEV
Canada - Quebec - Time-Weighted Average Exposure Values
50 ppm TWAEV; 233 mg/m3 TWAEV
United Kingdom - Occupational Exposure Standards - TWAs
50 ppm TWA; 240 mg/m3 TWA
Israel - Time Weighted Averages
50 ppm TWA; 233 mg/m3 TWA
Israel - Action Levels
25 ppm AL; 116.5 mg/m3 AL

Mexico - Instruction No. 10 - TWAs
50 ppm TWA; 235 mg/m3 TWA
Mexico - Instruction No. 10 - STELs
100 ppm STEL; 465 mg/m3 STEL
STATE LISTS
California - Exposure Limits - PELs
50 ppm PEL; 235 mg/m3 PEL
California - Directors List of Hazardous Substances (8 CCR 339)
[present]
Florida Hazardous Substance List
[present]
Minnesota Hazardous Substance List
[present]
NJ Right to Know List (Total)
sn 1279
Pennsylvania Right to Know List
[present]

2-METHYL-1,4-BENZOQUINONE   **553-97-9**
STATE LISTS
California - Directors List of Hazardous Substances (8 CCR 339)
[present]

4-(6-METHYL-2-BENZOTHIAZOLYL)-BENZENAMINE   **92-36-4**
HEALTH AND SAFETY LISTS
IARC - Group 3 (not classifiable)
[present]

METHYL PHENYL CARBINOL   **98-85-1**
HEALTH AND SAFETY LISTS
ACGIH 1995 - Time Weighted Averages
0.5 ppm TWA; 2.2 mg/m3 TWA
ACGIH 1995 - Skin Designations
skin - potential for cutaneous absorption
AIHA - Odor Threshold Values
no geometric mean air odor threshold
U.S. DOT - Substances From 49 CFR 172.101
regulated by DOT (UN2294)
U.S. DOT - Hazard Classes
DOT hazard class = 6.1
NIOSH 1990 - Pocket Guide - RELs
0.5 ppm TWA; 2 mg/m3 TWA
NIOSH 1990 - Pocket Guide - IDLHs
100 ppm IDLH
NIOSH 1990 - Pocket Guide - Target organs
respiratory system, liver, kidneys, blood
NIOSH 1990 - Pocket Guide - Skin list
Potential for dermal absorption
OSHA - Vacated PELs - Time Weighted Averages
0.5 ppm TWA; 2 mg/m3 TWA
OSHA - Vacated PELs - Skin Designation
Prevent or reduce skin absorption
OSHA - Final PELs - Time Weighted Averages
2 ppm TWA; 9 mg/m3 TWA
OSHA - Final PELs - Skin Notations
prevent or reduce skin absorption
ENVIRONMENTAL LISTS
CAA - HON Rule - SOCMI Chemicals
compliance by April 24, 1995
INTERNATIONAL LISTS
Australian Exposure Standards - Time Weighted Averages
0.5 ppm TWA; 2.2 mg/m3 TWA

Australian Exposure Standards - Skin Effects
skin absorption
Canada - WHMIS: Ingredient Disclosure
1% item 1017 (1116)
Canada - Alberta - 8 Hour Occupational Exposure Limit
0.5 ppm TWA; 2.2 mg/m3 TWA
Canada - Alberta - 15 Minute Occupational Exposure Limit
1 ppm STEL; 4.4 mg/m3 STEL
Canada - Alberta - Skin Designation
can be absorbed through the intact skin
Canada - British Columbia - 8 Hour Exposure Limits
2 ppm TWA; 9 mg/m3 TWA
Canada - British Columbia - 15 Minute Exposure Limits
4 ppm STEL; 18 mg/m3 STEL
Canada - British Columbia - Skin Notations
skin - potential for skin absorption
Canada - Ontario - OHSA - TWAEVs
0.5 ppm TWAEV; 2.2 mg/m3 TWAEV
Canada - Ontario - OHSA - Skin Notations
absorption through skin, eyes, or mucous membranes
Canada - Quebec - Time-Weighted Average Exposure Values
0.5 ppm TWAEV; 2.2 mg/m3 TWAEV
Canada - Quebec - Skin Designations
absorbed through the skin
United Kingdom - Occupational Exposure Standards - TWAs
0.5 ppm TWA; 2 mg/m3 TWA
United Kingdom - Occupational Exposure Standards - Notes
can be absorbed through skin
German (DFG) - MAK Values
0.5 ppm MAK; 2 mg/m3 MAK
German (DFG) - Peak Limitations
2 x normal MAK (30 min. average value); don't exceed 4 times during shift
German (DFG) - Skin/Sensitizers
danger of cutaneous absorption
Israel - Time Weighted Averages
0.5 ppm TWA 2.2 mg/m3 TWA
Israel - Action Levels
0.25 ppm AL; 1.1 mg/m3 AL
Mexico - Instruction No. 10 - TWAs
2 ppm TWA; 9 mg/m3 TWA
Mexico - Instruction No. 10 - Skin designation
skin - potential for cutaneous absorption
STATE LISTS
California - Exposure Limits - PELs
0.5 ppm PEL; 2 mg/m3 PEL
California - Exposure Limits - Skin Notation
material may be absorbed through the skin, eyes or mucous membrane
California - Directors List of Hazardous Substances (8 CCR 339)
[present]
Florida Hazardous Substance List
[present]
Massachusetts Right To Know List
[present]
Minnesota Hazardous Substance List
skin
NJ Right to Know List (Total)
sn 1229
Pennsylvania Right to Know List
[present]

P-METHYLBENZYL ALCOHOL   589-18-4

HEALTH AND SAFETY LISTS

NFPA - Flash Points
flash point > 212 degrees F (100 degrees C)

NFPA - Hazard Identification Ratings
health-0; flammability-1; reactivity-0

ALPHA-METHYLBENZYLAMINE   98-84-0

HEALTH AND SAFETY LISTS

IARC - Group 2B (sufficient animal data)
[present] (Overall evaluation based only on evidence of carcinogenicity in monograph (10, 1976) or in Supplement 4)

OSHA - Possible Select Carcinogens
[present]

STATE LISTS

California - Air Bill 2588 Appendix A-II
9/90

California - Prop. 65 - Cancer list
carcinogen - initial date 4/1/88

California - Directors List of Hazardous Substances (8 CCR 339)
[present]

Florida Hazardous Substance List
[present]

Massachusetts Right To Know List
carcinogen; extraordinarily hazardous

Pennsylvania Right to Know List
special hazardous substance

Pennsylvania RTK - Special Hazardous Substances
[present]

ALPHA-METHYLBENZYL DIMETHYL AMINE   2449-49-2

HEALTH AND SAFETY LISTS

IARC - Group 2B (sufficient animal data)
[present] (Overall evaluation based only on evidence of carcinogenicity in monograph (10, 1976) or in Supplement 4)

OSHA - Possible Select Carcinogens
[present]

INTERNATIONAL LISTS

Canada - WHMIS: Ingredient Disclosure
1% item 1018 (25)

STATE LISTS

California - Air Bill 2588 Appendix A-II
known or potential carcinogen: 9/89

California - Prop. 65 - Cancer list
carcinogen - initial date 4/1/88

California - Directors List of Hazardous Substances (8 CCR 339)
[present]

Florida Hazardous Substance List
[present]

Massachusetts Right To Know List
carcinogen; extraordinarily hazardous

Pennsylvania Right to Know List
special hazardous substance

Pennsylvania RTK - Special Hazardous Substances
[present]

1-METHYL-2-BENZYLHYDRAZINE   10309-79-2

HEALTH AND SAFETY LISTS

U.S. DOT - Substances From 49 CFR 172.101
regulated by DOT (UN2938)

U.S. DOT - Hazard Classes
DOT hazard class = 6.1

NFPA - Flash Points
flash point = 181 degrees F (83 degrees C)

NFPA - Hazard Identification Ratings
health-0; flammability-2; reactivity-0

NIOSH - Selected LD50s and LC50s
Oral, rat: LD50 = 1350 mg/kg

STATE LISTS

NJ Right to Know List (Total)
sn 1230

2-METHYLBIPHENYL   643-58-3

INTERNATIONAL LISTS

Canada - WHMIS: Ingredient Disclosure
1% item 1019 (1117)

4-METHYL BIPHENYL   644-08-6

HEALTH AND SAFETY LISTS

NTP Chemical Status Reports - Testing Status and NTIS Number
Prechronic studies for which toxicity technical reports were not prepared

4-METHYL-2,6-BIS(METHYLTHIO)-   102093-68-5

HEALTH AND SAFETY LISTS

U.S. DOT - Substances From 49 CFR 172.101
regulated by DOT (UN2937)

U.S. DOT - Hazard Classes
DOT hazard class = 6.1

NFPA - Flash Points
flash point = 200 degrees F (93 degrees C)

NFPA - Hazard Identification Ratings
health-0; flammability-2; reactivity-0

NIOSH - Selected LD50s and LC50s
Oral, rat: LD50 = 400 mg/kg Skin, rabbit: LD50 = 2500 mg/kg

NTP Chemical Status Reports - Testing Status and NTIS Number
Technical reports printed (PB90241092)

NTP Chemical Status Reports - Evidence of Carcinogenicity
male rat-some evidence; female rat-no evidence; male mice-no evidence; female mice-no evidence

ENVIRONMENTAL LISTS

CAA - HON Rule - SOCMI Chemicals
compliance by Jan. 23, 1995

INTERNATIONAL LISTS

Canada - WHMIS: Ingredient Disclosure
1% item 1020 (182)

STATE LISTS

NJ Right to Know List (Total)
sn 0049

METHYL BROMIDE (BROMOMETHANE)   74-83-9

HEALTH AND SAFETY LISTS

NIOSH - Selected LD50s and LC50s
Oral, rat: LD50 = 1400 mg/kg

METHYL BROMIDE AND ETHYLENE DIBROMIDE   RR-01300-3
MIXTURES, LIQUID

HEALTH AND SAFETY LISTS

NFPA - Flash Points
flash point = 175 degrees F (79 degrees C)

NFPA - Hazard Identification Ratings
health-2; flammability-2; reactivity-0

STATE LISTS

Florida Hazardous Substance List
[present]

Massachusetts Right To Know List
[present]

Pennsylvania Right to Know List
[present]

METHYL BROMOACETATE　　　　　　　　　96-32-2

HEALTH AND SAFETY LISTS

NFPA - Flash Points
flash point = 175 degrees F (79 degrees C)

NFPA - Hazard Identification Ratings
health-2; flammability-2; reactivity-0

STATE LISTS

Florida Hazardous Substance List
[present]

Massachusetts Right To Know List
[present]

Pennsylvania Right to Know List
[present]

1-METHYLBUTADIENE　　　　　　　　　　504-60-9

INTERNATIONAL LISTS

Canada - WHMIS: Ingredient Disclosure
0.1% item 1021 (1118)

3-METHYL-2-BUTANETHIOL　　　　　　　2084-18-6

HEALTH AND SAFETY LISTS

NFPA - Flash Points
flash point = 280 degrees F (137 degrees C)

NFPA - Hazard Identification Ratings
health-2; reactivity-0

STATE LISTS

Florida Hazardous Substance List
[present]

Massachusetts Right To Know List
[present]

Pennsylvania Right to Know List
[present]

2-METHYL-2-BUTANOL　　　　　　　　　　75-85-4

HEALTH AND SAFETY LISTS

NIOSH - Selected LD50s and LC50s
Oral, rat: LD50 = 2570 mg/kg

2-METHYL-1-BUTANOL　　　　　　　　　　137-32-6

ENVIRONMENTAL LISTS

TSCA - Code of Federal Regulations Citations
40 CFR 721.557

3-METHYL-2-BUTANOL　　　　　　　　　　598-75-4
SEE ALSO:
K132-HAZARDOUS WASTES
BUTYLDECALIN
F039-HAZARDOUS WASTES
K131-HAZARDOUS WASTES
AMYL OXALATE

HEALTH AND SAFETY LISTS

ACGIH 1995 - Time Weighted Averages
5 ppm TWA; 19 mg/m3 TWA

ACGIH 1995 - Skin Designations
skin - potential for cutaneous absorption

U.S. DOT - Substances From 49 CFR 172.101
regulated by DOT (UN1062)

U.S. DOT - Hazard Classes
DOT hazard class = 2.3

U.S. DOT - Substances Which Are Poisonous by Inhalation
gaseous hazardous material poisonous by inhalation (UN1062); liquid hazardous material poisonous by inhalation (with Ethylene dibromide mixtures) (UN1647)

U.S. DOT - Appendix A Table 1 - Hazardous Substances
final RQ = 1000 pounds (454 kg)

IARC - Group 3 (not classifiable)
[present]

NFPA - Flash Points
practically non-flammable

NFPA - Hazard Identification Ratings
health-3; flammability-1; reactivity-0

NIOSH - Selected LD50s and LC50s
Inhalation, rat: LC50 = 302 ppm 8 hr Oral, rat: LD50 = 214 mg/kg

NIOSH 1990 - Pocket Guide - RELs
Reduce exposure to lowest feasible concentration

NIOSH 1990 - Pocket Guide - IDLHs
2000 ppm IDLH (not considering carcinogenic effects)

NIOSH 1990 - Pocket Guide - Carcinogens
occupational carcinogen

NIOSH 1990 - Pocket Guide - Target organs
skin, eyes, CNS, respiratory system

NIOSH - Health Standards - Exposure Limits
reduce exposure to lowest feasible concentration (Listed under 'Monohalomethanes')

NIOSH - Health Standards - Health Effects and Precautions
has produced tumors of the kidney, forestomach, and lung in animals (Prevent skin contact) (Listed under 'Monohalomethanes')

NIOSH - Health Standards - Carcinogenic Chemicals
potential human carcinogen (Listed under 'Monohalomethanes')

NTP Chemical Status Reports - Testing Status and NTIS Number
Technical reports printed (PB92-189257); prechronic studies for which toxicity technical reports were not printed

NTP Chemical Status Reports - Evidence of Carcinogenicity
male mice: no evidence; female mice: no evidence

OSHA - Vacated PELs - Time Weighted Averages
5 ppm TWA; 20 mg/m3 TWA

OSHA - Vacated PELs - Skin Designation
Prevent or reduce skin absorption

OSHA - Final PELs - Ceiling Limits
C 20 ppm; C 80 mg/m3

OSHA - Final PELs - Skin Notations
prevent or reduce skin absorption

OSHA - List of Highly Hazardous Chemicals
threshhold quantity = 2500 pounds

ENVIRONMENTAL LISTS

CERCLA/SARA - Section 302 Extremely Hazardous Substances and TPQs
TPQ = 1000 pounds

CERCLA/SARA - Section 313 - Emission Reporting
form R reporting required for 1.0% de minimus concentration

CERCLA/SARA - Hazardous Substances and their Reportable Quantities
final RQ = 1000 pounds (454 kg)

Clean Air Act (1990) - List of Hazardous Air Contaminants
[present]

Class 1 Ozone Depletors
ozone depletion weight = 0.7

CAA - HON Rule - SOCMI Chemicals
compliance by July 24, 1995

CAA - HON Rule - Organic HAPs
[present]

Clean Water Act - Priority Pollutants
[present]

Safe Drinking Water Act - Monitoring
monitoring required
RCRA - U Series Wastes
waste number U029
RCRA - Hazardous Constituents-Appendix VIII
waste number U029
RCRA - Basis for Listing - Appendix VII
Included in waste streams: F039, K131, K132
RCRA - Substances Banned From Land Disposal
[present]
RCRA - TSD Facilities Ground Water Monitoring
TM 8010 = 20 ug/L PQL; TM 8240 = 10 ug/L PQL
RCRA - Universal Treatment Standards (LDR)
WW: 0.11 mg/l; NWW: 15 mg/kg
TSCA - Code of Federal Regulations Citations
40 CFR 716.120(a)
TSCA - PAIR - Reporting List
Reporting Date: June 13, 1989
TSCA - Health and Safety Reporting List
Effective Date: June 1, 1987

INTERNATIONAL LISTS

Australian Exposure Standards - Time Weighted Averages
5 ppm TWA; 19 mg/m3 TWA
Australian Exposure Standards - Skin Effects
skin absorption
Canada - NPRI (National Pollutant Release Inventory)
[present]
Canada - Alberta - 15 Minute Occupational Exposure Limit
15 ppm STEL; 58 mg/m3 STEL
Canada - Alberta - Designated Substances
designated substance - requires code of practice
Canada - British Columbia - 8 Hour Exposure Limits
15 ppm TWA; 60 mg/m3 TWA
Canada - British Columbia - Skin Notations
skin - potential for skin absorption
Canada - Ontario - OHSA - TWAEVs
5 ppm TWAEV; 19 mg/m3 TWAEV
Canada - Ontario - OHSA - Skin Notations
absorption through skin, eyes, or mucous membranes
Canada - Quebec - Time-Weighted Average Exposure Values
5 ppm TWAEV; 19 mg/m3 TWAEV
Canada - Quebec - Skin Designations
absorbed through the skin
United Kingdom - Occupational Exposure Standards - TWAs
5 ppm TWA; 20 mg/m3 TWA
United Kingdom - Occupational Exposure Standards - STELs
15 ppm STEL; 60 mg/m3 STEL
United Kingdom - Occupational Exposure Standards - Notes
can be absorbed through skin
German (DFG) - Skin/Sensitizers
danger of cutaneous absorption
German (DFG) - Carcinogens
suspected carcinogen
Israel - Time Weighted Averages
5 ppm TWA; 19 mg/m3 TWA
Israel - Action Levels
2.5 ppm AL; 9.5 mg/m3 AL
Mexico - Instruction No. 10 - TWAs
5 ppm TWA; 20 mg/m3 TWA
Mexico - Instruction No. 10 - STELs
15 ppm STEL; 60 mg/m3 STEL
Mexico - Instruction No. 10 - Skin designation
skin - potential for cutaneous absorption

Mexico - Wastewater - Organic Toxic Pollutants and Heavy Metals
Listed under [Halomethanes]

STATE LISTS

California - Air Bill 2588 Appendix A-I
[present]
California - Prop. 65 - Developmental Toxicity
as a strictural fumigant: developmental toxicity - initial date 1/1/93
California - Exposure Limits - PELs
5 ppm PEL; 20 mg/m3 PEL
California - Exposure Limits - STELs
25 ppm 5-minute STEL
California - Exposure Limits - Ceilings
C 50 ppm
California - Exposure Limits - Skin Notation
material may be absorbed through the skin, eyes or mucous membrane
California - Directors List of Hazardous Substances (8 CCR 339)
[present]
Florida Hazardous Substance List
[present]
Massachusetts Right To Know List
extraordinarily hazardous
Minnesota Hazardous Substance List
carcinogen; skin
NJ Right to Know List (Total)
sn 1231
Pennsylvania Right to Know List
environmental hazard

PROPOSED REGULATIONS

Safe Drinking Water Act - Priority list
[present]

2-METHYL-2-BUTENE   513-35-9

HEALTH AND SAFETY LISTS

U.S. DOT - Appendix B - Marine Pollutants
DOT regulated marine pollutant

3-METHYL-1-BUTENE   563-45-1

HEALTH AND SAFETY LISTS

U.S. DOT - Substances From 49 CFR 172.101
regulated by DOT (UN2643)
U.S. DOT - Hazard Classes
DOT hazard class = 6.1

INTERNATIONAL LISTS

Canada - WHMIS: Ingredient Disclosure
1% item 1023 (323)

STATE LISTS

NJ Right to Know List (Total)
sn 1232

METHYL BUTENE   26760-64-5

HEALTH AND SAFETY LISTS

U.S. DOT - Appendix A Table 1 - Hazardous Substances
final RQ = 100 pounds (45.4 kg)
NFPA - Flash Points
flash point = -20 degrees F (-29 degrees C)
NFPA - Hazard Identification Ratings
health-0; flammability-4; reactivity-2
NIOSH - Selected LD50s and LC50s
Inhalation, rat: LC50 = 140 gm/m3 2 hr

ENVIRONMENTAL LISTS

CERCLA/SARA - Hazardous Substances and their Reportable Quantities
final RQ = 100 pounds (45.4 kg)

CAA - Flammable Substances for Accidental Release Prevention
threshold quantity = 10,000 lbs

EPA - Master Testing List
[present]

RCRA - U Series Wastes
waste number U186 (Ignitable waste)

RCRA - Hazardous Constituents-Appendix VIII
waste number U186 (Ignitable waste)

RCRA - Substances Banned From Land Disposal
[present]

STATE LISTS

Florida Hazardous Substance List
[present]

Massachusetts Right To Know List
[present]

Pennsylvania Right to Know List
environmental hazard

2-METHYL-1-BUTENE (TECHNICAL)   563-46-2

HEALTH AND SAFETY LISTS

NFPA - Flash Points
flash point = 37 degrees F (3 degrees C)

NFPA - Hazard Identification Ratings
health-2; flammability-3; reactivity-0

STATE LISTS

Florida Hazardous Substance List
[present]

Massachusetts Right To Know List
[present]

Pennsylvania Right to Know List
[present]

N-METHYLBUTYLAMINE   110-68-9

HEALTH AND SAFETY LISTS

NFPA - Flash Points
flash point = 67 degrees F (19 degrees C)

NFPA - Hazard Identification Ratings
health-1; flammability-3; reactivity-0

NIOSH - Selected LD50s and LC50s
Oral, rat: LD50 = 1000 mg/kg

STATE LISTS

Florida Hazardous Substance List
[present]

Massachusetts Right To Know List
[present]

Pennsylvania Right to Know List
[present]

METHYL TERT-BUTYL ETHER   1634-04-4

HEALTH AND SAFETY LISTS

NFPA - Flash Points
flash point = 122 degrees F (50 degrees C)

NFPA - Hazard Identification Ratings
health-2; flammability-2; reactivity-0

NIOSH - Selected LD50s and LC50s
Oral, rat: LD50 = 4920 mg/kg Skin, rabbit: LD50 = 3540 mg/kg

STATE LISTS

Florida Hazardous Substance List
[present]

Massachusetts Right To Know List
[present]

Pennsylvania Right to Know List
[present]

4-METHYL-2-T-BUTYLPHENOL   2409-55-4

STATE LISTS

Florida Hazardous Substance List
[present]

Massachusetts Right To Know List
[present]

Pennsylvania Right to Know List
[present]

3-METHYL BUTYNOL   115-19-5

HEALTH AND SAFETY LISTS

U.S. DOT - Substances From 49 CFR 172.101
regulated by DOT (UN2460)

U.S. DOT - Hazard Classes
DOT hazard class = 3

NFPA - Flash Points
flash point < 20 degrees F (-7 degrees C)

NFPA - Hazard Identification Ratings
health-2; flammability-3; reactivity-0

STATE LISTS

Florida Hazardous Substance List
[present]

Massachusetts Right To Know List
[present]

Pennsylvania Right to Know List
[present]

METHYL BUTYRATE   623-42-7

HEALTH AND SAFETY LISTS

U.S. DOT - Substances From 49 CFR 172.101
regulated by DOT (UN2561)

U.S. DOT - Hazard Classes
DOT hazard class = 3

NFPA - Flash Points
flash point < 20 degrees F (-7 degrees C)

NFPA - Hazard Identification Ratings
health-2; flammability-4; reactivity-0

ENVIRONMENTAL LISTS

CAA - Flammable Substances for Accidental Release Prevention
threshold quantity = 10,000 lbs

STATE LISTS

Florida Hazardous Substance List
[present]

Massachusetts Right To Know List
[present]

Pennsylvania Right to Know List
[present]

METHYL CARBAMATE   598-55-0

STATE LISTS

NJ Right to Know List (Total)
sn 1233

NJ Special Hazardous Substances
(flammable - third degree; reactive - second degree)

METHYLCELLULOSE   9004-67-5

HEALTH AND SAFETY LISTS

U.S. DOT - Substances From 49 CFR 172.101
regulated by DOT (UN2459)

U.S. DOT - Hazard Classes
DOT hazard class = 3

NFPA - Flash Points
flash point < 20 degrees F (-7 degrees C)

NFPA - Hazard Identification Ratings
health-2; flammability-4; reactivity-0

ENVIRONMENTAL LISTS

CAA - Flammable Substances for Accidental Release Prevention
threshold quantity = 10,000 lbs

STATE LISTS

Florida Hazardous Substance List
[present]

Massachusetts Right To Know List
[present]

Pennsylvania Right to Know List
[present]

METHYL CHLORIDE   74-87-3

HEALTH AND SAFETY LISTS

U.S. DOT - Substances From 49 CFR 172.101
regulated by DOT (UN2945)

U.S. DOT - Hazard Classes
DOT hazard class = 3

NFPA - Flash Points
flash point = 55 degrees F (13 degrees C)

NFPA - Hazard Identification Ratings
health-3; flammability-3; reactivity-0

NIOSH - Selected LD50s and LC50s
Oral, rat: LD50 = 420 mg/kg Skin, rabbit: LD50 = 1260 mg/kg

INTERNATIONAL LISTS

Canada - WHMIS: Ingredient Disclosure
1% item 1025 (1121)

STATE LISTS

Florida Hazardous Substance List
[present]

Massachusetts Right To Know List
[present]

NJ Right to Know List (Total)
sn 1401

Pennsylvania Right to Know List
[present]

METHYL CHLOROACETATE   96-34-4

HEALTH AND SAFETY LISTS

ACGIH 1995 - Time Weighted Averages
(40) ppm TWA; (144) mg/m3 TWA

AIHA - WEEL - Time Weighted Averages
100 ppm TWA; 360 mg/m3 TWA

U.S. DOT - Substances From 49 CFR 172.101
regulated by DOT (UN2398)

U.S. DOT - Hazard Classes
DOT hazard class = 3

ENVIRONMENTAL LISTS

ATSDR Priority List
Rank (of 275): 234

CERCLA/SARA - Section 313 - Emission Reporting
form R reporting required for 1.0% de minimus concentration

CERCLA/SARA - Hazardous Substances and their Reportable Quantities
final RQ = 1 pound (.454 kg)

Clean Air Act (1990) - List of Hazardous Air Contaminants
[present]

CAA - HON Rule - SOCMI Chemicals
compliance by Oct. 23, 1995

CAA - HON Rule - Organic HAPs
[present]

EPA - Master Testing List
[present]

TSCA - Code of Federal Regulations Citations
40 CFR 712.30(t); 40 CFR 716.120(a); 40 CFR 799.5000

TSCA - PAIR - Reporting List
Reporting Date: February 12, 1987

TSCA - Health and Safety Reporting List
Effective Date: December 15, 1986

TSCA - Substances Subject to Testing Consent Orders
Test for: Health Effects

TSCA - Section 12(b) - Export Notification
export notification required - Section 4

INTERNATIONAL LISTS

Canada - NPRI (National Pollutant Release Inventory)
[present]

STATE LISTS

California - Air Bill 2588 Appendix A-I
6/91

Massachusetts Right To Know List
[present]

NJ Right to Know List (Total)
sn 1293

Pennsylvania Right to Know List
environmental hazard

PROPOSED REGULATIONS

ACGIH 1995 - Notice of Intended Changes
40 ppm TWA; 144 mg/m3 TWA; A3-animal carcinogen

Safe Drinking Water Act - Priority list
[present]

METHYL-2-CHLOROACRYLATE   80-63-7

HEALTH AND SAFETY LISTS

NIOSH - Selected LD50s and LC50s
Oral, rat: LD50 = 2390 mg/kg Skin, rabbit: LD50 = 2200 mg/kg

METHYLCHLOROBENZENES   RR-01301-4

HEALTH AND SAFETY LISTS

NFPA - Flash Points
flash point = 77 degrees F (25 degrees C)

NFPA - Hazard Identification Ratings
health-2; flammability-3; reactivity-0

NIOSH - Selected LD50s and LC50s
Oral, rat: LD50 = 1950 mg/kg

ENVIRONMENTAL LISTS

EPA - Master Testing List
[present]

List of Pesticide Product Inert Ingredients
[present]

STATE LISTS

Florida Hazardous Substance List
[present]

Massachusetts Right To Know List
[present]

Pennsylvania Right to Know List
[present]

METHYL CHLOROFORMATE   79-22-1

HEALTH AND SAFETY LISTS

U.S. DOT - Substances From 49 CFR 172.101
regulated by DOT (UN1237)

U.S. DOT - Hazard Classes
DOT hazard class = 3

NFPA - Flash Points
flash point = 57 degrees F (14 degrees C)

NFPA - Hazard Identification Ratings
health-2; flammability-3; reactivity-0

NIOSH - Selected LD50s and LC50s
Inhalation, mouse: LC50 = 18 gm/m3 2 hr Oral, rabbit: LD50 = 3380 mg/kg Skin, rabbit: LD50 = 3560 mg/kg

INTERNATIONAL LISTS

Canada - WHMIS: Ingredient Disclosure
1% item 1026 (375)

STATE LISTS

Florida Hazardous Substance List
[present]

Massachusetts Right To Know List
[present]

NJ Right to Know List (Total)
sn 1234

NJ Special Hazardous Substances
(flammable - third degree)

Pennsylvania Right to Know List
[present]

4-(2-METHYL-4-CHLOROPHENOXY) BUTYRIC ACID   94-81-5

HEALTH AND SAFETY LISTS

IARC - Group 3 (not classifiable)
[present]

NIOSH - Selected LD50s and LC50s
Oral, quail: LD50 = 21 mg/kg

NTP Chemical Status Reports - Testing Status and NTIS Number
Technical reports printed (PB88168570/AS)

NTP Chemical Status Reports - Evidence of Carcinogenicity
male rat-clear evidence; female rat-clear evidence; male mice-no evidence; female mice-no evidence

2-(2-METHYL-4-CHLOROPHENOXY) PROPIONIC   93-65-2
ACID

ENVIRONMENTAL LISTS

List of Pesticide Product Inert Ingredients
[present]

METHYL-2-CHLOROPROPIONATE   17639-93-9
SEE ALSO:
K149-HAZARDOUS WASTES
K009-HAZARDOUS WASTES
K150-HAZARDOUS WASTES
F025-HAZARDOUS WASTES
K010-HAZARDOUS WASTES
F039-HAZARDOUS WASTES
F024-HAZARDOUS WASTES
BIS(2,4-DIMETHYLBUTYL) MALEATE

HEALTH AND SAFETY LISTS

ACGIH 1995 - Time Weighted Averages
50 ppm TWA; 103 mg/m3 TWA

ACGIH 1995 - Short Term Exposure Limits
100 ppm STEL; 207 mg/m3 STEL

ACGIH 1995 - Skin Designations
skin - potential for cutaneous absorption

AIHA - Odor Threshold Values
no geometric mean air odor threshold

U.S. DOT - Substances From 49 CFR 172.101
regulated by DOT (UN1912, UN1063)

U.S. DOT - Hazard Classes
DOT hazard class = 2.3

U.S. DOT - Appendix A Table 1 - Hazardous Substances
final RQ = 100 pounds (45.4 kg)

IARC - Group 3 (not classifiable)
[present]

NFPA - Flash Points
flash point = -50 degrees F (-46 degrees C)

NFPA - Hazard Identification Ratings
health-1; flammability-4; reactivity-0

NIOSH - Selected LD50s and LC50s
Inhalation, rat: LC50 = 152000 mg/m3 (30 mn)

NIOSH 1990 - Pocket Guide - RELs
Reduce exposure to lowest feasible concentration

NIOSH 1990 - Pocket Guide - IDLHs
10,000 ppm IDLH (not considering carcinogenic effects)

NIOSH 1990 - Pocket Guide - Carcinogens
occupational carcinogen

NIOSH 1990 - Pocket Guide - Target organs
CNS, liver, skin, kidneys

NIOSH - Health Standards - Exposure Limits
reduce exposure to lowest feasible concentration (Listed under 'Mono-halomethanes')

NIOSH - Health Standards - Health Effects and Precautions
Potential teratogen; has produced tumors of the kidney, forestom-ach, and lung in animals (Prevent skin contact) (Listed under 'Mono-halomethanes')

NIOSH - Health Standards - Carcinogenic Chemicals
potential human carcinogen (Listed under 'Monohalomethanes')

OSHA - Vacated PELs - Time Weighted Averages
50 ppm TWA; 105 mg/m3 TWA

OSHA - Vacated PELs - Short Term Exposure Limits
100 ppm STEL; 210 mg/m3 STEL

OSHA - Final PELs - Time Weighted Averages
100 ppm TWA; C 200 ppm

OSHA - Final PELs - Ceiling Limits
C 200 ppm

OSHA - List of Highly Hazardous Chemicals
threshhold quantity = 15,000 pounds

ENVIRONMENTAL LISTS

ATSDR Priority List
Rank (of 275): 166

CERCLA/SARA - Section 313 - Emission Reporting
form R reporting required for 1.0% de minimus concentration

CERCLA/SARA - Hazardous Substances and their Reportable Quantities
final RQ = 100 pounds (45.4 kg)

Clean Air Act (1990) - List of Hazardous Air Contaminants
[present]

CAA -Toxic Substances for Accidental Release Prevention
threshold quantity = 10,000 lbs

CAA - HON Rule - SOCMI Chemicals
compliance by July 24, 1995

CAA - HON Rule - Organic HAPs
[present]

Clean Water Act - Priority Pollutants
[present]

Safe Drinking Water Act - Monitoring
monitoring required

EPA - Carcinogen Hazard Ranking for RQ Adjustment
Hazard ranking = Low

EPA - Master Testing List
[present]

RCRA - U Series Wastes
waste number U045 (Ignitable waste; Toxic waste)

RCRA - Hazardous Constituents-Appendix VIII
waste number U045

RCRA - Basis for Listing - Appendix VII
Included in waste streams: F024, F025, F039, K009, K010, K149, K150

RCRA - Substances Banned From Land Disposal
[present]

RCRA - TSD Facilities Ground Water Monitoring
TM 8010 = 1 ug/L PQL; TM 8240 = 5 ug/L PQL

RCRA - Universal Treatment Standards (LDR)
WW: 0.19 mg/l; NWW: 30 mg/kg

TSCA - Code of Federal Regulations Citations
40 CFR 712.30(d); 40 CFR 716.120(a)

TSCA - PAIR - Reporting List
Reporting Date: November 19, 1982

TSCA - Health and Safety Reporting List
Effective Date: October 4, 1982

TSCA - Multichemical Test Rules - Waste Constituents
hydrolysis testing required for chemical fate

TSCA - Section 12(b) - Export Notification
export notification required - Section 4

INTERNATIONAL LISTS

Australian Exposure Standards - Time Weighted Averages
50 ppm TWA; 103 mg/m3 TWA

Australian Exposure Standards - Short Term Exposure Limits
100 ppm STEL; 207 mg/m3 STEL

Canada - WHMIS: Ingredient Disclosure
1% item 1027 (507)

Canada - NPRI (National Pollutant Release Inventory)
[present]

Canada - Alberta - 8 Hour Occupational Exposure Limit
25 ppm TWA; 51 mg/m3 TWA

Canada - Alberta - 15 Minute Occupational Exposure Limit
50 ppm STEL; 102 mg/m3 STEL

Canada - British Columbia - 8 Hour Exposure Limits
100 ppm TWA; 210 mg/m3 TWA

Canada - British Columbia - 15 Minute Exposure Limits
125 ppm STEL; 260 mg/m3 STEL

Canada - Ontario - OHSA - TWAEVs
50 ppm TWAEV; 103 mg/m3 TWAEV

Canada - Ontario - OHSA - STEVs
100 ppm STEV; 205 mg/m3 STEV

Canada - Quebec - Time-Weighted Average Exposure Values
50 ppm TWAEV; 103 mg/m3 TWAEV

Canada - Quebec - Short-term Exposure Values
100 ppm STEV; 207 mg/m3 STEV

Canada - Quebec - Skin Designations
absorbed through the skin

United Kingdom - Occupational Exposure Standards - TWAs
50 ppm TWA; 105 mg/m3 TWA

United Kingdom - Occupational Exposure Standards - STELs
100 ppm STEL; 210 mg/m3 STEL

German (DFG) - MAK Values
50 ppm MAK; 105 mg/m3 MAK

German (DFG) - Peak Limitations
2 x normal MAK (30 min. average value); don't exceed 4 times during shift

German (DFG) - Carcinogens
suspected carcinogen

German (DFG) - Pregnancy
risk to embryo/fetus probable

Israel - Time Weighted Averages
50 ppm TWA; 103 mg/m3 TWA

Israel - Short Term Exposure Limits
100 ppm STEL; 207 mg/m3 STEL

Israel - Action Levels
25 ppm AL; 51.5 mg/m3 AL

Mexico - Instruction No. 10 - TWAs
50 ppm TWA; 105 mg/m3 TWA

Mexico - Instruction No. 10 - STELs
100 ppm STEL; 205 mg/m3 STEL

Mexico - Wastewater - Organic Toxic Pollutants and Heavy Metals
Listed under [Halomethanes]

Mexico - Drinking Water - Ecological Criteria
0.002 mg/l This level has been extrapolated by using a mathematic model

STATE LISTS

California - Air Bill 2588 Appendix A-I
6/91

California - Exposure Limits - PELs
5 ppm PEL; 105 mg/m3 PEL

California - Exposure Limits - STELs
100 ppm STEL; 210 mg/m3 STEL

California - Exposure Limits - Ceilings
C 300 ppm

California - Directors List of Hazardous Substances (8 CCR 339)
[present]

Florida Hazardous Substance List
[present]

Massachusetts Right To Know List
[present]

Minnesota Hazardous Substance List
carcinogen

NJ Right to Know List (Total)
sn 1235

NJ Special Hazardous Substances
(flammable - fourth degree)

Pennsylvania Right to Know List
environmental hazard

PROPOSED REGULATIONS

Safe Drinking Water Act - Priority list
[present]

METHYLCHLOROSILANE   993-00-0

HEALTH AND SAFETY LISTS

U.S. DOT - Substances From 49 CFR 172.101
regulated by DOT (UN2295)

U.S. DOT - Hazard Classes
DOT hazard class = 6.1

NFPA - Flash Points
flash point = 135 degrees F (57 degrees C)

NFPA - Hazard Identification Ratings
health-2; flammability-2; reactivity-1

NIOSH - Selected LD50s and LC50s
Inhalation, mouse: LC50 = 1 gm/m3 2 hr Oral, mouse: LD50 = 240 mg/kg

INTERNATIONAL LISTS

German (DFG) - MAK Values
1 ppm MAK; 5 mg/m3 MAK

German (DFG) - Peak Limitations
2 x normal MAK (5 min. momentary value) don't exceed 8 times during shift

German (DFG) - Skin/Sensitizers
danger of cutaneous absorption and danger of sensitization

German (DFG) - Pregnancy
classification not yet possible

STATE LISTS

Florida Hazardous Substance List
[present]

Massachusetts Right To Know List
[present]

NJ Right to Know List (Total)
sn 1236

Pennsylvania Right to Know List
[present]

3-METHYLCHOLANTHRENE   56-49-5

HEALTH AND SAFETY LISTS

NIOSH - Selected LD50s and LC50s
Inhalation, rat: LC50 = 500 mg/m3 2 hr

ENVIRONMENTAL LISTS

CERCLA/SARA - Section 302 Extremely Hazardous Substances and
TPQs
TPQ = 500 pounds

STATE LISTS

Florida Hazardous Substance List
effective March 13, 1992

Massachusetts Right To Know List
extraordinarily hazardous

NJ Right to Know List (Total)
sn 2556

Pennsylvania Right to Know List
environmental hazard

PROPOSED REGULATIONS

CERCLA/SARA - Proposed Hazardous Substance Additions
proposed RQ = 1 pound (.454 kg)

CERCLA/SARA - 1989 Proposed RQ Adjustments
proposed RQ = 100 pounds (45.4 kg)

6-METHYLCHRYSENE   1705-85-7

HEALTH AND SAFETY LISTS

U.S. DOT - Appendix B - Marine Pollutants
DOT regulated marine pollutant

1-METHYLCHRYSENE   3351-28-8

HEALTH AND SAFETY LISTS

U.S. DOT - Substances From 49 CFR 172.101
regulated by DOT (UN1238)

U.S. DOT - Hazard Classes
DOT hazard class = 6.1

U.S. DOT - Substances Which Are Poisonous by Inhalation
liquid hazardous material poisonous by inhalation (UN1238)

U.S. DOT - Appendix A Table 1 - Hazardous Substances
final RQ = 1000 pounds (454 kg)

NIOSH - Selected LD50s and LC50s
*Inhalation, rat: LC50 = 88 ppm 1 hr Oral, rat: LD50 = 60 mg/kg
Skin, mouse: LD50 = 1750 mg/kg*

OSHA - List of Highly Hazardous Chemicals
threshhold quantity = 500 pounds

ENVIRONMENTAL LISTS

CERCLA/SARA - Section 302 Extremely Hazardous Substances and
TPQs
TPQ = 500 pounds

CERCLA/SARA - Section 313 - Emission Reporting
form R reporting required

CERCLA/SARA - Hazardous Substances and their Reportable Quantities
final RQ = 1000 pounds (454 kg)

CAA -Toxic Substances for Accidental Release Prevention
threshold quantity = 5,000 lbs

RCRA - U Series Wastes
waste number U156 (Ignitable waste; Toxic waste)

RCRA - Hazardous Constituents-Appendix VIII
waste number U156

RCRA - Substances Banned From Land Disposal
[present]

INTERNATIONAL LISTS

Canada - WHMIS: Ingredient Disclosure
1% item 1028 (420)

STATE LISTS

Florida Hazardous Substance List
effective March 13, 1992

Massachusetts Right To Know List
extraordinarily hazardous

NJ Right to Know List (Total)
sn 1238

Pennsylvania Right to Know List
environmental hazard

4-METHYLCHRYSENE   3351-30-2

ENVIRONMENTAL LISTS

TSCA - HDD/HDF - Precursors Required for Reporting
[present]

3-METHYLCHRYSENE   3351-31-3

HEALTH AND SAFETY LISTS

NIOSH - Selected LD50s and LC50s
Oral, rat: LD50 = 650 mg/kg Skin, rabbit: LD50 = 900 mg/kg

ENVIRONMENTAL LISTS

CERCLA/SARA - Section 313 - Emission Reporting
form R reporting required

2-METHYLCHRYSENE   3351-32-4

HEALTH AND SAFETY LISTS

U.S. DOT - Substances From 49 CFR 172.101
regulated by DOT (UN2933)

U.S. DOT - Hazard Classes
DOT hazard class = 3

STATE LISTS

NJ Right to Know List (Total)
sn 1239

5-METHYLCHRYSENE   3697-24-3

HEALTH AND SAFETY LISTS

U.S. DOT - Substances From 49 CFR 172.101
regulated by DOT (UN2534)

U.S. DOT - Hazard Classes
DOT hazard class = 2.3

U.S. DOT - Substances Which Are Poisonous by Inhalation
gaseous hazardous material poisonous by inhalation (UN2534)

INTERNATIONAL LISTS

Canada - WHMIS: Ingredient Disclosure
1% item 1030 (1122)

STATE LISTS

NJ Right to Know List (Total)
sn 1240

METHYL CINNAMATE
103-26-4

SEE ALSO:
F039-HAZARDOUS WASTES

HEALTH AND SAFETY LISTS

U.S. DOT - Appendix A Table 1 - Hazardous Substances
final RQ = 10 pounds (4.54 kg)

ENVIRONMENTAL LISTS

CERCLA/SARA - Hazardous Substances and their Reportable Quantities
final RQ = 10 pounds (4.54 kg)

EPA - Carcinogen Hazard Ranking for RQ Adjustment
Hazard ranking = Medium

RCRA - U Series Wastes
waste number U157

RCRA - Hazardous Constituents-Appendix VIII
waste number U157

RCRA - Basis for Listing - Appendix VII
Included in waste stream: F039

RCRA - Substances Banned From Land Disposal
[present]

RCRA - TSD Facilities Ground Water Monitoring
TM 8270 = 10 ug/L PQL

RCRA - Universal Treatment Standards (LDR)
WW: 0.0055 mg/l; NWW: 15 mg/kg

TSCA - Chemicals with Significant New Use Rules
[present]

TSCA - Section 12(b) - Export Notification
export notification required - Section 5

INTERNATIONAL LISTS

Canada - WHMIS: Ingredient Disclosure
0.1% item 1031 (1123)

STATE LISTS

California - Air Bill 2588 Appendix A-I
known or potential carcinogen: 9/90

California - Prop. 65 - Cancer list
carcinogen - initial date 1/1/90

California - Prop. 65 - No Significant Risk Levels
no significant risk level = 0.03 ug/day

Florida Hazardous Substance List
[present]

Massachusetts Right To Know List
carcinogen; extraordinarily hazardous

NJ Special Hazardous Substances
(carcinogen; mutagen)

Pennsylvania Right to Know List
environmental hazard; special hazardous substance

Pennsylvania RTK - Special Hazardous Substances
[present]

6-METHYLCOUMARIN
92-48-8

HEALTH AND SAFETY LISTS

IARC - Group 3 (not classifiable)
[present]

STATE LISTS

California - Directors List of Hazardous Substances (8 CCR 339)
[present]

METHYL PARA-CRESOL
104-93-8

HEALTH AND SAFETY LISTS

IARC - Group 3 (not classifiable)
[present]

METHYL 2-CYANOACRYLATE
137-05-3

HEALTH AND SAFETY LISTS

IARC - Group 3 (not classifiable)
[present]

STATE LISTS

California - Directors List of Hazardous Substances (8 CCR 339)
[present]

METHYLCYCLOHEXANE
108-87-2

HEALTH AND SAFETY LISTS

IARC - Group 3 (not classifiable)
[present]

STATE LISTS

California - Directors List of Hazardous Substances (8 CCR 339)
[present]

2-METHYLCYCLOHEXANOL
583-59-5

HEALTH AND SAFETY LISTS

IARC - Group 3 (not classifiable)
[present]

STATE LISTS

California - Directors List of Hazardous Substances (8 CCR 339)
[present]

4-METHYLCYCLOHEXANOL
589-91-3

HEALTH AND SAFETY LISTS

IARC - Group 2B (sufficient animal data)
[present] (Overall evaluation based only on evidence of carcinogenicity in monograph (32, 1983) or in Supplement 4)

NTP Seventh Report - Suspect Carcinogens
suspect carcinogen (Listed under 'Polycyclic aromatic hydrocarbons')

OSHA - Possible Select Carcinogens
[present]

ENVIRONMENTAL LISTS

CERCLA/SARA - Section 313 - Emission Reporting
form R reporting required; (Listed under 'Polycyclic aromatic compounds')

INTERNATIONAL LISTS

Canada - WHMIS: Ingredient Disclosure
1% item 1032 (1124)

STATE LISTS

California - Air Bill 2588 Appendix A-I
known or potential carcinogen

California - Prop. 65 - Cancer list
carcinogen - initial date 4/1/88

California - Directors List of Hazardous Substances (8 CCR 339)
[present]

Florida Hazardous Substance List
effective March 13, 1992

Massachusetts Right To Know List
carcinogen; extraordinarily hazardous

Minnesota Hazardous Substance List
carcinogen

Pennsylvania Right to Know List
special hazardous substance

Pennsylvania RTK - Special Hazardous Substances
[present]

3-METHYLCYCLOHEXANOL
591-23-1

HEALTH AND SAFETY LISTS

NIOSH - Selected LD50s and LC50s
Oral, rat: LD50 = 2610 mg/kg

METHYLCYCLOHEXANOL   1331-23-3

HEALTH AND SAFETY LISTS

NIOSH - Selected LD50s and LC50s
Oral, rat: LD50 = 1680 mg/kg

NTP Chemical Status Reports - Testing Status and NTIS Number
Prechronic studies for which toxicity technical reports were not prepared

METHYLCYCLOHEXANOL   25639-42-3

HEALTH AND SAFETY LISTS

NFPA - Flash Points
flash point = 140 degrees F (60 degrees C)

NFPA - Hazard Identification Ratings
flammability-2; reactivity-0

METHYL CYCLOHEXANOLS   590-67-0

HEALTH AND SAFETY LISTS

ACGIH 1995 - Time Weighted Averages
2 ppm TWA; 9.1 mg/m3 TWA

ACGIH 1995 - Short Term Exposure Limits
4 ppm STEL; 18 mg/m3 STEL

OSHA - Vacated PELs - Time Weighted Averages
2 ppm TWA; 8 mg/m3 TWA

OSHA - Vacated PELs - Short Term Exposure Limits
4 ppm STEL; 16 mg/m3 STEL

ENVIRONMENTAL LISTS

TSCA - PAIR - Reporting List
Effective Date: January 26, 1994;   Reporting Date: March 28, 1994

TSCA - Health and Safety Reporting List
Effective Date: January 26, 1994;   Sunset Date: Jnauary 26, 2004

INTERNATIONAL LISTS

Australian Exposure Standards - Time Weighted Averages
2 ppm TWA; 9.1 mg/m3 TWA

Australian Exposure Standards - Short Term Exposure Limits
4 ppm STEL; 18 mg/m3 STEL

Canada - WHMIS: Ingredient Disclosure
1% item 1033 (585)

Canada - Alberta - 8 Hour Occupational Exposure Limit
2 ppm TWA; 9 mg/m3 TWA

Canada - Alberta - 15 Minute Occupational Exposure Limit
4 ppm STEL; 18 mg/m3 STEL

Canada - British Columbia - 8 Hour Exposure Limits
2 ppm TWA; 8 mg/m3 TWA

Canada - British Columbia - 15 Minute Exposure Limits
4 ppm STEL; 16 mg/m3 STEL

Canada - Ontario - OHSA - TWAEVs
2 ppm TWAEV; 9 mg/m3 TWAEV

Canada - Ontario - OHSA - STEVs
4 ppm STEV; 18 mg/m3 STEV

Canada - Quebec - Time-Weighted Average Exposure Values
2 ppm TWAEV; 9.1 mg/m3 TWAEV

Canada - Quebec - Short-term Exposure Values
4 ppm STEV; 18 mg/m3 STEV

United Kingdom - Occupational Exposure Standards - TWAs
2 ppm TWA; 8 mg/m3 TWA

United Kingdom - Occupational Exposure Standards - STELs
4 ppm STEL; 16 mg/m3 STEL

German (DFG) - MAK Values
2 ppm MAK; 8 mg/m3 MAK

German (DFG) - Skin/Sensitizers
danger of sensitization (skin or respiratory)

Israel - Time Weighted Averages
2 ppm TWA; 9.1 mg/m3 TWA

Israel - Short Term Exposure Limits
4 ppm STEL; 18 mg/m3 STEL

Israel - Action Levels
1 ppm AL; 4.55 mg/m3 AL

Mexico - Instruction No. 10 - TWAs
2 ppm TWA; 8 mg/m3 TWA

Mexico - Instruction No. 10 - STELs
4 ppm STEL; 16 mg/m3 STEL

STATE LISTS

California - Exposure Limits - PELs
2 ppm PEL; 8 mg/m3 PEL

California - Exposure Limits - STELs
4 ppm STEL; 16 mg/m3 STEL

California - Directors List of Hazardous Substances (8 CCR 339)
[present]

Florida Hazardous Substance List
[present]

Massachusetts Right To Know List
[present]

Minnesota Hazardous Substance List
[present]

Pennsylvania Right to Know List
[present]

O-METHYLCYCLOHEXANONE   583-60-8

HEALTH AND SAFETY LISTS

ACGIH 1995 - Time Weighted Averages
400 ppm TWA; 1610 mg/m3 TWA

U.S. DOT - Substances From 49 CFR 172.101
regulated by DOT (UN2296, UN2297)

U.S. DOT - Hazard Classes
DOT hazard class = 3

NFPA - Flash Points
flash point = 25 degrees F (-4 degrees C)

NFPA - Hazard Identification Ratings
health-2; flammability-3; reactivity-0

NIOSH - Selected LD50s and LC50s
Inhalation, mouse: LC50 = 41500 mg/m3 (2 hr) Oral, mouse: LD50 = 2250 mg/kg

NIOSH 1990 - Pocket Guide - RELs
400 ppm TWA; 1600 mg/m3 TWA

NIOSH 1990 - Pocket Guide - IDLHs
10,000 ppm IDLH

NIOSH 1990 - Pocket Guide - Target organs
skin, respiratory system

OSHA - Vacated PELs - Time Weighted Averages
400 ppm TWA; 1600 mg/m3 TWA

OSHA - Final PELs - Time Weighted Averages
500 ppm TWA; 2000 mg/m3 TWA

ENVIRONMENTAL LISTS

CAA - HON Rule - SOCMI Chemicals
compliance by April 24, 1995

TSCA - PAIR - Reporting List
Effective Date: January 26, 1994;   Reporting Date: March 28, 1994

TSCA - Health and Safety Reporting List
Effective Date: January 26, 1994;   Sunset Date: January 26, 2004

INTERNATIONAL LISTS

Australian Exposure Standards - Time Weighted Averages
400 ppm TWA; 1610 mg/m3 TWA

Canada - WHMIS: Ingredient Disclosure
1% item 1034 (1125)

Canada - Alberta - 8 Hour Occupational Exposure Limit
400 ppm TWA; 1600 mg/m3 TWA

Canada - Alberta - 15 Minute Occupational Exposure Limit
500 ppm STEL; 2000 mg/m3 STEL

Canada - British Columbia - 8 Hour Exposure Limits
400 ppm TWA; 1600 mg/m3 TWA

Canada - British Columbia - 15 Minute Exposure Limits
500 ppm STEL; 2000 mg/m3 STEL

Canada - Ontario - OHSA - TWAEVs
400 ppm TWAEV; 1600 mg/m3 TWAEV

Canada - Quebec - Time-Weighted Average Exposure Values
400 ppm TWAEV; 1610 mg/m3 TWAEV

United Kingdom - Occupational Exposure Standards - TWAs
400 ppm TWA; 1600 mg/m3 TWA

United Kingdom - Occupational Exposure Standards - STELs
500 ppm STEL; 2000 mg/m3 STEL

German (DFG) - MAK Values
500 ppm MAK; 2000 mg/m3 MAK

German (DFG) - Peak Limitations
2 x normal MAK (30 min. average value); don't exceed 4 times during shift

Israel - Time Weighted Averages
400 ppm TWA; 1610 mg/m3 TWA

Israel - Action Levels
200 ppm AL; 805 mg/m3 AL

Mexico - Instruction No. 10 - TWAs
400 ppm TWA; 1600 mg/m3 TWA

Mexico - Instruction No. 10 - STELs
500 ppm STEL; 2000 mg/m3 STEL

STATE LISTS

California - Exposure Limits - PELs
400 ppm PEL; 1600 mg/m3 PEL

California - Directors List of Hazardous Substances (8 CCR 339)
[present]

Florida Hazardous Substance List
[present]

Massachusetts Right To Know List
[present]

Minnesota Hazardous Substance List
[present]

NJ Right to Know List (Total)
sn 1242

NJ Special Hazardous Substances
(flammable - third degree)

Pennsylvania Right to Know List
[present]

PROPOSED REGULATIONS

TSCA - ITC 31st Report Priority Testing List
designated to be tested

METHYLCYCLOHEXANONE   1331-22-2

HEALTH AND SAFETY LISTS

NFPA - Flash Points
flash point = 149 degrees F (65 degrees C)

NFPA - Hazard Identification Ratings
flammability-2; reactivity-0

INTERNATIONAL LISTS

Canada - Ontario - OHSA - TWAEVs
50 ppm TWAEV; 233 mg/m3 TWAEV

STATE LISTS

California - Directors List of Hazardous Substances (8 CCR 339)
[present] (Listed under 'Methylcyclohexanol, all isomers')

METHYLCYCLOHEXANONE PEROXIDE   RR-01758-3

HEALTH AND SAFETY LISTS

NFPA - Flash Points
flash point = 158 degrees F (70 degrees C)

NFPA - Hazard Identification Ratings
flammability-2; reactivity-0

4-METHYLCYCLOHEXENE   591-47-9

HEALTH AND SAFETY LISTS

NFPA - Flash Points
flash point around 158 degrees F (70 degrees C)

NFPA - Hazard Identification Ratings
health-0; flammability-2; reactivity-0

STATE LISTS

California - Directors List of Hazardous Substances (8 CCR 339)
[present] (Listed under 'Methylcyclohexanol, all isomers')

N-METHYL CYCLOHEXYLAMINE   100-60-7

INTERNATIONAL LISTS

Canada - Alberta - 15 Minute Occupational Exposure Limit
75 ppm STEL; 350 mg/m3 STEL

Canada - British Columbia - 8 Hour Exposure Limits
50 ppm TWA; 235 mg/m3 TWA

Canada - British Columbia - 15 Minute Exposure Limits
75 ppm STEL; 350 mg/m3 STEL

Canada - Quebec - Time-Weighted Average Exposure Values
50 ppm TWAEV; 234 mg/m3 TWAEV

United Kingdom - Occupational Exposure Standards - TWAs
50 ppm TWA; 235 mg/m3 TWA

United Kingdom - Occupational Exposure Standards - STELs
75 ppm STEL; 350 mg/m3 STEL

Mexico - Instruction No. 10 - TWAs
50 ppm TWA; 235 mg/m3 TWA

Mexico - Instruction No. 10 - STELs
75 ppm STEL; 350 mg/m3 STEL

STATE LISTS

California - Directors List of Hazardous Substances (8 CCR 339)
[present] (Listed under 'Methylcyclohexanol, all isomers')

METHYLCYCLOHEXYL ACETATE   30232-11-2

HEALTH AND SAFETY LISTS

ACGIH 1995 - Time Weighted Averages
50 ppm TWA; 234 mg/m3 TWA

NIOSH - Selected LD50s and LC50s
Oral, rat: LD50 = 1660 mg/kg

NIOSH 1990 - Pocket Guide - RELs
50 ppm TWA; 235 mg/m3 TWA

NIOSH 1990 - Pocket Guide - IDLHs
10,000 ppm IDLH

NIOSH 1990 - Pocket Guide - Target organs
skin, eyes, respiratory system; in animals: CNS, liver, kidneys

OSHA - Vacated PELs - Time Weighted Averages
50 ppm TWA; 235 mg/m3 TWA

OSHA - Final PELs - Time Weighted Averages
100 ppm TWA; 470 mg/m3 TWA

ENVIRONMENTAL LISTS

CAA - HON Rule - SOCMI Chemicals
compliance by Oct. 23, 1995

INTERNATIONAL LISTS

Australian Exposure Standards - Time Weighted Averages
50 ppm TWA; 234 mg/m3 TWA

Canada - WHMIS: Ingredient Disclosure
1% item 1035 (1126)

Canada - Alberta - 8 Hour Occupational Exposure Limit
50 ppm TWA; 235 mg/m3 TWA

German (DFG) - MAK Values
50 ppm MAK; 235 mg/m3 MAK apply to all members of Glycidyl subcategory VII-B)

German (DFG) - Peak Limitations
2 x normal MAK (30 min. average value); don't exceed 4 times during shift

Israel - Time Weighted Averages
50 ppm TWA; 234 mg/m3 TWA

Israel - Action Levels
25 ppm AL; 117 mg/m3 AL

STATE LISTS

California - Exposure Limits - PELs
50 ppm PEL; 235 mg/m3 PEL

California - Directors List of Hazardous Substances (8 CCR 339)
[present] (Listed under 'Methylcyclohexanol, all isomers')

Florida Hazardous Substance List
[present]

Massachusetts Right To Know List
[present]

Minnesota Hazardous Substance List
[present]

NJ Right to Know List (Total)
sn 1243

Pennsylvania Right to Know List
[present]

METHYL CYCLOPENTADIENE   26472-00-4

HEALTH AND SAFETY LISTS

U.S. DOT - Substances From 49 CFR 172.101
regulated by DOT (UN2617)

U.S. DOT - Hazard Classes
DOT hazard class = 3

METHYLCYCLOPENTADIENYL MANGANESE TRICARBONYL   12108-13-3

HEALTH AND SAFETY LISTS

ACGIH 1995 - Time Weighted Averages
50 ppm TWA; 229 mg/m3 TWA

ACGIH 1995 - Short Term Exposure Limits
75 ppm STEL; 344 mg/m3 STEL

ACGIH 1995 - Skin Designations
skin - potential for cutaneous absorption

NIOSH - Selected LD50s and LC50s
Oral, rat: LD50 = 2140 mg/kg Skin, rabbit: LD50 = 1635 mg/kg

NIOSH 1990 - Pocket Guide - RELs
50 ppm TWA; 230 mg/m3 TWA; 75 ppm STEL; 345 mg/m3 STEL

NIOSH 1990 - Pocket Guide - IDLHs
2500 ppm IDLH

NIOSH 1990 - Pocket Guide - Target organs
in animals: respiratory system, liver, kidneys, skin

NIOSH 1990 - Pocket Guide - Skin list
Potential for dermal absorption

OSHA - Vacated PELs - Time Weighted Averages
50 ppm TWA; 230 mg/m3 TWA

OSHA - Vacated PELs - Short Term Exposure Limits
75 ppm STEL; 345 mg/m3 STEL

OSHA - Vacated PELs - Skin Designation
Prevent or reduce skin absorption

OSHA - Final PELs - Time Weighted Averages
100 ppm TWA; 460 mg/m3 TWA

OSHA - Final PELs - Skin Notations
prevent or reduce skin absorption

INTERNATIONAL LISTS

Australian Exposure Standards - Time Weighted Averages
50 ppm TWA; 229 mg/m3 TWA

Australian Exposure Standards - Short Term Exposure Limits
75 ppm STEL; 344 mg/m3 STEL

Australian Exposure Standards - Skin Effects
skin absorption

Canada - WHMIS: Ingredient Disclosure
1% item 1036 (1127)

Canada - Alberta - 8 Hour Occupational Exposure Limit
50 ppm TWA; 230 mg/m3 TWA

Canada - Alberta - 15 Minute Occupational Exposure Limit
75 ppm STEL; 345 mg/m3 STEL

Canada - Alberta - Skin Designation
can be absorbed through the intact skin

Canada - British Columbia - 8 Hour Exposure Limits
50 ppm TWA; 230 mg/m3 TWA

Canada - British Columbia - 15 Minute Exposure Limits
75 ppm STEL; 345 mg/m3 STEL

Canada - British Columbia - Skin Notations
skin - potential for skin absorption

Canada - Ontario - OHSA - TWAEVs
50 ppm TWAEV; 230 mg/m3 TWAEV

Canada - Ontario - OHSA - STEVs
75 ppm STEV; 345 mg/m3 STEV

Canada - Ontario - OHSA - Skin Notations
absorption through skin, eyes, or mucous membranes

Canada - Quebec - Time-Weighted Average Exposure Values
50 ppm TWAEV; 229 mg/m3 TWAEV

Canada - Quebec - Short-term Exposure Values
75 ppm STEV; 344 mg/m3 STEV

Canada - Quebec - Skin Designations
absorbed through the skin

United Kingdom - Occupational Exposure Standards - TWAs
50 ppm TWA; 230 mg/m3 TWA

United Kingdom - Occupational Exposure Standards - STELs
75 ppm STEL; 345 mg/m3 STEL

United Kingdom - Occupational Exposure Standards - Notes
can be absorbed through skin

German (DFG) - MAK Values
50 ppm MAK; 230 mg/m3 MAK

German (DFG) - Peak Limitations
2 x normal MAK (30 min. average value); don't exceed 4 times during shift

German (DFG) - Skin/Sensitizers
danger of cutaneous absorption

Israel - Time Weighted Averages
50 ppm TWA; 229 mg/m3 TWA

Israel - Short Term Exposure Limits
75 ppm STEL; 344 mg/m3 STEL

Israel - Action Levels
25 ppm AL; 114.5 mg/m3 AL

Mexico - Instruction No. 10 - TWAs
50 ppm TWA; 230 mg/m3 TWA

Mexico - Instruction No. 10 - STELs
75 ppm STEL; 345 mg/m3 STEL

Mexico - Instruction No. 10 - Skin designation
skin - potential for cutaneous absorption

STATE LISTS

California - Exposure Limits - PELs
50 ppm PEL; 230 mg/m3 PEL

California - Exposure Limits - STELs
75 ppm STEL;   345 mg/m3 STEL

California - Exposure Limits - Skin Notation
material may be absorbed through the skin, eyes or mucous membrane

California - Directors List of Hazardous Substances (8 CCR 339)
[present]

Florida Hazardous Substance List
[present]

Massachusetts Right To Know List
[present]

Minnesota Hazardous Substance List
skin

NJ Right to Know List (Total)
sn 1438

Pennsylvania Right to Know List
[present]

METHYLCYCLOPENTANE   96-37-7

HEALTH AND SAFETY LISTS

NFPA - Flash Points
flash point = 118 degrees F (48 degrees C)

NFPA - Hazard Identification Ratings
flammability-2;   reactivity-0

ENVIRONMENTAL LISTS

CAA - HON Rule - SOCMI Chemicals
compliance by April 24, 1995

2-METHYLDECANE   6975-98-0

HEALTH AND SAFETY LISTS

U.S. DOT - Organic Peroxides Table
Organic peroxide UN3115

METHYL DEMETON   8022-00-2

HEALTH AND SAFETY LISTS

NFPA - Flash Points
flash point = 30 degrees F (-1 degrees C)

NFPA - Hazard Identification Ratings
health-1;   flammability-3;   reactivity-0

STATE LISTS

Florida Hazardous Substance List
[present]

Massachusetts Right To Know List
[present]

Pennsylvania Right to Know List
[present]

METHYL DICHLOROACETATE   116-54-1

HEALTH AND SAFETY LISTS

NIOSH - Selected LD50s and LC50s
Inhalation, guinea pig: LC50 = 7000 mg/m3 (8 hr) Oral, rat: LD50 = 400 mg/kg

METHYL DICHLOROARSINE   593-89-5

HEALTH AND SAFETY LISTS

NFPA - Flash Points
flash point = 147 degrees F (64 degrees C)

NFPA - Hazard Identification Ratings
health-1;   flammability-2;   reactivity-0

METHYL DICHLOROSILANE   75-54-7

HEALTH AND SAFETY LISTS

NFPA - Flash Points
flash point = 120 degrees F (49 degrees C)

NFPA - Hazard Identification Ratings
health-1;   flammability-2;   reactivity-1

METHYLDIETHANOLAMINE   105-59-9
SEE ALSO:
MANGANESE

HEALTH AND SAFETY LISTS

ACGIH 1995 - Time Weighted Averages
as Mn: 0.2 mg/m3 TWA

ACGIH 1995 - Skin Designations
as Mn: skin - potential for cutaneous absorption

NIOSH - Selected LD50s and LC50s
Inhalation, rat: LC50 = 76 mg/m3 4 hr Oral, rat: LD50 = 50 mg/kg Skin, rabbit: LD50 = 140 mg/kg

OSHA - Vacated PELs - Time Weighted Averages
as Mn: 0.2 mg/m3 TWA

OSHA - Vacated PELs - Skin Designation
Prevent or reduce skin absorption

ENVIRONMENTAL LISTS

CERCLA/SARA - Section 302 Extremely Hazardous Substances and TPQs
TPQ = 100 pounds

INTERNATIONAL LISTS

Australian Exposure Standards - Time Weighted Averages
as Mn: 0.2 mg/m3 TWA

Australian Exposure Standards - Skin Effects
as Mn: skin absorption

Canada - WHMIS: Ingredient Disclosure
1% item 1037 (1078)

Canada - Alberta - 8 Hour Occupational Exposure Limit
as Mn: 0.2 mg/m3 TWA

Canada - Alberta - 15 Minute Occupational Exposure Limit
as Mn: 0.6 mg/m3 STEL

Canada - Alberta - Skin Designation
can be absorbed through the intact skin

Canada - British Columbia - 8 Hour Exposure Limits
as Mn: 0.1 ppm TWA;   0.2 mg/m3 TWA

Canada - British Columbia - 15 Minute Exposure Limits
as Mn: 0.3 ppm STEL;   0.6 mg/m3 STEL

Canada - British Columbia - Skin Notations
skin - potential for skin absorption

Canada - Ontario - OHSA - TWAEVs
as Mn: 0.2 mg/m3 TWAEV

Canada - Ontario - OHSA - Skin Notations
as Mn: absorption through skin, eyes, or mucous membranes

Canada - Quebec - Time-Weighted Average Exposure Values
0.2 mg/m3 TWAEV

United Kingdom - Occupational Exposure Standards - TWAs
as Mn: 0.2 mg/m3 TWA

United Kingdom - Occupational Exposure Standards - STELs
as Mn: 0.6 mg/m3 STEL

United Kingdom - Occupational Exposure Standards - Notes
can be absorbed through skin

Israel - Time Weighted Averages
as Mn: 0.2 mg/m3 TWA

Israel - Action Levels
as Mn: 0.1 mg/m3 AL

Mexico - Instruction No. 10 - TWAs
0.2 mg/m3 TWA

Mexico - Instruction No. 10 - STELs
0.6 mg/m3 STEL

Mexico - Instruction No. 10 - Skin designation
skin - potential for cutaneous absorption

STATE LISTS

California - Exposure Limits - PELs
as Mn: 0.2 mg/m3 PEL

California - Exposure Limits - Skin Notation
material may be absorbed through the skin, eyes or mucous membrane

California - Directors List of Hazardous Substances (8 CCR 339)
[present]

Florida Hazardous Substance List
[present]

Massachusetts Right To Know List
extraordinarily hazardous

Minnesota Hazardous Substance List
as Mn: skin

Pennsylvania Right to Know List
environmental hazard

PROPOSED REGULATIONS

CERCLA/SARA - Proposed Hazardous Substance Additions
proposed RQ = 1 pound (.454 kg)

CERCLA/SARA - 1989 Proposed RQ Adjustments
proposed RQ = 100 pounds (45.4 kg)

1-METHYL-3,5-DIETHYLBENZENE   2050-24-0

HEALTH AND SAFETY LISTS

U.S. DOT - Substances From 49 CFR 172.101
regulated by DOT (UN2298)

U.S. DOT - Hazard Classes
DOT hazard class = 3

NFPA - Flash Points
flash point < 20 degrees F (-7 degrees C)

NFPA - Hazard Identification Ratings
health-2; flammability-3; reactivity-0

ENVIRONMENTAL LISTS

TSCA - Code of Federal Regulations Citations
40 CFR 712.30(o); 40 CFR 716.120(a); 40 CFR 799.2155

TSCA - PAIR - Reporting List
Reporting Date: August 19, 1985

TSCA - Health and Safety Reporting List
Effective Date: June 20, 1985; Sunset Date: November 9, 1993

STATE LISTS

Florida Hazardous Substance List
[present]

Massachusetts Right To Know List
[present]

NJ Right to Know List (Total)
sn 1245

NJ Special Hazardous Substances
(flammable - third degree)

Pennsylvania Right to Know List
[present]

METHYL DIHYDROABIETATE   RR-00867-3

HEALTH AND SAFETY LISTS

NFPA - Hazard Identification Ratings
health-0; flammability-2; reactivity-0

N-METHYL-N,4-DINITROSOANILINE   99-80-9

HEALTH AND SAFETY LISTS

ACGIH 1995 - Time Weighted Averages
0.5 mg/m3 TWA

ACGIH 1995 - Skin Designations
skin - potential for cutaneous absorption

NIOSH - Selected LD50s and LC50s
Oral, rat: LD50 = 65 mg/kg Skin, rat: LD50 = 300 mg/kg

OSHA - Vacated PELs - Time Weighted Averages
0.5 mg/m3 TWA

OSHA - Vacated PELs - Skin Designation
Prevent or reduce skin absorption

INTERNATIONAL LISTS

Australian Exposure Standards - Time Weighted Averages
0.5 mg/m3 TWA

Australian Exposure Standards - Skin Effects
skin absorption

Canada - Alberta - 8 Hour Occupational Exposure Limit
0.5 mg/m3 TWA

Canada - Alberta - 15 Minute Occupational Exposure Limit
1.5 mg/m3 STEL

Canada - Alberta - Skin Designation
can be absorbed through the intact skin

Canada - British Columbia - 8 Hour Exposure Limits
0.5 mg/m3 TWA

Canada - British Columbia - 15 Minute Exposure Limits
1.5 mg/m3 STEL

Canada - British Columbia - Skin Notations
skin - potential for skin absorption

Canada - Ontario - OHSA - TWAEVs
0.5 mg/m3 TWAEV

Canada - Ontario - OHSA - Skin Notations
absorption through skin, eyes, or mucous membranes

Canada - Quebec - Time-Weighted Average Exposure Values
0.5 mg/m3 TWAEV

Canada - Quebec - Skin Designations
absorbed through the skin

German (DFG) - MAK Values
0.5 ppm MAK; 5 mg/m3 MAK

German (DFG) - Peak Limitations
10 x normal MAK (30 min average value); don't exceed during shift

German (DFG) - Skin/Sensitizers
danger of cutaneous absorption

Israel - Time Weighted Averages
0.5 mg/m3 TWA

Israel - Action Levels
0.25 mg/m3 AL

Mexico - Instruction No. 10 - TWAs
0.5 mg/m3 TWA

Mexico - Instruction No. 10 - STELs
1.5 mg/m3 STEL

Mexico - Instruction No. 10 - Skin designation
skin - potential for cutaneous absorption

STATE LISTS

California - Exposure Limits - PELs
0.5 mg/m3 PEL

California - Exposure Limits - Skin Notation
material may be absorbed through the skin, eyes or mucous membrane

California - Directors List of Hazardous Substances (8 CCR 339)
[present]

Florida Hazardous Substance List
[present]

Massachusetts Right To Know List
[present]

Minnesota Hazardous Substance List
skin

NJ Right to Know List (Total)
sn 1246

Pennsylvania Right to Know List
[present]

ALPHA-METHYLDOPA   555-30-6

HEALTH AND SAFETY LISTS

U.S. DOT - Substances From 49 CFR 172.101
regulated by DOT (UN2299)

U.S. DOT - Hazard Classes
DOT hazard class = 6.1

STATE LISTS

NJ Right to Know List (Total)
sn 1247

NJ Special Hazardous Substances
(corrosive)

METHYLDOPA SESQUIHYDRATE   41372-08-1
SEE ALSO:
ARSENIC

HEALTH AND SAFETY LISTS

U.S. DOT - Substances From 49 CFR 172.101
regulated by DOT (NA1556)

U.S. DOT - Hazard Classes
DOT hazard class = 6.1

U.S. DOT - Substances Which Are Poisonous by Inhalation
liquid hazardous material poisonous by inhalation (NA1556)

NIOSH - Selected LD50s and LC50s
Inhalation, mouse: LC50 = 2700 mg/m3 (10 mn)

STATE LISTS

NJ Right to Know List (Total)
sn 1248

4,4'-METHYLENEBIS(2-CHLOROANILINE) (MBOCA)   101-14-4

HEALTH AND SAFETY LISTS

U.S. DOT - Substances From 49 CFR 172.101
regulated by DOT (UN1242)

U.S. DOT - Hazard Classes
DOT hazard class = 4.3

NFPA - Flash Points
flash point = 15 degrees F (-9 degrees C)

NFPA - Hazard Identification Ratings
health-3; flammability-3; reactivity-2 (avoid use of water)

NIOSH - Selected LD50s and LC50s
Inhalation, rat: LC50 = 300 ppm 4 hr

ENVIRONMENTAL LISTS

EPA - Master Testing List
[present]

INTERNATIONAL LISTS

Canada - WHMIS: Ingredient Disclosure
1% item 1038 (1128)

STATE LISTS

Florida Hazardous Substance List
[present]

Massachusetts Right To Know List
[present]

NJ Right to Know List (Total)
sn 1249

NJ Special Hazardous Substances
(corrosive; flammable - third degree; reactive - second degree)

Pennsylvania Right to Know List
[present]

METHYLENE BIS(4-CYCLOHEXYLISOCYANATE)   5124-30-1

HEALTH AND SAFETY LISTS

NFPA - Flash Points
flash point = 260 degrees F (127 degrees C)

NFPA - Hazard Identification Ratings
health-1; flammability-1; reactivity-0

NIOSH - Selected LD50s and LC50s
Oral, rat: LD50 = 4780 mg/kg

4,4'-METHYLENEBIS(N,N-DIMETHYL)   101-61-1
BENZENAMINE

HEALTH AND SAFETY LISTS

NFPA - Hazard Identification Ratings
health-0; flammability-2; reactivity-0

2,2'-METHYLENEBIS(4-ETHYL-6-TERT-BUTYLPHE-   88-24-4
NOL)

HEALTH AND SAFETY LISTS

NFPA - Flash Points
flash point = 361 degrees F (183 degrees C)

NFPA - Hazard Identification Ratings
health-1; flammability-1; reactivity-0

1,1'-METHYLENEBIS (ISOCYANATO-) BENZENE   26447-40-5

HEALTH AND SAFETY LISTS

IARC - Group 3 (not classifiable)
[present]

STATE LISTS

California - Directors List of Hazardous Substances (8 CCR 339)
[present]

METHYLENEBIS(4-ISOCYANATOBENZENE), POLY-   RR-00218-6
MER WITH POLYCAPROLACTONE TRIOL AND
ALKOXYLATED ALKANEPOLYOL, HYDROX-
YALKYL METHACRYLATE ESTER

HEALTH AND SAFETY LISTS

NIOSH - Selected LD50s and LC50s
Oral, rat: LD50 = 5000 mg/kg

STATE LISTS

NJ Right to Know List (Total)
sn 0050

4,4'-METHYLENE BIS(2-METHYLANILINE)   838-88-0

HEALTH AND SAFETY LISTS

NTP Chemical Status Reports - Testing Status and NTIS Number
Technical reports printed (PB89216527/AS)

NTP Chemical Status Reports - Evidence of Carcinogenicity
male rat-no evidence; female rat-no evidence; male mice-equivocal evidence; female mice-no evidence

2,2'-METHYLENEBIS(4-METHYL-6-TERT-   119-47-1
BUTYLPHENOL)
SEE ALSO:
F039-HAZARDOUS WASTES

HEALTH AND SAFETY LISTS

ACGIH 1995 - Time Weighted Averages
0.01 ppm TWA; 0.11 mg/m3 TWA

ACGIH 1995 - Skin Designations
skin - potential for cutaneous absorption

ACGIH 1995 - Carcinogens
A2-suspected human carcinogen

U.S. DOT - Appendix A Table 1 - Hazardous Substances
final RQ = 10 pounds (4.54 kg)

IARC - Group 2A (limited human data)
[present] (Overall evaluation 2A and not 2B on the basis of supporting evidence from other relevant data)

NIOSH - Selected LD50s and LC50s
Oral, rat: LD50 = 2100 mg/kg

NIOSH - Health Standards - Exposure Limits
3 ug/m3 TWA (lowest detectable concentration)

NIOSH - Health Standards - Health Effects and Precautions
has produced liver and lung tumors in animals (Blood and urine testing and periodic chest X-ray required)

NIOSH - Health Standards - Carcinogenic Chemicals
potential human carcinogen

NTP Seventh Report - Suspect Carcinogens
suspect carcinogen

OSHA - Vacated PELs - Time Weighted Averages
0.02 ppm TWA; 0.22 mg/m3 TWA

OSHA - Vacated PELs - Skin Designation
Prevent or reduce skin absorption

OSHA - Possible Select Carcinogens
[present]

ENVIRONMENTAL LISTS

ATSDR Priority List
Rank (of 275): 132

CERCLA/SARA - Section 313 - Emission Reporting
form R reporting required for 0.1% de minimus concentration

CERCLA/SARA - Hazardous Substances and their Reportable Quantities
final RQ = 10 pounds (4.54 kg)

Clean Air Act (1990) - List of Hazardous Air Contaminants
[present]

EPA - Carcinogen Hazard Ranking for RQ Adjustment
Hazard ranking = Medium

RCRA - U Series Wastes
waste number U158

RCRA - Hazardous Constituents-Appendix VIII
waste number U158

RCRA - Basis for Listing - Appendix VII
Included in waste stream: F039

RCRA - Substances Banned From Land Disposal
[present]

RCRA - Universal Treatment Standards (LDR)
WW: 0.50 mg/l; NWW: 30 mg/kg

TSCA - Code of Federal Regulations Citations
40 CFR 704.175; 40 CFR 704.225(a)

TSCA - CAIR - Reporting List
reporting required by: distributor; processor

INTERNATIONAL LISTS

Australian Exposure Standards - Time Weighted Averages
0.02 ppm TWA; 0.22 mg/m3 TWA

Australian Exposure Standards - Skin Effects
skin absorption

Australian Exposure Standards - Carcinogens
probable carcinogen

Canada - WHMIS: Ingredient Disclosure
0.1% item 1039 (1129)

Canada - NPRI (National Pollutant Release Inventory)
[present]

Canada - Alberta - 8 Hour Occupational Exposure Limit
0.02 ppm TWA; 0.22 mg/m3 TWA

Canada - Alberta - 15 Minute Occupational Exposure Limit
0.06 ppm STEL; 0.66 mg/m3 STEL

Canada - Alberta - Skin Designation
can be absorbed through the intact skin

Canada - Alberta - Designated Substances
designated substance - requires code of practice

Canada - British Columbia - 8 Hour Exposure Limits
0.02 ppm TWA; 0.2 mg/m3 TWA

Canada - British Columbia - Skin Notations
skin - potential for skin absorption

Canada - British Columbia - Carcinogens
carcinogen - 0.02 ppm TWA; 0.2 mg/m3 TWA

Canada - Ontario - OHSA - TWAEVs
0.0005 ppm TWAEV; 0.005 mg/m3 TWAEV

Canada - Ontario - OHSA - Skin Notations
absorption through skin, eyes, or mucous membranes

Canada - Quebec - Time-Weighted Average Exposure Values
0.02 ppm TWAEV; 0.22 mg/m3 TWAEV (substance of which the recirculation is prohibited)

Canada - Quebec - Skin Designations
absorbed through the skin

Canada - Quebec - Carcinogens
C2 carcinogen: effect suspected in humans

United Kingdom - Maximum Exposure Limits - TWAs
0.005 mg/m3 TWA

United Kingdom - Maximum Exposure Limits - Notes
can be absorbed through skin

German (DFG) - Skin/Sensitizers
danger of cutaneous absorption

German (DFG) - Carcinogens
animal evidence of carcinogenicity

Israel - Time Weighted Averages
(0.02) ppm TWA; (0.22) mg/m3 TWA

Israel - Action Levels
0.01 ppm AL; 0.11 mg/m3 AL

Mexico - Instruction No. 10 - TWAs
0.02 ppm TWA; 0.22 mg/m3 TWA

Mexico - Instruction No. 10 - Skin designation
skin - potential for cutaneous absorption

Mexico - Instruction No. 10 - Carcinogens
potential carcinogen in humans - limited epidemiological evidence

STATE LISTS

California - Air Bill 2588 Appendix A-I
known or potential carcinogen

California - Prop. 65 - Cancer list
carcinogen - initial date 7/1/87

California - Prop. 65 - No Significant Risk Levels
no significant risk level = 0.5 ug/day

California - Exposure Limits - PELs
0.01 mg/m3 PEL; avoid inhalation of dusts and vapors; avoid skin contact; see also section 5215

California - Exposure Limits - Skin Notation
material may be absorbed through the skin, eyes or mucous membrane

California - Directors List of Hazardous Substances (8 CCR 339)
[present]

Florida Hazardous Substance List
[present]

Massachusetts Right To Know List
carcinogen; extraordinarily hazardous

Minnesota Hazardous Substance List
carcinogen; skin

NJ Right to Know List (Total)
sn 1250

NJ Special Hazardous Substances
(carcinogen)

Pennsylvania Right to Know List
environmental hazard; special hazardous substance

Pennsylvania RTK - Special Hazardous Substances
[present]

METHYLENE BISPHENOL ISOCYANATE (MDI)   101-68-8

HEALTH AND SAFETY LISTS

ACGIH 1995 - Time Weighted Averages
0.005 ppm TWA; 0.054 mg/m3 TWA

NIOSH - Selected LD50s and LC50s
Inhalation, rat: LC50 = 20 ppm 5 hr Oral, rat: LD50 = 9900 mg/kg

NIOSH - Health Standards - Exposure Limits
5 ppb TWA; 55 ug/m3 TWA; C (10 min) 20 ppb; C (10 min) 210 ug/m3 (Listed under 'Diisocyanates')

NIOSH - Health Standards - Health Effects and Precautions
Respiratory effects and sensitization; pulmonary irritation (Periodic chest X-ray and pulmonary function testing required) (Listed under 'Diisocyanates')

OSHA - Vacated PELs - Ceiling Limits
C 0.01 ppm; C 0.11 mg/m3

OSHA - Vacated PELs - Skin Designation
Prevent or reduce skin absorption

ENVIRONMENTAL LISTS

CERCLA/SARA - Section 313 - Emission Reporting
form R reporting required; (Listed under 'Diisocyanates')

TSCA - Code of Federal Regulations Citations
40 CFR 712.30(x); 40 CFR 716.120(a)

TSCA - PAIR - Reporting List
Reporting Date: December 27, 1990

TSCA - Health and Safety Reporting List
Effective Date: June 1, 1987

INTERNATIONAL LISTS

Canada - WHMIS: Ingredient Disclosure
0.1% item 1040 (1131)

Canada - Alberta - Ceiling Occupational Exposure Limit
C 0.01 ppm; C 0.1 mg/m3

Canada - British Columbia - Ceiling Exposure Limits
C 0.01 ppm; C 0.11 mg/m3

Canada - Quebec - Time-Weighted Average Exposure Values
0.005 ppm TWAEV; 0.054 mg/m3 TWAEV

Israel - Time Weighted Averages
0.005 ppm TWA; 0.054 mg/m3 TWA

Israel - Action Levels
0.0025 ppm AL; 0.027 mg/m3 AL

Mexico - Instruction No. 10 - TWAs
0.01 ppm TWA; 0.11 mg/m3 TWA

STATE LISTS

California - Exposure Limits - PELs
0.005 ppm PEL; 0.054 mg/m3 PEL

California - Directors List of Hazardous Substances (8 CCR 339)
[present]

Florida Hazardous Substance List
[present]

Massachusetts Right To Know List
[present]

Minnesota Hazardous Substance List
[present]

Pennsylvania Right to Know List
[present]

1,1'-[METHYLENEBIS(SULFONYL)] BIS-2-   41123-59-5
CHLOROETHANE

HEALTH AND SAFETY LISTS

IARC - Group 3 (not classifiable)
[present]

NIOSH - Selected LD50s and LC50s
Oral, mouse: LD50 = 3160 mg/kg

NTP Chemical Status Reports - Testing Status and NTIS Number
Technical reports printed (PB299856/AS)

NTP Chemical Status Reports - Evidence of Carcinogenicity
male rat-positive; female rat-positive; male mice-equivocal; female mice-positive

NTP Seventh Report - Suspect Carcinogens
suspect carcinogen

OSHA - Possible Select Carcinogens
[present]

ENVIRONMENTAL LISTS

CERCLA/SARA - Section 313 - Emission Reporting
form R reporting required for 0.1% de minimus concentration

INTERNATIONAL LISTS

Canada - WHMIS: Ingredient Disclosure
0.1% item 1041 (1130)

German (DFG) - Carcinogens
animal evidence of carcinogenicity

STATE LISTS

California - Air Bill 2588 Appendix A-II
known or potential carcinogen

California - Prop. 65 - Cancer list
carcinogen - initial date 10/1/89

California - Prop. 65 - No Significant Risk Levels
no significant risk level = 20 ug/day

California - Directors List of Hazardous Substances (8 CCR 339)
[present]

Florida Hazardous Substance List
[present]

Massachusetts Right To Know List
carcinogen; extraordinarily hazardous

Minnesota Hazardous Substance List
carcinogen

NJ Special Hazardous Substances
(carcinogen; mutagen)

Pennsylvania Right to Know List
environmental hazard; special hazardous substance

Pennsylvania RTK - Special Hazardous Substances
[present]

2,2'-[METHYLENEBIS(SULFONYL)] BISETHANOL   41123-69-7

ENVIRONMENTAL LISTS

List of Pesticide Product Inert Ingredients
[present]

1,1'-[METHYLENE BIS(SULFONYL)]BISETHENE   3278-22-6

ENVIRONMENTAL LISTS

TSCA - Code of Federal Regulations Citations
40 CFR 712.30(x); 40 CFR 716.120(a)

TSCA - PAIR - Reporting List
Reporting Date: December 27, 1990

TSCA - Health and Safety Reporting List
Effective Date: June 1, 1987

METHYLENE BIS(THIOCYANATE)   6317-18-6

ENVIRONMENTAL LISTS

TSCA - Chemicals with Significant New Use Rules
PMN number: P-89-749

METHYLENEBISTRISUBSTITUTED ANILINE   RR-00904-1

HEALTH AND SAFETY LISTS

IARC - Group 2B (sufficient animal data)
[present]

NIOSH - Selected LD50s and LC50s
Oral, rat: LD50 = 1490 mg/kg

OSHA - Possible Select Carcinogens
[present]

INTERNATIONAL LISTS

German (DFG) - Carcinogens
animal evidence of carcinogenicity

STATE LISTS

California - Air Bill 2588 Appendix A-II
known or potential carcinogen: 9/89

California - Prop. 65 - Cancer list
carcinogen - initial date 4/1/88

California - Prop. 65 - No Significant Risk Levels
no significant risk level = 0.8 ug/day

California - Directors List of Hazardous Substances (8 CCR 339)
[present]

Florida Hazardous Substance List
[present]

Massachusetts Right To Know List
carcinogen; extraordinarily hazardous

Minnesota Hazardous Substance List
carcinogen

Pennsylvania Right to Know List
special hazardous substance

Pennsylvania RTK - Special Hazardous Substances
[present]

METHYLENE BLUE   61-73-4

HEALTH AND SAFETY LISTS

NIOSH - Selected LD50s and LC50s
Oral, rat: LD50 = 4880 mg/kg

ENVIRONMENTAL LISTS

List of Pesticide Product Inert Ingredients
[present]

INTERNATIONAL LISTS

Canada - WHMIS: Ingredient Disclosure
1% item 1042 (1132)

METHYLENE BLUE TRIHYDRATE   7220-79-3

HEALTH AND SAFETY LISTS

ACGIH 1995 - Time Weighted Averages
0.005 ppm TWA; 0.051 mg/m3 TWA

AIHA - Odor Threshold Values
no geometric mean air odor threshold

U.S. DOT - Substances From 49 CFR 172.101
regulated by DOT (UN2489)

U.S. DOT - Hazard Classes
DOT hazard class = 6.1

IARC - Group 3 (not classifiable)
[present]

NIOSH - Selected LD50s and LC50s
Inhalation, rat: LC50 = 178 mg/m3 8 hr

NIOSH 1990 - Pocket Guide - RELs
0.05 mg/m3 TWA; 0.005 ppm TWA; C 0.2 mg/m3 (10 min); C 0.020 ppm (10 min)

NIOSH 1990 - Pocket Guide - IDLHs
100 mg/m3 IDLH

NIOSH 1990 - Pocket Guide - Target organs
eyes, respiratory system

NIOSH - Health Standards - Exposure Limits
5 ppb TWA; 50 ug/m3 TWA; C (10 min) 20 ppb; C (10 min) 200 ug/m3 (Listed under 'Diisocyanates')

NIOSH - Health Standards - Health Effects and Precautions
Respiratory effects and sensitization, pulmonary irritation (Periodic chest X-ray and pulmonary function testing required) (Listed under 'Diisocyanates')

OSHA - Vacated PELs - Ceiling Limits
C 0.02 ppm; C 0.2 mg/m3

OSHA - Final PELs - Ceiling Limits
C 0.02 ppm; C 0.2 mg/m3

ENVIRONMENTAL LISTS

CERCLA/SARA - Section 313 - Emission Reporting
form R reporting required for 1.0% de minimus concentration; (Listed under 'Diisocyanates')

CERCLA/SARA - Hazardous Substances and their Reportable Quantities
final RQ = 1 pound (.454 kg)

Clean Air Act (1990) - List of Hazardous Air Contaminants
[present]

CAA - HON Rule - SOCMI Chemicals
compliance by April 24, 1995

CAA - HON Rule - Organic HAPs
[present]

EPA - Master Testing List
[present]

TSCA - Code of Federal Regulations Citations
40 CFR 712.30(x); 40 CFR 716.120(a)

TSCA - PAIR - Reporting List
Reporting Date: December 27, 1990

TSCA - Health and Safety Reporting List
Effective Date: June 1, 1987

INTERNATIONAL LISTS

Canada - WHMIS: Ingredient Disclosure
0.1% item 663 (717)

Canada - NPRI (National Pollutant Release Inventory)
[present]

Canada - Alberta - 8 Hour Occupational Exposure Limit
0.005 ppm TWA; 0.057 mg/m3 TWA

Canada - Alberta - Ceiling Occupational Exposure Limit
C 0.02 ppm; C 0.2 mg/m3

Canada - British Columbia - 8 Hour Exposure Limits
0.02 ppm TWA; 0.2 mg/m3 TWA

Canada - British Columbia - Ceiling Exposure Limits
C 0.02 ppm; C 0.2 mg/m3

Canada - Ontario - OHSA - TWAEVs
0.005 ppm TWAEV; 0.2 micromoles/m3 TWAEV (designated substance regulation)

Canada - Ontario - OHSA - CEVs
0.02 ppm CEV; 0.8 micromoles/m3 CEV (designated substance regulation)

Canada - Ontario - OHSA - Designated Substances
0.005 ppm TWAEV; 0.2 micromoles/m3 TWAEV; See Ontario Reg. 842 for full information.

Canada - Quebec - Time-Weighted Average Exposure Values
0.005 ppm TWAEV; 0.051 mg/m3 TWAEV

German (DFG) - MAK Values
0.005 mg/m3 MAK; 0.05 mg/m3 MAK

German (DFG) - Peak Limitations
2 x normal MAK (5 min momentary value); don't exceed 8 times during shift

German (DFG) - Skin/Sensitizers
danger of sensitization

German (DFG) - Carcinogens
animal evidence of carcinogenicity

Israel - Time Weighted Averages
0.005 ppm TWA; 0.051 mg/m3 TWA

Israel - Action Levels
0.0025 ppm AL; 0.0255 mg/m3 AL

Mexico - Instruction No. 10 - Ceiling Limits
P 0.02 ppm; P 0.2 mg/m3

STATE LISTS

California - Air Bill 2588 Appendix A-I
6/91

California - Exposure Limits - PELs
0.005 ppm PEL; 0.051 mg/m3 PEL

California - Directors List of Hazardous Substances (8 CCR 339)
[present]

Florida Hazardous Substance List
[present]

Massachusetts Right To Know List
[present]

Minnesota Hazardous Substance List
[present]

NJ Right to Know List (Total)
sn 1253

Pennsylvania Right to Know List
environmental hazard

METHYLENE BROMIDE   74-95-3

ENVIRONMENTAL LISTS

TSCA - Code of Federal Regulations Citations
40 CFR 712.30(x); 40 CFR 716.120(d)

TSCA - PAIR - Reporting List
Reporting Date: November 27, 1991

TSCA - Health and Safety Reporting List
Effective Date: September 30, 1991

METHYLENE CHLORIDE   75-09-2

ENVIRONMENTAL LISTS

TSCA - Code of Federal Regulations Citations
40 CFR 712.30(x); 40 CFR 716.120(d)

TSCA - PAIR - Reporting List
Reporting Date: November 27, 1991

TSCA - Health and Safety Reporting List
Effective Date: September 30, 1991

METHYLENE DIACRYLAMIDE   110-26-9

ENVIRONMENTAL LISTS

TSCA - Code of Federal Regulations Citations
40 CFR 712.30(x); 40 CFR 716.120(d)

TSCA - PAIR - Reporting List
Reporting Date: November 27, 1991

TSCA - Health and Safety Reporting List
Effective Date: September 30, 1991

4,4'-METHYLENEDIANILINE   101-77-9

HEALTH AND SAFETY LISTS

NIOSH - Selected LD50s and LC50s
Oral, rat: LD50 = 161 mg/kg

NTP Chemical Status Reports - Testing Status and NTIS Number
Technical reports printed

4,4'-METHYLENEDIANILINE DIHYDROCHLORIDE   13552-44-8

ENVIRONMENTAL LISTS

TSCA - Chemicals with Significant New Use Rules
PMN number: P-87-90

METHYLENE DIISOCYANATE   RR-00870-8

HEALTH AND SAFETY LISTS

NIOSH - Selected LD50s and LC50s
Oral, rat: LD50 = 1180 mg/kg

ENVIRONMENTAL LISTS

List of Pesticide Product Inert Ingredients
[present]

INTERNATIONAL LISTS

Canada - WHMIS: Ingredient Disclosure
1% item 1043 (311)

3,4-METHYLENEDIOXYPHENYL-2-PROPANONE   4676-39-5

HEALTH AND SAFETY LISTS

NTP Chemical Status Reports - Testing Status and NTIS Number
Prechronic studies completed: in review for further evaluation

N,N'-(METHYLENEDI-4,1-PHENYLENE)BIS[2,2-   85095-61-0
DIMETHYL-

SEE ALSO:
F039-HAZARDOUS WASTES

HEALTH AND SAFETY LISTS

U.S. DOT - Substances From 49 CFR 172.101
regulated by DOT (UN2664)

U.S. DOT - Hazard Classes
DOT hazard class = 6.1

U.S. DOT - Appendix B - Marine Pollutants
DOT regulated marine pollutant

U.S. DOT - Appendix A Table 1 - Hazardous Substances
final RQ = 1000 pounds (454 kg)

NIOSH - Selected LD50s and LC50s
Inhalation, rat: LC50 = 40 gm/m3 2 hr

ENVIRONMENTAL LISTS

CERCLA/SARA - Section 313 - Emission Reporting
form R reporting required for 1.0% de minimus concentration

CERCLA/SARA - Hazardous Substances and their Reportable Quantities
final RQ = 1000 pounds (454 kg)

CAA - HON Rule - SOCMI Chemicals
compliance by Oct. 23, 1995

Safe Drinking Water Act - Monitoring
monitoring required

RCRA - U Series Wastes
waste number U068

RCRA - Hazardous Constituents-Appendix VIII
waste number U068

RCRA - Basis for Listing - Appendix VII
Included in waste stream: F039

RCRA - Substances Banned From Land Disposal
[present]

RCRA - TSD Facilities Ground Water Monitoring
TM 8010 = 15 ug/L PQL; TM 8240 = 5 ug/L PQL

RCRA - Universal Treatment Standards (LDR)
WW: 0.11 mg/l; NWW: 15 mg/kg

TSCA - Code of Federal Regulations Citations
40 CFR 716.120(a); 40 CFR 799.5055(e),(d)(2)

TSCA - Health and Safety Reporting List
Effective Date: June 1, 1987

TSCA - Multichemical Test Rules - Waste Constituents
hydrolysis testing for Chemical Fate

TSCA - Section 12(b) - Export Notification
export notification required - Section 4

INTERNATIONAL LISTS

Canada - WHMIS: Ingredient Disclosure
1% item 512 (644)

STATE LISTS

California - Air Bill 2588 Appendix A-II
6/91

Massachusetts Right To Know List
[present]

NJ Right to Know List (Total)
sn 1254

Pennsylvania Right to Know List
environmental hazard

PROPOSED REGULATIONS

Safe Drinking Water Act - Priority list
[present]

METHYLENE GLYCOL DINITRATE   38483-28-2
SEE ALSO:
F025-HAZARDOUS WASTES
BIS(2,4-DIMETHYLBUTYL) MALEATE
K010-HAZARDOUS WASTES
F001-HAZARDOUS WASTES
F039-HAZARDOUS WASTES
F002-HAZARDOUS WASTES
K009-HAZARDOUS WASTES
F024-HAZARDOUS WASTES

HEALTH AND SAFETY LISTS

ACGIH 1995 - Time Weighted Averages
50 ppm TWA;   174 mg/m3 TWA

ACGIH 1995 - Carcinogens
A2-suspected human carcinogen

AIHA - Odor Threshold Values
geometric mean air odor threshold = 160 ppm (detectable);   230 ppm (recognizable)

U.S. DOT - Substances From 49 CFR 172.101
regulated by DOT (UN1593)

U.S. DOT - Hazard Classes
DOT hazard class = 6.1

U.S. DOT - Substances Which Are Poisonous by Inhalation
liquid hazardous material poisonous by inhalation (when mixed with Aldicarb)

U.S. DOT - Appendix A Table 1 - Hazardous Substances
final RQ = 1000 pounds (454 kg)

IARC - Group 2B (sufficient animal data)
[present]

NFPA - Flash Points
no flash point

NFPA - Hazard Identification Ratings
health-2;   flammability-1;   reactivity-0

NIOSH - Selected LD50s and LC50s
Inhalation, rat: LC50 = 88000 mg/m3 (30 mn) Oral, rat: LD50 = 2136 mg/kg

NIOSH 1990 - Pocket Guide - RELs
Reduce exposure to lowest feasible concentration

NIOSH 1990 - Pocket Guide - IDLHs
5000 ppm IDLH (not considering carcinogenic effects)

NIOSH 1990 - Pocket Guide - Carcinogens
occupational carcinogen

NIOSH 1990 - Pocket Guide - Target organs
skin, CVS, eyes, CNS

NIOSH - Health Standards - Exposure Limits
reduce exposure to lowest feasible concentration

NIOSH - Health Standards - Health Effects and Precautions
has produced tumors of the lung, liver, salivary, and mammary glands in animals

NIOSH - Health Standards - Carcinogenic Chemicals
potential human carcinogen

NTP Chemical Status Reports - Testing Status and NTIS Number
Technical reports printed (PB86187903/AS);

NTP Chemical Status Reports - Evidence of Carcinogenicity
male rat-some evidence; female rat-clear evidence; male mice-clear evidence; female mice-clear evidence

NTP Seventh Report - Suspect Carcinogens
suspect carcinogen

OSHA - Vacated PELs - Time Weighted Averages
500 ppm TWA

OSHA - Vacated PELs - Short Term Exposure Limits
2000 ppm STEL (5 min in any 3 hrs)

OSHA - Vacated PELs - Ceiling Limits
C 1000 ppm

OSHA - Final PELs - Time Weighted Averages
500 ppm TWA;   C 1000 ppm

OSHA - Final PELs - Ceiling Limits
C 1000 ppm

OSHA - Possible Select Carcinogens
[present]

ENVIRONMENTAL LISTS

ATSDR Priority List
Rank (of 275): 065

CERCLA/SARA - Section 313 - Emission Reporting
form R reporting required for 0.1% de minimus concentration

CERCLA/SARA - Hazardous Substances and their Reportable Quantities
final RQ = 1000 pounds (454 kg)

Clean Air Act (1990) - List of Hazardous Air Contaminants
[present]

CAA - HON Rule - SOCMI Chemicals
compliance by Oct. 24, 1994

CAA - HON Rule - Organic HAPs
[present]

Clean Water Act - Priority Pollutants
[present]

Safe Drinking Water Act - MCLs
MCL = 0.005 mg/L

Safe Drinking Water Act - MCLGs
MCLG = Zero

Safe Drinking Water Act - Monitoring
monitoring required

List of Pesticide Product Inert Ingredients
[present]

RCRA - U Series Wastes
waste number U080

RCRA - Hazardous Constituents-Appendix VIII
waste number U080

RCRA - Basis for Listing - Appendix VII
Included in waste streams: F001, F002, F024, F025, F039, K009, K010

RCRA - Substances Banned From Land Disposal
[present]

RCRA - TSD Facilities Ground Water Monitoring
TM 8010 = 5 ug/L PQL;   TM 8240 = 5 ug/L PQL

RCRA - Universal Treatment Standards (LDR)
WW: 0.089 mg/l;   NWW: 30 mg/kg

TSCA - Code of Federal Regulations Citations
40 CFR 712.30(d),(x);   40 CFR 716.120(a)

TSCA - PAIR - Reporting List
Reporting Date: November 19, 1982

TSCA - Health and Safety Reporting List
Effective Date: October 4, 1982

INTERNATIONAL LISTS

Australian Exposure Standards - Time Weighted Averages
100 ppm TWA;   174 mg/m3 TWA

Australian Exposure Standards - Carcinogens
suspected carcinogen

Australian Exposure Standards - Under Review
exposure limits under review

Canada - WHMIS: Ingredient Disclosure
0.1% item 1044 (508)

Canada - NPRI (National Pollutant Release Inventory)
[present]

Canada - CEPA - Priority Substances List
estimated time for completion of assessment reports: 4 years

Canada - Drinking Water Quality - MACs
0.05 mg/L MAC

Canada - Alberta - 8 Hour Occupational Exposure Limit
50 ppm TWA; 173 mg/m3 TWA

Canada - Alberta - 15 Minute Occupational Exposure Limit
200 ppm STEL; 754 mg/m3 STEL

Canada - British Columbia - 8 Hour Exposure Limits
200 ppm TWA; 700 mg/m3 TWA

Canada - Ontario - OHSA - TWAEVs
50 ppm TWAEV; 175 mg/m3 TWAEV

Canada - Quebec - Time-Weighted Average Exposure Values
50 ppm TWAEV; 174 mg/m3 TWAEV

Canada - Quebec - Carcinogens
C2 carcinogen: effect suspected in humans

United Kingdom - Maximum Exposure Limits - TWAs
100 ppm TWA; 350 mg/m3 TWA

German (DFG) - MAK Values
100 ppm MAK; 360 mg/m3 MAK

German (DFG) - Peak Limitations
5 x normal MAK (30 min. average value); don't exceed 2 times during shift

German (DFG) - Carcinogens
suspected carcinogen

German (DFG) - Pregnancy
classification not yet possible

Israel - Time Weighted Averages
50 ppm TWA; 174 mg/m3 TWA

Israel - Action Levels
25 ppm AL; 87 mg/m3 AL

Mexico - Instruction No. 10 - TWAs
50 ppm TWA; 105 mg/m3 TWA

Mexico - Instruction No. 10 - STELs
100 ppm STEL; 205 mg/m3 STEL

Mexico - Wastewater - Organic Toxic Pollutants and Heavy Metals
Listed under [Halomethanes]

Mexico - Drinking Water - Ecological Criteria
0.002 mg/l This level has been extrapolated by using a mathematic model

STATE LISTS

California - Air Bill 2588 Appendix A-I
known or potential carcinogen

California - Prop. 65 - Cancer list
carcinogen - initial date 4/1/88

California - Prop. 65 - No Significant Risk Levels
no significant risk level = 50 ug/day

California - Exposure Limits - PELs
50 ppm PEL; 174 mg/m3 PEL

California - Directors List of Hazardous Substances (8 CCR 339)
[present]

Florida Hazardous Substance List
[present]

Massachusetts Right To Know List
carcinogen; extraordinarily hazardous

Minnesota Hazardous Substance List
carcinogen

Pennsylvania Right to Know List
environmental hazard; special hazardous substance

Pennsylvania RTK - Special Hazardous Substances
[present]

PROPOSED REGULATIONS

Proposed Amendments to OSHA Regulated Chemicals
proposed addition to specifically regulated chemicals in 29 CFR 1910.1052

N-METHYLEPHEDRINE   552-79-4

HEALTH AND SAFETY LISTS

NIOSH - Selected LD50s and LC50s
Oral, rat: LD50 = 390 mg/kg

N-METHYLEPHEDRINE SALTS, OPTICAL ISOMERS,   RR-01785-6
AND SALTS OF OPTICAL ISOMERS

HEALTH AND SAFETY LISTS

ACGIH 1995 - Time Weighted Averages
0.1 ppm TWA; 0.81 mg/m3 TWA

ACGIH 1995 - Skin Designations
skin - potential for cutaneous absorption

ACGIH 1995 - Carcinogens
A2-suspected human carcinogen

U.S. DOT - Substances From 49 CFR 172.101
regulated by DOT (UN2651)

U.S. DOT - Hazard Classes
DOT hazard class = 6.1

IARC - Group 2B (sufficient animal data)
[present] (Overall evaluation based only on evidence of carcinogenicity in monograph (39, 1986) or in Supplement 4)

NFPA - Flash Points
flash point = 428 degrees F (220 degrees C)

NFPA - Hazard Identification Ratings
health-3; flammability-1; reactivity-0

NIOSH - Selected LD50s and LC50s
Oral, rat: LD50 = 347 mg/kg

NIOSH - Health Standards - Exposure Limits
reduce exposure to lowest feasible concentration

NIOSH - Health Standards - Health Effects and Precautions
Bladder cancer; skin and liver effects (Prevent skin contact)

NIOSH - Health Standards - Carcinogenic Chemicals
potential human carcinogen

NTP Seventh Report - Suspect Carcinogens
suspect carcinogen

OSHA - 29 CFR 1910 Specifically Regulated Chemicals
10 ppb TWA; 100 ppb STEL; 5 ppb Action Level; Cancer suspect agent; Liver toxin (see 29 CFR 1910.1050)

OSHA - Select Carcinogens
[present]

OSHA - Possible Select Carcinogens
[present]

ENVIRONMENTAL LISTS

CERCLA/SARA - Section 313 - Emission Reporting
form R reporting required for 0.1% de minimus concentration

CERCLA/SARA - Hazardous Substances and their Reportable Quantities
final RQ = 1 pound (.454 kg)

Clean Air Act (1990) - List of Hazardous Air Contaminants
[present]

CAA - HON Rule - SOCMI Chemicals
compliance by Oct. 24, 1994

CAA - HON Rule - Organic HAPs
[present]

TSCA - Code of Federal Regulations Citations
40 CFR 712.30(d); 40 CFR 716.120(a)

TSCA - PAIR - Reporting List
Reporting Date: November 19, 1982

TSCA - Health and Safety Reporting List
Effective Date: October 4, 1982

INTERNATIONAL LISTS

Australian Exposure Standards - Time Weighted Averages
0.1 ppm TWA; 0.81 mg/m3 TWA

Australian Exposure Standards - Skin Effects
skin absorption

Australian Exposure Standards - Carcinogens
probable carcinogen

Canada - WHMIS: Ingredient Disclosure
0.1% item 489 (623)

Canada - NPRI (National Pollutant Release Inventory)
[present]

Canada - Alberta - 8 Hour Occupational Exposure Limit
0.1 ppm TWA; 0.8 mg/m3 TWA

Canada - Alberta - 15 Minute Occupational Exposure Limit
0.5 ppm STEL; 4 mg/m3 STEL

Canada - Alberta - Skin Designation
can be absorbed through the intact skin

Canada - Alberta - Designated Substances
designated substance - requires code of practice

Canada - Ontario - OHSA - TWAEVs
0.04 mg/m3 TWAEV

Canada - Ontario - OHSA - Skin Notations
absorption through skin, eyes, or mucous membranes

Canada - Quebec - Time-Weighted Average Exposure Values
0.1 ppm TWAEV; 0.81 mg/m3 TWAEV

Canada - Quebec - Skin Designations
absorbed through the skin

Canada - Quebec - Carcinogens
C2 carcinogen: effect suspected in humans

German (DFG) - Skin/Sensitizers
danger of cutaneous absorption; danger of sensitization (skin or respiratory)

German (DFG) - Carcinogens
animal evidence of carcinogenicity

Israel - Time Weighted Averages
0.1 ppm TWA; 0.81 mg/m3 TWA

Israel - Action Levels
0.05 ppm AL; 0.405 mg/m3 AL

STATE LISTS

California - Air Bill 2588 Appendix A-I
known or potential carcinogen

California - Prop. 65 - Cancer list
carcinogen - initial date 1/1/88

California - Prop. 65 - No Significant Risk Levels
no significant risk level = 0.4 ug/day

California - Exposure Limits - PELs
0.1 ppm PEL; 0.8 mg/m3 PEL

California - Exposure Limits - Skin Notation
material may be absorbed through the skin, eyes or mucous membrane

California - Directors List of Hazardous Substances (8 CCR 339)
[present]

Florida Hazardous Substance List
[present]

Massachusetts Right To Know List
carcinogen; extraordinarily hazardous

Minnesota Hazardous Substance List
carcinogen; skin (includes its dihydrochloride)

NJ Right to Know List (Total)
sn 1256

Pennsylvania Right to Know List
environmental hazard

METHYL EPOXYSTEARATE AND TETRAETHYLENE   68298-14-6
PENTAMINE(TEPA), REACTION PRODUCTS

HEALTH AND SAFETY LISTS

NTP Chemical Status Reports - Testing Status and NTIS Number
Technical reports printed (PB83238824)

NTP Chemical Status Reports - Evidence of Carcinogenicity
male rat-positive; female rat-positive; male mice-positive; female mice-positive

NTP Seventh Report - Suspect Carcinogens
suspect carcinogen

OSHA - Possible Select Carcinogens
[present]

STATE LISTS

California - Prop. 65 - Cancer list
carcinogen - initial date 1/1/88

California - Prop. 65 - No Significant Risk Levels
no significant risk level = 0.6 ug/day

California - Directors List of Hazardous Substances (8 CCR 339)
[present]

Pennsylvania Right to Know List
special hazardous substance

Pennsylvania RTK - Special Hazardous Substances
[present]

METHYL ESTER OCTANOIC ACID   111-11-5

HEALTH AND SAFETY LISTS

NFPA - Flash Points
flash point = 185 degrees F (85 degrees C)

NFPA - Hazard Identification Ratings
health-1; flammability-2; reactivity-1 (avoid use of water)

METHYL ESTER OF ROSIN, PARTIALLY   8050-13-3
HYDROGENATED

HEALTH AND SAFETY LISTS

FDA - Controlled Substances Act - Precursor chemicals
Threshold by base weight = 4 kilograms

METHYL ESTERS OF COTTONSEED OIL   61788-60-1

ENVIRONMENTAL LISTS

TSCA - Code of Federal Regulations Citations
40 CFR 721.1235

TSCA - Chemicals with Significant New Use Rules
PMN number: P-88-522

TSCA - Section 12(b) - Export Notification
P-88-522; export notification required - Section 5

N-METHYLETHANOLAMINE   109-83-1

HEALTH AND SAFETY LISTS

U.S. DOT - Hazard Classes
Forbidden from transport by the DOT

2-METHYL-2-ETHYL-1,3-DIOXOLANE   126-39-6

HEALTH AND SAFETY LISTS

FDA - Controlled Substances Act - Precursor chemicals
Threshold by base weight = 1 kilogram

METHYL ETHYLENE GLYCOL ETHERS AND   RR-01716-3
ESTERS

HEALTH AND SAFETY LISTS

FDA - Controlled Substances Act - Precursor chemicals
Threshold by base weight = 1 kilogram

METHYL ETHYL ETHER   540-67-0

ENVIRONMENTAL LISTS

List of Pesticide Product Inert Ingredients
[present]

2-METHYL-4-ETHYLHEXANE   3074-75-7

ENVIRONMENTAL LISTS

EPA - Master Testing List
[present]

PROPOSED REGULATIONS

TSCA - Proposed Substances for Developmental/ReproductiveTesting
proposed testing for: Developmental Toxicity - oral

3-METHYL-4-ETHYLHEXANE   3074-77-9

ENVIRONMENTAL LISTS

List of Pesticide Product Inert Ingredients
[present]

2,2'-[(1-METHYLETHYLIDENE)BIS[4,1-PHENY-LOXY[1-(BUTOXYMETHYL)-[2,1-ETHANEDIYL] OXYMETHYLENE]BISOXIRANE, REACTION PROD-UCT WITH A DIAMINE   RR-01221-5

ENVIRONMENTAL LISTS

List of Pesticide Product Inert Ingredients
[present]

METHYL ETHYL KETONE   78-93-3

HEALTH AND SAFETY LISTS

NFPA - Flash Points
flash point = 165 degrees F (74 degrees C)

NFPA - Hazard Identification Ratings
health-2; flammability-2; reactivity-0

NIOSH - Selected LD50s and LC50s
Oral, rat: LD50 = 2340 mg/kg

STATE LISTS

Florida Hazardous Substance List
[present]

Massachusetts Right To Know List
[present]

Pennsylvania Right to Know List
[present]

METHYL ETHYL KETONE PEROXIDE   1338-23-4

HEALTH AND SAFETY LISTS

NFPA - Flash Points
flash point = 74 degrees F (23 degrees C)

NFPA - Hazard Identification Ratings
health-2; flammability-3; reactivity-0

STATE LISTS

Florida Hazardous Substance List
[present]

Massachusetts Right To Know List
[present]

Pennsylvania Right to Know List
[present]

METHYL ETHYL KETOXIME   96-29-7

PROPOSED REGULATIONS

TSCA - ITC 33rd Report Priority Testing List
recommended for testing

TSCA - ITC 34th Report Priority Testing List
recommended for testing

2-METHYL-3-ETHYLPENTANE   609-26-7

HEALTH AND SAFETY LISTS

U.S. DOT - Substances From 49 CFR 172.101
regulated by DOT (UN1039)

U.S. DOT - Hazard Classes
DOT hazard class = 2.1

NFPA - Flash Points
flash point = -35 degrees F (-37 degrees C)

NFPA - Hazard Identification Ratings
health-1; flammability-4; reactivity-1

STATE LISTS

Florida Hazardous Substance List
[present]

Massachusetts Right To Know List
[present]

NJ Right to Know List (Total)
sn 1257

NJ Special Hazardous Substances
(flammable - fourth degree)

Pennsylvania Right to Know List
[present]

2-METHYL-5-ETHYLPIPERIDINE   104-89-2

HEALTH AND SAFETY LISTS

NFPA - Flash Points
flash point < 70 degrees F (21 degrees C)

NFPA - Hazard Identification Ratings
health-0; flammability-3; reactivity-0

STATE LISTS

Florida Hazardous Substance List
[present]

Massachusetts Right To Know List
[present]

Pennsylvania Right to Know List
[present]

2-METHYL-5-ETHYLPYRIDINE   104-90-5

HEALTH AND SAFETY LISTS

NFPA - Flash Points
flash point = 75 degrees F (24 degrees C)

NFPA - Hazard Identification Ratings
health-0; flammability-3; reactivity-0

STATE LISTS

Florida Hazardous Substance List
[present]

Massachusetts Right To Know List
[present]

Pennsylvania Right to Know List
[present]

METHYLEUGENOL   93-15-2

ENVIRONMENTAL LISTS

TSCA - Chemicals with Significant New Use Rules
PMN number: P-91-934

3-METHYLFLUORANTHENE   **1706-01-0**
SEE ALSO:
 KETONES, LIQUID, N.O.S.
 F039-HAZARDOUS WASTES
 VINYL ISOOCTYL ETHER
 F005-HAZARDOUS WASTES

HEALTH AND SAFETY LISTS

ACGIH 1995 - Time Weighted Averages
 200 ppm TWA; 590 mg/m3 TWA

ACGIH 1995 - Short Term Exposure Limits
 300 ppm STEL; 885 mg/m3 STEL

ACGIH 1995 - Biological Exposure Indices
 MEK in urine: 2 mg/L, end of shift

AIHA - Odor Threshold Values
 geometric mean air odor threshold = 16 ppm (detectable); 17 ppm (recognizable)

U.S. DOT - Substances From 49 CFR 172.101
 regulated by DOT (UN1193)

U.S. DOT - Hazard Classes
 DOT hazard class = 3

U.S. DOT - Appendix A Table 1 - Hazardous Substances
 final RQ = 5000 pounds (2270 kg)

FDA - Controlled Substances Act - Essential chemicals
 Import/Export threshold volume = 500 gallons; weight = 1,455 kilograms; Domestic threshold volume = 50 gallons; weight = 145 kilograms

NFPA - Flash Points
 flash point = 16 degrees F (-9 degrees C)

NFPA - Hazard Identification Ratings
 health-1; flammability-3; reactivity-0

NIOSH - Selected LD50s and LC50s
 Inhalation, mouse: LC50 = 40 gm/m3 2 hr Oral, rat: LD50 = 2737 mg/kg Skin, rabbit: LD50 = 13 gm/kg

NIOSH 1990 - Pocket Guide - RELs
 200 ppm TWA; 590 mg/m3 TWA; 300 ppm STEL; 885 mg/m3 STEL

NIOSH 1990 - Pocket Guide - IDLHs
 3000 ppm IDLH

NIOSH 1990 - Pocket Guide - Target organs
 CNS, lungs

NIOSH - Health Standards - Exposure Limits
 200 ppm TWA; 590 mg/m3 TWA (Listed under 'Ketones')

NIOSH - Health Standards - Health Effects and Precautions
 Irritation; liver, kidney, and nervous system effects (Urinalysis required) (Listed under 'Ketones')

OSHA - Vacated PELs - Time Weighted Averages
 200 ppm TWA; 590 mg/m3 TWA

OSHA - Vacated PELs - Short Term Exposure Limits
 300 ppm STEL; 885 mg/m3 STEL

OSHA - Final PELs - Time Weighted Averages
 200 ppm TWA; 590 mg/m3 TWA

ENVIRONMENTAL LISTS

ATSDR Priority List
 Rank (of 275): 168

CERCLA/SARA - Section 313 - Emission Reporting
 form R reporting required for 1.0% de minimus concentration

CERCLA/SARA - Hazardous Substances and their Reportable Quantities
 final RQ = 5000 pounds (2270 kg)

Clean Air Act (1990) - List of Hazardous Air Contaminants
 [present]

CAA - HON Rule - SOCMI Chemicals
 compliance by Oct. 23, 1995

CAA - HON Rule - Organic HAPs
 [present]

EPA - Master Testing List
 [present]

List of Pesticide Product Inert Ingredients
 [present]

RCRA - D Series - Maximum Concentration of Contaminants
 waste number D035; regulatory level = 200.0 mg/L

RCRA - D Series - Chronic Toxicity Reference Levels
 chronic toxicity reference level = 2 mg/L

RCRA - U Series Wastes
 waste number U159 (Ignitable waste; Toxic waste)

RCRA - Hazardous Constituents-Appendix VIII
 waste number U159

RCRA - Basis for Listing - Appendix VII
 Included in waste streams: F005, F039

RCRA - Substances Banned From Land Disposal
 [present]

RCRA - TSD Facilities Ground Water Monitoring
 TM 8015 = 10 ug/L PQL; TM 8249 = 100 ug/L PQL

RCRA - Universal Treatment Standards (LDR)
 WW: 0.28 mg/l; NWW: 36 mg/kg

TSCA - Code of Federal Regulations Citations
 40 CFR 712.30(d); 40 CFR 716.120(a)

TSCA - PAIR - Reporting List
 Reporting Date: November 19, 1982

TSCA - Health and Safety Reporting List
 Effective Date: October 4, 1982

INTERNATIONAL LISTS

Australian Exposure Standards - Time Weighted Averages
 150 ppm TWA; 445 mg/m3 TWA

Australian Exposure Standards - Short Term Exposure Limits
 300 ppm STEL; 890 mg/m3 STEL

Canada - WHMIS: Ingredient Disclosure
 1% item 1045 (1133)

Canada - NPRI (National Pollutant Release Inventory)
 [present]

Canada - Alberta - 8 Hour Occupational Exposure Limit
 200 ppm TWA; 590 mg/m3 TWA

Canada - Alberta - 15 Minute Occupational Exposure Limit
 300 ppm STEL; 885 mg/m3 STEL

Canada - British Columbia - 8 Hour Exposure Limits
 200 ppm TWA; 590 mg/m3 TWA

Canada - British Columbia - 15 Minute Exposure Limits
 300 ppm STEL; 885 mg/m3 STEL

Canada - Ontario - OHSA - TWAEVs
 200 ppm TWAEV; 590 mg/m3 TWAEV

Canada - Ontario - OHSA - STEVs
 300 ppm STEV; 885 mg/m3 STEV

Canada - Quebec - Time-Weighted Average Exposure Values
 50 ppm TWAEV; 150 mg/m3 TWAEV

Canada - Quebec - Short-term Exposure Values
 100 ppm STEV; 300 mg/m3 STEV

United Kingdom - Occupational Exposure Standards - TWAs
 200 ppm TWA; 590 mg/m3 TWA

United Kingdom - Occupational Exposure Standards - STELs
 300 ppm STEL; 885 mg/m3 STEL

German (DFG) - MAK Values
 200 ppm MAK; 590 mg/m3 MAK

German (DFG) - Peak Limitations
 2 x normal MAK (30 min. average value); don t exceed 4 times during shift

German (DFG) - Pregnancy
 classification not yet possible

Israel - Time Weighted Averages
 200 ppm TWA; 590 mg/m3 TWA

Israel - Short Term Exposure Limits
300 ppm STEL; 885 mg/m3 STEL

Israel - Action Levels
100 ppm AL; 295 mg/m3 AL

Mexico - Instruction No. 10 - TWAs
200 ppm TWA; 590 mg/m3 TWA

STATE LISTS

California - Air Bill 2588 Appendix A-I
6/91

California - Exposure Limits - PELs
200 ppm PEL; 590 mg/m3 PEL

California - Exposure Limits - STELs
300 ppm STEL; 885 mg/m3 STEL

California - Directors List of Hazardous Substances (8 CCR 339)
[present]

Florida Hazardous Substance List
[present]

Massachusetts Right To Know List
[present]

Minnesota Hazardous Substance List
[present]

NJ Right to Know List (Total)
sn 1258

Pennsylvania Right to Know List
environmental hazard

PROPOSED REGULATIONS

Safe Drinking Water Act - Priority list
[present]

Canada - Ontario - Proposed Occupational TWAEVs
50 ppm TWAEV; 150 mg/m3 TWAEV

Canada - Ontario - Proposed Occupational STEVs
100 ppm STEV; 300 mg/m3 STEV

2-METHYLFLUORANTHENE   33543-31-6

HEALTH AND SAFETY LISTS

ACGIH 1995 - Ceiling Limits
C 0.2 ppm; C 1.5 mg/m3

U.S. DOT - Hazard Classes
Forbidden from transport by the DOT

U.S. DOT - Appendix A Table 1 - Hazardous Substances
final RQ = 10 pounds (4.54 kg)

U.S. DOT - Organic Peroxides Table
Organic peroxide UN3101; UN3105; UN3107

NIOSH - Selected LD50s and LC50s
Inhalation, rat: LC50 = 200 ppm 4 hr Oral, rat: LD50 = 484 mg/kg

NTP Chemical Status Reports - Testing Status and NTIS Number
Technical reports printed (PB94-119278)

OSHA - Vacated PELs - Ceiling Limits
C 0.7 ppm; C 5 mg/m3

OSHA - List of Highly Hazardous Chemicals
concentration > 60%: threshhold quantity = 5000 pounds

ENVIRONMENTAL LISTS

CERCLA/SARA - Hazardous Substances and their Reportable Quantities
final RQ = 10 pounds (4.54 kg)

RCRA - U Series Wastes
waste number U160 (Reactive waste; Toxic waste)

RCRA - Hazardous Constituents-Appendix VIII
waste number U160

RCRA - Substances Banned From Land Disposal
[present]

INTERNATIONAL LISTS

Australian Exposure Standards - Time Weighted Averages
Peak Limitation: 0.2 ppm; 1.5 mg/m3

Canada - WHMIS: Ingredient Disclosure
1% item 1046 (1367)

Canada - Alberta - Ceiling Occupational Exposure Limit
C 0.2 ppm; C 1.4 mg/m3

Canada - British Columbia - Ceiling Exposure Limits
C 0.2 ppm; C 1.5 mg/m3

Canada - Ontario - OHSA - CEVs
0.1 ppm CEV; 1.5 mg/m3 CEV

Canada - Quebec - Ceiling Limits
P 0.2 ppm; P 1.5 mg/m3

United Kingdom - Occupational Exposure Standards - STELs
0.2 ppm STEL; 1.5 mg/m3 STEL

German (DFG) - Skin/Sensitizers
very strong effects on skin

Israel - Ceiling Exposure Limits
C 0.2 ppm; C 1.5 mg/m3

Mexico - Instruction No. 10 - TWAs
0.2 ppm TWA; 1.5 mg/m3 TWA

STATE LISTS

California - Exposure Limits - Ceilings
C 0.2 ppm; C 1.5 mg/m3

California - Directors List of Hazardous Substances (8 CCR 339)
[present]

Florida Hazardous Substance List
[present]

Massachusetts Right To Know List
[present]

Minnesota Hazardous Substance List
[present]

NJ Right to Know List (Total)
sn 1259

Pennsylvania Right to Know List
environmental hazard

METHYL FLUORIDE   593-53-3

HEALTH AND SAFETY LISTS

AIHA - WEEL - Time Weighted Averages
10 ppm TWA; 36 mg/m3 TWA

NFPA - Flash Points
flash point = 156 to 170 degrees F (69 to 77 degrees C)

NFPA - Hazard Identification Ratings
flammability-2; reactivity-0

NTP Chemical Status Reports - Testing Status and NTIS Number
Short term toxicity studies scheduled for peer review; project leader assigned/study in design

ENVIRONMENTAL LISTS

EPA - Master Testing List
[present]

List of Pesticide Product Inert Ingredients
[present]

TSCA - Code of Federal Regulations Citations
40 CFR 712.30(t); 40 CFR 716.120(a); 40 CFR 799.2700

TSCA - PAIR - Reporting List
Reporting Date: February 12, 1987

TSCA - Health and Safety Reporting List
Effective Date: December 15, 1986

TSCA - Chemical Test Rules
Testing required by: manufacturers; importers; processors (40 CFR 799.2700)

TSCA - Section 12(b) - Export Notification
export notification required - Section 4

METHYL FLUOROACETATE   453-18-9

HEALTH AND SAFETY LISTS
NFPA - Flash Points
flash point < 70 degrees F (21 degrees C)
NFPA - Hazard Identification Ratings
health-0; flammability-3; reactivity-0

STATE LISTS
Florida Hazardous Substance List
[present]
Massachusetts Right To Know List
[present]
Pennsylvania Right to Know List
[present]

METHYL FLUOROSULFATE   421-20-5

HEALTH AND SAFETY LISTS
NFPA - Flash Points
flash point = 126 degrees F (52 degrees C)
NFPA - Hazard Identification Ratings
health-2; flammability-2; reactivity-0

STATE LISTS
Florida Hazardous Substance List
[present]
Massachusetts Right To Know List
[present]
Pennsylvania Right to Know List
[present]

N-METHYLFORMAMIDE   123-39-7

HEALTH AND SAFETY LISTS
U.S. DOT - Substances From 49 CFR 172.101
regulated by DOT (UN2300)
U.S. DOT - Hazard Classes
DOT hazard class = 6.1
U.S. DOT - Appendix B - Marine Pollutants
DOT regulated marine pollutant
NFPA - Flash Points
flash point = 155 degrees F (68 degrees C)
NFPA - Hazard Identification Ratings
health-3; flammability-2; reactivity-0
NIOSH - Selected LD50s and LC50s
Oral, rat: LD50 = 368 mg/kg Skin, rabbit: LD50 = 1000 mg/kg

ENVIRONMENTAL LISTS
EPA - Master Testing List
[present]

INTERNATIONAL LISTS
Canada - WHMIS: Ingredient Disclosure
1% item 1047 (1134)

STATE LISTS
Florida Hazardous Substance List
[present]
Massachusetts Right To Know List
[present]
NJ Right to Know List (Total)
sn 1260
NJ Special Hazardous Substances
(corrosive; flammable - third degree)
Pennsylvania Right to Know List
[present]

METHYL FORMATE   107-31-3

HEALTH AND SAFETY LISTS
NFPA - Flash Points
flash point = 210 degrees F (99 degrees C)
NFPA - Hazard Identification Ratings
health-0; flammability-1; reactivity-0
NTP Chemical Status Reports - Testing Status and NTIS Number
Prechronic studies for which toxicity technical reports were not prepared; Two year studies in progress

2-METHYLFURAN   534-22-5

HEALTH AND SAFETY LISTS
IARC - Group 3 (not classifiable)
[present]

METHYL-2-FUROATE   611-13-2

HEALTH AND SAFETY LISTS
IARC - Group 3 (not classifiable)
[present]

STATE LISTS
California - Directors List of Hazardous Substances (8 CCR 339)
[present]

A-METHYLGLUCOSIDE TETRANITRATE   13225-10-0

HEALTH AND SAFETY LISTS
U.S. DOT - Substances From 49 CFR 172.101
regulated by DOT (UN2454)
U.S. DOT - Hazard Classes
DOT hazard class = 2.1

STATE LISTS
NJ Right to Know List (Total)
sn 1261

A-METHYLGLYCEROL TRINITRATE   84002-64-2

HEALTH AND SAFETY LISTS
OSHA - List of Highly Hazardous Chemicals
threshhold quantity = 100 pounds

N-METHYLGLYCINE, SODIUM SALT   4316-73-8

HEALTH AND SAFETY LISTS
OSHA - List of Highly Hazardous Chemicals
threshhold quantity = 100 pounds

16-METHYLHEPTADECANOIC ACID   2724-58-5

HEALTH AND SAFETY LISTS
NIOSH - Selected LD50s and LC50s
Oral, rat: LD50 = 4000 mg/kg

METHYL HEPTADECYL KETONE   RR-00868-4

HEALTH AND SAFETY LISTS
ACGIH 1995 - Time Weighted Averages
100 ppm TWA; 246 mg/m3 TWA
ACGIH 1995 - Short Term Exposure Limits
150 ppm STEL; 368 mg/m3 STEL
AIHA - Odor Threshold Values
geometric mean air odor threshold = 2000 ppm (detectable); 2800 ppm (recognizable)
U.S. DOT - Substances From 49 CFR 172.101
regulated by DOT (UN1243)
U.S. DOT - Hazard Classes
DOT hazard class = 3
NFPA - Flash Points
flash point = -2 degrees F (-19 degrees C)

NFPA - Hazard Identification Ratings
health-2; flammability-4; reactivity-0
NIOSH - Selected LD50s and LC50s
Oral, rabbit: LD50 = 1622 mg/kg
NIOSH 1990 - Pocket Guide - RELs
100 ppm TWA; 250 mg/m3 TWA; 150 ppm STEL; 375 mg/m3 STEL
NIOSH 1990 - Pocket Guide - IDLHs
5000 ppm IDLH
NIOSH 1990 - Pocket Guide - Target organs
eyes, CNS, respiratory system
OSHA - Vacated PELs - Time Weighted Averages
100 ppm TWA; 250 mg/m3 TWA
OSHA - Vacated PELs - Short Term Exposure Limits
150 ppm STEL; 375 mg/m3 STEL
OSHA - Final PELs - Time Weighted Averages
100 ppm TWA; 250 mg/m3 TWA

ENVIRONMENTAL LISTS
CAA - Flammable Substances for Accidental Release Prevention
threshold quantity = 10,000 lbs
CAA - HON Rule - SOCMI Chemicals
compliance by Jan. 23, 1995
TSCA - PAIR - Reporting List
Effective Date: January 26, 1994; Reporting Date: March 28, 1994
TSCA - Health and Safety Reporting List
Effective Date: January 26, 1994; Sunset Date: January 26, 2004

INTERNATIONAL LISTS
Australian Exposure Standards - Time Weighted Averages
100 ppm TWA; 246 mg/m3 TWA
Australian Exposure Standards - Short Term Exposure Limits
150 ppm STEL; 368 mg/m3 STEL
Canada - WHMIS: Ingredient Disclosure
1% item 1048 (923)
Canada - Alberta - 8 Hour Occupational Exposure Limit
100 ppm TWA; 246 mg/m3 TWA
Canada - Alberta - 15 Minute Occupational Exposure Limit
150 ppm STEL; 368 mg/m3 STEL
Canada - British Columbia - 8 Hour Exposure Limits
100 ppm TWA; 250 mg/m3 TWA
Canada - British Columbia - 15 Minute Exposure Limits
150 ppm STEL; 375 mg/m3 STEL
Canada - Ontario - OHSA - TWAEVs
100 ppm TWAEV; 245 mg/m3 TWAEV
Canada - Ontario - OHSA - STEVs
150 ppm STEV; 370 mg/m3 STEV
Canada - Quebec - Time-Weighted Average Exposure Values
100 ppm TWAEV; 246 mg/m3 TWAEV
Canada - Quebec - Short-term Exposure Values
150 ppm STEV; 369 mg/m3 STEV
United Kingdom - Occupational Exposure Standards - TWAs
100 ppm TWA; 250 mg/m3 TWA
United Kingdom - Occupational Exposure Standards - STELs
150 ppm STEL; 375 mg/m3 STEL
German (DFG) - MAK Values
100 ppm MAK; 250 mg/m3 MAK
German (DFG) - Peak Limitations
2 x normal MAK (5 min momentary value); don't exceed 8 times during shift
German (DFG) - Pregnancy
classification not yet possible
Israel - Time Weighted Averages
100 ppm TWA; 246 mg/m3 TWA
Israel - Short Term Exposure Limits
150 ppm STEL; 368 mg/m3 STEL

Israel - Action Levels
50 ppm AL; 123 mg/m3 AL
Mexico - Instruction No. 10 - TWAs
100 ppm TWA; 250 mg/m3 TWA
Mexico - Instruction No. 10 - STELs
150 ppm STEL; 375 mg/m3 STEL

STATE LISTS
California - Exposure Limits - PELs
100 ppm PEL; 250 mg/m3 PEL
California - Exposure Limits - STELs
150 ppm STEL; 375 mg/m3 STEL
California - Directors List of Hazardous Substances (8 CCR 339)
[present]
Florida Hazardous Substance List
[present]
Massachusetts Right To Know List
[present]
Minnesota Hazardous Substance List
[present]
NJ Right to Know List (Total)
sn 1262
NJ Special Hazardous Substances
(flammable - fourth degree)
Pennsylvania Right to Know List
[present]

PROPOSED REGULATIONS
TSCA - ITC 31st Report Priority Testing List
designated to be tested

METHYLHEPTENONE 409-02-9
HEALTH AND SAFETY LISTS
U.S. DOT - Substances From 49 CFR 172.101
regulated by DOT (UN2301)
U.S. DOT - Hazard Classes
DOT hazard class = 3
NFPA - Flash Points
flash point = -22 degrees F (-30 degrees C)
NFPA - Hazard Identification Ratings
health-2; flammability-3; reactivity-1
NIOSH - Selected LD50s and LC50s
Inhalation, rat: LC50 = 377 ppm 4 hr Oral, rat: LD50 = 167 mg/kg
INTERNATIONAL LISTS
Canada - WHMIS: Ingredient Disclosure
1% item 1049 (1135)
STATE LISTS
Florida Hazardous Substance List
[present]
Massachusetts Right To Know List
[present]
NJ Right to Know List (Total)
sn 1263
Pennsylvania Right to Know List
[present]

METHYL HEPTINE CARBONATE 111-12-6
INTERNATIONAL LISTS
Canada - WHMIS: Ingredient Disclosure
1% item 1050 (928)

6-METHYLHEPTYL GLYCIDYL ETHER 68134-07-6
HEALTH AND SAFETY LISTS
U.S. DOT - Hazard Classes
Forbidden from transport by the DOT

3-METHYLHEXANE   589-34-4

HEALTH AND SAFETY LISTS
U.S. DOT - Hazard Classes
Forbidden from transport by the DOT

METHYL HEXANOATE   106-70-7

ENVIRONMENTAL LISTS
List of Pesticide Product Inert Ingredients
[present]

METHYL HYDRAZINE   60-34-4

ENVIRONMENTAL LISTS
List of Pesticide Product Inert Ingredients
[present]

METHYL P-HYDROXYBENZOATE   99-76-3

HEALTH AND SAFETY LISTS
NFPA - Flash Points
flash point = 255 degrees F (124 degrees C)
NFPA - Hazard Identification Ratings
health-0; flammability-1; reactivity-0

METHYL-3-HYDROXYBUTYRATE   1487-49-6

HEALTH AND SAFETY LISTS
NFPA - Flash Points
flash point = 135 degrees F (57 degrees C)
NFPA - Hazard Identification Ratings
health-1; flammability-2; reactivity-0

1-METHYLIMIDAZOLE   616-47-7

HEALTH AND SAFETY LISTS
NFPA - Flash Points
flash point = 190 degrees F (88 degrees C)
NFPA - Hazard Identification Ratings
flammability-2; reactivity-0

2-METHYLIMIDAZOLE   693-98-1

ENVIRONMENTAL LISTS
EPA - Master Testing List
[present]

PROPOSED REGULATIONS
TSCA - Proposed Testing Rule for Glycidyl Ethers
member of Glycidyl subcategory I-A

4-METHYLIMIDAZOLE   822-36-6

HEALTH AND SAFETY LISTS
NFPA - Flash Points
flash point = 25 degrees F (-4 degrees C)
NFPA - Hazard Identification Ratings
health-0; flammability-3; reactivity-0

STATE LISTS
Florida Hazardous Substance List
[present]
Massachusetts Right To Know List
[present]
Pennsylvania Right to Know List
[present]

1,1'-(METHYLIMINO)BIS(3-CHLORO-2-PROPANOL)   68140-76-1
, POLYMER WITH N,N,N',N'-TETRAMETHYL-1,2-
ETHANEDIAMINE

HEALTH AND SAFETY LISTS
NIOSH - Selected LD50s and LC50s
Inhalation, mouse: LC50 = 14 gm/m3 2 hr

3-METHYLINDOLE   83-34-1

HEALTH AND SAFETY LISTS
ACGIH 1995 - Ceiling Limits
(C 0.2) ppm; (C 0.38) mg/m3
ACGIH 1995 - Skin Designations
skin - potential for cutaneous absorption
ACGIH 1995 - Carcinogens
(A2)-suspected human carcinogen
U.S. DOT - Substances From 49 CFR 172.101
regulated by DOT (UN1244)
U.S. DOT - Hazard Classes
DOT hazard class = 6.1
U.S. DOT - Substances Which Are Poisonous by Inhalation
liquid hazardous material poisonous by inhalation (UN1244)
U.S. DOT - Appendix A Table 1 - Hazardous Substances
final RQ = 10 pounds (4.54 kg)
NFPA - Flash Points
flash point = 17 degrees F (-8 degrees C)
NFPA - Hazard Identification Ratings
health-4; flammability-3; reactivity-2
NIOSH - Selected LD50s and LC50s
Inhalation, rat: LC50 = 34 ppm 4 hr Oral, rat: LD50 = 32 mg/kg
Skin, rat: LD50 = 183 mg/kg
NIOSH 1990 - Pocket Guide - RELs
C 0.04 ppm (2 hr); C 0.08 mg/m3 (2 hr)
NIOSH 1990 - Pocket Guide - IDLHs
50 ppm IDLH (not considering carcinogenic effects)
NIOSH 1990 - Pocket Guide - Carcinogens
occupational carcinogen
NIOSH 1990 - Pocket Guide - Target organs
CNS, respiratory system, liver, blood, CVS, eyes
NIOSH - Health Standards - Exposure Limits
C (2 hr) 0.04 ppm; C (2 hr) 0.08 mg/m3 (Listed under 'Hydrazines')
NIOSH - Health Standards - Health Effects and Precautions
has produced tumors of the lung, liver, blood vessels, and intestines in animals; blood, live, and skin effects (Blood and urine monitoring and periodic chest X-ray required; bowel examinations for workers above age 40) (Listed under 'Hydrazines')
NIOSH - Health Standards - Carcinogenic Chemicals
potential human carcinogen (Listed under 'Hydrazines')
OSHA - Vacated PELs - Ceiling Limits
C 0.2 ppm; C 0.35 mg/m3
OSHA - Vacated PELs - Skin Designation
Prevent or reduce skin absorption
OSHA - Final PELs - Ceiling Limits
C 0.2 ppm; C 0.35 mg/m3
OSHA - Final PELs - Skin Notations
prevent or reduce skin absorption
OSHA - List of Highly Hazardous Chemicals
threshhold quantity = 100 pounds

ENVIRONMENTAL LISTS
CERCLA/SARA - Section 302 Extremely Hazardous Substances and TPQs
TPQ = 500 pounds
CERCLA/SARA - Section 313 - Emission Reporting
form R reporting required for 1.0% de minimus concentration

CERCLA/SARA - Hazardous Substances and their Reportable Quantities
final RQ = 10 pounds (4.54 kg)

Clean Air Act (1990) - List of Hazardous Air Contaminants
[present]

CAA -Toxic Substances for Accidental Release Prevention
threshold quantity = 15,000 lbs

CAA - HON Rule - SOCMI Chemicals
compliance by July 24, 1995

CAA - HON Rule - Organic HAPs
[present]

RCRA - P Series Wastes
waste number P068

RCRA - Hazardous Constituents-Appendix VIII
waste number P068

RCRA - Substances Banned From Land Disposal
[present]

INTERNATIONAL LISTS

Australian Exposure Standards - Time Weighted Averages
Peak Limitation: 0.2 ppm; 0.38 mg/m3

Australian Exposure Standards - Skin Effects
skin absorption

Australian Exposure Standards - Carcinogens
suspected carcinogen

Australian Exposure Standards - Under Review
exposure limits under review

Canada - WHMIS: Ingredient Disclosure
0.1% item 1052 (1137)

Canada - Alberta - Ceiling Occupational Exposure Limit
C 0.2 ppm; C 0.38 mg/m3

Canada - Alberta - Skin Designation
can be absorbed through the intact skin

Canada - Alberta - Designated Substances
designated substance - requires code of practice

Canada - British Columbia - Ceiling Exposure Limits
C 0.2 ppm; C 0.35 mg/m3

Canada - British Columbia - Skin Notations
skin - potential for skin absorption

Canada - Ontario - OHSA - CEVs
0.1 ppm CEV; 0.38 mg/m3 CEV

Canada - Ontario - OHSA - Skin Notations
absorption through skin, eyes, or mucous membranes

Canada - Quebec - Ceiling Limits
P 0.2 ppm; P 0.38 mg/m3

Canada - Quebec - Skin Designations
absorbed through the skin

Canada - Quebec - Carcinogens
C2 carcinogen: effect suspected in humans

Israel - Ceiling Exposure Limits
C (0.2) ppm; C (0.38) mg/m3

Mexico - Instruction No. 10 - TWAs
0.35 mg/m3 TWA

Mexico - Instruction No. 10 - Ceiling Limits
P 0.2 ppm

Mexico - Instruction No. 10 - Skin designation
skin - potential for cutaneous absorption

Mexico - Instruction No. 10 - Carcinogens
potential carcinogen in humans - limited epidemiological evidence

STATE LISTS

California - Air Bill 2588 Appendix A-I
6/91

California - Prop. 65 - Cancer list
carcinogen (including Methylhydrazine salts) - initial date 7/1/92

California - Exposure Limits - Ceilings
C 0.2 ppm; C 0.35 mg/m3

California - Exposure Limits - Skin Notation
material may be absorbed through the skin, eyes or mucous membrane

California - Directors List of Hazardous Substances (8 CCR 339)
[present]

Florida Hazardous Substance List
[present]

Massachusetts Right To Know List
extraordinarily hazardous

Minnesota Hazardous Substance List
carcinogen; skin

NJ Right to Know List (Total)
sn 1265

NJ Special Hazardous Substances
(carcinogen; corrosive; flammable - third degree; mutagen)

Pennsylvania Right to Know List
environmental hazard

PROPOSED REGULATIONS

ACGIH 1995 - Notice of Intended Changes
(skin) 0.01 ppm TWA; 0.019 mg/m3 TWA; A3-animal carcinogen

METHYL IODIDE   74-88-4

HEALTH AND SAFETY LISTS

NIOSH - Selected LD50s and LC50s
Oral, dog: LD50 = 3000 mg/kg

ENVIRONMENTAL LISTS

List of Pesticide Product Inert Ingredients
[present]

METHYL IONONE   1335-94-0

HEALTH AND SAFETY LISTS

NFPA - Flash Points
flash point = 180 degrees F (82 degrees C)

NFPA - Hazard Identification Ratings
health-1; flammability-2; reactivity-0

METHYLIONONES (A-)   79-69-6

HEALTH AND SAFETY LISTS

NIOSH - Selected LD50s and LC50s
Oral, mouse: LD50 = 1400 mg/kg

METHYL ISOAMYL KETONE   110-12-3

HEALTH AND SAFETY LISTS

NIOSH - Selected LD50s and LC50s
Oral, mouse: LD50 = 1400 mg/kg

NTP Chemical Status Reports - Testing Status and NTIS Number
Prechronic studies completed; in review for further evaluation

ENVIRONMENTAL LISTS

List of Pesticide Product Inert Ingredients
[present]

METHYL ISOBUTYL CARBINOL   108-11-2

HEALTH AND SAFETY LISTS

NTP Chemical Status Reports - Testing Status and NTIS Number
Prechronic studies completed; in review for further evaluation

METHYL ISOBUTYL KETONE   108-10-1

ENVIRONMENTAL LISTS

List of Pesticide Product Inert Ingredients
[present]

METHYL ISOBUTYL KETONE PEROXIDE   37206-20-5

HEALTH AND SAFETY LISTS

NIOSH - Selected LD50s and LC50s
Oral, rat: LD50 = 3450 mg/kg

METHYL ISOBUTYL KETONE PEROXIDE   39811-34-2

SEE ALSO:
F039-HAZARDOUS WASTES

HEALTH AND SAFETY LISTS

ACGIH 1995 - Time Weighted Averages
2 ppm TWA; 12 mg/m3 TWA

ACGIH 1995 - Skin Designations
skin - potential for cutaneous absorption

ACGIH 1995 - Carcinogens
A2-suspected human carcinogen

U.S. DOT - Substances From 49 CFR 172.101
regulated by DOT (UN2644)

U.S. DOT - Hazard Classes
DOT hazard class = 6.1

U.S. DOT - Substances Which Are Poisonous by Inhalation
liquid hazardous material poisonous by inhalation (UN2644)

U.S. DOT - Appendix A Table 1 - Hazardous Substances
final RQ = 100 pounds (45.4 kg)

IARC - Group 3 (not classifiable)
[present]

NIOSH - Selected LD50s and LC50s
Inhalation, rat: LC50 = 1300 mg/m3 4 hr

NIOSH 1990 - Pocket Guide - RELs
2 ppm TWA; 10 mg/m3 TWA

NIOSH 1990 - Pocket Guide - IDLHs
800 ppm IDLH (not considering carcinogenic effects)

NIOSH 1990 - Pocket Guide - Carcinogens
occupational carcinogen

NIOSH 1990 - Pocket Guide - Target organs
CNS, skin, eyes,

NIOSH 1990 - Pocket Guide - Skin list
Potential for dermal absorption

NIOSH - Health Standards - Exposure Limits
reduce exposure to lowest feasible concentration (Listed under 'Monohalomethanes')

NIOSH - Health Standards - Health Effects and Precautions
has produced tumors of the kidney, forestomach, and lung in animals (Prevent skin contact) (Listed under 'Monohalomethanes')

NIOSH - Health Standards - Carcinogenic Chemicals
potential human carcinogen (Listed under 'Monohalomethanes')

OSHA - Vacated PELs - Time Weighted Averages
2 ppm TWA; 10 mg/m3 TWA

OSHA - Vacated PELs - Skin Designation
Prevent or reduce skin absorption

OSHA - Final PELs - Time Weighted Averages
5 ppm TWA; 28 mg/m3 TWA

OSHA - Final PELs - Skin Notations
prevent or reduce skin absorption

OSHA - List of Highly Hazardous Chemicals
threshhold quantity = 7500 pounds

ENVIRONMENTAL LISTS

CERCLA/SARA - Section 313 - Emission Reporting
form R reporting required for 0.1% de minimus concentration

CERCLA/SARA - Hazardous Substances and their Reportable Quantities
final RQ = 100 pounds (45.4 kg)

Clean Air Act (1990) - List of Hazardous Air Contaminants
[present]

EPA - Carcinogen Hazard Ranking for RQ Adjustment
Hazard ranking = Low

RCRA - U Series Wastes
waste number U138

RCRA - Hazardous Constituents-Appendix VIII
waste number U138

RCRA - Basis for Listing - Appendix VII
Included in waste stream: F039

RCRA - Substances Banned From Land Disposal
[present]

RCRA - TSD Facilities Ground Water Monitoring
TM 8010 = 40 ug/L PQL; TM 8240 = 5 ug/L PQL

RCRA - Universal Treatment Standards (LDR)
WW: 0.19 mg/l; NWW: 65 mg/kg

INTERNATIONAL LISTS

Australian Exposure Standards - Time Weighted Averages
2 ppm TWA; 12 mg/m3 TWA

Australian Exposure Standards - Skin Effects
skin absorption

Australian Exposure Standards - Carcinogens
suspected carcinogen

Canada - WHMIS: Ingredient Disclosure
0.1% item 1053 (1026)

Canada - NPRI (National Pollutant Release Inventory)
[present]

Canada - Alberta - 8 Hour Occupational Exposure Limit
2 ppm TWA; 12 mg/m3 TWA

Canada - Alberta - 15 Minute Occupational Exposure Limit
5 ppm STEL; 29 mg/m3 STEL

Canada - Alberta - Skin Designation
can be absorbed through the intact skin

Canada - Alberta - Designated Substances
designated substance - requires code of practice

Canada - British Columbia - 8 Hour Exposure Limits
5 ppm TWA; 28 mg/m3 TWA

Canada - British Columbia - 15 Minute Exposure Limits
10 ppm STEL; 56 mg/m3 STEL

Canada - British Columbia - Skin Notations
skin - potential for skin absorption

Canada - Ontario - OHSA - TWAEVs
2 ppm TWAEV; 12 mg/m3 TWAEV

Canada - Ontario - OHSA - Skin Notations
absorption through skin, eyes, or mucous membranes

Canada - Quebec - Time-Weighted Average Exposure Values
2 ppm TWAEV; 12 mg/m3 TWAEV

Canada - Quebec - Skin Designations
absorbed through the skin

Canada - Quebec - Carcinogens
C2 carcinogen: effect suspected in humans

German (DFG) - Carcinogens
animal evidence of carcinogenicity

Israel - Time Weighted Averages
2 ppm TWA; 12 mg/m3 TWA

Israel - Action Levels
1 ppm AL; 6 mg/m3 AL

Mexico - Instruction No. 10 - TWAs
2 ppm TWA; 10 mg/m3 TWA

Mexico - Instruction No. 10 - STELs
5 ppm STEL; 30 mg/m3 STEL

Mexico - Instruction No. 10 - Skin designation
skin - potential for cutaneous absorption

Mexico - Instruction No. 10 - Carcinogens
potential carcinogen in humans - limited epidemiological evidence

STATE LISTS

California - Air Bill 2588 Appendix A-I
known or potential carcinogen

California - Prop. 65 - Cancer list
carcinogen - initial date 4/1/88

California - Exposure Limits - PELs
2 ppm PEL; 10 mg/m3 PEL

California - Exposure Limits - Skin Notation
material may be absorbed through the skin, eyes or mucous membrane

California - Directors List of Hazardous Substances (8 CCR 339)
[present]

Florida Hazardous Substance List
[present]

Massachusetts Right To Know List
carcinogen; extraordinarily hazardous

Minnesota Hazardous Substance List
carcinogen; skin

NJ Right to Know List (Total)
sn 1266

NJ Special Hazardous Substances
(mutagen)

Pennsylvania Right to Know List
environmental hazard; special hazardous substance

Pennsylvania RTK - Special Hazardous Substances
[present]

PROPOSED REGULATIONS

Canada - Ontario - Proposed Occupational TWAEVs
1 ppm TWAEV; 6 mg/m3 TWAEV

Canada - Ontario - Proposed Occupational STEVs
5 ppm STEV; 30 mg/m3 STEV

METHYL ISOBUTYRATE   547-63-7

HEALTH AND SAFETY LISTS

NFPA - Flash Points
flash point > 212 degrees F (100 degrees C)

NFPA - Hazard Identification Ratings
health-0; flammability-1; reactivity-0

METHYL ISOCYANATE   624-83-9

ENVIRONMENTAL LISTS

CAA - HON Rule - SOCMI Chemicals
compliance by Oct. 23, 1995

METHYL ISO EUGENOL   93-16-3

HEALTH AND SAFETY LISTS

ACGIH 1995 - Time Weighted Averages
50 ppm TWA; 234 mg/m3 TWA

AIHA - Odor Threshold Values
geometric mean air odor threshold = 0.013 ppm (detectable); 0.049 ppm (recognizable)

U.S. DOT - Substances From 49 CFR 172.101
regulated by DOT (UN2302)

U.S. DOT - Hazard Classes
DOT hazard class = 3

NFPA - Flash Points
flash point = 96 degrees F (36 degrees C)

NFPA - Hazard Identification Ratings
health-1; flammability-2; reactivity-0

NIOSH - Selected LD50s and LC50s
Oral, rat: LD50 = 4760 mg/kg Skin, rabbit: LD50 = 10 gm/kg

NIOSH - Health Standards - Exposure Limits
50 ppm TWA; 230 mg/m3 TWA (Listed under 'Ketones')

NIOSH - Health Standards - Health Effects and Precautions
Irritation; liver, kidney, and nervous system effects (Urinalysis required) (Listed under 'Ketones')

OSHA - Vacated PELs - Time Weighted Averages
50 ppm TWA; 240 mg/m3 TWA

ENVIRONMENTAL LISTS

List of Pesticide Product Inert Ingredients
[present]

INTERNATIONAL LISTS

Australian Exposure Standards - Time Weighted Averages
50 ppm TWA; 234 mg/m3 TWA

Canada - WHMIS: Ingredient Disclosure
1% item 1051 (1136)

Canada - Alberta - 8 Hour Occupational Exposure Limit
50 ppm TWA; 233 mg/m3 TWA

Canada - Alberta - 15 Minute Occupational Exposure Limit
75 ppm STEL; 350 mg/m3 STEL

Canada - British Columbia - 8 Hour Exposure Limits
100 ppm TWA; 475 mg/m3 TWA

Canada - British Columbia - 15 Minute Exposure Limits
150 ppm STEL; 710 mg/m3 STEL

Canada - Ontario - OHSA - TWAEVs
50 ppm TWAEV; 233 mg/m3 TWAEV

Canada - Quebec - Time-Weighted Average Exposure Values
50 ppm TWAEV; 234 mg/m3 TWAEV

United Kingdom - Occupational Exposure Standards - TWAs
50 ppm TWA; 240 mg/m3 TWA

United Kingdom - Occupational Exposure Standards - STELs
75 ppm STEL; 360 mg/m3 STEL

Israel - Time Weighted Averages
50 ppm TWA; 234 mg/m3 TWA

Israel - Action Levels
25 ppm AL; 117 mg/m3 AL

Mexico - Instruction No. 10 - TWAs
100 ppm TWA; 475 mg/m3 TWA

STATE LISTS

California - Exposure Limits - PELs
50 ppm PEL; 234 mg/m3 PEL

California - Directors List of Hazardous Substances (8 CCR 339)
[present]

Florida Hazardous Substance List
[present]

Massachusetts Right To Know List
[present]

Minnesota Hazardous Substance List
[present]

NJ Right to Know List (Total)
sn 1267

Pennsylvania Right to Know List
[present]

METHYL ISOPROPENYL KETONE   814-78-8

HEALTH AND SAFETY LISTS

ACGIH 1995 - Time Weighted Averages
25 ppm TWA; 104 mg/m3 TWA

ACGIH 1995 - Short Term Exposure Limits
40 ppm STEL; 167 mg/m3 STEL

ACGIH 1995 - Skin Designations
skin - potential for cutaneous absorption

U.S. DOT - Substances From 49 CFR 172.101
regulated by DOT (UN2053)

U.S. DOT - Hazard Classes
DOT hazard class = 3

NFPA - Flash Points
flash point = 106 degrees F (41 degrees C)

NFPA - Hazard Identification Ratings
health-2; flammability-2; reactivity-0

NIOSH - Selected LD50s and LC50s
Oral, rat: LD50 = 2590 mg/kg Skin, rabbit: LD50 = 3560 mg/kg

NIOSH 1990 - Pocket Guide - RELs
25 ppm TWA; 100 mg/m3 TWA; 40 ppm STEL; 165 mg/m3 STEL

NIOSH 1990 - Pocket Guide - IDLHs
2000 ppm IDLH

NIOSH 1990 - Pocket Guide - Target organs
eyes, skin

NIOSH 1990 - Pocket Guide - Skin list
Potential for dermal absorption

OSHA - Vacated PELs - Time Weighted Averages
25 ppm TWA; 100 mg/m3 TWA

OSHA - Vacated PELs - Short Term Exposure Limits
40 ppm STEL; 165 mg/m3 STEL

OSHA - Vacated PELs - Skin Designation
Prevent or reduce skin absorption

OSHA - Final PELs - Time Weighted Averages
25 ppm TWA; 100 mg/m3 TWA

OSHA - Final PELs - Skin Notations
prevent or reduce skin absorption

ENVIRONMENTAL LISTS

CAA - HON Rule - SOCMI Chemicals
compliance by July 24, 1995

INTERNATIONAL LISTS

Australian Exposure Standards - Time Weighted Averages
25 ppm TWA; 104 mg/m3 TWA

Australian Exposure Standards - Short Term Exposure Limits
40 ppm STEL; 167 mg/m3 STEL

Australian Exposure Standards - Skin Effects
skin absorption

Canada - WHMIS: Ingredient Disclosure
1% item 1054 (1138)

Canada - Alberta - 8 Hour Occupational Exposure Limit
25 ppm TWA; 105 mg/m3 TWA

Canada - Alberta - 15 Minute Occupational Exposure Limit
40 ppm STEL; 167 mg/m3 STEL

Canada - Alberta - Skin Designation
can be absorbed through the intact skin

Canada - British Columbia - 8 Hour Exposure Limits
25 ppm TWA; 100 mg/m3 TWA

Canada - British Columbia - Skin Notations
skin - potential for skin absorption

Canada - Ontario - OHSA - TWAEVs
25 ppm TWAEV; 104 mg/m3 TWAEV

Canada - Ontario - OHSA - STEVs
40 ppm STEV; 167 mg/m3 STEV

Canada - Ontario - OHSA - Skin Notations
absorption through skin, eyes, or mucous membranes

Canada - Quebec - Time-Weighted Average Exposure Values
25 ppm TWAEV; 104 mg/m3 TWAEV

Canada - Quebec - Short-term Exposure Values
40 ppm STEV; 166 mg/m3 STEV

Canada - Quebec - Skin Designations
absorbed through the skin

United Kingdom - Occupational Exposure Standards - TWAs
25 ppm TWA; 100 mg/m3 TWA

United Kingdom - Occupational Exposure Standards - STELs
40 ppm STEL; 160 mg/m3 STEL

United Kingdom - Occupational Exposure Standards - Notes
can be absorbed through skin

German (DFG) - MAK Values
25 ppm MAK; 100 mg/m3 MAK

German (DFG) - Peak Limitations
5 x normal MAK (30 min. average value); don't exceed 2 times during shift

German (DFG) - Skin/Sensitizers
danger of cutaneous absorption

Israel - Time Weighted Averages
25 ppm TWA; 104 mg/m3 TWA

Israel - Short Term Exposure Limits
40 ppm STEL; 167 mg/m3 STEL

Israel - Action Levels
12.5 ppm AL; 52 mg/m3 AL

Mexico - Instruction No. 10 - TWAs
25 ppm TWA; 100 mg/m3 TWA

Mexico - Instruction No. 10 - STELs
40 ppm STEL; 165 mg/m3 STEL

Mexico - Instruction No. 10 - Skin designation
skin - potential for cutaneous absorption

STATE LISTS

California - Exposure Limits - PELs
25 ppm PEL; 100 mg/m3 PEL

California - Exposure Limits - STELs
40 ppm STEL; 165 mg/m3 STEL

California - Exposure Limits - Skin Notation
material may be absorbed through the skin, eyes or mucous membrane

California - Directors List of Hazardous Substances (8 CCR 339)
[present]

Florida Hazardous Substance List
[present]

Massachusetts Right To Know List
[present]

Minnesota Hazardous Substance List
skin

NJ Right to Know List (Total)
sn 1228

Pennsylvania Right to Know List
[present]

METHYL ISOPROPYL KETONE   563-80-4
SEE ALSO:
F039-HAZARDOUS WASTES
F003-HAZARDOUS WASTES

HEALTH AND SAFETY LISTS

ACGIH 1995 - Time Weighted Averages
50 ppm TWA; 205 mg/m3 TWA

ACGIH 1995 - Short Term Exposure Limits
75 ppm STEL; 307 mg/m3 STEL

ACGIH 1995 - Biological Exposure Indices
MIBK in urine: 2 mg/L, end of shift

AIHA - Odor Threshold Values
geometric mean air odor threshold = 0.88 ppm (detectable); 2.1 ppm (recognizable)

U.S. DOT - Substances From 49 CFR 172.101
regulated by DOT (UN1245)

U.S. DOT - Hazard Classes
DOT hazard class = 3

U.S. DOT - Appendix A Table 1 - Hazardous Substances
final RQ = 5000 pounds (2270 kg)

NFPA - Flash Points
flash point = 64 degrees F (18 degrees C)

NFPA - Hazard Identification Ratings
health-2; flammability-3; reactivity-1

NIOSH - Selected LD50s and LC50s
Inhalation, rat: LC50 = 8000 ppm 4 hr Oral, rat: LD50 = 2080 mg/kg

NIOSH 1990 - Pocket Guide - RELs
50 ppm TWA; 205 mg/m3 TWA; 75 ppm STEL; 300 mg/m3 STEL

NIOSH 1990 - Pocket Guide - IDLHs
3000 ppm IDLH

NIOSH 1990 - Pocket Guide - Target organs
eyes, skin, CNS, respiratory system

NIOSH - Health Standards - Exposure Limits
50 ppm TWA; 200 mg/m3 TWA (Listed under 'Ketones')

NIOSH - Health Standards - Health Effects and Precautions
Irritation, liver, kidney, and nervous system effects (Urinalysis required) (Listed under 'Ketones')

OSHA - Vacated PELs - Time Weighted Averages
50 ppm TWA; 205 mg/m3 TWA

OSHA - Vacated PELs - Short Term Exposure Limits
75 ppm STEL; 300 mg/m3 STEL

OSHA - Final PELs - Time Weighted Averages
100 ppm TWA; 410 mg/m3 TWA

ENVIRONMENTAL LISTS

ATSDR Priority List
Rank (of 275): 230

CERCLA/SARA - Section 313 - Emission Reporting
form R reporting required for 1.0% de minimus concentration

CERCLA/SARA - Hazardous Substances and their Reportable Quantities
final RQ = 5000 pounds (2270 kg)

Clean Air Act (1990) - List of Hazardous Air Contaminants
[present]

CAA - HON Rule - SOCMI Chemicals
compliance by July 24, 1995

CAA - HON Rule - Organic HAPs
[present]

EPA - Master Testing List
[present]

List of Pesticide Product Inert Ingredients
[present]

RCRA - U Series Wastes
waste number U161 (Ignitable waste)

RCRA - Hazardous Constituents-Appendix VIII
waste number U161 (Ignitable waste)

RCRA - Basis for Listing - Appendix VII
Included in waste stream: F039

RCRA - Substances Banned From Land Disposal
[present]

RCRA - TSD Facilities Ground Water Monitoring
TM 8015 = 5 ug/L PQL; TM 8240 = 50 ug/L PQL

RCRA - Universal Treatment Standards (LDR)
WW: 0.14 mg/l; NWW: 33 mg/kg

TSCA - Code of Federal Regulations Citations
40 CFR 712.30(d); 40 CFR 716.120(a)

TSCA - PAIR - Reporting List
Reporting Date: November 19, 1982

TSCA - Health and Safety Reporting List
Effective Date: October 4, 1982

TSCA - Multichemical Test Rules - Neurotoxicity
administrative stay for neurotoxicity tests effective June 27, 1994

INTERNATIONAL LISTS

Australian Exposure Standards - Time Weighted Averages
50 ppm TWA; 205 mg/m3 TWA

Australian Exposure Standards - Short Term Exposure Limits
75 ppm STEL; 307 mg/m3 STEL

Australian Exposure Standards - Under Review
exposure limits under review

Canada - WHMIS: Ingredient Disclosure
1% item 1055 (1139)

Canada - NPRI (National Pollutant Release Inventory)
[present]

Canada - Alberta - 8 Hour Occupational Exposure Limit
50 ppm TWA; 205 mg/m3 TWA

Canada - Alberta - 15 Minute Occupational Exposure Limit
75 ppm STEL; 300 mg/m3 STEL

Canada - Alberta - Skin Designation
can be absorbed through the intact skin

Canada - British Columbia - 8 Hour Exposure Limits
100 ppm TWA; 410 mg/m3 TWA

Canada - British Columbia - 15 Minute Exposure Limits
125 ppm STEL; 510 mg/m3 STEL

Canada - British Columbia - Skin Notations
skin - potential for skin absorption

Canada - Ontario - OHSA - TWAEVs
50 ppm TWAEV; 205 mg/m3 TWAEV

Canada - Ontario - OHSA - STEVs
75 ppm STEV; 305 mg/m3 STEV

Canada - Quebec - Time-Weighted Average Exposure Values
50 ppm TWAEV; 205 mg/m3 TWAEV

Canada - Quebec - Short-term Exposure Values
75 ppm STEV; 310 mg/m3 STEV

United Kingdom - Occupational Exposure Standards - TWAs
50 ppm TWA; 205 mg/m3 TWA

United Kingdom - Occupational Exposure Standards - STELs
75 ppm STEL; 300 mg/m3 STEL

United Kingdom - Occupational Exposure Standards - Notes
can be absorbed through skin

German (DFG) - MAK Values
100 ppm MAK; 400 mg/m3 MAK

German (DFG) - Peak Limitations
5 x normal MAK (30 min. average value); don't exceed 2 times during shift

Israel - Time Weighted Averages
50 ppm TWA; 205 mg/m3 TWA

Israel - Short Term Exposure Limits
75 ppm STEL; 307 mg/m3 STEL

Israel - Action Levels
25 ppm AL; 102.5 mg/m3 AL

Mexico - Instruction No. 10 - TWAs
50 ppm TWA; 203 mg/m3 TWA

Mexico - Instruction No. 10 - Skin designation
skin - potential for cutaneous absorption

STATE LISTS

California - Air Bill 2588 Appendix A-I
6/91

California - Exposure Limits - PELs
50 ppm PEL; 205 mg/m3 PEL

California - Exposure Limits - STELs
75 ppm STEL; 300 mg/m3 STEL

California - Directors List of Hazardous Substances (8 CCR 339)
[present]

Florida Hazardous Substance List
[present]

Massachusetts Right To Know List
[present]

Minnesota Hazardous Substance List
[present]

NJ Right to Know List (Total)
sn 1268

NJ Special Hazardous Substances
(flammable - third degree)

Pennsylvania Right to Know List
environmental hazard

PROPOSED REGULATIONS

Safe Drinking Water Act - Priority list
[present]

2-METHYL-3-ISOTHIAZOLONE   2682-20-4

HEALTH AND SAFETY LISTS

U.S. DOT - Hazard Classes
Forbidden from transport by the DOT

U.S. DOT - Organic Peroxides Table
Organic peroxide UN3105

STATE LISTS

NJ Right to Know List (Total)
sn 1269

METHYL ISOTHIOCYANATE   556-61-6

HEALTH AND SAFETY LISTS

U.S. DOT - Organic Peroxides Table
Organic peroxide UN3105

METHYL ISOVALERATE   556-24-1

HEALTH AND SAFETY LISTS

NIOSH - Selected LD50s and LC50s
Inhalation, mouse: LC50 = 25500 mg/kg (8 hr) Oral, rat: LD50 = 16000 mg/kg

METHYL LACTATE   547-64-8

HEALTH AND SAFETY LISTS

ACGIH 1995 - Time Weighted Averages
0.02 ppm TWA; 0.047 mg/m3 TWA

ACGIH 1995 - Skin Designations
skin - potential for cutaneous absorption

AIHA - Odor Threshold Values
no geometric mean air odor threshold

U.S. DOT - Substances From 49 CFR 172.101
regulated by DOT (UN2480)

U.S. DOT - Hazard Classes
DOT hazard class = 6.1

U.S. DOT - Substances Which Are Poisonous by Inhalation
liquid hazardous material poisonous by inhalation (UN2480)

U.S. DOT - Appendix A Table 1 - Hazardous Substances
final RQ = 10 pounds (4.54 kg)

NFPA - Flash Points
flash point = 19 degrees F (-7 degrees C)

NFPA - Hazard Identification Ratings
health-4; flammability-3; reactivity-2 (avoid use of water)

NIOSH - Selected LD50s and LC50s
Inhalation, rat: LC50 = 6100 ppb 6 hr Oral, rat: LD50 = 69 mg/kg Skin, rabbit: LD50 = 213 mg/kg

NIOSH 1990 - Pocket Guide - RELs
0.02 ppm TWA; 0.05 mg/m3 TWA

NIOSH 1990 - Pocket Guide - IDLHs
20 ppm IDLH

NIOSH 1990 - Pocket Guide - Target organs
skin, eyes, respiratory system

NIOSH 1990 - Pocket Guide - Skin list
Potential for dermal absorption

OSHA - Vacated PELs - Time Weighted Averages
0.02 ppm TWA; 0.05 mg/m3 TWA

OSHA - Vacated PELs - Skin Designation
Prevent or reduce skin absorption

OSHA - Final PELs - Time Weighted Averages
0.02 ppm TWA; 0.05 mg/m3 TWA

OSHA - Final PELs - Skin Notations
prevent or reduce skin absorption

OSHA - List of Highly Hazardous Chemicals
threshhold quantity = 250 pounds

ENVIRONMENTAL LISTS

CERCLA/SARA - Section 302 Extremely Hazardous Substances and TPQs
TPQ = 500 pounds

CERCLA/SARA - Section 313 - Emission Reporting
form R reporting required for 1.0% de minimus concentration

CERCLA/SARA - Hazardous Substances and their Reportable Quantities
final RQ = 10 pounds (4.54 kg)

Clean Air Act (1990) - List of Hazardous Air Contaminants
[present]

CAA -Toxic Substances for Accidental Release Prevention
threshold quantity = 10,000 lbs

CAA - HON Rule - SOCMI Chemicals
compliance by July 24, 1995

CAA - HON Rule - Organic HAPs
[present]

RCRA - P Series Wastes
waste number P064

RCRA - Hazardous Constituents-Appendix VIII
waste number P064

RCRA - Substances Banned From Land Disposal
[present]

TSCA - Code of Federal Regulations Citations
40 CFR 712.30(x); 40 CFR 716.120(d)

TSCA - PAIR - Reporting List
Reporting Date: December 27, 1990

TSCA - Health and Safety Reporting List
Effective Date: October 29, 1990

INTERNATIONAL LISTS

Canada - WHMIS: Ingredient Disclosure
0.1% item 1056 (1044)

Canada - Alberta - 8 Hour Occupational Exposure Limit
0.02 ppm TWA; 0.05 mg/m3 TWA

Canada - Alberta - 15 Minute Occupational Exposure Limit
0.06 ppm STEL; 0.15 mg/m3 STEL

Canada - Alberta - Skin Designation
can be absorbed through the intact skin

Canada - British Columbia - 8 Hour Exposure Limits
0.02 ppm TWA; 0.05 mg/m3 TWA

Canada - British Columbia - Skin Notations
skin - potential for skin absorption

Canada - Quebec - Time-Weighted Average Exposure Values
0.02 ppm TWAEV; 0.047 mg/m3 TWAEV

Canada - Quebec - Skin Designations
absorbed through the skin

German (DFG) - MAK Values
0.01 ppm MAK; 0.025 mg/m3 MAK

German (DFG) - Peak Limitations
2 x normal MAK (5 min momentary value); don't exceed 8 times during shift

German (DFG) - Skin/Sensitizers
danger of sensitization (skin or respiratory)

German (DFG) - Pregnancy
classification not yet possible

Israel - Time Weighted Averages
0.02 ppm TWA; 0.047 mg/m3 TWA

Israel - Action Levels
0.01 ppm AL; 0.0235 mg/m3 AL

Mexico - Instruction No. 10 - TWAs
0.02 ppm TWA; 0.05 mg/m3 TWA

Mexico - Instruction No. 10 - Skin designation
skin - potential for cutaneous absorption

STATE LISTS

California - Air Bill 2588 Appendix A-I
[present]

California - Exposure Limits - PELs
0.02 ppm PEL; 0.05 mg/m3 PEL

California - Exposure Limits - Skin Notation
material may be absorbed through the skin, eyes or mucous membrane

California - Directors List of Hazardous Substances (8 CCR 339)
[present]

Florida Hazardous Substance List
[present]

Massachusetts Right To Know List
extraordinarily hazardous

Minnesota Hazardous Substance List
skin

NJ Right to Know List (Total)
sn 1270

NJ Special Hazardous Substances
(flammable - third degree; reactive - third degree)

Pennsylvania Right to Know List
environmental hazard

METHYL MAGNESIUM BROMIDE   75-16-1

HEALTH AND SAFETY LISTS

NFPA - Flash Points
flash point > 212 degrees F (100 degrees C)

NFPA - Hazard Identification Ratings
health-0; flammability-1; reactivity-0

METHYL MERCAPTAN   74-93-1

HEALTH AND SAFETY LISTS

U.S. DOT - Substances From 49 CFR 172.101
regulated by DOT (UN1246)

U.S. DOT - Hazard Classes
DOT hazard class = 3

NFPA - Hazard Identification Ratings
health-2; reactivity-0

INTERNATIONAL LISTS

Canada - WHMIS: Ingredient Disclosure
1% item 1057 (1140)

STATE LISTS

Florida Hazardous Substance List
[present]

Massachusetts Right To Know List
[present]

NJ Right to Know List (Total)
sn 2557

Pennsylvania Right to Know List
[present]

6-METHYLMERCAPTOPURINE RIBONUCLEOSIDE 342-69-8

HEALTH AND SAFETY LISTS

ACGIH 1995 - Time Weighted Averages
200 ppm TWA; 705 mg/m3 TWA

AIHA - Odor Threshold Values
no geometric mean air odor threshold

U.S. DOT - Substances From 49 CFR 172.101
regulated by DOT (UN2397)

U.S. DOT - Hazard Classes
DOT hazard class = 3

NIOSH - Selected LD50s and LC50s
Oral, rat: LD50 = 148 mg/kg Skin, rabbit: LD50 = 6350 mg/kg

OSHA - Vacated PELs - Time Weighted Averages
200 ppm TWA; 705 mg/m3 TWA

INTERNATIONAL LISTS

Australian Exposure Standards - Time Weighted Averages
200 ppm TWA; 705 mg/m3 TWA

Canada - WHMIS: Ingredient Disclosure
1% item 1024 (1120)

Canada - Alberta - 8 Hour Occupational Exposure Limit
200 ppm TWA; 705 mg/m3 TWA

Canada - Alberta - 15 Minute Occupational Exposure Limit
250 ppm STEL; 881 mg/m3 STEL

Canada - Ontario - OHSA - TWAEVs
200 ppm TWAEV; 705 mg/m3 TWAEV

Canada - Quebec - Time-Weighted Average Exposure Values
200 ppm TWAEV; 705 mg/m3 TWAEV

Israel - Time Weighted Averages
200 ppm TWA; 705 mg/m3 TWA

Israel - Action Levels
100 ppm AL; 352.5 mg/m3 AL

Mexico - Instruction No. 10 - TWAs
200 ppm TWA; 700 mg/m3 TWA

STATE LISTS

California - Exposure Limits - PELs
200 ppm PEL; 705 mg/m3 PEL

California - Directors List of Hazardous Substances (8 CCR 339)
[present]

Florida Hazardous Substance List
[present]

Massachusetts Right To Know List
[present]

Minnesota Hazardous Substance List
[present]

NJ Right to Know List (Total)
sn 1271

Pennsylvania Right to Know List
[present]

METHYLMERCURY   22967-92-6

ENVIRONMENTAL LISTS

List of Pesticide Product Inert Ingredients
[present]

METHYL MERCURY DICYANDIAMIDE 502-39-6

HEALTH AND SAFETY LISTS

U.S. DOT - Substances From 49 CFR 172.101
regulated by DOT (UN2477)

U.S. DOT - Hazard Classes
DOT hazard class = 3

U.S. DOT - Substances Which Are Poisonous by Inhalation
liquid hazardous material poisonous by inhalation (UN2477)

NIOSH - Selected LD50s and LC50s
Oral, rat: LD50 = 97 mg/kg Skin, rat: LD50 = 2780 mg/kg

ENVIRONMENTAL LISTS

CERCLA/SARA - Section 302 Extremely Hazardous Substances and TPQs
TPQ = 500 pounds (This material is a reactive solid. The TPQ does not default to 10,000 pounds for non-powder, non-moleten, non-solution form)

CERCLA/SARA - Section 313 - Emission Reporting
form R reporting required

STATE LISTS

Florida Hazardous Substance List
effective March 13, 1992

Massachusetts Right To Know List
extraordinarily hazardous

NJ Right to Know List (Total)
sn 1272

Pennsylvania Right to Know List
environmental hazard

PROPOSED REGULATIONS

CERCLA/SARA - Proposed Hazardous Substance Additions
proposed RQ = 1 pound (.454 kg)

CERCLA/SARA - 1989 Proposed RQ Adjustments
proposed RQ = 100 pounds (45.4 kg)

METHYL METHACRYLATE   80-62-6

HEALTH AND SAFETY LISTS

U.S. DOT - Substances From 49 CFR 172.101
regulated by DOT (UN2400)

U.S. DOT - Hazard Classes
DOT hazard class = 3

NIOSH - Selected LD50s and LC50s
Inhalation, mouse: LC50 = 20250 mg/m3 (2 hr) Oral, rabbit: LD50 = 5693 mg/kg

STATE LISTS

NJ Right to Know List (Total)
sn 1273

METHYL METHACRYLATE-BUTYL ACRYLATE-   25035-69-2
METHACRYLIC ACID POLYMER

HEALTH AND SAFETY LISTS

NFPA - Flash Points
flash point = 121 degrees F (49 degrees C)

NFPA - Hazard Identification Ratings
health-1; flammability-2; reactivity-0

METHYL METHACRYLATE POLYMER   9011-14-7

HEALTH AND SAFETY LISTS

U.S. DOT - Substances From 49 CFR 172.101
regulated by DOT (with ethyl ether (UN1928)

U.S. DOT - Hazard Classes
DOT hazard class = 4.3

STATE LISTS

NJ Right to Know List (Total)
sn 1274

METHYL METHANESULFONATE   66-27-3

HEALTH AND SAFETY LISTS

ACGIH 1995 - Time Weighted Averages
0.5 ppm TWA; 0.98 mg/m3 TWA

AIHA - Odor Threshold Values
geometric mean air odor threshold = 0.00054 ppm (detectable); 0.0010 ppm (recognizable)

U.S. DOT - Substances From 49 CFR 172.101
regulated by DOT (UN1064)

U.S. DOT - Hazard Classes
DOT hazard class = 2.3

U.S. DOT - Substances Which Are Poisonous by Inhalation
gaseous hazardous material poisonous by inhalation (UN1064)

U.S. DOT - Appendix B - Marine Pollutants
DOT regulated marine pollutant

U.S. DOT - Appendix A Table 1 - Hazardous Substances
final RQ = 100 pounds (45.4 kg)

NFPA - Hazard Identification Ratings
health-4; flammability-4; reactivity-0

NIOSH - Selected LD50s and LC50s
Inhalation, rat: LC50 = 675 ppm 8 hr

NIOSH 1990 - Pocket Guide - RELs
C 0.5 ppm (15 min); C 1 mg/m3 (15 min)

NIOSH 1990 - Pocket Guide - IDLHs
400 ppm IDLH

NIOSH 1990 - Pocket Guide - Target organs
CNS, respiratory system

NIOSH - Health Standards - Exposure Limits
C (15 min) 0.5 ppm; C (15 min) 1.0 mg/m3 (Listed under 'Thiols')

NIOSH - Health Standards - Health Effects and Precautions
Irritation; eye, skin, blood, and nervous system effects (Blood and urine monitoring required; prevent skin contact) (Listed under 'Thiols')

OSHA - Vacated PELs - Time Weighted Averages
0.5 ppm TWA; 1 mg/m3 TWA

OSHA - Final PELs - Ceiling Limits
C 10 ppm; C 20 mg/m3

OSHA - List of Highly Hazardous Chemicals
threshhold quantity = 5000 pounds

ENVIRONMENTAL LISTS

CERCLA/SARA - Section 302 Extremely Hazardous Substances and TPQs
TPQ = 500 pounds

CERCLA/SARA - Section 313 - Emission Reporting
form R reporting required

CERCLA/SARA - Hazardous Substances and their Reportable Quantities
final RQ = 100 pounds (45.4 kg)

CAA -Toxic Substances for Accidental Release Prevention
threshold quantity = 10,000 lbs

CAA - HON Rule - SOCMI Chemicals
compliance by July 24, 1995

Clean Water Act - Hazardous Substances
[present]

RCRA - U Series Wastes
waste number U153 (Ignitable waste; Toxic waste)

RCRA - Hazardous Constituents-Appendix VIII
waste number U153

RCRA - Substances Banned From Land Disposal
[present]

TSCA - Code of Federal Regulations Citations
40 CFR 799.5055(c), (d)(1)

TSCA - Multichemical Test Rules - Waste Constituents
soil adsorption testing for Chemical Fate

TSCA - Section 12(b) - Export Notification
export notification required - Section 4

INTERNATIONAL LISTS

Australian Exposure Standards - Time Weighted Averages
0.5 ppm TWA; 0.98 mg/m3 TWA

Canada - WHMIS: Ingredient Disclosure
1% item 1058 (1142)

Canada - Alberta - 8 Hour Occupational Exposure Limit
0.5 ppm TWA; 1 mg/m3 TWA

Canada - Alberta - 15 Minute Occupational Exposure Limit
1.5 ppm STEL; 2.9 mg/m3 STEL

Canada - British Columbia - Ceiling Exposure Limits
C 3 ppm; C 5.9 mg/m3

Canada - Ontario - OHSA - TWAEVs
0.5 ppm TWAEV; 1 mg/m3 TWAEV

Canada - Quebec - Time-Weighted Average Exposure Values
0.5 ppm TWAEV; 0.98 mg/m3 TWAEV

United Kingdom - Occupational Exposure Standards - TWAs
0.5 ppm TWA; 1 mg/m3 TWA

German (DFG) - MAK Values
0.5 ppm MAK; 1 mg/m3 MAK

German (DFG) - Peak Limitations
2 x normal MAK (10 min momentary value); don't exceed 4 times per shift

Israel - Time Weighted Averages
0.5 ppm TWA; 0.98 mg/m3 TWA

Israel - Action Levels
0.25 ppm AL; 0.49 mg/m3 AL

Mexico - Instruction No. 10 - TWAs
0.5 ppm TWA; 1 mg/m3 TWA

STATE LISTS

California - Exposure Limits - PELs
0.5 ppm PEL; 1 mg/m3 PEL

California - Directors List of Hazardous Substances (8 CCR 339)
[present]

Florida Hazardous Substance List
[present]

Massachusetts Right To Know List
extraordinarily hazardous

Minnesota Hazardous Substance List
[present]

NJ Right to Know List (Total)
sn 1275

NJ Special Hazardous Substances
(flammable - fourth degree)

Pennsylvania Right to Know List
environmental hazard

METHYLMORPHOLINE   109-02-4

HEALTH AND SAFETY LISTS

NTP Chemical Status Reports - Testing Status and NTIS Number
Chronic studies exist for which technical reports were not prepared

1-METHYLNAPHTHALENE   90-12-0

ENVIRONMENTAL LISTS

ATSDR Priority List
Rank (of 275): 094

INTERNATIONAL LISTS

German (DFG) - MAK Values
total dust: 0.01 mg/m3 MAK

German (DFG) - Peak Limitations
10 x normal MAK (30 min average value); don't exceed during shift

German (DFG) - Skin/Sensitizers
danger of cutaneous absorption; danger of sensitization (skin or respiratory)

German (DFG) - Pregnancy
risk to embryo/fetus unequivocal

STATE LISTS

California - Prop. 65 - Developmental Toxicity
developmental toxicity - initial date 7/1/87

Massachusetts Right To Know List
teratogen

2-METHYLNAPHTHALENE   91-57-6

SEE ALSO:
MERCURY

HEALTH AND SAFETY LISTS

U.S. DOT - Substances From 49 CFR 172.101
regulated by DOT (UN1244)

U.S. DOT - Hazard Classes
DOT hazard class = 6.1

NIOSH - Selected LD50s and LC50s
Oral, mouse: LD50 = 20 mg/kg

ENVIRONMENTAL LISTS

CERCLA/SARA - Section 302 Extremely Hazardous Substances and TPQs
TPQ = 500/10,000 pounds

STATE LISTS

Florida Hazardous Substance List
[present]

Massachusetts Right To Know List
extraordinarily hazardous

NJ Right to Know List (Total)
sn 1276

NJ Special Hazardous Substances
(teratogen)

Pennsylvania Right to Know List
environmental hazard

PROPOSED REGULATIONS

CERCLA/SARA - Proposed Hazardous Substance Additions
proposed RQ = 1 pound (.454 kg)

CERCLA/SARA - 1989 Proposed RQ Adjustments
proposed RQ = 10 pounds (4.54 kg)

METHYLNAPHTHALENE   1321-94-4

SEE ALSO:
F039-HAZARDOUS WASTES

HEALTH AND SAFETY LISTS

ACGIH 1995 - Time Weighted Averages
100 ppm TWA; 410 mg/m3 TWA

AIHA - Odor Threshold Values
geometric mean air odor threshold = 0.049 ppm (detectable); 0.34 ppm (recognizable)

U.S. DOT - Substances From 49 CFR 172.101
regulated by DOT (UN1247)

U.S. DOT - Hazard Classes
DOT hazard class = 3

U.S. DOT - Appendix A Table 1 - Hazardous Substances
final RQ = 1000 pounds (454 kg)

IARC - Group 3 (not classifiable)
[present]

NFPA - Flash Points
flash point = 50 degrees F (10 degrees C)

NFPA - Hazard Identification Ratings
health-2; flammability-3; reactivity-2

NIOSH - Selected LD50s and LC50s
Inhalation, rat: LC50 = 3750 ppm 8 hr Oral, rat: LD50 = 7872 mg/kg

NIOSH 1990 - Pocket Guide - RELs
100 ppm TWA; 410 mg/m3 TWA

NIOSH 1990 - Pocket Guide - IDLHs
4000 ppm IDLH

NIOSH 1990 - Pocket Guide - Target organs
eyes, upper respiratory system, skin

NTP Chemical Status Reports - Testing Status and NTIS Number
Technical reports printed (PB87146742/AS)

NTP Chemical Status Reports - Evidence of Carcinogenicity
male rat-no evidence; female rat-no evidence; male mice-no evidence; female mice-no evidence

OSHA - Vacated PELs - Time Weighted Averages
100 ppm TWA; 410 mg/m3 TWA

OSHA - Final PELs - Time Weighted Averages
100 ppm TWA; 410 mg/m3 TWA

ENVIRONMENTAL LISTS

ATSDR Priority List
Rank (of 275): 270

CERCLA/SARA - Section 313 - Emission Reporting
form R reporting required for 1.0% de minimus concentration

CERCLA/SARA - Hazardous Substances and their Reportable Quantities
final RQ = 1000 pounds (454 kg)

Clean Air Act (1990) - List of Hazardous Air Contaminants
[present]

CAA - HON Rule - SOCMI Chemicals
compliance by July 24, 1995

CAA - HON Rule - Organic HAPs
[present]

Clean Water Act - Hazardous Substances
[present]

List of Pesticide Product Inert Ingredients
[present]

RCRA - U Series Wastes
waste number U162 (Ignitable waste; Toxic waste)

RCRA - Hazardous Constituents-Appendix VIII
waste number U162

RCRA - Basis for Listing - Appendix VII
Included in waste stream: F039

RCRA - Substances Banned From Land Disposal
[present]

RCRA - TSD Facilities Ground Water Monitoring
TM 8015 = 2 ug/L PQL; TM 8240 = 5 ug/L PQL

RCRA - Universal Treatment Standards (LDR)
WW: 0.14 mg/l; NWW: 160 mg/kg

TSCA - Code of Federal Regulations Citations
40 CFR 712.30(d); 40 CFR 716.120(a)

INTERNATIONAL LISTS

Australian Exposure Standards - Time Weighted Averages
100 ppm TWA; 410 mg/m3 TWA

Australian Exposure Standards - Skin Effects
sensitiser

Canada - WHMIS: Ingredient Disclosure
1% item 1059 (1096)

Canada - NPRI (National Pollutant Release Inventory)
[present]

Canada - Alberta - 8 Hour Occupational Exposure Limit
100 ppm TWA; 410 mg/m3 TWA

Canada - Alberta - 15 Minute Occupational Exposure Limit
125 ppm STEL; 510 mg/m3 STEL

Canada - British Columbia - 8 Hour Exposure Limits
100 ppm TWA; 410 mg/m3 TWA

Canada - British Columbia - 15 Minute Exposure Limits
125 ppm STEL; 510 mg/m3 STEL

Canada - Ontario - OHSA - TWAEVs
100 ppm TWAEV; 410 mg/m3 TWAEV

Canada - Quebec - Time-Weighted Average Exposure Values
100 ppm TWAEV; 410 mg/m3 TWAEV

United Kingdom - Occupational Exposure Standards - TWAs
100 ppm TWA; 410 mg/m3 TWA

United Kingdom - Occupational Exposure Standards - STELs
125 ppm STEL; 510 mg/m3 STEL

German (DFG) - MAK Values
50 ppm MAK; 210 mg/m3 MAK

German (DFG) - Peak Limitations
2 x normal MAK (5 min momentary value); don't exceed 8 times during shift

German (DFG) - Skin/Sensitizers
danger of sensitization (skin or respiratory)

German (DFG) - Pregnancy
no risk to embryo/fetus if exposure limits adhered to

Israel - Time Weighted Averages
100 ppm TWA; 410 mg/m3 TWA

Israel - Action Levels
50 ppm AL; 205 mg/m3 AL

Mexico - Instruction No. 10 - TWAs
100 ppm TWA; 410 mg/m3 TWA

Mexico - Instruction No. 10 - STELs
125 ppm STEL; 510 mg/m3 STEL

STATE LISTS

California - Air Bill 2588 Appendix A-I
[present]

California - Exposure Limits - PELs
100 ppm PEL; 410 mg/m3 PEL

California - Directors List of Hazardous Substances (8 CCR 339)
[present]

Florida Hazardous Substance List
[present]

Massachusetts Right To Know List
[present]

Minnesota Hazardous Substance List
[present]

NJ Right to Know List (Total)
sn 1277

NJ Special Hazardous Substances
(flammable - third degree; reactive - second degree)

Pennsylvania Right to Know List
environmental hazard

PROPOSED REGULATIONS

TSCA - ITC 32nd Report Priority Testing List
designated for dermal absorption testing

METHYL NITRAMINE   89460-82-2

ENVIRONMENTAL LISTS

List of Pesticide Product Inert Ingredients
[present]

METHYL NITRAMINE   RR-01440-4

HEALTH AND SAFETY LISTS

IARC - Group 3 (not classifiable)
[present]

ENVIRONMENTAL LISTS

List of Pesticide Product Inert Ingredients
[present]

METHYLNITRATE   598-58-3

HEALTH AND SAFETY LISTS

IARC - Group 2B (sufficient animal data)
[present]

NIOSH - Selected LD50s and LC50s
Oral, rat: LD50 = 225 mg/kg

NTP Seventh Report - Suspect Carcinogens
suspect carcinogen

OSHA - Possible Select Carcinogens
[present]

ENVIRONMENTAL LISTS

RCRA - Hazardous Constituents-Appendix VIII
hazardous constituent - no waste number

RCRA - Basis for Listing - Appendix VII
Included in waste stream: F039

RCRA - TSD Facilities Ground Water Monitoring
TM 8270 = 10 ug/L PQL

RCRA - Universal Treatment Standards (LDR)
WW: 0.018 mg/l; NWW: Not applicable

INTERNATIONAL LISTS

Canada - WHMIS: Ingredient Disclosure
1% item 1060 (1103)

STATE LISTS

California - Air Bill 2588 Appendix A-II
known or potential carcinogen

California - Prop. 65 - Cancer list
carcinogen - initial date 4/1/88

California - Prop. 65 - No Significant Risk Levels
no significant risk level = 7 ug/day

California - Directors List of Hazardous Substances (8 CCR 339)
[present]

Florida Hazardous Substance List
[present]

Massachusetts Right To Know List
carcinogen; extraordinarily hazardous

Minnesota Hazardous Substance List
carcinogen

NJ Special Hazardous Substances
(carcinogen; mutagen)

Pennsylvania Right to Know List
environmental hazard; special hazardous substance

Pennsylvania RTK - Special Hazardous Substances
[present]

METHYL NITRITE   624-91-9

HEALTH AND SAFETY LISTS

U.S. DOT - Substances From 49 CFR 172.101
regulated by DOT (UN2535)

U.S. DOT - Hazard Classes
DOT hazard class = 3

NFPA - Flash Points
flash point = 75 degrees F (24 degrees C)

NFPA - Hazard Identification Ratings
health-2; flammability-3; reactivity-0

NIOSH - Selected LD50s and LC50s
Inhalation, mouse: LC50 = 25200 mg/m3 (2 hr) Oral, rat: LD50 = 2720 mg/kg Skin, rabbit: LD50 = 1242 mg/kg

INTERNATIONAL LISTS

Canada - WHMIS: Ingredient Disclosure
1% item 1061 (1143)

STATE LISTS

Florida Hazardous Substance List
[present]

Massachusetts Right To Know List
[present]

NJ Right to Know List (Total)
sn 1278

NJ Special Hazardous Substances
(corrosive; flammable - third degree)

Pennsylvania Right to Know List
[present]

2-METHYL-1-NITROANTHRAQUINONE   129-15-7

HEALTH AND SAFETY LISTS

NFPA - Hazard Identification Ratings
health-2; flammability-2; reactivity-0

NIOSH - Selected LD50s and LC50s
Oral, rat: LD50 = 1840 mg/kg

ENVIRONMENTAL LISTS

ATSDR Priority List
Rank (of 275): 255

STATE LISTS

Florida Hazardous Substance List
[present]

Massachusetts Right To Know List
[present]

Pennsylvania Right to Know List
[present]

N-METHYL-N'-NITRO-N-NITROSOGUANIDINE   70-25-7

HEALTH AND SAFETY LISTS

NIOSH - Selected LD50s and LC50s
Oral, rat: LD50 = 1630 mg/kg

ENVIRONMENTAL LISTS

ATSDR Priority List
Rank (of 275): 141

RCRA - TSD Facilities Ground Water Monitoring
TM 8270 = 10 ug/L PQL

4-METHYL-2-NITRO-PHENOL   119-33-5

HEALTH AND SAFETY LISTS

U.S. DOT - Appendix B - Marine Pollutants
DOT regulated marine pollutant

ENVIRONMENTAL LISTS

ATSDR Priority List
Rank (of 275): 233

List of Pesticide Product Inert Ingredients
[present]

INTERNATIONAL LISTS

Canada - WHMIS: Ingredient Disclosure
1% item 1062 (1144)

3-METHYL-4-NITROPHENOL   2581-34-2

HEALTH AND SAFETY LISTS

U.S. DOT - Hazard Classes
Forbidden from transport by the DOT

METHYLNITROPHENOLS   RR-01302-5

STATE LISTS

NJ Right to Know List (Total)
sn 2458

3-METHYLNITROSOAMINOPROPIONITRILE   60153-49-3

HEALTH AND SAFETY LISTS

U.S. DOT - Hazard Classes
Forbidden from transport by the DOT

NIOSH - Selected LD50s and LC50s
Inhalation, rat: LC50 = 1275 ppm 4 hr Oral, rat: LD50 = 344 mg/kg

N-METHYL-N-NITROSO-P-TOLUENESULFONAMIDE   80-11-5

STATE LISTS

NJ Right to Know List (Total)
sn 1281

3-METHYLOCTANE   2216-33-3

HEALTH AND SAFETY LISTS

IARC - Group 2B (sufficient animal data)
[present] (Overall evaluation based only on evidence of carcinogenicity in monograph (27, 1982) or in Supplement 4)

NTP Chemical Status Reports - Testing Status and NTIS Number
Technical reports printed (PB277439/AS); chronic studies for which technical reports were not prepared

NTP Chemical Status Reports - Evidence of Carcinogenicity
male rat-positive; female rat-positive; male mice-positive; female mice-positive

OSHA - Possible Select Carcinogens
[present]

STATE LISTS

California - Air Bill 2588 Appendix A-II
known or potential carcinogen

California - Prop. 65 - Cancer list
carcinogen (of uncertain purity) - initial date 4/1/88

California - Prop. 65 - No Significant Risk Levels
no significant risk level = 0.2 ug/day

California - Directors List of Hazardous Substances (8 CCR 339)
[present]

Florida Hazardous Substance List
[present]

Massachusetts Right To Know List
carcinogen; extraordinarily hazardous

Minnesota Hazardous Substance List
[present]

Pennsylvania Right to Know List
special hazardous substance

Pennsylvania RTK - Special Hazardous Substances
[present]

4-METHYLOCTANE   2216-34-4

HEALTH AND SAFETY LISTS

U.S. DOT - Substances From 49 CFR 172.101
regulated by DOT (NA0473)

U.S. DOT - Hazard Classes
DOT hazard class = 1.1A

U.S. DOT - Appendix A Table 1 - Hazardous Substances
final RQ = 10 pounds (4.54 kg)

IARC - Group 2A (limited human data)
[present] (Other relevant data, as given in Supplement 7, influenced the making of the overall evaluation)

NIOSH - Selected LD50s and LC50s
Oral, rat: LD50 = 90 mg/kg

NTP Seventh Report - Suspect Carcinogens
suspect carcinogen

OSHA - Possible Select Carcinogens
[present]

ENVIRONMENTAL LISTS

CERCLA/SARA - Hazardous Substances and their Reportable Quantities
final RQ = 10 pounds (4.54 kg)

EPA - Carcinogen Hazard Ranking for RQ Adjustment
Hazard ranking = Medium

RCRA - U Series Wastes
waste number U163

RCRA - Hazardous Constituents-Appendix VIII
waste number U163

RCRA - Substances Banned From Land Disposal
[present]

TSCA - Chemicals with Significant New Use Rules
[present]

TSCA - Section 12(b) - Export Notification
export notification required - Section 5

INTERNATIONAL LISTS

Canada - WHMIS: Ingredient Disclosure
0.1% item 1063 (1145)

STATE LISTS

California - Air Bill 2588 Appendix A-II
known or potential carcinogen

California - Prop. 65 - Cancer list
carcinogen - initial date 4/1/88

California - Prop. 65 - No Significant Risk Levels
no significant risk level = 0.08 ug/day

California - Directors List of Hazardous Substances (8 CCR 339)
[present]

Florida Hazardous Substance List
[present]

Massachusetts Right To Know List
carcinogen; extraordinarily hazardous

Minnesota Hazardous Substance List
carcinogen

NJ Right to Know List (Total)
sn 0963

NJ Special Hazardous Substances
(carcinogen; mutagen)

Pennsylvania Right to Know List
environmental hazard; special hazardous substance

Pennsylvania RTK - Special Hazardous Substances
[present]

2-METHYLOCTANE   3221-61-2

HEALTH AND SAFETY LISTS

NIOSH - Selected LD50s and LC50s
Oral, rat: LD50 = 3360 mg/kg

ENVIRONMENTAL LISTS

TSCA - Code of Federal Regulations Citations
40 CFR 712.30(x); 40 CFR 716.120(d)

TSCA - PAIR - Reporting List
Reporting Date: November 27, 1991

TSCA - Health and Safety Reporting List
Effective Date: September 30, 1991

3-METHYL-3-OCTANOL   5340-36-3

HEALTH AND SAFETY LISTS

NIOSH - Selected LD50s and LC50s
Oral, rat: LD50 = 4410 mg/kg

ENVIRONMENTAL LISTS

EPA - Master Testing List
[present]

N-METHYLOLACRYLAMIDE   924-42-5

HEALTH AND SAFETY LISTS

U.S. DOT - Appendix B - Marine Pollutants
DOT regulated marine pollutant

2-METHYLOL-4,4'-ISOPROPYLIDENEDIPHENOL   3188-83-8
DIGLYCIDYL ETHER

HEALTH AND SAFETY LISTS

IARC - Group 2B (sufficient animal data)
[present] (Overall evaluation based only on evidence of carcinogenicity in monograph (37, 1985) or in supplement 4)

OSHA - Possible Select Carcinogens
[present]

STATE LISTS

California - Air Bill 2588 Appendix A-II
known or potential carcinogen

California - Prop. 65 - Cancer list
carcinogen - initial date 4/1/90

California - Directors List of Hazardous Substances (8 CCR 339)
[present]

Massachusetts Right To Know List
carcinogen; extraordinarily hazardous

Minnesota Hazardous Substance List
carcinogen

Pennsylvania Right to Know List
special hazardous substance

Pennsylvania RTK - Special Hazardous Substances
[present]

METHYLOLUREA   1000-82-4

HEALTH AND SAFETY LISTS

NIOSH - Selected LD50s and LC50s
Oral, rat: LD50 = 2700 mg/kg

METHYL ORTHOFORMATE   149-73-5

HEALTH AND SAFETY LISTS

NFPA - Hazard Identification Ratings
health-0; flammability-3; reactivity-0

STATE LISTS

Florida Hazardous Substance List
[present]

Massachusetts Right To Know List
[present]

Pennsylvania Right to Know List
[present]

METHYL OXIRANE POLYMER WITH OXIRANE, MONOOCTYL ETHER   37311-02-7

HEALTH AND SAFETY LISTS

NFPA - Hazard Identification Ratings
health-0; flammability-3; reactivity-0

STATE LISTS

Florida Hazardous Substance List
[present]

Massachusetts Right To Know List
[present]

Pennsylvania Right to Know List
[present]

METHYL OXIRANE POLYMER WITH OXIRANE, OCTYL ETHER   61827-84-7

HEALTH AND SAFETY LISTS

NFPA - Hazard Identification Ratings
health-0; flammability-3; reactivity-0

STATE LISTS

Florida Hazardous Substance List
[present]

Massachusetts Right To Know List
[present]

Pennsylvania Right to Know List
[present]

METHYL PALMITATE   112-39-0

INTERNATIONAL LISTS

Canada - WHMIS: Ingredient Disclosure
1% item 1064 (1147)

N-METHYL-N-PALMITOYLTAURINE, SODIUM SALT   3737-55-1

HEALTH AND SAFETY LISTS

NIOSH - Selected LD50s and LC50s
Oral, rat: LD50 = 474 mg/kg

NTP Chemical Status Reports - Testing Status and NTIS Number
Technical reports printed (PB90226374/AS)

NTP Chemical Status Reports - Evidence of Carcinogenicity
male rat-no evidence; female rat-no evidence; male mice-clear evidence; female mice-clear evidence

ENVIRONMENTAL LISTS

CERCLA/SARA - Section 313 - Emission Reporting
form R reporting required

STATE LISTS

California - Air Bill 2588 Appendix A-II
known or potential carcinogen: 9/90

California - Prop. 65 - Cancer list
carcinogen - initial date 7/1/90

METHYL PARATHION   298-00-0

ENVIRONMENTAL LISTS

EPA - Master Testing List
[present]

PROPOSED REGULATIONS

TSCA - Proposed Testing Rule for Glycidyl Ethers
member of Glycidyl subcategory VI-A

METHYL PENTADECYL KETONE   RR-00869-5

ENVIRONMENTAL LISTS

EPA - Master Testing List
[present]

TSCA - Code of Federal Regulations Citations
40 CFR 712.30(h); 40 CFR 716.120(a)

TSCA - PAIR - Reporting List
Reporting Date: September 20, 1983

TSCA - Health and Safety Reporting List
Effective Date: July 1, 1983

4-METHYL-1,3-PENTADIENE   926-56-7

HEALTH AND SAFETY LISTS

NIOSH - Selected LD50s and LC50s
Inhalation, rat: LC50 = 5000 ppm 4 hr Oral, rat: LD50 = 3130 mg/kg

INTERNATIONAL LISTS

Canada - WHMIS: Ingredient Disclosure
1% item 1065 (1290)

2-METHYL-1,3-PENTADIENE   1118-58-7

ENVIRONMENTAL LISTS

List of Pesticide Product Inert Ingredients
[present]

METHYLPENTADIENE   54363-49-4

ENVIRONMENTAL LISTS

List of Pesticide Product Inert Ingredients
[present]

METHYLPENTADIENES   39382-31-5

ENVIRONMENTAL LISTS

List of Pesticide Product Inert Ingredients
[present]

METHYLPENTALDEHYDE   73513-30-1

ENVIRONMENTAL LISTS

List of Pesticide Product Inert Ingredients
[present]

3-METHYLPENTANE   96-14-0
SEE ALSO:
F039-HAZARDOUS WASTES

HEALTH AND SAFETY LISTS

ACGIH 1995 - Time Weighted Averages
0.2 mg/m3 TWA

ACGIH 1995 - Skin Designations
skin - potential for cutaneous absorption

AIHA - Odor Threshold Values
no geometric mean air odor threshold

U.S. DOT - Substances From 49 CFR 172.101
regulated by DOT (NA2783, NA3018)

U.S. DOT - Hazard Classes
DOT hazard class = 6.1

U.S. DOT - Substances Which Are Poisonous by Inhalation
gaseous hazardous material poisonous by inhalation (when mixed with compressed gas) (UN1967)

U.S. DOT - Appendix B - Marine Pollutants
DOT regulated severe marine pollutant

U.S. DOT - Appendix A Table 1 - Hazardous Substances
final RQ = 100 pounds (45.4 kg)

IARC - Group 3 (not classifiable)
[present]

NIOSH - Selected LD50s and LC50s
Inhalation, rat: LC50 = 34 mg/m3 4 hr Oral, rat: LD50 = 6 mg/kg Skin, rat: LD50 = 63 mg/kg

NIOSH - Health Standards - Exposure Limits
0.2 mg/m3 TWA

NIOSH - Health Standards - Health Effects and Precautions
Central nervous system effects (Prevent skin contact; blood monitoring required)

NTP Chemical Status Reports - Testing Status and NTIS Number
Technical reports printed (PB295891/AS)

NTP Chemical Status Reports - Evidence of Carcinogenicity
male rat-negative; female rat-negative; male mice-negative; female mice-negative

OSHA - Vacated PELs - Time Weighted Averages
0.2 mg/m3 TWA

OSHA - Vacated PELs - Skin Designation
Prevent or reduce skin absorption

ENVIRONMENTAL LISTS

ATSDR Priority List
Rank (of 275): 248

CERCLA/SARA - Section 302 Extremely Hazardous Substances and TPQs
TPQ = 100/10,000 pounds

CERCLA/SARA - Section 313 - Emission Reporting
form R reporting required

CERCLA/SARA - Hazardous Substances and their Reportable Quantities
final RQ = 100 pounds (45.4 kg)

Clean Water Act - Hazardous Substances
[present]

RCRA - P Series Wastes
waste number P071

RCRA - Hazardous Constituents-Appendix VIII
waste number P071

RCRA - Basis for Listing - Appendix VII
Included in waste stream: F039

RCRA - Substances Banned From Land Disposal
[present]

RCRA - TSD Facilities Ground Water Monitoring
TM 8140 = 0.5 ug/L PQL;   TM 8270 = 10 ug/L PQL

RCRA - Universal Treatment Standards (LDR)
WW: 0.014 mg/l;   NWW: 4.6 mg/kg

INTERNATIONAL LISTS

Australian Exposure Standards - Time Weighted Averages
0.2 mg/m3 TWA

Australian Exposure Standards - Skin Effects
skin absorption

Canada - Alberta - 8 Hour Occupational Exposure Limit
0.2 mg/m3 TWA

Canada - Alberta - 15 Minute Occupational Exposure Limit
0.6 mg/m3 STEL

Canada - Alberta - Skin Designation
can be absorbed through the intact skin

Canada - British Columbia - 8 Hour Exposure Limits
0.2 mg/m3 TWA

Canada - British Columbia - 15 Minute Exposure Limits
0.6 mg/m3 STEL

Canada - British Columbia - Skin Notations
skin - potential for skin absorption

Canada - Ontario - OHSA - TWAEVs
0.2 mg/m3 TWAEV

Canada - Ontario - OHSA - Skin Notations
absorption through skin, eyes, or mucous membranes

Canada - Quebec - Time-Weighted Average Exposure Values
0.2 mg/m3 TWAEV

Canada - Quebec - Skin Designations
absorbed through the skin

United Kingdom - Occupational Exposure Standards - TWAs
0.2 mg/m3 TWA

United Kingdom - Occupational Exposure Standards - STELs
0.6 mg/m3 STEL

United Kingdom - Occupational Exposure Standards - Notes
can be absorbed through skin

Israel - Time Weighted Averages
0.2 mg/m3 TWA

Israel - Action Levels
0.1 mg/m3 AL

Mexico - Instruction No. 10 - TWAs
0.2 mg/m3 TWA

Mexico - Instruction No. 10 - STELs
0.6 mg/m3 STEL

Mexico - Instruction No. 10 - Skin designation
skin - potential for cutaneous absorption

STATE LISTS

California - Exposure Limits - PELs
0.2 mg/m3 PEL

California - Exposure Limits - Skin Notation
material may be absorbed through the skin, eyes or mucous membrane

California - Directors List of Hazardous Substances (8 CCR 339)
[present]

Florida Hazardous Substance List
[present]

Massachusetts Right To Know List
extraordinarily hazardous; neurotoxin

Minnesota Hazardous Substance List
skin

NJ Right to Know List (Total)
sn 1283

NJ Special Hazardous Substances
(mutagen)

Pennsylvania Right to Know List
environmental hazard

METHYLPENTANE   43133-95-5

HEALTH AND SAFETY LISTS

NFPA - Flash Points
flash point = 248 degrees F (120 degrees C)

NFPA - Hazard Identification Ratings
health-0; flammability-1; reactivity-0

2-METHYL-1,3-PENTANEDIOL   149-31-5

HEALTH AND SAFETY LISTS

NFPA - Flash Points
flash point = -30 degrees F (-34 degrees C)

NFPA - Hazard Identification Ratings
health-0; flammability-3; reactivity-1

STATE LISTS

Florida Hazardous Substance List
[present]

Massachusetts Right To Know List
[present]

Pennsylvania Right to Know List
[present]

2-METHYLPENTANOIC ACID   97-61-0

HEALTH AND SAFETY LISTS

NFPA - Flash Points
flash point < -4 degrees F (-20 degrees C)

NFPA - Hazard Identification Ratings
health-0; flammability-3; reactivity-0

STATE LISTS

Florida Hazardous Substance List
[present]

Massachusetts Right To Know List
[present]

Pennsylvania Right to Know List
[present]

2-METHYL-1-PENTANOL   105-30-6

STATE LISTS

NJ Right to Know List (Total)
sn 1284

2-METHYL-2-PENTANOL   590-36-3

HEALTH AND SAFETY LISTS

U.S. DOT - Substances From 49 CFR 172.101
regulated by DOT (UN2461)

U.S. DOT - Hazard Classes
DOT hazard class = 3

4-METHYL-2-PENTANOL   1320-98-5

HEALTH AND SAFETY LISTS

NFPA - Flash Points
flash point = 68 degrees F (20 degrees C)

NFPA - Hazard Identification Ratings
health-2; flammability-3; reactivity-1

STATE LISTS

Florida Hazardous Substance List
[present]

Massachusetts Right To Know List
[present]

Pennsylvania Right to Know List
[present]

2-METHYL-2-PENTENE   625-27-4

HEALTH AND SAFETY LISTS

NFPA - Flash Points
flash point < 20 degrees F (-7 degrees C)

NFPA - Hazard Identification Ratings
health-1; flammability-3; reactivity-0

STATE LISTS

Florida Hazardous Substance List
[present]

Massachusetts Right To Know List
[present]

Pennsylvania Right to Know List
[present]

4-METHYL-1-PENTENE   691-37-2

STATE LISTS

NJ Right to Know List (Total)
sn 1285

2-METHYL-1-PENTENE   763-29-1

HEALTH AND SAFETY LISTS

NFPA - Flash Points
flash point = 230 degrees F (110 degrees C)

NFPA - Hazard Identification Ratings
health-2; flammability-1; reactivity-0

STATE LISTS

Florida Hazardous Substance List
[present]

Massachusetts Right To Know List
[present]

Pennsylvania Right to Know List
[present]

4-METHYL-2-PENTENE   4461-48-7

HEALTH AND SAFETY LISTS

NFPA - Flash Points
flash point = 225 degrees F (107 degrees C)

NFPA - Hazard Identification Ratings
health-0; flammability-1; reactivity-0

1-METHYLPHENANTHRENE   832-69-9

HEALTH AND SAFETY LISTS

NFPA - Flash Points
flash point = 129 degrees F (54 degrees C)

NFPA - Hazard Identification Ratings
health-0; flammability-2; reactivity-0

NIOSH - Selected LD50s and LC50s
Oral, rat: LD50 = 1410 mg/kg Skin, rabbit: LD50 = 3560 mg/kg

INTERNATIONAL LISTS

Canada - WHMIS: Ingredient Disclosure
1% item 1067 (1149)

STATE LISTS

Massachusetts Right To Know List
[present]

METHYLPHENIDATE HYDROCHLORIDE   298-59-9

HEALTH AND SAFETY LISTS

U.S. DOT - Substances From 49 CFR 172.101
regulated by DOT (UN2560)

U.S. DOT - Hazard Classes
DOT hazard class = 3

STATE LISTS
NJ Right to Know List (Total)
sn 2558

METHYL PHENKAPTON   3735-23-7

HEALTH AND SAFETY LISTS
NIOSH - Selected LD50s and LC50s
Oral, rat: LD50 = 6500 mg/kg Skin, rabbit: LD50 = 3970 ug/kg

METHYL PHENYLACETATE   101-41-7

HEALTH AND SAFETY LISTS
NFPA - Flash Points
flash point < 20 degrees F (-7 degrees C)
NFPA - Hazard Identification Ratings
health-1; flammability-3; reactivity-0

STATE LISTS
Florida Hazardous Substance List
[present]
Massachusetts Right To Know List
[present]
Pennsylvania Right to Know List
[present]

METHYL PHENYL CARBINYL ACETATE   50373-55-2

HEALTH AND SAFETY LISTS
NFPA - Flash Points
flash point < 20 degrees F (-7 degrees C)
NFPA - Hazard Identification Ratings
health-1; flammability-3; reactivity-0

STATE LISTS
Florida Hazardous Substance List
[present]
Massachusetts Right To Know List
[present]
Pennsylvania Right to Know List
[present]

METHYLPHENYLDICHLOROSILANE   149-74-6

HEALTH AND SAFETY LISTS
NFPA - Flash Points
flash point < 20 degrees F (-7 degrees C)
NFPA - Hazard Identification Ratings
health-1; flammability-3; reactivity-0

STATE LISTS
Florida Hazardous Substance List
[present]
Massachusetts Right To Know List
[present]
Pennsylvania Right to Know List
[present]

2,2'-[(2-METHYLPHENYL)IMINO]BISETHANOL   28005-74-5

HEALTH AND SAFETY LISTS
NFPA - Flash Points
flash point < 20 degrees F (-7 degrees C)
NFPA - Hazard Identification Ratings
health-1; flammability-3; reactivity-0

STATE LISTS
Florida Hazardous Substance List
[present]
Massachusetts Right To Know List
[present]

Pennsylvania Right to Know List
[present]

3-METHYL-1-PHENYL-2-PYRAZOLIN-5-ONE   89-25-8

HEALTH AND SAFETY LISTS
IARC - Group 3 (not classifiable)
[present]

METHYL PHOSPHONIC DICHLORIDE   676-97-1

HEALTH AND SAFETY LISTS
NTP Chemical Status Reports - Testing Status and NTIS Number
Galley or camera copy technical reports in progress; prechronic studies for which toxicity technical reports were not prepared

METHYL PHOSPHONIC DIFLUORIDE   676-99-3

HEALTH AND SAFETY LISTS
NIOSH - Selected LD50s and LC50s
Oral, rat: LD50 = 220 mg/kg

ENVIRONMENTAL LISTS
CERCLA/SARA - Section 302 Extremely Hazardous Substances and TPQs
TPQ = 500 pounds

STATE LISTS
Massachusetts Right To Know List
extraordinarily hazardous
NJ Right to Know List (Total)
sn 2559
Pennsylvania Right to Know List
environmental hazard

PROPOSED REGULATIONS
CERCLA/SARA - Proposed Hazardous Substance Additions
proposed RQ = 1 pound (.454 kg)
CERCLA/SARA - 1989 Proposed RQ Adjustments
proposed RQ = 100 pounds (45.4 kg)

METHYL PHOSPHONOTHIOIC DICHLORIDE   676-98-2

HEALTH AND SAFETY LISTS
NFPA - Flash Points
flash point = 195 degrees F (91 degrees C)
NFPA - Hazard Identification Ratings
health-0; flammability-2; reactivity-0

METHYL PHOSPHONOUS DICHLORIDE   676-83-5

HEALTH AND SAFETY LISTS
NFPA - Flash Points
flash point = 195 degrees F (91 degrees C)
NFPA - Hazard Identification Ratings
health-0; flammability-2; reactivity-0

METHYL PHTHALYL ETHYL GLYCOLATE   85-71-2

HEALTH AND SAFETY LISTS
U.S. DOT - Substances From 49 CFR 172.101
regulated by DOT (UN2437)
U.S. DOT - Hazard Classes
DOT hazard class = 8

ENVIRONMENTAL LISTS
CERCLA/SARA - Section 302 Extremely Hazardous Substances and TPQs
TPQ = 1000 pounds

INTERNATIONAL LISTS
Canada - WHMIS: Ingredient Disclosure
1% item 1068 (1150)

STATE LISTS

Florida Hazardous Substance List
effective March 13, 1992

Massachusetts Right To Know List
extraordinarily hazardous

NJ Right to Know List (Total)
sn 1286

NJ Special Hazardous Substances
(corrosive)

Pennsylvania Right to Know List
environmental hazard

PROPOSED REGULATIONS

CERCLA/SARA - Proposed Hazardous Substance Additions
proposed RQ = 1 pound (.454 kg)

CERCLA/SARA - 1989 Proposed RQ Adjustments
proposed RQ = 100 pounds (45.4 kg)

1-METHYL PIPERAZINE   109-01-3

HEALTH AND SAFETY LISTS

NIOSH - Selected LD50s and LC50s
Oral, rat: LD50 = 2200 mg/kg Skin, rabbit: LD50 = 1000 mg/kg

1-METHYLPIPERIDINE   626-67-5

HEALTH AND SAFETY LISTS

NIOSH - Selected LD50s and LC50s
Oral, rat: LD50 = 3500 mg/kg

NTP Chemical Status Reports - Testing Status and NTIS Number
Technical reports printed (PB287122/AS)

NTP Chemical Status Reports - Evidence of Carcinogenicity
male rat-negative; female rat-negative; male mice-negative; female mice-negative

INTERNATIONAL LISTS

Canada - WHMIS: Ingredient Disclosure
1% item 1069 (1151)

METHYLPOLYCHLORO ALIPHATIC KETONE   RR-01573-6

HEALTH AND SAFETY LISTS

U.S. DOT - Substances From 49 CFR 172.101
regulated by DOT (NA9206)

U.S. DOT - Hazard Classes
DOT hazard class = 6.1

U.S. DOT - Substances Which Are Poisonous by Inhalation
liquid hazardous material poisonous by inhalation (NA9206)

NIOSH - Selected LD50s and LC50s
Inhalation, rat: LC50 = 26 ppm 4 hr

ENVIRONMENTAL LISTS

CERCLA/SARA - Section 302 Extremely Hazardous Substances and TPQs
TPQ = 100 pounds (This material is a reactive solid. The TPQ does not default to 10,000 pounds for non-powder, non-molten, non-solution form)

STATE LISTS

Florida Hazardous Substance List
effective March 13, 1992

Massachusetts Right To Know List
extraordinarily hazardous

NJ Right to Know List (Total)
sn 2560

Pennsylvania Right to Know List
environmental hazard

PROPOSED REGULATIONS

CERCLA/SARA - Proposed Hazardous Substance Additions
proposed RQ = 1 pound (.454 kg)

CERCLA/SARA - 1989 Proposed RQ Adjustments
proposed RQ = 100 pounds (45.4 kg)

METHYLPOLYSILOXANE   9004-73-3

HEALTH AND SAFETY LISTS

U.S. DOT - Substances Which Are Poisonous by Inhalation
liquid hazardous material poisonous by inhalation

2-METHYL-2-PROPANETHIOL   75-66-1

INTERNATIONAL LISTS

Canada - WHMIS: Ingredient Disclosure
1% item 1070 (510)

STATE LISTS

NJ Right to Know List (Total)
sn 1287

METHYL PROPIONATE   554-12-1

HEALTH AND SAFETY LISTS

U.S. DOT - Substances From 49 CFR 172.101
regulated by DOT (NA2845)

U.S. DOT - Hazard Classes
DOT hazard class = 6.1

U.S. DOT - Substances Which Are Poisonous by Inhalation
liquid hazardous material poisonous by inhalation (NA2845)

INTERNATIONAL LISTS

Canada - WHMIS: Ingredient Disclosure
1% item 1071 (509)

STATE LISTS

NJ Right to Know List (Total)
sn 1288

NJ Special Hazardous Substances
(corrosive)

2-(1-METHYLPROPOXY) ETHANOL   7795-91-7

HEALTH AND SAFETY LISTS

NFPA - Flash Points
flash point = 380 degrees F (193 degrees C)

NFPA - Hazard Identification Ratings
health-2; flammability-1; reactivity-0

NIOSH - Selected LD50s and LC50s
Oral, rat: LD50 = 3200 mg/kg

STATE LISTS

Florida Hazardous Substance List
[present]

Massachusetts Right To Know List
[present]

Pennsylvania Right to Know List
[present]

METHYL PROPYL ACETYLENE   764-35-2

HEALTH AND SAFETY LISTS

NFPA - Flash Points
flash point = 108 degrees F (42 degrees C)

NFPA - Hazard Identification Ratings
health-2; flammability-2; reactivity-0

STATE LISTS

Florida Hazardous Substance List
[present]

Massachusetts Right To Know List
[present]

Pennsylvania Right to Know List
[present]

METHYL PROPYL ETHER   557-17-5

HEALTH AND SAFETY LISTS

U.S. DOT - Substances From 49 CFR 172.101
regulated by DOT (UN2399)

U.S. DOT - Hazard Classes
DOT hazard class = 3

STATE LISTS

NJ Right to Know List (Total)
sn 1289

METHYL PROPYL KETONE   107-87-9

ENVIRONMENTAL LISTS

TSCA - Chemicals with Significant New Use Rules
PMN number: P-91-1321

N-METHYLPSEUDOEPHEDRINE   14222-20-9

ENVIRONMENTAL LISTS

TSCA - PAIR - Reporting List
Effective Date: October 12, 1993;   Reporting Date: February 28, 1994
TSCA - Health and Safety Reporting List
Effective Date: October 12, 1993;   Sunset Date: October 12, 2003

N-METHYLPSEUDOEPHEDRINE SALTS, OPTICAL   RR-01787-8
ISOMERS, AND SALTS OFOPTICAL ISOMERS

HEALTH AND SAFETY LISTS

NFPA - Flash Points
flash point < -20 degrees F (-29 degrees C)

NFPA - Hazard Identification Ratings
health-2; flammability-3; reactivity-0

NIOSH - Selected LD50s and LC50s
Inhalation, rat: LC50 = 32200 ppm 4 hr Oral, rat: LD50 = 4729 mg/kg

STATE LISTS

Florida Hazardous Substance List
[present]

Massachusetts Right To Know List
[present]

Pennsylvania Right to Know List
[present]

2-METHYLPYRAZINE   109-08-0

HEALTH AND SAFETY LISTS

U.S. DOT - Substances From 49 CFR 172.101
regulated by DOT (UN1248)

U.S. DOT - Hazard Classes
DOT hazard class = 3

NFPA - Flash Points
flash point = 28 degrees F (-2 degrees C)

NFPA - Hazard Identification Ratings
health-1; flammability-3; reactivity-0

NIOSH - Selected LD50s and LC50s
Inhalation, mouse: LC50 = 27 gm/m3 8 hr Oral, rat: LD50 = 5 gm/kg

INTERNATIONAL LISTS

Canada - WHMIS: Ingredient Disclosure
1% item 1073 (1449)

STATE LISTS

Florida Hazardous Substance List
[present]

Massachusetts Right To Know List
[present]

NJ Right to Know List (Total)
sn 1290

NJ Special Hazardous Substances
(flammable - third degree)

Pennsylvania Right to Know List
[present]

3-METHYLPYRIDINE   108-99-6

HEALTH AND SAFETY LISTS

NIOSH - Selected LD50s and LC50s
Oral, rat: LD50 = 790 mg/kg

7-METHYLPYRIDO[3,4-C]PSORALEN   RR-01536-1

HEALTH AND SAFETY LISTS

NFPA - Flash Points
flash point < 14 degrees F (-10 degrees C)

NFPA - Hazard Identification Ratings
flammability-3

STATE LISTS

Florida Hazardous Substance List
[present]

Massachusetts Right To Know List
[present]

Pennsylvania Right to Know List
[present]

1-METHYLPYRROLE   96-54-8

HEALTH AND SAFETY LISTS

U.S. DOT - Substances From 49 CFR 172.101
regulated by DOT (UN2612)

U.S. DOT - Hazard Classes
DOT hazard class = 3

NFPA - Flash Points
flash point < -4 degrees F (-20 degrees C)

NFPA - Hazard Identification Ratings
health-0; flammability-3; reactivity-0

STATE LISTS

Florida Hazardous Substance List
[present]

Massachusetts Right To Know List
[present]

NJ Right to Know List (Total)
sn 1291

NJ Special Hazardous Substances
(flammable - third degree)

Pennsylvania Right to Know List
[present]

METHYLPYRROLIDINE   120-94-5

HEALTH AND SAFETY LISTS

ACGIH 1995 - Time Weighted Averages
200 ppm TWA; 705 mg/m3 TWA

ACGIH 1995 - Short Term Exposure Limits
250 ppm STEL; 881 mg/m3 STEL

AIHA - Odor Threshold Values
geometric mean air odor threshold = 7.7 ppm (detectable); 14 ppm (recognizable)

U.S. DOT - Substances From 49 CFR 172.101
regulated by DOT (UN1249)

U.S. DOT - Hazard Classes
DOT hazard class = 3

NFPA - Flash Points
flash point = 45 degrees F (7 degrees C)

NFPA - Hazard Identification Ratings
health-2; flammability-3; reactivity-0

NIOSH - Selected LD50s and LC50s
Oral, rat: LD50 = 3730 mg/kg Skin, rabbit: LD50 = 6500 mg/kg

NIOSH 1990 - Pocket Guide - RELs
150 ppm TWA; 530 mg/m3 TWA

NIOSH 1990 - Pocket Guide - IDLHs
5000 ppm IDLH

NIOSH 1990 - Pocket Guide - Target organs
skin, eyes, CNS, respiratory system

NIOSH - Health Standards - Exposure Limits
150 ppm TWA; 530 mg/m3 TWA (Listed under 'Ketones')

NIOSH - Health Standards - Health Effects and Precautions
Irritation; liver, kidney, and nervous system effects (Urinalysis required) (Listed under 'Ketones')

OSHA - Vacated PELs - Time Weighted Averages
200 ppm TWA; 700 mg/m3 TWA

OSHA - Vacated PELs - Short Term Exposure Limits
250 ppm STEL; 875 mg/m3 STEL

OSHA - Final PELs - Time Weighted Averages
200 ppm TWA; 700 mg/m3 TWA

ENVIRONMENTAL LISTS

List of Pesticide Product Inert Ingredients
[present]

INTERNATIONAL LISTS

Australian Exposure Standards - Time Weighted Averages
200 ppm TWA; 705 mg/m3 TWA

Australian Exposure Standards - Short Term Exposure Limits
250 ppm STEL; 881 mg/m3 STEL

Canada - WHMIS: Ingredient Disclosure
1% item 1074 (1152)

Canada - Alberta - 8 Hour Occupational Exposure Limit
200 ppm TWA; 704 mg/m3 TWA

Canada - Alberta - 15 Minute Occupational Exposure Limit
250 ppm STEL; 879 mg/m3 STEL

Canada - British Columbia - 8 Hour Exposure Limits
200 ppm TWA; 700 mg/m3 TWA

Canada - British Columbia - 15 Minute Exposure Limits
250 ppm STEL; 875 mg/m3 STEL

Canada - Ontario - OHSA - TWAEVs
200 ppm TWAEV; 700 mg/m3 TWAEV

Canada - Ontario - OHSA - STEVs
250 ppm STEV; 880 mg/m3 STEV

Canada - Quebec - Time-Weighted Average Exposure Values
150 ppm TWAEV; 530 mg/m3 TWAEV

United Kingdom - Occupational Exposure Standards - TWAs
200 ppm TWA; 700 mg/m3 TWA

United Kingdom - Occupational Exposure Standards - STELs
250 ppm STEL; 875 mg/m3 STEL

German (DFG) - MAK Values
200 ppm MAK; 700 mg/m3 MAK

German (DFG) - Peak Limitations
2 x normal MAK (30 min. average value); don't exceed 4 times during shift

Israel - Time Weighted Averages
200 ppm TWA; 705 mg/m3 TWA

Israel - Short Term Exposure Limits
250 ppm STEL; 881 mg/m3 STEL

Israel - Action Levels
100 ppm AL; 352.5 mg/m3 AL

Mexico - Instruction No. 10 - TWAs
200 ppm TWA; 700 mg/m3 TWA

STATE LISTS

California - Exposure Limits - PELs
200 ppm PEL; 700 mg/m3 PEL

California - Exposure Limits - STELs
250 ppm STEL; 875 mg/m3 STEL

California - Directors List of Hazardous Substances (8 CCR 339)
[present]

Florida Hazardous Substance List
[present]

Massachusetts Right To Know List
[present]

Minnesota Hazardous Substance List
[present]

NJ Right to Know List (Total)
sn 1292

NJ Special Hazardous Substances
(flammable - third degree)

Pennsylvania Right to Know List
[present]

1-METHYL-2-PYRROLIDONE   872-50-4

HEALTH AND SAFETY LISTS

FDA - Controlled Substances Act - Precursor chemicals
Threshold by base weight = 1 kilogram

METHYL RED   493-52-7

HEALTH AND SAFETY LISTS

FDA - Controlled Substances Act - Precursor chemicals
Threshold by base weight = 1 kilogram

METHYL SALICYLATE   119-36-8

HEALTH AND SAFETY LISTS

NFPA - Flash Points
flash point = 122 degrees F (50 degrees C)

NFPA - Hazard Identification Ratings
health-2; flammability-2; reactivity-0

NIOSH - Selected LD50s and LC50s
Oral, rat: LD50 = 1800 mg/kg

STATE LISTS

Florida Hazardous Substance List
[present]

Massachusetts Right To Know List
[present]

Pennsylvania Right to Know List
environmental hazard

METHYL SELENAC   144-34-3

HEALTH AND SAFETY LISTS

AIHA - WEEL - Time Weighted Averages
2 ppm TWA; 7.6 mg/m3 TWA

AIHA - WEEL - Ceilings or Short Term Time Weighted Averages
5 ppm STEL; 19 mg/m3 STEL

AIHA - WEEL - Skin Absorption Designations
skin absorber

U.S. DOT - Appendix B - Marine Pollutants
DOT regulated marine pollutant

NIOSH - Selected LD50s and LC50s
Oral, bird: LD50 = 1 gm/kg

ENVIRONMENTAL LISTS

CAA - HON Rule - SOCMI Chemicals
compliance by Jan. 23, 1995

EPA - Master Testing List
[present]

TSCA - Code of Federal Regulations Citations
40 CFR 716.120(a)

TSCA - Health and Safety Reporting List
Effective Date: September 10, 1984

METHYL SELENIDE   593-79-3

HEALTH AND SAFETY LISTS

IARC - Group 3 (not classifiable)
[present]

METHYL SILICATE   681-84-5

HEALTH AND SAFETY LISTS

NFPA - Flash Points
flash point = 61 degrees F (16 degrees C)

NFPA - Hazard Identification Ratings
health-2; flammability-3; reactivity-1

STATE LISTS

Florida Hazardous Substance List
[present]

Massachusetts Right To Know List
[present]

METHYL STEARATE   112-61-8

HEALTH AND SAFETY LISTS

NFPA - Flash Points
flash point = 7 degrees F (-14 degrees C)

NFPA - Hazard Identification Ratings
health-2; flammability-3; reactivity-1

ENVIRONMENTAL LISTS

List of Pesticide Product Inert Ingredients
[present]

STATE LISTS

Florida Hazardous Substance List
[present]

Massachusetts Right To Know List
[present]

Pennsylvania Right to Know List
[present]

ALPHA-METHYL STYRENE   98-83-9

HEALTH AND SAFETY LISTS

NFPA - Flash Points
flash point = 204 degrees F (96 degrees C)

NFPA - Hazard Identification Ratings
health-2; flammability-1; reactivity-0

NIOSH - Selected LD50s and LC50s
Oral, rat: LD50 = 3914 mg/kg Skin, rabbit: LD50 = 8000 mg/kg

ENVIRONMENTAL LISTS

CERCLA/SARA - Section 313 - Emission Reporting
form R reporting required

EPA - Master Testing List
[present]

List of Pesticide Product Inert Ingredients
[present]

TSCA - Substances Subject to Testing Consent Orders
Test for: Health Effects

TSCA - Section 12(b) - Export Notification
export notification required - Section 4

INTERNATIONAL LISTS

Australian Exposure Standards - Under Review
exposure limits under review

United Kingdom - Occupational Exposure Standards - TWAs
100 ppm TWA; 400 mg/m3 TWA

German (DFG) - MAK Values
vapor: 20 ppm MAK; 80 mg/m3 MAK

German (DFG) - Peak Limitations
5 x normal MAK (30 min., average value) don't exceed 2 times during shift

German (DFG) - Skin/Sensitizers
danger of cutaneous absorption

German (DFG) - Pregnancy
no risk to embryo/fetus if exposure limits adhered to

STATE LISTS

Florida Hazardous Substance List
[present]

Massachusetts Right To Know List
[present]

Pennsylvania Right to Know List
[present]

METHYL SULFATE, SODIUM SALT   512-42-5

HEALTH AND SAFETY LISTS

IARC - Group 3 (not classifiable)
[present]

6-(METHYLSULFONYL)-2-BENZOTHIAZOLAMINE   17557-67-4

HEALTH AND SAFETY LISTS

U.S. DOT - Appendix B - Marine Pollutants
DOT regulated marine pollutant

NFPA - Flash Points
flash point = 205 degrees F (96 degrees C)

NFPA - Hazard Identification Ratings
health-1; flammability-1; reactivity-0

NIOSH - Selected LD50s and LC50s
Oral, rat: LD50 = 887 mg/kg

ENVIRONMENTAL LISTS

List of Pesticide Product Inert Ingredients
[present]

INTERNATIONAL LISTS

Canada - WHMIS: Ingredient Disclosure
0.1% item 1075 (1474)

STATE LISTS

Pennsylvania Right to Know List
[present]

METHYL TALLATE   74499-22-2

HEALTH AND SAFETY LISTS

IARC - Group 3 (not classifiable)
[present]

METHYL, TALLOW DIETHYLENETRIAMINE CON-   68413-04-7
DENSATE, PROPOXYLATED, METHYL SULFATE
SEE ALSO:
SELENIUM

INTERNATIONAL LISTS

Canada - WHMIS: Ingredient Disclosure
1% item 1076 (1481)

METHYLTESTOSTERONE   58-18-4

HEALTH AND SAFETY LISTS

ACGIH 1995 - Time Weighted Averages
1 ppm TWA; 6 mg/m3 TWA

U.S. DOT - Substances From 49 CFR 172.101
regulated by DOT (UN2606)

U.S. DOT - Hazard Classes
DOT hazard class = 3

U.S. DOT - Substances Which Are Poisonous by Inhalation
liquid hazardous material poisonous by inhalation (UN2606)
NIOSH - Selected LD50s and LC50s
Skin, rabbit: LD50 = 17 gm/kg
OSHA - Vacated PELs - Time Weighted Averages
1 ppm TWA; 6 mg/m3 TWA

INTERNATIONAL LISTS

Australian Exposure Standards - Time Weighted Averages
1 ppm TWA; 6 mg/m3 TWA
Canada - WHMIS: Ingredient Disclosure
1% item 1077 (1486)
Canada - Alberta - 8 Hour Occupational Exposure Limit
1 ppm TWA; 6 mg/m3 TWA
Canada - Alberta - 15 Minute Occupational Exposure Limit
5 ppm STEL; 31 mg/m3 STEL
Canada - British Columbia - Ceiling Exposure Limits
C 5 ppm; C 30 mg/m3
Canada - Ontario - OHSA - TWAEVs
1 ppm TWAEV; 6.2 mg/m3 TWAEV
Canada - Quebec - Time-Weighted Average Exposure Values
1 ppm TWAEV; 6 mg/m3 TWAEV
United Kingdom - Occupational Exposure Standards - TWAs
1 ppm TWA; 6 mg/m3 TWA
United Kingdom - Occupational Exposure Standards - STELs
5 ppm STEL; 30 mg/m3 STEL
Israel - Time Weighted Averages
1 ppm TWA; 6 mg/m3 TWA
Israel - Action Levels
0.5 ppm AL; 3 mg/m3 AL
Mexico - Instruction No. 10 - TWAs
1 ppm TWA; 6 mg/m3 TWA
Mexico - Instruction No. 10 - STELs
5 ppm STEL; 30 mg/m3 STEL

STATE LISTS

California - Exposure Limits - PELs
1 ppm PEL; 6 mg/m3 PEL
California - Directors List of Hazardous Substances (8 CCR 339)
[present]
Florida Hazardous Substance List
[present]
Massachusetts Right To Know List
[present]
Minnesota Hazardous Substance List
[present]
NJ Right to Know List (Total)
sn 1282
Pennsylvania Right to Know List
[present]

METHYL TETRADECANOATE   124-10-7

HEALTH AND SAFETY LISTS

NFPA - Flash Points
flash point = 307 degrees F (153 degrees C)
NFPA - Hazard Identification Ratings
health-0; flammability-1; reactivity-0

ENVIRONMENTAL LISTS

List of Pesticide Product Inert Ingredients
[present]

METHYLTETRAHYDROFURAN   96-47-9

HEALTH AND SAFETY LISTS

ACGIH 1995 - Time Weighted Averages
50 ppm TWA; 242 mg/m3 TWA

ACGIH 1995 - Short Term Exposure Limits
100 ppm STEL; 483 mg/m3 STEL
U.S. DOT - Substances From 49 CFR 172.101
regulated by DOT (UN2303)
U.S. DOT - Hazard Classes
DOT hazard class = 3
U.S. DOT - Appendix B - Marine Pollutants
DOT regulated marine pollutant
NFPA - Flash Points
flash point = 129 degrees F (54 degrees C)
NFPA - Hazard Identification Ratings
health-1; flammability-2; reactivity-1
NIOSH - Selected LD50s and LC50s
Oral, rat: LD50 = 4900 mg/kg
NIOSH 1990 - Pocket Guide - RELs
50 ppm TWA; 240 mg/m3 TWA; 100 ppm STEL; 485 mg/m3 STEL
NIOSH 1990 - Pocket Guide - IDLHs
5000 ppm IDLH
NIOSH 1990 - Pocket Guide - Target organs
eyes, skin, respiratory system
OSHA - Vacated PELs - Time Weighted Averages
50 ppm TWA; 240 mg/m3 TWA
OSHA - Vacated PELs - Short Term Exposure Limits
100 ppm STEL; 485 mg/m3 STEL
OSHA - Final PELs - Ceiling Limits
C 100 ppm; C 480 mg/m3

ENVIRONMENTAL LISTS

CAA - HON Rule - SOCMI Chemicals
compliance by Oct. 24, 1994
TSCA - Code of Federal Regulations Citations
40 CFR 716.120(a)
TSCA - Health and Safety Reporting List
Effective Date: June 1, 1987; Sunset Date: November 9, 1993

INTERNATIONAL LISTS

Australian Exposure Standards - Time Weighted Averages
50 ppm TWA; 242 mg/m3 TWA
Australian Exposure Standards - Short Term Exposure Limits
100 ppm STEL; 483 mg/m3 STEL
Canada - WHMIS: Ingredient Disclosure
1% item 1078 (1153)
Canada - Alberta - 8 Hour Occupational Exposure Limit
50 ppm TWA; 243 mg/m3 TWA
Canada - Alberta - 15 Minute Occupational Exposure Limit
100 ppm STEL; 486 mg/m3 STEL
Canada - British Columbia - Ceiling Exposure Limits
C 100 ppm; C 480 mg/m3
Canada - Ontario - OHSA - TWAEVs
50 ppm TWAEV; 241 mg/m3 TWAEV
Canada - Ontario - OHSA - STEVs
100 ppm STEV; 482 mg/m3 STEV
Canada - Quebec - Time-Weighted Average Exposure Values
50 ppm TWAEV; 242 mg/m3 TWAEV
Canada - Quebec - Short-term Exposure Values
100 ppm STEV; 484 mg/m3 STEV
United Kingdom - Occupational Exposure Standards - STELs
100 ppm STEL; 480 mg/m3 STEL
German (DFG) - MAK Values
100 ppm MAK; 480 mg/m3 MAK
Israel - Time Weighted Averages
50 ppm TWA; 242 mg/m3 TWA
Israel - Short Term Exposure Limits
100 ppm STEL; 483 mg/m3 STEL

Israel - Action Levels
25 ppm AL;   121 mg/m3 AL
Mexico - Instruction No. 10 - TWAs
50 ppm TWA;   240 mg/m3 TWA
Mexico - Instruction No. 10 - STELs
100 ppm STEL;   485 mg/m3 STEL

STATE LISTS

California - Exposure Limits - PELs
50 ppm PEL;   240 mg/m3 PEL
California - Exposure Limits - STELs
100 ppm STEL;   485 mg/m3 STEL
California - Directors List of Hazardous Substances (8 CCR 339)
[present]
Florida Hazardous Substance List
[present]
Massachusetts Right To Know List
[present]
Minnesota Hazardous Substance List
[present]
NJ Right to Know List (Total)
sn 1072

METHYL TETRAHYDROFURAN   25265-68-3

ENVIRONMENTAL LISTS

List of Pesticide Product Inert Ingredients
[present]

METHYL THIOCYANATE   556-64-9

ENVIRONMENTAL LISTS

TSCA - Code of Federal Regulations Citations
40 CFR 712.30(x);   40 CFR 716.120(d)
TSCA - PAIR - Reporting List
Reporting Date: November 27, 1991
TSCA - Health and Safety Reporting List
Effective Date: September 30, 1991

METHYLTHIOURACIL   56-04-2

ENVIRONMENTAL LISTS

List of Pesticide Product Inert Ingredients
[present]

METHYL-P-TOLUENESULFONATE   80-48-8

ENVIRONMENTAL LISTS

TSCA - Code of Federal Regulations Citations
40 CFR 712.30(w);   40 CFR 716.120(a)
TSCA - PAIR - Reporting List
Reporting Date: August 18, 1988

METHYLTRICAPRYLAMMONIUM CHLORIDE   5137-55-3

STATE LISTS

California - Air Bill 2588 Appendix A-II
9/90
California - Prop. 65 - Developmental Toxicity
developmental toxicity - initial date 4/1/90
California - Directors List of Hazardous Substances (8 CCR 339)
[present]
Massachusetts Right To Know List
teratogen

METHYL TRICHLOROACETATE   598-99-2

ENVIRONMENTAL LISTS

List of Pesticide Product Inert Ingredients
[present]

METHYL TRICHLOROSILANE   75-79-6

HEALTH AND SAFETY LISTS

NFPA - Flash Points
flash point = .12 degrees F (-11 degrees C)
NFPA - Hazard Identification Ratings
health-2;   flammability-3;   reactivity-0
NIOSH - Selected LD50s and LC50s
Inhalation, rat: LC50 = 6000 ppm 4 hr Oral, rat: LD50 = 5720 mg/kg Skin, rabbit: LD50 = 4500 mg/kg

INTERNATIONAL LISTS

Canada - WHMIS: Ingredient Disclosure
1% item 1079 (1154)

STATE LISTS

Florida Hazardous Substance List
[present]
Massachusetts Right To Know List
[present]
Pennsylvania Right to Know List
[present]

METHYLTRIETHOXYSILANE   2031-67-6

HEALTH AND SAFETY LISTS

U.S. DOT - Substances From 49 CFR 172.101
regulated by DOT (UN2536)
U.S. DOT - Hazard Classes
DOT hazard class = 3

STATE LISTS

NJ Right to Know List (Total)
sn 1294
NJ Special Hazardous Substances
(flammable - third degree)

METHYLTRIMETHOXYSILANE   1185-55-3

ENVIRONMENTAL LISTS

CERCLA/SARA - Section 302 Extremely Hazardous Substances and TPQs
TPQ = 10,000 pounds
CAA -Toxic Substances for Accidental Release Prevention
threshold quantity = 20,000 lbs

STATE LISTS

Florida Hazardous Substance List
effective March 13, 1992
Massachusetts Right To Know List
extraordinarily hazardous
NJ Right to Know List (Total)
sn 2562
Pennsylvania Right to Know List
environmental hazard

PROPOSED REGULATIONS

CERCLA/SARA - Proposed Hazardous Substance Additions
proposed RQ = 1 pound (.454 kg)
CERCLA/SARA - 1989 Proposed RQ Adjustments
proposed RQ = 1000 pounds (454 kg)

METHYL TRIMETHYLOL METHANE TRINITRATE   RR-01442-6

HEALTH AND SAFETY LISTS

U.S. DOT - Appendix A Table 1 - Hazardous Substances
final RQ = 10 pounds (4.54 kg)
IARC - Group 2B (sufficient animal data)
[present] (Overall evaluation based only on evidence of carcinogenicity in monograph (7, 1974) or in Supplement 4)

NIOSH - Selected LD50s and LC50s
Oral, rat: LD50 = 2790 mg/kg

OSHA - Possible Select Carcinogens
[present]

ENVIRONMENTAL LISTS

CERCLA/SARA - Hazardous Substances and their Reportable Quantities
final RQ = 10 pounds (4.54 kg)

EPA - Carcinogen Hazard Ranking for RQ Adjustment
Hazard ranking = Medium

RCRA - U Series Wastes
waste number U164

RCRA - Hazardous Constituents-Appendix VIII
waste number U164

RCRA - Substances Banned From Land Disposal
[present]

TSCA - Chemicals with Significant New Use Rules
[present]

TSCA - Section 12(b) - Export Notification
export notification required - Section 5

STATE LISTS

California - Air Bill 2588 Appendix A-II
known or potential carcinogen

California - Prop. 65 - Cancer list
carcinogen - initial date 10/1/89

California - Prop. 65 - No Significant Risk Levels
no significant risk level = 2 ug/day

California - Directors List of Hazardous Substances (8 CCR 339)
[present]

Florida Hazardous Substance List
[present]

Massachusetts Right To Know List
carcinogen; extraordinarily hazardous

Minnesota Hazardous Substance List
carcinogen

NJ Right to Know List (Total)
sn 1967

NJ Special Hazardous Substances
(carcinogen)

Pennsylvania Right to Know List
environmental hazard; special hazardous substance

Pennsylvania RTK - Special Hazardous Substances
[present]

METHYLTRITHION   7332-32-3

HEALTH AND SAFETY LISTS

NFPA - Flash Points
flash point = 306 degrees F (152 degrees C)

NFPA - Hazard Identification Ratings
health-2; flammability-1; reactivity-0

NIOSH - Selected LD50s and LC50s
Oral, rat: LD50 = 341 mg/kg

INTERNATIONAL LISTS

Canada - WHMIS: Ingredient Disclosure
1% item 1080 (1631)

STATE LISTS

Florida Hazardous Substance List
[present]

Massachusetts Right To Know List
[present]

Pennsylvania Right to Know List
[present]

METHYL UNDECYLENATE   5760-50-9

HEALTH AND SAFETY LISTS

NIOSH - Selected LD50s and LC50s
Oral, rat: LD50 = 223 mg/kg

ENVIRONMENTAL LISTS

List of Pesticide Product Inert Ingredients
[present]

METHYL UNDECYL KETONE   593-08-8

HEALTH AND SAFETY LISTS

U.S. DOT - Substances From 49 CFR 172.101
regulated by DOT (UN2533)

U.S. DOT - Hazard Classes
DOT hazard class = 6.1

STATE LISTS

NJ Right to Know List (Total)
sn 1295

2-METHYL VALERALDEHYDE   123-15-9

HEALTH AND SAFETY LISTS

AIHA - WEEL - Ceilings or Short Term Time Weighted Averages
1 ppm STEL; 6 mg/m3 STEL

U.S. DOT - Substances From 49 CFR 172.101
regulated by DOT (UN1250)

U.S. DOT - Hazard Classes
DOT hazard class = 3

NFPA - Flash Points
flash point = 15 degrees F (-9 degrees C)

NFPA - Hazard Identification Ratings
health-3; flammability-3; reactivity-2 (avoid use of water)

NIOSH - Selected LD50s and LC50s
Inhalation, rat: LC50 = 450 ppm 4 hr

OSHA - List of Highly Hazardous Chemicals
threshhold quantity = 500 pounds

ENVIRONMENTAL LISTS

CERCLA/SARA - Section 302 Extremely Hazardous Substances and TPQs
TPQ = 500 pounds

CERCLA/SARA - Section 313 - Emission Reporting
form R reporting required

CAA -Toxic Substances for Accidental Release Prevention
threshold quantity = 5,000 lbs

EPA - Master Testing List
[present]

INTERNATIONAL LISTS

Canada - WHMIS: Ingredient Disclosure
1% item 1081 (1155)

STATE LISTS

Florida Hazardous Substance List
[present]

Massachusetts Right To Know List
extraordinarily hazardous

NJ Right to Know List (Total)
sn 1296

NJ Special Hazardous Substances
(corrosive; flammable - third degree; reactive - second degree)

Pennsylvania Right to Know List
environmental hazard

PROPOSED REGULATIONS

CERCLA/SARA - Proposed Hazardous Substance Additions
proposed RQ = 1 pound (.454 kg)

CERCLA/SARA - 1989 Proposed RQ Adjustments
proposed RQ = 100 pounds (45.4 kg)

4-METHYL VALERALDEHYDE   1119-16-0

HEALTH AND SAFETY LISTS

NIOSH - Selected LD50s and LC50s
Oral, rat: LD50 = 15700 mg/kg

ENVIRONMENTAL LISTS

List of Pesticide Product Inert Ingredients
[present]

INTERNATIONAL LISTS

Canada - WHMIS: Ingredient Disclosure
1% item 1082 (1156)

METHYL VALERATE   624-24-8

HEALTH AND SAFETY LISTS

NIOSH - Selected LD50s and LC50s
Oral, rat: LD50 = 12500 mg/kg

INTERNATIONAL LISTS

Canada - WHMIS: Ingredient Disclosure
1% item 1083 (1157)

METHYLVINYLBENZENES   RR-01303-6

HEALTH AND SAFETY LISTS

U.S. DOT - Hazard Classes
Forbidden from transport by the DOT

METHYLVINYLCYCLOSILOXANE   2554-06-5

HEALTH AND SAFETY LISTS

U.S. DOT - Appendix B - Marine Pollutants
DOT regulated marine pollutant

METHYL VINYL KETONE   78-94-4

INTERNATIONAL LISTS

Canada - WHMIS: Ingredient Disclosure
1% item 1084 (1705)

METHYL VIOLET   1325-82-2

HEALTH AND SAFETY LISTS

NFPA - Flash Points
flash point = 225 degrees F (107 degrees C)
NFPA - Hazard Identification Ratings
health-1; flammability-1; reactivity-0

METHYL VIOLET   8004-87-3

HEALTH AND SAFETY LISTS

U.S. DOT - Substances From 49 CFR 172.101
regulated by DOT (UN2367)
U.S. DOT - Hazard Classes
DOT hazard class = 3
NFPA - Flash Points
flash point = 62 degrees F (17 degrees C)
NFPA - Hazard Identification Ratings
health-1; flammability-3; reactivity-0

STATE LISTS

Florida Hazardous Substance List
[present]
Massachusetts Right To Know List
[present]
NJ Right to Know List (Total)
sn 1299
NJ Special Hazardous Substances
(flammable - third degree)

Pennsylvania Right to Know List
[present]

METIRAM   9006-42-2

HEALTH AND SAFETY LISTS

NIOSH - Selected LD50s and LC50s
Oral, rat: LD50 = 5660 mg/kg Skin, rabbit: LD50 = 4460 mg/kg

METOLACHLOR   51218-45-2

HEALTH AND SAFETY LISTS

NIOSH - Selected LD50s and LC50s
Inhalation, mouse: LC50 = 6600 mg/m3 (2 hr)

METOLCARB   1129-41-5

HEALTH AND SAFETY LISTS

U.S. DOT - Appendix B - Marine Pollutants
inhibited form: DOT regulated marine pollutant

METRIBUZIN   21087-64-9

ENVIRONMENTAL LISTS

TSCA - PAIR - Reporting List
Effective Date: October 12, 1993; Reporting Date: February 28, 1994
TSCA - Health and Safety Reporting List
Effective Date: October 12, 1993; Sunset Date: October 12, 2003

METRONIDAZOLE   443-48-1

HEALTH AND SAFETY LISTS

U.S. DOT - Substances From 49 CFR 172.101
regulated by DOT (UN1251)
U.S. DOT - Hazard Classes
DOT hazard class = 3
U.S. DOT - Substances Which Are Poisonous by Inhalation
liquid hazardous material poisonous by inhalation (UN1251)
NFPA - Flash Points
flash point = 20 degrees F (-7 degrees C)
NFPA - Hazard Identification Ratings
health-4; flammability-3; reactivity-2
NIOSH - Selected LD50s and LC50s
Inhalation, rat: LC50 = 7 mg/m3 4 hr Oral, rat: LD50 = 30 mg/kg
NTP Chemical Status Reports - Testing Status and NTIS Number
Project leader assigned/study in design
OSHA - List of Highly Hazardous Chemicals
threshhold quantity = 100 pounds

ENVIRONMENTAL LISTS

CERCLA/SARA - Section 302 Extremely Hazardous Substances and TPQs
TPQ = 10 pounds

INTERNATIONAL LISTS

Canada - WHMIS: Ingredient Disclosure
1% item 1085 (1158)
German (DFG) - Skin/Sensitizers
danger of cutaneous absorption and danger of sensitization

STATE LISTS

Florida Hazardous Substance List
[present]
Massachusetts Right To Know List
extraordinarily hazardous
NJ Right to Know List (Total)
sn 1301
NJ Special Hazardous Substances
(flammable - third degree; reactive - second degree)
Pennsylvania Right to Know List
environmental hazard

PROPOSED REGULATIONS

CERCLA/SARA - Proposed Hazardous Substance Additions
proposed RQ = 1 pound (.454 kg)

CERCLA/SARA - 1989 Proposed RQ Adjustments
proposed RQ = 10 pounds (4.54 kg)

MEVINPHOS   7786-34-7

INTERNATIONAL LISTS

Canada - WHMIS: Ingredient Disclosure
1% item 1087 (1720)

MEXACARBATE   315-18-4

HEALTH AND SAFETY LISTS

NIOSH - Selected LD50s and LC50s
Oral, rat: LD50 = 413 mg/kg

ENVIRONMENTAL LISTS

List of Pesticide Product Inert Ingredients
[present]

MICA   12001-26-2

ENVIRONMENTAL LISTS

CERCLA/SARA - Section 313 - Emission Reporting
form R reporting required

STATE LISTS

California - Air Bill 2588 Appendix A-II
9/90

California - Prop. 65 - Cancer list
carcinogen - initial date 1/1/90

MICHLER'S KETONE   90-94-8

INTERNATIONAL LISTS

Canada - Drinking Water Quality - IMACs
0.05 mg/L IMAC

PROPOSED REGULATIONS

Safe Drinking Water Act - Priority list
[present]

MICROCRYSTALLINE WAX   63231-60-7

HEALTH AND SAFETY LISTS

NIOSH - Selected LD50s and LC50s
*Inhalation, rat: LC50 = 475 mg/m3 8 hr Oral, rat: LD50 = 268
mg/kg Skin, rat: LD50 = 6000 mg/kg*

ENVIRONMENTAL LISTS

CERCLA/SARA - Section 302 Extremely Hazardous Substances and
TPQs
TPQ = 100/10,000 pounds

STATE LISTS

Florida Hazardous Substance List
effective March 13, 1992

Massachusetts Right To Know List
extraordinarily hazardous

NJ Right to Know List (Total)
sn 2563

Pennsylvania Right to Know List
environmental hazard

PROPOSED REGULATIONS

CERCLA/SARA - Proposed Hazardous Substance Additions
proposed RQ = 1 pound (.454 kg)

CERCLA/SARA - 1989 Proposed RQ Adjustments
proposed RQ = 100 pounds (45.4 kg)

MIDAZOLAM HYDROCHLORIDE   59467-96-8

HEALTH AND SAFETY LISTS

ACGIH 1995 - Time Weighted Averages
5 mg/m3 TWA

NIOSH - Selected LD50s and LC50s
Oral, rat: LD50 = 1100 mg/kg

OSHA - Vacated PELs - Time Weighted Averages
5 mg/m3 TWA

ENVIRONMENTAL LISTS

CERCLA/SARA - Section 313 - Emission Reporting
form R reporting required

INTERNATIONAL LISTS

Australian Exposure Standards - Time Weighted Averages
5 mg/m3 TWA

Canada - Drinking Water Quality - MACs
0.08 mg/L MAC

Canada - Alberta - 8 Hour Occupational Exposure Limit
5 mg/m3 TWA

Canada - Alberta - 15 Minute Occupational Exposure Limit
10 mg/m3 STEL

Canada - Ontario - OHSA - TWAEVs
5 mg/m3 TWAEV

Canada - Quebec - Time-Weighted Average Exposure Values
5 mg/m3 TWAEV

Israel - Time Weighted Averages
5 mg/m3 TWA

Israel - Action Levels
2.5 mg/m3 AL

STATE LISTS

California - Exposure Limits - PELs
5 mg/m3 PEL

Florida Hazardous Substance List
effective March 13, 1992

Massachusetts Right To Know List
[present]

Minnesota Hazardous Substance List
[present]

NJ Right to Know List (Total)
sn 1302

Pennsylvania Right to Know List
[present]

PROPOSED REGULATIONS

Safe Drinking Water Act - Priority list
[present]

MILK, HYDROLYZED   68514-61-4

HEALTH AND SAFETY LISTS

IARC - Group 2B (sufficient animal data)
[present]

NIOSH - Selected LD50s and LC50s
Oral, rat: LD50 = 3 gm/kg

NTP Seventh Report - Suspect Carcinogens
suspect carcinogen

OSHA - Possible Select Carcinogens
[present]

STATE LISTS

California - Air Bill 2588 Appendix A-I
known or potential carcinogen

California - Prop. 65 - Cancer list
carcinogen - initial date 1/1/88

California - Directors List of Hazardous Substances (8 CCR 339)
[present]

Florida Hazardous Substance List
[present]

Massachusetts Right To Know List
carcinogen; extraordinarily hazardous; teratogen

Minnesota Hazardous Substance List
carcinogen

NJ Right to Know List (Total)
sn 1303

NJ Special Hazardous Substances
(carcinogen; mutagen)

Pennsylvania Right to Know List
special hazardous substance

Pennsylvania RTK - Special Hazardous Substances
[present]

MILLET SEED   RR-01095-7

HEALTH AND SAFETY LISTS

ACGIH 1995 - Time Weighted Averages
0.01 ppm TWA; 0.092 mg/m3 TWA

ACGIH 1995 - Short Term Exposure Limits
0.03 ppm STEL; 0.27 mg/m3 STEL

ACGIH 1995 - Skin Designations
skin - potential for cutaneous absorption

U.S. DOT - Appendix B - Marine Pollutants
DOT regulated severe marine pollutant

U.S. DOT - Appendix A Table 1 - Hazardous Substances
final RQ = 10 pounds (4.54 kg)

NIOSH - Selected LD50s and LC50s
Inhalation, rat: LC50 = 14 ppm 1 hr Oral, rat: LD50 = 3 mg/kg Skin, rat: LD50 = 4200 ug/kg

NIOSH 1990 - Pocket Guide - RELs
0.01 ppm TWA; 0.1 mg/m3 TWA; 0.03 ppm STEL; 0.3 mg/m3 STEL

NIOSH 1990 - Pocket Guide - IDLHs
4 ppm IDLH

NIOSH 1990 - Pocket Guide - Target organs
respiratory system, CNS, CVS, skin, blood cholinesterase

NIOSH 1990 - Pocket Guide - Skin list
Potential for dermal absorption

OSHA - Vacated PELs - Time Weighted Averages
0.01 ppm TWA; 0.1 mg/m3 TWA

OSHA - Vacated PELs - Short Term Exposure Limits
0.03 ppm STEL; 0.3 mg/m3 STEL

OSHA - Vacated PELs - Skin Designation
Prevent or reduce skin absorption

OSHA - Final PELs - Time Weighted Averages
0.1 mg/m3 TWA

OSHA - Final PELs - Skin Notations
prevent or reduce skin absorption

ENVIRONMENTAL LISTS

CERCLA/SARA - Section 302 Extremely Hazardous Substances and TPQs
TPQ = 500 pounds

CERCLA/SARA - Section 313 - Emission Reporting
form R reporting required

CERCLA/SARA - Hazardous Substances and their Reportable Quantities
final RQ = 10 pounds (4.54 kg)

Clean Water Act - Hazardous Substances
[present]

INTERNATIONAL LISTS

Australian Exposure Standards - Time Weighted Averages
0.01 ppm TWA; 0.092 mg/m3 TWA

Australian Exposure Standards - Short Term Exposure Limits
0.03 ppm STEL; 0.27 mg/m3 STEL

Australian Exposure Standards - Skin Effects
skin absorption

Canada - Alberta - 8 Hour Occupational Exposure Limit
0.01 ppm TWA; 0.092 mg/m3 TWA

Canada - Alberta - 15 Minute Occupational Exposure Limit
0.03 ppm STEL; 0.27 mg/m3 STEL

Canada - Alberta - Skin Designation
can be absorbed through the intact skin

Canada - British Columbia - 8 Hour Exposure Limits
0.01 ppm TWA; 0.1 mg/m3 TWA

Canada - British Columbia - 15 Minute Exposure Limits
0.03 ppm STEL; 0.3 mg/m3 STEL

Canada - British Columbia - Skin Notations
skin - potential for skin absorption

Canada - Ontario - OHSA - TWAEVs
0.01 ppm TWAEV; 0.1 mg/m3 TWAEV

Canada - Ontario - OHSA - STEVs
0.03 ppm STEV; 0.3 mg/m3 STEV

Canada - Ontario - OHSA - Skin Notations
absorption through skin, eyes, or mucous membranes

Canada - Quebec - Time-Weighted Average Exposure Values
0.01 ppm TWAEV; 0.09 mg/m3 TWAEV

Canada - Quebec - Short-term Exposure Values
0.03 ppm STEV; 0.27 mg/m3 STEV

Canada - Quebec - Skin Designations
absorbed through the skin

United Kingdom - Occupational Exposure Standards - TWAs
0.01 ppm TWA; 0.1 mg/m3 TWA

United Kingdom - Occupational Exposure Standards - STELs
0.03 ppm STEL; 0.3 mg/m3 STEL

United Kingdom - Occupational Exposure Standards - Notes
can be absorbed through skin

German (DFG) - MAK Values
0.01 ppm MAK; 0.1 mg/m3 MAK

German (DFG) - Skin/Sensitizers
danger of cutaneous absorption

Israel - Time Weighted Averages
0.01 ppm TWA; 0.092 mg/m3 TWA

Israel - Short Term Exposure Limits
0.03 ppm STEL; 0.27 mg/m3 STEL

Israel - Action Levels
0.005 ppm AL; 0.045 mg/m3 AL

Mexico - Instruction No. 10 - TWAs
0.01 ppm TWA; 0.1 mg/m3 TWA

Mexico - Instruction No. 10 - STELs
0.03 ppm STEL; 0.3 mg/m3 STEL

Mexico - Instruction No. 10 - Skin designation
skin - potential for cutaneous absorption

STATE LISTS

California - Exposure Limits - PELs
0.01 ppm PEL; 0.1 mg/m3 PEL

California - Exposure Limits - STELs
0.03 ppm STEL; 0.3 mg/m3 STEL

California - Exposure Limits - Skin Notation
material may be absorbed through the skin, eyes or mucous membrane

California - Directors List of Hazardous Substances (8 CCR 339)
[present]

Florida Hazardous Substance List
[present]

Massachusetts Right To Know List
extraordinarily hazardous; neurotoxin

Minnesota Hazardous Substance List
skin

NJ Right to Know List (Total)
sn 1509

Pennsylvania Right to Know List
environmental hazard

MILO   RR-01096-8

HEALTH AND SAFETY LISTS

U.S. DOT - Appendix B - Marine Pollutants
DOT regulated marine pollutant

U.S. DOT - Appendix A Table 1 - Hazardous Substances
final RQ = 1000 pounds (454 kg)

IARC - Group 3 (not classifiable)
[present]

NIOSH - Selected LD50s and LC50s
Oral, rat: LD50 = 14 mg/kg Skin, rat: LD50 = 1500 mg/kg

NTP Chemical Status Reports - Testing Status and NTIS Number
Technical reports printed (PB287471/AS)

NTP Chemical Status Reports - Evidence of Carcinogenicity
male rat-negative; female rat-negative; male mice-negative; female mice-negative

ENVIRONMENTAL LISTS

CERCLA/SARA - Section 302 Extremely Hazardous Substances and TPQs
TPQ = 500/10,000 pounds

CERCLA/SARA - Hazardous Substances and their Reportable Quantities
final RQ = 1000 pounds (454 kg)

Clean Water Act - Hazardous Substances
[present]

STATE LISTS

California - Directors List of Hazardous Substances (8 CCR 339)
[present]

Florida Hazardous Substance List
effective March 13, 1992

Massachusetts Right To Know List
extraordinarily hazardous

NJ Right to Know List (Total)
sn 1304

Pennsylvania Right to Know List
environmental hazard

MILORGANITE   8049-99-8

HEALTH AND SAFETY LISTS

ACGIH 1995 - Time Weighted Averages
3 mg/m3 TWA (this TLV is for the respirable fraction of dust for Mica)

NIOSH 1990 - Pocket Guide - RELs
respirable dust: 3 mg/m3 TWA

NIOSH 1990 - Pocket Guide - Target organs
lungs

OSHA - Vacated PELs - Time Weighted Averages
respirable dust (less than 1% crystalline silica): 3 mg/m3 TWA (Listed under 'Silicates')

OSHA - Final PELs - Time Weighted Averages
see Table Z-3

INTERNATIONAL LISTS

Australian Exposure Standards - Time Weighted Averages
inspirable dust: 2.5 mg/m3 TWA

Australian Exposure Standards - Under Review
exposure limits under review

Canada - WHMIS: Ingredient Disclosure
1% item 1088 (1160)

Canada - Alberta - 8 Hour Occupational Exposure Limit
respirable mass: 3 mg/m3 TWA; total mass: 6 mg/m3 TWA

Canada - Ontario - OHSA - TWAEVs
total dust: 6 mg/m3 TWAEV; respirable dust: 3 mg/m3 TWAEV (listed as mineral dust)

Canada - Quebec - Time-Weighted Average Exposure Values
3 mg/m3 TWAEV

United Kingdom - Occupational Exposure Standards - TWAs
total inhalable dust: 10 mg/m3 TWA; respirable dust: 1 mg/m3 TWA

Israel - Time Weighted Averages
3 mg/m3 TWA (this TLV is for the respirable fraction of dust for Mica)

Israel - Action Levels
respirable dust: 1.5 mg/m3 AL

STATE LISTS

California - Exposure Limits - PELs
respirable dust (<1% crystalline silica): 3 mg/m3 PEL (Listed under 'Silicates')

California - Directors List of Hazardous Substances (8 CCR 339)
[present] (exempt except when inhalable dust is present or can be generated)

Florida Hazardous Substance List
[present]

Massachusetts Right To Know List
[present]

Minnesota Hazardous Substance List
[present]

Pennsylvania Right to Know List
[present]

MINERAL FIBERS   RR-01501-0

HEALTH AND SAFETY LISTS

NIOSH - Selected LD50s and LC50s
Oral, bird: LD50 = 100 mg/kg

NTP Chemical Status Reports - Testing Status and NTIS Number
Technical reports printed (PB299855/AS)

NTP Chemical Status Reports - Evidence of Carcinogenicity
male rat-positive; female rat-positive; male mice-positive; female mice-positive

NTP Seventh Report - Suspect Carcinogens
suspect carcinogen

OSHA - Possible Select Carcinogens
[present]

ENVIRONMENTAL LISTS

CERCLA/SARA - Section 313 - Emission Reporting
form R reporting required for 0.1% de minimus concentration

INTERNATIONAL LISTS

Canada - WHMIS: Ingredient Disclosure
0.1% item 1089 (398)

Canada - NPRI (National Pollutant Release Inventory)
[present]

German (DFG) - Carcinogens
suspected carcinogen

STATE LISTS

California - Air Bill 2588 Appendix A-I
known or potential carcinogen

California - Prop. 65 - Cancer list
carcinogen - initial date 1/1/88

California - Prop. 65 - No Significant Risk Levels
no significant risk level = 0.8 ug/day

Florida Hazardous Substance List
[present]

Massachusetts Right To Know List
carcinogen; extraordinarily hazardous

NJ Right to Know List (Total)
sn 1305

NJ Special Hazardous Substances
(carcinogen)

Pennsylvania Right to Know List
environmental hazard; special hazardous substance

Pennsylvania RTK - Special Hazardous Substances
[present]

MINERAL FIBERS (OTHER THAN MANMADE)   RR-01739-0

HEALTH AND SAFETY LISTS

NFPA - Flash Points
flash point > 400 degrees F (204 degrees C)

NFPA - Hazard Identification Ratings
health-0; flammability-1; reactivity-0

MINERAL WOOL FIBER   RR-00015-7

STATE LISTS

California - Air Bill 2588 Appendix A-II
9/90

California - Prop. 65 - Developmental Toxicity
developmental toxicity - initial date 7/1/90

MINING REAGENT, LIQUID   RR-01502-1

ENVIRONMENTAL LISTS

List of Pesticide Product Inert Ingredients
[present]

MINK OIL   8023-74-3

ENVIRONMENTAL LISTS

List of Pesticide Product Inert Ingredients
[present]

MINOCYCLINE HYDROCHLORIDE   13614-98-7

ENVIRONMENTAL LISTS

List of Pesticide Product Inert Ingredients
[present]

MIREX   2385-85-5

ENVIRONMENTAL LISTS

List of Pesticide Product Inert Ingredients
[present]

MISOPROSTOL   62015-39-8

INTERNATIONAL LISTS

Canada - CEPA - Priority Substances List
estimated time for completion of assessment reports: 5 years

STATE LISTS

California - Air Bill 2588 Appendix A-I
manmade and of respirable size: known or potential carcinogen: 6/91

MITOMYCIN C   50-07-7

STATE LISTS

California - Air Bill 2588 Appendix A-I
[present]

MITOXANTRONE HYDROCHLORIDE   70476-82-3

HEALTH AND SAFETY LISTS

ACGIH 1995 - Time Weighted Averages
10 mg/m3 TWA (The value is for total dust containing no asbestos and < 1% crystalline silica)

U.S. DOT - Appendix A Table 1 - Hazardous Substances
final RQ = 1 pound (.454 kg)

IARC - Group 2B (sufficient animal data)
[present]

NTP Seventh Report - Suspect Carcinogens
suspect carcinogen

OSHA - Possible Select Carcinogens
[present]

ENVIRONMENTAL LISTS

CERCLA/SARA - Hazardous Substances and their Reportable Quantities
final RQ = 1 pound (.454 kg)

Clean Air Act (1990) - List of Hazardous Air Contaminants
[present] (includes mineral fiber emissions from facilities manufacturing or processing glass, rock, or slag fibers of average diameter 1 micrometer or less)

INTERNATIONAL LISTS

Australian Exposure Standards - Time Weighted Averages
respirable dust: 0.5 fibres per ml of air; non-respirable dust: 2 mg/m3 TWA

Canada - Alberta - 8 Hour Occupational Exposure Limit
1 f/cm3 TWA

Canada - British Columbia - 8 Hour Exposure Limits
nuisance dust, respirable mass: 10 mg/m3 TWA

Canada - Ontario - OHSA - TWAEVs
total dust: 10 mg/m3 TWAEV (listed as nuisance particulates)

Canada - Quebec - Time-Weighted Average Exposure Values
2 mg/m3 TWAEV

Canada - Quebec - Carcinogens
C3 carcinogen: effect detected in animals

United Kingdom - Maximum Exposure Limits - TWAs
5 mg/m3 TWA

German (DFG) - Carcinogens
as fibrous dust: suspected carcinogen

Israel - Time Weighted Averages
10 mg/m3 TWA (The value is for total dust containing no asbestos and < 1% crystalline silica)

Israel - Action Levels
fiber: 5 mg/m3 AL

STATE LISTS

California - Air Bill 2588 Appendix A-I
known or potential carcinogen: 9/89

California - Prop. 65 - Cancer list
carcinogen (airborne particles of respirable size) - initial date 7/1/90

California - Directors List of Hazardous Substances (8 CCR 339)
[present]

Minnesota Hazardous Substance List
[present] (includes inert or nuisance dust)

Pennsylvania Right to Know List
[present]

PROPOSED REGULATIONS

Canada - Ontario - Proposed Occupational TWAEVs
total dust: 5 mg/m3 TWAEV

MIXED C10-16-ALKYL SULFATES   68611-55-2

STATE LISTS

NJ Right to Know List (Total)
sn 2564

MIXED FATTY ACIDS   67254-79-9

ENVIRONMENTAL LISTS

List of Pesticide Product Inert Ingredients
[present]

MIXED FATTY AND ROSIN ACIDS   RR-01097-9

STATE LISTS

California - Prop. 65 - Developmental Toxicity
internal use: developmental toxicity - initial date 1/1/92

MIXTURE OF 1,3-BENZENEDIAMINE, 2-METHYL-4,   106264-79-3
6-BIS(METHYLTHIO)AND 1,3-BENZENEDIAMINE, 4-
METHYL-2,6-BIS(METHYLTHIO)

HEALTH AND SAFETY LISTS

U.S. DOT - Appendix B - Marine Pollutants
DOT regulated marine pollutant

IARC - Group 2B (sufficient animal data)
*[present] (Overall evaluation based only on evidence of carcinogenicity
in monograph (20, 1979) or in Supplement 4)*

NIOSH - Selected LD50s and LC50s
*Inhalation, bird: LC50 = 1400 ppm 8 hr Oral, rat: LD50 = 235
mg/kg Skin, rabbit: LD50 = 800 mg/kg*

NTP Chemical Status Reports - Testing Status and NTIS Number
Technical reports printed (PB90241084)

NTP Chemical Status Reports - Evidence of Carcinogenicity
male rat-clear evidence; female rat-clear evidence

NTP Seventh Report - Suspect Carcinogens
suspect carcinogen

OSHA - Possible Select Carcinogens
[present]

INTERNATIONAL LISTS

Canada - CEPA Schedule I - Toxic Substances
prohibited commercial, manufacturing or processing uses

Canada - CEPA Schedule II Part I - Prohibited Substances (Export)
[present]

STATE LISTS

California - Air Bill 2588 Appendix A-II
known or potential carcinogen

California - Prop. 65 - Cancer list
carcinogen - initial date 1/1/88

California - Prop. 65 - No Significant Risk Levels
no significant risk level = 0.04 ug/day

California - Directors List of Hazardous Substances (8 CCR 339)
[present]

Florida Hazardous Substance List
[present]

Massachusetts Right To Know List
carcinogen; extraordinarily hazardous

Minnesota Hazardous Substance List
carcinogen

NJ Right to Know List (Total)
sn 1306

NJ Special Hazardous Substances
(carcinogen)

Pennsylvania Right to Know List
special hazardous substance

Pennsylvania RTK - Special Hazardous Substances
[present]

MOCIC ACID   526-99-8

STATE LISTS

California - Air Bill 2588 Appendix A-II
9/90

California - Prop. 65 - Developmental Toxicity
developmental toxicity - initial date 4/1/90

MODACRYLIC FIBRES   RR-01537-2

HEALTH AND SAFETY LISTS

U.S. DOT - Appendix A Table 1 - Hazardous Substances
final RQ = 10 pounds (4.54 kg)

IARC - Group 2B (sufficient animal data)
*[present] (Overall evaluation based only on evidence of carcinogenicity
in monograph (10, 1976) or in Supplement 4)*

NIOSH - Selected LD50s and LC50s
Oral, rat: LD50 = 30 mg/kg

NTP Chemical Status Reports - Testing Status and NTIS Number
Chronic studies exist for which technical reports were not prepared

OSHA - Possible Select Carcinogens
[present]

ENVIRONMENTAL LISTS

CERCLA/SARA - Section 302 Extremely Hazardous Substances and
TPQs
TPQ = 500/10,000 pounds

CERCLA/SARA - Hazardous Substances and their Reportable Quantities
final RQ = 10 pounds (4.54 kg)

EPA - Carcinogen Hazard Ranking for RQ Adjustment
Hazard ranking = Medium

RCRA - U Series Wastes
waste number U010

RCRA - Hazardous Constituents-Appendix VIII
waste number U010

RCRA - Substances Banned From Land Disposal
[present]

TSCA - Chemicals with Significant New Use Rules
[present]

TSCA - Section 12(b) - Export Notification
export notification required - Section 5

STATE LISTS

California - Air Bill 2588 Appendix A-II
known or potential carcinogen

California - Prop. 65 - Cancer list
carcinogen - initial date 4/1/88

California - Prop. 65 - No Significant Risk Levels
no significant risk level = 0.00009 ug/day

California - Directors List of Hazardous Substances (8 CCR 339)
[present]

Florida Hazardous Substance List
[present]

Massachusetts Right To Know List
carcinogen; extraordinarily hazardous

Minnesota Hazardous Substance List
carcinogen

NJ Right to Know List (Total)
sn 1307

Pennsylvania Right to Know List
environmental hazard; special hazardous substance

Pennsylvania RTK - Special Hazardous Substances
[present]

MODIFIED ACRYLIC ESTER   RR-00928-9

STATE LISTS

California - Air Bill 2588 Appendix A-II
9/90

California - Prop. 65 - Developmental Toxicity
developmental toxicity - initial date 7/1/90

MODIFIED HYDROCARBON RESIN   RR-01262-4
ENVIRONMENTAL LISTS
List of Pesticide Product Inert Ingredients
[present]

MOLASSES   8052-35-5
ENVIRONMENTAL LISTS
List of Pesticide Product Inert Ingredients
[present]

MOLASSES   68476-78-8
ENVIRONMENTAL LISTS
List of Pesticide Product Inert Ingredients
[present]

MOLYBDATE ORANGE (LEAD CHROMATE PIG-   12656-85-8
MENT)
ENVIRONMENTAL LISTS
TSCA - Section 12(b) - Export Notification
P-86-1322; export notification required - Section 5

MOLYBDATES   RR-01621-7
HEALTH AND SAFETY LISTS
NIOSH - Selected LD50s and LC50s
Oral, mouse: LD50 = 8000 mg/kg

MOLYBDENUM   7439-98-7
HEALTH AND SAFETY LISTS
IARC - Group 3 (not classifiable)
[present]

MOLYBDENUM 90   15690-77-4
ENVIRONMENTAL LISTS
TSCA - Chemicals with Significant New Use Rules
PMN number: P-86-387

MOLYBDENUM 93   14119-13-2
ENVIRONMENTAL LISTS
TSCA - Chemicals with Significant New Use Rules
PMN number: P-91-1418

MOLYBDENUM 93M   RR-00402-4
ENVIRONMENTAL LISTS
List of Pesticide Product Inert Ingredients
[present]

MOLYBDENUM 99   14119-15-4
ENVIRONMENTAL LISTS
List of Pesticide Product Inert Ingredients
[present]

MOLYBDENUM 101   14191-83-4
SEE ALSO:
LEAD
CHROMIUM (VI) COMPOUNDS
LEAD COMPOUNDS
CHROMIUM COMPOUNDS
LEAD
ENVIRONMENTAL LISTS
List of Pesticide Product Inert Ingredients
[present]
INTERNATIONAL LISTS
Canada - WHMIS: Ingredient Disclosure
0.1% item 1090 (1288)

MOLYBDENUM COMPOUNDS, N.O.S.   RR-00603-1
INTERNATIONAL LISTS
Canada - Ontario - OHSA - TWAEVs
as Mo: 5 mg/m3 TWAEV

MOLYBDENUM INSOLUBLE COMPOUNDS   RR-00037-3
HEALTH AND SAFETY LISTS
ACGIH 1995 - Time Weighted Averages
soluble compounds as Mo: 5 mg/m3 TWA; insoluble compounds as Mo: 10 mg/m3 TWA
NIOSH 1990 - Pocket Guide - RELs
as Mo: See Appendix D (Listed under 'Molybdenum (insoluble compounds)')
NIOSH 1990 - Pocket Guide - Target organs
none known in humans
OSHA - Vacated PELs - Time Weighted Averages
soluble compounds, as Mo: 5 mg/m3 TWA; insoluble compounds (total dust), as Mo: 10 mg/m3 TWA
OSHA - Final PELs - Time Weighted Averages
soluble compounds, as Mo: 5 mg/m3 TWA; insoluble compounds (total dust), as Mo: 15 mg/m3 TWA
ENVIRONMENTAL LISTS
ATSDR Priority List
Rank (of 275): 227
INTERNATIONAL LISTS
Australian Exposure Standards - Time Weighted Averages
insoluble compounds, as Mo: 10 mg/m3 TWA; soluble compounds, as Mo: 5 mg/m3 TWA
Canada - WHMIS: Ingredient Disclosure
1% item 1092 (1167)
Canada - British Columbia - 8 Hour Exposure Limits
as Mo: 10 mg/m3 TWA
Canada - British Columbia - 15 Minute Exposure Limits
as Mo: 20 mg/m3 STEL
Canada - Ontario - OHSA - TWAEVs
as Mo: 10 mg/m3 TWAEV
German (DFG) - MAK Values
insoluble compounds-total dust, as Mo: 15 mg/m3 MAK; soluble compounds-total dust, as Mo: 5 mg/m3 MAK
German (DFG) - Peak Limitations
10 x normal MAK (30 min average value); don't exceed during shift
Israel - Time Weighted Averages
soluble compounds, as Mo: 5 mg/m3 TWA; insoluble compounds, as Mo: 10 mg/m3 TWA
Israel - Action Levels
soluble compounds, as Mo: 2.5 mg/m3 AL; insoluble compounds, as Mo: 5 mg/m3 AL
Mexico - Wastewater - Organic Toxic Pollutants and Heavy Metals
Listed under [Heavy Metals]
STATE LISTS
California - Directors List of Hazardous Substances (8 CCR 339)
[present]
Florida Hazardous Substance List
[present]
Massachusetts Right To Know List
[present]
Minnesota Hazardous Substance List
[present] as Mo (includes soluble and insoluble compounds)
Pennsylvania Right to Know List
[present]
PROPOSED REGULATIONS
Safe Drinking Water Act - Priority list
[present]

MOLYBDENUM OXIDE   18868-43-4

HEALTH AND SAFETY LISTS

U.S. DOT - Appendix A Table 2 - Radionuclides
final RQ = 100 curies (3.7E 12 Bq)

ENVIRONMENTAL LISTS

CERCLA/SARA List of Radionuclides (Appendix B) and Their Reportable Quantities
final RQ = 100 curies (3.7E 12 Bq)

MOLYBDENUM PENTACHLORIDE   10241-05-1

HEALTH AND SAFETY LISTS

U.S. DOT - Appendix A Table 2 - Radionuclides
final RQ = 100 curies (3.7E 12 Bq)

ENVIRONMENTAL LISTS

CERCLA/SARA List of Radionuclides (Appendix B) and Their Reportable Quantities
final RQ = 100 curies (3.7E 12 Bq)

MOLYBDENUM SOLUBLE COMPOUNDS   RR-00036-2

HEALTH AND SAFETY LISTS

U.S. DOT - Appendix A Table 2 - Radionuclides
final RQ = 10 curies (3.7E 11 Bq)

ENVIRONMENTAL LISTS

CERCLA/SARA List of Radionuclides (Appendix B) and Their Reportable Quantities
final RQ = 10 curies (3.7E 11 Bq)

MOLYBDENUM (IV) SULFIDE   1317-33-5

HEALTH AND SAFETY LISTS

U.S. DOT - Appendix A Table 2 - Radionuclides
final RQ = 100 curies (3.7E 12 Bq)

ENVIRONMENTAL LISTS

CERCLA/SARA List of Radionuclides (Appendix B) and Their Reportable Quantities
final RQ = 100 curies (3.7E 12 Bq)

MOLYBDENUM TRIOXIDE   1313-27-5

HEALTH AND SAFETY LISTS

U.S. DOT - Appendix A Table 2 - Radionuclides
final RQ = 1000 curies (3.7E 13 Bq)

ENVIRONMENTAL LISTS

CERCLA/SARA List of Radionuclides (Appendix B) and Their Reportable Quantities
final RQ = 1000 curies (3.7E 13 Bq)

MONOACRYLATE   RR-01001-5

INTERNATIONAL LISTS

Canada - WHMIS: Ingredient Disclosure
1% item 1091 (1160)

United Kingdom - Occupational Exposure Standards - TWAs
soluble compounds, as Mo: 5 mg/m3 TWA; insoluble compounds, as Mo: 10 mg/m3 TWA

United Kingdom - Occupational Exposure Standards - STELs
soluble compounds, as Mo: 10 mg/m3 STEL; insoluble compounds, as Mo: 20 mg/m3 STEL

STATE LISTS

California - Directors List of Hazardous Substances (8 CCR 339)
[present]

MONOAMMONIUM PHOSPHATE   7722-76-1

INTERNATIONAL LISTS

Canada - Alberta - 8 Hour Occupational Exposure Limit
as Mo: 10 mg/m3 TWA

Canada - Alberta - 15 Minute Occupational Exposure Limit
20 mg/m3 STEL

Canada - Quebec - Time-Weighted Average Exposure Values
as Mo: 10 mg/m3 TWAEV

Mexico - Instruction No. 10 - TWAs
10 mg/m3 TWA

Mexico - Instruction No. 10 - STELs
20 mg/m3 STEL

STATE LISTS

California - Exposure Limits - PELs
as Mo: 10 mg/m3 PEL

MONOBUTYLTIN TRIS (ISOOCTYL) MERCAPTO-   25852-70-4
ACETATE

INTERNATIONAL LISTS

Canada - Ontario - OHSA - TWAEVs
as Mo: 10 mg/m3 TWAEV

MONOCROTALINE   315-22-0

HEALTH AND SAFETY LISTS

U.S. DOT - Substances From 49 CFR 172.101
regulated by DOT (UN2508)

U.S. DOT - Hazard Classes
DOT hazard class = 8

STATE LISTS

NJ Right to Know List (Total)
sn 1311

NJ Special Hazardous Substances
(corrosive)

MONOCROTOPHOS   6923-22-4

HEALTH AND SAFETY LISTS

NIOSH 1990 - Pocket Guide - RELs
as Mo: See Appendix D

NIOSH 1990 - Pocket Guide - Target organs
respiratory system; in animals: kidneys, blood

INTERNATIONAL LISTS

Canada - Alberta - 8 Hour Occupational Exposure Limit
as Mo: 5 mg/m3 TWA

Canada - Alberta - 15 Minute Occupational Exposure Limit
as Mo: 10 mg/m3 STEL

Canada - British Columbia - 8 Hour Exposure Limits
as Mo: 5 mg/m3 TWA

Canada - British Columbia - 15 Minute Exposure Limits
as Mo: 10 mg/m3 STEL

Canada - Quebec - Time-Weighted Average Exposure Values
as Mo: 5 mg/m3 TWAEV

Mexico - Instruction No. 10 - TWAs
5 mg/m3 TWA

Mexico - Instruction No. 10 - STELs
10 mg/m3 STEL

STATE LISTS

California - Exposure Limits - PELs
as Mo: 5 mg/m3 PEL

MONODECYL ACID PHOSPHATE   3921-30-0
SEE ALSO:
MOLYBDENUM

ENVIRONMENTAL LISTS

List of Pesticide Product Inert Ingredients
[present]

INTERNATIONAL LISTS

Canada - Ontario - OHSA - TWAEVs
as Mo: 10 mg/m3 TWAEV

MONOETHANOLAMINE ALKYL(C8-C24) BENZENE SULFONATE   RR-01098-0

SEE ALSO:
MOLYBDENUM

HEALTH AND SAFETY LISTS

NIOSH - Selected LD50s and LC50s
Oral, rat: LD50 = 125 mg/kg

NTP Chemical Status Reports - Testing Status and NTIS Number
Two year studies: pathology quality assessment in progress; prechronic studies for which toxicity technical reports were not prepared

ENVIRONMENTAL LISTS

CERCLA/SARA - Section 313 - Emission Reporting
form R reporting required for 1.0% de minimus concentration

List of Pesticide Product Inert Ingredients
[present]

TSCA - Code of Federal Regulations Citations
40 CFR 712.30(w)

TSCA - PAIR - Reporting List
Reporting Date: July 13, 1988

INTERNATIONAL LISTS

Canada - WHMIS: Ingredient Disclosure
1% item 1093 (1694)

Canada - NPRI (National Pollutant Release Inventory)
[present]

Canada - Ontario - OHSA - TWAEVs
as Mo: 5 mg/m3 TWAEV

STATE LISTS

California - Air Bill 2588 Appendix A-I
6/91

Massachusetts Right To Know List
[present]

NJ Right to Know List (Total)
sn 1312

Pennsylvania Right to Know List
environmental hazard

MONOFLUOROPHOSPHORIC ACID   13537-32-1

ENVIRONMENTAL LISTS

TSCA - Chemicals with Significant New Use Rules
PMN number: P-85-415

MONOMETHOXY NEOPENTYL GLYCOL PROPOXYLATE MONOACRYLATE   RR-00980-3

ENVIRONMENTAL LISTS

List of Pesticide Product Inert Ingredients
[present]

MONOMETHYLTIN TRIS(ISOOOCTYL MERCAPTOACETATE)   54849-38-6

ENVIRONMENTAL LISTS

TSCA - Code of Federal Regulations Citations
40 CFR 712.30(g); 40 CFR 716.120(a)

TSCA - PAIR - Reporting List
Reporting Date: October 8, 1984

TSCA - Health and Safety Reporting List
Effective Date: January 3, 1983

MONONITRO-M-TOLUIDINE   RR-00604-2

HEALTH AND SAFETY LISTS

IARC - Group 2B (sufficient animal data)
[present] (Overall evaluation based only on evidence of carcinogenicity in monograph (10, 1976) or in Supplement 4)

NIOSH - Selected LD50s and LC50s
Oral, rat: LD50 = 66 mg/kg

OSHA - Possible Select Carcinogens
[present]

STATE LISTS

California - Air Bill 2588 Appendix A-II
known or potential carcinogen

California - Prop. 65 - Cancer list
carcinogen - initial date 4/1/88

California - Prop. 65 - No Significant Risk Levels
no significant risk level = 0.07 ug/day

California - Directors List of Hazardous Substances (8 CCR 339)
[present]

Florida Hazardous Substance List
[present]

Massachusetts Right To Know List
carcinogen; extraordinarily hazardous

Minnesota Hazardous Substance List
carcinogen

Pennsylvania Right to Know List
special hazardous substance

Pennsylvania RTK - Special Hazardous Substances
[present]

MONONITRO-O-TOLUIDINE   RR-00605-3

HEALTH AND SAFETY LISTS

ACGIH 1995 - Time Weighted Averages
0.25 mg/m3 TWA

ACGIH 1995 - Skin Designations
skin - potential for cutaneous absorption

U.S. DOT - Appendix B - Marine Pollutants
DOT regulated marine pollutant

NIOSH - Selected LD50s and LC50s
Inhalation, rat: LC50 = 63 mg/m3 4 hr Oral, rat: LD50 = 8 mg/kg Skin, rat: LD50 = 112 mg/kg

OSHA - Vacated PELs - Time Weighted Averages
0.25 mg/m3 TWA

ENVIRONMENTAL LISTS

CERCLA/SARA - Section 302 Extremely Hazardous Substances and TPQs
TPQ = 10/10,000 pounds

INTERNATIONAL LISTS

Australian Exposure Standards - Time Weighted Averages
0.25 mg/m3 TWA

Canada - Alberta - 8 Hour Occupational Exposure Limit
0.25 mg/m3 TWA

Canada - Alberta - 15 Minute Occupational Exposure Limit
0.75 mg/m3 STEL

Canada - British Columbia - 8 Hour Exposure Limits
0.25 mg/m3 TWA

Canada - Ontario - OHSA - TWAEVs
0.25 mg/m3 TWAEV

Canada - Quebec - Time-Weighted Average Exposure Values
0.25 mg/m3 TWAEV

Canada - Quebec - Skin Designations
absorbed through the skin

Israel - Time Weighted Averages
0.25 mg/m3 TWA

Israel - Action Levels
0.125 mg/m3 AL

Mexico - Instruction No. 10 - TWAs
0.25 mg/m3 TWA

Mexico - Instruction No. 10 - Skin designation
skin - potential for cutaneous absorption

STATE LISTS

California - Exposure Limits - PELs
0.25 mg/m3 PEL

California - Directors List of Hazardous Substances (8 CCR 339)
[present]

Florida Hazardous Substance List
[present]

Massachusetts Right To Know List
extraordinarily hazardous; neurotoxin

Minnesota Hazardous Substance List
[present]

NJ Right to Know List (Total)
sn 1313

Pennsylvania Right to Know List
environmental hazard

PROPOSED REGULATIONS

CERCLA/SARA - Proposed Hazardous Substance Additions
proposed RQ = 1 pound (.454 kg)

CERCLA/SARA - 1989 Proposed RQ Adjustments
proposed RQ = 10 pounds (4.54 kg)

MONONITRO-P-TOLUIDINE   RR-00606-4

ENVIRONMENTAL LISTS

List of Pesticide Product Inert Ingredients
[present]

MONOOCTYLTIN OXIDE   13356-20-2

ENVIRONMENTAL LISTS

List of Pesticide Product Inert Ingredients
[present]

MONOOCTYLTIN TRICHLORIDE   3091-25-6

INTERNATIONAL LISTS

Canada - WHMIS: Ingredient Disclosure
1% item 1097 (107)

STATE LISTS

NJ Right to Know List (Total)
sn 0940

NJ Special Hazardous Substances
(corrosive)

MONOOCTYLTIN TRIS(2-ETHYLHEXYL THIOGLY-COLATE)   27107-89-7

ENVIRONMENTAL LISTS

TSCA - Chemicals with Significant New Use Rules
PMN number: P-88-1690

MONOOCTYLTIN TRIS(ISOOCTYL THIOGLYCO-LATE)   26401-86-5

SEE ALSO:
ALKYLTIN COMPOUNDS

ENVIRONMENTAL LISTS

TSCA - Code of Federal Regulations Citations
40 CFR 712.30(g); 40 CFR 716.120

TSCA - PAIR - Reporting List
Reporting Date: October 8, 1984

MONOPOTASSIUM PEROXYMONOSULFATE   10058-23-8

INTERNATIONAL LISTS

Canada - WHMIS: Ingredient Disclosure
1% item 1098 (1171)

MONOSUBSTITUTED ALKOXYAMINOTRIZINES   RR-00162-7

INTERNATIONAL LISTS

Canada - WHMIS: Ingredient Disclosure
1% item 1099 (1172)

MONO-(TRICHLORO)TETRA(MONO-POTASSIUM DICHLORO)-PENTA-S-TRIAZINE-TRIONE   34651-95-1

INTERNATIONAL LISTS

Canada - WHMIS: Ingredient Disclosure
1% item 1100 (1173)

MONURON   150-68-5

INTERNATIONAL LISTS

German (DFG) - MAK Values
total dust: 0.1 mg/m3 MAK (Listed under 'Tin compounds, organic')

German (DFG) - Peak Limitations
2 x normal MAK (30 min. average value); don't exceed 4 times during shift (Listed under 'Tin compounds, organic')

German (DFG) - Skin/Sensitizers
danger of cutaneous absorption (Listed under 'Tin compounds, organic')

MOPP   RR-00053-3

INTERNATIONAL LISTS

German (DFG) - MAK Values
total dust: 0.1 mg/m3 MAK (Listed under 'Tin compounds, organic')

German (DFG) - Peak Limitations
2 x normal MAK (30 min. average value); don't exceed 4 times during shift (Listed under 'Tin compounds, organic')

German (DFG) - Skin/Sensitizers
danger of cutaneous absorption (Listed under 'Tin compounds, organic')

MORPHINAN-3,6-DIOL, 7,8-DIDEHYDRO-4,5-EPOXY-17-METHYL-(5A,6A)-, (Z)-9-OCTADECENOATE (SALT)   6033-05-2

INTERNATIONAL LISTS

German (DFG) - MAK Values
total dust: 0.1 mg/m3 MAK (Listed under 'Tin compounds, organic')

German (DFG) - Peak Limitations
2 x normal MAK (30 min. average value); don't exceed 4 times during shift (Listed under 'Tin compounds, organic')

German (DFG) - Skin/Sensitizers
danger of cutaneous absorption (Listed under 'Tin compounds, organic')

MORPHOLINE   110-91-8

INTERNATIONAL LISTS

German (DFG) - MAK Values
total dust: 0.1 mg/m3 MAK (Listed under 'Tin compounds, organic')

German (DFG) - Peak Limitations
2 x normal MAK (30 min. average value); don't exceed 4 times during shift (Listed under 'Tin compounds, organic')

German (DFG) - Skin/Sensitizers
danger of cutaneous absorption (Listed under 'Tin compounds, organic')

MORPHOLINE, 4-ACETYL-   1696-20-4

ENVIRONMENTAL LISTS

List of Pesticide Product Inert Ingredients
[present]

MORPHOLINE ETHANOL 622-40-2
ENVIRONMENTAL LISTS
TSCA - Chemicals with Significant New Use Rules
PMN numbers: P-86-1043; P-86-1044

MORPHOLINE OLEATE 1095-66-5
STATE LISTS
Massachusetts Right To Know List
[present]

Pennsylvania Right to Know List
[present]

5-(MORPHOLINOMETHYL)-3-[(5-NITRO-FUR-FURYLIDENE)-AMINO]-2-OXAZOLIDINONE 139-91-3
HEALTH AND SAFETY LISTS
IARC - Group 3 (not classifiable)
[present]

NIOSH - Selected LD50s and LC50s
Oral, rat: LD50 = 1053 mg/kg

NTP Chemical Status Reports - Testing Status and NTIS Number
Technical reports printed (PB89109615/AS)

NTP Chemical Status Reports - Evidence of Carcinogenicity
male rat-clear evidence; female rat-no evidence; male mice-no evidence; female mice-no evidence
ENVIRONMENTAL LISTS
CERCLA/SARA - Section 313 - Emission Reporting
form R reporting required
STATE LISTS
California - Directors List of Hazardous Substances (8 CCR 339)
[present]

5-MORPHOLINOMETHYL-3((5-NITROFUR-FURYLIDENE)AMINO)-2-OXAZOLIDINONE HYDROCHLORIDE 3759-92-0
HEALTH AND SAFETY LISTS
IARC - Group 1 (carcinogenic to humans)
[present] (Combined therapy with nitrogen mustard, vincristine, procarbazine and prednisone)
OSHA - Select Carcinogens
[present]
STATE LISTS
California - Directors List of Hazardous Substances (8 CCR 339)
[present]

Pennsylvania Right to Know List
special hazardous substance

Pennsylvania RTK - Special Hazardous Substances
[present]

5-(MORPHOLINOMETHYL)-3-((5-NITROFURFURYLI-DENE)AMINO)-2-OXAZOLIDINONE 3795-88-8
ENVIRONMENTAL LISTS
TSCA - Code of Federal Regulations Citations
40 CFR 712.30(d)

TSCA - PAIR - Reporting List
Reporting Date: November 19, 1982

DI-5-(MORPHOLINOMETHYL)-3((5-NITROFUR-FURYLIDENE)AMINO)-2-OXAZOLIDINONE HYDROCHLORIDE 13146-28-6
HEALTH AND SAFETY LISTS
ACGIH 1995 - Time Weighted Averages
20 ppm TWA; 71 mg/m3 TWA

ACGIH 1995 - Skin Designations
skin - potential for cutaneous absorption

AIHA - Odor Threshold Values
geometric mean air odor threshold = 0.011 ppm (detectable); 0.070 ppm (recognizable)

U.S. DOT - Substances From 49 CFR 172.101
regulated by DOT (UN2054)

U.S. DOT - Hazard Classes
DOT hazard class = 3

IARC - Group 3 (not classifiable)
[present]

NFPA - Flash Points
flash point = 98 degrees F (37 degrees C)

NFPA - Hazard Identification Ratings
health-3; flammability-3; reactivity-0

NIOSH - Selected LD50s and LC50s
Inhalation, rat: LC50 = 8000 ppm 8 hr Oral, rat: LD50 = 1050 mg/kg Skin, rabbit: LD50 = 500 mg/kg

NIOSH 1990 - Pocket Guide - RELs
20 ppm TWA; 70 mg/m3 TWA; 30 ppm STEL; 105 mg/m3 STEL

NIOSH 1990 - Pocket Guide - IDLHs
8000 ppm IDLH

NIOSH 1990 - Pocket Guide - Target organs
skin, eyes, respiratory system

NIOSH 1990 - Pocket Guide - Skin list
Potential for dermal absorption

OSHA - Vacated PELs - Time Weighted Averages
20 ppm TWA; 70 mg/m3 TWA

OSHA - Vacated PELs - Short Term Exposure Limits
30 ppm STEL; 105 mg/m3 STEL

OSHA - Vacated PELs - Skin Designation
Prevent or reduce skin absorption

OSHA - Final PELs - Time Weighted Averages
20 ppm TWA; 70 mg/m3 TWA

OSHA - Final PELs - Skin Notations
prevent or reduce skin absorption
ENVIRONMENTAL LISTS
EPA - Master Testing List
[present]

List of Pesticide Product Inert Ingredients
[present]
INTERNATIONAL LISTS
Australian Exposure Standards - Time Weighted Averages
20 ppm TWA; 71 mg/m3 TWA

Australian Exposure Standards - Short Term Exposure Limits
30 ppm STEL; 107 mg/m3 STEL

Australian Exposure Standards - Skin Effects
skin absorption

Canada - WHMIS: Ingredient Disclosure
1% item 1101 (1177)

Canada - Alberta - 8 Hour Occupational Exposure Limit
20 ppm TWA; 70 mg/m3 TWA

Canada - Alberta - 15 Minute Occupational Exposure Limit
30 ppm STEL; 105 mg/m3 STEL

Canada - Alberta - Skin Designation
can be absorbed through the intact skin

Canada - British Columbia - 8 Hour Exposure Limits
20 ppm TWA; 70 mg/m3 TWA

Canada - British Columbia - 15 Minute Exposure Limits
30 ppm STEL; 105 mg/m3 STEL

Canada - British Columbia - Skin Notations
skin - potential for skin absorption

Canada - Ontario - OHSA - TWAEVs
20 ppm TWAEV; 70 mg/m3 TWAEV

Canada - Ontario - OHSA - STEVs
30 ppm STEV; 105 mg/m3 STEV

Canada - Ontario - OHSA - Skin Notations
absorption through skin, eyes, or mucous membranes
Canada - Quebec - Time-Weighted Average Exposure Values
20 ppm TWAEV; 71 mg/m3 TWAEV
Canada - Quebec - Skin Designations
absorbed through the skin
United Kingdom - Occupational Exposure Standards - TWAs
20 ppm TWA; 70 mg/m3 TWA
United Kingdom - Occupational Exposure Standards - STELs
30 ppm STEL; 105 mg/m3 STEL
United Kingdom - Occupational Exposure Standards - Notes
can be absorbed through skin
German (DFG) - MAK Values
20 ppm MAK; 70 mg/m3 MAK
German (DFG) - Peak Limitations
2 x normal MAK (5 min momentary value); don't exceed 8 times during shift
German (DFG) - Skin/Sensitizers
danger of cutaneous absorption
Israel - Time Weighted Averages
(20) ppm TWA; (71) mg/m3 TWA
Israel - Short Term Exposure Limits
(30) ppm STEL; (107) mg/m3 STEL
Israel - Action Levels
10 ppm AL; 35.5 mg/m3 AL
Mexico - Instruction No. 10 - TWAs
20 ppm TWA; 70 mg/m3 TWA
Mexico - Instruction No. 10 - STELs
30 ppm STEL; 105 mg/m3 STEL
Mexico - Instruction No. 10 - Skin designation
skin - potential for cutaneous absorption

STATE LISTS

California - Exposure Limits - PELs
20 ppm PEL; 70 mg/m3 PEL
California - Exposure Limits - STELs
30 ppm STEL; 105 mg/m3 STEL
California - Exposure Limits - Skin Notation
material may be absorbed through the skin, eyes or mucous membrane
California - Directors List of Hazardous Substances (8 CCR 339)
[present]
Florida Hazardous Substance List
[present]
Massachusetts Right To Know List
[present]
Minnesota Hazardous Substance List
skin
NJ Right to Know List (Total)
sn 1315
NJ Special Hazardous Substances
(corrosive; flammable - third degree)
Pennsylvania Right to Know List
[present]

2-(4-MORPHOLINYLDITHIO)-BENZOTHIAZOLE   95-32-9

HEALTH AND SAFETY LISTS

NFPA - Flash Points
flash point = 235 degrees F (113 degrees F)
NFPA - Hazard Identification Ratings
health-2; flammability-1; reactivity-1

STATE LISTS

Florida Hazardous Substance List
[present]

Massachusetts Right To Know List
[present]
Pennsylvania Right to Know List
[present]

MORTMORILLONITE   1318-93-0

HEALTH AND SAFETY LISTS

NFPA - Flash Points
flash point = 210 degrees F (99 degrees C)
NFPA - Hazard Identification Ratings
health-2; flammability-1; reactivity-0

INTERNATIONAL LISTS

Canada - WHMIS: Ingredient Disclosure
1% item 1102 (1178)

STATE LISTS

Florida Hazardous Substance List
[present]
Massachusetts Right To Know List
[present]
Pennsylvania Right to Know List
[present]

MULTI SOURCE LEACHATE   RR-00071-5

ENVIRONMENTAL LISTS

List of Pesticide Product Inert Ingredients
[present]

MUSCIMOL   2763-96-4

STATE LISTS

California - Air Bill 2588 Appendix A-II
known or potential carcinogen
California - Prop. 65 - Cancer list
carcinogen - initial date 4/1/88
California - Directors List of Hazardous Substances (8 CCR 339)
[present]
Florida Hazardous Substance List
[present]
Massachusetts Right To Know List
carcinogen; extraordinarily hazardous
Pennsylvania Right to Know List
special hazardous substance
Pennsylvania RTK - Special Hazardous Substances
[present]

MUSK XYLENE   81-15-2

HEALTH AND SAFETY LISTS

IARC - Group 2B (sufficient animal data)
[present] (Overall evaluation based only on evidence of carcinogenicity in monograph (7, 1974) or in Supplement 4)
OSHA - Possible Select Carcinogens
[present]

MUSTARD GAS   505-60-2

STATE LISTS

Massachusetts Right To Know List
carcinogen; extraordinarily hazardous
Minnesota Hazardous Substance List
carcinogen

MYCLOBUTANIL [.ALPHA.-BUTYL-.ALPHA.-(4-CHLOROPHENYL)-1H-1,2,4-TRIAZOLE-1-PROPANETRILE]   88671-89-0

STATE LISTS

Massachusetts Right To Know List
carcinogen; extraordinarily hazardous

MYLAR   25038-59-9

HEALTH AND SAFETY LISTS

NIOSH - Selected LD50s and LC50s
Oral, rat: LD50 = 11500 mg/kg

ENVIRONMENTAL LISTS

TSCA - Code of Federal Regulations Citations
40 CFR 712.30(x); 40 CFR 716.120(d)

TSCA - PAIR - Reporting List
Reporting Date: November 27, 1991

TSCA - Health and Safety Reporting List
Effective Date: September 30, 1991

MYRCENE   123-35-3

ENVIRONMENTAL LISTS

List of Pesticide Product Inert Ingredients
[present]

MYRISTIC ACID   544-63-8

HEALTH AND SAFETY LISTS

U.S. DOT - Appendix A Table 1 - Hazardous Substances
final RQ = 1 pound (0.454 kg)

ENVIRONMENTAL LISTS

CERCLA/SARA - Hazardous Substances and their Reportable Quantities
final RQ = 1 pound (0.454 kg)

MYRISTIC ALCOHOL   112-72-1

HEALTH AND SAFETY LISTS

U.S. DOT - Appendix A Table 1 - Hazardous Substances
final RQ = 1000 pounds (454 kg)

NIOSH - Selected LD50s and LC50s
Oral, rat: LD50 = 45 mg/kg

ENVIRONMENTAL LISTS

CERCLA/SARA - Section 302 Extremely Hazardous Substances and TPQs
TPQ = 500/10,000 pounds

CERCLA/SARA - Hazardous Substances and their Reportable Quantities
final RQ = 1000 pounds (454 kg)

RCRA - P Series Wastes
waste number P007

RCRA - Hazardous Constituents-Appendix VIII
waste number P007

RCRA - Substances Banned From Land Disposal
[present]

TSCA - Code of Federal Regulations Citations
40 CFR 716.120(a)

TSCA - Health and Safety Reporting List
Effective Date: March 7, 1986

STATE LISTS

Florida Hazardous Substance List
effective March 13, 1992

Massachusetts Right To Know List
extraordinarily hazardous

NJ Right to Know List (Total)
sn 2571

Pennsylvania Right to Know List
environmental hazard

MYRISTIC ALDEHYDE   5780-07-4

HEALTH AND SAFETY LISTS

U.S. DOT - Substances From 49 CFR 172.101
regulated by DOT (UN2956)

U.S. DOT - Hazard Classes
DOT hazard class = 4.1

NAFENOPIN   3771-19-5

HEALTH AND SAFETY LISTS

U.S. DOT - Hazard Classes
Forbidden from transport by the DOT

IARC - Group 1 (carcinogenic to humans)
[present]

NIOSH - Selected LD50s and LC50s
Inhalation, rat: LC50 = 100 mg/m3 10 mn Skin, rat: LD50 = 5 mg/kg

NTP Seventh Report - Known Carcinogens
known carcinogen

OSHA - Select Carcinogens
[present]

ENVIRONMENTAL LISTS

CERCLA/SARA - Section 302 Extremely Hazardous Substances and TPQs
TPQ = 500 pounds

CERCLA/SARA - Section 313 - Emission Reporting
form R reporting required for 0.1% de minimus concentration

RCRA - Hazardous Constituents-Appendix VIII
hazardous constituent - no waste number

INTERNATIONAL LISTS

Canada - WHMIS: Ingredient Disclosure
0.1% item 1103 (1730)

German (DFG) - Carcinogens
proven carcinogen

STATE LISTS

California - Air Bill 2588 Appendix A-II
known or potential carcinogen

California - Prop. 65 - Cancer list
carcinogen - initial date 2/27/87

California - Directors List of Hazardous Substances (8 CCR 339)
[present]

Florida Hazardous Substance List
[present]

Massachusetts Right To Know List
carcinogen; extraordinarily hazardous

Minnesota Hazardous Substance List
carcinogen

NJ Right to Know List (Total)
sn 1319

NJ Special Hazardous Substances
(carcinogen; mutagen)

Pennsylvania Right to Know List
environmental hazard; special hazardous substance

Pennsylvania RTK - Special Hazardous Substances
[present]

PROPOSED REGULATIONS

CERCLA/SARA - Proposed Hazardous Substance Additions
proposed RQ = 1 pound (.454 kg)

CERCLA/SARA - 1989 Proposed RQ Adjustments
proposed RQ = 1 pounds (.454 kg)

NAFRELIN ACETATE   86220-42-0

ENVIRONMENTAL LISTS

CERCLA/SARA - Section 313 - Emission Reporting
form R reporting required

NALED   300-76-5

ENVIRONMENTAL LISTS

List of Pesticide Product Inert Ingredients
[present]

NALIDIXIC ACID   389-08-2

INTERNATIONAL LISTS

Canada - WHMIS: Ingredient Disclosure
0.1% item 1104 (1179)

NAPHTHA 49 DEGREE BE-COAL TAR TYPE   8030-31-7

ENVIRONMENTAL LISTS

List of Pesticide Product Inert Ingredients
[present]

INTERNATIONAL LISTS

Canada - WHMIS: Ingredient Disclosure
1% item 1105 (108)

5,12-NAPHTHACENEDIONE, 10-[(3-AMINO-2,3,6-TRIDEOXY-.ALPHA.-L-LYXO-HEXOPYRANOSYL)OXY]-7,8,9,10-TETRAHYDRO-6,8,11-TRIHYDROXY-8-(HYDROXYACETYL)-1-METHOXY-, HYDROCHLORIDE, (8S-CIS)-   25316-40-9

ENVIRONMENTAL LISTS

EPA - Master Testing List
[present]

INTERNATIONAL LISTS

Canada - WHMIS: Ingredient Disclosure
1% item 1106 (184)

NAPHTHA (COAL TAR)   65996-79-4

ENVIRONMENTAL LISTS

TSCA - Code of Federal Regulations Citations
40 CFR 712.30(x); 40 CFR 716.120(d)

TSCA - PAIR - Reporting List
Reporting Date: November 27, 1991

TSCA - Health and Safety Reporting List
Effective Date: September 30, 1991

INTERNATIONAL LISTS

Canada - WHMIS: Ingredient Disclosure
1% item 1107 (190)

NAPHTHALENE   91-20-3

HEALTH AND SAFETY LISTS

IARC - Group 2B (sufficient animal data)
[present] (Overall evaluation based only on evidence of carcinogenicity in monograph (24, 1980) or in Supplement 4)

OSHA - Possible Select Carcinogens
[present]

STATE LISTS

California - Air Bill 2588 Appendix A-II
known or potential carcinogen

California - Prop. 65 - Cancer list
carcinogen - initial date 4/1/88

California - Directors List of Hazardous Substances (8 CCR 339)
[present]

Florida Hazardous Substance List
[present]

Massachusetts Right To Know List
carcinogen; extraordinarily hazardous

Pennsylvania Right to Know List
special hazardous substance

Pennsylvania RTK - Special Hazardous Substances
[present]

NAPHTHALENE   RR-01304-7

STATE LISTS

California - Air Bill 2588 Appendix A-II
9/90

California - Prop. 65 - Developmental Toxicity
developmental toxicity - initial date 4/1/90

NAPHTHALENE SULFONIC ACID (A-)   85-47-2

HEALTH AND SAFETY LISTS

ACGIH 1995 - Time Weighted Averages
3 mg/m3 TWA

ACGIH 1995 - Skin Designations
skin - potential for cutaneous absorption

U.S. DOT - Appendix B - Marine Pollutants
DOT regulated marine pollutant

U.S. DOT - Appendix A Table 1 - Hazardous Substances
final RQ = 10 pounds (4.54 kg)

NIOSH - Selected LD50s and LC50s
Inhalation, rat: LC50 = 7700 ug/kg 8 hr Oral, rat: LD50 = 250 mg/kg Skin, rat: LD50 = 800 mg/kg

NIOSH 1990 - Pocket Guide - RELs
3 mg/m3 TWA

NIOSH 1990 - Pocket Guide - IDLHs
1800 mg/m3 IDLH

NIOSH 1990 - Pocket Guide - Target organs
respiratory system, CNS, CVS, skin, eyes, blood cholinesterase

NIOSH 1990 - Pocket Guide - Skin list
Potential for dermal absorption

OSHA - Vacated PELs - Time Weighted Averages
3 mg/m3 TWA

OSHA - Vacated PELs - Skin Designation
Prevent or reduce skin absorption

OSHA - Final PELs - Time Weighted Averages
3 mg/m3 TWA

ENVIRONMENTAL LISTS

ATSDR Priority List
Rank (of 275): 215

CERCLA/SARA - Section 313 - Emission Reporting
form R reporting required

CERCLA/SARA - Hazardous Substances and their Reportable Quantities
final RQ = 10 pounds (4.54 kg)

Clean Water Act - Hazardous Substances
[present]

INTERNATIONAL LISTS

Australian Exposure Standards - Time Weighted Averages
3 mg/m3 TWA

Australian Exposure Standards - Skin Effects
skin absorption

Canada - Alberta - 8 Hour Occupational Exposure Limit
3 mg/m3 TWA

Canada - Alberta - 15 Minute Occupational Exposure Limit
6 mg/m3 STEL

Canada - British Columbia - 8 Hour Exposure Limits
3 mg/m3 TWA

Canada - British Columbia - 15 Minute Exposure Limits
6 mg/m3 STEL

Canada - Ontario - OHSA - TWAEVs
3 mg/m3 TWAEV

Canada - Ontario - OHSA - Skin Notations
absorption through skin, eyes, or mucous membranes

Canada - Quebec - Time-Weighted Average Exposure Values
3 mg/m3 TWAEV

Canada - Quebec - Skin Designations
absorbed through the skin

United Kingdom - Occupational Exposure Standards - TWAs
3 mg/m3 TWA

United Kingdom - Occupational Exposure Standards - STELs
6 mg/m3 STEL

German (DFG) - MAK Values
total dust: 3 mg/m3 MAK

German (DFG) - Peak Limitations
10 x normal MAK (30 min average value); don't exceed during shift

Israel - Time Weighted Averages
3 mg/m3 TWA

Israel - Action Levels
1.5 mg/m3 AL

Mexico - Instruction No. 10 - TWAs
3 mg/m3 TWA

STATE LISTS

California - Exposure Limits - PELs
3 mg/m3 PEL

California - Exposure Limits - Skin Notation
material may be absorbed through the skin, eyes or mucous membrane

California - Directors List of Hazardous Substances (8 CCR 339)
[present]

Florida Hazardous Substance List
[present]

Massachusetts Right To Know List
neurotoxin

Minnesota Hazardous Substance List
[present]

NJ Right to Know List (Total)
sn 0751

Pennsylvania Right to Know List
environmental hazard

1-NAPHTHALENEACETIC ACID   86-87-3

HEALTH AND SAFETY LISTS

NIOSH - Selected LD50s and LC50s
Oral, rat: LD50 = 1160 mg/kg

NTP Chemical Status Reports - Testing Status and NTIS Number
Technical reports printed (PB90256389)

NTP Chemical Status Reports - Evidence of Carcinogenicity
male rat-clear evidence; female rat-clear evidence; male mice-equivocal evidence; female mice-no evidence

1-NAPHTHALENECARBOXALDEHYDE   66-77-3

HEALTH AND SAFETY LISTS

NFPA - Flash Points
flash point = 107 degrees F (42 degrees C)

NFPA - Hazard Identification Ratings
health-2; flammability-2; reactivity-0

STATE LISTS

California - Exposure Limits - PELs
100 ppm PEL; 400 mg/m3 PEL

California - Directors List of Hazardous Substances (8 CCR 339)
[present]

Massachusetts Right To Know List
[present]

Pennsylvania Right to Know List
[present]

NAPHTHALENE, 1-CHLORO-   90-13-1

STATE LISTS

Pennsylvania Right to Know List
special hazardous substance

Pennsylvania RTK - Special Hazardous Substances
[present]

NAPHTHALENE, CHLORO-   25586-43-0

STATE LISTS

Minnesota Hazardous Substance List
[present]

NJ Right to Know List (Total)
sn 0518

NAPHTHALENE, CHLORO DERIVATIVES   70776-03-3
SEE ALSO:
K001-HAZARDOUS WASTES
K087-HAZARDOUS WASTES
K035-HAZARDOUS WASTES
K060-HAZARDOUS WASTES
K147-HAZARDOUS WASTES
BIS(2,4-DIMETHYLBUTYL) MALEATE
F024-HAZARDOUS WASTES
F039-HAZARDOUS WASTES
F025-HAZARDOUS WASTES
K141-HAZARDOUS WASTES
K142-HAZARDOUS WASTES
K143-HAZARDOUS WASTES
K144-HAZARDOUS WASTES
K145-HAZARDOUS WASTES

HEALTH AND SAFETY LISTS

ACGIH 1995 - Time Weighted Averages
10 ppm TWA; 52 mg/m3 TWA

ACGIH 1995 - Short Term Exposure Limits
15 ppm STEL; 79 mg/m3 STEL

AIHA - Odor Threshold Values
geometric mean air odor threshold = 0.038 ppm (detectable)

U.S. DOT - Substances From 49 CFR 172.101
regulated by DOT (UN1334, 2304)

U.S. DOT - Hazard Classes
DOT hazard class = 4.1

U.S. DOT - Appendix A Table 1 - Hazardous Substances
final RQ = 100 pounds (45.4 kg)

NFPA - Flash Points
flash point = 174 degrees F (79 degrees C)

NFPA - Hazard Identification Ratings
health-2; flammability-2; reactivity-0

NIOSH - Selected LD50s and LC50s
Oral, rat: LD50 = 1250 mg/kg

NIOSH 1990 - Pocket Guide - RELs
10 ppm TWA; 50 mg/m3 TWA; 15 ppm STEL; 75 mg/m3 STEL

NIOSH 1990 - Pocket Guide - IDLHs
500 ppm IDLH

NIOSH 1990 - Pocket Guide - Target organs
eyes, blood, liver, kidneys, skin, RBC, CNS

NTP Chemical Status Reports - Testing Status and NTIS Number
Technical reports printed (PB92-224260/AS); approved for toxicology/carcinogenesis study

NTP Chemical Status Reports - Evidence of Carcinogenicity
male mice: no evidence; female mice: some evidence

OSHA - Vacated PELs - Time Weighted Averages
10 ppm TWA; 50 mg/m3 TWA

OSHA - Vacated PELs - Short Term Exposure Limits
15 ppm STEL; 75 mg/m3 STEL
OSHA - Final PELs - Time Weighted Averages
10 ppm TWA; 50 mg/m3 TWA

ENVIRONMENTAL LISTS

ATSDR Priority List
Rank (of 275): 070
CERCLA/SARA - Section 313 - Emission Reporting
form R reporting required for 1.0% de minimus concentration
CERCLA/SARA - Hazardous Substances and their Reportable Quantities
final RQ = 100 pounds (45.4 kg)
Clean Air Act (1990) - List of Hazardous Air Contaminants
[present]
CAA - HON Rule - SOCMI Chemicals
compliance by July 24, 1995
CAA - HON Rule - Organic HAPs
[present]
Clean Water Act - Hazardous Substances
[present]
Clean Water Act - Priority Pollutants
[present]
Clean Water Act - Toxic Pollutants
[present]
Safe Drinking Water Act - Monitoring
monitoring required at discretion of the state
List of Pesticide Product Inert Ingredients
[present]
RCRA - U Series Wastes
waste number U165
RCRA - Hazardous Constituents-Appendix VIII
waste number U165
RCRA - Basis for Listing - Appendix VII
Included in waste streams: F024, F025, F039, K001, K035, K060, K087, K141, K142, K143, K144, K145, K147
RCRA - Substances Banned From Land Disposal
[present]
RCRA - TSD Facilities Ground Water Monitoring
TM 8100 = 200 ug/L PQL; TM 8270 = 10 ug/L PQL
RCRA - Universal Treatment Standards (LDR)
WW: 0.059 mg/l; NWW: 5.6 mg/kg
TSCA - Code of Federal Regulations Citations
40 CFR 716.120(a)
TSCA - Health and Safety Reporting List
Effective Date: June 1, 1987

INTERNATIONAL LISTS

Australian Exposure Standards - Time Weighted Averages
10 ppm TWA; 52 mg/m3 TWA
Australian Exposure Standards - Short Term Exposure Limits
15 ppm STEL; 79 mg/m3 STEL
Canada - WHMIS: Ingredient Disclosure
1% item 1108 (1181)
Canada - NPRI (National Pollutant Release Inventory)
[present]
Canada - Alberta - 8 Hour Occupational Exposure Limit
10 ppm TWA; 52 mg/m3 TWA
Canada - Alberta - 15 Minute Occupational Exposure Limit
15 ppm STEL; 79 mg/m3 STEL
Canada - British Columbia - 8 Hour Exposure Limits
10 ppm TWA; 50 mg/m3 TWA
Canada - British Columbia - 15 Minute Exposure Limits
15 ppm STEL; 75 mg/m3 STEL
Canada - Ontario - OHSA - TWAEVs
10 ppm TWAEV; 52 mg/m3 TWAEV

Canada - Ontario - OHSA - STEVs
15 ppm STEV; 78 mg/m3 STEV
Canada - Quebec - Time-Weighted Average Exposure Values
10 ppm TWAEV; 52 mg/m3 TWAEV
Canada - Quebec - Short-term Exposure Values
15 ppm STEV; 79 mg/m3 STEV
United Kingdom - Occupational Exposure Standards - TWAs
10 ppm TWA; 50 mg/m3 TWA
United Kingdom - Occupational Exposure Standards - STELs
15 ppm STEL; 75 mg/m3 STEL
German (DFG) - MAK Values
10 ppm MAK; 50 mg/m3 MAK
Israel - Time Weighted Averages
10 ppm TWA; 52 mg/m3 TWA
Israel - Short Term Exposure Limits
15 ppm STEL; 79 mg/m3 STEL
Israel - Action Levels
5 ppm AL; 26 mg/m3 AL
Mexico - Instruction No. 10 - TWAs
10 ppm TWA; 50 mg/m3 TWA
Mexico - Instruction No. 10 - STELs
15 ppm STEL; 75 mg/m3 STEL
Mexico - Wastewater - Organic Toxic Pollutants and Heavy Metals
Listed under [Aromatic Hydrocarbons]
Mexico - Drinking Water - Ecological Criteria
None given.

STATE LISTS

California - Air Bill 2588 Appendix A-I
known or potential carcinogen
California - Exposure Limits - PELs
10 ppm PEL; 50 mg/m3 PEL
California - Exposure Limits - STELs
15 ppm STEL; 75 mg/m3 STEL
California - Directors List of Hazardous Substances (8 CCR 339)
[present]
Florida Hazardous Substance List
[present]
Massachusetts Right To Know List
[present]
Minnesota Hazardous Substance List
[present]
NJ Right to Know List (Total)
sn 1322
Pennsylvania Right to Know List
environmental hazard

PROPOSED REGULATIONS

Safe Drinking Water Act - Priority list
[present]

1,5-NAPHTHALENEDIAMINE   2243-62-1

HEALTH AND SAFETY LISTS
U.S. DOT - Appendix B - Marine Pollutants
DOT regulated marine pollutant

NAPHTHALENEDICARBOXYLIC ANHYDRIDE   81-84-5

ENVIRONMENTAL LISTS
CAA - HON Rule - SOCMI Chemicals
compliance by July 24, 1995

NAPHTHALENE, 1,5-DICHLORO-   1825-30-5

HEALTH AND SAFETY LISTS
NIOSH - Selected LD50s and LC50s
Oral, rat: LD50 = 1000 mg/kg

NAPHTHALENE, 1,4-DICHLORO-   1825-31-6

ENVIRONMENTAL LISTS

TSCA - Code of Federal Regulations Citations
40 CFR 712.30(x); 40 CFR 716.120(d)

TSCA - PAIR - Reporting List
Reporting Date: November 27, 1991

TSCA - Health and Safety Reporting List
Effective Date: September 30, 1991

NAPHTHALENE, 1,2-DICHLORO-   2050-69-3
SEE ALSO:
CHLORINATED NAPHTHALENES

HEALTH AND SAFETY LISTS

NFPA - Flash Points
flash point = 250 degrees F (121 degrees C)

NFPA - Hazard Identification Ratings
health-1; flammability-1; reactivity-0

NIOSH - Selected LD50s and LC50s
Oral, rat: LD50 = 1540 mg/kg

ENVIRONMENTAL LISTS

TSCA - Code of Federal Regulations Citations
40 CFR 704.83; 40 CFR 712.30(d); 40 CFR 716.120(c)

TSCA - PAIR - Reporting List
Reporting Date: November 19, 1982

NAPHTHALENE, 1,6-DICHLORO-   2050-72-8
SEE ALSO:
CHLORINATED NAPHTHALENES

ENVIRONMENTAL LISTS

CAA - HON Rule - SOCMI Chemicals
compliance by July 24, 1995

TSCA - Code of Federal Regulations Citations
40 CFR 704.83; 40 CFR 716.120(c)

INTERNATIONAL LISTS

Mexico - Wastewater - Organic Toxic Pollutants and Heavy Metals
Listed under [Organic Toxic Pollutants]

Mexico - Drinking Water - Ecological Criteria
None given.

NAPHTHALENE, 1,7-DICHLORO-   2050-73-9
SEE ALSO:
CHLORINATED NAPHTHALENES

ENVIRONMENTAL LISTS

Clean Water Act - Toxic Pollutants
[present]

RCRA - Hazardous Constituents-Appendix VIII
hazardous constituent - no waste number

TSCA - Code of Federal Regulations Citations
40 CFR 704.83; 40 CFR 716.120(c)

STATE LISTS

California - Directors List of Hazardous Substances (8 CCR 339)
[present]

Pennsylvania Right to Know List
environmental hazard

NAPHTHALENE, 1,8-DICHLORO-   2050-74-0

HEALTH AND SAFETY LISTS

IARC - Group 3 (not classifiable)
[present]

NTP Chemical Status Reports - Testing Status and NTIS Number
Technical reports printed (PB287646/AS)

NTP Chemical Status Reports - Evidence of Carcinogenicity
male rat-negative; female rat-positive; male mice-positive; female mice-positive

STATE LISTS

California - Directors List of Hazardous Substances (8 CCR 339)
[present]

Massachusetts Right To Know List
carcinogen; extraordinarily hazardous

NAPHTHALENE, 2,3-DICHLORO-   2050-75-1

ENVIRONMENTAL LISTS

TSCA - Code of Federal Regulations Citations
40 CFR 712.30(x); 40 CFR 716.120(d)

TSCA - PAIR - Reporting List
Reporting Date: November 27, 1991

TSCA - Health and Safety Reporting List
Effective Date: September 30, 1991

NAPHTHALENE, 2,6-DICHLORO-   2065-70-5

ENVIRONMENTAL LISTS

TSCA - Code of Federal Regulations Citations
40 CFR 704.83; 40 CFR 716.120(a)

TSCA - Health and Safety Reporting List
Effective Date: October 4, 1982

NAPHTHALENE, 1,3-DICHLORO-   2198-75-6

ENVIRONMENTAL LISTS

TSCA - Code of Federal Regulations Citations
40 CFR 704.83; 40 CFR 716.120(a)

TSCA - Health and Safety Reporting List
Effective Date: October 4, 1982

NAPHTHALENE, 2,7-DICHLORO-   2198-77-8

ENVIRONMENTAL LISTS

TSCA - Code of Federal Regulations Citations
40 CFR 704.83; 40 CFR 716.120(a)

TSCA - Health and Safety Reporting List
Effective Date: October 4, 1982

NAPHTHALENE DIISOCYANATE   25551-28-4

ENVIRONMENTAL LISTS

TSCA - Code of Federal Regulations Citations
40 CFR 704.83; 40 CFR 716.120(a)

TSCA - Health and Safety Reporting List
Effective Date: October 4, 1982

NAPHTHALENE, 1,5-DIISOCYANATO-   3173-72-6

ENVIRONMENTAL LISTS

TSCA - Code of Federal Regulations Citations
40 CFR 704.83; 40 CFR 716.120(a)

TSCA - Health and Safety Reporting List
Effective Date: October 4, 1982

1,4-NAPHTHALENEDIONE   130-15-4

ENVIRONMENTAL LISTS

TSCA - Code of Federal Regulations Citations
40 CFR 704.83; 40 CFR 716.120(a)

TSCA - Health and Safety Reporting List
Effective Date: October 4, 1982

NAPHTHALENE DIOZONIDE   RR-01448-2

ENVIRONMENTAL LISTS

TSCA - Code of Federal Regulations Citations
40 CFR 704.83; 40 CFR 716.120(a)

TSCA - Health and Safety Reporting List
Effective Date: October 4, 1982

1-NAPHTHALENEDISULFONIC ACID, 3-[[4'-[(6-AMINO-1-HYDROXY-3-SULFO-2-NAPHTHALENYL)AZO]-3,3'-DIMETHOXY[1,1'-BIPHENYL]-4-YL]AZO]-4-HYDROXY-, DISODIUM SALT   6449-35-0

ENVIRONMENTAL LISTS

TSCA - Code of Federal Regulations Citations
40 CFR 704.83; 40 CFR 716.120(a)

TSCA - Health and Safety Reporting List
Effective Date: October 4, 1982

NAPHTHALENE, HEPTACHLORO-   32241-08-0

ENVIRONMENTAL LISTS

TSCA - Code of Federal Regulations Citations
40 CFR 704.83; 40 CFR 716.120(a)

TSCA - Health and Safety Reporting List
Effective Date: October 4, 1982

NAPHTHALENE, PENTYL-   1320-27-0

ENVIRONMENTAL LISTS

TSCA - Code of Federal Regulations Citations
40 CFR 704.83; 40 CFR 716.120(a)

TSCA - Health and Safety Reporting List
Effective Date: October 4, 1982

2-NAPHTHALENESULFONIC ACID   120-18-3

HEALTH AND SAFETY LISTS

NIOSH - Health Standards - Exposure Limits
5 ppb TWA; 40 ug/m3 TWA; C (10 min) 20 ppb; C (10 min) 170 ug/m3 (Listed under 'Diisocyanates')

NIOSH - Health Standards - Health Effects and Precautions
Respiratory effects and sensitization, pulmonary irritation (Periodic chest X-ray and pulmonary function testing required) (Listed under 'Diisocyanates')

STATE LISTS

California - Exposure Limits - Ceilings
C 0.01 ppm; C 0.085 mg/m3

California - Directors List of Hazardous Substances (8 CCR 339)
[present]

Minnesota Hazardous Substance List
[present]

NJ Right to Know List (Total)
sn 1323

2-NAPHTHALENESULFONIC ACID, 6-AMINO-   93-00-5

HEALTH AND SAFETY LISTS

IARC - Group 3 (not classifiable)
[present]

ENVIRONMENTAL LISTS

CERCLA/SARA - Section 313 - Emission Reporting
form R reporting required; (Listed under 'Diisocyanates')

TSCA - Code of Federal Regulations Citations
40 CFR 716.120(a)

TSCA - Health and Safety Reporting List
Effective Date: June 1, 1987

INTERNATIONAL LISTS

German (DFG) - MAK Values
0.01 ppm MAK; 0.09 mg/m3 MAK

German (DFG) - Peak Limitations
2 x normal MAK (5 min momentary value); don't exceed 8 times during shift

German (DFG) - Skin/Sensitizers
danger of sensitization (skin or respiratory)

NAPHTHALENESULFONIC ACID, METHYL-, SODIUM SALT   26264-58-4
SEE ALSO:
K024-HAZARDOUS WASTES

HEALTH AND SAFETY LISTS

U.S. DOT - Appendix A Table 1 - Hazardous Substances
final RQ = 5000 pounds (2270 kg)

NIOSH - Selected LD50s and LC50s
Oral, rat: LD50 = 190 mg/kg

ENVIRONMENTAL LISTS

CERCLA/SARA - Hazardous Substances and their Reportable Quantities
final RQ = 5000 pounds (2270 kg)

RCRA - U Series Wastes
waste number U166

RCRA - Hazardous Constituents-Appendix VIII
waste number U166

RCRA - Basis for Listing - Appendix VII
Included in waste stream: K024

RCRA - Substances Banned From Land Disposal
[present]

RCRA - TSD Facilities Ground Water Monitoring
TM 8270 = 10 ug/L PQL

STATE LISTS

Massachusetts Right To Know List
[present]

Pennsylvania Right to Know List
environmental hazard

NAPHTHALENESULFONIC ACID, METHLENEBIS-, SODIUM SALT   26545-58-4

HEALTH AND SAFETY LISTS

U.S. DOT - Hazard Classes
Forbidden from transport by the DOT

NAPHTHALENESULFONIC ACID/PARAMETALDEHYDE CONDENSATE, AMMONIUM SALT   RR-01099-1
SEE ALSO:
BISAZOBIPHENYL DYES

ENVIRONMENTAL LISTS

TSCA - Code of Federal Regulations Citations
40 CFR 716.120(c)

NAPHTHALENESULFONIC ACID, POLYMER WITH FORMALDEHYDE   9017-33-8
SEE ALSO:
CHLORINATED NAPHTHALENES

ENVIRONMENTAL LISTS

TSCA - Code of Federal Regulations Citations
40 CFR 704.83; 40 CFR 716.120(c)

2-NAPHTHALENESULFONIC ACID, SODIUM SALT   532-02-5

STATE LISTS

Pennsylvania Right to Know List
[present]

1,4,5,8-NAPHTHALENE TETRACARBOXYLIC DIANHYDRIDE   81-30-1

HEALTH AND SAFETY LISTS

NIOSH - Selected LD50s and LC50s
Oral, rat: LD50 = 400 mg/kg

ENVIRONMENTAL LISTS
  CAA - HON Rule - SOCMI Chemicals
  compliance by July 24, 1995

NAPHTHALENE, 1,2,3,4-TETRAHYDRO(1-   RR-00165-0
PHENYLETHYL)
  ENVIRONMENTAL LISTS
    TSCA - Code of Federal Regulations Citations
    40 CFR 704.225(a)
    TSCA - CAIR - Reporting List
    reporting required by: manufacturer; importer

2-NAPHTHALENETHIOL   91-60-1
  ENVIRONMENTAL LISTS
    List of Pesticide Product Inert Ingredients
    [present]

2-NAPHTHALENOL, 1-[[2-METHYL-4-[(2-   85-83-6
METHYLPHENYL)AZO]PHENY
  ENVIRONMENTAL LISTS
    List of Pesticide Product Inert Ingredients
    [present]

NAPHTHA (PETROLEUM), HEAVY AROMATIC   64742-94-5
  ENVIRONMENTAL LISTS
    List of Pesticide Product Inert Ingredients
    [present]

NAPHTHA (PETROLEUM), LIGHT STEAM-   68527-23-1
CRACKED AROMATIC
  ENVIRONMENTAL LISTS
    List of Pesticide Product Inert Ingredients
    [present]

NAPHTHENIC ACID SOAP OF N-ALKYL (C16-C18)   RR-01100-7
TRIMETHYLENEDIAMINE
  ENVIRONMENTAL LISTS
    List of Pesticide Product Inert Ingredients
    [present]

NAPHTHENIC ACIDS, ZINC SALTS   12001-85-3
  HEALTH AND SAFETY LISTS
    NIOSH - Selected LD50s and LC50s
    Oral, rat: LD50 = 7400 mg/kg

NAPHTHENIC OILS (PETROLEUM), CATALYTIC   64742-68-3
DEWAXED HEAVY
  ENVIRONMENTAL LISTS
    TSCA - Chemicals with Significant New Use Rules
    PMN number: P-85-1331

NAPHTHENIC OILS (PETROLEUM), CATALYTIC   64742-69-4
DEWAXED LIGHT
  HEALTH AND SAFETY LISTS
    NIOSH - Selected LD50s and LC50s
    Oral, mouse: LD50 = 385 mg/kg

ALPHA-NAPHTHOL   90-15-3
  HEALTH AND SAFETY LISTS
    IARC - Group 3 (not classifiable)
    [present]

2-NAPHTHOL   135-19-3
  ENVIRONMENTAL LISTS
    List of Pesticide Product Inert Ingredients
    [present]

BETA-NAPHTHOL ETHYL ETHER   93-18-5
  ENVIRONMENTAL LISTS
    List of Pesticide Product Inert Ingredients
    [present]

NAPHTHOLSULFONIC ACID (1-)   567-18-0
  ENVIRONMENTAL LISTS
    List of Pesticide Product Inert Ingredients
    [present]

BETA-NAPHTHYLAMINE   91-59-8
  SEE ALSO:
    ZINC COMPOUNDS
    ZINC
    ZINC

  HEALTH AND SAFETY LISTS
    NIOSH - Selected LD50s and LC50s
    Oral, rat: LD50 = 4920 mg/kg

  ENVIRONMENTAL LISTS
    List of Pesticide Product Inert Ingredients
    [present]
    TSCA - Code of Federal Regulations Citations
    40 CFR 716.120(a)
    TSCA - Health and Safety Reporting List
    Effective Date: June 1, 1987

ALPHA-NAPHTHYLAMINE   134-32-7
  STATE LISTS
    Massachusetts Right To Know List
    carcinogen; extraordinarily hazardous

NAPHTHYLAMINE SULFONIC ACID (1,4-)   84-86-6
  STATE LISTS
    Massachusetts Right To Know List
    carcinogen; extraordinarily hazardous

1-NAPHTHYLAMINE-6-SULFONIC ACID   119-79-9
  HEALTH AND SAFETY LISTS
    NIOSH - Selected LD50s and LC50s
    Oral, rat: LD50 = 2400 mg/kg Skin, rabbit: LD50 = 880 mg/kg

  ENVIRONMENTAL LISTS
    CAA - HON Rule - SOCMI Chemicals
    compliance by July 24, 1995
    TSCA - Code of Federal Regulations Citations
    40 CFR 712.30(x); 40 CFR 716.120(d)
    TSCA - PAIR - Reporting List
    Reporting Date: November 27, 1991
    TSCA - Health and Safety Reporting List
    Effective Date: September 30, 1991

  INTERNATIONAL LISTS
    Canada - WHMIS: Ingredient Disclosure
    1% item 1109 (1184)

NAPHTHYL AMINEPERCHLORATE   RR-01449-3
  HEALTH AND SAFETY LISTS
    NFPA - Flash Points
    flash point = 307 degrees F (153 degrees C)
    NFPA - Hazard Identification Ratings
    flammability-1; reactivity-0
    NIOSH - Selected LD50s and LC50s
    Oral, rat: LD50 = 1960 mg/kg

ENVIRONMENTAL LISTS

CAA - HON Rule - SOCMI Chemicals
compliance by July 24, 1995

EPA - Master Testing List
[present]

List of Pesticide Product Inert Ingredients
[present]

INTERNATIONAL LISTS

Canada - WHMIS: Ingredient Disclosure
1% item 1110 (1185)

N-(1-NAPHTHYL)ETHYLENEDIAMINE   **1465-25-4**
DIHYDROCHLORIDE

HEALTH AND SAFETY LISTS

NIOSH - Selected LD50s and LC50s
Oral, rat: LD50 = 3110 mg/kg

INTERNATIONAL LISTS

Canada - WHMIS: Ingredient Disclosure
1% item 1111 (828)

1-NAPHTHYL METHYLNITROSOCARBONATE   **7090-25-7**

ENVIRONMENTAL LISTS

CAA - HON Rule - SOCMI Chemicals
compliance by Oct. 23, 1995

2-(1-NAPHTHYL)THIOACETAMIDE   **17518-47-7**
SEE ALSO:
F039-HAZARDOUS WASTES

HEALTH AND SAFETY LISTS

ACGIH 1995 - Carcinogens
A1-confirmed human carcinogen

U.S. DOT - Substances From 49 CFR 172.101
regulated by DOT (UN1650)

U.S. DOT - Hazard Classes
DOT hazard class = 6.1

U.S. DOT - Appendix A Table 1 - Hazardous Substances
final RQ = 10 pounds (4.54 kg)

IARC - Group 1 (carcinogenic to humans)
[present]

NIOSH - Selected LD50s and LC50s
Oral, rat: LD50 = 727 mg/kg

NIOSH 1990 - Pocket Guide - Carcinogens
occupational carcinogen

NIOSH 1990 - Pocket Guide - Target organs
bladder, skin

NIOSH - Health Standards - Exposure Limits
use 29 CFR 1910.1009

NIOSH - Health Standards - Health Effects and Precautions
Bladder cancer

NIOSH - Health Standards - Carcinogenic Chemicals
potential human carcinogen

NTP Seventh Report - Known Carcinogens
known carcinogen

OSHA - 29 CFR 1910 Specifically Regulated Chemicals
Cancer suspect agent (see 29 CFR 1910.1009)

OSHA - Select Carcinogens
[present]

ENVIRONMENTAL LISTS

CERCLA/SARA - Section 313 - Emission Reporting
form R reporting required for 0.1% de minimus concentration

CERCLA/SARA - Hazardous Substances and their Reportable Quantities
final RQ = 10 pounds (4.54 kg)

CAA - HON Rule - SOCMI Chemicals
compliance by Oct. 23, 1995

EPA - Carcinogen Hazard Ranking for RQ Adjustment
Hazard ranking = High

RCRA - U Series Wastes
waste number U168

RCRA - Hazardous Constituents-Appendix VIII
waste number U168

RCRA - Basis for Listing - Appendix VII
Included in waste stream: F039

RCRA - Substances Banned From Land Disposal
[present]

RCRA - TSD Facilities Ground Water Monitoring
TM 8270 = 10 ug/L PQL

RCRA - Universal Treatment Standards (LDR)
WW: 0.52 mg/l; NWW: Not applicable

INTERNATIONAL LISTS

Australian Exposure Standards - Time Weighted Averages
prohibition recommended

Australian Exposure Standards - Carcinogens
confirmed carcinogen

Canada - WHMIS: Ingredient Disclosure
0.1% item 1113 (1187)

Canada - Alberta - Designated Substances
designated substance - requires code of practice

Canada - British Columbia - 8 Hour Exposure Limits
carcinogen with no established permitted concentration

Canada - British Columbia - Carcinogens
carcinogen with no established permitted concentration

Canada - Quebec - Time-Weighted Average Exposure Values
substance of which the recirculation is prohibited

Canada - Quebec - Carcinogens
C1 carcinogen: effect detected in humans

German (DFG) - Skin/Sensitizers
danger of cutaneous absorption

German (DFG) - Carcinogens
proven carcinogen

Mexico - Instruction No. 10 - Carcinogens
carcinogen in humans

STATE LISTS

California - Air Bill 2588 Appendix A-II
known or potential carcinogen

California - Prop. 65 - Cancer list
carcinogen - initial date 2/27/87

California - Prop. 65 - No Significant Risk Levels
no significant risk level = 0.4 ug/day

California - Exposure Limits - Carcinogens
cancer-suspect agent (at a concentration >= 0.1%) (does not apply to materials in operations involving the destructive distillation of carbonaceous materials)

California - Directors List of Hazardous Substances (8 CCR 339)
[present] (refers to any mixture containing 0.1% or greater)

Florida Hazardous Substance List
[present]

Massachusetts Right To Know List
carcinogen; extraordinarily hazardous

Minnesota Hazardous Substance List
carcinogen

NJ Right to Know List (Total)
sn 1324

NJ Special Hazardous Substances
(carcinogen; mutagen)

Pennsylvania Right to Know List
environmental hazard; special hazardous substance

Pennsylvania RTK - Special Hazardous Substances
[present]

NAPHTHYLTHIOUREA   30553-04-9

HEALTH AND SAFETY LISTS

U.S. DOT - Substances From 49 CFR 172.101
regulated by DOT (UN2077)

U.S. DOT - Hazard Classes
DOT hazard class = 6.1

U.S. DOT - Appendix A Table 1 - Hazardous Substances
final RQ = 100 pounds (45.4 kg)

IARC - Group 3 (not classifiable)
[present]

NFPA - Flash Points
flash point = 315 degrees F (157 degrees C)

NFPA - Hazard Identification Ratings
health-2; flammability-1; reactivity-0

NIOSH - Selected LD50s and LC50s
Oral, rat: LD50 = 779 mg/kg

NIOSH 1990 - Pocket Guide - Carcinogens
occupational carcinogen

NIOSH 1990 - Pocket Guide - Target organs
bladder, skin

NIOSH - Health Standards - Exposure Limits
use 29 CFR 1910.1004

NIOSH - Health Standards - Health Effects and Precautions
Bladder cancer

NIOSH - Health Standards - Carcinogenic Chemicals
potential human carcinogen

OSHA - 29 CFR 1910 Specifically Regulated Chemicals
Cancer suspect agent (see 29 CFR 1910.1004)

OSHA - Select Carcinogens
[present]

ENVIRONMENTAL LISTS

CERCLA/SARA - Section 313 - Emission Reporting
form R reporting required for 0.1% de minimus concentration

CERCLA/SARA - Hazardous Substances and their Reportable Quantities
final RQ = 100 pounds (45.4 kg)

CAA - HON Rule - SOCMI Chemicals
compliance by Oct. 23, 1995

EPA - Carcinogen Hazard Ranking for RQ Adjustment
Hazard ranking = Low

RCRA - U Series Wastes
waste number U167

RCRA - Hazardous Constituents-Appendix VIII
waste number U167

RCRA - Substances Banned From Land Disposal
[present]

RCRA - TSD Facilities Ground Water Monitoring
TM 8270 = 10 ug/L PQL

INTERNATIONAL LISTS

Canada - WHMIS: Ingredient Disclosure
1% item 1112 (1186)

STATE LISTS

California - Air Bill 2588 Appendix A-II
known or potential carcinogen: 9/90

California - Prop. 65 - Cancer list
carcinogen - initial date 10/1/89

California - Exposure Limits - Carcinogens
cancer-suspect agent (at a concentration >= 1.0%) (does not apply to materials involved in the distillation of carbonaceous materials)

California - Directors List of Hazardous Substances (8 CCR 339)
[present]

Florida Hazardous Substance List
[present]

Massachusetts Right To Know List
carcinogen; extraordinarily hazardous

Minnesota Hazardous Substance List
carcinogen

NJ Right to Know List (Total)
sn 1325

NJ Special Hazardous Substances
(carcinogen; mutagen)

Pennsylvania Right to Know List
environmental hazard

1-NAPHTHYLUREA   6950-84-1

ENVIRONMENTAL LISTS

CAA - HON Rule - SOCMI Chemicals
compliance by Oct. 23, 1995

2-NAPHTHYLUREA   13114-62-0

INTERNATIONAL LISTS

Canada - WHMIS: Ingredient Disclosure
1% item 1114 (109)

NAPTHENIC ACID   1338-24-5

HEALTH AND SAFETY LISTS

U.S. DOT - Hazard Classes
Forbidden from transport by the DOT

NATURAL GAS   8006-14-2

HEALTH AND SAFETY LISTS

NTP Chemical Status Reports - Testing Status and NTIS Number
Technical reports printed (PB289733/AS)

NTP Chemical Status Reports - Evidence of Carcinogenicity
male rat-negative; female rat-negative; male mice-negative; female mice-negative

NATURAL RUBBERS   9006-04-6

INTERNATIONAL LISTS

Canada - WHMIS: Ingredient Disclosure
1% item 1115 (1146)

NATURAL YELLOW 10   117-39-5

INTERNATIONAL LISTS

Canada - WHMIS: Ingredient Disclosure
1% item 1116 (1188)

NAUGAUHITE   7786-17-6

HEALTH AND SAFETY LISTS

U.S. DOT - Substances From 49 CFR 172.101
regulated by DOT (UN1651)

U.S. DOT - Hazard Classes
DOT hazard class = 6.1

STATE LISTS

NJ Right to Know List (Total)
sn 1326

NEATSFOOT OIL   8002-64-0

HEALTH AND SAFETY LISTS

U.S. DOT - Substances From 49 CFR 172.101
regulated by DOT (UN1652)

U.S. DOT - Hazard Classes
DOT hazard class = 6.1

STATE LISTS

NJ Right to Know List (Total)
sn 1327

NEMALITE/BRUCITE   1317-43-7

STATE LISTS

NJ Right to Know List (Total)
sn 1328

NEODECANEPEROXOIC ACID, 1,1,3,3-TETRAM-   51240-95-0
ETHYLBUTYL ESTER

HEALTH AND SAFETY LISTS

U.S. DOT - Appendix B - Marine Pollutants
DOT regulated marine pollutant

U.S. DOT - Appendix A Table 1 - Hazardous Substances
final RQ = 100 pounds (45.4 kg)

NIOSH - Selected LD50s and LC50s
Oral, rat: LD50 = 3000 mg/kg

ENVIRONMENTAL LISTS

CERCLA/SARA - Hazardous Substances and their Reportable Quantities
final RQ = 100 pounds (45.4 kg)

Clean Water Act - Hazardous Substances
[present]

List of Pesticide Product Inert Ingredients
[present]

STATE LISTS

California - Directors List of Hazardous Substances (8 CCR 339)
[present]

Massachusetts Right To Know List
[present]

NJ Right to Know List (Total)
sn 2576

Pennsylvania Right to Know List
environmental hazard

NEODECANOIC ACID   26896-20-8

HEALTH AND SAFETY LISTS

NFPA - Hazard Identification Ratings
health-1; flammability-4; reactivity-0

STATE LISTS

Massachusetts Right To Know List
[present]

Pennsylvania Right to Know List
[present]

NEODECANOIC ACID, OXIRANYLMETHYL ESTER   26761-45-5

HEALTH AND SAFETY LISTS

U.S. DOT - Substances From 49 CFR 172.101
regulated by DOT (UN1287)

U.S. DOT - Hazard Classes
DOT hazard class = 3

ENVIRONMENTAL LISTS

List of Pesticide Product Inert Ingredients
[present]

INTERNATIONAL LISTS

Australian Exposure Standards - Under Review
fumes and dust: exposure limits under review

United Kingdom - Maximum Exposure Limits - TWAs
fume: 0.6 mg/m3 TWA

STATE LISTS

NJ Right to Know List (Total)
solution: sn 2755; scrap, powdered or granulated: sn 2754

NEODECANOIC ACID, ZIRCONIUM SALT   39049-04-2

HEALTH AND SAFETY LISTS

IARC - Group 3 (not classifiable)
[present]

NTP Chemical Status Reports - Testing Status and NTIS Number
Technical reports printed (PB83165761)

NTP Chemical Status Reports - Evidence of Carcinogenicity
male rat: some evidence; female rat: no evidence

INTERNATIONAL LISTS

Canada - WHMIS: Ingredient Disclosure
1% item 1117 (1061)

STATE LISTS

California - Directors List of Hazardous Substances (8 CCR 339)
[present]

NEODYMIUM 136   22095-52-9

HEALTH AND SAFETY LISTS

NIOSH - Selected LD50s and LC50s
Oral, rat: LD50 = 33 gm/kg

NEODYMIUM 138   15700-34-2

HEALTH AND SAFETY LISTS

NFPA - Flash Points
flash point = 470 degrees F (243 degrees C)

NFPA - Hazard Identification Ratings
health-0; flammability-1; reactivity-0

NEODYMIUM 139   18411-36-4

INTERNATIONAL LISTS

German (DFG) - Carcinogens
suspected carcinogen

NEODYMIUM 139M   RR-00401-3

ENVIRONMENTAL LISTS

TSCA - Chemicals with Significant New Use Rules
PMN number: P-89-764

TSCA - Section 12(b) - Export Notification
P-89-764; export notification required - Section 5

NEODYMIUM 141   14877-64-6

HEALTH AND SAFETY LISTS

NIOSH - Selected LD50s and LC50s
Oral, mammal: LD50 = 3400 mg/kg

NEODYMIUM 147   14269-74-0
SEE ALSO:
GLYCIDOL (OXIRANEMETHANOL) AND ITS DERIVATIVES

ENVIRONMENTAL LISTS

EPA - Master Testing List
[present]

TSCA - Code of Federal Regulations Citations
40 CFR 712.30(d); 40 CFR 716.120(c)

TSCA - PAIR - Reporting List
Reporting Date: November 19, 1982

PROPOSED REGULATIONS

TSCA - Proposed Testing Rule for Glycidyl Ethers
subject to screening subcategory testing (results apply to all members of Glycidyl subcategory VII-A)

NEODYMIUM 149   15749-81-2
  SEE ALSO:
    ZIRCONIUM
  ENVIRONMENTAL LISTS
    List of Pesticide Product Inert Ingredients
    [present]

NEODYMIUM 151   15690-82-1
  HEALTH AND SAFETY LISTS
    U.S. DOT - Appendix A Table 2 - Radionuclides
    final RQ = 1000 curies (3.7E 13 Bq)
  ENVIRONMENTAL LISTS
    CERCLA/SARA List of Radionuclides (Appendix B) and Their Re-
    portable Quantities
    final RQ = 1000 curies (3.7E 13 Bq)

NEODYMIUM CHLORIDE   10024-93-8
  HEALTH AND SAFETY LISTS
    U.S. DOT - Appendix A Table 2 - Radionuclides
    final RQ = 1000 curies (3.7E 13 Bq)
  ENVIRONMENTAL LISTS
    CERCLA/SARA List of Radionuclides (Appendix B) and Their Re-
    portable Quantities
    final RQ = 1000 curies (3.7E 13 Bq)

NEOHEXANE   75-83-2
  HEALTH AND SAFETY LISTS
    U.S. DOT - Appendix A Table 2 - Radionuclides
    final RQ = 1000 curies (3.7E 13 Bq)
  ENVIRONMENTAL LISTS
    CERCLA/SARA List of Radionuclides (Appendix B) and Their Re-
    portable Quantities
    final RQ = 1000 curies (3.7E 13 Bq)

NEOMYCIN SULFATE   1405-10-3
  HEALTH AND SAFETY LISTS
    U.S. DOT - Appendix A Table 2 - Radionuclides
    final RQ = 100 curies (3.7E 12 Bq)
  ENVIRONMENTAL LISTS
    CERCLA/SARA List of Radionuclides (Appendix B) and Their Re-
    portable Quantities
    final RQ = 100 curies (3.7E 12 Bq)

NEON   7440-01-9
  HEALTH AND SAFETY LISTS
    U.S. DOT - Appendix A Table 2 - Radionuclides
    final RQ = 1000 curies (3.7E 13 Bq)
  ENVIRONMENTAL LISTS
    CERCLA/SARA List of Radionuclides (Appendix B) and Their Re-
    portable Quantities
    final RQ = 1000 curies (3.7E 13 Bq)

NEONONANOIC ACID, ETHENYL ESTER   RR-01651-3
  HEALTH AND SAFETY LISTS
    U.S. DOT - Appendix A Table 2 - Radionuclides
    final RQ = 10 curies (3.7E 11 Bq)
  ENVIRONMENTAL LISTS
    CERCLA/SARA List of Radionuclides (Appendix B) and Their Re-
    portable Quantities
    final RQ = 10 curies (3.7E 11 Bq)

NEOPENTYL GLYCOL DIACRYLATE   2223-82-7
  HEALTH AND SAFETY LISTS
    U.S. DOT - Appendix A Table 2 - Radionuclides
    final RQ = 100 curies (3.7E 12 Bq)
  ENVIRONMENTAL LISTS
    CERCLA/SARA List of Radionuclides (Appendix B) and Their Re-
    portable Quantities
    final RQ = 100 curies (3.7E 12 Bq)

NEOPRENE RUBBER   9010-98-4
  HEALTH AND SAFETY LISTS
    U.S. DOT - Appendix A Table 2 - Radionuclides
    final RQ = 1000 curies (3.7E 13 Bq)
  ENVIRONMENTAL LISTS
    CERCLA/SARA List of Radionuclides (Appendix B) and Their Re-
    portable Quantities
    final RQ = 1000 curies (3.7E 13 Bq)

NEOPYNAMIN   7696-12-0
  HEALTH AND SAFETY LISTS
    NIOSH - Selected LD50s and LC50s
    Oral, mouse: LD50 = 3692 mg/kg

NEPTUNIUM 232   29687-52-3
  HEALTH AND SAFETY LISTS
    NFPA - Flash Points
    flash point = -54 degrees F (-48 degrees C)
    NFPA - Hazard Identification Ratings
    health-1; flammability-3; reactivity-0
  INTERNATIONAL LISTS
    Canada - WHMIS: Ingredient Disclosure
    1% item 1118 (1189)
  STATE LISTS
    Florida Hazardous Substance List
    [present]
    Massachusetts Right To Know List
    [present]
    Minnesota Hazardous Substance List
    [present]
    NJ Right to Know List (Total)
    sn 1335
    NJ Special Hazardous Substances
    (flammable - third degree)
    Pennsylvania Right to Know List
    [present]

NEPTUNIUM 233   15832-46-9
  STATE LISTS
    California - Air Bill 2588 Appendix A-II
    9/90
    California - Prop. 65 - Developmental Toxicity
    internal use: developmental toxicity - initial date 10/1/92

NEPTUNIUM 234   15116-90-2
  HEALTH AND SAFETY LISTS
    ACGIH 1995 - Time Weighted Averages
    simple asphyxiant
    U.S. DOT - Substances From 49 CFR 172.101
    regulated by DOT (UN1913, UN1065)
    U.S. DOT - Hazard Classes
    DOT hazard class = 2.2

INTERNATIONAL LISTS
  Australian Exposure Standards - Time Weighted Averages
    Asphyxiant at < 18% oxygen by volume; explosion hazard
  Canada - British Columbia - 8 Hour Exposure Limits
    asphyxiant substance
  Canada - Ontario - OHSA - TWAEVs
    simple asphyxiant
  Canada - Quebec - Time-Weighted Average Exposure Values
    simple asphyxiant
  Israel - Time Weighted Averages
    Asphyxiant
  Mexico - Instruction No. 10 - TWAs
    simple asphyxiant
STATE LISTS
  California - Exposure Limits - PELs
    asphyxiant (limit depends on level of oxygen)
  Florida Hazardous Substance List
    [present]
  Massachusetts Right To Know List
    [present]
  Minnesota Hazardous Substance List
    [present]
  NJ Right to Know List (Total)
    sn 1336
  Pennsylvania Right to Know List
    [present]

NEPTUNIUM 235   15700-37-5
  ENVIRONMENTAL LISTS
    TSCA - Chemicals with Significant New Use Rules
      PMN number: P-92-129

NEPTUNIUM 236   15700-36-4
  ENVIRONMENTAL LISTS
    TSCA - Code of Federal Regulations Citations
      40 CFR 712.30(d)
    TSCA - PAIR - Reporting List
      Reporting Date: November 19, 1982
  INTERNATIONAL LISTS
    Canada - WHMIS: Ingredient Disclosure
      1% item 1119 (620)

NEPTUNIUM 237   13994-20-2
  HEALTH AND SAFETY LISTS
    IARC - Group 3 (not classifiable)
      [present]

NEPTUNIUM 238   15766-25-3
  HEALTH AND SAFETY LISTS
    NIOSH - Selected LD50s and LC50s
      Oral, rat: LD50 = 20000 mg/kg
  ENVIRONMENTAL LISTS
    CERCLA/SARA - Section 313 - Emission Reporting
      form R reporting required

NEPTUNIUM 239   13968-59-7
  HEALTH AND SAFETY LISTS
    U.S. DOT - Appendix A Table 2 - Radionuclides
      final RQ = 1000 curies (3.7E 13 Bq)
  ENVIRONMENTAL LISTS
    CERCLA/SARA List of Radionuclides (Appendix B) and Their Reportable Quantities
      final RQ = 1000 curies (3.7E 13 Bq)

NEPTUNIUM 240   15690-84-3
  HEALTH AND SAFETY LISTS
    U.S. DOT - Appendix A Table 2 - Radionuclides
      final RQ = 1000 curies (3.7E 13 Bq)
  ENVIRONMENTAL LISTS
    CERCLA/SARA List of Radionuclides (Appendix B) and Their Reportable Quantities
      final RQ = 1000 curies (3.7E 13 Bq)

NETILMICIN SULFATE   56391-57-2
  HEALTH AND SAFETY LISTS
    U.S. DOT - Appendix A Table 2 - Radionuclides
      final RQ = 10 curies (3.7E 11 Bq)
  ENVIRONMENTAL LISTS
    CERCLA/SARA List of Radionuclides (Appendix B) and Their Reportable Quantities
      final RQ = 10 curies (3.7E 11 Bq)

NIAX A 1   3033-62-3
  HEALTH AND SAFETY LISTS
    U.S. DOT - Appendix A Table 2 - Radionuclides
      final RQ = 1000 curies (3.7E 13 Bq)
  ENVIRONMENTAL LISTS
    CERCLA/SARA List of Radionuclides (Appendix B) and Their Reportable Quantities
      final RQ = 1000 curies (3.7E 13 Bq)

NIAX CATALYST ESN   62765-43-9
  HEALTH AND SAFETY LISTS
    U.S. DOT - Appendix A Table 2 - Radionuclides
      120,000 year half-life: Final RQ = 0.1 curies (3.7E 9 Bq); 22.5 hour half-life: Final RQ = 100 curies (3.7E 12 Bq)
  ENVIRONMENTAL LISTS
    CERCLA/SARA List of Radionuclides (Appendix B) and Their Reportable Quantities
      120,000 year half-life: Final RQ = 0.1 curies (3.7E 9 Bq); 22.5 hour half-life: Final RQ = 100 curies (3.7E 12 Bq)

NIAX CATALYST ESN   62765-93-9
  HEALTH AND SAFETY LISTS
    U.S. DOT - Appendix A Table 2 - Radionuclides
      final RQ = 0.01 curies (3.7E 8 Bq)
  ENVIRONMENTAL LISTS
    CERCLA/SARA List of Radionuclides (Appendix B) and Their Reportable Quantities
      final RQ = 0.01 curies (3.7E 8 Bq)

NICKEL   7440-02-0
  HEALTH AND SAFETY LISTS
    U.S. DOT - Appendix A Table 2 - Radionuclides
      final RQ = 10 curies (3.7E 11 Bq)
  ENVIRONMENTAL LISTS
    CERCLA/SARA List of Radionuclides (Appendix B) and Their Reportable Quantities
      final RQ = 10 curies (3.7E 11 Bq)

NICKEL 56   14932-64-0
  HEALTH AND SAFETY LISTS
    U.S. DOT - Appendix A Table 2 - Radionuclides
      final RQ = 100 curies (3.7E 12 Bq)

ENVIRONMENTAL LISTS

CERCLA/SARA List of Radionuclides (Appendix B) and Their Reportable Quantities
final RQ = 100 curies (3.7E 12 Bq)

NICKEL 57   13981-99-2

HEALTH AND SAFETY LISTS

U.S. DOT - Appendix A Table 2 - Radionuclides
final RQ = 100 curies (3.7E 12 Bq)

ENVIRONMENTAL LISTS

CERCLA/SARA List of Radionuclides (Appendix B) and Their Reportable Quantities
final RQ = 100 curies (3.7E 12 Bq)

NICKEL 59   14336-70-0

STATE LISTS

California - Air Bill 2588 Appendix A-II
9/90

California - Prop. 65 - Developmental Toxicity
developmental toxicity - initial date 7/1/90

NICKEL 63   13981-37-8

HEALTH AND SAFETY LISTS

NIOSH - Selected LD50s and LC50s
Oral, rat: LD50 = 1070 mg/kg Skin, rabbit: LD50 = 280 mg/kg

NICKEL 65   14833-49-9

HEALTH AND SAFETY LISTS

NIOSH - Health Standards - Exposure Limits
minimize exposure to Niax Catalyst ESN and its components (Dimethylaminopropionitrile and Bis(2-(di-methylamino) ethyl ether)
NIOSH - Health Standards - Health Effects and Precautions
Urological disorders; nervous system effects (Use work practice and engineering controls to reduce exposure)

NICKEL 66   15766-33-3

STATE LISTS

Minnesota Hazardous Substance List
[present]

NICKEL(II) ACETATE   373-02-4
SEE ALSO:
F039-HAZARDOUS WASTES
F006-HAZARDOUS WASTES
NICKEL COMPOUNDS

HEALTH AND SAFETY LISTS

ACGIH 1995 - Time Weighted Averages
metal: (1) mg/m3 TWA; insoluble compounds as Ni: (1) mg/m3 TWA; soluble compounds as Ni: (0.1) mg/m3 TWA
U.S. DOT - Appendix A Table 1 - Hazardous Substances
final RQ = 100 pounds (45.4 kg) (no reporting of releases of this hazardous substance is required if the diameter of the pieces of the solid metal released is equal to or exceeds 0.004 inches)
IARC - Group 2B (sufficient animal data)
[present]
NIOSH 1990 - Pocket Guide - RELs
as Ni: 0.015 mg/m3 TWA
NIOSH 1990 - Pocket Guide - Carcinogens
occupational carcinogen
NIOSH 1990 - Pocket Guide - Target organs
lungs, paranasal sinus, CNS
NTP Seventh Report - Suspect Carcinogens
suspect carcinogen (Listed under 'Nickel and certain nickel compounds')

OSHA - Vacated PELs - Time Weighted Averages
metal and insoluble compounds, as Ni: 1 mg/m3 TWA; soluble compounds, as Ni: 0.1 mg/m3 TWA
OSHA - Final PELs - Time Weighted Averages
metal and insoluble compounds, as Ni: 1 mg/m3 TWA; soluble compounds, as Ni: 1 mg/m3 TWA
OSHA - Possible Select Carcinogens
[present]

ENVIRONMENTAL LISTS

ATSDR Priority List
Rank (of 275): 047
CERCLA/SARA - Section 313 - Emission Reporting
form R reporting required for 0.1% de minimus concentration
CERCLA/SARA - Hazardous Substances and their Reportable Quantities
final RQ = 100 pounds (45.4 kg) (no reporting of releases of this hazardous substance is required if the diameter of the pieces of the solid metal released is equal to or exceeds 0.004 inches)
Clean Water Act - Priority Pollutants
[present]
Clean Water Act - Toxic Pollutants
[present] (Listed under 'Nickel and compounds')
Safe Drinking Water Act - MCLs
MCL = 0.1 mg/L
Safe Drinking Water Act - MCLGs
MCLG = 0.1 mg/L
EPA - Carcinogen Hazard Ranking for RQ Adjustment
Hazard ranking = Low
RCRA - Hazardous Constituents-Appendix VIII
hazardous constituent - no waste number
RCRA - Basis for Listing - Appendix VII
Included in waste streams: F006, F039
RCRA - TSD Facilities Ground Water Monitoring
TM 6010 = 50 ug/L PQL; TM 7520 = 400 ug/L PQL (all species in the ground water that contain this element are included)
RCRA - Universal Treatment Standards (LDR)
WW: 3.98 mg/l; NWW: 5.0 mg/l TCLP

INTERNATIONAL LISTS

Australian Exposure Standards - Time Weighted Averages
metal: 1 mg/m3 TWA; soluble compounds as Ni: 0.1 mg/m3 TWA
Australian Exposure Standards - Skin Effects
metal and soluble compounds, as Ni: sensitiser
Australian Exposure Standards - Under Review
exposure limits under review; soluble compounds as Ni: exposure limits under review
Canada - WHMIS: Ingredient Disclosure
0.1% item 1126 (1193)
Canada - NPRI (National Pollutant Release Inventory)
[present]
Canada - CEPA - Priority Substances List
estimated time for completion of assessment reports: 4 years
Canada - Alberta - 8 Hour Occupational Exposure Limit
1 mg/m3 TWA
Canada - Alberta - 15 Minute Occupational Exposure Limit
2 mg/m3 STEL
Canada - British Columbia - 8 Hour Exposure Limits
as Ni: 0.05 mg/m3 TWA
Canada - British Columbia - Skin Notations
skin - potential for skin absorption
Canada - British Columbia - Carcinogens
carcinogen - 0.05 mg/m3 TWA
Canada - Ontario - OHSA - TWAEVs
as Ni: 1 mg/m3 TWAEV
Canada - Quebec - Time-Weighted Average Exposure Values
1 mg/m3 TWAEV

United Kingdom - Maximum Exposure Limits - TWAs
0.5 mg/m3 TWA

United Kingdom - Occupational Exposure Standards - TWAs
organic compounds, as Ni: 1 mg/m3 TWA

United Kingdom - Occupational Exposure Standards - STELs
inorganic compounds, as Ni: 3 mg/m3 STEL

German (DFG) - Skin/Sensitizers
danger of sensitization (skin or respiratory) (as inspirable dust/aerosols from nickel metal, nickel sulfide and sulfidic ores, nickel oxide and nickel carbonate arising in production and processing)

German (DFG) - Carcinogens
proven carcinogen (as inspirable dust/aerosols from nickel metal, nickel sulfide and sulfidic ores, nickel oxide and nickel carbonate arising in production and processing)

Israel - Time Weighted Averages
metal: (1) mg/m3 TWA; Insoluble compounds, as Ni: (1) mg/m3 TWA; soluble compounds, as Ni: (0.1) mg/m3 TWA

Israel - Action Levels
metal: .5 mg/m3 AL; insoluble compounds, as Ni: .5 mg/m3 AL; soluble compounds, as Ni: 0.05 mg/m3 AL

Mexico - Instruction No. 10 - TWAs
1 mg/m3 TWA

Mexico - Wastewater - Organic Toxic Pollutants and Heavy Metals
Listed under [Heavy Metals]

Mexico - Drinking Water - Ecological Criteria
0.01 mg/l

STATE LISTS

California - Air Bill 2588 Appendix A-I
known or potential carcinogen

California - Exposure Limits - PELs
metal, as Ni: 1 mg/m3 PEL; insoluble compounds, as Ni: 1 mg/m3 PEL; soluble compounds, as Ni: 0.1 mg/m3 PEL

California - Directors List of Hazardous Substances (8 CCR 339)
[present]

Florida Hazardous Substance List
[present]

Massachusetts Right To Know List
carcinogen; extraordinarily hazardous

Minnesota Hazardous Substance List
as Ni: carcinogen

NJ Right to Know List (Total)
sn 1341

NJ Special Hazardous Substances
(carcinogen)

Pennsylvania Right to Know List
environmental hazard; special hazardous substance (any compound of this substance is also an environmental hazard)

Pennsylvania RTK - Special Hazardous Substances
[present]

PROPOSED REGULATIONS

ACGIH 1995 - Notice of Intended Changes
as Ni: 0.05 mg/m3 TWA; A1-confirmed human carcinogen

Canada - Ontario - Proposed Occupational TWAEVs
as Ni: 0.5 mg/m3 TWAEV

NICKEL ACRYLATE COMPLEX   RR-01007-1

HEALTH AND SAFETY LISTS

U.S. DOT - Appendix A Table 2 - Radionuclides
final RQ = 10 curies (3.7E 11 Bq)

ENVIRONMENTAL LISTS

CERCLA/SARA List of Radionuclides (Appendix B) and Their Reportable Quantities
final RQ = 10 curies (3.7E 11 Bq)

NICKEL AMMONIUM SULFATE   15699-18-0

HEALTH AND SAFETY LISTS

U.S. DOT - Appendix A Table 2 - Radionuclides
final RQ = 10 curies (3.7E 11 Bq)

ENVIRONMENTAL LISTS

CERCLA/SARA List of Radionuclides (Appendix B) and Their Reportable Quantities
final RQ = 10 curies (3.7E 11 Bq)

NICKELATE(6-), [[[1,2-ETHANEDIYLBIS[NITRILOBIS(METHYLENE)]TETRAKIS[PHOSPHONATO](8-)], PENTAAMMONIUM HYDROGEN, (OC-6-21)-   68958-86-1

HEALTH AND SAFETY LISTS

U.S. DOT - Appendix A Table 2 - Radionuclides
final RQ = 100 curies (3.7E 12 Bq)

ENVIRONMENTAL LISTS

CERCLA/SARA List of Radionuclides (Appendix B) and Their Reportable Quantities
final RQ = 100 curies (3.7E 12 Bq)

NICKELATE(6-), [[[1,2-ETHANEDIYLBIS[NITRILOBIS(METHYLENE)]TETRAKIS[PHOSPHONATO](8-)], PENTAPOTASSIUM HYDROGEN, (OC-6-21)-   68958-87-2

HEALTH AND SAFETY LISTS

U.S. DOT - Appendix A Table 2 - Radionuclides
final RQ = 100 curies (3.7E 12 Bq)

ENVIRONMENTAL LISTS

CERCLA/SARA List of Radionuclides (Appendix B) and Their Reportable Quantities
final RQ = 100 curies (3.7E 12 Bq)

NICKELATE(6-), [[[1,2-ETHANEDIYLBIS[NITRILOBIS(METHYLENE)]TETRAKIS[PHOSPHONATO](8-)], PENTASODIUM HYDROGEN, (OC-6-21)-   68958-88-3

HEALTH AND SAFETY LISTS

U.S. DOT - Appendix A Table 2 - Radionuclides
final RQ = 100 curies (3.7E 12 Bq)

ENVIRONMENTAL LISTS

CERCLA/SARA List of Radionuclides (Appendix B) and Their Reportable Quantities
final RQ = 100 curies (3.7E 12 Bq)

NICKEL CARBONATE   3333-67-3

HEALTH AND SAFETY LISTS

U.S. DOT - Appendix A Table 2 - Radionuclides
final RQ = 10 curies (3.7E 11 Bq)

ENVIRONMENTAL LISTS

CERCLA/SARA List of Radionuclides (Appendix B) and Their Reportable Quantities
final RQ = 10 curies (3.7E 11 Bq)

NICKEL CARBONYL   13463-39-3
SEE ALSO:
NICKEL

HEALTH AND SAFETY LISTS

NIOSH - Selected LD50s and LC50s
Oral, rat: LD50 = 350 mg/kg

NTP Seventh Report - Suspect Carcinogens
suspect carcinogen (Listed under 'Nickel and certain nickel compounds')

OSHA - Possible Select Carcinogens
[present]

INTERNATIONAL LISTS

Canada - WHMIS: Ingredient Disclosure
0.1% item 1120 (27)

STATE LISTS

California - Air Bill 2588 Appendix A-I
known or potential carcinogen: 6/91

NICKEL CARBONYL (VAN)　　　　　　　12612-55-4

ENVIRONMENTAL LISTS

TSCA - Chemicals with Significant New Use Rules
PMN number: P-85-1034

NICKEL CATALYST　　　　　　　　　　RR-00376-9

SEE ALSO:
NICKEL

HEALTH AND SAFETY LISTS

U.S. DOT - Appendix A Table 1 - Hazardous Substances
final RQ = 100 pounds (45.4 kg)

NIOSH - Selected LD50s and LC50s
Oral, rat: LD50 = 400 mg/kg

ENVIRONMENTAL LISTS

CERCLA/SARA - Hazardous Substances and their Reportable Quantities
final RQ = 100 pounds (45.4 kg)

Clean Water Act - Hazardous Substances
[present]

EPA - Carcinogen Hazard Ranking for RQ Adjustment
Hazard ranking = Low

INTERNATIONAL LISTS

Canada - WHMIS: Ingredient Disclosure
0.1% item 1121 (1526)

STATE LISTS

Florida Hazardous Substance List
[present]

Massachusetts Right To Know List
[present]

NJ Right to Know List (Total)
sn 1342

Pennsylvania Right to Know List
environmental hazard

NICKEL(II) CHLORIDE　　　　　　　　7718-54-9

ENVIRONMENTAL LISTS

TSCA - Code of Federal Regulations Citations
40 CFR 704.95

NICKEL CHLORIDE　　　　　　　　　　37211-05-5

ENVIRONMENTAL LISTS

TSCA - Code of Federal Regulations Citations
40 CFR 704.95

NICKEL(II) CHLORIDE HEXAHYDRATE (1:2:6)　　7791-20-0

ENVIRONMENTAL LISTS

TSCA - Code of Federal Regulations Citations
40 CFR 704.95

NICKEL COMPLEX OF DIMETHYLAMINO METHYL　RR-01101-8
OCTYL PHENOL

SEE ALSO:
NICKEL COMPOUNDS
NICKEL
NICKEL

HEALTH AND SAFETY LISTS

NIOSH - Selected LD50s and LC50s
Oral, rat: LD50 = 840 mg/kg

NTP Seventh Report - Suspect Carcinogens
suspect carcinogen (Listed under 'Nickel and certain nickel compounds')

OSHA - Possible Select Carcinogens
[present]

INTERNATIONAL LISTS

Canada - WHMIS: Ingredient Disclosure
0.1% item 1122 (388)

STATE LISTS

California - Air Bill 2588 Appendix A-I
known or potential carcinogen: 6/91

Florida Hazardous Substance List
[present]

Massachusetts Right To Know List
carcinogen; extraordinarily hazardous

Pennsylvania Right to Know List
environmental hazard; special hazardous substance

Pennsylvania RTK - Special Hazardous Substances
[present]

NICKEL SALT OF AN ORGANO COMPOUND CON-　RR-01673-9
TAINING NITROGEN

SEE ALSO:
NICKEL

HEALTH AND SAFETY LISTS

ACGIH 1995 - Time Weighted Averages
as Ni: (0.05) ppm TWA; (0.12) mg/m3 TWA

AIHA - Odor Threshold Values
no geometric mean air odor threshold

U.S. DOT - Appendix B - Marine Pollutants
DOT regulated severe marine pollutant

U.S. DOT - Appendix A Table 1 - Hazardous Substances
final RQ = 10 pounds (4.54 kg)

NFPA - Flash Points
flash point < -4 degrees F (-24 degrees C)

NFPA - Hazard Identification Ratings
health-4; flammability-3; reactivity-3

NIOSH - Selected LD50s and LC50s
Inhalation, rat: LC50 = 35 ppm 30 mn

NIOSH 1990 - Pocket Guide - RELs
as Ni: 0.001 ppm TWA; 0.007 mg/m3 TWA

NIOSH 1990 - Pocket Guide - IDLHs
as Ni: 7 ppm IDLH (not considering carcinogenic effects)

NIOSH 1990 - Pocket Guide - Carcinogens
occupational carcinogen

NIOSH 1990 - Pocket Guide - Target organs
nasal cavities, lungs, skin

NIOSH - Health Standards - Exposure Limits
1 ppb TWA; 7 ug/m3 TWA (lowest detectable concentration)

NIOSH - Health Standards - Health Effects and Precautions
Lung and nasal cancer (Periodic chest X-ray, pulmonary function testing, and urine monitoring required)

NIOSH - Health Standards - Carcinogenic Chemicals
potential human carcinogen

NTP Seventh Report - Suspect Carcinogens
suspect carcinogen (Listed under 'Nickel and certain nickel compounds')
OSHA - Vacated PELs - Time Weighted Averages
as Ni: 0.001 ppm TWA; 0.007 mg/m3 TWA
OSHA - Final PELs - Time Weighted Averages
as Ni: 0.001 ppm TWA; 0.007 mg/m3 TWA
OSHA - Possible Select Carcinogens
[present]
OSHA - List of Highly Hazardous Chemicals
threshhold quantity = 150 pounds

ENVIRONMENTAL LISTS
CERCLA/SARA - Section 302 Extremely Hazardous Substances and TPQs
TPQ = 1 pound
CERCLA/SARA - Hazardous Substances and their Reportable Quantities
final RQ = 10 pounds (4.54 kg)
CAA -Toxic Substances for Accidental Release Prevention
threshold quantity = 1,000 lbs
EPA - Carcinogen Hazard Ranking for RQ Adjustment
Hazard ranking = Medium
RCRA - P Series Wastes
waste number P073
RCRA - Hazardous Constituents-Appendix VIII
waste number P073
RCRA - Substances Banned From Land Disposal
[present]

INTERNATIONAL LISTS
Australian Exposure Standards - Time Weighted Averages
as Ni: 0.05 ppm TWA; 0.12 mg/m3 TWA
Canada - WHMIS: Ingredient Disclosure
0.1% item 1123 (1190)
Canada - Ontario - OHSA - TWAEVs
as Ni: 0.05 ppm TWAEV; 0.35 mg/m3 TWAEV
United Kingdom - Occupational Exposure Standards - STELs
as Ni: 0.1 ppm STEL; 0.24 mg/m3 STEL
German (DFG) - Skin/Sensitizers
danger of cutaneous absorption
Israel - Time Weighted Averages
as Ni: (0.05) ppm TWA; (0.12) mg/m3 TWA
Israel - Action Levels
as Ni: 0.025 ppm AL; 0.06 mg/m3 AL

STATE LISTS
California - Air Bill 2588 Appendix A-I
known or potential carcinogen
California - Prop. 65 - Cancer list
carcinogen - initial date 10/1/87
California - Exposure Limits - PELs
0.001 ppm PEL; 0.007 mg/m3 PEL
Florida Hazardous Substance List
[present]
Massachusetts Right To Know List
carcinogen; extraordinarily hazardous
Minnesota Hazardous Substance List
as Ni: carcinogen
NJ Right to Know List (Total)
sn 1343
NJ Special Hazardous Substances
(carcinogen; flammable - third degree; reactive - third degree)
Pennsylvania Right to Know List
environmental hazard; special hazardous substance

Pennsylvania RTK - Special Hazardous Substances
[present]

NICKEL COMPOUNDS   RR-00800-4

HEALTH AND SAFETY LISTS
U.S. DOT - Substances From 49 CFR 172.101
regulated by DOT (UN1259)
U.S. DOT - Hazard Classes
DOT hazard class = 6.1
U.S. DOT - Substances Which Are Poisonous by Inhalation
liquid hazardous material poisonous by inhalation (UN1259)
U.S. DOT - Appendix B - Marine Pollutants
DOT regulated severe marine pollutant

INTERNATIONAL LISTS
Canada - Alberta - 8 Hour Occupational Exposure Limit
as Ni: 0.05 ppm TWA; 0.12 mg/m3 TWA
Canada - Alberta - 15 Minute Occupational Exposure Limit
as Ni: 0.15 ppm STEL; 0.36 mg/m3 STEL
Canada - British Columbia - 8 Hour Exposure Limits
0.05 ppm TWA; 0.35 mg/m3 TWA
Canada - Quebec - Time-Weighted Average Exposure Values
0.001 ppm TWAEV; 0.007 mg/m3 TWAEV
Mexico - Instruction No. 10 - TWAs
0.05 ppm TWA; 0.35 mg/m3 TWA

NICKEL CYANIDE   557-19-7

STATE LISTS
NJ Right to Know List (Total)
sn 2580; sn 2581
Pennsylvania Right to Know List
environmental hazard

NICKEL DIMETHYLDITHIOCARBAMATE   15521-65-0
SEE ALSO:
NICKEL

HEALTH AND SAFETY LISTS
U.S. DOT - Appendix A Table 1 - Hazardous Substances
final RQ = 100 pounds (45.4 kg)
NIOSH - Selected LD50s and LC50s
Oral, rat: LD50 = 105 mg/kg

ENVIRONMENTAL LISTS
CERCLA/SARA - Hazardous Substances and their Reportable Quantities
final RQ = 100 pounds (45.4 kg)
Clean Water Act - Hazardous Substances
[present]
EPA - Carcinogen Hazard Ranking for RQ Adjustment
Hazard ranking = Low

INTERNATIONAL LISTS
Canada - WHMIS: Ingredient Disclosure
0.1% item 1124 (511)

STATE LISTS
Massachusetts Right To Know List
[present]
NJ Right to Know List (Total)
sn 1344
Pennsylvania Right to Know List
environmental hazard

NICKEL(II) FLUOBORATE   14708-14-6

HEALTH AND SAFETY LISTS
U.S. DOT - Appendix A Table 1 - Hazardous Substances
final RQ = 100 pounds (45.4 kg)

ENVIRONMENTAL LISTS
CERCLA/SARA - Hazardous Substances and their Reportable Quantities
final RQ = 100 pounds (45.4 kg)
Clean Water Act - Hazardous Substances
[present]

STATE LISTS
Massachusetts Right To Know List
[present]
Pennsylvania Right to Know List
environmental hazard

NICKEL(II) FLUOSILICATE   26043-11-8
SEE ALSO:
NICKEL

HEALTH AND SAFETY LISTS
NIOSH - Selected LD50s and LC50s
Oral, rat: LD50 = 175 mg/kg

NICKEL FORMATE   15843-02-4

ENVIRONMENTAL LISTS
List of Pesticide Product Inert Ingredients
[present]

NICKEL HYDROXIDE   12054-48-7

ENVIRONMENTAL LISTS
TSCA - Chemicals with Significant New Use Rules
PMN number: P-92-686

NICKEL HYDROXIDE   12125-56-3

HEALTH AND SAFETY LISTS
IARC - Group 1 (carcinogenic to humans)
[present] (This evaluation applies to the group of chemicals as a whole and not necessarily to all individual chemicals within the group)
OSHA - Select Carcinogens
[present]

ENVIRONMENTAL LISTS
CERCLA/SARA - Section 313 - Emission Reporting
form R reporting required for 0.1% de minimus concentration
Clean Air Act (1990) - List of Hazardous Air Contaminants
[present] (includes any unique chemical substance that contains Nickel as part of that chemical's infrastructure)
Clean Water Act - Toxic Pollutants
[present] (Listed under 'Nickel and compounds')
RCRA - Hazardous Constituents-Appendix VIII
hazardous constituent - no waste number

INTERNATIONAL LISTS
Canada - NPRI (National Pollutant Release Inventory)
[present]
Canada - CEPA - Priority Substances List
estimated time for completion of assessment reports: 4 years

STATE LISTS
California - Air Bill 2588 Appendix A-I
[present]
California - Directors List of Hazardous Substances (8 CCR 339)
[present]
NJ Right to Know List (Total)
sn 3142

NICKEL HYDROXIDE (VAN)   11113-74-9
SEE ALSO:
CYANIDE ANION
NICKEL

HEALTH AND SAFETY LISTS
U.S. DOT - Substances From 49 CFR 172.101
regulated by DOT (UN1653)
U.S. DOT - Hazard Classes
DOT hazard class = 6.1
U.S. DOT - Appendix B - Marine Pollutants
DOT regulated severe marine pollutant
U.S. DOT - Appendix A Table 1 - Hazardous Substances
final RQ = 10 pounds (4.54 kg)

ENVIRONMENTAL LISTS
CERCLA/SARA - Hazardous Substances and their Reportable Quantities
final RQ = 10 pounds (4.54 kg)
EPA - Carcinogen Hazard Ranking for RQ Adjustment
Hazard ranking = Low
RCRA - P Series Wastes
waste number P074
RCRA - Hazardous Constituents-Appendix VIII
waste number P074
RCRA - Substances Banned From Land Disposal
[present]

INTERNATIONAL LISTS
Canada - WHMIS: Ingredient Disclosure
1% item 1125 (594)

STATE LISTS
Massachusetts Right To Know List
[present]
NJ Right to Know List (Total)
sn 1345
Pennsylvania Right to Know List
environmental hazard

NICKEL, INORGANIC COMPOUNDS   RR-00571-0
SEE ALSO:
NICKEL

HEALTH AND SAFETY LISTS
NIOSH - Selected LD50s and LC50s
Oral, rat: LD50 = 17 gm/kg

NICKEL INSOLUBLE COMPOUNDS   RR-00522-1
SEE ALSO:
NICKEL

INTERNATIONAL LISTS
Canada - WHMIS: Ingredient Disclosure
1% item 1127 (888)

NICKEL(II) IODIDE   13462-90-3
SEE ALSO:
NICKEL

INTERNATIONAL LISTS
Canada - WHMIS: Ingredient Disclosure
1% item 1128 (917)

NICKEL NITRATE   14216-75-2

INTERNATIONAL LISTS
Canada - WHMIS: Ingredient Disclosure
0.1% item 1129 (924)

NICKEL(II) NITRATE, HEXAHYDRATE (1:2:6)   13478-00-7

SEE ALSO:
NICKEL

HEALTH AND SAFETY LISTS

U.S. DOT - Appendix A Table 1 - Hazardous Substances
final RQ - 10 pounds (4.54 kg)

IARC - Group Unspecified
[present] (Listed under 'Nickel and nickel compounds')

NIOSH - Selected LD50s and LC50s
Oral, rat: LD50 - 1500 mg/kg

NTP Seventh Report - Suspect Carcinogens
suspect carcinogen (Listed under 'Nickel and certain nickel compounds')

OSHA - Possible Select Carcinogens
[present]

ENVIRONMENTAL LISTS

CERCLA/SARA - Hazardous Substances and their Reportable Quantities
final RQ - 10 pounds (4.54 kg)

Clean Water Act - Hazardous Substances
[present]

EPA - Carcinogen Hazard Ranking for RQ Adjustment
Hazard ranking - Low

INTERNATIONAL LISTS

Canada - WHMIS: Ingredient Disclosure
0.1% item 1130 (994)

STATE LISTS

California - Air Bill 2588 Appendix A-I
known or potential carcinogen: 6/91

Massachusetts Right To Know List
carcinogen; extraordinarily hazardous

NJ Right to Know List (Total)
sn 1346

Pennsylvania Right to Know List
environmental hazard

NICKEL NITRATE (2+ SALT)   13138-45-9

HEALTH AND SAFETY LISTS

NTP Seventh Report - Suspect Carcinogens
suspect carcinogen (Listed under 'Nickel and certain nickel compounds')

OSHA - Possible Select Carcinogens
[present]

NICKEL NITRITE   17861-62-0

STATE LISTS

Pennsylvania Right to Know List
environmental hazard

NICKELOCENE   1271-28-9

HEALTH AND SAFETY LISTS

NIOSH - Health Standards - Exposure Limits
15 ug Ni/m3 TWA

NIOSH - Health Standards - Health Effects and Precautions
Lung and nasal cancer; skin effects (Periodic chest X-ray and pulmonary function testing required)

NIOSH - Health Standards - Carcinogenic Chemicals
potential human carcinogen

STATE LISTS

NJ Right to Know List (Total)
sn 2875

NICKEL OXIDE   1313-99-1

INTERNATIONAL LISTS

Canada - WHMIS: Ingredient Disclosure
1% item 1143 (1192)

Canada - British Columbia - 8 Hour Exposure Limits
as Ni: 0.05 mg/m3 TWA

Canada - British Columbia - Carcinogens
carcinogen - 0.05 mg/m3 TWA

Canada - Quebec - Time-Weighted Average Exposure Values
1 mg/m3 TWAEV

United Kingdom - Maximum Exposure Limits - TWAs
as Ni: 0.5 mg/m3 TWA

PROPOSED REGULATIONS

ACGIH 1995 - Notice of Intended Changes
as Ni: 0.05 mg/m3 TWA; A1-confirmed human carcinogen

NICKEL OXIDE   11099-02-8

SEE ALSO:
NICKEL

INTERNATIONAL LISTS

Canada - WHMIS: Ingredient Disclosure
0.1% item 1131 (1027)

NICKEL PERCHLORATE   13637-71-3

HEALTH AND SAFETY LISTS

U.S. DOT - Appendix A Table 1 - Hazardous Substances
final RQ - 100 pounds (45.4 kg)

ENVIRONMENTAL LISTS

CERCLA/SARA - Hazardous Substances and their Reportable Quantities
final RQ - 100 pounds (45.4 kg)

Clean Water Act - Hazardous Substances
[present]

EPA - Carcinogen Hazard Ranking for RQ Adjustment
Hazard ranking - Low

STATE LISTS

Massachusetts Right To Know List
[present]

Pennsylvania Right to Know List
environmental hazard

NICKEL PICRATE   RR-01450-6

SEE ALSO:
NICKEL

HEALTH AND SAFETY LISTS

NIOSH - Selected LD50s and LC50s
Oral, rat: LD50 - 1620 mg/kg

NICKEL POTASSIUM CYANIDE   14220-17-8

SEE ALSO:
NICKEL COMPOUNDS
NICKEL NITRATE
NICKEL
NICKEL

HEALTH AND SAFETY LISTS

U.S. DOT - Substances From 49 CFR 172.101
regulated by DOT (UN2725)

U.S. DOT - Hazard Classes
DOT hazard class - 5.1

INTERNATIONAL LISTS

Canada - WHMIS: Ingredient Disclosure
0.1% item 1132 (1208)

STATE LISTS
Florida Hazardous Substance List
[present]
Massachusetts Right To Know List
[present]
NJ Right to Know List (Total)
sn 1347
Pennsylvania Right to Know List
environmental hazard

NICKEL REFINERY DUST FROM THE PYROMET-   RR-00084-0
ALLURGICAL PROCESS
HEALTH AND SAFETY LISTS
U.S. DOT - Substances From 49 CFR 172.101
regulated by DOT (UN2726)
U.S. DOT - Hazard Classes
DOT hazard class = 5.1
STATE LISTS
NJ Right to Know List (Total)
sn 2582

NICKEL SOLUBLE COMPOUNDS   RR-00038-4
SEE ALSO:
NICKEL
HEALTH AND SAFETY LISTS
NIOSH - Selected LD50s and LC50s
Oral, rat: LD50 = 490 mg/kg
NTP Seventh Report - Suspect Carcinogens
suspect carcinogen (Listed under 'Nickel and certain nickel compounds')
OSHA - Possible Select Carcinogens
[present]
INTERNATIONAL LISTS
Canada - WHMIS: Ingredient Disclosure
0.1% item 1133 (1194)
STATE LISTS
California - Air Bill 2588 Appendix A-I
known or potential carcinogen: 6/91
Florida Hazardous Substance List
[present]
Massachusetts Right To Know List
carcinogen; extraordinarily hazardous
Pennsylvania Right to Know List
special hazardous substance
Pennsylvania RTK - Special Hazardous Substances
[present]

NICKEL SUBSULFIDE   12035-72-2
SEE ALSO:
NICKEL
NICKEL
NICKEL COMPOUNDS
HEALTH AND SAFETY LISTS
IARC - Group Unspecified
[present] (Listed under 'Nickel and nickel compounds')
NTP Chemical Status Reports - Testing Status and NTIS Number
Two year studies scheduled for peer review
NTP Seventh Report - Suspect Carcinogens
suspect carcinogen (Listed under 'Nickel and certain nickel compounds')
OSHA - Possible Select Carcinogens
[present]

INTERNATIONAL LISTS
Canada - WHMIS: Ingredient Disclosure
0.1% item 1134 (1316)
Canada - Ontario - OHSA - TWAEVs
as Ni: 1 mg/m3 TWAEV
STATE LISTS
California - Air Bill 2588 Appendix A-I
known or potential carcinogen: 6/91
Florida Hazardous Substance List
[present]
Massachusetts Right To Know List
carcinogen; extraordinarily hazardous
Pennsylvania Right to Know List
environmental hazard; special hazardous substance
Pennsylvania RTK - Special Hazardous Substances
[present]
PROPOSED REGULATIONS
Canada - Ontario - Proposed Occupational TWAEVs
as Ni: 0.5 mg/m3 TWAEV

NICKEL SULFATE   7786-81-4
STATE LISTS
Pennsylvania Right to Know List
environmental hazard; special hazardous substance
Pennsylvania RTK - Special Hazardous Substances
[present]

NICKEL(II) SULFATE HEXAHYDRATE (1:1:6)   10101-97-0
SEE ALSO:
NICKEL
INTERNATIONAL LISTS
Canada - WHMIS: Ingredient Disclosure
0.1% item 1136 (1355)

NICKEL(II) SULFIDE   11113-75-0
HEALTH AND SAFETY LISTS
U.S. DOT - Hazard Classes
Forbidden from transport by the DOT

NICKEL SULFIDE ROASTING   RR-00798-7
SEE ALSO:
NICKEL
INTERNATIONAL LISTS
Canada - WHMIS: Ingredient Disclosure
0.1% item 1137 (595)

NICKEL SULFIDES, CRYSTALLINE   RR-01204-4
STATE LISTS
California - Air Bill 2588 Appendix A-I
known or potential carcinogen: 9/89
California - Prop. 65 - Cancer list
carcinogen - initial date 10/1/87
California - Prop. 65 - No Significant Risk Levels
no significant risk level = 0.8 ug/day
Pennsylvania Right to Know List
environmental hazard; special hazardous substance
Pennsylvania RTK - Special Hazardous Substances
[present]

NICKEL TELLURIDE   12142-88-0
INTERNATIONAL LISTS
Canada - WHMIS: Ingredient Disclosure
1% item 1144 (1191)

Canada - Alberta - 8 Hour Occupational Exposure Limit
as Ni: 0.1 mg/m3 TWA

Canada - Alberta - 15 Minute Occupational Exposure Limit
as Ni: 0.3 mg/m3 STEL

Canada - British Columbia - 8 Hour Exposure Limits
as Ni: 0.1 mg/m3 TWA

Canada - British Columbia - 15 Minute Exposure Limits
as Ni: 0.3 mg/m3 STEL

Canada - British Columbia - Carcinogens
carcinogen - 0.05 mg/m3 TWA

Canada - Ontario - OHSA - TWAEVs
as Ni: 0.1 mg/m3 TWAEV

Canada - Quebec - Time-Weighted Average Exposure Values
as Ni: 0.1 mg/m3 TWAEV

United Kingdom - Maximum Exposure Limits - TWAs
as Ni: 0.1 mg/m3 TWA

Mexico - Instruction No. 10 - TWAs
0.1 mg/m3 TWA

Mexico - Instruction No. 10 - STELs
0.3 mg/m3 STEL

STATE LISTS

California - Prop. 65 - Cancer list
carcinogen - initial date 10/1/89

PROPOSED REGULATIONS

ACGIH 1995 - Notice of Intended Changes
as Ni: 0.05 mg/m3 TWA; A1-confirmed human carcinogen

NICKEL TITANIUM OXIDE   12035-39-1
SEE ALSO:
NICKEL

HEALTH AND SAFETY LISTS

NTP Chemical Status Reports - Testing Status and NTIS Number
Two year studies: pathology working group in progress

NTP Seventh Report - Suspect Carcinogens
suspect carcinogen (Listed under 'Nickel and certain nickel compounds')

OSHA - Possible Select Carcinogens
[present]

INTERNATIONAL LISTS

Canada - WHMIS: Ingredient Disclosure
0.1% item 1138 (1501)

Canada - Ontario - OHSA - TWAEVs
as Ni: 1 mg/m3 TWAEV

STATE LISTS

California - Air Bill 2588 Appendix A-I
known or potential carcinogen

California - Prop. 65 - Cancer list
carcinogen - initial date 10/1/87

California - Prop. 65 - No Significant Risk Levels
no significant risk level = 0.4 ug/day

Florida Hazardous Substance List
[present]

Massachusetts Right To Know List
carcinogen; extraordinarily hazardous

Pennsylvania Right to Know List
environmental hazard; special hazardous substance

Pennsylvania RTK - Special Hazardous Substances
[present]

PROPOSED REGULATIONS

Canada - Ontario - Proposed Occupational TWAEVs
as Ni: 0.5 mg/m3 TWAEV

NICKEL TRIOXIDE   1314-06-3
SEE ALSO:
NICKEL
NICKEL COMPOUNDS
NICKEL

HEALTH AND SAFETY LISTS

U.S. DOT - Appendix A Table 1 - Hazardous Substances
final RQ = 100 pounds (45.4 kg)

IARC - Group Unspecified
[present] (Listed under 'Nickel and nickel compounds')

ENVIRONMENTAL LISTS

CERCLA/SARA - Hazardous Substances and their Reportable Quantities
final RQ = 100 pounds (45.4 kg)

Clean Water Act - Hazardous Substances
[present]

EPA - Carcinogen Hazard Ranking for RQ Adjustment
Hazard ranking = Low

INTERNATIONAL LISTS

Canada - WHMIS: Ingredient Disclosure
0.1% item 1139 (1525)

STATE LISTS

Massachusetts Right To Know List
[present]

NJ Right to Know List (Total)
sn 1348

Pennsylvania Right to Know List
environmental hazard

NICOTINAMIDE   98-92-0
SEE ALSO:
NICKEL
NICKEL SOLUBLE COMPOUNDS

HEALTH AND SAFETY LISTS

NIOSH - Selected LD50s and LC50s
Oral, rat: LD50 = 275 mg/kg

NTP Chemical Status Reports - Testing Status and NTIS Number
Two year studies scheduled for peer review

NICOTINE   54-11-5
SEE ALSO:
NICKEL

INTERNATIONAL LISTS

Canada - WHMIS: Ingredient Disclosure
0.1% item 1140 (1551)

Canada - Ontario - OHSA - TWAEVs
as Ni: 1 mg/m3 TWAEV

PROPOSED REGULATIONS

Canada - Ontario - Proposed Occupational TWAEVs
as Ni: 0.5 mg/m3 TWAEV

NICOTINE, COMPOUNDS AND PREPARATIONS, N.O.S.   RR-00378-1

HEALTH AND SAFETY LISTS

ACGIH 1995 - Time Weighted Averages
(1) mg/m3 TWA

ACGIH 1995 - Carcinogens
as Ni: A1-confirmed human carcinogen

INTERNATIONAL LISTS

Australian Exposure Standards - Time Weighted Averages
fume and dust, as Ni: 1 mg/m3 TWA

Australian Exposure Standards - Skin Effects
fume and dust, as Ni: sensitiser

Australian Exposure Standards - Carcinogens
fume and dust, as Ni: confirmed carcinogen

Canada - Alberta - 8 Hour Occupational Exposure Limit
as Ni: 1 mg/m3 TWA

Canada - Alberta - 15 Minute Occupational Exposure Limit
as Ni: 3 mg/m3 STEL

Canada - Alberta - Designated Substances
designated substance - requires code of practice

Canada - British Columbia - 8 Hour Exposure Limits
as Ni: 1.0 mg/m3 TWA

Canada - British Columbia - Carcinogens
carcinogen - as Ni: 1.0 mg/m3 TWA

Canada - Quebec - Time-Weighted Average Exposure Values
as Ni: 1 mg/m3 TWAEV (substance of which the recirculation is prohibited)

Canada - Quebec - Carcinogens
as Ni: C1 carcinogen: effect detected in humans

Israel - Time Weighted Averages
as Ni: (1) mg/m3 TWA

Israel - Action Levels
as Ni: 0.5 mg/m3 AL

Mexico - Instruction No. 10 - TWAs
1 mg/m3 TWA

Mexico - Instruction No. 10 - Carcinogens
carcinogen in humans

STATE LISTS

Minnesota Hazardous Substance List
as Ni: carcinogen

NICOTINE HYDROCHLORIDE   2820-51-1

HEALTH AND SAFETY LISTS

IARC - Group Unspecified
[present] (Listed under 'Nickel and nickel compounds')

NICOTINE SALICYLATE   29790-52-1
SEE ALSO:
  TELLURIUM
  NICKEL

INTERNATIONAL LISTS

Canada - WHMIS: Ingredient Disclosure
0.1% item 1141 (1560)

NICOTINE SULFATE   65-30-5
SEE ALSO:
  NICKEL

INTERNATIONAL LISTS

Canada - WHMIS: Ingredient Disclosure
0.1% item 1142 (1318)

NICOTINE SULFATE   6505-86-8
SEE ALSO:
  NICKEL

INTERNATIONAL LISTS

Canada - WHMIS: Ingredient Disclosure
1% item 1135 (1317)

Canada - Ontario - OHSA - TWAEVs
as Ni: 1 mg/m3 TWAEV

STATE LISTS

Pennsylvania Right to Know List
environmental hazard; special hazardous substance

Pennsylvania RTK - Special Hazardous Substances
[present]

PROPOSED REGULATIONS

Canada - Ontario - Proposed Occupational TWAEVs
as Ni: 0.5 mg/m3 TWAEV

NICOTINE TARTRATE   3275-73-8

HEALTH AND SAFETY LISTS

NIOSH - Selected LD50s and LC50s
Oral, mouse: LD50 = 2500 mg/kg

NIOBIUM   7440-03-1

HEALTH AND SAFETY LISTS

ACGIH 1995 - Time Weighted Averages
0.5 mg/m3 TWA

ACGIH 1995 - Skin Designations
skin - potential for cutaneous absorption

U.S. DOT - Substances From 49 CFR 172.101
regulated by DOT (UN1654)

U.S. DOT - Hazard Classes
DOT hazard class = 6.1

U.S. DOT - Appendix A Table 1 - Hazardous Substances
final RQ = 100 pounds (45.4 kg)

NFPA - Hazard Identification Ratings
health-4; flammability-1; reactivity-0

NIOSH - Selected LD50s and LC50s
Oral, rat: LD50 = 50 mg/kg Skin, rat: LD50 = 140 mg/kg

NIOSH 1990 - Pocket Guide - RELs
0.5 mg/m3 TWA

NIOSH 1990 - Pocket Guide - IDLHs
35 mg/m3 IDLH

NIOSH 1990 - Pocket Guide - Target organs
CNS, CVS, lungs, GI tract

NIOSH 1990 - Pocket Guide - Skin list
Potential for dermal absorption

OSHA - Vacated PELs - Time Weighted Averages
0.5 mg/m3 TWA

OSHA - Vacated PELs - Skin Designation
Prevent or reduce skin absorption

OSHA - Final PELs - Time Weighted Averages
0.5 mg/m3 TWA

OSHA - Final PELs - Skin Notations
prevent or reduce skin absorption

ENVIRONMENTAL LISTS

CERCLA/SARA - Section 302 Extremely Hazardous Substances and TPQs
TPQ = 100 pounds

CERCLA/SARA - Section 313 - Emission Reporting
form R reporting required

CERCLA/SARA - Hazardous Substances and their Reportable Quantities
final RQ = 100 pounds (45.4 kg)

RCRA - P Series Wastes
waste number P075

RCRA - Hazardous Constituents-Appendix VIII
waste number P075

RCRA - Substances Banned From Land Disposal
[present]

INTERNATIONAL LISTS

Australian Exposure Standards - Time Weighted Averages
0.5 mg/m3 TWA

Australian Exposure Standards - Skin Effects
skin absorption

Canada - WHMIS: Ingredient Disclosure
1% item 1145 (1195)

Canada - Alberta - 8 Hour Occupational Exposure Limit
0.5 mg/m3 TWA

Canada - Alberta - 15 Minute Occupational Exposure Limit
1.5 mg/m3 STEL

Canada - Alberta - Skin Designation
can be absorbed through the intact skin

Canada - British Columbia - 8 Hour Exposure Limits
0.5 mg/m3 TWA

Canada - British Columbia - 15 Minute Exposure Limits
1.5 mg/m3 STEL

Canada - British Columbia - Skin Notations
skin - potential for skin absorption

Canada - Ontario - OHSA - TWAEVs
0.5 mg/m3 TWAEV

Canada - Ontario - OHSA - Skin Notations
absorption through skin, eyes, or mucous membranes

Canada - Quebec - Time-Weighted Average Exposure Values
0.5 mg/m3 TWAEV

Canada - Quebec - Skin Designations
absorbed through the skin

United Kingdom - Occupational Exposure Standards - TWAs
0.5 mg/m3 TWA

United Kingdom - Occupational Exposure Standards - STELs
1.5 mg/m3 STEL

United Kingdom - Occupational Exposure Standards - Notes
can be absorbed through skin

German (DFG) - MAK Values
0.07 ppm MAK; 0.5 mg/m3 MAK

German (DFG) - Peak Limitations
2 x normal MAK (30 min. average value); don't exceed 4 times during shift

German (DFG) - Skin/Sensitizers
danger of cutaneous absorption

Israel - Time Weighted Averages
0.5 mg/m3 TWA

Israel - Action Levels
0.25 mg/m3 AL

Mexico - Instruction No. 10 - TWAs
0.5 mg/m3 TWA

Mexico - Instruction No. 10 - STELs
1.5 mg/m3 STEL

Mexico - Instruction No. 10 - Skin designation
skin - potential for cutaneous absorption

STATE LISTS

California - Air Bill 2588 Appendix A-II
9/90

California - Prop. 65 - Developmental Toxicity
developmental toxicity - initial date 4/1/90

California - Exposure Limits - PELs
0.075 ppm PEL; 0.5 mg/m3 PEL

California - Exposure Limits - Skin Notation
material may be absorbed through the skin, eyes or mucous membrane

California - Directors List of Hazardous Substances (8 CCR 339)
[present]

Florida Hazardous Substance List
[present]

Massachusetts Right To Know List
extraordinarily hazardous

Minnesota Hazardous Substance List
skin

NJ Right to Know List (Total)
sn 1349

Pennsylvania Right to Know List
environmental hazard

NIOBIUM 88   14681-74-4

HEALTH AND SAFETY LISTS

U.S. DOT - Substances From 49 CFR 172.101
regulated by DOT (UN1655, UN3144)

U.S. DOT - Hazard Classes
DOT hazard class = 6.1

ENVIRONMENTAL LISTS

CERCLA/SARA - Section 313 - Emission Reporting
form R reporting required

STATE LISTS

NJ Right to Know List (Total)
sn 2583

NIOBIUM 89   15700-40-0

HEALTH AND SAFETY LISTS

U.S. DOT - Substances From 49 CFR 172.101
regulated by DOT (UN1656)

U.S. DOT - Hazard Classes
DOT hazard class = 6.1

STATE LISTS

NJ Right to Know List (Total)
sn 1350

NIOBIUM 90   14681-65-3

HEALTH AND SAFETY LISTS

U.S. DOT - Substances From 49 CFR 172.101
regulated by DOT (UN1657)

U.S. DOT - Hazard Classes
DOT hazard class = 6.1

STATE LISTS

NJ Right to Know List (Total)
sn 1351

NIOBIUM 93M   RR-00400-2

HEALTH AND SAFETY LISTS

NIOSH - Selected LD50s and LC50s
Oral, rat: LD50 = 50 mg/kg Skin, rat: LD50 = 285 mg/kg

ENVIRONMENTAL LISTS

CERCLA/SARA - Section 302 Extremely Hazardous Substances and TPQs
TPQ = 100/10,000 pounds

STATE LISTS

Florida Hazardous Substance List
effective March 13, 1992

Massachusetts Right To Know List
extraordinarily hazardous

NJ Right to Know List (Total)
sn 1352

Pennsylvania Right to Know List
environmental hazard

PROPOSED REGULATIONS

CERCLA/SARA - Proposed Hazardous Substance Additions
proposed RQ = 1 pound (.454 kg)

CERCLA/SARA - 1989 Proposed RQ Adjustments
proposed RQ = 10 pounds (4.54 kg)

NIOBIUM 94   14681-63-1

HEALTH AND SAFETY LISTS

U.S. DOT - Substances From 49 CFR 172.101
regulated by DOT (UN1658)

U.S. DOT - Hazard Classes
DOT hazard class = 6.1

NIOBIUM 95   13967-76-5

HEALTH AND SAFETY LISTS

U.S. DOT - Substances From 49 CFR 172.101
regulated by DOT (UN1659)

U.S. DOT - Hazard Classes
DOT hazard class = 6.1

STATE LISTS

NJ Right to Know List (Total)
sn 1353

NIOBIUM 95M   RR-00398-5

INTERNATIONAL LISTS

Australian Exposure Standards - Under Review
exposure limits under review

NIOBIUM 96   15832-32-3

HEALTH AND SAFETY LISTS

U.S. DOT - Appendix A Table 2 - Radionuclides
final RQ = 100 curies (3.7E 12 Bq)

ENVIRONMENTAL LISTS

CERCLA/SARA List of Radionuclides (Appendix B) and Their Reportable Quantities
final RQ = 100 curies (3.7E 12 Bq)

NIOBIUM 97   18496-04-3

HEALTH AND SAFETY LISTS

U.S. DOT - Appendix A Table 2 - Radionuclides
66 minute half-life: Final RQ = 100 curies (3.7E 12 Bq); 122 minute half-life: Final RQ = 100 curies (3.7E 12 Bq)

ENVIRONMENTAL LISTS

CERCLA/SARA List of Radionuclides (Appendix B) and Their Reportable Quantities
66 minute half-life: Final RQ = 100 curies (3.7E 12 Bq); 122 minute half-life: Final RQ = 100 curies (3.7E 12 Bq)

NIOBIUM 98   15700-41-1

HEALTH AND SAFETY LISTS

U.S. DOT - Appendix A Table 2 - Radionuclides
final RQ = 10 curies (3.7E 11 Bq)

ENVIRONMENTAL LISTS

CERCLA/SARA List of Radionuclides (Appendix B) and Their Reportable Quantities
final RQ = 10 curies (3.7E 11 Bq)

NIOBIUM CHLORIDE   10026-12-7

HEALTH AND SAFETY LISTS

U.S. DOT - Appendix A Table 2 - Radionuclides
final RQ = 100 curies (3.7E 12 Bq)

ENVIRONMENTAL LISTS

CERCLA/SARA List of Radionuclides (Appendix B) and Their Reportable Quantities
final RQ = 100 curies (3.7E 12 Bq)

NIRIDAZOLE   61-57-4

HEALTH AND SAFETY LISTS

U.S. DOT - Appendix A Table 2 - Radionuclides
final RQ = 10 curies (3.7E 11 Bq)

ENVIRONMENTAL LISTS

CERCLA/SARA List of Radionuclides (Appendix B) and Their Reportable Quantities
final RQ = 10 curies (3.7E 11 Bq)

NITHIAZIDE   139-94-6

HEALTH AND SAFETY LISTS

U.S. DOT - Appendix A Table 2 - Radionuclides
final RQ = 10 curies (3.7E 11 Bq)

ENVIRONMENTAL LISTS

CERCLA/SARA List of Radionuclides (Appendix B) and Their Reportable Quantities
final RQ = 10 curies (3.7E 11 Bq)

NITRAPYRIN   1929-82-4

HEALTH AND SAFETY LISTS

U.S. DOT - Appendix A Table 2 - Radionuclides
final RQ = 100 curies (3.7E 12 Bq)

ENVIRONMENTAL LISTS

CERCLA/SARA List of Radionuclides (Appendix B) and Their Reportable Quantities
final RQ = 100 curies (3.7E 12 Bq)

NITRATE   14797-55-8

HEALTH AND SAFETY LISTS

U.S. DOT - Appendix A Table 2 - Radionuclides
final RQ = 10 curies (3.7E 11 Bq)

ENVIRONMENTAL LISTS

CERCLA/SARA List of Radionuclides (Appendix B) and Their Reportable Quantities
final RQ = 10 curies (3.7E 11 Bq)

NITRATE COMPOUNDS   RR-01770-9

HEALTH AND SAFETY LISTS

U.S. DOT - Appendix A Table 2 - Radionuclides
final RQ = 100 curies (3.7E 12 Bq)

ENVIRONMENTAL LISTS

CERCLA/SARA List of Radionuclides (Appendix B) and Their Reportable Quantities
final RQ = 100 curies (3.7E 12 Bq)

NITRATED PAPER   RR-01451-7

HEALTH AND SAFETY LISTS

U.S. DOT - Appendix A Table 2 - Radionuclides
final RQ = 1000 curies (3.7E 13 Bq)

ENVIRONMENTAL LISTS

CERCLA/SARA List of Radionuclides (Appendix B) and Their Reportable Quantities
final RQ = 1000 curies (3.7E 13 Bq)

NITRATE POLYETHER POLYOL   RR-00936-9

HEALTH AND SAFETY LISTS

NIOSH - Selected LD50s and LC50s
Oral, rat: LD50 = 1400 mg/kg

NITRATES, INORGANIC, N.O.S.   RR-00379-2

HEALTH AND SAFETY LISTS

IARC - Group 2B (sufficient animal data)
[present] (Overall evaluation based only on evidence of carcinogenicity in monograph (13, 1977) or in Supplement 4)
NIOSH - Selected LD50s and LC50s
Oral, rat: LD50 = 900 mg/kg

OSHA - Possible Select Carcinogens
[present]

STATE LISTS

California - Air Bill 2588 Appendix A-I
known or potential carcinogen

California - Prop. 65 - Cancer list
carcinogen - initial date 4/1/88

California - Directors List of Hazardous Substances (8 CCR 339)
[present]

Florida Hazardous Substance List
[present]

Massachusetts Right To Know List
carcinogen; extraordinarily hazardous

Minnesota Hazardous Substance List
carcinogen

Pennsylvania Right to Know List
special hazardous substance

Pennsylvania RTK - Special Hazardous Substances
[present]

NITRATING ACID MIXTURES   RR-01345-6

HEALTH AND SAFETY LISTS

IARC - Group 3 (not classifiable)
[present]

NTP Chemical Status Reports - Testing Status and NTIS Number
Technical reports printed (PB295897/AS)

NTP Chemical Status Reports - Evidence of Carcinogenicity
male rat-negative; female rat-positive; male mice-positive; female mice-equivocal

STATE LISTS

California - Directors List of Hazardous Substances (8 CCR 339)
[present]

Massachusetts Right To Know List
carcinogen; extraordinarily hazardous

NITRIC ACID   7697-37-2

HEALTH AND SAFETY LISTS

ACGIH 1995 - Time Weighted Averages
10 mg/m3 TWA

ACGIH 1995 - Short Term Exposure Limits
20 mg/m3 STEL

NIOSH - Selected LD50s and LC50s
Oral, rat: LD50 = 940 mg/kg Skin, rabbit: LD50 = 850 mg/kg

OSHA - Vacated PELs - Time Weighted Averages
total dust: 15 mg/m3 TWA; respirable fraction: 5 mg/m3 TWA

OSHA - Final PELs - Time Weighted Averages
total dust: 15 mg/m3 TWA; respirable fraction: 5 mg/m3 TWA

ENVIRONMENTAL LISTS

CERCLA/SARA - Section 313 - Emission Reporting
form R reporting required

INTERNATIONAL LISTS

Australian Exposure Standards - Time Weighted Averages
10 mg/m3 TWA

Australian Exposure Standards - Short Term Exposure Limits
20 mg/m3 STEL

Canada - Alberta - 8 Hour Occupational Exposure Limit
10 mg/m3 TWA

Canada - Alberta - 15 Minute Occupational Exposure Limit
20 mg/m3 STEL

Canada - British Columbia - 8 Hour Exposure Limits
10 mg/m3 TWA

Canada - British Columbia - 15 Minute Exposure Limits
20 mg/m3 STEL

Canada - Ontario - OHSA - TWAEVs
10 mg/m3 TWAEV

Canada - Ontario - OHSA - STEVs
20 mg/m3 STEV

Canada - Quebec - Time-Weighted Average Exposure Values
10 mg/m3 TWAEV

Canada - Quebec - Short-term Exposure Values
20 mg/m3 STEV

United Kingdom - Occupational Exposure Standards - TWAs
10 mg/m3 TWA

United Kingdom - Occupational Exposure Standards - STELs
20 mg/m3 STEL

Israel - Time Weighted Averages
10 mg/m3 TWA

Israel - Short Term Exposure Limits
20 mg/m3 STEL

Israel - Action Levels
5 mg/m3 AL

Mexico - Instruction No. 10 - TWAs
20 ppm TWA; 100 mg/m3 TWA

STATE LISTS

California - Exposure Limits - PELs
total dust: 10 mg/m3 PEL; respirable fraction: 5 mg/m3 PEL

California - Directors List of Hazardous Substances (8 CCR 339)
[present]

Florida Hazardous Substance List
[present]

Massachusetts Right To Know List
[present]

Minnesota Hazardous Substance List
[present]

Pennsylvania Right to Know List
[present]

NITRIC ACID, URANIUM SALT   15905-86-9

ENVIRONMENTAL LISTS

ATSDR Priority List
Rank (of 275): 214

Safe Drinking Water Act - MCLs
as N: MCL = 10 mg/L (or 10 mg/L total with Nitrite)

Safe Drinking Water Act - MCLGs
as N: MCLG = 10 mg/L (or 10 mg/L total with Nitrite)

INTERNATIONAL LISTS

Canada - Drinking Water Quality - MACs
45.0 mg/L MAC

Mexico - Drinking Water - Ecological Criteria
5.0 mg/l

NITRIC OXIDE   10102-43-9

ENVIRONMENTAL LISTS

CERCLA/SARA - Section 313 - Emission Reporting
form R reporting required; (water dissociable; reportable only when in aqueous solution)

NITRILOTRIACETIC ACID (NTA)   139-13-9

HEALTH AND SAFETY LISTS

U.S. DOT - Hazard Classes
Forbidden from transport by the DOT

NITRILOTRIACETIC ACID (SALTS)   RR-01745-8

ENVIRONMENTAL LISTS

TSCA - Chemicals with Significant New Use Rules
PMN number: P-88-2540

NITRILOTRIACETIC ACID DISODIUM SALT   15467-20-6

HEALTH AND SAFETY LISTS

U.S. DOT - Substances From 49 CFR 172.101
regulated by DOT (UN1477)

U.S. DOT - Hazard Classes
DOT hazard class = 5.1

U.S. DOT - Appendix B - Marine Pollutants
DOT regulated marine pollutant

STATE LISTS

NJ Right to Know List (Total)
sn 2584

NITRILOTRIACETIC ACID TRISODIUM SALT   5064-31-3

HEALTH AND SAFETY LISTS

U.S. DOT - Substances From 49 CFR 172.101
regulated by DOT (UN1796, UN1826)

U.S. DOT - Hazard Classes
DOT hazard class = 8

STATE LISTS

NJ Right to Know List (Total)
sn 2060

2,2',2"-NITRILOTRISETHANOL POLYMER WITH 1,   66507-71-9
4-DICHLORO-2-BUTENE AND N,N,N',N'-TETRAM-
ETHYL-2-BUTENE-1,4-DIAMINE

HEALTH AND SAFETY LISTS

ACGIH 1995 - Time Weighted Averages
2 ppm TWA; 5.2 mg/m3 TWA

ACGIH 1995 - Short Term Exposure Limits
4 ppm STEL; 10 mg/m3 STEL

AIHA - Odor Threshold Values
no geometric mean air odor threshold

U.S. DOT - Substances From 49 CFR 172.101
regulated by DOT (UN2032, UN2031)

U.S. DOT - Hazard Classes
DOT hazard class = 8

U.S. DOT - Substances Which Are Poisonous by Inhalation
liquid hazardous material poisonous by inhalation (red fuming form)
(UN2032)

U.S. DOT - Appendix A Table 1 - Hazardous Substances
final RQ = 1000 pounds (454 kg)

NIOSH - Selected LD50s and LC50s
Inhalation, rat: LC50 = 67 ppm 4 hr

NIOSH 1990 - Pocket Guide - RELs
2 ppm TWA; 5 mg/m3 TWA; 4 ppm STEL; 10 mg/m3 STEL

NIOSH 1990 - Pocket Guide - IDLHs
100 ppm IDLH

NIOSH 1990 - Pocket Guide - Target organs
eyes, skin, teeth, respiratory system

NIOSH - Health Standards - Exposure Limits
2 ppm TWA; 5 mg/m3 TWA

NIOSH - Health Standards - Health Effects and Precautions
Dental erosion; nasal/lung irritation (Prevent skin and eye contact;
periodic chest X-ray required)

OSHA - Vacated PELs - Time Weighted Averages
2 ppm TWA; 5 mg/m3 TWA

OSHA - Vacated PELs - Short Term Exposure Limits
4 ppm STEL; 10 mg/m3 STEL

OSHA - Final PELs - Time Weighted Averages
2 ppm TWA; 5 mg/m3 TWA

OSHA - List of Highly Hazardous Chemicals
94.5% by weight or greater: threshhold quantity = 500 pounds

ENVIRONMENTAL LISTS

CERCLA/SARA - Section 302 Extremely Hazardous Substances and
TPQs
TPQ = 1000 pounds

CERCLA/SARA - Section 313 - Emission Reporting
form R reporting required for 1.0% de minimus concentration

CERCLA/SARA - Hazardous Substances and their Reportable Quantities
final RQ = 1000 pounds (454 kg)

CAA -Toxic Substances for Accidental Release Prevention
conc. 80% or greater: threshold quantity = 15,000 lbs

Clean Water Act - Hazardous Substances
[present]

INTERNATIONAL LISTS

Australian Exposure Standards - Time Weighted Averages
2 ppm TWA; 5.2 mg/m3 TWA

Australian Exposure Standards - Short Term Exposure Limits
4 ppm STEL; 10 mg/m3 STEL

Canada - WHMIS: Ingredient Disclosure
1% item 1146 (111)

Canada - NPRI (National Pollutant Release Inventory)
[present]

Canada - Alberta - 8 Hour Occupational Exposure Limit
2 ppm TWA; 5.2 mg/m3 TWA

Canada - Alberta - 15 Minute Occupational Exposure Limit
4 ppm STEL; 10 mg/m3 STEL

Canada - British Columbia - 8 Hour Exposure Limits
2 ppm TWA; 5 mg/m3 TWA

Canada - British Columbia - 15 Minute Exposure Limits
4 ppm STEL; 10 mg/m3 STEL

Canada - Ontario - OHSA - TWAEVs
2 ppm TWAEV; 5 mg/m3 TWAEV

Canada - Ontario - OHSA - STEVs
4 ppm STEV; 10 mg/m3 STEV

Canada - Quebec - Time-Weighted Average Exposure Values
2 ppm TWAEV; 5.2 mg/m3 TWAEV

Canada - Quebec - Short-term Exposure Values
4 ppm STEV; 10 mg/m3 STEV

United Kingdom - Occupational Exposure Standards - TWAs
2 ppm TWA; 5 mg/m3 TWA

United Kingdom - Occupational Exposure Standards - STELs
4 ppm STEL; 10 mg/m3 STEL

German (DFG) - MAK Values
2 ppm MAK; 5 mg/m3 MAK

German (DFG) - Peak Limitations
2 x normal MAK (5 min momentary value); don't exceed 8 times
during shift

Israel - Time Weighted Averages
2 ppm TWA; 5.2 mg/m3 TWA

Israel - Short Term Exposure Limits
4 ppm STEL; 10 mg/m3 STEL

Israel - Action Levels
1 ppm AL; 2.6 mg/m3 AL

Mexico - Instruction No. 10 - TWAs
2 ppm TWA; 5 mg/m3 TWA

Mexico - Instruction No. 10 - STELs
4 ppm STEL; 10 mg/m3 STEL

STATE LISTS

California - Air Bill 2588 Appendix A-I
6/91

California - Exposure Limits - PELs
2 ppm PEL; 5 mg/m3 PEL

California - Exposure Limits - STELs
4 ppm STEL; 10 mg/m3 STEL

California - Directors List of Hazardous Substances (8 CCR 339)
[present]

Florida Hazardous Substance List
[present]

Massachusetts Right To Know List
extraordinarily hazardous

Minnesota Hazardous Substance List
[present]

NJ Right to Know List (Total)
sn 1356

NJ Special Hazardous Substances
(corrosive)

Pennsylvania Right to Know List
environmental hazard

NITRITE   14797-65-0

STATE LISTS

Pennsylvania Right to Know List
environmental hazard

NITRITES, INORGANIC, N.O.S.   RR-00380-5

HEALTH AND SAFETY LISTS

ACGIH 1995 - Time Weighted Averages
25 ppm TWA; 31 mg/m3 TWA

U.S. DOT - Substances From 49 CFR 172.101
regulated by DOT (UN1660)

U.S. DOT - Hazard Classes
DOT hazard class = 2.3

U.S. DOT - Substances Which Are Poisonous by Inhalation
gaseous hazardous material poisonous by inhalation (alone or mixed with Nitrogen dioxide) (UN1660, UN1975)

U.S. DOT - Appendix A Table 1 - Hazardous Substances
final RQ = 10 pounds (4.54 kg)

NIOSH - Selected LD50s and LC50s
Inhalation, rat: LC50 = 1068 mg/m3 8 hr

NIOSH 1990 - Pocket Guide - RELs
25 ppm TWA; 30 mg/m3 TWA

NIOSH 1990 - Pocket Guide - IDLHs
100 ppm IDLH

NIOSH 1990 - Pocket Guide - Target organs
respiratory system

OSHA - Vacated PELs - Time Weighted Averages
25 ppm TWA; 30 mg/m3 TWA

OSHA - Final PELs - Time Weighted Averages
25 ppm TWA; 30 mg/m3 TWA

OSHA - List of Highly Hazardous Chemicals
threshhold quantity = 250 pounds

ENVIRONMENTAL LISTS

CERCLA/SARA - Section 302 Extremely Hazardous Substances and TPQs
TPQ = 100 pounds

CERCLA/SARA - Hazardous Substances and their Reportable Quantities
final RQ = 10 pounds (4.54 kg)

CAA -Toxic Substances for Accidental Release Prevention
threshold quantity = 10,000 lbs

RCRA - P Series Wastes
waste number P076

RCRA - Hazardous Constituents-Appendix VIII
waste number P076

RCRA - Substances Banned From Land Disposal
[present]

INTERNATIONAL LISTS

Australian Exposure Standards - Time Weighted Averages
25 ppm TWA; 31 mg/m3 TWA

Canada - WHMIS: Ingredient Disclosure
1% item 1147 (1330)

Canada - Alberta - 8 Hour Occupational Exposure Limit
25 ppm TWA; 31 mg/m3 TWA

Canada - Alberta - 15 Minute Occupational Exposure Limit
35 ppm STEL; 43 mg/m3 STEL

Canada - British Columbia - 8 Hour Exposure Limits
25 ppm TWA; 30 mg/m3 TWA

Canada - British Columbia - 15 Minute Exposure Limits
35 ppm STEL; 45 mg/m3 STEL

Canada - Ontario - OHSA - TWAEVs
25 ppm TWAEV; 31 mg/m3 TWAEV

Canada - Quebec - Time-Weighted Average Exposure Values
25 ppm TWAEV; 31 mg/m3 TWAEV

Israel - Time Weighted Averages
25 ppm TWA; 31 mg/m3 TWA

Israel - Action Levels
12.5 ppm AL; 15.5 mg/m3 AL

Mexico - Instruction No. 10 - TWAs
25 ppm TWA; 30 mg/m3 TWA

Mexico - Instruction No. 10 - STELs
35 ppm STEL; 45 mg/m3 STEL

STATE LISTS

California - Exposure Limits - PELs
25 ppm PEL; 30 mg/m3 PEL

California - Directors List of Hazardous Substances (8 CCR 339)
[present]

Florida Hazardous Substance List
[present]

Massachusetts Right To Know List
extraordinarily hazardous

Minnesota Hazardous Substance List
[present]

NJ Right to Know List (Total)
sn 1357

Pennsylvania Right to Know List
environmental hazard

5-NITROACENAPHTHENE   602-87-9

HEALTH AND SAFETY LISTS

IARC - Group 2B (sufficient animal data)
[present] (includes its sodium salts)

NIOSH - Selected LD50s and LC50s
Oral, rat: LD50 = 1470 mg/kg

NTP Chemical Status Reports - Testing Status and NTIS Number
Technical reports printed (PB266177/AS)

NTP Chemical Status Reports - Evidence of Carcinogenicity
male rat-positive; female rat-positive; male mice-positive; female mice-positive

NTP Seventh Report - Suspect Carcinogens
suspect carcinogen

OSHA - Possible Select Carcinogens
[present]

ENVIRONMENTAL LISTS

CERCLA/SARA - Section 313 - Emission Reporting
form R reporting required for 0.1% de minimus concentration

List of Pesticide Product Inert Ingredients
[present]

INTERNATIONAL LISTS

Canada - WHMIS: Ingredient Disclosure
0.1% item 1148 (110)

Canada - NPRI (National Pollutant Release Inventory)
[present]

Canada - Drinking Water Quality - MACs
0.4 mg/L MAC

STATE LISTS

California - Air Bill 2588 Appendix A-I
known or potential carcinogen

California - Prop. 65 - Cancer list
carcinogen - initial date 1/1/88

California - Prop. 65 - No Significant Risk Levels
no significant risk level = 100 ug/day

California - Directors List of Hazardous Substances (8 CCR 339)
[present]

Florida Hazardous Substance List
[present]

Massachusetts Right To Know List
carcinogen; extraordinarily hazardous

Minnesota Hazardous Substance List
carcinogen

NJ Right to Know List (Total)
sn 1358

NJ Special Hazardous Substances
(carcinogen)

Pennsylvania Right to Know List
environmental hazard; special hazardous substance

Pennsylvania RTK - Special Hazardous Substances
[present]

3-NITRO-P-ACETOPHENETIDE   1777-84-0

STATE LISTS

California - Air Bill 2588 Appendix A-II
known or potential carcinogen: 6/91

3'-NITROACETOPHENONE   121-89-1

HEALTH AND SAFETY LISTS

NIOSH - Selected LD50s and LC50s
Oral, rat: LD50 = 1460 mg/kg

M-NITROANILINE   99-09-2

HEALTH AND SAFETY LISTS

NIOSH - Selected LD50s and LC50s
Oral, rat: LD50 = 1100 mg/kg

ENVIRONMENTAL LISTS

List of Pesticide Product Inert Ingredients
[present]

STATE LISTS

Massachusetts Right To Know List
carcinogen; extraordinarily hazardous

P-NITROANILINE   100-01-6

ENVIRONMENTAL LISTS

List of Pesticide Product Inert Ingredients
[present]

N-NITROANILINE   645-55-6

ENVIRONMENTAL LISTS

ATSDR Priority List
Rank (of 275): 228

Safe Drinking Water Act - MCLs
as N: MCL = 1 mg/L (or 10 mg/L total with Nitrate)

Safe Drinking Water Act - MCLGs
as N: MCLG = 1 mg/L (or 10 mg/L total with Nitrate)

INTERNATIONAL LISTS

Mexico - Drinking Water - Ecological Criteria
0.05 mg/l

N-NITROANILINE   29757-24-2

HEALTH AND SAFETY LISTS

U.S. DOT - Substances From 49 CFR 172.101
regulated by DOT (UN2627)

U.S. DOT - Hazard Classes
DOT hazard class = 5.1

U.S. DOT - Appendix B - Marine Pollutants
DOT regulated marine pollutant

STATE LISTS

NJ Right to Know List (Total)
sn 2587

5-NITRO-O-ANISIDINE   99-59-2

HEALTH AND SAFETY LISTS

IARC - Group 2B (sufficient animal data)
[present] (Overall evaluation based only on evidence of carcinogenicity in monograph (16, 1978) or in Supplement 4)

NTP Chemical Status Reports - Testing Status and NTIS Number
Technical reports printed (PB287347/AS)

NTP Chemical Status Reports - Evidence of Carcinogenicity
male rat-positive; female rat-positive; male mice-negative; female mice-positive

OSHA - Possible Select Carcinogens
[present]

INTERNATIONAL LISTS

Canada - WHMIS: Ingredient Disclosure
1% item 1149 (1219)

German (DFG) - Carcinogens
animal evidence of carcinogenicity

STATE LISTS

California - Air Bill 2588 Appendix A-II
known or potential carcinogen

California - Prop. 65 - Cancer list
carcinogen - initial date 4/1/88

California - Prop. 65 - No Significant Risk Levels
no significant risk level = 6 ug/day

California - Directors List of Hazardous Substances (8 CCR 339)
[present]

Florida Hazardous Substance List
[present]

Massachusetts Right To Know List
carcinogen; extraordinarily hazardous

Minnesota Hazardous Substance List
carcinogen

Pennsylvania Right to Know List
special hazardous substance

Pennsylvania RTK - Special Hazardous Substances
[present]

2-NITROANISOLE   91-23-6

HEALTH AND SAFETY LISTS

NTP Chemical Status Reports - Testing Status and NTIS Number
Technical reports printed (PB299857/AS)

NTP Chemical Status Reports - Evidence of Carcinogenicity
male rat-negative; female rat-negative; male mice-positive; female mice-negative

STATE LISTS

Massachusetts Right To Know List
carcinogen; extraordinarily hazardous

4-NITROANISOLE   100-17-4

HEALTH AND SAFETY LISTS

NIOSH - Selected LD50s and LC50s
Oral, rat: LD50 = 3250 mg/kg Skin, rabbit: LD50 = 3 gm/kg

3-NITROANISOLE   555-03-3
SEE ALSO:
ANILINE AND CHLORO-, BROMO-, AND/OR NITROANILINES

HEALTH AND SAFETY LISTS

NIOSH - Selected LD50s and LC50s
Oral, rat: LD50 = 535 mg/kg

ENVIRONMENTAL LISTS

CAA - HON Rule - SOCMI Chemicals
compliance by Jan. 23, 1995

EPA - Master Testing List
[present]

RCRA - TSD Facilities Ground Water Monitoring
TM 8270 = 50 ug/L PQL

TSCA - Code of Federal Regulations Citations
40 CFR 712.30(d); 40 CFR 716.120(a)

TSCA - PAIR - Reporting List
Reporting Date: November 19, 1982

INTERNATIONAL LISTS

Canada - WHMIS: Ingredient Disclosure
1% item 1150 (1220)

9-NITROANTHRACENE   602-60-8
SEE ALSO:
COCO SHELL FLOUR
ANILINE AND CHLORO-, BROMO-, AND/OR NITROANILINES
F039-HAZARDOUS WASTES

HEALTH AND SAFETY LISTS

ACGIH 1995 - Time Weighted Averages
3 mg/m3 TWA

ACGIH 1995 - Skin Designations
skin - potential for cutaneous absorption

U.S. DOT - Appendix A Table 1 - Hazardous Substances
final RQ = 5000 pounds (2270 kg)

NFPA - Flash Points
flash point = 390 degrees F (199 degrees C)

NFPA - Hazard Identification Ratings
health-3; flammability-1; reactivity-3

NIOSH - Selected LD50s and LC50s
Oral, rat: LD50 = 750 mg/kg

NIOSH 1990 - Pocket Guide - RELs
3 mg/m3 TWA

NIOSH 1990 - Pocket Guide - IDLHs
300 mg/m3 IDLH

NIOSH 1990 - Pocket Guide - Target organs
blood, heart, lungs, liver

NIOSH 1990 - Pocket Guide - Skin list
Potential for dermal absorption

NTP Chemical Status Reports - Testing Status and NTIS Number
Prechronic studies for which toxicity technical reports were not prepared; technical reports printed (PB94-104528)

OSHA - Vacated PELs - Time Weighted Averages
3 mg/m3 TWA

OSHA - Vacated PELs - Skin Designation
Prevent or reduce skin absorption

OSHA - Final PELs - Time Weighted Averages
1 ppm TWA; 6 mg/m3 TWA

OSHA - Final PELs - Skin Notations
prevent or reduce skin absorption

OSHA - List of Highly Hazardous Chemicals
threshhold quantity = 5000 pounds

ENVIRONMENTAL LISTS

CERCLA/SARA - Section 313 - Emission Reporting
form R reporting required

CERCLA/SARA - Hazardous Substances and their Reportable Quantities
final RQ = 5000 pounds (2270 kg)

RCRA - P Series Wastes
waste number P077

RCRA - Hazardous Constituents-Appendix VIII
waste number P077

RCRA - Basis for Listing - Appendix VII
Included in waste stream: F039

RCRA - Substances Banned From Land Disposal
[present]

RCRA - TSD Facilities Ground Water Monitoring
TM 8270 = 50 ug/L PQL

RCRA - Universal Treatment Standards (LDR)
WW: 0.028 mg/l; NWW: 28 mg/kg

TSCA - Code of Federal Regulations Citations
40 CFR 712.30(d); 40 CFR 716.120(c); 40 CFR 799.5000

TSCA - PAIR - Reporting List
Reporting Date: November 19, 1982

TSCA - Health and Safety Reporting List
Effective Date: March 11, 1994; Sunset Date: March 11, 2004

TSCA - Substances Subject to Testing Consent Orders
Test for: Health Effects

TSCA - Section 12(b) - Export Notification
export notification required - Section 4

INTERNATIONAL LISTS

Australian Exposure Standards - Time Weighted Averages
3 mg/m3 TWA

Australian Exposure Standards - Skin Effects
skin absorption

Canada - WHMIS: Ingredient Disclosure
1% item 1152 (1222)

Canada - Alberta - 8 Hour Occupational Exposure Limit
3 mg/m3 TWA

Canada - Alberta - 15 Minute Occupational Exposure Limit
6 mg/m3 STEL

Canada - Alberta - Skin Designation
can be absorbed through the intact skin

Canada - British Columbia - 8 Hour Exposure Limits
1 ppm TWA; 6 mg/m3 TWA

Canada - British Columbia - 15 Minute Exposure Limits
2 ppm STEL; 12 mg/m3 STEL

Canada - British Columbia - Skin Notations
skin - potential for skin absorption

Canada - Ontario - OHSA - TWAEVs
3 mg/m3 TWAEV

Canada - Ontario - OHSA - Skin Notations
absorption through skin, eyes, or mucous membranes

Canada - Quebec - Short-term Exposure Values
3 ppm STEV

Canada - Quebec - Skin Designations
absorbed through the skin

United Kingdom - Occupational Exposure Standards - TWAs
6 mg/m3 TWA

United Kingdom - Occupational Exposure Standards - Notes
can be absorbed through skin

German (DFG) - MAK Values
1 ppm MAK; 6 mg/m3 MAK

German (DFG) - Skin/Sensitizers
danger of cutaneous absorption

Israel - Time Weighted Averages
3 mg/m3 TWA

Israel - Action Levels
1.5 mg/m3 AL

Mexico - Instruction No. 10 - TWAs
1 ppm TWA; 6 mg/m3 TWA

Mexico - Instruction No. 10 - Skin designation
skin - potential for cutaneous absorption

STATE LISTS

California - Exposure Limits - PELs
3 mg/m3 PEL

California - Exposure Limits - Skin Notation
material may be absorbed through the skin, eyes or mucous membrane

California - Directors List of Hazardous Substances (8 CCR 339)
[present]

Florida Hazardous Substance List
[present]

Massachusetts Right To Know List
[present]

Minnesota Hazardous Substance List
skin

NJ Right to Know List (Total)
sn 1548

NJ Special Hazardous Substances
(mutagen)

Pennsylvania Right to Know List
environmental hazard

PROPOSED REGULATIONS

TSCA - ITC 32nd Report Priority Testing List
designated for dermal absorption testing

4-NITROANTHRANILIC ACID   619-17-0

HEALTH AND SAFETY LISTS

U.S. DOT - Hazard Classes
Forbidden from transport by the DOT

P-NITROBENZALDEHYDE   555-16-8

HEALTH AND SAFETY LISTS

U.S. DOT - Substances From 49 CFR 172.101
regulated by DOT (UN1661)

U.S. DOT - Hazard Classes
DOT hazard class = 6.1

STATE LISTS

NJ Right to Know List (Total)
sn 1359

NJ Special Hazardous Substances
(reactive - third degree)

NITROBENZENE   98-95-3

HEALTH AND SAFETY LISTS

IARC - Group 3 (not classifiable)
[present]

NIOSH - Selected LD50s and LC50s
Oral, rat: LD50 = 704 mg/kg

NTP Chemical Status Reports - Testing Status and NTIS Number
Technical reports printed (PB287411/AS)

NTP Chemical Status Reports - Evidence of Carcinogenicity
male rat-positive; female rat-positive; male mice-equivocal; female mice-positive

ENVIRONMENTAL LISTS

CERCLA/SARA - Section 313 - Emission Reporting
form R reporting required for 0.1% de minimus concentration

INTERNATIONAL LISTS

Canada - WHMIS: Ingredient Disclosure
0.1% item 1153 (1223)

STATE LISTS

California - Air Bill 2588 Appendix A-II
known or potential carcinogen

California - Prop. 65 - Cancer list
carcinogen - initial date 10/1/89

California - Prop. 65 - No Significant Risk Levels
no significant risk level = 10 ug/day

California - Directors List of Hazardous Substances (8 CCR 339)
[present]

Florida Hazardous Substance List
[present]

Massachusetts Right To Know List
carcinogen; extraordinarily hazardous

Minnesota Hazardous Substance List
carcinogen

NJ Right to Know List (Total)
sn 1388

NJ Special Hazardous Substances
(carcinogen)

Pennsylvania Right to Know List
environmental hazard; special hazardous substance

Pennsylvania RTK - Special Hazardous Substances
[present]

M-NITROBENZENE DIAZONIUM PERCHLORATE   22751-24-2

HEALTH AND SAFETY LISTS

NIOSH - Selected LD50s and LC50s
Oral, rat: LD50 = 1980 mg/kg

NTP Chemical Status Reports - Testing Status and NTIS Number
Technical reports printed (no NTIS number given); prechronic studies for which toxicity technical reports were not prepared

ENVIRONMENTAL LISTS

CAA - HON Rule - SOCMI Chemicals
compliance by April 24, 1995

INTERNATIONAL LISTS

Canada - WHMIS: Ingredient Disclosure
1% item 1154 (1224)

German (DFG) - Carcinogens
animal evidence of carcinogenicity

STATE LISTS

California - Prop. 65 - Cancer list
carcinogen - initial date 10/1/92

PROPOSED REGULATIONS

NTP - Proposed Additions to Annual Report on Carcinogens
proposed as a suspect carcinogen for the NTP 9th report

M-NITROBENZENESULFONIC ACID   98-47-5

HEALTH AND SAFETY LISTS

U.S. DOT - Substances From 49 CFR 172.101
regulated by DOT (UN2730)

U.S. DOT - Hazard Classes
DOT hazard class = 6.1

NIOSH - Selected LD50s and LC50s
Oral, rat: LD50 = 2600 mg/kg

ENVIRONMENTAL LISTS

CAA - HON Rule - SOCMI Chemicals
compliance by April 24, 1995

INTERNATIONAL LISTS

Canada - WHMIS: Ingredient Disclosure
1% item 1156 (1226)

STATE LISTS

NJ Right to Know List (Total)
sn 1360

M-NITROBENZENESULFONIC ACID, SODIUM SALT   127-68-4

INTERNATIONAL LISTS

Canada - WHMIS: Ingredient Disclosure
1% item 1155 (1225)

NITROBENZENESULPHONIC ACID   31212-28-9

HEALTH AND SAFETY LISTS

IARC - Group 3 (not classifiable)
[present]

6-NITROBENZIMIDAZOLE   94-52-0

HEALTH AND SAFETY LISTS

NTP Chemical Status Reports - Testing Status and NTIS Number
Technical reports printed (PB286942/AS)

NTP Chemical Status Reports - Evidence of Carcinogenicity
male rat-negative; female rat-negative; male mice-negative; female mice-negative

7-NITROBENZO[A]ANTHRACENE   20268-51-3

HEALTH AND SAFETY LISTS

NIOSH - Selected LD50s and LC50s
Oral, rat: LD50 = 4700 mg/kg Skin, rat: LD50 = 16000 mg/kg

6-NITROBENZO(A)PYRENE   63041-90-7
SEE ALSO:
F039-HAZARDOUS WASTES
K141-HAZARDOUS WASTES
K103-HAZARDOUS WASTES
F004-HAZARDOUS WASTES
K104-HAZARDOUS WASTES
K083-HAZARDOUS WASTES

HEALTH AND SAFETY LISTS

ACGIH 1995 - Time Weighted Averages
1 ppm TWA; 5 mg/m3 TWA

ACGIH 1995 - Skin Designations
skin - potential for cutaneous absorption

ACGIH 1995 - Biological Exposure Indices
Total p-nitrophenol in urine: 5 mg/g creatinine, end of shift and week (Ns); Methemoglobin in blood: 1.5% of hemoglobin, end of shift (B, Ns, Sq)

AIHA - Odor Threshold Values
geometric mean air odor threshold = 0.37 ppm (detectable)

U.S. DOT - Substances From 49 CFR 172.101
regulated by DOT (UN1662)

U.S. DOT - Hazard Classes
DOT hazard class = 6.1

U.S. DOT - Appendix A Table 1 - Hazardous Substances
final RQ = 1000 pounds (454 kg)

NFPA - Flash Points
flash point = 190 degrees F (88 degrees C)

NFPA - Hazard Identification Ratings
health-3; flammability-2; reactivity-1

NIOSH - Selected LD50s and LC50s
Oral, rat: LD50 = 489 mg/kg Skin, rat: LD50 = 2100 mg/kg

NIOSH 1990 - Pocket Guide - RELs
1 ppm TWA; 5 mg/m3 TWA

NIOSH 1990 - Pocket Guide - IDLHs
200 ppm IDLH

NIOSH 1990 - Pocket Guide - Target organs
blood, liver, kidneys, skin, CVS

NIOSH 1990 - Pocket Guide - Skin list
Potential for dermal absorption

NTP Chemical Status Reports - Testing Status and NTIS Number
Prechronic studies for which toxicity technical reports were not prepared

OSHA - Vacated PELs - Time Weighted Averages
1 ppm TWA; 5 mg/m3 TWA

OSHA - Vacated PELs - Skin Designation
Prevent or reduce skin absorption

OSHA - Final PELs - Time Weighted Averages
1 ppm TWA; 5 mg/m3 TWA

OSHA - Final PELs - Skin Notations
prevent or reduce skin absorption

ENVIRONMENTAL LISTS

ATSDR Priority List
Rank (of 275): 274

CERCLA/SARA - Section 302 Extremely Hazardous Substances and TPQs
TPQ = 10,000 pounds

CERCLA/SARA - Section 313 - Emission Reporting
form R reporting required for 1.0% de minimus concentration

CERCLA/SARA - Hazardous Substances and their Reportable Quantities
final RQ = 1000 pounds (454 kg)

Clean Air Act (1990) - List of Hazardous Air Contaminants
[present]

CAA - HON Rule - SOCMI Chemicals
compliance by Oct. 24, 1994

CAA - HON Rule - Organic HAPs
[present]

Clean Water Act - Hazardous Substances
[present]

Clean Water Act - Priority Pollutants
[present]

Clean Water Act - Toxic Pollutants
[present]

RCRA - D Series - Maximum Concentration of Contaminants
waste number D036; regulatory level = 2.0 mg/L

RCRA - D Series - Chronic Toxicity Reference Levels
chronic toxicity reference level = 0.02 mg/L

RCRA - U Series Wastes
waste number U169 (Ignitable waste; Toxic waste)

RCRA - Hazardous Constituents-Appendix VIII
waste number U169

RCRA - Basis for Listing - Appendix VII
Included in waste streams: F004, F039, K083, K103, K104

RCRA - Substances Banned From Land Disposal
[present]

RCRA - TSD Facilities Ground Water Monitoring
TM 8090 = 40 ug/L PQL; TM 8270 = 10 ug/L PQL

RCRA - Universal Treatment Standards (LDR)
WW: 0.068 mg/l; NWW: 14 mg/kg

TSCA - Code of Federal Regulations Citations
40 CFR 712.30(d); 40 CFR 716.120(a)

TSCA - PAIR - Reporting List
Reporting Date: November 19, 1982

TSCA - Health and Safety Reporting List
Effective Date: October 4, 1982

INTERNATIONAL LISTS

Australian Exposure Standards - Time Weighted Averages
1 ppm TWA; 5 mg/m3 TWA

Australian Exposure Standards - Skin Effects
skin absorption

Canada - WHMIS: Ingredient Disclosure
1% item 1157 (1227)

Canada - NPRI (National Pollutant Release Inventory)
[present]

Canada - Alberta - 8 Hour Occupational Exposure Limit
1 ppm TWA; 5 mg/m3 TWA

Canada - Alberta - 15 Minute Occupational Exposure Limit
2 ppm STEL; 10 mg/m3 STEL

Canada - Alberta - Skin Designation
can be absorbed through the intact skin

Canada - British Columbia - 8 Hour Exposure Limits
1 ppm TWA; 5 mg/m3 TWA

Canada - British Columbia - 15 Minute Exposure Limits
2 ppm STEL; 10 mg/m3 STEL

Canada - British Columbia - Skin Notations
skin - potential for skin absorption

Canada - Ontario - OHSA - TWAEVs
1 ppm TWAEV; 5 mg/m3 TWAEV

Canada - Ontario - OHSA - Skin Notations
absorption through skin, eyes, or mucous membranes

Canada - Quebec - Time-Weighted Average Exposure Values
1 mg/m3 TWAEV

Canada - Quebec - Short-term Exposure Values
5 ppm STEV

Canada - Quebec - Skin Designations
absorbed through the skin

United Kingdom - Occupational Exposure Standards - TWAs
1 ppm TWA; 5 mg/m3 TWA

United Kingdom - Occupational Exposure Standards - STELs
2 ppm STEL; 10 mg/m3 STEL

United Kingdom - Occupational Exposure Standards - Notes
can be absorbed through skin

German (DFG) - MAK Values
1 ppm MAK; 5 mg/m3 MAK

German (DFG) - Peak Limitations
2 x normal MAK (30 min. average value); don't exceed 4 times during shift

German (DFG) - Skin/Sensitizers
danger of cutaneous absorption

German (DFG) - Pregnancy
classification not yet possible

Israel - Time Weighted Averages
1 ppm TWA; 5 mg/m3 TWA

Israel - Action Levels
0.5 ppm AL; 2.5 mg/m3 AL

Mexico - Instruction No. 10 - TWAs
1 ppm TWA; 5 mg/m3 TWA

Mexico - Instruction No. 10 - STELs
2 ppm STEL; 10 mg/m3 STEL

Mexico - Instruction No. 10 - Skin designation
skin - potential for cutaneous absorption

Mexico - Wastewater - Organic Toxic Pollutants and Heavy Metals
Listed under [Organic Toxic Pollutants]

Mexico - Drinking Water - Ecological Criteria
20.0 mg/l

STATE LISTS

California - Air Bill 2588 Appendix A-I
[present]

California - Exposure Limits - PELs
1 ppm PEL; 5 mg/m3 PEL

California - Exposure Limits - Skin Notation
material may be absorbed through the skin, eyes or mucous membrane

California - Directors List of Hazardous Substances (8 CCR 339)
[present]

Florida Hazardous Substance List
[present]

Massachusetts Right To Know List
extraordinarily hazardous

Minnesota Hazardous Substance List
skin

NJ Right to Know List (Total)
sn 1361

Pennsylvania Right to Know List
environmental hazard

PROPOSED REGULATIONS

Safe Drinking Water Act - Priority list
[present]

P-NITROBENZOIC ACID   62-23-7

HEALTH AND SAFETY LISTS

U.S. DOT - Hazard Classes
Forbidden from transport by the DOT

M-NITROBENZOIC ACID   121-92-6

INTERNATIONAL LISTS

Canada - WHMIS: Ingredient Disclosure
1% item 1159 (113)

NITROBENZOIC ACID OCTYL ESTER   RR-01687-5

HEALTH AND SAFETY LISTS

NIOSH - Selected LD50s and LC50s
Oral, rat: LD50 = 11 gm/kg

INTERNATIONAL LISTS

Canada - WHMIS: Ingredient Disclosure
1% item 1160 (114)

6-NITRO-2(3H)-BENZOOXAZOLONE   RR-00198-9

HEALTH AND SAFETY LISTS

U.S. DOT - Substances From 49 CFR 172.101
regulated by DOT (UN2305)

U.S. DOT - Hazard Classes
DOT hazard class = 8

INTERNATIONAL LISTS

Canada - WHMIS: Ingredient Disclosure
1% item 1158 (112)

STATE LISTS

NJ Right to Know List (Total)
sn 1362

NJ Special Hazardous Substances
(corrosive)

5-NITROBENZOTRIAZOL   2338-12-7

HEALTH AND SAFETY LISTS

NTP Chemical Status Reports - Testing Status and NTIS Number
Technical reports printed (PB293834/AS)

NTP Chemical Status Reports - Evidence of Carcinogenicity
male rat-negative; female rat-negative; male mice-positive; female mice-positive

STATE LISTS
Massachusetts Right To Know List
carcinogen; extraordinarily hazardous

3-NITROBENZOTRIFLUORIDE — 98-46-4
HEALTH AND SAFETY LISTS
IARC - Group 3 (not classifiable)
[present]

O-NITROBENZOTRIFLUORIDE — 384-22-5
HEALTH AND SAFETY LISTS
IARC - Group 3 (not classifiable)
[present]

P-NITROBENZOTRIFLUORIDE — 402-54-0
HEALTH AND SAFETY LISTS
NIOSH - Selected LD50s and LC50s
Oral, rat: LD50 - 1960 mg/kg
NTP Chemical Status Reports - Testing Status and NTIS Number
Galley or camera copy technical reports in progress; prechronic studies for which toxicity technical reports were not prepared
INTERNATIONAL LISTS
Canada - WHMIS: Ingredient Disclosure
1% item 1161 (115)

NITROBENZOTRIFLUORIDES — RR-01305-8
HEALTH AND SAFETY LISTS
NTP Chemical Status Reports - Testing Status and NTIS Number
Prechronic studies for which toxicity technical reports were not prepared

2-NITROBIPHENYL — 86-00-0
ENVIRONMENTAL LISTS
TSCA - Chemicals with Significant New Use Rules
PMN number: P-93-343

4-NITROBIPHENYL — 92-93-3
ENVIRONMENTAL LISTS
TSCA - Chemicals with Significant New Use Rules
PMN number: P-84-963

NITROBIPHENYL — 28984-85-2
HEALTH AND SAFETY LISTS
U.S. DOT - Substances From 49 CFR 172.101
regulated by DOT (UN0385)
U.S. DOT - Hazard Classes
DOT hazard class - 1.1D
STATE LISTS
NJ Right to Know List (Total)
sn 1363

O-NITROBROMOBENZENE — 577-19-5
HEALTH AND SAFETY LISTS
NFPA - Flash Points
flash point - 217 degrees F (103 degrees C)
NFPA - Hazard Identification Ratings
flammability-1
NIOSH - Selected LD50s and LC50s
Inhalation, rat: LC50 - 870 mg/kg 8 hr Oral, rat: LD50 - 610 mg/kg
INTERNATIONAL LISTS
Canada - WHMIS: Ingredient Disclosure
1% item 1162 (1228)

STATE LISTS
NJ Right to Know List (Total)
sn 1364

M-NITROBROMOBENZENE — 585-79-5
INTERNATIONAL LISTS
Canada - WHMIS: Ingredient Disclosure
1% item 1163 (1229)

P-NITROBROMOBENZENE — 586-78-7
INTERNATIONAL LISTS
Canada - WHMIS: Ingredient Disclosure
1% item 1164 (1230)

NITROBROMOBENZENES — RR-01355-8
HEALTH AND SAFETY LISTS
U.S. DOT - Substances From 49 CFR 172.101
regulated by DOT (UN2306)
U.S. DOT - Hazard Classes
DOT hazard class - 6.1
U.S. DOT - Appendix B - Marine Pollutants
DOT regulated marine pollutant

4-(2-NITROBUTYL)MORPHOLINE — 2224-44-4
HEALTH AND SAFETY LISTS
NIOSH - Selected LD50s and LC50s
Oral, rat: LD50 - 1230 mg/kg

NITROCELLULOSE — 9004-70-0
HEALTH AND SAFETY LISTS
ACGIH 1995 - Skin Designations
skin - potential for cutaneous absorption
ACGIH 1995 - Carcinogens
A1-confirmed human carcinogen
IARC - Group 3 (not classifiable)
[present]
NIOSH - Selected LD50s and LC50s
Oral, rat: LD50 - 2230 mg/kg
NIOSH 1990 - Pocket Guide - Carcinogens
occupational carcinogen
NIOSH 1990 - Pocket Guide - Target organs
bladder, blood
NIOSH - Health Standards - Exposure Limits
use 29 CFR 1910.1003
NIOSH - Health Standards - Health Effects and Precautions
has produced bladder tumors in animals
NIOSH - Health Standards - Carcinogenic Chemicals
potential human carcinogen
OSHA - 29 CFR 1910 Specifically Regulated Chemicals
Cancer suspect agent (see 29 CFR 1910.1003)
OSHA - Select Carcinogens
[present]
ENVIRONMENTAL LISTS
CERCLA/SARA - Section 313 - Emission Reporting
form R reporting required for 0.1% de minimus concentration
CERCLA/SARA - Hazardous Substances and their Reportable Quantities
final RQ - 1 pound (.454 kg)
Clean Air Act (1990) - List of Hazardous Air Contaminants
[present]
INTERNATIONAL LISTS
Australian Exposure Standards - Time Weighted Averages
prohibition recommended

Australian Exposure Standards - Carcinogens
confirmed carcinogen

Canada - WHMIS: Ingredient Disclosure
0.1% item 1165 (1231)

Canada - Alberta - Designated Substances
designated substance - requires code of practice

Canada - British Columbia - 8 Hour Exposure Limits
carcinogen with no permitted exposure or contact by any route

Canada - British Columbia - Carcinogens
carcinogen with no permitted exposure or contact by any route

Canada - Quebec - Time-Weighted Average Exposure Values
substance of which the recirculation is prohibited

Canada - Quebec - Skin Designations
absorbed through the skin

Canada - Quebec - Carcinogens
C1 carcinogen: effect detected in humans

German (DFG) - Skin/Sensitizers
danger of cutaneous absorption

German (DFG) - Carcinogens
animal evidence of carcinogenicity

Mexico - Instruction No. 10 - Carcinogens
carcinogen in humans

STATE LISTS

California - Air Bill 2588 Appendix A-I
known or potential carcinogen: 9/89

California - Prop. 65 - Cancer list
carcinogen - initial date 4/1/88

California - Exposure Limits - Carcinogens
cancer-suspect agent (at a concentration >= 0.1%)

California - Directors List of Hazardous Substances (8 CCR 339)
[present] (refers to any mixture containing 0.1% or greater)

Florida Hazardous Substance List
[present]

Massachusetts Right To Know List
carcinogen; extraordinarily hazardous

Minnesota Hazardous Substance List
carcinogen

NJ Right to Know List (Total)
sn 0229

Pennsylvania Right to Know List
environmental hazard

O-NITROCHLOROBENZENE   88-73-3

HEALTH AND SAFETY LISTS

NFPA - Flash Points
flash point = 290 degrees F (143 degrees C)

NFPA - Hazard Identification Ratings
health-2; flammability-1; reactivity-0

STATE LISTS

Florida Hazardous Substance List
[present]

Massachusetts Right To Know List
[present]

Pennsylvania Right to Know List
[present]

P-NITROCHLOROBENZENE   100-00-5

INTERNATIONAL LISTS

Canada - WHMIS: Ingredient Disclosure
1% item 1167 (1233)

M-NITROCHLOROBENZENE   121-73-3

INTERNATIONAL LISTS

Canada - WHMIS: Ingredient Disclosure
1% item 1166 (1232)

1-NITRO-4-CHLOROBENZO-2-TRIFLUORIDE   118-83-2

HEALTH AND SAFETY LISTS

U.S. DOT - Hazard Classes
Forbidden from transport by the DOT

INTERNATIONAL LISTS

Canada - WHMIS: Ingredient Disclosure
1% item 1168 (1234)

STATE LISTS

NJ Right to Know List (Total)
sn 1365

2-NITRO-1-CHLOROBENZO-4-TRIFLUORIDE   121-17-5

HEALTH AND SAFETY LISTS

U.S. DOT - Substances From 49 CFR 172.101
regulated by DOT (UN2732)

U.S. DOT - Hazard Classes
DOT hazard class = 6.1

4-NITRO-1-CHLOROBENZO-2-TRIFLUORIDE   777-37-7

INTERNATIONAL LISTS

German (DFG) - MAK Values
0.5 ppm MAK; 0.6 mg/m3 MAK

German (DFG) - Peak Limitations
2 x normal MAK (5 minute momentary value); don't exceed 8 times during shift

German (DFG) - Skin/Sensitizers
danger of sensitization

German (DFG) - Pregnancy
no risk to embryo/fetus if exposure limits adhered to

2-NITRO-4-CHLOROBENZO-1-TRIFLUORIDE   25889-38-7

HEALTH AND SAFETY LISTS

U.S. DOT - Substances From 49 CFR 172.101
regulated by DOT (UN2555, UN2556, UN2557, UN2059, UN0341, UN0342, UN0343, UN0340)

U.S. DOT - Hazard Classes
DOT hazard class = 1.1D; DOT hazard class = 1.3C; DOT hazard class = 3; DOT hazard class = 4.1 (depending on the percentage of nitrogen, alcohol, water and/or plasticizing agents)

NFPA - Flash Points
solution of nitrated cellulose in ether-alcohol: flash point < 0 degrees F (-18 degrees C)

NFPA - Hazard Identification Ratings
solution of nitrated cellulose in ether-alcohol: health-1; flammability-4; reactivity-0

OSHA - List of Highly Hazardous Chemicals
concentration > 12.6% nitrogen: threshhold quanitity = 2500 pounds

ENVIRONMENTAL LISTS

List of Pesticide Product Inert Ingredients
[present]

STATE LISTS

Florida Hazardous Substance List
[present]

Massachusetts Right To Know List
[present]

NJ Right to Know List (Total)
sn 1366; as Collodion: sn 2264

NJ Special Hazardous Substances
(flammable - fourth degree; reactive - third degree)

Pennsylvania Right to Know List
[present]

1-NITRO-2-CHLOROBENZO-3-TRIFLUORIDE   39974-35-1

HEALTH AND SAFETY LISTS

NIOSH - Selected LD50s and LC50s
Oral, rat: LD50 = 288 mg/kg

NTP Chemical Status Reports - Testing Status and NTIS Number
Technical reports printed (PB94-118262)

ENVIRONMENTAL LISTS

CAA - HON Rule - SOCMI Chemicals
compliance by Oct. 24, 1994

INTERNATIONAL LISTS

Canada - WHMIS: Ingredient Disclosure
1% item 1170 (1236)

German (DFG) - Skin/Sensitizers
danger of cutaneous absorption

German (DFG) - Carcinogens
suspected carcinogen

STATE LISTS

Florida Hazardous Substance List
[present]

Massachusetts Right To Know List
[present]

Pennsylvania Right to Know List
[present]

6-NITROCHRYSENE   7496-02-8

HEALTH AND SAFETY LISTS

ACGIH 1995 - Time Weighted Averages
0.1 ppm TWA; 0.64 mg/m3 TWA

ACGIH 1995 - Skin Designations
skin - potential for cutaneous absorption

NFPA - Flash Points
flash point = 261 degrees F (127 degrees C)

NFPA - Hazard Identification Ratings
health-2; flammability-1; reactivity-3

NIOSH - Selected LD50s and LC50s
Oral, rat: LD50 = 420 mg/kg Skin, rat: LD50 = 16 gm/kg

NIOSH 1990 - Pocket Guide - IDLHs
1000 mg/m3 IDLH (not considering carcinogenic effects)

NIOSH 1990 - Pocket Guide - Carcinogens
occupational carcinogen

NIOSH 1990 - Pocket Guide - Target organs
blood, liver, kidneys, CVS

NIOSH 1990 - Pocket Guide - Skin list
Potential for dermal absorption

NTP Chemical Status Reports - Testing Status and NTIS Number
Technical reports printed (PB94118262)

OSHA - Vacated PELs - Time Weighted Averages
1 mg/m3 TWA

OSHA - Vacated PELs - Skin Designation
Prevent or reduce skin absorption

OSHA - Final PELs - Time Weighted Averages
1 mg/m3 TWA

OSHA - Final PELs - Skin Notations
prevent or reduce skin absorption

ENVIRONMENTAL LISTS

CAA - HON Rule - SOCMI Chemicals
compliance by Oct. 24, 1994

TSCA - Health and Safety Reporting List
Effective Date: March 11, 1994; Sunset Date: March 11, 2004

INTERNATIONAL LISTS

Australian Exposure Standards - Time Weighted Averages
0.1 ppm TWA; 0.64 mg/m3 TWA

Australian Exposure Standards - Skin Effects
skin absorption

Canada - WHMIS: Ingredient Disclosure
1% item 1171 (1237)

Canada - Alberta - 8 Hour Occupational Exposure Limit
1 mg/m3 TWA

Canada - Alberta - 15 Minute Occupational Exposure Limit
2 mg/m3 STEL

Canada - Alberta - Skin Designation
can be absorbed through the intact skin

Canada - British Columbia - 8 Hour Exposure Limits
1 mg/m3 TWA

Canada - British Columbia - 15 Minute Exposure Limits
2 mg/m3 STEL

Canada - British Columbia - Skin Notations
skin - potential for skin absorption

Canada - Ontario - OHSA - TWAEVs
0.1 ppm TWAEV; 0.6 mg/m3 TWAEV

Canada - Quebec - Time-Weighted Average Exposure Values
0.1 ppm TWAEV; 0.64 mg/m3 TWAEV

Canada - Quebec - Skin Designations
absorbed through the skin

United Kingdom - Occupational Exposure Standards - TWAs
1 mg/m3 TWA

United Kingdom - Occupational Exposure Standards - STELs
2 mg/m3 STEL

United Kingdom - Occupational Exposure Standards - Notes
can be absorbed through skin

German (DFG) - MAK Values
total dust: 1 mg/m3 MAK

German (DFG) - Peak Limitations
2 x normal MAK (30 min. average value); don't exceed 4 times during shift

German (DFG) - Skin/Sensitizers
danger of cutaneous absorption

German (DFG) - Carcinogens
suspected carcinogen

Israel - Time Weighted Averages
0.1 ppm TWA; 0.64 mg/m3 TWA

Israel - Action Levels
0.05 ppm AL; 0.32 mg/m3 AL

Mexico - Instruction No. 10 - TWAs
1 mg/m3 TWA

Mexico - Instruction No. 10 - STELs
2 mg/m3 STEL

Mexico - Instruction No. 10 - Skin designation
skin - potential for cutaneous absorption

STATE LISTS

California - Exposure Limits - PELs
0.1 ppm PEL; 0.64 mg/m3 PEL

California - Exposure Limits - Skin Notation
material may be absorbed through the skin, eyes or mucous membrane

California - Directors List of Hazardous Substances (8 CCR 339)
[present]

Florida Hazardous Substance List
[present]

Massachusetts Right To Know List
[present]

Minnesota Hazardous Substance List
[present]

NJ Right to Know List (Total)
sn 1549

NJ Special Hazardous Substances
(corrosive)

Pennsylvania Right to Know List
[present]
PROPOSED REGULATIONS
TSCA - ITC 32nd Report Priority Testing List
designated for dermal absorption testing

NITROCRESOLS   12167-20-3
HEALTH AND SAFETY LISTS
NIOSH - Selected LD50s and LC50s
Oral, rat: LD50 = 470 mg/kg
ENVIRONMENTAL LISTS
CAA - HON Rule - SOCMI Chemicals
compliance by Oct. 24, 1994
INTERNATIONAL LISTS
Canada - WHMIS: Ingredient Disclosure
1% item 1169 (1235)
STATE LISTS
Florida Hazardous Substance List
[present]
Massachusetts Right To Know List
[present]
Pennsylvania Right to Know List
[present]

NITROCYCLOHEXANE   1122-60-7
STATE LISTS
NJ Right to Know List (Total)
sn 1368

6-NITRO-4-DIAZOTOLUENE-3-SULFONIC ACID   RR-01389-8
HEALTH AND SAFETY LISTS
U.S. DOT - Substances From 49 CFR 172.101
regulated by DOT (UN2307)
U.S. DOT - Hazard Classes
DOT hazard class = 6.1
U.S. DOT - Appendix B - Marine Pollutants
DOT regulated marine pollutant
NIOSH - Selected LD50s and LC50s
Oral, rat: LD50 = 1075 mg/kg
INTERNATIONAL LISTS
Canada - WHMIS: Ingredient Disclosure
1% item 1172 (1238)
STATE LISTS
NJ Right to Know List (Total)
sn 1367

P-NITRODIPHENYLAMINE   836-30-6
HEALTH AND SAFETY LISTS
NFPA - Flash Points
flash point = 275 degrees F (135 degrees C)
NFPA - Hazard Identification Ratings
flammability-1; reactivity-3
STATE LISTS
Florida Hazardous Substance List
[present]
Massachusetts Right To Know List
[present]
NJ Right to Know List (Total)
sn 1370
Pennsylvania Right to Know List
[present]

NITROETHANE   79-24-3
STATE LISTS
NJ Right to Know List (Total)
sn 1371

NITROETHYLENE POLYMER   26618-70-2
STATE LISTS
NJ Right to Know List (Total)
sn 1369

NITROETHYL NITRATE   4528-34-1
HEALTH AND SAFETY LISTS
IARC - Group 2B (sufficient animal data)
[present]
OSHA - Possible Select Carcinogens
[present]
STATE LISTS
California - Air Bill 2588 Appendix A-I
known or potential carcinogen: 6/91
California - Prop. 65 - Cancer list
carcinogen - initial date 10/1/90
California - Directors List of Hazardous Substances (8 CCR 339)
[present]
PROPOSED REGULATIONS
NTP - Proposed Additions to Annual Report on Carcinogens
proposed as a suspect carcinogen for NTP 8th report

NITROFEN   1836-75-5
HEALTH AND SAFETY LISTS
U.S. DOT - Substances From 49 CFR 172.101
regulated by DOT (UN2446)
U.S. DOT - Hazard Classes
DOT hazard class = 6.1
U.S. DOT - Appendix B - Marine Pollutants
DOT regulated marine pollutant
STATE LISTS
NJ Right to Know List (Total)
sn 1372

3-NITROFLUORANTHENE   892-21-7
HEALTH AND SAFETY LISTS
NFPA - Flash Points
flash point = 190 degrees F (88 degrees C)
NFPA - Hazard Identification Ratings
health-2; flammability-2; reactivity-3
NIOSH - Selected LD50s and LC50s
Inhalation, rat: LC50 = 150 mg/m3 4 hr Oral, mouse: LD50 = 250 mg/kg
ENVIRONMENTAL LISTS
CERCLA/SARA - Section 302 Extremely Hazardous Substances and TPQs
TPQ = 500 pounds
STATE LISTS
Florida Hazardous Substance List
[present]
Massachusetts Right To Know List
extraordinarily hazardous
NJ Right to Know List (Total)
sn 2588
Pennsylvania Right to Know List
environmental hazard

PROPOSED REGULATIONS

CERCLA/SARA - Proposed Hazardous Substance Additions
proposed RQ = 1 pound (.454 kg)

CERCLA/SARA - 1989 Proposed RQ Adjustments
proposed RQ = 100 pounds (45.4 kg)

2-NITROFLUORENE 607-57-8

HEALTH AND SAFETY LISTS

U.S. DOT - Hazard Classes
Forbidden from transport by the DOT

NITROFURANTOIN 67-20-9

ENVIRONMENTAL LISTS

EPA - Master Testing List
[present]

INTERNATIONAL LISTS

Canada - WHMIS: Ingredient Disclosure
1% item 1173 (1239)

NITROFURAZONE 59-87-0

HEALTH AND SAFETY LISTS

ACGIH 1995 - Time Weighted Averages
100 ppm TWA; 307 mg/m3 TWA

U.S. DOT - Substances From 49 CFR 172.101
regulated by DOT (UN2842)

U.S. DOT - Hazard Classes
DOT hazard class = 3

NFPA - Flash Points
flash point = 82 degrees F (28 degrees C)

NFPA - Hazard Identification Ratings
health-1; flammability-3; reactivity-3 (explodes on heating)

NIOSH - Selected LD50s and LC50s
Oral, rat: LD50 = 1100 mg/kg

NIOSH 1990 - Pocket Guide - RELs
100 ppm TWA; 310 mg/m3 TWA

NIOSH 1990 - Pocket Guide - IDLHs
1000 ppm IDLH

NIOSH 1990 - Pocket Guide - Target organs
skin

OSHA - Vacated PELs - Time Weighted Averages
100 ppm TWA; 310 mg/m3 TWA

OSHA - Final PELs - Time Weighted Averages
100 ppm TWA; 310 mg/m3 TWA

ENVIRONMENTAL LISTS

List of Pesticide Product Inert Ingredients
[present]

TSCA - Code of Federal Regulations Citations
40 CFR 712.30(w); 40 CFR 716.120(a)

TSCA - PAIR - Reporting List
Reporting Date: June 13, 1989

TSCA - Health and Safety Reporting List
Effective Date: April 13, 1989

INTERNATIONAL LISTS

Australian Exposure Standards - Time Weighted Averages
100 ppm TWA; 307 mg/m3 TWA

Canada - WHMIS: Ingredient Disclosure
1% item 1174 (1240)

Canada - Alberta - 8 Hour Occupational Exposure Limit
100 ppm TWA; 307 mg/m3 TWA

Canada - Alberta - 15 Minute Occupational Exposure Limit
150 ppm STEL; 460 mg/m3 STEL

Canada - British Columbia - 8 Hour Exposure Limits
100 ppm TWA; 310 mg/m3 TWA

Canada - British Columbia - 15 Minute Exposure Limits
150 ppm STEL; 465 mg/m3 STEL

Canada - Ontario - OHSA - TWAEVs
100 ppm TWAEV; 306 mg/m3 TWAEV

Canada - Quebec - Time-Weighted Average Exposure Values
100 ppm TWAEV; 307 mg/m3 TWAEV

United Kingdom - Occupational Exposure Standards - TWAs
100 ppm TWA; 310 mg/m3 TWA

German (DFG) - MAK Values
100 ppm MAK; 310 mg/m3 MAK

Israel - Time Weighted Averages
100 ppm TWA; 307 mg/m3 TWA

Israel - Action Levels
50 ppm AL; 153.5 mg/m3 AL

Mexico - Instruction No. 10 - TWAs
100 ppm TWA; 310 mg/m3 TWA

Mexico - Instruction No. 10 - STELs
150 ppm STEL; 465 mg/m3 STEL

STATE LISTS

California - Exposure Limits - PELs
100 ppm PEL; 310 mg/m3 PEL

California - Directors List of Hazardous Substances (8 CCR 339)
[present]

Florida Hazardous Substance List
[present]

Massachusetts Right To Know List
[present]

Minnesota Hazardous Substance List
[present]

NJ Right to Know List (Total)
sn 1373

NJ Special Hazardous Substances
(flammable - third degree; reactive - third degree)

Pennsylvania Right to Know List
[present]

1-((5-NITROFURFURYLIDENE)AMINO)-2-IMIDAZOLIDINONE 555-84-0

HEALTH AND SAFETY LISTS

U.S. DOT - Hazard Classes
Forbidden from transport by the DOT

N-[4-(5-NITRO-2-FURYL)-2-THIAZOLYL]ACETAMIDE 531-82-8

HEALTH AND SAFETY LISTS

U.S. DOT - Hazard Classes
Forbidden from transport by the DOT

NITROGEN 7727-37-9

HEALTH AND SAFETY LISTS

IARC - Group 2B (sufficient animal data)
[present] (Overall evaluation based only on evidence of carcinogenicity in monograph (30, 1983) or in Supplement 4)

NIOSH - Selected LD50s and LC50s
Oral, rat: LD50 = 740 mg/kg Skin, rat: LD50 = 5000 mg/kg

NTP Chemical Status Reports - Testing Status and NTIS Number
Technical reports printed (PB296038/AS) (PB277440/AS)

NTP Chemical Status Reports - Evidence of Carcinogenicity
PB296038/AS: male rat-negative; female rat-negative; male mice-positive; female mice-positive; PB277440/AS: male rat-inadequate; female rat-positive; male mice-positive; female mice-positive

NTP Seventh Report - Suspect Carcinogens
suspect carcinogen

OSHA - Possible Select Carcinogens
[present]

ENVIRONMENTAL LISTS
  CERCLA/SARA - Section 313 - Emission Reporting
  form R reporting required for 0.1% de minimus concentration
STATE LISTS
  California - Air Bill 2588 Appendix A-II
  known or potential carcinogen
  California - Prop. 65 - Cancer list
  carcinogen - initial date 1/1/88
  California - Prop. 65 - No Significant Risk Levels
  no significant risk level = 9 ug/day
  California - Directors List of Hazardous Substances (8 CCR 339)
  [present]
  Florida Hazardous Substance List
  [present]
  Massachusetts Right To Know List
  carcinogen; extraordinarily hazardous
  Minnesota Hazardous Substance List
  carcinogen
  NJ Right to Know List (Total)
  sn 1374
  NJ Special Hazardous Substances
  (carcinogen)
  Pennsylvania Right to Know List
  environmental hazard; special hazardous substance
  Pennsylvania RTK - Special Hazardous Substances
  [present]

NITROGEN COMPOUNDS, INORGANIC   RR-01026-4
HEALTH AND SAFETY LISTS
  IARC - Group 3 (not classifiable)
  [present]

NITROGEN DIOXIDE   10102-44-0
HEALTH AND SAFETY LISTS
  IARC - Group 2B (sufficient animal data)
  [present]
  OSHA - Possible Select Carcinogens
  [present]
STATE LISTS
  California - Air Bill 2588 Appendix A-I
  known or potential carcinogen: 6/91
  California - Prop. 65 - Cancer list
  carcinogen - initial date 10/1/90
  California - Directors List of Hazardous Substances (8 CCR 339)
  [present]

NITROGEN FERTILIZER SOLUTION   RR-00381-6
HEALTH AND SAFETY LISTS
  IARC - Group 3 (not classifiable)
  [present]
  NIOSH - Selected LD50s and LC50s
  Oral, rat: LD50 = 604 mg/kg
  NTP Chemical Status Reports - Testing Status and NTIS Number
  Technical reports printed (PB90197930/AS)
  NTP Chemical Status Reports - Evidence of Carcinogenicity
  male rat-sufficient evidence; female rat-no evidence; male mice-no evidence; female mice-clear evidence
STATE LISTS
  California - Air Bill 2588 Appendix A-II
  known or potential carcinogen: 6/91
  California - Prop. 65 - Reproductive - Male
  male reproductive toxicity - initial date 4/1/91

NITROGEN FIXING BACTERIA   RR-01102-9
HEALTH AND SAFETY LISTS
  IARC - Group 3 (not classifiable)
  [present]
  NTP Chemical Status Reports - Testing Status and NTIS Number
  Technical reports printed (PB89102388/AS)
  NTP Chemical Status Reports - Evidence of Carcinogenicity
  male rat-equivocal evidence; female rat-clear evidence; male mice-no evidence; female mice-clear evidence
STATE LISTS
  California - Air Bill 2588 Appendix A-II
  known or potential carcinogen: 9/90
  California - Prop. 65 - Cancer list
  carcinogen - initial date 1/1/90
  California - Prop. 65 - No Significant Risk Levels
  no significant risk level = 0.5 ug/day
  California - Directors List of Hazardous Substances (8 CCR 339)
  [present]

NITROGEN FLUORIDE OXIDE   RR-01168-7
HEALTH AND SAFETY LISTS
  IARC - Group 2B (sufficient animal data)
  [present] (Overall evaluation based only on evidence of carcinogenicity in monograph (7, 1974) or in Supplement 4)
  OSHA - Possible Select Carcinogens
  [present]
STATE LISTS
  California - Air Bill 2588 Appendix A-II
  known or potential carcinogen
  California - Prop. 65 - Cancer list
  carcinogen - initial date 4/1/88
  California - Prop. 65 - No Significant Risk Levels
  no significant risk level = 0.4 ug/day
  California - Directors List of Hazardous Substances (8 CCR 339)
  [present]
  Florida Hazardous Substance List
  [present]
  Massachusetts Right To Know List
  carcinogen; extraordinarily hazardous
  Minnesota Hazardous Substance List
  carcinogen
  Pennsylvania Right to Know List
  special hazardous substance
  Pennsylvania RTK - Special Hazardous Substances
  [present]

NITROGEN MUSTARD   51-75-2
HEALTH AND SAFETY LISTS
  IARC - Group 2B (sufficient animal data)
  [present] (Overall evaluation based only on evidence of carcinogenicity in monograph (7, 1974) or in Supplement 4)
  OSHA - Possible Select Carcinogens
  [present]
STATE LISTS
  California - Air Bill 2588 Appendix A-II
  known or potential carcinogen
  California - Prop. 65 - Cancer list
  carcinogen - initial date 4/1/88
  California - Prop. 65 - No Significant Risk Levels
  no significant risk level = 0.5 ug/day
  California - Directors List of Hazardous Substances (8 CCR 339)
  [present]

Florida Hazardous Substance List
[present]

Massachusetts Right To Know List
carcinogen; extraordinarily hazardous

Minnesota Hazardous Substance List
carcinogen

Pennsylvania Right to Know List
special hazardous substance

Pennsylvania RTK - Special Hazardous Substances
[present]

NITROGEN MUSTARD HYDROCHLORIDE 55-86-7

HEALTH AND SAFETY LISTS

ACGIH 1995 - Time Weighted Averages
simple asphyxiant

U.S. DOT - Substances From 49 CFR 172.101
regulated by DOT (UN1977, UN1066)

U.S. DOT - Hazard Classes
DOT hazard class = 2.2

ENVIRONMENTAL LISTS

List of Pesticide Product Inert Ingredients
[present]

INTERNATIONAL LISTS

Australian Exposure Standards - Time Weighted Averages
Asphyxiant at < 18% oxygen by volume

Canada - British Columbia - 8 Hour Exposure Limits
asphyxiant substance

Canada - Ontario - OHSA - TWAEVs
simple asphyxiant

Canada - Quebec - Time-Weighted Average Exposure Values
simple asphyxiant

Israel - Time Weighted Averages
Asphyxiant

STATE LISTS

California - Exposure Limits - PELs
asphyxiant (limit depends on level of oxygen)

Florida Hazardous Substance List
[present]

Massachusetts Right To Know List
[present]

NJ Right to Know List (Total)
sn 1375

Pennsylvania Right to Know List
[present]

NITROGEN MUSTARD N-OXIDE 126-85-2

STATE LISTS

California - Directors List of Hazardous Substances (8 CCR 339)
[present]

NITROGEN MUSTARD N-OXIDE HYDROCHLORIDE 302-70-5

HEALTH AND SAFETY LISTS

ACGIH 1995 - Time Weighted Averages
3 ppm TWA; 5.6 mg/m3 TWA

ACGIH 1995 - Short Term Exposure Limits
5 ppm STEL; 9.4 mg/m3 STEL

AIHA - Odor Threshold Values
no geometric mean air odor threshold

U.S. DOT - Substances From 49 CFR 172.101
regulated by DOT (UN1067)

U.S. DOT - Hazard Classes
DOT hazard class = 2.3

U.S. DOT - Appendix A Table 1 - Hazardous Substances
final RQ = 10 pounds (4.54 kg)

NIOSH - Selected LD50s and LC50s
Inhalation, rat: LC50 = 88 ppm 4 hr

NIOSH 1990 - Pocket Guide - RELs
1 ppm STEL; 1.8 mg/m3 STEL

NIOSH 1990 - Pocket Guide - IDLHs
50 ppm IDLH

NIOSH 1990 - Pocket Guide - Target organs
CVS, respiratory system

NIOSH - Health Standards - Exposure Limits
C (15 min) 1 ppm; C (15 min) 1.8 mg/m3 (Listed under 'Nitrogen, oxides of')

NIOSH - Health Standards - Health Effects and Precautions
Respiratory effects; blood effects (Pulmonary function testing required) (Listed under 'Nitrogen, oxides of')

OSHA - Vacated PELs - Short Term Exposure Limits
1 ppm STEL; 1.8 mg/m3 STEL

OSHA - Final PELs - Ceiling Limits
C 5 ppm; C 9 mg/m3

OSHA - List of Highly Hazardous Chemicals
threshhold quantity = 250 pounds

ENVIRONMENTAL LISTS

CERCLA/SARA - Section 302 Extremely Hazardous Substances and TPQs
TPQ = 100 pounds

CERCLA/SARA - Hazardous Substances and their Reportable Quantities
final RQ = 10 pounds (4.54 kg)

Clean Water Act - Hazardous Substances
[present]

RCRA - P Series Wastes
waste number P078

RCRA - Hazardous Constituents-Appendix VIII
waste number P078

RCRA - Substances Banned From Land Disposal
[present]

INTERNATIONAL LISTS

Australian Exposure Standards - Time Weighted Averages
3 ppm TWA; 5.6 mg/m3 TWA

Australian Exposure Standards - Short Term Exposure Limits
5 ppm STEL; 9.4 mg/m3 STEL

Australian Exposure Standards - Under Review
exposure limits under review

Canada - WHMIS: Ingredient Disclosure
1% item 1175 (769)

Canada - National Air Quality Objectives - Schedule I
desirable limits: 0-60 ug/m3 year; acceptable limits: 0-100 ug/m3 year; 0-200 ug/m3 day; 0-400 ug/m3 hour; tolerable limits: 200-300 ug/m3 day; 400-1000 ug/mg hour;

Canada - Alberta - 8 Hour Occupational Exposure Limit
3 ppm TWA; 6 mg/m3 TWA

Canada - Alberta - 15 Minute Occupational Exposure Limit
5 ppm STEL; 9.4 mg/m3 STEL

Canada - British Columbia - 8 Hour Exposure Limits
1 ppm TWA

Canada - British Columbia - Ceiling Exposure Limits
C 5 ppm; C 9 mg/m3

Canada - Ontario - OHSA - TWAEVs
3 ppm TWAEV; 5.6 mg/m3 TWAEV

Canada - Ontario - OHSA - STEVs
5 ppm STEV; 9.4 mg/m3 STEV

Canada - Quebec - Time-Weighted Average Exposure Values
3 ppm TWAEV; 5.6 mg/m3 TWAEV

United Kingdom - Occupational Exposure Standards - TWAs
3 ppm TWA; 5 mg/m3 TWA

United Kingdom - Occupational Exposure Standards - STELs
5 ppm STEL; 9 mg/m3 STEL

German (DFG) - MAK Values
5 ppm MAK; 9 mg/m3 MAK

German (DFG) - Peak Limitations
2 x normal MAK (5 min momentary value); don't exceed 8 times during shift

Israel - Time Weighted Averages
3 ppm TWA; 5.6 mg/m3 TWA

Israel - Short Term Exposure Limits
5 ppm STEL; 9.4 mg/m3 STEL

Israel - Action Levels
1.5 ppm AL; 2.8 mg/m3 AL

Mexico - Instruction No. 10 - TWAs
3 ppm TWA; 6 mg/m3 TWA

Mexico - Instruction No. 10 - STELs
5 ppm STEL; 10 mg/m3 STEL

STATE LISTS

California - Exposure Limits - STELs
1 ppm STEL; 1.8 mg/m3 STEL

California - Directors List of Hazardous Substances (8 CCR 339)
[present]

Florida Hazardous Substance List
[present]

Massachusetts Right To Know List
extraordinarily hazardous

Minnesota Hazardous Substance List
[present]

NJ Right to Know List (Total)
sn 1376

Pennsylvania Right to Know List
environmental hazard

PROPOSED REGULATIONS

Canada - Ontario - Proposed Occupational TWAEVs
1 ppm TWAEV; 2 mg/m3 TWAEV

NITROGEN TETROXIDE   10544-72-6

STATE LISTS

NJ Right to Know List (Total)
sn 2589

NITROGEN TRICHLORIDE   10025-85-1

ENVIRONMENTAL LISTS

List of Pesticide Product Inert Ingredients
[present]

NITROGEN TRIFLUORIDE   7783-54-2

HEALTH AND SAFETY LISTS

U.S. DOT - Substances Which Are Poisonous by Inhalation
gaseous hazardous material poisonous by inhalation

NITROGEN TRIIODIDE   13444-85-4

HEALTH AND SAFETY LISTS

IARC - Group 2A (limited human data)
[present]

NIOSH - Selected LD50s and LC50s
Inhalation, rat: LC50 = 600 mg/m3 2 mn Oral, rat: LD50 = 10 mg/kg Skin, rat: LD50 = 12 mg/kg

OSHA - Possible Select Carcinogens
[present]

ENVIRONMENTAL LISTS

CERCLA/SARA - Section 302 Extremely Hazardous Substances and TPQs
TPQ = 10 pounds

CERCLA/SARA - Section 313 - Emission Reporting
form R reporting required for 0.1% de minimus concentration

RCRA - Hazardous Constituents-Appendix VIII
hazardous constituent - no waste number

INTERNATIONAL LISTS

Canada - WHMIS: Ingredient Disclosure
1% item 1022 (1119)

German (DFG) - Skin/Sensitizers
danger of cutaneous absorption; danger of sensitization (skin or respiratory)

German (DFG) - Carcinogens
proven carcinogen

STATE LISTS

California - Air Bill 2588 Appendix A-II
known or potential carcinogen

California - Prop. 65 - Cancer list
carcinogen - initial date 1/1/88

California - Prop. 65 - Developmental Toxicity
developmental toxicity - initial date 1/1/89

California - Directors List of Hazardous Substances (8 CCR 339)
[present]

Florida Hazardous Substance List
[present]

Massachusetts Right To Know List
carcinogen; extraordinarily hazardous; teratogen

Minnesota Hazardous Substance List
carcinogen

NJ Right to Know List (Total)
sn 1377

NJ Special Hazardous Substances
(carcinogen; mutagen; teratogen)

Pennsylvania Right to Know List
environmental hazard

PROPOSED REGULATIONS

CERCLA/SARA - Proposed Hazardous Substance Additions
proposed RQ = 1 pound (.454 kg)

CERCLA/SARA - 1989 Proposed RQ Adjustments
proposed RQ = 10 pounds (4.54 kg)

NITROGEN TRIIODIDE MONOAMINE   RR-01452-8

HEALTH AND SAFETY LISTS

NIOSH - Selected LD50s and LC50s
Oral, rat: LD50 = 10 mg/kg

NTP Seventh Report - Suspect Carcinogens
suspect carcinogen

OSHA - Possible Select Carcinogens
[present]

ENVIRONMENTAL LISTS

RCRA - Hazardous Constituents-Appendix VIII
hazardous constituent - no waste number

STATE LISTS

California - Air Bill 2588 Appendix A-II
known or potential carcinogen: 9/89

California - Prop. 65 - Cancer list
carcinogen - initial date 4/1/88

California - Prop. 65 - Developmental Toxicity
developmental toxicity - initial date 7/1/90

California - Directors List of Hazardous Substances (8 CCR 339)
[present]

Florida Hazardous Substance List
[present]

Massachusetts Right To Know List
carcinogen; extraordinarily hazardous

Pennsylvania Right to Know List
environmental hazard; special hazardous substance

Pennsylvania RTK - Special Hazardous Substances
[present]

NITROGEN TRIOXIDE 10544-73-7

HEALTH AND SAFETY LISTS

IARC - Group 2B (sufficient animal data)
[present] (Overall evaluation based only on evidence of carcinogenicity in monograph [9,1975] or in Supplement 4)

OSHA - Possible Select Carcinogens
[present]

ENVIRONMENTAL LISTS

RCRA - Hazardous Constituents-Appendix VIII
hazardous constituent - no waste number

STATE LISTS

California - Prop. 65 - Cancer list
carcinogen - initial date 4/1/88

California - Directors List of Hazardous Substances (8 CCR 339)
[present]

Florida Hazardous Substance List
[present]

Massachusetts Right To Know List
carcinogen; extraordinarily hazardous

NITROGEN TRIOXIDE 12033-49-7

HEALTH AND SAFETY LISTS

NIOSH - Selected LD50s and LC50s
Oral, rat: LD50 = 60 mg/kg

ENVIRONMENTAL LISTS

RCRA - Hazardous Constituents-Appendix VIII
hazardous constituent - no waste number

STATE LISTS

California - Air Bill 2588 Appendix A-I
known or potential carcinogen

California - Prop. 65 - Cancer list
carcinogen - initial date 4/1/88

California - Directors List of Hazardous Substances (8 CCR 339)
[present]

Florida Hazardous Substance List
[present]

Massachusetts Right To Know List
carcinogen; extraordinarily hazardous

NJ Special Hazardous Substances
(carcinogen)

Pennsylvania Right to Know List
environmental hazard; special hazardous substance

Pennsylvania RTK - Special Hazardous Substances
[present]

NITROGLYCERIN 55-63-0

HEALTH AND SAFETY LISTS

U.S. DOT - Substances From 49 CFR 172.101
regulated by DOT (UN1975)

U.S. DOT - Hazard Classes
DOT hazard class = 2.3

U.S. DOT - Substances Which Are Poisonous by Inhalation
gaseous hazardous material poisonous by inhalation (UN1067)

U.S. DOT - Appendix A Table 1 - Hazardous Substances
final RQ = 10 pounds (4.54 kg)

NIOSH - Selected LD50s and LC50s
Inhalation, rabbit: LC50 = 315 ppm (15 mn)

OSHA - List of Highly Hazardous Chemicals
threshhold quantity = 250 pounds

ENVIRONMENTAL LISTS

CERCLA/SARA - Hazardous Substances and their Reportable Quantities
final RQ = 10 pounds (4.54 kg)

INTERNATIONAL LISTS

Canada - WHMIS: Ingredient Disclosure
1% item 1176 (1602)

STATE LISTS

Florida Hazardous Substance List
[present]

Massachusetts Right To Know List
[present]

NJ Right to Know List (Total)
sn 1379

Pennsylvania Right to Know List
environmental hazard

1-NITROGUANIDINE 556-88-7

HEALTH AND SAFETY LISTS

U.S. DOT - Hazard Classes
Forbidden from transport by the DOT

NITROGUANIDINE NITRATE RR-01453-9

HEALTH AND SAFETY LISTS

ACGIH 1995 - Time Weighted Averages
10 ppm TWA; 29 mg/m3 TWA

U.S. DOT - Substances From 49 CFR 172.101
regulated by DOT (UN2451)

U.S. DOT - Hazard Classes
DOT hazard class = 2.2

NIOSH - Selected LD50s and LC50s
Inhalation, rat: LC50 = 6700 ppm 1 hr

NIOSH 1990 - Pocket Guide - RELs
10 ppm TWA; 29 mg/m3 TWA

NIOSH 1990 - Pocket Guide - IDLHs
2000 ppm IDLH

NIOSH 1990 - Pocket Guide - Target organs
in animals: blood

OSHA - Vacated PELs - Time Weighted Averages
10 ppm TWA; 29 mg/m3 TWA

OSHA - Final PELs - Time Weighted Averages
10 ppm TWA; 29 mg/m3 TWA

OSHA - List of Highly Hazardous Chemicals
threshhold quantity = 5000 pounds

INTERNATIONAL LISTS

Australian Exposure Standards - Time Weighted Averages
10 ppm TWA; 29 mg/m3 TWA

Canada - WHMIS: Ingredient Disclosure
1% item 1177 (1673)

Canada - Alberta - 8 Hour Occupational Exposure Limit
10 ppm TWA; 29 mg/m3 TWA

Canada - Alberta - 15 Minute Occupational Exposure Limit
15 ppm STEL; 44 mg/m3 STEL

Canada - British Columbia - 8 Hour Exposure Limits
10 ppm TWA; 29 mg/m3 TWA

Canada - British Columbia - 15 Minute Exposure Limits
15 ppm STEL; 45 mg/m3 STEL

Canada - Ontario - OHSA - TWAEVs
10 ppm TWAEV; 29 mg/m3 TWAEV

Canada - Quebec - Time-Weighted Average Exposure Values
10 ppm TWAEV; 29 mg/m3 TWAEV

United Kingdom - Occupational Exposure Standards - TWAs
10 ppm TWA; 30 mg/m3 TWA

United Kingdom - Occupational Exposure Standards - STELs
15 ppm STEL; 45 mg/m3 STEL

Israel - Time Weighted Averages
10 ppm TWA; 29 mg/m3 TWA

Israel - Action Levels
5 ppm AL; 14.5 mg/m3 AL

Mexico - Instruction No. 10 - TWAs
10 ppm TWA; 30 mg/m3 TWA

Mexico - Instruction No. 10 - STELs
15 ppm STEL; 45 mg/m3 STEL

STATE LISTS

California - Exposure Limits - PELs
10 ppm PEL; 29 mg/m3 PEL

California - Directors List of Hazardous Substances (8 CCR 339)
[present]

Florida Hazardous Substance List
[present]

Massachusetts Right To Know List
[present]

Minnesota Hazardous Substance List
[present]

NJ Right to Know List (Total)
sn 1380

Pennsylvania Right to Know List
[present]

1-NITROHYDANTOIN   2825-15-2

HEALTH AND SAFETY LISTS

U.S. DOT - Hazard Classes
Forbidden from transport by the DOT

NITROHYDROCHLORIC ACID (AQUA REGIA)   8007-56-5

HEALTH AND SAFETY LISTS

U.S. DOT - Hazard Classes
Forbidden from transport by the DOT

NITRO ISOBUTANE TRIOL TRINITRATE   20820-44-4

HEALTH AND SAFETY LISTS

U.S. DOT - Substances From 49 CFR 172.101
regulated by DOT (UN2421)

U.S. DOT - Hazard Classes
DOT hazard class = 2.3

U.S. DOT - Substances Which Are Poisonous by Inhalation
gaseous hazardous material poisonous by inhalation (UN2421)

OSHA - List of Highly Hazardous Chemicals
threshhold quantity = 250 pounds

INTERNATIONAL LISTS

Canada - WHMIS: Ingredient Disclosure
1% item 1178 (1693)

STATE LISTS

Florida Hazardous Substance List
[present]

Massachusetts Right To Know List
[present]

NJ Right to Know List (Total)
sn 1381

Pennsylvania Right to Know List
[present]

NITROMETHANE   75-52-5

STATE LISTS

NJ Right to Know List (Total)
sn 1382

Pennsylvania Right to Know List
[present]

N-NITRO-N-METHYLGLYCOLAMIDE NITRATE   RR-01447-1

HEALTH AND SAFETY LISTS

ACGIH 1995 - Time Weighted Averages
0.05 ppm TWA; 0.46 mg/m3 TWA

ACGIH 1995 - Skin Designations
skin - potential for cutaneous absorption

U.S. DOT - Substances From 49 CFR 172.101
regulated by DOT (UN1204, UN3064, UN0143, NA0144)

U.S. DOT - Hazard Classes
Forbidden from transport by the DOT

U.S. DOT - Appendix A Table 1 - Hazardous Substances
final RQ = 10 pounds (4.54 kg)

NFPA - Flash Points
explodes

NFPA - Hazard Identification Ratings
health-2; flammability-2; reactivity-4

NIOSH - Selected LD50s and LC50s
Oral, rat: LD50 = 105 mg/kg

NIOSH 1990 - Pocket Guide - RELs
0.1 mg/m3 STEL

NIOSH 1990 - Pocket Guide - IDLHs
500 mg/m3 IDLH

NIOSH 1990 - Pocket Guide - Target organs
CVS, blood, skin

NIOSH 1990 - Pocket Guide - Skin list
Potential for dermal absorption

NIOSH - Health Standards - Exposure Limits
C (20 min) 0.1 mg/m3 (recommended limit for substance alone or with Ethylene glycol dinitrate)

NIOSH - Health Standards - Health Effects and Precautions
Circulatory system effects (Prevent skin contact)

OSHA - Vacated PELs - Short Term Exposure Limits
0.1 mg/m3 STEL (not in effect as a result of reconsideration)

OSHA - Vacated PELs - Skin Designation
Prevent or reduce skin absorption

OSHA - Final PELs - Ceiling Limits
C 0.2 ppm; C 2 mg/m3

OSHA - Final PELs - Skin Notations
prevent or reduce skin absorption

ENVIRONMENTAL LISTS

CERCLA/SARA - Section 313 - Emission Reporting
form R reporting required for 1.0% de minimus concentration

CERCLA/SARA - Hazardous Substances and their Reportable Quantities
final RQ = 10 pounds (4.54 kg)

RCRA - P Series Wastes
waste number P081 (Reactive waste)

RCRA - Hazardous Constituents-Appendix VIII
waste number P081

RCRA - Substances Banned From Land Disposal
[present]

INTERNATIONAL LISTS

Australian Exposure Standards - Time Weighted Averages
0.05 ppm TWA; 0.46 mg/m3 TWA

Australian Exposure Standards - Skin Effects
skin absorption
Canada - NPRI (National Pollutant Release Inventory)
[present]
Canada - Alberta - 8 Hour Occupational Exposure Limit
0.02 ppm TWA; 0.19 mg/m3 TWA
Canada - Alberta - 15 Minute Occupational Exposure Limit
0.05 ppm STEL; 0.48 mg/m3 STEL
Canada - Alberta - Skin Designation
can be absorbed through the intact skin
Canada - British Columbia - Ceiling Exposure Limits
C 0.2 ppm; C 2 mg/m3
Canada - British Columbia - Skin Notations
skin - potential for skin absorption
Canada - Ontario - OHSA - TWAEVs
0.05 ppm TWAEV; 0.5 mg/m3 TWAEV
Canada - Ontario - OHSA - Skin Notations
absorption through skin, eyes, or mucous membranes
Canada - Quebec - Ceiling Limits
P 0.2 ppm; P 1.86 mg/m3
Canada - Quebec - Skin Designations
absorbed through the skin
United Kingdom - Occupational Exposure Standards - TWAs
0.2 ppm TWA; 2 mg/m3 TWA
United Kingdom - Occupational Exposure Standards - STELs
0.2 ppm STEL; 2 mg/m3 STEL
United Kingdom - Occupational Exposure Standards - Notes
can be absorbed through skin
German (DFG) - MAK Values
0.05 ppm MAK; 0.5 mg/m3 MAK
German (DFG) - Peak Limitations
2 x normal MAK (30 min. average value); don't exceed 4 times during shift
German (DFG) - Skin/Sensitizers
danger of cutaneous absorption
Israel - Time Weighted Averages
0.05 ppm TWA 0.46 mg/m3 TWA
Israel - Action Levels
0.025 ppm AL; 0.23 mg/m3 AL
Mexico - Instruction No. 10 - TWAs
0.05 ppm TWA; 0.5 mg/m3 TWA
Mexico - Instruction No. 10 - STELs
0.1 ppm STEL

STATE LISTS
California - Air Bill 2588 Appendix A-II
6/91
California - Exposure Limits - PELs
when mixed with Ethylene glycol dinitrate: 0.05 mg/m3 PEL
California - Exposure Limits - STELs
0.1 mg/m3 STEL
California - Exposure Limits - Skin Notation
material may be absorbed through the skin, eyes or mucous membrane
California - Directors List of Hazardous Substances (8 CCR 339)
[present]
Florida Hazardous Substance List
[present]
Massachusetts Right To Know List
[present]
Minnesota Hazardous Substance List
skin
NJ Right to Know List (Total)
sn 1383
NJ Special Hazardous Substances
(reactive - fourth degree)

Pennsylvania Right to Know List
environmental hazard
PROPOSED REGULATIONS
Canada - Ontario - Proposed Occupational TWAEVs
0.03 ppm TWAEV; 0.3 mg/m3 TWAEV
Canada - Ontario - Proposed Occupational STEVs
0.1 ppm STEV; 0.9 mg/m3 STEV

2-NITRO-2-METHYLPROPANOL NITRATE   24884-69-3
HEALTH AND SAFETY LISTS
U.S. DOT - Substances From 49 CFR 172.101
regulated by DOT (UN0282, UN1336)
U.S. DOT - Hazard Classes
with less than 20% water: DOT hazard class = 1.1D; with 20% or more water: DOT hazard class = 4.1
INTERNATIONAL LISTS
Canada - WHMIS: Ingredient Disclosure
1% item 1179 (1241)
STATE LISTS
NJ Right to Know List (Total)
sn 1384

1-NITRONAPHTHALENE   86-57-7
HEALTH AND SAFETY LISTS
U.S. DOT - Hazard Classes
Forbidden from transport by the DOT

2-NITRONAPHTHALENE   581-89-5
HEALTH AND SAFETY LISTS
U.S. DOT - Hazard Classes
Forbidden from transport by the DOT

NITRONAPHTHALENE   27254-36-0
HEALTH AND SAFETY LISTS
U.S. DOT - Substances From 49 CFR 172.101
regulated by DOT (UN1798)
U.S. DOT - Hazard Classes
DOT hazard class = 8

INTERNATIONAL LISTS
Canada - WHMIS: Ingredient Disclosure
1% item 1180 (116)
STATE LISTS
NJ Right to Know List (Total)
sn 1385
NJ Special Hazardous Substances
(corrosive)

3-NITROPERYLENE   20539-63-3
HEALTH AND SAFETY LISTS
U.S. DOT - Hazard Classes
Forbidden from transport by the DOT

4-NITROPHENETOLE   100-29-8
HEALTH AND SAFETY LISTS
ACGIH 1995 - Time Weighted Averages
20 ppm TWA; 50 mg/m3 TWA
U.S. DOT - Substances From 49 CFR 172.101
regulated by DOT (UN1261)
U.S. DOT - Hazard Classes
DOT hazard class = 3
NFPA - Flash Points
flash point = 95 degrees F (35 degrees C)

NFPA - Hazard Identification Ratings
health-1; flammability-3; reactivity-4
NIOSH - Selected LD50s and LC50s
Oral, rat: LD50 = 940 mg/kg
NIOSH 1990 - Pocket Guide - RELs
See Appendix D
NIOSH 1990 - Pocket Guide - IDLHs
1000 ppm IDLH
NIOSH 1990 - Pocket Guide - Target organs
skin
NTP Chemical Status Reports - Testing Status and NTIS Number
Assigned to laboratory for toxicology/carcinogenesis study; two year studies: pathology working group in progress
OSHA - Vacated PELs - Time Weighted Averages
100 ppm TWA; 250 mg/m3 TWA
OSHA - Final PELs - Time Weighted Averages
100 ppm TWA; 250 mg/m3 TWA
OSHA - List of Highly Hazardous Chemicals
threshhold quantity = 2500 pounds

ENVIRONMENTAL LISTS
List of Pesticide Product Inert Ingredients
[present]
TSCA - Code of Federal Regulations Citations
40 CFR 712.30(w); 40 CFR 716.120(a)
TSCA - PAIR - Reporting List
Reporting Date: June 13, 1989
TSCA - Health and Safety Reporting List
Effective Date: April 13, 1989

INTERNATIONAL LISTS
Australian Exposure Standards - Time Weighted Averages
100 ppm TWA; 250 mg/m3 TWA
Canada - WHMIS: Ingredient Disclosure
1% item 1181 (1242)
Canada - Alberta - 8 Hour Occupational Exposure Limit
100 ppm TWA; 250 mg/m3 TWA
Canada - Alberta - 15 Minute Occupational Exposure Limit
150 ppm STEL; 375 mg/m3 STEL
Canada - British Columbia - 8 Hour Exposure Limits
100 ppm TWA; 250 mg/m3 TWA
Canada - British Columbia - 15 Minute Exposure Limits
150 ppm STEL; 375 mg/m3 STEL
Canada - Ontario - OHSA - TWAEVs
100 ppm TWAEV; 250 mg/m3 TWAEV
Canada - Quebec - Time-Weighted Average Exposure Values
100 ppm TWAEV; 250 mg/m3 TWAEV
United Kingdom - Occupational Exposure Standards - TWAs
100 ppm TWA; 250 mg/m3 TWA
United Kingdom - Occupational Exposure Standards - STELs
150 ppm STEL; 375 mg/m3 STEL
German (DFG) - MAK Values
100 ppm MAK; 250 mg/m3 MAK
Israel - Time Weighted Averages
(100) ppm TWA; (250) mg/m3 TWA
Israel - Action Levels
50 ppm AL; 125 mg/m3 AL
Mexico - Instruction No. 10 - TWAs
100 ppm TWA; 250 mg/m3 TWA
Mexico - Instruction No. 10 - STELs
150 ppm STEL; 375 mg/m3 STEL

STATE LISTS
California - Exposure Limits - PELs
100 ppm PEL; 250 mg/m3 PEL

California - Directors List of Hazardous Substances (8 CCR 339)
[present]
Florida Hazardous Substance List
[present]
Massachusetts Right To Know List
[present]
Minnesota Hazardous Substance List
[present]
NJ Right to Know List (Total)
sn 1386
NJ Special Hazardous Substances
(flammable - third degree; reactive - fourth degree)
Pennsylvania Right to Know List
[present]

O-NITROPHENOL   88-75-5
HEALTH AND SAFETY LISTS
U.S. DOT - Hazard Classes
Forbidden from transport by the DOT

P-NITROPHENOL   100-02-7
HEALTH AND SAFETY LISTS
U.S. DOT - Hazard Classes
Forbidden from transport by the DOT

M-NITROPHENOL   554-84-7
HEALTH AND SAFETY LISTS
IARC - Group 3 (not classifiable)
[present]
NFPA - Flash Points
flash point = 327 degrees F (164 degrees C)
NFPA - Hazard Identification Ratings
health-1; flammability-1; reactivity-0
NIOSH - Selected LD50s and LC50s
Oral, rat: LD50 = 120 mg/kg
NTP Chemical Status Reports - Testing Status and NTIS Number
Technical reports printed (PB282310/AS)
NTP Chemical Status Reports - Evidence of Carcinogenicity
male rat-negative; female rat-negative; male mice-negative; female mice-negative
ENVIRONMENTAL LISTS
CAA - HON Rule - SOCMI Chemicals
compliance by July 24, 1995
INTERNATIONAL LISTS
German (DFG) - Carcinogens
suspected carcinogen

NITROPHENOL (MIXED ISOMERS)   25154-55-6
HEALTH AND SAFETY LISTS
IARC - Group 3 (not classifiable)
[present]
NIOSH - Selected LD50s and LC50s
Oral, rat: LD50 = 4400 mg/kg
NIOSH - Health Standards - Exposure Limits
reduce exposure to lowest feasible concentration
NIOSH - Health Standards - Health Effects and Precautions
Bladder cancer
NIOSH - Health Standards - Carcinogenic Chemicals
potential human carcinogen
INTERNATIONAL LISTS
German (DFG) - Carcinogens
animal evidence of carcinogenicity

STATE LISTS

Minnesota Hazardous Substance List
carcinogen

NITROPHENOXYALKANOIC ACID SUBSTITUTED   RR-00932-5
THIAZINO HYDRAZIDE

HEALTH AND SAFETY LISTS

U.S. DOT - Substances From 49 CFR 172.101
regulated by DOT (UN2538)

U.S. DOT - Hazard Classes
DOT hazard class = 4.1

STATE LISTS

NJ Right to Know List (Total)
sn 1387

M-NITROPHENYLDINITRO METHANE   RR-01434-6

HEALTH AND SAFETY LISTS

IARC - Group 3 (not classifiable)
[present]

2-NITRO-P-PHENYLENEDIAMINE   5307-14-2

ENVIRONMENTAL LISTS

TSCA - Code of Federal Regulations Citations
40 CFR 712.30(x); 40 CFR 716.120(d)

TSCA - PAIR - Reporting List
Reporting Date: November 27, 1991

TSCA - Health and Safety Reporting List
Effective Date: September 30, 1991

INTERNATIONAL LISTS

Canada - WHMIS: Ingredient Disclosure
1% item 1182 (1243)

2-[(3-NITROPHENYL)SULFONYL] ETHANOL   41687-30-3

HEALTH AND SAFETY LISTS

U.S. DOT - Appendix A Table 1 - Hazardous Substances
final RQ = 100 pounds (45.4 kg) (Listed under 'Nitrophenol (mixed)')

NIOSH - Selected LD50s and LC50s
Oral, rat: LD50 = 334 mg/kg

ENVIRONMENTAL LISTS

ATSDR Priority List
Rank (of 275): 251

CERCLA/SARA - Section 313 - Emission Reporting
form R reporting required for 1.0% de minimus concentration

CERCLA/SARA - Hazardous Substances and their Reportable Quantities
final RQ = 100 pounds (45.4 kg) (Listed under 'Nitrophenol (mixed)')

CAA - HON Rule - SOCMI Chemicals
compliance by April 24, 1995

Clean Water Act - Hazardous Substances
[present] (Listed under 'Nitrophenol (mixed)')

Clean Water Act - Priority Pollutants
[present]

RCRA - TSD Facilities Ground Water Monitoring
TM 8040 = 5 ug/L PQL; TM 8270 = 10 ug/L PQL

RCRA - Universal Treatment Standards (LDR)
WW: 0.028 mg/l; NWW: 13 mg/kg

INTERNATIONAL LISTS

Canada - WHMIS: Ingredient Disclosure
1% item 1184 (1245)

Mexico - Wastewater - Organic Toxic Pollutants and Heavy Metals
Listed under [Nitrophenols]

Mexico - Drinking Water - Ecological Criteria
0.07 mg/l (combined with 4-Nitrophenol)

STATE LISTS

California - Air Bill 2588 Appendix A-II
6/91

California - Directors List of Hazardous Substances (8 CCR 339)
[present] (Listed under 'Nitrophenols, all isomers')

Massachusetts Right To Know List
[present]

NJ Right to Know List (Total)
sn 1391

Pennsylvania Right to Know List
environmental hazard

2-NITROPROPANE   79-46-9
SEE ALSO:
F039-HAZARDOUS WASTES

HEALTH AND SAFETY LISTS

U.S. DOT - Appendix A Table 1 - Hazardous Substances
final RQ = 100 pounds (45.4 kg) (Listed under 'Nitrophenol (mixed)')

NIOSH - Selected LD50s and LC50s
Oral, rat: LD50 = 250 mg/kg Skin, mammal: LD50 = 920 mg/kg

NTP Chemical Status Reports - Testing Status and NTIS Number
Technical reports printed (no NTIS number given)

NTP Chemical Status Reports - Evidence of Carcinogenicity
male mice: no evidence; female mice: no evidence

ENVIRONMENTAL LISTS

ATSDR Priority List
Rank (of 275): 216

CERCLA/SARA - Section 313 - Emission Reporting
form R reporting required for 1.0% de minimus concentration

CERCLA/SARA - Hazardous Substances and their Reportable Quantities
final RQ = 100 pounds (45.4 kg) (Listed under 'Nitrophenol (mixed)')

Clean Air Act (1990) - List of Hazardous Air Contaminants
[present]

CAA - HON Rule - SOCMI Chemicals
compliance by April 24, 1995

CAA - HON Rule - Organic HAPs
[present]

Clean Water Act - Hazardous Substances
[present] (Listed under 'Nitrophenol (mixed)')

Clean Water Act - Priority Pollutants
[present]

List of Pesticide Product Inert Ingredients
[present]

RCRA - U Series Wastes
waste number U170

RCRA - Hazardous Constituents-Appendix VIII
waste number U170

RCRA - Basis for Listing - Appendix VII
Included in waste stream: F039

RCRA - Substances Banned From Land Disposal
[present]

RCRA - TSD Facilities Ground Water Monitoring
TM 8040 = 10 ug/L PQL; TM 8270 = 50 ug/L PQL

RCRA - Universal Treatment Standards (LDR)
WW: 0.12 mg/l; NWW: 29 mg/kg

TSCA - Code of Federal Regulations Citations
40 CFR 712.30(w); 40 CFR 716.120(a); 40 CFR 799.5055(c), (e) (1)

TSCA - PAIR - Reporting List
Reporting Date: June 13, 1989

TSCA - Health and Safety Reporting List
Effective Date: April 13, 1989

TSCA - Multichemical Test Rules - Waste Constituents
subchronic toxicity testing for Health Effects

TSCA - Section 12(b) - Export Notification
export notification required - Section 4

INTERNATIONAL LISTS

Canada - WHMIS: Ingredient Disclosure
1% item 1185 (1246)

Canada - NPRI (National Pollutant Release Inventory)
[present]

Mexico - Wastewater - Organic Toxic Pollutants and Heavy Metals
Listed under [Nitrophenols]

Mexico - Drinking Water - Ecological Criteria
0.07 mg/l (combined with 2-Nitrophenol)

STATE LISTS

California - Air Bill 2588 Appendix A-I
6/91

California - Directors List of Hazardous Substances (8 CCR 339)
[present] (Listed under 'Nitrophenols, all isomers')

Florida Hazardous Substance List
[present]

Massachusetts Right To Know List
[present]

NJ Right to Know List (Total)
sn 1390

Pennsylvania Right to Know List
environmental hazard

PROPOSED REGULATIONS

Safe Drinking Water Act - Priority list
[present]

1-NITROPROPANE   108-03-2

HEALTH AND SAFETY LISTS

U.S. DOT - Appendix A Table 1 - Hazardous Substances
final RQ = 100 pounds (45.4 kg) (Listed under 'Nitrophenol (mixed)')

NIOSH - Selected LD50s and LC50s
Oral, rat: LD50 = 328 mg/kg Skin, mammal: LD50 = 543 mg/kg

ENVIRONMENTAL LISTS

CERCLA/SARA - Hazardous Substances and their Reportable Quantities
final RQ = 100 pounds (45.4 kg) (Listed under 'Nitrophenol (mixed)')

Clean Water Act - Hazardous Substances
[present] (Listed under 'Nitrophenol (mixed)')

INTERNATIONAL LISTS

Canada - WHMIS: Ingredient Disclosure
1% item 1183 (1244)

STATE LISTS

California - Directors List of Hazardous Substances (8 CCR 339)
[present] (Listed under 'Nitrophenols, all isomers')

Massachusetts Right To Know List
[present]

NJ Right to Know List (Total)
sn 1389

Pennsylvania Right to Know List
environmental hazard

NITROPROPANE   25322-01-4

HEALTH AND SAFETY LISTS

U.S. DOT - Substances From 49 CFR 172.101
regulated by DOT (UN1663)

U.S. DOT - Hazard Classes
DOT hazard class = 6.1

U.S. DOT - Appendix A Table 1 - Hazardous Substances
final RQ = 100 pounds (45.4 kg)

ENVIRONMENTAL LISTS

CERCLA/SARA - Hazardous Substances and their Reportable Quantities
final RQ = 100 pounds (45.4 kg)

Clean Water Act - Hazardous Substances
[present]

Clean Water Act - Toxic Pollutants
[present]

STATE LISTS

California - Directors List of Hazardous Substances (8 CCR 339)
[present] (Listed under 'Nitrophenols, all isomers')

Massachusetts Right To Know List
[present]

NJ Right to Know List (Total)
sn 3001

Pennsylvania Right to Know List
environmental hazard

3-NITROPROPIONIC ACID   504-88-1

ENVIRONMENTAL LISTS

TSCA - Chemicals with Significant New Use Rules
PMN number: P-88-270

2-NITROPYRENE   789-07-1

HEALTH AND SAFETY LISTS

U.S. DOT - Hazard Classes
Forbidden from transport by the DOT

1-NITROPYRENE   5522-43-0
SEE ALSO:
2-NITRO-P-PHENYLENEDIAMINE

HEALTH AND SAFETY LISTS

IARC - Group 3 (not classifiable)
[present]

NIOSH - Selected LD50s and LC50s
Oral, rat: LD50 = 3080 mg/kg

NTP Chemical Status Reports - Testing Status and NTIS Number
Technical reports printed (PB290304/AS)

NTP Chemical Status Reports - Evidence of Carcinogenicity
male rat-negative; female rat-negative; male mice-negative; female mice-positive

ENVIRONMENTAL LISTS

TSCA - Code of Federal Regulations Citations
40 CFR 712.30(f); 40 CFR 716.120

TSCA - PAIR - Reporting List
Reporting Date: August 17, 1983

INTERNATIONAL LISTS

Canada - WHMIS: Ingredient Disclosure
1% item 76 (216)

German (DFG) - Skin/Sensitizers
danger of cutaneous absorption

German (DFG) - Carcinogens
suspected carcinogen

STATE LISTS

Massachusetts Right To Know List
carcinogen; extraordinarily hazardous

4-NITROPYRENE   57835-92-4

ENVIRONMENTAL LISTS

TSCA - Code of Federal Regulations Citations
40 CFR 712.30(x); 40 CFR 716.120(d)

TSCA - PAIR - Reporting List
Reporting Date: November 27, 1991

TSCA - Health and Safety Reporting List
Effective Date: September 30, 1991

NITROPYRENES (MONO-, DI-, TRI-, AND TETRA-)   RR-00815-1
SEE ALSO:
F005-HAZARDOUS WASTES

HEALTH AND SAFETY LISTS
ACGIH 1995 - Time Weighted Averages
10 ppm TWA; 36 mg/m3 TWA

ACGIH 1995 - Carcinogens
A2-suspected human carcinogen

AIHA - Odor Threshold Values
no geometric mean air odor threshold

U.S. DOT - Appendix A Table 1 - Hazardous Substances
final RQ = 10 pounds (4.54 kg)

IARC - Group 2B (sufficient animal data)
[present] (Overall evaluation based only on evidence of carcinogenicity in monograph (29, 1982) or in Supplement 4)

NFPA - Flash Points
flash point = 75 degrees F (24 degrees C)

NFPA - Hazard Identification Ratings
health-2; flammability-3; reactivity-2 (may explode on heating)

NIOSH - Selected LD50s and LC50s
Inhalation, rat: LC50 = 400 ppm 6 hr Oral, rat: LD50 = 720 mg/kg

NIOSH 1990 - Pocket Guide - RELs
Reduce exposure to lowest feasible concentration

NIOSH 1990 - Pocket Guide - IDLHs
2300 ppm IDLH (not considering carcinogenic effects)

NIOSH 1990 - Pocket Guide - Carcinogens
occupational carcinogen

NIOSH 1990 - Pocket Guide - Target organs
respiratory system, CNS

NIOSH - Health Standards - Exposure Limits
reduce exposure to lowest feasible concentration

NIOSH - Health Standards - Health Effects and Precautions
has produced liver tumors in rats (Conduct medical monitoring with emphasis on liver function tests)

NIOSH - Health Standards - Carcinogenic Chemicals
potential human carcinogen

NTP Seventh Report - Suspect Carcinogens
suspect carcinogen

OSHA - Vacated PELs - Time Weighted Averages
10 ppm TWA; 35 mg/m3 TWA

OSHA - Final PELs - Time Weighted Averages
25 ppm TWA; 90 mg/m3 TWA

OSHA - Possible Select Carcinogens
[present]

ENVIRONMENTAL LISTS
CERCLA/SARA - Section 313 - Emission Reporting
form R reporting required for 0.1% de minimus concentration

CERCLA/SARA - Hazardous Substances and their Reportable Quantities
final RQ = 10 pounds (4.54 kg)

Clean Air Act (1990) - List of Hazardous Air Contaminants
[present]

CAA - HON Rule - SOCMI Chemicals
compliance by Jan. 23, 1995

CAA - HON Rule - Organic HAPs
[present]

EPA - Carcinogen Hazard Ranking for RQ Adjustment
Hazard ranking = Medium

RCRA - U Series Wastes
waste number U171 (Ignitable waste; Toxic waste)

RCRA - Hazardous Constituents-Appendix VIII
waste number U171

RCRA - Basis for Listing - Appendix VII
Included in waste stream: F005

RCRA - Substances Banned From Land Disposal
[present]

TSCA - Health and Safety Reporting List
Effective Date: March 11, 1994; Sunset Date: March 11, 2004

INTERNATIONAL LISTS
Australian Exposure Standards - Time Weighted Averages
10 ppm TWA; 36 mg/m3 TWA

Australian Exposure Standards - Carcinogens
probable carcinogen

Canada - WHMIS: Ingredient Disclosure
0.1% item 1187 (1248)

Canada - NPRI (National Pollutant Release Inventory)
[present]

Canada - Alberta - 8 Hour Occupational Exposure Limit
10 ppm TWA; 36 mg/m3 TWA

Canada - Alberta - 15 Minute Occupational Exposure Limit
20 ppm STEL; 72 mg/m3 STEL

Canada - Alberta - Ceiling Occupational Exposure Limit
C 25 ppm; C 91 mg/m3

Canada - Alberta - Designated Substances
designated substance - requires code of practice

Canada - British Columbia - 8 Hour Exposure Limits
25 ppm TWA; 90 mg/m3 TWA

Canada - British Columbia - Carcinogens
carcinogen - 25 ppm TWA; 90 mg/m3 TWA

Canada - Ontario - OHSA - TWAEVs
10 ppm TWAEV; 35 mg/m3 TWAEV

Canada - Ontario - OHSA - STEVs
20 ppm STEV; 70 mg/m3 STEV

Canada - Quebec - Time-Weighted Average Exposure Values
10 ppm TWAEV; 36 mg/m3 TWAEV

Canada - Quebec - Carcinogens
C2 carcinogen: effect suspected in humans

German (DFG) - Carcinogens
animal evidence of carcinogenicity

Israel - Time Weighted Averages
10 ppm TWA 36 mg/m3 TWA

Israel - Action Levels
5 ppm AL; 18 mg/m3 AL

Mexico - Instruction No. 10 - TWAs
25 ppm TWA; 90 mg/m3 TWA

Mexico - Instruction No. 10 - Carcinogens
potential carcinogen in humans - limited epidemiological evidence

STATE LISTS
California - Air Bill 2588 Appendix A-I
known or potential carcinogen

California - Prop. 65 - Cancer list
carcinogen - initial date 1/1/88

California - Exposure Limits - PELs
10 ppm PEL; 35 mg/m3 PEL

California - Directors List of Hazardous Substances (8 CCR 339)
[present] (Listed under 'Nitropropanes')

Florida Hazardous Substance List
[present]

Massachusetts Right To Know List
carcinogen; extraordinarily hazardous

Minnesota Hazardous Substance List
carcinogen

NJ Right to Know List (Total)
sn 1392

NJ Special Hazardous Substances
(reactive - first degree; flammable - third degree)

Pennsylvania Right to Know List
environmental hazard; special hazardous substance

Pennsylvania RTK - Special Hazardous Substances
[present]

PROPOSED REGULATIONS

TSCA - ITC 32nd Report Priority Testing List
designated for dermal absorption testing

Canada - Ontario - Proposed Occupational TWAEVs
1 ppm TWAEV;   3.6 mg/m3 TWAEV

4-NITROQUINOLINE-1-OXIDE   56-57-5

HEALTH AND SAFETY LISTS

ACGIH 1995 - Time Weighted Averages
25 ppm TWA;   91 mg/m3 TWA

AIHA - Odor Threshold Values
geometric mean air odor threshold = 140 ppm (detectable)

NFPA - Flash Points
flash point = 96 degrees F (36 degrees C)

NFPA - Hazard Identification Ratings
health-1;   flammability-3;   reactivity-2 (may explode on heating)

NIOSH - Selected LD50s and LC50s
Inhalation, rat: LC50 = 3100 ppm 8 hr Oral, rat: LD50 = 455 mg/kg

NIOSH 1990 - Pocket Guide - RELs
25 ppm TWA;   90 mg/m3 TWA

NIOSH 1990 - Pocket Guide - IDLHs
2300 ppm IDLH

NIOSH 1990 - Pocket Guide - Target organs
CNS, eyes

OSHA - Vacated PELs - Time Weighted Averages
25 ppm TWA;   90 mg/m3 TWA

OSHA - Final PELs - Time Weighted Averages
25 ppm TWA;   90 mg/m3 TWA

ENVIRONMENTAL LISTS

List of Pesticide Product Inert Ingredients
[present]

TSCA - PAIR - Reporting List
Effective Date: January 26, 1994;   Reporting Date: March 28, 1994

TSCA - Health and Safety Reporting List
Effective Date: January 26, 1994;   Sunset Date: January 26, 2004

INTERNATIONAL LISTS

Australian Exposure Standards - Time Weighted Averages
25 ppm TWA;   91 mg/m3 TWA

Canada - WHMIS: Ingredient Disclosure
1% item 1186 (1247)

Canada - Alberta - 8 Hour Occupational Exposure Limit
25 ppm TWA;   91 mg/m3 TWA

Canada - Alberta - 15 Minute Occupational Exposure Limit
35 ppm STEL;   128 mg/m3 STEL

Canada - British Columbia - 8 Hour Exposure Limits
25 ppm TWA;   90 mg/m3 TWA

Canada - British Columbia - 15 Minute Exposure Limits
35 ppm STEL;   135 mg/m3 STEL

Canada - Ontario - OHSA - TWAEVs
25 ppm TWAEV;   90 mg/m3 TWAEV

Canada - Quebec - Time-Weighted Average Exposure Values
25 ppm TWAEV;   91 mg/m3 TWAEV

United Kingdom - Occupational Exposure Standards - TWAs
25 ppm TWA;   90 mg/m3 TWA

German (DFG) - MAK Values
25 ppm MAK;   90 mg/m3 MAK

Israel - Time Weighted Averages
25 ppm TWA;   91 mg/m3 TWA

Israel - Action Levels
12.5 ppm AL;   45.5 mg/m3 AL

Mexico - Instruction No. 10 - TWAs
25 ppm TWA;   90 mg/m3 TWA

Mexico - Instruction No. 10 - STELs
35 ppm STEL;   135 mg/m3 STEL

STATE LISTS

California - Exposure Limits - PELs
25 ppm PEL;   90 mg/m3 PEL

California - Directors List of Hazardous Substances (8 CCR 339)
[present] (Listed under 'Nitropropanes')

Florida Hazardous Substance List
[present]

Massachusetts Right To Know List
[present]

Minnesota Hazardous Substance List
[present]

NJ Right to Know List (Total)
sn 1394

NJ Special Hazardous Substances
(reactive - second degree)

Pennsylvania Right to Know List
[present]

PROPOSED REGULATIONS

TSCA - ITC 31st Report Priority Testing List
designated to be tested

4-NITROQUINOLINE-1-OXIDE   3565-26-2

HEALTH AND SAFETY LISTS

U.S. DOT - Substances From 49 CFR 172.101
regulated by DOT (UN2608)

U.S. DOT - Hazard Classes
DOT hazard class = 3

NITROSAMINES   RR-00382-7

HEALTH AND SAFETY LISTS

NTP Chemical Status Reports - Testing Status and NTIS Number
Technical reports printed (PB281102/AS)

NTP Chemical Status Reports - Evidence of Carcinogenicity
male rat-equivocal;   female rat-negative;   male mice-negative;   female mice-negative

STATE LISTS

Massachusetts Right To Know List
carcinogen;   extraordinarily hazardous

N'-NITROSOANABASINE   1133-64-8

HEALTH AND SAFETY LISTS

IARC - Group 3 (not classifiable)
[present]

N'-NITROSOANATABINE   71267-22-6

HEALTH AND SAFETY LISTS

IARC - Group 2B (sufficient animal data)
[present]

NTP Chemical Status Reports - Testing Status and NTIS Number
Post peer review technical reports in progress

OSHA - Possible Select Carcinogens
[present]

ENVIRONMENTAL LISTS

CERCLA/SARA - Section 313 - Emission Reporting
form R reporting required; (Listed under 'Polycyclic aromatic compounds')

STATE LISTS

California - Air Bill 2588 Appendix A-I
known or potential carcinogen: 6/91

California - Prop. 65 - Cancer list
carcinogen - initial date 10/1/90

California - Directors List of Hazardous Substances (8 CCR 339)
[present]

PROPOSED REGULATIONS

NTP - Proposed Additions to Annual Report on Carcinogens
proposed as a suspect carcinogen for NTP 8th report

N-NITROSO-N-BUTYL-N-(3-CARBOXYPROPYL) AMINE   38252-74-3

HEALTH AND SAFETY LISTS

IARC - Group 2B (sufficient animal data)
[present]

OSHA - Possible Select Carcinogens
[present]

STATE LISTS

California - Air Bill 2588 Appendix A-II
known or potential carcinogen: 6/91

California - Prop. 65 - Cancer list
carcinogen - initial date 10/1/90

California - Directors List of Hazardous Substances (8 CCR 339)
[present]

PROPOSED REGULATIONS

NTP - Proposed Additions to Annual Report on Carcinogens
proposed as a suspect carcinogen for NTP 8th report

N-NITROSO-N-BUTYL-N-(4-HYDROXYBUTYL) AMINE   3817-11-6

INTERNATIONAL LISTS

German (DFG) - Carcinogens
suspected carcinogen

N-NITROSODI-N-BUTYLAMINE   924-16-3

ENVIRONMENTAL LISTS

RCRA - TSD Facilities Ground Water Monitoring
TM 8270 = 10 ug/L PQL

INTERNATIONAL LISTS

Canada - WHMIS: Ingredient Disclosure
1% item 1188 (1249)

STATE LISTS

NJ Right to Know List (Total)
sn 1629

NJ Special Hazardous Substances
(carcinogen; mutagen)

Pennsylvania Right to Know List
environmental hazard

N-NITROSODIETHANOLAMINE   1116-54-7

STATE LISTS

Florida Hazardous Substance List
[present]

Massachusetts Right To Know List
carcinogen; extraordinarily hazardous

N-NITROSODIETHYLAMINE   55-18-5

ENVIRONMENTAL LISTS

Clean Water Act - Toxic Pollutants
[present]

RCRA - Hazardous Constituents-Appendix VIII
hazardous constituent - no waste number

STATE LISTS

California - Directors List of Hazardous Substances (8 CCR 339)
[present]

Pennsylvania Right to Know List
environmental hazard

N-NITROSODIMETHYLAMINE   62-75-9

HEALTH AND SAFETY LISTS

IARC - Group 3 (not classifiable)
[present]

STATE LISTS

California - Directors List of Hazardous Substances (8 CCR 339)
[present]

P-NITROSODIMETHYLANILINE   138-89-6

HEALTH AND SAFETY LISTS

IARC - Group 3 (not classifiable)
[present]

P-NITROSODIPHENYLAMINE   156-10-5

HEALTH AND SAFETY LISTS

NTP Seventh Report - Suspect Carcinogens
suspect carcinogen (Listed under 'N-nitrosodi-n-butylamine')

OSHA - Possible Select Carcinogens
[present]

N-NITROSODI-I-PROPYLAMINE   601-77-4

HEALTH AND SAFETY LISTS

NIOSH - Selected LD50s and LC50s
Oral, rat: LD50 = 1800 mg/kg

NTP Seventh Report - Suspect Carcinogens
suspect carcinogen

N-NITROSODI-N-PROPYLAMINE   621-64-7
SEE ALSO:
NITROSAMINES
F039-HAZARDOUS WASTES

HEALTH AND SAFETY LISTS

U.S. DOT - Appendix A Table 1 - Hazardous Substances
final RQ = 1 pound (0.454 kg)

IARC - Group 2B (sufficient animal data)
[present] (Overall evaluation based only on evidence of carcinogenicity in monograph (17, 1978) or in Supplement 4)

NIOSH - Selected LD50s and LC50s
Oral, rat: LD50 = 1200 mg/kg

NTP Seventh Report - Suspect Carcinogens
suspect carcinogen

OSHA - Possible Select Carcinogens
[present]

ENVIRONMENTAL LISTS

CERCLA/SARA - Section 313 - Emission Reporting
form R reporting required for 0.1% de minimus concentration

CERCLA/SARA - Hazardous Substances and their Reportable Quantities
final RQ = 1 pound (0.454 kg)

EPA - Carcinogen Hazard Ranking for RQ Adjustment
Hazard ranking = Medium

RCRA - U Series Wastes
waste number U172

RCRA - Hazardous Constituents-Appendix VIII
waste number U172

RCRA - Basis for Listing - Appendix VII
Included in waste stream: F039

RCRA - Substances Banned From Land Disposal
[present]

RCRA - TSD Facilities Ground Water Monitoring
TM 8270 = 10 ug/L PQL

RCRA - Universal Treatment Standards (LDR)
WW: 0.40 mg/l;   NWW: 17 mg/kg

INTERNATIONAL LISTS

Canada - WHMIS: Ingredient Disclosure
0.1% item 1189 (1250)

German (DFG) - Carcinogens
animal evidence of carcinogenicity

STATE LISTS

California - Air Bill 2588 Appendix A-I
known or potential carcinogen

California - Prop. 65 - Cancer list
carcinogen - initial date 10/1/87

California - Prop. 65 - No Significant Risk Levels
no significant risk level = 0.06 ug/day

California - Directors List of Hazardous Substances (8 CCR 339)
[present]

Florida Hazardous Substance List
[present]

Massachusetts Right To Know List
carcinogen; extraordinarily hazardous

Minnesota Hazardous Substance List
carcinogen

NJ Right to Know List (Total)
sn 1406

NJ Special Hazardous Substances
(carcinogen; mutagen)

Pennsylvania Right to Know List
environmental hazard; special hazardous substance

Pennsylvania RTK - Special Hazardous Substances
[present]

N-NITROSOETHYLPHENYLAMINE   612-64-6

HEALTH AND SAFETY LISTS

U.S. DOT - Appendix A Table 1 - Hazardous Substances
final RQ = 1 pound (0.454 kg)

IARC - Group 2B (sufficient animal data)
[present] (Overall evaluation based only on evidence of carcinogenicity in monograph (17, 1978) or in Supplement 4)

NIOSH - Selected LD50s and LC50s
Oral, rat: LD50 = 7500 mg/kg

NTP Chemical Status Reports - Testing Status and NTIS Number
Chronic studies exist for which technical reports were not prepared

NTP Seventh Report - Suspect Carcinogens
suspect carcinogen

OSHA - Possible Select Carcinogens
[present]

ENVIRONMENTAL LISTS

CERCLA/SARA - Hazardous Substances and their Reportable Quantities
final RQ = 1 pound (0.454 kg)

EPA - Carcinogen Hazard Ranking for RQ Adjustment
Hazard ranking = High

RCRA - U Series Wastes
waste number U173

RCRA - Hazardous Constituents-Appendix VIII
waste number U173

RCRA - Substances Banned From Land Disposal
[present]

TSCA - Chemicals with Significant New Use Rules
[present]

TSCA - Section 12(b) - Export Notification
export notification required - Section 5

INTERNATIONAL LISTS

Canada - WHMIS: Ingredient Disclosure
0.1% item 1190 (1251)

German (DFG) - Carcinogens
animal evidence of carcinogenicity

STATE LISTS

California - Air Bill 2588 Appendix A-I
known or potential carcinogen

California - Prop. 65 - Cancer list
carcinogen - initial date 1/1/88

California - Prop. 65 - No Significant Risk Levels
no significant risk level = 0.3 ug/day

California - Directors List of Hazardous Substances (8 CCR 339)
[present]

Florida Hazardous Substance List
[present]

Massachusetts Right To Know List
carcinogen; extraordinarily hazardous

Minnesota Hazardous Substance List
carcinogen

NJ Special Hazardous Substances
(carcinogen)

Pennsylvania Right to Know List
environmental hazard; special hazardous substance

Pennsylvania RTK - Special Hazardous Substances
[present]

N-NITROSO-N-ETHYLUREA   759-73-9

HEALTH AND SAFETY LISTS

U.S. DOT - Appendix A Table 1 - Hazardous Substances
final RQ = 1 pound (0.454 kg)

IARC - Group 2A (limited human data)
[present] (Overall evaluation based only on evidence of carcinogenicity in monograph [17, 1978] or in Supplement 4) (Other relevant data, as given in Supplement 7 or in the monograph, influenced the making of the overall evaluation)

NIOSH - Selected LD50s and LC50s
Oral, rat: LD50 = 280 mg/kg

NTP Seventh Report - Suspect Carcinogens
suspect carcinogen

OSHA - Possible Select Carcinogens
[present]

ENVIRONMENTAL LISTS

CERCLA/SARA - Section 313 - Emission Reporting
form R reporting required for 0.1% de minimus concentration

CERCLA/SARA - Hazardous Substances and their Reportable Quantities
final RQ = 1 pound (0.454 kg)

EPA - Carcinogen Hazard Ranking for RQ Adjustment
Hazard ranking = High

RCRA - U Series Wastes
waste number U174

RCRA - Hazardous Constituents-Appendix VIII
waste number U174

RCRA - Basis for Listing - Appendix VII
Included in waste stream: F039

RCRA - Substances Banned From Land Disposal
[present]

RCRA - TSD Facilities Ground Water Monitoring
TM 8270 = 10 ug/L PQL

RCRA - Universal Treatment Standards (LDR)
WW: 0.40 mg/l; NWW: 28 mg/kg

INTERNATIONAL LISTS

Canada - WHMIS: Ingredient Disclosure
0.1% item 1191 (1252)

German (DFG) - Carcinogens
animal evidence of carcinogenicity

STATE LISTS

California - Air Bill 2588 Appendix A-I
known or potential carcinogen

California - Prop. 65 - Cancer list
carcinogen - initial date 10/1/87

California - Prop. 65 - No Significant Risk Levels
no significant risk level = 0.02 ug/day

California - Directors List of Hazardous Substances (8 CCR 339)
[present]

Florida Hazardous Substance List
[present]

Massachusetts Right To Know List
carcinogen; extraordinarily hazardous

Minnesota Hazardous Substance List
carcinogen

NJ Special Hazardous Substances
(carcinogen; mutagen)

Pennsylvania Right to Know List
environmental hazard; special hazardous substance

Pennsylvania RTK - Special Hazardous Substances
[present]

N-NITROSOFOLIC ACID   29291-35-8
SEE ALSO:
NITROSAMINES
F039-HAZARDOUS WASTES

HEALTH AND SAFETY LISTS

ACGIH 1995 - Skin Designations
skin - potential for cutaneous absorption

ACGIH 1995 - Carcinogens
A2-suspected human carcinogen

AIHA - Odor Threshold Values
no geometric mean air odor threshold

U.S. DOT - Appendix A Table 1 - Hazardous Substances
final RQ = 10 pounds (4.54 kg)

IARC - Group 2A (limited human data)
[present] (Overall evaluation based only on evidence of carcinogenicity in monograph [17, 1978] or in Supplement 4) (Other relevant data, as given in Supplement 7 or in the monograph, influenced the making of the overall evaluation)

NIOSH - Selected LD50s and LC50s
Inhalation, rat: LC50 = 78 ppm 4 hr Oral, hamster: LD50 = 28 mg/kg

NIOSH 1990 - Pocket Guide - Carcinogens
occupational carcinogen

NIOSH 1990 - Pocket Guide - Target organs
liver, kidneys, lungs

NIOSH - Health Standards - Exposure Limits
use 29 CFR 1910.1016

NIOSH - Health Standards - Health Effects and Precautions
has produced tumors of the liver, kidney, lung, and nasal cavity in animals

NIOSH - Health Standards - Carcinogenic Chemicals
potential human carcinogen

NTP Seventh Report - Suspect Carcinogens
suspect carcinogen

OSHA - 29 CFR 1910 Specifically Regulated Chemicals
Cancer suspect agent (see 29 CFR 1910.1016)

OSHA - Select Carcinogens
[present]

OSHA - Possible Select Carcinogens
[present]

ENVIRONMENTAL LISTS

ATSDR Priority List
Rank (of 275): 206

CERCLA/SARA - Section 302 Extremely Hazardous Substances and TPQs
TPQ = 1000 pounds

CERCLA/SARA - Section 313 - Emission Reporting
form R reporting required for 0.1% de minimus concentration

CERCLA/SARA - Hazardous Substances and their Reportable Quantities
final RQ = 10 pounds (4.54 kg)

Clean Air Act (1990) - List of Hazardous Air Contaminants
[present]

Clean Water Act - Priority Pollutants
[present]

EPA - Carcinogen Hazard Ranking for RQ Adjustment
Hazard ranking = Medium

RCRA - P Series Wastes
waste number P082

RCRA - Hazardous Constituents-Appendix VIII
waste number P082

RCRA - Basis for Listing - Appendix VII
Included in waste stream: F039

RCRA - Substances Banned From Land Disposal
[present]

RCRA - TSD Facilities Ground Water Monitoring
TM 8270 = 10 ug/L PQL

RCRA - Universal Treatment Standards (LDR)
WW: 0.40 mg/l; NWW: 2.3 mg/kg

INTERNATIONAL LISTS

Australian Exposure Standards - Time Weighted Averages
control to the lowest practical level

Australian Exposure Standards - Skin Effects
skin absorption

Australian Exposure Standards - Carcinogens
probable carcinogen

Canada - WHMIS: Ingredient Disclosure
0.1% item 1192 (1253)

Canada - Alberta - Skin Designation
can be absorbed through the intact skin

Canada - Alberta - Designated Substances
designated substance - requires code of practice

Canada - British Columbia - 8 Hour Exposure Limits
carcinogen with no established permitted concentration

Canada - British Columbia - Skin Notations
skin - potential for skin absorption

Canada - British Columbia - Carcinogens
carcinogen with no established permitted concentration

Canada - Quebec - Time-Weighted Average Exposure Values
substance of which the recirculation is prohibited

Canada - Quebec - Skin Designations
absorbed through the skin

Canada - Quebec - Carcinogens
C2 carcinogen: effect suspected in humans

German (DFG) - Carcinogens
animal evidence of carcinogenicity

Mexico - Wastewater - Organic Toxic Pollutants and Heavy Metals
Listed under [Nitrosamines]

Mexico - Drinking Water - Ecological Criteria
0.00001 mg/l Substance presents persistence, bioaccumulations or risk of cancer, reduce human exposure to a minimum; This level has been extrapolated by using a mathematic model

STATE LISTS

California - Air Bill 2588 Appendix A-I
known or potential carcinogen

California - Prop. 65 - Cancer list
carcinogen - initial date 10/1/87

California - Prop. 65 - No Significant Risk Levels
no significant risk level = 0.04 ug/day

California - Exposure Limits - Carcinogens
cancer-suspect agent (at a concentration >= 1.0%)

California - Directors List of Hazardous Substances (8 CCR 339)
[present]

Florida Hazardous Substance List
[present]

Massachusetts Right To Know List
carcinogen; extraordinarily hazardous

Minnesota Hazardous Substance List
carcinogen

NJ Right to Know List (Total)
sn 1405

NJ Special Hazardous Substances
(carcinogen; mutagen)

Pennsylvania Right to Know List
environmental hazard; special hazardous substance

Pennsylvania RTK - Special Hazardous Substances
[present]

N-NITROSOGUVACINE   55557-01-2

HEALTH AND SAFETY LISTS

U.S. DOT - Substances From 49 CFR 172.101
regulated by DOT (UN1369)

U.S. DOT - Hazard Classes
DOT hazard class = 4.2

NIOSH - Selected LD50s and LC50s
Oral, rat: LD50 = 65 mg/kg

STATE LISTS

NJ Right to Know List (Total)
sn 1550

N-NITROSOGUVACOLINE   55557-02-3

HEALTH AND SAFETY LISTS

IARC - Group 3 (not classifiable)
[present]

NIOSH - Selected LD50s and LC50s
Oral, rat: LD50 = 2140 mg/kg

NTP Chemical Status Reports - Testing Status and NTIS Number
Technical reports printed (PB295100/AS)

NTP Chemical Status Reports - Evidence of Carcinogenicity
male rat-positive; female rat-negative; male mice-positive; female mice-negative

NTP Seventh Report - Suspect Carcinogens
suspect carcinogen

OSHA - Possible Select Carcinogens
[present]

ENVIRONMENTAL LISTS

CERCLA/SARA - Section 313 - Emission Reporting
form R reporting required for 0.1% de minimus concentration

INTERNATIONAL LISTS

Canada - WHMIS: Ingredient Disclosure
1% item 1194 (1255)

STATE LISTS

California - Air Bill 2588 Appendix A-I
known or potential carcinogen

California - Prop. 65 - Cancer list
carcinogen - initial date 1/1/88

California - Prop. 65 - No Significant Risk Levels
no significant risk level = 30 ug/day

Florida Hazardous Substance List
[present]

Massachusetts Right To Know List
carcinogen; extraordinarily hazardous

Minnesota Hazardous Substance List
carcinogen

NJ Right to Know List (Total)
sn 1551

NJ Special Hazardous Substances
(carcinogen)

Pennsylvania Right to Know List
environmental hazard; special hazardous substance

Pennsylvania RTK - Special Hazardous Substances
[present]

N-NITROSOHYDROXYPROLINE   30310-80-6

INTERNATIONAL LISTS

German (DFG) - Carcinogens
animal evidence of carcinogenicity

3-(N-NITROSOMETHYLAMINO)   85502-23-4
PROPIONALDEHDYE

SEE ALSO:
NITROSAMINES
F039-HAZARDOUS WASTES

HEALTH AND SAFETY LISTS

U.S. DOT - Appendix A Table 1 - Hazardous Substances
final RQ = 10 pounds (4.54 kg)

IARC - Group 2B (sufficient animal data)
[present] (Overall evaluation based only on evidence of carcinogenicity in monograph (17, 1978) or in Supplement 4)

NIOSH - Selected LD50s and LC50s
Oral, rat: LD50 = 480 mg/kg

NTP Seventh Report - Suspect Carcinogens
suspect carcinogen

OSHA - Possible Select Carcinogens
[present]

ENVIRONMENTAL LISTS

ATSDR Priority List
Rank (of 275): 109

CERCLA/SARA - Section 313 - Emission Reporting
form R reporting required for 0.1% de minimus concentration

CERCLA/SARA - Hazardous Substances and their Reportable Quantities
final RQ = 10 pounds (4.54 kg)

Clean Water Act - Priority Pollutants
[present]

EPA - Carcinogen Hazard Ranking for RQ Adjustment
Hazard ranking = Medium

RCRA - U Series Wastes
waste number U111

RCRA - Hazardous Constituents-Appendix VIII
waste number U111

RCRA - Basis for Listing - Appendix VII
Included in waste stream: F039

RCRA - Substances Banned From Land Disposal
[present]

RCRA - TSD Facilities Ground Water Monitoring
TM 8270 = 10 ug/L PQL

RCRA - Universal Treatment Standards (LDR)
WW: 0.40 mg/l; NWW: 14 mg/kg

INTERNATIONAL LISTS

Canada - WHMIS: Ingredient Disclosure
0.1% item 1195 (1256)

German (DFG) - Carcinogens
animal evidence of carcinogenicity

Mexico - Wastewater - Organic Toxic Pollutants and Heavy Metals
Listed under [Nitrosamines]

Mexico - Drinking Water - Ecological Criteria
None given.

STATE LISTS

California - Air Bill 2588 Appendix A-I
known or potential carcinogen

California - Prop. 65 - Cancer list
carcinogen - initial date 1/1/88

California - Prop. 65 - No Significant Risk Levels
no significant risk level = 0.1 ug/day

California - Directors List of Hazardous Substances (8 CCR 339)
[present]

Florida Hazardous Substance List
[present]

Massachusetts Right To Know List
carcinogen; extraordinarily hazardous

Minnesota Hazardous Substance List
carcinogen

NJ Right to Know List (Total)
sn 1407

NJ Special Hazardous Substances
(carcinogen; mutagen)

Pennsylvania Right to Know List
environmental hazard; special hazardous substance

Pennsylvania RTK - Special Hazardous Substances
[present]

4-(N-NITROSOMETHYLAMINO)-1-(3-PYRIDYL)-1-BUTANONE   64091-91-4

INTERNATIONAL LISTS

German (DFG) - Carcinogens
animal evidence of carcinogenicity

3-(N-NITROSOMETHYLAMINO)-4-(3-PYRIDYL)-1-BUTANAL (NNA)   64091-90-3
SEE ALSO:
NITROSAMINES

HEALTH AND SAFETY LISTS

U.S. DOT - Appendix A Table 1 - Hazardous Substances
final RQ = 1 pound (0.454 kg)

IARC - Group 2A (limited human data)
[present] (Overall evaluation based only on evidence of carcinogenicity in monograph (17, 1978) or in Supplement 4) (Other relevant data, as given in Supplement 7 or in the monograph, influenced the making of the overall evaluation)

NIOSH - Selected LD50s and LC50s
Oral, rat: LD50 = 300 mg/kg

NTP Seventh Report - Suspect Carcinogens
suspect carcinogen

OSHA - Possible Select Carcinogens
[present]

ENVIRONMENTAL LISTS

CERCLA/SARA - Section 313 - Emission Reporting
form R reporting required for 0.1% de minimus concentration

CERCLA/SARA - Hazardous Substances and their Reportable Quantities
final RQ = 1 pound (0.454 kg)

EPA - Carcinogen Hazard Ranking for RQ Adjustment
Hazard ranking = High

RCRA - U Series Wastes
waste number U176

RCRA - Hazardous Constituents-Appendix VIII
waste number U176

RCRA - Substances Banned From Land Disposal
[present]

INTERNATIONAL LISTS

Canada - WHMIS: Ingredient Disclosure
0.1% item 1196 (1257)

STATE LISTS

California - Air Bill 2588 Appendix A-II
known or potential carcinogen

California - Prop. 65 - Cancer list
carcinogen - initial date 10/1/87

California - Prop. 65 - No Significant Risk Levels
no significant risk level = 0.03 ug/day

California - Directors List of Hazardous Substances (8 CCR 339)
[present]

Florida Hazardous Substance List
[present]

Massachusetts Right To Know List
carcinogen; extraordinarily hazardous

Minnesota Hazardous Substance List
carcinogen

NJ Right to Know List (Total)
sn 1410

NJ Special Hazardous Substances
(carcinogen; mutagen; teratogen)

Pennsylvania Right to Know List
environmental hazard; special hazardous substance

Pennsylvania RTK - Special Hazardous Substances
[present]

N-NITROSOMETHYLETHYLAMINE   10595-95-6

HEALTH AND SAFETY LISTS

IARC - Group 3 (not classifiable)
[present]

N-NITROSOMETHYLPHENYLAMINE   614-00-6

HEALTH AND SAFETY LISTS

IARC - Group 3 (not classifiable)
[present]

N-NITROSO-N-METHYLUREA   684-93-5

HEALTH AND SAFETY LISTS

IARC - Group 3 (not classifiable)
[present]

N-NITROSO-N-METHYLURETHANE   615-53-2

HEALTH AND SAFETY LISTS

IARC - Group 3 (not classifiable)
[present]

N-NITROSOMETHYLVINYLAMINE   4549-40-0

HEALTH AND SAFETY LISTS

IARC - Group 3 (not classifiable)
[present]

N-NITROSOMORPHOLINE   59-89-2

HEALTH AND SAFETY LISTS

IARC - Group 2B (sufficient animal data)
[present] (Overall evaluation based only on evidence of carcinogenicity in monograph (37, 1985) or in Supplement 4)
NTP Seventh Report - Suspect Carcinogens
suspect carcinogen
OSHA - Possible Select Carcinogens
[present]

STATE LISTS

California - Air Bill 2588 Appendix A-II
known or potential carcinogen
California - Prop. 65 - Cancer list
carcinogen - initial date 4/1/90
California - Directors List of Hazardous Substances (8 CCR 339)
[present]
Massachusetts Right To Know List
carcinogen; extraordinarily hazardous
Minnesota Hazardous Substance List
carcinogen

N-NITROSONORNICOTINE   16543-55-8

HEALTH AND SAFETY LISTS

IARC - Group 3 (not classifiable)
[present]

N-NITROSOPIPERIDINE   100-75-4

HEALTH AND SAFETY LISTS

IARC - Group 2B (sufficient animal data)
[present] (Overall evaluation based only on evidence of carcinogenicity in monograph (17, 1978) or in Supplement 4)
NIOSH - Selected LD50s and LC50s
Oral, rat: LD50 = 90 mg/kg
OSHA - Possible Select Carcinogens
[present]

ENVIRONMENTAL LISTS

RCRA - Hazardous Constituents-Appendix VIII
hazardous constituent - no waste number
RCRA - Basis for Listing - Appendix VII
Included in waste stream: F039
RCRA - TSD Facilities Ground Water Monitoring
TM 8270 = 10 ug/L PQL
RCRA - Universal Treatment Standards (LDR)
WW: 0.40 mg/l; NWW: 2.3 mg/kg

INTERNATIONAL LISTS

German (DFG) - Carcinogens
animal evidence of carcinogenicity

STATE LISTS

California - Air Bill 2588 Appendix A-I
known or potential carcinogen
California - Prop. 65 - Cancer list
carcinogen - initial date 10/1/89

California - Prop. 65 - No Significant Risk Levels
no significant risk level = 0.03 ug/day
California - Directors List of Hazardous Substances (8 CCR 339)
[present]
Florida Hazardous Substance List
[present]
Massachusetts Right To Know List
carcinogen; extraordinarily hazardous
Minnesota Hazardous Substance List
carcinogen
Pennsylvania Right to Know List
environmental hazard; special hazardous substance
Pennsylvania RTK - Special Hazardous Substances
[present]

N-NITROSOPROLINE   RR-01543-0

INTERNATIONAL LISTS

German (DFG) - Carcinogens
animal evidence of carcinogenicity

N-NITROSOPYRROLIDINE   930-55-2
SEE ALSO:
NITROSAMINES

HEALTH AND SAFETY LISTS

U.S. DOT - Appendix A Table 1 - Hazardous Substances
final RQ = 1 pound (0.454 kg)
IARC - Group 2A (limited human data)
[present] (Overall evaluation based only on evidence of carcinogenicity in monograph (17, 1978) or in Supplement 4) (Other relevant data, as given in Supplement 7 or in the monograph, influenced the making of the overall evaluation)
NIOSH - Selected LD50s and LC50s
Oral, rat: LD50 = 110 mg/kg
NTP Seventh Report - Suspect Carcinogens
suspect carcinogen
OSHA - Possible Select Carcinogens
[present]

ENVIRONMENTAL LISTS

CERCLA/SARA - Section 313 - Emission Reporting
form R reporting required for 0.1% de minimus concentration
CERCLA/SARA - Hazardous Substances and their Reportable Quantities
final RQ = 1 pound (0.454 kg)
Clean Air Act (1990) - List of Hazardous Air Contaminants
[present]
EPA - Carcinogen Hazard Ranking for RQ Adjustment
Hazard ranking = High
RCRA - U Series Wastes
waste number U177
RCRA - Hazardous Constituents-Appendix VIII
waste number U177
RCRA - Substances Banned From Land Disposal
[present]

INTERNATIONAL LISTS

Canada - WHMIS: Ingredient Disclosure
0.1% item 1197 (1258)

STATE LISTS

California - Air Bill 2588 Appendix A-I
known or potential carcinogen
California - Prop. 65 - Cancer list
carcinogen - initial date 10/1/87
California - Prop. 65 - No Significant Risk Levels
no significant risk level = 0.006 ug/day

California - Directors List of Hazardous Substances (8 CCR 339)
[present]

Florida Hazardous Substance List
[present]

Massachusetts Right To Know List
carcinogen; extraordinarily hazardous

Minnesota Hazardous Substance List
carcinogen

NJ Right to Know List (Total)
sn 1411

NJ Special Hazardous Substances
(carcinogen)

Pennsylvania Right to Know List
environmental hazard; special hazardous substance

Pennsylvania RTK - Special Hazardous Substances
[present]

N-NITROSOSARCOSINE   13256-22-9

HEALTH AND SAFETY LISTS

U.S. DOT - Appendix A Table 1 - Hazardous Substances
final RQ = 1 pound (0.454 kg)

IARC - Group 2B (sufficient animal data)
[present] (Overall evaluation based only on evidence of carcinogenicity in monograph (4, 1974) or in Supplement 4)

NIOSH - Selected LD50s and LC50s
Oral, rat: LD50 = 180 mg/kg

OSHA - Possible Select Carcinogens
[present]

ENVIRONMENTAL LISTS

CERCLA/SARA - Hazardous Substances and their Reportable Quantities
final RQ = 1 pound (0.454 kg)

EPA - Carcinogen Hazard Ranking for RQ Adjustment
Hazard ranking = High

RCRA - U Series Wastes
waste number U178

RCRA - Hazardous Constituents-Appendix VIII
waste number U178

RCRA - Substances Banned From Land Disposal
[present]

TSCA - Chemicals with Significant New Use Rules
[present]

TSCA - Section 12(b) - Export Notification
export notification required - Section 5

INTERNATIONAL LISTS

Canada - WHMIS: Ingredient Disclosure
1% item 1198 (1259)

STATE LISTS

California - Air Bill 2588 Appendix A-II
known or potential carcinogen

California - Prop. 65 - Cancer list
carcinogen - initial date 4/1/88

California - Prop. 65 - No Significant Risk Levels
no significant risk level = 0.006 ug/day

California - Directors List of Hazardous Substances (8 CCR 339)
[present]

Florida Hazardous Substance List
[present]

Massachusetts Right To Know List
carcinogen; extraordinarily hazardous

Minnesota Hazardous Substance List
carcinogen

NJ Special Hazardous Substances
(carcinogen; mutagen)

Pennsylvania Right to Know List
environmental hazard; special hazardous substance

Pennsylvania RTK - Special Hazardous Substances
[present]

NITROSTARCH   9056-38-6
SEE ALSO:
NITROSAMINES

HEALTH AND SAFETY LISTS

U.S. DOT - Appendix A Table 1 - Hazardous Substances
final RQ = 10 pounds (4.54 kg)

IARC - Group 2B (sufficient animal data)
[present] (Overall evaluation based only on evidence of carcinogenicity in monograph (17, 1978) or in Supplement 4)

NIOSH - Selected LD50s and LC50s
Inhalation, rat: LD50 = 22 mg/kg 8 hr Oral, rat: LD50 = 24 mg/kg

NTP Seventh Report - Suspect Carcinogens
suspect carcinogen

OSHA - Possible Select Carcinogens
[present]

ENVIRONMENTAL LISTS

CERCLA/SARA - Section 313 - Emission Reporting
form R reporting required for 0.1% de minimus concentration

CERCLA/SARA - Hazardous Substances and their Reportable Quantities
final RQ = 10 pounds (4.54 kg)

EPA - Carcinogen Hazard Ranking for RQ Adjustment
Hazard ranking = Medium

RCRA - P Series Wastes
waste number P084

RCRA - Hazardous Constituents-Appendix VIII
waste number P084

RCRA - Substances Banned From Land Disposal
[present]

INTERNATIONAL LISTS

Canada - WHMIS: Ingredient Disclosure
0.1% item 1199 (1260)

STATE LISTS

California - Air Bill 2588 Appendix A-II
known or potential carcinogen

California - Prop. 65 - Cancer list
carcinogen - initial date 1/1/88

California - Directors List of Hazardous Substances (8 CCR 339)
[present]

Florida Hazardous Substance List
[present]

Massachusetts Right To Know List
carcinogen; extraordinarily hazardous

Minnesota Hazardous Substance List
carcinogen

NJ Right to Know List (Total)
sn 2907

Pennsylvania Right to Know List
environmental hazard; special hazardous substance

Pennsylvania RTK - Special Hazardous Substances
[present]

BETA-NITROSTYRENE   102-96-5
SEE ALSO:
NITROSAMINES
F039-HAZARDOUS WASTES

HEALTH AND SAFETY LISTS

IARC - Group 2B (sufficient animal data)
[present] (Overall evaluation based only on evidence of carcinogenicity in monograph (17, 1978) or in Supplement 4)

NIOSH - Selected LD50s and LC50s
Oral, rat: LD50 = 282 mg/kg

NTP Seventh Report - Suspect Carcinogens
suspect carcinogen

OSHA - Possible Select Carcinogens
[present]

ENVIRONMENTAL LISTS

CERCLA/SARA - Section 313 - Emission Reporting
form R reporting required for 0.1% de minimus concentration

CERCLA/SARA - Hazardous Substances and their Reportable Quantities
final RQ = 1 pound (.454 kg)

Clean Air Act (1990) - List of Hazardous Air Contaminants
[present]

RCRA - Hazardous Constituents-Appendix VIII
hazardous constituent - no waste number

RCRA - Basis for Listing - Appendix VII
Included in waste stream: F039

RCRA - TSD Facilities Ground Water Monitoring
TM 8270 = 10 ug/L PQL

RCRA - Universal Treatment Standards (LDR)
WW: 0.40 mg/l; NWW: 2.3 mg/kg

INTERNATIONAL LISTS

Canada - WHMIS: Ingredient Disclosure
0.1% item 1200 (1261)

German (DFG) - Carcinogens
animal evidence of carcinogenicity

STATE LISTS

California - Air Bill 2588 Appendix A-I
known or potential carcinogen

California - Prop. 65 - Cancer list
carcinogen - initial date 1/1/88

California - Prop. 65 - No Significant Risk Levels
no significant risk level = 0.1 ug/day

California - Directors List of Hazardous Substances (8 CCR 339)
[present]

Florida Hazardous Substance List
[present]

Massachusetts Right To Know List
carcinogen; extraordinarily hazardous

Minnesota Hazardous Substance List
carcinogen

NJ Right to Know List (Total)
sn 1409

NJ Special Hazardous Substances
(carcinogen; mutagen)

Pennsylvania Right to Know List
environmental hazard; special hazardous substance

Pennsylvania RTK - Special Hazardous Substances
[present]

NITROSUGARS   RR-01454-0
SEE ALSO:
NITROSAMINES

HEALTH AND SAFETY LISTS

IARC - Group 2B (sufficient animal data)
[present] (Overall evaluation based only on evidence of carcinogenicity in monograph (37, 1985) or in Supplement 4)

NTP Seventh Report - Suspect Carcinogens
suspect carcinogen

OSHA - Possible Select Carcinogens
[present]

ENVIRONMENTAL LISTS

CERCLA/SARA - Section 313 - Emission Reporting
form R reporting required for 0.1% de minimus concentration

RCRA - Hazardous Constituents-Appendix VIII
hazardous constituent - no waste number

INTERNATIONAL LISTS

Canada - WHMIS: Ingredient Disclosure
0.1% item 1201 (1262)

STATE LISTS

California - Air Bill 2588 Appendix A-II
known or potential carcinogen

California - Prop. 65 - Cancer list
carcinogen - initial date 1/1/88

California - Prop. 65 - No Significant Risk Levels
no significant risk level = 0.5 ug/day

California - Directors List of Hazardous Substances (8 CCR 339)
[present]

Florida Hazardous Substance List
[present]

Massachusetts Right To Know List
carcinogen; extraordinarily hazardous

Minnesota Hazardous Substance List
carcinogen

NJ Right to Know List (Total)
sn 2900

Pennsylvania Right to Know List
environmental hazard; special hazardous substance

Pennsylvania RTK - Special Hazardous Substances
[present]

NITROSYL CHLORIDE   2696-92-6
SEE ALSO:
F039-HAZARDOUS WASTES
NITROSAMINES

HEALTH AND SAFETY LISTS

U.S. DOT - Appendix A Table 1 - Hazardous Substances
final RQ = 10 pounds (4.54 kg)

IARC - Group 2B (sufficient animal data)
[present] (Overall evaluation based only on evidence of carcinogenicity in monograph (17, 1978) or in Supplement 4)

NIOSH - Selected LD50s and LC50s
Oral, rat: LD50 = 200 mg/kg

NTP Seventh Report - Suspect Carcinogens
suspect carcinogen

OSHA - Possible Select Carcinogens
[present]

ENVIRONMENTAL LISTS

CERCLA/SARA - Section 313 - Emission Reporting
form R reporting required for 0.1% de minimus concentration

CERCLA/SARA - Hazardous Substances and their Reportable Quantities
final RQ = 10 pounds (4.54 kg)

EPA - Carcinogen Hazard Ranking for RQ Adjustment
Hazard ranking = Medium

RCRA - U Series Wastes
waste number U179

RCRA - Hazardous Constituents-Appendix VIII
waste number U179

RCRA - Basis for Listing - Appendix VII
Included in waste stream: F039

RCRA - Substances Banned From Land Disposal
[present]

RCRA - TSD Facilities Ground Water Monitoring
TM 8270 = 10 ug/L PQL

RCRA - Universal Treatment Standards (LDR)
WW: 0.013 mg/l; NWW: 35 mg/kg

INTERNATIONAL LISTS

Canada - WHMIS: Ingredient Disclosure
0.1% item 1202 (1263)

German (DFG) - Carcinogens
animal evidence of carcinogenicity

STATE LISTS

California - Air Bill 2588 Appendix A-I
known or potential carcinogen

California - Prop. 65 - Cancer list
carcinogen - initial date 1/1/88

California - Prop. 65 - No Significant Risk Levels
no significant risk level = 0.07 ug/day

California - Directors List of Hazardous Substances (8 CCR 339)
[present]

Florida Hazardous Substance List
[present]

Massachusetts Right To Know List
carcinogen; extraordinarily hazardous

Minnesota Hazardous Substance List
carcinogen

NJ Right to Know List (Total)
sn 1412

NJ Special Hazardous Substances
(carcinogen; mutagen)

Pennsylvania Right to Know List
environmental hazard; special hazardous substance

Pennsylvania RTK - Special Hazardous Substances
[present]

NITROSYLSULPHURIC ACID 7782-78-7

HEALTH AND SAFETY LISTS

IARC - Group 3 (not classifiable)
[present]

O-NITROTOLUENE 88-72-2
SEE ALSO:
NITROSAMINES
F039-HAZARDOUS WASTES

HEALTH AND SAFETY LISTS

U.S. DOT - Appendix A Table 1 - Hazardous Substances
final RQ = 1 pound (0.454 kg)

IARC - Group 2B (sufficient animal data)
[present] (Overall evaluation based only on evidence of carcinogenicity in monograph (17, 1978) or in Supplement 4)

NIOSH - Selected LD50s and LC50s
Oral, rat: LD50 = 900 mg/kg

NTP Seventh Report - Suspect Carcinogens
suspect carcinogen

OSHA - Possible Select Carcinogens
[present]

ENVIRONMENTAL LISTS

CERCLA/SARA - Hazardous Substances and their Reportable Quantities
final RQ = 1 pound (0.454 kg)

EPA - Carcinogen Hazard Ranking for RQ Adjustment
Hazard ranking = High

RCRA - U Series Wastes
waste number U180

RCRA - Hazardous Constituents-Appendix VIII
waste number U180

RCRA - Basis for Listing - Appendix VII
Included in waste stream: F039

RCRA - Substances Banned From Land Disposal
[present]

RCRA - TSD Facilities Ground Water Monitoring
TM 8270 = 10 ug/L PQL

RCRA - Universal Treatment Standards (LDR)
WW: 0.013 mg/l; NWW: 35 mg/kg

TSCA - Chemicals with Significant New Use Rules
[present]

TSCA - Section 12(b) - Export Notification
export notification required - Section 5

INTERNATIONAL LISTS

Canada - WHMIS: Ingredient Disclosure
0.1% item 1203 (1264)

German (DFG) - Carcinogens
animal evidence of carcinogenicity

STATE LISTS

California - Air Bill 2588 Appendix A-I
known or potential carcinogen

California - Prop. 65 - Cancer list
carcinogen - initial date 10/1/87

California - Prop. 65 - No Significant Risk Levels
no significant risk level = 0.3 ug/day

California - Directors List of Hazardous Substances (8 CCR 339)
[present]

Florida Hazardous Substance List
[present]

Massachusetts Right To Know List
carcinogen; extraordinarily hazardous

Minnesota Hazardous Substance List
carcinogen

NJ Right to Know List (Total)
sn 3000

Pennsylvania Right to Know List
environmental hazard; special hazardous substance

Pennsylvania RTK - Special Hazardous Substances
[present]

M-NITROTOLUENE 99-08-1
SEE ALSO:
NITROSAMINES

HEALTH AND SAFETY LISTS

IARC - Group 2B (sufficient animal data)
[present] (Overall evaluation based only on evidence of carcinogenicity in monograph (17, 1978) or in Supplement 4)

NIOSH - Selected LD50s and LC50s
Oral, rat: LD50 = 5000 mg/kg

NTP Seventh Report - Suspect Carcinogens
suspect carcinogen

OSHA - Possible Select Carcinogens
[present]

ENVIRONMENTAL LISTS

RCRA - Hazardous Constituents-Appendix VIII
hazardous constituent - no waste number

INTERNATIONAL LISTS

Canada - WHMIS: Ingredient Disclosure
0.1% item 1204 (1265)

STATE LISTS

California - Air Bill 2588 Appendix A-II
known or potential carcinogen

California - Prop. 65 - Cancer list
carcinogen - initial date 1/1/88

California - Directors List of Hazardous Substances (8 CCR 339)
[present]

Florida Hazardous Substance List
[present]

Massachusetts Right To Know List
carcinogen; extraordinarily hazardous

Minnesota Hazardous Substance List
carcinogen

NJ Right to Know List (Total)
sn 1413

NJ Special Hazardous Substances
(carcinogen)

Pennsylvania Right to Know List
environmental hazard; special hazardous substance

Pennsylvania RTK - Special Hazardous Substances
[present]

P-NITROTOLUENE   99-99-0

HEALTH AND SAFETY LISTS

U.S. DOT - Substances From 49 CFR 172.101
regulated by DOT (UN0146, UN1337)

U.S. DOT - Hazard Classes
DOT hazard class = 4.1

STATE LISTS

NJ Right to Know List (Total)
sn 1395

NITROTOLUENE (MIXED ISOMERS)   1321-12-6

HEALTH AND SAFETY LISTS

NTP Chemical Status Reports - Testing Status and NTIS Number
Technical reports printed (PB300949/AS)

NTP Chemical Status Reports - Evidence of Carcinogenicity
male rat-negative; female rat-negative; male mice-negative; female mice-negative

2-NITRO-P-TOLUIDINE   89-62-3

HEALTH AND SAFETY LISTS

U.S. DOT - Hazard Classes
Forbidden from transport by the DOT

5-NITRO-O-TOLUIDINE   99-55-8

HEALTH AND SAFETY LISTS

U.S. DOT - Substances From 49 CFR 172.101
regulated by DOT (UN1069)

U.S. DOT - Hazard Classes
DOT hazard class = 2.3

U.S. DOT - Substances Which Are Poisonous by Inhalation
gaseous hazardous material poisonous by inhalation (UN1069)

INTERNATIONAL LISTS

Canada - WHMIS: Ingredient Disclosure
1% item 1205 (512)

STATE LISTS

NJ Right to Know List (Total)
sn 1396

NITROTRIAZOLONE   RR-01334-3

HEALTH AND SAFETY LISTS

U.S. DOT - Substances From 49 CFR 172.101
regulated by DOT (UN2308)

U.S. DOT - Hazard Classes
DOT hazard class = 8

INTERNATIONAL LISTS

Canada - WHMIS: Ingredient Disclosure
1% item 1206 (117)

STATE LISTS

NJ Right to Know List (Total)
sn 1397

NJ Special Hazardous Substances
(corrosive)

NITRO UREA   556-89-8

HEALTH AND SAFETY LISTS

ACGIH 1995 - Time Weighted Averages
2 ppm TWA; 11 mg/m3 TWA

ACGIH 1995 - Skin Designations
skin - potential for cutaneous absorption

U.S. DOT - Appendix A Table 1 - Hazardous Substances
final RQ = 1000 pounds (454 kg) (Listed under 'Nitrotoluene')

NFPA - Flash Points
flash point = 223 degrees F (106 degrees C)

NFPA - Hazard Identification Ratings
health-3; flammability-1; reactivity-1

NIOSH - Selected LD50s and LC50s
Oral, rat: LD50 = 891 mg/kg

NIOSH 1990 - Pocket Guide - RELs
2 ppm TWA; 11 mg/m3 TWA (Listed under 'Nitrotoluene')

NIOSH 1990 - Pocket Guide - IDLHs
200 ppm IDLH (Listed under 'Nitrotoluene')

NIOSH 1990 - Pocket Guide - Target organs
blood, CNS, CVS, skin, GI tract (Listed under 'Nitrotoluene')

NIOSH 1990 - Pocket Guide - Skin list
Potential for dermal absorption (Listed under 'Nitrotoluene')

NTP Chemical Status Reports - Testing Status and NTIS Number
Technical reports printed (PB93-150092); Approved for toxicology/ carcinogenesis study; Prechronic studies completed: in review for further evaluation

OSHA - Vacated PELs - Time Weighted Averages
2 ppm TWA; 11 mg/m3 TWA (Listed under 'Nitrotoluene')

OSHA - Vacated PELs - Skin Designation
Prevent or reduce skin absorption (Listed under 'Nitrotoluene')

OSHA - Final PELs - Time Weighted Averages
5 ppm TWA; 30 mg/m3 TWA (Listed under 'Nitrotoluene')

OSHA - Final PELs - Skin Notations
prevent or reduce skin absorption

ENVIRONMENTAL LISTS

CERCLA/SARA - Hazardous Substances and their Reportable Quantities
final RQ = 1000 pounds (454 kg) (Listed under 'Nitrotoluene')

CAA - HON Rule - SOCMI Chemicals
compliance by April 24, 1995

Clean Water Act - Hazardous Substances
[present] (Listed under 'Nitrotoluene')

EPA - Master Testing List
[present]

TSCA - Health and Safety Reporting List
Effective Date: March 11, 1994; Sunset Date: March 11, 2004

INTERNATIONAL LISTS

Australian Exposure Standards - Time Weighted Averages
2 ppm TWA; 11 mg/m3 TWA

Australian Exposure Standards - Skin Effects
skin absorption

Canada - WHMIS: Ingredient Disclosure
1% item 1208 (1267)

German (DFG) - Skin/Sensitizers
danger of cutaneous absorption

German (DFG) - Carcinogens
animal evidence of carcinogenicity

Israel - Time Weighted Averages
2 ppm TWA; 11 mg/m3 TWA

Israel - Action Levels
1 ppm AL; 5.5 mg/m3 AL

STATE LISTS

California - Exposure Limits - PELs
2 ppm PEL; 11 mg/m3 PEL

California - Exposure Limits - Skin Notation
material may be absorbed through the skin, eyes or mucous membrane

California - Directors List of Hazardous Substances (8 CCR 339)
[present] (Listed under 'Nitrotoluenes')

Florida Hazardous Substance List
[present]

Massachusetts Right To Know List
[present]

Pennsylvania Right to Know List
environmental hazard

PROPOSED REGULATIONS

TSCA - ITC 32nd Report Priority Testing List
designated for dermal absorption testing

NITROUS ACID, LEAD (2+) SALT   13826-65-8

HEALTH AND SAFETY LISTS

ACGIH 1995 - Time Weighted Averages
2 ppm TWA; 11 mg/m3 TWA

ACGIH 1995 - Skin Designations
skin - potential for cutaneous absorption

U.S. DOT - Appendix A Table 1 - Hazardous Substances
final RQ = 1000 pounds (454 kg) (Listed under 'Nitrotoluene')

NFPA - Flash Points
flash point = 223 degrees F (106 degrees C)

NFPA - Hazard Identification Ratings
health-3; flammability-1; reactivity-1

NIOSH - Selected LD50s and LC50s
Oral, rat: LD50 = 1072 mg/kg

NIOSH 1990 - Pocket Guide - RELs
2 ppm TWA; 11 mg/m3 TWA (Listed under 'Nitrotoluene')

NIOSH 1990 - Pocket Guide - IDLHs
200 ppm IDLH (Listed under 'Nitrotoluene')

NIOSH 1990 - Pocket Guide - Target organs
blood, CNS, CVS, skin, GI tract (Listed under 'Nitrotoluene')

NIOSH 1990 - Pocket Guide - Skin list
Potential for dermal absorption (Listed under 'Nitrotoluene')

NTP Chemical Status Reports - Testing Status and NTIS Number
Technical reports printed (PB93-150092); project leader assigned/ study in design

OSHA - Vacated PELs - Time Weighted Averages
2 ppm TWA; 11 mg/m3 TWA (Listed under 'Nitrotoluene')

OSHA - Vacated PELs - Skin Designation
Prevent or reduce skin absorption (Listed under 'Nitrotoluene')

OSHA - Final PELs - Time Weighted Averages
5 ppm TWA; 30 mg/m3 TWA (Listed under 'Nitrotoluene')

OSHA - Final PELs - Skin Notations
prevent or reduce skin absorption

ENVIRONMENTAL LISTS

CERCLA/SARA - Hazardous Substances and their Reportable Quantities
final RQ = 1000 pounds (454 kg) (Listed under 'Nitrotoluene')

CAA - HON Rule - SOCMI Chemicals
compliance by April 24, 1995

Clean Water Act - Hazardous Substances
[present] (Listed under 'Nitrotoluene')

INTERNATIONAL LISTS

Australian Exposure Standards - Time Weighted Averages
2 ppm TWA; 11 mg/m3 TWA

Australian Exposure Standards - Skin Effects
skin absorption

Canada - WHMIS: Ingredient Disclosure
1% item 1207 (1266)

Israel - Time Weighted Averages
2 ppm TWA; 11 mg/m3 TWA

Israel - Action Levels
1 ppm AL; 5.5 mg/m3 AL

STATE LISTS

California - Exposure Limits - PELs
2 ppm PEL; 11 mg/m3 PEL

California - Exposure Limits - Skin Notation
material may be absorbed through the skin, eyes or mucous membrane

California - Directors List of Hazardous Substances (8 CCR 339)
[present] (Listed under 'Nitrotoluenes')

Florida Hazardous Substance List
[present]

Massachusetts Right To Know List
[present]

Minnesota Hazardous Substance List
skin

Pennsylvania Right to Know List
environmental hazard

NITROUS OXIDE   10024-97-2

HEALTH AND SAFETY LISTS

ACGIH 1995 - Time Weighted Averages
2 ppm TWA; 11 mg/m3 TWA

ACGIH 1995 - Skin Designations
skin - potential for cutaneous absorption

U.S. DOT - Appendix A Table 1 - Hazardous Substances
final RQ = 1000 pounds (454 kg) (Listed under 'Nitrotoluene')

NFPA - Flash Points
flash point = 223 degrees F (106 degrees C)

NFPA - Hazard Identification Ratings
health-3; flammability-1; reactivity-1

NIOSH - Selected LD50s and LC50s
Oral, rat: LD50 = 1960 mg/kg Skin, rat: LD50 = 16000 mg/kg

NIOSH 1990 - Pocket Guide - RELs
2 ppm TWA; 11 mg/m3 TWA (Listed under 'Nitrotoluene')

NIOSH 1990 - Pocket Guide - IDLHs
200 ppm IDLH (Listed under 'Nitrotoluene')

NIOSH 1990 - Pocket Guide - Target organs
blood, CNS, CVS, skin, GI tract (Listed under 'Nitrotoluene')

NIOSH 1990 - Pocket Guide - Skin list
Potential for dermal absorption (Listed under 'Nitrotoluene')

NTP Chemical Status Reports - Testing Status and NTIS Number
Technical reports printed (PB93-150092); project leader assigned/ study in design

OSHA - Vacated PELs - Time Weighted Averages
2 ppm TWA; 11 mg/m3 TWA (Listed under 'Nitrotoluene')

OSHA - Vacated PELs - Skin Designation
Prevent or reduce skin absorption (Listed under 'Nitrotoluene')

OSHA - Final PELs - Time Weighted Averages
5 ppm TWA; 30 mg/m3 TWA (Listed under 'Nitrotoluene')

OSHA - Final PELs - Skin Notations
prevent or reduce skin absorption

ENVIRONMENTAL LISTS

CERCLA/SARA - Hazardous Substances and their Reportable Quantities
final RQ = 1000 pounds (454 kg) (Listed under 'Nitrotoluene')

CAA - HON Rule - SOCMI Chemicals
compliance by April 24, 1995

Clean Water Act - Hazardous Substances
[present] (Listed under 'Nitrotoluene')

INTERNATIONAL LISTS

Australian Exposure Standards - Time Weighted Averages
2 ppm TWA; 11 mg/m3 TWA

Australian Exposure Standards - Skin Effects
skin absorption

Canada - WHMIS: Ingredient Disclosure
1% item 1209 (1268)

Israel - Time Weighted Averages
2 ppm TWA; 11 mg/m3 TWA

Israel - Action Levels
1 ppm AL; 5.5 mg/m3 AL

STATE LISTS

California - Exposure Limits - PELs
2 ppm PEL; 11 mg/m3 PEL

California - Exposure Limits - Skin Notation
material may be absorbed through the skin, eyes or mucous membrane

California - Directors List of Hazardous Substances (8 CCR 339)
[present] (Listed under 'Nitrotoluenes')

Florida Hazardous Substance List
[present]

Massachusetts Right To Know List
[present]

Pennsylvania Right to Know List
environmental hazard

NITROVIN   804-36-4

HEALTH AND SAFETY LISTS

U.S. DOT - Substances From 49 CFR 172.101
regulated by DOT (UN1664)

U.S. DOT - Hazard Classes
DOT hazard class = 6.1

U.S. DOT - Appendix A Table 1 - Hazardous Substances
final RQ = 1000 pounds (454 kg)

ENVIRONMENTAL LISTS

CERCLA/SARA - Hazardous Substances and their Reportable Quantities
final RQ = 1000 pounds (454 kg)

CAA - HON Rule - SOCMI Chemicals
compliance by April 24, 1995

Clean Water Act - Hazardous Substances
[present] (Listed under 'Nitrotoluene')

INTERNATIONAL LISTS

Canada - Alberta - 8 Hour Occupational Exposure Limit
2 ppm TWA; 11 mg/m3 TWA

Canada - Alberta - 15 Minute Occupational Exposure Limit
4 ppm STEL; 22 mg/m3 STEL

Canada - Alberta - Skin Designation
can be absorbed through the intact skin

Canada - British Columbia - 8 Hour Exposure Limits
5 ppm TWA; 30 mg/m3 TWA

Canada - British Columbia - 15 Minute Exposure Limits
10 ppm STEL; 60 mg/m3 STEL

Canada - British Columbia - Skin Notations
skin - potential for skin absorption

Canada - Ontario - OHSA - TWAEVs
2 ppm TWAEV; 11 mg/m3 TWAEV

Canada - Ontario - OHSA - Skin Notations
absorption through skin, eyes, or mucous membranes

Canada - Quebec - Time-Weighted Average Exposure Values
2 ppm TWAEV; 11 mg/m3 TWAEV

Canada - Quebec - Skin Designations
absorbed through the skin

United Kingdom - Occupational Exposure Standards - TWAs
5 ppm TWA; 30 mg/m3 TWA

United Kingdom - Occupational Exposure Standards - STELs
10 ppm STEL; 60 mg/m3 STEL

United Kingdom - Occupational Exposure Standards - Notes
can be absorbed through skin

German (DFG) - MAK Values
5 ppm MAK; 30 mg/m3 MAK

German (DFG) - Peak Limitations
2 x normal MAK (30 min. average value); don't exceed 4 times during shift

German (DFG) - Skin/Sensitizers
danger of cutaneous absorption

Mexico - Instruction No. 10 - TWAs
5 ppm TWA; 30 mg/m3 TWA

Mexico - Instruction No. 10 - STELs
10 ppm STEL; 60 mg/m3 STEL

STATE LISTS

California - Exposure Limits - PELs
2 ppm PEL; 11 mg/m3 PEL

California - Exposure Limits - Skin Notation
material may be absorbed through the skin, eyes or mucous membrane

California - Directors List of Hazardous Substances (8 CCR 339)
[present] (Listed under 'Nitrotoluenes')

Massachusetts Right To Know List
[present]

NJ Right to Know List (Total)
sn 1398

NJ Special Hazardous Substances
(reactive - fourth degree)

Pennsylvania Right to Know List
environmental hazard

PROPOSED REGULATIONS

Canada - Ontario - Proposed Occupational TWAEVs
1 ppm TWAEV; 5.5 mg/m3 TWAEV

M-NITROXYLENE   99-51-4

HEALTH AND SAFETY LISTS

NFPA - Flash Points
flash point = 315 degrees F (157 degrees C)

NFPA - Hazard Identification Ratings
health-2; flammability-1; reactivity-4

STATE LISTS

Florida Hazardous Substance List
[present]

Massachusetts Right To Know List
[present]

Pennsylvania Right to Know List
[present]

NITROXYLENE, ALL ISOMERS   RR-00607-5
SEE ALSO:
F039-HAZARDOUS WASTES

HEALTH AND SAFETY LISTS

U.S. DOT - Appendix A Table 1 - Hazardous Substances
final RQ = 100 pounds (45.4 kg)

IARC - Group 3 (not classifiable)
[present]

NIOSH - Selected LD50s and LC50s
Oral, rat: LD50 = 574 mg/kg

NTP Chemical Status Reports - Testing Status and NTIS Number
Technical reports printed (PB285872/AS)

NTP Chemical Status Reports - Evidence of Carcinogenicity
male rat-negative; female rat-negative; male mice-positive; female mice-positive

ENVIRONMENTAL LISTS

CERCLA/SARA - Section 313 - Emission Reporting
form R reporting required

CERCLA/SARA - Hazardous Substances and their Reportable Quantities
final RQ = 100 pounds (45.4 kg)

EPA - Carcinogen Hazard Ranking for RQ Adjustment
Hazard ranking = Low

RCRA - U Series Wastes
waste number U181

RCRA - Hazardous Constituents-Appendix VIII
waste number U181

RCRA - Basis for Listing - Appendix VII
Included in waste stream: F039

RCRA - Substances Banned From Land Disposal
[present]

RCRA - TSD Facilities Ground Water Monitoring
TM 8270 = 10 ug/L PQL

RCRA - Universal Treatment Standards (LDR)
WW: 0.32 mg/l; NWW: 28 mg/kg

INTERNATIONAL LISTS

Canada - WHMIS: Ingredient Disclosure
1% item 1210 (1269)

German (DFG) - Carcinogens
animal evidence of carcinogenicity

STATE LISTS

Florida Hazardous Substance List
[present]

Massachusetts Right To Know List
carcinogen; extraordinarily hazardous

NJ Right to Know List (Total)
sn 1444

NJ Special Hazardous Substances
(carcinogen)

Pennsylvania Right to Know List
environmental hazard

NITROXYLOL   25168-04-1

HEALTH AND SAFETY LISTS

U.S. DOT - Substances From 49 CFR 172.101
regulated by DOT (UN0490)

U.S. DOT - Hazard Classes
DOT hazard class = 1.1D

5-(4-NITROPHENYL)-2,4-PENTADIEN-1-AL   2608-48-2

HEALTH AND SAFETY LISTS

U.S. DOT - Substances From 49 CFR 172.101
regulated by DOT (UN0147)

U.S. DOT - Hazard Classes
DOT hazard class = 1.1D

NIVALENOL   23282-20-4

STATE LISTS

Massachusetts Right To Know List
[present]

NONADECANE   18435-45-5

HEALTH AND SAFETY LISTS

ACGIH 1995 - Time Weighted Averages
50 ppm TWA; 90 mg/m3 TWA

U.S. DOT - Substances From 49 CFR 172.101
regulated by DOT (UN1070, UN2201)

U.S. DOT - Hazard Classes
DOT hazard class = 2.2

NIOSH - Selected LD50s and LC50s
Inhalation, rat: LC50 = 1068 mg/m3 4 hr

NIOSH - Health Standards - Exposure Limits
25 ppm TWA; 30 mg/m3 TWA (Listed under 'Nitrogen, oxides of')

NIOSH - Health Standards - Health Effects and Precautions
Respiratory effects; blood effects (Pulmonary function testing required) (Listed under 'Nitrogen, oxides of')

ENVIRONMENTAL LISTS

List of Pesticide Product Inert Ingredients
[present]

INTERNATIONAL LISTS

Australian Exposure Standards - Time Weighted Averages
25 ppm TWA; 45 mg/m3 TWA

Canada - WHMIS: Ingredient Disclosure
0.1% item 1211 (1329)

Canada - Alberta - 8 Hour Occupational Exposure Limit
100 ppm TWA; 180 mg/m3 TWA

Canada - Alberta - 15 Minute Occupational Exposure Limit
200 ppm STEL; 360 mg/m3 STEL

Canada - British Columbia - 8 Hour Exposure Limits
25 ppm TWA

Canada - Ontario - OHSA - TWAEVs
25 ppm TWAEV; 45 mg/m3 TWAEV

Canada - Quebec - Time-Weighted Average Exposure Values
50 ppm TWAEV; 90 mg/m3 TWAEV

United Kingdom - Occupational Exposure Standards - TWAs
25 ppm TWA; 30 mg/m3 TWA

United Kingdom - Occupational Exposure Standards - STELs
35 ppm STEL; 45 mg/m3 STEL

German (DFG) - MAK Values
100 ppm MAK; 200 mg/m3 MAK

German (DFG) - Peak Limitations
2 x normal MAK (30 minute average value); don't exceed 4 times during shift

German (DFG) - Pregnancy
classification not yet possible

Israel - Time Weighted Averages
50 ppm TWA; 90 mg/m3 TWA

Israel - Action Levels
25 ppm AL; 45 mg/m3 AL

STATE LISTS

California - Exposure Limits - PELs
50 ppm PEL; 90 mg/m3 PEL

California - Directors List of Hazardous Substances (8 CCR 339)
[present]

Florida Hazardous Substance List
effective March 13, 1992

Minnesota Hazardous Substance List
[present]

NJ Right to Know List (Total)
sn 1399

Pennsylvania Right to Know List
[present]

NONANAL   124-19-6

HEALTH AND SAFETY LISTS

IARC - Group 3 (not classifiable)
[present]

NONANE   111-84-2

STATE LISTS

NJ Right to Know List (Total)
sn 1308

1-NONANETHIOL   1455-21-6

HEALTH AND SAFETY LISTS

U.S. DOT - Substances From 49 CFR 172.101
regulated by DOT (UN1665)

U.S. DOT - Hazard Classes
DOT hazard class = 6.1

U.S. DOT - Appendix B - Marine Pollutants
DOT regulated marine pollutant

INTERNATIONAL LISTS

Canada - WHMIS: Ingredient Disclosure
1% item 1212 (1270)

1-NONANOL   143-08-8

HEALTH AND SAFETY LISTS

NIOSH - Selected LD50s and LC50s
Oral, rat: LD50 = 2440 mg/kg

ENVIRONMENTAL LISTS

CAA - HON Rule - SOCMI Chemicals
compliance by Oct. 23, 1995

STATE LISTS

NJ Right to Know List (Total)
sn 1400

4-NONANOL, 2,6,8-TRIMETHYL-   123-17-1

HEALTH AND SAFETY LISTS

NTP Chemical Status Reports - Testing Status and NTIS Number
Prechronic studies for which toxicity technical reports were not prepared

2-NONANONE   821-55-6

HEALTH AND SAFETY LISTS

IARC - Group Unspecified
[present] (Listed under 'Toxins derived from Fusarium graminearum, F. culmorum and F. crookwellense')

NONENE   27215-95-8

HEALTH AND SAFETY LISTS

NFPA - Flash Points
flash point > 212 degrees F (100 degrees C)

NFPA - Hazard Identification Ratings
health-0; flammability-1; reactivity-0

NONFLAMMABLE GAS, N.O.S.   RR-00383-8

ENVIRONMENTAL LISTS

TSCA - Code of Federal Regulations Citations
40 CFR 712.30(x); 40 CFR 716.120(d)

TSCA - PAIR - Reporting List
Reporting Date: November 27, 1991

TSCA - Health and Safety Reporting List
Effective Date: September 30, 1991

NONYL ACETATE   143-13-5

HEALTH AND SAFETY LISTS

ACGIH 1995 - Time Weighted Averages
200 ppm TWA; 1050 mg/m3 TWA

AIHA - Odor Threshold Values
no geometric mean air odor threshold

U.S. DOT - Substances From 49 CFR 172.101
regulated by DOT (UN1920)

U.S. DOT - Hazard Classes
DOT hazard class = 3

NFPA - Flash Points
flash point = 88 degrees F (31 degrees C)

NFPA - Hazard Identification Ratings
health-0; flammability-3; reactivity-0

NIOSH - Selected LD50s and LC50s
Inhalation, rat: LC50 = 3200 ppm 4 hr

OSHA - Vacated PELs - Time Weighted Averages
200 ppm TWA; 1050 mg/m3 TWA

ENVIRONMENTAL LISTS

TSCA - PAIR - Reporting List
Effective Date: January 26, 1994; Reporting Date: March 28, 1994

TSCA - Health and Safety Reporting List
Effective Date: January 26, 1994; Sunset Date: January 26, 2004

INTERNATIONAL LISTS

Australian Exposure Standards - Time Weighted Averages
200 ppm TWA; 1050 mg/m3 TWA

Canada - WHMIS: Ingredient Disclosure
1% item 1213 (1274)

Canada - Alberta - 8 Hour Occupational Exposure Limit
200 ppm TWA; 1049 mg/m3 TWA

Canada - Alberta - 15 Minute Occupational Exposure Limit
250 ppm STEL; 1311 mg/m3 STEL

Canada - British Columbia - 8 Hour Exposure Limits
200 ppm TWA; 1050 mg/m3 TWA

Canada - British Columbia - 15 Minute Exposure Limits
250 ppm STEL; 1300 mg/m3 STEL

Canada - Ontario - OHSA - TWAEVs
200 ppm TWAEV; 1050 mg/m3 TWAEV

Canada - Quebec - Time-Weighted Average Exposure Values
200 ppm TWAEV; 1050 mg/m3 TWAEV

Israel - Time Weighted Averages
200 ppm TWA; 1050 mg/m3 TWA

Israel - Action Levels
100 ppm AL; 525 mg/m3 AL

Mexico - Instruction No. 10 - TWAs
200 ppm TWA; 1050 mg/m3 TWA

Mexico - Instruction No. 10 - STELs
250 ppm STEL; 1300 mg/m3 STEL

STATE LISTS

California - Exposure Limits - PELs
200 ppm PEL; 1050 mg/m3 PEL

California - Directors List of Hazardous Substances (8 CCR 339)
[present]

Florida Hazardous Substance List
[present]

Massachusetts Right To Know List
[present]

Minnesota Hazardous Substance List
[present]

NJ Right to Know List (Total)
sn 1414

NJ Special Hazardous Substances
(flammable - third degree)

Pennsylvania Right to Know List
[present]

PROPOSED REGULATIONS

TSCA - ITC 31st Report Priority Testing List
designated to be tested

Canada - Ontario - Proposed Occupational TWAEVs
100 ppm TWAEV; 525 mg/m3 TWAEV

NONYLBENZENE   1081-77-2

HEALTH AND SAFETY LISTS

NIOSH - Health Standards - Exposure Limits
C (15 min) 0.5 ppm; C (15 min) 3.3 mg/m3 (Listed under 'Thiols')

NIOSH - Health Standards - Health Effects and Precautions
Irritation; eye, skin, blood, and nervous system effects (Blood and urine monitoring required; prevent skin contact) (Listed under 'Thiols')

TERT-NONYL MERCAPTAN   25360-10-5

HEALTH AND SAFETY LISTS

NIOSH - Selected LD50s and LC50s
Inhalation, mouse: LC50 = 5500 mg/m3 (2 hr) Skin, rabbit: LD50 = 5660 mg/kg

ENVIRONMENTAL LISTS

List of Pesticide Product Inert Ingredients
[present]

NONYLNAPHTHALENE   27193-93-7

HEALTH AND SAFETY LISTS

NFPA - Flash Points
flash point = 200 degrees F (93 degrees C)

NFPA - Hazard Identification Ratings
health-2; flammability-2; reactivity-0

ENVIRONMENTAL LISTS

List of Pesticide Product Inert Ingredients
[present]

STATE LISTS

Florida Hazardous Substance List
[present]

Massachusetts Right To Know List
[present]

Pennsylvania Right to Know List
[present]

N-(NONYLOXYPROPYL)-1,3-PROPANEDIAMINE   68877-34-9

HEALTH AND SAFETY LISTS

NFPA - Flash Points
flash point = 140 degrees F (60 degrees C)

NFPA - Hazard Identification Ratings
health-0; flammability-2; reactivity-0

NIOSH - Selected LD50s and LC50s
Oral, rat: LD50 = 3200 mg/kg

NONYLPHENOL   25154-52-3

HEALTH AND SAFETY LISTS

NFPA - Flash Points
flash point = 78 degrees F (26 degrees C)

NFPA - Hazard Identification Ratings
health-0; flammability-3; reactivity-0

STATE LISTS

Florida Hazardous Substance List
[present]

Massachusetts Right To Know List
[present]

Pennsylvania Right to Know List
[present]

4-NONYLPHENOL, BRANCHED   84852-15-3

STATE LISTS

NJ Right to Know List (Total)
sn 2590

NONYLPHENOL, ETHOXYLATED AND   51811-79-1
PHOSPHATED

HEALTH AND SAFETY LISTS

NFPA - Flash Points
flash point = 155 degrees F (68 degrees C)

NFPA - Hazard Identification Ratings
health-1; flammability-2; reactivity-0

P-NONYLPHENOL, ETHOXYLATED AND   51609-41-7
PHOSPHATED

HEALTH AND SAFETY LISTS

NFPA - Flash Points
flash point = 210 degrees F (99 degrees C)

NFPA - Hazard Identification Ratings
health-0; flammability-1; reactivity-0

ENVIRONMENTAL LISTS

CAA - HON Rule - SOCMI Chemicals
compliance by Oct. 23, 1995

NONYLPHENOL, ETHOXYLATED AND SULFATED   9081-17-8

HEALTH AND SAFETY LISTS

NFPA - Flash Points
flash point = 154 degrees F (68 degrees C)

NFPA - Hazard Identification Ratings
health-2; flammability-2; reactivity-0

STATE LISTS

Florida Hazardous Substance List
[present]

Massachusetts Right To Know List
[present]

Pennsylvania Right to Know List
[present]

P-NONYLPHENOL, ETHOXYLATED, PHOSPHATED,   RR-01103-0
CALCIUM SALT

HEALTH AND SAFETY LISTS

NFPA - Flash Points
flash point < 200 degrees F (93 degrees C)

NFPA - Hazard Identification Ratings
health-0; flammability-2; reactivity-0

P-NONYLPHENOL, ETHOXYLATED, PHOSPHATED,   106151-63-7
DIPOTASSIUM SALT

ENVIRONMENTAL LISTS

List of Pesticide Product Inert Ingredients
[present]

NONYLPHENOL, ETHOXYLATED, PHOSPHATED,   67922-57-0
MAGNESIUM SALT

HEALTH AND SAFETY LISTS

U.S. DOT - Appendix B - Marine Pollutants
DOT regulated marine pollutant

NFPA - Flash Points
flash point = 285 degrees F (141 degrees C)

NFPA - Hazard Identification Ratings
health-2; flammability-1; reactivity-0

NIOSH - Selected LD50s and LC50s
Oral, rat: LD50 = 1620 mg/kg Skin, rabbit: LD50 = 2140 mg/kg

ENVIRONMENTAL LISTS

CAA - HON Rule - SOCMI Chemicals
compliance by Oct. 23, 1995

INTERNATIONAL LISTS

Canada - WHMIS: Ingredient Disclosure
1% item 1214 (1276)

STATE LISTS

Florida Hazardous Substance List
[present]

Massachusetts Right To Know List
[present]

Pennsylvania Right to Know List
[present]

NONYLPHENOL, ETHOXYLATED, PHOSPHATED,   59139-23-0
MONOETHANOLAMINE SALT

ENVIRONMENTAL LISTS

EPA - Master Testing List
[present]

TSCA - Code of Federal Regulations Citations
40 CFR 799.5000

TSCA - Substances Subject to Testing Consent Orders
Test for: Environmental Effects and Chemical Fate

TSCA - Section 12(b) - Export Notification
export notification required - Section 4

NONYLPHENOL, ETHOXYLATED, PHOSPHATED,   52503-15-8
POTASSIUM SALT

ENVIRONMENTAL LISTS

List of Pesticide Product Inert Ingredients
[present]

NONYLPHENOL, ETHOXYLATED, SULFATED,   57451-03-3
TRIETHANOLAMINE SALT

ENVIRONMENTAL LISTS

List of Pesticide Product Inert Ingredients
[present]

NONYLPHENOL PHOSPHATE   30526-26-2

ENVIRONMENTAL LISTS

List of Pesticide Product Inert Ingredients
[present]

NONYLPHENOXYPOLY(ETHYLENEOXY) ETHYL   37340-60-6
PHOSPHATE, SODIUM SALT

ENVIRONMENTAL LISTS

List of Pesticide Product Inert Ingredients
[present]

NONYLPHENYLPOLYOXYETHYLENE   54612-36-1
SULFOSUCCINATE

ENVIRONMENTAL LISTS

List of Pesticide Product Inert Ingredients
[present]

NONYLTRICHLOROSILANE   5283-67-0

ENVIRONMENTAL LISTS

List of Pesticide Product Inert Ingredients
[present]

NORBORMIDE   991-42-4

ENVIRONMENTAL LISTS

List of Pesticide Product Inert Ingredients
[present]

NORETHISTERONE   68-22-4

ENVIRONMENTAL LISTS

List of Pesticide Product Inert Ingredients
[present]

NORETHISTERONE ACETATE (NORETHINDRONE   51-98-9
ACETATE)

ENVIRONMENTAL LISTS

List of Pesticide Product Inert Ingredients
[present]

NORETHISTERONE AND OESTROGENS   RR-00039-5

ENVIRONMENTAL LISTS

List of Pesticide Product Inert Ingredients
[present]

NORETHYNODREL   68-23-5

ENVIRONMENTAL LISTS

List of Pesticide Product Inert Ingredients
[present]

NORETHYNODREL AND OESTROGENS   RR-00550-5

ENVIRONMENTAL LISTS

List of Pesticide Product Inert Ingredients
[present]

NORFLURAZON [4-CHLORO-5-METHYLAMINO)-2-   27314-13-2
[3-(TRIFLUOROMETHYL)PHENYL]-3(2H)-PYRIDAZI-
NONE]

HEALTH AND SAFETY LISTS

U.S. DOT - Substances From 49 CFR 172.101
regulated by DOT (UN1799)

U.S. DOT - Hazard Classes
DOT hazard class = 8

INTERNATIONAL LISTS

Canada - WHMIS: Ingredient Disclosure
1% item 1215 (1277)

STATE LISTS

Massachusetts Right To Know List
[present]

NJ Right to Know List (Total)
sn 1416

NJ Special Hazardous Substances
(corrosive)

NORGESTREL   6533-00-2

HEALTH AND SAFETY LISTS

NIOSH - Selected LD50s and LC50s
Oral, rat: LD50 = 3800 ug/kg

ENVIRONMENTAL LISTS
  CERCLA/SARA - Section 302 Extremely Hazardous Substances and
  TPQs
    TPQ = 100/10,000 pounds

STATE LISTS
  Florida Hazardous Substance List
    effective March 13, 1992
  Massachusetts Right To Know List
    extraordinarily hazardous
  NJ Right to Know List (Total)
    sn 2591
  Pennsylvania Right to Know List
    environmental hazard

PROPOSED REGULATIONS
  CERCLA/SARA - Proposed Hazardous Substance Additions
    proposed RQ = 1 pound (.454 kg)
  CERCLA/SARA - 1989 Proposed RQ Adjustments
    proposed RQ = 100 pounds (45.4 kg)

NORGESTREL AND OESTROGENS   RR-00034-0

HEALTH AND SAFETY LISTS
  IARC - Group Unspecified
    [present] (Listed under 'Progestins')
  NIOSH - Selected LD50s and LC50s
    Oral, mouse: LD50 = 12 gm/kg
  NTP Seventh Report - Suspect Carcinogens
    suspect carcinogen
  OSHA - Possible Select Carcinogens
    [present]

STATE LISTS
  California - Air Bill 2588 Appendix A-II
    known or potential carcinogen
  California - Prop. 65 - Cancer list
    carcinogen - initial date 10/1/89
  California - Prop. 65 - Developmental Toxicity
    developmental toxicity - initial date 4/1/90
  California - Directors List of Hazardous Substances (8 CCR 339)
    [present]
  Florida Hazardous Substance List
    [present]
  Massachusetts Right To Know List
    carcinogen; extraordinarily hazardous; teratogen
  Minnesota Hazardous Substance List
    carcinogen
  NJ Right to Know List (Total)
    sn 1417
  NJ Special Hazardous Substances
    (carcinogen; mutagen)
  Pennsylvania Right to Know List
    special hazardous substance
  Pennsylvania RTK - Special Hazardous Substances
    [present]

NORPOR 10   116712-07-3

STATE LISTS
  California - Prop. 65 - Developmental Toxicity
    developmental toxicity - initial date 10/1/91
  California - Directors List of Hazardous Substances (8 CCR 339)
    [present]

NORPSEUDOEPHEDRINE   36393-56-3

HEALTH AND SAFETY LISTS
  IARC - Group Unspecified
    [present] (Listed under 'Combined oral contraceptives')

NORPSEUDOEPHEDRINE SALTS, OPTICAL ISO-MERS, AND SALTS OF OPTICAL ISOMERS   RR-01777-6

HEALTH AND SAFETY LISTS
  IARC - Group Unspecified
    [present] (Listed under 'Progestins')

NOVOBIOCIN   303-81-1

HEALTH AND SAFETY LISTS
  IARC - Group Unspecified
    [present] (Listed under 'Combined oral contraceptives')

NUISANCE PARTICULATES   RR-00072-6

ENVIRONMENTAL LISTS
  CERCLA/SARA - Section 313 - Emission Reporting
    form R reporting required

NYLON   9008-75-7

HEALTH AND SAFETY LISTS
  IARC - Group Unspecified
    [present] (Listed under 'Progestins')

STATE LISTS
  California - Air Bill 2588 Appendix A-II
    9/90
  California - Prop. 65 - Developmental Toxicity
    developmental toxicity - initial date 4/1/90

NYLON 6   25038-54-4

HEALTH AND SAFETY LISTS
  IARC - Group Unspecified
    [present] (Listed under 'Combined oral contraceptives')

O-NITROANILINE   88-74-4

ENVIRONMENTAL LISTS
  List of Pesticide Product Inert Ingredients
    [present]

OATS AND OATMEAL   RR-01104-1

HEALTH AND SAFETY LISTS
  FDA - Controlled Substances Act - Precursor chemicals
    Threshold by base weight = 2.5 kilograms

OCHRATOXIN A   303-47-9

HEALTH AND SAFETY LISTS
  FDA - Controlled Substances Act - Precursor chemicals
    Threshold by base weight = 2.5 kilograms

OCOTEA CYMBARUM OIL   68917-09-9

HEALTH AND SAFETY LISTS
  NIOSH - Selected LD50s and LC50s
    Oral, mouse: LD50 = 1500 mg/kg

OCTABROMOBIPHENYL   27858-07-7

HEALTH AND SAFETY LISTS
  ACGIH 1995 - Time Weighted Averages
    *10 mg/m3 TWA (The value is for total dust containing no asbestos
    and <1% crystalline silica)*
  OSHA - Vacated PELs - Time Weighted Averages
    total dust: 15 mg/m3 TWA; respirable fraction: 5 mg/m3 TWA

OSHA - Final PELs - Time Weighted Averages
total dust: 15 mg/m3 TWA; respirable fraction: 5 mg/m3 TWA

INTERNATIONAL LISTS

Canada - Alberta - 8 Hour Occupational Exposure Limit
respirable mass: 5 mg/m3 TWA; total mass: 10 mg/m3 TWA

Canada - Quebec - Time-Weighted Average Exposure Values
total dust: 10 mg/m3 TWAEV; respirable dust: 5 mg/m3 TWAEV

Israel - Time Weighted Averages
10 mg/m3 TWA (The value is for total dust containing no asbestos and <1% crystalline silica)

Israel - Action Levels
5 mg/m3 AL

STATE LISTS

California - Exposure Limits - PELs
total dust: 10 mg/m3 PEL; respirable fraction: 5 mg/m3 PEL

PROPOSED REGULATIONS

ACGIH 1995 - Notice of Intended Changes
inhalable particulate: 10 mg/m3 TWA; respirable particulate: 3 mg/m3 TWA These values are for total dust containing no asbestos and <1% crystalline silica.

OCTABROMODIPHENYLOXIDE   32536-52-0

ENVIRONMENTAL LISTS

List of Pesticide Product Inert Ingredients
[present]

OCTACHLORODIBENZO-P-DIOXIN   3268-87-9

HEALTH AND SAFETY LISTS

IARC - Group 3 (not classifiable)
[present]

NIOSH - Selected LD50s and LC50s
Inhalation, mouse: LC50 = 11 gm/m3 (30 mn) Oral, rat: LD50 = 3200 mg/kg

ENVIRONMENTAL LISTS

List of Pesticide Product Inert Ingredients
[present]

OCTACHLORODIBENZOFURAN   39001-02-0

SEE ALSO:
ANILINE AND CHLORO-, BROMO-, AND/OR NITROANILINES

HEALTH AND SAFETY LISTS

NIOSH - Selected LD50s and LC50s
Oral, rat: LD50 = 1600 mg/kg Skin, rabbit: LD50 = 20 gm/kg

ENVIRONMENTAL LISTS

CAA - HON Rule - SOCMI Chemicals
compliance by Oct. 24, 1994

RCRA - TSD Facilities Ground Water Monitoring
TM 8270 = 50 ug/L PQL

RCRA - Universal Treatment Standards (LDR)
WW: 0.27 mg/l; NWW: 14 mg/kg

TSCA - Code of Federal Regulations Citations
40 CFR 712.30(d); 40 CFR 716.120(c); 40 CFR 799.5000

TSCA - PAIR - Reporting List
Reporting Date: November 19, 1982

TSCA - Substances Subject to Testing Consent Orders
Test for: Health Effects

TSCA - Section 12(b) - Export Notification
export notification required - Section 4

INTERNATIONAL LISTS

Canada - WHMIS: Ingredient Disclosure
1% item 1151 (1221)

OCTACHLORONAPHTHALENE   2234-13-1

ENVIRONMENTAL LISTS

List of Pesticide Product Inert Ingredients
[present]

OCTACOSAMETHYLTRIDECASILOXANE   2471-09-2

HEALTH AND SAFETY LISTS

IARC - Group 2B (sufficient animal data)
[present]

IARC - Group 3 (not classifiable)
[present]

NTP Chemical Status Reports - Testing Status and NTIS Number
Technical reports printed (PB90219478/AS)

NTP Chemical Status Reports - Evidence of Carcinogenicity
male rat-clear evidence; female rat-clear evidence

NTP Seventh Report - Suspect Carcinogens
suspect carcinogen

OSHA - Possible Select Carcinogens
[present]

STATE LISTS

California - Air Bill 2588 Appendix A-II
known or potential carcinogen: 9/90

California - Prop. 65 - Cancer list
carcinogen - initial date 7/1/90

California - Directors List of Hazardous Substances (8 CCR 339)
[present]

OCTADECAMETHYLCYCLONONASILOXANE   556-71-8

ENVIRONMENTAL LISTS

List of Pesticide Product Inert Ingredients
[present]

OCTADECAMETHYLOCTASILOXANE   556-69-4

ENVIRONMENTAL LISTS

TSCA - Code of Federal Regulations Citations
40 CFR 721.600

TSCA - Chemicals with Significant New Use Rules
[present]

TSCA - Section 12(b) - Export Notification
export notification required - Section 5

STATE LISTS

Massachusetts Right To Know List
carcinogen; extraordinarily hazardous

OCTADECANE   593-45-3

ENVIRONMENTAL LISTS

EPA - Master Testing List
[present]

TSCA - Code of Federal Regulations Citations
40 CFR 712.30(w); 40 CFR 716.120(a); 40 CFR 766.35

TSCA - PAIR - Reporting List
Reporting Date: March 12, 1990

TSCA - Health and Safety Reporting List
Effective Date: January 11, 1990

TSCA - HDD/HDF - Chemicals Required for Testing
[present]

TSCA - Section 12(b) - Export Notification
export notification required - Section 4

OCTADECANE, 1-ISOCYANATO-   112-96-9

ENVIRONMENTAL LISTS

ATSDR Priority List
Rank (of 275): 104

1-OCTADECANETHIOL   **2885-00-9**

ENVIRONMENTAL LISTS

ATSDR Priority List
Rank (of 275): 102

9-OCTADECANOIC ACID (Z)-, MONOESTER WITH   **1330-80-9**
1,2-PROPANEDIOL

HEALTH AND SAFETY LISTS

ACGIH 1995 - Time Weighted Averages
0.1 mg/m3 TWA

ACGIH 1995 - Short Term Exposure Limits
0.3 mg/m3 STEL

ACGIH 1995 - Skin Designations
skin - potential for cutaneous absorption

NIOSH 1990 - Pocket Guide - RELs
0.1 mg/m3 TWA; 0.3 mg/m3 STEL

NIOSH 1990 - Pocket Guide - Target organs
skin, liver

NIOSH 1990 - Pocket Guide - Skin list
Potential for dermal absorption

OSHA - Vacated PELs - Time Weighted Averages
0.1 mg/m3 TWA

OSHA - Vacated PELs - Short Term Exposure Limits
0.3 mg/m3 STEL

OSHA - Vacated PELs - Skin Designation
Prevent or reduce skin absorption

OSHA - Final PELs - Time Weighted Averages
0.1 mg/m3 TWA

OSHA - Final PELs - Skin Notations
prevent or reduce skin absorption

ENVIRONMENTAL LISTS

CERCLA/SARA - Section 313 - Emission Reporting
form R reporting required for 1.0% de minimus concentration

TSCA - Code of Federal Regulations Citations
40 CFR 704.83; 40 CFR 712.30(d); 40 CFR 716.120(a)

TSCA - PAIR - Reporting List
Reporting Date: November 19, 1982

TSCA - Health and Safety Reporting List
Effective Date: October 4, 1982

INTERNATIONAL LISTS

Australian Exposure Standards - Time Weighted Averages
0.1 mg/m3 TWA

Australian Exposure Standards - Short Term Exposure Limits
0.3 mg/m3 STEL

Australian Exposure Standards - Skin Effects
skin absorption

Canada - WHMIS: Ingredient Disclosure
1% item 1216 (1278)

Canada - Alberta - 8 Hour Occupational Exposure Limit
0.1 mg/m3 TWA

Canada - Alberta - 15 Minute Occupational Exposure Limit
0.3 mg/m3 STEL

Canada - Alberta - Skin Designation
can be absorbed through the intact skin

Canada - British Columbia - 8 Hour Exposure Limits
0.1 mg/m3 TWA

Canada - British Columbia - 15 Minute Exposure Limits
0.3 mg/m3 STEL

Canada - British Columbia - Skin Notations
skin - potential for skin absorption

Canada - Ontario - OHSA - TWAEVs
0.1 mg/m3 TWAEV

Canada - Ontario - OHSA - STEVs
0.3 mg/m3 STEV

Canada - Ontario - OHSA - Skin Notations
absorption through skin, eyes, or mucous membranes

Canada - Quebec - Time-Weighted Average Exposure Values
0.1 mg/m3 TWAEV

Canada - Quebec - Short-term Exposure Values
0.3 ppm STEV

Canada - Quebec - Skin Designations
absorbed through the skin

United Kingdom - Occupational Exposure Standards - TWAs
0.1 mg/m3 TWA

United Kingdom - Occupational Exposure Standards - STELs
0.3 mg/m3 STEL

United Kingdom - Occupational Exposure Standards - Notes
can be absorbed through skin

Israel - Time Weighted Averages
0.1 mg/m3 TWA

Israel - Short Term Exposure Limits
0.3 mg/m3 STEL

Israel - Action Levels
0.05 mg/m3 AL

Mexico - Instruction No. 10 - TWAs
0.1 mg/m3 TWA

Mexico - Instruction No. 10 - STELs
0.3 mg/m3 STEL

Mexico - Instruction No. 10 - Skin designation
skin - potential for cutaneous absorption

STATE LISTS

California - Air Bill 2588 Appendix A-II
6/91

California - Exposure Limits - PELs
0.1 mg/m3 PEL

California - Exposure Limits - STELs
0.3 mg/m3 STEL

California - Exposure Limits - Skin Notation
material may be absorbed through the skin, eyes or mucous membrane

California - Directors List of Hazardous Substances (8 CCR 339)
[present]

Florida Hazardous Substance List
[present]

Massachusetts Right To Know List
[present]

Minnesota Hazardous Substance List
skin

NJ Right to Know List (Total)
sn 1427

Pennsylvania Right to Know List
environmental hazard

9-OCTADECENOIC ACID (Z)-, SULFONATED,   **68443-05-0**
SODIUM SALT

ENVIRONMENTAL LISTS

TSCA - PAIR - Reporting List
Effective Date: October 12, 1993; Reporting Date: February 28, 1994

TSCA - Health and Safety Reporting List
Effective Date: October 12, 1993; Sunset Date: October 12, 2003

1-OCTADECANOL   **112-92-5**

ENVIRONMENTAL LISTS

TSCA - PAIR - Reporting List
Effective Date: October 12, 1993; Reporting Date: February 28, 1994

TSCA - Health and Safety Reporting List
Effective Date: October 12, 1993; Sunset Date: October 12, 2003

9-OCTADECENOIC ACID (Z)-, SULFONATED   68988-76-1

ENVIRONMENTAL LISTS

TSCA - PAIR - Reporting List
Effective Date: October 12, 1993; Reporting Date: February 28, 1994

TSCA - Health and Safety Reporting List
Effective Date: October 12, 1993; Sunset Date: October 12, 2003

CIS-9-OCTADECEN-1-OL   143-28-2

HEALTH AND SAFETY LISTS

NFPA - Flash Points
flash point > 212 degrees F (100 degrees C)

NFPA - Hazard Identification Ratings
health-0; flammability-1; reactivity-0

N-CIS-9-OCTADECENYL-1,3-PROPANEDIAMINE   7173-62-8

ENVIRONMENTAL LISTS

TSCA - Code of Federal Regulations Citations
40 CFR 712.30(x); 40 CFR 716.120(d)

TSCA - PAIR - Reporting List
Reporting Date: December 27, 1990

TSCA - Health and Safety Reporting List
Effective Date: October 29, 1990; Sunset Date: November 9, 1993

CIS-9-OCTADECENYL SULFATE, SODIUM SALT   1847-55-8

HEALTH AND SAFETY LISTS

NIOSH - Health Standards - Exposure Limits
C (15 min) 0.5 ppm; C (15 min) 5.9 mg/m3 (Listed under 'Thiols')

NIOSH - Health Standards - Health Effects and Precautions
Irritation; eye, skin, blood, and nervous system effects (Blood and urine monitoring required; prevent skin contact) (Listed under 'Thiols')

N-OCTADECYLAMINE   124-30-1

ENVIRONMENTAL LISTS

List of Pesticide Product Inert Ingredients
[present]

OCTADECYLAMINE ACETATE   2190-04-7

ENVIRONMENTAL LISTS

List of Pesticide Product Inert Ingredients
[present]

OCTADECYL 3-(3',5'-DI-TERT-BUTYL-4'-HYDROX-YPHENYL)PROPIONATE   2082-79-3

HEALTH AND SAFETY LISTS

NFPA - Hazard Identification Ratings
health-0; reactivity-0

NIOSH - Selected LD50s and LC50s
Oral, rat: LD50 = 20 gm/kg

ENVIRONMENTAL LISTS

EPA - Master Testing List
[present]

List of Pesticide Product Inert Ingredients
[present]

OCTADECYLENE, ALPHA-   112-88-9

ENVIRONMENTAL LISTS

List of Pesticide Product Inert Ingredients
[present]

ALPHA-(OCTADECYLPHENYL)-OMEGA-HYDROXY-POLY(OXY-1,2-ETHANEDIYL)   51617-79-9

ENVIRONMENTAL LISTS

List of Pesticide Product Inert Ingredients
[present]

OCTADECYL SULFATE, SODIUM SALT   1120-04-3

ENVIRONMENTAL LISTS

List of Pesticide Product Inert Ingredients
[present]

OCTADECYLTRICHLOROSILANE   112-04-9

ENVIRONMENTAL LISTS

List of Pesticide Product Inert Ingredients
[present]

2,6-OCTADIENAL, 3,7-DIMETHYL-, (Z)-   106-26-3

INTERNATIONAL LISTS

Canada - WHMIS: Ingredient Disclosure
1% item 1217 (1279)

2,6-OCTADIENAL, 3,7-DIMETHYL-,(E)-   141-27-5

HEALTH AND SAFETY LISTS

NIOSH - Selected LD50s and LC50s
Oral, mouse: LD50 = 1000 mg/kg

OCTADIENE   63597-41-1

ENVIRONMENTAL LISTS

List of Pesticide Product Inert Ingredients
[present]

1,7-OCTADINE-3,5-DIYNE-1,8-DIMETHOXY-9-OC-TADECYNOIC ACID   RR-01369-4

HEALTH AND SAFETY LISTS

NFPA - Flash Points
flash point > 212 degrees F (100 degrees C)

NFPA - Hazard Identification Ratings
health-0; flammability-1; reactivity-0

OCTAFLUOROBUT-2-ENE   360-89-4

ENVIRONMENTAL LISTS

List of Pesticide Product Inert Ingredients
[present]

OCTAFLUOROCYCLOBUTANE   115-25-3

ENVIRONMENTAL LISTS

List of Pesticide Product Inert Ingredients
[present]

OCTAFLUOROPROPANE   76-19-7

HEALTH AND SAFETY LISTS

U.S. DOT - Substances From 49 CFR 172.101
regulated by DOT (UN1800)

U.S. DOT - Hazard Classes
DOT hazard class = 8

NFPA - Flash Points
flash point = 193 degrees F (89 degrees C)

NFPA - Hazard Identification Ratings
health-3; flammability-2; reactivity-2

INTERNATIONAL LISTS

Canada - WHMIS: Ingredient Disclosure
1% item 1218 (1280)

STATE LISTS

Florida Hazardous Substance List
[present]

Massachusetts Right To Know List
[present]

NJ Right to Know List (Total)
sn 1429

NJ Special Hazardous Substances
(corrosive)

Pennsylvania Right to Know List
[present]

OCTAMETHYLCYCLOTETRASILOXANE 556-57-0

ENVIRONMENTAL LISTS

TSCA - Code of Federal Regulations Citations
40 CFR 712.30(x); 40 CFR 716.120(d)

TSCA - PAIR - Reporting List
Reporting Date: November 27, 1991

TSCA - Health and Safety Reporting List
Effective Date: September 30, 1991

OCTAMETHYLCYCLOTETRASILOXANE 556-67-2

ENVIRONMENTAL LISTS

TSCA - Code of Federal Regulations Citations
40 CFR 712.30(x); 40 CFR 716.120(d)

TSCA - PAIR - Reporting List
Reporting Date: November 27, 1991

TSCA - Health and Safety Reporting List
Effective Date: September 30, 1991

OCTAMETHYLPYROPHOSPHORAMIDE 152-16-9

HEALTH AND SAFETY LISTS

U.S. DOT - Substances From 49 CFR 172.101
regulated by DOT (UN2309)

U.S. DOT - Hazard Classes
DOT hazard class = 3

STATE LISTS

NJ Right to Know List (Total)
sn 1430

NJ Special Hazardous Substances
(corrosive)

OCTAMETHYLTRISILOXANE 107-51-7

HEALTH AND SAFETY LISTS

U.S. DOT - Hazard Classes
Forbidden from transport by the DOT

OCTANAL, 3,7-DIMETHYL- 5988-91-0

HEALTH AND SAFETY LISTS

U.S. DOT - Substances From 49 CFR 172.101
regulated by DOT (UN2422)

U.S. DOT - Hazard Classes
DOT hazard class = 2.2

STATE LISTS

NJ Right to Know List (Total)
sn 1431

OCTANAL, 7-HYDROXY-3,7-DIMETHYL- 107-75-5

HEALTH AND SAFETY LISTS

U.S. DOT - Substances From 49 CFR 172.101
regulated by DOT (UN1976)

U.S. DOT - Hazard Classes
DOT hazard class = 2.2

STATE LISTS

NJ Right to Know List (Total)
sn 1432

OCTANAL, 7-METHOXY-3,7-DIMETHYL- 3613-30-7

HEALTH AND SAFETY LISTS

U.S. DOT - Substances From 49 CFR 172.101
regulated by DOT (UN2424)

U.S. DOT - Hazard Classes
DOT hazard class = 2.2

STATE LISTS

NJ Right to Know List (Total)
sn 1433

OCTANAL, 2-(PHENYLMETHYLENE)- 101-86-0

ENVIRONMENTAL LISTS

TSCA - Substances Subject to Testing Consent Orders
Test for: Chemical Fate and Environmental Effects

1-OCTANAMINE 111-86-4

ENVIRONMENTAL LISTS

EPA - Master Testing List
[present]

TSCA - Code of Federal Regulations Citations
40 CFR 712.30(m); 40 CFR 716.120(a); 40 CFR 799.5000

TSCA - PAIR - Reporting List
Effective Date: October 12, 1993; Reporting Date: February 28, 1994

TSCA - Health and Safety Reporting List
Effective Date: December 28, 1984

TSCA - Section 12(b) - Export Notification
export notification required - Section 4

1-OCTANAMINE, N-OCTYL- 1120-48-5

HEALTH AND SAFETY LISTS

U.S. DOT - Appendix A Table 1 - Hazardous Substances
final RQ = 100 pounds (45.4 kg)

NIOSH - Selected LD50s and LC50s
Oral, rat: LD50 = 5 mg/kg Skin, rat: LD50 = 15 mg/kg

ENVIRONMENTAL LISTS

CERCLA/SARA - Section 302 Extremely Hazardous Substances and TPQs
TPQ = 100 pounds

CERCLA/SARA - Hazardous Substances and their Reportable Quantities
final RQ = 100 pounds (45.4 kg)

RCRA - P Series Wastes
waste number P085

RCRA - Hazardous Constituents-Appendix VIII
waste number P085

RCRA - Substances Banned From Land Disposal
[present]

STATE LISTS

California - Directors List of Hazardous Substances (8 CCR 339)
[present]

Florida Hazardous Substance List
effective March 13, 1992

Massachusetts Right To Know List
extraordinarily hazardous

NJ Right to Know List (Total)
sn 2357

Pennsylvania Right to Know List
environmental hazard

OCTANE   111-65-9

ENVIRONMENTAL LISTS

TSCA - PAIR - Reporting List
Effective Date: October 12, 1993; Reporting Date: February 28, 1994

TSCA - Health and Safety Reporting List
Effective Date: October 12, 1993; Sunset Date: October 12, 2003

1-OCTANETHIOL   111-88-6

ENVIRONMENTAL LISTS

TSCA - Code of Federal Regulations Citations
40 CFR 712.30(x); 40 CFR 716.120(d)

TSCA - PAIR - Reporting List
Reporting Date: November 27, 1991

TSCA - Health and Safety Reporting List
Effective Date: September 30, 1991

OCTANOIC ACID   124-07-2

HEALTH AND SAFETY LISTS

NFPA - Flash Points
flash point > 212 degrees F (100 degrees C)

NFPA - Hazard Identification Ratings
flammability-1; reactivity-0

ENVIRONMENTAL LISTS

TSCA - Code of Federal Regulations Citations
40 CFR 712.30(x); 40 CFR 716.120(d)

TSCA - PAIR - Reporting List
Reporting Date: November 27, 1991

TSCA - Health and Safety Reporting List
Effective Date: September 30, 1991

OCTANOIC ACID, CADMIUM SALT (2:1)   2191-10-8

ENVIRONMENTAL LISTS

TSCA - Code of Federal Regulations Citations
40 CFR 712.30(x); 40 CFR 716.120(d)

TSCA - PAIR - Reporting List
Reporting Date: November 27, 1991

TSCA - Health and Safety Reporting List
Effective Date: September 30, 1991

OCTANOIC ACID, HYDRAZIDE   6304-39-8

HEALTH AND SAFETY LISTS

NFPA - Flash Points
flash point > 212 degrees F (100 degrees C)

NFPA - Hazard Identification Ratings
flammability-1; reactivity-0

ENVIRONMENTAL LISTS

TSCA - Code of Federal Regulations Citations
40 CFR 712.30(x); 40 CFR 716.120(d)

TSCA - PAIR - Reporting List
Reporting Date: November 27, 1991

TSCA - Health and Safety Reporting List
Effective Date: September 30, 1991

OCTANOIC ACID, ZIRCONIUM SALT   18312-04-4

HEALTH AND SAFETY LISTS

NFPA - Flash Points
flash point = 140 degrees F (60 degrees C)

NFPA - Hazard Identification Ratings
health-2; flammability-2; reactivity-0

STATE LISTS

Florida Hazardous Substance List
[present]

Massachusetts Right To Know List
[present]

Pennsylvania Right to Know List
[present]

1-OCTANOL   111-87-5

STATE LISTS

Pennsylvania Right to Know List
[present]

3-OCTANOL   589-98-0

HEALTH AND SAFETY LISTS

ACGIH 1995 - Time Weighted Averages
300 ppm TWA; 1400 mg/m3 TWA

ACGIH 1995 - Short Term Exposure Limits
375 ppm STEL; 1750 mg/m3 STEL

AIHA - Odor Threshold Values
geometric mean air odor threshold = 150 ppm (detectable); 240 ppm (recognizable)

U.S. DOT - Substances From 49 CFR 172.101
regulated by DOT (UN1262)

U.S. DOT - Hazard Classes
DOT hazard class = 3

NFPA - Flash Points
flash point = 56 degrees F (13 degrees C)

NFPA - Hazard Identification Ratings
health-0; flammability-3; reactivity-0

NIOSH 1990 - Pocket Guide - RELs
75 ppm TWA; 350 mg/m3 TWA; C 385 ppm (15 min); C 1800 mg/m3 (15 min)

NIOSH 1990 - Pocket Guide - IDLHs
5000 ppm IDLH

NIOSH 1990 - Pocket Guide - Target organs
skin, eyes, respiratory system

NIOSH - Health Standards - Exposure Limits
75 ppm TWA; 350 mg/m3 TWA; C (15 min) 385 ppm; C (15 min) 1800 mg/m3 (Listed under 'Alkanes (C5-C8)')

NIOSH - Health Standards - Health Effects and Precautions
Skin and nervous system effects (Listed under 'Alkanes (C5-C8)')

OSHA - Vacated PELs - Time Weighted Averages
300 ppm TWA; 1450 mg/m3 TWA

OSHA - Vacated PELs - Short Term Exposure Limits
375 ppm STEL; 1800 mg/m3 STEL

OSHA - Final PELs - Time Weighted Averages
500 ppm TWA; 2350 mg/m3 TWA

INTERNATIONAL LISTS

Australian Exposure Standards - Time Weighted Averages
300 ppm TWA; 1400 mg/m3 TWA

Australian Exposure Standards - Short Term Exposure Limits
375 ppm STEL; 1750 mg/m3 STEL

Canada - WHMIS: Ingredient Disclosure
1% item 1219 (1281)

Canada - Alberta - 8 Hour Occupational Exposure Limit
300 ppm TWA; 1402 mg/m3 TWA

Canada - Alberta - 15 Minute Occupational Exposure Limit
10 mg/m3 STEL

Canada - British Columbia - 8 Hour Exposure Limits
300 ppm TWA; 1450 mg/m3 TWA

Canada - British Columbia - 15 Minute Exposure Limits
375 ppm STEL; 1800 mg/m3 STEL

Canada - Ontario - OHSA - TWAEVs
300 ppm TWAEV; 1400 mg/m3 TWAEV

Canada - Ontario - OHSA - STEVs
375 ppm STEV; 1750 mg/m3 STEV

Canada - Quebec - Time-Weighted Average Exposure Values
300 ppm TWAEV; 1400 mg/m3 TWAEV

Canada - Quebec - Short-term Exposure Values
375 ppm STEV; 1750 mg/m3 STEV

United Kingdom - Occupational Exposure Standards - TWAs
300 ppm TWA; 1450 mg/m3 TWA

United Kingdom - Occupational Exposure Standards - STELs
375 ppm STEL; 1800 mg/m3 STEL

German (DFG) - MAK Values
500 ppm MAK; 2350 mg/m3 MAK

German (DFG) - Peak Limitations
2 x normal MAK (30 min. average value); don't exceed 4 times during shift

Israel - Time Weighted Averages
300 ppm TWA; 1400 mg/m3 TWA

Israel - Short Term Exposure Limits
375 ppm STEL; 1750 mg/m3 STEL

Israel - Action Levels
150 ppm AL; 700 mg/m3 AL

Mexico - Instruction No. 10 - TWAs
300 ppm TWA; 1450 mg/m3 TWA

Mexico - Instruction No. 10 - STELs
375 ppm STEL; 1800 mg/m3 STEL

STATE LISTS

California - Exposure Limits - PELs
300 ppm PEL; 1450 mg/m3 PEL

California - Exposure Limits - STELs
375 ppm STEL; 1800 mg/m3 STEL

California - Directors List of Hazardous Substances (8 CCR 339)
[present]

Florida Hazardous Substance List
[present]

Massachusetts Right To Know List
[present]

Minnesota Hazardous Substance List
[present]

NJ Right to Know List (Total)
sn 1434

NJ Special Hazardous Substances
(flammable - third degree)

Pennsylvania Right to Know List
[present]

PROPOSED REGULATIONS

Canada - Ontario - Proposed Occupational TWAEVs
150 ppm TWAEV; 725 mg/m3 TWAEV

2-OCTANOL   5978-70-1

HEALTH AND SAFETY LISTS

NFPA - Flash Points
flash point = 156 degrees F (69 degrees C)

NFPA - Hazard Identification Ratings
health-2; flammability-2; reactivity-0

NIOSH - Health Standards - Exposure Limits
C (15 min) 0.5 ppm; C (15 min) 3.0 mg/m3 (Listed under 'Thiols')

NIOSH - Health Standards - Health Effects and Precautions
Irritation; eye, skin, blood, and nervous system effects (Blood and urine monitoring required; prevent skin contact) (Listed under 'Thiols')

STATE LISTS

Florida Hazardous Substance List
[present]

Massachusetts Right To Know List
[present]

Pennsylvania Right to Know List
[present]

2-OCTANONE   111-13-7

HEALTH AND SAFETY LISTS

NIOSH - Selected LD50s and LC50s
Oral, rat: LD50 = 10080 mg/kg

ENVIRONMENTAL LISTS

List of Pesticide Product Inert Ingredients
[present]

INTERNATIONAL LISTS

Canada - WHMIS: Ingredient Disclosure
1% item 1220 (118)

OCTAPHENYLCYCLOTETRASILOXANE   546-56-5
SEE ALSO:
CADMIUM

HEALTH AND SAFETY LISTS

NIOSH - Selected LD50s and LC50s
Oral, rat: LD50 = 950 mg/kg

OCTATRIACONTAMETHYLOCTADECASILOXANE   36938-52-0

ENVIRONMENTAL LISTS

TSCA - Chemicals with Significant New Use Rules
PMN number: P-92-1086

TSCA - Section 12(b) - Export Notification
P-92-1086; export notification required - Section 5

6-OCTENAL, 3,7-DIMETHYL-   106-23-0

ENVIRONMENTAL LISTS

List of Pesticide Product Inert Ingredients
[present]

6-OCTENAL, 3,7-DIMETHYL-, (S)-   5949-05-3

HEALTH AND SAFETY LISTS

AIHA - WEEL - Time Weighted Averages
50 ppm TWA; 265 mg/m3 TWA

NFPA - Flash Points
flash point = 178 degrees F (81 degrees C)

NFPA - Hazard Identification Ratings
health-1; flammability-2; reactivity-0

NIOSH - Selected LD50s and LC50s
Oral, mouse: LD50 = 1790 mg/kg

ENVIRONMENTAL LISTS

List of Pesticide Product Inert Ingredients
[present]

STATE LISTS

Minnesota Hazardous Substance List
[present]

Pennsylvania Right to Know List
[present]

1-OCTENE   111-66-0

INTERNATIONAL LISTS

Canada - WHMIS: Ingredient Disclosure
1% item 1221 (1282)

2-OCTENE   111-67-1

HEALTH AND SAFETY LISTS

NFPA - Flash Points
flash point = 190 degrees F (88 degrees C)

NFPA - Hazard Identification Ratings
health-1; flammability-2; reactivity-0

OCTENE (MIXED ISOMERS)   25377-83-7

HEALTH AND SAFETY LISTS

NFPA - Flash Points
flash point = 125 degrees F (52 degrees C)

NFPA - Hazard Identification Ratings
health-0; flammability-2; reactivity-0

NIOSH - Selected LD50s and LC50s
Oral, mouse: LD50 = 3824 mg/kg

6-OCTEN-1-YN-3-OL, 3,7-DIMETHYL-   29171-20-8

ENVIRONMENTAL LISTS

TSCA - PAIR - Reporting List
Effective Date: October 12, 1993; Reporting Date: February 28, 1994

TSCA - Health and Safety Reporting List
Effective Date: October 12, 1993; Sunset Date: October 12, 2003

OCTOLITE   RR-01330-9

ENVIRONMENTAL LISTS

TSCA - PAIR - Reporting List
Effective Date: October 12, 1993; Reporting Date: February 28, 1994

TSCA - Health and Safety Reporting List
Effective Date: October 12, 1993; Sunset Date: October 12, 2003

OCTYL ALCOHOL BOTTOMS   68526-82-9

ENVIRONMENTAL LISTS

TSCA - Code of Federal Regulations Citations
40 CFR 712.30(x); 40 CFR 716.120(d)

TSCA - PAIR - Reporting List
Reporting Date: November 27, 1991

TSCA - Health and Safety Reporting List
Effective Date: September 30, 1991

OCTYL ALDEHYDES   RR-01335-4

ENVIRONMENTAL LISTS

TSCA - Code of Federal Regulations Citations
40 CFR 712.30(x); 40 CFR 716.120(d)

TSCA - PAIR - Reporting List
Reporting Date: November 27, 1991

TSCA - Health and Safety Reporting List
Effective Date: September 30, 1991

TERT-OCTYLAMINE   107-45-9

HEALTH AND SAFETY LISTS

NFPA - Flash Points
flash point = 70 degrees F (21 degrees C)

NFPA - Hazard Identification Ratings
health-1; flammability-3; reactivity-0

ENVIRONMENTAL LISTS

CAA - HON Rule - SOCMI Chemicals
compliance by Oct. 24, 1994

EPA - Master Testing List
[present]

STATE LISTS

Florida Hazardous Substance List
[present]

Massachusetts Right To Know List
[present]

Pennsylvania Right to Know List
[present]

N-OCTYL BICYCLOHEPTENEDICARBOXIMIDE   113-48-4

STATE LISTS

Massachusetts Right To Know List
[present]

Pennsylvania Right to Know List
[present]

GAMMA-N-OCTYL-GAMMA-N-BUTYROLACTONE   2305-05-7

HEALTH AND SAFETY LISTS

NFPA - Flash Points
flash point = 70 degrees F (21 degrees C)

NFPA - Hazard Identification Ratings
health-1; flammability-3; reactivity-0

STATE LISTS

Florida Hazardous Substance List
[present]

Pennsylvania Right to Know List
[present]

OCTYL CHLORIDE   111-85-3

ENVIRONMENTAL LISTS

EPA - Master Testing List
[present]

OCTYL DIPHENYLAMINE   4175-37-5

HEALTH AND SAFETY LISTS

U.S. DOT - Substances From 49 CFR 172.101
regulated by DOT (UN0266)

U.S. DOT - Hazard Classes
DOT hazard class = 1.1D

2-OCTYLDODECANOL   533-42-6

ENVIRONMENTAL LISTS

List of Pesticide Product Inert Ingredients
[present]

OCTYL EPOXYTALLATES   61788-72-5

HEALTH AND SAFETY LISTS

U.S. DOT - Substances From 49 CFR 172.101
regulated by DOT (UN1191)

U.S. DOT - Hazard Classes
DOT hazard class = 3

TERT-OCTYL HYDROPEROXIDE   RR-01522-5

HEALTH AND SAFETY LISTS

NFPA - Flash Points
flash point = 91 degrees F (33 degrees C)

NFPA - Hazard Identification Ratings
flammability-3; reactivity-0

STATE LISTS

Florida Hazardous Substance List
[present]

Massachusetts Right To Know List
[present]

Pennsylvania Right to Know List
[present]

OCTYL 12-HYDROXYSTEARATE   29710-25-6

HEALTH AND SAFETY LISTS

NIOSH - Selected LD50s and LC50s
Oral, rat: LD50 = 2800 mg/kg Skin, rat: LD50 = 470 mg/kg

TERT-OCTYL MERCAPTAN   141-59-3

INTERNATIONAL LISTS

Canada - WHMIS: Ingredient Disclosure
1% item 1222 (1283)

OCTYLOXYPOLY(ETHYLENEOXY)ETHYL PHOSPHATE   31800-88-1

HEALTH AND SAFETY LISTS

U.S. DOT - Appendix B - Marine Pollutants
DOT regulated marine pollutant

NFPA - Flash Points
flash point = 158 degrees F (70 degrees C)

NFPA - Hazard Identification Ratings
health-1; flammability-2; reactivity-0

TERT-OCTYL PEROXY-2-ETHYLHEXANOATE   RR-00750-1

ENVIRONMENTAL LISTS

List of Pesticide Product Inert Ingredients
[present]

OCTYLPHENOL   27193-28-8

ENVIRONMENTAL LISTS

List of Pesticide Product Inert Ingredients
[present]

2-(OCTYLPHENOXY)-ETHANOL   1322-97-0

ENVIRONMENTAL LISTS

List of Pesticide Product Inert Ingredients
[present]

P-OCTYLPHENYL SALICYLATE   2553-08-4

STATE LISTS

NJ Right to Know List (Total)
sn 2592

N-OCTYL SALICYLATE   6969-49-9

ENVIRONMENTAL LISTS

List of Pesticide Product Inert Ingredients
[present]

OCTYL TRICHLOROSILANE   5283-66-9

HEALTH AND SAFETY LISTS

U.S. DOT - Substances From 49 CFR 172.101
regulated by DOT (UN3023)

U.S. DOT - Hazard Classes
DOT hazard class = 6.1

U.S. DOT - Substances Which Are Poisonous by Inhalation
liquid hazardous material poisonous by inhalation (UN3023)

NFPA - Flash Points
flash point = 115 degrees F (46 degrees C)

NFPA - Hazard Identification Ratings
health-2; flammability-2; reactivity-0

STATE LISTS

Florida Hazardous Substance List
[present]

Massachusetts Right To Know List
[present]

Pennsylvania Right to Know List
[present]

4-OCTYNE-3,6-DIOL, 3,6-DIMETHYL-   78-66-0

ENVIRONMENTAL LISTS

List of Pesticide Product Inert Ingredients
[present]

OESTRADIOL-17B AND ESTERS   RR-00549-2

STATE LISTS

NJ Right to Know List (Total)
sn 2594

OESTROGEN-PROGESTIN COMBIN., SEQUENTIAL ORAL CONTRACEPTIVES   RR-00542-5

ENVIRONMENTAL LISTS

CAA - HON Rule - SOCMI Chemicals
compliance by April 24, 1995

OESTROGEN-PROGESTIN REPLACEMENT THERAPY   RR-01544-1

HEALTH AND SAFETY LISTS

NIOSH - Selected LD50s and LC50s
Oral, rat: LD50 = 7100 mg/kg

OESTROGEN REPLACEMENT THERAPY   RR-00541-4

HEALTH AND SAFETY LISTS

NFPA - Flash Points
flash point = 420 degrees F (216 degrees C)

NFPA - Hazard Identification Ratings
health-1; flammability-1; reactivity-0

OESTROGENS, NONSTEROIDAL   RR-00051-1

ENVIRONMENTAL LISTS

List of Pesticide Product Inert Ingredients
[present]

OESTROGENS, STEROIDAL   RR-00052-2

HEALTH AND SAFETY LISTS

U.S. DOT - Substances From 49 CFR 172.101
regulated by DOT (UN1801)

U.S. DOT - Hazard Classes
DOT hazard class = 8

INTERNATIONAL LISTS

Canada - WHMIS: Ingredient Disclosure
1% item 1223 (1284)

STATE LISTS

NJ Right to Know List (Total)
sn 1436

NJ Special Hazardous Substances
(corrosive; reactive - second degree)

OIL MIST, MINERAL   8012-95-1

ENVIRONMENTAL LISTS

List of Pesticide Product Inert Ingredients
[present]

OIL OF ANISE   8007-70-3

HEALTH AND SAFETY LISTS

IARC - Group Unspecified
[present] (Listed under 'Steroidal oestrogens')

OIL OF BITTER ALMOND   8013-76-1

HEALTH AND SAFETY LISTS

IARC - Group 1 (carcinogenic to humans)
[present] (Listed under 'Oestrogen-progestin combinations')

OSHA - Select Carcinogens
[present]

STATE LISTS

California - Prop. 65 - Cancer list
carcinogen - initial date 10/1/89

Minnesota Hazardous Substance List
carcinogen

OIL ORANGE SS   2646-17-5
HEALTH AND SAFETY LISTS
IARC - Group 3 (not classifiable)
[present]

OILS, LARD   8016-28-2
HEALTH AND SAFETY LISTS
IARC - Group 1 (carcinogenic to humans)
[present] (Listed under 'Steroidal oestrogens')
OSHA - Select Carcinogens
[present]

OILS, MENHADEN, POLYMERIZED   68213-57-0
HEALTH AND SAFETY LISTS
IARC - Group 1 (carcinogenic to humans)
*[present] (This evaluation applies to the group of chemicals as a whole
and not necessarily to all individual chemicals within the group) (Listed
under 'Oestrogens')*
OSHA - Select Carcinogens
[present]
STATE LISTS
California - Directors List of Hazardous Substances (8 CCR 339)
[present]

OILS, MUSTARD   8007-40-7
HEALTH AND SAFETY LISTS
IARC - Group 1 (carcinogenic to humans)
*[present] (This evaluation applies to the group of chemicals as a whole
and not necessarily to all individual chemicals within the group) (Listed
under 'Oestrogens')*
OSHA - Select Carcinogens
[present]
STATE LISTS
California - Directors List of Hazardous Substances (8 CCR 339)
[present]

OILS, PINE NEEDLE   8000-26-8
SEE ALSO:
PETROLEUM DISTILLATES (NAPHTHA)
HEALTH AND SAFETY LISTS
ACGIH 1995 - Time Weighted Averages
5 mg/m3 TWA (as sampled by a method that does not collect vapor)
ACGIH 1995 - Short Term Exposure Limits
*(10) mg/m3 STEL (as sampled by a method that does not collect
vapor)*
IARC - Group 3 (not classifiable)
[present] (highly refined oils)
NFPA - Flash Points
flash point = 380 degrees F (193 degrees C)
NFPA - Hazard Identification Ratings
health-0; flammability-1; reactivity-0
NIOSH - Selected LD50s and LC50s
Oral, mouse: LD50 = 22 gm/kg
NIOSH 1990 - Pocket Guide - RELs
5 mg/m3 TWA; 10 mg/m3 STEL
NIOSH 1990 - Pocket Guide - Target organs
skin, respiratory system
OSHA - Vacated PELs - Time Weighted Averages
5 mg/m3 TWA
OSHA - Final PELs - Time Weighted Averages
5 mg/m3 TWA

INTERNATIONAL LISTS
Australian Exposure Standards - Time Weighted Averages
5 mg/m3 TWA
Australian Exposure Standards - Short Term Exposure Limits
10 mg/m3 STEL
Canada - WHMIS: Ingredient Disclosure
1% item 1224 (977)
Canada - Alberta - 8 Hour Occupational Exposure Limit
5 mg/m3 TWA
Canada - Alberta - 15 Minute Occupational Exposure Limit
10 mg/m3 STEL
Canada - British Columbia - 8 Hour Exposure Limits
5 mg/m3 TWA
Canada - British Columbia - 15 Minute Exposure Limits
10 mg/m3 STEL
Canada - Ontario - OHSA - TWAEVs
5 mg/m3 TWAEV (listed as an agent of variable composition)
Canada - Ontario - OHSA - STEVs
10 mg/m3 STEV (listed as 'Agent of variable composition')
Canada - Quebec - Time-Weighted Average Exposure Values
5 mg/m3 TWAEV
Canada - Quebec - Short-term Exposure Values
10 mg/m3 STEV
United Kingdom - Occupational Exposure Standards - TWAs
5 mg/m3 TWA
United Kingdom - Occupational Exposure Standards - STELs
10 mg/m3 STEL
Israel - Time Weighted Averages
5 mg/m3 TWA (As sampled by method that does not collect vapor)
Israel - Short Term Exposure Limits
10 mg/m3 STEL (As sampled by method that does not collect vapor)
Israel - Action Levels
2.5 mg/m3 AL
Mexico - Instruction No. 10 - TWAs
5 mg/m3 TWA
Mexico - Instruction No. 10 - STELs
10 mg/m3 STEL
STATE LISTS
California - Exposure Limits - PELs
*particulate: 5 mg/m3 PEL (as sampled by a method that does not
collect vapor)*
California - Directors List of Hazardous Substances (8 CCR 339)
*[present] (refers to particulate forms which are not generated in the
ordinary use of the product)*
Massachusetts Right To Know List
[present]
Minnesota Hazardous Substance List
carcinogen
Pennsylvania Right to Know List
[present]
PROPOSED REGULATIONS
ACGIH 1995 - Notice of Intended Changes
*severely refined: 5 mg/m3 TWA as sampled by method that does not
collect vapor; mildly refined, as cyclohexane soluble particulate containing polynuclear aromatic hydrocarbons (PNAs): 0.2 mg/m3 TWA
as sampled by method that does not collect vapor; A1-confirmed human carcinogen*
Canada - Ontario - Proposed Occupational TWAEVs
1 mg/m3 TWAEV
Canada - Ontario - Proposed Occupational STEVs
3 mg/m3 STEV

OILS, TOBACCO   8037-19-2

ENVIRONMENTAL LISTS
  List of Pesticide Product Inert Ingredients
  [present]

OILS, VEGETABLE, HYDROGENATED   68334-28-1

ENVIRONMENTAL LISTS
  List of Pesticide Product Inert Ingredients
  [present]

OLAQUINDOX (N-(2-HYDROXYETHYL)-3-METHYL-   23696-28-8
2-QUINOXALINECARBOXAMIDE 1,4-DIOXIDE)

HEALTH AND SAFETY LISTS
  IARC - Group 2B (sufficient animal data)
  *[present] (Overall evaluation based only on evidence of carcinogenicity
  in monograph (8, 1975) or in Supplement 4)*
  OSHA - Possible Select Carcinogens
  [present]

STATE LISTS
  California - Air Bill 2588 Appendix A-II
  known or potential carcinogen
  California - Prop. 65 - Cancer list
  carcinogen - initial date 4/1/88
  California - Directors List of Hazardous Substances (8 CCR 339)
  [present]
  Florida Hazardous Substance List
  [present]
  Massachusetts Right To Know List
  carcinogen; extraordinarily hazardous
  NJ Right to Know List (Total)
  sn 3143
  Pennsylvania Right to Know List
  special hazardous substance
  Pennsylvania RTK - Special Hazardous Substances
  [present]

ALPHA-OLEFINS   69898-00-6

HEALTH AND SAFETY LISTS
  NFPA - Flash Points
  *commercial or animal: flash point = 395 degrees F (202 degrees C);
  no. 1: flash point = 440 degrees F (227 degrees C); pure: flash point
  = 500 degrees F (260 degrees C); no. 2: flash point = 419 degrees F
  (215 degrees C); mineral: flash point = 404 degrees F (207 degrees
  C)*
  NFPA - Hazard Identification Ratings
  health-0; flammability-1; reactivity-0

STATE LISTS
  Pennsylvania Right to Know List
  [present]

ALPHA-OLEFIN SULFONATE, POTASSIUM SALT   RR-00963-2

ENVIRONMENTAL LISTS
  List of Pesticide Product Inert Ingredients
  [present]

ALPHA-OLEFIN SULFONATE, SODIUM SALTS   RR-01234-0

STATE LISTS
  Pennsylvania Right to Know List
  [present]

OLEIC ACID   112-80-1

HEALTH AND SAFETY LISTS
  NIOSH - Selected LD50s and LC50s
  Oral, rat: LD50 = 6880 mg/kg

OLEIC ACID DIETHANOLAMIDE   93-83-4

STATE LISTS
  Massachusetts Right To Know List
  carcinogen; extraordinarily hazardous

OLEIC ACID DIETHANOLAMINE   13961-86-9

STATE LISTS
  Pennsylvania Right to Know List
  [present]

OLEIC ACID ESTER OF TETRA(HYDROXYETHYL)   RR-01105-2
ETHYLENEDIAMINE

INTERNATIONAL LISTS
  German (DFG) - Skin/Sensitizers
  danger of photo-contact sensitization
  German (DFG) - Carcinogens
  suspected carcinogen

OLEIC ACID ESTER WITH 1,2,3-PROPANETRIOL   37220-82-9

ENVIRONMENTAL LISTS
  List of Pesticide Product Inert Ingredients
  [present]

OLEIC ACID, 2-(2-(2-(2-HYDROXYETHOXY)   71012-10-7
ETHOXY)ETHOXY)ETHYL ESTER

ENVIRONMENTAL LISTS
  TSCA - Chemicals with Significant New Use Rules
  PMN number: P-91-100

OLEOYL IMIDAZOLINE   7347-29-7

ENVIRONMENTAL LISTS
  TSCA - Chemicals with Significant New Use Rules
  PMN number: P-88-2210

OLEO OIL   RR-00792-1

HEALTH AND SAFETY LISTS
  NFPA - Flash Points
  *flash point = 372 degrees F (189 degrees C); distilled: flash point =
  364 degrees F (184 degrees C)*
  NFPA - Hazard Identification Ratings
  health-0; flammability-1; reactivity-0
  NIOSH - Selected LD50s and LC50s
  Oral, rat: LD50 = 74 gm/kg

ENVIRONMENTAL LISTS
  List of Pesticide Product Inert Ingredients
  [present]

INTERNATIONAL LISTS
  Canada - WHMIS: Ingredient Disclosure
  1% item 1225 (119)

STATE LISTS
  Pennsylvania Right to Know List
  [present]

OLEOYL SARCOSINE   110-25-8

HEALTH AND SAFETY LISTS
  NTP Chemical Status Reports - Testing Status and NTIS Number
  Two year studies in progress

ENVIRONMENTAL LISTS
  List of Pesticide Product Inert Ingredients
  [present]

OLEYLAMINE   112-90-3

ENVIRONMENTAL LISTS

List of Pesticide Product Inert Ingredients
[present]

OLEYL BASED IMIDAZOLINE 21652-27-7
ENVIRONMENTAL LISTS
List of Pesticide Product Inert Ingredients
[present]

OLEYL ETHER PHOSPHATE (ACID) 7722-71-6
ENVIRONMENTAL LISTS
List of Pesticide Product Inert Ingredients
[present]

OLEYL ETHER PHOSPHATE (NEUTRAL) 37310-83-1
ENVIRONMENTAL LISTS
List of Pesticide Product Inert Ingredients
[present]

OLIGOESTER DERIVFED BY CONDENSATION OF RR-01106-3
ADIPIC ACID, PHTHALIC ANHYDRIDE, ETHY-
LENE GLYCOL, N-OCTYL ALCOHOL AND N-DECYL
ALCOHOL
ENVIRONMENTAL LISTS
List of Pesticide Product Inert Ingredients
[present]

OLIGOMETRIC SILICIC ACID ESTER COMPOUND RR-01211-3
WITH AN HYDROXYLALKYLAMINE
HEALTH AND SAFETY LISTS
NFPA - Flash Points
flash point = 450 degrees F (232 degrees C)
NFPA - Hazard Identification Ratings
health-0; flammability-1; reactivity-0
STATE LISTS
Pennsylvania Right to Know List
[present]

OLIVE OIL 8001-25-0
ENVIRONMENTAL LISTS
List of Pesticide Product Inert Ingredients
[present]

OLIVINE 1317-71-1
ENVIRONMENTAL LISTS
EPA - Master Testing List
[present]
List of Pesticide Product Inert Ingredients
[present]
TSCA - Code of Federal Regulations Citations
40 CFR 712.30(j); 40 CFR 716.120(a); 40 CFR 799.3175
TSCA - PAIR - Reporting List
Reporting Date: March 13, 1984
TSCA - Health and Safety Reporting List
Effective Date: January 13, 1984
TSCA - Chemical Test Rules
Testing required by: manufacturers; processors (40 CFR 799.3175)
TSCA - Section 12(b) - Export Notification
export notification required - Section 4

OMADINE 1121-30-8
ENVIRONMENTAL LISTS
List of Pesticide Product Inert Ingredients
[present]

ONIONS, OIL 8002-72-0
ENVIRONMENTAL LISTS
List of Pesticide Product Inert Ingredients
[present]

ORANGE I 523-44-4
ENVIRONMENTAL LISTS
List of Pesticide Product Inert Ingredients
[present]

ORANGE OIL 8008-57-9
ENVIRONMENTAL LISTS
List of Pesticide Product Inert Ingredients
[present]

ORDRAM (MOLINATE) 2212-67-1
ENVIRONMENTAL LISTS
TSCA - Chemicals with Significant New Use Rules
PMN number: P-91-118

ORGANIC SYNTHETIC FIBRES (CARBON AND RR-01807-5
GRAPHITE FIBRES)
HEALTH AND SAFETY LISTS
NFPA - Flash Points
flash point = 437 degrees F (225 degrees C)
NFPA - Hazard Identification Ratings
health-0; flammability-1; reactivity-0
ENVIRONMENTAL LISTS
List of Pesticide Product Inert Ingredients
[present]

ORGANIC PEROXIDES, N.O.S. RR-00385-0
INTERNATIONAL LISTS
Canada - British Columbia - 8 Hour Exposure Limits
respirable mass: 3.3 mg/m3 TWA

ORGANIC PHOSPHORUS COMPOUND, MIXED RR-00390-7
WITH COMPRESSED GAS
HEALTH AND SAFETY LISTS
NIOSH - Selected LD50s and LC50s
Oral, mouse: LD50 = 535 mg/kg

ORGANOCHLORINE PESTICIDES, N.O.S. RR-00391-8
ENVIRONMENTAL LISTS
List of Pesticide Product Inert Ingredients
[present]

ORGANOHALOGEN COMPOUNDS RR-01603-5
HEALTH AND SAFETY LISTS
IARC - Group 3 (not classifiable)
[present]

ORGANOPHOSPHOROUS PESTICIDES, N.O.S. RR-00395-2
ENVIRONMENTAL LISTS
List of Pesticide Product Inert Ingredients
[present]

ORGANORHODIUM COMPLEX RR-00069-1
ENVIRONMENTAL LISTS
CERCLA/SARA - Section 313 - Emission Reporting
form R reporting required

STATE LISTS

California - Directors List of Hazardous Substances (8 CCR 339)
[present]

ORGANOSILICON COMPOUNDS   RR-01609-1

INTERNATIONAL LISTS

Canada - Quebec - Time-Weighted Average Exposure Values
total dust: 10 mg/m3;   respirable dust: 5 mg/m3

ORGANOTIN CATALYSTS   RR-01661-5

HEALTH AND SAFETY LISTS

U.S. DOT - Substances From 49 CFR 172.101
regulated by DOT (UN3101, UN3102, UN3103, UN3104, UN3105, UN3106, UN3107, UN3108, UN3109, UN3110, UN3111, UN3112, UN3113, UN3114, UN3115, UN3116, UN3117, UN3118, UN3119, UN3120)
U.S. DOT - Hazard Classes
DOT hazard class = 5.2

STATE LISTS

NJ Right to Know List (Total)
sn 2596;   sn 2597;   sn 2598;   sn 2599;   sn 2600

ORGANOTIN COMPOUNDS, N.O.S.   RR-00134-3

STATE LISTS

NJ Right to Know List (Total)
sn 2602

ORGANOTIN LITHIUM COMPOUNDS   RR-01725-4

HEALTH AND SAFETY LISTS

U.S. DOT - Substances From 49 CFR 172.101
regulated by DOT (UN2761, UN2762, UN2995, UN2996)
U.S. DOT - Hazard Classes
toxic or toxic, flammable: DOT hazard class = 6.1;   flammable, toxic: DOT hazard class = 3

STATE LISTS

NJ Right to Know List (Total)
sn 2603;   sn 2604;   sn 2605;   sn 2606

ORGANOTIN PESTICIDES, N.O.S.   RR-00399-6

INTERNATIONAL LISTS

Canada - CEPA Schedule III Part I - Prohibited Substances (Ocean Dumping)
[present]

L-ORNITHINE HCL   3184-13-2

HEALTH AND SAFETY LISTS

ACGIH 1995 - Biological Exposure Indices
Cholinesterase activity in red cells: 70% of individual's baseline, discretionary (B, Ns, Sq)
U.S. DOT - Substances From 49 CFR 172.101
regulated by DOT (UN2783, UN2784, UN3017, UN3018)
U.S. DOT - Hazard Classes
toxic or toxic, flammable: DOT hazard class = 6.1;   flammable, toxic: DOT hazard class = 3

STATE LISTS

NJ Right to Know List (Total)
sn 2607;   sn 2608;   sn 2609;   sn 2610

OROTIC ACID   65-86-1

ENVIRONMENTAL LISTS

CERCLA/SARA - Section 302 Extremely Hazardous Substances and TPQs
TPQ = 10/10,000 pounds

STATE LISTS

Massachusetts Right To Know List
extraordinarily hazardous
NJ Right to Know List (Total)
sn 2611
Pennsylvania Right to Know List
environmental hazard

PROPOSED REGULATIONS

CERCLA/SARA - Proposed Hazardous Substance Additions
proposed RQ = 1 pound (.454 kg)
CERCLA/SARA - 1989 Proposed RQ Adjustments
proposed RQ = 10 pounds (4.54 kg)

ORTHOARSENIC ACID   RR-01306-9

INTERNATIONAL LISTS

Canada - CEPA Schedule III Part II - Restricted Substances (Ocean Dumping)
[present]

ORYZALIN [4-(DIPROPYLAMINO)-3,5-DINITROBEN-ZENESULFONAMIDE]   19044-88-3

ENVIRONMENTAL LISTS

TSCA - Chemicals with Significant New Use Rules
PMN numbers: P-93-853 through P-93-858

OSMIUM 180   14993-35-2

HEALTH AND SAFETY LISTS

U.S. DOT - Substances From 49 CFR 172.101
regulated by DOT (UN2788, UN3146)
U.S. DOT - Hazard Classes
DOT hazard class = 6.1
U.S. DOT - Appendix B - Marine Pollutants
DOT regulated severe marine pollutant
NIOSH - Health Standards - Exposure Limits
0.1 mg Sn/m3 TWA
NIOSH - Health Standards - Health Effects and Precautions
Eye, skin, liver, nervous system, and heart effects (Periodic chest X-ray, blood and urine monitoring, eye tests, heart examination, and nervous system testing required;   prevent skin and eye contact)

INTERNATIONAL LISTS

Canada - CEPA - Priority Substances List
estimated time for completion of assessment reports: 5 years

STATE LISTS

Minnesota Hazardous Substance List
[present]

OSMIUM 181   14993-64-7

ENVIRONMENTAL LISTS

TSCA - Chemicals with Significant New Use Rules
PMN number: P-93-1119

OSMIUM 182   14993-36-3

HEALTH AND SAFETY LISTS

U.S. DOT - Substances From 49 CFR 172.101
regulated by DOT (UN2786, UN2787, UN3019, UN3020)
U.S. DOT - Hazard Classes
toxic or toxic, flammable: DOT hazard class = 6.1;   flammable, toxic: DOT hazard class = 3
U.S. DOT - Appendix B - Marine Pollutants
DOT regulated severe marine pollutant

STATE LISTS

NJ Right to Know List (Total)
sn 2612;   sn 2613;   sn 2614;   sn 2615

OSMIUM 185   15766-50-4

HEALTH AND SAFETY LISTS

NIOSH - Selected LD50s and LC50s
Oral, rat: LD50 = 10 gm/kg

OSMIUM 189M   RR-00397-4

HEALTH AND SAFETY LISTS

NIOSH - Selected LD50s and LC50s
Oral, mouse: LD50 = 2 gm/kg

STATE LISTS

Massachusetts Right To Know List
[present]

OSMIUM 191   14119-24-5

HEALTH AND SAFETY LISTS

U.S. DOT - Appendix B - Marine Pollutants
DOT regulated marine pollutant

OSMIUM 191M   RR-00396-3

ENVIRONMENTAL LISTS

CERCLA/SARA - Section 313 - Emission Reporting
form R reporting required

OSMIUM 193   16057-77-5

HEALTH AND SAFETY LISTS

U.S. DOT - Appendix A Table 2 - Radionuclides
final RQ = 1000 curies (3.7E 13 Bq)

ENVIRONMENTAL LISTS

CERCLA/SARA List of Radionuclides (Appendix B) and Their Reportable Quantities
final RQ = 1000 curies (3.7E 13 Bq)

OSMIUM 194   15766-57-1

HEALTH AND SAFETY LISTS

U.S. DOT - Appendix A Table 2 - Radionuclides
final RQ = 100 curies (3.7E 12 Bq)

ENVIRONMENTAL LISTS

CERCLA/SARA List of Radionuclides (Appendix B) and Their Reportable Quantities
final RQ = 100 curies (3.7E 12 Bq)

OSMIUM, TETRAAMINETETRACHLORO-   RR-00608-6

HEALTH AND SAFETY LISTS

U.S. DOT - Appendix A Table 2 - Radionuclides
final RQ = 100 curies (3.7E 12 Bq)

ENVIRONMENTAL LISTS

CERCLA/SARA List of Radionuclides (Appendix B) and Their Reportable Quantities
final RQ = 100 curies (3.7E 12 Bq)

OSMIUM TETROXIDE   20816-12-0

HEALTH AND SAFETY LISTS

U.S. DOT - Appendix A Table 2 - Radionuclides
final RQ = 10 curies (3.7E 11 Bq)

ENVIRONMENTAL LISTS

CERCLA/SARA List of Radionuclides (Appendix B) and Their Reportable Quantities
final RQ = 10 curies (3.7E 11 Bq)

OUABAIN   630-60-4

HEALTH AND SAFETY LISTS

U.S. DOT - Appendix A Table 2 - Radionuclides
final RQ = 1000 curies (3.7E 13 Bq)

ENVIRONMENTAL LISTS

CERCLA/SARA List of Radionuclides (Appendix B) and Their Reportable Quantities
final RQ = 1000 curies (3.7E 13 Bq)

1-OXA-4-AZASPIRO(4,5)DECANE, 4-   71526-07-3
(DICHLOROACETYL)-

HEALTH AND SAFETY LISTS

U.S. DOT - Appendix A Table 2 - Radionuclides
final RQ = 100 curies (3.7E 12 Bq)

ENVIRONMENTAL LISTS

CERCLA/SARA List of Radionuclides (Appendix B) and Their Reportable Quantities
final RQ = 100 curies (3.7E 12 Bq)

3-OXA-9-AZONIATRICYCLO[3.3.1.02,4]NONANE,   155-41-9
7-(3-HYDROXY-1-OXO-2-PHENYLPROPOXY)-9,9-
DIMETHYL-, BROMIDE, [7(S)-(1-ALPHA, 2-BETA,4-
BETA,5-ALPHA,7-BETA)]-

HEALTH AND SAFETY LISTS

U.S. DOT - Appendix A Table 2 - Radionuclides
final RQ = 1000 curies (3.7E 13 Bq)

ENVIRONMENTAL LISTS

CERCLA/SARA List of Radionuclides (Appendix B) and Their Reportable Quantities
final RQ = 1000 curies (3.7E 13 Bq)

7-OXABICYCLO[4.1.0]HEPTANE   286-20-4

HEALTH AND SAFETY LISTS

U.S. DOT - Appendix A Table 2 - Radionuclides
final RQ = 100 curies (3.7E 12 Bq)

ENVIRONMENTAL LISTS

CERCLA/SARA List of Radionuclides (Appendix B) and Their Reportable Quantities
final RQ = 100 curies (3.7E 12 Bq)

OXABICYCLO[4.1.0]HEPTANE, 3-ETHENYL, HO-   RR-01015-1
MOPOLYMER, ETHER WITH2-ETHYL-2-(HYDROX-
YMETHYL)-1,3-PROPANEDIOL (3:1), EPOXIZED

HEALTH AND SAFETY LISTS

U.S. DOT - Appendix A Table 2 - Radionuclides
final RQ = 1 curie (3.7E 10 Bq)

ENVIRONMENTAL LISTS

CERCLA/SARA List of Radionuclides (Appendix B) and Their Reportable Quantities
final RQ = 1 curie (3.7E 10 Bq)

7-OXABICYCLO[2.2.1]HEPTANE, 1-METHYL-4-(1-   470-67-7
METHYLETHYL)-

INTERNATIONAL LISTS

Canada - WHMIS: Ingredient Disclosure
1% item 1226 (1599)

7-OXABICYCLO[4.1.0]HEPTANE, 1-METHYL-4-(2-   96-08-2
METHYLOXIRANYL)-

HEALTH AND SAFETY LISTS

ACGIH 1995 - Time Weighted Averages
as Os: 0.0002 ppm TWA; 0.0016 mg/m3 TWA

ACGIH 1995 - Short Term Exposure Limits
as Os: 0.0006 ppm STEL; 0.0047 mg/m3 STEL

U.S. DOT - Substances From 49 CFR 172.101
regulated by DOT (UN2472)

U.S. DOT - Hazard Classes
DOT hazard class = 6.1

U.S. DOT - Appendix B - Marine Pollutants
DOT regulated marine pollutant

U.S. DOT - Appendix A Table 1 - Hazardous Substances
final RQ = 1000 pounds (454 kg)

NIOSH - Selected LD50s and LC50s
Oral, mouse: LD50 = 162 mg/kg

NIOSH 1990 - Pocket Guide - RELs
as Os: 0.002 mg/m3 TWA; 0.0002 ppm TWA; 0.006 mg/m3 STEL; 0.0006 ppm STEL

NIOSH 1990 - Pocket Guide - IDLHs
as Os: 1 mg/m3 IDLH

NIOSH 1990 - Pocket Guide - Target organs
skin, eyes, respiratory system

OSHA - Vacated PELs - Time Weighted Averages
as Os: 0.0002 ppm TWA; 0.002 mg/m3 TWA

OSHA - Vacated PELs - Short Term Exposure Limits
as Os: 0.0006 ppm STEL; 0.006 mg/m3 STEL

OSHA - Final PELs - Time Weighted Averages
as Os: 0.002 mg/m3 TWA

OSHA - List of Highly Hazardous Chemicals
threshhold quantity = 100 pounds

ENVIRONMENTAL LISTS

CERCLA/SARA - Section 313 - Emission Reporting
form R reporting required for 1.0% de minimus concentration

CERCLA/SARA - Hazardous Substances and their Reportable Quantities
final RQ = 1000 pounds (454 kg)

RCRA - P Series Wastes
waste number P087

RCRA - Hazardous Constituents-Appendix VIII
waste number P087

RCRA - Substances Banned From Land Disposal
[present]

INTERNATIONAL LISTS

Australian Exposure Standards - Time Weighted Averages
as Os: 0.0002 ppm TWA; 0.0016 mg/m3 TWA

Australian Exposure Standards - Short Term Exposure Limits
as Os: 0.0006 ppm STEL; 0.0047 mg/m3 STEL

Canada - WHMIS: Ingredient Disclosure
0.1% item 1227 (1603)

Canada - Alberta - 8 Hour Occupational Exposure Limit
as Os: 0.0002 ppm TWA; 0.0016 mg/m3 TWA

Canada - Alberta - 15 Minute Occupational Exposure Limit
as Os: 0.0006 ppm STEL; 0.0047 mg/m3 STEL

Canada - British Columbia - 8 Hour Exposure Limits
as Os: 0.0002 ppm TWA; 0.002 mg/m3 TWA

Canada - British Columbia - 15 Minute Exposure Limits
as Os: 0.0006 ppm STEL; 0.006 mg/m3 STEL

Canada - Ontario - OHSA - TWAEVs
as Osmium: 0.0002 ppm TWAEV; 0.002 mg/m3 TWAEV

Canada - Ontario - OHSA - STEVs
as Os: 0.0006 ppm STEV; 0.006 mg/m3 STEV

Canada - Quebec - Time-Weighted Average Exposure Values
as Os: 0.0002 ppm TWAEV; 0.0016 mg/m3 TWAEV

Canada - Quebec - Short-term Exposure Values
as Os: 0.0006 ppm STEV; 0.0048 mg/m3 STEV

United Kingdom - Occupational Exposure Standards - TWAs
as Os: 0.0002 ppm TWA; 0.002 mg/m3 TWA

United Kingdom - Occupational Exposure Standards - STELs
as Os: 0.0006 ppm STEL; 0.006 mg/m3 STEL

German (DFG) - MAK Values
0.0002 ppm MAK; 0.002 mg/m3 MAK

German (DFG) - Peak Limitations
2 x normal MAK (5 min momentary value); don't exceed 8 times during shift

Israel - Time Weighted Averages
as Os: 0.0002 ppm TWA; 0.0016 mg/m3 TWA

Israel - Short Term Exposure Limits
as Os: 0.0006 ppm STEL; 0.0047 mg/m3 STEL

Israel - Action Levels
as Os: 0.0001 ppm AL; 0.0008 mg/m3 AL

Mexico - Instruction No. 10 - TWAs
0.0002 ppm TWA; 0.002 mg/m3 TWA

Mexico - Instruction No. 10 - STELs
0.0006 ppm STEL; 0.006 mg/m3 STEL

STATE LISTS

California - Air Bill 2588 Appendix A-II
6/91

California - Exposure Limits - PELs
as Os: 0.0002 ppm PEL; 0.002 mg/m3 PEL

California - Exposure Limits - STELs
as Os: 0.0006 ppm STEL; 0.006 mg/m3 STEL

California - Directors List of Hazardous Substances (8 CCR 339)
[present]

Florida Hazardous Substance List
[present]

Massachusetts Right To Know List
[present]

Minnesota Hazardous Substance List
[present] as Os

NJ Right to Know List (Total)
sn 1441

Pennsylvania Right to Know List
environmental hazard

6-OXABICYCLO[3.1.0]HEXANE   285-67-6

HEALTH AND SAFETY LISTS

NIOSH - Selected LD50s and LC50s
Oral, guinea pig: LD50 = 8280 ug/kg

ENVIRONMENTAL LISTS

CERCLA/SARA - Section 302 Extremely Hazardous Substances and TPQs
TPQ = 100/10,000 pounds

STATE LISTS

Florida Hazardous Substance List
effective March 13, 1992

Massachusetts Right To Know List
extraordinarily hazardous

NJ Right to Know List (Total)
sn 2617

Pennsylvania Right to Know List
environmental hazard

PROPOSED REGULATIONS

CERCLA/SARA - Proposed Hazardous Substance Additions
proposed RQ = 1 pound (.454 kg)

CERCLA/SARA - 1989 Proposed RQ Adjustments
proposed RQ = 100 pounds (45.4 kg)

OXADIAZON   19666-30-9

ENVIRONMENTAL LISTS

List of Pesticide Product Inert Ingredients
[present]

TSCA - Chemicals with Significant New Use Rules
PMN number: P-86-1648

TSCA - Section 12(b) - Export Notification
P-86-1648; export notification required - Section 5

OXALATES   RR-01351-4
ENVIRONMENTAL LISTS

TSCA - Code of Federal Regulations Citations
40 CFR 712.30(d)

TSCA - PAIR - Reporting List
Reporting Date: November 19, 1982

OXALIC ACID   144-62-7
HEALTH AND SAFETY LISTS

NIOSH - Selected LD50s and LC50s
Oral, rat: LD50 = 1090 mg/kg Skin, rabbit: LD50 = 630 mg/kg

NTP Chemical Status Reports - Testing Status and NTIS Number
Approved for toxicology/carcinogenesis study

ENVIRONMENTAL LISTS

TSCA - Code of Federal Regulations Citations
40 CFR 712.30(d)

TSCA - PAIR - Reporting List
Reporting Date: Nobember 19, 1982

OXALIC ACID, AMMONIUM IRON (3+) SALT (3:3:1)   2944-67-4
ENVIRONMENTAL LISTS

TSCA - Chemicals with Significant New Use Rules
PMN number: P-88-1898

OXALIC ACID, AMMONIUM SALT   14258-49-2
ENVIRONMENTAL LISTS

TSCA - Code of Federal Regulations Citations
40 CFR 712.30(d)

TSCA - PAIR - Reporting List
Reporting Date: November 19, 1982

OXAMIDE   471-46-5
HEALTH AND SAFETY LISTS

NIOSH - Selected LD50s and LC50s
Oral, rat: LD50 = 5630 mg/kg Skin, rabbit: LD50 = 1770 mg/kg

ENVIRONMENTAL LISTS

TSCA - Code of Federal Regulations Citations
40 CFR 712.30(d)

TSCA - PAIR - Reporting List
Reporting Date: November 19, 1982

OXAMYL   23135-22-0
ENVIRONMENTAL LISTS

TSCA - Code of Federal Regulations Citations
40 CFR 712.30(d)

TSCA - PAIR - Reporting List
Reporting Date: November 19, 1982

1,4-OXATHIANE   15980-15-1
ENVIRONMENTAL LISTS

CERCLA/SARA - Section 313 - Emission Reporting
form R reporting required

STATE LISTS

California - Prop. 65 - Cancer list
carcinogen - initial date 7/1/91

5-OXATRICYCLO[8.2.0.04,6]DODECANE, 4,12,12-TRIMETHYL-9-METHYLENE-, [1R-(1R*,4R*,6R*,10S*)]-   1139-30-6
HEALTH AND SAFETY LISTS

U.S. DOT - Substances From 49 CFR 172.101
regulated by DOT (UN2449)

U.S. DOT - Hazard Classes
DOT hazard class = 6.1

3-OXAURACIL   34314-63-1
HEALTH AND SAFETY LISTS

ACGIH 1995 - Time Weighted Averages
1 mg/m3 TWA

ACGIH 1995 - Short Term Exposure Limits
2 mg/m3 STEL

NIOSH - Selected LD50s and LC50s
Oral, rat: LD50 = 375 mg/kg Skin, rabbit: LD50 = 20 gm/kg

NIOSH 1990 - Pocket Guide - RELs
1 mg/m3 TWA; 2 mg/m3 STEL

NIOSH 1990 - Pocket Guide - IDLHs
500 mg/m3 IDLH

NIOSH 1990 - Pocket Guide - Target organs
respiratory system, skin, eyes, kidneys

OSHA - Vacated PELs - Time Weighted Averages
1 mg/m3 TWA

OSHA - Vacated PELs - Short Term Exposure Limits
2 mg/m3 STEL

OSHA - Final PELs - Time Weighted Averages
1 mg/m3 TWA

ENVIRONMENTAL LISTS

List of Pesticide Product Inert Ingredients
[present]

INTERNATIONAL LISTS

Australian Exposure Standards - Time Weighted Averages
1 mg/m3 TWA

Australian Exposure Standards - Short Term Exposure Limits
2 mg/m3 STEL

Canada - WHMIS: Ingredient Disclosure
0.1% item 1228 (120)

Canada - Alberta - 8 Hour Occupational Exposure Limit
1 mg/m3 TWA

Canada - Alberta - 15 Minute Occupational Exposure Limit
2 mg/m3 STEL

Canada - British Columbia - 8 Hour Exposure Limits
1 mg/m3 TWA

Canada - British Columbia - 15 Minute Exposure Limits
2 mg/m3 STEL

Canada - Ontario - OHSA - TWAEVs
1 mg/m3 TWAEV

Canada - Ontario - OHSA - STEVs
2 mg/m3 STEV

Canada - Quebec - Time-Weighted Average Exposure Values
1 mg/m3 TWAEV

Canada - Quebec - Short-term Exposure Values
2 mg/m3 STEV

United Kingdom - Occupational Exposure Standards - TWAs
1 mg/m3 TWA

United Kingdom - Occupational Exposure Standards - STELs
2 mg/m3 STEL

Israel - Time Weighted Averages
1 mg/m3 TWA

Israel - Short Term Exposure Limits
2 mg/m3 STEL

Israel - Action Levels
0.5 mg/m3 AL

Mexico - Instruction No. 10 - TWAs
1 mg/m3 TWA

Mexico - Instruction No. 10 - STELs
2 mg/m3 STEL

STATE LISTS

California - Exposure Limits - PELs
1 mg/m3 PEL

California - Exposure Limits - STELs
2 mg/m3 STEL

California - Directors List of Hazardous Substances (8 CCR 339)
[present]

Florida Hazardous Substance List
[present]

Massachusetts Right To Know List
[present]

Minnesota Hazardous Substance List
[present]

NJ Right to Know List (Total)
sn 1445

NJ Special Hazardous Substances
(corrosive)

Pennsylvania Right to Know List
[present]

OXAZEPAM   604-75-1

HEALTH AND SAFETY LISTS

U.S. DOT - Appendix A Table 1 - Hazardous Substances
final RQ = 1000 pounds (454 kg)

ENVIRONMENTAL LISTS

CERCLA/SARA - Hazardous Substances and their Reportable Quantities
final RQ = 1000 pounds (454 kg)

Clean Water Act - Hazardous Substances
[present]

STATE LISTS

Massachusetts Right To Know List
[present]

Pennsylvania Right to Know List
environmental hazard

OXAZOLODINE-1,3   129-20-4

HEALTH AND SAFETY LISTS

U.S. DOT - Appendix A Table 1 - Hazardous Substances
final RQ = 5000 pounds (2270 kg)

ENVIRONMENTAL LISTS

CERCLA/SARA - Hazardous Substances and their Reportable Quantities
final RQ = 5000 pounds (2270 kg)

Clean Water Act - Hazardous Substances
[present] (Listed under 'Ammonium oxalate')

STATE LISTS

Massachusetts Right To Know List
[present]

Pennsylvania Right to Know List
environmental hazard

1H,3H,5H-OXAZOLO(3,4-C)OXAZOLE-7A,(7H)-   6542-37-6
METHANOL

HEALTH AND SAFETY LISTS

NIOSH - Selected LD50s and LC50s
Oral, rat: LD50 = 447 mg/kg

INTERNATIONAL LISTS

Canada - WHMIS: Ingredient Disclosure
1% item 1229 (1294)

1H,3H,5H-OXAZOLO(3,4-C)OXAZOLE, METHANOL   59720-42-2
DERIVATIVE

HEALTH AND SAFETY LISTS

U.S. DOT - Appendix B - Marine Pollutants
DOT regulated marine pollutant

NIOSH - Selected LD50s and LC50s
Inhalation, rat: LC50 = 170 mg/m3 1 hr Oral, rat: LD50 = 2500 ug/kg Skin, rabbit: LD50 = 740 mg/kg

ENVIRONMENTAL LISTS

CERCLA/SARA - Section 302 Extremely Hazardous Substances and TPQs
TPQ = 100/10,000 pounds

Safe Drinking Water Act - MCLs
MCL = 0.2 mg/L

Safe Drinking Water Act - MCLGs
MCLG = 0.2 mg/L

STATE LISTS

Florida Hazardous Substance List
effective March 13, 1992

Massachusetts Right To Know List
extraordinarily hazardous

NJ Right to Know List (Total)
sn 2618

Pennsylvania Right to Know List
environmental hazard

PROPOSED REGULATIONS

CERCLA/SARA - Proposed Hazardous Substance Additions
proposed RQ = 1 pound (.454 kg)

CERCLA/SARA - 1989 Proposed RQ Adjustments
proposed RQ = 100 pounds (45.4 kg)

1H,3H,5H-OXAZOLO(3,4-C)OXAZOLE, POLY   56709-13-8
(OXYMETHYLENE)DERIVATIVE

HEALTH AND SAFETY LISTS

NFPA - Flash Points
flash point = 108 degrees F (42 degrees C)

NFPA - Hazard Identification Ratings
health-2; flammability-2; reactivity-0

STATE LISTS

Florida Hazardous Substance List
[present]

Massachusetts Right To Know List
[present]

Pennsylvania Right to Know List
[present]

2-OXEPANONE, POLYMER WITH 4,4'-(1-   RR-00915-4
METHYLETHYLIDENE)BISPHENOLAND 2,2-
[(1-METHYLETHYLIDENE)BIS(4,1-PHENYLE-
NEOXYMETHYLENE)]BISOXIRANE, GRAFT

ENVIRONMENTAL LISTS

TSCA - Code of Federal Regulations Citations
40 CFR 712.30(d)

TSCA - PAIR - Reporting List
Reporting Date: November 19, 1982

OXETANE, 3,3-BIS(CHLOROMETHYL)-   78-71-7

HEALTH AND SAFETY LISTS

NIOSH - Selected LD50s and LC50s
Oral, rat: LD50 = 850 mg/kg

OXIDANTS (OZONE)   RR-01556-5

HEALTH AND SAFETY LISTS

IARC - Group 3 (not classifiable)
[present]

NIOSH - Selected LD50s and LC50s
Oral, mouse: LD50 = 525 mg/kg

NTP Chemical Status Reports - Testing Status and NTIS Number
Technical reports printed; two year studies: pathology quality assessment in progress

STATE LISTS

California - Prop. 65 - Cancer list
carcinogen - initial date 10/01/94

California - Prop. 65 - Developmental Toxicity
developmental toxicity - initial date 10/1/92

California - Directors List of Hazardous Substances (8 CCR 339)
[present]

NJ Right to Know List (Total)
sn 1447

NJ Special Hazardous Substances
(teratogen)

OXIDIZERS, N.O.S.   RR-00403-5

HEALTH AND SAFETY LISTS

IARC - Group 3 (not classifiable)
[present]

NIOSH - Selected LD50s and LC50s
Oral, rat: LD50 = 350 mg/kg

OXIRANE, 2,2'-[1,4-BUTANEDIYLBIS(OXYMETHYLENE)]BIS-   2425-79-8

ENVIRONMENTAL LISTS

List of Pesticide Product Inert Ingredients
[present]

OXIRANE,2,2'-4-BUTYLIDENEBISPHENYLENEOXYMETHYLENE (DGEBA)   25085-99-8

ENVIRONMENTAL LISTS

List of Pesticide Product Inert Ingredients
[present]

OXIRANECARBOXYLIC ACID, 3-METHYL-3-PHENYL-, ETHYL ESTER   77-83-8

ENVIRONMENTAL LISTS

List of Pesticide Product Inert Ingredients
[present]

OXIRANECARBOXYLIC ACID, 3-PHENYL-, ETHYL ESTER   121-39-1

ENVIRONMENTAL LISTS

TSCA - Chemicals with Significant New Use Rules
PMN number: P-88-2582

OXIRANE, 2,2'-[1,4-CYCLOHEXANEDILBIS (METHYLENEOXYMETHYLENE)]BIS-   14228-73-0

HEALTH AND SAFETY LISTS

NIOSH - Selected LD50s and LC50s
Inhalation, mouse: LC50 = 200 mg/m3 (2 hr) Oral, mouse: LD50 = 420 mg/kg

ENVIRONMENTAL LISTS

CERCLA/SARA - Section 302 Extremely Hazardous Substances and TPQs
TPQ = 500 pounds

STATE LISTS

Florida Hazardous Substance List
effective March 13, 1992

Massachusetts Right To Know List
extraordinarily hazardous

Pennsylvania Right to Know List
environmental hazard

PROPOSED REGULATIONS

CERCLA/SARA - Proposed Hazardous Substance Additions
proposed RQ = 1 pound (.454 kg)

CERCLA/SARA - 1989 Proposed RQ Adjustments
proposed RQ = 100 pounds (45.4 kg)

OXIRANE, DECYL-   2855-19-8

INTERNATIONAL LISTS

Canada - National Air Quality Objectives - Schedule I
desirable limits: 0-30 ug/m3 day; 0-100 ug/m3 hour; acceptable limits: 0-30 ug/m3 year; 30-50 ug/m3 day; 100-160 ug/m3 hour; tolerable limits: 160-300 ug/m3 hour

OXIRANE, [(2,4-DIBROMOPHENOXY)METHYL]-   20217-01-0

HEALTH AND SAFETY LISTS

U.S. DOT - Substances From 49 CFR 172.101
regulated by DOT (UN1479, UN3085, UN3087, UN3098, UN3099, UN3100, UN3121, UN3139)
U.S. DOT - Hazard Classes
DOT hazard class = 5.1

STATE LISTS

NJ Right to Know List (Total)
sn 2620; sn 2621; sn 2622; sn 2623; sn 2624

OXIRANE, [(1,2-DIBROMOPROPOXY)METHYL]-   35243-89-1
SEE ALSO:
GLYCIDOL (OXIRANEMETHANOL) AND ITS DERIVATIVES

HEALTH AND SAFETY LISTS

NIOSH - Selected LD50s and LC50s
Oral, rat: LD50 = 2980 mg/kg Skin, rabbit: LD50 = 1130 mg/kg

ENVIRONMENTAL LISTS

EPA - Master Testing List
[present]

TSCA - Code of Federal Regulations Citations
40 CFR 712.30(d); 40 CFR 716.120(c)

TSCA - PAIR - Reporting List
Reporting Date: November 19, 1982

PROPOSED REGULATIONS

TSCA - Proposed Testing Rule for Glycidyl Ethers
subject to subchronic toxicity, neurotoxicity and mutagenicity testing (results apply to all members of Glycidyl subcategory V-A)

OXIRANE, 2,2-DIMETHYL-   558-30-5
SEE ALSO:
GLYCIDOL (OXIRANEMETHANOL) AND ITS DERIVATIVES

ENVIRONMENTAL LISTS

TSCA - Code of Federal Regulations Citations
40 CFR 712.30(d); 40 CFR 716.120(c)

TSCA - PAIR - Reporting List
Reporting Date: November 19, 1982

OXIRANE, 2,3-DIMETHYL-   3266-23-7
SEE ALSO:
GLYCIDOL (OXIRANEMETHANOL) AND ITS DERIVATIVES

HEALTH AND SAFETY LISTS

NIOSH - Selected LD50s and LC50s
Oral, rat: LD50 = 5470 mg/kg

ENVIRONMENTAL LISTS

TSCA - Code of Federal Regulations Citations
40 CFR 712.30(d); 40 CFR 716.120(c)

TSCA - PAIR - Reporting List
Reporting Date: November 19, 1982

**OXIRANE, 2,2'-[(2,2-DIMETHYL-1,3-PROPANEDIYL)   17557-23-2
BIS(OXYMETHYLENE)]BIS-**

ENVIRONMENTAL LISTS

TSCA - Code of Federal Regulations Citations
40 CFR 712.30(d); 40 CFR 716.120(c)

TSCA - PAIR - Reporting List
Reporting Date: November 19, 1982

OXIRANE, DODECYL-   3234-28-4
SEE ALSO:
GLYCIDOL (OXIRANEMETHANOL) AND ITS DERIVATIVES

ENVIRONMENTAL LISTS

EPA - Master Testing List
[present]

TSCA - Code of Federal Regulations Citations
40 CFR 716.120(c)

PROPOSED REGULATIONS

TSCA - Proposed Testing Rule for Glycidyl Ethers
member of Glycidyl subcategory V-A

OXIRANE, [(DODECYLOXY)METHYL]-   2461-18-9
SEE ALSO:
ALKYL EPOXIDES

ENVIRONMENTAL LISTS

TSCA - Code of Federal Regulations Citations
40 CFR 712.30(d); 40 CFR 716.120(c)

TSCA - PAIR - Reporting List
Reporting Date: November 19, 1982

**OXIRANE, 2,2'-[1,2-ETHANEDIYLBIS (OXYMETHY-   2224-15-9
LENE)]BIS-**
SEE ALSO:
GLYCIDOL (OXIRANEMETHANOL) AND ITS DERIVATIVES

ENVIRONMENTAL LISTS

EPA - Master Testing List
[present]

TSCA - Code of Federal Regulations Citations
40 CFR 716.120(c)

PROPOSED REGULATIONS

TSCA - Proposed Testing Rule for Glycidyl Ethers
member of Glycidyl subcategory IV-C

**OXIRANE, 2,2',2",2'''-[1,2-ETHANEDIYLIDENETE-   7328-97-4
TRAKIS-(4,1-PHENYLENEOXYMETHYLENE)]TE-
TRAKIS-**
SEE ALSO:
GLYCIDOL (OXIRANEMETHANOL) AND ITS DERIVATIVES

ENVIRONMENTAL LISTS

EPA - Master Testing List
[present]

TSCA - Code of Federal Regulations Citations
40 CFR 716.120

PROPOSED REGULATIONS

TSCA - Proposed Testing Rule for Glycidyl Ethers
member of Glycidyl subcategory I-B

OXIRANE, [[(2-ETHYLHEXYL)OXY]METHYL]-   2461-15-6
SEE ALSO:
ALKYL EPOXIDES

ENVIRONMENTAL LISTS

TSCA - Code of Federal Regulations Citations
40 CFR 716.120(c)

OXIRANE, HEPTADECYL-   67860-04-2
SEE ALSO:
ALKYL EPOXIDES

HEALTH AND SAFETY LISTS

NFPA - Flash Points
flash point = 5 degrees F (-15 degrees C)

NFPA - Hazard Identification Ratings
health-2; flammability-3; reactivity-2

ENVIRONMENTAL LISTS

TSCA - Code of Federal Regulations Citations
40 CFR 716.120(c)

OXIRANE, HEXADECYL-   7390-81-0
SEE ALSO:
GLYCIDOL (OXIRANEMETHANOL) AND ITS DERIVATIVES

ENVIRONMENTAL LISTS

EPA - Master Testing List
[present]

TSCA - Code of Federal Regulations Citations
40 CFR 716.120(c)

PROPOSED REGULATIONS

TSCA - Proposed Testing Rule for Glycidyl Ethers
subject to subchronic toxicity testing (results apply to all members of Glycidyl subcategory V-A)

OXIRANE, [(HEXADECYLOXY)METHYL]-   15965-99-8
SEE ALSO:
ALKYL EPOXIDES

ENVIRONMENTAL LISTS

TSCA - Code of Federal Regulations Citations
40 CFR 716.120(c)

**OXIRANE, 2,2'-(1,6-HEXANEDIYLBIS(OXYMETHY-   RR-00982-5
LENE)) BIS-**
SEE ALSO:
GLYCIDOL (OXIRANEMETHANOL) AND ITS DERIVATIVES

ENVIRONMENTAL LISTS

EPA - Master Testing List
[present]

TSCA - Code of Federal Regulations Citations
40 CFR 712.30(d); 40 CFR 716.120(c)

TSCA - PAIR - Reporting List
Reporting Date: November 19, 1982

PROPOSED REGULATIONS

TSCA - Proposed Testing Rule for Glycidyl Ethers
member of Glycidyl subcategory II-A

**OXIRANE, 2,2',2"-[1,2,6-HEXANETRIYLTRIS-   68959-23-9
(OXYMETHYLENE)]TRIS-**
SEE ALSO:
GLYCIDOL (OXIRANEMETHANOL) AND ITS DERIVATIVES

ENVIRONMENTAL LISTS

EPA - Master Testing List
[present]

TSCA - Code of Federal Regulations Citations
40 CFR 716.120(c)

PROPOSED REGULATIONS

TSCA - Proposed Testing Rule for Glycidyl Ethers
member of Glycidyl subcategory V-A

**OXIRANEMETHANAMINE, N,N'-[METHYLENEBIS[2-   130728-76-6
ETHYL-4,1-PHENYLENE]BIS[N-OXIRANYLMETHYL]**

SEE ALSO:
GLYCIDOL (OXIRANEMETHANOL) AND ITS DERIVATIVES

ENVIRONMENTAL LISTS

EPA - Master Testing List
[present]

TSCA - Code of Federal Regulations Citations
40 CFR 716.120(c)

PROPOSED REGULATIONS

TSCA - Proposed Testing Rule for Glycidyl Ethers
member of Glycidyl subcategory VI-A

**OXIRANEMETHANAMINE, N-[4-(OXIRANYL-   5026-74-4
METHOXY)PHENYL]-N-(OXIRANYLMETHYL)-**
SEE ALSO:
GLYCIDOL (OXIRANEMETHANOL) AND ITS DERIVATIVES

ENVIRONMENTAL LISTS

EPA - Master Testing List
[present]

TSCA - Code of Federal Regulations Citations
40 CFR 716.120(c)

PROPOSED REGULATIONS

TSCA - Proposed Testing Rule for Glycidyl Ethers
subject to mutagenicity testing (results apply to all members of Glycidyl subcategory I-D)

OXIRANE, METHOXYMETHYL-   930-37-0
SEE ALSO:
ALKYL EPOXIDES

ENVIRONMENTAL LISTS

TSCA - Code of Federal Regulations Citations
40 CFR 716.120(c)

**OXIRANE, 2,2'-[METHYLENEBIS(PHENYLE-   39817-09-9
NEOXYMETHYLENE)]BIS-**

ENVIRONMENTAL LISTS

TSCA - Code of Federal Regulations Citations
40 CFR 716.120(c)

**OXIRANE, 2,2'-[METHYLENEBIS(2,1-PHENYLE-   54208-63-8
NEOXYMETHYLENE)]BIS-**
SEE ALSO:
GLYCIDOL (OXIRANEMETHANOL) AND ITS DERIVATIVES

ENVIRONMENTAL LISTS

EPA - Master Testing List
[present]

TSCA - Code of Federal Regulations Citations
40 CFR 716.120(c)

PROPOSED REGULATIONS

TSCA - Proposed Testing Rule for Glycidyl Ethers
member of Glycidyl subcategory II-A

**OXIRANE, 2,2'-[(1-METHYLETHYLIDINE)BIS[4,   71033-08-4
1-PHENYLENEOXY[1-(BUTOXYMETHYL)-2,1-
ETHANEDIYL]OXYMETHYLENE]BIS-**
ENVIRONMENTAL LISTS

TSCA - Code of Federal Regulations Citations
40 CFR 721.1502

TSCA - Chemicals with Significant New Use Rules
PMN numbers: P-88-2179; P-89-539

TSCA - Section 12(b) - Export Notification
P-88-2179, P-89-539; export notification required - Section 5

**OXIRANE, 2,2'-[(1-METHYLETHYLIDINE)BIS[4,   72319-24-5
1-PHENYLENEOXY-3,1-PROPANEDIYLOXY-4,
1-PHENYLENE(1-METHYLETHYLIDENE)-4,1-
PHENYLENEOXYMETHYLENE]BIS-**
SEE ALSO:
GLYCIDOL (OXIRANEMETHANOL) AND ITS DERIVATIVES

ENVIRONMENTAL LISTS

EPA - Master Testing List
[present]

TSCA - Code of Federal Regulations Citations
40 CFR 716.120(c)

PROPOSED REGULATIONS

TSCA - Proposed Testing Rule for Glycidyl Ethers
member of Glycidyl subcategory V-B

OXIRANE, [(2-METHYLPHENOXY)METHYL]-   2210-79-9

ENVIRONMENTAL LISTS

TSCA - Chemicals with Significant New Use Rules
PMN number: P-91-411

TSCA - Section 12(b) - Export Notification
P-91-411; export notification required - Section 5

**OXIRANE, [[4-(1-METHYL-1-PHENYLETHYL)PHE-   61578-04-9
NOXY]METHYL]-**
ENVIRONMENTAL LISTS

EPA - Master Testing List
[present]

TSCA - Code of Federal Regulations Citations
40 CFR 716.120

PROPOSED REGULATIONS

TSCA - Proposed Testing Rule for Glycidyl Ethers
subject to mutagenicity testing (results apply to all members of Glycidyl subcategory VI-C)

**OXIRANE, METHYL-, POLYMER WITH OXIRANE,   9038-29-3
DECYL ETHER**
SEE ALSO:
GLYCIDOL (OXIRANEMETHANOL) AND ITS DERIVATIVES

ENVIRONMENTAL LISTS

EPA - Master Testing List
[present]

TSCA - Code of Federal Regulations Citations
40 CFR 712.30(d); 40 CFR 716.120(c)

TSCA - PAIR - Reporting List
Reporting Date: November 30, 1982

PROPOSED REGULATIONS

TSCA - Proposed Testing Rule for Glycidyl Ethers
member of Glycidyl subcategory I-A

**OXIRANE METHYL-, POLYMER WITH OXIRANE,   61725-89-1
TRIDECYL ETHER**
SEE ALSO:
GLYCIDOL (OXIRANEMETHANOL) AND ITS DERIVATIVES

ENVIRONMENTAL LISTS

TSCA - Code of Federal Regulations Citations
40 CFR 716.120(c)

OXIRANE, MONO[C(6-12)-ALKYLOXY)METHYL]   68987-80-4
DERIVATIVES
SEE ALSO:
   GLYCIDOL (OXIRANEMETHANOL) AND ITS DERIVATIVES

ENVIRONMENTAL LISTS
   EPA - Master Testing List
   [present]

   TSCA - Code of Federal Regulations Citations
   40 CFR 716.120

PROPOSED REGULATIONS
   TSCA - Proposed Testing Rule for Glycidyl Ethers
   member of Glycidyl subcategory VI-A

OXIRANE, MONO[(C10-16-ALKYLOXY)METHYL]   68081-84-5
DERIVATIVES
SEE ALSO:
   GLYCIDOL (OXIRANEMETHANOL) AND ITS DERIVATIVES

ENVIRONMENTAL LISTS
   EPA - Master Testing List
   [present]

   TSCA - Code of Federal Regulations Citations
   40 CFR 716.120(c)

PROPOSED REGULATIONS
   TSCA - Proposed Testing Rule for Glycidyl Ethers
   member of Glycidyl subcategory VI-A

OXIRANE, [(4-NITROPHENOXY)METHYL]-   5255-75-4
SEE ALSO:
   GLYCIDOL (OXIRANEMETHANOL) AND ITS DERIVATIVES

ENVIRONMENTAL LISTS
   EPA - Master Testing List
   [present]

   TSCA - Code of Federal Regulations Citations
   40 CFR 716.120(c)

PROPOSED REGULATIONS
   TSCA - Proposed Testing Rule for Glycidyl Ethers
   member of Glycidyl subcategory VI-A

OXIRANE, [(4-NONYLPHENOXY)METHYL]-   6178-32-1
SEE ALSO:
   GLYCIDOL (OXIRANEMETHANOL) AND ITS DERIVATIVES

ENVIRONMENTAL LISTS
   EPA - Master Testing List
   [present]

   TSCA - Code of Federal Regulations Citations
   40 CFR 716.120(c)

PROPOSED REGULATIONS
   TSCA - Proposed Testing Rule for Glycidyl Ethers
   subject to neurotoxicity, screening subcategory and mutagenicity testing (results apply to all members of Glycidyl subcategory IV-A)

OXIRANE, [(9-OCTADECENYLOXY)METHYL]-, (Z)-   60501-41-9
SEE ALSO:
   GLYCIDOL (OXIRANEMETHANOL) AND ITS DERIVATIVES

ENVIRONMENTAL LISTS
   EPA - Master Testing List
   [present]

   TSCA - Code of Federal Regulations Citations
   40 CFR 716.120(c)

PROPOSED REGULATIONS
   TSCA - Proposed Testing Rule for Glycidyl Ethers
   member of Glycidyl subcategory IV-A

OXIRANE, [(OCTADECYLOXY)METHYL]-   16245-97-9

ENVIRONMENTAL LISTS
   List of Pesticide Product Inert Ingredients
   [present]

OXIRANEOCTANOIC ACID, 3-OCTYL-, BUTYL   106-83-2
ESTER

ENVIRONMENTAL LISTS
   List of Pesticide Product Inert Ingredients
   [present]

OXIRANEOCTANOIC ACID, 3-OCTYL-, 2-ETHYL-   141-38-8
HEXYL ESTER
SEE ALSO:
   GLYCIDOL (OXIRANEMETHANOL) AND ITS DERIVATIVES

ENVIRONMENTAL LISTS
   EPA - Master Testing List
   [present]

   TSCA - Code of Federal Regulations Citations
   40 CFR 716.120(c)

PROPOSED REGULATIONS
   TSCA - Proposed Testing Rule for Glycidyl Ethers
   member of Glycidyl subcategory I-A

OXIRANEOCTANOIC ACID, 3-OCTYL-, OCTYL   106-84-3
ESTER
SEE ALSO:
   GLYCIDOL (OXIRANEMETHANOL) AND ITS DERIVATIVES

ENVIRONMENTAL LISTS
   EPA - Master Testing List
   [present]

   TSCA - Code of Federal Regulations Citations
   40 CFR 716.120(a)

   TSCA - Health and Safety Reporting List
   Effective Date: October 4, 1982

PROPOSED REGULATIONS
   TSCA - Proposed Testing Rule for Glycidyl Ethers
   subject to subchronic toxicity testing (results apply to all members of Glycidyl subcategory II-A)

OXIRANE, OCTYL-   2404-44-6
SEE ALSO:
   GLYCIDOL (OXIRANEMETHANOL) AND ITS DERIVATIVES

ENVIRONMENTAL LISTS
   EPA - Master Testing List
   [present]

   TSCA - Code of Federal Regulations Citations
   40 CFR 716.120

PROPOSED REGULATIONS
   TSCA - Proposed Testing Rule for Glycidyl Ethers
   member of Glycidyl subcategory IV-B

OXIRANE, 2,2'-(OXIRANYLMETHOXY)-1,3-PHENY-   13561-08-5
LENE]BIS(METHYLENE)]BIS-
SEE ALSO:
   GLYCIDOL (OXIRANEMETHANOL) AND ITS DERIVATIVES

ENVIRONMENTAL LISTS
   EPA - Master Testing List
   [present]

   TSCA - Code of Federal Regulations Citations
   40 CFR 716.120

PROPOSED REGULATIONS
   TSCA - Proposed Testing Rule for Glycidyl Ethers
   member of Glycidyl subcategory IV-A

OXIRANE, 2,2'-[[[(2-OXIRANYLMETHOXY)PHENYL]   67786-03-2
METHYLENE]BIS(4,1-PHENYLENEOXYMETHY-
LENE)]BIS-
SEE ALSO:
   GLYCIDOL (OXIRANEMETHANOL) AND ITS DERIVATIVES

ENVIRONMENTAL LISTS

  EPA - Master Testing List
  [present]

  TSCA - Code of Federal Regulations Citations
  40 CFR 712.30(e); 40 CFR 716.120(c)

PROPOSED REGULATIONS

  TSCA - Proposed Testing Rule for Glycidyl Ethers
  member of Glycidyl subcategory II-B

OXIRANE, PENTADECYL-   22092-38-2
SEE ALSO:
   GLYCIDOL (OXIRANEMETHANOL) AND ITS DERIVATIVES

ENVIRONMENTAL LISTS

  EPA - Master Testing List
  [present]

  TSCA - Code of Federal Regulations Citations
  40 CFR 716.120(c)

PROPOSED REGULATIONS

  TSCA - Proposed Testing Rule for Glycidyl Ethers
  member of Glycidyl subcategory II-A

OXIRANE, 2,2'-[1,4-PHENYLENEBIS (OXYMETHY-   2425-01-6
LENE)]BIS-
ENVIRONMENTAL LISTS

  TSCA - Code of Federal Regulations Citations
  40 CFR 712.30(d)

  TSCA - PAIR - Reporting List
  Reporting Date: November 19, 1982

OXIRANE, 2,2',2''-[1,2,3-PROPANETRIYL TRIS   13236-02-7
(OXYMETHYLENE)]TRIS-
HEALTH AND SAFETY LISTS

  NIOSH - Selected LD50s and LC50s
  Oral, rat: LD50 = 30800 mg/kg

ENVIRONMENTAL LISTS

  TSCA - Code of Federal Regulations Citations
  40 CFR 712.30(d)

  TSCA - PAIR - Reporting List
  Reporting Date: November 19, 1982

OXIRANE, 2,2',2''-[PROPYLIDYNETRIS(4,1-   68517-02-2
PHENYLENEOXYMETHYLENE)]TRIS-
ENVIRONMENTAL LISTS

  TSCA - Code of Federal Regulations Citations
  40 CFR 712.30(d)

  TSCA - PAIR - Reporting List
  Reporting Date: November 19, 1982

OXIRANE, TETRADECYL-   7320-37-8
SEE ALSO:
  ALKYL EPOXIDES

ENVIRONMENTAL LISTS

  TSCA - Code of Federal Regulations Citations
  40 CFR 716.120(c)

OXIRANE, [(TETRADECYLOXY)METHYL]-   38954-75-5
SEE ALSO:
  GLYCIDOL (OXIRANEMETHANOL) AND ITS DERIVATIVES

ENVIRONMENTAL LISTS

EPA - Master Testing List
[present]

TSCA - Code of Federal Regulations Citations
40 CFR 712.30(d); 40 CFR 716.120(c)

TSCA - PAIR - Reporting List
Reporting Date: November 19, 1982

PROPOSED REGULATIONS

TSCA - Proposed Testing Rule for Glycidyl Ethers
member of Glycidyl subcategory VI-C

OXIRANE, (2,2,3,3,4,4,5,5,6,6,7,7,7-TRIDECAFLUORO-   38565-52-5
HEPTYL)-
SEE ALSO:
  GLYCIDOL (OXIRANEMETHANOL) AND ITS DERIVATIVES

ENVIRONMENTAL LISTS

EPA - Master Testing List
[present]

TSCA - Code of Federal Regulations Citations
40 CFR 716.120

PROPOSED REGULATIONS

TSCA - Proposed Testing Rule for Glycidyl Ethers
member of Glycidyl subcategory VI-A

OXIRANE, TRIDECYL-   18633-25-5
SEE ALSO:
  ALKYL EPOXIDES

ENVIRONMENTAL LISTS

TSCA - Code of Federal Regulations Citations
40 CFR 716.120(c)

OXO ALCOHOL STILL BOTTOMS (C8-18 ALCO-   68551-07-5
HOLS)
SEE ALSO:
  GLYCIDOL (OXIRANEMETHANOL) AND ITS DERIVATIVES

ENVIRONMENTAL LISTS

EPA - Master Testing List
[present]

TSCA - Code of Federal Regulations Citations
40 CFR 716.120(a)

PROPOSED REGULATIONS

TSCA - Proposed Testing Rule for Glycidyl Ethers
member of Glycidyl subcategory VI-C

OXO ALCOHOL STILL BOTTOMS, SULFATED,   68130-43-8
SODIUM SALT
SEE ALSO:
  GLYCIDOL (OXIRANEMETHANOL) AND ITS DERIVATIVES

ENVIRONMENTAL LISTS

EPA - Master Testing List
[present]

TSCA - Code of Federal Regulations Citations
40 CFR 712.30(d); 40 CFR 716.120(c)

TSCA - PAIR - Reporting List
Reporting Date: November 19, 1982

PROPOSED REGULATIONS

TSCA - Proposed Testing Rule for Glycidyl Ethers
member of Glycidyl subcategory V-A

OXOSUBSTITUTED AMINOALKANOIC ACID   RR-01568-9
DERIVATIVE
SEE ALSO:
  GLYCIDOL (OXIRANEMETHANOL) AND ITS DERIVATIVES

ENVIRONMENTAL LISTS

EPA - Master Testing List
[present]

TSCA - Code of Federal Regulations Citations
40 CFR 716.120(c)

PROPOSED REGULATIONS

TSCA - Proposed Testing Rule for Glycidyl Ethers
member of Glycidyl subcategory VI-A

OXYALKANEPOLYOL POLYACRYLATE RR-00985-8

ENVIRONMENTAL LISTS

TSCA - Code of Federal Regulations Citations
40 CFR 712.30(d); 40 CFR 716.120(a)

TSCA - PAIR - Reporting List
Reporting Date: November 19, 1982

TSCA - Health and Safety Reporting List
Effective Date: October 4, 1982

OXYBIS(DIBUTYL(2,4,5-TRICHLOROPHENOXY)TIN) 74007-80-0
SEE ALSO:
GLYCIDOL (OXIRANEMETHANOL) AND ITS DERIVATIVES

ENVIRONMENTAL LISTS

EPA - Master Testing List
[present]

TSCA - Code of Federal Regulations Citations
40 CFR 716.120(c)

PROPOSED REGULATIONS

TSCA - Proposed Testing Rule for Glycidyl Ethers
member of Glycidyl subcategory II-A

2,2-OXYBISETHANE BIS(4-METHYLBENZENESUL-FONATE) 7460-82-4
SEE ALSO:
HALOGENATED ALKYL EPOXIDES

ENVIRONMENTAL LISTS

TSCA - Code of Federal Regulations Citations
40 CFR 716.120(c)

1,1'-[OXYBIS(METHYLENESULFONYL)] BIS-2-CHLOROETHANE 53061-10-2

ENVIRONMENTAL LISTS

TSCA - Code of Federal Regulations Citations
40 CFR 716.120(a)

TSCA - Health and Safety Reporting List
Effective Date: October 4, 1982

1,1'-[OXYBIS(METHYLENESULFONYL)] BISETHENE 26750-50-5

ENVIRONMENTAL LISTS

List of Pesticide Product Inert Ingredients
[present]

2,2'-[OXYBIS(METHYLENESULFONYL)] BISETHANOL 36724-43-3

ENVIRONMENTAL LISTS

List of Pesticide Product Inert Ingredients
[present]

10,10'-OXYBISPHENOXARSINE 58-36-6

ENVIRONMENTAL LISTS

TSCA - Chemicals with Significant New Use Rules
PMN number: P-92-692

OXYCHLORDANE 27304-13-8

ENVIRONMENTAL LISTS

TSCA - Chemicals with Significant New Use Rules
PMN number: P-89-1072

OXYDEMETON-METHYL 301-12-2

INTERNATIONAL LISTS

Canada - WHMIS: Ingredient Disclosure
1% item 1230 (1295)

OXYDI-2,1-ETHANEDIYL TETRAKIS(2-CHLOROETHYL) PHOSPHATE 53461-82-8

ENVIRONMENTAL LISTS

TSCA - Chemicals with Significant New Use Rules
PMN number: P-93-1194

TSCA - Section 12(b) - Export Notification
P-93-1194; export notification required - Section 5

N-(OXYDIETHYLENE)BENZOTHIAZOLE-2-SULFENAMIDE 102-77-2

ENVIRONMENTAL LISTS

TSCA - Code of Federal Regulations Citations
40 CFR 712.30(x); 40 CFR 716.120(d)

TSCA - PAIR - Reporting List
Reporting Date: November 27, 1991

TSCA - Health and Safety Reporting List
Effective Date: September 30, 1991

3,3'-OXYDIPROPIONITRILE 1656-48-0

ENVIRONMENTAL LISTS

TSCA - Code of Federal Regulations Citations
40 CFR 712.30(x); 40 CFR 716.120(d)

TSCA - PAIR - Reporting List
Reporting Date: November 27, 1991

TSCA - Health and Safety Reporting List
Effective Date: September 30, 1991

OXYDISULFOTON 2497-07-6

ENVIRONMENTAL LISTS

TSCA - Code of Federal Regulations Citations
40 CFR 712.30(x); 40 CFR 716.120(d)

TSCA - PAIR - Reporting List
Reporting Date: November 27, 1991

TSCA - Health and Safety Reporting List
Effective Date: September 30, 1991

1-OXYETHYL-2-STEARIC IMIDAZOLINE 31866-76-9
SEE ALSO:
ARSENIC

HEALTH AND SAFETY LISTS

NIOSH - Selected LD50s and LC50s
Oral, rat: LD50 = 40 mg/kg

ENVIRONMENTAL LISTS

CERCLA/SARA - Section 302 Extremely Hazardous Substances and TPQs
TPQ = 500/10,000 pounds

STATE LISTS

Massachusetts Right To Know List
extraordinarily hazardous

NJ Right to Know List (Total)
sn 2653

Pennsylvania Right to Know List
environmental hazard

PROPOSED REGULATIONS

CERCLA/SARA - Proposed Hazardous Substance Additions
proposed RQ = 1 pound (.454 kg)

CERCLA/SARA - 1989 Proposed RQ Adjustments
proposed RQ = 100 pounds (45.4 kg)

OXYFLUORFEN   42874-03-3
ENVIRONMENTAL LISTS
ATSDR Priority List
Rank (of 275): 121

OXYGEN   7782-44-7
ENVIRONMENTAL LISTS
CERCLA/SARA - Section 313 - Emission Reporting
form R reporting required
STATE LISTS
California - Directors List of Hazardous Substances (8 CCR 339)
[present]

Massachusetts Right To Know List
[present]

OXYGEN DIFLUORIDE   7783-41-7
ENVIRONMENTAL LISTS
TSCA - PAIR - Reporting List
Effective Date: June 14, 1993
TSCA - Health and Safety Reporting List
Effective Date: June 14, 1993

OXYMETHOLONE   434-07-1
HEALTH AND SAFETY LISTS
NIOSH - Selected LD50s and LC50s
Oral, mouse: LD50 = 1870 mg/kg

OXYTETRACYCLINE   79-57-2
HEALTH AND SAFETY LISTS
NIOSH - Selected LD50s and LC50s
Oral, rat: LD50 = 2830 mg/kg

OXYTETRACYCLINE HYDROCHLORIDE   2058-46-0
HEALTH AND SAFETY LISTS
U.S. DOT - Appendix B - Marine Pollutants
DOT regulated marine pollutant
NIOSH - Selected LD50s and LC50s
Oral, rat: LD50 = 3500 ug/kg Skin, rat: LD50 = 192 mg/kg
ENVIRONMENTAL LISTS
CERCLA/SARA - Section 302 Extremely Hazardous Substances and TPQs
TPQ = 500 pounds
STATE LISTS
Florida Hazardous Substance List
effective March 13, 1992
Massachusetts Right To Know List
extraordinarily hazardous
NJ Right to Know List (Total)
sn 2625
Pennsylvania Right to Know List
environmental hazard
PROPOSED REGULATIONS
CERCLA/SARA - Proposed Hazardous Substance Additions
proposed RQ = 1 pound (.454 kg)
CERCLA/SARA - 1989 Proposed RQ Adjustments
proposed RQ = 100 pounds (45.4 kg)

OYSTER SHELLS   RR-01107-4
ENVIRONMENTAL LISTS
List of Pesticide Product Inert Ingredients
[present]

OZONE   10028-15-6
ENVIRONMENTAL LISTS
CERCLA/SARA - Section 313 - Emission Reporting
form R reporting required

OZONE BY-PRODUCTS   RR-00110-5
HEALTH AND SAFETY LISTS
U.S. DOT - Substances From 49 CFR 172.101
regulated by DOT (UN1073, UN1072, UN1014)
U.S. DOT - Hazard Classes
DOT hazard class = 2.2
ENVIRONMENTAL LISTS
List of Pesticide Product Inert Ingredients
[present]
INTERNATIONAL LISTS
Mexico - Drinking Water - Ecological Criteria
4.0 mg/l; The established levels shall be considered minimums
STATE LISTS
Florida Hazardous Substance List
[present]
Massachusetts Right To Know List
[present]
NJ Right to Know List (Total)
sn 1448
Pennsylvania Right to Know List
[present]

PAINT   RR-01518-9
HEALTH AND SAFETY LISTS
ACGIH 1995 - Ceiling Limits
C 0.05 ppm; C 0.11 mg/m3
AIHA - Odor Threshold Values
no geometric mean air odor threshold
U.S. DOT - Substances From 49 CFR 172.101
regulated by DOT (UN2190)
U.S. DOT - Hazard Classes
DOT hazard class = 2.3
U.S. DOT - Substances Which Are Poisonous by Inhalation
gaseous hazardous material poisonous by inhalation (UN2190)
NIOSH - Selected LD50s and LC50s
Inhalation, rat: LC50 = 136 ppm 1 hr
NIOSH 1990 - Pocket Guide - RELs
C 0.05 ppm; C 0.1 mg/m3
NIOSH 1990 - Pocket Guide - IDLHs
0.5 ppm IDLH
NIOSH 1990 - Pocket Guide - Target organs
lungs, eyes
OSHA - Vacated PELs - Ceiling Limits
C 0.05 ppm; C 0.1 mg/m3 (Enforcement indefinitely stayed)
OSHA - Final PELs - Time Weighted Averages
0.05 ppm TWA; 0.1 mg/m3 TWA
OSHA - List of Highly Hazardous Chemicals
threshhold quantity = 100 pounds
INTERNATIONAL LISTS
Australian Exposure Standards - Time Weighted Averages
Peak Limitation: 0.05 ppm; 0.11 mg/m3
Canada - WHMIS: Ingredient Disclosure
1% item 1231 (710)
Canada - Alberta - Ceiling Occupational Exposure Limit
C 0.05 ppm; C 0.11 mg/m3
Canada - British Columbia - 8 Hour Exposure Limits
0.05 ppm TWA; 0.1 mg/m3 TWA

Canada - British Columbia - 15 Minute Exposure Limits
0.15 ppm STEL; 0.3 mg/m3 STEL

Canada - Ontario - OHSA - CEVs
0.05 ppm CEV; 0.1 mg/m3 CEV

Canada - Quebec - Ceiling Limits
P 0.05 ppm; P 0.11 mg/m3

Israel - Ceiling Exposure Limits
C 0.05 ppm; C 0.11 mg/m3

Mexico - Instruction No. 10 - TWAs
0.05 ppm TWA; 0.1 mg/m3 TWA

Mexico - Instruction No. 10 - STELs
0.15 ppm STEL; 0.3 mg/m3 STEL

STATE LISTS

California - Exposure Limits - Ceilings
C 0.05 ppm; C 0.1 mg/m3

California - Directors List of Hazardous Substances (8 CCR 339)
[present]

Florida Hazardous Substance List
[present]

Massachusetts Right To Know List
[present]

Minnesota Hazardous Substance List
[present]

NJ Right to Know List (Total)
sn 1449

Pennsylvania Right to Know List
[present]

PAINT MANUFACTURE AND PAINTING   RR-00316-7

HEALTH AND SAFETY LISTS

IARC - Group Unspecified
[present] (Listed under 'Androgenic (anabolic) steroids')

NTP Chemical Status Reports - Testing Status and NTIS Number
Two year studies in progress; short term toxicity studies scheduled for peer review; prechronic studies for which toxicity technical reports were not prepared

NTP Seventh Report - Suspect Carcinogens
suspect carcinogen

OSHA - Possible Select Carcinogens
[present]

STATE LISTS

California - Air Bill 2588 Appendix A-II
known or potential carcinogen

California - Prop. 65 - Cancer list
carcinogen - initial date 4/1/88

Florida Hazardous Substance List
[present]

Massachusetts Right To Know List
carcinogen; extraordinarily hazardous

Minnesota Hazardous Substance List
carcinogen

NJ Right to Know List (Total)
sn 1450

NJ Special Hazardous Substances
(carcinogen)

Pennsylvania Right to Know List
special hazardous substance

Pennsylvania RTK - Special Hazardous Substances
[present]

PALLADIUM   7440-05-3

STATE LISTS

California - Air Bill 2588 Appendix A-II
6/91

California - Prop. 65 - Developmental Toxicity
developmental toxicity (internal use) - initial date 1/1/91

PALLADIUM 100   15690-69-4

HEALTH AND SAFETY LISTS

NTP Chemical Status Reports - Testing Status and NTIS Number
Technical reports printed (PB87204103/AS)

NTP Chemical Status Reports - Evidence of Carcinogenicity
male rat-equivocal evidence; female rat-equivocal evidence; male mice-no evidence; female mice-no evidence

STATE LISTS

California - Prop. 65 - Developmental Toxicity
developmental toxicity (internal use) - initial date 10/1/91

PALLADIUM 101   15749-54-9

ENVIRONMENTAL LISTS

List of Pesticide Product Inert Ingredients
[present]

PALLADIUM 103   14967-68-1

HEALTH AND SAFETY LISTS

ACGIH 1995 - Time Weighted Averages
(-) ppm TWA; (-) mg/m3 TWA

ACGIH 1995 - Ceiling Limits
(C 0.1) ppm; (C 0.20) mg/m3

AIHA - Odor Threshold Values
no geometric mean air odor threshold

NIOSH - Selected LD50s and LC50s
Inhalation, rat: LC50 = 4800 ppb 4 hr

NIOSH 1990 - Pocket Guide - RELs
C 0.1 ppm; C 0.2 mg/m3

NIOSH 1990 - Pocket Guide - IDLHs
10 ppm IDLH

NIOSH 1990 - Pocket Guide - Target organs
eyes, respiratory system

NTP Chemical Status Reports - Testing Status and NTIS Number
with or without NNK: Galley or camera copy technical reports in progress

OSHA - Vacated PELs - Time Weighted Averages
0.1 ppm TWA; 0.2 mg/m3 TWA

OSHA - Vacated PELs - Short Term Exposure Limits
0.3 ppm STEL; 0.6 mg/m3 STEL

OSHA - Final PELs - Time Weighted Averages
0.1 ppm TWA; 0.2 mg/m3 TWA

OSHA - List of Highly Hazardous Chemicals
threshhold quantity = 100 pounds

ENVIRONMENTAL LISTS

CERCLA/SARA - Section 302 Extremely Hazardous Substances and TPQs
TPQ = 100 pounds

CERCLA/SARA - Section 313 - Emission Reporting
form R reporting required

INTERNATIONAL LISTS

Australian Exposure Standards - Time Weighted Averages
0.1 ppm TWA; 0.2 mg/m3 TWA

Canada - WHMIS: Ingredient Disclosure
1% item 1232 (1332)

Canada - Alberta - 8 Hour Occupational Exposure Limit
0.1 ppm TWA; 0.2 mg/m3 TWA

Canada - Alberta - 15 Minute Occupational Exposure Limit
0.3 ppm STEL; 0.59 mg/m3 STEL

Canada - British Columbia - 8 Hour Exposure Limits
0.1 ppm TWA; 0.2 mg/m3 TWA

Canada - British Columbia - 15 Minute Exposure Limits
0.3 ppm STEL; 0.6 mg/m3 STEL
Canada - Ontario - OHSA - TWAEVs
0.1 ppm TWAEV; 0.2 mg/m3 TWAEV
Canada - Ontario - OHSA - STEVs
0.3 ppm STEV; 0.6 mg/m3 STEV
Canada - Quebec - Ceiling Limits
P 0.1 ppm; P 0.2 mg/m3
United Kingdom - Occupational Exposure Standards - TWAs
0.1 ppm TWA; 0.2 mg/m3 TWA
United Kingdom - Occupational Exposure Standards - STELs
0.3 ppm STEL; 0.6 mg/m3 STEL
German (DFG) - MAK Values
0.1 ppm MAK; 0.2 mg/m3 MAK
German (DFG) - Peak Limitations
2 x normal MAK (5 min momentary value);   don't exceed 8 times during shift
Israel - Ceiling Exposure Limits
C 0.1 ppm; C 0.20 mg/m3
Mexico - Instruction No. 10 - TWAs
0.1 ppm TWA; 0.2 mg/m3 TWA
Mexico - Instruction No. 10 - STELs
0.3 ppm STEL; 0.6 mg/m3 STEL

STATE LISTS

California - Exposure Limits - PELs
0.1 ppm PEL; 0.2 mg/m3 PEL
California - Exposure Limits - STELs
0.3 ppm STEL; 0.6 mg/m3 STEL
California - Directors List of Hazardous Substances (8 CCR 339)
[present] (see documentation for criteria of ozone)
Florida Hazardous Substance List
[present]
Massachusetts Right To Know List
extraordinarily hazardous
Minnesota Hazardous Substance List
[present]
NJ Right to Know List (Total)
sn 1451
Pennsylvania Right to Know List
environmental hazard

PROPOSED REGULATIONS

ACGIH 1995 - Notice of Intended Changes
0.05 ppm TWA; 0.1 mg/m3 TWA; 0.2 ppm STEL; 0.4 mg/m3 STEL
CERCLA/SARA - Proposed Hazardous Substance Additions
proposed RQ = 1 pound (.454 kg)
CERCLA/SARA - 1989 Proposed RQ Adjustments
proposed RQ = 10 pounds (4.54 kg)

PALLADIUM 107   17637-99-9

PROPOSED REGULATIONS

Safe Drinking Water Act - Priority list
[present]

PALLADIUM 109   14981-64-7

STATE LISTS

NJ Right to Know List (Total)
flammable liquid, enamels, lacquers, stains: sn 2628; corrosive liquid: sn 26 27

PALLADIUM(II) CHLORIDE   7647-10-1

HEALTH AND SAFETY LISTS

IARC - Group 1 (carcinogenic to humans)
[present] (occupational exposures as a painter)

OSHA - Select Carcinogens
[present]

PALLADIUM DINITRATE   10102-05-3

STATE LISTS

California - Directors List of Hazardous Substances (8 CCR 339)
[present]

PALMITIC ACID   57-10-3

HEALTH AND SAFETY LISTS

U.S. DOT - Appendix A Table 2 - Radionuclides
final RQ = 100 curies (3.7E 12 Bq)

ENVIRONMENTAL LISTS

CERCLA/SARA List of Radionuclides (Appendix B) and Their Reportable Quantities
final RQ = 100 curies (3.7E 12 Bq)

PALM KERNEL OIL   8023-79-8

HEALTH AND SAFETY LISTS

U.S. DOT - Appendix A Table 2 - Radionuclides
final RQ = 100 curies (3.7E 12 Bq)

ENVIRONMENTAL LISTS

CERCLA/SARA List of Radionuclides (Appendix B) and Their Reportable Quantities
final RQ = 100 curies (3.7E 12 Bq)

PALM OIL   8002-75-3

HEALTH AND SAFETY LISTS

U.S. DOT - Appendix A Table 2 - Radionuclides
final RQ = 100 curies (3.7E 12 Bq)

ENVIRONMENTAL LISTS

CERCLA/SARA List of Radionuclides (Appendix B) and Their Reportable Quantities
final RQ = 100 curies (3.7E 12 Bq)

PANCREATIN   8049-47-6

HEALTH AND SAFETY LISTS

U.S. DOT - Appendix A Table 2 - Radionuclides
final RQ = 100 curies (3.7E 12 Bq)

ENVIRONMENTAL LISTS

CERCLA/SARA List of Radionuclides (Appendix B) and Their Reportable Quantities
final RQ = 100 curies (3.7E 12 Bq)

PANFURAN S   794-93-4

HEALTH AND SAFETY LISTS

U.S. DOT - Appendix A Table 2 - Radionuclides
final RQ = 1000 curies (3.7E 13 Bq)

ENVIRONMENTAL LISTS

CERCLA/SARA List of Radionuclides (Appendix B) and Their Reportable Quantities
final RQ = 1000 curies (3.7E 13 Bq)

PAPAIN   9001-73-4

HEALTH AND SAFETY LISTS

NIOSH - Selected LD50s and LC50s
Oral, rat: LD50 = 2704 mg/kg

ENVIRONMENTAL LISTS

TSCA - Code of Federal Regulations Citations
40 CFR 712.30(w)
TSCA - PAIR - Reporting List
Reporting Date: July 13, 1988

INTERNATIONAL LISTS

Canada - WHMIS: Ingredient Disclosure
1% item 1233 (513)

PAPER   RR-01108-5

INTERNATIONAL LISTS

Canada - WHMIS: Ingredient Disclosure
1% item 1234 (750)

PAPER, UNSATURATED OIL TREATED, INCOM-   RR-01503-2
PLETELY DRY

ENVIRONMENTAL LISTS

EPA - Master Testing List
[present]

List of Pesticide Product Inert Ingredients
[present]

PROPOSED REGULATIONS

TSCA - Proposed Substances for Developmental/Reproductive Testing
proposed testing for: Developmental Toxicity - oral

PAPRIKA   68991-42-4

HEALTH AND SAFETY LISTS

NFPA - Flash Points
flash point = 398 degrees F (203 degrees C)

NFPA - Hazard Identification Ratings
health-0; flammability-1; reactivity-0

PARA-ARAMIDE FIBRES (KEVLAR, TWARON)   RR-01805-3

HEALTH AND SAFETY LISTS

NFPA - Flash Points
flash point = 323 degrees F (162 degrees C)

NFPA - Hazard Identification Ratings
health-0; flammability-1; reactivity-0

PARAFFINIC DISTILLATE SOLVENT EXTRACT   64742-04-7

STATE LISTS

NJ Right to Know List (Total)
sn 1452

PARAFFIN OIL   8012-45-1

HEALTH AND SAFETY LISTS

IARC - Group 2B (sufficient animal data)
*[present] (Overall evaluation based only on evidence of carcinogenicity
in monograph (24, 1980) or in Supplement 4)*
NIOSH - Selected LD50s and LC50s
Oral, mouse: LD50 = 2690 mg/kg
OSHA - Possible Select Carcinogens
[present]

STATE LISTS

California - Air Bill 2588 Appendix A-II
known or potential carcinogen
California - Prop. 65 - Cancer list
carcinogen - initial date 1/1/88
California - Directors List of Hazardous Substances (8 CCR 339)
[present]
Florida Hazardous Substance List
[present]
Massachusetts Right To Know List
carcinogen; extraordinarily hazardous
Minnesota Hazardous Substance List
carcinogen
Pennsylvania Right to Know List
special hazardous substance

Pennsylvania RTK - Special Hazardous Substances
[present]

PARAFFIN OILS, CATALYTIC DEWAXED LIGHT   64742-71-8

ENVIRONMENTAL LISTS

List of Pesticide Product Inert Ingredients
[present]

PARAFFIN OILS (PETROLEUM), CATALYTIC DE-   64742-70-7
WAXED HEAVY

HEALTH AND SAFETY LISTS

U.S. DOT - Substances From 49 CFR 172.101
regulated by DOT (UN1379)
U.S. DOT - Hazard Classes
DOT hazard class = 4.2

ENVIRONMENTAL LISTS

List of Pesticide Product Inert Ingredients
[present]

PARAFFIN WAXES AND HYDROCARBON WAXES   8002-74-2

STATE LISTS

NJ Right to Know List (Total)
sn 2629

PARAFORMALDEHYDE   30325-89-4

ENVIRONMENTAL LISTS

List of Pesticide Product Inert Ingredients
[present]

PARAFORMALDEHYDE   30525-89-4

INTERNATIONAL LISTS

Canada - Quebec - Time-Weighted Average Exposure Values
1 fibre/cm3 TWAEV

PARALDEHYDE   123-63-7

STATE LISTS

Massachusetts Right To Know List
carcinogen; extraordinarily hazardous

PARAMETHADIONE   115-67-3

HEALTH AND SAFETY LISTS

NFPA - Flash Points
flash point = 300 to 450 degrees F (149 to 232 degrees C)
NFPA - Hazard Identification Ratings
health-0; flammability-1; reactivity-0

PARAQUAT   1910-42-5

STATE LISTS

Massachusetts Right To Know List
carcinogen; extraordinarily hazardous

PARAQUAT   4685-14-7

STATE LISTS

Massachusetts Right To Know List
carcinogen; extraordinarily hazardous

PARAQUAT METHOSULFATE   2074-50-2

HEALTH AND SAFETY LISTS

ACGIH 1995 - Time Weighted Averages
2 mg/m3 TWA
NFPA - Flash Points
flash point = 390 degrees F (199 degrees C)
NFPA - Hazard Identification Ratings
health-0; flammability-1; reactivity-0

OSHA - Vacated PELs - Time Weighted Averages
2 mg/m3 TWA

ENVIRONMENTAL LISTS

List of Pesticide Product Inert Ingredients
[present]

INTERNATIONAL LISTS

Australian Exposure Standards - Time Weighted Averages
2 mg/m3 TWA

Canada - Alberta - 8 Hour Occupational Exposure Limit
2 mg/m3 TWA

Canada - Alberta - 15 Minute Occupational Exposure Limit
6 mg/m3 STEL

Canada - British Columbia - 8 Hour Exposure Limits
2 mg/m3 TWA

Canada - British Columbia - 15 Minute Exposure Limits
6 mg/m3 STEL

Canada - Ontario - OHSA - TWAEVs
2 mg/m3 TWAEV (listed as an agent of variable composition)

Canada - Quebec - Time-Weighted Average Exposure Values
2 mg/m3 TWAEV

United Kingdom - Occupational Exposure Standards - TWAs
fume: 2 mg/m3 TWA

United Kingdom - Occupational Exposure Standards - STELs
fume: 6 mg/m3 STEL

Israel - Time Weighted Averages
2 mg/m3 TWA

Israel - Action Levels
1 mg/m3 AL

Mexico - Instruction No. 10 - TWAs
2 mg/m3 TWA

Mexico - Instruction No. 10 - STELs
6 mg/m3 STEL

STATE LISTS

California - Exposure Limits - PELs
fume: 2 mg/m3 PEL

California - Directors List of Hazardous Substances (8 CCR 339)
[present]

Florida Hazardous Substance List
[present]

Massachusetts Right To Know List
[present]

Minnesota Hazardous Substance List
[present]

Pennsylvania Right to Know List
[present]

PARASORBIC ACID   10048-32-5

ENVIRONMENTAL LISTS

List of Pesticide Product Inert Ingredients
[present]

PARATHION   56-38-2

HEALTH AND SAFETY LISTS

U.S. DOT - Substances From 49 CFR 172.101
regulated by DOT (UN2213)

U.S. DOT - Hazard Classes
DOT hazard class = 4.1

U.S. DOT - Appendix A Table 1 - Hazardous Substances
final RQ = 1000 pounds (454 kg)

NFPA - Flash Points
flash point = 158 degrees F (70 degrees C)

NFPA - Hazard Identification Ratings
health-3; flammability-1; reactivity-0

NIOSH - Selected LD50s and LC50s
Oral, rat: LD50 = 800 mg/kg

ENVIRONMENTAL LISTS

CERCLA/SARA - Hazardous Substances and their Reportable Quantities
final RQ = 1000 pounds (454 kg)

CAA - HON Rule - SOCMI Chemicals
compliance by Oct. 24, 1994

Clean Water Act - Hazardous Substances
[present]

INTERNATIONAL LISTS

Canada - WHMIS: Ingredient Disclosure
1% item 1235 (1333)

STATE LISTS

California - Directors List of Hazardous Substances (8 CCR 339)
[present]

Massachusetts Right To Know List
[present]

NJ Right to Know List (Total)
sn 1454

Pennsylvania Right to Know List
environmental hazard

PARATHION AND COMPRESSED GAS MIXTURE   RR-00411-5
SEE ALSO:
   K009-HAZARDOUS WASTES
   K010-HAZARDOUS WASTES
   K026-HAZARDOUS WASTES

HEALTH AND SAFETY LISTS

U.S. DOT - Substances From 49 CFR 172.101
regulated by DOT (UN1264)

U.S. DOT - Hazard Classes
DOT hazard class = 3

U.S. DOT - Appendix A Table 1 - Hazardous Substances
final RQ = 1000 pounds (454 kg)

NFPA - Flash Points
flash point = 96 degrees F (36 degrees C)

NFPA - Hazard Identification Ratings
health-2; flammability-3; reactivity-1

NIOSH - Selected LD50s and LC50s
Oral, rat: LD50 = 1530 mg/kg Skin, rabbit: LD50 = 14 gm/kg

ENVIRONMENTAL LISTS

CERCLA/SARA - Section 313 - Emission Reporting
form R reporting required

CERCLA/SARA - Hazardous Substances and their Reportable Quantities
final RQ = 1000 pounds (454 kg)

CAA - HON Rule - SOCMI Chemicals
compliance by Jan. 23, 1995

RCRA - U Series Wastes
waste number U182

RCRA - Hazardous Constituents-Appendix VIII
waste number U182

RCRA - Basis for Listing - Appendix VII
Included in waste streams: K009, K010, K026

RCRA - Substances Banned From Land Disposal
[present]

INTERNATIONAL LISTS

Canada - WHMIS: Ingredient Disclosure
0.1% item 1236 (1334)

STATE LISTS

Florida Hazardous Substance List
[present]

Massachusetts Right To Know List
[present]
NJ Right to Know List (Total)
sn 1455
NJ Special Hazardous Substances
(flammable - third degree)
Pennsylvania Right to Know List
environmental hazard

PARSLEY APIOLE   8000-68-8

STATE LISTS

California - Air Bill 2588 Appendix A-II
9/90

California - Prop. 65 - Developmental Toxicity
developmental toxicity - initial date 7/1/90

Massachusetts Right To Know List
teratogen

PATULIN   149-29-1
SEE ALSO:
   PARAQUAT

HEALTH AND SAFETY LISTS

NIOSH - Selected LD50s and LC50s
Oral, rat: LD50 = 57 mg/kg Skin, rat: LD50 = 80 mg/kg

NIOSH 1990 - Pocket Guide - RELs
respirable dust: 0.1 mg/m3 TWA

NIOSH 1990 - Pocket Guide - IDLHs
1.5 mg/m3 IDLH

NIOSH 1990 - Pocket Guide - Target organs
eyes, respiratory system, heart, liver, kidneys, GI tract

NIOSH 1990 - Pocket Guide - Skin list
Potential for dermal absorption

OSHA - Vacated PELs - Time Weighted Averages
respirable dust: 0.1 mg/m3 TWA (Listed under "Paraquat")

OSHA - Vacated PELs - Skin Designation
Prevent or reduce skin absorption (Listed under 'Paraquat')

OSHA - Final PELs - Time Weighted Averages
0.5 mg/m3 TWA (Listed under 'Paraquat')

OSHA - Final PELs - Skin Notations
prevent or reduce skin absorption

ENVIRONMENTAL LISTS

CERCLA/SARA - Section 302 Extremely Hazardous Substances and TPQs
TPQ = 10/10,000 pounds

CERCLA/SARA - Section 313 - Emission Reporting
form R reporting required

INTERNATIONAL LISTS

Canada - Drinking Water Quality - IMACs
0.01 mg/L IMAC

Canada - Alberta - 8 Hour Occupational Exposure Limit
0.1 mg/m3 TWA

Canada - Alberta - 15 Minute Occupational Exposure Limit
0.3 mg/m3 STEL

Canada - British Columbia - 8 Hour Exposure Limits
respirable sizes: 0.1 mg/m3 TWA

Canada - British Columbia - 15 Minute Exposure Limits
respirable sizes: 0.3 mg/m3 STEL

Canada - Ontario - OHSA - TWAEVs
0.1 mg/m3 TWAEV

Canada - Quebec - Time-Weighted Average Exposure Values
0.1 mg/m3 TWAEV

United Kingdom - Occupational Exposure Standards - TWAs
respirable dust: 0.1 mg/m3 TWA

German (DFG) - MAK Values
total dust: 0.1 mg/m3 MAK

German (DFG) - Peak Limitations
2 x normal MAK (5 min momentary value); don't exceed 8 times during shift

German (DFG) - Skin/Sensitizers
danger of cutaneous absorption

Mexico - Instruction No. 10 - TWAs
0.1 mg/m3 TWA

STATE LISTS

California - Exposure Limits - PELs
total particulates: 0.5 mg/m3 PEL; respirable sizes: 0.1 mg/m3 PEL

California - Exposure Limits - Skin Notation
material may be absorbed through the skin, eyes or mucous membrane

California - Directors List of Hazardous Substances (8 CCR 339)
[present]

Florida Hazardous Substance List
[present]

Massachusetts Right To Know List
extraordinarily hazardous

Minnesota Hazardous Substance List
skin

NJ Right to Know List (Total)
sn 1458

Pennsylvania Right to Know List
environmental hazard

PROPOSED REGULATIONS

CERCLA/SARA - Proposed Hazardous Substance Additions
proposed RQ = 1 pound (.454 kg)

CERCLA/SARA - 1989 Proposed RQ Adjustments
proposed RQ = 10 pounds (4.54 kg)

PEANUT BUTTER   RR-01109-6

HEALTH AND SAFETY LISTS

ACGIH 1995 - Time Weighted Averages
total dust: 0.5 mg/m3 TWA; respirable fraction: 0.1 mg/m3 TWA

NIOSH - Selected LD50s and LC50s
Oral, rat: LD50 = 150 mg/kg

OSHA - Vacated PELs - Time Weighted Averages
respirable dust: 0.1 mg/m3 TWA

OSHA - Vacated PELs - Skin Designation
Prevent or reduce skin absorption

OSHA - Final PELs - Time Weighted Averages
0.5 mg/m3 TWA

OSHA - Final PELs - Skin Notations
prevent or reduce skin absorption

INTERNATIONAL LISTS

Australian Exposure Standards - Time Weighted Averages
respirable sizes: 0.1 mg/m3 TWA

Australian Exposure Standards - Under Review
exposure limits under review

Israel - Time Weighted Averages
respirable sizes: 0.1 mg/m3 TWA

Israel - Action Levels
respirable sizes: 0.05 mg/m3 AL

STATE LISTS

Massachusetts Right To Know List
[present]

Minnesota Hazardous Substance List
[present]

Pennsylvania Right to Know List
[present]

PEANUT OIL

SEE ALSO:

PARAQUAT

8002-03-7

HEALTH AND SAFETY LISTS

NIOSH - Selected LD50s and LC50s
Oral, rat: LD50 = 100 mg/kg

OSHA - Vacated PELs - Time Weighted Averages
respirable dust: 0.1 mg/m3 TWA (Listed under 'Paraquat')

OSHA - Vacated PELs - Skin Designation
Prevent or reduce skin absorption (Listed under 'Paraquat')

OSHA - Final PELs - Time Weighted Averages
0.5 mg/m3 TWA (Listed under 'Paraquat')

OSHA - Final PELs - Skin Notations
prevent or reduce skin absorption

ENVIRONMENTAL LISTS

CERCLA/SARA - Section 302 Extremely Hazardous Substances and TPQs
TPQ = 10/10,000 pounds

STATE LISTS

California - Exposure Limits - PELs
total particulates: 0.5 mg/m3 PEL; respirable sizes: 0.1 mg/m3 PEL (Listed under 'Paraquat')

California - Exposure Limits - Skin Notation
material may be absorbed through the skin, eyes or mucous membrane (Listed under 'Paraquat')

California - Directors List of Hazardous Substances (8 CCR 339)
[present] (Listed under 'Paraquat')

Florida Hazardous Substance List
effective March 13, 1992

Massachusetts Right To Know List
extraordinarily hazardous

NJ Right to Know List (Total)
sn 2630

Pennsylvania Right to Know List
environmental hazard

PROPOSED REGULATIONS

CERCLA/SARA - Proposed Hazardous Substance Additions
proposed RQ = 1 pound (.454 kg)

CERCLA/SARA - 1989 Proposed RQ Adjustments
proposed RQ = 10 pounds (4.54 kg)

PEANUTS

68476-82-4

HEALTH AND SAFETY LISTS

IARC - Group 3 (not classifiable)
[present]

STATE LISTS

California - Directors List of Hazardous Substances (8 CCR 339)
[present]

PEANUT SHELLS

SEE ALSO:

F039-HAZARDOUS WASTES

RR-01110-9

HEALTH AND SAFETY LISTS

ACGIH 1995 - Time Weighted Averages
0.1 mg/m3 TWA

ACGIH 1995 - Skin Designations
skin - potential for cutaneous absorption

ACGIH 1995 - Biological Exposure Indices
Total p-Nitrophenol in urine: 0.5 mg/g creatinine, end of shift (Ns); Chlolinesterase activity in red cells: 70% of an individual's baseline, discretionary (B, Ns, Sq)

U.S. DOT - Substances From 49 CFR 172.101
regulated by DOT (NA2783)

U.S. DOT - Hazard Classes
DOT hazard class = 6.1

U.S. DOT - Substances Which Are Poisonous by Inhalation
gaseous hazardous material poisonous by inhalation (when mixed with compressed gas) (NA1967)

U.S. DOT - Appendix B - Marine Pollutants
DOT regulated severe marine pollutant

U.S. DOT - Appendix A Table 1 - Hazardous Substances
final RQ = 10 pounds (4.54 kg)

IARC - Group 3 (not classifiable)
[present]

NIOSH - Selected LD50s and LC50s
Inhalation, rat: LC50 = 84 mg/m3 4 hr Oral, rat: LD50 = 2 mg/kg Skin, rat: LD50 = 6800 ug/kg

NIOSH 1990 - Pocket Guide - RELs
0.05 mg/m3 TWA

NIOSH 1990 - Pocket Guide - IDLHs
20 mg/m3 IDLH

NIOSH 1990 - Pocket Guide - Target organs
respiratory system, CNS, CVS, eyes, skin, blood cholinesterase

NIOSH 1990 - Pocket Guide - Skin list
Potential for dermal absorption

NIOSH - Health Standards - Exposure Limits
0.05 mg/m3 TWA

NIOSH - Health Standards - Health Effects and Precautions
Nervous system effects (Prevent skin contact; blood monitoring required)

NTP Chemical Status Reports - Testing Status and NTIS Number
Technical reports printed (PB288803/AS)

NTP Chemical Status Reports - Evidence of Carcinogenicity
male rat-equivocal; female rat-equivocal; male mice-negative; female mice-negative

OSHA - Vacated PELs - Time Weighted Averages
0.1 mg/m3 TWA

OSHA - Vacated PELs - Skin Designation
Prevent or reduce skin absorption

OSHA - Final PELs - Time Weighted Averages
0.1 mg/m3 TWA

OSHA - Final PELs - Skin Notations
prevent or reduce skin absorption

ENVIRONMENTAL LISTS

ATSDR Priority List
Rank (of 275): 138

CERCLA/SARA - Section 302 Extremely Hazardous Substances and TPQs
TPQ = 100 pounds

CERCLA/SARA - Section 313 - Emission Reporting
form R reporting required for 1.0% de minimus concentration

CERCLA/SARA - Hazardous Substances and their Reportable Quantities
final RQ = 10 pounds (4.54 kg)

Clean Air Act (1990) - List of Hazardous Air Contaminants
[present]

Clean Water Act - Hazardous Substances
[present]

RCRA - P Series Wastes
waste number P089

RCRA - Hazardous Constituents-Appendix VIII
waste number P089

RCRA - Basis for Listing - Appendix VII
Included in waste stream: F039

RCRA - Substances Banned From Land Disposal
[present]

RCRA - TSD Facilities Ground Water Monitoring
TM 8270 = 10 ug/L PQL

RCRA - Universal Treatment Standards (LDR)
WW: 0.014 mg/l; NWW: 4.6 mg/kg

INTERNATIONAL LISTS

Australian Exposure Standards - Time Weighted Averages
0.1 mg/m3 TWA

Australian Exposure Standards - Skin Effects
skin absorption

Canada - Drinking Water Quality - MACs
0.05 mg/L MAC

Canada - Alberta - 8 Hour Occupational Exposure Limit
0.1 mg/m3 TWA

Canada - Alberta - 15 Minute Occupational Exposure Limit
0.3 mg/m3 STEL

Canada - Alberta - Skin Designation
can be absorbed through the intact skin

Canada - British Columbia - 8 Hour Exposure Limits
0.1 mg/m3 TWA

Canada - British Columbia - 15 Minute Exposure Limits
0.3 mg/m3 STEL

Canada - British Columbia - Skin Notations
skin - potential for skin absorption

Canada - Ontario - OHSA - TWAEVs
0.05 mg/m3 TWAEV

Canada - Ontario - OHSA - Skin Notations
absorption through skin, eyes, or mucous membranes

Canada - Quebec - Time-Weighted Average Exposure Values
0.1 mg/m3 TWAEV

Canada - Quebec - Skin Designations
absorbed through the skin

United Kingdom - Occupational Exposure Standards - TWAs
0.1 mg/m3 TWA

United Kingdom - Occupational Exposure Standards - STELs
0.3 mg/m3 STEL

United Kingdom - Occupational Exposure Standards - Notes
can be absorbed through skin

German (DFG) - MAK Values
total dust: 0.1 mg/m3 MAK

German (DFG) - Skin/Sensitizers
danger of cutaneous absorption

German (DFG) - Pregnancy
classification not yet possible

Israel - Time Weighted Averages
0.1 mg/m3 TWA

Israel - Action Levels
0.05 mg/m3 AL

Mexico - Instruction No. 10 - TWAs
0.1 mg/m3 TWA

Mexico - Instruction No. 10 - STELs
0.3 mg/m3 STEL

Mexico - Instruction No. 10 - Skin designation
skin - potential for cutaneous absorption

Mexico - Drinking Water - Ecological Criteria
0.00003 mg/l

STATE LISTS

California - Air Bill 2588 Appendix A-I
6/91

California - Exposure Limits - PELs
0.1 mg/m3 PEL

California - Exposure Limits - Skin Notation
material may be absorbed through the skin, eyes or mucous membrane

California - Directors List of Hazardous Substances (8 CCR 339)
[present]

Florida Hazardous Substance List
[present]

Massachusetts Right To Know List
extraordinarily hazardous; neurotoxin

Minnesota Hazardous Substance List
skin

NJ Right to Know List (Total)
sn 1459

Pennsylvania Right to Know List
environmental hazard

PEAT MOSS   RR-01111-0

HEALTH AND SAFETY LISTS

U.S. DOT - Substances From 49 CFR 172.101
regulated by DOT (NA1967)

U.S. DOT - Hazard Classes
DOT hazard class = 2.3

STATE LISTS

NJ Right to Know List (Total)
sn 2631

PEBULATE [BUTYLETHYLCARBAMOTHIOIC ACID   1114-71-2
S-PROPYL ESTER]

ENVIRONMENTAL LISTS

List of Pesticide Product Inert Ingredients
[present]

PECAN SHELL FLOUR   RR-01112-1

HEALTH AND SAFETY LISTS

IARC - Group 3 (not classifiable)
[present]

PECTIN   9000-69-5

ENVIRONMENTAL LISTS

List of Pesticide Product Inert Ingredients
[present]

PEG (400) DIOLEATE   9005-07-6

HEALTH AND SAFETY LISTS

NFPA - Flash Points
flash point = 540 degrees F (282 degrees C)

NFPA - Hazard Identification Ratings
health-0; flammability-1; reactivity-0

ENVIRONMENTAL LISTS

List of Pesticide Product Inert Ingredients
[present]

STATE LISTS

Pennsylvania Right to Know List
[present]

PELARGONIC ACID   112-05-0

ENVIRONMENTAL LISTS

List of Pesticide Product Inert Ingredients
[present]

PENDIMETHALIN [N-(1-ETHYLPROPYL)-3,4-   40487-42-1
DIMETHYL-2,6-DINITROBENZENEAMINE]

ENVIRONMENTAL LISTS

List of Pesticide Product Inert Ingredients
[present]

D-PENICILLAMINE   52-67-5

ENVIRONMENTAL LISTS

List of Pesticide Product Inert Ingredients
[present]

PENICILLIC ACID   90-65-3

ENVIRONMENTAL LISTS

CERCLA/SARA - Section 313 - Emission Reporting
form R reporting required

PENICILLIN VK   132-98-9

ENVIRONMENTAL LISTS

List of Pesticide Product Inert Ingredients
[present]

PENTABORANE   19624-22-7

ENVIRONMENTAL LISTS

List of Pesticide Product Inert Ingredients
[present]

1,2,3,4,5-PENTABROMO-6-CHLORO-CYCLOHEXANE   87-84-3

ENVIRONMENTAL LISTS

List of Pesticide Product Inert Ingredients
[present]

PENTABROMODIPHENYLOXIDE   32534-81-9

HEALTH AND SAFETY LISTS

NIOSH - Selected LD50s and LC50s
Oral, mouse: LD50 = 15 gm/kg

ENVIRONMENTAL LISTS

List of Pesticide Product Inert Ingredients
[present]

INTERNATIONAL LISTS

Canada - WHMIS: Ingredient Disclosure
1% item 1238 (121)

PENTABROMOETHANE   75-95-6

ENVIRONMENTAL LISTS

CERCLA/SARA - Section 313 - Emission Reporting
form R reporting required

PENTABROMOETHYLBENZENE   85-22-3

STATE LISTS

California - Air Bill 2588 Appendix A-II
6/91

California - Prop. 65 - Developmental Toxicity
developmental toxicity - initial date 1/1/91

Massachusetts Right To Know List
teratogen

PENTABROMOPHENOL   608-71-9

HEALTH AND SAFETY LISTS

IARC - Group 3 (not classifiable)
[present]

STATE LISTS

California - Directors List of Hazardous Substances (8 CCR 339)
[present]

PENTACARBOXYMETHYL DIETHYLENETRIAMINE   67-43-6

HEALTH AND SAFETY LISTS

NTP Chemical Status Reports - Testing Status and NTIS Number
Technical reports printed (PB89128615/AS)

NTP Chemical Status Reports - Evidence of Carcinogenicity
male rat-no evidence; female rat-no evidence; male mice-no evidence; female mice-no evidence

PENT-ACETATE   RR-01761-8

HEALTH AND SAFETY LISTS

ACGIH 1995 - Time Weighted Averages
0.005 ppm TWA; 0.013 mg/m3 TWA

ACGIH 1995 - Short Term Exposure Limits
0.015 ppm STEL; 0.039 mg/m3 STEL

AIHA - Odor Threshold Values
no geometric mean air odor threshold

U.S. DOT - Substances From 49 CFR 172.101
regulated by DOT (UN1380)

U.S. DOT - Hazard Classes
DOT hazard class = 4.2

U.S. DOT - Substances Which Are Poisonous by Inhalation
liquid hazardous material poisonous by inhalation (UN1380)

NFPA - Flash Points
ignites spontaneously in air

NFPA - Hazard Identification Ratings
health-4; flammability-4; reactivity-2 (reacts violently with halogenated extinguishing agents)

NIOSH - Selected LD50s and LC50s
Inhalation, rat: LC50 = 6 ppm 4 hr

NIOSH 1990 - Pocket Guide - RELs
0.005 ppm TWA; 0.01 mg/m3 TWA; 0.015 ppm STEL; 0.03 mg/m3 STEL

NIOSH 1990 - Pocket Guide - IDLHs
3 ppm IDLH

NIOSH 1990 - Pocket Guide - Target organs
CNS, eyes, skin

OSHA - Vacated PELs - Time Weighted Averages
0.005 ppm TWA; 0.01 mg/m3 TWA

OSHA - Vacated PELs - Short Term Exposure Limits
0.015 ppm STEL; 0.03 mg/m3 STEL

OSHA - Final PELs - Time Weighted Averages
0.005 ppm TWA; 0.01 mg/m3 TWA

OSHA - List of Highly Hazardous Chemicals
threshhold quantity = 100 pounds

ENVIRONMENTAL LISTS

CERCLA/SARA - Section 302 Extremely Hazardous Substances and TPQs
TPQ = 500 pounds

INTERNATIONAL LISTS

Australian Exposure Standards - Time Weighted Averages
0.005 ppm TWA; 0.013 mg/m3 TWA

Australian Exposure Standards - Short Term Exposure Limits
0.015 ppm STEL; 0.039 mg/m3 STEL

Canada - WHMIS: Ingredient Disclosure
1% item 1239 (1336)

Canada - Alberta - 8 Hour Occupational Exposure Limit
0.005 ppm TWA; 0.013 mg/m3 TWA

Canada - Alberta - 15 Minute Occupational Exposure Limit
0.015 ppm STEL; 0.039 mg/m3 STEL

Canada - British Columbia - 8 Hour Exposure Limits
0.005 ppm TWA; 0.01 mg/m3 TWA

Canada - British Columbia - 15 Minute Exposure Limits
0.015 ppm STEL; 0.03 mg/m3 STEL

Canada - Ontario - OHSA - TWAEVs
0.005 ppm TWAEV; 0.013 mg/m3 TWAEV

Canada - Ontario - OHSA - STEVs
0.015 ppm STEV; 0.039 mg/m3 STEV

Canada - Quebec - Time-Weighted Average Exposure Values
0.005 ppm TWAEV; 0.013 mg/m3 TWAEV

Canada - Quebec - Short-term Exposure Values
0.015 ppm STEV; 0.039 mg/m3 STEV

German (DFG) - MAK Values
0.005 ppm MAK; 0.01 mg/m3 MAK

German (DFG) - Peak Limitations
2 x normal MAK (5 min momentary value); don't exceed 8 times during shift

Israel - Time Weighted Averages
0.005 ppm TWA; 0.013 mg/m3 TWA

Israel - Short Term Exposure Limits
0.015 ppm STEL; 0.039 mg/m3 STEL

Israel - Action Levels
0.0025 ppm AL; 0.0065 mg/m3 AL

Mexico - Instruction No. 10 - TWAs
0.005 ppm TWA; 0.01 mg/m3 TWA

Mexico - Instruction No. 10 - STELs
0.015 ppm STEL; 0.03 mg/m3 STEL

STATE LISTS

California - Exposure Limits - PELs
0.005 ppm PEL; 0.01 mg/m3 PEL

California - Exposure Limits - STELs
0.015 ppm STEL; 0.03 mg/m3 STEL

California - Directors List of Hazardous Substances (8 CCR 339)
[present]

Florida Hazardous Substance List
[present]

Massachusetts Right To Know List
extraordinarily hazardous

Minnesota Hazardous Substance List
[present]

NJ Right to Know List (Total)
sn 1470

NJ Special Hazardous Substances
(flammable - third degree; reactive - second degree)

Pennsylvania Right to Know List
environmental hazard

PROPOSED REGULATIONS

CERCLA/SARA - Proposed Hazardous Substance Additions
proposed RQ = 1 pound (.454 kg)

CERCLA/SARA - 1989 Proposed RQ Adjustments
proposed RQ = 10 pounds (4.54 kg)

PENTACHLOROANISOLE   1825-21-4

ENVIRONMENTAL LISTS

TSCA - Code of Federal Regulations Citations
40 CFR 712.30(x); 40 CFR 716.120(d)

TSCA - PAIR - Reporting List
Reporting Date: December 27, 1990

TSCA - Health and Safety Reporting List
Effective Date: October 29, 1990

TSCA - HDD/HDF - Precursors Required for Reporting
[present]

PENTACHLOROBENZENE   608-93-5

ENVIRONMENTAL LISTS

EPA - Master Testing List
[present]

TSCA - Code of Federal Regulations Citations
40 CFR 712.30(w); 40 CFR 716.120(a); 40 CFR 766.35

TSCA - PAIR - Reporting List
Reporting Date: March 12, 1990

TSCA - Health and Safety Reporting List
Effective Date: January 11, 1990

TSCA - HDD/HDF - Chemicals Required for Testing
[present]

TSCA - Section 12(b) - Export Notification
export notification required - Section 4

1,2,3,7,8-PENTACHLORODIBENZO-P-DIOXIN   36088-22-9

HEALTH AND SAFETY LISTS

NTP Chemical Status Reports - Testing Status and NTIS Number
Prechronic studies completed: in review for further evaluation

1,2,3,7,8-PENTACHLORODIBENZO-P-DIOXIN   40321-76-4

ENVIRONMENTAL LISTS

TSCA - Code of Federal Regulations Citations
40 CFR 712.30(m); 40 CFR 716.120(a)

TSCA - PAIR - Reporting List
Reporting Date: February 26, 1985

TSCA - Health and Safety Reporting List
Effective Date: December 28, 1984

TSCA - Chemicals with Significant New Use Rules
[present]

TSCA - HDD/HDF - Precursors Required for Reporting
[present]

TSCA - Section 12(b) - Export Notification
export notification required - Section 5

PENTACHLORODIBENZO-P-DIOXINS   RR-00511-8

ENVIRONMENTAL LISTS

TSCA - HDD/HDF - Chemicals Required for Testing
[present]

TSCA - Section 12(b) - Export Notification
export notification required - Section 4

PENTACHLORODIBENZOFURAN   30402-15-4

HEALTH AND SAFETY LISTS

NIOSH - Selected LD50s and LC50s
Oral, mouse: LD50 = 4840 mg/kg

1,2,3,7,8-PENTACHLORO DIBENZOFURAN   57117-41-6

HEALTH AND SAFETY LISTS

NFPA - Flash Points
flash point = 98 degrees F (37 degrees C)

NFPA - Hazard Identification Ratings
health-2; flammability-3; reactivity-0

2,3,4,7,8-PENTACHLORO DIBENZOFURANS   57117-31-4

HEALTH AND SAFETY LISTS

NTP Chemical Status Reports - Testing Status and NTIS Number
Technical reports printed (no NTIS number given)

PENTACHLORODIBENZOFURANS   RR-00507-2

SEE ALSO:
F026-HAZARDOUS WASTES
F024-HAZARDOUS WASTES
K150-HAZARDOUS WASTES
K151-HAZARDOUS WASTES
K149-HAZARDOUS WASTES
K085-HAZARDOUS WASTES
F025-HAZARDOUS WASTES
F039-HAZARDOUS WASTES
F022-HAZARDOUS WASTES

HEALTH AND SAFETY LISTS

U.S. DOT - Appendix A Table 1 - Hazardous Substances
final RQ = 10 pounds (4.54 kg)

NIOSH - Selected LD50s and LC50s
Oral, rat: LD50 = 1080 mg/kg

NTP Chemical Status Reports - Testing Status and NTIS Number
Technical reports printed (PB91185983)

ENVIRONMENTAL LISTS

ATSDR Priority List
Rank (of 275): 131

CERCLA/SARA - Hazardous Substances and their Reportable Quantities
final RQ = 10 pounds (4.54 kg)

RCRA - U Series Wastes
waste number U183

RCRA - Hazardous Constituents-Appendix VIII
waste number U183

RCRA - Basis for Listing - Appendix VII
Included in waste streams: F024, F025, F039, K085, K149, K150, K151

RCRA - Substances Banned From Land Disposal
[present]

RCRA - TSD Facilities Ground Water Monitoring
TM 8270 = 10 ug/L PQL

RCRA - Universal Treatment Standards (LDR)
WW: 0.055 mg/l; NWW: 10 mg/kg

TSCA - Code of Federal Regulations Citations
40 CFR 712.30(d); 40 CFR 716.120(c); 40 CFR 799.5055(c),(d)(2)

TSCA - PAIR - Reporting List
Reporting Date: November 19, 1982

TSCA - Chemicals with Significant New Use Rules
[present]

TSCA - Multichemical Test Rules - Waste Constituents
hydrolysis testing for Chemical Fate

TSCA - Section 12(b) - Export Notification
export notification required - Section 5

INTERNATIONAL LISTS

Canada - CEPA - Priority Substances List
estimated time for completion of assessment reports: 4 years

STATE LISTS

Massachusetts Right To Know List
[present]

Pennsylvania Right to Know List
environmental hazard

PENTACHLORODIFLUOROPROPANE   134237-36-8

ENVIRONMENTAL LISTS

ATSDR Priority List
Rank (of 275): 153

PENTACHLOROETHANE   76-01-7

STATE LISTS

California - Air Bill 2588 Appendix A-I
known or potential carcinogen

PENTACHLOROFLUOROETHANE   354-56-3
SEE ALSO:
F039-HAZARDOUS WASTES

ENVIRONMENTAL LISTS

RCRA - Hazardous Constituents-Appendix VIII
hazardous constituent - no waste number

RCRA - Basis for Listing - Appendix VII
Included in waste streams: F020, F021, F022, F023, F026, F027, F028, F039

RCRA - Universal Treatment Standards (LDR)
WW: 0.000063 mg/l; NWW: 0.001 mg/kg

PENTACHLOROFLUOROPROPANE   134190-48-0

ENVIRONMENTAL LISTS

ATSDR Priority List
Rank (of 275): 155

PENTACHLORONAPHTHALENE   1321-64-8

STATE LISTS

California - Air Bill 2588 Appendix A-I
known or potential carcinogen

PENTACHLORONITROBENZENE   82-68-8

STATE LISTS

California - Air Bill 2588 Appendix A-I
known or potential carcinogen

PENTACHLOROPHENOL   87-86-5
SEE ALSO:
F039-HAZARDOUS WASTES

ENVIRONMENTAL LISTS

RCRA - Hazardous Constituents-Appendix VIII
hazardous constituent - no waste number

RCRA - Basis for Listing - Appendix VII
Included in waste streams: F020, F021, F022, F023, F026, F027, F028, F039

RCRA - Universal Treatment Standards (LDR)
WW: 0.000035 mg/l; NWW: 0.001 mg/kg

PENTACHLOROPHENOL DERIVATIVES   RR-00512-9

ENVIRONMENTAL LISTS

Class 2 Ozone Depletors
ozone depletion weight reserved

PENTACHLOROPHENYL LAURATE   3772-94-9
SEE ALSO:
F025-HAZARDOUS WASTES
F024-HAZARDOUS WASTES

HEALTH AND SAFETY LISTS

U.S. DOT - Substances From 49 CFR 172.101
regulated by DOT (UN1669)

U.S. DOT - Hazard Classes
DOT hazard class = 6.1

U.S. DOT - Appendix B - Marine Pollutants
DOT regulated marine pollutant

U.S. DOT - Appendix A Table 1 - Hazardous Substances
final RQ = 10 pounds (4.54 kg)

IARC - Group 3 (not classifiable)
[present]

NIOSH - Health Standards - Exposure Limits
Handle with caution in the workplace

NIOSH - Health Standards - Health Effects and Precautions
Central nervous system effects; possible liver and kidney effects

NTP Chemical Status Reports - Testing Status and NTIS Number
Technical reports printed (PB83206748); Prechronic studies completed: in review for further evaluation

ENVIRONMENTAL LISTS

ATSDR Priority List
Rank (of 275): 273

CERCLA/SARA - Section 313 - Emission Reporting
form R reporting required

CERCLA/SARA - Hazardous Substances and their Reportable Quantities
final RQ = 10 pounds (4.54 kg)

EPA - Carcinogen Hazard Ranking for RQ Adjustment
Hazard ranking = Low

RCRA - U Series Wastes
waste number U184

RCRA - Hazardous Constituents-Appendix VIII
waste number U184

RCRA - Basis for Listing - Appendix VII
Included in waste streams: F024, F025

RCRA - Substances Banned From Land Disposal
[present]

RCRA - TSD Facilities Ground Water Monitoring
TM 8240 = 5 ug/L PQL; TM 8270 = 10 ug/L PQL

RCRA - Universal Treatment Standards (LDR)
WW: 0.055 mg/l; NWW: 6.0 mg/kg

TSCA - Code of Federal Regulations Citations
40 CFR 721.1525; 40 CFR 799.5055(c)

TSCA - Chemicals with Significant New Use Rules
[present]

TSCA - Multichemical Test Rules - Waste Constituents
hydrolysis testing for Chemical Fate

TSCA - Section 12(b) - Export Notification
export notification required - Section 4

INTERNATIONAL LISTS

Canada - WHMIS: Ingredient Disclosure
1% item 1240 (1338)

German (DFG) - MAK Values
5 ppm MAK; 40 mg/m3 MAK

German (DFG) - Peak Limitations
2 x normal MAK (30 min. average value); don't exceed 4 times during shift

STATE LISTS

California - Directors List of Hazardous Substances (8 CCR 339)
[present]

Massachusetts Right To Know List
[present]

Minnesota Hazardous Substance List
[present]

NJ Right to Know List (Total)
sn 1471

Pennsylvania Right to Know List
environmental hazard

PENTACHLOROTHIOPHENOL   133-49-3

ENVIRONMENTAL LISTS

Class 1 Ozone Depletors
ozone depletion potential = 1.0

PENTACHLOROTRIFLUOROPROPANE   134237-31-3

ENVIRONMENTAL LISTS

Class 2 Ozone Depletors
ozone depletion weight reserved

PENTADECYLAMINE   2570-26-5

HEALTH AND SAFETY LISTS

ACGIH 1995 - Time Weighted Averages
0.5 mg/m3 TWA

ACGIH 1995 - Skin Designations
skin - potential for cutaneous absorption

NIOSH 1990 - Pocket Guide - RELs
0.5 mg/m3 TWA

NIOSH 1990 - Pocket Guide - Target organs
skin, liver, CNS

NIOSH 1990 - Pocket Guide - Skin list
Potential for dermal absorption

OSHA - Vacated PELs - Time Weighted Averages
0.5 mg/m3 TWA

OSHA - Vacated PELs - Skin Designation
Prevent or reduce skin absorption

OSHA - Final PELs - Time Weighted Averages
0.5 mg/m3 TWA

OSHA - Final PELs - Skin Notations
prevent or reduce skin absorption

ENVIRONMENTAL LISTS

TSCA - Code of Federal Regulations Citations
40 CFR 704.83; 40 CFR 712.30(d); 40 CFR 716.120(a)

TSCA - PAIR - Reporting List
Reporting Date: November 19, 1982

TSCA - Health and Safety Reporting List
Effective Date: October 4, 1982

INTERNATIONAL LISTS

Australian Exposure Standards - Time Weighted Averages
0.5 mg/m3 TWA

Canada - WHMIS: Ingredient Disclosure
1% item 1241 (1339)

Canada - Alberta - 8 Hour Occupational Exposure Limit
0.5 mg/m3 TWA

Canada - Alberta - 15 Minute Occupational Exposure Limit
2 mg/m3 STEL

Canada - British Columbia - 8 Hour Exposure Limits
0.5 mg/m3 TWA

Canada - British Columbia - 15 Minute Exposure Limits
2 mg/m3 STEL

Canada - Ontario - OHSA - TWAEVs
0.5 mg/m3 TWAEV

Canada - Quebec - Time-Weighted Average Exposure Values
0.5 mg/m3 TWAEV

Canada - Quebec - Skin Designations
absorbed through the skin

German (DFG) - MAK Values
total dust: 0.5 mg/m3 MAK

German (DFG) - Peak Limitations
5 x normal MAK (30 min. average value); don't exceed 2 times during shift

German (DFG) - Skin/Sensitizers
danger of cutaneous absorption

Israel - Time Weighted Averages
0.5 mg/m3 TWA

Israel - Action Levels
0.25 mg/m3 AL

Mexico - Instruction No. 10 - TWAs
0.5 mg/m3 TWA

Mexico - Instruction No. 10 - STELs
2 mg/m3 STEL

STATE LISTS

California - Exposure Limits - PELs
0.5 mg/m3 PEL

California - Exposure Limits - Skin Notation
material may be absorbed through the skin, eyes or mucous membrane

California - Directors List of Hazardous Substances (8 CCR 339)
[present]

Florida Hazardous Substance List
[present]

Massachusetts Right To Know List
[present]

Minnesota Hazardous Substance List
[present]

Pennsylvania Right to Know List
[present]

PENTAERYTHRITE TETRANITRATE   78-11-5
SEE ALSO:
F039-HAZARDOUS WASTES

HEALTH AND SAFETY LISTS

ACGIH 1995 - Time Weighted Averages
0.5 mg/m3 TWA

U.S. DOT - Appendix A Table 1 - Hazardous Substances
final RQ = 100 pounds (45.4 kg)

IARC - Group 3 (not classifiable)
[present]

NIOSH - Selected LD50s and LC50s
Inhalation, rat: LC50 = 1400 mg/m3 8 hr Oral, rat: LD50 = 1100 mg/kg

NTP Chemical Status Reports - Testing Status and NTIS Number
Technical reports printed (PB281732/AS) (PB87208633/AS)

NTP Chemical Status Reports - Evidence of Carcinogenicity
PB87208633/AS: male mice-no evidence; female mice-no evidence; PB281732/AS: male rat-negative; female rat-negative; male mice-negative; female mice-negative

ENVIRONMENTAL LISTS

CERCLA/SARA - Section 313 - Emission Reporting
form R reporting required for 1.0% de minimus concentration

CERCLA/SARA - Hazardous Substances and their Reportable Quantities
final RQ = 100 pounds (45.4 kg)

Clean Air Act (1990) - List of Hazardous Air Contaminants
[present]

EPA - Carcinogen Hazard Ranking for RQ Adjustment
Hazard ranking = Low

RCRA - U Series Wastes
waste number U185

RCRA - Hazardous Constituents-Appendix VIII
waste number U185

RCRA - Basis for Listing - Appendix VII
Included in waste stream: F039

RCRA - Substances Banned From Land Disposal
[present]

RCRA - TSD Facilities Ground Water Monitoring
TM 8270 = 10 ug/L PQL

RCRA - Universal Treatment Standards (LDR)
WW: 0.055 mg/l; NWW: 4.8 mg/kg

STATE LISTS

California - Air Bill 2588 Appendix A-I
6/91

California - Directors List of Hazardous Substances (8 CCR 339)
[present]

Florida Hazardous Substance List
[present]

Massachusetts Right To Know List
carcinogen; extraordinarily hazardous

NJ Right to Know List (Total)
sn 1630

NJ Special Hazardous Substances
(carcinogen)

Pennsylvania Right to Know List
environmental hazard

PENTAERYTHRITOL   115-77-5
SEE ALSO:
F039-HAZARDOUS WASTES
K142-HAZARDOUS WASTES
F021-HAZARDOUS WASTES
F027-HAZARDOUS WASTES
K001-HAZARDOUS WASTES
TETRACHLORODIBENZO-P-DIOXINS
F028-HAZARDOUS WASTES

HEALTH AND SAFETY LISTS

ACGIH 1995 - Time Weighted Averages
0.5 mg/m3 TWA

ACGIH 1995 - Skin Designations
skin - potential for cutaneous absorption

ACGIH 1995 - Biological Exposure Indices
Total PCP in urine: 2 mg/g creatinine, prior to the last shift of workweek (B); Free PCP in plasma: 5 mg/L creatinine, end of shift (B)

U.S. DOT - Appendix B - Marine Pollutants
DOT regulated severe marine pollutant

U.S. DOT - Appendix A Table 1 - Hazardous Substances
final RQ = 10 pounds (4.54 kg)

IARC - Group 2B (sufficient animal data)
[present]

NIOSH - Selected LD50s and LC50s
Inhalation, rat: LC50 =355 mg/m3 8 hr Oral, rat: LD50 =27 mg/kg Skin, rat: LD50 = 96 mg/kg

NIOSH 1990 - Pocket Guide - RELs
0.5 mg/m3 TWA

NIOSH 1990 - Pocket Guide - IDLHs
150 mg/m3 IDLH

NIOSH 1990 - Pocket Guide - Target organs
respiratory system, CVS, eyes, liver, kidneys, skin, CNS

NIOSH 1990 - Pocket Guide - Skin list
Potential for dermal absorption

NTP Chemical Status Reports - Testing Status and NTIS Number
dowicide EC-7 and technical grade: Technical reports printed (PB89216536); two year studies in progress

NTP Chemical Status Reports - Evidence of Carcinogenicity
PB89216536/AS (dowicide EC-7): male mice-clear evidence; female mice-clear evidence; PB89216536/AS (technical): male mice-clear evidence; female mice-some evidence

OSHA - Vacated PELs - Time Weighted Averages
0.5 mg/m3 TWA

OSHA - Vacated PELs - Skin Designation
Prevent or reduce skin absorption

OSHA - Final PELs - Time Weighted Averages
0.5 mg/m3 TWA

OSHA - Final PELs - Skin Notations
prevent or reduce skin absorption

OSHA - Possible Select Carcinogens
[present]

ENVIRONMENTAL LISTS

ATSDR Priority List
Rank (of 275): 051

CERCLA/SARA - Section 313 - Emission Reporting
form R reporting required for 1.0% de minimus concentration

CERCLA/SARA - Hazardous Substances and their Reportable Quantities
final RQ = 10 pounds (4.54 kg)

Clean Air Act (1990) - List of Hazardous Air Contaminants
[present]

CAA - HON Rule - SOCMI Chemicals
compliance by April 24, 1995

Clean Water Act - Hazardous Substances
[present]

Clean Water Act - Priority Pollutants
[present]

Clean Water Act - Toxic Pollutants
[present]

Safe Drinking Water Act - MCLs
MCL = 0.001 mg/L

Safe Drinking Water Act - MCLGs
MCLG = Zero

RCRA - D Series - Maximum Concentration of Contaminants
waste number D037; regulatory level = 100.0 mg/L

RCRA - D Series - Chronic Toxicity Reference Levels
chronic toxicity reference level = 1 mg/L

RCRA - Hazardous Constituents-Appendix VIII
contained in F027 waste stream

RCRA - Basis for Listing - Appendix VII
Included in waste streams: F021, F027, F028, F039, K001

RCRA - TSD Facilities Ground Water Monitoring
TM 8040 = 5 ug/L PQL; TM 8270 = 50 ug/L PQL

RCRA - Universal Treatment Standards (LDR)
WW: 0.089 mg/l; NWW: 7.4 mg/kg

INTERNATIONAL LISTS

Australian Exposure Standards - Time Weighted Averages
0.5 mg/m3 TWA

Australian Exposure Standards - Skin Effects
skin absorption

Canada - Drinking Water Quality - MACs
0.06 mg/L MAC

Canada - Drinking Water Quality - AOs
<= 0.03 mg/L AO

Canada - Alberta - 8 Hour Occupational Exposure Limit
0.5 mg/m3 TWA

Canada - Alberta - 15 Minute Occupational Exposure Limit
1.5 mg/m3 STEL

Canada - Alberta - Skin Designation
can be absorbed through the intact skin

Canada - British Columbia - 8 Hour Exposure Limits
0.5 mg/m3 TWA

Canada - British Columbia - 15 Minute Exposure Limits
1.5 mg/m3 STEL

Canada - British Columbia - Skin Notations
skin - potential for skin absorption

Canada - Ontario - OHSA - TWAEVs
0.5 mg/m3 TWAEV

Canada - Ontario - OHSA - Skin Notations
absorption through skin, eyes, or mucous membranes

Canada - Quebec - Time-Weighted Average Exposure Values
0.5 mg/m3 TWAEV

Canada - Quebec - Skin Designations
absorbed through the skin

Canada - Quebec - Carcinogens
C2 carcinogen: effect suspected in humans

United Kingdom - Occupational Exposure Standards - TWAs
0.5 mg/m3 TWA

United Kingdom - Occupational Exposure Standards - STELs
1.5 mg/m3 STEL

United Kingdom - Occupational Exposure Standards - Notes
can be absorbed through skin

German (DFG) - Skin/Sensitizers
danger of cutaneous absorption

German (DFG) - Carcinogens
animal evidence of carcinogenicity

Israel - Time Weighted Averages
0.5 mg/m3 TWA

Israel - Action Levels
0.25 mg/m3 AL

Mexico - Instruction No. 10 - TWAs
0.5 mg/m3 TWA

Mexico - Instruction No. 10 - STELs
1.5 mg/m3 STEL

Mexico - Instruction No. 10 - Skin designation
skin - potential for cutaneous absorption

Mexico - Wastewater - Organic Toxic Pollutants and Heavy Metals
Listed under [Chlorinated Phenols]

Mexico - Drinking Water - Ecological Criteria
0.03 mg/l

STATE LISTS

California - Air Bill 2588 Appendix A-I
known or potential carcinogen: 9/90

California - Prop. 65 - Cancer list
carcinogen - initial date 1/1/90

California - Prop. 65 - No Significant Risk Levels
no significant risk level = 40 ug/day

California - Exposure Limits - PELs
0.5 mg/m3 PEL

California - Exposure Limits - Skin Notation
material may be absorbed through the skin, eyes or mucous membrane

California - Directors List of Hazardous Substances (8 CCR 339)
[present]

Florida Hazardous Substance List
[present]

Massachusetts Right To Know List
[present]

Minnesota Hazardous Substance List
[present]

NJ Right to Know List (Total)
sn 1473

Pennsylvania Right to Know List
environmental hazard

PENTAERYTHRITOL, MIXED ESTERS WITH CARBOXYLIC ACIDS   RR-01260-2

ENVIRONMENTAL LISTS

RCRA - Basis for Listing - Appendix VII
Included in waste stream: F021

PENTAERYTHRITOL, PHTHALIC ANHYDRIDE AND TALL-OIL POLYMER WITH ROSIN   68554-37-0

ENVIRONMENTAL LISTS

TSCA - HDD/HDF - Chemicals Required for Testing
[present]

TSCA - Section 12(b) - Export Notification
export notification required - Section 4

PENTAERYTHRITOL TETRABENZOATE   4196-86-5

INTERNATIONAL LISTS

Canada - WHMIS: Ingredient Disclosure
1% item 1242 (1340)

PENTAERYTHRITOL TRIACRYLATE   3524-68-3

ENVIRONMENTAL LISTS

Class 1 Ozone Depletors
ozone depletion potential = 1.0

PENTAERYTHRITOL TRISTEARATE   28188-24-1

HEALTH AND SAFETY LISTS

NIOSH - Selected LD50s and LC50s
Inhalation, rat: LC50 = 900 mg/m3 4 hr Oral, rat: LD50 = 660 mg/kg

ENVIRONMENTAL LISTS

CERCLA/SARA - Section 302 Extremely Hazardous Substances and TPQs
TPQ = 100/10,000 pounds

STATE LISTS

Florida Hazardous Substance List
effective March 13, 1992

Massachusetts Right To Know List
extraordinarily hazardous

NJ Right to Know List (Total)
sn 2634

Pennsylvania Right to Know List
environmental hazard

PROPOSED REGULATIONS

CERCLA/SARA - Proposed Hazardous Substance Additions
proposed RQ = 1 pound (.454 kg)

CERCLA/SARA - 1989 Proposed RQ Adjustments
proposed RQ = 100 pounds (45.4 kg)

1,2,3,4,5-PENTAMETHYL BENZENE   700-12-9

HEALTH AND SAFETY LISTS

U.S. DOT - Substances From 49 CFR 172.101
regulated by DOT (UN0411, NA0150)

U.S. DOT - Hazard Classes
DOT hazard class = 1.1D

NTP Chemical Status Reports - Testing Status and NTIS Number
Technical reports printed (PB90219452/AS)

NTP Chemical Status Reports - Evidence of Carcinogenicity
male rat-equivocal evidence; female rat-equivocal evidence; male mice-no evidence; female mice-no evidence

STATE LISTS

NJ Right to Know List (Total)
sn 1474

PENTAMETHYLENE OXIDE   142-68-7

HEALTH AND SAFETY LISTS

ACGIH 1995 - Time Weighted Averages
10 mg/m3 TWA

NIOSH - Selected LD50s and LC50s
Oral, mouse: LD50 = 25500 mg/kg

OSHA - Vacated PELs - Time Weighted Averages
total dust: 10 mg/m3 TWA; respirable fraction: 5 mg/m3 TWA

OSHA - Final PELs - Time Weighted Averages
total dust: 15 mg/m3 TWA; respirable fraction: 5 mg/m3 TWA

ENVIRONMENTAL LISTS

CAA - HON Rule - SOCMI Chemicals
compliance by Oct. 24, 1994

List of Pesticide Product Inert Ingredients
[present]

INTERNATIONAL LISTS

Australian Exposure Standards - Time Weighted Averages
10 mg/m3 TWA

Canada - Alberta - 8 Hour Occupational Exposure Limit
respirable mass: 5 mg/m3 TWA; total mass: 10 mg/m3 TWA

Canada - British Columbia - 8 Hour Exposure Limits
nuisance dust, mists, and fumes: 10 mg/m3 TWA

Canada - British Columbia - 15 Minute Exposure Limits
20 mg/m3 STEL

Canada - Ontario - OHSA - TWAEVs
total dust: 10 mg/m3 TWAEV (listed as a nuisance particulate)

Canada - Quebec - Time-Weighted Average Exposure Values
10 mg/m3 TWAEV

United Kingdom - Occupational Exposure Standards - TWAs
total inhalable dust: 10 mg/m3 TWA; respirable dust: 5 mg/m3 TWA

United Kingdom - Occupational Exposure Standards - STELs
total inhalable dust: 20 mg/m3 STEL

Israel - Time Weighted Averages
10 mg/m3 TWA

Israel - Action Levels
5 mg/m3 AL

Mexico - Instruction No. 10 - TWAs
10 mg/m3 TWA; (nuisance particulate)

Mexico - Instruction No. 10 - STELs
20 mg/m3 STEL

STATE LISTS

Massachusetts Right To Know List
[present]

Minnesota Hazardous Substance List
[present] (includes inert or nuisance dust)

Pennsylvania Right to Know List
[present]

PENTAMETHYLHEPTANE   30586-18-6

ENVIRONMENTAL LISTS

TSCA - Chemicals with Significant New Use Rules
PMN number: P-91-1250

SEC-PENTAMINE   41444-43-3

ENVIRONMENTAL LISTS

List of Pesticide Product Inert Ingredients
[present]

PENTANAL, 2,3-DIMETHYL-   32749-94-3

HEALTH AND SAFETY LISTS

NIOSH - Selected LD50s and LC50s
Oral, rat: LD50 = 10 gm/kg

1-PENTANAMINE, N,N-DIPENTYL-   621-77-2

HEALTH AND SAFETY LISTS

AIHA - WEEL - Time Weighted Averages
1 mg/m3 TWA

NIOSH - Selected LD50s and LC50s
Oral, rat: LD50 = 2460 mg/kg Skin, rabbit: LD50 = 4000 mg/kg

NTP Chemical Status Reports - Testing Status and NTIS Number
Approved for toxicology/carcinogenesis study

STATE LISTS

Minnesota Hazardous Substance List
[present]

PENTANE   109-66-0

ENVIRONMENTAL LISTS

List of Pesticide Product Inert Ingredients
[present]

1,5-PENTANEDIOL   111-29-5

HEALTH AND SAFETY LISTS

NFPA - Flash Points
95%: flash point = 200 degrees F (93 degrees C)

NFPA - Hazard Identification Ratings
95%: flammability-2; reactivity-0

2,4-PENTANEDIOL   625-69-4

HEALTH AND SAFETY LISTS

NFPA - Flash Points
flash point = -4 degrees F (-20 degrees C)

NFPA - Hazard Identification Ratings
health-2; flammability-3; reactivity-1

STATE LISTS

Florida Hazardous Substance List
[present]

Massachusetts Right To Know List
[present]

Pennsylvania Right to Know List
[present]

2,4-PENTANEDIONE   123-54-6

HEALTH AND SAFETY LISTS

U.S. DOT - Substances From 49 CFR 172.101
regulated by DOT (UN2286)

U.S. DOT - Hazard Classes
DOT hazard class = 3

STATE LISTS

NJ Right to Know List (Total)
sn 1475

2,3-PENTANEDIONE   600-14-6

STATE LISTS

Pennsylvania Right to Know List
[present]

PENTANENITRILE, 3-AMINO-   75405-06-0

STATE LISTS

Florida Hazardous Substance List
[present]

Massachusetts Right To Know List
[present]

Pennsylvania Right to Know List
[present]

PENTANE, 2,2,3,4-TETRAMETHYL-   1186-53-4

HEALTH AND SAFETY LISTS

NFPA - Flash Points
flash point = 215 degrees F (102 degrees C)

NFPA - Hazard Identification Ratings
health-2; flammability-1; reactivity-0

STATE LISTS

Florida Hazardous Substance List
[present]

Massachusetts Right To Know List
[present]

Pennsylvania Right to Know List
[present]

PENTANE, 2,2,3,3-TETRAMETHYL-   7154-79-2

HEALTH AND SAFETY LISTS

ACGIH 1995 - Time Weighted Averages
600 ppm TWA; 1770 mg/m3 TWA

ACGIH 1995 - Short Term Exposure Limits
750 ppm STEL; 2210 mg/m3 STEL

AIHA - Odor Threshold Values
no geometric mean air odor threshold

U.S. DOT - Substances From 49 CFR 172.101
regulated by DOT (UN1265)

U.S. DOT - Hazard Classes
DOT hazard class = 3

NFPA - Flash Points
flash point < -40 degrees F (-40 degrees C)

NFPA - Hazard Identification Ratings
health-1; flammability-4; reactivity-0

NIOSH 1990 - Pocket Guide - RELs
120 ppm TWA; 350 mg/m3 TWA; C 610 ppm (15 min); C 1800 mg/m3 (15 min)

NIOSH 1990 - Pocket Guide - IDLHs
15,000 ppm IDLH (lower explosive level)

NIOSH 1990 - Pocket Guide - Target organs
skin, eyes, respiratory system

NIOSH - Health Standards - Exposure Limits
120 ppm TWA; 350 mg/m3 TWA; C (15 min) 610 ppm; C (15 min) 1800 mg/m3 (Listed under 'Alkanes (C5-C8)')

NIOSH - Health Standards - Health Effects and Precautions
Skin and nervous system effects (Listed under 'Alkanes (C5-C8)')

OSHA - Vacated PELs - Time Weighted Averages
600 ppm TWA; 1800 mg/m3 TWA

OSHA - Vacated PELs - Short Term Exposure Limits
750 ppm STEL; 2250 mg/m3 STEL

OSHA - Final PELs - Time Weighted Averages
1000 ppm TWA; 2950 mg/m3 TWA

ENVIRONMENTAL LISTS

CAA - Flammable Substances for Accidental Release Prevention
threshold quantity = 10,000 lbs

List of Pesticide Product Inert Ingredients
[present]

TSCA - PAIR - Reporting List
Effective Date: January 26, 1994; Reporting Date: March 28, 1994

TSCA - Health and Safety Reporting List
Effective Date: January 26, 1994; Sunset Date: January 26, 2004

INTERNATIONAL LISTS

Australian Exposure Standards - Time Weighted Averages
600 ppm TWA; 1770 mg/m3 TWA

Australian Exposure Standards - Short Term Exposure Limits
750 ppm STEL; 2210 mg/m3 STEL

Canada - WHMIS: Ingredient Disclosure
1% item 1243 (1348)

Canada - Alberta - 8 Hour Occupational Exposure Limit
600 ppm TWA; 1771 mg/m3 TWA

Canada - Alberta - 15 Minute Occupational Exposure Limit
750 ppm STEL; 2213 mg/m3 STEL

Canada - British Columbia - 8 Hour Exposure Limits
600 ppm TWA; 1800 mg/m3 TWA

Canada - British Columbia - 15 Minute Exposure Limits
750 ppm STEL; 2250 mg/m3 STEL

Canada - Ontario - OHSA - TWAEVs
600 ppm TWAEV; 1770 mg/m3 TWAEV

Canada - Ontario - OHSA - STEVs
750 ppm STEV; 2210 mg/m3 STEV

Canada - Quebec - Time-Weighted Average Exposure Values
120 ppm TWAEV; 350 mg/m3 TWAEV

United Kingdom - Occupational Exposure Standards - TWAs
600 ppm TWA; 1800 mg/m3 TWA

United Kingdom - Occupational Exposure Standards - STELs
750 ppm STEL; 2250 mg/m3 STEL

German (DFG) - MAK Values
1000 ppm MAK; 2950 mg/m3 MAK (Listed under 'Pentane')

German (DFG) - Peak Limitations
2 x normal MAK (1 hour momentary value); don't exceed 3 times per shift (Listed under 'Pentane')

Israel - Time Weighted Averages
600 ppm TWA; 1770 mg/m3 TWA

Israel - Short Term Exposure Limits
750 ppm STEL; 2210 mg/m3 STEL

Israel - Action Levels
300 ppm AL; 885 mg/m3 AL

Mexico - Instruction No. 10 - TWAs
600 ppm TWA; 1800 mg/m3 TWA

Mexico - Instruction No. 10 - STELs
760 ppm STEL; 2250 mg/m3 STEL

STATE LISTS

California - Exposure Limits - PELs
600 ppm PEL; 1800 mg/m3 PEL

California - Exposure Limits - STELs
750 ppm STEL; 2250 mg/m3 STEL

California - Directors List of Hazardous Substances (8 CCR 339)
[present]

Florida Hazardous Substance List
[present]

Massachusetts Right To Know List
[present]

Minnesota Hazardous Substance List
[present]

NJ Right to Know List (Total)
sn 1476

NJ Special Hazardous Substances
(flammable - fourth degree)

Pennsylvania Right to Know List
[present]

PROPOSED REGULATIONS

TSCA - ITC 31st Report Priority Testing List
designated to be tested

Canada - Ontario - Proposed Occupational TWAEVs
250 ppm TWAEV; 750 mg/m3 TWAEV

PENTANE, 1,1'-THIOBIS-   872-10-6

HEALTH AND SAFETY LISTS

NFPA - Flash Points
flash point = 265 degrees F (129 degrees C)

NFPA - Hazard Identification Ratings
health-1; flammability-1; reactivity-0

NIOSH - Selected LD50s and LC50s
Oral, rat: LD50 = 5890 mg/kg

PENTANETHIOLS   RR-01307-0

HEALTH AND SAFETY LISTS

NIOSH - Selected LD50s and LC50s
Oral, rat: LD50 = 6860 mg/kg Skin, rabbit: LD50 = 14100 mg/kg

PENTANE, 2,3,3-TRIMETHYL-   560-21-4

HEALTH AND SAFETY LISTS

U.S. DOT - Substances From 49 CFR 172.101
regulated by DOT (UN2310)

U.S. DOT - Hazard Classes
DOT hazard class = 3

NFPA - Flash Points
flash point = 93 degrees F (34 degrees C)

NFPA - Hazard Identification Ratings
health-2; flammability-2; reactivity-0

NIOSH - Selected LD50s and LC50s
Oral, rat: LD50 = 1000 mg/kg Skin, rabbit: LD50 = 5000 mg/kg

ENVIRONMENTAL LISTS

TSCA - Code of Federal Regulations Citations
40 CFR 712.30(x); 40 CFR 716.120(d)

TSCA - PAIR - Reporting List
Reporting Date: November 27, 1991

TSCA - Health and Safety Reporting List
Effective Date: September 30, 1991

TSCA - Section 12(b) - Export Notification
export notification required (because of proposed action) - Section 5

INTERNATIONAL LISTS

Canada - WHMIS: Ingredient Disclosure
1% item 1244 (1349)

STATE LISTS

Florida Hazardous Substance List
[present]

Massachusetts Right To Know List
[present]

NJ Right to Know List (Total)
sn 1477

Pennsylvania Right to Know List
[present]

PENTANE, 2,2,3-TRIMETHYL-   564-02-3

HEALTH AND SAFETY LISTS

NIOSH - Selected LD50s and LC50s
Oral, rat: LD50 = 3000 mg/kg

PENTANITROANILINE   21985-87-5

ENVIRONMENTAL LISTS

TSCA - Chemicals with Significant New Use Rules
PMN number: P-91-222

TSCA - Section 12(b) - Export Notification
P-91-222; export notification required - Section 5

3-PENTANOL   584-02-1

HEALTH AND SAFETY LISTS

NFPA - Flash Points
flash point < 70 degrees F (21 degrees C)

NFPA - Hazard Identification Ratings
health-0; flammability-3; reactivity-0

STATE LISTS

Florida Hazardous Substance List
[present]

Massachusetts Right To Know List
[present]

Pennsylvania Right to Know List
[present]

1-PENTANOL, METHYL-   54972-97-3

HEALTH AND SAFETY LISTS

NFPA - Flash Points
flash point < 70 degrees F (21 degrees C)

NFPA - Hazard Identification Ratings
health-0; flammability-3; reactivity-0

STATE LISTS

Florida Hazardous Substance List
[present]

Massachusetts Right To Know List
[present]

Pennsylvania Right to Know List
[present]

2,5,8,10,13-PENTAOXAHEXADEC-15-ENOIC ACID, 9,14-DIOXO-2-[(1-OXO-2-PROPENYL)OXY]ETHYL ESTER — 31206-94-7

HEALTH AND SAFETY LISTS

NFPA - Flash Points
flash point = 185 degrees F (85 degrees C)

NFPA - Hazard Identification Ratings
health-2; flammability-2; reactivity-0

STATE LISTS

Florida Hazardous Substance List
[present]

Massachusetts Right To Know List
[present]

Pennsylvania Right to Know List
[present]

2,5,8,10,13-PENTAOXAHEXADEC-15-ENOIC ACID, 9,14-DIOXO-2-[(1-OXO-2-PROPENYL)OXY]ETHYL ESTER — RR-01647-7

HEALTH AND SAFETY LISTS

U.S. DOT - Appendix B - Marine Pollutants
DOT regulated marine pollutant

3,6,9,12,16-PENTAOXANONADEC-18-ENE-1,14-DIOL, BIS(4-METHYLBENZENSULFONATE) — RR-01729-8

HEALTH AND SAFETY LISTS

NFPA - Flash Points
flash point < 70 degrees F (21 degrees C)

NFPA - Hazard Identification Ratings
health-0; flammability-3; reactivity-0

STATE LISTS

Florida Hazardous Substance List
[present]

Massachusetts Right To Know List
[present]

Pennsylvania Right to Know List
[present]

2,5,8,11,14-PENTAOXAPENTADECANE — 143-24-8

HEALTH AND SAFETY LISTS

NFPA - Flash Points
flash point < 70 degrees F (21 degrees C)

NFPA - Hazard Identification Ratings
health-0; flammability-3; reactivity-0

STATE LISTS

Florida Hazardous Substance List
[present]

Massachusetts Right To Know List
[present]

Pennsylvania Right to Know List
[present]

PENTASODIUM TRIPHOSPHATE — 7758-29-4

HEALTH AND SAFETY LISTS

U.S. DOT - Hazard Classes
Forbidden from transport by the DOT

PENTASODIUM TRIPHOSPHATE — 7758-79-4

HEALTH AND SAFETY LISTS

NFPA - Flash Points
flash point = 105 degrees F (41 degrees C)

NFPA - Hazard Identification Ratings
health-1; flammability-2; reactivity-0

NIOSH - Selected LD50s and LC50s
Oral, rat: LD50 = 1870 mg/kg Skin, rabbit: LD50 = 2520 mg/kg

INTERNATIONAL LISTS

Canada - WHMIS: Ingredient Disclosure
1% item 577 (699)

STATE LISTS

NJ Right to Know List (Total)
sn 0696

NJ Special Hazardous Substances
(flammable - third degree)

Pennsylvania Right to Know List
[present]

1-PENTENE — 109-67-1

STATE LISTS

Pennsylvania Right to Know List
[present]

1-PENTENE, 2,4,4-TRIMETHYL- — 107-39-1

ENVIRONMENTAL LISTS

TSCA - Section 12(b) - Export Notification
P-91-548; export notification required - Section 5

2-PENTENE, 2,4,4-TRIMETHYL- — 107-40-4

ENVIRONMENTAL LISTS

TSCA - Chemicals with Significant New Use Rules
PMN number: P-91-548

1-PENTENE, 2,3,4-TRIMETHYL- — 565-76-4

ENVIRONMENTAL LISTS

TSCA - Chemicals with Significant New Use Rules
PMN number: P-93-1202

PENTOBARBITAL SODIUM — 57-33-0

HEALTH AND SAFETY LISTS

NFPA - Flash Points
flash point = 285 degrees F (141 degrees C)

NFPA - Hazard Identification Ratings
health-1; flammability-1; reactivity-0

NIOSH - Selected LD50s and LC50s
Oral, rat: LD50 = 5140 mg/kg

PENTOL — 12772-47-3

HEALTH AND SAFETY LISTS

U.S. DOT - Appendix A Table 1 - Hazardous Substances
final RQ = 5000 pounds (2270 kg) (Listed under 'Sodium phosphate, tribasic')

ENVIRONMENTAL LISTS

CERCLA/SARA - Hazardous Substances and their Reportable Quantities
final RQ = 5000 pounds (2270 kg) (Listed under 'Sodium phosphate, tribasic')

Clean Water Act - Hazardous Substances
[present] (Listed under 'Sodium phosphate, tribasic')

List of Pesticide Product Inert Ingredients
[present]

STATE LISTS

California - Directors List of Hazardous Substances (8 CCR 339)
[present] (Listed under 'Sodium phosphate, tribasic')

Massachusetts Right To Know List
[present]

NJ Right to Know List (Total)
sn 3042

Pennsylvania Right to Know List
environmental hazard

PENTOLITE 8066-33-9
STATE LISTS
California - Directors List of Hazardous Substances (8 CCR 339)
[present]

P-TERT-PENTYLPHENOL 80-46-6
HEALTH AND SAFETY LISTS
NFPA - Flash Points
flash point = 0 degrees F (-18 degrees C)
NFPA - Hazard Identification Ratings
health-1; flammability-4; reactivity-0
ENVIRONMENTAL LISTS
CAA - Flammable Substances for Accidental Release Prevention
threshold quantity = 10,000 lbs
STATE LISTS
Florida Hazardous Substance List
[present]
Massachusetts Right To Know List
[present]
Pennsylvania Right to Know List
[present]

PENTYL PHENYL ACID PHOSPHATE 69867-71-6
HEALTH AND SAFETY LISTS
NFPA - Flash Points
flash point = 23 degrees F (-5 degrees C)
NFPA - Hazard Identification Ratings
health-2; flammability-3; reactivity-0
STATE LISTS
Florida Hazardous Substance List
[present]
Massachusetts Right To Know List
[present]
Pennsylvania Right to Know List
[present]

1-PENTYNE 627-19-0
HEALTH AND SAFETY LISTS
NFPA - Flash Points
flash point = 35 degrees F (2 degrees C)
NFPA - Hazard Identification Ratings
health-2; flammability-3; reactivity-0
STATE LISTS
Florida Hazardous Substance List
[present]
Massachusetts Right To Know List
[present]
Pennsylvania Right to Know List
[present]

1-PENTYN-3-OL, 3-METHYL- 77-75-8
HEALTH AND SAFETY LISTS
NFPA - Flash Points
flash point < 70 degrees F (21 degrees C)
NFPA - Hazard Identification Ratings
health-0; flammability-3; reactivity-0
STATE LISTS
Florida Hazardous Substance List
[present]

Massachusetts Right To Know List
[present]
Pennsylvania Right to Know List
[present]

PERACETIC ACID DILUTED WITH 60% OF ACETIC ACID RR-00797-6
ENVIRONMENTAL LISTS
CERCLA/SARA - Section 313 - Emission Reporting
form R reporting required
STATE LISTS
California - Air Bill 2588 Appendix A-II
9/90
California - Prop. 65 - Developmental Toxicity
developmental toxicity - initial date 7/1/90

PERCHLORATE COMPOUNDS, N.O.S. RR-00609-7
HEALTH AND SAFETY LISTS
U.S. DOT - Substances From 49 CFR 172.101
regulated by DOT (UN2705)
U.S. DOT - Hazard Classes
DOT hazard class = 8
STATE LISTS
NJ Right to Know List (Total)
sn 1478

PERCHLORATES, INORGANIC, N.O.S. RR-00413-7
HEALTH AND SAFETY LISTS
U.S. DOT - Substances From 49 CFR 172.101
regulated by DOT (UN0151)
U.S. DOT - Hazard Classes
DOT hazard class = 1.1D
STATE LISTS
NJ Right to Know List (Total)
sn 1479

PERCHLORIC ACID 7601-90-3
HEALTH AND SAFETY LISTS
NFPA - Flash Points
flash point = 232 degrees F (111 degrees C)
NFPA - Hazard Identification Ratings
health-2; flammability-1; reactivity-0
NIOSH - Selected LD50s and LC50s
Oral, rat: LD50 = 1830 mg/kg Skin, rabbit: LD50 = 2000 mg/kg
INTERNATIONAL LISTS
Canada - WHMIS: Ingredient Disclosure
1% item 1246 (1353)
STATE LISTS
Florida Hazardous Substance List
[present]
Massachusetts Right To Know List
[present]
Pennsylvania Right to Know List
[present]

PERCHLORYL FLUORIDE 7616-94-6
ENVIRONMENTAL LISTS
List of Pesticide Product Inert Ingredients
[present]

PERFLUOROALKYL AROMATIC CARBAMATE   RR-01231-7
MODIFIED ALKYL METHACRYLATE COPOLYMER

HEALTH AND SAFETY LISTS

NFPA - Flash Points
flash point < -4 degrees F (-20 degrees C)

NFPA - Hazard Identification Ratings
flammability-3; reactivity-3

STATE LISTS

Florida Hazardous Substance List
[present]

Massachusetts Right To Know List
[present]

Pennsylvania Right to Know List
[present]

PERFLUOROALKYL EPOXIDE   RR-00190-1

HEALTH AND SAFETY LISTS

NFPA - Flash Points
flash point = 101 degrees F (38 degrees C)

NFPA - Hazard Identification Ratings
health-1; flammability-2; reactivity-0

NIOSH - Selected LD50s and LC50s
Oral, rat: LD50 = 300 mg/kg

ENVIRONMENTAL LISTS

CAA - HON Rule - SOCMI Chemicals
compliance by Oct. 23, 1995

PERFLUOROETHYLVINYL ETHER   RR-01363-8

STATE LISTS

Pennsylvania Right to Know List
[present]

PERFLUORO-N-HEXANE   355-42-0

INTERNATIONAL LISTS

Canada - WHMIS: Ingredient Disclosure
1% item 1247 (1354)

PERFLUOROISOBUTYLENE   382-21-8

HEALTH AND SAFETY LISTS

U.S. DOT - Substances From 49 CFR 172.101
regulated by DOT (UN1481)

U.S. DOT - Hazard Classes
DOT hazard class = 5.1

STATE LISTS

NJ Right to Know List (Total)
sn 2636

PERFLUOROMETHYLVINYL ETHER   RR-01362-7

HEALTH AND SAFETY LISTS

U.S. DOT - Substances From 49 CFR 172.101
regulated by DOT (UN1802, UN1873)

U.S. DOT - Hazard Classes
DOT hazard class = 8

NIOSH - Selected LD50s and LC50s
Oral, rat: LD50 = 1100 mg/kg

OSHA - List of Highly Hazardous Chemicals
concentration > 60%: threshhold quantity = 5000 pounds

INTERNATIONAL LISTS

Canada - WHMIS: Ingredient Disclosure
1% item 1248 (122)

STATE LISTS

Florida Hazardous Substance List
[present]

Massachusetts Right To Know List
[present]

NJ Right to Know List (Total)
sn 2637; sn 2638

Pennsylvania Right to Know List
[present]

PERFLUOROTRIBUTYLAMINE   311-89-7

HEALTH AND SAFETY LISTS

ACGIH 1995 - Time Weighted Averages
3 ppm TWA; 13 mg/m3 TWA

ACGIH 1995 - Short Term Exposure Limits
6 ppm STEL; 25 mg/m3 STEL

U.S. DOT - Substances From 49 CFR 172.101
regulated by DOT (UN3083)

U.S. DOT - Hazard Classes
DOT hazard class = 2.3

U.S. DOT - Substances Which Are Poisonous by Inhalation
gaseous hazardous material poisonous by inhalation (UN3083)

NIOSH - Selected LD50s and LC50s
Inhalation, mouse: LC50 = 630 ppm 4 hr

NIOSH 1990 - Pocket Guide - RELs
3 ppm TWA; 14 mg/m3 TWA; 6 ppm STEL; 28 mg/m3 STEL

NIOSH 1990 - Pocket Guide - IDLHs
385 ppm IDLH

NIOSH 1990 - Pocket Guide - Target organs
skin, blood, respiratory system

OSHA - Vacated PELs - Time Weighted Averages
3 ppm TWA; 14 mg/m3 TWA

OSHA - Vacated PELs - Short Term Exposure Limits
6 ppm STEL; 28 mg/m3 STEL

OSHA - Final PELs - Time Weighted Averages
3 ppm TWA; 13.5 mg/m3 TWA

OSHA - List of Highly Hazardous Chemicals
threshhold quantity = 5000 pounds

INTERNATIONAL LISTS

Australian Exposure Standards - Time Weighted Averages
3 ppm TWA; 13 mg/m3 TWA

Australian Exposure Standards - Short Term Exposure Limits
6 ppm STEL; 25 mg/m3 STEL

Canada - WHMIS: Ingredient Disclosure
1% item 1251 (909)

Canada - Alberta - 8 Hour Occupational Exposure Limit
3 ppm TWA; 13 mg/m3 TWA

Canada - Alberta - 15 Minute Occupational Exposure Limit
6 ppm STEL; 25 mg/m3 STEL

Canada - British Columbia - 8 Hour Exposure Limits
3 ppm TWA; 14 mg/m3 TWA

Canada - British Columbia - 15 Minute Exposure Limits
6 ppm STEL; 28 mg/m3 STEL

Canada - Ontario - OHSA - TWAEVs
3 ppm TWAEV; 13 mg/m3 TWAEV

Canada - Ontario - OHSA - STEVs
6 ppm STEV; 25 mg/m3 STEV

Canada - Quebec - Time-Weighted Average Exposure Values
3 ppm TWAEV; 13 mg/m3 TWAEV

Canada - Quebec - Short-term Exposure Values
6 ppm STEV; 25 mg/m3 STEV

United Kingdom - Occupational Exposure Standards - TWAs
3 ppm TWA; 14 mg/m3 TWA

United Kingdom - Occupational Exposure Standards - STELs
6 ppm STEL; 28 mg/m3 STEL

Israel - Time Weighted Averages
3 ppm TWA; 13 mg/m3 TWA

Israel - Short Term Exposure Limits
6 ppm STEL; 25 mg/m3 STEL

Israel - Action Levels
1.5 ppm AL; 6.5 mg/m3 AL

Mexico - Instruction No. 10 - TWAs
3 ppm TWA; 14 mg/m3 TWA

Mexico - Instruction No. 10 - STELs
6 ppm STEL; 28 mg/m3 STEL

STATE LISTS

California - Exposure Limits - PELs
3 ppm PEL; 14 mg/m3 PEL

California - Exposure Limits - STELs
6 ppm STEL; 28 mg/m3 STEL

California - Directors List of Hazardous Substances (8 CCR 339)
[present]

Florida Hazardous Substance List
[present]

Massachusetts Right To Know List
[present]

Minnesota Hazardous Substance List
[present]

NJ Right to Know List (Total)
sn 1481

Pennsylvania Right to Know List
[present]

PERFUMERY PRODUCTS, WITH FLAMMABLE SOLVENT   RR-01519-0

ENVIRONMENTAL LISTS

TSCA - Chemicals with Significant New Use Rules
PMN number: P-87-1555

PERFUMES AND ESSENCES, JASMIN   8024-43-9

ENVIRONMENTAL LISTS

TSCA - Chemicals with Significant New Use Rules
PMN number: P-86-562

PERHALO ALKOXY ETHER   RR-00943-8

HEALTH AND SAFETY LISTS

U.S. DOT - Substances From 49 CFR 172.101
regulated by DOT (UN3154)

U.S. DOT - Hazard Classes
DOT hazard class = 2.1

PERHYDROPHENANTHRENE   5743-97-5

ENVIRONMENTAL LISTS

TSCA - Code of Federal Regulations Citations
40 CFR 712.30(x); 40 CFR 716.120(d)

TSCA - PAIR - Reporting List
Reporting Date: November 27, 1991

TSCA - Health and Safety Reporting List
Effective Date: September 30, 1991

PERILLA ALCOHOL   536-59-4

HEALTH AND SAFETY LISTS

ACGIH 1995 - Ceiling Limits
C 0.01 ppm; C 0.082 mg/m3

PERILLA OIL   68132-21-8

HEALTH AND SAFETY LISTS

U.S. DOT - Substances From 49 CFR 172.101
regulated by DOT (UN3153)

U.S. DOT - Hazard Classes
DOT hazard class = 2.1

PERLITE   93763-70-3

ENVIRONMENTAL LISTS

TSCA - Code of Federal Regulations Citations
40 CFR 712.30(x); 40 CFR 716.120(d)

TSCA - PAIR - Reporting List
Reporting Date: November 27, 1991

TSCA - Health and Safety Reporting List
Effective Date: September 30, 1991

PERLITE   130885-09-5

STATE LISTS

NJ Right to Know List (Total)
sn 2639

PERMANGANATES, INORGANIC, N.O.S.   RR-00416-0

ENVIRONMENTAL LISTS

List of Pesticide Product Inert Ingredients
[present]

PERMETHRIN   52645-53-1

ENVIRONMENTAL LISTS

TSCA - Chemicals with Significant New Use Rules
PMN number: P-83-1227

PEROXIDES, INORGANIC, N.O.S.   RR-00417-1

HEALTH AND SAFETY LISTS

NFPA - Hazard Identification Ratings
reactivity-0

PEROXISOME PROJECT (GEMFIBROZIL)   25812-30-0

INTERNATIONAL LISTS

Canada - WHMIS: Ingredient Disclosure
1% item 1252 (1358)

PEROXISOME PROJECT (WY-14643)   50892-23-4

HEALTH AND SAFETY LISTS

NFPA - Flash Points
flash point = 522 degrees F (272 degrees C)

NFPA - Hazard Identification Ratings
health-0; flammability-1; reactivity-0

PEROXYACETIC ACID   79-21-0

HEALTH AND SAFETY LISTS

ACGIH 1995 - Time Weighted Averages
10 mg/m3 TWA (The value is for total dust containing no asbestos and <1% crystalline silica)

OSHA - Vacated PELs - Time Weighted Averages
total dust: 15 mg/m3 TWA; respirable fraction: 5 mg/m3 TWA

OSHA - Final PELs - Time Weighted Averages
total dust: 15 mg/m3 TWA; respirable fraction: 5 mg/m3 TWA

ENVIRONMENTAL LISTS

List of Pesticide Product Inert Ingredients
[present]

INTERNATIONAL LISTS

Australian Exposure Standards - Time Weighted Averages
10 mg/m3 TWA

Canada - Alberta - 8 Hour Occupational Exposure Limit
respirable mass: 5 mg/m3 TWA; total mass: 10 mg/m3 TWA

Canada - Ontario - OHSA - TWAEVs
total dust: 10 mg/m3 TWAEV (listed as mineral dust)
Canada - Quebec - Time-Weighted Average Exposure Values
total dust: 10 mg/m3 TWAEV; respirable dust: 5 mg/m3 TWAEV
Israel - Time Weighted Averages
10 mg/m3 TWA (The value is for total dust containing no asbestos and <1% crystalline silica)
Israel - Action Levels
5 mg/m3 AL

STATE LISTS

California - Exposure Limits - PELs
total dust: 10 mg/m3 PEL; respirable fraction: 5 mg/m3 PEL
Florida Hazardous Substance List
[present] (as a dust only)
Massachusetts Right To Know List
[present]
Minnesota Hazardous Substance List
[present] (includes inert or nuisance dust)
Pennsylvania Right to Know List
[present]

PERSISTENT SYNTHETIC MATERIALS (INCLUDING PERSISTENT PLASTICS)   RR-01604-6

ENVIRONMENTAL LISTS

List of Pesticide Product Inert Ingredients
[present]

PERSULFATE COMPOUNDS   RR-00610-0

HEALTH AND SAFETY LISTS

U.S. DOT - Substances From 49 CFR 172.101
regulated by DOT (UN1482)
U.S. DOT - Hazard Classes
DOT hazard class = 5.1

STATE LISTS

NJ Right to Know List (Total)
sn 2640

PERYLENE   198-55-0

HEALTH AND SAFETY LISTS

IARC - Group 3 (not classifiable)
[present]

ENVIRONMENTAL LISTS

CERCLA/SARA - Section 313 - Emission Reporting
form R reporting required

STATE LISTS

Massachusetts Right To Know List
[present]

PESTICIDES, N.O.S.   RR-00418-2

HEALTH AND SAFETY LISTS

U.S. DOT - Substances From 49 CFR 172.101
regulated by DOT (UN1483)
U.S. DOT - Hazard Classes
DOT hazard class = 5.1

STATE LISTS

NJ Right to Know List (Total)
sn 2641

PETASITENINE   60102-37-6

HEALTH AND SAFETY LISTS

NTP Chemical Status Reports - Testing Status and NTIS Number
Prechronic studies in progress

PETROLATUM   8009-03-8

HEALTH AND SAFETY LISTS

NTP Chemical Status Reports - Testing Status and NTIS Number
Prechronic studies in progress

PETROLEUM ETHER   RR-01763-0

HEALTH AND SAFETY LISTS

U.S. DOT - Hazard Classes
Forbidden from transport by the DOT
U.S. DOT - Organic Peroxides Table
Organic peroxide UN3105;   UN3107;   UN3109
NFPA - Flash Points
diluted with 60% of acetic acid: flash point = 105 degrees F (41 degrees C)
NFPA - Hazard Identification Ratings
diluted with 60% of acetic acid: health-3;   flammability-2;   reactivity-4 (explodes on heating)
NIOSH - Selected LD50s and LC50s
Inhalation, rat: LC50 = 450 mg/m3 8 hr Oral, rat: LD50 = 1540 mg/kg Skin, rabbit: LD50 = 1410 mg/kg
OSHA - List of Highly Hazardous Chemicals
concentration > 60%: threshhold quantity = 1000 pounds

ENVIRONMENTAL LISTS

CERCLA/SARA - Section 302 Extremely Hazardous Substances and TPQs
TPQ = 500 pounds
CERCLA/SARA - Section 313 - Emission Reporting
form R reporting required for 1.0% de minimus concentration
CAA -Toxic Substances for Accidental Release Prevention
threshold quantity = 10,000 lbs
CAA - HON Rule - SOCMI Chemicals
compliance by Jan. 23, 1995

INTERNATIONAL LISTS

Canada - WHMIS: Ingredient Disclosure
1% item 1253 (123)
Canada - NPRI (National Pollutant Release Inventory)
[present]
German (DFG) - Skin/Sensitizers
very strong effects on skin
German (DFG) - Carcinogens
suspected carcinogen

STATE LISTS

California - Air Bill 2588 Appendix A-I
6/91
Florida Hazardous Substance List
[present]
Massachusetts Right To Know List
extraordinarily hazardous
NJ Right to Know List (Total)
sn 1482
NJ Special Hazardous Substances
(corrosive;   reactive - fourth degree)
Pennsylvania Right to Know List
environmental hazard

PROPOSED REGULATIONS

CERCLA/SARA - Proposed Hazardous Substance Additions
proposed RQ = 1 pound (.454 kg)
CERCLA/SARA - 1989 Proposed RQ Adjustments
proposed RQ = 10 pounds (4.54 kg)

PETROLEUM DISTILLATE RESIDUES   RR-01605-7

INTERNATIONAL LISTS

Canada - CEPA Schedule III Part I - Prohibited Substances (Ocean Dumping)
[present]

PETROLEUM DISTILLATES, ACID TREATED,   64742-18-3
HEAVY NAPHTHENIC

INTERNATIONAL LISTS

Canada - WHMIS: Ingredient Disclosure
0.1% item 1254 (1369)

Canada - Alberta - 8 Hour Occupational Exposure Limit
as S2O8: 2 mg/m3 TWA

Canada - Alberta - 15 Minute Occupational Exposure Limit
as S2O8: 4 mg/m3 STEL

Canada - Ontario - OHSA - TWAEVs
5 mg/m3 TWAEV

PROPOSED REGULATIONS

Canada - Ontario - Proposed Occupational TWAEVs
2 mg/m3 TWAEV

PETROLEUM DISTILLATES, ACID TREATED,   64742-20-7
HEAVY PARAFFINIC

HEALTH AND SAFETY LISTS

IARC - Group 3 (not classifiable)
[present]

ENVIRONMENTAL LISTS

ATSDR Priority List
Rank (of 275): 164

INTERNATIONAL LISTS

Canada - WHMIS: Ingredient Disclosure
1% item 1255 (1370)

PETROLEUM DISTILLATES, ACID TREATED,   64742-19-4
LIGHT NAPHTHENIC

HEALTH AND SAFETY LISTS

U.S. DOT - Substances From 49 CFR 172.101
regulated by DOT (UN2588, UN2902, UN2903, UN3021)

U.S. DOT - Hazard Classes
toxic or toxic, flammable: DOT hazard class = 6.1; flammable, toxic: DOT hazard class = 3

INTERNATIONAL LISTS

Canada - CEPA Schedule III Part II - Restricted Substances (Ocean Dumping)
[present]

STATE LISTS

NJ Right to Know List (Total)
sn 2642; sn 2643; sn 2644; sn 2645

PETROLEUM DISTILLATES, ACID TREATED,   64742-21-8
LIGHT PARAFFINIC

HEALTH AND SAFETY LISTS

IARC - Group 3 (not classifiable)
[present]

STATE LISTS

California - Directors List of Hazardous Substances (8 CCR 339)
[present]

PETROLEUM DISTILLATES, CATALYTIC, RE-   68477-31-6
FORMER FRACTIONATOR RESIDUE, LOW-
BOILING

ENVIRONMENTAL LISTS

List of Pesticide Product Inert Ingredients
[present]

PETROLEUM DISTILLATES, HEAVY CATALYTIC   64741-61-3
CRACKED

HEALTH AND SAFETY LISTS

NFPA - Flash Points
flash point < 0 degrees F (-18 degrees C)

NFPA - Hazard Identification Ratings
health-1; flammability-4; reactivity-0

PETROLEUM DISTILLATES, HEAVY NAPHTHENIC   64741-53-3

INTERNATIONAL LISTS

Canada - CEPA Schedule III Part I - Prohibited Substances (Ocean Dumping)
[present]

PETROLEUM DISTILLATES, HEAVY PARAFFINIC   64741-51-1

STATE LISTS

Massachusetts Right To Know List
carcinogen; extraordinarily hazardous

PETROLEUM DISTILLATES, HYDROTREATED   64742-53-6
LIGHT NAPHTHENIC

STATE LISTS

Massachusetts Right To Know List
carcinogen; extraordinarily hazardous

PETROLEUM DISTILLATES, HYDROTREATED   64742-55-8
LIGHT PARAFFINIC

STATE LISTS

Massachusetts Right To Know List
carcinogen; extraordinarily hazardous

PETROLEUM DISTILLATES, LIGHT CATALYTIC   64741-59-9
CRACKED

STATE LISTS

Massachusetts Right To Know List
carcinogen; extraordinarily hazardous

PETROLEUM DISTILLATES, LIGHT PARAFFINIC   64741-50-0

ENVIRONMENTAL LISTS

List of Pesticide Product Inert Ingredients
[present]

PETROLEUM DISTILLATES (NAPHTHA)   8002-05-9

HEALTH AND SAFETY LISTS

IARC - Group Unspecified
[present] (Listed under 'Occupational expsosures in petroleum refining')

PETROLEUM DISTILLATES, N.O.S.   RR-00422-8

ENVIRONMENTAL LISTS

List of Pesticide Product Inert Ingredients
[present]

STATE LISTS

Massachusetts Right To Know List
carcinogen; extraordinarily hazardous

PETROLEUM DISTILLATES, SOLVENT DEWAXED   64742-56-9
LIGHT PARAFFINIC

ENVIRONMENTAL LISTS

List of Pesticide Product Inert Ingredients
[present]

STATE LISTS

Massachusetts Right To Know List
carcinogen; extraordinarily hazardous

PETROLEUM DISTILLATES, SOLVENT-REFINED LIGHT PARAFFINIC   64741-89-5

STATE LISTS

Massachusetts Right To Know List
carcinogen; extraordinarily hazardous

PETROLEUM DISTILLATES, VACUUM   70592-78-8

STATE LISTS

Massachusetts Right To Know List
carcinogen; extraordinarily hazardous

PETROLEUM GASES, LIQUIFIED, SWEETENED   68476-86-8

HEALTH AND SAFETY LISTS

IARC - Group Unspecified
[present] (Listed under 'Occupational exposures in petroleum refining')

PETROLEUM NAPHTHA, LIGHT AROMATIC   64742-95-6

ENVIRONMENTAL LISTS

List of Pesticide Product Inert Ingredients
[present]

STATE LISTS

Massachusetts Right To Know List
carcinogen; extraordinarily hazardous

PETROLEUM OIL   RR-00423-9

HEALTH AND SAFETY LISTS

U.S. DOT - Substances From 49 CFR 172.101
regulated by DOT (UN1267, UN1270)

U.S. DOT - Hazard Classes
DOT hazard class = 3

IARC - Group 2A (limited human data)
[present] (refers to the occupational exposure during petroleum refining)

IARC - Group 3 (not classifiable)
[present] (as crude oil)

NFPA - Flash Points
sweet and sour: flash point = 20 to 90 degrees F (-7 to 32 degrees C)

NFPA - Hazard Identification Ratings
sweet: health-1; flammability-3; reactivity-0; sour: health-2; flammability-3; reactivity-0

NIOSH 1990 - Pocket Guide - RELs
350 mg/m3 TWA; C 1800 mg/m3 (15 min)

NIOSH 1990 - Pocket Guide - IDLHs
10,000 ppm IDLH

NIOSH 1990 - Pocket Guide - Target organs
skin, eyes, CNS, respiratory system

NIOSH - Health Standards - Exposure Limits
350 mg/m3 TWA; C (15 min) 1800 mg/m3 (Listed under 'Refined petroleum solvents')

NIOSH - Health Standards - Health Effects and Precautions
Eye, nose, and throat irritation; dermatitis; nervous system effects (Blood and urine monitoring required; prevent skin contact) (Listed under 'Refined petroleum solvents')

OSHA - Vacated PELs - Time Weighted Averages
400 ppm TWA; 1600 mg/m3 TWA

OSHA - Final PELs - Time Weighted Averages
500 ppm TWA; 2000 mg/m3 TWA

OSHA - Possible Select Carcinogens
[present]

INTERNATIONAL LISTS

Canada - CEPA Schedule III Part I - Prohibited Substances (Ocean Dumping)
[present]

STATE LISTS

California - Air Bill 2588 Appendix A-II
known or potential carcinogen

Florida Hazardous Substance List
[present]

Massachusetts Right To Know List
[present]

Minnesota Hazardous Substance List
carcinogen (includes lubricants, base oils, and derived products)

NJ Right to Know List (Total)
sn 2648

Pennsylvania Right to Know List
[present]

PETROLEUM REFINING RESIDUES, POLYMERIZED   68476-87-9

HEALTH AND SAFETY LISTS

U.S. DOT - Substances From 49 CFR 172.101
regulated by DOT (UN1268)

U.S. DOT - Hazard Classes
DOT hazard class = 3

STATE LISTS

Pennsylvania Right to Know List
[present]

PETROLEUM RESINS   64742-16-1

STATE LISTS

Massachusetts Right To Know List
carcinogen; extraordinarily hazardous

PETROLEUM SOLVENTS   RR-01546-3

STATE LISTS

Massachusetts Right To Know List
carcinogen; extraordinarily hazardous

PETROLEUM SULFONATE   RR-00877-5

HEALTH AND SAFETY LISTS

IARC - Group Unspecified
[present] (Listed under 'Occupational exposures in petroleum refining')

PHARMAMEDIA   12751-36-9

HEALTH AND SAFETY LISTS

U.S. DOT - Substances From 49 CFR 172.101
regulated by DOT (UN1075)

U.S. DOT - Hazard Classes
DOT hazard class = 2.1

BETA-PHELLANDRENE   555-10-2

ENVIRONMENTAL LISTS

List of Pesticide Product Inert Ingredients
[present]

TSCA - Code of Federal Regulations Citations
40 CFR 716.120

TSCA - Health and Safety Reporting List
Effective Date: February 13, 1984

PHENACEMIDE   63-98-9

STATE LISTS

NJ Right to Know List (Total)
sn 2651

PHENACETIN   62-44-2

HEALTH AND SAFETY LISTS

IARC - Group Unspecified
[present] (Listed under 'Occupational exposures in petroleum refining')

PHENACYL BROMIDE   70-11-1
ENVIRONMENTAL LISTS
List of Pesticide Product Inert Ingredients
[present]

PHENANTHRENE   85-01-8
HEALTH AND SAFETY LISTS
IARC - Group 3 (not classifiable)
[present]

1,10-PHENANTHROLINE   66-71-7
HEALTH AND SAFETY LISTS
NFPA - Flash Points
flash point = 400 degrees F (204 degrees C)
NFPA - Hazard Identification Ratings
health-0; flammability-1; reactivity-0

PHENAZOPYRIDINE   94-78-0
ENVIRONMENTAL LISTS
List of Pesticide Product Inert Ingredients
[present]

PHENAZOPYRIDINE HYDROCHLORIDE   136-40-3
HEALTH AND SAFETY LISTS
NFPA - Flash Points
flash point = 120 degrees F (49 degrees C)
NFPA - Hazard Identification Ratings
health-0; flammability-2; reactivity-0

PHENELZINE SULPHATE   156-51-4
STATE LISTS
California - Air Bill 2588 Appendix A-II
9/90
California - Prop. 65 - Developmental Toxicity
developmental toxicity - initial date 7/1/90

PHENESTERIN   3546-10-9
SEE ALSO:
F039-HAZARDOUS WASTES
HEALTH AND SAFETY LISTS
U.S. DOT - Appendix A Table 1 - Hazardous Substances
final RQ = 100 pounds (45.4 kg)
IARC - Group 2A (limited human data)
[present]
NIOSH - Selected LD50s and LC50s
Oral, rat: LD50 = 1500 mg/kg
NTP Seventh Report - Suspect Carcinogens
suspect carcinogen
OSHA - Possible Select Carcinogens
[present]
ENVIRONMENTAL LISTS
CERCLA/SARA - Hazardous Substances and their Reportable Quantities
final RQ = 100 pounds (45.4 kg)
EPA - Carcinogen Hazard Ranking for RQ Adjustment
Hazard ranking = Low
RCRA - U Series Wastes
waste number U187
RCRA - Hazardous Constituents-Appendix VIII
waste number U187
RCRA - Basis for Listing - Appendix VII
Included in waste stream: F039

RCRA - Substances Banned From Land Disposal
[present]
RCRA - TSD Facilities Ground Water Monitoring
TM 8270 = 10 ug/L PQL
RCRA - Universal Treatment Standards (LDR)
WW: 0.081 mg/l; NWW: 16 mg/kg
TSCA - Chemicals with Significant New Use Rules
[present]
TSCA - Section 12(b) - Export Notification
export notification required - Section 5
STATE LISTS
California - Air Bill 2588 Appendix A-II
known or potential carcinogen
California - Prop. 65 - Cancer list
carcinogen - initial date 10/1/89
California - Prop. 65 - No Significant Risk Levels
no significant risk level = 300 ug/day
California - Directors List of Hazardous Substances (8 CCR 339)
[present]
Florida Hazardous Substance List
[present]
Massachusetts Right To Know List
carcinogen; extraordinarily hazardous
Minnesota Hazardous Substance List
carcinogen
NJ Right to Know List (Total)
sn 1483
NJ Special Hazardous Substances
(carcinogen)
Pennsylvania Right to Know List
environmental hazard; special hazardous substance
Pennsylvania RTK - Special Hazardous Substances
[present]

PHENETHYL ALCOHOL   60-12-8
HEALTH AND SAFETY LISTS
U.S. DOT - Substances From 49 CFR 172.101
regulated by DOT (UN2645)
U.S. DOT - Hazard Classes
DOT hazard class = 6.1
INTERNATIONAL LISTS
Canada - WHMIS: Ingredient Disclosure
1% item 1256 (339)
STATE LISTS
NJ Right to Know List (Total)
sn 1484

PHENETHYLAMINE   64-04-0
SEE ALSO:
F039-HAZARDOUS WASTES
COAL TAR PITCHES
HEALTH AND SAFETY LISTS
U.S. DOT - Appendix A Table 1 - Hazardous Substances
final RQ = 5000 pounds (2270 kg)
IARC - Group 3 (not classifiable)
[present]
NFPA - Flash Points
flash point = 340 degrees F (171 degrees C)
NFPA - Hazard Identification Ratings
flammability-1; reactivity-0
NIOSH - Selected LD50s and LC50s
Oral, mouse: LD50 = 700 mg/kg

ENVIRONMENTAL LISTS
ATSDR Priority List
Rank (of 275): 188
CERCLA/SARA - Section 313 - Emission Reporting
form R reporting required
CERCLA/SARA - Hazardous Substances and their Reportable Quantities
final RQ = 5000 pounds (2270 kg)
CAA - HON Rule - SOCMI Chemicals
compliance by Oct. 23, 1995
Clean Water Act - Priority Pollutants
[present]
RCRA - Basis for Listing - Appendix VII
Included in waste stream: F039
RCRA - TSD Facilities Ground Water Monitoring
TM 8100 = 200 ug/L PQL; TM 8270 = 10 ug/L PQL
RCRA - Universal Treatment Standards (LDR)
WW: 0.059 mg/l; NWW: 5.6 mg/kg
TSCA - Code of Federal Regulations Citations
40 CFR 704.225(a)
TSCA - CAIR - Reporting List
reporting required by: manufacturer; distributor; importer; processor
INTERNATIONAL LISTS
Canada - WHMIS: Ingredient Disclosure
1% item 1257 (1371)
Mexico - Wastewater - Organic Toxic Pollutants and Heavy Metals
Listed under [Aromatic Hydrocarbons]
STATE LISTS
California - Directors List of Hazardous Substances (8 CCR 339)
[present]
Massachusetts Right To Know List
[present]
Pennsylvania Right to Know List
environmental hazard

O-PHENETIDINE   94-70-2
HEALTH AND SAFETY LISTS
NIOSH - Selected LD50s and LC50s
Oral, rat: LD50 = 132 mg/kg
ENVIRONMENTAL LISTS
List of Pesticide Product Inert Ingredients
[present]

P-PHENETIDINE   156-43-4
STATE LISTS
California - Prop. 65 - Cancer list
carcinogen - initial date 1/1/88
California - Prop. 65 - No Significant Risk Levels
no significant risk level = 4 ug/day
California - Directors List of Hazardous Substances (8 CCR 339)
[present]
Florida Hazardous Substance List
[present]
Massachusetts Right To Know List
carcinogen; extraordinarily hazardous
Minnesota Hazardous Substance List
carcinogen
Pennsylvania Right to Know List
special hazardous substance
Pennsylvania RTK - Special Hazardous Substances
[present]

PHENETIDINE   1321-31-9
HEALTH AND SAFETY LISTS
IARC - Group 2B (sufficient animal data)
[present]
NIOSH - Selected LD50s and LC50s
Oral, rat: LD50 = 403 mg/kg
NTP Chemical Status Reports - Testing Status and NTIS Number
Technical reports printed (PB286207/AS)
NTP Chemical Status Reports - Evidence of Carcinogenicity
male rat-positive; female rat-positive; male mice-negative; female mice-positive
NTP Seventh Report - Suspect Carcinogens
suspect carcinogen
OSHA - Possible Select Carcinogens
[present]
STATE LISTS
California - Air Bill 2588 Appendix A-II
known or potential carcinogen
California - Prop. 65 - Cancer list
carcinogen - initial date 1/1/88
California - Prop. 65 - No Significant Risk Levels
no significant risk level = 5 ug/day
California - Directors List of Hazardous Substances (8 CCR 339)
[present]
Florida Hazardous Substance List
[present]
Massachusetts Right To Know List
carcinogen; extraordinarily hazardous
Minnesota Hazardous Substance List
carcinogen
NJ Special Hazardous Substances
(carcinogen)
Pennsylvania Right to Know List
special hazardous substance
Pennsylvania RTK - Special Hazardous Substances
[present]

PHENETOLE   103-73-1
HEALTH AND SAFETY LISTS
IARC - Group 3 (not classifiable)
[present]
STATE LISTS
California - Directors List of Hazardous Substances (8 CCR 339)
[present]

PHENFORMIN HYDROCHLORIDE   834-28-6
HEALTH AND SAFETY LISTS
NTP Chemical Status Reports - Testing Status and NTIS Number
Technical reports printed (PB283361/AS)
NTP Chemical Status Reports - Evidence of Carcinogenicity
male rat-negative; female rat-positive; male mice-positive; female mice-positive
STATE LISTS
California - Air Bill 2588 Appendix A-II
known or potential carcinogen: 9/89
California - Prop. 65 - Cancer list
carcinogen - initial date 7/1/89
California - Prop. 65 - No Significant Risk Levels
no significant risk level = 0.005 ug/day
Massachusetts Right To Know List
carcinogen; extraordinarily hazardous

PHENICARBAZIDE 103-03-7

HEALTH AND SAFETY LISTS

NFPA - Flash Points
flash point = 205 degrees F (96 degrees C)

NFPA - Hazard Identification Ratings
health-1; flammability-1; reactivity-0

NIOSH - Selected LD50s and LC50s
Oral, rat: LD50 = 1790 mg/kg Skin, guinea pig: LD50 = 5000 mg/kg

INTERNATIONAL LISTS

Canada - WHMIS: Ingredient Disclosure
1% item 1258 (185)

PHENOBARBITAL 50-06-6

HEALTH AND SAFETY LISTS

NIOSH - Selected LD50s and LC50s
Oral, mouse: LD50 = 400 mg/kg

PHENOL 108-95-2

HEALTH AND SAFETY LISTS

NFPA - Flash Points
flash point = 239 degrees F (115 degrees C)

NFPA - Hazard Identification Ratings
health-2; flammability-1; reactivity-0

INTERNATIONAL LISTS

Canada - WHMIS: Ingredient Disclosure
1% item 1259 (1372)

STATE LISTS

Florida Hazardous Substance List
[present]

Massachusetts Right To Know List
[present]

Pennsylvania Right to Know List
[present]

PHENOL, [[(2-AMINOETHYL)AMINO]METHYL]- 53894-28-3

HEALTH AND SAFETY LISTS

NFPA - Flash Points
flash point = 241 degrees F (116 degrees C)

NFPA - Hazard Identification Ratings
health-2; flammability-1; reactivity-0

ENVIRONMENTAL LISTS

CAA - HON Rule - SOCMI Chemicals
compliance by April 24, 1995

EPA - Master Testing List
[present]

INTERNATIONAL LISTS

Canada - WHMIS: Ingredient Disclosure
1% item 1260 (1373)

STATE LISTS

Florida Hazardous Substance List
[present]

Massachusetts Right To Know List
[present]

Pennsylvania Right to Know List
[present]

PHENOL, 3-[2-CHLORO-4-(TRIFLUOROMETHYL) PHENOXY]-, ACETATE 50594-77-9

HEALTH AND SAFETY LISTS

U.S. DOT - Substances From 49 CFR 172.101
regulated by DOT (UN2311)

U.S. DOT - Hazard Classes
DOT hazard class = 6.1

STATE LISTS

NJ Right to Know List (Total)
sn 1486

PHENOL, 2,4-DIAMINO-6-METHYL- 15872-73-8

HEALTH AND SAFETY LISTS

NFPA - Flash Points
flash point = 145 degrees F (63 degrees C)

NFPA - Hazard Identification Ratings
health-0; flammability-2; reactivity-0

NIOSH - Selected LD50s and LC50s
Oral, mouse: LD50 = 2200 mg/kg

PHENOL, 2,4-DIAMINO-6-METHYL-, HYDROCHLORIDE 65879-44-9

HEALTH AND SAFETY LISTS

NTP Chemical Status Reports - Testing Status and NTIS Number
Technical reports printed (PB266176/AS)

NTP Chemical Status Reports - Evidence of Carcinogenicity
male rat-negative; female rat-negative; male mice-negative; female mice-negative

PHENOL, 2,4(OR 2,6)-DIBROMO-, HOMOPOLYMER 69882-11-7

HEALTH AND SAFETY LISTS

IARC - Group 3 (not classifiable)
[present]

STATE LISTS

California - Directors List of Hazardous Substances (8 CCR 339)
[present]

PHENOL, 4-(3,4-DIHYDRO-2,2,4-TRIMETHYL-2H-1-BENZOPYRAN-4-YL)- 472-41-3

HEALTH AND SAFETY LISTS

IARC - Group 2B (sufficient animal data)
[present]

NIOSH - Selected LD50s and LC50s
Oral, rat: LD50 = 162 mg/kg

OSHA - Possible Select Carcinogens
[present]

STATE LISTS

California - Air Bill 2588 Appendix A-I
known or potential carcinogen

California - Prop. 65 - Cancer list
carcinogen - initial date 1/1/90

California - Prop. 65 - No Significant Risk Levels
no significant risk level = 2 ug/day

California - Directors List of Hazardous Substances (8 CCR 339)
[present]

Massachusetts Right To Know List
teratogen

Minnesota Hazardous Substance List
carcinogen

PHENOL, 4-(1,1-DIMETHYLETHYL)-, PHOSPHATE   78-33-1
(3:1)

SEE ALSO:
- K001-HAZARDOUS WASTES
- K087-HAZARDOUS WASTES
- K022-HAZARDOUS WASTES
- F039-HAZARDOUS WASTES

HEALTH AND SAFETY LISTS

ACGIH 1995 - Time Weighted Averages
5 ppm TWA; 19 mg/m3 TWA

ACGIH 1995 - Skin Designations
skin - potential for cutaneous absorption

ACGIH 1995 - Biological Exposure Indices
Total phenol in urine: 250 mg/g creatinine, end of shift (B, Ns)

AIHA - Odor Threshold Values
geometric mean air odor threshold = 0.060 ppm (detectable)

U.S. DOT - Substances From 49 CFR 172.101
regulated by DOT (UN2312, UN1671)

U.S. DOT - Hazard Classes
DOT hazard class = 6.1

U.S. DOT - Appendix A Table 1 - Hazardous Substances
final RQ = 1000 pounds (454 kg)

IARC - Group 3 (not classifiable)
[present]

NFPA - Flash Points
flash point = 175 degrees F (79 degrees C)

NFPA - Hazard Identification Ratings
health-4; flammability-2; reactivity-0

NIOSH - Selected LD50s and LC50s
Inhalation, rat: LC50 = 316 mg/m3 8 hr Oral, rat: LD50 = 317 mg/kg Skin, rat: LD50 = 669 mg/kg

NIOSH 1990 - Pocket Guide - RELs
5 ppm TWA; 19 mg/m3 TWA; C 15.6 ppm (15 min); C 60 mg/m3 (15 min)

NIOSH 1990 - Pocket Guide - IDLHs
250 ppm IDLH

NIOSH 1990 - Pocket Guide - Target organs
liver, kidneys, skin

NIOSH 1990 - Pocket Guide - Skin list
Potential for dermal absorption

NIOSH - Health Standards - Exposure Limits
5.2 ppm TWA; 20 mg/m3 TWA; C (15 min) 15.6 ppm; C (15 min) 60 mg/m3

NIOSH - Health Standards - Health Effects and Precautions
Skin, eye, central nervous system, liver, and kidney effects (Prevent skin and eye contact)

NTP Chemical Status Reports - Testing Status and NTIS Number
Technical reports printed (PB80217946)

NTP Chemical Status Reports - Evidence of Carcinogenicity
male rat-negative; female rat-negative; male mice-negative; female mice-negative

OSHA - Vacated PELs - Time Weighted Averages
5 ppm TWA; 19 mg/m3 TWA

OSHA - Vacated PELs - Skin Designation
Prevent or reduce skin absorption

OSHA - Final PELs - Time Weighted Averages
5 ppm TWA; 19 mg/m3 TWA

OSHA - Final PELs - Skin Notations
prevent or reduce skin absorption

ENVIRONMENTAL LISTS

ATSDR Priority List
Rank (of 275): 129

CERCLA/SARA - Section 302 Extremely Hazardous Substances and TPQs
TPQ = 500/10,000 pounds

CERCLA/SARA - Section 313 - Emission Reporting
form R reporting required for 1.0% de minimus concentration

CERCLA/SARA - Hazardous Substances and their Reportable Quantities
final RQ = 1000 pounds (454 kg)

Clean Air Act (1990) - List of Hazardous Air Contaminants
[present]

CAA - HON Rule - SOCMI Chemicals
compliance by April 24, 1995

CAA - HON Rule - Organic HAPs
[present]

Clean Water Act - Hazardous Substances
[present]

Clean Water Act - Priority Pollutants
[present]

Clean Water Act - Toxic Pollutants
[present]

EPA - Master Testing List
[present]

RCRA - U Series Wastes
waste number U188

RCRA - Hazardous Constituents-Appendix VIII
waste number U188

RCRA - Basis for Listing - Appendix VII
Included in waste streams: F039, K001, K022, K087

RCRA - Substances Banned From Land Disposal
[present]

RCRA - TSD Facilities Ground Water Monitoring
TM 8040 = 1 ug/L PQL; TM 8270 = 10 ug/L PQL

RCRA - Universal Treatment Standards (LDR)
WW: 0.039 mg/l; NWW: 6.2 mg/kg

TSCA - Code of Federal Regulations Citations
40 CFR 716.120(a)

INTERNATIONAL LISTS

Australian Exposure Standards - Time Weighted Averages
5 ppm TWA; 19 mg/m3 TWA

Australian Exposure Standards - Skin Effects
skin absorption

Canada - WHMIS: Ingredient Disclosure
1% item 1261 (1374)

Canada - NPRI (National Pollutant Release Inventory)
[present]

Canada - Alberta - 8 Hour Occupational Exposure Limit
5 ppm TWA; 19 mg/m3 TWA

Canada - Alberta - 15 Minute Occupational Exposure Limit
10 ppm STEL; 38 mg/m3 STEL

Canada - Alberta - Skin Designation
can be absorbed through the intact skin

Canada - British Columbia - 8 Hour Exposure Limits
5 ppm TWA; 19 mg/m3 TWA

Canada - British Columbia - 15 Minute Exposure Limits
10 ppm STEL; 38 mg/m3 STEL

Canada - British Columbia - Skin Notations
skin - potential for skin absorption

Canada - Ontario - OHSA - TWAEVs
5 ppm TWAEV; 19 mg/m3 TWAEV

Canada - Ontario - OHSA - Skin Notations
absorption through skin, eyes, or mucous membranes

Canada - Quebec - Time-Weighted Average Exposure Values
5 ppm TWAEV; 19 mg/m3 TWAEV

Canada - Quebec - Skin Designations
absorbed through the skin

United Kingdom - Occupational Exposure Standards - TWAs
5 ppm TWA; 19 mg/m3 TWA

United Kingdom - Occupational Exposure Standards - STELs
10 ppm STEL; 38 mg/m3 STEL

United Kingdom - Occupational Exposure Standards - Notes
can be absorbed through skin

German (DFG) - MAK Values
5 ppm MAK; 19 mg/m3 MAK

German (DFG) - Peak Limitations
2 x normal MAK (5 min momentary value); don't exceed 8 times during shift

German (DFG) - Skin/Sensitizers
danger of cutaneous absorption

Israel - Time Weighted Averages
5 ppm TWA; 19 mg/m3 TWA

Israel - Action Levels
2.5 ppm AL; 9.5 mg/m3 AL

Mexico - Instruction No. 10 - TWAs
5 ppm TWA; 19 mg/m3 TWA

Mexico - Instruction No. 10 - STELs
10 ppm STEL; 38 mg/m3 STEL

Mexico - Instruction No. 10 - Skin designation
skin - potential for cutaneous absorption

Mexico - Wastewater - Organic Toxic Pollutants and Heavy Metals
Listed under [Organic Toxic Pollutants]

Mexico - Drinking Water - Ecological Criteria
0.3 mg/l

STATE LISTS

California - Air Bill 2588 Appendix A-I
[present]

California - Exposure Limits - PELs
5 ppm PEL; 19 mg/m3 PEL

California - Exposure Limits - Skin Notation
material may be absorbed through the skin, eyes or mucous membrane

California - Directors List of Hazardous Substances (8 CCR 339)
[present]

Florida Hazardous Substance List
[present]

Massachusetts Right To Know List
extraordinarily hazardous

Minnesota Hazardous Substance List
skin

NJ Right to Know List (Total)
sn 1487

NJ Special Hazardous Substances
(mutagen)

Pennsylvania Right to Know List
environmental hazard

PROPOSED REGULATIONS

TSCA - ITC 33rd Report Priority Testing List
designated for testing

Canada - Ontario - Proposed Occupational TWAEVs
1 ppm TWAEV; 4 mg/m3 TWAEV

Canada - Ontario - Proposed Occupational STEVs
2 ppm STEV; 8 mg/m3 STEV

PHENOL, 2,4-DIPENTYL-   138-00-1

STATE LISTS

Pennsylvania Right to Know List
special hazardous substance

Pennsylvania RTK - Special Hazardous Substances
[present]

PHENOL, 4-ETHYL-   123-07-9

ENVIRONMENTAL LISTS

TSCA - PAIR - Reporting List
Reporting Date: June 10, 1993

TSCA - Health and Safety Reporting List
Effective Date: April 12, 1993

PHENOLIC COMPOUNDS   RR-00498-8
SEE ALSO:
PHENOL, 2,4-DIAMINO-6-METHYL-

ENVIRONMENTAL LISTS

TSCA - Code of Federal Regulations Citations
40 CFR 712.30(f); 40 CFR 716.120(c)

TSCA - PAIR - Reporting List
Reporting Date: August 17, 1983

PHENOL, 4-ISOCYANATO-, PHOSPHOROTHIOATE   4151-51-3
(3:1) (ESTER)
SEE ALSO:
PHENOL, 2,4-DIAMINO-6-METHYL-, HYDROCHLORIDE

ENVIRONMENTAL LISTS

TSCA - Code of Federal Regulations Citations
40 CFR 712.30(f); 40 CFR 716.120

TSCA - PAIR - Reporting List
Reporting Date: August 17, 1983

PHENOL, 4,4'-METHYLENEBIS(2,6-DIMETHYL)-   5384-21-4

ENVIRONMENTAL LISTS

TSCA - Code of Federal Regulations Citations
40 CFR 712.30(x); 40 CFR 716.120(d)

TSCA - PAIR - Reporting List
Reporting Date: December 27, 1990

TSCA - Health and Safety Reporting List
Effective Date: October 29, 1990

PHENOL, 4,4'-[METHYLENEBIS(OXY-2,1-   93589-69-6
ETHANEDIYLTHIO)]BIS-

ENVIRONMENTAL LISTS

TSCA - Code of Federal Regulations Citations
40 CFR 716.120(a)

TSCA - Health and Safety Reporting List
Effective Date: June 1, 1987

PHENOL, 3-(1-METHYLETHYL)-,   64-00-6
METHYLCARBAMATE
SEE ALSO:
BISAZOBIPHENYL DYES

ENVIRONMENTAL LISTS

EPA - Master Testing List
[present]

TSCA - Code of Federal Regulations Citations
40 CFR 712.30(d); 40 CFR 716.120(c)

TSCA - PAIR - Reporting List
Reporting Date: November 19, 1982

PHENOL, 4,4'-(OXYBIS(2,1-ETHANEDIYLTHIO))BIS-   90884-29-0

HEALTH AND SAFETY LISTS

NFPA - Flash Points
flash point = 260 degrees F (127 degrees C)

NFPA - Hazard Identification Ratings
health-2; flammability-1; reactivity-0

STATE LISTS

Florida Hazardous Substance List
[present]

Massachusetts Right To Know List
[present]
Pennsylvania Right to Know List
[present]

PHENOLPHTHALEIN   77-09-8

HEALTH AND SAFETY LISTS

NFPA - Flash Points
flash point = 219 degrees F (104 degrees C)
NFPA - Hazard Identification Ratings
health-2; flammability-1; reactivity-0

STATE LISTS

Florida Hazardous Substance List
[present]
Massachusetts Right To Know List
[present]
Pennsylvania Right to Know List
[present]

PHENOL RED   143-74-8

SEE ALSO:
K060-HAZARDOUS WASTES

ENVIRONMENTAL LISTS

RCRA - Basis for Listing - Appendix VII
Included in waste stream: K060

STATE LISTS

California - Directors List of Hazardous Substances (8 CCR 339)
[present]

PHENOLSULPHONIC ACID   98-11-3

ENVIRONMENTAL LISTS

TSCA - Code of Federal Regulations Citations
40 CFR 712.30(x); 40 CFR 716.120(d)
TSCA - PAIR - Reporting List
Reporting Date: December 27, 1990
TSCA - Health and Safety Reporting List
Effective Date: October 29, 1990; Sunset Date: November 9, 1993

PHENOLSULPHONIC ACID   1333-39-7

ENVIRONMENTAL LISTS

TSCA - Code of Federal Regulations Citations
40 CFR 721.1537
TSCA - Chemicals with Significant New Use Rules
PMN number: P-88-864
TSCA - Section 12(b) - Export Notification
P-88-864; export notification required - Section 5

PHENOL, 4-(1,1,3,3-TETRAMETHYLBUTYL)-   140-66-9

ENVIRONMENTAL LISTS

TSCA - Code of Federal Regulations Citations
40 CFR 721.1540
TSCA - Chemicals with Significant New Use Rules
PMN number: P-87-1760
TSCA - Section 12(b) - Export Notification
P-87-1760; export notification required - Section 5

PHENOL, 2,2'-THIOBIS(4-CHLORO-6-METHYL)-   4418-66-0

HEALTH AND SAFETY LISTS

NIOSH - Selected LD50s and LC50s
Oral, rat: LD50 = 29 mg/kg Skin, rat: LD50 = 113 mg/kg

ENVIRONMENTAL LISTS

CERCLA/SARA - Section 302 Extremely Hazardous Substances and
TPQs
TPQ = 500/10,000 pounds

STATE LISTS

Florida Hazardous Substance List

]effective March 13, 1992
Massachusetts Right To Know List
extraordinarily hazardous
NJ Right to Know List (Total)
sn 2654
Pennsylvania Right to Know List
environmental hazard

PROPOSED REGULATIONS

CERCLA/SARA - Proposed Hazardous Substance Additions
proposed RQ = 1 pound (.454 kg)
CERCLA/SARA - 1989 Proposed RQ Adjustments
proposed RQ = 100 pounds (45.4 kg)

PHENOL, 2,2'-THIOBIS(4,6-DICHLORO)-   97-18-7

ENVIRONMENTAL LISTS

TSCA - Code of Federal Regulations Citations
40 CFR 721.1538
TSCA - Chemicals with Significant New Use Rules
PMN number: P-89-651
TSCA - Section 12(b) - Export Notification
P-89-651; export notification required - Section 5

PHENOL, 2,4,6-TRIBROMO-   118-79-6

HEALTH AND SAFETY LISTS

NTP Chemical Status Reports - Testing Status and NTIS Number
*Two year studies: pathology quality assessment in progress;
prechronic studies for which toxicity technical reports were not prepared*

ENVIRONMENTAL LISTS

CAA - HON Rule - SOCMI Chemicals
compliance by April 24, 1995

PHENOL, 2,4,6-TRIMETHYL-   527-60-6

ENVIRONMENTAL LISTS

List of Pesticide Product Inert Ingredients
[present]

PHENOLSULFONIC ACID - FORMALDEHYDE -   ·52277-29-9
UREA CONDENSATE, SODIUM SALT

HEALTH AND SAFETY LISTS

U.S. DOT - Substances From 49 CFR 172.101
regulated by DOT (UN1803)
U.S. DOT - Hazard Classes
DOT hazard class = 8
NIOSH - Selected LD50s and LC50s
Oral, rat: LD50 = 890 mg/kg

ENVIRONMENTAL LISTS

CAA - HON Rule - SOCMI Chemicals
compliance by Oct. 24, 1994

STATE LISTS

NJ Right to Know List (Total)
sn 1488
NJ Special Hazardous Substances
(corrosive)

PHENOTHIAZINE   92-84-2

ENVIRONMENTAL LISTS

CAA - HON Rule - SOCMI Chemicals
compliance by April 24, 1995

INTERNATIONAL LISTS

Canada - WHMIS: Ingredient Disclosure
1% item 1262 (124)

PHENOTHRIN [2,2-DIMETHYL-3-(2-METHYL-1-   26002-80-2
PROPENYL)CYCLOPROPANECARBOXYLIC ACID
(3-PHENOXYPHENYL)METHYL ESTER]

ENVIRONMENTAL LISTS

EPA - Master Testing List
[present]

TSCA - Code of Federal Regulations Citations
40 CFR 712.30(g);   40 CFR 716.120(a)

TSCA - PAIR - Reporting List
Reporting Date: October 8, 1984

TSCA - Health and Safety Reporting List
Effective Date: January 30, 1983

PHENOXYACETIC ACID   122-59-8

HEALTH AND SAFETY LISTS

NIOSH - Selected LD50s and LC50s
Oral, rat: LD50 = 1300 ug/kg

ENVIRONMENTAL LISTS

CERCLA/SARA - Section 302 Extremely Hazardous Substances and TPQs
TPQ = 100/10,000 pounds

STATE LISTS

Florida Hazardous Substance List
effective March 13, 1992

Massachusetts Right To Know List
extraordinarily hazardous

Pennsylvania Right to Know List
environmental hazard

PROPOSED REGULATIONS

CERCLA/SARA - Proposed Hazardous Substance Additions
proposed RQ = 1 pound (.454 kg)

CERCLA/SARA - 1989 Proposed RQ Adjustments
proposed RQ = 100 pounds (45.4 kg)

PHENOXYACETIC ACID HERBICIDES   RR-00776-1

HEALTH AND SAFETY LISTS

NIOSH - Selected LD50s and LC50s
Oral, rat: LD50 = 7 mg/kg

ENVIRONMENTAL LISTS

TSCA - Code of Federal Regulations Citations
40 CFR 716.120

TSCA - Health and Safety Reporting List
Effective Date: June 1, 1987

INTERNATIONAL LISTS

German (DFG) - Skin/Sensitizers
danger of photo-contact sensitization

STATE LISTS

Massachusetts Right To Know List
extraordinarily hazardous

NJ Right to Know List (Total)
sn 2817

Pennsylvania Right to Know List
environmental hazard

PROPOSED REGULATIONS

CERCLA/SARA - Proposed Hazardous Substance Additions
proposed RQ = 1 pound (.454 kg)

CERCLA/SARA - 1989 Proposed RQ Adjustments
proposed RQ = 100 pounds (45.4 kg)

PHENOXYBENZAMINE   59-96-1

ENVIRONMENTAL LISTS

EPA - Master Testing List
[present]

TSCA - Code of Federal Regulations Citations
40 CFR 712.30(d),(w);   40 CFR 716.120(a);   40 CFR 766.35

TSCA - PAIR - Reporting List
Reporting Dates: November 19, 1982;   March 12, 1990

TSCA - Health and Safety Reporting List
Effective Date: January 11, 1990

TSCA - HDD/HDF - Chemicals Required for Testing
[present]

TSCA - Section 12(b) - Export Notification
export notification required - Section 4

PHENOXYBENZAMINE HYDROCHLORIDE   63-92-3

ENVIRONMENTAL LISTS

EPA - Master Testing List
[present]

N-(2-PHENOXYETHYL) ANILINE   RR-00874-2

ENVIRONMENTAL LISTS

List of Pesticide Product Inert Ingredients
[present]

PHENOXY PESTICIDES, N.O.S.   RR-00424-0

HEALTH AND SAFETY LISTS

ACGIH 1995 - Time Weighted Averages
5 mg/m3 TWA

ACGIH 1995 - Skin Designations
skin - potential for cutaneous absorption

NIOSH - Selected LD50s and LC50s
Oral, mouse: LD50 = 5000 mg/kg

OSHA - Vacated PELs - Time Weighted Averages
5 mg/m3 TWA

OSHA - Vacated PELs - Skin Designation
Prevent or reduce skin absorption

ENVIRONMENTAL LISTS

TSCA - Code of Federal Regulations Citations
40 CFR 716.120(a)

TSCA - Health and Safety Reporting List
Effective Date: June 1, 1987

INTERNATIONAL LISTS

Australian Exposure Standards - Time Weighted Averages
5 mg/m3 TWA

Australian Exposure Standards - Skin Effects
skin absorption

Canada - WHMIS: Ingredient Disclosure
1% item 1263 (1376)

Canada - Alberta - 8 Hour Occupational Exposure Limit
5 mg/m3 TWA

Canada - Alberta - 15 Minute Occupational Exposure Limit
10 mg/m3 STEL

Canada - Alberta - Skin Designation
can be absorbed through the intact skin

Canada - British Columbia - 8 Hour Exposure Limits
5 mg/m3 TWA

Canada - British Columbia - 15 Minute Exposure Limits
10 mg/m3 STEL
Canada - British Columbia - Skin Notations
skin - potential for skin absorption
Canada - Ontario - OHSA - TWAEVs
5 mg/m3 TWAEV
Canada - Ontario - OHSA - Skin Notations
absorption through skin, eyes, or mucous membranes
Canada - Quebec - Time-Weighted Average Exposure Values
5 mg/m3 TWAEV
Canada - Quebec - Skin Designations
absorbed through the skin
Israel - Time Weighted Averages
5 mg/m3 TWA
Israel - Action Levels
2.5 mg/m3 AL
Mexico - Instruction No. 10 - TWAs
5 mg/m3 TWA
Mexico - Instruction No. 10 - STELs
10 mg/m3 STEL

STATE LISTS
California - Exposure Limits - PELs
5 mg/m3 PEL
California - Exposure Limits - Skin Notation
material may be absorbed through the skin, eyes or mucous membrane
California - Directors List of Hazardous Substances (8 CCR 339)
[present]
Florida Hazardous Substance List
[present]
Massachusetts Right To Know List
[present]
Minnesota Hazardous Substance List
skin
Pennsylvania Right to Know List
[present]

PHENPROCOUMON   435-97-2
ENVIRONMENTAL LISTS
CERCLA/SARA - Section 313 - Emission Reporting
form R reporting required

PHENTHOATE   RR-01308-1
HEALTH AND SAFETY LISTS
NIOSH - Selected LD50s and LC50s
Oral, rat: LD50 = 3700 mg/kg

PHENYL(DISUBSTITUTEDPOLYCYCLIC)   RR-01654-6
STATE LISTS
Minnesota Hazardous Substance List
carcinogen
Pennsylvania Right to Know List
special hazardous substance
Pennsylvania RTK - Special Hazardous Substances
[present]

PHENYL ACETATE   122-79-2
HEALTH AND SAFETY LISTS
NIOSH - Selected LD50s and LC50s
Oral, rat: LD50 = 2500 mg/kg
STATE LISTS
California - Air Bill 2588 Appendix A-II
known or potential carcinogen: 9/89

California - Prop. 65 - Cancer list
carcinogen - initial date 4/1/88
California - Prop. 65 - No Significant Risk Levels
no significant risk level = 0.2 ug/day
California - Directors List of Hazardous Substances (8 CCR 339)
[present]
Florida Hazardous Substance List
[present]
Massachusetts Right To Know List
carcinogen; extraordinarily hazardous
Pennsylvania Right to Know List
special hazardous substance
Pennsylvania RTK - Special Hazardous Substances
[present]

PHENYLACETIC ACID   103-82-2
HEALTH AND SAFETY LISTS
IARC - Group 2B (sufficient animal data)
[present] (Overall evaluation based only on evidence of carcinogenicity in monograph (24, 1980) or in Supplement 4)
NIOSH - Selected LD50s and LC50s
Oral, mouse: LD50 = 900 mg/kg
NTP Chemical Status Reports - Testing Status and NTIS Number
Technical reports printed (PB285095/AS)
NTP Chemical Status Reports - Evidence of Carcinogenicity
male rat-positive; female rat-positive; male mice-positive; female mice-positive
NTP Seventh Report - Suspect Carcinogens
suspect carcinogen
OSHA - Possible Select Carcinogens
[present]
STATE LISTS
California - Air Bill 2588 Appendix A-II
known or potential carcinogen: 9/90
California - Prop. 65 - Cancer list
carcinogen - initial date 4/1/88
California - Prop. 65 - No Significant Risk Levels
no significant risk level = 0.3 ug/day
California - Directors List of Hazardous Substances (8 CCR 339)
[present]
Florida Hazardous Substance List
[present]
Massachusetts Right To Know List
carcinogen; extraordinarily hazardous
Minnesota Hazardous Substance List
carcinogen
Pennsylvania Right to Know List
special hazardous substance
Pennsylvania RTK - Special Hazardous Substances
[present]

PHENYLACETIC ACID SALTS   RR-01778-7
HEALTH AND SAFETY LISTS
NFPA - Flash Points
flash point = 338 degrees F (170 degrees C)
NFPA - Hazard Identification Ratings
health-1; flammability-1; reactivity-0

PHENYLACETYL CHLORIDE   103-80-0
HEALTH AND SAFETY LISTS
U.S. DOT - Substances From 49 CFR 172.101
regulated by DOT (UN2765, UN2766, UN2999, UN3000)

U.S. DOT - Hazard Classes
toxic or toxic, flammable: DOT hazard class = 6.1; flammable, toxic: DOT hazard class = 3

STATE LISTS

NJ Right to Know List (Total)
sn 2655; sn 2656; sn 2657; sn 2658

PHENYL ANTHRANILIC ACID (ALL ISOMERS)   91-40-7

STATE LISTS

California - Prop. 65 - Developmental Toxicity
developmental toxicity - initial date 10/1/92

PHENYL BENZOATE   93-99-2

HEALTH AND SAFETY LISTS

U.S. DOT - Appendix B - Marine Pollutants
DOT regulated severe marine pollutant

PHENYLBUTAZONE   50-33-9

ENVIRONMENTAL LISTS

TSCA - Chemicals with Significant New Use Rules
PMN number: P-92-1337

1-PHENYL-2-BUTENE   1560-06-1

HEALTH AND SAFETY LISTS

NFPA - Flash Points
flash point = 176 degrees F (80 degrees C)

NFPA - Hazard Identification Ratings
health-1; flammability-2; reactivity-0

NIOSH - Selected LD50s and LC50s
Oral, rat: LD50 = 1630 mg/kg Skin, rabbit: LD50 = 8000 mg/kg

PHENYLCARBYLAMINE CHLORIDE   622-44-6

HEALTH AND SAFETY LISTS

FDA - Controlled Substances Act - Precursor chemicals
Threshold by base weight = 1 kilogram

NFPA - Flash Points
flash point > 212 degrees F (100 degrees C)

NFPA - Hazard Identification Ratings
health-1; flammability-1; reactivity-0

NIOSH - Selected LD50s and LC50s
Oral, rat: LD50 = 2250 mg/kg

PHENYLCHLOROFORMATE   1885-14-9

HEALTH AND SAFETY LISTS

FDA - Controlled Substances Act - Precursor chemicals
Threshold by base weight = 1 kilogram

PHENYL DIDECYL PHOSPHITE   1254-78-0

HEALTH AND SAFETY LISTS

U.S. DOT - Substances From 49 CFR 172.101
regulated by DOT (UN2577)

U.S. DOT - Hazard Classes
DOT hazard class = 8

INTERNATIONAL LISTS

Canada - WHMIS: Ingredient Disclosure
1% item 1265 (516)

STATE LISTS

NJ Right to Know List (Total)
sn 1491

NJ Special Hazardous Substances
(corrosive)

N-PHENYL DIETHANOLAMINE   120-07-0

ENVIRONMENTAL LISTS

CAA - HON Rule - SOCMI Chemicals
compliance by April 24, 1995

3-PHENYL-1,1-DIMETHYLUREA   101-42-8

HEALTH AND SAFETY LISTS

NIOSH - Selected LD50s and LC50s
Oral, mouse: LD50 = 1225 mg/kg

PHENYL DISELENIDE   1666-13-3

HEALTH AND SAFETY LISTS

IARC - Group 3 (not classifiable)
[present]

NTP Chemical Status Reports - Testing Status and NTIS Number
Technical reports printed (PB90258765)

NTP Chemical Status Reports - Evidence of Carcinogenicity
male rat-equivocal evidence; female rat-some evidence; male mice-some evidence; female mice-no evidence

PHENYL DI-O-XENYL PHOSPHATE   RR-00878-6

HEALTH AND SAFETY LISTS

NFPA - Flash Points
flash point = 160 degrees F (71 degrees C)

NFPA - Hazard Identification Ratings
flammability-2; reactivity-0

O-PHENYLENEDIAMINE   95-54-5

HEALTH AND SAFETY LISTS

U.S. DOT - Substances From 49 CFR 172.101
regulated by DOT (UN1672)

U.S. DOT - Hazard Classes
DOT hazard class = 6.1

U.S. DOT - Substances Which Are Poisonous by Inhalation
liquid hazardous material poisonous by inhalation (UN1672)

INTERNATIONAL LISTS

Canada - WHMIS: Ingredient Disclosure
1% item 1267 (517)

STATE LISTS

NJ Right to Know List (Total)
sn 1492

P-PHENYLENE DIAMINE   106-50-3

HEALTH AND SAFETY LISTS

U.S. DOT - Substances From 49 CFR 172.101
regulated by DOT (UN2746)

U.S. DOT - Hazard Classes
DOT hazard class = 6.1

NIOSH - Selected LD50s and LC50s
Oral, rat: LD50 = 490 mg/kg Skin, rabbit: LD50 = 3970 mg/kg

INTERNATIONAL LISTS

Canada - WHMIS: Ingredient Disclosure
1% item 1268 (439)

STATE LISTS

NJ Right to Know List (Total)
sn 1493

NJ Special Hazardous Substances
(corrosive)

M-PHENYLENEDIAMINE   108-45-2

HEALTH AND SAFETY LISTS

NFPA - Flash Points
flash point = 425 degrees F (218 degrees C)

NFPA - Hazard Identification Ratings
health-0; flammability-1; reactivity-0

ENVIRONMENTAL LISTS

List of Pesticide Product Inert Ingredients
[present]

P-PHENYLENEDIAMINE DIHYDROCHLORIDE   624-18-0

HEALTH AND SAFETY LISTS

NFPA - Flash Points
flash point = 385 degrees F (196 degrees C)

NFPA - Hazard Identification Ratings
health-1; flammability-1; reactivity-0

NIOSH - Selected LD50s and LC50s
Oral, rat: LD50 = 980 mg/kg

INTERNATIONAL LISTS

Canada - WHMIS: Ingredient Disclosure
1% item 1269 (1378)

M-PHENYLENE DIAMINEDIPERCHLORATE   RR-01435-7

HEALTH AND SAFETY LISTS

NIOSH - Selected LD50s and LC50s
Oral, rat: LD50 = 6400 mg/kg

STATE LISTS

California - Directors List of Hazardous Substances (8 CCR 339)
[present]

PHENYLENEDIAMINES   25265-76-3

SEE ALSO:
SELENIUM

INTERNATIONAL LISTS

Canada - WHMIS: Ingredient Disclosure
1% item 1270 (784)

M-PHENYLENEDIAMINE, SULFATE SALT   54-17-1

HEALTH AND SAFETY LISTS

NFPA - Flash Points
flash point = 482 degrees F (250 degrees C)

NFPA - Hazard Identification Ratings
health-0; flammability-1; reactivity-1

PHENYLEPHRINE HYDROCHLORIDE   61-76-7

SEE ALSO:
O-PHENYLENEDIAMINE

HEALTH AND SAFETY LISTS

ACGIH 1995 - Time Weighted Averages
0.1 mg/m3 TWA

ACGIH 1995 - Carcinogens
A2-suspected human carcinogen

NFPA - Flash Points
flash point = 313 degrees F (156 degrees C)

NFPA - Hazard Identification Ratings
flammability-1; reactivity-0

NIOSH - Selected LD50s and LC50s
Oral, rat: LD50 = 1070 mg/kg

ENVIRONMENTAL LISTS

CERCLA/SARA - Section 313 - Emission Reporting
form R reporting required

EPA - Master Testing List
[present]

TSCA - Code of Federal Regulations Citations
40 CFR 712.30(d); 40 CFR 716.120(c); 40 CFR 799.3300

TSCA - PAIR - Reporting List
Reporting Date: November 19, 1982

TSCA - Chemical Test Rules
Testing required by: manufacturers; importers; processors (40 CFR 799.3300) (Listed under 'Unsubstituted phenylenediamines')

TSCA - Section 12(b) - Export Notification
export notification required - Section 4

INTERNATIONAL LISTS

Canada - WHMIS: Ingredient Disclosure
0.1% item 1272 (1380)

German (DFG) - Carcinogens
animal evidence of carcinogenicity

STATE LISTS

NJ Right to Know List (Total)
sn 1495

N-PHENYL ETHANOLAMINE   122-98-5

HEALTH AND SAFETY LISTS

ACGIH 1995 - Time Weighted Averages
0.1 mg/m3 TWA

IARC - Group 3 (not classifiable)
[present]

NIOSH - Selected LD50s and LC50s
Oral, rat: LD50 = 80 mg/kg

NIOSH 1990 - Pocket Guide - RELs
0.1 mg/m3 TWA

NIOSH 1990 - Pocket Guide - Target organs
skin, respiratory system

NIOSH 1990 - Pocket Guide - Skin list
Potential for dermal absorption

OSHA - Vacated PELs - Time Weighted Averages
0.1 mg/m3 TWA

OSHA - Vacated PELs - Skin Designation
Prevent or reduce skin absorption

OSHA - Final PELs - Time Weighted Averages
0.1 mg/m3 TWA

OSHA - Final PELs - Skin Notations
prevent or reduce skin absorption

ENVIRONMENTAL LISTS

CERCLA/SARA - Section 313 - Emission Reporting
form R reporting required for 1.0% de minimus concentration

CERCLA/SARA - Hazardous Substances and their Reportable Quantities
final RQ = 1 pound (.454 kg)

Clean Air Act (1990) - List of Hazardous Air Contaminants
[present]

CAA - HON Rule - SOCMI Chemicals
compliance by Oct. 24, 1994

CAA - HON Rule - Organic HAPs
[present]

EPA - Master Testing List
[present]

RCRA - TSD Facilities Ground Water Monitoring
TM 8270 = 10 ug/L PQL

TSCA - Code of Federal Regulations Citations
40 CFR 712.30(d); 40 CFR 716.120(a); 40 CFR 79.3300

TSCA - PAIR - Reporting List
Reporting Date: November 19, 1982

TSCA - Health and Safety Reporting List
Effective Date: October 4, 1982

TSCA - Chemical Test Rules
Testing required by: manufacturers; importers; processors (40 CFR 799.3300) (Listed under 'Unsubstituted phenylenediamines')

TSCA - Section 12(b) - Export Notification
export notification required - Section 4

INTERNATIONAL LISTS

Australian Exposure Standards - Time Weighted Averages
0.1 mg/m3 TWA

Australian Exposure Standards - Skin Effects
skin absorption; sensitiser

Canada - WHMIS: Ingredient Disclosure
0.1% item 1273 (1381)

Canada - NPRI (National Pollutant Release Inventory)
[present]

Canada - Alberta - 8 Hour Occupational Exposure Limit
0.1 mg/m3 TWA

Canada - Alberta - 15 Minute Occupational Exposure Limit
0.3 mg/m3 STEL

Canada - Alberta - Skin Designation
can be absorbed through the intact skin

Canada - British Columbia - 8 Hour Exposure Limits
0.1 mg/m3 TWA

Canada - British Columbia - Skin Notations
skin - potential for skin absorption

Canada - Ontario - OHSA - TWAEVs
0.1 mg/m3 TWAEV

Canada - Ontario - OHSA - Skin Notations
absorption through skin, eyes, or mucous membranes

United Kingdom - Occupational Exposure Standards - TWAs
0.1 mg/m3 TWA

United Kingdom - Occupational Exposure Standards - Notes
can be absorbed through skin

German (DFG) - MAK Values
total dust: 0.1 mg/m3 MAK

German (DFG) - Peak Limitations
2 x normal MAK (30 min. average value); don't exceed 4 times during shift

German (DFG) - Skin/Sensitizers
danger of cutaneous absorption; danger of sensitization (skin or respiratory)

German (DFG) - Carcinogens
suspected carcinogen

Israel - Time Weighted Averages
(0.1) mg/m3 TWA

Israel - Action Levels
0.05 mg/m3 AL

Mexico - Instruction No. 10 - TWAs
0.1 mg/m3 TWA

Mexico - Instruction No. 10 - Skin designation
skin - potential for cutaneous absorption

STATE LISTS

California - Air Bill 2588 Appendix A-I
6/91

California - Exposure Limits - PELs
0.1 mg/m3 PEL

California - Exposure Limits - Skin Notation
material may be absorbed through the skin, eyes or mucous membrane

California - Directors List of Hazardous Substances (8 CCR 339)
[present]

Florida Hazardous Substance List
[present]

Massachusetts Right To Know List
[present]

Minnesota Hazardous Substance List
skin

NJ Right to Know List (Total)
sn 1586

Pennsylvania Right to Know List
environmental hazard

PHENYL ETHER   101-84-8
SEE ALSO:
M-PHENYLENEDIAMINE

HEALTH AND SAFETY LISTS

ACGIH 1995 - Time Weighted Averages
0.1 mg/m3 TWA

IARC - Group 3 (not classifiable)
[present]

NIOSH - Selected LD50s and LC50s
Oral, rat: LD50 = 650 mg/kg

ENVIRONMENTAL LISTS

CERCLA/SARA - Section 313 - Emission Reporting
form R reporting required

EPA - Master Testing List
[present]

TSCA - Code of Federal Regulations Citations
40 CFR 712.30(d); 40 CFR 716.120(a); 40 CFR 799.3300

TSCA - PAIR - Reporting List
Reporting Date: November 19, 1982

TSCA - Chemical Test Rules
Testing required by: manufacturers; importers; processors (40 CFR 799.3300) (Listed under 'Unsubstituted phenylenediamines')

TSCA - Section 12(b) - Export Notification
export notification required - Section 4

INTERNATIONAL LISTS

Canada - WHMIS: Ingredient Disclosure
0.1% item 1271 (1379)

German (DFG) - Skin/Sensitizers
danger of cutaneous absorption

German (DFG) - Carcinogens
suspected carcinogen

PHENYL ETHER-BIPHENYL MIXTURE VAPOR   8004-13-5
SEE ALSO:
P-PHENYLENEDIAMINE DIHYDROCHLORIDE

HEALTH AND SAFETY LISTS

NTP Chemical Status Reports - Testing Status and NTIS Number
Technical reports printed (PB290124/AS)

NTP Chemical Status Reports - Evidence of Carcinogenicity
male rat-negative; female rat-negative; male mice-negative; female mice-negative

ENVIRONMENTAL LISTS

CERCLA/SARA - Section 313 - Emission Reporting
form R reporting required

TSCA - Code of Federal Regulations Citations
40 CFR 712.30(d); 40 CFR 716.120(c)

TSCA - PAIR - Reporting List
Reporting Date: November 19, 1982

2-PHYENYLETHYL ACETATE   103-45-7

HEALTH AND SAFETY LISTS
U.S. DOT - Hazard Classes
Forbidden from transport by the DOT

PHENYL GLYCIDYL ETHER   **122-60-1**
SEE ALSO:
K103-HAZARDOUS WASTES
K104-HAZARDOUS WASTES
K083-HAZARDOUS WASTES

HEALTH AND SAFETY LISTS

U.S. DOT - Substances From 49 CFR 172.101
regulated by DOT (UN1673)

U.S. DOT - Hazard Classes
DOT hazard class = 6.1

ENVIRONMENTAL LISTS

RCRA - Hazardous Constituents-Appendix VIII
hazardous constituent - no waste number

RCRA - Basis for Listing - Appendix VII
Included in waste streams: K083, K103, K104

TSCA - Health and Safety Reporting List
Effective Date: April 29, 1983 (This category is defined as all nitrogen unsubstituted phenylenediamines and their salts with zero to two substitutents on the ring selected from the same of different members of the group of halo, nitro, hydroxy, hydroxy-lower alkoxy, lower-alkyl, and lower alkoxy)

INTERNATIONAL LISTS

Canada - Quebec - Time-Weighted Average Exposure Values
0.1 mg/m3 TWAEV

Canada - Quebec - Skin Designations
absorbed through the skin

PHENYL GLYCOL ETHER   **93-56-1**

ENVIRONMENTAL LISTS

TSCA - Code of Federal Regulations Citations
40 CFR 799.3300

PHENYLHYDRAZINE   **100-63-0**

HEALTH AND SAFETY LISTS

NTP Chemical Status Reports - Testing Status and NTIS Number
Technical reports printed (PB87208609/AS)

NTP Chemical Status Reports - Evidence of Carcinogenicity
male rat-no evidence; female rat-no evidence; male mice-no evidence; female mice-no evidence

PHENYLHYDRAZINE HYDROCHLORIDE   **59-88-1**

HEALTH AND SAFETY LISTS

NFPA - Flash Points
flash point = 305 degrees F (152 degrees C)

NFPA - Hazard Identification Ratings
health-2; flammability-1; reactivity-0

NIOSH - Selected LD50s and LC50s
Oral, rat: LD50 = 2230 mg/kg Skin, rabbit: LD50 = 63 mg/kg

INTERNATIONAL LISTS

Canada - WHMIS: Ingredient Disclosure
1% item 1274 (1382)

STATE LISTS

Florida Hazardous Substance List
[present]

Massachusetts Right To Know List
[present]

Pennsylvania Right to Know List
[present]

PHENYL ISOCYANATE   **103-71-9**

HEALTH AND SAFETY LISTS

ACGIH 1995 - Time Weighted Averages
vapor: 1 ppm TWA; 7 mg/m3 TWA

ACGIH 1995 - Short Term Exposure Limits
vapor: 2 ppm STEL; 14 mg/m3 STEL

U.S. DOT - Appendix B - Marine Pollutants
DOT regulated marine pollutant

NFPA - Flash Points
flash point = 239 degrees F (115 degrees C)

NFPA - Hazard Identification Ratings
health-1; flammability-1; reactivity-0

NIOSH - Selected LD50s and LC50s
Oral, rat: LD50 = 3370 mg/kg

NIOSH 1990 - Pocket Guide - RELs
1 ppm TWA; 7 mg/m3 TWA

NIOSH 1990 - Pocket Guide - Target organs
eyes, skin, respiratory system

OSHA - Vacated PELs - Time Weighted Averages
vapor: 1 ppm TWA; 7 mg/m3 TWA

OSHA - Final PELs - Time Weighted Averages
vapor: 1 ppm TWA; 7 mg/m3 TWA

ENVIRONMENTAL LISTS

CAA - HON Rule - SOCMI Chemicals
compliance by Oct. 24, 1994

TSCA - Code of Federal Regulations Citations
40 CFR 712.30(w); 40 CFR 716.120(a)

TSCA - PAIR - Reporting List
Reporting Date: June 13, 1989

TSCA - Health and Safety Reporting List
Effective Date: April 13, 1989

INTERNATIONAL LISTS

Australian Exposure Standards - Time Weighted Averages
vapour: 1 ppm TWA; 7 mg/m3 TWA

Australian Exposure Standards - Short Term Exposure Limits
vapour: 2 ppm STEL; 14 mg/m3 STEL

Canada - WHMIS: Ingredient Disclosure
1% item 661 (820)

Canada - Alberta - 8 Hour Occupational Exposure Limit
as vapor: 1 ppm TWA; 7 mg/m3 TWA

Canada - Alberta - 15 Minute Occupational Exposure Limit
2 ppm STEL; 14 mg/m3 STEL

Canada - British Columbia - 8 Hour Exposure Limits
1 ppm TWA; 7 mg/m3 TWA

Canada - British Columbia - 15 Minute Exposure Limits
2 ppm STEL; 14 mg/m3 STEL

Canada - Ontario - OHSA - TWAEVs
1 ppm TWAEV; 7 mg/m3 TWAEV

Canada - Ontario - OHSA - STEVs
2 ppm STEV; 14 mg/m3 STEV

Canada - Quebec - Time-Weighted Average Exposure Values
1 ppm TWAEV; 7 mg/m3 TWAEV

Canada - Quebec - Short-term Exposure Values
2 ppm STEV; 14 mg/m3 STEV

United Kingdom - Occupational Exposure Standards - TWAs
vapour: 1 ppm TWA; 7 mg/m3 TWA

German (DFG) - MAK Values
vapor: 1 ppm MAK; 7 mg/m3 MAK

Israel - Time Weighted Averages
1 ppm TWA; 7 mg/m3 TWA

Israel - Short Term Exposure Limits
2 ppm STEL; 14 mg/m3 STEL

Israel - Action Levels
0.5 ppm AL; 3.5 mg/m3 AL

Mexico - Instruction No. 10 - TWAs
1 ppm TWA; 7 mg/m3 TWA

Mexico - Instruction No. 10 - STELs
2 ppm STEL;   14 mg/m3 STEL

STATE LISTS

California - Exposure Limits - PELs
vapor: 1 ppm PEL;   7 mg/m3 PEL

California - Directors List of Hazardous Substances (8 CCR 339)
[present] (exempt when vapors or particulates can be formed due to work practices or procedures)

Florida Hazardous Substance List
[present]

Massachusetts Right To Know List
[present]

Minnesota Hazardous Substance List
[present]

NJ Right to Know List (Total)
sn 1496

Pennsylvania Right to Know List
[present]

PHENYLMERCURIC ACETATE   62-38-4

HEALTH AND SAFETY LISTS

NIOSH 1990 - Pocket Guide - RELs
1 ppm TWA;   7 mg/m3 TWA

NIOSH 1990 - Pocket Guide - Target organs
eyes, skin, respiratory system

OSHA - Vacated PELs - Time Weighted Averages
1 ppm TWA;   7 mg/m3 TWA

OSHA - Final PELs - Time Weighted Averages
1 ppm TWA;   7 mg/m3 TWA

INTERNATIONAL LISTS

Canada - Alberta - 8 Hour Occupational Exposure Limit
0.5 ppm TWA;   4 mg/m3 TWA

Canada - Alberta - 15 Minute Occupational Exposure Limit
2 ppm STEL;   16 mg/m3 STEL

Canada - British Columbia - 8 Hour Exposure Limits
1 ppm TWA;   7 mg/m3 TWA

Canada - British Columbia - 15 Minute Exposure Limits
2 ppm STEL;   14 mg/m3 STEL

STATE LISTS

Florida Hazardous Substance List
[present]

Massachusetts Right To Know List
[present]

Minnesota Hazardous Substance List
[present]

Pennsylvania Right to Know List
[present]

PHENYLMERCURIC COMPOUND, SOLID, N.O.S.   RR-00428-4

HEALTH AND SAFETY LISTS

NFPA - Flash Points
flash point = 230 degrees F (110 degrees C)

NFPA - Hazard Identification Ratings
health-0; flammability-1; reactivity-0

NIOSH - Selected LD50s and LC50s
Oral, rat: LD50 = 3670 mg/kg Skin, rabbit: LD50 = 6210 mg/kg

PHENYLMERCURIC HYDROXIDE   100-57-2
SEE ALSO:
GLYCIDOL (OXIRANEMETHANOL) AND ITS DERIVATIVES

HEALTH AND SAFETY LISTS

ACGIH 1995 - Time Weighted Averages
0.1 ppm TWA 0.6 mg/m3 TWA

ACGIH 1995 - Skin Designations
skin - potential for cutaneous absorption

ACGIH 1995 - Carcinogens
A3-animal carcinogen

IARC - Group 2B (sufficient animal data)
[present]

NIOSH - Selected LD50s and LC50s
Oral, rat: LD50 = 2150 mg/kg Skin, rabbit: LD50 = 1500 mg/kg

NIOSH 1990 - Pocket Guide - RELs
C 1 ppm (15 min);   C 6 mg/m3 (15 min)

NIOSH 1990 - Pocket Guide - Carcinogens
occupational carcinogen

NIOSH 1990 - Pocket Guide - Target organs
skin, eyes, CNS

NIOSH - Health Standards - Exposure Limits
C (15 min) 1 ppm;   C (15 min) 5 mg/m3 (Listed under 'Glycidyl ethers')

NIOSH - Health Standards - Health Effects and Precautions
Skin and mucous membrane effects; sensitization potential; possible hematopoietic and reproductive system effects (Medical monitoring required) (Listed under 'Glycidyl ethers')

OSHA - Vacated PELs - Time Weighted Averages
1 ppm TWA;   6 mg/m3 TWA

OSHA - Final PELs - Time Weighted Averages
10 ppm TWA;   60 mg/m3 TWA

OSHA - Possible Select Carcinogens
[present]

ENVIRONMENTAL LISTS

EPA - Master Testing List
[present]

TSCA - Code of Federal Regulations Citations
40 CFR 712.30(d);   40 CFR 716.120(c)

TSCA - PAIR - Reporting List
Reporting Date: November 19, 1982

INTERNATIONAL LISTS

Australian Exposure Standards - Time Weighted Averages
1 ppm TWA;   6.1 mg/m3 TWA

Australian Exposure Standards - Skin Effects
sensitiser

Canada - WHMIS: Ingredient Disclosure
0.1% item 1275 (829)

Canada - Alberta - 8 Hour Occupational Exposure Limit
1 ppm TWA;   6 mg/m3 TWA

Canada - Alberta - 15 Minute Occupational Exposure Limit
2 ppm STEL;   12 mg/m3 STEL

Canada - British Columbia - 8 Hour Exposure Limits
10 ppm TWA;   60 mg/m3 TWA

Canada - British Columbia - 15 Minute Exposure Limits
15 ppm STEL;   90 mg/m3 STEL

Canada - Ontario - OHSA - TWAEVs
1 ppm TWAEV;   6 mg/m3 TWAEV

Canada - Quebec - Time-Weighted Average Exposure Values
1 ppm TWAEV;   6.1 mg/m3 TWAEV

Canada - Quebec - Carcinogens
C3 carcinogen: effect detected in animals

United Kingdom - Occupational Exposure Standards - TWAs
1 ppm TWA;   6 mg/m3 TWA

German (DFG) - Peak Limitations
2 x normal MAK (5 min momentary value);   don't exceed 8 times during shift

German (DFG) - Skin/Sensitizers
danger of cutaneous absorption;   danger of sensitization (skin or respiratory)

German (DFG) - Carcinogens
animal evidence of carcinogenicity

Israel - Time Weighted Averages
1 ppm TWA; 6.1 mg/m3 TWA

Israel - Action Levels
0.5 ppm AL; 3.05 mg/m3 AL

Mexico - Instruction No. 10 - TWAs
10 ppm TWA; 60 mg/m3 TWA

STATE LISTS

California - Air Bill 2588 Appendix A-II
known or potential carcinogen: 9/90

California - Prop. 65 - Cancer list
carcinogen - initial date 10/1/90

California - Exposure Limits - PELs
1 ppm PEL; 6 mg/m3 PEL

California - Directors List of Hazardous Substances (8 CCR 339)
[present] (exempt when part of a cured epoxy or rubber)

Florida Hazardous Substance List
[present]

Massachusetts Right To Know List
[present]

Minnesota Hazardous Substance List
[present]

Pennsylvania Right to Know List
[present]

PROPOSED REGULATIONS

TSCA - Proposed Testing Rule for Glycidyl Ethers
subject to neurotoxicity, reproductive and fertility effects, and screening subcategory testing (results apply to all members of Glycidyl subcategory IV-A)

PHENYLMERCURIC NITRATE   55-68-5

ENVIRONMENTAL LISTS

List of Pesticide Product Inert Ingredients
[present]

4-[[4-(PHENYLMETHOXY)PHENYL]SULFONYL] PHENOL   63134-33-8

HEALTH AND SAFETY LISTS

ACGIH 1995 - Time Weighted Averages
0.1 ppm TWA; 0.44 mg/m3 TWA

ACGIH 1995 - Skin Designations
skin - potential for cutaneous absorption

ACGIH 1995 - Carcinogens
A2-suspected human carcinogen

U.S. DOT - Substances From 49 CFR 172.101
regulated by DOT (UN2572)

U.S. DOT - Hazard Classes
DOT hazard class = 6.1

NFPA - Flash Points
flash point = 190 degrees F (88 degrees C)

NFPA - Hazard Identification Ratings
health-3; flammability-2; reactivity-0

NIOSH - Selected LD50s and LC50s
Oral, rat: LD50 = 188 mg/kg

NIOSH 1990 - Pocket Guide - RELs
C 0.14 ppm (2 hr); C 0.6 mg/m3 (2 hr)

NIOSH 1990 - Pocket Guide - IDLHs
295 ppm IDLH (not considering carcinogenic effects)

NIOSH 1990 - Pocket Guide - Carcinogens
occupational carcinogen

NIOSH 1990 - Pocket Guide - Target organs
blood, respiratory system, liver, kidneys, skin

NIOSH - Health Standards - Exposure Limits
C (2 hr) 0.14 ppm; C (2 hr) 0.6 mg/m3 (Listed under 'Hydrazines')

NIOSH - Health Standards - Health Effects and Precautions
has produced tumors of the lung, liver, blood vessels, and intestines in animals; blood, liver, and skin effects (Blood and urine monitoring and periodic chest X-ray required; bowel examination for workers above age 40) (Listed under 'Hydrazines')

NIOSH - Health Standards - Carcinogenic Chemicals
potential human carcinogen (Listed under 'Hydrazines')

OSHA - Vacated PELs - Time Weighted Averages
5 ppm TWA; 20 mg/m3 TWA

OSHA - Vacated PELs - Short Term Exposure Limits
10 ppm STEL; 45 mg/m3 STEL

OSHA - Vacated PELs - Skin Designation
Prevent or reduce skin absorption

OSHA - Final PELs - Time Weighted Averages
5 ppm TWA; 22 mg/m3 TWA

OSHA - Final PELs - Skin Notations
prevent or reduce skin absorption

ENVIRONMENTAL LISTS

TSCA - Health and Safety Reporting List
Effective Date: March 11, 1994; Sunset Date: March 11, 2004

INTERNATIONAL LISTS

Australian Exposure Standards - Time Weighted Averages
5 ppm TWA; 22 mg/m3 TWA

Australian Exposure Standards - Short Term Exposure Limits
10 ppm STEL; 44 mg/m3 STEL

Australian Exposure Standards - Skin Effects
skin absorption; sensitiser

Australian Exposure Standards - Carcinogens
suspected carcinogen

Canada - WHMIS: Ingredient Disclosure
0.1% item 1276 (1383)

Canada - Alberta - 8 Hour Occupational Exposure Limit
5 ppm TWA; 22 mg/m3 TWA

Canada - Alberta - 15 Minute Occupational Exposure Limit
10 ppm STEL; 44 mg/m3 STEL

Canada - Alberta - Skin Designation
can be absorbed through the intact skin

Canada - Alberta - Designated Substances
designated substance - requires code of practice

Canada - British Columbia - 8 Hour Exposure Limits
5 ppm TWA; 22 mg/m3 TWA

Canada - British Columbia - 15 Minute Exposure Limits
10 ppm STEL; 44 mg/m3 STEL

Canada - British Columbia - Skin Notations
skin - potential for skin absorption

Canada - Ontario - OHSA - TWAEVs
5 ppm TWAEV; 22 mg/m3 TWAEV

Canada - Ontario - OHSA - STEVs
10 ppm STEV; 44 mg/m3 STEV

Canada - Ontario - OHSA - Skin Notations
absorption through skin, eyes, or mucous membranes

Canada - Quebec - Time-Weighted Average Exposure Values
0.1 ppm TWAEV; 0.44 mg/m3 TWAEV

Canada - Quebec - Skin Designations
absorbed through the skin

Canada - Quebec - Carcinogens
C2 carcinogen: effect suspected in humans

German (DFG) - MAK Values
5 ppm MAK; 22 mg/m3 MAK

German (DFG) - Skin/Sensitizers
danger of cutaneous absorption; danger of sensitization (skin or respiratory)

German (DFG) - Carcinogens
suspected carcinogen

Israel - Time Weighted Averages
(5) ppm TWA; (22) mg/m3 TWA

Israel - Short Term Exposure Limits
10 ppm STEL; 44 mg/m3 STEL

Israel - Action Levels
2.5 ppm AL; 11 mg/m3 AL

Mexico - Instruction No. 10 - TWAs
5 ppm TWA; 20 mg/m3 TWA

Mexico - Instruction No. 10 - STELs
10 ppm STEL; 45 mg/m3 STEL

Mexico - Instruction No. 10 - Skin designation
skin - potential for cutaneous absorption

STATE LISTS

California - Prop. 65 - Cancer list
carcinogen (includes Phenylhydrazine salts) - initial date 7/1/92

California - Exposure Limits - PELs
5 ppm PEL; 20 mg/m3 PEL

California - Exposure Limits - STELs
10 ppm STEL; 45 mg/m3 STEL

California - Exposure Limits - Skin Notation
material may be absorbed through the skin, eyes or mucous membrane

California - Directors List of Hazardous Substances (8 CCR 339)
[present]

Florida Hazardous Substance List
[present]

Massachusetts Right To Know List
[present]

Minnesota Hazardous Substance List
carcinogen; skin

NJ Right to Know List (Total)
sn 1500

Pennsylvania Right to Know List
[present]

PROPOSED REGULATIONS

TSCA - ITC 32nd Report Priority Testing List
designated for dermal absorption testing

Canada - Ontario - Proposed Occupational TWAEVs
0.6 mg/m3 TWAEV

4-PHENYLMORPHOLINE   92-53-5

ENVIRONMENTAL LISTS

CERCLA/SARA - Section 302 Extremely Hazardous Substances and TPQs
TPQ = 1000/10,000 pounds

STATE LISTS

Florida Hazardous Substance List
effective March 13, 1992

Massachusetts Right To Know List
extraordinarily hazardous

NJ Right to Know List (Total)
sn 2659

Pennsylvania Right to Know List
environmental hazard

PROPOSED REGULATIONS

CERCLA/SARA - Proposed Hazardous Substance Additions
proposed RQ = 1 pound (454 kg)

CERCLA/SARA - 1989 Proposed RQ Adjustments
proposed RQ = 100 pounds (45.4 kg)

N-PHENYL-1-NAPHTHYLAMINE   90-30-2

HEALTH AND SAFETY LISTS

U.S. DOT - Substances From 49 CFR 172.101
regulated by DOT (UN2487)

U.S. DOT - Hazard Classes
DOT hazard class = 6.1

U.S. DOT - Substances Which Are Poisonous by Inhalation
liquid hazardous material poisonous by inhalation (UN2487)

NIOSH - Selected LD50s and LC50s
Oral, rat: LD50 = 940 mg/kg Skin, rabbit: LD50 = 7130 mg/kg

ENVIRONMENTAL LISTS

TSCA - Code of Federal Regulations Citations
40 CFR 712.30(d),(x)

TSCA - PAIR - Reporting List
Reporting Date: November 19, 1982, December 27, 1990

TSCA - Health and Safety Reporting List
Effective Date: October 29, 1990

INTERNATIONAL LISTS

Canada - WHMIS: Ingredient Disclosure
0.1% item 1277 (1045)

STATE LISTS

NJ Right to Know List (Total)
sn 1501

N-PHENYL-BETA-NAPHTHYLAMINE   135-88-6
SEE ALSO:
  MERCURY
  MERCURY
  MERCURY COMPOUNDS

HEALTH AND SAFETY LISTS

U.S. DOT - Substances From 49 CFR 172.101
regulated by DOT (UN1674)

U.S. DOT - Hazard Classes
DOT hazard class = 6.1

U.S. DOT - Appendix B - Marine Pollutants
DOT regulated severe marine pollutant

U.S. DOT - Appendix A Table 1 - Hazardous Substances
final RQ = 100 pounds (45.4 kg)

NIOSH - Selected LD50s and LC50s
Oral, rat: LD50 = 22 mg/kg

ENVIRONMENTAL LISTS

CERCLA/SARA - Section 302 Extremely Hazardous Substances and TPQs
TPQ = 500/10,000 pounds

CERCLA/SARA - Hazardous Substances and their Reportable Quantities
final RQ = 100 pounds (45.4 kg)

RCRA - P Series Wastes
waste number P092

RCRA - Hazardous Constituents-Appendix VIII
waste number P092

RCRA - Substances Banned From Land Disposal
[present]

INTERNATIONAL LISTS

Canada - WHMIS: Ingredient Disclosure
0.1% item 1279 (29)

STATE LISTS

Florida Hazardous Substance List
[present]

Massachusetts Right To Know List
extraordinarily hazardous

NJ Right to Know List (Total)
sn 1502

Pennsylvania Right to Know List
environmental hazard

2-PHENYLPHENOL   90-43-7

HEALTH AND SAFETY LISTS

U.S. DOT - Substances From 49 CFR 172.101
regulated by DOT (UN2026)

U.S. DOT - Hazard Classes
DOT hazard class = 6.1

U.S. DOT - Appendix B - Marine Pollutants
DOT regulated severe marine pollutant

STATE LISTS

NJ Right to Know List (Total)
sn 2660

PHENYLPHOSPHINE   638-21-1

SEE ALSO:
MERCURY

HEALTH AND SAFETY LISTS

U.S. DOT - Substances From 49 CFR 172.101
regulated by DOT (UN1894)

U.S. DOT - Hazard Classes
DOT hazard class = 6.1

U.S. DOT - Appendix B - Marine Pollutants
DOT regulated severe marine pollutant

INTERNATIONAL LISTS

Canada - WHMIS: Ingredient Disclosure
1% item 1280 (995)

STATE LISTS

NJ Right to Know List (Total)
sn 1503

NJ Special Hazardous Substances
(teratogen)

PHENYL PHOSPHORUS DICHLORIDE   644-97-3

SEE ALSO:
MERCURY

HEALTH AND SAFETY LISTS

U.S. DOT - Substances From 49 CFR 172.101
regulated by DOT (UN1895)

U.S. DOT - Hazard Classes
DOT hazard class = 6.1

U.S. DOT - Appendix B - Marine Pollutants
DOT regulated severe marine pollutant

STATE LISTS

NJ Right to Know List (Total)
sn 1504

NJ Special Hazardous Substances
(teratogen)

PHENYL PHOSPHORUS THIODICHLORIDE   3497-00-5

ENVIRONMENTAL LISTS

TSCA - Code of Federal Regulations Citations
40 CFR 712.30(x); 40 CFR 716.120(d)

TSCA - PAIR - Reporting List
Reporting Date: November 27, 1991

TSCA - Health and Safety Reporting List
Effective Date: September 30, 1991

3-PHENYL-1-PROPANOL   122-97-4

HEALTH AND SAFETY LISTS

NFPA - Flash Points
flash point = 220 degrees F (104 degrees C)

NFPA - Hazard Identification Ratings
health-2; flammability-1; reactivity-0

STATE LISTS

Florida Hazardous Substance List
[present]

Massachusetts Right To Know List
[present]

Pennsylvania Right to Know List
[present]

PHENYLPROPANOLAMINE   14838-15-4

HEALTH AND SAFETY LISTS

NIOSH - Selected LD50s and LC50s
Oral, rat: LD50 = 1625 mg/kg

ENVIRONMENTAL LISTS

TSCA - Code of Federal Regulations Citations
40 CFR 712.30(w); 40 CFR 716.120(a)

TSCA - PAIR - Reporting List
Reporting Date: November 27, 1991

TSCA - Health and Safety Reporting List
Effective Date: September 30, 1991

PROPOSED REGULATIONS

TSCA - ITC 33rd Report Priority Testing List
recommended for testing

PHENYLPROPANOLAMINE SALTS, OPTICAL ISO-   RR-01779-8
MERS, AND SALTS OF OPTICAL ISOMERS

HEALTH AND SAFETY LISTS

ACGIH 1995 - Carcinogens
A2-suspected human carcinogen

IARC - Group 3 (not classifiable)
[present]

NIOSH - Selected LD50s and LC50s
Oral, rat: LD50 = 8730 mg/kg

NIOSH - Health Standards - Exposure Limits
reduce exposure to lowest feasible concentration

NIOSH - Health Standards - Health Effects and Precautions
Bladder cancer

NIOSH - Health Standards - Carcinogenic Chemicals
potential human carcinogen

NTP Chemical Status Reports - Testing Status and NTIS Number
Technical reports printed (PB88216270/AS)

NTP Chemical Status Reports - Evidence of Carcinogenicity
male rat-no evidence; female rat-no evidence; male mice-no evidence; female mice-equivocal evidence

INTERNATIONAL LISTS

Australian Exposure Standards - Time Weighted Averages
control to the lowest practical level

Australian Exposure Standards - Carcinogens
probable carcinogen

Australian Exposure Standards - Under Review
exposure limits under review

Canada - WHMIS: Ingredient Disclosure
0.1% item 1281 (1385)

Canada - Alberta - Designated Substances
designated substance - requires code of practice

Canada - Quebec - Time-Weighted Average Exposure Values
substance of which the recirculation is prohibited

Canada - Quebec - Carcinogens
C2 carcinogen: effect suspected in humans

German (DFG) - Carcinogens
suspected carcinogen

STATE LISTS

California - Directors List of Hazardous Substances (8 CCR 339)
[present]

Florida Hazardous Substance List
[present]

Massachusetts Right To Know List
[present]

Minnesota Hazardous Substance List
carcinogen

Pennsylvania Right to Know List
[present]

PHENYL PROPYL ALDEHYDE   1335-10-0

HEALTH AND SAFETY LISTS

IARC - Group 3 (not classifiable)
[present]

NFPA - Flash Points
flash point = 255 degrees F (124 degrees C)

NFPA - Hazard Identification Ratings
health-1; flammability-1; reactivity-0

NIOSH - Selected LD50s and LC50s
Oral, rat: LD50 = 2000 mg/kg

NTP Chemical Status Reports - Testing Status and NTIS Number
Technical reports printed (PB86217239/AS)

NTP Chemical Status Reports - Evidence of Carcinogenicity
male mice-no evidence; female mice-no evidence

ENVIRONMENTAL LISTS

CERCLA/SARA - Section 313 - Emission Reporting
form R reporting required for 1.0% de minimus concentration

INTERNATIONAL LISTS

Canada - WHMIS: Ingredient Disclosure
1% item 1282 (1386)

Canada - NPRI (National Pollutant Release Inventory)
[present]

STATE LISTS

California - Air Bill 2588 Appendix A-I
6/91

Massachusetts Right To Know List
[present]

NJ Right to Know List (Total)
sn 1439

Pennsylvania Right to Know List
environmental hazard

PHENYL SALICYLATE   118-55-8

HEALTH AND SAFETY LISTS

ACGIH 1995 - Ceiling Limits
C 0.05 ppm; C 0.23 mg/m3

NIOSH - Selected LD50s and LC50s
Inhalation, rat: LC50 = 38 ppm 4 hr

OSHA - Vacated PELs - Ceiling Limits
C 0.05 ppm; C 0.25 mg/m3 (Enforcement indefinitely stayed)

INTERNATIONAL LISTS

Australian Exposure Standards - Time Weighted Averages
Peak Limitation: 0.05 ppm; 0.23 mg/m3

Canada - WHMIS: Ingredient Disclosure
1% item 1283 (1387)

Canada - Alberta - Ceiling Occupational Exposure Limit
C 0.05 ppm; C 0.23 mg/m3

Canada - British Columbia - Ceiling Exposure Limits
C 0.05 ppm; C 0.25 mg/m3

Canada - Ontario - OHSA - CEVs
0.05 ppm CEV; 0.23 mg/m3 CEV

Canada - Quebec - Ceiling Limits
P 0.05 ppm; P 0.23 mg/m3

Israel - Ceiling Exposure Limits
C 0.05 ppm; C 0.23 mg/m3

Mexico - Instruction No. 10 - TWAs
0.05 ppm TWA; 0.25 mg/m3 TWA

STATE LISTS

California - Exposure Limits - Ceilings
C 0.05 ppm; C 0.25 mg/m3

California - Directors List of Hazardous Substances (8 CCR 339)
[present]

Florida Hazardous Substance List
[present]

Massachusetts Right To Know List
[present]

Minnesota Hazardous Substance List
[present]

Pennsylvania Right to Know List
[present]

PHENYLSILATRANE   2097-19-0

HEALTH AND SAFETY LISTS

U.S. DOT - Substances From 49 CFR 172.101
regulated by DOT (UN2798)

U.S. DOT - Hazard Classes
Forbidden from transport by the DOT

INTERNATIONAL LISTS

Canada - WHMIS: Ingredient Disclosure
1% item 1284 (518)

STATE LISTS

NJ Right to Know List (Total)
sn 0200; sn 2661

NJ Special Hazardous Substances
(corrosive)

PHENYL SULFIDE   139-66-2

HEALTH AND SAFETY LISTS

U.S. DOT - Substances From 49 CFR 172.101
regulated by DOT (UN2799)

U.S. DOT - Hazard Classes
DOT hazard class = 8

INTERNATIONAL LISTS

Canada - WHMIS: Ingredient Disclosure
1% item 1285 (1612)

STATE LISTS

NJ Right to Know List (Total)
sn 2662

4-PHENYLTHIOMORPHOLINE, 1,1-DIOXIDE-   17688-68-5

HEALTH AND SAFETY LISTS

NFPA - Flash Points
flash point = 212 degrees F (100 degrees C)

NFPA - Hazard Identification Ratings
health-0; flammability-1; reactivity-0

NIOSH - Selected LD50s and LC50s
Oral, rat: LD50 = 2300 mg/kg Skin, rabbit: LD50 = 5000 mg/kg

PHENYLTHIOUREA   103-85-5

HEALTH AND SAFETY LISTS
FDA - Controlled Substances Act - Precursor chemicals
Threshold by base weight = 2.5 kilograms

PHENYLTOLUENE   28652-72-4

HEALTH AND SAFETY LISTS
FDA - Controlled Substances Act - Precursor chemicals
Threshold by base weight = 2.5 kilograms

PHENYL TRICHLOROSILANE   98-13-5

HEALTH AND SAFETY LISTS
NFPA - Flash Points
flash point = 205 degrees F (96 degrees C)
NFPA - Hazard Identification Ratings
flammability-1; reactivity-0

PHENYL TRIMETHYL SILOXANE   2116-84-9

HEALTH AND SAFETY LISTS
NIOSH - Selected LD50s and LC50s
Oral, rat: LD50 = 3 gm/kg

PHENYLUREA   64-10-8

HEALTH AND SAFETY LISTS
NIOSH - Selected LD50s and LC50s
Oral, rat: LD50 = 1 mg/kg

ENVIRONMENTAL LISTS
CERCLA/SARA - Section 302 Extremely Hazardous Substances and
TPQs
TPQ = 100/10,000 pounds

STATE LISTS
Florida Hazardous Substance List
effective March 13, 1992
Massachusetts Right To Know List
extraordinarily hazardous
Pennsylvania Right to Know List
environmental hazard

PROPOSED REGULATIONS
CERCLA/SARA - Proposed Hazardous Substance Additions
proposed RQ = 1 pound (.454 kg)
CERCLA/SARA - 1989 Proposed RQ Adjustments
proposed RQ = 100 pounds (45.4 kg)

PHENYL UREA PESTICIDES, N.O.S.   RR-00429-5

HEALTH AND SAFETY LISTS
NIOSH - Selected LD50s and LC50s
Oral, rat: LD50 = 490 mg/kg Skin, rabbit: LD50 = 11300 mg/kg

INTERNATIONAL LISTS
Canada - WHMIS: Ingredient Disclosure
1% item 1286 (1552)

PHENYTOIN   57-41-0

ENVIRONMENTAL LISTS
TSCA - Code of Federal Regulations Citations
40 CFR 712.30(x); 40 CFR 716.120(d)
TSCA - PAIR - Reporting List
Reporting Date: November 27, 1991
TSCA - Health and Safety Reporting List
Effective Date: September 30, 1991

**PHIP (2-AMINO-1-METHYL-6-PHENYLIMIDAZO-[4,5-   105650-23-5
B]PYRIDINE)**

HEALTH AND SAFETY LISTS
U.S. DOT - Appendix A Table 1 - Hazardous Substances
final RQ = 100 pounds (45.4 kg)
NIOSH - Selected LD50s and LC50s
Oral, rat: LD50 = 3 mg/kg
NTP Chemical Status Reports - Testing Status and NTIS Number
Technical reports printed (PB287357/AS)
NTP Chemical Status Reports - Evidence of Carcinogenicity
*male rat-negative; female rat-negative; male mice-negative; female
mice-negative*

ENVIRONMENTAL LISTS
CERCLA/SARA - Section 302 Extremely Hazardous Substances and
TPQs
TPQ = 100/10,000 pounds
CERCLA/SARA - Hazardous Substances and their Reportable Quan-
tities
final RQ = 100 pounds (45.4 kg)
RCRA - P Series Wastes
waste number P093
RCRA - Hazardous Constituents-Appendix VIII
waste number P093
RCRA - Substances Banned From Land Disposal
[present]

INTERNATIONAL LISTS
Canada - WHMIS: Ingredient Disclosure
1% item 1287 (1388)

STATE LISTS
Florida Hazardous Substance List
effective March 13, 1992
Massachusetts Right To Know List
extraordinarily hazardous
Pennsylvania Right to Know List
environmental hazard

PHLOROGLUCINOL   108-73-6

HEALTH AND SAFETY LISTS
NFPA - Flash Points
flash point > 212 degrees F (> 100 degrees C)
NFPA - Hazard Identification Ratings
flammability-1; reactivity-0

PHORATE   298-02-2

HEALTH AND SAFETY LISTS
U.S. DOT - Substances From 49 CFR 172.101
regulated by DOT (UN1804)
U.S. DOT - Hazard Classes
DOT hazard class = 8
NFPA - Flash Points
flash point = 196 degrees F (91 degrees C)
NFPA - Hazard Identification Ratings
health-3; flammability-2; reactivity-0
NIOSH - Selected LD50s and LC50s
*Inhalation, mouse: LC50 = 330 mg/m3 (2 hr) Oral, rat: LD50 =
2390 mg/kg Skin, rabbit: LD50 = 890 mg/kg*

ENVIRONMENTAL LISTS
CERCLA/SARA - Section 302 Extremely Hazardous Substances and
TPQs
TPQ = 500 pounds

INTERNATIONAL LISTS
Canada - WHMIS: Ingredient Disclosure
1% item 1288 (1389)

STATE LISTS

Florida Hazardous Substance List
[present]

Massachusetts Right To Know List
extraordinarily hazardous

NJ Right to Know List (Total)
sn 1506

NJ Special Hazardous Substances
(corrosive)

Pennsylvania Right to Know List
environmental hazard

PROPOSED REGULATIONS

CERCLA/SARA - Proposed Hazardous Substance Additions
proposed RQ = 1 pound (.454 kg)

CERCLA/SARA - 1989 Proposed RQ Adjustments
proposed RQ = 100 pounds (45.4 kg)

PHORONE   504-20-1

ENVIRONMENTAL LISTS

List of Pesticide Product Inert Ingredients
[present]

PHOSACETIM   4104-14-7

HEALTH AND SAFETY LISTS

NIOSH - Selected LD50s and LC50s
Oral, rat: LD50 = 2000 mg/kg

PHOSALONE   2310-17-0

HEALTH AND SAFETY LISTS

U.S. DOT - Substances From 49 CFR 172.101
regulated by DOT (UN2767, UN2768, UN3001, UN3002)

U.S. DOT - Hazard Classes
toxic or toxic, flammable: DOT hazard class = 6.1; flammable, toxic: DOT hazard class = 3

STATE LISTS

NJ Right to Know List (Total)
sn 2665; sn 2666; sn 2667; sn 2668

PHOSFOLAN   947-02-4

HEALTH AND SAFETY LISTS

IARC - Group 2B (sufficient animal data)
[present]

NIOSH - Selected LD50s and LC50s
Oral, rat: LD50 = 2195 mg/kg

NTP Chemical Status Reports - Testing Status and NTIS Number
Technical reports printed

NTP Seventh Report - Suspect Carcinogens
suspect carcinogen

OSHA - Possible Select Carcinogens
[present]

ENVIRONMENTAL LISTS

CERCLA/SARA - Section 313 - Emission Reporting
form R reporting required

STATE LISTS

California - Air Bill 2588 Appendix A-II
known or potential carcinogen

California - Prop. 65 - Cancer list
carcinogen - initial date 1/1/88

California - Prop. 65 - Developmental Toxicity
developmental toxicity - initial date 7/1/87

California - Directors List of Hazardous Substances (8 CCR 339)
[present]

Florida Hazardous Substance List
[present]

Massachusetts Right To Know List
carcinogen; extraordinarily hazardous; teratogen

Minnesota Hazardous Substance List
carcinogen (includes its sodium salts)

NJ Right to Know List (Total)
sn 1507

NJ Special Hazardous Substances
(carcinogen; teratogen)

Pennsylvania Right to Know List
special hazardous substance

Pennsylvania RTK - Special Hazardous Substances
[present]

PHOSGENE   75-44-5

HEALTH AND SAFETY LISTS

IARC - Group 2B (sufficient animal data)
[present]

OSHA - Possible Select Carcinogens
[present]

STATE LISTS

California - Prop. 65 - Cancer list
carcinogen - initial date 10/01/94

PHOSMET   732-11-6

HEALTH AND SAFETY LISTS

NIOSH - Selected LD50s and LC50s
Oral, rat: LD50 = 5200 mg/kg

ENVIRONMENTAL LISTS

CAA - HON Rule - SOCMI Chemicals
compliance by April 24, 1995

9-PHOSPHABICYCLONONANE   13396-80-0
SEE ALSO:
K038-HAZARDOUS WASTES
K040-HAZARDOUS WASTES
F039-HAZARDOUS WASTES

HEALTH AND SAFETY LISTS

ACGIH 1995 - Time Weighted Averages
0.05 mg/m3 TWA

ACGIH 1995 - Short Term Exposure Limits
0.2 mg/m3 STEL

ACGIH 1995 - Skin Designations
skin - potential for cutaneous absorption

U.S. DOT - Appendix B - Marine Pollutants
DOT regulated severe marine pollutant

U.S. DOT - Appendix A Table 1 - Hazardous Substances
final RQ = 10 pounds (4.54 kg)

NIOSH - Selected LD50s and LC50s
Inhalation, rat: LC50 = 11 mg/m3 1 hr Oral, rat: LD50 = 1 mg/kg Skin, rat: LD50 = 2500 ug/kg

OSHA - Vacated PELs - Time Weighted Averages
0.05 mg/m3 TWA

OSHA - Vacated PELs - Short Term Exposure Limits
0.2 mg/m3 STEL

OSHA - Vacated PELs - Skin Designation
Prevent or reduce skin absorption

ENVIRONMENTAL LISTS

CERCLA/SARA - Section 302 Extremely Hazardous Substances and TPQs
TPQ = 10 pounds

CERCLA/SARA - Hazardous Substances and their Reportable Quantities
final RQ = 10 pounds (4.54 kg)

RCRA - P Series Wastes
waste number P094

RCRA - Hazardous Constituents-Appendix VIII
waste number P094

RCRA - Basis for Listing - Appendix VII
Included in waste streams: F039, K038, K040

RCRA - Substances Banned From Land Disposal
[present]

RCRA - TSD Facilities Ground Water Monitoring
TM 8140 = 2 ug/L PQL; TM 8270 = 10 ug/L PQL

RCRA - Universal Treatment Standards (LDR)
WW: 0.021 mg/l; NWW: 4.6 mg/kg

INTERNATIONAL LISTS

Australian Exposure Standards - Time Weighted Averages
0.05 mg/m3 TWA

Australian Exposure Standards - Short Term Exposure Limits
0.2 mg/m3 STEL

Australian Exposure Standards - Skin Effects
skin absorption

Canada - Drinking Water Quality - IMACs
0.002 mg/L IMAC

Canada - Alberta - 8 Hour Occupational Exposure Limit
0.05 mg/m3 TWA

Canada - Alberta - 15 Minute Occupational Exposure Limit
0.2 mg/m3 STEL

Canada - Alberta - Skin Designation
can be absorbed through the intact skin

Canada - British Columbia - 8 Hour Exposure Limits
0.05 mg/m3 TWA

Canada - British Columbia - 15 Minute Exposure Limits
0.2 mg/m3 STEL

Canada - British Columbia - Skin Notations
skin - potential for skin absorption

Canada - Ontario - OHSA - TWAEVs
0.05 mg/m3 TWAEV

Canada - Ontario - OHSA - STEVs
0.2 mg/m3 STEV

Canada - Ontario - OHSA - Skin Notations
absorption through skin, eyes, or mucous membranes

Canada - Quebec - Time-Weighted Average Exposure Values
0.05 mg/m3 TWAEV

Canada - Quebec - Short-term Exposure Values
0.2 mg/m3 STEV

Canada - Quebec - Skin Designations
absorbed through the skin

United Kingdom - Occupational Exposure Standards - TWAs
0.05 mg/m3 TWA

United Kingdom - Occupational Exposure Standards - STELs
0.2 mg/m3 STEL

United Kingdom - Occupational Exposure Standards - Notes
can be absorbed through skin

Israel - Time Weighted Averages
0.05 mg/m3 TWA

Israel - Short Term Exposure Limits
0.2 mg/m3 STEL

Israel - Action Levels
0.025 mg/m3 AL

Mexico - Instruction No. 10 - TWAs
0.05 mg/m3 TWA

Mexico - Instruction No. 10 - STELs
0.2 mg/m3 STEL

Mexico - Instruction No. 10 - Skin designation
skin - potential for cutaneous absorption

STATE LISTS

California - Exposure Limits - PELs
0.05 mg/m3 PEL

California - Exposure Limits - STELs
0.2 mg/m3 STEL

California - Exposure Limits - Skin Notation
material may be absorbed through the skin, eyes or mucous membrane

California - Directors List of Hazardous Substances (8 CCR 339)
[present]

Florida Hazardous Substance List
[present]

Massachusetts Right To Know List
extraordinarily hazardous; neurotoxin

Minnesota Hazardous Substance List
skin

NJ Right to Know List (Total)
sn 1508

Pennsylvania Right to Know List
environmental hazard

PHOSPHAMIDON   13171-21-6

HEALTH AND SAFETY LISTS

NFPA - Flash Points
flash point = 185 degrees F (85 degrees C)

NFPA - Hazard Identification Ratings
health-2; flammability-2; reactivity-0

STATE LISTS

Florida Hazardous Substance List
[present]

Massachusetts Right To Know List
[present]

Pennsylvania Right to Know List
[present]

PHOSPHATE   14265-44-2

HEALTH AND SAFETY LISTS

NIOSH - Selected LD50s and LC50s
Oral, rat: LD50 = 3700 ug/kg Skin, rat: LD50 = 25 mg/kg

ENVIRONMENTAL LISTS

CERCLA/SARA - Section 302 Extremely Hazardous Substances and TPQs
TPQ = 100/10,000 pounds

STATE LISTS

California - Directors List of Hazardous Substances (8 CCR 339)
[present]

Florida Hazardous Substance List
effective March 13, 1992

Massachusetts Right To Know List
extraordinarily hazardous

NJ Right to Know List (Total)
sn 2669

Pennsylvania Right to Know List
environmental hazard

PROPOSED REGULATIONS

CERCLA/SARA - Proposed Hazardous Substance Additions
proposed RQ = 1 pound (.454 kg)

CERCLA/SARA - 1989 Proposed RQ Adjustments
proposed RQ = 100 pounds (45.4 kg)

PHOSPHATE, BIS(TERT-BUTYLPHENYL) PHENYL   **65652-41-7**
HEALTH AND SAFETY LISTS
U.S. DOT - Appendix B - Marine Pollutants
DOT regulated severe marine pollutant

PHOSPHATE, DIISODECYL PHENYL   **51363-64-5**
HEALTH AND SAFETY LISTS
NIOSH - Selected LD50s and LC50s
Oral, rat: LD50 = 9 mg/kg Skin, guinea pig: LD50 = 54 mg/kg
ENVIRONMENTAL LISTS
CERCLA/SARA - Section 302 Extremely Hazardous Substances and TPQs
TPQ = 100/10,000 pounds
STATE LISTS
Florida Hazardous Substance List
effective March 13, 1992
Massachusetts Right To Know List
extraordinarily hazardous
NJ Right to Know List (Total)
sn 2760
Pennsylvania Right to Know List
environmental hazard
PROPOSED REGULATIONS
CERCLA/SARA - Proposed Hazardous Substance Additions
proposed RQ = 1 pound (.454 kg)
CERCLA/SARA - 1989 Proposed RQ Adjustments
proposed RQ = 100 pounds (45.4 kg)

**PHOSPHATED POLYARYLPHENOL ETHOXYLATE,   RR-01731-2
POTASSIUM SALT**
SEE ALSO:
K116-HAZARDOUS WASTES
HEALTH AND SAFETY LISTS
ACGIH 1995 - Time Weighted Averages
0.1 ppm TWA; 0.40 mg/m3 TWA
AIHA - Odor Threshold Values
no geometric mean air odor threshold
U.S. DOT - Substances From 49 CFR 172.101
regulated by DOT (UN1076)
U.S. DOT - Hazard Classes
DOT hazard class = 2.3
U.S. DOT - Substances Which Are Poisonous by Inhalation
gaseous hazardous material poisonous by inhalation (UN1076)
U.S. DOT - Appendix A Table 1 - Hazardous Substances
final RQ = 10 pounds (4.54 kg)
NIOSH - Selected LD50s and LC50s
Inhalation, rat: LC50 = 1400 mg/m3 (30 mn)
NIOSH 1990 - Pocket Guide - RELs
0.1 ppm TWA; 0.4 mg/m3 TWA; C 0.2 ppm (15 min); C 0.8 mg/m3 (15 min)
NIOSH 1990 - Pocket Guide - IDLHs
2 ppm IDLH
NIOSH 1990 - Pocket Guide - Target organs
skin, eyes, respiratory system
NIOSH - Health Standards - Exposure Limits
0.1 ppm TWA; 0.4 mg/m3 TWA; C (15 min) 0.2 ppm; C (15 min) 0.8 mg/m3
NIOSH - Health Standards - Health Effects and Precautions
Respiratory effects (Periodic chest X-ray and pulmonary function testing required)
OSHA - Vacated PELs - Time Weighted Averages
0.1 ppm TWA; 0.4 mg/m3 TWA
OSHA - Final PELs - Time Weighted Averages
0.1 ppm TWA; 0.4 mg/m3 TWA

OSHA - List of Highly Hazardous Chemicals
threshhold quantity = 100 pounds
ENVIRONMENTAL LISTS
CERCLA/SARA - Section 302 Extremely Hazardous Substances and TPQs
TPQ = 10 pounds
CERCLA/SARA - Section 313 - Emission Reporting
form R reporting required for 1.0% de minimus concentration
CERCLA/SARA - Hazardous Substances and their Reportable Quantities
final RQ = 10 pounds (4.54 kg)
Clean Air Act (1990) - List of Hazardous Air Contaminants
[present]
CAA -Toxic Substances for Accidental Release Prevention
threshold quantity = 500 lbs
CAA - HON Rule - SOCMI Chemicals
compliance by July 24, 1995
CAA - HON Rule - Organic HAPs
[present]
Clean Water Act - Hazardous Substances
[present]
RCRA - P Series Wastes
waste number P095
RCRA - Hazardous Constituents-Appendix VIII
waste number P095
RCRA - Basis for Listing - Appendix VII
Included in waste stream: K116
RCRA - Substances Banned From Land Disposal
[present]
INTERNATIONAL LISTS
Australian Exposure Standards - Time Weighted Averages
0.1 ppm TWA; 0.4 mg/m3 TWA
Canada - WHMIS: Ingredient Disclosure
1% item 1289 (1390)
Canada - NPRI (National Pollutant Release Inventory)
[present]
Canada - Alberta - 8 Hour Occupational Exposure Limit
0.1 ppm TWA; 0.4 mg/m3 TWA
Canada - Alberta - 15 Minute Occupational Exposure Limit
0.3 ppm STEL; 1.2 mg/m3 STEL
Canada - British Columbia - 8 Hour Exposure Limits
0.1 ppm TWA; 0.4 mg/m3 TWA
Canada - Ontario - OHSA - TWAEVs
0.1 ppm TWAEV; 0.4 mg/m3 TWAEV
Canada - Quebec - Time-Weighted Average Exposure Values
0.1 ppm TWAEV; 0.4 mg/m3 TWAEV
United Kingdom - Occupational Exposure Standards - TWAs
0.1 ppm TWA; 0.4 mg/m3 TWA
German (DFG) - MAK Values
0.1 ppm MAK; 0.4 mg/m3 MAK
German (DFG) - Peak Limitations
2 x normal MAK (30 min. average value); don't exceed 4 times during shift
Israel - Time Weighted Averages
0.1 ppm TWA; 0.40 mg/m3 TWA
Israel - Action Levels
0.05 ppm AL; 0.2 mg/m3 AL
Mexico - Instruction No. 10 - TWAs
0.1 ppm TWA; 0.4 mg/m3 TWA
STATE LISTS
California - Air Bill 2588 Appendix A-I
[present]

California - Exposure Limits - PELs
0.1 ppm PEL; 0.4 mg/m3 PEL

California - Directors List of Hazardous Substances (8 CCR 339)
[present]

Florida Hazardous Substance List
[present]

Massachusetts Right To Know List
extraordinarily hazardous

Minnesota Hazardous Substance List
[present]

NJ Right to Know List (Total)
sn 1510

Pennsylvania Right to Know List
environmental hazard

PHOSPHATE, ISODECYL DIPHENYL   29761-21-5

HEALTH AND SAFETY LISTS

U.S. DOT - Appendix B - Marine Pollutants
DOT regulated marine pollutant

NIOSH - Selected LD50s and LC50s
Inhalation, rat: LC50 = 54 mg/m3 4 hr Oral, rat: LD50 = 113 mg/kg
Skin, rat: LD50 = 1550 mg/kg

ENVIRONMENTAL LISTS

CERCLA/SARA - Section 302 Extremely Hazardous Substances and TPQs
TPQ = 10/10,000 pounds

STATE LISTS

Florida Hazardous Substance List
effective March 13, 1992

Massachusetts Right To Know List
extraordinarily hazardous

NJ Right to Know List (Total)
sn 0603

Pennsylvania Right to Know List
environmental hazard

PROPOSED REGULATIONS

CERCLA/SARA - Proposed Hazardous Substance Additions
proposed RQ = 1 pound (.454 kg)

CERCLA/SARA - 1989 Proposed RQ Adjustments
proposed RQ = 10 pounds (4.54 kg)

PHOSPHATE, ISOPROPYLPHENYL DIPHENYL   28108-99-8

HEALTH AND SAFETY LISTS

U.S. DOT - Substances From 49 CFR 172.101
regulated by DOT (UN2940)

U.S. DOT - Hazard Classes
DOT hazard class = 4.2

STATE LISTS

NJ Right to Know List (Total)
sn 1511

NJ Special Hazardous Substances
(corrosive)

PHOSPHATE, TRIS(ISOPROPYLPHENYL)   26967-76-0

HEALTH AND SAFETY LISTS

U.S. DOT - Appendix B - Marine Pollutants
DOT regulated severe marine pollutant

NIOSH - Selected LD50s and LC50s
Inhalation, rat: LC50 = 135 mg/m3 4 hr Oral, rat: LD50 = 10900 ug/kg Skin, rat: LD50 = 125 mg/kg

NTP Chemical Status Reports - Testing Status and NTIS Number
Technical reports printed (PB288800/AS)

NTP Chemical Status Reports - Evidence of Carcinogenicity
male rat-equivocal; female rat-equivocal; male mice-negative; female mice-negative

ENVIRONMENTAL LISTS

CERCLA/SARA - Section 302 Extremely Hazardous Substances and TPQs
TPQ = 100 pounds

INTERNATIONAL LISTS

Canada - CEPA Schedule II Part I - Prohibited Substances (Export)
[present]

STATE LISTS

California - Directors List of Hazardous Substances (8 CCR 339)
[present]

Florida Hazardous Substance List
effective March 13, 1992

Massachusetts Right To Know List
extraordinarily hazardous

Minnesota Hazardous Substance List
[present]

NJ Right to Know List (Total)
sn 1513

Pennsylvania Right to Know List
environmental hazard

PROPOSED REGULATIONS

CERCLA/SARA - Proposed Hazardous Substance Additions
proposed RQ = 1 pound (.454 kg)

CERCLA/SARA - 1989 Proposed RQ Adjustments
proposed RQ = 100 pounds (45.4 kg)

PHOSPHATE, TRIXYLYL   25155-23-1

HEALTH AND SAFETY LISTS

U.S. DOT - Substances From 49 CFR 172.101
regulated by DOT (NA1955)

U.S. DOT - Hazard Classes
DOT hazard class = 2.3

U.S. DOT - Substances Which Are Poisonous by Inhalation
gaseous hazardous material poisonous by inhalation (when mixed with compressed gas) (NA1955)

INTERNATIONAL LISTS

Mexico - Drinking Water - Ecological Criteria
0.1 mg/l

STATE LISTS

NJ Right to Know List (Total)
sn 2601

PHOSPHINE   7803-51-2

ENVIRONMENTAL LISTS

EPA - Master Testing List
[present]

PHOSPHINE, TRIPHENYL-   603-35-0
SEE ALSO:
BISAZOBIPHENYL DYES

ENVIRONMENTAL LISTS

TSCA - Code of Federal Regulations Citations
40 CFR 712.30(d); 40 CFR 716.120(c)

TSCA - PAIR - Reporting List
Reporting Date: November 19, 1982

PHOSPHONIC ACID, [1,2-ETHANEDIYLBIS[NI-   68901-17-7
TRILOBIS(METHYLENE)]TETRAKIS-, OCTAAMMO-
NIUM SALT

ENVIRONMENTAL LISTS

TSCA - Chemicals with Significant New Use Rules
PMN number: P-93-1222

PHOSPHONIC ACID, (1-HYDROXYETHYLIDINE)BIS-   14860-53-8
TETRAPOTASSIUMSALT
SEE ALSO:
BISAZOBIPHENYL DYES

HEALTH AND SAFETY LISTS

U.S. DOT - Appendix B - Marine Pollutants
DOT regulated marine pollutant

ENVIRONMENTAL LISTS

EPA - Master Testing List
[present]

TSCA - Code of Federal Regulations Citations
40 CFR 712.30(d);   40 CFR 716.120(c)

TSCA - PAIR - Reporting List
Reporting Date: November 19, 1982

PHOSPHONIC ACID, (1-HYDROXYETHYLIDINE)BIS-   3794-83-0
, TETRASODIUM SALT
SEE ALSO:
BISAZOBIPHENYL DYES

ENVIRONMENTAL LISTS

EPA - Master Testing List
[present]

TSCA - Code of Federal Regulations Citations
40 CFR 712.30(d)

PHOSPHONIUM SALT   RR-00175-2

ENVIRONMENTAL LISTS

EPA - Master Testing List
[present]

2-PHOSPHONO-1,2-4-BUTANETRICARBOXYLIC   37971-36-1
ACID
SEE ALSO:
BISAZOBIPHENYL DYES

HEALTH AND SAFETY LISTS

U.S. DOT - Appendix B - Marine Pollutants
DOT regulated marine pollutant

NIOSH - Selected LD50s and LC50s
Oral, mouse: LD50 = 11800 mg/kg

ENVIRONMENTAL LISTS

EPA - Master Testing List
[present]

TSCA - Code of Federal Regulations Citations
40 CFR 712.30(d);   40 CFR 716.120(c)

TSCA - PAIR - Reporting List
Reporting Date: November 19, 1982

2-PHOSPHONO-1,2,4-BUTANETRICARBOXYLIC   40372-66-5
ACID, SODIUM SALT

HEALTH AND SAFETY LISTS

ACGIH 1995 - Time Weighted Averages
0.3 ppm TWA;   0.42 mg/m3 TWA

ACGIH 1995 - Short Term Exposure Limits
1 ppm STEL;   1.4 mg/m3 STEL

AIHA - Odor Threshold Values
geometric mean air odor threshold = 0.14 ppm (recognizable)

U.S. DOT - Substances From 49 CFR 172.101
regulated by DOT (UN2199)

U.S. DOT - Hazard Classes
DOT hazard class = 2.3

U.S. DOT - Substances Which Are Poisonous by Inhalation
gaseous hazardous material poisonous by inhalation (UN2199)

U.S. DOT - Appendix A Table 1 - Hazardous Substances
final RQ = 100 pounds (45.4 kg)

NFPA - Flash Points
gas (no flash point given)

NFPA - Hazard Identification Ratings
health-4;   flammability-4;   reactivity-2

NIOSH - Selected LD50s and LC50s
Inhalation, rat: LC50 = 11 ppm 4 hr

NIOSH 1990 - Pocket Guide - RELs
0.3 ppm TWA;   0.4 mg/m3 TWA;   1 ppm STEL;   1 mg/m3 STEL

NIOSH 1990 - Pocket Guide - IDLHs
200 ppm IDLH

NIOSH 1990 - Pocket Guide - Target organs
respiratory system

NTP Chemical Status Reports - Testing Status and NTIS Number
Prechronic studies for which toxicity technical reports were not pre-pared

OSHA - Vacated PELs - Time Weighted Averages
0.3 ppm TWA;   0.4 mg/m3 TWA

OSHA - Vacated PELs - Short Term Exposure Limits
1 ppm STEL;   1 mg/m3 STEL

OSHA - Final PELs - Time Weighted Averages
0.3 ppm TWA;   0.4 mg/m3 TWA

OSHA - List of Highly Hazardous Chemicals
threshhold quantity = 100 pounds

ENVIRONMENTAL LISTS

CERCLA/SARA - Section 302 Extremely Hazardous Substances and TPQs
TPQ = 500 pounds

CERCLA/SARA - Section 313 - Emission Reporting
form R reporting required

CERCLA/SARA - Hazardous Substances and their Reportable Quan-tities
final RQ = 100 pounds (45.4 kg)

Clean Air Act (1990) - List of Hazardous Air Contaminants
[present]

CAA -Toxic Substances for Accidental Release Prevention
threshold quantity = 5,000 lbs

RCRA - P Series Wastes
waste number P096

RCRA - Hazardous Constituents-Appendix VIII
waste number P096

RCRA - Substances Banned From Land Disposal
[present]

INTERNATIONAL LISTS

Australian Exposure Standards - Time Weighted Averages
0.3 ppm TWA;   0.42 mg/m3 TWA

Australian Exposure Standards - Short Term Exposure Limits
1 ppm STEL;   1.4 mg/m3 STEL

Australian Exposure Standards - Under Review
exposure limits under review

Canada - WHMIS: Ingredient Disclosure
1% item 1290 (1402)

Canada - Alberta - 8 Hour Occupational Exposure Limit
0.3 ppm TWA;   0.42 mg/m3 TWA

Canada - Alberta - 15 Minute Occupational Exposure Limit
1 ppm STEL;   1.3 mg/m3 STEL

Canada - British Columbia - 8 Hour Exposure Limits
0.3 ppm TWA;   0.4 mg/m3 TWA

Canada - British Columbia - 15 Minute Exposure Limits
1 ppm STEL;   1 mg/m3 STEL

Canada - Ontario - OHSA - TWAEVs
0.3 ppm TWAEV;   0.4 mg/m3 TWAEV

Canada - Ontario - OHSA - STEVs
1 ppm STEV; 1.4 mg/m3 STEV
Canada - Quebec - Time-Weighted Average Exposure Values
0.3 ppm TWAEV; 0.42 mg/m3 TWAEV
Canada - Quebec - Short-term Exposure Values
1 ppm STEV; 1.4 mg/m3 STEV
United Kingdom - Occupational Exposure Standards - STELs
0.3 ppm STEL; 0.4 mg/m3 STEL
German (DFG) - MAK Values
0.1 ppm MAK; 0.15 mg/m3 MAK
German (DFG) - Peak Limitations
2 x normal MAK (5 min momentary value); don't exceed 8 times during shift
Israel - Time Weighted Averages
0.3 ppm TWA; 0.42 mg/m3 TWA
Israel - Short Term Exposure Limits
1 ppm STEL; 1.4 mg/m3 STEL
Israel - Action Levels
0.15 ppm AL; 0.21 mg/m3 AL
Mexico - Instruction No. 10 - TWAs
0.3 ppm TWA; 0.4 mg/m3 TWA
Mexico - Instruction No. 10 - STELs
1 ppm STEL; 1 mg/m3 STEL

STATE LISTS

California - Air Bill 2588 Appendix A-I
[present]
California - Exposure Limits - PELs
0.3 ppm PEL; 0.4 mg/m3 PEL
California - Exposure Limits - STELs
1 ppm STEL; 1 mg/m3 STEL
California - Directors List of Hazardous Substances (8 CCR 339)
[present]
Florida Hazardous Substance List
[present]
Massachusetts Right To Know List
extraordinarily hazardous
Minnesota Hazardous Substance List
[present]
NJ Right to Know List (Total)
sn 1514
NJ Special Hazardous Substances
(flammable - fourth degree)
Pennsylvania Right to Know List
environmental hazard

PHOSPHONOCARBOXYLATE SALTS RR-01658-0

HEALTH AND SAFETY LISTS

NFPA - Flash Points
flash point = 356 degrees F (180 degrees C)
NFPA - Hazard Identification Ratings
health-0; flammability-1; reactivity-0
NIOSH - Selected LD50s and LC50s
Inhalation, rat: LC50 = 1135 ppm 4 hr Oral, rat: LD50 = 700 mg/kg

ENVIRONMENTAL LISTS

TSCA - Code of Federal Regulations Citations
40 CFR 712.30(d)
TSCA - PAIR - Reporting List
Reporting Date: November 19, 1982

PHOSPHONOTHIOIC ACID, METHYL-, S-[2-[BIS(1- 50782-69-9 METHYLETHYL)AMINOETHYL]O-ETHYL ESTER

ENVIRONMENTAL LISTS

TSCA - Code of Federal Regulations Citations
40 CFR 704.95

PHOSPHONOTHIOIC ACID, METHYL-,O-ETHYL O- 2703-13-1 (4-(METHYLTHIO)PHENYL) ESTER

HEALTH AND SAFETY LISTS

NIOSH - Selected LD50s and LC50s
Oral, rat: LD50 = 520 mg/kg

PHOSPHONOTHIOIC ACID, METHYL-,O-(4-NITRO- 2665-30-7 PHENYL) O-PHENYL ESTER

HEALTH AND SAFETY LISTS

NIOSH - Selected LD50s and LC50s
Oral, rat: LD50 = 990 mg/kg

ALPHA-PHOSPHONO-OMEGA-(TRIDECYLOXY) 26915-70-8 POLY(OXY-1,2-ETHANEDIYL)

ENVIRONMENTAL LISTS

TSCA - Chemicals with Significant New Use Rules
PMN number: P-84-820

PHOSPHORAMIDE RR-01666-0

ENVIRONMENTAL LISTS

EPA - Master Testing List
[present]
List of Pesticide Product Inert Ingredients
[present]

PHOSPHORIC ACID 7664-38-2

ENVIRONMENTAL LISTS

List of Pesticide Product Inert Ingredients
[present]

PHOSPHORIC ACID, BERYLLIUM SALT (2:3) 13598-26-0

ENVIRONMENTAL LISTS

TSCA - Chemicals with Significant New Use Rules
PMN number: P-93-722, P-93-723, P-93-724

PHOSPHORIC ACID, 2,2-BIS(BROMOMETHYL) 66108-37-0 -3-CHLOROPROPYL BIS[2-CHLORO-1- (CHLOROMETHYL)ETHYL] ESTER

ENVIRONMENTAL LISTS

CERCLA/SARA - Section 302 Extremely Hazardous Substances and TPQs
TPQ = 100 pounds

STATE LISTS

Florida Hazardous Substance List
effective March 13, 1992
Massachusetts Right To Know List
extraordinarily hazardous
NJ Right to Know List (Total)
sn 2673
Pennsylvania Right to Know List
environmental hazard

PROPOSED REGULATIONS

CERCLA/SARA - Proposed Hazardous Substance Additions
proposed RQ = 1 pound (.454 kg)
CERCLA/SARA - 1989 Proposed RQ Adjustments
proposed RQ = 10 pounds (4.54 kg)

PHOSPHORIC ACID, BIS(2-ETHYLHEXYL)ESTER, 141-65-1 SODIUM SALT

ENVIRONMENTAL LISTS

CERCLA/SARA - Section 302 Extremely Hazardous Substances and TPQs
TPQ = 500 pounds

STATE LISTS

Florida Hazardous Substance List
effective March 13, 1992

Massachusetts Right To Know List
extraordinarily hazardous

NJ Right to Know List (Total)
sn 2671

Pennsylvania Right to Know List
environmental hazard

PROPOSED REGULATIONS

CERCLA/SARA - Proposed Hazardous Substance Additions
proposed RQ = 1 pound (.454 kg)

CERCLA/SARA - 1989 Proposed RQ Adjustments
proposed RQ = 10 pounds (4.54 kg)

PHOSPHORIC ACID, BIS(2-ETHYLHEXYL) PHENYL ESTER   16368-97-1

HEALTH AND SAFETY LISTS

NIOSH - Selected LD50s and LC50s
Oral, rat: LD50 = 8 mg/kg

ENVIRONMENTAL LISTS

CERCLA/SARA - Section 302 Extremely Hazardous Substances and TPQs
TPQ = 500 pounds

STATE LISTS

Florida Hazardous Substance List
effective March 13, 1992

Massachusetts Right To Know List
extraordinarily hazardous

NJ Right to Know List (Total)
sn 2672

Pennsylvania Right to Know List
environmental hazard

PROPOSED REGULATIONS

CERCLA/SARA - Proposed Hazardous Substance Additions
proposed RQ = 1 pound (.454 kg)

CERCLA/SARA - 1989 Proposed RQ Adjustments
proposed RQ = 100 pounds (45.4 kg)

PHOSPHORIC ACID, BUTYL ESTER, MANGANESE (2+) SALT   69011-04-7

ENVIRONMENTAL LISTS

List of Pesticide Product Inert Ingredients
[present]

PHOSPHORIC ACID, C(6-12)-ALKYL ESTERS, COMPOUND WITH 2-(DIBUTYLAMIONO) ETHANOL   129733-59-1

ENVIRONMENTAL LISTS

TSCA - Chemicals with Significant New Use Rules
PMN number: P-89-538

PHOSPHORIC ACID, DIDODECYL ESTER   7057-92-3

HEALTH AND SAFETY LISTS

ACGIH 1995 - Time Weighted Averages
1 mg/m3 TWA

ACGIH 1995 - Short Term Exposure Limits
3 mg/m3 STEL

U.S. DOT - Substances From 49 CFR 172.101
regulated by DOT (UN1805)

U.S. DOT - Hazard Classes
DOT hazard class = 8

U.S. DOT - Appendix A Table 1 - Hazardous Substances
final RQ = 5000 pounds (2270 kg)

NIOSH - Selected LD50s and LC50s
Oral, rat: LD50 = 1530 mg/kg Skin, rabbit: LD50 = 2740 mg/kg

NIOSH 1990 - Pocket Guide - RELs
1 mg/m3 TWA; 3 mg/m3 STEL

NIOSH 1990 - Pocket Guide - IDLHs
10,000 mg/m3 IDLH

NIOSH 1990 - Pocket Guide - Target organs
skin, eyes, respiratory system

OSHA - Vacated PELs - Time Weighted Averages
1 mg/m3 TWA

OSHA - Vacated PELs - Short Term Exposure Limits
3 mg/m3 STEL

OSHA - Final PELs - Time Weighted Averages
1 mg/m3 TWA

ENVIRONMENTAL LISTS

CERCLA/SARA - Section 313 - Emission Reporting
form R reporting required for 1.0% de minimus concentration

CERCLA/SARA - Hazardous Substances and their Reportable Quantities
final RQ = 5000 pounds (2270 kg)

Clean Water Act - Hazardous Substances
[present]

List of Pesticide Product Inert Ingredients
[present]

INTERNATIONAL LISTS

Australian Exposure Standards - Time Weighted Averages
1 mg/m3 TWA

Australian Exposure Standards - Short Term Exposure Limits
3 mg/m3 STEL

Canada - WHMIS: Ingredient Disclosure
1% item 1291 (127)

Canada - NPRI (National Pollutant Release Inventory)
[present]

Canada - Alberta - 8 Hour Occupational Exposure Limit
1 mg/m3 TWA

Canada - Alberta - 15 Minute Occupational Exposure Limit
3 mg/m3 STEL

Canada - British Columbia - 8 Hour Exposure Limits
1 mg/m3 TWA

Canada - British Columbia - 15 Minute Exposure Limits
3 mg/m3 STEL

Canada - Ontario - OHSA - TWAEVs
1 mg/m3 TWAEV

Canada - Ontario - OHSA - STEVs
3 mg/m3 STEV

Canada - Quebec - Time-Weighted Average Exposure Values
1 mg/m3 TWAEV

Canada - Quebec - Short-term Exposure Values
3 mg/m3 STEV

United Kingdom - Occupational Exposure Standards - TWAs
1 mg/m3 TWA

United Kingdom - Occupational Exposure Standards - STELs
3 mg/m3 STEL

Israel - Time Weighted Averages
1 mg/m3 TWA

Israel - Short Term Exposure Limits
3 mg/m3 STEL

Israel - Action Levels
0.5 mg/m3 AL

Mexico - Instruction No. 10 - TWAs
1 mg/m3 TWA

Mexico - Instruction No. 10 - STELs
3 mg/m3 STEL

STATE LISTS

California - Air Bill 2588 Appendix A-I
9/89

California - Exposure Limits - PELs
1 mg/m3 PEL

California - Exposure Limits - STELs
3 mg/m3 STEL

California - Directors List of Hazardous Substances (8 CCR 339)
[present]

Florida Hazardous Substance List
[present]

Massachusetts Right To Know List
[present]

Minnesota Hazardous Substance List
[present]

NJ Right to Know List (Total)
sn 1516

NJ Special Hazardous Substances
(corrosive)

Pennsylvania Right to Know List
environmental hazard

PHOSPHORIC ACID, DIMETHYL 4-(METHYLTHIO) PHENYL ESTER   3254-63-5

STATE LISTS

Pennsylvania Right to Know List
environmental hazard; special hazardous substance

Pennsylvania RTK - Special Hazardous Substances
[present]

PHOSPHORIC ACID, DISODIUM SALT, DODECAHYDRATE   10039-32-4

ENVIRONMENTAL LISTS

TSCA - Code of Federal Regulations Citations
40 CFR 712.30(d)

TSCA - PAIR - Reporting List
Reporting Date: November 19, 1982

PHOSPHORIC ACID, DISODIUM SALT, HYDRATE   10140-65-5

INTERNATIONAL LISTS

Canada - WHMIS: Ingredient Disclosure
1% item 1292 (1401)

PHOSPHORIC ACID, DODECYL ESTER   12751-23-4

HEALTH AND SAFETY LISTS

NIOSH - Selected LD50s and LC50s
Inhalation, mouse: LC50 = 5 gm/m3 8 hr Oral, mouse: LD50 = 9333 mg/kg

PHOSPHORIC ACID, [1,2-ETHANEDIYLBIS[NI-TRILOBIS(METHYLENE)]TETRAKIS-, HEXASODIUM SALT   15142-96-8

ENVIRONMENTAL LISTS

List of Pesticide Product Inert Ingredients
[present]

PHOSPHORIC ACID, [1,2-ETHANEDIYLBIS[NI-TRILOBIS(METHYLENE)]TETRAKIS-, POTASSIUM SALT   34274-30-1

ENVIRONMENTAL LISTS

TSCA - Code of Federal Regulations Citations
40 CFR 721.1610

TSCA - Chemicals with Significant New Use Rules
PMN number: P-90-384

TSCA - Section 12(b) - Export Notification
P-90-384; export notification required - Section 5

PHOSPHORIC ACID, [1,2-ETHANEDIYLBIS[NI-TRILOBIS(METHYLENE)]TETRAKIS-, AMMONIUM SALT   57011-27-5

ENVIRONMENTAL LISTS

TSCA - Code of Federal Regulations Citations
40 CFR 712.30(x); 40 CFR 716.120(d)

TSCA - PAIR - Reporting List
Reporting Date: December 27, 1990

TSCA - Health and Safety Reporting List
Effective Date: October 29, 1990; Sunset Date: November 9, 1993

PHOSPHORIC ACID, [1,2-ETHANEDIYLBIS[NITRILOBIS(METHYLENE)]TETRAKIS-, TE-TRAPOTASSIUM SALT   68188-96-5

HEALTH AND SAFETY LISTS

NIOSH - Selected LD50s and LC50s
Oral, rat: LD50 = 7 mg/kg Skin, rabbit: LD50 = 48 mg/kg

ENVIRONMENTAL LISTS

CERCLA/SARA - Section 302 Extremely Hazardous Substances and TPQs
TPQ = 500 pounds

STATE LISTS

Florida Hazardous Substance List
effective March 13, 1992

Massachusetts Right To Know List
extraordinarily hazardous

NJ Right to Know List (Total)
sn 2674

Pennsylvania Right to Know List
environmental hazard

PROPOSED REGULATIONS

CERCLA/SARA - Proposed Hazardous Substance Additions
proposed RQ = 1 pound (.454 kg)

CERCLA/SARA - 1989 Proposed RQ Adjustments
proposed RQ = 100 pounds (45.4 kg)

PHOSPHORIC ACID, 1,2-ETHANEDIYL TETRAKIS(2-CHLORO-1-METHYLETHYL) ESTER   RR-00901-8
SEE ALSO:
SODIUM PHOSPHATE DIBASIC

HEALTH AND SAFETY LISTS

U.S. DOT - Appendix A Table 1 - Hazardous Substances
final RQ = 5000 pounds (2270 kg) (Listed under 'Sodium phosphate, dibasic')

ENVIRONMENTAL LISTS

CERCLA/SARA - Hazardous Substances and their Reportable Quantities
final RQ = 5000 pounds (2270 kg) (Listed under 'Sodium phosphate, dibasic')

Clean Water Act - Hazardous Substances
[present] (Listed under 'Sodium phosphate, dibasic')

STATE LISTS

California - Directors List of Hazardous Substances (8 CCR 339)
[present] (Listed under 'Sodium phosphate, dibasic')

Massachusetts Right To Know List
[present]

NJ Right to Know List (Total)
sn 3040

Pennsylvania Right to Know List
environmental hazard

PHOSPHORIC ACID, 2-ETHYLHEXYL DIPHENYL ESTER   1241-94-7

SEE ALSO:
SODIUM PHOSPHATE DIBASIC

HEALTH AND SAFETY LISTS

U.S. DOT - Appendix A Table 1 - Hazardous Substances
final RQ = 5000 pounds (2270 kg) (Listed under 'Sodium phosphate, dibasic')

ENVIRONMENTAL LISTS

CERCLA/SARA - Hazardous Substances and their Reportable Quantities
final RQ = 5000 pounds (2270 kg) (Listed under 'Sodium phosphate, dibasic')

Clean Water Act - Hazardous Substances
[present] (Listed under 'Sodium phosphate, dibasic')

STATE LISTS

California - Directors List of Hazardous Substances (8 CCR 339)
[present] (Listed under 'Sodium phosphate, dibasic')

Massachusetts Right To Know List
[present]

NJ Right to Know List (Total)
sn 3041

Pennsylvania Right to Know List
environmental hazard

PHOSPHORIC ACID, 2-ETHYLHEXYL ESTER   12645-31-7

ENVIRONMENTAL LISTS

TSCA - Code of Federal Regulations Citations
40 CFR 712.30(x); 40 CFR 716.120(d)

TSCA - PAIR - Reporting List
Reporting Date: December 27, 1990

TSCA - Health and Safety Reporting List
Effective Date: October 29, 1990; Sunset Date: November 9, 1993

PHOSPHORIC ACID, METHYLPHENYL DIPHENYL ESTER   26444-49-5

ENVIRONMENTAL LISTS

TSCA - Code of Federal Regulations Citations
40 CFR 704.95

PHOSPHORIC ACID, (1-METHYL-1-PHENYLETHYL) PHENYL DIPHENYLESTER   34364-42-6

ENVIRONMENTAL LISTS

TSCA - Code of Federal Regulations Citations
40 CFR 704.95

PHOSPHORIC ACID, MONOBUTYL ESTER   1623-15-0

ENVIRONMENTAL LISTS

TSCA - Code of Federal Regulations Citations
40 CFR 704.95

PHOSPHORIC ACID MONO(C8-C10)ALKYL SODIUM SALTS   68909-59-1

ENVIRONMENTAL LISTS

TSCA - Code of Federal Regulations Citations
40 CFR 704.95

PHOSPHORIC ACID, MONO(2-ETHYLHEXYL) ESTER   1070-03-7

ENVIRONMENTAL LISTS

TSCA - Chemicals with Significant New Use Rules
PMN number: P-86-1263

PHOSPHORIC ACID, MONOHEXYL ESTER   3900-04-7

SEE ALSO:
BISAZOBIPHENYL DYES

ENVIRONMENTAL LISTS

EPA - Master Testing List
[present]

TSCA - Code of Federal Regulations Citations
40 CFR 712.30(d); 40 CFR 716.120(c)

TSCA - PAIR - Reporting List
Reporting Date: November 19, 1982

PHOSPHORIC ACID, MONO(2-METHYLPHENYL) ESTER   18351-85-4

ENVIRONMENTAL LISTS

TSCA - Code of Federal Regulations Citations
40 CFR 712.30(x); 40 CFR 716.120(d)

TSCA - PAIR - Reporting List
Reporting Date: December 27, 1990

TSCA - Health and Safety Reporting List
Effective Date: October 29, 1990; Sunset Date: November 9, 1993

PHOSPHORIC ACID, MONOMETHYL ESTER   812-00-0

SEE ALSO:
BISAZOBIPHENYL DYES

HEALTH AND SAFETY LISTS

NIOSH - Selected LD50s and LC50s
Oral, rat: LD50 = 6400 mg/kg

ENVIRONMENTAL LISTS

EPA - Master Testing List
[present]

TSCA - Code of Federal Regulations Citations
40 CFR 712.30(d); 40 CFR 716.120(c)

TSCA - PAIR - Reporting List
Reporting Date: November 19, 1982

PHOSPHORIC ACID, MONOOCTADECYL ESTER   2958-09-0

SEE ALSO:
BISAZOBIPHENYL DYES

ENVIRONMENTAL LISTS

TSCA - Code of Federal Regulations Citations
40 CFR 716.120(c)

PHOSPHORIC ACID, MONOOCTYL ESTER   3991-73-9

ENVIRONMENTAL LISTS

TSCA - Code of Federal Regulations Citations
40 CFR 712.30(x); 40 CFR 716.120(d)

TSCA - PAIR - Reporting List
Reporting Date: December 27, 1990

TSCA - Health and Safety Reporting List
Effective Date: October 29, 1990; Sunset Date: November 9, 1993

PHOSPHORIC ACID, TRIS(4-METHYLPHENYL) ESTER   78-32-0

ENVIRONMENTAL LISTS

List of Pesticide Product Inert Ingredients
[present]

PHOSPHORIC ACID, TRIS(3-METHYLPHENYL) ESTER   563-04-2

ENVIRONMENTAL LISTS

List of Pesticide Product Inert Ingredients
[present]

TSCA - Code of Federal Regulations Citations
40 CFR 712.30(x); 40 CFR 716.120(d)

TSCA - PAIR - Reporting List
Reporting Date: December 27, 1990

TSCA - Health and Safety Reporting List
Effective Date: October 29, 1990; Sunset Date: November 9, 1993

PHOSPHORIC ACID, TRISODIUM SALT, DECAHYDRATE   10361-89-4

ENVIRONMENTAL LISTS

TSCA - Code of Federal Regulations Citations
40 CFR 712.30(x); 40 CFR 716.120(d)

TSCA - PAIR - Reporting List
Reporting Date: December 27, 1990

TSCA - Health and Safety Reporting List
Effective Date: October 29, 1990; Sunset Date: November 9, 1993

PHOSPHORODITHIOIC ACID, O,O-DIISOOCTYL   28629-66-5

STATE LISTS

Pennsylvania Right to Know List
[present]

PHOSPHORODITHIOIC AND PHOSPHOTHIOIC ACID ESTERS   RR-00497-7

ENVIRONMENTAL LISTS

TSCA - Code of Federal Regulations Citations
40 CFR 712.30(x); 40 CFR 716.120(d)

TSCA - PAIR - Reporting List
Reporting Date: December 27, 1990

TSCA - Health and Safety Reporting List
Effective Date: October 29, 1990; Sunset Date: November 9, 1993

PHOSPHOROTHIOIC ACID, O-(5-CHLORO-1-(1-METHYLETHYL)-1H-1,2,4-TRIAZOL-3-YL) O,O-DIETHYL ESTER   42509-80-8

ENVIRONMENTAL LISTS

TSCA - Code of Federal Regulations Citations
40 CFR 712.30(x); 40 CFR 716.120(d)

TSCA - PAIR - Reporting List
Reporting Date: December 27, 1990

TSCA - Health and Safety Reporting List
Effective Date: October 29, 1990; Sunset Date: November 9, 1993

PHOSPHOROTHIOIC ACID, O,O-DIMETHYL-S-(2-METHYLTHIO)ETHYL ESTER   2587-90-8

ENVIRONMENTAL LISTS

TSCA - Code of Federal Regulations Citations
40 CFR 712.30(x); 40 CFR 716.120(d)

TSCA - PAIR - Reporting List
Reporting Date: December 27, 1990

TSCA - Health and Safety Reporting List
Effective Date: October 29, 1990; Sunset Date: November 9, 1993

PHOSPHOROUS ACID   13598-36-2

SEE ALSO:
BISAZOBIPHENYL DYES

ENVIRONMENTAL LISTS

TSCA - Code of Federal Regulations Citations
40 CFR 712.30(d); 40 CFR 716.120(c)

TSCA - PAIR - Reporting List
Reporting Date: November 19, 1982

PHOSPHOROUS ACID, ORTHO   10294-56-1

SEE ALSO:
BISAZOBIPHENYL DYES

ENVIRONMENTAL LISTS

TSCA - Code of Federal Regulations Citations
40 CFR 716.120(c)

PHOSPHOROUS ACID, DIBUTYL ESTER   109-47-7

HEALTH AND SAFETY LISTS

U.S. DOT - Appendix A Table 1 - Hazardous Substances
final RQ = 5000 pounds (2270 kg) (Listed under 'Sodium phosphate, tribasic')

ENVIRONMENTAL LISTS

CERCLA/SARA - Hazardous Substances and their Reportable Quantities
final RQ = 5000 pounds (2270 kg) (Listed under 'Sodium phosphate, tribasic')

Clean Water Act - Hazardous Substances
[present] (Listed under 'Sodium phosphate, tribasic')

STATE LISTS

California - Directors List of Hazardous Substances (8 CCR 339)
[present] (Listed under 'Sodium phosphate, tribasic')

Massachusetts Right To Know List
[present]

NJ Right to Know List (Total)
sn 3046

Pennsylvania Right to Know List
environmental hazard

PHOSPHOROUS ACID, (DIISODECYL)PHENYL ETHER   25550-98-5

ENVIRONMENTAL LISTS

EPA - Master Testing List
[present]

PHOSPHOROUS ACID, TRIBUTYL ESTER   102-85-2

SEE ALSO:
K039-HAZARDOUS WASTES
K036-HAZARDOUS WASTES
K038-HAZARDOUS WASTES
K040-HAZARDOUS WASTES

ENVIRONMENTAL LISTS

RCRA - Basis for Listing - Appendix VII
Included in waste streams: K036, K037, K038, K039, K040

PHOSPHORUS   7723-14-0

STATE LISTS

Massachusetts Right To Know List
[present]

PHOSPHORUS 32   14596-37-3

ENVIRONMENTAL LISTS

CERCLA/SARA - Section 302 Extremely Hazardous Substances and TPQs
TPQ = 500 pounds

STATE LISTS

Florida Hazardous Substance List
effective March 13, 1992

Massachusetts Right To Know List
extraordinarily hazardous

Pennsylvania Right to Know List
environmental hazard

PROPOSED REGULATIONS

CERCLA/SARA - Proposed Hazardous Substance Additions
proposed RQ = 1 pound (.454 kg)

CERCLA/SARA - 1989 Proposed RQ Adjustments
proposed RQ = 100 pounds (45.4 kg)

PHOSPHORUS 33   15749-66-3

INTERNATIONAL LISTS

Canada - WHMIS: Ingredient Disclosure
1% item 1294 (126)

STATE LISTS

NJ Right to Know List (Total)
sn 1519

PHOSPHORUS, AMORPHOUS   RR-01336-5

HEALTH AND SAFETY LISTS

U.S. DOT - Substances From 49 CFR 172.101
regulated by DOT (UN2834)

U.S. DOT - Hazard Classes
DOT hazard class = 8

ENVIRONMENTAL LISTS

List of Pesticide Product Inert Ingredients
[present]

PHOSPHORUS COMPOUNDS, INORGANIC   RR-01027-5

STATE LISTS

Pennsylvania Right to Know List
[present]

PHOSPHORUS HEPTASULPHIDE   12037-82-0

ENVIRONMENTAL LISTS

List of Pesticide Product Inert Ingredients
[present]

TSCA - Code of Federal Regulations Citations
40 CFR 712.30(p);   40 CFR 716.120(a);   40 CFR 799.5000

TSCA - PAIR - Reporting List
Reporting Date: February 18, 1986

TSCA - Health and Safety Reporting List
Effective Date: December 19, 1985

TSCA - Substances Subject to Testing Consent Orders
Test for: Neurotoxic Effects

TSCA - Section 12(b) - Export Notification
export notification required - Section 4

PHOSPHORUS OXYBROMIDE   7789-59-5

HEALTH AND SAFETY LISTS

NFPA - Flash Points
flash point = 248 degrees F (120 degrees C)

NFPA - Hazard Identification Ratings
health-2;   flammability-1;   reactivity-1

NIOSH - Selected LD50s and LC50s
Oral, rat: LD50 = 3000 mg/kg

STATE LISTS

Florida Hazardous Substance List
[present]

Massachusetts Right To Know List
[present]

Pennsylvania Right to Know List
[present]

PHOSPHORUS OXYCHLORIDE   10025-87-3

HEALTH AND SAFETY LISTS

ACGIH 1995 - Time Weighted Averages
0.02 ppm TWA;   0.1 mg/m3 TWA

U.S. DOT - Substances From 49 CFR 172.101
regulated by DOT (UN1381, UN2447)

U.S. DOT - Hazard Classes
DOT hazard class = 4.2

U.S. DOT - Appendix B - Marine Pollutants
DOT regulated severe marine pollutant

U.S. DOT - Appendix A Table 1 - Hazardous Substances
final RQ = 1 pound (0.454 kg)

NIOSH - Selected LD50s and LC50s
Oral, rat: LD50 = 3030 ug/kg

NIOSH 1990 - Pocket Guide - RELs
0.1 mg/m3 TWA

NIOSH 1990 - Pocket Guide - Target organs
respiratory system, liver, kidneys, jaw, teeth, blood, eyes, skin

OSHA - Vacated PELs - Time Weighted Averages
0.1 mg/m3 TWA

OSHA - Final PELs - Time Weighted Averages
0.1 mg/m3 TWA

ENVIRONMENTAL LISTS

ATSDR Priority List
Rank (of 275): 034

CERCLA/SARA - Section 302 Extremely Hazardous Substances and TPQs
TPQ = 100 pounds (This material is a reactive solid. The TPQ does not default to 10,000 pounds for non-powder, non-molten, non-solution form)

CERCLA/SARA - Section 313 - Emission Reporting
form R reporting required for 1.0% de minimus concentration

CERCLA/SARA - Hazardous Substances and their Reportable Quantities
final RQ = 1 pound (0.454 kg)

Clean Air Act (1990) - List of Hazardous Air Contaminants
[present]

Clean Water Act - Hazardous Substances
[present]

TSCA - Code of Federal Regulations Citations
40 CFR 712.30(w)

TSCA - PAIR - Reporting List
Reporting Date: July 13, 1988

INTERNATIONAL LISTS

Australian Exposure Standards - Time Weighted Averages
0.1 mg/m3 TWA

Canada - WHMIS: Ingredient Disclosure
1% item 1295 (1407)

Canada - NPRI (National Pollutant Release Inventory)
[present]

Canada - Alberta - 8 Hour Occupational Exposure Limit
0.1 mg/m3 TWA

Canada - Alberta - 15 Minute Occupational Exposure Limit
0.3 mg/m3 STEL

Canada - British Columbia - 8 Hour Exposure Limits
0.1 mg/m3 TWA

Canada - British Columbia - 15 Minute Exposure Limits
0.3 mg/m3 STEL

Canada - Ontario - OHSA - TWAEVs
0.1 mg/m3 TWAEV

Canada - Quebec - Time-Weighted Average Exposure Values
0.1 ppm TWAEV

United Kingdom - Occupational Exposure Standards - TWAs
0.1 mg/m3 TWA

United Kingdom - Occupational Exposure Standards - STELs
0.3 mg/m3 STEL

German (DFG) - MAK Values
total dust: 0.1 mg/m3 MAK

German (DFG) - Peak Limitations
2 x normal MAK (5 min momentary value);   don't exceed 8 times during shift

German (DFG) - Pregnancy
classification not yet possible
Israel - Time Weighted Averages
0.1 mg/m3 TWA
Israel - Action Levels
0.05 mg/m3 AL
Mexico - Instruction No. 10 - TWAs
0.1 mg/m3 TWA
Mexico - Instruction No. 10 - STELs
0.3 mg/m3 STEL
Mexico - Drinking Water - Ecological Criteria
None given.

STATE LISTS

California - Air Bill 2588 Appendix A-I
[present]
California - Exposure Limits - PELs
0.1 mg/m3 PEL
California - Directors List of Hazardous Substances (8 CCR 339)
[present]
Florida Hazardous Substance List
[present]
Massachusetts Right To Know List
extraordinarily hazardous
Minnesota Hazardous Substance List
[present]
NJ Right to Know List (Total)
sn 1520; yellow: sn 1534
NJ Special Hazardous Substances
(flammable - third degree)
Pennsylvania Right to Know List
environmental hazard

PROPOSED REGULATIONS

TSCA - ITC 33rd Report Priority Testing List
recommended for testing
TSCA - ITC 34th Report Priority Testing List
designated for testing

PHOSPHORUS PENTABROMIDE   7789-69-7

HEALTH AND SAFETY LISTS

U.S. DOT - Appendix A Table 2 - Radionuclides
final RQ = 0.1 curies (3.7E 9 Bq)

ENVIRONMENTAL LISTS

CERCLA/SARA List of Radionuclides (Appendix B) and Their Reportable Quantities
final RQ = 0.1 curies (3.7E 9 Bq)

PHOSPHORUS PENTACHLORIDE   10026-13-8

HEALTH AND SAFETY LISTS

U.S. DOT - Appendix A Table 2 - Radionuclides
final RQ = 1 curie (3.7E 10 Bq)

ENVIRONMENTAL LISTS

CERCLA/SARA List of Radionuclides (Appendix B) and Their Reportable Quantities
final RQ = 1 curie (3.7E 10 Bq)

PHOSPHORUS PENTAFLUORIDE   7647-19-0

HEALTH AND SAFETY LISTS

U.S. DOT - Substances From 49 CFR 172.101
regulated by DOT (UN1338)
U.S. DOT - Hazard Classes
DOT hazard class = 4.1

PHOSPHORUS PENTASULFIDE   1314-80-3

STATE LISTS

California - Air Bill 2588 Appendix A-I
9/89
California - Directors List of Hazardous Substances (8 CCR 339)
[present]

PHOSPHORUS PENTOXIDE   1314-56-3

HEALTH AND SAFETY LISTS

U.S. DOT - Substances From 49 CFR 172.101
regulated by DOT (UN1339)
U.S. DOT - Hazard Classes
DOT hazard class = 4.1

STATE LISTS

NJ Right to Know List (Total)
sn 1521

PHOSPHORUS SESQUISULFIDE   1314-85-8

HEALTH AND SAFETY LISTS

U.S. DOT - Substances From 49 CFR 172.101
regulated by DOT (UN1939, UN2576)
U.S. DOT - Hazard Classes
DOT hazard class = 8

INTERNATIONAL LISTS

Canada - WHMIS: Ingredient Disclosure
1% item 1296 (1296)

STATE LISTS

NJ Right to Know List (Total)
sn 1522
NJ Special Hazardous Substances
(corrosive)

PHOSPHORUS TRIBROMIDE   7789-60-8

HEALTH AND SAFETY LISTS

ACGIH 1995 - Time Weighted Averages
0.1 ppm TWA; 0.63 mg/m3 TWA
U.S. DOT - Substances From 49 CFR 172.101
regulated by DOT (UN1810)
U.S. DOT - Hazard Classes
DOT hazard class = 8
U.S. DOT - Substances Which Are Poisonous by Inhalation
liquid hazardous material poisonous by inhalation (UN1810)
U.S. DOT - Appendix A Table 1 - Hazardous Substances
final RQ = 1000 pounds (454 kg)
NIOSH - Selected LD50s and LC50s
Inhalation, rat: LC50 = 48 ppm 4 hr Oral, rat: LD50 = 380 mg/kg
OSHA - Vacated PELs - Time Weighted Averages
0.1 ppm TWA; 0.6 mg/m3 TWA
OSHA - List of Highly Hazardous Chemicals
threshhold quantity = 1000 pounds

ENVIRONMENTAL LISTS

CERCLA/SARA - Section 302 Extremely Hazardous Substances and TPQs
TPQ = 500 pounds
CERCLA/SARA - Hazardous Substances and their Reportable Quantities
final RQ = 1000 pounds (454 kg)
CAA -Toxic Substances for Accidental Release Prevention
threshhold quantity = 5,000 lbs
Clean Water Act - Hazardous Substances
[present]

INTERNATIONAL LISTS

Australian Exposure Standards - Time Weighted Averages
0.1 ppm TWA; 0.63 mg/m3 TWA

Canada - WHMIS: Ingredient Disclosure
1% item 1297 (1297)

Canada - Alberta - 8 Hour Occupational Exposure Limit
0.1 ppm TWA; 0.6 mg/m3 TWA

Canada - Alberta - 15 Minute Occupational Exposure Limit
0.5 ppm STEL; 3.1 mg/m3 STEL

Canada - Ontario - OHSA - TWAEVs
0.1 ppm TWAEV; 0.6 mg/m3 TWAEV

Canada - Ontario - OHSA - STEVs
0.5 ppm STEV; 3 mg/m3 STEV

Canada - Quebec - Time-Weighted Average Exposure Values
0.1 ppm TWAEV; 0.63 mg/m3 TWAEV

United Kingdom - Occupational Exposure Standards - TWAs
0.2 ppm TWA; 1.2 mg/m3 TWA

United Kingdom - Occupational Exposure Standards - STELs
0.6 ppm STEL; 3.6 mg/m3 STEL

German (DFG) - MAK Values
0.2 ppm MAK; 1 mg/m3 MAK

German (DFG) - Peak Limitations
2 x normal MAK (30 min. average value); don't exceed 4 times during shift

Israel - Time Weighted Averages
0.1 ppm TWA; 0.63 mg/m3 TWA

Israel - Action Levels
0.05 ppm AL; 0.315 mg/m3 AL

STATE LISTS

California - Air Bill 2588 Appendix A-I
9/89

California - Exposure Limits - PELs
0.1 ppm PEL; 0.6 mg/m3 PEL

California - Directors List of Hazardous Substances (8 CCR 339)
[present]

Florida Hazardous Substance List
[present]

Massachusetts Right To Know List
extraordinarily hazardous

Minnesota Hazardous Substance List
[present]

NJ Right to Know List (Total)
sn 1523

Pennsylvania Right to Know List
environmental hazard

PHOSPHORUS TRICHLORIDE   7719-12-2

HEALTH AND SAFETY LISTS

U.S. DOT - Substances From 49 CFR 172.101
regulated by DOT (UN2691)

U.S. DOT - Hazard Classes
DOT hazard class = 8

INTERNATIONAL LISTS

Canada - WHMIS: Ingredient Disclosure
1% item 1298 (1337)

STATE LISTS

NJ Right to Know List (Total)
sn 1524

NJ Special Hazardous Substances
(corrosive)

PHOSPHORUS TRIFLUORIDE   7783-55-3

HEALTH AND SAFETY LISTS

ACGIH 1995 - Time Weighted Averages
0.1 ppm TWA; 0.85 mg/m3 TWA

U.S. DOT - Substances From 49 CFR 172.101
regulated by DOT (UN1806)

U.S. DOT - Hazard Classes
DOT hazard class = 8

NIOSH - Selected LD50s and LC50s
Inhalation, rat: LC50 = 205 mg/m3 8 hr Oral, rat: LD50 = 660 mg/kg

NIOSH 1990 - Pocket Guide - RELs
1 mg/m3 TWA

NIOSH 1990 - Pocket Guide - IDLHs
200 mg/m3 IDLH

NIOSH 1990 - Pocket Guide - Target organs
skin, eyes, respiratory system

OSHA - Vacated PELs - Time Weighted Averages
1 mg/m3 TWA

OSHA - Final PELs - Time Weighted Averages
1 mg/m3 TWA

ENVIRONMENTAL LISTS

CERCLA/SARA - Section 302 Extremely Hazardous Substances and TPQs
TPQ = 500 pounds (This material is a reactive solid. The TPQ does not default to 10,000 pounds for non-powder, non-molten, non-solution form)

INTERNATIONAL LISTS

Australian Exposure Standards - Time Weighted Averages
0.1 ppm TWA; 0.85 mg/m3 TWA

Canada - WHMIS: Ingredient Disclosure
1% item 1299 (1341)

Canada - Alberta - 8 Hour Occupational Exposure Limit
0.1 ppm TWA; 0.85 mg/m3 TWA

Canada - Alberta - 15 Minute Occupational Exposure Limit
0.3 ppm STEL; 2.56 mg/m3 STEL

Canada - British Columbia - 8 Hour Exposure Limits
1 mg/m3 TWA

Canada - British Columbia - 15 Minute Exposure Limits
3 mg/m3 STEL

Canada - Ontario - OHSA - TWAEVs
0.1 ppm TWAEV; 0.85 mg/m3 TWAEV

Canada - Quebec - Time-Weighted Average Exposure Values
0.1 ppm TWAEV; 0.85 mg/m3 TWAEV

United Kingdom - Occupational Exposure Standards - TWAs
0.1 ppm TWA; 1 mg/m3 TWA

German (DFG) - MAK Values
total dust: 1 mg/m3 MAK

German (DFG) - Peak Limitations
2 x normal MAK (5 min momentary value); don't exceed 8 times during shift

Israel - Time Weighted Averages
0.1 ppm TWA; 0.85 mg/m3 TWA

Israel - Action Levels
0.05 ppm AL; 0.425 mg/m3 AL

Mexico - Instruction No. 10 - TWAs
0.1 ppm TWA; 1 mg/m3 TWA

STATE LISTS

California - Air Bill 2588 Appendix A-I
9/89

California - Exposure Limits - PELs
0.1 ppm PEL; 1 mg/m3 PEL

California - Directors List of Hazardous Substances (8 CCR 339)
[present]

Florida Hazardous Substance List
[present]

Massachusetts Right To Know List
extraordinarily hazardous

Minnesota Hazardous Substance List
[present]

NJ Right to Know List (Total)
sn 1525

NJ Special Hazardous Substances
(corrosive)

Pennsylvania Right to Know List
environmental hazard

PROPOSED REGULATIONS

CERCLA/SARA - Proposed Hazardous Substance Additions
proposed RQ = 1 pound (.454 kg)

CERCLA/SARA - 1989 Proposed RQ Adjustments
proposed RQ = 100 pounds (45.4 kg)

PHOSPHORUS TRIOXIDE   1314-24-5

HEALTH AND SAFETY LISTS

U.S. DOT - Substances From 49 CFR 172.101
regulated by DOT (UN2198)

U.S. DOT - Hazard Classes
DOT hazard class = 2.3

U.S. DOT - Substances Which Are Poisonous by Inhalation
gaseous hazardous material poisonous by inhalation (UN2198)

INTERNATIONAL LISTS

Canada - WHMIS: Ingredient Disclosure
1% item 1300 (1346)

STATE LISTS

NJ Right to Know List (Total)
sn 1526

PHOSPHORUS TRISULFIDE   12165-69-4

HEALTH AND SAFETY LISTS

ACGIH 1995 - Time Weighted Averages
1 mg/m3 TWA

ACGIH 1995 - Short Term Exposure Limits
3 mg/m3 STEL

U.S. DOT - Substances From 49 CFR 172.101
regulated by DOT (UN1340)

U.S. DOT - Hazard Classes
DOT hazard class = 4.3

U.S. DOT - Appendix A Table 1 - Hazardous Substances
final RQ = 100 pounds (45.4 kg)

NIOSH - Selected LD50s and LC50s
Oral, rat: LD50 = 389 mg/kg

NIOSH 1990 - Pocket Guide - RELs
1 mg/m3 TWA; 3 mg/m3 STEL

NIOSH 1990 - Pocket Guide - IDLHs
750 mg/m3 IDLH

NIOSH 1990 - Pocket Guide - Target organs
CNS, eyes, skin, respiratory system

OSHA - Vacated PELs - Time Weighted Averages
1 mg/m3 TWA

OSHA - Vacated PELs - Short Term Exposure Limits
3 mg/m3 STEL

OSHA - Final PELs - Time Weighted Averages
1 mg/m3 TWA

ENVIRONMENTAL LISTS

CERCLA/SARA - Hazardous Substances and their Reportable Quantities
final RQ = 100 pounds (45.4 kg)

Clean Water Act - Hazardous Substances
[present]

RCRA - U Series Wastes
waste number U189 (Reactive waste)

RCRA - Hazardous Constituents-Appendix VIII
waste number U189

RCRA - Substances Banned From Land Disposal
[present]

INTERNATIONAL LISTS

Australian Exposure Standards - Time Weighted Averages
1 mg/m3 TWA

Australian Exposure Standards - Short Term Exposure Limits
3 mg/m3 STEL

Canada - WHMIS: Ingredient Disclosure
1% item 1301 (1350)

Canada - Alberta - 8 Hour Occupational Exposure Limit
1 mg/m3 TWA

Canada - Alberta - 15 Minute Occupational Exposure Limit
3 mg/m3 STEL

Canada - British Columbia - 8 Hour Exposure Limits
1 mg/m3 TWA

Canada - British Columbia - 15 Minute Exposure Limits
3 mg/m3 STEL

Canada - Ontario - OHSA - TWAEVs
1 mg/m3 TWAEV

Canada - Ontario - OHSA - STEVs
3 mg/m3 STEV

Canada - Quebec - Time-Weighted Average Exposure Values
1 mg/m3 TWAEV

Canada - Quebec - Short-term Exposure Values
3 mg/m3 STEV

United Kingdom - Occupational Exposure Standards - TWAs
1 mg/m3 TWA

United Kingdom - Occupational Exposure Standards - STELs
3 mg/m3 STEL

German (DFG) - MAK Values
total dust: 1 mg/m3 MAK

German (DFG) - Peak Limitations
2 x normal MAK (5 min momentary value); don't exceed 8 times during shift

Israel - Time Weighted Averages
1 mg/m3 TWA

Israel - Short Term Exposure Limits
3 mg/m3 STEL

Israel - Action Levels
0.5 mg/m3 AL

Mexico - Instruction No. 10 - TWAs
1 mg/m3 TWA

Mexico - Instruction No. 10 - STELs
3 mg/m3 STEL

STATE LISTS

California - Exposure Limits - PELs
1 mg/m3 PEL

California - Exposure Limits - STELs
3 mg/m3 STEL

California - Directors List of Hazardous Substances (8 CCR 339)
[present]

Florida Hazardous Substance List
[present]

Massachusetts Right To Know List
[present]

Minnesota Hazardous Substance List
[present]

NJ Right to Know List (Total)
sn 1527

Pennsylvania Right to Know List
environmental hazard

PHOSPHORYLATED CAPROLACTONE, ALKYLOX-   RR-00941-6
OHETEROMONO-CYCLE AND POLYALKYLENE
POLYOL ALKYL ETHER

HEALTH AND SAFETY LISTS

U.S. DOT - Substances From 49 CFR 172.101
regulated by DOT (UN1807)

U.S. DOT - Hazard Classes
DOT hazard class = 8

NIOSH - Selected LD50s and LC50s
Inhalation, rat: LC50 = 1217 mg/m3 1 hr

INTERNATIONAL LISTS

Canada - WHMIS: Ingredient Disclosure
1% item 1293 (232)

German (DFG) - MAK Values
total dust: 1 mg/m3 MAK

German (DFG) - Peak Limitations
2 x normal MAK (5 min momentary value); don't exceed 8 times during shift

German (DFG) - Pregnancy
no risk to embryo/fetus if exposure limits adhered to

STATE LISTS

California - Air Bill 2588 Appendix A-I
9/89

Massachusetts Right To Know List
extraordinarily hazardous

NJ Right to Know List (Total)
sn 1517

NJ Special Hazardous Substances
(corrosive)

Pennsylvania Right to Know List
environmental hazard

PROPOSED REGULATIONS

CERCLA/SARA - Proposed Hazardous Substance Additions
proposed RQ = 1 pound (.454 kg)

CERCLA/SARA - 1989 Proposed RQ Adjustments
proposed RQ = 10 pounds (4.54 kg)

PHOSPHORYLATED OXOHETEROMONOCYCLE   RR-00940-5
POLYOXYETHYLENE ALKYL ETHER

HEALTH AND SAFETY LISTS

U.S. DOT - Substances From 49 CFR 172.101
regulated by DOT (UN1341)

U.S. DOT - Hazard Classes
DOT hazard class = 4.1

STATE LISTS

Florida Hazardous Substance List
[present]

Massachusetts Right To Know List
[present]

NJ Right to Know List (Total)
sn 1528

Pennsylvania Right to Know List
[present]

PHOSPHORYL FLUORIDE   13478-20-1

HEALTH AND SAFETY LISTS

U.S. DOT - Substances From 49 CFR 172.101
regulated by DOT (UN1808)

U.S. DOT - Hazard Classes
DOT hazard class = 8

INTERNATIONAL LISTS

Canada - WHMIS: Ingredient Disclosure
1% item 1302 (1637)

STATE LISTS

NJ Right to Know List (Total)
sn 1529

NJ Special Hazardous Substances
(corrosive)

PHOSPHOTUNGSTIC ACIDS   RR-01637-5

HEALTH AND SAFETY LISTS

ACGIH 1995 - Time Weighted Averages
0.2 ppm TWA; 1.1 mg/m3 TWA

ACGIH 1995 - Short Term Exposure Limits
0.5 ppm STEL; 2.8 mg/m3 STEL

U.S. DOT - Substances From 49 CFR 172.101
regulated by DOT (UN1809)

U.S. DOT - Hazard Classes
DOT hazard class = 8

U.S. DOT - Substances Which Are Poisonous by Inhalation
liquid hazardous material poisonous by inhalation (UN1809)

U.S. DOT - Appendix A Table 1 - Hazardous Substances
final RQ = 1000 pounds (454 kg)

NIOSH - Selected LD50s and LC50s
Inhalation, rat: LC50 = 104 ppm 4 hr Oral, rat: LD50 = 550 mg/kg

NIOSH 1990 - Pocket Guide - RELs
0.2 ppm TWA; 1.5 mg/m3 TWA; 0.5 ppm STEL; 3 mg/m3 STEL

NIOSH 1990 - Pocket Guide - IDLHs
50 ppm IDLH

NIOSH 1990 - Pocket Guide - Target organs
skin, eyes, respiratory system

OSHA - Vacated PELs - Time Weighted Averages
0.2 ppm TWA; 1.5 mg/m3 TWA

OSHA - Vacated PELs - Short Term Exposure Limits
0.5 ppm STEL; 3 mg/m3 STEL

OSHA - Final PELs - Time Weighted Averages
0.5 ppm TWA; 3 mg/m3 TWA

OSHA - List of Highly Hazardous Chemicals
threshhold quantity = 1000 pounds

ENVIRONMENTAL LISTS

CERCLA/SARA - Section 302 Extremely Hazardous Substances and TPQs
TPQ = 1000 pounds

CERCLA/SARA - Hazardous Substances and their Reportable Quantities
final RQ = 1000 pounds (454 kg)

CAA -Toxic Substances for Accidental Release Prevention
threshold quantity = 15,000 lbs

Clean Water Act - Hazardous Substances
[present]

INTERNATIONAL LISTS

Australian Exposure Standards - Time Weighted Averages
0.2 ppm TWA; 1.1 mg/m3 TWA

Australian Exposure Standards - Short Term Exposure Limits
0.5 ppm STEL; 2.8 mg/m3 STEL

Canada - WHMIS: Ingredient Disclosure
1% item 1303 (1658)

Canada - Alberta - 8 Hour Occupational Exposure Limit
0.2 ppm TWA; 1.1 mg/m3 TWA

Canada - Alberta - 15 Minute Occupational Exposure Limit
0.5 ppm STEL; 2.8 mg/m3 STEL

Canada - British Columbia - 8 Hour Exposure Limits
0.5 ppm TWA; 3 mg/m3 TWA

Canada - Ontario - OHSA - TWAEVs
0.2 ppm TWAEV; 1.1 mg/m3 TWAEV

Canada - Ontario - OHSA - STEVs
0.5 ppm STEV; 2.8 mg/m3 STEV

Canada - Quebec - Time-Weighted Average Exposure Values
0.2 ppm TWAEV; 1.1 mg/m3 TWAEV

Canada - Quebec - Short-term Exposure Values
0.5 ppm STEV; 2.8 mg/m3 STEV

United Kingdom - Occupational Exposure Standards - TWAs
0.2 ppm TWA; 1.5 mg/m3 TWA

United Kingdom - Occupational Exposure Standards - STELs
0.5 ppm STEL; 3 mg/m3 STEL

German (DFG) - MAK Values
0.5 ppm MAK; 3 mg/m3 MAK

German (DFG) - Peak Limitations
2 x normal MAK (5 min momentary value); don't exceed 8 times during shift

Israel - Time Weighted Averages
0.2 ppm TWA; 1.1 mg/m3 TWA

Israel - Short Term Exposure Limits
0.5 ppm STEL; 2.8 mg/m3 STEL

Israel - Action Levels
0.1 ppm AL; 0.55 mg/m3 AL

Mexico - Instruction No. 10 - TWAs
0.5 ppm TWA; 5 mg/m3 TWA

STATE LISTS

California - Air Bill 2588 Appendix A-I
9/89

California - Exposure Limits - PELs
0.2 ppm PEL; 1.5 mg/m3 PEL

California - Exposure Limits - STELs
0.5 ppm STEL; 3 mg/m3 STEL

California - Directors List of Hazardous Substances (8 CCR 339)
[present]

Florida Hazardous Substance List
[present]

Massachusetts Right To Know List
extraordinarily hazardous

Minnesota Hazardous Substance List
[present]

NJ Right to Know List (Total)
sn 1530

NJ Special Hazardous Substances
(corrosive; reactive - second degree)

Pennsylvania Right to Know List
environmental hazard

PHOTODIELDRIN   13366-73-9

HEALTH AND SAFETY LISTS

U.S. DOT - Substances Which Are Poisonous by Inhalation
gaseous hazardous material poisonous by inhalation

INTERNATIONAL LISTS

Canada - WHMIS: Ingredient Disclosure
1% item 1304 (1679)

PHTHALAMIDE   88-96-0

HEALTH AND SAFETY LISTS

U.S. DOT - Substances From 49 CFR 172.101
regulated by DOT (UN2578)

U.S. DOT - Hazard Classes
DOT hazard class = 8

INTERNATIONAL LISTS

Canada - WHMIS: Ingredient Disclosure
1% item 1305 (1695)

STATE LISTS

NJ Right to Know List (Total)
sn 1532

NJ Special Hazardous Substances
(corrosive)

PHTHALATE ESTERS   RR-00433-1

HEALTH AND SAFETY LISTS

U.S. DOT - Substances From 49 CFR 172.101
regulated by DOT (UN1343)

U.S. DOT - Hazard Classes
DOT hazard class = 4.1

STATE LISTS

NJ Right to Know List (Total)
sn 1533

PHTHALIC ACID   88-99-3

ENVIRONMENTAL LISTS

TSCA - Chemicals with Significant New Use Rules
PMN number: P-89-837

PHTHALIC ACID, DIALKYL (C7) ESTER   68515-44-6

ENVIRONMENTAL LISTS

TSCA - Chemicals with Significant New Use Rules
PMN number: P-89-836

PHTHALIC ACID, DIALKYL (C7-C11) ESTER   68515-42-4

HEALTH AND SAFETY LISTS

U.S. DOT - Substances From 49 CFR 172.101
regulated by DOT (UN1776)

U.S. DOT - Hazard Classes
DOT hazard class = 8

PHTHALIC ACID, DIALKYL (C9) ESTER   68515-45-7

INTERNATIONAL LISTS

Canada - Ontario - OHSA - TWAEVs
as W: 1 mg/m3 TWAEV

Canada - Ontario - OHSA - STEVs
as W: 3 mg/m3 STEV

PHTHALIC ACID, POLYMER WITH 1,3-DIHYDRO-   64382-04-3
1,3-DIOXO-5-ISOBENZOFURANCARBOXYLIC ACID
AND TRIMETHYLOLPROPANE, LINOLEATE

HEALTH AND SAFETY LISTS

NTP Chemical Status Reports - Testing Status and NTIS Number
Technical reports printed (PB274393/AS)

NTP Chemical Status Reports - Evidence of Carcinogenicity
male rat-negative; female rat-negative; male mice-negative; female mice-negative

PHTHALIC ANHYDRIDE   85-44-9

HEALTH AND SAFETY LISTS

NTP Chemical Status Reports - Testing Status and NTIS Number
Technical reports printed (PB293831/AS)

NTP Chemical Status Reports - Evidence of Carcinogenicity
male rat-negative; female rat-negative; male mice-negative; female mice-negative

PHTHALIC DIMALEIMIDE   3006-93-7

ENVIRONMENTAL LISTS

Clean Water Act - Toxic Pollutants
[present]

RCRA - Hazardous Constituents-Appendix VIII
hazardous constituent - no waste number

STATE LISTS

California - Directors List of Hazardous Substances (8 CCR 339)
[present] (except butyl benzyl phthalates)

Pennsylvania Right to Know List
environmental hazard

PHTHALIMIDE   85-41-6

HEALTH AND SAFETY LISTS

NFPA - Flash Points
flash point = 334 degrees F (168 degrees C)

NFPA - Hazard Identification Ratings
health-0; flammability-1; reactivity-1 (dust explosion hazard)

NIOSH - Selected LD50s and LC50s
Oral, rat: LD50 = 7900 mg/kg

ENVIRONMENTAL LISTS

CAA - HON Rule - SOCMI Chemicals
compliance by April 24, 1995

PHTHALIMIDE DERIVATIVE PESTICIDES, N.O.S.   RR-00434-2

ENVIRONMENTAL LISTS

TSCA - Code of Federal Regulations Citations
40 CFR 799.5025

PHTHALOCYANINE BLUE   147-14-8

ENVIRONMENTAL LISTS

TSCA - Code of Federal Regulations Citations
40 CFR 799.5000

TSCA - Section 12(b) - Export Notification
Section 4 - Notification is required for specific substances listed in 55 FR 55038

O-PHTHALODINITRILE   91-15-6

ENVIRONMENTAL LISTS

TSCA - Code of Federal Regulations Citations
40 CFR 799.5025

M-PHTHALODINITRILE   626-17-5

ENVIRONMENTAL LISTS

List of Pesticide Product Inert Ingredients
[present]

PHYLLOQUINONE   84-80-0

SEE ALSO:

K093-HAZARDOUS WASTES
K094-HAZARDOUS WASTES
K024-HAZARDOUS WASTES
K023-HAZARDOUS WASTES

HEALTH AND SAFETY LISTS

ACGIH 1995 - Time Weighted Averages
1 ppm TWA; 6.1 mg/m3 TWA

AIHA - Odor Threshold Values
no geometric mean air odor threshold

U.S. DOT - Substances From 49 CFR 172.101
regulated by DOT (UN2214)

U.S. DOT - Hazard Classes
DOT hazard class = 8

U.S. DOT - Appendix A Table 1 - Hazardous Substances
final RQ = 5000 pounds (2270 kg)

NFPA - Flash Points
flash point = 305 degrees F (152 degrees C)

NFPA - Hazard Identification Ratings
health-3; flammability-1; reactivity-0

NIOSH - Selected LD50s and LC50s
Oral, rat: LD50 = 4020 mg/kg

NIOSH 1990 - Pocket Guide - RELs
6 mg/m3 TWA; 1 ppm TWA

NIOSH 1990 - Pocket Guide - IDLHs
10,000 mg/m3 IDLH

NIOSH 1990 - Pocket Guide - Target organs
respiratory system, skin, eyes, liver, kidneys

NTP Chemical Status Reports - Testing Status and NTIS Number
Technical reports printed (PB293594/AS)

NTP Chemical Status Reports - Evidence of Carcinogenicity
male rat-negative; female rat-negative; male mice-negative; female mice-negative

OSHA - Vacated PELs - Time Weighted Averages
1 ppm TWA; 6 mg/m3 TWA

OSHA - Final PELs - Time Weighted Averages
2 ppm TWA; 12 mg/m3 TWA

ENVIRONMENTAL LISTS

CERCLA/SARA - Section 313 - Emission Reporting
form R reporting required for 1.0% de minimus concentration

CERCLA/SARA - Hazardous Substances and their Reportable Quantities
final RQ = 5000 pounds (2270 kg)

Clean Air Act (1990) - List of Hazardous Air Contaminants
[present]

CAA - HON Rule - SOCMI Chemicals
compliance by April 24, 1995

CAA - HON Rule - Organic HAPs
[present]

List of Pesticide Product Inert Ingredients
[present]

RCRA - U Series Wastes
waste number U190

RCRA - Hazardous Constituents-Appendix VIII
waste number U190

RCRA - Basis for Listing - Appendix VII
Included in waste streams: K023, K024, K093, K094

RCRA - Universal Treatment Standards (LDR)
WW: 0.055 mg/l; NWW: 28 mg/kg

INTERNATIONAL LISTS

Australian Exposure Standards - Time Weighted Averages
1 ppm TWA; 6.1 mg/m3 TWA

Australian Exposure Standards - Skin Effects
sensitiser

Canada - WHMIS: Ingredient Disclosure
0.1% item 1306 (233)

Canada - NPRI (National Pollutant Release Inventory)
[present]

Canada - Alberta - 8 Hour Occupational Exposure Limit
1 ppm TWA; 6 mg/m3 TWA

Canada - Alberta - 15 Minute Occupational Exposure Limit
4 ppm STEL; 24 mg/m3 STEL

Canada - British Columbia - 8 Hour Exposure Limits
1 ppm TWA; 6 mg/m3 TWA

Canada - British Columbia - 15 Minute Exposure Limits
4 ppm STEL; 24 mg/m3 STEL

Canada - Ontario - OHSA - TWAEVs
1 ppm TWAEV; 6 mg/m3 TWAEV
Canada - Quebec - Time-Weighted Average Exposure Values
1 ppm TWAEV; 6.1 mg/m3 TWAEV
German (DFG) - MAK Values
total dust: 5 mg/m3 MAK
German (DFG) - Peak Limitations
2 x normal MAK (5 min momentary value); don't exceed 8 times during shift
German (DFG) - Skin/Sensitizers
danger of sensitization (skin or respiratory)
German (DFG) - Pregnancy
classification not yet possible
Israel - Time Weighted Averages
1 ppm TWA; 6.1 mg/m3 TWA
Israel - Action Levels
0.5 ppm AL; 3.55 mg/m3 AL
Mexico - Instruction No. 10 - TWAs
1 ppm TWA; 6 mg/m3 TWA
Mexico - Instruction No. 10 - STELs
4 ppm STEL; 24 mg/m3 STEL

STATE LISTS
California - Air Bill 2588 Appendix A-I
known or potential carcinogen
California - Exposure Limits - PELs
1 ppm PEL; 6 mg/m3 PEL
California - Directors List of Hazardous Substances (8 CCR 339)
[present]
Florida Hazardous Substance List
[present]
Massachusetts Right To Know List
[present]
Minnesota Hazardous Substance List
[present]
NJ Right to Know List (Total)
sn 1535
Pennsylvania Right to Know List
environmental hazard

PHYSOSTIGMINE 57-47-6

HEALTH AND SAFETY LISTS
NIOSH - Selected LD50s and LC50s
Oral, rat: LD50 = 1370 mg/kg

PHYSOSTIGMINE SALICYLATE (1:1) 57-64-7

HEALTH AND SAFETY LISTS
NIOSH - Selected LD50s and LC50s
Oral, mouse: LD50 = 5000 mg/kg

ENVIRONMENTAL LISTS
CAA - HON Rule - SOCMI Chemicals
compliance by April 24, 1995
List of Pesticide Product Inert Ingredients
[present]

PICKLED VEGETABLES, TRADITIONAL ASIAN RR-01705-0

HEALTH AND SAFETY LISTS
U.S. DOT - Substances From 49 CFR 172.101
regulated by DOT (UN2773, UN2774, UN3007, UN3008)
U.S. DOT - Hazard Classes
toxic or toxic, flammable: DOT hazard class = 6.1; flammable, toxic: DOT hazard class = 3
STATE LISTS
NJ Right to Know List (Total)
sn 2676; sn 2677; sn 2678; sn 2679

PICLORAM 1918-02-1
SEE ALSO:
COPPER

ENVIRONMENTAL LISTS
EPA - Master Testing List
[present]
List of Pesticide Product Inert Ingredients
[present]

P-PICOLINE 108-89-4

HEALTH AND SAFETY LISTS
NIOSH - Selected LD50s and LC50s
Oral, mouse: LD50 = 65 mg/kg

ENVIRONMENTAL LISTS
CAA - HON Rule - SOCMI Chemicals
compliance by April 24, 1995

2-PICOLINE 109-06-8

HEALTH AND SAFETY LISTS
ACGIH 1995 - Time Weighted Averages
5 mg/m3 TWA
NIOSH - Selected LD50s and LC50s
Oral, rat: LD50 = 1860 mg/kg
OSHA - Vacated PELs - Time Weighted Averages
5 mg/m3 TWA

ENVIRONMENTAL LISTS
TSCA - Code of Federal Regulations Citations
40 CFR 712.30(x); 40 CFR 716.120(d)
TSCA - PAIR - Reporting List
Reporting Date: November 27, 1991
TSCA - Health and Safety Reporting List
Effective Date: September 30, 1991

INTERNATIONAL LISTS
Australian Exposure Standards - Time Weighted Averages
5 mg/m3 TWA
Canada - WHMIS: Ingredient Disclosure
1% item 1307 (1424)
Canada - Alberta - 8 Hour Occupational Exposure Limit
5 mg/m3 TWA
Canada - Alberta - 15 Minute Occupational Exposure Limit
10 mg/m3 STEL
Canada - British Columbia - 8 Hour Exposure Limits
5 mg/m3 TWA
Canada - Ontario - OHSA - TWAEVs
5 mg/m3 TWAEV
Canada - Quebec - Time-Weighted Average Exposure Values
5 mg/m3 TWAEV
Israel - Time Weighted Averages
5 mg/m3 TWA
Israel - Action Levels
2.5 mg/m3 AL
Mexico - Instruction No. 10 - TWAs
5 mg/m3 TWA

STATE LISTS
California - Exposure Limits - PELs
5 mg/m3 PEL
California - Directors List of Hazardous Substances (8 CCR 339)
[present]
Florida Hazardous Substance List
[present]
Massachusetts Right To Know List
[present]

Minnesota Hazardous Substance List
[present]

Pennsylvania Right to Know List
[present]

PROPOSED REGULATIONS

TSCA - ITC 32nd Report Priority Testing List
designated for dermal absorption testing

PICOLINES   1333-41-1

HEALTH AND SAFETY LISTS

NIOSH - Selected LD50s and LC50s
Oral, mouse: LD50 = 25 gm/kg

STATE LISTS

Massachusetts Right To Know List
[present]

NJ Right to Know List (Total)
sn 2680

PICOLINIC ACID   98-98-6

HEALTH AND SAFETY LISTS

NIOSH - Selected LD50s and LC50s
Oral, mouse: LD50 = 4500 ug/kg

ENVIRONMENTAL LISTS

CERCLA/SARA - Section 302 Extremely Hazardous Substances and
TPQs
TPQ = 100/10,000 pounds

STATE LISTS

Florida Hazardous Substance List
effective March 13, 1992

Massachusetts Right To Know List
extraordinarily hazardous

NJ Right to Know List (Total)
sn 2681

Pennsylvania Right to Know List
environmental hazard

PROPOSED REGULATIONS

CERCLA/SARA - Proposed Hazardous Substance Additions
proposed RQ = 1 pound (.454 kg)

CERCLA/SARA - 1989 Proposed RQ Adjustments
proposed RQ = 100 pounds (45.4 kg)

PICRIC ACID   88-89-1

HEALTH AND SAFETY LISTS

NIOSH - Selected LD50s and LC50s
Oral, mouse: LD50 = 2500 ug/kg

ENVIRONMENTAL LISTS

CERCLA/SARA - Section 302 Extremely Hazardous Substances and
TPQs
TPQ = 100/10,000 pounds

STATE LISTS

Florida Hazardous Substance List
effective March 13, 1992

Massachusetts Right To Know List
extraordinarily hazardous

NJ Right to Know List (Total)
sn 2682

Pennsylvania Right to Know List
environmental hazard

PROPOSED REGULATIONS

CERCLA/SARA - Proposed Hazardous Substance Additions
proposed RQ = 1 pound (.454 kg)

CERCLA/SARA - 1989 Proposed RQ Adjustments
proposed RQ = 100 pounds (45.4 kg)

PICRIC ACID, SODIUM SALT   3324-58-1

HEALTH AND SAFETY LISTS

IARC - Group 2B (sufficient animal data)
[present]

OSHA - Possible Select Carcinogens
[present]

PICRITE   RR-01520-3

HEALTH AND SAFETY LISTS

ACGIH 1995 - Time Weighted Averages
10 mg/m3 TWA

IARC - Group 3 (not classifiable)
[present]

NIOSH - Selected LD50s and LC50s
Oral, rat: LD50 = 2898 mg/kg

NTP Chemical Status Reports - Testing Status and NTIS Number
Technical reports printed (PB276471/AS)

NTP Chemical Status Reports - Evidence of Carcinogenicity
*male rat-negative; female rat-equivocal; male mice-negative; female
mice-negative*

OSHA - Vacated PELs - Time Weighted Averages
total dust: 10 mg/m3 TWA; respirable fraction: 5 mg/m3 TWA

OSHA - Final PELs - Time Weighted Averages
total dust: 15 mg/m3 TWA; respirable fraction: 5 mg/m3 TWA

ENVIRONMENTAL LISTS

CERCLA/SARA - Section 313 - Emission Reporting
form R reporting required

Safe Drinking Water Act - MCLs
MCL = 0.5 mg/L

Safe Drinking Water Act - MCLGs
MCLG = 0.5 mg/L

INTERNATIONAL LISTS

Australian Exposure Standards - Time Weighted Averages
10 mg/m3 TWA

Canada - Drinking Water Quality - IMACs
0.19 mg/L IMAC

Canada - Alberta - 8 Hour Occupational Exposure Limit
10 mg/m3 TWA

Canada - Alberta - 15 Minute Occupational Exposure Limit
20 mg/m3 STEL

Canada - British Columbia - 8 Hour Exposure Limits
10 mg/m3 TWA

Canada - British Columbia - 15 Minute Exposure Limits
20 mg/m3 STEL

Canada - Ontario - OHSA - TWAEVs
10 mg/m3 TWAEV

Canada - Ontario - OHSA - STEVs
20 mg/m3 STEV

Canada - Quebec - Time-Weighted Average Exposure Values
10 mg/m3 TWAEV

United Kingdom - Occupational Exposure Standards - TWAs
10 mg/m3 TWA

United Kingdom - Occupational Exposure Standards - STELs
20 mg/m3 STEL

Israel - Time Weighted Averages
10 mg/m3 TWA

Israel - Action Levels
5 mg/m3 AL

Mexico - Instruction No. 10 - TWAs
10 mg/m3 TWA

Mexico - Instruction No. 10 - STELs
20 mg/m3 STEL

STATE LISTS

California - Exposure Limits - PELs
total dust: 10 mg/m3 PEL; respirable fraction: 5 mg/m3 PEL

California - Directors List of Hazardous Substances (8 CCR 339)
[present]

Florida Hazardous Substance List
[present]

Massachusetts Right To Know List
[present]

Minnesota Hazardous Substance List
[present]

NJ Right to Know List (Total)
sn 1536

Pennsylvania Right to Know List
[present]

PICROTOXIN   124-87-8

HEALTH AND SAFETY LISTS

AIHA - WEEL - Time Weighted Averages
2 ppm TWA; 7.6 mg/m3 TWA

AIHA - WEEL - Ceilings or Short Term Time Weighted Averages
5 ppm STEL; 19 mg/m3 STEL

AIHA - WEEL - Skin Absorption Designations
skin absorber

NFPA - Flash Points
flash point = 134 degrees F (57 degrees C)

NFPA - Hazard Identification Ratings
health-2; flammability-2; reactivity-0

NIOSH - Selected LD50s and LC50s
Oral, rat: LD50 = 1290 mg/kg Skin, rabbit: LD50 = 270 mg/kg

ENVIRONMENTAL LISTS

EPA - Master Testing List
[present]

TSCA - Code of Federal Regulations Citations
40 CFR 716.120(a)

TSCA - Health and Safety Reporting List
Effective Date: September 10, 1984

INTERNATIONAL LISTS

Canada - WHMIS: Ingredient Disclosure
1% item 1309 (1426)

STATE LISTS

Florida Hazardous Substance List
[present]

Massachusetts Right To Know List
[present]

Pennsylvania Right to Know List
[present]

PIGMENT RED 28 (PERMANENT RED)   3564-21-4
SEE ALSO:
K026-HAZARDOUS WASTES

HEALTH AND SAFETY LISTS

AIHA - WEEL - Time Weighted Averages
2 ppm TWA; 7.6 mg/m3 TWA

AIHA - WEEL - Ceilings or Short Term Time Weighted Averages
5 ppm STEL; 19 mg/m3 STEL

AIHA - WEEL - Skin Absorption Designations
skin absorber

U.S. DOT - Appendix A Table 1 - Hazardous Substances
final RQ = 5000 pounds (2270 kg)

NFPA - Flash Points
flash point = 102 degrees F (39 degrees C)

NFPA - Hazard Identification Ratings
health-2; flammability-2; reactivity-0

NIOSH - Selected LD50s and LC50s
Oral, rat: LD50 = 790 mg/kg Skin, rabbit: LD50 = 410 mg/kg

ENVIRONMENTAL LISTS

CERCLA/SARA - Section 313 - Emission Reporting
form R reporting required

CERCLA/SARA - Hazardous Substances and their Reportable Quantities
final RQ = 5000 pounds (2270 kg)

EPA - Master Testing List
[present]

RCRA - U Series Wastes
waste number U191

RCRA - Hazardous Constituents-Appendix VIII
waste number U191

RCRA - Basis for Listing - Appendix VII
Included in waste stream: K026

RCRA - Substances Banned From Land Disposal
[present]

RCRA - TSD Facilities Ground Water Monitoring
TM 8240 = 5 ug/L PQL; TM 8270 = 10 ug/L PQL

TSCA - Code of Federal Regulations Citations
40 CFR 716.120(a)

TSCA - Health and Safety Reporting List
Effective Date: September 10, 1984

INTERNATIONAL LISTS

Canada - WHMIS: Ingredient Disclosure
1% item 1308 (1425)

STATE LISTS

Florida Hazardous Substance List
[present]

Massachusetts Right To Know List
[present]

NJ Right to Know List (Total)
sn 2955

Pennsylvania Right to Know List
environmental hazard

PIGMENT RED 48, CI NO. 15865   5858-82-2

HEALTH AND SAFETY LISTS

U.S. DOT - Substances From 49 CFR 172.101
regulated by DOT (UN2313)

U.S. DOT - Hazard Classes
DOT hazard class = 3

ENVIRONMENTAL LISTS

TSCA - Code of Federal Regulations Citations
40 CFR 716.120(a)

TSCA - Health and Safety Reporting List
Effective Date: September 10, 1984

STATE LISTS

Minnesota Hazardous Substance List
skin

NJ Right to Know List (Total)
sn 1537

PIGMENT YELLOW 16   5979-28-2

HEALTH AND SAFETY LISTS

NIOSH - Selected LD50s and LC50s
Oral, bird: LD50 = 178 mg/kg

PIGMENT YELLOW 74   6358-31-2
HEALTH AND SAFETY LISTS
ACGIH 1995 - Time Weighted Averages
0.1 mg/m3 TWA

U.S. DOT - Substances From 49 CFR 172.101
regulated by DOT (NA1344, UN0154, UN1344)

U.S. DOT - Hazard Classes
with less than 30% water: DOT hazard class = 1.1D; with 30% or more water: DOT hazard class = 4.1

NIOSH 1990 - Pocket Guide - RELs
0.1 mg/m3 TWA; 0.3 mg/m3 STEL

NIOSH 1990 - Pocket Guide - IDLHs
100 mg/m3 IDLH

NIOSH 1990 - Pocket Guide - Target organs
kidneys, liver, blood, skin, eyes

NIOSH 1990 - Pocket Guide - Skin list
Potential for dermal absorption

OSHA - Vacated PELs - Time Weighted Averages
0.1 mg/m3 TWA

OSHA - Vacated PELs - Skin Designation
Prevent or reduce skin absorption

OSHA - Final PELs - Time Weighted Averages
0.1 mg/m3 TWA

OSHA - Final PELs - Skin Notations
prevent or reduce skin absorption

ENVIRONMENTAL LISTS
ATSDR Priority List
Rank (of 275): 272

CERCLA/SARA - Section 313 - Emission Reporting
form R reporting required for 1.0% de minimus concentration

List of Pesticide Product Inert Ingredients
[present]

INTERNATIONAL LISTS
Australian Exposure Standards - Time Weighted Averages
0.1 mg/m3 TWA

Australian Exposure Standards - Skin Effects
skin absorption

Canada - WHMIS: Ingredient Disclosure
1% item 1310 (128)

Canada - Alberta - 8 Hour Occupational Exposure Limit
0.1 mg/m3 TWA

Canada - Alberta - 15 Minute Occupational Exposure Limit
0.3 mg/m3 STEL

Canada - Alberta - Skin Designation
can be absorbed through the intact skin

Canada - British Columbia - 8 Hour Exposure Limits
0.1 mg/m3 TWA

Canada - British Columbia - 15 Minute Exposure Limits
0.3 mg/m3 STEL

Canada - British Columbia - Skin Notations
skin - potential for skin absorption

Canada - Ontario - OHSA - TWAEVs
0.1 mg/m3 TWAEV

Canada - Ontario - OHSA - STEVs
0.3 mg/m3 STEV

Canada - Ontario - OHSA - Skin Notations
absorption through skin, eyes, or mucous membranes

Canada - Quebec - Time-Weighted Average Exposure Values
0.1 mg/m3 TWAEV

Canada - Quebec - Skin Designations
absorbed through the skin

United Kingdom - Occupational Exposure Standards - TWAs
0.1 mg/m3 TWA

United Kingdom - Occupational Exposure Standards - STELs
0.3 mg/m3 STEL

United Kingdom - Occupational Exposure Standards - Notes
can be absorbed through skin

German (DFG) - MAK Values
total dust: 0.1 mg/m3 MAK

German (DFG) - Peak Limitations
2 x normal MAK (5 min momentary value); don't exceed 8 times during shift

German (DFG) - Skin/Sensitizers
danger of cutaneous absorption

Israel - Time Weighted Averages
0.1 mg/m3 TWA

Israel - Action Levels
0.2 mg/m3 AL

Mexico - Instruction No. 10 - TWAs
0.1 mg/m3 TWA

Mexico - Instruction No. 10 - STELs
0.3 mg/m3 STEL

Mexico - Instruction No. 10 - Skin designation
skin - potential for cutaneous absorption

STATE LISTS
California - Air Bill 2588 Appendix A-II
6/91

California - Exposure Limits - PELs
0.1 mg/m3 PEL

California - Exposure Limits - Skin Notation
material may be absorbed through the skin, eyes or mucous membrane

California - Directors List of Hazardous Substances (8 CCR 339)
[present]

Florida Hazardous Substance List
[present]

Massachusetts Right To Know List
[present]

Minnesota Hazardous Substance List
skin

NJ Right to Know List (Total)
sn 1946

Pennsylvania Right to Know List
environmental hazard

PIGMENT YELLOW 97, CI 11767   12225-18-2
ENVIRONMENTAL LISTS
List of Pesticide Product Inert Ingredients
[present]

PIMARICIN   7681-93-8
STATE LISTS
NJ Right to Know List (Total)
with > 20% water: sn 2683

PINACOL   76-09-5
HEALTH AND SAFETY LISTS
U.S. DOT - Substances From 49 CFR 172.101
regulated by DOT (UN1584)

U.S. DOT - Hazard Classes
DOT hazard class = 6.1

U.S. DOT - Appendix B - Marine Pollutants
DOT regulated marine pollutant

NIOSH - Selected LD50s and LC50s
Oral, mouse: LD50 = 15 mg/kg

ENVIRONMENTAL LISTS
   CERCLA/SARA - Section 302 Extremely Hazardous Substances and TPQs
     TPQ = 500/10,000 pounds
STATE LISTS
   Florida Hazardous Substance List
     effective March 13, 1992
   Massachusetts Right To Know List
     extraordinarily hazardous
   NJ Right to Know List (Total)
     sn 0526
   Pennsylvania Right to Know List
     environmental hazard
PROPOSED REGULATIONS
   CERCLA/SARA - Proposed Hazardous Substance Additions
     proposed RQ = 1 pound (.454 kg)
   CERCLA/SARA - 1989 Proposed RQ Adjustments
     proposed RQ = 100 pounds (45.4 kg)

PINANE   473-55-2
ENVIRONMENTAL LISTS
   List of Pesticide Product Inert Ingredients
     [present]

PINANE HYDROPEROXIDE   28324-52-9
ENVIRONMENTAL LISTS
   List of Pesticide Product Inert Ingredients
     [present]

PINANE HYDROPEROXIDE   29240-17-3
ENVIRONMENTAL LISTS
   List of Pesticide Product Inert Ingredients
     [present]

PINANE HYDROPEROXIDE   30580-75-7
ENVIRONMENTAL LISTS
   List of Pesticide Product Inert Ingredients
     [present]

PINDONE   83-26-1
ENVIRONMENTAL LISTS
   List of Pesticide Product Inert Ingredients
     [present]

ALPHA-PINENE   80-56-8
HEALTH AND SAFETY LISTS
   NIOSH - Selected LD50s and LC50s
     Oral, rat: LD50 = 2730 mg/kg

BETA-PINENE   127-91-3
HEALTH AND SAFETY LISTS
   NIOSH - Selected LD50s and LC50s
     Oral, mouse: LD50 = 3380 mg/kg
INTERNATIONAL LISTS
   Canada - WHMIS: Ingredient Disclosure
     1% item 1311 (1427)

PINENE   1330-16-1
HEALTH AND SAFETY LISTS
   NFPA - Hazard Identification Ratings
     health-0; reactivity-0

BETA-PINENE POLYMER   25719-60-2
HEALTH AND SAFETY LISTS
   U.S. DOT - Organic Peroxides Table
     Organic peroxide UN3105; UN3109
STATE LISTS
   NJ Right to Know List (Total)
     sn 1538

PINE OIL   8002-09-3
HEALTH AND SAFETY LISTS
   U.S. DOT - Organic Peroxides Table
     Organic peroxide UN3113

PINE PITCH   RR-00879-7
HEALTH AND SAFETY LISTS
   U.S. DOT - Organic Peroxides Table
     Organic peroxide UN3105

PINE TAR   8011-48-1
HEALTH AND SAFETY LISTS
   ACGIH 1995 - Time Weighted Averages
     0.1 mg/m3 TWA
   U.S. DOT - Appendix B - Marine Pollutants
     DOT regulated marine pollutant (includes its salts)
   NIOSH - Selected LD50s and LC50s
     Oral, rat: LD50 = 280 mg/kg
   NIOSH 1990 - Pocket Guide - RELs
     0.1 mg/m3 TWA
   NIOSH 1990 - Pocket Guide - IDLHs
     200 mg/m3 IDLH
   NIOSH 1990 - Pocket Guide - Target organs
     blood prothrombin
   OSHA - Vacated PELs - Time Weighted Averages
     0.1 mg/m3 TWA
   OSHA - Final PELs - Time Weighted Averages
     0.1 mg/m3 TWA
INTERNATIONAL LISTS
   Australian Exposure Standards - Time Weighted Averages
     0.1 mg/m3 TWA
   Canada - Alberta - 8 Hour Occupational Exposure Limit
     0.1 mg/m3 TWA
   Canada - Alberta - 15 Minute Occupational Exposure Limit
     0.3 mg/m3 STEL
   Canada - British Columbia - 8 Hour Exposure Limits
     0.1 mg/m3 TWA
   Canada - British Columbia - 15 Minute Exposure Limits
     0.3 mg/m3 STEL
   Canada - Ontario - OHSA - TWAEVs
     0.1 mg/m3 TWAEV
   Canada - Quebec - Time-Weighted Average Exposure Values
     0.1 mg/m3 TWAEV
   Israel - Time Weighted Averages
     0.1 mg/m3 TWA
   Israel - Action Levels
     0.05 mg/m3 AL
   Mexico - Instruction No. 10 - TWAs
     0.1 mg/m3 TWA
   Mexico - Instruction No. 10 - STELs
     0.3 mg/m3 STEL
STATE LISTS
   California - Exposure Limits - PELs
     0.1 mg/m3 PEL

California - Directors List of Hazardous Substances (8 CCR 339)
[present]

Florida Hazardous Substance List
[present]

Massachusetts Right To Know List
[present]

Minnesota Hazardous Substance List
[present]

NJ Right to Know List (Total)
sn 1546

Pennsylvania Right to Know List
[present]

PIPERAZINE   110-85-0

HEALTH AND SAFETY LISTS

U.S. DOT - Substances From 49 CFR 172.101
regulated by DOT (UN2368)

U.S. DOT - Hazard Classes
DOT hazard class = 3

U.S. DOT - Appendix B - Marine Pollutants
DOT regulated marine pollutant

NFPA - Flash Points
flash point = 91 degrees F (33 degrees C)

NFPA - Hazard Identification Ratings
health-1; flammability-3; reactivity-0

NIOSH - Selected LD50s and LC50s
Oral, rat: LD50 = 3700 mg/kg

ENVIRONMENTAL LISTS

List of Pesticide Product Inert Ingredients
[present]

INTERNATIONAL LISTS

Canada - WHMIS: Ingredient Disclosure
1% item 1312 (1428)

STATE LISTS

Florida Hazardous Substance List
[present]

Massachusetts Right To Know List
[present]

NJ Right to Know List (Total)
sn 0052

Pennsylvania Right to Know List
[present]

PIPERAZINE DIHYDROCHLORIDE   142-64-3

INTERNATIONAL LISTS

Canada - WHMIS: Ingredient Disclosure
1% item 1313 (1429)

PIPERAZINE ESTRONE SULFATE   7280-37-7

STATE LISTS

NJ Right to Know List (Total)
sn 1539

NJ Special Hazardous Substances
(flammable - third degree)

1-PIPERAZINEETHANOL   103-76-4

ENVIRONMENTAL LISTS

List of Pesticide Product Inert Ingredients
[present]

PIPERAZINONE, 1,1',1"-[1,3,5-TRIAZINE-2, 4,6-TRIYLTRIS[[CYCLOHEXYLIMINO]-2,1-ETHANEDIYL]TRIS-[3,3,4,5,5-PENTAMETHYL]-,   130277-45-1

HEALTH AND SAFETY LISTS

U.S. DOT - Substances From 49 CFR 172.101
regulated by DOT (UN1272)

U.S. DOT - Hazard Classes
DOT hazard class = 3

NFPA - Flash Points
flash point = 172 degrees F (78 degrees C);   steam distilled: flash point = 138 degrees F (59 degrees C)

NFPA - Hazard Identification Ratings
health-0; flammability-2; reactivity-0

NIOSH - Selected LD50s and LC50s
Oral, rat: LD50 = 3200 mg/kg Skin, rabbit: LD50 = 5 gm/kg

ENVIRONMENTAL LISTS

List of Pesticide Product Inert Ingredients
[present]

STATE LISTS

NJ Right to Know List (Total)
sn 2684

PIPERIDINE   110-89-4

HEALTH AND SAFETY LISTS

NFPA - Flash Points
flash point = 285 degrees F (141 degrees C)

NFPA - Hazard Identification Ratings
health-0; flammability-1; reactivity-0

1-PIPERIDINECARBOXALDEHYDE   2591-86-8

HEALTH AND SAFETY LISTS

NFPA - Flash Points
flash point = 130 degrees F (54 degrees C)

NFPA - Hazard Identification Ratings
health-0; flammability-2; reactivity-0

ENVIRONMENTAL LISTS

List of Pesticide Product Inert Ingredients
[present]

PIPERIDINE SALTS   RR-01780-1

HEALTH AND SAFETY LISTS

U.S. DOT - Substances From 49 CFR 172.101
regulated by DOT (UN2579)

U.S. DOT - Hazard Classes
DOT hazard class = 8

NFPA - Flash Points
flash point = 178 degrees F (81 degrees C)

NFPA - Hazard Identification Ratings
health-2; flammability-2; reactivity-0

NIOSH - Selected LD50s and LC50s
Inhalation, mouse: LC50 = 5400 mg/m3 (2 hr) Oral, rat: LD50 = 1900 mg/kg Skin, rabbit: LD50 = 4000 mg/kg

ENVIRONMENTAL LISTS

CAA - HON Rule - SOCMI Chemicals
compliance by Oct. 24, 1994

INTERNATIONAL LISTS

Canada - WHMIS: Ingredient Disclosure
0.1% item 1314 (1430)

German (DFG) - Skin/Sensitizers
danger of sensitization

STATE LISTS

Florida Hazardous Substance List
[present]

Massachusetts Right To Know List
[present]

NJ Right to Know List (Total)
sn 1540

NJ Special Hazardous Substances
(corrosive)

Pennsylvania Right to Know List
[present]

PIPERONYL BUTOXIDE 51-03-6

HEALTH AND SAFETY LISTS

ACGIH 1995 - Time Weighted Averages
5 mg/m3 TWA

NIOSH - Selected LD50s and LC50s
Oral, rat: LD50 = 4900 mg/kg

OSHA - Vacated PELs - Time Weighted Averages
5 mg/m3 TWA

INTERNATIONAL LISTS

Australian Exposure Standards - Time Weighted Averages
5 mg/m3 TWA

Canada - WHMIS: Ingredient Disclosure
0.1% item 1315 (1431)

Canada - Alberta - 8 Hour Occupational Exposure Limit
5 mg/m3 TWA

Canada - Alberta - 15 Minute Occupational Exposure Limit
10 mg/m3 STEL

Canada - Ontario - OHSA - TWAEVs
5 mg/m3 TWAEV

Canada - Quebec - Time-Weighted Average Exposure Values
5 mg/m3 TWAEV

United Kingdom - Occupational Exposure Standards - TWAs
5 mg/m3 TWA

Israel - Time Weighted Averages
5 mg/m3 TWA

Israel - Action Levels
2.5 mg/m3 AL

STATE LISTS

California - Exposure Limits - PELs
5 mg/m3 PEL

Florida Hazardous Substance List
[present]

Massachusetts Right To Know List
[present]

Minnesota Hazardous Substance List
[present]

Pennsylvania Right to Know List
[present]

PIPERONYL SULFOXIDE 120-62-7

HEALTH AND SAFETY LISTS

NTP Seventh Report - Known Carcinogens
known carcinogen (Listed under 'Conjugated estrogens')

OSHA - Select Carcinogens
[present]

STATE LISTS

Pennsylvania Right to Know List
special hazardous substance

Pennsylvania RTK - Special Hazardous Substances
[present]

PIPOBROMANE 54-91-1

HEALTH AND SAFETY LISTS

NFPA - Flash Points
flash point = 255 degrees F (124 degrees C)

NFPA - Hazard Identification Ratings
health-0; flammability-1; reactivity-0

NIOSH - Selected LD50s and LC50s
Oral, rat: LD50 = 4920 mg/kg

PIPROTAL 5281-13-0

ENVIRONMENTAL LISTS

TSCA - Chemicals with Significant New Use Rules
PMN number: P-89-589

TSCA - Section 12(b) - Export Notification
P-89-589; export notification required - Section 5

PIRIMICARB 23103-98-2

HEALTH AND SAFETY LISTS

AIHA - WEEL - Time Weighted Averages
1 ppm TWA; 3.5 mg/m3 TWA

U.S. DOT - Substances From 49 CFR 172.101
regulated by DOT (UN2401)

U.S. DOT - Hazard Classes
DOT hazard class = 3

FDA - Controlled Substances Act - Precursor chemicals
Threshold by base weight = 500 grams

NFPA - Flash Points
flash point = 61 degrees F (16 degrees C)

NFPA - Hazard Identification Ratings
health-3; flammability-3; reactivity-0

NIOSH - Selected LD50s and LC50s
Inhalation, mouse: LC50 = 6000 mg/m3 (2 hr) Oral, rat: LD50 = 400 mg/kg Skin, rabbit: LD50 = 320 mg/kg

ENVIRONMENTAL LISTS

CERCLA/SARA - Section 302 Extremely Hazardous Substances and TPQs
TPQ = 1000 pounds

CAA -Toxic Substances for Accidental Release Prevention
threshold quantity = 15,000 lbs

INTERNATIONAL LISTS

United Kingdom - Occupational Exposure Standards - TWAs
1 ppm TWA; 3.5 mg/m3 TWA

United Kingdom - Occupational Exposure Standards - Notes
can be absorbed through skin

STATE LISTS

Florida Hazardous Substance List
[present]

Massachusetts Right To Know List
extraordinarily hazardous

Minnesota Hazardous Substance List
[present]

NJ Right to Know List (Total)
sn 1543

NJ Special Hazardous Substances
(flammable - third degree; reactive - third degree)

Pennsylvania Right to Know List
environmental hazard

PROPOSED REGULATIONS

CERCLA/SARA - Proposed Hazardous Substance Additions
proposed RQ = 1 pound (.454 kg)

CERCLA/SARA - 1989 Proposed RQ Adjustments
proposed RQ = 100 pounds (45.4 kg)

PIRIMIPHOS-ETHYL   RR-01309-2

ENVIRONMENTAL LISTS

TSCA - Code of Federal Regulations Citations
40 CFR 712.30(x); 40 CFR 716.120(d)

TSCA - PAIR - Reporting List
Reporting Date: November 27, 1991

TSCA - Health and Safety Reporting List
Effective Date: September 30, 1991

PIRIMIPHOS METHYL [O-(2-(DIETHYLAMINO)-6-   29232-93-7
METHYL-4-PYRIMIDINYL)-O,O-DIMETHYLPHOS-
PHOROTHIOATE]

HEALTH AND SAFETY LISTS

FDA - Controlled Substances Act - Precursor chemicals
Threshold by base weight = 500 grams

PIRIMPHOS-ETHYL   23505-41-1

HEALTH AND SAFETY LISTS

IARC - Group 3 (not classifiable)
[present]

NIOSH - Selected LD50s and LC50s
Oral, rat: LD50 = 6150 mg/kg Skin, rabbit: LD50 = 200 mg/kg

NTP Chemical Status Reports - Testing Status and NTIS Number
Technical reports printed (PB288753/AS)

NTP Chemical Status Reports - Evidence of Carcinogenicity
male rat-negative; female rat-negative; male mice-negative; female mice-negative

ENVIRONMENTAL LISTS

CERCLA/SARA - Section 313 - Emission Reporting
form R reporting required

PIVALOLACTONE   1955-45-9

HEALTH AND SAFETY LISTS

NIOSH - Selected LD50s and LC50s
Oral, rat: LD50 = 2000 mg/kg Skin, rabbit: LD50 = 9000 mg/kg

NTP Chemical Status Reports - Testing Status and NTIS Number
Technical reports printed (PB288778/AS)

NTP Chemical Status Reports - Evidence of Carcinogenicity
male rat-negative; female rat-negative; male mice-positive; female mice-negative

STATE LISTS

Massachusetts Right To Know List
carcinogen; extraordinarily hazardous

PIVALOYL CHLORIDE   3282-30-2

STATE LISTS

California - Air Bill 2588 Appendix A-II
9/90

California - Prop. 65 - Developmental Toxicity
developmental toxicity - initial date 7/1/90

PLASTER OF PARIS   26499-65-0

HEALTH AND SAFETY LISTS

NIOSH - Selected LD50s and LC50s
Oral, rat: LD50 = 4400 ug/kg

STATE LISTS

Massachusetts Right To Know List
extraordinarily hazardous

NJ Right to Know List (Total)
sn 2685

Pennsylvania Right to Know List
environmental hazard

PROPOSED REGULATIONS

CERCLA/SARA - Proposed Hazardous Substance Additions
proposed RQ = 1 pound (.454 kg)

CERCLA/SARA - 1989 Proposed RQ Adjustments
proposed RQ = 100 pounds (45.4 kg)

PLATINIUM(II) SULFATE   53231-79-1

HEALTH AND SAFETY LISTS

U.S. DOT - Appendix B - Marine Pollutants
DOT regulated marine pollutant

NIOSH - Selected LD50s and LC50s
Oral, rat: LD50 = 147 mg/kg

STATE LISTS

NJ Right to Know List (Total)
sn 1544

PLATINOUS CHLORIDE   10025-65-7

HEALTH AND SAFETY LISTS

U.S. DOT - Appendix B - Marine Pollutants
DOT regulated severe marine pollutant

PLATINUM   7440-06-4

ENVIRONMENTAL LISTS

CERCLA/SARA - Section 313 - Emission Reporting
form R reporting required

PLATINUM 186   14993-39-6

HEALTH AND SAFETY LISTS

NIOSH - Selected LD50s and LC50s
Oral, rat: LD50 = 140 mg/kg Skin, rat: LD50 = 1000 mg/kg

ENVIRONMENTAL LISTS

CERCLA/SARA - Section 302 Extremely Hazardous Substances and TPQs
TPQ = 1000 pounds

STATE LISTS

Florida Hazardous Substance List
effective March 13, 1992

Massachusetts Right To Know List
extraordinarily hazardous

NJ Right to Know List (Total)
sn 1545

Pennsylvania Right to Know List
environmental hazard

PROPOSED REGULATIONS

CERCLA/SARA - Proposed Hazardous Substance Additions
proposed RQ = 1 pound (.454 kg)

CERCLA/SARA - 1989 Proposed RQ Adjustments
proposed RQ = 1000 pounds (454 kg)

PLATINUM 188   14922-70-4

HEALTH AND SAFETY LISTS

NTP Chemical Status Reports - Testing Status and NTIS Number
Technical reports printed (PB287645/AS)

NTP Chemical Status Reports - Evidence of Carcinogenicity
male rat-positive; female rat-positive; male mice-negative; female mice-negative

STATE LISTS

Massachusetts Right To Know List
carcinogen; extraordinarily hazardous

PLATINUM 189   15055-30-8

HEALTH AND SAFETY LISTS

U.S. DOT - Substances From 49 CFR 172.101
regulated by DOT (UN2438)

U.S. DOT - Hazard Classes
DOT hazard class = 8

U.S. DOT - Substances Which Are Poisonous by Inhalation
liquid hazardous material poisonous by inhalation (UN2438)

INTERNATIONAL LISTS

Canada - WHMIS: Ingredient Disclosure
1% item 1316 (519)

STATE LISTS

NJ Right to Know List (Total)
sn 1925

NJ Special Hazardous Substances
(corrosive)

PLATINUM 191   15706-36-2

HEALTH AND SAFETY LISTS

OSHA - Vacated PELs - Time Weighted Averages
total dust: 15 mg/m3 TWA; respirable fraction: 5 mg/m3 TWA

OSHA - Final PELs - Time Weighted Averages
total dust: 15 mg/m3 TWA; respirable fraction: 5 mg/m3 TWA

INTERNATIONAL LISTS

Canada - Alberta - 8 Hour Occupational Exposure Limit
respirable mass: 5 mg/m3 TWA; total mass: 10 mg/m3 TWA

Canada - British Columbia - 8 Hour Exposure Limits
nuisance dust: 10 mg/m3 TWA

Canada - British Columbia - 15 Minute Exposure Limits
20 mg/m3 STEL

Canada - Quebec - Time-Weighted Average Exposure Values
total dust: 10 mg/m3 TWAEV; respirable dust: 5 mg/m3 TWAEV

United Kingdom - Occupational Exposure Standards - TWAs
total inhalable dust: 10 mg/m3 TWA; respirable dust: 5 mg/m3 TWA

Mexico - Instruction No. 10 - TWAs
10 mg/m3 TWA; (nuisance particulate)

Mexico - Instruction No. 10 - STELs
20 mg/m3 STEL

STATE LISTS

Pennsylvania Right to Know List
[present]

PLATINUM 193   15735-70-3

INTERNATIONAL LISTS

Canada - WHMIS: Ingredient Disclosure
0.1% item 1319 (1527)

PLATINUM 193M   RR-00394-1

SEE ALSO:
PLATINUM

HEALTH AND SAFETY LISTS

NIOSH - Selected LD50s and LC50s
Oral, rat: LD50 = 3423 mg/m3

INTERNATIONAL LISTS

Canada - WHMIS: Ingredient Disclosure
0.1% item 1317 (520)

STATE LISTS

Massachusetts Right To Know List
[present]

PLATINUM 195M   RR-00393-0

HEALTH AND SAFETY LISTS

ACGIH 1995 - Time Weighted Averages
metal: 1 mg/m3 TWA; soluble salts, as Pt: 0.002 mg/m3 TWA

OSHA - Vacated PELs - Time Weighted Averages
metal, as Pt: 1 mg/m3 TWA; soluble salts, as Pt: 0.002 mg/m3 TWA

OSHA - Final PELs - Time Weighted Averages
soluble salts, as Pt: 0.002 mg/m3 TWA

INTERNATIONAL LISTS

Australian Exposure Standards - Time Weighted Averages
metal: 1 mg/m3 TWA; soluble salts, as Pt: 0.002 mg/m3 TWA

Australian Exposure Standards - Skin Effects
soluble salts, as Pt: sensitiser

Canada - WHMIS: Ingredient Disclosure
1% item 1318 (1432)

Canada - Alberta - 8 Hour Occupational Exposure Limit
1 mg/m3 TWA

Canada - Alberta - 15 Minute Occupational Exposure Limit
2 mg/m3 STEL

Canada - Ontario - OHSA - TWAEVs
1 mg/m3 TWAEV

Canada - Quebec - Time-Weighted Average Exposure Values
1 mg/m3 TWAEV

United Kingdom - Occupational Exposure Standards - TWAs
5 mg/m3 TWA

German (DFG) - MAK Values
total dust, as Pt: 0.002 mg/m3 MAK

German (DFG) - Skin/Sensitizers
danger of sensitization (skin or respiratory)

Israel - Time Weighted Averages
metal: 1 mg/m3 TWA; soluble salts, as Pt: 0.002 mg/m3 TWA

Israel - Action Levels
metal: 0.5 mg/m3 AL; soluble salts, as Pt: 0.001 mg/m3 AL

Mexico - Instruction No. 10 - TWAs
0.002 mg/m3 TWA

STATE LISTS

California - Exposure Limits - PELs
metal: 1 mg/m3 PEL; soluble salts, as Pt: 0.002 mg/m3 PEL

California - Directors List of Hazardous Substances (8 CCR 339)
[present]

Florida Hazardous Substance List
[present]

Massachusetts Right To Know List
[present]

Minnesota Hazardous Substance List
[present] (includes soluble salts)

Pennsylvania Right to Know List
[present]

PLATINUM 197   15735-74-7

HEALTH AND SAFETY LISTS

U.S. DOT - Appendix A Table 2 - Radionuclides
final RQ = 100 curies (3.7E 12 Bq)

ENVIRONMENTAL LISTS

CERCLA/SARA List of Radionuclides (Appendix B) and Their Reportable Quantities
final RQ = 100 curies (3.7E 12 Bq)

PLATINUM 197M   RR-00392-9

HEALTH AND SAFETY LISTS

U.S. DOT - Appendix A Table 2 - Radionuclides
final RQ = 100 curies (3.7E 12 Bq)

ENVIRONMENTAL LISTS

CERCLA/SARA List of Radionuclides (Appendix B) and Their Reportable Quantities
final RQ = 100 curies (3.7E 12 Bq)

PLATINUM 199   15706-54-4

HEALTH AND SAFETY LISTS

U.S. DOT - Appendix A Table 2 - Radionuclides
final RQ = 100 curies (3.7E 12 Bq)

ENVIRONMENTAL LISTS

CERCLA/SARA List of Radionuclides (Appendix B) and Their Reportable Quantities
final RQ = 100 curies (3.7E 12 Bq)

PLATINUM 200   29687-31-8

HEALTH AND SAFETY LISTS

U.S. DOT - Appendix A Table 2 - Radionuclides
final RQ = 100 curies (3.7E 12 Bq)

ENVIRONMENTAL LISTS

CERCLA/SARA List of Radionuclides (Appendix B) and Their Reportable Quantities
final RQ = 100 curies (3.7E 12 Bq)

PLATINUM SOLUBLE SALTS   RR-00046-4

HEALTH AND SAFETY LISTS

U.S. DOT - Appendix A Table 2 - Radionuclides
final RQ = 1000 curies (3.7E 13 Bq)

ENVIRONMENTAL LISTS

CERCLA/SARA List of Radionuclides (Appendix B) and Their Reportable Quantities
final RQ = 1000 curies (3.7E 13 Bq)

PLATINUM TETRACHLORIDE   13454-96-1

HEALTH AND SAFETY LISTS

U.S. DOT - Appendix A Table 2 - Radionuclides
final RQ = 100 curies (3.7E 12 Bq)

ENVIRONMENTAL LISTS

CERCLA/SARA List of Radionuclides (Appendix B) and Their Reportable Quantities
final RQ = 100 curies (3.7E 12 Bq)

PLICAMYCIN   18378-89-7

HEALTH AND SAFETY LISTS

U.S. DOT - Appendix A Table 2 - Radionuclides
final RQ = 100 curies (3.7E 12 Bq)

ENVIRONMENTAL LISTS

CERCLA/SARA List of Radionuclides (Appendix B) and Their Reportable Quantities
final RQ = 100 curies (3.7E 12 Bq)

PLUTONIUM   7440-07-5

HEALTH AND SAFETY LISTS

U.S. DOT - Appendix A Table 2 - Radionuclides
final RQ = 1000 curies (3.7E 13 Bq)

ENVIRONMENTAL LISTS

CERCLA/SARA List of Radionuclides (Appendix B) and Their Reportable Quantities
final RQ = 1000 curies (3.7E 13 Bq)

PLUTONIUM 234   34018-47-8

HEALTH AND SAFETY LISTS

U.S. DOT - Appendix A Table 2 - Radionuclides
final RQ = 1000 curies (3.7E 13 Bq)

ENVIRONMENTAL LISTS

CERCLA/SARA List of Radionuclides (Appendix B) and Their Reportable Quantities
final RQ = 1000 curies (3.7E 13 Bq)

PLUTONIUM 235   14928-39-3

HEALTH AND SAFETY LISTS

U.S. DOT - Appendix A Table 2 - Radionuclides
final RQ = 1000 curies (3.7E 13 Bq)

ENVIRONMENTAL LISTS

CERCLA/SARA List of Radionuclides (Appendix B) and Their Reportable Quantities
final RQ = 1000 curies (3.7E 13 Bq)

PLUTONIUM 236   15411-92-4

HEALTH AND SAFETY LISTS

U.S. DOT - Appendix A Table 2 - Radionuclides
final RQ = 100 curies (3.7E 12 Bq)

ENVIRONMENTAL LISTS

CERCLA/SARA List of Radionuclides (Appendix B) and Their Reportable Quantities
final RQ = 100 curies (3.7E 12 Bq)

PLUTONIUM 237   15411-93-5

HEALTH AND SAFETY LISTS

NIOSH 1990 - Pocket Guide - RELs
as Pt: 0.002 mg/m3 TWA

NIOSH 1990 - Pocket Guide - Target organs
skin, eyes, respiratory system

INTERNATIONAL LISTS

Canada - WHMIS: Ingredient Disclosure
0.1% item 1321 (1433)

Canada - Alberta - 8 Hour Occupational Exposure Limit
as Pt: 0.002 mg/m3 TWA

Canada - Alberta - 15 Minute Occupational Exposure Limit
as Pt: 0.006 mg/m3 STEL

Canada - British Columbia - 8 Hour Exposure Limits
as Pt: 0.002 mg/m3 TWA

Canada - Ontario - OHSA - TWAEVs
as Pt: 0.002 mg/m3 TWAEV

Canada - Quebec - Time-Weighted Average Exposure Values
0.002 mg/m3 TWAEV

United Kingdom - Occupational Exposure Standards - TWAs
as Pt: 0.002 mg/m3 TWA

United Kingdom - Occupational Exposure Standards - Notes
sensitisation

Mexico - Instruction No. 10 - TWAs
0.002 mg/m3 TWA

STATE LISTS

California - Directors List of Hazardous Substances (8 CCR 339)
[present]

Pennsylvania Right to Know List
[present]

PLUTONIUM 238   13981-16-3

SEE ALSO:
PLATINUM

HEALTH AND SAFETY LISTS

NIOSH - Selected LD50s and LC50s
Oral, rat: LD50 = 276 mg/kg

INTERNATIONAL LISTS

Canada - WHMIS: Ingredient Disclosure
0.1% item 1320 (1581)

STATE LISTS
Massachusetts Right To Know List
[present]

PLUTONIUM 239   15117-48-3

STATE LISTS
California - Air Bill 2588 Appendix A-II
9/90
California - Prop. 65 - Developmental Toxicity
developmental toxicity - initial date 4/1/90

PLUTONIUM 240   14119-33-6

ENVIRONMENTAL LISTS
ATSDR Priority List
Rank (of 275): 192

PLUTONIUM 241   14119-32-5

HEALTH AND SAFETY LISTS
U.S. DOT - Appendix A Table 2 - Radionuclides
final RQ = 1000 curies (3.7E 13 Bq)

ENVIRONMENTAL LISTS
CERCLA/SARA List of Radionuclides (Appendix B) and Their Reportable Quantities
final RQ = 1000 curies (3.7E 13 Bq)

PLUTONIUM 242   13982-10-0

HEALTH AND SAFETY LISTS
U.S. DOT - Appendix A Table 2 - Radionuclides
final RQ = 1000 curies (3.7E 13 Bq)

ENVIRONMENTAL LISTS
CERCLA/SARA List of Radionuclides (Appendix B) and Their Reportable Quantities
final RQ = 1000 curies (3.7E 13 Bq)

PLUTONIUM 243   15706-37-3

HEALTH AND SAFETY LISTS
U.S. DOT - Appendix A Table 2 - Radionuclides
final RQ = 0.1 curies (3.7E 9 Bq)

ENVIRONMENTAL LISTS
CERCLA/SARA List of Radionuclides (Appendix B) and Their Reportable Quantities
final RQ = 0.1 curies (3.7E 9 Bq)

PLUTONIUM 244   14119-34-7

HEALTH AND SAFETY LISTS
U.S. DOT - Appendix A Table 2 - Radionuclides
final RQ = 1000 curies (3.7E 13 Bq)

ENVIRONMENTAL LISTS
CERCLA/SARA List of Radionuclides (Appendix B) and Their Reportable Quantities
final RQ = 1000 curies (3.7E 13 Bq)

PLUTONIUM 245   18784-52-6

HEALTH AND SAFETY LISTS
U.S. DOT - Appendix A Table 2 - Radionuclides
final RQ = 0.01 curies (3.7E 8 Bq)

ENVIRONMENTAL LISTS
ATSDR Priority List
Rank (of 275): 185
CERCLA/SARA List of Radionuclides (Appendix B) and Their Reportable Quantities
final RQ = 0.01 curies (3.7E 8 Bq)

POE SORBITAN DITALLATE   RR-01113-2

HEALTH AND SAFETY LISTS
U.S. DOT - Appendix A Table 2 - Radionuclides
final RQ = 0.01 curies (3.7E 8 Bq)

ENVIRONMENTAL LISTS
ATSDR Priority List
Rank (of 275): 189
CERCLA/SARA List of Radionuclides (Appendix B) and Their Reportable Quantities
final RQ = 0.01 curies (3.7E 8 Bq)

POISONOUS LIQUIDS, N.O.S.   RR-00440-0

HEALTH AND SAFETY LISTS
U.S. DOT - Appendix A Table 2 - Radionuclides
final RQ = 0.01 curies (3.7E 8 Bq)

ENVIRONMENTAL LISTS
CERCLA/SARA List of Radionuclides (Appendix B) and Their Reportable Quantities
final RQ = 0.01 curies (3.7E 8 Bq)

POISONOUS SOLIDS, N.O.S.   RR-00445-5

HEALTH AND SAFETY LISTS
U.S. DOT - Appendix A Table 2 - Radionuclides
final RQ = 1 curie (3.7E 10 Bq)

ENVIRONMENTAL LISTS
CERCLA/SARA List of Radionuclides (Appendix B) and Their Reportable Quantities
final RQ = 1 curie (3.7E 10 Bq)

POLISH   RR-01504-3

HEALTH AND SAFETY LISTS
U.S. DOT - Appendix A Table 2 - Radionuclides
final RQ = 0.01 curies (3.7E 8 Bq)

ENVIRONMENTAL LISTS
CERCLA/SARA List of Radionuclides (Appendix B) and Their Reportable Quantities
final RQ = 0.01 curies (3.7E 8 Bq)

POLONIUM 203   16729-74-1

HEALTH AND SAFETY LISTS
U.S. DOT - Appendix A Table 2 - Radionuclides
final RQ = 1000 curies (3.7E 13 Bq)

ENVIRONMENTAL LISTS
CERCLA/SARA List of Radionuclides (Appendix B) and Their Reportable Quantities
final RQ = 1000 curies (3.7E 13 Bq)

POLONIUM 205   16729-76-3

HEALTH AND SAFETY LISTS
U.S. DOT - Appendix A Table 2 - Radionuclides
final RQ = 0.01 curies (3.7E 8 Bq)

ENVIRONMENTAL LISTS
CERCLA/SARA List of Radionuclides (Appendix B) and Their Reportable Quantities
final RQ = 0.01 curies (3.7E 8 Bq)

POLONIUM 207   15720-45-3

HEALTH AND SAFETY LISTS
U.S. DOT - Appendix A Table 2 - Radionuclides
final RQ = 100 curies (3.7E 12 Bq)

ENVIRONMENTAL LISTS

CERCLA/SARA List of Radionuclides (Appendix B) and Their Reportable Quantities
final RQ = 100 curies (3.7E 12 Bq)

POLONIUM 210   13981-52-7

ENVIRONMENTAL LISTS

List of Pesticide Product Inert Ingredients
[present]

POLYACRYLAMIDES   9003-05-8

HEALTH AND SAFETY LISTS

U.S. DOT - Substances From 49 CFR 172.101
regulated by DOT (UN2927, UN2929, UN2810, UN3122, UN3123)

U.S. DOT - Hazard Classes
DOT hazard class = 6.1

U.S. DOT - Substances Which Are Poisonous by Inhalation
gaseous hazardous material poisonous by inhalation (zone A or B) (UN2810, UN2927, UN2929, UN3122)

STATE LISTS

NJ Right to Know List (Total)
sn 2690; sn 2691; sn 2692; sn 2693; sn 2694

POLYACRYLONITRILE, OXIDIZED   68908-35-0

HEALTH AND SAFETY LISTS

U.S. DOT - Substances From 49 CFR 172.101
regulated by DOT (UN2928, UN2930, UN2811, UN3086, UN3124, UN3125)

U.S. DOT - Hazard Classes
DOT hazard class = 6.1

STATE LISTS

NJ Right to Know List (Total)
sn 2695; sn 2696; sn 2697

POLYALKYLENE GLYCOL ALKYL ETHER ACRYLATE   RR-00981-4

STATE LISTS

NJ Right to Know List (Total)
liquid: sn 2698; metal, stove, and furniture: sn 2699

POLYALKYLENE GLYCOL SUBSTITUTED ACETATE   RR-01261-3

HEALTH AND SAFETY LISTS

U.S. DOT - Appendix A Table 2 - Radionuclides
final RQ = 100 curies (3.7E 12 Bq)

ENVIRONMENTAL LISTS

CERCLA/SARA List of Radionuclides (Appendix B) and Their Reportable Quantities
final RQ = 100 curies (3.7E 12 Bq)

POLYALKYLENE POLYAMINE   RR-01559-8

HEALTH AND SAFETY LISTS

U.S. DOT - Appendix A Table 2 - Radionuclides
final RQ = 100 curies (3.7E 12 Bq)

ENVIRONMENTAL LISTS

CERCLA/SARA List of Radionuclides (Appendix B) and Their Reportable Quantities
final RQ = 100 curies (3.7E 12 Bq)

POLYALKYLENEPOLYOL ALKYLAMINE   RR-00916-5

HEALTH AND SAFETY LISTS

U.S. DOT - Appendix A Table 2 - Radionuclides
final RQ = 10 curies (3.7E 11 Bq)

ENVIRONMENTAL LISTS

CERCLA/SARA List of Radionuclides (Appendix B) and Their Reportable Quantities
final RQ = 10 curies (3.7E 11 Bq)

POLYALKYLPOLYSILAZANE, BIS(SUBSTITUTED ACRYLATE)   RR-00226-6

HEALTH AND SAFETY LISTS

U.S. DOT - Appendix A Table 2 - Radionuclides
final RQ = 0.01 curies (3.7E 8 Bq)

ENVIRONMENTAL LISTS

ATSDR Priority List
Rank (of 275): 179

CERCLA/SARA List of Radionuclides (Appendix B) and Their Reportable Quantities
final RQ = 0.01 curies (3.7E 8 Bq)

POLYAMINE DITHIOCARBAMATE   RR-01574-7

ENVIRONMENTAL LISTS

List of Pesticide Product Inert Ingredients
[present]

POLYAMINOPOLYACID   RR-01565-6

ENVIRONMENTAL LISTS

List of Pesticide Product Inert Ingredients
[present]

POLYAMYL NAPTHALENE MIXTURE OF POLYMERS   RR-00140-1

ENVIRONMENTAL LISTS

TSCA - Chemicals with Significant New Use Rules
PMN number: P-88-1691

POLYBROMINATED BIPHENYLS (PBB)(GENERIC)   RR-00086-2

ENVIRONMENTAL LISTS

TSCA - Chemicals with Significant New Use Rules
PMN number: P-91-1269

POLYBUTADIENE RESIN   9003-17-2

ENVIRONMENTAL LISTS

TSCA - Chemicals with Significant New Use Rules
PMN number: P-89-963

POLYBUTENE   9003-29-6

ENVIRONMENTAL LISTS

TSCA - Chemicals with Significant New Use Rules
PMN number: P-89-483

POLYBUTYL ACRYLATE   9003-49-0

ENVIRONMENTAL LISTS

TSCA - Chemicals with Significant New Use Rules
PMN number: P-89-423

POLYBUTYLENE   9003-28-5

ENVIRONMENTAL LISTS

TSCA - Chemicals with Significant New Use Rules
PMN number: P-91-1328

POLYCHLORINATED ALKANES   RR-01771-0

ENVIRONMENTAL LISTS

TSCA - Chemicals with Significant New Use Rules
PMN number: P-92-491

POLYCHLORINATED BIPHENYLS   1336-36-3

HEALTH AND SAFETY LISTS

NFPA - Flash Points
flash point = 360 degrees F (182 degrees C)

NFPA - Hazard Identification Ratings
health-0; flammability-1; reactivity-0

POLYCHLORINATED DIBENZO-P-DIOXINS   RR-00296-0

ENVIRONMENTAL LISTS

CERCLA/SARA - Section 313 - Emission Reporting
form R reporting required for 0.1% de minimus concentration

INTERNATIONAL LISTS

Canada - CEPA Schedule I - Toxic Substances
that have the molecular formula C12H10-nBrn in which 'n' is greater than 2: prohibited commercial, manufacturing or processing uses

Canada - CEPA Schedule II Part I - Prohibited Substances (Export)
molecular formula C12H10-nBrn in which "n" is greater than 2:
[present]

STATE LISTS

California - Air Bill 2588 Appendix A-II
known or potential carcinogen

California - Prop. 65 - Cancer list
carcinogen - initial date 1/1/88

California - Prop. 65 - Developmental Toxicity
developmental toxicity - initial date 10/01/94

California - Prop. 65 - No Significant Risk Levels
no significant risk level = 0.02 ug/day

California - Directors List of Hazardous Substances (8 CCR 339)
[present]

Pennsylvania Right to Know List
environmental hazard

POLYCHLORINATED DIBENZOFURANS   RR-00295-9

ENVIRONMENTAL LISTS

List of Pesticide Product Inert Ingredients
[present]

POLYCHLORINATED O-TERPHENYL   11126-42-4

ENVIRONMENTAL LISTS

List of Pesticide Product Inert Ingredients
[present]

POLYCHLORINATED TERPHENYLS   RR-01611-5

ENVIRONMENTAL LISTS

List of Pesticide Product Inert Ingredients
[present]

POLYCHLORINATED TRIPHENYLS   12642-23-8

ENVIRONMENTAL LISTS

List of Pesticide Product Inert Ingredients
[present]

POLYCYCLIC AROMATIC HYDROCARBON   RR-01740-3
DERIVATIVES

ENVIRONMENTAL LISTS

CERCLA/SARA - Section 313 - Emission Reporting
form R reporting required; (Includes those chemicals defined by the following formula: CxH2x-yCly where x = 10 to 13 and y = 3 to 12; and where the average chlorine content ranges from 40-70% with the limitng molecular formulas C10H19Cl13 and C13H16Cl12)

POLYCYCLIC AROMATIC HYDROCARBONS   RR-01523-6

HEALTH AND SAFETY LISTS

U.S. DOT - Substances From 49 CFR 172.101
regulated by DOT (UN2315)

U.S. DOT - Hazard Classes
DOT hazard class = 9

U.S. DOT - Appendix B - Marine Pollutants
DOT regulated severe marine pollutant

U.S. DOT - Appendix A Table 1 - Hazardous Substances
final RQ = 1 pound (0.454 kg)

IARC - Group 2A (limited human data)
[present]

NIOSH - Selected LD50s and LC50s
Oral, mouse: LD50 = 1900 mg/kg

NIOSH - Health Standards - Exposure Limits
1 ug/m3 TWA (the minimum reliably detectable concentration)

NIOSH - Health Standards - Health Effects and Precautions
has produced tumors of the liver and pituitary gland and leukemias in animals; skin, liver, and reproductive system effects (Blood testing required); prevent skin contact; warn women of potential reproductive hazards)

NIOSH - Health Standards - Carcinogenic Chemicals
potential human carcinogen

NTP Seventh Report - Suspect Carcinogens
suspect carcinogen

OSHA - Possible Select Carcinogens
[present]

ENVIRONMENTAL LISTS

ATSDR Priority List
Rank (of 275): 007

CERCLA/SARA - Section 313 - Emission Reporting
form R reporting required for 0.1% de minimus concentration

CERCLA/SARA - Hazardous Substances and their Reportable Quantities
final RQ = 1 pound (0.454 kg)

Clean Air Act (1990) - List of Hazardous Air Contaminants
[present]

Clean Water Act - Hazardous Substances
[present]

Clean Water Act - Toxic Pollutants
[present]

Safe Drinking Water Act - MCLs
as decachlorobiphenyl: MCL = 0.0005 mg/L

Safe Drinking Water Act - MCLGs
as decachlorobiphenyl: MCLG = Zero

EPA - Carcinogen Hazard Ranking for RQ Adjustment
Hazard ranking = Medium

RCRA - Hazardous Constituents-Appendix VIII
hazardous constituent - no waste number

RCRA - TSD Facilities Ground Water Monitoring
TM 8080 = 50 ug/L PQL; TM 8250 = 100 ug/L PQL (These values are averages for PCB congeners (Aroclor nos. 1016, 1221, 1232, 1242, 1248, 1254, and 1260))

RCRA - Universal Treatment Standards (LDR)
WW: 0.10 mg/l; NWW: 10 mg/kg

TSCA - Section 12(b) - Export Notification
export notification required - Section 6

INTERNATIONAL LISTS

Canada - WHMIS: Ingredient Disclosure
0.1% item 1323 (1440)

Canada - Ontario - OHSA - TWAEVs
0.05 mg/m3 TWAEV (listed as an agent of variable composition) (As sum of components assayed by chromatographic procedure with reference to the bulk sample.)

Canada - Ontario - OHSA - Skin Notations
absorption through skin, eyes, or mucous membranes (listed under 'Agents of Variable Composition')
Mexico - Drinking Water - Ecological Criteria
0.0000008 mg/l Substance presents persistence, bioaccumulations or risk of cancer, reduce human exposure to a minimum; This level has been extrapolated by using a mathematic model

STATE LISTS

California - Air Bill 2588 Appendix A-I
known or potential carcinogen
California - Prop. 65 - Cancer list
carcinogen - initial date 10/1/89
California - Prop. 65 - Developmental Toxicity
developmental toxicity - initial date 1/1/91
California - Prop. 65 - No Significant Risk Levels
no significant risk level = 0.09 ug/day
California - Directors List of Hazardous Substances (8 CCR 339)
[present]
Florida Hazardous Substance List
[present]
Massachusetts Right To Know List
carcinogen; extraordinarily hazardous; teratogen
Minnesota Hazardous Substance List
carcinogen
NJ Right to Know List (Total)
sn 1554
NJ Special Hazardous Substances
(carcinogen; teratogen)
Pennsylvania Right to Know List
environmental hazard; special hazardous substance
Pennsylvania RTK - Special Hazardous Substances
[present]

PROPOSED REGULATIONS

Canada - Ontario - Proposed Occupational TWAEVs
0.01 mg/m3 TWAEV
Canada - Ontario - Proposed Occupational STEVs
0.03 mg/m3 STEV

POLYCYCLIC ORGANIC MATTER   RR-00079-3

ENVIRONMENTAL LISTS

RCRA - TSD Facilities Ground Water Monitoring
TM 8280 = 0.01 ug/L PQL (this is an average value for PCDF congeners (hexa-, penta-, and tetrachlorodibenzo-p-dioxins))

INTERNATIONAL LISTS

Canada - CEPA Schedule I - Toxic Substances
that have the molecular formula C12H(8-n)O2Cl2 where 'n' is greater than 2: (a) maximum concentrations in products; (b) maximum concentrations that may be released into the environment

STATE LISTS

California - Air Bill 2588 Appendix A-I
known or potential carcinogen
California - Prop. 65 - Cancer list
carcinogen - initial date 10/1/92

POLYDIMETHYLDIPHENYL SILOXANE   68083-14-7
COPOLYMER

ENVIRONMENTAL LISTS

RCRA - TSD Facilities Ground Water Monitoring
TM 8280 = 0.01 ug/L PQL (this value is an average for PCDF congeners (hexa-, penta-, and tetrachlorodibenzofurans))

INTERNATIONAL LISTS

Canada - CEPA Schedule I - Toxic Substances
that have the molecular formula C12H(8-n)OCln where 'n' is greater than 2: (a) maximum concentrations in products; (b) maximum concentrations that may be released into the environment

STATE LISTS

California - Air Bill 2588 Appendix A-I
known or potential carcinogen
California - Prop. 65 - Cancer list
carcinogen - initial date 10/1/92

POLY(DIMETHYLSILOXANE)   63148-62-9

HEALTH AND SAFETY LISTS

NIOSH - Selected LD50s and LC50s
Oral, rat: LD50 = 19200 mg/kg

POLYEPOXYPOLYOL   RR-01723-2

INTERNATIONAL LISTS

Canada - CEPA Schedule I - Toxic Substances
that have a molecular formula C18H14-nCln in which 'n' is greater than 2: prohibited commercial, manufacturing or processing uses
Canada - CEPA Schedule II Part I - Prohibited Substances (Export)
that have the molecular formula C18H14-nCln in which 'n' is greater than 2: [present]

POLYESTER POLYURETHANE ACRYLATE   RR-01692-2

STATE LISTS

NJ Right to Know List (Total)
sn 3146

POLYETHER-SULPHONE   25667-42-9

STATE LISTS

California - Air Bill 2588 Appendix A-I
6/91

POLYETHOXYAMINE HK   37271-20-8

INTERNATIONAL LISTS

Canada - CEPA - Priority Substances List
estimated time for completion of assessment reports: 3 years
Mexico - Instruction No. 10 - TWAs
0.2 mg/m3 TWA
Mexico - Instruction No. 10 - Carcinogens
potentially carcinogenic contaminant

STATE LISTS

California - Air Bill 2588 Appendix A-I
9/89

POLYETHOXYLATED POLYPROPOXYLATED C12-15   68551-13-3
ALCOHOLS

HEALTH AND SAFETY LISTS

U.S. DOT - Appendix A Table 1 - Hazardous Substances
final RQ = 1 pound (.454 kg)

ENVIRONMENTAL LISTS

CERCLA/SARA - Hazardous Substances and their Reportable Quantities
final RQ = 1 pound (.454 kg)
Clean Air Act (1990) - List of Hazardous Air Contaminants
[present] (includes organic compounds with more than one benzene ring, and which have a boiling point greater than or equal to 100 degrees C)
CAA - HON Rule - Organic HAPs
Includes organic compounds with more than one benzene ring, and which have a boiling point greater than or equal to 100 deg. C.

INTERNATIONAL LISTS
  Canada - British Columbia - 8 Hour Exposure Limits
    as benzene solubles: 0.2 mg/m3 TWA
  Canada - British Columbia - Carcinogens
    carcinogen - (as benzene solubles) 0.2 mg/m3 TWA
STATE LISTS
  California - Air Bill 2588 Appendix A-I
    9/89
  Pennsylvania Right to Know List
    [present]

POLYETHOXYLATED PRIMARY AMINE (C14-18)   68755-33-9

ENVIRONMENTAL LISTS
  List of Pesticide Product Inert Ingredients
    [present]
  TSCA - PAIR - Reporting List
    Effective Date: October 12, 1993;   Reporting Date: February 28, 1994
  TSCA - Health and Safety Reporting List
    Effective Date: October 12, 1993;   Sunset Date: October 12, 2003

POLYETHOXY POLYPROPOXY OLYETHOXY ETHANOL-IODINE COMPLEX   11096-42-7

ENVIRONMENTAL LISTS
  List of Pesticide Product Inert Ingredients
    [present]
  TSCA - PAIR - Reporting List
    Effective Date: October 12, 1993;   Reporting Date: February 28, 1994
  TSCA - Health and Safety Reporting List
    Effective Date: October 12, 1993;   Sunset Date: October 12, 2003

POLYETHYLENE   9002-88-4

ENVIRONMENTAL LISTS
  TSCA - Chemicals with Significant New Use Rules
    PMN number: P-93-364

POLYETHYLENE GLYCOL   25322-68-3

ENVIRONMENTAL LISTS
  TSCA - Chemicals with Significant New Use Rules
    PMN number: P-93-498

POLYETHYLENE GLYCOL BRANCHED NONYLPHENYL ETHER   68412-54-4

HEALTH AND SAFETY LISTS
  NIOSH - Selected LD50s and LC50s
    Oral, mouse: LD50 − 1300 mg/kg

POLYETHYLENE GLYCOL METHYL ETHER   9004-74-4

ENVIRONMENTAL LISTS
  List of Pesticide Product Inert Ingredients
    [present]

POLYETHYLENE GLYCOL, MONOBUTYL ETHER   9004-77-7

ENVIRONMENTAL LISTS
  List of Pesticide Product Inert Ingredients
    [present]

POLYETHYLENE GLYCOL MONOLAURATE   9004-81-3

ENVIRONMENTAL LISTS
  List of Pesticide Product Inert Ingredients
    [present]

POLYETHYLENE GLYCOL MONOLAURYL ETHER SULFATE AMMONIUM SALT   32612-48-9

HEALTH AND SAFETY LISTS
  NIOSH - Selected LD50s and LC50s
    Oral, rat: LD50 = 2100 mg/kg
ENVIRONMENTAL LISTS
  List of Pesticide Product Inert Ingredients
    [present]

POLYETHYLENE GLYCOL MONOOLEATE   9004-96-0

HEALTH AND SAFETY LISTS
  IARC - Group 3 (not classifiable)
    [present]
  NIOSH - Selected LD50s and LC50s
    Inhalation, mouse: LC50 = 12 gm/m3 (30 mn)
ENVIRONMENTAL LISTS
  List of Pesticide Product Inert Ingredients
    [present]

POLYETHYLENE GLYCOL MONOOLEYL ETHER   9004-98-2

HEALTH AND SAFETY LISTS
  AIHA - WEEL - Time Weighted Averages
    10 mg/m3 TWA
  NFPA - Flash Points
    flash point = 360 to 550 degrees F (182 to 287 degrees C)
  NFPA - Hazard Identification Ratings
    health-0;   flammability-1;   reactivity-0
  NIOSH - Selected LD50s and LC50s
    Oral, rat: LD50 = 33750 mg/kg
ENVIRONMENTAL LISTS
  CAA - HON Rule - SOCMI Chemicals
    compliance by Oct. 23, 1995
  List of Pesticide Product Inert Ingredients
    [present]
STATE LISTS
  Minnesota Hazardous Substance List
    [present]

POLYETHYLENE GLYCOL MONOETHER WITH OLEIC ACID MONO-ETHANOLAMIDE   26027-37-2

ENVIRONMENTAL LISTS
  List of Pesticide Product Inert Ingredients
    [present]

POLYETHYLENE GLYCOL NONYLPHENYL ETHER SODIUM SULFATE   9014-90-8

HEALTH AND SAFETY LISTS
  NIOSH - Selected LD50s and LC50s
    Oral, rat: LD50 = 22 gm/kg

POLYETHYLENE GLYCOL OCTYLPHENOL ETHER   9002-93-1

ENVIRONMENTAL LISTS
  List of Pesticide Product Inert Ingredients
    [present]

POLYETHYLENE GLYCOL, OLEIC ACID DIESTER   55069-68-6

ENVIRONMENTAL LISTS
  List of Pesticide Product Inert Ingredients
    [present]

POLYETHYLENE GLYCOL STEARATE 9004-99-3

HEALTH AND SAFETY LISTS

NIOSH - Selected LD50s and LC50s
Oral, rat: LD50 = 1600 mg/kg

ENVIRONMENTAL LISTS

List of Pesticide Product Inert Ingredients
[present]

INTERNATIONAL LISTS

Canada - WHMIS: Ingredient Disclosure
1% item 1445 (845)

POLYETHYLENEIMINE 9002-98-6

ENVIRONMENTAL LISTS

List of Pesticide Product Inert Ingredients
[present]

POLYETHYLENE-POLYPROPYLENE GLYCOL 9003-11-6

HEALTH AND SAFETY LISTS

NIOSH - Selected LD50s and LC50s
Oral, rat: LD50 = 25800 mg/kg

ENVIRONMENTAL LISTS

List of Pesticide Product Inert Ingredients
[present]

INTERNATIONAL LISTS

Canada - WHMIS: Ingredient Disclosure
1% item 1324 (830)

POLYETHYLENE TEREPHTHALATE - POLYETHYLENE ISOPHTHALATE FILM 24938-04-3

ENVIRONMENTAL LISTS

List of Pesticide Product Inert Ingredients
[present]

POLYFLUOROSULFONIC ACID SALT RR-00976-7

ENVIRONMENTAL LISTS

List of Pesticide Product Inert Ingredients
[present]

POLYGEENAN 53973-98-1

HEALTH AND SAFETY LISTS

NIOSH - Selected LD50s and LC50s
Oral, rat: LD50 = 1800 mg/kg

ENVIRONMENTAL LISTS

List of Pesticide Product Inert Ingredients
[present]

INTERNATIONAL LISTS

Canada - WHMIS: Ingredient Disclosure
1% item 1325 (831)

POLYGLYCERINE 25618-55-7

ENVIRONMENTAL LISTS

List of Pesticide Product Inert Ingredients
[present]

POLYGLYCEROL ESTER OF OLEIC ACID 9007-48-1

HEALTH AND SAFETY LISTS

NIOSH - Selected LD50s and LC50s
Oral, rat: LD50 = 53 gm/kg

ENVIRONMENTAL LISTS

List of Pesticide Product Inert Ingredients
[present]

INTERNATIONAL LISTS

Canada - WHMIS: Ingredient Disclosure
1% item 1327 (1503)

POLYGLYCERYL PHTHALATE ESTER OF COCONUT OIL FATTY ACID 66070-87-9

HEALTH AND SAFETY LISTS

NIOSH - Selected LD50s and LC50s
Oral, rat: LD50 = 1350 mg/kg

POLYGLYCIDOL 25722-70-7

HEALTH AND SAFETY LISTS

NIOSH - Selected LD50s and LC50s
Oral, rat: LD50 = 9380 mg/kg

ENVIRONMENTAL LISTS

List of Pesticide Product Inert Ingredients
[present]

POLYHALOGENATED BIPHENYLS RR-01310-5

ENVIRONMENTAL LISTS

List of Pesticide Product Inert Ingredients
[present]

POLYMER RR-01008-2

ENVIRONMENTAL LISTS

TSCA - Chemicals with Significant New Use Rules
PMN number: P-90-587

POLYMERIZED FATTY ACID PHOSPHATE (100% C12-C18) RR-01117-6

STATE LISTS

California - Air Bill 2588 Appendix A-II
known or potential carcinogen: 9/89

California - Prop. 65 - Cancer list
carcinogen - initial date 1/1/88

POLYMER OF ADIPIC ACID, ALKANEPOLYOL, ALKYLDIISOCYANATOCARBOMONOCYCLE, HYDROXYALKYL ACRYLATE ESTER RR-01017-3

ENVIRONMENTAL LISTS

List of Pesticide Product Inert Ingredients
[present]

POLYMER OF ALKANEDIOIC ACID, METHYLENEBISCARBOMONOCYCLIC DIISOCYANATE, AND ALKYLENE GLYCOLS, HYDROXYALKYL ACRYLATE ESTER RR-01254-4

ENVIRONMENTAL LISTS

List of Pesticide Product Inert Ingredients
[present]

POLYMER OF ALKANEPOLYOL AND POLYALKYLPOLYISOCYANATOCARBOMONOCYCLE, ACETONE OXIME-BLOCKED RR-00188-7

ENVIRONMENTAL LISTS

List of Pesticide Product Inert Ingredients
[present]

POLYMER OF ALKENOIC ACID, SUBSTITUTED ALKYLACRYLATE, SODIUMSALT RR-00189-8

ENVIRONMENTAL LISTS

List of Pesticide Product Inert Ingredients
[present]

POLYMER OF ALKYL CARBOMONOCYCLE   RR-00227-7
DIISOCYANATE WITH ALKANEPOLYOL
POLYACRYLATES
  HEALTH AND SAFETY LISTS
    U.S. DOT - Substances From 49 CFR 172.101
      regulated by DOT (UN3151, UN3152)
    U.S. DOT - Hazard Classes
      DOT hazard class = 9
    U.S. DOT - Appendix B - Marine Pollutants
      DOT regulated severe marine pollutant

POLYMER OF BISPHENOL A DIGLYCIDYL ETHER,   RR-00970-1
SUBSTITUTED ALKENES, AND BUTADIENE
  ENVIRONMENTAL LISTS
    TSCA - Chemicals with Significant New Use Rules
      PMN number: P-86-164

POLYMER OF N-BUTYL ACRYLATE, METHYL   RR-01116-5
METHACRYLATE, METHACRYLIC ACID AND
AMINOPROPYL METHACRYLATE
  ENVIRONMENTAL LISTS
    List of Pesticide Product Inert Ingredients
      [present]

POLYMER OF DISODIUM MALEATE, ALLYL   RR-01257-7
ETHER, AND ETHYLENE OXIDE
  ENVIRONMENTAL LISTS
    TSCA - Chemicals with Significant New Use Rules
      PMN number: P-89-726

POLYMER OF DISUBSTITUTED PHTHALATE, DIOX-   RR-01179-0
OHETEROPOLYCYCLE, AND METHACRYLIC ACID
  ENVIRONMENTAL LISTS
    TSCA - Chemicals with Significant New Use Rules
      PMN number: P-91-505

POLYMER OF HYDROXYETHYL ACRYLATE AND   RR-00999-4
POLYISOCYANATE
  ENVIRONMENTAL LISTS
    TSCA - Chemicals with Significant New Use Rules
      PMN number: P-88-1658

POLYMER OF ISOPHORONE DIISO-   RR-00962-1
CYANATE, TRIMETHYLOLPROPANE,
POLYALKYLENEPOLYOL, DISUBSTITUTED ALKA-
NES AND HYDROXYETHYL ACRYLATE
  ENVIRONMENTAL LISTS
    TSCA - Chemicals with Significant New Use Rules
      PMN number: P-88-854

POLYMER OF POLYETHYLENEPOLYAMINE AND   RR-00949-4
ALKANEDIOL DIGLYCIDYL ETHER
  ENVIRONMENTAL LISTS
    TSCA - Chemicals with Significant New Use Rules
      PMN number: P-89-73

POLYMER OF STYRENE, SUBSTITUTED ALKYL   RR-01014-0
METHACRYLATES, 2-ETHYLHEXYL ACRY-
LATE, METHACRYLIC ACID AND SUBSTITUTED
BISBENZENE
  ENVIRONMENTAL LISTS
    TSCA - Chemicals with Significant New Use Rules
      PMN numbers: P-90-244; P-90-245

POLYMER OF SUBSTITUTED ALKYLPHENOL   RR-00911-0
FORMALDEHYDE AND PHTHALICANHYDRIDE,
ACRYLATE
  ENVIRONMENTAL LISTS
    List of Pesticide Product Inert Ingredients
      [present]

POLYMER OF SUBSTITUTED ARYL OLEFIN   RR-01175-6
  ENVIRONMENTAL LISTS
    TSCA - Chemicals with Significant New Use Rules
      PMN number: P-91-1086

POLYMER OF SUBSTITUTED PHENOL,   RR-00987-0
FORMALDEHYDE, EPICHLOROHYDRIN, AND DIS-
UBSTITUTED BENZENE
  ENVIRONMENTAL LISTS
    TSCA - Chemicals with Significant New Use Rules
      PMN number: P-91-937

POLYMER OF VINYL ACETATE, N-BUTYL ACRY-   30938-41-1
LATE, VINYL CHLORIDE,AND ACRYLIC ACID
  ENVIRONMENTAL LISTS
    TSCA - Chemicals with Significant New Use Rules
      PMN number: P-84-938

POLYMETHYLCARBOMONOCYCLE, REACTION   RR-01240-8
PRODUCT WITH 2-HYDROXYETHYL ACRYLATE
  ENVIRONMENTAL LISTS
    TSCA - Chemicals with Significant New Use Rules
      PMN number: P-91-11

POLYMETHYLENE POLYPHENYLENE   9016-87-9
ISOCYANATE
  ENVIRONMENTAL LISTS
    TSCA - Chemicals with Significant New Use Rules
      PMN number: P-89-810

POLYMETHYLOCTADECYLSILOXANE   RR-01190-5
  ENVIRONMENTAL LISTS
    TSCA - Chemicals with Significant New Use Rules
      PMN number: P-87-730

POLYMETHYLOCTADECYLSILOXANE   RR-01191-6
  ENVIRONMENTAL LISTS
    TSCA - Chemicals with Significant New Use Rules
      PMN number: P-88-1616

POLYNUCLEAR AROMATIC HYDROCARBIDES   RR-01810-0
  ENVIRONMENTAL LISTS
    TSCA - Chemicals with Significant New Use Rules
      PMN number: P-85-612

POLYNUCLEAR AROMATIC HYDROCARBONS   RR-00450-2
  ENVIRONMENTAL LISTS
    TSCA - Chemicals with Significant New Use Rules
      PMN number: P-89-1104

POLYOL CARBOXYLATE ESTER   RR-00245-9
  ENVIRONMENTAL LISTS
    List of Pesticide Product Inert Ingredients
      [present]

POLYOLEFINES FIBRES   RR-01806-4
  ENVIRONMENTAL LISTS

TSCA - Chemicals with Significant New Use Rules
PMN number: P-90-1338

POLY-OXO ALUMINUM STEARATE RR-01115-4

HEALTH AND SAFETY LISTS

IARC - Group 3 (not classifiable)
[present]

ENVIRONMENTAL LISTS

CERCLA/SARA - Section 313 - Emission Reporting
form R reporting required; (Listed under 'Diisocyanates')

List of Pesticide Product Inert Ingredients
[present]

TSCA - Code of Federal Regulations Citations
40 CFR 716.120(a)

TSCA - Health and Safety Reporting List
Effective Date: June 1, 1987

INTERNATIONAL LISTS

Canada - Alberta - 8 Hour Occupational Exposure Limit
0.005 ppm TWA; 0.07 mg/m3 TWA

Canada - Alberta - Ceiling Occupational Exposure Limit
C 0.02 ppm; C 0.28 mg/m3

German (DFG) - Carcinogens
suspected carcinogen

POLYOXYALKYLENE GLYCOL RR-01118-7

ENVIRONMENTAL LISTS

TSCA - PAIR - Reporting List
Effective Date: October 12, 1993; Reporting Date: February 28, 1994

TSCA - Health and Safety Reporting List
Effective Date: October 12, 1993; Sunset Date: October 12, 2003

POLYOXY ALKYLENE GLYCOL AMINE RR-01180-3

ENVIRONMENTAL LISTS

TSCA - Health and Safety Reporting List
Effective Date: June 14, 1993

POLYOXYALKYLENE SILOXANE RR-01119-8

INTERNATIONAL LISTS

Mexico - Drinking Water - Ecological Criteria
0.00003 mg/l Substance presents persistence, bioaccumulations or risk of cancer, reduce human exposure to a minimum; This level has been extrapolated by using a mathematic model

POLYOXYALKYLENE SUBSTITUTED AROMATIC RR-01677-3
AZO COLORANT

ENVIRONMENTAL LISTS

Clean Water Act - Toxic Pollutants
[present]

STATE LISTS

California - Directors List of Hazardous Substances (8 CCR 339)
[present] (including benzopyrenes, benzofluoranthrene, chrysenes, dibenzanthracenes and indenopyrenes)

Pennsylvania Right to Know List
environmental hazard

POLY(OXY-1,4-BUTANEDIYL), ALPHA-(1-OXO-2- 52277-33-5
PROPENYL)-X-[(1-OXO-2-PROPENYL)OXY]-

ENVIRONMENTAL LISTS

TSCA - Chemicals with Significant New Use Rules
PMN number: P-84-27

POLY[OXY(DIMETHYLSILYLENE)] 9016-00-6

INTERNATIONAL LISTS

Canada - Quebec - Time-Weighted Average Exposure Values
total dust: 10 mg/m3 TWAEV

POLY(OXY-1,2-ETHANEDIYL), ALPHA-(3-(3-(2H- 104810-48-2
BENZOTRIAZOL-2-YL)-5-(1,1-DIMETHYLETHYL)
-4-HYDROXYPHENYL)-1-OXOPROPYL)-OMEGA-
HYDROXY-

ENVIRONMENTAL LISTS

List of Pesticide Product Inert Ingredients
[present]

POLY(OXY-1,2-ETHANEDIYL), ALPHA-[2-[BIS 68389-88-8
(2-AMINOETHYL)METHYLAMMONIO]ETHYL]-
OMEGA-HYDROXY-, N,N'-DICOCO ACYL DERIVA-
TIVES, METHYL SULFATES (SALTS)

ENVIRONMENTAL LISTS

List of Pesticide Product Inert Ingredients
[present]

POLY(OXY-1,2-ETHANEDIYL), ALPHA-[2-[BIS(2- 68389-89-9
AMINOETHYL)METHYLAMMONIO]ETHYL-OMEGA]
-HYDROXY-, N,N'-BIS(HYDROGENATED TALLOW
ACYL) DERIVATIVES, METHYL SULFATES (SALTS)

ENVIRONMENTAL LISTS

TSCA - Chemicals with Significant New Use Rules
PMN number: P-91-1372

POLY(OXY-1,2-ETHANEDIYL), ALPHA-[2-[BIS 68410-69-5
(2-AMINOETHYL)METHYLAMMONIO]ETHYL]
-OMEGA-HYDROXY-, N,N'-DITALLOW ACYL
DERIVATIVES, METHYL SULFATES (SALTS)

ENVIRONMENTAL LISTS

List of Pesticide Product Inert Ingredients
[present]

POLY(OXY-1,2-ETHANEDIYL), ALPHA-[3-[BIS(2- 68554-06-3
AMINOETHYL)METHYLAMMONIO]-2-HYDROX-
YPROPYL]-OMEGA-HYDROXY-, N-COCO ACYL
DERIVATIVES, METHYL SULFATES (SALTS)

ENVIRONMENTAL LISTS

TSCA - Chemicals with Significant New Use Rules
PMN number: P-92-1131

POLY(OXY-1,2-ETHANEDIYL), ALPHA-[2-[BIS 70914-09-9
(2-AMINOETHYL)METHYLAMMONIO]ETHYL]
-OMEGA-HYDROXY-, N,N'-DI-C(14-18) ACYL
DERIVAT IVES, METHYL SULFATES (SALTS)

ENVIRONMENTAL LISTS

TSCA - Code of Federal Regulations Citations
40 CFR 721.1770

TSCA - Chemicals with Significant New Use Rules
PMN number: P-84-274

TSCA - Section 12(b) - Export Notification
P-84-274; export notification required - Section 5

POLY(OXY-1,2-ETHANEDIYL), ALPHA-(CAR- 70632-06-3
BOXYMETHYL)-OMEGA-HYDROXY-, C12-15-ALKYL
ETHERS, SODIUM SALTS

ENVIRONMENTAL LISTS

List of Pesticide Product Inert Ingredients
[present]

TSCA - PAIR - Reporting List
Effective Date: October 12, 1993; Reporting Date: February 28, 1994

TSCA - Health and Safety Reporting List
Effective Date: June 14, 1993

POLY[OXY-1,2-ETHANEDIYL(DIMETHYLIM-INO)-1,3-PROPANEDIYLIMINOCARBONYLIM-INO-1,3-PROPANEDIYL(DIMETHYLIMINO)-1,2-ETHANEDIYL DICHLORIDE], ALPHA-(2-CHLOROETHYL)-OMEGA-(2-CHLOROETHOXY)-   130547-87-4

ENVIRONMENTAL LISTS

    List of Pesticide Product Inert Ingredients
    [present]

POLY(OXY-1,2-ETHANEDIYL), .ALPHA.-HYDRO-.W.-HYDROXY-, ETHERWITH 2-ETHYL-2-(HY-DROXYMETHYL)-1,3-PROPANEDIOL (3:1) DI-2-PROPENOATE, METHYL ETHER   106158-22-9

ENVIRONMENTAL LISTS

    TSCA - Code of Federal Regulations Citations
    40 CFR 712.30(w); 40 CFR 716.120(a)

    TSCA - PAIR - Reporting List
    Reporting Date: August 18, 1988

    TSCA - Health and Safety Reporting List
    Effective Date: June 20, 1988; Sunset Date: November 9, 1993

POLY(OXY-1,2-ETHANEDIYL), ALPHA-HYDRO-3-(OXIRANYLMETHYL)-1,3-PROPANEDIOL (3:1)   52495-71-3

ENVIRONMENTAL LISTS

    TSCA - Code of Federal Regulations Citations
    40 CFR 712.30(w); 40 CFR 716.120(a)

    TSCA - PAIR - Reporting List
    Reporting Date: August 18, 1988

    TSCA - Health and Safety Reporting List
    Effective Date: June 20, 1988; Sunset Date: November 9, 1993

POLY(OXY-1,2-ETHANEDIYL), ALPHA-(2-METHYL-1-OXO-2-PROPENYL)-X-HYDROXY-, C10-16-ALKYL ETHERS   RR-00235-7

ENVIRONMENTAL LISTS

    List of Pesticide Product Inert Ingredients
    [present]

    TSCA - Code of Federal Regulations Citations
    40 CFR 712.30(w); 40 CFR 716.120(a)

    TSCA - PAIR - Reporting List
    Reporting Date: August 18, 1988

    TSCA - Health and Safety Reporting List
    Effective Date: June 20, 1988

POLY(OXY-1,2-ETHANEDIYL), ALPHA-[4-OXI-RANYLMETHOXY)BENZOYL]-OMEGA-[[(4-OXI-RANYLMETHOXY)BENZOYL]OXY]-   69943-75-5

ENVIRONMENTAL LISTS

    TSCA - Code of Federal Regulations Citations
    40 CFR 712.30(w); 40 CFR 716.120(a)

    TSCA - PAIR - Reporting List
    Reporting Date: August 18, 1988

    TSCA - Health and Safety Reporting List
    Effective Date: June 20, 1988; Sunset Date: November 9, 1993

POLY(OXY-1,2-ETHANEDIYL), ALPHA-(1-OXO-2-PROPENYL)-OMEGA-HYDROXY-, C(10-16)-ALKYL ETHERS   125304-11-2

ENVIRONMENTAL LISTS

    TSCA - Code of Federal Regulations Citations
    40 CFR 712.30(w); 40 CFR 716.120(a)

    TSCA - PAIR - Reporting List
    Reporting Date: August 18, 1988

TSCA - Health and Safety Reporting List
Effective Date: June 20, 1988

POLYOXYETHYLENE AMYLPHENOL - FORMALDE-HYDE RESIN   RR-01120-1

ENVIRONMENTAL LISTS

    List of Pesticide Product Inert Ingredients
    [present]

POLYOXYETHYLENE P-TERT-BUTYLPHENOL - FORMALDEHYDE RESIN   30704-63-3

ENVIRONMENTAL LISTS

    List of Pesticide Product Inert Ingredients
    [present]

POLYOXYETHYLENE C4-18-ALKYL ESTER OF PHOSPHORIC ACID, MONOETHANOLAMINE SALTS   70247-86-8

ENVIRONMENTAL LISTS

    TSCA - Code of Federal Regulations Citations
    40 CFR 721.1702

    TSCA - Chemicals with Significant New Use Rules
    PMN number: P-88-1211

    TSCA - Section 12(b) - Export Notification
    P-88-1211; export notification required - Section 5

POLYOXYETHYLENE C8-18 AND C18 UNSATU-RATED, ALKYLAMINES   68439-72-5

ENVIRONMENTAL LISTS

    TSCA - Code of Federal Regulations Citations
    40 CFR 721.1708

    TSCA - Chemicals with Significant New Use Rules
    PMN number: P-88-2188

    TSCA - Section 12(b) - Export Notification
    P-88-2188; export notification required - Section 5

POLYOXYETHYLENE C12-13-ALKYL ESTER OF PHOSPHORIC ACID, MONOETHANOLAMINE SALTS   71549-82-1

ENVIRONMENTAL LISTS

    TSCA - Chemicals with Significant New Use Rules
    PMN number: P-86-588

POLYOXYETHYLENE CETYL AND OLEYL ALCOHOLS   68155-01-1
SEE ALSO:
    GLYCIDOL (OXIRANEMETHANOL) AND ITS DERIVATIVES

ENVIRONMENTAL LISTS

    TSCA - Code of Federal Regulations Citations
    40 CFR 716.120(c)

POLY[OXYETHYLENE (DIMETHYLIMINO) ETHY-LENE (DIMETHYLIMINO) ETHYLENE DICHLO-RIDE]   31512-74-0

ENVIRONMENTAL LISTS

    TSCA - Code of Federal Regulations Citations
    40 CFR 721.1780

    TSCA - Chemicals with Significant New Use Rules
    PMN number: P-86-554

POLYOXYETHYLENE DINONYLPHENOL   9014-93-1

ENVIRONMENTAL LISTS

    List of Pesticide Product Inert Ingredients
    [present]

POLYOXYETHYLENE DISTEARATE   9005-08-7

ENVIRONMENTAL LISTS

List of Pesticide Product Inert Ingredients
[present]

POLYOXYETHYLENE DOCOSYL ETHER   26636-40-8
ENVIRONMENTAL LISTS
List of Pesticide Product Inert Ingredients
[present]

POLY(OXYETHYLENE) 10-DODECYLMERCAPTAN   13081-34-0
ENVIRONMENTAL LISTS
List of Pesticide Product Inert Ingredients
[present]

POLYOXYETHYLENE ESTER OF ROSIN   8050-33-7
ENVIRONMENTAL LISTS
List of Pesticide Product Inert Ingredients
[present]

POLYOXYETHYLENE ESTERS OF MONO & DICAR-   RR-01121-2
BOXYLIC ACID & OIL SOLUBLE SULFONATES
ENVIRONMENTAL LISTS
List of Pesticide Product Inert Ingredients
[present]

POLYOXYETHYLENE ISOPROPYLIDENEDIPHENOL   32492-61-8
HEALTH AND SAFETY LISTS
NIOSH - Selected LD50s and LC50s
Oral, rat: LD50 - 1850 mg/kg

POLYOXYETHYLENE LANOLIN ALCOHOL   68648-38-4
ENVIRONMENTAL LISTS
List of Pesticide Product Inert Ingredients
[present]

POLYOXYETHYLENE LINEAR PRIMARY C8-16   71243-46-4
ALCOHOLS
ENVIRONMENTAL LISTS
List of Pesticide Product Inert Ingredients
[present]

POLY(OXYETHYLENE)METHYLENEBIS(DI-   60874-89-7
AMYLPHENOL)
ENVIRONMENTAL LISTS
List of Pesticide Product Inert Ingredients
[present]

POLY(OXYETHYLENE)METHYLENEBIS   41928-09-0
(OCTYLPHENOL)
ENVIRONMENTAL LISTS
List of Pesticide Product Inert Ingredients
[present]

POLYOXYETHYLENE MONOHEXADECYL ETHER   9004-95-9
ENVIRONMENTAL LISTS
List of Pesticide Product Inert Ingredients
[present]

POLYOXYETHYLENE MONOOCTADECYL ETHER   9005-00-9
ENVIRONMENTAL LISTS
List of Pesticide Product Inert Ingredients
[present]

POLYOXYETHYLENE MONOTETRADECYL ETHER   27306-79-2
ENVIRONMENTAL LISTS

List of Pesticide Product Inert Ingredients
[present]

POLYOXYETHYLENE POLYOXYPROPYLENE DI-   69029-39-6
SEC-BUTYLPHENOL
ENVIRONMENTAL LISTS
List of Pesticide Product Inert Ingredients
[present]

POLYOXYETHYLENE POLYOXYPROPYLENE TERT-   RR-01123-4
C12-13-ALKYL AMINE
ENVIRONMENTAL LISTS
List of Pesticide Product Inert Ingredients
[present]

POLYOXYETHYLENE POLYOXYPROPYLENE   RR-01122-3
MONOISOPROPANOLAMIDE OF MIXED CAPRYLIC
AND CAPRIC ACIS
ENVIRONMENTAL LISTS
List of Pesticide Product Inert Ingredients
[present]

POLYOXYETHYLENE POLYOXYPROPYLENE   37251-69-7
NONYLPHENOL
ENVIRONMENTAL LISTS
List of Pesticide Product Inert Ingredients
[present]

POLYOXYETHYLENE POLYOXYPROPYLENE   37280-82-3
PHOSPHATE
ENVIRONMENTAL LISTS
List of Pesticide Product Inert Ingredients
[present]

POLYOXYETHYLENE 20 SORBITAN MONOOLEATE   9005-65-6
ENVIRONMENTAL LISTS
List of Pesticide Product Inert Ingredients
[present]

POLYOXYETHYLENE SORBITAN HEPTAOLEATE   54846-79-6
ENVIRONMENTAL LISTS
List of Pesticide Product Inert Ingredients
[present]

POLYOXYETHYLENE SORBITAN   9005-66-7
MONOPALMITATE
ENVIRONMENTAL LISTS
List of Pesticide Product Inert Ingredients
[present]

POLYOXYETHYLENE SORBITAN MONOSTEARATE   9005-67-8
ENVIRONMENTAL LISTS
List of Pesticide Product Inert Ingredients
[present]

POLYOXYETHYLENE SORBITAN MONOTALLATE   61790-86-1
ENVIRONMENTAL LISTS
List of Pesticide Product Inert Ingredients
[present]

POLYOXYETHYLENE SORBITAN TETRATALLATE   RR-01124-5
ENVIRONMENTAL LISTS
List of Pesticide Product Inert Ingredients
[present]

POLYOXYETHYLENE SORBITAN TRIOLEATE   9005-70-3

ENVIRONMENTAL LISTS

    List of Pesticide Product Inert Ingredients
      [present]

POLYOXYETHYLENE SORBITAN TRISTEARATE   9005-71-4

HEALTH AND SAFETY LISTS

    NIOSH - Selected LD50s and LC50s
      Oral, mouse: LD50 = 25 gm/kg

    NTP Chemical Status Reports - Testing Status and NTIS Number
      Technical reports printed (PB92-189331/AS)

    NTP Chemical Status Reports - Evidence of Carcinogenicity
      male rat: equivocal evidence; female rat: no evidence; male mice: no evidence; female mice: no evidence

ENVIRONMENTAL LISTS

    List of Pesticide Product Inert Ingredients
      [present]

POLYOXYETHYLENE SORBITOL   RR-01125-6

ENVIRONMENTAL LISTS

    List of Pesticide Product Inert Ingredients
      [present]

POLYOXYETHYLENE SORBITOL HEXAOLEATE   57171-56-9

ENVIRONMENTAL LISTS

    List of Pesticide Product Inert Ingredients
      [present]

POLYOXYETHYLENE SORBITOL TETRAOLEATE   63089-86-1

ENVIRONMENTAL LISTS

    List of Pesticide Product Inert Ingredients
      [present]

POLYOXYETHYLENE SOYA ACID ESTERS   61791-23-9

ENVIRONMENTAL LISTS

    List of Pesticide Product Inert Ingredients
      [present]

POLYOXYETHYLENE 2,4,7,9-TETRAMETHYL-5-DECYNE-4,7-DIOL   9014-85-1

ENVIRONMENTAL LISTS

    List of Pesticide Product Inert Ingredients
      [present]

POLYOXYETHYLENE TRIMETHYLDECYL ALCOHOL   69011-36-5

ENVIRONMENTAL LISTS

    List of Pesticide Product Inert Ingredients
      [present]

POLY(OXYMETHYLENE)   9002-81-7

ENVIRONMENTAL LISTS

    List of Pesticide Product Inert Ingredients
      [present]

POLY(OXYMETHYLENE), .ALPHA.-HYDRO-.OMEGA.-HYDROXY-   9015-98-9

ENVIRONMENTAL LISTS

    List of Pesticide Product Inert Ingredients
      [present]

POLY[OXY(METHYL-1,2-ETHANEDIYL)], ALPHA, ALPHA'-(2,2-DIMETHYL-1,3-PROPANEDIYL(BIS[3-(OXIRANYLMETHOXY)-   RR-00983-6

ENVIRONMENTAL LISTS

    List of Pesticide Product Inert Ingredients
      [present]

POLY(OXY(METHYL-1,2-ETHANEDIYL)), ALPHA-(2-(DIETHYLMETHYLAMMONIO)ETHYL-OMEGA-HYDROXY-, CHLORIDE   RR-01114-3

ENVIRONMENTAL LISTS

    List of Pesticide Product Inert Ingredients
      [present]

POLY[OXY(METHYL-1,2-ETHANEDIYL)], ALPHA, ALPHA'-[(1-METHYLETHYLIDENE)DI-4,1-PHENYLENE]BIS[3-(OXIRANYLMETHOXY)-   54140-64-6

ENVIRONMENTAL LISTS

    List of Pesticide Product Inert Ingredients
      [present]

POLY[OXY[METHYL(3,3,3-TRIFLUOROPROPYL) SILYLENE], ALPHA-(TRIMETHYLSILYL)-OMEGA-[(TRIMETHYLSILYL)OXY]-   42557-13-1

ENVIRONMENTAL LISTS

    List of Pesticide Product Inert Ingredients
      [present]

POLYOXY(1-OXO-1,6-HEXANEDIYL), ALPHA-HYDRO-OMEGA-HYDROXY-ESTER WITH 3-HYDROXY-2,2-DIMETHYLPROPYL 3-HYDROXY-2,2-DIMETHYLPROPANOATE (2:1), DI-2-[PROPANOATE   96915-49-0

ENVIRONMENTAL LISTS

    List of Pesticide Product Inert Ingredients
      [present]

POLYOXY(1-OXO-1,6-HEXANEDIYL), ALPHA-HYDRO-OMEGA-HYDROXY-ESTER WITH 2,2'-(OXYBIS(METHYLENE)BIS(2-HYDROXYMETHYL)-1,3-PROPANEDIOL)-2-PROPENOATE, 2-PROP   89800-10-2

ENVIRONMENTAL LISTS

    List of Pesticide Product Inert Ingredients
      [present]

STATE LISTS

    Pennsylvania Right to Know List
      [present]

POLYOXY[OXY(1-OXO-1,6 HEXANEDIYL)], ALPHA-(1-OXO-2-PROPENYL)-OMEGA-((TETRAHYDRO-2-FURANYL)METHOXY)-   96915-50-3

STATE LISTS

    Pennsylvania Right to Know List
      [present]

POLYOXYPROPYLENE DITALL-OIL ESTER   68648-12-4

ENVIRONMENTAL LISTS

    TSCA - Chemicals with Significant New Use Rules
      PMN number: P-88-2180

POLYOXYPROPYLENE OLEATE BUTYL ETHER   37281-78-0

ENVIRONMENTAL LISTS

    List of Pesticide Product Inert Ingredients
      [present]

POLY(OXYPROPYLENE) TRIOL   25791-96-2

ENVIRONMENTAL LISTS

    TSCA - Code of Federal Regulations Citations
      40 CFR 721.1706

    TSCA - Chemicals with Significant New Use Rules
      PMN number: P-88-2181

TSCA - Section 12(b) - Export Notification
P-88-2181; export notification required - Section 5

POLYPHOSPHORIC ACIDS, ESTERS WITH   68458-49-1
POLYETHYLENEGLYCOL-NONYLPHENYL ETHER
ENVIRONMENTAL LISTS
List of Pesticide Product Inert Ingredients
[present]

POLYPIPERIDINOL-ACRYLATE METHACRYLATE   RR-01718-5
ENVIRONMENTAL LISTS
TSCA - Section 12(b) - Export Notification
P-84-341; export notification required - Section 5

POLYPROPYLENE   9003-07-0
ENVIRONMENTAL LISTS
TSCA - Section 12(b) - Export Notification
P-84-343; export notification required - Section 5

POLYPROPYLENE GLYCOL   25322-69-4
ENVIRONMENTAL LISTS
TSCA - Section 12(b) - Export Notification
P-88-342; export notification required - Section 5

POLYPROPYLENE GLYCOL METHYL ETHER   37286-64-9
ENVIRONMENTAL LISTS
List of Pesticide Product Inert Ingredients
[present]

POLYPROPYLENE GLYCOL ALPHA-METHYL GLU-   52673-60-6
COSIDE ETHER (4:1)
ENVIRONMENTAL LISTS
List of Pesticide Product Inert Ingredients
[present]

POLYPROPYLENE GLYCOL BETA-METHYL GLU-   61849-72-7
COSIDE ETHER (4:1)
HEALTH AND SAFETY LISTS
NIOSH - Selected LD50s and LC50s
Oral, rat: LD50 = 2830 mg/kg
ENVIRONMENTAL LISTS
CAA - HON Rule - SOCMI Chemicals
compliance by Jan. 23, 1995
List of Pesticide Product Inert Ingredients
[present]

POLYPROPYLENE GLYCOL, MONOBUTYL ETHER   9003-13-8
ENVIRONMENTAL LISTS
List of Pesticide Product Inert Ingredients
[present]

POLYPROPYLENE STEARYL ETHER   25231-21-4
ENVIRONMENTAL LISTS
TSCA - Chemicals with Significant New Use Rules
PMN number: P-88-1304
INTERNATIONAL LISTS
Canada - British Columbia - 8 Hour Exposure Limits
2.5 mg/m3 TWA
Canada - British Columbia - 15 Minute Exposure Limits
5 mg/m3 STEL

POLYSORBATE 20   9005-64-5
HEALTH AND SAFETY LISTS
IARC - Group 3 (not classifiable)
[present]
ENVIRONMENTAL LISTS
List of Pesticide Product Inert Ingredients
[present]

POLYSTYRENE   9003-53-6
HEALTH AND SAFETY LISTS
AIHA - WEEL - Time Weighted Averages
10 mg/m3 TWA
NFPA - Flash Points
flash point = 365 degrees F (185 degrees C)
NFPA - Hazard Identification Ratings
health-0; flammability-1; reactivity-0
NIOSH - Selected LD50s and LC50s
Oral, rat: LD50 = 4190 mg/kg Skin, rabbit: LD50 = 20 gm/kg
ENVIRONMENTAL LISTS
CAA - HON Rule - SOCMI Chemicals
compliance by Oct. 23, 1995
List of Pesticide Product Inert Ingredients
[present]
STATE LISTS
Minnesota Hazardous Substance List
[present]

POLYSUBSTITUTED PHENYLAZOPOLYSUBSTI-   RR-01657-9
TUTED PHENYL DYE
ENVIRONMENTAL LISTS
List of Pesticide Product Inert Ingredients
[present]

POLYSUBSTITUTED POLYOL   RR-00242-6
ENVIRONMENTAL LISTS
List of Pesticide Product Inert Ingredients
[present]

POLY(SUBSTITUTED TRIAZINYL) PIPERAZINE   RR-00933-6
ENVIRONMENTAL LISTS
List of Pesticide Product Inert Ingredients
[present]

POLYTERPENE RESINS, SYNTHETIC   70750-53-7
HEALTH AND SAFETY LISTS
NIOSH - Selected LD50s and LC50s
Oral, rat: LD50 = 9100 mg/kg Skin, rabbit: LD50 = 21 gm/kg
ENVIRONMENTAL LISTS
List of Pesticide Product Inert Ingredients
[present]

POLYTETRAFLUOROETHYLENE   9002-84-0
ENVIRONMENTAL LISTS
List of Pesticide Product Inert Ingredients
[present]

POLYTETRAFLUORETHYLENE DECOMPOSITION   RR-00073-7
PRODUCTS
HEALTH AND SAFETY LISTS
NIOSH - Selected LD50s and LC50s
Oral, rat: LD50 = 37 gm/kg

ENVIRONMENTAL LISTS
  List of Pesticide Product Inert Ingredients
  [present]

POLYTRIMETHYLHYDROSILYSILICONE   68988-56-7

HEALTH AND SAFETY LISTS
  IARC - Group 3 (not classifiable)
  [present]

ENVIRONMENTAL LISTS
  List of Pesticide Product Inert Ingredients
  [present]

STATE LISTS
  NJ Right to Know List (Total)
  sn 2700

POLYURETHANE   9017-09-8

ENVIRONMENTAL LISTS
  TSCA - Chemicals with Significant New Use Rules
  PMN number: P-93-658

POLYURETHANE   68400-67-9

ENVIRONMENTAL LISTS
  TSCA - Chemicals with Significant New Use Rules
  PMN number: P-84-814

POLYVINYL ACETATE   RR-01538-3

ENVIRONMENTAL LISTS
  TSCA - Chemicals with Significant New Use Rules
  PMN number: P-88-436

POLYVINYL ALCOHOL   9002-89-5

ENVIRONMENTAL LISTS
  List of Pesticide Product Inert Ingredients
  [present]

POLYVINYL BUTYRAL   63148-65-2

HEALTH AND SAFETY LISTS
  IARC - Group 3 (not classifiable)
  [present]

INTERNATIONAL LISTS
  Canada - British Columbia - 8 Hour Exposure Limits
  as fluorine: 2.5 mg/m3 TWA
  Canada - Quebec - Time-Weighted Average Exposure Values
  Determine quantitatively the decomposition products in the air and express the results as fluorides (see "fluorides" standards).

STATE LISTS
  California - Exposure Limits - PELs
  keep decomposition products below the analytical detection level
  Pennsylvania Right to Know List
  [present]

POLYVINYL BUTYRATE   24991-31-9

HEALTH AND SAFETY LISTS
  NIOSH - Health Standards - Exposure Limits
  Use good work practices, engineering controls, and medical management
  NIOSH - Health Standards - Health Effects and Precautions
  Lung effects; polymer fume fever (Monitor workroom air for inorganic fluorides and Hydrogen fluoride)

INTERNATIONAL LISTS
  Israel - Time Weighted Averages
  air concentration should be kept to a minimum

STATE LISTS
  California - Directors List of Hazardous Substances (8 CCR 339)
  [present]
  Minnesota Hazardous Substance List
  [present]

POLYVINYL CHLORIDE   9002-86-2

ENVIRONMENTAL LISTS
  List of Pesticide Product Inert Ingredients
  [present]

POLY(VINYL ETHYL ETHER)   25104-37-4

ENVIRONMENTAL LISTS
  List of Pesticide Product Inert Ingredients
  [present]

POLYVINYL PYRROLIDONE   9003-39-8

HEALTH AND SAFETY LISTS
  IARC - Group 3 (not classifiable)
  [present] (in foam form)

ENVIRONMENTAL LISTS
  TSCA - Chemicals with Significant New Use Rules
  PMN number: P-85-118

POLYVINYLPYRROLIDONE - VINYL ACETATE   25086-89-9
COPOLYMER

HEALTH AND SAFETY LISTS
  IARC - Group 3 (not classifiable)
  [present]

PONCEAU 3R   3564-09-8

HEALTH AND SAFETY LISTS
  IARC - Group 3 (not classifiable)
  [present]
  NFPA - Flash Points
  mixture of polymers: flash point = 175 degrees F (79 degrees C)
  NFPA - Hazard Identification Ratings
  mixture of polymers: health-0; flammability-2; reactivity-0
  NTP Chemical Status Reports - Testing Status and NTIS Number
  Two year studies: pathology quality assessment in progress

ENVIRONMENTAL LISTS
  List of Pesticide Product Inert Ingredients
  [present]

PONCEAU MX   3761-53-3

ENVIRONMENTAL LISTS
  List of Pesticide Product Inert Ingredients
  [present]

POP NONYLPHENOL - FORMALDEHYDE RESIN   37523-33-4

ENVIRONMENTAL LISTS
  List of Pesticide Product Inert Ingredients
  [present]

POPPY SEED OIL   RR-00880-0

HEALTH AND SAFETY LISTS
  IARC - Group 3 (not classifiable)
  [present]

ENVIRONMENTAL LISTS
  List of Pesticide Product Inert Ingredients
  [present]

INTERNATIONAL LISTS

United Kingdom - Occupational Exposure Standards - TWAs
total inhalable dust: 10 mg/m3 TWA; respirable dust: 5 mg/m3 TWA

German (DFG) - MAK Values
fine dust: 5 mg/m3 MAK

PORTLAND CEMENT   65997-15-1

ENVIRONMENTAL LISTS

List of Pesticide Product Inert Ingredients
[present]

POTASSIUM   7440-09-7

HEALTH AND SAFETY LISTS

IARC - Group 3 (not classifiable)
[present]

ENVIRONMENTAL LISTS

List of Pesticide Product Inert Ingredients
[present]

STATE LISTS

California - Directors List of Hazardous Substances (8 CCR 339)
[present]

POTASSIUM 40   13966-00-2

ENVIRONMENTAL LISTS

List of Pesticide Product Inert Ingredients
[present]

POTASSIUM 42   14378-21-3

HEALTH AND SAFETY LISTS

IARC - Group 2B (sufficient animal data)
[present] (Overall evaluation based only on evidence of carcinogenicity in monograph (8, 1975) or in Supplement 4)

OSHA - Possible Select Carcinogens
[present]

STATE LISTS

California - Air Bill 2588 Appendix A-II
known or potential carcinogen

California - Prop. 65 - Cancer list
carcinogen - initial date 4/1/88

California - Prop. 65 - No Significant Risk Levels
no significant risk level = 40 ug/day

California - Directors List of Hazardous Substances (8 CCR 339)
[present]

Florida Hazardous Substance List
[present]

Massachusetts Right To Know List
carcinogen; extraordinarily hazardous

Minnesota Hazardous Substance List
carcinogen

Pennsylvania Right to Know List
special hazardous substance

Pennsylvania RTK - Special Hazardous Substances
[present]

POTASSIUM 43   14903-02-7

HEALTH AND SAFETY LISTS

IARC - Group 2B (sufficient animal data)
[present] (Overall evaluation based only on evidence of carcinogenicity in monograph (8, 1975) or in Supplement 4)

OSHA - Possible Select Carcinogens
[present]

ENVIRONMENTAL LISTS

CERCLA/SARA - Section 313 - Emission Reporting
form R reporting required for 0.1% de minimus concentration

INTERNATIONAL LISTS

Canada - WHMIS: Ingredient Disclosure
1% item 867 (1008)

STATE LISTS

California - Air Bill 2588 Appendix A-II
known or potential carcinogen

California - Prop. 65 - Cancer list
carcinogen - initial date 4/1/88

California - Prop. 65 - No Significant Risk Levels
no significant risk level = 200 ug/day

California - Directors List of Hazardous Substances (8 CCR 339)
[present]

Florida Hazardous Substance List
[present]

Massachusetts Right To Know List
carcinogen; extraordinarily hazardous

Minnesota Hazardous Substance List
carcinogen

NJ Right to Know List (Total)
sn 0504

Pennsylvania Right to Know List
environmental hazard; special hazardous substance

Pennsylvania RTK - Special Hazardous Substances
[present]

POTASSIUM 44   14378-22-4

ENVIRONMENTAL LISTS

List of Pesticide Product Inert Ingredients
[present]

POTASSIUM 45   15706-41-9

HEALTH AND SAFETY LISTS

NFPA - Flash Points
flash point = 491 degrees F (255 degrees C)

NFPA - Hazard Identification Ratings
health-0; flammability-1; reactivity-0

POTASSIUM ACETATE   127-08-2

HEALTH AND SAFETY LISTS

ACGIH 1995 - Time Weighted Averages
10 mg/m3 TWA (The value is for total dust containing no asbestos and <1% crystalline silica)

NIOSH 1990 - Pocket Guide - RELs
total: 10 mg/m3 TWA; respirable dust: 5 mg/m3 TWA

NIOSH 1990 - Pocket Guide - Target organs
skin, eyes, respiratory system

OSHA - Vacated PELs - Time Weighted Averages
total dust: 10 mg/m3 TWA; respirable fraction: 5 mg/m3 TWA

OSHA - Final PELs - Time Weighted Averages
total dust: 15 mg/m3 TWA; respirable fraction: 5 mg/m3 TWA

INTERNATIONAL LISTS

Australian Exposure Standards - Time Weighted Averages
10 mg/m3 TWA

Canada - Alberta - 8 Hour Occupational Exposure Limit
respirable mass: 5 mg/m3 TWA; total mass: 10 mg/m3 TWA

Canada - British Columbia - 8 Hour Exposure Limits
nuisance dust: 10 mg/m3 TWA

Canada - British Columbia - 15 Minute Exposure Limits
20 mg/m3 STEL

Canada - Ontario - OHSA - TWAEVs
total dust: 10 mg/m3 TWAEV (listed as a nuisance particulate)
Canada - Quebec - Time-Weighted Average Exposure Values
total dust: 10 mg/m3 TWAEV; respirable dust: 5 mg/m3 TWAEV
German (DFG) - MAK Values
dusts: 5 mg/m3 total dust
Israel - Time Weighted Averages
10 mg/m3 TWA (The value is for total dust containing no asbestos and <1% crystalline silica)
Israel - Action Levels
5 mg/m3 AL

STATE LISTS

Minnesota Hazardous Substance List
[present] (includes inert or nuisance dust)
Pennsylvania Right to Know List
[present]

POTASSIUM ALUMINUM SULFATE   10043-67-1

HEALTH AND SAFETY LISTS

U.S. DOT - Substances From 49 CFR 172.101
regulated by DOT (UN2257)
U.S. DOT - Hazard Classes
DOT hazard class = 4.3

STATE LISTS

California - Directors List of Hazardous Substances (8 CCR 339)
[present]
Florida Hazardous Substance List
[present]
Massachusetts Right To Know List
[present]
NJ Right to Know List (Total)
sn 1555
NJ Special Hazardous Substances
(reactive - second degree)
Pennsylvania Right to Know List
[present]

POTASSIUM ARSENATE   7784-41-0

HEALTH AND SAFETY LISTS

U.S. DOT - Appendix A Table 2 - Radionuclides
final RQ = 1 curie (3.7E 10 Bq)

ENVIRONMENTAL LISTS

ATSDR Priority List
Rank (of 275): 124
CERCLA/SARA List of Radionuclides (Appendix B) and Their Reportable Quantities
final RQ = 1 curie (3.7E 10 Bq)

POTASSIUM ARSENITE   10124-50-2

HEALTH AND SAFETY LISTS

U.S. DOT - Appendix A Table 2 - Radionuclides
final RQ = 100 curies (3.7E 12 Bq)

ENVIRONMENTAL LISTS

CERCLA/SARA List of Radionuclides (Appendix B) and Their Reportable Quantities
final RQ = 100 curies (3.7E 12 Bq)

POTASSIUM ARSENITE   13464-35-2

HEALTH AND SAFETY LISTS

U.S. DOT - Appendix A Table 2 - Radionuclides
final RQ = 10 curies (3.7E 11 Bq)

ENVIRONMENTAL LISTS

CERCLA/SARA List of Radionuclides (Appendix B) and Their Reportable Quantities
final RQ = 10 curies (3.7E 11 Bq)

POTASSIUM BICARBONATE   298-14-6

HEALTH AND SAFETY LISTS

U.S. DOT - Appendix A Table 2 - Radionuclides
final RQ = 100 curies (3.7E 12 Bq)

ENVIRONMENTAL LISTS

CERCLA/SARA List of Radionuclides (Appendix B) and Their Reportable Quantities
final RQ = 100 curies (3.7E 12 Bq)

POTASSIUM BIFLUORIDE   7789-29-9

HEALTH AND SAFETY LISTS

U.S. DOT - Appendix A Table 2 - Radionuclides
final RQ = 1000 curies (3.7E 13 Bq)

ENVIRONMENTAL LISTS

CERCLA/SARA List of Radionuclides (Appendix B) and Their Reportable Quantities
final RQ = 1000 curies (3.7E 13 Bq)

POTASSIUM BIS(2-HYDROXYETHYL)   23746-34-1
DITHIOCARBAMATE

HEALTH AND SAFETY LISTS

NIOSH - Selected LD50s and LC50s
Oral, rat: LD50 = 3250 mg/kg

ENVIRONMENTAL LISTS

List of Pesticide Product Inert Ingredients
[present]

POTASSIUM BORATE   1332-77-0

ENVIRONMENTAL LISTS

List of Pesticide Product Inert Ingredients
[present]

POTASSIUM BORATE TETRAHYDRATE   12045-78-2
SEE ALSO:
ARSENIC

HEALTH AND SAFETY LISTS

U.S. DOT - Substances From 49 CFR 172.101
regulated by DOT (UN1677)
U.S. DOT - Hazard Classes
DOT hazard class = 6.1
U.S. DOT - Appendix B - Marine Pollutants
DOT regulated marine pollutant
U.S. DOT - Appendix A Table 1 - Hazardous Substances
final RQ = 1 pound (0.454 kg)
NTP Seventh Report - Known Carcinogens
known carcinogen (Listed under 'Arsenic and certain arsenic compounds')
OSHA - Select Carcinogens
[present]

ENVIRONMENTAL LISTS

CERCLA/SARA - Hazardous Substances and their Reportable Quantities
final RQ = 1 pound (0.454 kg)
Clean Water Act - Hazardous Substances
[present]
EPA - Carcinogen Hazard Ranking for RQ Adjustment
Hazard ranking = High

INTERNATIONAL LISTS
Canada - WHMIS: Ingredient Disclosure
1% item 1332 (264)

STATE LISTS
Massachusetts Right To Know List
[present]
NJ Right to Know List (Total)
sn 1556
NJ Special Hazardous Substances
(carcinogen)
Pennsylvania Right to Know List
environmental hazard; special hazardous substance
Pennsylvania RTK - Special Hazardous Substances
[present]

POTASSIUM BOROHYDRIDE   13762-51-1
SEE ALSO:
ARSENIC

HEALTH AND SAFETY LISTS
U.S. DOT - Substances From 49 CFR 172.101
regulated by DOT (UN1678)
U.S. DOT - Hazard Classes
DOT hazard class = 6.1
U.S. DOT - Appendix A Table 1 - Hazardous Substances
final RQ = 1 pound (0.454 kg)
NIOSH - Selected LD50s and LC50s
Oral, rat: LD50 = 14 mg/kg Skin, rat: LD50 = 150 mg/kg

ENVIRONMENTAL LISTS
CERCLA/SARA - Section 302 Extremely Hazardous Substances and TPQs
TPQ = 500/10,000 pounds
CERCLA/SARA - Hazardous Substances and their Reportable Quantities
final RQ = 1 pound (0.454 kg)
Clean Water Act - Hazardous Substances
[present]
EPA - Carcinogen Hazard Ranking for RQ Adjustment
Hazard ranking = High

STATE LISTS
Massachusetts Right To Know List
extraordinarily hazardous
NJ Right to Know List (Total)
sn 1557
NJ Special Hazardous Substances
(carcinogen)
Pennsylvania Right to Know List
environmental hazard

POTASSIUM BROMATE   7758-01-2
HEALTH AND SAFETY LISTS
NTP Seventh Report - Known Carcinogens
known carcinogen (Listed under 'Arsenic and certain arsenic compounds')
OSHA - Select Carcinogens
[present]

POTASSIUM BROMIDE   7758-02-3
ENVIRONMENTAL LISTS
List of Pesticide Product Inert Ingredients
[present]

POTASSIUM CARBONATE   584-08-7
HEALTH AND SAFETY LISTS
U.S. DOT - Substances From 49 CFR 172.101
regulated by DOT (UN1811, UN1812)
U.S. DOT - Hazard Classes
DOT hazard class = 8

STATE LISTS
NJ Right to Know List (Total)
sn 1568
NJ Special Hazardous Substances
(corrosive)

POTASSIUM CARBONYL   12397-35-2
HEALTH AND SAFETY LISTS
IARC - Group 3 (not classifiable)
[present]

STATE LISTS
California - Directors List of Hazardous Substances (8 CCR 339)
[present]

POTASSIUM CHLORATE   3811-04-9
ENVIRONMENTAL LISTS
List of Pesticide Product Inert Ingredients
[present]

POTASSIUM CHLORIDE   7447-40-7
ENVIRONMENTAL LISTS
List of Pesticide Product Inert Ingredients
[present]

POTASSIUM CHROMATE   7789-00-6
HEALTH AND SAFETY LISTS
U.S. DOT - Substances From 49 CFR 172.101
regulated by DOT (UN1870)
U.S. DOT - Hazard Classes
DOT hazard class = 4.3

STATE LISTS
NJ Right to Know List (Total)
sn 1558

POTASSIUM N,N-BIS (HYDROXYETHYL) CO-   85712-26-1
COAMINE OXIDE PHOSPHATE
HEALTH AND SAFETY LISTS
AIHA - WEEL - Time Weighted Averages
0.1 mg/m3 TWA
U.S. DOT - Substances From 49 CFR 172.101
regulated by DOT (UN1484)
U.S. DOT - Hazard Classes
DOT hazard class = 5.1
IARC - Group 2B (sufficient animal data)
[present] (Overall evaluation based only on evidence of carcinogenicity in monograph (40, 1986) or in Supplement 4)
NIOSH - Selected LD50s and LC50s
Oral, rat: LD50 = 321 mg/kg
OSHA - Possible Select Carcinogens
[present]

ENVIRONMENTAL LISTS
CERCLA/SARA - Section 313 - Emission Reporting
form R reporting required
List of Pesticide Product Inert Ingredients
[present]

STATE LISTS

California - Air Bill 2588 Appendix A-I
[present]

California - Prop. 65 - Cancer list
carcinogen - initial date 1/1/90

California - Prop. 65 - No Significant Risk Levels
no significant risk level = 1 ug/day

California - Directors List of Hazardous Substances (8 CCR 339)
[present]

Florida Hazardous Substance List
[present]

Massachusetts Right To Know List
carcinogen; extraordinarily hazardous

Minnesota Hazardous Substance List
[present]

NJ Right to Know List (Total)
sn 1559

Pennsylvania Right to Know List
[present]

POTASSIUM N-COCO-N-HYDROXYETHYL AMINO-3-ETHOXY PROPANE SULFONATE   RR-01127-8

INTERNATIONAL LISTS

Canada - WHMIS: Ingredient Disclosure
0.1% item 1328 (340)

POTASSIUM COCONUT FATTY OIL SOAP   61789-31-9

HEALTH AND SAFETY LISTS

NIOSH - Selected LD50s and LC50s
Oral, rat: LD50 = 1870 mg/kg

ENVIRONMENTAL LISTS

List of Pesticide Product Inert Ingredients
[present]

INTERNATIONAL LISTS

Canada - WHMIS: Ingredient Disclosure
1% item 1329 (390)

POTASSIUM CRESYL E5 PHOSPHATE   RR-01126-7

HEALTH AND SAFETY LISTS

U.S. DOT - Hazard Classes
Forbidden from transport by the DOT

POTASSIUM CUPROCYANIDE   13682-73-0

HEALTH AND SAFETY LISTS

U.S. DOT - Substances From 49 CFR 172.101
regulated by DOT (UN1485, UN2427)

U.S. DOT - Hazard Classes
DOT hazard class = 5.1

NIOSH - Selected LD50s and LC50s
Oral, rat: LD50 = 1870 mg/kg

ENVIRONMENTAL LISTS

List of Pesticide Product Inert Ingredients
[present]

STATE LISTS

Florida Hazardous Substance List
[present]

Massachusetts Right To Know List
[present]

NJ Right to Know List (Total)
sn 1560

Pennsylvania Right to Know List
[present]

POTASSIUM CYANATE   590-28-3

HEALTH AND SAFETY LISTS

NIOSH - Selected LD50s and LC50s
Oral, rat: LD50 = 2600 mg/kg

ENVIRONMENTAL LISTS

List of Pesticide Product Inert Ingredients
[present]

POTASSIUM CYANIDE   151-50-8
SEE ALSO:
CHROMIUM (VI) COMPOUNDS- WATER SOLUBLE
CHROMIUM

HEALTH AND SAFETY LISTS

U.S. DOT - Appendix A Table 1 - Hazardous Substances
final RQ = 10 pounds (4.54 kg)

OSHA - Select Carcinogens
[present]

ENVIRONMENTAL LISTS

CERCLA/SARA - Hazardous Substances and their Reportable Quantities
final RQ = 10 pounds (4.54 kg)

Clean Water Act - Hazardous Substances
[present]

EPA - Carcinogen Hazard Ranking for RQ Adjustment
Hazard ranking = High

List of Pesticide Product Inert Ingredients
[present]

INTERNATIONAL LISTS

Canada - WHMIS: Ingredient Disclosure
1% item 1330 (551)

STATE LISTS

Massachusetts Right To Know List
[present]

NJ Right to Know List (Total)
sn 1561

NJ Special Hazardous Substances
(carcinogen)

Pennsylvania Right to Know List
environmental hazard

POTASSIUM DICHLOROISOCYANURATE   2244-21-5

ENVIRONMENTAL LISTS

TSCA - Code of Federal Regulations Citations
40 CFR 721.1750

TSCA - Chemicals with Significant New Use Rules
[present]

TSCA - Section 12(b) - Export Notification
P-82-400; export notification required - Section 5

POTASSIUM DICHROMATE   7778-50-9

ENVIRONMENTAL LISTS

List of Pesticide Product Inert Ingredients
[present]

POTASSIUM DIMETHYLDITHIOCARBAMATE   128-03-0

ENVIRONMENTAL LISTS

List of Pesticide Product Inert Ingredients
[present]

POTASSIUM DITHIONITE   RR-00752-3

ENVIRONMENTAL LISTS

List of Pesticide Product Inert Ingredients
[present]

POTASSIUM 2-ETHYLHEXANOATE   3164-85-0

HEALTH AND SAFETY LISTS

U.S. DOT - Substances From 49 CFR 172.101
regulated by DOT (UN1679)

U.S. DOT - Hazard Classes
DOT hazard class = 6.1

U.S. DOT - Appendix B - Marine Pollutants
DOT regulated severe marine pollutant

STATE LISTS

NJ Right to Know List (Total)
sn 2702

POTASSIUM FLUORIDE   7789-23-3

HEALTH AND SAFETY LISTS

NIOSH - Selected LD50s and LC50s
Oral, mouse: LD50 = 841 mg/kg

POTASSIUM FLUOROACETATE   23745-86-0
SEE ALSO:
CYANIDE ANION
CYANIDE ANION

HEALTH AND SAFETY LISTS

ACGIH 1995 - Ceiling Limits
C 5 mg/m3

ACGIH 1995 - Skin Designations
skin - potential for cutaneous absorption

U.S. DOT - Substances From 49 CFR 172.101
regulated by DOT (UN1680)

U.S. DOT - Hazard Classes
DOT hazard class = 6.1

U.S. DOT - Appendix B - Marine Pollutants
DOT regulated marine pollutant

U.S. DOT - Appendix A Table 1 - Hazardous Substances
final RQ = 10 pounds (4.54 kg)

NIOSH - Selected LD50s and LC50s
Oral, rat: LD50 = 10 mg/kg

NIOSH 1990 - Pocket Guide - RELs
as CN: C 5 mg/m3 (10 min);   C 4.7 ppm (10 min) (Listed under 'Cyanides')

NIOSH 1990 - Pocket Guide - IDLHs
as CN: 50 mg/m3 IDLH (Listed under 'Cyanides')

NIOSH 1990 - Pocket Guide - Target organs
CVS, CNS, liver, kidneys, skin (Listed under "Cyanides")

ENVIRONMENTAL LISTS

CERCLA/SARA - Section 302 Extremely Hazardous Substances and TPQs
TPQ = 100 pounds (This material is a reactive solid. The TPQ does not default to 10,000 pounds for non-powder, non-molten, non-solution form)

CERCLA/SARA - Hazardous Substances and their Reportable Quantities
final RQ = 10 pounds (4.54 kg)

Clean Water Act - Hazardous Substances
[present]

RCRA - P Series Wastes
waste number P098

RCRA - Hazardous Constituents-Appendix VIII
waste number P098

RCRA - Substances Banned From Land Disposal
[present]

INTERNATIONAL LISTS

Australian Exposure Standards - Time Weighted Averages
as CN: 5 mg/m3 TWA

Australian Exposure Standards - Skin Effects
as CN: skin absorption

Canada - Ontario - OHSA - TWAEVs
as CN: 5 mg/m3 TWAEV

Canada - Ontario - OHSA - Skin Notations
as CN: absorption through skin, eyes, or mucous membranes

Israel - Time Weighted Averages
as CN: 5 mg/m3 TWA

Israel - Action Levels
as CN: 2.5 mg/m3 AL

STATE LISTS

Florida Hazardous Substance List
[present]

Massachusetts Right To Know List
extraordinarily hazardous

Minnesota Hazardous Substance List
as Cn: skin

NJ Right to Know List (Total)
sn 1562

Pennsylvania Right to Know List
environmental hazard

POTASSIUM HEPTAFLUOROTANTALATE   16924-00-8

INTERNATIONAL LISTS

Canada - WHMIS: Ingredient Disclosure
1% item 548 (666)

STATE LISTS

Florida Hazardous Substance List
[present]

Massachusetts Right To Know List
[present]

NJ Right to Know List (Total)
sn 1563

Pennsylvania Right to Know List
[present]

POTASSIUM HEXACHLOROPLATINATE(IV)   16921-30-5
SEE ALSO:
CHROMIUM (VI) COMPOUNDS- WATER SOLUBLE
CHROMIUM COMPOUNDS
CHROMIUM (VI) COMPOUNDS
CHROMIUM

HEALTH AND SAFETY LISTS

U.S. DOT - Appendix A Table 1 - Hazardous Substances
final RQ = 10 pounds (4.54 kg)

NIOSH - Selected LD50s and LC50s
Oral, mouse: LD50 = 190 mg/kg

OSHA - Select Carcinogens
[present]

ENVIRONMENTAL LISTS

CERCLA/SARA - Hazardous Substances and their Reportable Quantities
final RQ = 10 pounds (4.54 kg)

Clean Water Act - Hazardous Substances
[present]

EPA - Carcinogen Hazard Ranking for RQ Adjustment
Hazard ranking = High

List of Pesticide Product Inert Ingredients
[present]

INTERNATIONAL LISTS

Canada - WHMIS: Ingredient Disclosure
0.1% item 1331 (687)

STATE LISTS

Florida Hazardous Substance List
[present]

Massachusetts Right To Know List
[present]

NJ Right to Know List (Total)
sn 1564

NJ Special Hazardous Substances
(carcinogen)

Pennsylvania Right to Know List
environmental hazard

POTASSIUM HYDROGEN SULFATE   7646-93-7

ENVIRONMENTAL LISTS

CERCLA/SARA - Section 313 - Emission Reporting
form R reporting required

POTASSIUM HYDROSULFITE   14293-73-3

HEALTH AND SAFETY LISTS

U.S. DOT - Substances From 49 CFR 172.101
regulated by DOT (UN1929)

U.S. DOT - Hazard Classes
DOT hazard class = 4.2

POTASSIUM HYDROXIDE   1310-58-3

ENVIRONMENTAL LISTS

List of Pesticide Product Inert Ingredients
[present]

POTASSIUM HYPOCHLORITE   7778-66-7

HEALTH AND SAFETY LISTS

U.S. DOT - Substances From 49 CFR 172.101
regulated by DOT (UN1812)

U.S. DOT - Hazard Classes
DOT hazard class = 6.1

NIOSH - Selected LD50s and LC50s
Oral, rat: LD50 = 245 mg/kg

STATE LISTS

NJ Right to Know List (Total)
sn 1565

NJ Special Hazardous Substances
(corrosive)

POTASSIUM IODIDE   7681-11-0

HEALTH AND SAFETY LISTS

U.S. DOT - Substances From 49 CFR 172.101
regulated by DOT (UN2628)

U.S. DOT - Hazard Classes
DOT hazard class = 6.1

STATE LISTS

NJ Right to Know List (Total)
sn 1566

POTASSIUM METABISULFITE   16731-55-8

HEALTH AND SAFETY LISTS

NIOSH - Selected LD50s and LC50s
Oral, rat: LD50 = 2500 mg/kg

POTASSIUM, METAL ALLOYS   RR-01337-6

INTERNATIONAL LISTS

Canada - WHMIS: Ingredient Disclosure
0.1% item 1333 (952)

POTASSIUM METAVANADATE   13769-43-2

HEALTH AND SAFETY LISTS

U.S. DOT - Substances From 49 CFR 172.101
regulated by DOT (UN2509)

U.S. DOT - Hazard Classes
DOT hazard class = 8

NIOSH - Selected LD50s and LC50s
Oral, rat: LD50 = 2340 mg/kg

INTERNATIONAL LISTS

Canada - WHMIS: Ingredient Disclosure
1% item 1334 (1513)

STATE LISTS

NJ Right to Know List (Total)
sn 1569

NJ Special Hazardous Substances
(corrosive)

POTASSIUM N-METHYLDITHIOCARBAMATE   137-41-7

STATE LISTS

NJ Right to Know List (Total)
sn 1570

POTASSIUM NITRATE   7757-79-1

HEALTH AND SAFETY LISTS

ACGIH 1995 - Ceiling Limits
C 2 mg/m3

U.S. DOT - Substances From 49 CFR 172.101
regulated by DOT (UN1814, UN1813)

U.S. DOT - Hazard Classes
DOT hazard class = 8

U.S. DOT - Appendix A Table 1 - Hazardous Substances
final RQ = 1000 pounds (454 kg)

NIOSH - Selected LD50s and LC50s
Oral, rat: LD50 = 365 mg/kg

OSHA - Vacated PELs - Ceiling Limits
C 2 mg/m3

ENVIRONMENTAL LISTS

CERCLA/SARA - Hazardous Substances and their Reportable Quantities
final RQ = 1000 pounds (454 kg)

Clean Water Act - Hazardous Substances
[present]

List of Pesticide Product Inert Ingredients
[present]

INTERNATIONAL LISTS

Australian Exposure Standards - Time Weighted Averages
Peak Limitation: 2 mg/m3

Canada - WHMIS: Ingredient Disclosure
1% item 1335 (996)

Canada - Alberta - Ceiling Occupational Exposure Limit
C 2 mg/m3

Canada - British Columbia - Ceiling Exposure Limits
C 2 mg/m3

Canada - Ontario - OHSA - CEVs
2 mg/m3 CEV

Canada - Quebec - Ceiling Limits
P 2 mg/m3

United Kingdom - Occupational Exposure Standards - STELs
2 mg/m3 STEL

Israel - Ceiling Exposure Limits
C 2 mg/m3

STATE LISTS

California - Exposure Limits - Ceilings
C 2 mg/m3

California - Directors List of Hazardous Substances (8 CCR 339)
[present]

Florida Hazardous Substance List
[present]

Massachusetts Right To Know List
[present]

Minnesota Hazardous Substance List
[present]

NJ Right to Know List (Total)
sn 1571

NJ Special Hazardous Substances
(corrosive)

Pennsylvania Right to Know List
environmental hazard

POTASSIUM NITRATE AND SODIUM NITRITE, MIXTURE   RR-00452-4

INTERNATIONAL LISTS

Canada - WHMIS: Ingredient Disclosure
1% item 1336 (1012)

STATE LISTS

NJ Right to Know List (Total)
sn 1572

POTASSIUM NITRITE   7758-09-0

INTERNATIONAL LISTS

Canada - WHMIS: Ingredient Disclosure
1% item 1337 (1028)

POTASSIUM OLEATE   143-18-0

INTERNATIONAL LISTS

Canada - WHMIS: Ingredient Disclosure
1% item 1338 (1082)

POTASSIUM OXIDE   12136-45-7

HEALTH AND SAFETY LISTS

U.S. DOT - Substances From 49 CFR 172.101
regulated by DOT (UN1420)

U.S. DOT - Hazard Classes
DOT hazard class = 4.3

POTASSIUM PALLADIUM CHLORIDE   10025-98-6

HEALTH AND SAFETY LISTS

U.S. DOT - Substances From 49 CFR 172.101
regulated by DOT (UN2864)

U.S. DOT - Hazard Classes
DOT hazard class = 6.1

STATE LISTS

NJ Right to Know List (Total)
sn 1573

POTASSIUM PENTACHLOROPHENATE   7778-73-6

ENVIRONMENTAL LISTS

CERCLA/SARA - Section 313 - Emission Reporting
form R reporting required

POTASSIUM PERCHLORATE   7778-74-7

HEALTH AND SAFETY LISTS

U.S. DOT - Substances From 49 CFR 172.101
regulated by DOT (UN1486)

U.S. DOT - Hazard Classes
DOT hazard class = 5.1

NIOSH - Selected LD50s and LC50s
Oral, rat: LD50 = 3750 mg/kg

ENVIRONMENTAL LISTS

List of Pesticide Product Inert Ingredients
[present]

STATE LISTS

Florida Hazardous Substance List
[present]

Massachusetts Right To Know List
[present]

NJ Right to Know List (Total)
sn 1574

Pennsylvania Right to Know List
[present]

POTASSIUM PERMANGANATE   7722-64-7

HEALTH AND SAFETY LISTS

U.S. DOT - Substances From 49 CFR 172.101
regulated by DOT (UN1487)

U.S. DOT - Hazard Classes
DOT hazard class = 5.1

STATE LISTS

NJ Right to Know List (Total)
sn 2704

POTASSIUM PEROXIDE   17014-71-0

HEALTH AND SAFETY LISTS

U.S. DOT - Substances From 49 CFR 172.101
regulated by DOT (UN1488)

U.S. DOT - Hazard Classes
DOT hazard class = 5.1

NIOSH - Selected LD50s and LC50s
Oral, rabbit: LD50 = 200 mg/kg

ENVIRONMENTAL LISTS

List of Pesticide Product Inert Ingredients
[present]

INTERNATIONAL LISTS

Canada - WHMIS: Ingredient Disclosure
0.1% item 1339 (1217)

STATE LISTS

NJ Right to Know List (Total)
sn 1575

POTASSIUM PERSULFATE   7727-21-1

ENVIRONMENTAL LISTS

List of Pesticide Product Inert Ingredients
[present]

POTASSIUM PHOSPHIDE   20770-41-6

HEALTH AND SAFETY LISTS

U.S. DOT - Substances From 49 CFR 172.101
regulated by DOT (UN2033)

U.S. DOT - Hazard Classes
DOT hazard class = 8

STATE LISTS

NJ Right to Know List (Total)
sn 1576

NJ Special Hazardous Substances
(corrosive)

POTASSIUM PYROPHOSPHATE 7320-34-5

INTERNATIONAL LISTS
Canada - WHMIS: Ingredient Disclosure
1% item 1340 (448)

POTASSIUM PYROSULFATE 7790-62-7

ENVIRONMENTAL LISTS
RCRA - Hazardous Constituents-Appendix VIII
hazardous constituent - no waste number

STATE LISTS
Massachusetts Right To Know List
[present]

POTASSIUM RICINOLEATE 7492-30-0

HEALTH AND SAFETY LISTS
U.S. DOT - Substances From 49 CFR 172.101
regulated by DOT (UN1489)
U.S. DOT - Hazard Classes
DOT hazard class = 5.1

STATE LISTS
Florida Hazardous Substance List
[present]
Massachusetts Right To Know List
[present]
NJ Right to Know List (Total)
sn 1577
NJ Special Hazardous Substances
(reactive - second degree)
Pennsylvania Right to Know List
[present]

POTASSIUM SALT OF NAPHTHALENE SULFONIC ACID - FORMALDEHYDE CONDENSATE RR-01128-9
SEE ALSO:
MANGANESE
MANGANESE COMPOUNDS, N.O.S.
MANGANESE

HEALTH AND SAFETY LISTS
U.S. DOT - Substances From 49 CFR 172.101
regulated by DOT (UN1490)
U.S. DOT - Hazard Classes
DOT hazard class = 5.1
U.S. DOT - Appendix A Table 1 - Hazardous Substances
final RQ = 100 pounds (45.4 kg)
FDA - Controlled Substances Act - Essential chemicals
Import/Export threshold volume = N/A; weight = 500 kilograms; Domestic threshold volume = N/A; weight = 55 kilograms
NIOSH - Selected LD50s and LC50s
Oral, rat: LD50 = 1090 mg/kg

ENVIRONMENTAL LISTS
CERCLA/SARA - Hazardous Substances and their Reportable Quantities
final RQ = 100 pounds (45.4 kg)
Clean Water Act - Hazardous Substances
[present]
List of Pesticide Product Inert Ingredients
[present]

INTERNATIONAL LISTS
Canada - WHMIS: Ingredient Disclosure
1% item 1341 (1359)

STATE LISTS
California - Directors List of Hazardous Substances (8 CCR 339)
[present]

Florida Hazardous Substance List
[present]
Massachusetts Right To Know List
[present]
NJ Right to Know List (Total)
sn 1578
Pennsylvania Right to Know List
environmental hazard

POTASSIUM SALTS OF FATTY ACIDS (C12-C20) 69669-25-6

HEALTH AND SAFETY LISTS
U.S. DOT - Substances From 49 CFR 172.101
regulated by DOT (UN1491)
U.S. DOT - Hazard Classes
DOT hazard class = 5.1

STATE LISTS
Florida Hazardous Substance List
[present]
Massachusetts Right To Know List
[present]
NJ Right to Know List (Total)
sn 1579
NJ Special Hazardous Substances
(reactive - second degree)
Pennsylvania Right to Know List
[present]

POTASSIUM SALTS OF NITRO-AROMATIC DERIVATIVES, EXPLOSIVE RR-00453-5

HEALTH AND SAFETY LISTS
U.S. DOT - Substances From 49 CFR 172.101
regulated by DOT (UN1492)
U.S. DOT - Hazard Classes
DOT hazard class = 5.1

INTERNATIONAL LISTS
Canada - Ontario - OHSA - TWAEVs
as S2O8: 5 mg/m3 TWAEV
United Kingdom - Occupational Exposure Standards - TWAs
measured as S2O8: 1 mg/m3 TWA

STATE LISTS
Florida Hazardous Substance List
[present]
Massachusetts Right To Know List
[present]
NJ Right to Know List (Total)
sn 1580
Pennsylvania Right to Know List
[present]

PROPOSED REGULATIONS
Canada - Ontario - Proposed Occupational TWAEVs
2 mg/m3 TWAEV

POTASSIUM SELENATE 7790-59-2

HEALTH AND SAFETY LISTS
U.S. DOT - Substances From 49 CFR 172.101
regulated by DOT (UN2012)
U.S. DOT - Hazard Classes
DOT hazard class = 4.3

INTERNATIONAL LISTS
Canada - WHMIS: Ingredient Disclosure
1% item 1342 (1412)

STATE LISTS
  NJ Right to Know List (Total)
    sn 1581

POTASSIUM SILICATE   1312-76-1
  ENVIRONMENTAL LISTS
    List of Pesticide Product Inert Ingredients
    [present]

POTASSIUM SILICOFLUORIDE   16871-90-2
  ENVIRONMENTAL LISTS
    List of Pesticide Product Inert Ingredients
    [present]

POTASSIUM SILVER CYANIDE   506-61-6
  ENVIRONMENTAL LISTS
    List of Pesticide Product Inert Ingredients
    [present]

POTASSIUM-SODIUM ALLOY   11135-81-2
  ENVIRONMENTAL LISTS
    List of Pesticide Product Inert Ingredients
    [present]

POTASSIUM SORBATE   24634-61-5
  ENVIRONMENTAL LISTS
    List of Pesticide Product Inert Ingredients
    [present]

POTASSIUM STEARATE   593-29-3
  HEALTH AND SAFETY LISTS
    U.S. DOT - Substances From 49 CFR 172.101
    regulated by DOT (UN0158)
    U.S. DOT - Hazard Classes
    DOT hazard class = 1.3C
  STATE LISTS
    NJ Right to Know List (Total)
    sn 2705

POTASSIUM SULFATORICINOLEATE   67785-93-7
  SEE ALSO:
    SELENIUM
  INTERNATIONAL LISTS
    Canada - WHMIS: Ingredient Disclosure
    0.1% item 1343 (1476)

POTASSIUM SULFIDE   1312-73-8
  ENVIRONMENTAL LISTS
    List of Pesticide Product Inert Ingredients
    [present]

POTASSIUM SULFIDE (VAN)   37248-34-3
  HEALTH AND SAFETY LISTS
    U.S. DOT - Substances From 49 CFR 172.101
    regulated by DOT (UN2655)
    U.S. DOT - Hazard Classes
    DOT hazard class = 6.1
    NIOSH - Selected LD50s and LC50s
    Oral, guinea pig: LD50 = 500 mg/kg
  STATE LISTS
    NJ Right to Know List (Total)
    sn 1582

POTASSIUM SULFITE   10117-38-1
  SEE ALSO:
    SILVER
    CYANIDE ANION
  HEALTH AND SAFETY LISTS
    U.S. DOT - Appendix A Table 1 - Hazardous Substances
    final RQ = 1 pound (0.454 kg)
    NIOSH - Selected LD50s and LC50s
    Oral, rat: LD50 = 20900 ug/kg
  ENVIRONMENTAL LISTS
    CERCLA/SARA - Section 302 Extremely Hazardous Substances and
    TPQs
    *TPQ = 500 pounds (This material is a reactive solid. The TPQ does not
    default to 10,000 pounds for non-powder, non-molten, non-solution
    form)*
    CERCLA/SARA - Hazardous Substances and their Reportable Quan-
    tities
    final RQ = 1 pound (0.454 kg)
    RCRA - P Series Wastes
    waste number P099
    RCRA - Hazardous Constituents-Appendix VIII
    waste number P099
    RCRA - Substances Banned From Land Disposal
    [present]
  INTERNATIONAL LISTS
    Canada - WHMIS: Ingredient Disclosure
    1% item 1418 (588)
  STATE LISTS
    Florida Hazardous Substance List
    effective March 13, 1992
    Massachusetts Right To Know List
    extraordinarily hazardous
    NJ Right to Know List (Total)
    sn 2708
    Pennsylvania Right to Know List
    environmental hazard

POTASSIUM SUPEROXIDE   12030-88-5
  HEALTH AND SAFETY LISTS
    U.S. DOT - Hazard Classes
    DOT hazard class = 4.3
  STATE LISTS
    Florida Hazardous Substance List
    [present]
    Massachusetts Right To Know List
    [present]
    NJ Right to Know List (Total)
    sn 2709
    Pennsylvania Right to Know List
    [present]

POTASSIUM N,N'-BIS(HYDROXYETHYL) TAL-   85712-27-2
LOWAMINE OXIDE PHOSPHATE
  ENVIRONMENTAL LISTS
    List of Pesticide Product Inert Ingredients
    [present]

POTASSIUM TETRACHLOROPHENATE   53535-27-6
  ENVIRONMENTAL LISTS
    List of Pesticide Product Inert Ingredients
    [present]

STATE LISTS
California - Exposure Limits - PELs
10 mg/m3 PEL

POTASSIUM TETRACHLOROPLATINATE(II)   10025-99-7

ENVIRONMENTAL LISTS
List of Pesticide Product Inert Ingredients
[present]

POTASSIUM THIOCYANATE   333-20-0

HEALTH AND SAFETY LISTS
U.S. DOT - Substances From 49 CFR 172.101
regulated by DOT (UN1382, UN1847)
U.S. DOT - Hazard Classes
DOT hazard class = 4.2

STATE LISTS
Florida Hazardous Substance List
[present]
Massachusetts Right To Know List
[present]
NJ Right to Know List (Total)
sn 1583
NJ Special Hazardous Substances
(corrosive)
Pennsylvania Right to Know List
[present]

POTASSIUM TITANATE   12030-97-6

STATE LISTS
Pennsylvania Right to Know List
[present]

POTASSIUM TITANATE (K2TI2O5)   12056-46-1

ENVIRONMENTAL LISTS
List of Pesticide Product Inert Ingredients
[present]

POTASSIUM TITANATE (K2TI4O9)   12056-49-4

HEALTH AND SAFETY LISTS
U.S. DOT - Substances From 49 CFR 172.101
regulated by DOT (UN2466)
U.S. DOT - Hazard Classes
DOT hazard class = 5.1

STATE LISTS
NJ Right to Know List (Total)
sn 1584

POTASSIUM TITANATE (K2TI8O17)   59766-31-3

ENVIRONMENTAL LISTS
TSCA - Code of Federal Regulations Citations
40 CFR 721.1750
TSCA - Chemicals with Significant New Use Rules
[present]
TSCA - Section 12(b) - Export Notification
P-82-409; export notification required - Section 5

POTASSIUM TOLUENESULFONATE   30526-22-8

ENVIRONMENTAL LISTS
RCRA - Hazardous Constituents-Appendix VIII
hazardous constituent - no waste number

STATE LISTS
Massachusetts Right To Know List
[present]

POTASSIUM TRIPOLYPHOSPHATE   13845-36-8

INTERNATIONAL LISTS
Canada - WHMIS: Ingredient Disclosure
0.1% item 1344 (1577)

POTATOES   RR-01129-0

HEALTH AND SAFETY LISTS
NIOSH - Selected LD50s and LC50s
Oral, rat: LD50 = 854 mg/kg

PRASEODYMIUM 136   22095-53-0

INTERNATIONAL LISTS
German (DFG) - Carcinogens
as fibrous dust: animal evidence of carcinogenicity

PRASEODYMIUM 137   15125-66-3

INTERNATIONAL LISTS
German (DFG) - Carcinogens
as fibrous dust: animal evidence of carcinogenicity

PRASEODYMIUM 138M   RR-00389-4

INTERNATIONAL LISTS
German (DFG) - Carcinogens
as fibrous dust: animal evidence of carcinogenicity

PRASEODYMIUM 139   14191-76-5

INTERNATIONAL LISTS
German (DFG) - Carcinogens
as fibrous dust: animal evidence of carcinogenicity

PRASEODYMIUM 142   14191-64-1

ENVIRONMENTAL LISTS
List of Pesticide Product Inert Ingredients
[present]

PRASEODYMIUM 142M   RR-00388-3

ENVIRONMENTAL LISTS
List of Pesticide Product Inert Ingredients
[present]

PRASEODYMIUM 143   14981-79-4

ENVIRONMENTAL LISTS
List of Pesticide Product Inert Ingredients
[present]

PRASEODYMIUM 144   14119-05-2

HEALTH AND SAFETY LISTS
U.S. DOT - Appendix A Table 2 - Radionuclides
final RQ = 1000 curies (3.7E 13 Bq)

ENVIRONMENTAL LISTS
CERCLA/SARA List of Radionuclides (Appendix B) and Their Reportable Quantities
final RQ = 1000 curies (3.7E 13 Bq)

PRASEODYMIUM 145   15765-23-8

HEALTH AND SAFETY LISTS
U.S. DOT - Appendix A Table 2 - Radionuclides
final RQ = 1000 curies (3.7E 13 Bq)

ENVIRONMENTAL LISTS
CERCLA/SARA List of Radionuclides (Appendix B) and Their Reportable Quantities
final RQ = 1000 curies (3.7E 13 Bq)

PRASEODYMIUM 147   15765-24-9

HEALTH AND SAFETY LISTS

U.S. DOT - Appendix A Table 2 - Radionuclides
final RQ = 100 curies (3.7E 12 Bq)

ENVIRONMENTAL LISTS

CERCLA/SARA List of Radionuclides (Appendix B) and Their Reportable Quantities
final RQ = 100 curies (3.7E 12 Bq)

PRASEODYMIUM CHLORIDE   10361-79-2

HEALTH AND SAFETY LISTS

U.S. DOT - Appendix A Table 2 - Radionuclides
final RQ = 1000 curies (3.7E 13 Bq)

ENVIRONMENTAL LISTS

CERCLA/SARA List of Radionuclides (Appendix B) and Their Reportable Quantities
final RQ = 1000 curies (3.7E 13 Bq)

PRASEODYMIUM NITRATE   10361-80-5

HEALTH AND SAFETY LISTS

U.S. DOT - Appendix A Table 2 - Radionuclides
final RQ = 100 curies (3.7E 12 Bq)

ENVIRONMENTAL LISTS

CERCLA/SARA List of Radionuclides (Appendix B) and Their Reportable Quantities
final RQ = 100 curies (3.7E 12 Bq)

PRAZEPAM   2955-38-6

HEALTH AND SAFETY LISTS

U.S. DOT - Appendix A Table 2 - Radionuclides
final RQ = 1000 curies (3.7E 13 Bq)

ENVIRONMENTAL LISTS

CERCLA/SARA List of Radionuclides (Appendix B) and Their Reportable Quantities
final RQ = 1000 curies (3.7E 13 Bq)

PREDNIMUSTINE   29069-24-7

HEALTH AND SAFETY LISTS

U.S. DOT - Appendix A Table 2 - Radionuclides
final RQ = 10 curies (3.7E 11 Bq)

ENVIRONMENTAL LISTS

CERCLA/SARA List of Radionuclides (Appendix B) and Their Reportable Quantities
final RQ = 10 curies (3.7E 11 Bq)

PREDNISOLONE   50-24-8

HEALTH AND SAFETY LISTS

U.S. DOT - Appendix A Table 2 - Radionuclides
final RQ = 1000 curies (3.7E 13 Bq)

ENVIRONMENTAL LISTS

CERCLA/SARA List of Radionuclides (Appendix B) and Their Reportable Quantities
final RQ = 1000 curies (3.7E 13 Bq)

PREDNISONE   53-03-2

HEALTH AND SAFETY LISTS

U.S. DOT - Appendix A Table 2 - Radionuclides
final RQ = 1000 curies (3.7E 13 Bq)

ENVIRONMENTAL LISTS

CERCLA/SARA List of Radionuclides (Appendix B) and Their Reportable Quantities
final RQ = 1000 curies (3.7E 13 Bq)

PRIMACLONE   125-33-7

HEALTH AND SAFETY LISTS

U.S. DOT - Appendix A Table 2 - Radionuclides
final RQ = 1000 curies (3.7E 13 Bq)

ENVIRONMENTAL LISTS

CERCLA/SARA List of Radionuclides (Appendix B) and Their Reportable Quantities
final RQ = 1000 curies (3.7E 13 Bq)

PROBENECID   57-66-9

HEALTH AND SAFETY LISTS

NIOSH - Selected LD50s and LC50s
Oral, mouse: LD50 = 2987 mg/kg

PROCARBAZINE   671-16-9

HEALTH AND SAFETY LISTS

NIOSH - Selected LD50s and LC50s
Oral, rat: LD50 = 1859 mg/kg

PROCARBAZINE HYDROCHLORIDE   366-70-1

HEALTH AND SAFETY LISTS

NIOSH - Selected LD50s and LC50s
Oral, mouse: LD50 = 2300 mg/kg

STATE LISTS

NJ Right to Know List (Total)
sn 1587

NJ Special Hazardous Substances
(teratogen)

PROCYMIDONE   32809-16-8

HEALTH AND SAFETY LISTS

IARC - Group 3 (not classifiable)
[present]

PROFLAVINE SALTS   RR-01539-4

HEALTH AND SAFETY LISTS

NIOSH - Selected LD50s and LC50s
Oral, mouse: LD50 = 1680 mg/kg

STATE LISTS

NJ Right to Know List (Total)
sn 1588

PROFLAVIN HYDROCHLORIDE   952-23-8

HEALTH AND SAFETY LISTS

IARC - Group 3 (not classifiable)
[present]

NTP Chemical Status Reports - Testing Status and NTIS Number
Chronic studies exist for which technical reports were not prepared

STATE LISTS

NJ Right to Know List (Total)
sn 1589

PROGESTERONE   57-83-0

HEALTH AND SAFETY LISTS

NTP Chemical Status Reports - Testing Status and NTIS Number
Two year studies: laboratory study report in preparation

PROGESTERONE AND OESTROGENS   RR-00018-0

HEALTH AND SAFETY LISTS

NTP Chemical Status Reports - Testing Status and NTIS Number
Technical reports printed (PB92129584/AS)

NTP Chemical Status Reports - Evidence of Carcinogenicity
male rat: no evidence; female rat: no evidence; male mice: no evidence; female mice: some evidence

PROGESTINS   RR-00033-9

STATE LISTS

California - Prop. 65 - Cancer list
carcinogen - initial date 1/1/88

California - Prop. 65 - No Significant Risk Levels
no significant risk level = 0.05 ug/day

Florida Hazardous Substance List
[present]

Massachusetts Right To Know List
carcinogen; extraordinarily hazardous; teratogen

Minnesota Hazardous Substance List
carcinogen

NJ Special Hazardous Substances
(carcinogen; mutagen)

Pennsylvania Right to Know List
special hazardous substance

Pennsylvania RTK - Special Hazardous Substances
[present]

PROMECARB   2631-37-0

HEALTH AND SAFETY LISTS

IARC - Group 2A (limited human data)
[present] (Other relevant data, as given in Supplement 7, influenced the making of the overall evaluation)

NIOSH - Selected LD50s and LC50s
Oral, rat: LD50 = 570 mg/kg

NTP Chemical Status Reports - Testing Status and NTIS Number
Technical reports printed (PB299902/AS); Chronic studies exist for which technical reports were not prepared

NTP Chemical Status Reports - Evidence of Carcinogenicity
male rat-positive; female rat-positive; male mice-positive; female mice-positive

NTP Seventh Report - Suspect Carcinogens
suspect carcinogen

OSHA - Possible Select Carcinogens
[present]

STATE LISTS

California - Air Bill 2588 Appendix A-II
known or potential carcinogen

California - Prop. 65 - Cancer list
carcinogen - initial date 1/1/88

California - Prop. 65 - Developmental Toxicity
developmental toxicity - initial date 7/1/90

California - Prop. 65 - No Significant Risk Levels
no significant risk level = 0.06 ug/day

California - Directors List of Hazardous Substances (8 CCR 339)
[present]

Florida Hazardous Substance List
[present]

Massachusetts Right To Know List
carcinogen; extraordinarily hazardous; teratogen

Minnesota Hazardous Substance List
carcinogen

NJ Right to Know List (Total)
sn 1590

NJ Special Hazardous Substances
(carcinogen; mutagen; teratogen)

Pennsylvania Right to Know List
special hazardous substance

Pennsylvania RTK - Special Hazardous Substances
[present]

PROMETHAZINE HYDROCHLORIDE   58-33-3

STATE LISTS

California - Prop. 65 - Cancer list
carcinogen - initial date 10/01/94

PROMETHIUM 141   14952-27-3

HEALTH AND SAFETY LISTS

IARC - Group 3 (not classifiable)
[present]

PROMETHIUM 143   14834-72-1

HEALTH AND SAFETY LISTS

NTP Chemical Status Reports - Testing Status and NTIS Number
Technical reports printed (PB268553/AS)

NTP Chemical Status Reports - Evidence of Carcinogenicity
male rat-equivocal; female rat-negative; male mice-equivocal; female mice-equivocal

STATE LISTS

Massachusetts Right To Know List
[present]

PROMETHIUM 144   14834-73-2

HEALTH AND SAFETY LISTS

IARC - Group Unspecified
[present] (Listed under 'Progestins')

NTP Seventh Report - Suspect Carcinogens
suspect carcinogen

OSHA - Possible Select Carcinogens
[present]

STATE LISTS

California - Air Bill 2588 Appendix A-I
known or potential carcinogen

California - Prop. 65 - Cancer list
carcinogen - initial date 1/1/88

California - Directors List of Hazardous Substances (8 CCR 339)
[present]

Florida Hazardous Substance List
[present]

Massachusetts Right To Know List
carcinogen; extraordinarily hazardous

Minnesota Hazardous Substance List
carcinogen

NJ Right to Know List (Total)
sn 1591

NJ Special Hazardous Substances
(carcinogen; mutagen)

Pennsylvania Right to Know List
special hazardous substance

Pennsylvania RTK - Special Hazardous Substances
[present]

PROMETHIUM 145   15706-44-2

HEALTH AND SAFETY LISTS

IARC - Group Unspecified
[present] (Listed under 'Combined oral contraceptives')

PROMETHIUM 146   14834-74-3

HEALTH AND SAFETY LISTS

IARC - Group 2B (sufficient animal data)
[present] (Listed under 'Oestrogens, progestins, and combinations')

OSHA - Possible Select Carcinogens
[present]

STATE LISTS

California - Air Bill 2588 Appendix A-II
known or potential carcinogen

California - Directors List of Hazardous Substances (8 CCR 339)
[present]

Minnesota Hazardous Substance List
carcinogen

PROMETHIUM 147   14380-75-7

HEALTH AND SAFETY LISTS

U.S. DOT - Appendix B - Marine Pollutants
DOT regulated marine pollutant

NIOSH - Selected LD50s and LC50s
Oral, rat: LD50 = 60 mg/kg Skin, rat: LD50 = 450 mg/kg

ENVIRONMENTAL LISTS

CERCLA/SARA - Section 302 Extremely Hazardous Substances and TPQs
TPQ = 500/10,000 pounds

STATE LISTS

Florida Hazardous Substance List
effective March 13, 1992

Massachusetts Right To Know List
extraordinarily hazardous

NJ Right to Know List (Total)
sn 2710

Pennsylvania Right to Know List
environmental hazard

PROPOSED REGULATIONS

CERCLA/SARA - Proposed Hazardous Substance Additions
proposed RQ = 1 pound (.454 kg)

CERCLA/SARA - 1989 Proposed RQ Adjustments
proposed RQ = 100 pounds (45.4 kg)

PROMETHIUM 148   14683-19-3

HEALTH AND SAFETY LISTS

NTP Chemical Status Reports - Testing Status and NTIS Number
Technical reports printed; prechronic studies for which toxicity technical reports were not prepared

PROMETHIUM 148M   RR-00387-2

HEALTH AND SAFETY LISTS

U.S. DOT - Appendix A Table 2 - Radionuclides
final RQ = 1000 curies (3.7E 13 Bq)

ENVIRONMENTAL LISTS

CERCLA/SARA List of Radionuclides (Appendix B) and Their Reportable Quantities
final RQ = 1000 curies (3.7E 13 Bq)

PROMETHIUM 149   15765-31-8

HEALTH AND SAFETY LISTS

U.S. DOT - Appendix A Table 2 - Radionuclides
final RQ = 100 curies (3.7E 12 Bq)

ENVIRONMENTAL LISTS

CERCLA/SARA List of Radionuclides (Appendix B) and Their Reportable Quantities
final RQ = 100 curies (3.7E 12 Bq)

PROMETHIUM 150   15720-47-5

HEALTH AND SAFETY LISTS

U.S. DOT - Appendix A Table 2 - Radionuclides
final RQ = 10 curies (3.7E 11 Bq)

ENVIRONMENTAL LISTS

CERCLA/SARA List of Radionuclides (Appendix B) and Their Reportable Quantities
final RQ = 10 curies (3.7E 11 Bq)

PROMETHIUM 151   15766-03-7

HEALTH AND SAFETY LISTS

U.S. DOT - Appendix A Table 2 - Radionuclides
final RQ = 100 curies (3.7E 12 Bq)

ENVIRONMENTAL LISTS

CERCLA/SARA List of Radionuclides (Appendix B) and Their Reportable Quantities
final RQ = 100 curies (3.7E 12 Bq)

PROMETON   1610-18-0

HEALTH AND SAFETY LISTS

U.S. DOT - Appendix A Table 2 - Radionuclides
final RQ = 10 curies (3.7E 11 Bq)

ENVIRONMENTAL LISTS

CERCLA/SARA List of Radionuclides (Appendix B) and Their Reportable Quantities
final RQ = 10 curies (3.7E 11 Bq)

PROMETRYN   7287-19-6

HEALTH AND SAFETY LISTS

U.S. DOT - Appendix A Table 2 - Radionuclides
final RQ = 10 curies (3.7E 11 Bq)

ENVIRONMENTAL LISTS

CERCLA/SARA List of Radionuclides (Appendix B) and Their Reportable Quantities
final RQ = 10 curies (3.7E 11 Bq)

PRONAMIDE   23950-58-5

HEALTH AND SAFETY LISTS

U.S. DOT - Appendix A Table 2 - Radionuclides
final RQ = 10 curies (3.7E 11 Bq)

ENVIRONMENTAL LISTS

CERCLA/SARA List of Radionuclides (Appendix B) and Their Reportable Quantities
final RQ = 10 curies (3.7E 11 Bq)

PRONETALOL HYDROCHLORIDE   51-02-5

HEALTH AND SAFETY LISTS

U.S. DOT - Appendix A Table 2 - Radionuclides
final RQ = 10 curies (3.7E 11 Bq)

ENVIRONMENTAL LISTS

CERCLA/SARA List of Radionuclides (Appendix B) and Their Reportable Quantities
final RQ = 10 curies (3.7E 11 Bq)

PROPANTHELINE BROMIDE   50-34-0

HEALTH AND SAFETY LISTS

U.S. DOT - Appendix A Table 2 - Radionuclides
final RQ = 100 curies (3.7E 12 Bq)

ENVIRONMENTAL LISTS

CERCLA/SARA List of Radionuclides (Appendix B) and Their Reportable Quantities
final RQ = 100 curies (3.7E 12 Bq)

PROPACHLOR   1918-16-7

HEALTH AND SAFETY LISTS

U.S. DOT - Appendix A Table 2 - Radionuclides
final RQ = 100 curies (3.7E 12 Bq)

ENVIRONMENTAL LISTS

CERCLA/SARA List of Radionuclides (Appendix B) and Their Reportable Quantities
final RQ = 100 curies (3.7E 12 Bq)

PROPADIENE   463-49-0

HEALTH AND SAFETY LISTS

U.S. DOT - Appendix A Table 2 - Radionuclides
final RQ = 100 curies (3.7E 12 Bq)

ENVIRONMENTAL LISTS

CERCLA/SARA List of Radionuclides (Appendix B) and Their Reportable Quantities
final RQ = 100 curies (3.7E 12 Bq)

PROPANAL, 3-HYDROXY-2,2-DIMETHYL-   597-31-9

HEALTH AND SAFETY LISTS

NIOSH - Selected LD50s and LC50s
Inhalation, rat: LC50 = 3260 mg/m3 4 hr Oral, rat: LD50 = 503 mg/kg Skin, rabbit: LD50 = 2200 mg/kg

PROPOSED REGULATIONS

Safe Drinking Water Act - Priority list
[present]

1-PROPANAMINE   107-10-8

ENVIRONMENTAL LISTS

CERCLA/SARA - Section 313 - Emission Reporting
form R reporting required

2-PROPANAMINE, 1-CHLORO-N,N-DIMETHYL-, HYDROCHLORIDE   17256-39-2

SEE ALSO:
F039-HAZARDOUS WASTES

HEALTH AND SAFETY LISTS

U.S. DOT - Appendix A Table 1 - Hazardous Substances
final RQ = 5000 pounds (2270 kg)

NIOSH - Selected LD50s and LC50s
Oral, rat: LD50 = 5620 mg/kg

ENVIRONMENTAL LISTS

CERCLA/SARA - Section 313 - Emission Reporting
form R reporting required

CERCLA/SARA - Hazardous Substances and their Reportable Quantities
final RQ = 5000 pounds (2270 kg)

RCRA - U Series Wastes
waste number U192

RCRA - Hazardous Constituents-Appendix VIII
waste number U192

RCRA - Basis for Listing - Appendix VII
Included in waste stream: F039

RCRA - Substances Banned From Land Disposal
[present]

RCRA - TSD Facilities Ground Water Monitoring
TM 8270 = 10 ug/L PQL

RCRA - Universal Treatment Standards (LDR)
WW: 0.093 mg/l; NWW: 1.5 mg/kg

STATE LISTS

Florida Hazardous Substance List
[present]

Massachusetts Right To Know List
carcinogen; extraordinarily hazardous

NJ Special Hazardous Substances
(carcinogen)

Pennsylvania Right to Know List
environmental hazard

PROPANAMINIUM, N-(3-AMINOPROPYL)-2-HYDROXY-N,N-DIMETHYL-3-SULFO-, N-COCO ACYL DERIVATIVES, HYDROXIDES, INNER SALTS   68139-30-0

HEALTH AND SAFETY LISTS

IARC - Group 3 (not classifiable)
[present]

STATE LISTS

California - Directors List of Hazardous Substances (8 CCR 339)
[present]

PROPANE   74-98-6

HEALTH AND SAFETY LISTS

NTP Chemical Status Reports - Testing Status and NTIS Number
Prechronic studies for which toxicity technical reports were not prepared

1,3-PROPANEDIAMINE   109-76-2

ENVIRONMENTAL LISTS

CERCLA/SARA - Section 313 - Emission Reporting
form R reporting required

PROPANE, 2,2-DICHLORO-   594-20-7

HEALTH AND SAFETY LISTS

U.S. DOT - Substances From 49 CFR 172.101
regulated by DOT (UN2200)

U.S. DOT - Hazard Classes
DOT hazard class = 2.1

ENVIRONMENTAL LISTS

CAA - Flammable Substances for Accidental Release Prevention
threshold quantity = 10,000 lbs

STATE LISTS

NJ Right to Know List (Total)
sn 1593

PROPANE, 1,1-DICHLORO-1-NITRO-   595-44-8

ENVIRONMENTAL LISTS

TSCA - Code of Federal Regulations Citations
40 CFR 712.30(x); 40 CFR 716.120(d)

TSCA - PAIR - Reporting List
Reporting Date: November 27, 1991

TSCA - Health and Safety Reporting List
Effective Date: September 30, 1991

1,3-PROPANEDIOL   504-63-2

HEALTH AND SAFETY LISTS

U.S. DOT - Substances From 49 CFR 172.101
regulated by DOT (UN1277)

U.S. DOT - Hazard Classes
DOT hazard class = 3

U.S. DOT - Appendix A Table 1 - Hazardous Substances
final RQ = 5000 pounds (2270 kg)

NFPA - Flash Points
flash point = -35 degrees F (-37 degrees C)

NFPA - Hazard Identification Ratings
health-3; flammability-3; reactivity-0

NIOSH - Selected LD50s and LC50s
Inhalation, rat: LC50 = 2310 ppm 4 hr Oral, rat: LD50 = 570 mg/kg Skin, rabbit: LD50 = 560 mg/kg

ENVIRONMENTAL LISTS

CERCLA/SARA - Hazardous Substances and their Reportable Quantities
final RQ = 5000 pounds (2270 kg)

RCRA - U Series Wastes
waste number U194 (Ignitable waste; Toxic waste)

RCRA - Hazardous Constituents-Appendix VIII
waste number U194

RCRA - Substances Banned From Land Disposal
[present]

TSCA - Code of Federal Regulations Citations
40 CFR 716.120(a)

TSCA - Health and Safety Reporting List
Effective Date: March 7, 1986

INTERNATIONAL LISTS

Canada - WHMIS: Ingredient Disclosure
1% item 1356 (1452)

STATE LISTS

Florida Hazardous Substance List
[present]

Massachusetts Right To Know List
[present]

NJ Right to Know List (Total)
sn 1606

NJ Special Hazardous Substances
(flammable - third degree)

Pennsylvania Right to Know List
environmental hazard

1,2-PROPANEDIOL DIACETATE   623-84-7

ENVIRONMENTAL LISTS

TSCA - Code of Federal Regulations Citations
40 CFR 712.30(d)

TSCA - PAIR - Reporting List
Reporting Date: November 19, 1982

1,3-PROPANEDIOL, 2,2-DIMETHYL-   126-30-7

ENVIRONMENTAL LISTS

List of Pesticide Product Inert Ingredients
[present]

1,3-PROPANEDIOL, 2-ETHYL-2-(HYDROXYMETHYL)   77-99-6

HEALTH AND SAFETY LISTS

ACGIH 1995 - Time Weighted Averages
simple asphyxiant

AIHA - Odor Threshold Values
no geometric mean air odor threshold

U.S. DOT - Substances From 49 CFR 172.101
regulated by DOT (UN1978)

U.S. DOT - Hazard Classes
DOT hazard class = 2.1

NFPA - Flash Points
gas (no flash point given)

NFPA - Hazard Identification Ratings
health-1; flammability-4; reactivity-0

NIOSH 1990 - Pocket Guide - RELs
1000 ppm TWA; 1800 mg/m3 TWA

NIOSH 1990 - Pocket Guide - IDLHs
20,000 ppm IDLH (lower explosive level)

NIOSH 1990 - Pocket Guide - Target organs
CNS

OSHA - Vacated PELs - Time Weighted Averages
1000 ppm TWA; 1800 mg/m3 TWA

OSHA - Final PELs - Time Weighted Averages
1000 ppm TWA; 1800 mg/m3 TWA

ENVIRONMENTAL LISTS

CAA - Flammable Substances for Accidental Release Prevention
threshold quantity = 10,000 lbs

List of Pesticide Product Inert Ingredients
[present]

INTERNATIONAL LISTS

Australian Exposure Standards - Time Weighted Averages
Asphyxiant at < 18% oxygen by volume; explosion hazard

Canada - British Columbia - 8 Hour Exposure Limits
asphyxiant substance

Canada - Ontario - OHSA - TWAEVs
simple asphyxiant

Canada - Quebec - Time-Weighted Average Exposure Values
1000 ppm TWAEV; 1800 mg/m3 TWAEV

German (DFG) - MAK Values
1000 ppm MAK; 1800 mg/m3 MAK

German (DFG) - Peak Limitations
2 x normal MAK (1 hour momentary value); don't exceed 3 times per shift

Israel - Time Weighted Averages
Asphyxiant

Mexico - Instruction No. 10 - TWAs
simple asphyxiant

STATE LISTS

California - Exposure Limits - PELs
1000 ppm PEL; 1800 mg/m3 PEL

Massachusetts Right To Know List
[present]

Minnesota Hazardous Substance List
[present]

NJ Right to Know List (Total)
sn 1594

NJ Special Hazardous Substances
(flammable - fourth degree)

Pennsylvania Right to Know List
[present]

1,3-PROPANEDIOL, 2-HYDROXYMETHYL-2-NITRO-   126-11-4

HEALTH AND SAFETY LISTS

NFPA - Flash Points
flash point = 75 degrees F (24 degrees C)

NFPA - Hazard Identification Ratings
health-2; flammability-3; reactivity-0

NIOSH - Selected LD50s and LC50s
Oral, rat: LD50 = 350 mg/kg Skin, rabbit: LD50 = 200 mg/kg

INTERNATIONAL LISTS

Canada - WHMIS: Ingredient Disclosure
1% item 491 (625)

STATE LISTS

Florida Hazardous Substance List
[present]

Massachusetts Right To Know List
[present]

Pennsylvania Right to Know List
[present]

1,2-PROPANEDIOL, MONOISOPROPYL ETHER   29387-84-6

ENVIRONMENTAL LISTS

Safe Drinking Water Act - Monitoring
monitoring required

TSCA - Code of Federal Regulations Citations
40 CFR 716.120(a)

TSCA - Health and Safety Reporting List
Effective Date: March 7, 1986

STATE LISTS

Massachusetts Right To Know List
[present]

PROPOSED REGULATIONS

Safe Drinking Water Act - Priority list
[present]

1,3-PROPANEDIOL, 2,2'-[OXYBIS(METHYLENE)]   126-58-9

HEALTH AND SAFETY LISTS

NFPA - Flash Points
flash point = 151 degrees F (66 degrees C)

NFPA - Hazard Identification Ratings
health-2; flammability-2; reactivity-3

STATE LISTS

Florida Hazardous Substance List
[present]

Massachusetts Right To Know List
[present]

Pennsylvania Right to Know List
[present]

1,2-PROPANEDIOL,3-(2-PROPENYLOXY)-,BIS(4-METHYLBENZENESULFONATE)   114719-19-6

HEALTH AND SAFETY LISTS

NFPA - Hazard Identification Ratings
health-1; reactivity-0

NIOSH - Selected LD50s and LC50s
Oral, mouse: LD50 = 4773 mg/kg

PROPANE, 2-(ETHENYLOXY)-   926-65-8

HEALTH AND SAFETY LISTS

NIOSH - Selected LD50s and LC50s
Oral, rat: LD50 = 13530 mg/kg

PROPANE, 1,1,1,2,3,3,3-HEPTAFLUORO-   431-89-0

HEALTH AND SAFETY LISTS

NFPA - Flash Points
flash point = 265 degrees F (129 degrees C)

NFPA - Hazard Identification Ratings
health-1; flammability-1; reactivity-0

ENVIRONMENTAL LISTS

EPA - Master Testing List
[present]

List of Pesticide Product Inert Ingredients
[present]

PROPANENITRILE, 3-METHOXY-   110-67-8

HEALTH AND SAFETY LISTS

NIOSH - Selected LD50s and LC50s
Oral, rat: LD50 = 14100 mg/kg

ENVIRONMENTAL LISTS

CAA - HON Rule - SOCMI Chemicals
compliance by Oct. 24, 1994

EPA - Master Testing List
[present]

List of Pesticide Product Inert Ingredients
[present]

PROPANE, 2,2'-OXYBIS[DICHLORO-   63283-80-7

HEALTH AND SAFETY LISTS

NIOSH - Selected LD50s and LC50s
Oral, rat: LD50 = 1900 mg/kg

1,3-PROPANE SULTONE   1120-71-4

ENVIRONMENTAL LISTS

List of Pesticide Product Inert Ingredients
[present]

PROPANE, 1,1,1,2-TETRACHLORO-   812-03-3

ENVIRONMENTAL LISTS

EPA - Master Testing List
[present]

PROPANE, 1,1,1,3-TETRACHLORO-   1070-78-6

ENVIRONMENTAL LISTS

TSCA - Chemicals with Significant New Use Rules
PMN number: P-93-1198

TSCA - Section 12(b) - Export Notification
P-93-1198; export notification required - Section 5

PROPANE, 1,1,2,3-TETRACHLORO-   18495-30-2

HEALTH AND SAFETY LISTS

NFPA - Flash Points
flash point = -26 degrees F (-32 degrees C)

NFPA - Hazard Identification Ratings
health-2; flammability-4; reactivity-2

STATE LISTS

Florida Hazardous Substance List
[present]

Massachusetts Right To Know List
[present]

Pennsylvania Right to Know List
[present]

1-PROPANETHIOL   107-03-9

ENVIRONMENTAL LISTS

TSCA - Chemicals with Significant New Use Rules
PMN number: P-91-831

TSCA - Section 12(b) - Export Notification
P-91-831; export notification required - Section 5

PROPANETHIOL   79869-58-2

HEALTH AND SAFETY LISTS

NFPA - Flash Points
flash point = 149 degrees F (65 degrees C)

NFPA - Hazard Identification Ratings
health-4; flammability-2; reactivity-1

NIOSH - Selected LD50s and LC50s
Oral, rat: LD50 = 4390 mg/kg

STATE LISTS

Florida Hazardous Substance List
[present]

Massachusetts Right To Know List
[present]

Pennsylvania Right to Know List
[present]

1,2,3-PROPANETRICARBOXYLIC ACID, 2-(ACETYLOXY)-, TRIBUTYL ESTER   77-90-7

STATE LISTS

Pennsylvania Right to Know List
[present]

1,2,3-PROPANETRIYL ESTER OF 12-(OXIRANYL-   74398-71-3
METHOXY)-9-OCTADECANOIC ACID

HEALTH AND SAFETY LISTS

ACGIH 1995 - Carcinogens
A2-suspected human carcinogen

U.S. DOT - Appendix A Table 1 - Hazardous Substances
final RQ = 10 pounds (4.54 kg)

IARC - Group 2B (sufficient animal data)
[present] (Overall evaluation based only on evidence of carcinogenicity in monograph (4, 1974) or in Supplement 4)

NTP Seventh Report - Suspect Carcinogens
suspect carcinogen

OSHA - Possible Select Carcinogens
[present]

ENVIRONMENTAL LISTS

CERCLA/SARA - Section 313 - Emission Reporting
form R reporting required for 0.1% de minimus concentration

CERCLA/SARA - Hazardous Substances and their Reportable Quantities
final RQ = 10 pounds (4.54 kg)

Clean Air Act (1990) - List of Hazardous Air Contaminants
[present]

EPA - Carcinogen Hazard Ranking for RQ Adjustment
Hazard ranking = Medium

RCRA - U Series Wastes
waste number U193

RCRA - Hazardous Constituents-Appendix VIII
waste number U193

RCRA - Substances Banned From Land Disposal
[present]

INTERNATIONAL LISTS

Australian Exposure Standards - Time Weighted Averages
control to the lowest practical level

Australian Exposure Standards - Carcinogens
probable carcinogen

Canada - WHMIS: Ingredient Disclosure
0.1% item 1345 (1442)

Canada - Alberta - Designated Substances
designated substance - requires code of practice

Canada - Quebec - Time-Weighted Average Exposure Values
substance of which the recirculation is prohibited

Canada - Quebec - Carcinogens
C2 carcinogen: effect suspected in humans

German (DFG) - Skin/Sensitizers
danger of cutaneous absorption

German (DFG) - Carcinogens
animal evidence of carcinogenicity

STATE LISTS

California - Air Bill 2588 Appendix A-I
known or potential carcinogen

California - Prop. 65 - Cancer list
carcinogen - initial date 1/1/88

California - Prop. 65 - No Significant Risk Levels
no significant risk level = 0.3 ug/day

California - Directors List of Hazardous Substances (8 CCR 339)
[present]

Florida Hazardous Substance List
[present]

Massachusetts Right To Know List
carcinogen; extraordinarily hazardous

Minnesota Hazardous Substance List
carcinogen

NJ Special Hazardous Substances
(carcinogen; mutagen)

Pennsylvania Right to Know List
environmental hazard; special hazardous substance

Pennsylvania RTK - Special Hazardous Substances
[present]

PROPANIL   709-98-8

ENVIRONMENTAL LISTS

TSCA - Code of Federal Regulations Citations
40 CFR 716.120(a)

TSCA - Health and Safety Reporting List
Effective Date: June 1, 1987

PROPANOIC ACID, 2,2-DIMETHYL-, ETHENYL   RR-01643-3
ESTER

ENVIRONMENTAL LISTS

TSCA - Code of Federal Regulations Citations
40 CFR 716.120(a)

TSCA - Health and Safety Reporting List
Effective Date: June 1, 1987

PROPANOIC ACID, ETHENYL ESTER   105-38-4

ENVIRONMENTAL LISTS

TSCA - Code of Federal Regulations Citations
40 CFR 716.120(a)

TSCA - Health and Safety Reporting List
Effective Date: June 1, 1987

PROPANOIC ACID, 3-ETHOXY-   4324-38-3

HEALTH AND SAFETY LISTS

NIOSH - Selected LD50s and LC50s
Inhalation, rat: LC50 = 7300 ppm 4 hr Oral, rat: LD50 = 1790 mg/kg

NIOSH - Health Standards - Exposure Limits
C (15 min) 0.5 ppm; C (15 min) 1.6 mg/m3 (Listed under 'Thiols')

NIOSH - Health Standards - Health Effects and Precautions
Irritation; eye, skin, blood, and nervous system effects (Blood and urine monitoring required; prevent skin contact) (Listed under 'Thiols')

INTERNATIONAL LISTS

Canada - WHMIS: Ingredient Disclosure
1% item 1346 (1443)

STATE LISTS

Massachusetts Right To Know List
[present]

Minnesota Hazardous Substance List
[present]

NJ Right to Know List (Total)
sn 1618

2-PROPANOL,1-[2-[2-[[(4-METHYLPHENYL)SUL-   124028-99-5
FONYL]OXY]ETHOXY]ETHOXY]-3-(2-PROPENY-
LOXY)-,4-METHYLBENZENESULFONATE

HEALTH AND SAFETY LISTS

U.S. DOT - Substances From 49 CFR 172.101
regulated by DOT (UN2402)

U.S. DOT - Hazard Classes
DOT hazard class = 3

STATE LISTS

NJ Right to Know List (Total)
sn 1595

PROPANOLAMINE   156-87-6
ENVIRONMENTAL LISTS
List of Pesticide Product Inert Ingredients
[present]

2-PROPANOL, 1-(2-BUTOXYETHOXY)-   124-16-3
ENVIRONMENTAL LISTS
EPA - Master Testing List
[present]
PROPOSED REGULATIONS
TSCA - Proposed Testing Rule for Glycidyl Ethers
subject to screening subcategory testing (results apply to all members of Glycidyl subcategory V-B)

2-PROPANOL, 1-(2-BUTOXY-1-METHYLETHOXY)-   29911-28-2
ENVIRONMENTAL LISTS
CERCLA/SARA - Section 313 - Emission Reporting
form R reporting required
STATE LISTS
California - Directors List of Hazardous Substances (8 CCR 339)
[present]

1-PROPANOL, 2-CHLORO-, PHOSPHATE (3:1)   6145-73-9
ENVIRONMENTAL LISTS
TSCA - Chemicals with Significant New Use Rules
PMN number: P-89-1058

2-PROPANOL, 1-CHLORO-, PHOSPHATE (3:1)   13674-84-5
HEALTH AND SAFETY LISTS
NFPA - Flash Points
flash point = 34 degrees F (1 degree C)
NFPA - Hazard Identification Ratings
health-2; flammability-3; reactivity-2
STATE LISTS
Florida Hazardous Substance List
[present]
Massachusetts Right To Know List
[present]
Pennsylvania Right to Know List
[present]

1-PROPANOL, 2,3-DICHLORO-   616-23-9
STATE LISTS
Pennsylvania Right to Know List
[present]

1-PROPANOL, 2,2-DIMETHYL-   75-84-3
ENVIRONMENTAL LISTS
TSCA - Chemicals with Significant New Use Rules
PMN number: P-93-1200
TSCA - Section 12(b) - Export Notification
P-93-1200; export notification required - Section 5

1-PROPANOL, 2,2-DIMETHYL-, TRIBROMO DERIVATIVE   36483-57-5
HEALTH AND SAFETY LISTS
NFPA - Flash Points
flash point = 175 degrees F (80 degrees C)
NFPA - Hazard Identification Ratings
health-3; flammability-2; reactivity-0
NIOSH - Selected LD50s and LC50s
Oral, rat: LD50 = 2830 mg/kg Skin, rabbit: LD50 = 1250 mg/kg

INTERNATIONAL LISTS
Canada - WHMIS: Ingredient Disclosure
1% item 1347 (1444)
STATE LISTS
Florida Hazardous Substance List
[present]
Massachusetts Right To Know List
[present]
Pennsylvania Right to Know List
[present]

1-PROPANOL, 3-MERCAPTO   19721-22-3
HEALTH AND SAFETY LISTS
NFPA - Flash Points
flash point = 250 degrees F (121 degrees C)
NFPA - Hazard Identification Ratings
health-2; flammability-1; reactivity-0
NIOSH - Selected LD50s and LC50s
Oral, rat: LD50 = 4000 mg/kg Skin, rabbit: LD50 = 2830 mg/kg
ENVIRONMENTAL LISTS
List of Pesticide Product Inert Ingredients
[present]
TSCA - Code of Federal Regulations Citations
40 CFR 712.30(w); 40 CFR 716.120(a)
TSCA - PAIR - Reporting List
Reporting Date: June 13, 1989
TSCA - Health and Safety Reporting List
Effective Date: April 13, 1989
STATE LISTS
Florida Hazardous Substance List
[present]
Massachusetts Right To Know List
[present]
Pennsylvania Right to Know List
[present]

2-PROPANOL,1-[2-[[(4-METHYLPHENYL)SUL-FONYL]OXY]ETHOXY]-3-(2-PROPENYLOXY)-4-METHYLBENZENESULFONATE   124213-39-4
ENVIRONMENTAL LISTS
TSCA - PAIR - Reporting List
Effective Date: January 26, 1994; Reporting Date: March 28, 1994
TSCA - Health and Safety Reporting List
Effective Date: January 26, 1994; Sunset Date: January 26, 2004
PROPOSED REGULATIONS
TSCA - ITC 31st Report Priority Testing List
recommended for testing

1-PROPANOL, METHOXY-   28677-93-2
ENVIRONMENTAL LISTS
TSCA - Code of Federal Regulations Citations
40 CFR 712.30(w); 40 CFR 716.120(a)
TSCA - PAIR - Reporting List
Reporting Date: February 14, 1989
TSCA - Health and Safety Reporting List
Effective Date: December 16, 1988; Sunset Date: November 9, 1993
PROPOSED REGULATIONS
TSCA - ITC 34th Report Priority Testing List
recommended with intent to designate

PROPANOL 1 (OR 2)-2-METHOXYMETHYLETHOXY,   **88917-22-0**
ACETATE

ENVIRONMENTAL LISTS

EPA - Master Testing List
[present]

TSCA - Code of Federal Regulations Citations
40 CFR 712.30(w); 40 CFR 716.120(a)

TSCA - PAIR - Reporting List
Reporting Date: February 14, 1989

TSCA - Health and Safety Reporting List
Effective Date: December 16, 1988; Sunset Date: November 9, 1993

PROPOSED REGULATIONS

TSCA - ITC 33rd Report Priority Testing List
recommended with intent to designate

TSCA - ITC 34th Report Priority Testing List
recommended with intent to designate

PROPANOL, [(1-METHYL-1,2-ETHANEDIYL)BIS   **24800-44-0**
(OXY)]BIS-

HEALTH AND SAFETY LISTS

NFPA - Flash Points
flash point = 200 degrees F (93 degrees C)

NFPA - Hazard Identification Ratings
health-2; flammability-1; reactivity-0

NIOSH - Selected LD50s and LC50s
Oral, rat: LD50 = 90 mg/kg Skin, rabbit: LD50 = 200 mg/kg

ENVIRONMENTAL LISTS

TSCA - Code of Federal Regulations Citations
40 CFR 716.120(a)

TSCA - Health and Safety Reporting List
Effective Date: March 7, 1986

STATE LISTS

Pennsylvania Right to Know List
[present]

2-PROPANOL, 1-(2-METHYLPROPOXY)-   **23436-19-3**

HEALTH AND SAFETY LISTS

NFPA - Flash Points
flash point = 98 degrees F (37 degrees C)

NFPA - Hazard Identification Ratings
health-2; flammability-3; reactivity-0

STATE LISTS

Florida Hazardous Substance List
[present]

Massachusetts Right To Know List
[present]

Pennsylvania Right to Know List
[present]

2-PROPANOL, 1,1'-OXYBIS   **110-98-5**

ENVIRONMENTAL LISTS

TSCA - Code of Federal Regulations Citations
40 CFR 712.30(x); 40 CFR 716.120(d)

TSCA - PAIR - Reporting List
Reporting Date: December 27, 1990

TSCA - Health and Safety Reporting List
Effective Date: October 29, 1990

1-PROPANOL, 3,3'-OXYBIS[2,3-BIS(BROMOMETHYL)   **RR-01230-6**
]-

ENVIRONMENTAL LISTS

TSCA - Section 12(b) - Export Notification
P-85-433; export notification required (because of proposed action)
- Section 5

2-PROPANOL, 1-PHENOXY-   **770-35-4**

ENVIRONMENTAL LISTS

TSCA - Chemicals with Significant New Use Rules
PMN number: P-93-1199

TSCA - Section 12(b) - Export Notification
P-93-1199; export notification required - Section 5

2-PROPANOL, 1-PROPOXY-   **1569-01-3**

ENVIRONMENTAL LISTS

TSCA - PAIR - Reporting List
Effective Date: January 26, 1994; Reporting Date: March 28, 1994

TSCA - Health and Safety Reporting List
Effective Date: January 26, 1994; Sunset Date: January 26, 2004

PROPOSED REGULATIONS

TSCA - ITC 31st Report Priority Testing List
recommended for testing

PROPAPHOS   **7292-16-2**

ENVIRONMENTAL LISTS

TSCA - PAIR - Reporting List
Effective Date: January 26, 1994; Reporting Date: March 28, 1994

TSCA - Health and Safety Reporting List
Effective Date: January 26, 1994; Sunset Date: January 26, 2004

PROPOSED REGULATIONS

TSCA - ITC 31st Report Priority Testing List
recommended for testing

PROPARGITE   **2312-35-8**

ENVIRONMENTAL LISTS

CAA - HON Rule - SOCMI Chemicals
compliance by Oct. 23, 1995

EPA - Master Testing List
[present]

PROPARGYL ALCOHOL   **107-19-7**

HEALTH AND SAFETY LISTS

NIOSH - Selected LD50s and LC50s
Oral, rat: LD50 = 4290 mg/kg Skin, rabbit: LD50 = 8000 mg/kg

PROPARGYL ALCOHOL   **707-19-7**

HEALTH AND SAFETY LISTS

NIOSH - Selected LD50s and LC50s
Oral, rat: LD50 = 14850 mg/kg

ENVIRONMENTAL LISTS

CAA - HON Rule - SOCMI Chemicals
compliance by Oct. 24, 1994

TSCA - PAIR - Reporting List
Effective Date: January 26, 1994; Reporting Date: March 28, 1994

TSCA - Health and Safety Reporting List
Effective Date: January 26, 1994; Sunset Date: January 26, 2004

PROPOSED REGULATIONS

TSCA - ITC 31st Report Priority Testing List
recommended for testing

2-PROPENAL, 3,4-(1,1-DIMETHYLETHYL)PHENYL-2-   **13586-68-0**
METHYL-

ENVIRONMENTAL LISTS

TSCA - Chemicals with Significant New Use Rules
PMN number: P-87-1273

2-PROPENAL, 3-(2-METHOXYPHENYL)-   1504-74-1

ENVIRONMENTAL LISTS

TSCA - PAIR - Reporting List
Effective Date: January 26, 1994; Reporting Date: March 28, 1994

TSCA - Health and Safety Reporting List
Effective Date: January 26, 1994; Sunset Date: January 26, 2004

PROPOSED REGULATIONS

TSCA - ITC 31st Report Priority Testing List
recommended for testing

2-PROPENAL, 2-METHYL-3-PHENYL-   101-39-3

HEALTH AND SAFETY LISTS

NIOSH - Selected LD50s and LC50s
Oral, rat: LD50 = 3250 mg/kg Skin, rabbit: LD50 = 3560 mg/kg

ENVIRONMENTAL LISTS

List of Pesticide Product Inert Ingredients
[present]

2-PROPENAL, 3-PHENYL-, MONOPENTYL   1331-92-6
DERIVATIVE

HEALTH AND SAFETY LISTS

U.S. DOT - Appendix B - Marine Pollutants
DOT regulated marine pollutant

2-PROPENAMIDE, N-[3-(DIMETHYLAMINO)   RR-00910-9
PROPYL]-

HEALTH AND SAFETY LISTS

U.S. DOT - Appendix A Table 1 - Hazardous Substances
final RQ = 10 pounds (4.54 kg)

NIOSH - Selected LD50s and LC50s
Oral, rat: LD50 = 1480 mg/kg Skin, rat: LD50 = 250 mg/kg

ENVIRONMENTAL LISTS

CERCLA/SARA - Section 313 - Emission Reporting
form R reporting required

CERCLA/SARA - Hazardous Substances and their Reportable Quantities
final RQ = 10 pounds (4.54 kg)

Clean Water Act - Hazardous Substances
[present]

STATE LISTS

California - Prop. 65 - Cancer list
carcinogen - initial date 10/01/94

California - Directors List of Hazardous Substances (8 CCR 339)
[present]

Massachusetts Right To Know List
[present]

NJ Right to Know List (Total)
sn 1596

Pennsylvania Right to Know List
environmental hazard

1-PROPENE, 1,2-DICHLORO-   563-54-2

HEALTH AND SAFETY LISTS

ACGIH 1995 - Time Weighted Averages
1 ppm TWA; 2.3 mg/m3 TWA

ACGIH 1995 - Skin Designations
skin - potential for cutaneous absorption

U.S. DOT - Substances From 49 CFR 172.101
regulated by DOT (NA1986)

U.S. DOT - Hazard Classes
DOT hazard class = 3

U.S. DOT - Appendix A Table 1 - Hazardous Substances
final RQ = 1000 pounds (454 kg)

NFPA - Flash Points
flash point = 97 degrees F (36 degrees C)

NFPA - Hazard Identification Ratings
health-4; flammability-3; reactivity-3

NIOSH - Selected LD50s and LC50s
Inhalation, rat: LC50 = 2000 mg/m3 2 hr Oral, rat: LD50 = 55 mg/kg Skin, rabbit: LD50 = 88 mg/kg

OSHA - Vacated PELs - Time Weighted Averages
1 ppm TWA; 2 mg/m3 TWA

OSHA - Vacated PELs - Skin Designation
Prevent or reduce skin absorption

ENVIRONMENTAL LISTS

CERCLA/SARA - Section 313 - Emission Reporting
form R reporting required

CERCLA/SARA - Hazardous Substances and their Reportable Quantities
final RQ = 1000 pounds (454 kg)

RCRA - P Series Wastes
waste number P102

RCRA - Hazardous Constituents-Appendix VIII
waste number P102

RCRA - Substances Banned From Land Disposal
[present]

TSCA - Code of Federal Regulations Citations
40 CFR 716.120(a)

TSCA - Health and Safety Reporting List
Effective Date: March 7, 1986

INTERNATIONAL LISTS

Australian Exposure Standards - Time Weighted Averages
1 ppm TWA; 2.3 mg/m3 TWA

Australian Exposure Standards - Skin Effects
skin absorption

Canada - WHMIS: Ingredient Disclosure
1% item 1348 (186)

Canada - Alberta - 8 Hour Occupational Exposure Limit
1 ppm TWA; 2.3 mg/m3 TWA

Canada - Alberta - 15 Minute Occupational Exposure Limit
3 ppm STEL; 6.9 mg/m3 STEL

Canada - Alberta - Skin Designation
can be absorbed through the intact skin

Canada - British Columbia - 8 Hour Exposure Limits
1 ppm TWA; 2 mg/m3 TWA

Canada - British Columbia - 15 Minute Exposure Limits
3 ppm STEL; 6 mg/m3 STEL

Canada - British Columbia - Skin Notations
skin - potential for skin absorption

Canada - Ontario - OHSA - TWAEVs
1 ppm TWAEV; 2.3 mg/m3 TWAEV

Canada - Ontario - OHSA - Skin Notations
absorption through skin, eyes, or mucous membranes

Canada - Quebec - Time-Weighted Average Exposure Values
1 ppm TWAEV; 2.3 mg/m3 TWAEV

Canada - Quebec - Skin Designations
absorbed through the skin

United Kingdom - Occupational Exposure Standards - TWAs
1 ppm TWA; 2 mg/m3 TWA

United Kingdom - Occupational Exposure Standards - STELs
3 ppm STEL; 6 mg/m3 STEL

United Kingdom - Occupational Exposure Standards - Notes
can be absorbed through skin

German (DFG) - MAK Values
2 ppm MAK; 5 mg/m3 MAK

German (DFG) - Skin/Sensitizers
danger of cutaneous absorption

Israel - Time Weighted Averages
1 ppm TWA; 2.3 mg/m3 TWA
Israel - Action Levels
0.5 ppm AL; 1.15 mg/m3 AL
STATE LISTS
California - Exposure Limits - PELs
1 ppm PEL; 2 mg/m3 PEL
California - Exposure Limits - Skin Notation
material may be absorbed through the skin, eyes or mucous membrane
California - Directors List of Hazardous Substances (8 CCR 339)
[present]
Florida Hazardous Substance List
[present]
Massachusetts Right To Know List
[present]
Minnesota Hazardous Substance List
skin
NJ Right to Know List (Total)
sn 1597
NJ Special Hazardous Substances
(flammable - third degree; reactive - third degree)
Pennsylvania Right to Know List
environmental hazard

1-PROPENE, 3-(ETHENYLOXY)-   3917-15-5

ENVIRONMENTAL LISTS
List of Pesticide Product Inert Ingredients
[present]

INTERNATIONAL LISTS
Canada - Alberta - 8 Hour Occupational Exposure Limit
1 ppm TWA
Canada - Alberta - 15 Minute Occupational Exposure Limit
3 ppm STEL
Canada - Alberta - Skin Designation
can be absorbed through the intact skin

1-PROPENE, 1-ETHOXY-   928-55-2

ENVIRONMENTAL LISTS
TSCA - Code of Federal Regulations Citations
40 CFR 712.30(x); 40 CFR 716.120(d)
TSCA - PAIR - Reporting List
Reporting Date: November 27, 1991
TSCA - Health and Safety Reporting List
Effective Date: September 30, 1991

PROPENE, 3-ISOCYANATO-   1476-23-9

ENVIRONMENTAL LISTS
TSCA - Code of Federal Regulations Citations
40 CFR 712.30(x); 40 CFR 716.120(d)
TSCA - PAIR - Reporting List
Reporting Date: November 27, 1991
TSCA - Health and Safety Reporting List
Effective Date: September 30, 1991

2-PROPENENITRILE POLYMER WITH   9003-54-7
ETHENYLBENZENE

ENVIRONMENTAL LISTS
TSCA - Code of Federal Regulations Citations
40 CFR 712.30(x); 40 CFR 716.120(d)
TSCA - PAIR - Reporting List
Reporting Date: November 27, 1991
TSCA - Health and Safety Reporting List
Effective Date: September 30, 1991

2-PROPENENITRILE, POLYMER WITH 1,3-BUTA-   RR-00997-2
DIENE, 3-CARBOXY-1-CYANO-1-METHYLPROPYL-
TERMINATED, POLYMERS WITH BISPHENOL A,
EPICHLOROHYDRIN AND 4,4'-(1-METHYLETHYLI-
DENE) BIS(2,6-DIBROMOPHENOL

ENVIRONMENTAL LISTS
TSCA - Code of Federal Regulations Citations
40 CFR 712.30(x); 40 CFR 716.120(d)
TSCA - PAIR - Reporting List
Reporting Date: November 27, 1991
TSCA - Health and Safety Reporting List
Effective Date: September 30, 1991

2-PROPENENITRILE, POLYMER WITH 1,3-BUTA-   RR-00996-1
DIENE, 3-CARBOXY-1-CYANO-1-METHYLPROPYL-
TERMINATED, POLYMERS WITH EPICHLOROHY-
DRIN, FORMALDEHYDE, 4,4'-(1-METHYL ETHYLI-
DENE)BIS(2,6-DI

ENVIRONMENTAL LISTS
TSCA - Chemicals with Significant New Use Rules
PMN number: P-86-1602

1-PROPENE, 3,3,3-TRIFLUORO-   677-21-4

ENVIRONMENTAL LISTS
TSCA - Code of Federal Regulations Citations
40 CFR 716.120(a)
TSCA - Health and Safety Reporting List
Effective Date: March 7, 1986

PROPENOATE-TERMINATED ALKYL SUBSTITUTED   RR-01177-8
SILYL ESTER

HEALTH AND SAFETY LISTS
NFPA - Flash Points
flash point < 68 degrees F (20 degrees C)
NFPA - Hazard Identification Ratings
health-2; flammability-3; reactivity-2
STATE LISTS
Florida Hazardous Substance List
[present]
Massachusetts Right To Know List
[present]
Pennsylvania Right to Know List
[present]

2-PROPENOIC ACID 3-(TRIMETHOXYSILYL)   RR-01732-3
PROPYL ESTER

HEALTH AND SAFETY LISTS
NFPA - Flash Points
flash point < 20 degrees F (-7 degrees C)
NFPA - Hazard Identification Ratings
health-2; flammability-3; reactivity-0
STATE LISTS
Florida Hazardous Substance List
[present]
Massachusetts Right To Know List
[present]
Pennsylvania Right to Know List
[present]

2-PROPENOIC ACID, BICYCLO[2.2.1]HEPT-5-EN-2-   95-39-6
YL-, METHYL ESTER

ENVIRONMENTAL LISTS
TSCA - Code of Federal Regulations Citations
40 CFR 712.30(x); 40 CFR 716.120(d)

TSCA - PAIR - Reporting List
Reporting Date: December 27, 1990

TSCA - Health and Safety Reporting List
Effective Date: October 29, 1990 Sunset Date: November 9, 1993

2-PROPENOIC ACID, C-18-26 AND C>20 ALKYL ESTERS   RR-01685-3

HEALTH AND SAFETY LISTS

IARC - Group 3 (not classifiable)
[present]

NIOSH - Selected LD50s and LC50s
Oral, rat: LD50 = 1800 mg/kg

2-PROPENOIC ACID, 2-CYANO-, BUTYL ESTER   6606-65-1

ENVIRONMENTAL LISTS

TSCA - Chemicals with Significant New Use Rules
PMN number: P-90-1393

2-PROPENOIC ACID, 2-CYANO-3,3-DIPHENYL-, 2-ETHYLHEXYL ESTER   6197-30-4

ENVIRONMENTAL LISTS

TSCA - Chemicals with Significant New Use Rules
PMN number: P-90-668

2-PROPENOIC ACID, 2-CYANO-, ETHOXY ETHYL ESTER   21982-43-4

ENVIRONMENTAL LISTS

TSCA - Code of Federal Regulations Citations
40 CFR 712.30(f); 40 CFR 716.120(a)

TSCA - PAIR - Reporting List
Reporting Date: August 17, 1983

TSCA - Health and Safety Reporting List
Effective Date: April 29, 1983

2-PROPENOIC ACID, 2-CYANO-, ISOBUTYL ESTER   1069-55-2

ENVIRONMENTAL LISTS

TSCA - Chemicals with Significant New Use Rules
PMN number: P-91-74

2-PROPENOIC ACID, 2-CYANO-, 2-METHOXYETHYL ESTER   27816-23-5

ENVIRONMENTAL LISTS

TSCA - Chemicals with Significant New Use Rules
PMN number: P-93-1235

2-PROPENOIC ACID, 2-CYANO-, 1-METHYLETHYL ESTER   10586-17-1

ENVIRONMENTAL LISTS

TSCA - Code of Federal Regulations Citations
40 CFR 712.30(d)

TSCA - PAIR - Reporting List
Reporting Date: November 19, 1982

2-PROPENOIC ACID, 2-CYANO-, 2-PROPENYL ESTER   7324-02-9

ENVIRONMENTAL LISTS

TSCA - Chemicals with Significant New Use Rules
PMN numbers: P-93-37, P-93-38

2-PROPENOIC ACID, 2-CYANO-, 2,2,2-TRIFLUO-ROMETHYL ESTER   23023-91-8

ENVIRONMENTAL LISTS

TSCA - PAIR - Reporting List
Effective Date: January 26, 1994; Reporting Date: March 28, 1994

TSCA - Health and Safety Reporting List
Effective Date: January 26, 1994; Sunset Date: January 26, 2004

2-PROPENOIC ACID, 2,3-DIBROMOPROPYL ESTER   19660-16-3

ENVIRONMENTAL LISTS

List of Pesticide Product Inert Ingredients
[present]

TSCA - PAIR - Reporting List
Effective Date: January 26, 1994; Reporting Date: March 28, 1994

TSCA - Health and Safety Reporting List
Effective Date: January 26, 1994; Sunset Date: January 26, 2004

2-PROPENOIC ACID, [2-[1,1-DIMETHYL-2-[(1-OXO-2-PROPENYL)OXY]ETHYL]-5-ETHYL-1,3-DIOXAN-5-YL)METHYL ESTER   87320-05-6

ENVIRONMENTAL LISTS

TSCA - PAIR - Reporting List
Effective Date: January 26, 1994; Reporting Date: March 28, 1994

TSCA - Health and Safety Reporting List
Effective Date: January 26, 1994; Sunset Date: January 26, 2004

2-PROPENOIC ACID, 1,1-DIMETHYLETHYL ESTER   1663-39-4

ENVIRONMENTAL LISTS

TSCA - PAIR - Reporting List
Effective Date: January 26, 1994; Reporting Date: March 28, 1994

TSCA - Health and Safety Reporting List
Effective Date: January 26, 1994; Sunset Date: January 26, 2004

2-PROPENOIC ACID, 2,2-DINITROPROPYL ESTER   17977-09-2

ENVIRONMENTAL LISTS

TSCA - PAIR - Reporting List
Effective Date: January 26, 1994; Reporting Date: March 28, 1994

TSCA - Health and Safety Reporting List
Effective Date: January 26, 1994; Sunset Date: January 26, 2004

2-PROPENOIC ACID, DOCOSYL ESTER   RR-01684-2

ENVIRONMENTAL LISTS

TSCA - PAIR - Reporting List
Effective Date: January 26, 1994; Reporting Date: March 28, 1994

TSCA - Health and Safety Reporting List
Effective Date: January 26, 1994; Sunset Date: Januray 26, 2004

2-PROPENOIC ACID, 3A,4,5,6,7,7A-HEXAHYDRO-4,7-METHANO-1H-INDENYL ESTER   33791-58-1

ENVIRONMENTAL LISTS

TSCA - PAIR - Reporting List
Effective Date: January 26, 1994

TSCA - Health and Safety Reporting List
Effective Date: January 26, 1994; Sunset Date: January 26, 2004

2-PROPENOIC ACID, HEXYL ESTER   2499-95-8

ENVIRONMENTAL LISTS

TSCA - PAIR - Reporting List
Effective Date: January 26, 1994; Reporting Date: March 28, 1994

TSCA - Health and Safety Reporting List
Effective Date: January 26, 1994; Sunset Date: January 26, 2004

2-PROPENOIC ACID, 2-HYDROXYBUTYL ESTER   2421-27-4

ENVIRONMENTAL LISTS

TSCA - Code of Federal Regulations Citations
40 CFR 716.120(a)

TSCA - Health and Safety Reporting List
Effective Date: June 1, 1987

2-PROPENOIC ACID, 1-[HYDROXYMETHYL] PROPYL ESTER   RR-00234-6

ENVIRONMENTAL LISTS

TSCA - Section 12(b) - Export Notification
P-84-344;   export notification required - Section 5

2-PROPENOIC ACID, ISODECYL ESTER   **1330-61-6**

ENVIRONMENTAL LISTS

TSCA - Code of Federal Regulations Citations
40 CFR 712.30(d)

TSCA - PAIR - Reporting List
Reporting Date: November 19, 1982

2-PROPENOIC ACID, ISOOCTYL ESTER   **29590-42-9**

ENVIRONMENTAL LISTS

TSCA - Code of Federal Regulations Citations
40 CFR 712.30(d)

TSCA - PAIR - Reporting List
Reporting Date: November 19, 1982

2-PROPENOIC ACID, 3-(DIMETHYLAMINO)-2,2-   **RR-01004-8**
DIMETHYLPROPYL ESTER

ENVIRONMENTAL LISTS

TSCA - Chemicals with Significant New Use Rules
PMN number: P-93-36

2-PROPENOIC ACID, 2-METHYL-, 2-[3-(2H-BENZO-   **96478-09-0**
TRIAZOL-2-YL)-4-HYDROXYPHENYL]ETHYL ESTER

ENVIRONMENTAL LISTS

TSCA - Code of Federal Regulations Citations
40 CFR 712.30(d)

TSCA - PAIR - Reporting List
Reporting Date: November 19, 1982

2-PROPENOIC ACID, 2-METHYL-, 2-[BIS(1-   **16715-83-6**
METHYLETHYL)AMINO]ETHYL ESTER

ENVIRONMENTAL LISTS

TSCA - Code of Federal Regulations Citations
40 CFR 712.30(d)

TSCA - PAIR - Reporting List
Reporting Date: November 19, 1982

2-PROPENOIC ACID, 2-METHYL-, CYCLOHEXYL   **101-43-9**
ESTER

ENVIRONMENTAL LISTS

TSCA - Code of Federal Regulations Citations
40 CFR 721.1810

TSCA - Chemicals with Significant New Use Rules
PMN number: P-87-930

2-PROPENOIC ACID, 2-METHYL-, 1,1-   **585-07-9**
DIMETHYLETHYL ESTER

ENVIRONMENTAL LISTS

TSCA - Chemicals with Significant New Use Rules
PMN number: P-87-931

2-PROPENOIC ACID, 2-METHYL-, 1,2-ETHANEDIYL   **97-90-5**
ESTER

HEALTH AND SAFETY LISTS

U.S. DOT - Appendix B - Marine Pollutants
DOT regulated marine pollutant

ENVIRONMENTAL LISTS

TSCA - Code of Federal Regulations Citations
40 CFR 712.30(d)

TSCA - PAIR - Reporting List
Reporting Date: November 19, 1982

2-PROPENOIC ACID, 1-METHYLETHYL ESTER   **689-12-3**

ENVIRONMENTAL LISTS

EPA - Master Testing List
[present]

2-PROPENOIC ACID, 2-METHYL-, 2-HYDROX-   **923-26-2**
YPROPYL ESTER

ENVIRONMENTAL LISTS

TSCA - Chemicals with Significant New Use Rules
PMN number: P-85-545

2-PROPENOIC ACID, 2-METHYL-, 2-[[[[[5-ISO-   **73597-26-9**
CYANATO-1,3,3-TRIMETHYLCYCLOHEXYL]
METHYL]AMINO]CARBONYL]OXY]ETHYL ESTER

ENVIRONMENTAL LISTS

TSCA - Code of Federal Regulations Citations
40 CFR 721.1817

TSCA - Section 12(b) - Export Notification
P-90-333;   export notification required - Section 5

2-PROPENOIC ACID, 2-METHYL-, OCTADECYL   **32360-05-7**
ESTER

ENVIRONMENTAL LISTS

TSCA - Code of Federal Regulations Citations
40 CFR 712.30(d)

TSCA - PAIR - Reporting List
Reporting Date: November 19, 1982

2-PROPENOIC ACID, 2-METHYL-, 7-OXABICYCLO   **82428-30-6**
[4.1.0]HEPT-3-YLMETHYL ESTER

ENVIRONMENTAL LISTS

TSCA - Code of Federal Regulations Citations
40 CFR 712.30(d)

TSCA - PAIR - Reporting List
Reporting Date: November 19, 1982

2-PROPENOIC ACID, 2-METHYL-, PROPYL ESTER   **2210-28-8**

ENVIRONMENTAL LISTS

TSCA - Code of Federal Regulations Citations
40 CFR 721.1822

TSCA - Chemicals with Significant New Use Rules
PMN number: P-89-422

2-PROPENOIC ACID, 2-METHYL-, SODIUM SALT   **5536-61-8**

HEALTH AND SAFETY LISTS

NIOSH - Selected LD50s and LC50s
Oral, rat: LD50 = 3300 mg/kg

ENVIRONMENTAL LISTS

TSCA - Code of Federal Regulations Citations
40 CFR 712.30(d)

TSCA - PAIR - Reporting List
Reporting Date: November 19, 1982

2-PROPENOIC ACID, 2-METHYL-, 1,7,7-   **7534-94-3**
TRIMETHYLBICYCLO[2.2.1]HEPT-2-YL ESTER,
EXO-

ENVIRONMENTAL LISTS

TSCA - Code of Federal Regulations Citations
40 CFR 712.30(d)

TSCA - PAIR - Reporting List
Reporting Date: November 19, 1982

2-PROPENOIC ACID, 2-METHYL-, 7,7,9-TRIMETHYL-4,13-DIOXO-3,14-DIOXO-5,12-DIAZA-HEXADECANE, 1,16-DIYL ESTER   RR-01003-7

HEALTH AND SAFETY LISTS

NIOSH - Selected LD50s and LC50s
Oral, mouse: LD50 = 7964 mg/kg

2-PROPENOIC ACID, 2-METHYL-, 3,3,5-TRIMETHYL CYCLOHEXYL ESTER   RR-01005-9

ENVIRONMENTAL LISTS

TSCA - Code of Federal Regulations Citations
40 CFR 712.30(x); 40 CFR 716.120(d)

TSCA - PAIR - Reporting List
Reporting Date: December 27, 1990

TSCA - Health and Safety Reporting List
Effective Date: October 29, 1990; Sunset Date: November 9, 1993

2-PROPENOIC ACID, MONOESTER WITH 1,2-PROPANEDIOL   25584-83-2

ENVIRONMENTAL LISTS

TSCA - Code of Federal Regulations Citations
40 CFR 712.30(d)

TSCA - PAIR - Reporting List
Reporting Date: November 19, 1982

2-PROPENOIC ACID, 2-[[(1-METHYLETHOXY) CARBONYL]AMINO]ETHYLESTER   RR-01646-6

ENVIRONMENTAL LISTS

TSCA - Code of Federal Regulations Citations
40 CFR 721.1822

TSCA - Chemicals with Significant New Use Rules
PMN number: P-89-30

TSCA - Section 12(b) - Export Notification
P-89-30; export notification required - Section 5

2-PROPENOIC ACID, OCTAHYDRO-4,7-METHANO-1H-INDENYL ESTER   79637-74-4

ENVIRONMENTAL LISTS

TSCA - Code of Federal Regulations Citations
40 CFR 712.30(d)

TSCA - PAIR - Reporting List
Reporting Date: November 19, 1982

2-PROPENOIC ACID [OCTAHYDRO-4,7-METHANO-1H-INDENE-1,5(1,6 OR2,5)-DIYL]BIS(METHYLENE) ESTER   RR-00989-2

ENVIRONMENTAL LISTS

List of Pesticide Product Inert Ingredients
[present]

TSCA - Code of Federal Regulations Citations
40 CFR 712.30(d)

TSCA - PAIR - Reporting List
Reporting Date: November 19, 1982

2-PROPENOIC ACID, OCTYL ESTER   2499-59-4

ENVIRONMENTAL LISTS

TSCA - Code of Federal Regulations Citations
40 CFR 712.30(d)

TSCA - PAIR - Reporting List
Reporting Date: November 19, 1982

2-PROPENOIC ACID, 7-OXABICYCLO[4.1.0]HEPT-3-YLMETHYL ESTER   64630-63-3

ENVIRONMENTAL LISTS

TSCA - Chemicals with Significant New Use Rules
PMN number: P-85-544

2-PROPENOIC ACID, 2-(2-OXO-3-OXAZOLIDINYL) ETHYL ESTER   115965-75-8

ENVIRONMENTAL LISTS

TSCA - Chemicals with Significant New Use Rules
PMN number: P-85-546

2-PROPENOIC ACID, POLYMER WITH 2-PROPENAMIDE, SODIUM SALT   25987-30-8

ENVIRONMENTAL LISTS

TSCA - Code of Federal Regulations Citations
40 CFR 712.30(d)

TSCA - PAIR - Reporting List
Reporting Date: November 19, 1982

2-PROPENOIC ACID, PROPYL ESTER   925-60-0

ENVIRONMENTAL LISTS

TSCA - Chemicals with Significant New Use Rules
PMN number: P-91-503

2-PROPENOIC ACID, REACTION PRODUCT WITH 2-OXEPROPANONE AND ALKYL TRIOL   RR-01683-1

ENVIRONMENTAL LISTS

TSCA - Code of Federal Regulations Citations
40 CFR 721.1832

TSCA - Chemicals with Significant New Use Rules
PMN number: P-90-1285

TSCA - Section 12(b) - Export Notification
P-90-1285; export notification required - Section 5

2-PROPENOIC ACID, TRIDECYL ESTER   3076-04-8

ENVIRONMENTAL LISTS

TSCA - Chemicals with Significant New Use Rules
PMN number: P-90-333

2-PROPENOIC ACID, 3,3,5-TRIMETHYLCYCLO-HEXYL ESTER   RR-01006-0

STATE LISTS

Pennsylvania Right to Know List
[present]

2-PROPENOIC ACID, 2-[[[[[1,3,3-TRIMETHYL-5-[[[2-[(1-OXO-2-PROPENYL)OXY]ETHOXY]CARBONYL]AMINO]CYCLOHEXYL]METHYL]AMINO] CARBONYL]OXY]ETHYL ESTER   42404-50-2

ENVIRONMENTAL LISTS

TSCA - Code of Federal Regulations Citations
40 CFR 721.1815

TSCA - Chemicals with Significant New Use Rules
PMN number: P-89-31

TSCA - Section 12(b) - Export Notification
P-89-31; export notification required - Section 5

PROPETAMPHOS   31218-83-4

ENVIRONMENTAL LISTS

TSCA - Chemicals with Significant New Use Rules
PMN number: P-91-391

TSCA - Section 12(b) - Export Notification
P-91-391; export notification required - Section 5

PROPHAM   122-42-9

ENVIRONMENTAL LISTS

List of Pesticide Product Inert Ingredients
[present]

PROPICONAZOLE [1-[2-(2,4-DICHLOROPHENYL)-4-   **60207-90-1**
PROPYL-1,3-DIOXOLAN-2-YL]-METHYL-1H-1,2,4,-
TRIAZOLE]

ENVIRONMENTAL LISTS

TSCA - Code of Federal Regulations Citations
40 CFR 712.30(d)

TSCA - PAIR - Reporting List
Reporting Date: November 19, 1982

BETA-PROPIOLACTONE   **57-57-8**

ENVIRONMENTAL LISTS

TSCA - Chemicals with Significant New Use Rules
PMN number: P-92-1447

PROPIONALDEHYDE   **123-38-6**

HEALTH AND SAFETY LISTS

NFPA - Flash Points
flash point = 270 degrees F (132 degrees C)

NFPA - Hazard Identification Ratings
health-1; flammability-1; reactivity-0

ENVIRONMENTAL LISTS

TSCA - Code of Federal Regulations Citations
40 CFR 712.30(d)

TSCA - PAIR - Reporting List
Reporting Date: November 19, 1982

PROPIONIC ACID   **79-09-4**

ENVIRONMENTAL LISTS

TSCA - Chemicals with Significant New Use Rules
PMN number: P-85-547

PROPIONIC ACID, 2-(2,4,5-TRICHLOROPHENOXY)-,   **73826-29-6**
P-CHLOROPHENACYL ESTER

ENVIRONMENTAL LISTS

TSCA - Chemicals with Significant New Use Rules
PMN number: P-90-1825

TSCA - Section 12(b) - Export Notification
P-90-1825; export notification required - Section 5

PROPIONIC ANHYDRIDE   **123-62-6**

ENVIRONMENTAL LISTS

CERCLA/SARA - Section 313 - Emission Reporting
form R reporting required

STATE LISTS

Massachusetts Right To Know List
neurotoxin

PROPIONITRILE   **107-12-0**

HEALTH AND SAFETY LISTS

IARC - Group 3 (not classifiable)
[present]

STATE LISTS

California - Directors List of Hazardous Substances (8 CCR 339)
[present]

PROPIONYL CHLORIDE   **79-03-8**

ENVIRONMENTAL LISTS

CERCLA/SARA - Section 313 - Emission Reporting
form R reporting required

PROPIONYL PEROXIDE   **3248-28-0**

HEALTH AND SAFETY LISTS

ACGIH 1995 - Time Weighted Averages
0.5 ppm TWA;   1.5 mg/m3 TWA

ACGIH 1995 - Carcinogens
A2-suspected human carcinogen

IARC - Group 2B (sufficient animal data)
[present] (Overall evaluation based only on evidence of carcinogenicity in monograph (4, 1974) or in Supplement 4)

NFPA - Flash Points
flash point = 165 degrees F (74 degrees C)

NFPA - Hazard Identification Ratings
health-0; flammability-2; reactivity-0

NIOSH - Selected LD50s and LC50s
Inhalation, rat: LC50 = 25 ppm 6 hr

NIOSH 1990 - Pocket Guide - Carcinogens
occupational carcinogen

NIOSH 1990 - Pocket Guide - Target organs
kidneys, skin, lungs, eyes

NIOSH - Health Standards - Exposure Limits
use 29 CFR 1910.1013

NIOSH - Health Standards - Health Effects and Precautions
has produced tumors of the liver, skin, and stomach in animals

NIOSH - Health Standards - Carcinogenic Chemicals
potential human carcinogen

NTP Seventh Report - Suspect Carcinogens
suspect carcinogen

OSHA - 29 CFR 1910 Specifically Regulated Chemicals
Cancer suspect agent (see 29 CFR 1910.1013)

OSHA - Select Carcinogens
[present]

OSHA - Possible Select Carcinogens
[present]

ENVIRONMENTAL LISTS

CERCLA/SARA - Section 302 Extremely Hazardous Substances and TPQs
TPQ = 500 pounds

CERCLA/SARA - Section 313 - Emission Reporting
form R reporting required for 0.1% de minimus concentration

CERCLA/SARA - Hazardous Substances and their Reportable Quantities
final RQ = 1 pound (.454 kg)

Clean Air Act (1990) - List of Hazardous Air Contaminants
[present]

CAA - HON Rule - SOCMI Chemicals
compliance by Oct. 24, 1994

CAA - HON Rule - Organic HAPs
[present]

INTERNATIONAL LISTS

Australian Exposure Standards - Time Weighted Averages
0.5 ppm TWA; 1.5 mg/m3 TWA

Australian Exposure Standards - Carcinogens
probable carcinogen

Canada - WHMIS: Ingredient Disclosure
0.1% item 1349 (1445)

Canada - Alberta - 8 Hour Occupational Exposure Limit
0.5 ppm TWA; 1.5 mg/m3 TWA

Canada - Alberta - 15 Minute Occupational Exposure Limit
1 ppm STEL; 3 mg/m3 STEL

Canada - Alberta - Designated Substances
designated substance - requires code of practice

Canada - British Columbia - 8 Hour Exposure Limits
carcinogen with no established permitted concentration

Canada - British Columbia - Carcinogens
carcinogen with no established permitted concentration
Canada - Ontario - OHSA - TWAEVs
0.5 ppm TWAEV; 1.5 mg/m3 TWAEV
Canada - Quebec - Time-Weighted Average Exposure Values
0.5 ppm TWAEV; 1.5 mg/m3 TWAEV
Canada - Quebec - Carcinogens
C2 carcinogen: effect suspected in humans
German (DFG) - Carcinogens
animal evidence of carcinogenicity
Israel - Time Weighted Averages
0.5 ppm TWA; 1.5 mg/m3 TWA
Israel - Action Levels
0.25 ppm AL; 0.75 mg/m3 AL

STATE LISTS

California - Air Bill 2588 Appendix A-I
known or potential carcinogen
California - Prop. 65 - Cancer list
carcinogen - initial date 1/1/88
California - Prop. 65 - No Significant Risk Levels
no significant risk level - 0.05 ug/day
California - Exposure Limits - PELs
0.5 ppm PEL; 1.5 mg/m3 PEL
California - Exposure Limits - Carcinogens
cancer-suspect agent (at a concentration >= 1.0%)
California - Directors List of Hazardous Substances (8 CCR 339)
[present]
Florida Hazardous Substance List
[present]
Massachusetts Right To Know List
carcinogen; extraordinarily hazardous
Minnesota Hazardous Substance List
carcinogen
NJ Right to Know List (Total)
sn 0228
NJ Special Hazardous Substances
(carcinogen; corrosive; mutagen)
Pennsylvania Right to Know List
environmental hazard; special hazardous substance
Pennsylvania RTK - Special Hazardous Substances
[present]

PROPOSED REGULATIONS

CERCLA/SARA - Proposed Hazardous Substance Additions
proposed RQ - 1 pound (.454 kg)
CERCLA/SARA - 1989 Proposed RQ Adjustments
proposed RQ - 10 pounds (4.54 kg)

PROPIOPHENONE, 4-AMINO-   70-69-9

HEALTH AND SAFETY LISTS

U.S. DOT - Substances From 49 CFR 172.101
regulated by DOT (UN1275)
U.S. DOT - Hazard Classes
DOT hazard class - 3
NFPA - Flash Points
flash point - -22 degrees F (-30 degrees C)
NFPA - Hazard Identification Ratings
health-2; flammability-3; reactivity-2
NIOSH - Selected LD50s and LC50s
Inhalation, mouse: LD50 - 21800 mg/m3 (2 hr) Oral, rat: LD50 - 1410 mg/kg Skin, rabbit: LD50 - 5040 mg/kg

ENVIRONMENTAL LISTS

CERCLA/SARA - Section 313 - Emission Reporting
form R reporting required for 1.0% de minimus concentration

CERCLA/SARA - Hazardous Substances and their Reportable Quantities
final RQ - 1 pound (.454 kg)
Clean Air Act (1990) - List of Hazardous Air Contaminants
[present]
CAA - HON Rule - SOCMI Chemicals
compliance by July 24, 1995
CAA - HON Rule - Organic HAPs
[present]
EPA - Master Testing List
[present]
List of Pesticide Product Inert Ingredients
[present]
TSCA - Code of Federal Regulations Citations
40 CFR 712.30(x); 40 CFR 716.120(d)
TSCA - PAIR - Reporting List
Reporting Date: November 27, 1991
TSCA - Health and Safety Reporting List
Effective Date: September 30, 1991

INTERNATIONAL LISTS

Canada - WHMIS: Ingredient Disclosure
1% item 1350 (1446)
Canada - NPRI (National Pollutant Release Inventory)
[present]

STATE LISTS

California - Air Bill 2588 Appendix A-I
6/91
Florida Hazardous Substance List
[present]
Massachusetts Right To Know List
[present]
NJ Right to Know List (Total)
sn 1598
NJ Special Hazardous Substances
(flammable - third degree)
Pennsylvania Right to Know List
environmental hazard

PROPOXUR   114-26-1

HEALTH AND SAFETY LISTS

ACGIH 1995 - Time Weighted Averages
10 ppm TWA; 30 mg/m3 TWA
AIHA - Odor Threshold Values
geoemtric mean air odor threshold - 0.066 ppm (detectable); 0.033 ppm (recognizable)
U.S. DOT - Substances From 49 CFR 172.101
regulated by DOT (UN1848)
U.S. DOT - Hazard Classes
DOT hazard class - 8
U.S. DOT - Appendix A Table 1 - Hazardous Substances
final RQ - 5000 pounds (2270 kg)
NFPA - Flash Points
flash point - 126 degrees F (52 degrees C)
NFPA - Hazard Identification Ratings
health-3; flammability-2; reactivity-0
NIOSH - Selected LD50s and LC50s
Oral, rat: LD50 - 3500 mg/kg Skin, rabbit: LD50 - 500 mg/kg
OSHA - Vacated PELs - Time Weighted Averages
10 ppm TWA; 30 mg/m3 TWA

ENVIRONMENTAL LISTS

CERCLA/SARA - Hazardous Substances and their Reportable Quantities
final RQ - 5000 pounds (2270 kg)

CAA - HON Rule - SOCMI Chemicals
compliance by Oct. 24, 1994

Clean Water Act - Hazardous Substances
[present]

List of Pesticide Product Inert Ingredients
[present]

INTERNATIONAL LISTS

Australian Exposure Standards - Time Weighted Averages
10 ppm TWA; 30 mg/m3 TWA

Canada - WHMIS: Ingredient Disclosure
1% item 1351 (129)

Canada - Alberta - 8 Hour Occupational Exposure Limit
10 ppm TWA; 30 mg/m3 TWA

Canada - Alberta - 15 Minute Occupational Exposure Limit
15 ppm STEL; 45 mg/m3 STEL

Canada - Ontario - OHSA - TWAEVs
10 ppm TWAEV; 30 mg/m3 TWAEV

Canada - Quebec - Time-Weighted Average Exposure Values
10 ppm TWAEV; 30 mg/m3 TWAEV

United Kingdom - Occupational Exposure Standards - TWAs
10 ppm TWA; 30 mg/m3 TWA

United Kingdom - Occupational Exposure Standards - STELs
15 ppm STEL; 45 mg/m3 STEL

German (DFG) - MAK Values
10 ppm MAK; 30 mg/m3 MAK

German (DFG) - Peak Limitations
2 x normal MAK (5 min momentary value); don't exceed 8 times during shift

Israel - Time Weighted Averages
10 ppm TWA; 30 mg/m3 TWA

Israel - Action Levels
5 ppm AL; 15 mg/m3 AL

STATE LISTS

California - Exposure Limits - PELs
10 ppm PEL; 30 mg/m3 PEL

California - Directors List of Hazardous Substances (8 CCR 339)
[present]

Florida Hazardous Substance List
[present]

Massachusetts Right To Know List
[present]

Minnesota Hazardous Substance List
[present]

NJ Right to Know List (Total)
sn 1599

NJ Special Hazardous Substances
(corrosive)

Pennsylvania Right to Know List
environmental hazard

2-[2-(PROPOXYETHOXY)ETHOXY]-ETHANOL   23305-64-8

STATE LISTS

NJ Right to Know List (Total)
sn 1898

N-PROPYL ACETATE   109-60-4

HEALTH AND SAFETY LISTS

U.S. DOT - Substances From 49 CFR 172.101
regulated by DOT (UN2496)

U.S. DOT - Hazard Classes
DOT hazard class = 8

U.S. DOT - Appendix A Table 1 - Hazardous Substances
final RQ = 5000 pounds (2270 kg)

FDA - Controlled Substances Act - Precursor chemicals
Threshold by base weight = 1 gram

NFPA - Flash Points
flash point = 145 degrees F (63 degrees C)

NFPA - Hazard Identification Ratings
health-3; flammability-2; reactivity-1 (decomposes in water)

NIOSH - Selected LD50s and LC50s
Oral, rat: LD50 = 2360 mg/kg Skin, rabbit: LD50 = 10 gm/kg

ENVIRONMENTAL LISTS

CERCLA/SARA - Hazardous Substances and their Reportable Quantities
final RQ = 5000 pounds (2270 kg)

Clean Water Act - Hazardous Substances
[present]

TSCA - Code of Federal Regulations Citations
40 CFR 712.30(x); 40 CFR 716.120(d)

TSCA - PAIR - Reporting List
Reporting Date: November 27, 1991

TSCA - Health and Safety Reporting List
Effective Date: September 30, 1991

INTERNATIONAL LISTS

Canada - WHMIS: Ingredient Disclosure
1% item 1352 (234)

STATE LISTS

California - Directors List of Hazardous Substances (8 CCR 339)
[present]

Florida Hazardous Substance List
[present]

Massachusetts Right To Know List
[present]

NJ Right to Know List (Total)
sn 1600

NJ Special Hazardous Substances
(corrosive)

Pennsylvania Right to Know List
environmental hazard

N-PROPYL ALCOHOL   71-23-8
SEE ALSO:
F039-HAZARDOUS WASTES

HEALTH AND SAFETY LISTS

U.S. DOT - Substances From 49 CFR 172.101
regulated by DOT (UN2404)

U.S. DOT - Hazard Classes
DOT hazard class = 3

U.S. DOT - Appendix A Table 1 - Hazardous Substances
final RQ = 10 pounds (4.54 kg)

NFPA - Flash Points
flash point = 36 degrees F (2 degrees C)

NFPA - Hazard Identification Ratings
health-4; flammability-3; reactivity-1

NIOSH - Selected LD50s and LC50s
Inhalation, mouse: LC50 = 163 ppm 1 hr Oral, rat: LD50 = 39 mg/kg Skin, rabbit: LD50 = 210 mg/kg

NIOSH - Health Standards - Exposure Limits
6 ppm TWA; 14 mg/m3 TWA (Listed under 'Nitriles')

NIOSH - Health Standards - Health Effects and Precautions
Hepatic, renal, respiratory, cardiovascular, gastrointestinal, and nervous system effects (Periodic chest X-ray and pulmonary function testing required; prevent skin and eye contact; make first-aid kits and personnel available during use) (Listed under 'Nitriles')

ENVIRONMENTAL LISTS

CERCLA/SARA - Section 302 Extremely Hazardous Substances and TPQs
TPQ = 500 pounds

CERCLA/SARA - Hazardous Substances and their Reportable Quantities
final RQ = 10 pounds (4.54 kg)

CAA -Toxic Substances for Accidental Release Prevention
threshold quantity = 10,000 lbs

RCRA - P Series Wastes
waste number P101

RCRA - Hazardous Constituents-Appendix VIII
waste number P101

RCRA - Basis for Listing - Appendix VII
Included in waste stream: F039

RCRA - Substances Banned From Land Disposal
[present]

RCRA - TSD Facilities Ground Water Monitoring
TM 8015 = 60 ug/L PQL; TM 8240 = 5 ug/L PQL

RCRA - Universal Treatment Standards (LDR)
WW: 0.24 mg/l; NWW: 360 mg/kg

INTERNATIONAL LISTS

Canada - WHMIS: Ingredient Disclosure
1% item 710 (592)

STATE LISTS

Florida Hazardous Substance List
[present]

Massachusetts Right To Know List
carcinogen; extraordinarily hazardous

NJ Right to Know List (Total)
sn 1601

NJ Special Hazardous Substances
(flammable - third degree)

Pennsylvania Right to Know List
environmental hazard

PROPYL BENZENE   103-65-1

HEALTH AND SAFETY LISTS

U.S. DOT - Substances From 49 CFR 172.101
regulated by DOT (UN1815)

U.S. DOT - Hazard Classes
DOT hazard class = 3

NFPA - Flash Points
flash point = 54 degrees F (12 degrees C)

NFPA - Hazard Identification Ratings
health-3; flammability-3; reactivity-1 (decomposes in water)

INTERNATIONAL LISTS

Canada - WHMIS: Ingredient Disclosure
1% item 1353 (522)

STATE LISTS

Florida Hazardous Substance List
[present]

Massachusetts Right To Know List
[present]

NJ Right to Know List (Total)
sn 1602

NJ Special Hazardous Substances
(corrosive; flammable - third degree)

Pennsylvania Right to Know List
[present]

2-PROPYLBIPHENYL   RR-00261-9

HEALTH AND SAFETY LISTS

U.S. DOT - Hazard Classes
Forbidden from transport by the DOT

U.S. DOT - Organic Peroxides Table
Organic peroxide UN3117

STATE LISTS

NJ Right to Know List (Total)
sn 1603

N-PROPYL BROMIDE   106-94-5

ENVIRONMENTAL LISTS

CERCLA/SARA - Section 302 Extremely Hazardous Substances and TPQs
TPQ = 100/10,000 pounds

STATE LISTS

Florida Hazardous Substance List
effective March 13, 1992

Massachusetts Right To Know List
extraordinarily hazardous

Pennsylvania Right to Know List
environmental hazard

PROPOSED REGULATIONS

CERCLA/SARA - Proposed Hazardous Substance Additions
proposed RQ = 1 pound (.454 kg)

CERCLA/SARA - 1989 Proposed RQ Adjustments
proposed RQ = 100 pounds (45.4 kg)

N-PROPYL CARBAMATE   627-12-3

HEALTH AND SAFETY LISTS

ACGIH 1995 - Time Weighted Averages
0.5 mg/m3 TWA

U.S. DOT - Appendix B - Marine Pollutants
DOT regulated marine pollutant

NIOSH - Selected LD50s and LC50s
Inhalation, rat: LC50 = 1440 mg/m3 1 hr Oral, rat: LD50 = 70 mg/kg Skin, rat: LD50 = 800 mg/kg

OSHA - Vacated PELs - Time Weighted Averages
0.5 mg/m3 TWA

ENVIRONMENTAL LISTS

CERCLA/SARA - Section 313 - Emission Reporting
form R reporting required for 1.0% de minimus concentration

CERCLA/SARA - Hazardous Substances and their Reportable Quantities
final RQ = 1 pound (.454 kg)

Clean Air Act (1990) - List of Hazardous Air Contaminants
[present]

INTERNATIONAL LISTS

Australian Exposure Standards - Time Weighted Averages
0.5 mg/m3 TWA

Canada - Alberta - 8 Hour Occupational Exposure Limit
0.5 mg/m3 TWA

Canada - Alberta - 15 Minute Occupational Exposure Limit
2 mg/m3 STEL

Canada - British Columbia - 8 Hour Exposure Limits
0.5 mg/m3 TWA

Canada - British Columbia - 15 Minute Exposure Limits
2 mg/m3 STEL

Canada - Ontario - OHSA - TWAEVs
0.5 mg/m3 TWAEV

Canada - Quebec - Time-Weighted Average Exposure Values
0.5 mg/m3 TWAEV

United Kingdom - Occupational Exposure Standards - TWAs
0.5 mg/m3 TWA

United Kingdom - Occupational Exposure Standards - STELs
2 mg/m3 STEL

German (DFG) - MAK Values
total dust: 2 mg/m3 MAK

Israel - Time Weighted Averages
0.5 mg/m3 TWA

Israel - Action Levels
0.25 mg/m3 AL

STATE LISTS

California - Air Bill 2588 Appendix A-I
6/91

California - Exposure Limits - PELs
0.5 mg/m3 PEL

California - Directors List of Hazardous Substances (8 CCR 339)
[present]

Florida Hazardous Substance List
[present]

Massachusetts Right To Know List
neurotoxin

Minnesota Hazardous Substance List
[present]

NJ Right to Know List (Total)
sn 1604

Pennsylvania Right to Know List
environmental hazard

PROPYL CHLORIDE   540-54-5

ENVIRONMENTAL LISTS

TSCA - Section 12(b) - Export Notification
P-94-138; export notification required

PROPYLCHLOROFORMATE   109-61-5

HEALTH AND SAFETY LISTS

ACGIH 1995 - Time Weighted Averages
200 ppm TWA; 835 mg/m3 TWA

ACGIH 1995 - Short Term Exposure Limits
250 ppm STEL; 1040 mg/m3 STEL

AIHA - Odor Threshold Values
geometric mean air odor threshold = 0.18 ppm (detectable); 1.9 ppm (recognizable)

U.S. DOT - Substances From 49 CFR 172.101
regulated by DOT (UN1276)

U.S. DOT - Hazard Classes
DOT hazard class = 3

NFPA - Flash Points
flash point = 55 degrees F (13 degrees C)

NFPA - Hazard Identification Ratings
health-1; flammability-3; reactivity-0

NIOSH - Selected LD50s and LC50s
Oral, rat: LD50 = 9370 mg/kg

NIOSH 1990 - Pocket Guide - RELs
200 ppm TWA; 840 mg/m3 TWA; 250 ppm STEL; 1050 mg/m3 STEL

NIOSH 1990 - Pocket Guide - IDLHs
8000 ppm IDLH

NIOSH 1990 - Pocket Guide - Target organs
skin, eyes, CNS, respiratory system

OSHA - Vacated PELs - Time Weighted Averages
200 ppm TWA; 840 mg/m3 TWA

OSHA - Vacated PELs - Short Term Exposure Limits
250 ppm STEL; 1050 mg/m3 STEL

OSHA - Final PELs - Time Weighted Averages
200 ppm TWA; 840 mg/m3 TWA

INTERNATIONAL LISTS

Australian Exposure Standards - Time Weighted Averages
200 ppm TWA; 835 mg/m3 TWA

Australian Exposure Standards - Short Term Exposure Limits
250 ppm STEL; 1040 mg/m3 STEL

Canada - WHMIS: Ingredient Disclosure
1% item 1354 (32)

Canada - Alberta - 8 Hour Occupational Exposure Limit
200 ppm TWA; 835 mg/m3 TWA

Canada - Alberta - 15 Minute Occupational Exposure Limit
250 ppm STEL; 1040 mg/m3 STEL

Canada - British Columbia - 8 Hour Exposure Limits
200 ppm TWA; 840 mg/m3 TWA

Canada - British Columbia - 15 Minute Exposure Limits
250 ppm STEL; 1050 mg/m3 STEL

Canada - Ontario - OHSA - TWAEVs
200 ppm TWAEV; 830 mg/m3 TWAEV

Canada - Ontario - OHSA - STEVs
250 ppm STEV; 1040 mg/m3 STEV

Canada - Quebec - Time-Weighted Average Exposure Values
200 ppm TWAEV; 835 mg/m3 TWAEV

Canada - Quebec - Short-term Exposure Values
250 ppm STEV; 1040 mg/m3 STEV

United Kingdom - Occupational Exposure Standards - TWAs
200 ppm TWA; 840 mg/m3 TWA

United Kingdom - Occupational Exposure Standards - STELs
250 ppm STEL; 1050 mg/m3 STEL

German (DFG) - MAK Values
200 ppm MAK; 840 mg/m3 MAK

German (DFG) - Peak Limitations
2 x normal MAK (5 min momentary value); don't exceed 8 times per shift

Israel - Time Weighted Averages
200 ppm TWA; 835 mg/m3 TWA

Israel - Short Term Exposure Limits
250 ppm STEL; 1040 mg/m3 STEL

Israel - Action Levels
100 ppm AL; 417.5 mg/m3 AL

Mexico - Instruction No. 10 - TWAs
200 ppm TWA; 840 mg/m3 TWA

Mexico - Instruction No. 10 - STELs
250 ppm STEL; 1050 mg/m3 STEL

STATE LISTS

California - Exposure Limits - PELs
200 ppm PEL; 840 mg/m3 PEL

California - Exposure Limits - STELs
250 ppm STEL; 1050 mg/m3 STEL

California - Directors List of Hazardous Substances (8 CCR 339)
[present]

Florida Hazardous Substance List
[present]

Massachusetts Right To Know List
[present]

Minnesota Hazardous Substance List
[present]

NJ Right to Know List (Total)
sn 1419

NJ Special Hazardous Substances
(flammable - third degree)

Pennsylvania Right to Know List
[present]

PROPYLCYCLOHEXANE   RR-00881-1

HEALTH AND SAFETY LISTS

ACGIH 1995 - Time Weighted Averages
200 ppm TWA; 492 mg/m3 TWA

ACGIH 1995 - Short Term Exposure Limits
250 ppm STEL; 614 mg/m3 STEL

ACGIH 1995 - Skin Designations
skin - potential for cutaneous absorption

AIHA - Odor Threshold Values
geometric mean air odor threshold = 5.3 ppm (detectable); 11 ppm (recognizable)

U.S. DOT - Substances From 49 CFR 172.101
regulated by DOT (UN1274)

U.S. DOT - Hazard Classes
DOT hazard class = 3

NFPA - Flash Points
flash point = 74 degrees F (23 degrees C)

NFPA - Hazard Identification Ratings
health-1; flammability-3; reactivity-0

NIOSH - Selected LD50s and LC50s
Inhalation, mouse: LC50 = 48 gm/m3 8 hr Oral, rat: LD50 = 1870 mg/kg Skin, rabbit: LD50 = 5040 mg/kg

NIOSH 1990 - Pocket Guide - RELs
200 ppm TWA; 500 mg/m3 TWA; 250 ppm STEL; 625 mg/m3 STEL

NIOSH 1990 - Pocket Guide - IDLHs
4000 ppm IDLH

NIOSH 1990 - Pocket Guide - Target organs
skin, eyes, respiratory system, GI tract

NIOSH 1990 - Pocket Guide - Skin list
Potential for dermal absorption

OSHA - Vacated PELs - Time Weighted Averages
200 ppm TWA; 500 mg/m3 TWA

OSHA - Vacated PELs - Short Term Exposure Limits
250 ppm STEL; 625 mg/m3 STEL

OSHA - Final PELs - Time Weighted Averages
200 ppm TWA; 500 mg/m3 TWA

ENVIRONMENTAL LISTS

List of Pesticide Product Inert Ingredients
[present]

INTERNATIONAL LISTS

Australian Exposure Standards - Time Weighted Averages
200 ppm TWA; 492 mg/m3 TWA

Australian Exposure Standards - Short Term Exposure Limits
250 ppm STEL; 614 mg/m3 STEL

Australian Exposure Standards - Skin Effects
skin absorption

Canada - WHMIS: Ingredient Disclosure
1% item 1355 (187)

Canada - Alberta - 8 Hour Occupational Exposure Limit
200 ppm TWA; 491 mg/m3 TWA

Canada - Alberta - 15 Minute Occupational Exposure Limit
250 ppm STEL; 615 mg/m3 STEL

Canada - Alberta - Skin Designation
can be absorbed through the intact skin

Canada - British Columbia - 8 Hour Exposure Limits
200 ppm TWA; 500 mg/m3 TWA

Canada - British Columbia - 15 Minute Exposure Limits
250 ppm STEL; 625 mg/m3 STEL

Canada - British Columbia - Skin Notations
skin - potential for skin absorption

Canada - Ontario - OHSA - TWAEVs
200 ppm TWAEV; 490 mg/m3 TWAEV

Canada - Ontario - OHSA - STEVs
250 ppm STEV; 615 mg/m3 STEV

Canada - Ontario - OHSA - Skin Notations
absorption through skin, eyes, or mucous membranes

Canada - Quebec - Time-Weighted Average Exposure Values
200 ppm TWAEV; 492 mg/m3 TWAEV

Canada - Quebec - Short-term Exposure Values
250 ppm STEV; 615 mg/m3 STEV

Canada - Quebec - Skin Designations
absorbed through the skin

United Kingdom - Occupational Exposure Standards - TWAs
200 ppm TWA; 500 mg/m3 TWA

United Kingdom - Occupational Exposure Standards - STELs
250 ppm STEL; 625 mg/m3 STEL

United Kingdom - Occupational Exposure Standards - Notes
can be absorbed through skin

Israel - Time Weighted Averages
200 ppm TWA; 492 mg/m3 TWA

Israel - Short Term Exposure Limits
250 ppm STEL; 615 mg/m3 STEL

Israel - Action Levels
100 ppm AL; 246 mg/m3 AL

Mexico - Instruction No. 10 - TWAs
200 ppm TWA; 500 mg/m3 TWA

Mexico - Instruction No. 10 - STELs
250 ppm STEL; 625 mg/m3 STEL

STATE LISTS

California - Exposure Limits - PELs
200 ppm PEL; 500 mg/m3 PEL

California - Exposure Limits - STELs
250 ppm STEL; 625 mg/m3 STEL

California - Exposure Limits - Skin Notation
material may be absorbed through the skin, eyes or mucous membrane

California - Directors List of Hazardous Substances (8 CCR 339)
[present]

Florida Hazardous Substance List
[present]

Massachusetts Right To Know List
[present]

Minnesota Hazardous Substance List
skin

NJ Right to Know List (Total)
sn 1605

NJ Special Hazardous Substances
(flammable - third degree)

Pennsylvania Right to Know List
[present]

PROPYLCYCLOPENTANE   2040-96-2

HEALTH AND SAFETY LISTS

U.S. DOT - Substances From 49 CFR 172.101
regulated by DOT (UN2364)

U.S. DOT - Hazard Classes
DOT hazard class = 3

NFPA - Flash Points
flash point = 86 degrees F (30 degrees C)

NFPA - Hazard Identification Ratings
health-2; flammability-3; reactivity-0

NIOSH - Selected LD50s and LC50s
Oral, rat: LD50 = 6040 mg/kg

ENVIRONMENTAL LISTS

Safe Drinking Water Act - Monitoring
monitoring required at discretion of the state

EPA - Master Testing List
[present]

TSCA - Section 12(b) - Export Notification
export notification required - Section 4

STATE LISTS

Florida Hazardous Substance List
[present]

Massachusetts Right To Know List
[present]

NJ Right to Know List (Total)
sn 1607

NJ Special Hazardous Substances
(flammable - third degree)

Pennsylvania Right to Know List
[present]

PROPYLENE   115-07-1

HEALTH AND SAFETY LISTS

NFPA - Flash Points
flash point > 212 degrees F (100 degrees C)

NFPA - Hazard Identification Ratings
health-0; flammability-1; reactivity-0

PROPYLENE CARBONATE   108-32-7

HEALTH AND SAFETY LISTS

NFPA - Hazard Identification Ratings
health-2; flammability-3; reactivity-0

NIOSH - Selected LD50s and LC50s
Inhalation, rat: LC50 = 253000 mg/m3 (30 mn)

STATE LISTS

Florida Hazardous Substance List
[present]

Massachusetts Right To Know List
[present]

Pennsylvania Right to Know List
[present]

PROPYLENE CHLOROHYDRIN   78-89-7

HEALTH AND SAFETY LISTS

IARC - Group 3 (not classifiable)
[present]

STATE LISTS

California - Directors List of Hazardous Substances (8 CCR 339)
[present]

PROPYLENE DIAMINE   78-90-0

HEALTH AND SAFETY LISTS

U.S. DOT - Substances From 49 CFR 172.101
regulated by DOT (UN1278)

U.S. DOT - Hazard Classes
DOT hazard class = 3

NFPA - Flash Points
flash point < 0 degrees F (-18 degrees C)

NFPA - Hazard Identification Ratings
health-2; flammability-3; reactivity-0

ENVIRONMENTAL LISTS

TSCA - Code of Federal Regulations Citations
40 CFR 716.120(a)

TSCA - Health and Safety Reporting List
Effective Date: June 1, 1987

STATE LISTS

Florida Hazardous Substance List
[present]

Massachusetts Right To Know List
[present]

NJ Right to Know List (Total)
sn 2062

Pennsylvania Right to Know List
[present]

PROPYLENE-ETHYLENE THIOETHER   RR-01131-4

HEALTH AND SAFETY LISTS

U.S. DOT - Substances From 49 CFR 172.101
regulated by DOT (UN2740)

U.S. DOT - Hazard Classes
DOT hazard class = 6.1

U.S. DOT - Substances Which Are Poisonous by Inhalation
liquid hazardous material poisonous by inhalation (UN2740)

NIOSH - Selected LD50s and LC50s
Inhalation, mouse: LC50 = 319 ppm 1 hr Oral, mouse: LD50 = 650 mg/kg Skin, mouse: LD50 = 10 mg/kg

ENVIRONMENTAL LISTS

CERCLA/SARA - Section 302 Extremely Hazardous Substances and TPQs
TPQ = 500 pounds

CAA -Toxic Substances for Accidental Release Prevention
threshold quantity = 15,000 lbs

INTERNATIONAL LISTS

Canada - WHMIS: Ingredient Disclosure
1% item 1357 (440)

STATE LISTS

Florida Hazardous Substance List
effective March 13, 1992

Massachusetts Right To Know List
extraordinarily hazardous

NJ Right to Know List (Total)
sn 1608

NJ Special Hazardous Substances
(corrosive)

Pennsylvania Right to Know List
environmental hazard

PROPOSED REGULATIONS

CERCLA/SARA - Proposed Hazardous Substance Additions
proposed RQ = 1 pound (.454 kg)

CERCLA/SARA - 1989 Proposed RQ Adjustments
proposed RQ = 100 pounds (45.4 kg)

1,2-PROPYLENE GLYCOL   57-55-6

HEALTH AND SAFETY LISTS

NFPA - Hazard Identification Ratings
health-0; reactivity-0

PROPYLENE GLYCOL ALGINATE   9005-37-2

HEALTH AND SAFETY LISTS

NFPA - Hazard Identification Ratings
health-0; reactivity-0

PROPYLENE GLYCOL, ALLYL ETHER   1331-17-5

HEALTH AND SAFETY LISTS

ACGIH 1995 - Time Weighted Averages
simple asphyxiant

AIHA - Odor Threshold Values
geometric mean air odor threshold = 23 ppm (detectable); 68 ppm (recognizable)
U.S. DOT - Substances From 49 CFR 172.101
regulated by DOT (UN1077)
U.S. DOT - Hazard Classes
DOT hazard class = 2.1
IARC - Group 3 (not classifiable)
[present]
NFPA - Flash Points
gas (no flash point given)
NFPA - Hazard Identification Ratings
health-1; flammability-4; reactivity-1
NTP Chemical Status Reports - Testing Status and NTIS Number
Technical reports printed (PB86145521/AS)
NTP Chemical Status Reports - Evidence of Carcinogenicity
male rat-no evidence; female rat-no evidence; male mice-no evidence; female mice-no evidence

ENVIRONMENTAL LISTS
CERCLA/SARA - Section 313 - Emission Reporting
form R reporting required for 1.0% de minimus concentration
CAA - Flammable Substances for Accidental Release Prevention
threshold quantity = 10,000 lbs

INTERNATIONAL LISTS
Australian Exposure Standards - Time Weighted Averages
Asphyxiant at < 18% oxygen by volume; explosion hazard
Canada - NPRI (National Pollutant Release Inventory)
[present]
Canada - British Columbia - 8 Hour Exposure Limits
asphyxiant substance
Canada - Ontario - OHSA - TWAEVs
simple asphyxiant
Canada - Quebec - Time-Weighted Average Exposure Values
simple asphyxiant
Israel - Time Weighted Averages
Asphyxiant
Mexico - Instruction No. 10 - TWAs
simple asphyxiant

STATE LISTS
California - Air Bill 2588 Appendix A-I
[present]
California - Exposure Limits - PELs
asphyxiant (limit depends on level of oxygen)
California - Directors List of Hazardous Substances (8 CCR 339)
[present]
Florida Hazardous Substance List
[present]
Massachusetts Right To Know List
[present]
Minnesota Hazardous Substance List
[present]
NJ Right to Know List (Total)
sn 1609
NJ Special Hazardous Substances
(flammable - fourth degree)
Pennsylvania Right to Know List
environmental hazard

PROPYLENE GLYCOL T-BUTYL ETHER   **57018-52-7**
HEALTH AND SAFETY LISTS
NFPA - Flash Points
flash point = 275 degrees F (135 degrees C)

NFPA - Hazard Identification Ratings
health-1; flammability-1; reactivity-0
NIOSH - Selected LD50s and LC50s
Oral, rat: LD50 = 29 gm/kg

ENVIRONMENTAL LISTS
CAA - HON Rule - SOCMI Chemicals
compliance by Oct. 23, 1995
List of Pesticide Product Inert Ingredients
[present]

INTERNATIONAL LISTS
Canada - WHMIS: Ingredient Disclosure
1% item 1358 (391)

PROPYLENE GLYCOL DINITRATE   **6423-43-4**
HEALTH AND SAFETY LISTS
U.S. DOT - Substances From 49 CFR 172.101
regulated by DOT (UN2611)
U.S. DOT - Hazard Classes
DOT hazard class = 6.1
NFPA - Flash Points
flash point = 125 degrees F (52 degrees C)
NFPA - Hazard Identification Ratings
health-2; flammability-2; reactivity-0
NIOSH - Selected LD50s and LC50s
Oral, rat: LD50 = 218 mg/kg Skin, rabbit: LD50 = 529 mg/kg

INTERNATIONAL LISTS
Canada - WHMIS: Ingredient Disclosure
1% item 1359 (442)

STATE LISTS
Florida Hazardous Substance List
[present]
Massachusetts Right To Know List
[present]
NJ Right to Know List (Total)
sn 1610
Pennsylvania Right to Know List
[present]

PROPYLENE GLYCOL ETHERS AND ESTERS   **RR-01715-2**
HEALTH AND SAFETY LISTS
U.S. DOT - Substances From 49 CFR 172.101
regulated by DOT (UN2258)
U.S. DOT - Hazard Classes
DOT hazard class = 8
NFPA - Flash Points
flash point = 92 degrees F (33 degrees C)
NFPA - Hazard Identification Ratings
health-2; flammability-3; reactivity-0
NIOSH - Selected LD50s and LC50s
Oral, rat: LD50 = 2230 mg/kg Skin, rabbit: LD50 = 500 mg/kg

INTERNATIONAL LISTS
Canada - WHMIS: Ingredient Disclosure
1% item 1360 (1453)

STATE LISTS
Florida Hazardous Substance List
[present]
Massachusetts Right To Know List
[present]
NJ Right to Know List (Total)
sn 1611
NJ Special Hazardous Substances
(corrosive; flammable - third degree)

Pennsylvania Right to Know List
[present]

PROPYLENE GLYCOL ETHYL ETHER   1569-02-4

ENVIRONMENTAL LISTS

List of Pesticide Product Inert Ingredients
[present]

PROPYLENE GLYCOL ISOBUTYL AND HIGHER HOMOLOGS   RR-01130-3

HEALTH AND SAFETY LISTS

AIHA - WEEL - Time Weighted Averages
total: 50 ppm TWA; aerosol only: 10 mg/m3 TWA

NFPA - Flash Points
flash point = 210 degrees F (99 degrees C)

NFPA - Hazard Identification Ratings
health-0; flammability-1; reactivity-0

NIOSH - Selected LD50s and LC50s
Oral, rat: LD50 = 20 gm/kg Skin, rabbit: LD50 = 20800 mg/kg

ENVIRONMENTAL LISTS

CAA - HON Rule - SOCMI Chemicals
compliance by Oct. 24, 1994

List of Pesticide Product Inert Ingredients
[present]

INTERNATIONAL LISTS

Canada - WHMIS: Ingredient Disclosure
1% item 1362 (1454)

United Kingdom - Occupational Exposure Standards - TWAs
total (vapour and particulates): 150 ppm TWA; 470 mg/m3 TWA; particulates: 10 mg/m3 TWA

STATE LISTS

Minnesota Hazardous Substance List
[present]

Pennsylvania Right to Know List
[present]

PROPYLENE GLYCOL ISOPROPYL ETHER   RR-01169-8

ENVIRONMENTAL LISTS

List of Pesticide Product Inert Ingredients
[present]

PROPYLENE GLYCOL MONOBUTYL ETHER   29387-86-8

HEALTH AND SAFETY LISTS

NIOSH - Selected LD50s and LC50s
Oral, rat: LD50 = 510 mg/kg Skin, rabbit: LD50 = 1100 mg/kg

STATE LISTS

Massachusetts Right To Know List
[present]

PROPYLENE GLYCOL MONOMETHYL ETHER   107-98-2

HEALTH AND SAFETY LISTS

NTP Chemical Status Reports - Testing Status and NTIS Number
Approved for toxicology/carcinogenesis study

ENVIRONMENTAL LISTS

TSCA - PAIR - Reporting List
Effective Date: January 26, 1994; Reporting Date: March 28, 1994

TSCA - Health and Safety Reporting List
Effective Date: January 26, 1994; Sunset Date: January 26, 2004

PROPOSED REGULATIONS

TSCA - ITC 31st Report Priority Testing List
recommended for testing

PROPYLENE GLYCOL MONOMETHYL ETHER ACETATE   108-65-6

HEALTH AND SAFETY LISTS

ACGIH 1995 - Time Weighted Averages
0.05 ppm TWA; 0.34 mg/m3 TWA

ACGIH 1995 - Skin Designations
skin - potential for cutaneous absorption

AIHA - Odor Threshold Values
geometric mean air odor threshold = 0.24 ppm (detectable)

NIOSH - Selected LD50s and LC50s
Oral, rat: LD50 = 250 mg/kg

OSHA - Vacated PELs - Time Weighted Averages
0.05 ppm TWA; 0.3 mg/m3 TWA

ENVIRONMENTAL LISTS

TSCA - Health and Safety Reporting List
Effective Date: March 11, 1994; Sunset Date: March 11, 2004

INTERNATIONAL LISTS

Australian Exposure Standards - Time Weighted Averages
0.05 ppm TWA; 0.34 mg/m3 TWA

Australian Exposure Standards - Skin Effects
skin absorption

Canada - WHMIS: Ingredient Disclosure
1% item 1363 (751)

Canada - Alberta - 8 Hour Occupational Exposure Limit
0.02 ppm TWA; 0.14 mg/m3 TWA

Canada - Alberta - 15 Minute Occupational Exposure Limit
0.05 ppm STEL; 0.34 mg/m3 STEL

Canada - Alberta - Skin Designation
can be absorbed through the intact skin

Canada - British Columbia - Ceiling Exposure Limits
C 0.2 ppm; C 2 mg/m3

Canada - British Columbia - Skin Notations
skin - potential for skin absorption

Canada - Ontario - OHSA - TWAEVs
0.05 ppm TWAEV; 0.34 mg/m3 TWAEV

Canada - Ontario - OHSA - Skin Notations
absorption through skin, eyes, or mucous membranes

Canada - Quebec - Time-Weighted Average Exposure Values
0.05 ppm TWAEV; 0.34 mg/m3 TWAEV

Canada - Quebec - Skin Designations
absorbed through the skin

United Kingdom - Occupational Exposure Standards - TWAs
0.2 ppm TWA; 1.2 mg/m3 TWA

United Kingdom - Occupational Exposure Standards - STELs
0.2 ppm STEL; 1.2 mg/m3 STEL

United Kingdom - Occupational Exposure Standards - Notes
can be absorbed through skin

German (DFG) - MAK Values
0.05 ppm MAK; 0.3 mg/m3 MAK (when skin contact does not occur)

German (DFG) - Skin/Sensitizers
danger of cutaneous absorption

Israel - Time Weighted Averages
0.05 ppm TWA; 0.34 mg/m3 TWA

Israel - Action Levels
0.025 ppm AL; 0.17 mg/m3 AL

STATE LISTS

California - Exposure Limits - PELs
0.05 ppm PEL; 0.3 mg/m3 PEL

California - Exposure Limits - Skin Notation
material may be absorbed through the skin, eyes or mucous membrane

California - Directors List of Hazardous Substances (8 CCR 339)
[present]

Florida Hazardous Substance List
[present]

Massachusetts Right To Know List
[present]

Minnesota Hazardous Substance List
skin

Pennsylvania Right to Know List
[present]

PROPOSED REGULATIONS

TSCA - ITC 32nd Report Priority Testing List
designated for dermal absorption testing

PROPYLENE GLYCOL MONOMETHYL ETHER ACETATE   84540-57-8

PROPOSED REGULATIONS

TSCA - ITC 33rd Report Priority Testing List
recommended for testing

TSCA - ITC 34th Report Priority Testing List
recommended for testing

PROPYLENE GLYCOL MONOPROPYL ETHER   30136-13-1

HEALTH AND SAFETY LISTS

NIOSH - Selected LD50s and LC50s
Oral, rat: LD50 = 4400 mg/kg Skin, rabbit: LD50 = 8100 mg/kg

PROPYLENEIMINE   75-55-8

ENVIRONMENTAL LISTS

List of Pesticide Product Inert Ingredients
[present]

PROPYLENE OXIDE   75-56-9

HEALTH AND SAFETY LISTS

NFPA - Flash Points
flash point = 110 degrees F (43 degrees C)

PROPYLENE SULFIDE   1072-43-1

ENVIRONMENTAL LISTS

TSCA - PAIR - Reporting List
Effective Date: January 26, 1994; Reporting Date: March 28, 1994

TSCA - Health and Safety Reporting List
Effective Date: January 26, 1994; Sunset Date: January 26, 2004

PROPOSED REGULATIONS

TSCA - ITC 31st Report Priority Testing List
recommended for testing

PROPYLENE TETRAMER   6842-15-5

HEALTH AND SAFETY LISTS

ACGIH 1995 - Time Weighted Averages
100 ppm TWA; 369 mg/m3 TWA

ACGIH 1995 - Short Term Exposure Limits
150 ppm STEL; 553 mg/m3 STEL

AIHA - Odor Threshold Values
no geometric mean air odor threshold

U.S. DOT - Substances From 49 CFR 172.101
regulated by DOT (UN3092)

U.S. DOT - Hazard Classes
DOT hazard class = 3

NFPA - Flash Points
flash point = 90 degrees F (32 degrees C)

NFPA - Hazard Identification Ratings
health-0; flammability-3; reactivity-0

NIOSH - Selected LD50s and LC50s
Oral, rat: LD50 = 5660 mg/kg Skin, rabbit: LD50 = 13 gm/kg

OSHA - Vacated PELs - Time Weighted Averages
100 ppm TWA; 360 mg/m3 TWA

OSHA - Vacated PELs - Short Term Exposure Limits
150 ppm STEL; 540 mg/m3 STEL

ENVIRONMENTAL LISTS

CAA - HON Rule - SOCMI Chemicals
compliance by Oct. 24, 1994

List of Pesticide Product Inert Ingredients
[present]

TSCA - Code of Federal Regulations Citations
40 CFR 712.30(w); 40 CFR 716.120

TSCA - PAIR - Reporting List
Reporting Date: June 13, 1989

TSCA - Health and Safety Reporting List
Effective Date: April 13, 1989

INTERNATIONAL LISTS

Australian Exposure Standards - Time Weighted Averages
100 ppm TWA; 369 mg/m3 TWA

Australian Exposure Standards - Short Term Exposure Limits
150 ppm STEL; 553 mg/m3 STEL

Australian Exposure Standards - Under Review
exposure limits under review

Canada - WHMIS: Ingredient Disclosure
1% item 1364 (833)

Canada - Alberta - 8 Hour Occupational Exposure Limit
100 ppm TWA; 360 mg/m3 TWA

Canada - Alberta - 15 Minute Occupational Exposure Limit
150 ppm STEL; 540 mg/m3 STEL

Canada - British Columbia - 8 Hour Exposure Limits
100 ppm TWA; 360 mg/m3 TWA

Canada - British Columbia - 15 Minute Exposure Limits
150 ppm STEL; 540 mg/m3 STEL

Canada - Ontario - OHSA - TWAEVs
100 ppm TWAEV; 365 mg/m3 TWAEV

Canada - Ontario - OHSA - STEVs
150 ppm STEV; 550 mg/m3 STEV

Canada - Quebec - Time-Weighted Average Exposure Values
100 ppm TWAEV; 369 mg/m3 TWAEV

Canada - Quebec - Short-term Exposure Values
150 ppm STEV; 553 mg/m3 STEV

United Kingdom - Occupational Exposure Standards - TWAs
100 ppm TWA; 360 mg/m3 TWA

United Kingdom - Occupational Exposure Standards - STELs
300 ppm STEL; 1080 mg/m3 STEL

United Kingdom - Occupational Exposure Standards - Notes
can be absorbed through skin

German (DFG) - MAK Values
100 ppm MAK; 375 mg/m3 MAK

German (DFG) - Peak Limitations
2 x normal MAK (5 min momentary value); don't exceed 8 times during shift

German (DFG) - Pregnancy
no risk to embryo/fetus if exposure limits adhered to

Israel - Time Weighted Averages
100 ppm TWA; 369 mg/m3 TWA

Israel - Short Term Exposure Limits
150 ppm STEL; 553 mg/m3 STEL

Israel - Action Levels
50 ppm AL; 184.5 mg/m3 AL

STATE LISTS

California - Air Bill 2588 Appendix A-I
9/90

California - Exposure Limits - PELs
100 ppm PEL;   360 mg/m3 PEL

California - Exposure Limits - STELs
150 ppm STEL;   540 mg/m3 STEL

California - Directors List of Hazardous Substances (8 CCR 339)
[present]

Florida Hazardous Substance List
[present]

Massachusetts Right To Know List
[present]

Minnesota Hazardous Substance List
[present]

Pennsylvania Right to Know List
[present]

PROPYL FORMATE   110-74-7

HEALTH AND SAFETY LISTS

AIHA - WEEL - Time Weighted Averages
commercial grade: 100 ppm TWA;   541 mg/m3 TWA

NFPA - Flash Points
99% pure: flash point = 108 degrees F (42 degrees C)

NFPA - Hazard Identification Ratings
health-0;   flammability-2;   reactivity-0

ENVIRONMENTAL LISTS

List of Pesticide Product Inert Ingredients
[present]

TSCA - PAIR - Reporting List
Effective Date: January 26, 1994;   Reporting Date: March 28, 1994

TSCA - Health and Safety Reporting List
Effective Date: January 26, 1994;   Sunset Date: January 26, 2004

INTERNATIONAL LISTS

German (DFG) - MAK Values
50 ppm MAK;   275 mg/m3 MAK

German (DFG) - Peak Limitations
2 x normal MAK (5 min momentary value);   don't exceed 8 times during shift

German (DFG) - Pregnancy
no risk to embryo/fetus is exposure limits are adhered to

STATE LISTS

California - Air Bill 2588 Appendix A-I
9/90

PROPOSED REGULATIONS

TSCA - ITC 31st Report Priority Testing List
recommended for testing

PROPYL FORMATES   RR-00137-6

ENVIRONMENTAL LISTS

List of Pesticide Product Inert Ingredients
[present]

PROPYL GALLATE   121-79-9

HEALTH AND SAFETY LISTS

NIOSH - Selected LD50s and LC50s
Oral, rat: LD50 = 3250 mg/kg Skin, rabbit: LD50 = 4000 mg/kg

4-PROPYLGUAIACOL   2785-87-7

HEALTH AND SAFETY LISTS

ACGIH 1995 - Time Weighted Averages
2 ppm TWA;   4.7 mg/m3 TWA

ACGIH 1995 - Skin Designations
skin - potential for cutaneous absorption

ACGIH 1995 - Carcinogens
A2-suspected human carcinogen

U.S. DOT - Substances From 49 CFR 172.101
regulated by DOT (UN1921)

U.S. DOT - Hazard Classes
DOT hazard class = 3

U.S. DOT - Appendix A Table 1 - Hazardous Substances
final RQ = 1 pound (0.454 kg)

IARC - Group 2B (sufficient animal data)
[present] (Overall evaluation based only on evidence of carcinogenicity in monograph (9, 1975) or in Supplement 4)

NIOSH - Selected LD50s and LC50s
Oral, rat: LD50 = 19 mg/kg Skin, guinea pig: LD50 = 43 mg/kg

NIOSH 1990 - Pocket Guide - RELs
2 ppm TWA;   5 mg/m3 TWA

NIOSH 1990 - Pocket Guide - IDLHs
500 ppm IDLH (not considering carcinogenic effects)

NIOSH 1990 - Pocket Guide - Carcinogens
occupational carcinogen

NIOSH 1990 - Pocket Guide - Target organs
eyes, skin

NIOSH 1990 - Pocket Guide - Skin list
Potential for dermal absorption

NTP Seventh Report - Suspect Carcinogens
suspect carcinogen

OSHA - Vacated PELs - Time Weighted Averages
2 ppm TWA;   5 mg/m3 TWA

OSHA - Vacated PELs - Skin Designation
Prevent or reduce skin absorption

OSHA - Final PELs - Time Weighted Averages
2 ppm TWA;   5 mg/m3 TWA

OSHA - Final PELs - Skin Notations
prevent or reduce skin absorption

OSHA - Possible Select Carcinogens
[present]

ENVIRONMENTAL LISTS

CERCLA/SARA - Section 302 Extremely Hazardous Substances and TPQs
TPQ = 10,000 pounds

CERCLA/SARA - Section 313 - Emission Reporting
form R reporting required for 0.1% de minimus concentration

CERCLA/SARA - Hazardous Substances and their Reportable Quantities
final RQ = 1 pound (0.454 kg)

Clean Air Act (1990) - List of Hazardous Air Contaminants
[present]

CAA -Toxic Substances for Accidental Release Prevention
threshold quantity = 10,000 lbs

EPA - Carcinogen Hazard Ranking for RQ Adjustment
Hazard ranking = High

RCRA - P Series Wastes
waste number P067

RCRA - Hazardous Constituents-Appendix VIII
waste number P067

RCRA - Substances Banned From Land Disposal
[present]

INTERNATIONAL LISTS

Australian Exposure Standards - Time Weighted Averages
2 ppm TWA;   4.7 mg/m3 TWA

Australian Exposure Standards - Skin Effects
skin absorption

Australian Exposure Standards - Carcinogens
probable carcinogen

Canada - WHMIS: Ingredient Disclosure
0.1% item 1367 (1455)

Canada - Alberta - 8 Hour Occupational Exposure Limit
2 ppm TWA; 4.7 mg/m3 TWA

Canada - Alberta - 15 Minute Occupational Exposure Limit
4 ppm STEL; 9.3 mg/m3 STEL

Canada - Alberta - Skin Designation
can be absorbed through the intact skin

Canada - Alberta - Designated Substances
designated substance - requires code of practice

Canada - British Columbia - 8 Hour Exposure Limits
2 ppm TWA; 5 mg/m3 TWA

Canada - British Columbia - Skin Notations
skin - potential for skin absorption

Canada - Ontario - OHSA - TWAEVs
2 ppm TWAEV; 4.7 mg/m3 TWAEV

Canada - Ontario - OHSA - Skin Notations
absorption through skin, eyes, or mucous membranes

Canada - Quebec - Time-Weighted Average Exposure Values
2 ppm TWAEV; 4.7 mg/m3 TWAEV

Canada - Quebec - Skin Designations
absorbed through the skin

Canada - Quebec - Carcinogens
C2 carcinogen: effect suspected in humans

German (DFG) - Skin/Sensitizers
danger of cutaneous absorption

German (DFG) - Carcinogens
animal evidence of carcinogenicity

Israel - Time Weighted Averages
2 ppm TWA; 4.7 mg/m3 TWA

Israel - Action Levels
1 ppm AL; 2.35 mg/m3 AL

Mexico - Instruction No. 10 - TWAs
2 ppm TWA; 5 mg/m3 TWA

Mexico - Instruction No. 10 - Carcinogens
potential carcinogen in humans - limited epidemiological evidence

STATE LISTS

California - Air Bill 2588 Appendix A-I
known or potential carcinogen

California - Prop. 65 - Cancer list
carcinogen - initial date 1/1/88

California - Exposure Limits - PELs
2 ppm PEL; 5 mg/m3 PEL

California - Exposure Limits - Skin Notation
material may be absorbed through the skin, eyes or mucous membrane

California - Directors List of Hazardous Substances (8 CCR 339)
[present]

Florida Hazardous Substance List
[present]

Massachusetts Right To Know List
carcinogen; extraordinarily hazardous

Minnesota Hazardous Substance List
carcinogen

Pennsylvania Right to Know List
environmental hazard; special hazardous substance

Pennsylvania RTK - Special Hazardous Substances
[present]

PROPYL-P-HYDROXYBENZOATE   94-13-3

HEALTH AND SAFETY LISTS

ACGIH 1995 - Time Weighted Averages
20 ppm TWA; 48 mg/m3 TWA

AIHA - Odor Threshold Values
geometric mean air odor threshold = 45 ppm (detectable); 35 ppm (recognizable)

U.S. DOT - Substances From 49 CFR 172.101
regulated by DOT (UN1280)

U.S. DOT - Hazard Classes
DOT hazard class = 3

U.S. DOT - Appendix A Table 1 - Hazardous Substances
final RQ = 100 pounds (45.4 kg)

IARC - Group 2A (limited human data)
[present] (Other relevant data, as given in Supplement 7, influenced the making of the overall evaluation)

NFPA - Flash Points
flash point = -35 degrees F (-37 degrees C)

NFPA - Hazard Identification Ratings
health-3; flammability-4; reactivity-2

NIOSH - Selected LD50s and LC50s
Inhalation, mouse: LC50 = 1740 ppm 4 hr Oral, rat: LD50 = 520 mg/kg Skin, rabbit: LD50 = 1245 mg/kg

NIOSH 1990 - Pocket Guide - IDLHs
2000 ppm IDLH (not considering carcinogenic effects)

NIOSH 1990 - Pocket Guide - Carcinogens
occupational carcinogen

NIOSH 1990 - Pocket Guide - Target organs
eyes, skin, respiratory system

NTP Chemical Status Reports - Testing Status and NTIS Number
Technical reports printed (PB85179653/AS)

NTP Chemical Status Reports - Evidence of Carcinogenicity
male rat-some evidence; female rat-some evidence; male mice-clear evidence; female mice-clear evidence

NTP Seventh Report - Suspect Carcinogens
suspect carcinogen

OSHA - Vacated PELs - Time Weighted Averages
20 ppm TWA; 50 mg/m3 TWA

OSHA - Final PELs - Time Weighted Averages
100 ppm TWA; 240 mg/m3 TWA

OSHA - Possible Select Carcinogens
[present]

ENVIRONMENTAL LISTS

CERCLA/SARA - Section 302 Extremely Hazardous Substances and TPQs
TPQ = 10,000 pounds

CERCLA/SARA - Section 313 - Emission Reporting
form R reporting required for 0.1% de minimus concentration

CERCLA/SARA - Hazardous Substances and their Reportable Quantities
final RQ = 100 pounds (45.4 kg)

Clean Air Act (1990) - List of Hazardous Air Contaminants
[present]

CAA -Toxic Substances for Accidental Release Prevention
threshold quantity = 10,000 lbs

CAA - HON Rule - SOCMI Chemicals
compliance by Oct. 24, 1994

CAA - HON Rule - Organic HAPs
[present]

Clean Water Act - Hazardous Substances
[present]

TSCA - Code of Federal Regulations Citations
40 CFR 712.30(d); 40 CFR 716.120(a); 40 CFR 799.3450

TSCA - PAIR - Reporting List
Reporting Date: November 19, 1982

TSCA - Health and Safety Reporting List
Effective Date: October 4, 1982

TSCA - Chemical Test Rules
Testing required by: manufacturers; processors (40 CFR 799.3450)

TSCA - Section 12(b) - Export Notification
export notification required - Section 4

INTERNATIONAL LISTS

Australian Exposure Standards - Time Weighted Averages
20 ppm TWA; 48 mg/m3 TWA

Australian Exposure Standards - Carcinogens
probable carcinogen

Australian Exposure Standards - Under Review
exposure limits under review

Canada - WHMIS: Ingredient Disclosure
1% item 1365 (1319)

Canada - NPRI (National Pollutant Release Inventory)
[present]

Canada - Alberta - 8 Hour Occupational Exposure Limit
20 ppm TWA; 47 mg/m3 TWA

Canada - Alberta - 15 Minute Occupational Exposure Limit
30 ppm STEL; 71 mg/m3 STEL

Canada - British Columbia - 8 Hour Exposure Limits
100 ppm TWA; 240 mg/m3 TWA

Canada - British Columbia - 15 Minute Exposure Limits
150 ppm STEL; 360 mg/m3 STEL

Canada - Ontario - OHSA - TWAEVs
20 ppm TWAEV; 47 mg/m3 TWAEV

Canada - Quebec - Time-Weighted Average Exposure Values
20 ppm TWAEV; 48 mg/m3 TWAEV

Canada - Quebec - Carcinogens
C2 carcinogen: effect suspected in humans

German (DFG) - Carcinogens
animal evidence of carcinogenicity

Israel - Time Weighted Averages
20 ppm TWA; 48 mg/m3 TWA

Israel - Action Levels
10 ppm AL; 24 mg/m3 AL

Mexico - Instruction No. 10 - TWAs
20 ppm TWA; 50 mg/m3 TWA

STATE LISTS

California - Air Bill 2588 Appendix A-I
known or potential carcinogen

California - Prop. 65 - Cancer list
carcinogen - initial date 10/1/88

California - Exposure Limits - PELs
20 ppm PEL; 50 mg/m3 PEL

California - Directors List of Hazardous Substances (8 CCR 339)
[present] (exempt when part of a cured epoxy or rubber)

Florida Hazardous Substance List
[present]

Massachusetts Right To Know List
carcinogen; extraordinarily hazardous

Minnesota Hazardous Substance List
[present]

NJ Right to Know List (Total)
sn 1615

NJ Special Hazardous Substances
(flammable - fourth degree; mutagen; reactive - second degree)

Pennsylvania Right to Know List
environmental hazard; special hazardous substance

Pennsylvania RTK - Special Hazardous Substances
[present]

PROPOSED REGULATIONS

Canada - Ontario - Proposed Occupational TWAEVs
1 ppm TWAEV; 2 mg/m3 TWAEV

PROPYLIDENE DICHLORIDE   RR-01311-6

INTERNATIONAL LISTS

Canada - WHMIS: Ingredient Disclosure
1% item 1366 (1553)

N-PROPYL ISOCYANATE   110-78-1

HEALTH AND SAFETY LISTS

U.S. DOT - Substances From 49 CFR 172.101
regulated by DOT (UN2850)

U.S. DOT - Hazard Classes
DOT hazard class = 3

NFPA - Flash Points
flash point < 212 degrees F (100 degrees C)

NFPA - Hazard Identification Ratings
health-0; flammability-1; reactivity-0

STATE LISTS

NJ Right to Know List (Total)
sn 2713

N-PROPYL NITRATE   627-13-4

HEALTH AND SAFETY LISTS

NFPA - Flash Points
flash point = 27 degrees F (-3 degrees C)

NFPA - Hazard Identification Ratings
health-2; flammability-3; reactivity-0

NIOSH - Selected LD50s and LC50s
Oral, rat: LD50 = 3980 mg/kg

STATE LISTS

Florida Hazardous Substance List
[present]

Massachusetts Right To Know List
[present]

NJ Right to Know List (Total)
sn 1616

NJ Special Hazardous Substances
(flammable - third degree)

Pennsylvania Right to Know List
[present]

P-PROPYLPHENOL   645-56-7

HEALTH AND SAFETY LISTS

U.S. DOT - Substances From 49 CFR 172.101
regulated by DOT (UN1281)

U.S. DOT - Hazard Classes
DOT hazard class = 3

PROPYL PROPIONATE   106-36-5

HEALTH AND SAFETY LISTS

NTP Chemical Status Reports - Testing Status and NTIS Number
Technical reports printed (PB83180042)

NTP Chemical Status Reports - Evidence of Carcinogenicity
male rat-equivocal; female rat-negative; male mice-equivocal; female mice-negative

ENVIRONMENTAL LISTS

List of Pesticide Product Inert Ingredients
[present]

PROPYLTHIOURACIL   51-52-5

INTERNATIONAL LISTS

Canada - WHMIS: Ingredient Disclosure
1% item 1368 (1456)

PROPYL TRICHLOROSILANE   141-57-1

HEALTH AND SAFETY LISTS

NIOSH - Selected LD50s and LC50s
Oral, dog: LD50 = 6000 mg/kg

ENVIRONMENTAL LISTS

List of Pesticide Product Inert Ingredients
[present]

PROTACTINIUM 227   29901-97-1

HEALTH AND SAFETY LISTS

U.S. DOT - Appendix B - Marine Pollutants
DOT regulated marine pollutant

PROTACTINIUM 228   15766-09-3

HEALTH AND SAFETY LISTS

U.S. DOT - Substances From 49 CFR 172.101
regulated by DOT (UN2482)

U.S. DOT - Hazard Classes
DOT hazard class = 3

U.S. DOT - Substances Which Are Poisonous by Inhalation
liquid hazardous material poisonous by inhalation (UN2482)

ENVIRONMENTAL LISTS

TSCA - Code of Federal Regulations Citations
40 CFR 712.30(x); 40 CFR 716.120(d)

TSCA - PAIR - Reporting List
Reporting Date: December 27, 1990

TSCA - Health and Safety Reporting List
Effective Date: October 29, 1990; Sunset Date: November 9, 1993

INTERNATIONAL LISTS

Canada - WHMIS: Ingredient Disclosure
0.1% item 1369 (1046)

STATE LISTS

NJ Right to Know List (Total)
sn 1617

PROTACTINIUM 230   15766-10-6

HEALTH AND SAFETY LISTS

ACGIH 1995 - Time Weighted Averages
25 ppm TWA; 107 mg/m3 TWA

ACGIH 1995 - Short Term Exposure Limits
40 ppm STEL; 172 mg/m3 STEL

U.S. DOT - Substances From 49 CFR 172.101
regulated by DOT (UN1865)

U.S. DOT - Hazard Classes
DOT hazard class = 3

NFPA - Flash Points
flash point = 68 degrees F (20 degrees C)

NFPA - Hazard Identification Ratings
health-2; flammability-3; reactivity-3 (oxidizing properties) (may explode on heating)

NIOSH 1990 - Pocket Guide - RELs
25 ppm TWA; 105 mg/m3 TWA; 40 ppm STEL; 170 mg/m3 STEL

NIOSH 1990 - Pocket Guide - IDLHs
2000 ppm IDLH

NIOSH 1990 - Pocket Guide - Target organs
none known

OSHA - Vacated PELs - Time Weighted Averages
25 ppm TWA; 105 mg/m3 TWA

OSHA - Vacated PELs - Short Term Exposure Limits
40 ppm STEL; 170 mg/m3 STEL

OSHA - Final PELs - Time Weighted Averages
25 ppm TWA; 110 mg/m3 TWA

OSHA - List of Highly Hazardous Chemicals
threshhold quantity = 2500 pounds

INTERNATIONAL LISTS

Australian Exposure Standards - Time Weighted Averages
25 ppm TWA; 107 mg/m3 TWA

Australian Exposure Standards - Short Term Exposure Limits
40 ppm STEL; 172 mg/m3 STEL

Canada - WHMIS: Ingredient Disclosure
1% item 1370 (1210)

Canada - Alberta - 8 Hour Occupational Exposure Limit
25 ppm TWA; 107 mg/m3 TWA

Canada - Alberta - 15 Minute Occupational Exposure Limit
40 ppm STEL; 172 mg/m3 STEL

Canada - British Columbia - 8 Hour Exposure Limits
25 ppm TWA; 105 mg/m3 TWA

Canada - British Columbia - 15 Minute Exposure Limits
40 ppm STEL; 470 mg/m3 STEL

Canada - Ontario - OHSA - TWAEVs
25 ppm TWAEV; 105 mg/m3 TWAEV

Canada - Ontario - OHSA - STEVs
40 ppm STEV; 170 mg/m3 STEV

Canada - Quebec - Time-Weighted Average Exposure Values
25 ppm TWAEV; 107 mg/m3 TWAEV

Canada - Quebec - Short-term Exposure Values
40 ppm STEV; 172 mg/m3 STEV

German (DFG) - MAK Values
25 ppm MAK; 110 mg/m3 MAK

Israel - Time Weighted Averages
25 ppm TWA; 107 mg/m3 TWA

Israel - Short Term Exposure Limits
40 ppm STEL; 172 mg/m3 STEL

Israel - Action Levels
12.5 ppm AL; 53.5 mg/m3 AL

Mexico - Instruction No. 10 - TWAs
25 ppm TWA; 105 mg/m3 TWA

Mexico - Instruction No. 10 - STELs
40 ppm STEL; 170 mg/m3 STEL

STATE LISTS

California - Exposure Limits - PELs
25 ppm PEL; 107 mg/m3 PEL

California - Exposure Limits - STELs
40 ppm STEL; 170 mg/m3 STEL

California - Directors List of Hazardous Substances (8 CCR 339)
[present]

Florida Hazardous Substance List
[present]

Massachusetts Right To Know List
[present]

Minnesota Hazardous Substance List
[present]

NJ Right to Know List (Total)
sn 1420

NJ Special Hazardous Substances
(flammable - third degree; reactive - third degree)

Pennsylvania Right to Know List
[present]

PROTACTINIUM 231   14331-85-2

HEALTH AND SAFETY LISTS

NIOSH - Selected LD50s and LC50s
Oral, rat: LD50 = 500 mg/kg Skin, mammal: LD50 = 2150 mg/kg

PROTACTINIUM 232   15766-06-0

HEALTH AND SAFETY LISTS

NFPA - Flash Points
flash point = 175 degrees F (79 degrees C)

NFPA - Hazard Identification Ratings
health-1; flammability-3; reactivity-0

NIOSH - Selected LD50s and LC50s
Inhalation, mouse: LC50 = 24 gm/m3 2 hr Oral, rabbit: LD50 = 3950 mg/kg

INTERNATIONAL LISTS

Canada - WHMIS: Ingredient Disclosure
1% item 1371 (1450)

STATE LISTS

Florida Hazardous Substance List
[present]

Massachusetts Right To Know List
[present]

Pennsylvania Right to Know List
[present]

PROTACTINIUM 233   13981-14-1

HEALTH AND SAFETY LISTS

IARC - Group 2B (sufficient animal data)
[present]

NIOSH - Selected LD50s and LC50s
Oral, rat: LD50 = 1980 mg/kg

NTP Seventh Report - Suspect Carcinogens
suspect carcinogen

OSHA - Possible Select Carcinogens
[present]

ENVIRONMENTAL LISTS

RCRA - Hazardous Constituents-Appendix VIII
hazardous constituent - no waste number

STATE LISTS

California - Air Bill 2588 Appendix A-II
known or potential carcinogen

California - Prop. 65 - Cancer list
carcinogen - initial date 1/1/88

California - Prop. 65 - Developmental Toxicity
developmental toxicity - initial date 7/1/90

California - Prop. 65 - No Significant Risk Levels
no significant risk level = 0.7 ug/day

California - Directors List of Hazardous Substances (8 CCR 339)
[present]

Florida Hazardous Substance List
[present]

Massachusetts Right To Know List
carcinogen; extraordinarily hazardous; teratogen

Minnesota Hazardous Substance List
carcinogen

NJ Special Hazardous Substances
(carcinogen; mutagen)

Pennsylvania Right to Know List
environmental hazard; special hazardous substance

Pennsylvania RTK - Special Hazardous Substances
[present]

PROTACTINIUM 234   15100-28-4

HEALTH AND SAFETY LISTS

U.S. DOT - Substances From 49 CFR 172.101
regulated by DOT (UN1816)

U.S. DOT - Hazard Classes
DOT hazard class = 8

NFPA - Flash Points
flash point = 98 degrees F (37 degrees C)

NFPA - Hazard Identification Ratings
health-3; flammability-3; reactivity-1

INTERNATIONAL LISTS

Canada - WHMIS: Ingredient Disclosure
1% item 1372 (1457)

STATE LISTS

Florida Hazardous Substance List
[present]

Massachusetts Right To Know List
[present]

NJ Right to Know List (Total)
sn 1619

NJ Special Hazardous Substances
(corrosive; flammable - third degree; reactive - second degree)

Pennsylvania Right to Know List
[present]

PROTEIN COLLOID   RR-01132-5

HEALTH AND SAFETY LISTS

U.S. DOT - Appendix A Table 2 - Radionuclides
final RQ = 100 curies (3.7E 12 Bq)

ENVIRONMENTAL LISTS

CERCLA/SARA List of Radionuclides (Appendix B) and Their Reportable Quantities
final RQ = 100 curies (3.7E 12 Bq)

PROTHOATE   2275-18-5

HEALTH AND SAFETY LISTS

U.S. DOT - Appendix A Table 2 - Radionuclides
final RQ = 10 curies (3.7E 11 Bq)

ENVIRONMENTAL LISTS

CERCLA/SARA List of Radionuclides (Appendix B) and Their Reportable Quantities
final RQ = 10 curies (3.7E 11 Bq)

PSEUDOCUMENE   95-63-6

HEALTH AND SAFETY LISTS

U.S. DOT - Appendix A Table 2 - Radionuclides
final RQ = 10 curies (3.7E 11 Bq)

ENVIRONMENTAL LISTS

CERCLA/SARA List of Radionuclides (Appendix B) and Their Reportable Quantities
final RQ = 10 curies (3.7E 11 Bq)

PSEUDOEPHEDRINE   90-82-4

HEALTH AND SAFETY LISTS

U.S. DOT - Appendix A Table 2 - Radionuclides
final RQ = 0.01 curies (3.7E 8 Bq)

ENVIRONMENTAL LISTS

CERCLA/SARA List of Radionuclides (Appendix B) and Their Reportable Quantities
final RQ = 0.01 curies (3.7E 8 Bq)

PSEUDOEPHEDRINE SALTS, OPTICAL ISOMERS, AND SALTS OF OPTICALISOMERS   RR-01781-2

HEALTH AND SAFETY LISTS

U.S. DOT - Appendix A Table 2 - Radionuclides
final RQ = 10 curies (3.7E 11 Bq)

ENVIRONMENTAL LISTS
CERCLA/SARA List of Radionuclides (Appendix B) and Their Reportable Quantities
final RQ = 10 curies (3.7E 11 Bq)

PTAQUILOSIDE   87625-62-5
HEALTH AND SAFETY LISTS
U.S. DOT - Appendix A Table 2 - Radionuclides
final RQ = 100 curies (3.7E 12 Bq)

ENVIRONMENTAL LISTS
CERCLA/SARA List of Radionuclides (Appendix B) and Their Reportable Quantities
final RQ = 100 curies (3.7E 12 Bq)

PUMICE   1332-09-8
HEALTH AND SAFETY LISTS
U.S. DOT - Appendix A Table 2 - Radionuclides
final RQ = 10 curies (3.7E 11 Bq)

ENVIRONMENTAL LISTS
CERCLA/SARA List of Radionuclides (Appendix B) and Their Reportable Quantities
final RQ = 10 curies (3.7E 11 Bq)

2H-PYRAN, 3,6-DIHYDRO-   3174-74-1
ENVIRONMENTAL LISTS
List of Pesticide Product Inert Ingredients
[present]

PYRAZINAMIDE   98-96-4
HEALTH AND SAFETY LISTS
U.S. DOT - Appendix B - Marine Pollutants
DOT regulated marine pollutant
NIOSH - Selected LD50s and LC50s
Inhalation, rat: LC50 = 165 mg/m3 4 hr Oral, rat: LD50 = 8 mg/kg
Skin, rat: LD50 = 100 mg/kg

ENVIRONMENTAL LISTS
CERCLA/SARA - Section 302 Extremely Hazardous Substances and TPQs
TPQ = 100/10,000 pounds

STATE LISTS
Florida Hazardous Substance List
effective March 13, 1992
Massachusetts Right To Know List
extraordinarily hazardous
NJ Right to Know List (Total)
sn 2715
Pennsylvania Right to Know List
environmental hazard

PROPOSED REGULATIONS
CERCLA/SARA - Proposed Hazardous Substance Additions
proposed RQ = 1 pound (.454 kg)
CERCLA/SARA - 1989 Proposed RQ Adjustments
proposed RQ = 100 pounds (45.4 kg)

3-PYRAZOLIDINONE, 1-PHENYL-   92-43-3
SEE ALSO:
AROMATIC C9 FRACTION FROM PETROLEUM REFINING
HEALTH AND SAFETY LISTS
AIHA - Odor Threshold Values
geometric mean air odor threshold = 2.4 ppm (detectable)
NFPA - Flash Points
flash point = 112 degrees F (44 degrees C)

NFPA - Hazard Identification Ratings
health-0; flammability-2; reactivity-0
NIOSH - Selected LD50s and LC50s
Inhalation, rat: LC50 = 18 gm/m3 4 hr

ENVIRONMENTAL LISTS
CERCLA/SARA - Section 313 - Emission Reporting
form R reporting required for 1.0% de minimus concentration
Safe Drinking Water Act - Monitoring
monitoring required at discretion of the state
TSCA - Code of Federal Regulations Citations
40 CFR 712.30(d); 40 CFR 716.120(b); 40 CFR 799.2175
TSCA - PAIR - Reporting List
Reporting Date: November 19, 1982
TSCA - Health and Safety Reporting List
Effective Date: April 29, 1983

INTERNATIONAL LISTS
Canada - WHMIS: Ingredient Disclosure
0.1% item 1640 (1684)
Canada - NPRI (National Pollutant Release Inventory)
[present]

STATE LISTS
California - Air Bill 2588 Appendix A-I
6/91
Massachusetts Right To Know List
[present]
NJ Right to Know List (Total)
sn 2716
Pennsylvania Right to Know List
environmental hazard

PYRAZOPHOS   13457-18-6
HEALTH AND SAFETY LISTS
FDA - Controlled Substances Act - Precursor chemicals
Threshold by base weight = 1 kilogram

PYRENE   129-00-0
HEALTH AND SAFETY LISTS
FDA - Controlled Substances Act - Precursor chemicals
Threshold by base weight = 1 kilogram

PYRETHRIN I   121-21-1
HEALTH AND SAFETY LISTS
IARC - Group 3 (not classifiable)
[present]

STATE LISTS
California - Directors List of Hazardous Substances (8 CCR 339)
[present]

PYRETHRIN II   121-29-9
ENVIRONMENTAL LISTS
List of Pesticide Product Inert Ingredients
[present]

PYRETHRUM   8003-34-7
STATE LISTS
Pennsylvania Right to Know List
[present]

PYRIDINE   110-86-1
HEALTH AND SAFETY LISTS
NTP Chemical Status Reports - Testing Status and NTIS Number
Technical reports printed (PB280251/AS)

NTP Chemical Status Reports - Evidence of Carcinogenicity
male rat-negative; female rat-negative; male mice-negative; female mice-inadequate

4-PYRIDINECARBONITRILE   100-48-1

HEALTH AND SAFETY LISTS

NIOSH - Selected LD50s and LC50s
Oral, rat: LD50 = 200 mg/kg

3-PYRIDINECARBONITRILE   100-54-9

HEALTH AND SAFETY LISTS

U.S. DOT - Appendix B - Marine Pollutants
DOT regulated severe marine pollutant

2-PYRIDINECARBONITRILE   100-70-9
SEE ALSO:
F039-HAZARDOUS WASTES
COAL TAR PITCHES

HEALTH AND SAFETY LISTS

U.S. DOT - Appendix A Table 1 - Hazardous Substances
final RQ = 5000 pounds (2270 kg)

IARC - Group 3 (not classifiable)
[present]

NIOSH - Selected LD50s and LC50s
Inhalation, rat: LC50 = 170 mg/m3 8 hr Oral, rat: LD50 = 2700 mg/kg

ENVIRONMENTAL LISTS

ATSDR Priority List
Rank (of 275): 201

CERCLA/SARA - Section 302 Extremely Hazardous Substances and TPQs
TPQ = 1000/10,000 pounds

CERCLA/SARA - Hazardous Substances and their Reportable Quantities
final RQ = 5000 pounds (2270 kg)

CAA - HON Rule - SOCMI Chemicals
compliance by Oct. 23, 1995

Clean Water Act - Priority Pollutants
[present]

RCRA - Basis for Listing - Appendix VII
Included in waste stream: F039

RCRA - TSD Facilities Ground Water Monitoring
TM 8100 = 200 ug/L PQL; TM 8270 = 10 ug/L PQL

RCRA - Universal Treatment Standards (LDR)
WW: 0.067 mg/l; NWW: 8.2 mg/kg

TSCA - Code of Federal Regulations Citations
40 CFR 716.120(a); 40 CFR 704.225(a)

TSCA - CAIR - Reporting List
reporting required by: manufacturer; importer

TSCA - Health and Safety Reporting List
Effective Date: June 1, 1987

INTERNATIONAL LISTS

Canada - WHMIS: Ingredient Disclosure
1% item 1373 (1458)

Mexico - Wastewater - Organic Toxic Pollutants and Heavy Metals
Listed under [Aromatic Hydrocarbons]

STATE LISTS

California - Directors List of Hazardous Substances (8 CCR 339)
[present]

Massachusetts Right To Know List
extraordinarily hazardous

NJ Special Hazardous Substances
(carcinogen)

Pennsylvania Right to Know List
environmental hazard

2-PYRIDINECARBOXALDEHYDE   1121-60-4
SEE ALSO:
PYRETHRIN II

HEALTH AND SAFETY LISTS

U.S. DOT - Appendix A Table 1 - Hazardous Substances
final RQ = 1 pound (0.454 kg) (Listed under 'Pyrethrins')

ENVIRONMENTAL LISTS

CERCLA/SARA - Hazardous Substances and their Reportable Quantities
final RQ = 1 pound (0.454 kg) (Listed under 'Pyrethrins')

Clean Water Act - Hazardous Substances
[present] (Listed under 'Pyrethrin')

STATE LISTS

California - Directors List of Hazardous Substances (8 CCR 339)
[present] (Listed under 'Pyrethrins')

Massachusetts Right To Know List
[present]

NJ Right to Know List (Total)
sn 3064

Pennsylvania Right to Know List
environmental hazard

3-PYRIDINECARBOXYLIC ACID   59-67-6

HEALTH AND SAFETY LISTS

U.S. DOT - Appendix A Table 1 - Hazardous Substances
final RQ = 1 pound (0.454 kg)

NIOSH - Selected LD50s and LC50s
Oral, rat: LD50 = 200 mg/kg

ENVIRONMENTAL LISTS

CERCLA/SARA - Hazardous Substances and their Reportable Quantities
final RQ = 1 pound (0.454 kg)

Clean Water Act - Hazardous Substances
[present] (Listed under 'Pyrethrin')

STATE LISTS

California - Directors List of Hazardous Substances (8 CCR 339)
[present] (Listed under 'Pyrethrins')

Massachusetts Right To Know List
[present]

NJ Right to Know List (Total)
sn 3065

Pennsylvania Right to Know List
environmental hazard

2,6-PYRIDINEDICARBOXYLIC ACID   499-83-2
SEE ALSO:
PYRETHRIN II

HEALTH AND SAFETY LISTS

ACGIH 1995 - Time Weighted Averages
5 mg/m3 TWA

U.S. DOT - Appendix A Table 1 - Hazardous Substances
final RQ = 1 pound (0.454 kg) (Listed under 'Pyrethrins')

NIOSH - Selected LD50s and LC50s
Oral, rat: LD50 = 200 mg/kg

NIOSH 1990 - Pocket Guide - RELs
5 mg/m3 TWA

NIOSH 1990 - Pocket Guide - IDLHs
5000 mg/m3 IDLH

NIOSH 1990 - Pocket Guide - Target organs
CNS, skin, respiratory system

OSHA - Vacated PELs - Time Weighted Averages
5 mg/m3 TWA

OSHA - Final PELs - Time Weighted Averages
5 mg/m3 TWA

ENVIRONMENTAL LISTS

CERCLA/SARA - Hazardous Substances and their Reportable Quantities
final RQ = 1 pound (0.454 kg) (Listed under 'Pyrethrins')

List of Pesticide Product Inert Ingredients
[present] (does not refer to pyrethrins)

INTERNATIONAL LISTS

Australian Exposure Standards - Time Weighted Averages
5 mg/m3 TWA

Australian Exposure Standards - Skin Effects
sensitiser

Canada - Alberta - 8 Hour Occupational Exposure Limit
5 mg/m3 TWA

Canada - Alberta - 15 Minute Occupational Exposure Limit
10 mg/m3 STEL

Canada - British Columbia - 8 Hour Exposure Limits
5 mg/m3 TWA

Canada - British Columbia - 15 Minute Exposure Limits
10 mg/m3 STEL

Canada - Ontario - OHSA - TWAEVs
5 mg/m3 TWAEV

Canada - Quebec - Time-Weighted Average Exposure Values
5 mg/m3 TWAEV

United Kingdom - Occupational Exposure Standards - TWAs
5 mg/m3 TWA

United Kingdom - Occupational Exposure Standards - STELs
10 mg/m3 STEL

German (DFG) - MAK Values
total dust: 5 mg/m3 MAK

German (DFG) - Peak Limitations
10 x normal MAK (30 min average value); don't exceed during shift

German (DFG) - Skin/Sensitizers
danger of sensitization (skin or respiratory) (does not apply to constituents of insecticides)

Israel - Time Weighted Averages
5 mg/m3 TWA

Israel - Action Levels
2.5 mg/m3 AL

Mexico - Instruction No. 10 - TWAs
5 mg/m3 TWA

Mexico - Instruction No. 10 - STELs
10 mg/m3 STEL

STATE LISTS

California - Exposure Limits - PELs
5 mg/m3 PEL

California - Directors List of Hazardous Substances (8 CCR 339)
[present]

Florida Hazardous Substance List
[present]

Massachusetts Right To Know List
[present]

Minnesota Hazardous Substance List
[present]

NJ Right to Know List (Total)
sn 1623

Pennsylvania Right to Know List
environmental hazard

PYRIDINE, 2-ETHENYL-5-ETHYL-   5408-74-2
SEE ALSO:
F039-HAZARDOUS WASTES
K143-HAZARDOUS WASTES
K026-HAZARDOUS WASTES
F005-HAZARDOUS WASTES

HEALTH AND SAFETY LISTS

ACGIH 1995 - Time Weighted Averages
5 ppm TWA; 16 mg/m3 TWA

AIHA - Odor Threshold Values
geometric mean air odor threshold = 0.66 ppm (detectable); 0.74 ppm (recognizable)

U.S. DOT - Substances From 49 CFR 172.101
regulated by DOT (UN1282)

U.S. DOT - Hazard Classes
DOT hazard class = 3

U.S. DOT - Appendix A Table 1 - Hazardous Substances
final RQ = 1000 pounds (454 kg)

NFPA - Flash Points
flash point = 68 degrees F (20 degrees C)

NFPA - Hazard Identification Ratings
health-3; flammability-3; reactivity-0

NIOSH - Selected LD50s and LC50s
Oral, rat: LD50 = 891 mg/kg Skin, rabbit: LD50 = 1121 mg/kg

NIOSH 1990 - Pocket Guide - RELs
5 ppm TWA; 15 mg/m3 TWA

NIOSH 1990 - Pocket Guide - IDLHs
3600 ppm IDLH

NIOSH 1990 - Pocket Guide - Target organs
CNS, liver, kidneys, skin, GI tract

NTP Chemical Status Reports - Testing Status and NTIS Number
Two year studies: pathology quality assessment in progress

OSHA - Vacated PELs - Time Weighted Averages
5 ppm TWA; 15 mg/m3 TWA

OSHA - Final PELs - Time Weighted Averages
5 ppm TWA; 15 mg/m3 TWA

ENVIRONMENTAL LISTS

CERCLA/SARA - Section 313 - Emission Reporting
form R reporting required for 1.0% de minimus concentration

CERCLA/SARA - Hazardous Substances and their Reportable Quantities
final RQ = 1000 pounds (454 kg)

CAA - HON Rule - SOCMI Chemicals
compliance by Jan. 23, 1995

RCRA - D Series - Maximum Concentration of Contaminants
waste number D038; regulatory level = 5.0 mg/L

RCRA - D Series - Chronic Toxicity Reference Levels
chronic toxicity reference level = 0.04 mg/L

RCRA - U Series Wastes
waste number U196

RCRA - Hazardous Constituents-Appendix VIII
waste number U196

RCRA - Basis for Listing - Appendix VII
Included in waste streams: F005, F039, K026

RCRA - Substances Banned From Land Disposal
[present]

RCRA - TSD Facilities Ground Water Monitoring
TM 8240 = 5 ug/L PQL; TM 8270 = 10 ug/L PQL

RCRA - Universal Treatment Standards (LDR)
WW: 0.014 mg/l; NWW: 16 mg/kg

TSCA - Code of Federal Regulations Citations
40 CFR 712.30(d); 40 CFR 716.120(a)

TSCA - PAIR - Reporting List
Reporting Date: November 19, 1982

TSCA - Health and Safety Reporting List
Effective Date: October 4, 1982

INTERNATIONAL LISTS

Australian Exposure Standards - Time Weighted Averages
5 ppm TWA; 16 mg/m3 TWA

Canada - WHMIS: Ingredient Disclosure
1% item 1374 (1459)

Canada - NPRI (National Pollutant Release Inventory)
[present]

Canada - Alberta - 8 Hour Occupational Exposure Limit
5 ppm TWA; 16 mg/m3 TWA

Canada - Alberta - 15 Minute Occupational Exposure Limit
10 ppm STEL; 32 mg/m3 STEL

Canada - British Columbia - 8 Hour Exposure Limits
5 ppm TWA; 15 mg/m3 TWA

Canada - British Columbia - 15 Minute Exposure Limits
10 ppm STEL; 30 mg/m3 STEL

Canada - Ontario - OHSA - TWAEVs
5 ppm TWAEV; 16 mg/m3 TWAEV

Canada - Quebec - Time-Weighted Average Exposure Values
5 ppm TWAEV; 16 mg/m3 TWAEV

United Kingdom - Occupational Exposure Standards - TWAs
5 ppm TWA; 15 mg/m3 TWA

United Kingdom - Occupational Exposure Standards - STELs
10 ppm STEL; 30 mg/m3 STEL

German (DFG) - MAK Values
5 ppm MAK; 15 mg/m3 MAK

German (DFG) - Peak Limitations
2 x normal MAK (30 min. average value); don't exceed 4 times during shift

Israel - Time Weighted Averages
5 ppm TWA; 16 mg/m3 TWA

Israel - Action Levels
2.5 ppm AL; 8 mg/m3 AL

Mexico - Instruction No. 10 - TWAs
5 ppm TWA; 15 mg/m3 TWA

Mexico - Instruction No. 10 - STELs
10 ppm STEL; 30 mg/m3 STEL

STATE LISTS

California - Air Bill 2588 Appendix A-I
6/91

California - Exposure Limits - PELs
5 ppm PEL; 15 mg/m3 PEL

California - Directors List of Hazardous Substances (8 CCR 339)
[present]

Florida Hazardous Substance List
[present]

Massachusetts Right To Know List
[present]

Minnesota Hazardous Substance List
[present]

NJ Right to Know List (Total)
sn 1624

NJ Special Hazardous Substances
(flammable - third degree)

Pennsylvania Right to Know List
environmental hazard

PYRIDINE, 2-METHYL-5-VINYL-   140-76-1

ENVIRONMENTAL LISTS

TSCA - Code of Federal Regulations Citations
40 CFR 716.120(a)

TSCA - Health and Safety Reporting List
Effective Date: June 1, 1987

PYRIDINE, 4-NITRO-, 1-OXIDE-   1124-33-0

HEALTH AND SAFETY LISTS

NIOSH - Selected LD50s and LC50s
Oral, rat: LD50 = 1185 mg/kg

ENVIRONMENTAL LISTS

TSCA - Code of Federal Regulations Citations
40 CFR 716.120(a)

TSCA - Health and Safety Reporting List
Effective Date: June 1, 1987

PYRIDINE, 3-(1-NITROSO-2-PYRROLIDINYL)   80508-23-2

ENVIRONMENTAL LISTS

TSCA - Code of Federal Regulations Citations
40 CFR 716.120(a)

TSCA - Health and Safety Reporting List
Effective Date: June 1, 1987; Sunset Date: November 9, 1993

PYRIDINE PERCHLORATE   15598-34-2

ENVIRONMENTAL LISTS

TSCA - Code of Federal Regulations Citations
40 CFR 712.30(x); 40 CFR 716.120(d)

TSCA - PAIR - Reporting List
Reporting Date: November 27, 1991

TSCA - Health and Safety Reporting List
Effective Date: September 30, 1991

PYRIDINE, 2,3,5,6-TETRACHLORO-   2402-79-1

HEALTH AND SAFETY LISTS

NIOSH - Selected LD50s and LC50s
Oral, rat: LD50 = 7000 mg/kg

ENVIRONMENTAL LISTS

EPA - Master Testing List
[present]

List of Pesticide Product Inert Ingredients
[present]

PYRIDO[3,4-C]PSORALEN   RR-01540-7

ENVIRONMENTAL LISTS

List of Pesticide Product Inert Ingredients
[present]

PYRIDOXAL-S'-PHOSPHATE   41468-25-1

HEALTH AND SAFETY LISTS

NFPA - Flash Points
flash point = 200 degrees F (93 degrees C)

NFPA - Hazard Identification Ratings
health-2; flammability-2; reactivity-2

STATE LISTS

Florida Hazardous Substance List
[present]

Massachusetts Right To Know List
[present]

Pennsylvania Right to Know List
[present]

PYRILAMINE   91-84-9

HEALTH AND SAFETY LISTS

NIOSH - Selected LD50s and LC50s
Inhalation, rat: LC50 = 189 mg/m3 2 hr Oral, rat: LD50 = 1167 mg/kg Skin, rabbit: LD50 = 718 mg/kg

ENVIRONMENTAL LISTS

CERCLA/SARA - Section 302 Extremely Hazardous Substances and TPQs
TPQ = 500 pounds

INTERNATIONAL LISTS

Canada - WHMIS: Ingredient Disclosure
1% item 1086 (1159)

STATE LISTS

Florida Hazardous Substance List
effective March 13, 1992

Massachusetts Right To Know List
extraordinarily hazardous

NJ Right to Know List (Total)
sn 2717

Pennsylvania Right to Know List
environmental hazard

PROPOSED REGULATIONS

CERCLA/SARA - Proposed Hazardous Substance Additions
proposed RQ = 1 pound (.454 kg)

CERCLA/SARA - 1989 Proposed RQ Adjustments
proposed RQ = 100 pounds (45.4 kg)

PYRIMETHAMINE   58-14-0

ENVIRONMENTAL LISTS

CERCLA/SARA - Section 302 Extremely Hazardous Substances and TPQs
TPQ = 500/10,000 pounds

STATE LISTS

Florida Hazardous Substance List
effective March 13, 1992

Massachusetts Right To Know List
extraordinarily hazardous

Pennsylvania Right to Know List
environmental hazard

PROPOSED REGULATIONS

CERCLA/SARA - Proposed Hazardous Substance Additions
proposed RQ = 1 pound (.454 kg)

CERCLA/SARA - 1989 Proposed RQ Adjustments
proposed RQ = 100 pounds (45.4 kg)

PYRIMINIL   53558-25-1

STATE LISTS

Massachusetts Right To Know List
carcinogen; extraordinarily hazardous

PYROGALLOL   87-66-1

HEALTH AND SAFETY LISTS

U.S. DOT - Hazard Classes
Forbidden from transport by the DOT

1H-PYROLE-2,5-DIONE, 1-(2,4,6-TRIBRO-   59789-51-4
MOPHENYL)-

ENVIRONMENTAL LISTS

EPA - Master Testing List
[present]

PYROPHORIC LIQUIDS, N.O.S.   RR-00454-6

HEALTH AND SAFETY LISTS

IARC - Group 3 (not classifiable)
[present]

PYROPHORIC METALS OR ALLOYS, N.O.S.   RR-00455-7

HEALTH AND SAFETY LISTS

NIOSH - Selected LD50s and LC50s
Oral, rat: LD50 = 5900 mg/kg

PYROPHORIC SOLIDS, N.O.S.   RR-00456-8

HEALTH AND SAFETY LISTS

NTP Chemical Status Reports - Testing Status and NTIS Number
Technical reports printed (no NTIS number)

PYROPHYLLITE   12269-78-2

HEALTH AND SAFETY LISTS

IARC - Group 3 (not classifiable)
[present]

NTP Chemical Status Reports - Testing Status and NTIS Number
Technical reports printed (PB282608/AS)

NTP Chemical Status Reports - Evidence of Carcinogenicity
male rat-negative; female rat-negative; male mice-inadequate; female mice-negative

STATE LISTS

California - Directors List of Hazardous Substances (8 CCR 339)
[present]

PYROSULFURIC ACID   RR-00457-9

ENVIRONMENTAL LISTS

CERCLA/SARA - Section 302 Extremely Hazardous Substances and TPQs
TPQ = 100/10,000 pounds

STATE LISTS

Florida Hazardous Substance List
effective March 13, 1992

Massachusetts Right To Know List
extraordinarily hazardous

NJ Right to Know List (Total)
sn 2719

Pennsylvania Right to Know List
environmental hazard

PROPOSED REGULATIONS

CERCLA/SARA - Proposed Hazardous Substance Additions
proposed RQ = 1 pound (.454 kg)

CERCLA/SARA - 1989 Proposed RQ Adjustments
proposed RQ = 100 pounds (45.4 kg)

PYROSULFURYL CHLORIDE   7791-27-7

HEALTH AND SAFETY LISTS

NIOSH - Selected LD50s and LC50s
Oral, rat: LD50 = 789 mg/kg

INTERNATIONAL LISTS

Canada - WHMIS: Ingredient Disclosure
1% item 1375 (1461)

PYROXYLIN SOLUTION   RR-01160-9

ENVIRONMENTAL LISTS

TSCA - Chemicals with Significant New Use Rules
PMN number: P-90-159

TSCA - Section 12(b) - Export Notification
P-90-159; export notification required - Section 5

1H-PYRROLE   109-97-7

HEALTH AND SAFETY LISTS

U.S. DOT - Substances From 49 CFR 172.101
regulated by DOT (UN2845)

U.S. DOT - Hazard Classes
DOT hazard class = 4.2

STATE LISTS
  NJ Right to Know List (Total)
  sn 2721; sn 2445

1H-PYRROLE, METHYL-   27417-39-6

HEALTH AND SAFETY LISTS
  U.S. DOT - Substances From 49 CFR 172.101
  regulated by DOT (UN1383)
  U.S. DOT - Hazard Classes
  DOT hazard class = 4.2

STATE LISTS
  NJ Right to Know List (Total)
  sn 2722

PYRROLIDINE   123-75-1

HEALTH AND SAFETY LISTS
  U.S. DOT - Substances From 49 CFR 172.101
  regulated by DOT (UN2846)
  U.S. DOT - Hazard Classes
  DOT hazard class = 4.2

STATE LISTS
  NJ Right to Know List (Total)
  sn 2723

2-PYRROLIDINONE   616-45-5

ENVIRONMENTAL LISTS
  List of Pesticide Product Inert Ingredients
  [present]

PYRUVIC ALDEHYDE   78-98-8

STATE LISTS
  NJ Right to Know List (Total)
  sn 2884

QUARTZ   14808-60-7

HEALTH AND SAFETY LISTS
  U.S. DOT - Substances From 49 CFR 172.101
  regulated by DOT (UN1817)
  U.S. DOT - Hazard Classes
  DOT hazard class = 8

INTERNATIONAL LISTS
  Canada - WHMIS: Ingredient Disclosure
  1% item 1376 (523)

STATE LISTS
  NJ Right to Know List (Total)
  sn 1625
  NJ Special Hazardous Substances
  (corrosive)

QUASSIA, EXTRACT   68915-32-2

HEALTH AND SAFETY LISTS
  NFPA - Flash Points
  flash point = 80 degrees F (27 degrees C)
  NFPA - Hazard Identification Ratings
  health-1; flammability-3; reactivity-0

QUATERNARY AMMONIUM COMPOUNDS, COCO ALKYLBIS(HYDROXYETHYL)METHYL, ETHOXYLATED, CHLORIDES   61791-10-4

HEALTH AND SAFETY LISTS
  NFPA - Flash Points
  flash point = 102 degrees F (39 degrees C)

NFPA - Hazard Identification Ratings
  health-2; flammability-2; reactivity-0

STATE LISTS
  Florida Hazardous Substance List
  [present]
  Massachusetts Right To Know List
  [present]
  Pennsylvania Right to Know List
  [present]

QUATERNARY AMMONIUM COMPOUNDS, (HYDROGENATED TALLOW ALKYL) BIS(HYDROXYETHYL) METHYL, CHLORIDES   68607-27-2

STATE LISTS
  Pennsylvania Right to Know List
  [present]

QUATERNARY AMMONIUM COMPOUNDS, (HYDROGENATED TALLOW ALKYL) BIS(HYDROXYETHYL) METHYL, ETHOXYLATED, CHLORIDES   68187-69-9

HEALTH AND SAFETY LISTS
  U.S. DOT - Substances From 49 CFR 172.101
  regulated by DOT (UN1922)
  U.S. DOT - Hazard Classes
  DOT hazard class = 3
  NFPA - Flash Points
  flash point = 37 degrees F (3 degrees C)
  NFPA - Hazard Identification Ratings
  health-2; flammability-3; reactivity-1
  NIOSH - Selected LD50s and LC50s
  Inhalation, mouse: LC50 = 1300 mg/m3 (2 hr) Oral, rat: LD50 = 300 mg/kg

STATE LISTS
  Florida Hazardous Substance List
  [present]
  Massachusetts Right To Know List
  [present]
  NJ Right to Know List (Total)
  sn 1626
  NJ Special Hazardous Substances
  (flammable - third degree)
  Pennsylvania Right to Know List
  [present]

QUATERNARY AMMONIUM SALT OF FLUORINATED ALKYLARYL AMIDE   RR-01567-8

HEALTH AND SAFETY LISTS
  NFPA - Flash Points
  flash point = 265 degrees F (129 degrees C)
  NFPA - Hazard Identification Ratings
  health-2; flammability-1; reactivity-0
  NIOSH - Selected LD50s and LC50s
  Oral, rat: LD50 = 6500 mg/kg

STATE LISTS
  Florida Hazardous Substance List
  [present]
  Massachusetts Right To Know List
  [present]
  Pennsylvania Right to Know List
  [present]

QUEBRACHITOL PENTANITRATE   RR-01457-3

HEALTH AND SAFETY LISTS

IARC - Group 3 (not classifiable)
[present]

NIOSH - Selected LD50s and LC50s
Oral, rat: LD50 = 1165 mg/kg

QUENCHING OIL   RR-00793-2
SEE ALSO:
SILICA, CRYSTALLINE (GENERAL FORM)

HEALTH AND SAFETY LISTS

ACGIH 1995 - Time Weighted Averages
0.1 mg/m3 TWA (this TLV is for the respirable fraction of dust)

NIOSH 1990 - Pocket Guide - RELs
as respirable dust: 0.05 mg/m3 TWA

NIOSH 1990 - Pocket Guide - Carcinogens
occupational carcinogen

NIOSH 1990 - Pocket Guide - Target organs
respiratory system

NTP Seventh Report - Suspect Carcinogens
respirable dust: suspect carcinogen (Listed under 'Silica, crystalline')

OSHA - Vacated PELs - Time Weighted Averages
respirable dust: 0.1 mg/m3 TWA

OSHA - Final PELs - Time Weighted Averages
see Table Z-3

OSHA - Possible Select Carcinogens
[present]

INTERNATIONAL LISTS

Canada - WHMIS: Ingredient Disclosure
1% item 1406 (1491)

Canada - Alberta - 8 Hour Occupational Exposure Limit
respirable mass: 0.1 mg/m3 TWA; total mass: 0.3 mg/m3 TWA (See additional requirements in Part 5)

Canada - Quebec - Time-Weighted Average Exposure Values
respirable dust: 0.1 mg/m3 TWAEV

Canada - Quebec - Carcinogens
C2 carcinogen: effect suspected in humans

United Kingdom - Maximum Exposure Limits - TWAs
respirable dust: 0.4 mg/m3 TWA

German (DFG) - MAK Values
fine dust: 0.15 mg/m3 MAK

German (DFG) - Pregnancy
no risk to embryo/fetus if exposure limits adhered to

Israel - Time Weighted Averages
total dust: 0.3 mg/m3 TWA; respirable dust: 0.1 mg/m3 TWA (Listed under 'Silica - Crystalline')

Israel - Action Levels
total dust: 0.075 mg/m3 AL; respirable dust: 0.025 mg/m3 AL

STATE LISTS

California - Exposure Limits - PELs
total dust: 0.3 mg/m3 PEL; respirable dust: 0.1 mg/m3 PEL

Florida Hazardous Substance List
[present]

Massachusetts Right To Know List
carcinogen; extraordinarily hazardous

Minnesota Hazardous Substance List
[present]

Pennsylvania Right to Know List
[present]

QUILLAJA (SAPONIN)   1393-03-9

ENVIRONMENTAL LISTS

List of Pesticide Product Inert Ingredients
[present]

QUINACRIDONE   1047-16-1

ENVIRONMENTAL LISTS

List of Pesticide Product Inert Ingredients
[present]

QUINACRINE HYDROCHLORIDE   69-05-6

ENVIRONMENTAL LISTS

List of Pesticide Product Inert Ingredients
[present]

QUINALDINE   91-63-4

ENVIRONMENTAL LISTS

List of Pesticide Product Inert Ingredients
[present]

QUINALPHOS   13593-03-8

ENVIRONMENTAL LISTS

TSCA - Chemicals with Significant New Use Rules
PMN number: P-92-688

QUININE   130-95-0

HEALTH AND SAFETY LISTS

U.S. DOT - Hazard Classes
Forbidden from transport by the DOT

QUINOLINE   91-22-5

HEALTH AND SAFETY LISTS

NFPA - Flash Points
flash point = 365 degrees F (185 degrees C)

NFPA - Hazard Identification Ratings
health-0; flammability-1; reactivity-0

STATE LISTS

Pennsylvania Right to Know List
[present]

QUINOLINE YELLOW   8004-92-0

ENVIRONMENTAL LISTS

List of Pesticide Product Inert Ingredients
[present]

QUINONE   106-51-4

ENVIRONMENTAL LISTS

TSCA - Code of Federal Regulations Citations
40 CFR 712.30(x); 40 CFR 716.120(d)

TSCA - PAIR - Reporting List
Reporting Date: November 27, 1991

TSCA - Health and Safety Reporting List
Effective Date: September 30, 1991

N-(2-QUINOXALINYL)-SULFANILIDE   59-40-5

HEALTH AND SAFETY LISTS

NIOSH - Selected LD50s and LC50s
Oral, rat: LD50 = 660 mg/kg

QUIZALOFOP-ETHYL [2-[4-[(6-CHLORO-2-QUINOX-ALINYL)OXY]PHENOXY]PROPANOIC ACID ETHYL ESTER]   76578-14-8

HEALTH AND SAFETY LISTS

NIOSH - Selected LD50s and LC50s
Oral, rat: LD50 = 1230 mg/kg Skin, rabbit: LD50 = 1870 mg/kg

INTERNATIONAL LISTS

Canada - WHMIS: Ingredient Disclosure
1% item 1377 (1463)

RADIOACTIVE MATERIALS, N.O.S.   RR-00459-1

HEALTH AND SAFETY LISTS

U.S. DOT - Appendix B - Marine Pollutants
DOT regulated marine pollutant

RADIONUCLIDES   RR-00066-8

STATE LISTS

Massachusetts Right To Know List
teratogen

RADIUM   7440-14-4

HEALTH AND SAFETY LISTS

AIHA - WEEL - Time Weighted Averages
0.1 ppm TWA; 0.5 mg/m3 TWA

AIHA - WEEL - Skin Absorption Designations
skin absorber

U.S. DOT - Substances From 49 CFR 172.101
regulated by DOT (UN2656)

U.S. DOT - Hazard Classes
DOT hazard class = 6.1

U.S. DOT - Appendix A Table 1 - Hazardous Substances
final RQ = 5000 pounds (2270 kg)

NFPA - Hazard Identification Ratings
health-2; flammability-1; reactivity-0

NIOSH - Selected LD50s and LC50s
Oral, rat: LD50 = 331 mg/kg Skin, rabbit: LD50 = 540 mg/kg

ENVIRONMENTAL LISTS

CERCLA/SARA - Section 313 - Emission Reporting
form R reporting required for 1.0% de minimus concentration

CERCLA/SARA - Hazardous Substances and their Reportable Quantities
final RQ = 5000 pounds (2270 kg)

Clean Air Act (1990) - List of Hazardous Air Contaminants
[present]

Clean Water Act - Hazardous Substances
[present]

INTERNATIONAL LISTS

Canada - WHMIS: Ingredient Disclosure
1% item 1378 (1464)

Canada - NPRI (National Pollutant Release Inventory)
[present]

STATE LISTS

California - Air Bill 2588 Appendix A-I
6/91

California - Directors List of Hazardous Substances (8 CCR 339)
[present]

Florida Hazardous Substance List
[present]

Massachusetts Right To Know List
[present]

Minnesota Hazardous Substance List
skin

NJ Right to Know List (Total)
sn 1628

NJ Special Hazardous Substances
(mutagen)

Pennsylvania Right to Know List
environmental hazard

RADIUM 223   15623-45-7

HEALTH AND SAFETY LISTS

NIOSH - Selected LD50s and LC50s
Oral, rat: LD50 = 2 gm/kg

ENVIRONMENTAL LISTS

List of Pesticide Product Inert Ingredients
[present]

RADIUM 224   13233-32-4

HEALTH AND SAFETY LISTS

ACGIH 1995 - Time Weighted Averages
0.1 ppm TWA; 0.44 mg/m3 TWA

U.S. DOT - Substances From 49 CFR 172.101
regulated by DOT (UN2587)

U.S. DOT - Hazard Classes
DOT hazard class = 6.1

U.S. DOT - Appendix A Table 1 - Hazardous Substances
final RQ = 10 pounds (4.54 kg)

IARC - Group 3 (not classifiable)
[present]

NFPA - Flash Points
flash point = 100 to 200 degrees F (38 to 93 degrees C)

NFPA - Hazard Identification Ratings
health-1; flammability-2; reactivity-1

NIOSH - Selected LD50s and LC50s
Oral, rat: LD50 = 130 mg/kg

NIOSH 1990 - Pocket Guide - RELs
0.4 mg/m3 TWA; 0.1 ppm TWA

NIOSH 1990 - Pocket Guide - IDLHs
300 mg/m3 IDLH

NIOSH 1990 - Pocket Guide - Target organs
eyes, skin

OSHA - Vacated PELs - Time Weighted Averages
0.1 ppm TWA; 0.4 mg/m3 TWA

OSHA - Final PELs - Time Weighted Averages
0.1 ppm TWA; 0.4 mg/m3 TWA

ENVIRONMENTAL LISTS

CERCLA/SARA - Section 313 - Emission Reporting
form R reporting required for 1.0% de minimus concentration

CERCLA/SARA - Hazardous Substances and their Reportable Quantities
final RQ = 10 pounds (4.54 kg)

Clean Air Act (1990) - List of Hazardous Air Contaminants
[present]

CAA - HON Rule - SOCMI Chemicals
compliance by April 24, 1995

CAA - HON Rule - Organic HAPs
[present]

RCRA - U Series Wastes
waste number U197

RCRA - Hazardous Constituents-Appendix VIII
waste number U197

RCRA - Substances Banned From Land Disposal
[present]

TSCA - Code of Federal Regulations Citations
40 CFR 712.30(d); 40 CFR 716.120(a)

TSCA - PAIR - Reporting List
Reporting Date: November 19, 1982

TSCA - Health and Safety Reporting List
Effective Date: October 4, 1982

INTERNATIONAL LISTS

Australian Exposure Standards - Time Weighted Averages
0.1 ppm TWA; 0.44 mg/m3 TWA

Canada - WHMIS: Ingredient Disclosure
1% item 165 (289)

Canada - NPRI (National Pollutant Release Inventory)
[present]

Canada - Alberta - 8 Hour Occupational Exposure Limit
0.1 ppm TWA; 0.42 mg/m3 TWA

Canada - Alberta - 15 Minute Occupational Exposure Limit
0.3 ppm STEL; 1.3 mg/m3 STEL

Canada - British Columbia - 8 Hour Exposure Limits
0.1 ppm TWA; 0.4 mg/m3 TWA

Canada - British Columbia - 15 Minute Exposure Limits
0.3 ppm STEL; 2 mg/m3 STEL

Canada - Ontario - OHSA - TWAEVs
0.1 ppm TWAEV; 0.44 mg/m3 TWAEV

Canada - Quebec - Time-Weighted Average Exposure Values
0.1 ppm TWAEV; 0.44 mg/m3 TWAEV

United Kingdom - Occupational Exposure Standards - TWAs
0.1 ppm TWA; 0.4 mg/m3 TWA

United Kingdom - Occupational Exposure Standards - STELs
0.3 ppm STEL; 1.2 mg/m3 STEL

German (DFG) - MAK Values
0.1 ppm MAK; 0.4 mg/m3 MAK

German (DFG) - Peak Limitations
2 x normal MAK (5 min momentary value); don't exceed 8 times during shift

Israel - Time Weighted Averages
0.1 ppm TWA; 0.44 mg/m3 TWA

Israel - Action Levels
0.05 ppm AL; 0.22 mg/m3 AL

Mexico - Instruction No. 10 - TWAs
0.1 ppm TWA; 0.4 mg/m3 TWA

STATE LISTS

California - Air Bill 2588 Appendix A-I
6/91

California - Exposure Limits - PELs
0.1 ppm PEL; 0.4 mg/m3 PEL

California - Directors List of Hazardous Substances (8 CCR 339)
[present]

Florida Hazardous Substance List
[present]

Massachusetts Right To Know List
[present]

Minnesota Hazardous Substance List
[present]

NJ Right to Know List (Total)
sn 1460

Pennsylvania Right to Know List
environmental hazard

RADIUM 225   13981-53-8

ENVIRONMENTAL LISTS

List of Pesticide Product Inert Ingredients
[present]

RADIUM 226   13982-63-3

ENVIRONMENTAL LISTS

CERCLA/SARA - Section 313 - Emission Reporting
form R reporting required

RADIUM 227   15743-84-7

INTERNATIONAL LISTS

Canada - CEPA Schedule III Part I - Prohibited Substances (Ocean Dumping)
[present]

STATE LISTS

NJ Right to Know List (Total)
sn 2725; sn 2726; sn 2728; sn 2729; sn 2730; sn 2732; sn 2733; sn 2735; sn 2724

RADIUM 228   15262-20-1

HEALTH AND SAFETY LISTS

U.S. DOT - Appendix A Table 2 - Radionuclides
final RQ = 1 curie (3.7E 12 Bq) (for all radionuclides not cited individualy)

ENVIRONMENTAL LISTS

CERCLA/SARA List of Radionuclides (Appendix B) and Their Reportable Quantities
final RQ = 1 curie (3.7E 12 Bq) (for all radionuclides not cited individualy)

Clean Air Act (1990) - List of Hazardous Air Contaminants
[present]

STATE LISTS

California - Air Bill 2588 Appendix A-I
known or potential carcinogen

California - Prop. 65 - Cancer list
carcinogen - initial date 7/1/89

California - Directors List of Hazardous Substances (8 CCR 339)
[present]

Pennsylvania Right to Know List
environmental hazard

RADON   10043-92-2

ENVIRONMENTAL LISTS

ATSDR Priority List
Rank (of 275): 078

RADON 220   22481-48-7

HEALTH AND SAFETY LISTS

U.S. DOT - Appendix A Table 2 - Radionuclides
final RQ = 1 curie (3.7E 10 Bq)

ENVIRONMENTAL LISTS

CERCLA/SARA List of Radionuclides (Appendix B) and Their Reportable Quantities
final RQ = 1 curie (3.7E 10 Bq)

RADON 222   14859-67-7

HEALTH AND SAFETY LISTS

U.S. DOT - Appendix A Table 2 - Radionuclides
final RQ = 10 curies (3.7E 11 Bq)

ENVIRONMENTAL LISTS

ATSDR Priority List
Rank (of 275): 116

CERCLA/SARA List of Radionuclides (Appendix B) and Their Reportable Quantities
final RQ = 10 curies (3.7E 11 Bq)

RAISINS   RR-01133-6

HEALTH AND SAFETY LISTS

U.S. DOT - Appendix A Table 2 - Radionuclides
final RQ = 1 curie (3.7E 10 Bq)

ENVIRONMENTAL LISTS

CERCLA/SARA List of Radionuclides (Appendix B) and Their Reportable Quantities
final RQ = 1 curie (3.7E 10 Bq)

RAPESEED OIL   8002-13-9

HEALTH AND SAFETY LISTS

U.S. DOT - Appendix A Table 2 - Radionuclides
final RQ = 0.1 curies (3.7E 9 Bq) (notification requirements for releases of mixtures or solutions can be found in 40 CFR 302.6(b))

ENVIRONMENTAL LISTS

ATSDR Priority List
Rank (of 275): 081

CERCLA/SARA List of Radionuclides (Appendix B) and Their Reportable Quantities
final RQ = 0.1 curies (3.7E 9 Bq) (notification requirements for releases of mixtures or solutions can be found in 40 CFR 302.6(b))

Safe Drinking Water Act - MCLs
combined with Radium 228: MCL = 5 pCi/l;   gross alpha particle activity: MCL = 15 pCi/l

STATE LISTS

California - Directors List of Hazardous Substances (8 CCR 339)
[present]

RARE GASES, MIXTURES   RR-01505-4

HEALTH AND SAFETY LISTS

U.S. DOT - Appendix A Table 2 - Radionuclides
final RQ = 1000 curies (3.7E 13 Bq)

ENVIRONMENTAL LISTS

CERCLA/SARA List of Radionuclides (Appendix B) and Their Reportable Quantities
final RQ = 1000 curies (3.7E 13 Bq)

RAW UMBER   12713-03-0

HEALTH AND SAFETY LISTS

U.S. DOT - Appendix A Table 2 - Radionuclides
final RQ = 0.1 curies (3.7E 9 Bq)

ENVIRONMENTAL LISTS

ATSDR Priority List
Rank (of 275): 089

CERCLA/SARA List of Radionuclides (Appendix B) and Their Reportable Quantities
final RQ = 0.1 curies (3.7E 9 Bq)

Safe Drinking Water Act - MCLs
combined with Radium 226: MCL = 5 pCi/l;   gross alpha particle activity: MCL = 15 pCi/l

REACTION PRODUCT OF ALKANEDIOL AND   RR-00948-3
EPICHLOROHYDRIN

HEALTH AND SAFETY LISTS

NTP Seventh Report - Known Carcinogens
known carcinogen

ENVIRONMENTAL LISTS

ATSDR Priority List
Rank (of 275): 087

STATE LISTS

California - Air Bill 2588 Appendix A-I
known or potential carcinogen: 9/89

California - Directors List of Hazardous Substances (8 CCR 339)
[present]

REACTION PRODUCT OF ALKYL CARBOXYLIC   RR-00986-9
ACIDS, ALKANE POLYOLS, ALKYL ACRYLATE,
AND ISOPHORONE DIISOCYANATE

HEALTH AND SAFETY LISTS

U.S. DOT - Appendix A Table 2 - Radionuclides
final RQ = 0.1 curies (3.7E 9 Bq)

NTP Seventh Report - Known Carcinogens
known carcinogen

ENVIRONMENTAL LISTS

ATSDR Priority List
Rank (of 275): 096

CERCLA/SARA List of Radionuclides (Appendix B) and Their Reportable Quantities
final RQ = 0.1 curies (3.7E 9 Bq)

REACTION PRODUCT OF ALKYLPHENOL, TE-   RR-00975-6
TRALKYL TITANATE, AND TINCOMPLEX

HEALTH AND SAFETY LISTS

U.S. DOT - Appendix A Table 2 - Radionuclides
final RQ = 0.1 curies (3.7E 9 Bq)

NTP Seventh Report - Known Carcinogens
known carcinogen

ENVIRONMENTAL LISTS

ATSDR Priority List
Rank (of 275): 094

CERCLA/SARA List of Radionuclides (Appendix B) and Their Reportable Quantities
final RQ = 0.1 curies (3.7E 9 Bq)

REACTION PRODUCT OF AMINOKETONES AND   RR-01134-7
FORMALDEHYDE

ENVIRONMENTAL LISTS

List of Pesticide Product Inert Ingredients
[present]

REACTION PRODUCT OF A MONOALKYL SUC-   RR-00232-4
CINIC ANHYDRIDE WITH ANX-HYDROXY
METHACRYLATE

HEALTH AND SAFETY LISTS

NFPA - Flash Points
flash point = 325 degrees F (163 degrees C)

NFPA - Hazard Identification Ratings
health-0; flammability-1; reactivity-0

REACTION PRODUCT OF HYDROXYETHYL ACRY-   60857-97-8
LATE AND METHYL OXIRANE

STATE LISTS

NJ Right to Know List (Total)
sn 2739; with nitrogen: sn 2740; with oxygen: sn 2741

REACTION PRODUCT OF ETHOXYLATED FATTY   RR-01263-5
ACID OILS AND A PHENOLIC PENTAERYTHRITOL
TETRAESTER

SEE ALSO:
MANGANESE COMPOUNDS, N.O.S.
MANGANESE

ENVIRONMENTAL LISTS

List of Pesticide Product Inert Ingredients
[present]

REACTION PRODUCTS OF PHENOLIC PENTAERY-   RR-01259-9
THRITOL TETRAESTER WITH FATTY ACID OILS
AND ESTERS, AND GLYCERIDE TRIESTERS

ENVIRONMENTAL LISTS

TSCA - Chemicals with Significant New Use Rules
PMN number: P-89-760

REACTION PROD-   **RR-01249-7**
UCTS OF SUBSTITUTED HYDROXYLALKANES AND
POLYALKYLPOLYISOCYANATOCARBOMONOCYCLE
ENVIRONMENTAL LISTS
TSCA - Chemicals with Significant New Use Rules
PMN number: P-89-1081

RECOVERED METAL HYDROXIDE   **RR-01256-6**
ENVIRONMENTAL LISTS
TSCA - Chemicals with Significant New Use Rules
PMN number: P-90-583

RED SQUILL   **RR-00753-4**
ENVIRONMENTAL LISTS
List of Pesticide Product Inert Ingredients
[present]

REDUCING LIQUID   **RR-01506-5**
ENVIRONMENTAL LISTS
TSCA - Chemicals with Significant New Use Rules
PMN number: P-88-701

REFRACTORY FIBRES (CERAMIC OR OTHERS)   **RR-01792-5**
ENVIRONMENTAL LISTS
TSCA - Code of Federal Regulations Citations
40 CFR 721.1500
TSCA - Chemicals with Significant New Use Rules
PMN number: P-86-832

REFRACTORY CERAMIC FIBERS   **142844-00-6**
ENVIRONMENTAL LISTS
TSCA - Chemicals with Significant New Use Rules
PMN number: P-92-63

REFRIGERANT GASES, N.O.S.   **RR-00472-8**
ENVIRONMENTAL LISTS
TSCA - Chemicals with Significant New Use Rules
PMN numbers: P-91-1231;   P-91-1232;   P-91-1233;   P-91-1234;
P-91-1235

RESERPINE   **50-55-5**
ENVIRONMENTAL LISTS
TSCA - Chemicals with Significant New Use Rules
PMN number: P-91-75

RESIDUAL (HEAVY) FUEL OILS   **RR-01741-4**
ENVIRONMENTAL LISTS
TSCA - Chemicals with Significant New Use Rules
PMN number: P-91-809

RESIDUAL HEAVY POLYMER (PETROLEUM)   **64741-71-5**
STATE LISTS
NJ Right to Know List (Total)
sn 1631

RESIDUAL OIL SOVENT EXTRACT (PETROLEUM)   **64742-10-5**
STATE LISTS
NJ Right to Know List (Total)
sn 2423

RESIN ACIDS AND ROSIN ACIDS, POTASSIUM   **61790-50-9**
SALT
INTERNATIONAL LISTS
Canada - Quebec - Time-Weighted Average Exposure Values
1 fibre/cm3 TWAEV
Canada - Quebec - Carcinogens
C3 carcinogen: effect detected in animals

RESIN ACIDS AND ROSIN ACIDS, SODIUM SALTS   **61790-51-0**
HEALTH AND SAFETY LISTS
IARC - Group 2B (sufficient animal data)
[present]
NTP Seventh Report - Suspect Carcinogens
suspect carcinogen
OSHA - Possible Select Carcinogens
[present]
ENVIRONMENTAL LISTS
EPA - Master Testing List
[present]
TSCA - Substances Subject to Testing Consent Orders
Test for: Health Effects
TSCA - Section 12(b) - Export Notification
export notification required - Section 4, proposed Section 5
INTERNATIONAL LISTS
Australian Exposure Standards - Time Weighted Averages
respirable dust: 0.5 fibres per ml of air; non-respirable dust: 2 mg/m3
TWA
Canada - Alberta - 8 Hour Occupational Exposure Limit
0.5 f/cm3 TWA
German (DFG) - Carcinogens
as fibrous dust: animal evidence of carcinogenicity
STATE LISTS
California - Air Bill 2588 Appendix A-I
known or potential carcinogen: 9/89
California - Prop. 65 - Cancer list
carcinogen (airborne particles of respirable size) - initial date 7/1/90

RESIN FAST BLACK WP   **RR-01507-6**
HEALTH AND SAFETY LISTS
U.S. DOT - Substances From 49 CFR 172.101
regulated by DOT (NA1954, UN1078)
U.S. DOT - Hazard Classes
flammable: DOT hazard class = 2.1; nonflammable: DOT hazard
class = 2.2
STATE LISTS
NJ Right to Know List (Total)
sn 2744; sn 2745

RESIN OIL   **RR-00475-1**
HEALTH AND SAFETY LISTS
U.S. DOT - Appendix A Table 1 - Hazardous Substances
final RQ = 5000 pounds (2270 kg)
IARC - Group 3 (not classifiable)
[present]
NIOSH - Selected LD50s and LC50s
Oral, rat: LD50 = 420 mg/kg
NTP Chemical Status Reports - Testing Status and NTIS Number
Technical reports printed (PB83165761); prechronic studies for
which toxicity technical reports were not prepared
NTP Chemical Status Reports - Evidence of Carcinogenicity
male rat-positive; female rat-negative; male mice-positive; female
mice-positive
NTP Seventh Report - Suspect Carcinogens
suspect carcinogen

OSHA - Possible Select Carcinogens
[present]

ENVIRONMENTAL LISTS

CERCLA/SARA - Hazardous Substances and their Reportable Quantities
final RQ = 5000 pounds (2270 kg)

RCRA - U Series Wastes
waste number U200

RCRA - Hazardous Constituents-Appendix VIII
waste number U200

RCRA - Substances Banned From Land Disposal
[present]

TSCA - Chemicals with Significant New Use Rules
[present]

TSCA - Section 12(b) - Export Notification
export notification required - Section 5

STATE LISTS

California - Air Bill 2588 Appendix A-I
known or potential carcinogen

California - Prop. 65 - Cancer list
carcinogen - initial date 10/1/89

California - Prop. 65 - No Significant Risk Levels
no significant risk level = 0.06 ug/day

California - Directors List of Hazardous Substances (8 CCR 339)
[present]

Florida Hazardous Substance List
[present]

Massachusetts Right To Know List
carcinogen; extraordinarily hazardous

Minnesota Hazardous Substance List
carcinogen

NJ Right to Know List (Total)
sn 1632

NJ Special Hazardous Substances
(carcinogen; mutagen)

Pennsylvania Right to Know List
environmental hazard; special hazardous substance

Pennsylvania RTK - Special Hazardous Substances
[present]

RESIN SOLUTION   RR-00476-2

STATE LISTS

California - Air Bill 2588 Appendix A-I
known or potential carcinogen: 6/91

California - Prop. 65 - Cancer list
carcinogen - initial date 10/1/90

RESMETHRIN   10453-86-8

HEALTH AND SAFETY LISTS

IARC - Group 2B (sufficient animal data)
[present] (Listed under 'Fuel oils')

OSHA - Possible Select Carcinogens
[present]

STATE LISTS

California - Prop. 65 - Cancer list
carcinogen - initial date 2/27/87

RESORCINOL   108-46-3

STATE LISTS

Massachusetts Right To Know List
carcinogen; extraordinarily hazardous

RESORCINOL, FORMALDEHYDE, SUBSTITUTED   RR-01018-4
CARBOMONOCYCLE RESIN

ENVIRONMENTAL LISTS

List of Pesticide Product Inert Ingredients
[present]

ALL-TRANS RETINOIC ACID   302-79-4

ENVIRONMENTAL LISTS

List of Pesticide Product Inert Ingredients
[present]

RETINOID PROJECT 2 (4-HYDROXYPHENYL)   65646-68-6

STATE LISTS

NJ Right to Know List (Total)
sn 1633

NJ Special Hazardous Substances
(carcinogen)

RETINOL, ACETATE   127-47-9

STATE LISTS

NJ Right to Know List (Total)
sn 2748

RETRORSINE   480-54-6

HEALTH AND SAFETY LISTS

U.S. DOT - Substances From 49 CFR 172.101
regulated by DOT (UN1866)

U.S. DOT - Hazard Classes
DOT hazard class = 3

STATE LISTS

NJ Right to Know List (Total)
sn 2749; sn 2751

RHAMSAN GUM   96949-21-2

HEALTH AND SAFETY LISTS

NIOSH - Selected LD50s and LC50s
Inhalation, mouse: LD50 = 99 mg/kg 8 hr Oral, rat: LD50 = 1244 mg/kg Skin, rabbit: LD50 = 2500 mg/kg

ENVIRONMENTAL LISTS

CERCLA/SARA - Section 313 - Emission Reporting
form R reporting required

RHENIUM 177   18853-09-3

HEALTH AND SAFETY LISTS

ACGIH 1995 - Time Weighted Averages
10 ppm TWA; 45 mg/m3 TWA

ACGIH 1995 - Short Term Exposure Limits
20 ppm STEL; 90 mg/m3 STEL

U.S. DOT - Substances From 49 CFR 172.101
regulated by DOT (UN2876)

U.S. DOT - Hazard Classes
DOT hazard class = 6.1

U.S. DOT - Appendix A Table 1 - Hazardous Substances
final RQ = 5000 pounds (2270 kg)

IARC - Group 3 (not classifiable)
[present]

NFPA - Flash Points
flash point = 261 degrees F (127 degrees C)

NFPA - Hazard Identification Ratings
flammability-1; reactivity-0

NIOSH - Selected LD50s and LC50s
Oral, rat: LD50 = 301 mg/kg Skin, rabbit: LD50 = 3360 mg/kg

NTP Chemical Status Reports - Testing Status and NTIS Number
Technical reports printed (PB93-126381)

NTP Chemical Status Reports - Evidence of Carcinogenicity
male rat: no evidence; female rat: no evidence; male mice: no evidence; female mice: no evidence

OSHA - Vacated PELs - Time Weighted Averages
10 ppm TWA; 45 mg/m3 TWA

OSHA - Vacated PELs - Short Term Exposure Limits
20 ppm STEL; 90 mg/m3 STEL

ENVIRONMENTAL LISTS

CERCLA/SARA - Hazardous Substances and their Reportable Quantities
final RQ = 5000 pounds (2270 kg)

CAA - HON Rule - SOCMI Chemicals
compliance by Oct. 24, 1994

Clean Water Act - Hazardous Substances
[present]

List of Pesticide Product Inert Ingredients
[present]

RCRA - U Series Wastes
waste number U201

RCRA - Hazardous Constituents-Appendix VIII
waste number U201

RCRA - Substances Banned From Land Disposal
[present]

INTERNATIONAL LISTS

Australian Exposure Standards - Time Weighted Averages
10 ppm TWA; 45 mg/m3 TWA

Australian Exposure Standards - Short Term Exposure Limits
20 ppm STEL; 90 mg/m3 STEL

Canada - WHMIS: Ingredient Disclosure
1% item 1379 (1465)

Canada - Alberta - 8 Hour Occupational Exposure Limit
10 ppm TWA; 45 mg/m3 TWA

Canada - Alberta - 15 Minute Occupational Exposure Limit
20 ppm STEL; 90 mg/m3 STEL

Canada - British Columbia - 8 Hour Exposure Limits
10 ppm TWA; 45 mg/m3 TWA

Canada - British Columbia - 15 Minute Exposure Limits
20 ppm STEL; 90 mg/m3 STEL

Canada - Ontario - OHSA - TWAEVs
10 ppm TWAEV; 45 mg/m3 TWAEV

Canada - Ontario - OHSA - STEVs
20 ppm STEV; 90 mg/m3 STEV

Canada - Quebec - Time-Weighted Average Exposure Values
10 ppm TWAEV; 45 mg/m3 TWAEV

Canada - Quebec - Short-term Exposure Values
20 ppm STEV; 90 mg/m3 STEV

United Kingdom - Occupational Exposure Standards - TWAs
10 ppm TWA; 45 mg/m3 TWA

United Kingdom - Occupational Exposure Standards - STELs
20 ppm STEL; 90 mg/m3 STEL

Israel - Time Weighted Averages
10 ppm TWA; 45 mg/m3 TWA

Israel - Short Term Exposure Limits
20 ppm STEL; 90 mg/m3 STEL

Israel - Action Levels
5 ppm AL; 22.5 mg/m3 AL

Mexico - Instruction No. 10 - TWAs
10 ppm TWA; 45 mg/m3 TWA

Mexico - Instruction No. 10 - STELs
20 ppm STEL; 90 mg/m3 STEL

STATE LISTS

California - Exposure Limits - PELs
10 ppm PEL; 45 mg/m3 PEL

California - Exposure Limits - STELs
20 ppm STEL; 90 mg/m3 STEL

California - Directors List of Hazardous Substances (8 CCR 339)
[present]

Florida Hazardous Substance List
[present]

Massachusetts Right To Know List
[present]

Minnesota Hazardous Substance List
[present]

NJ Right to Know List (Total)
sn 1634

Pennsylvania Right to Know List
environmental hazard

RHENIUM 178   18853-08-2

ENVIRONMENTAL LISTS

TSCA - Chemicals with Significant New Use Rules
PMN number: P-89-769

RHENIUM 181   14993-65-8

STATE LISTS

California - Air Bill 2588 Appendix A-II
9/89

California - Prop. 65 - Developmental Toxicity
developmental toxicity - initial date 1/1/89

RHENIUM 182   21459-71-2

HEALTH AND SAFETY LISTS

NTP Chemical Status Reports - Testing Status and NTIS Number
Prechronic studies in progress

RHENIUM 184   14983-46-1

ENVIRONMENTAL LISTS

List of Pesticide Product Inert Ingredients
[present]

RHENIUM 184M   RR-00386-1

HEALTH AND SAFETY LISTS

IARC - Group 3 (not classifiable)
[present]

STATE LISTS

California - Directors List of Hazardous Substances (8 CCR 339)
[present]

RHENIUM 186   14998-63-1

ENVIRONMENTAL LISTS

List of Pesticide Product Inert Ingredients
[present]

RHENIUM 186M   RR-00377-0

HEALTH AND SAFETY LISTS

U.S. DOT - Appendix A Table 2 - Radionuclides
final RQ = 1000 curies (3.7E 13 Bq)

ENVIRONMENTAL LISTS

CERCLA/SARA List of Radionuclides (Appendix B) and Their Reportable Quantities
final RQ = 1000 curies (3.7E 13 Bq)

RHENIUM 187   **14391-29-8**

HEALTH AND SAFETY LISTS

U.S. DOT - Appendix A Table 2 - Radionuclides
final RQ = 1000 curies (3.7E 13 Bq)

ENVIRONMENTAL LISTS

CERCLA/SARA List of Radionuclides (Appendix B) and Their Reportable Quantities
final RQ = 1000 curies (3.7E 13 Bq)

RHENIUM 188   **14378-26-8**

HEALTH AND SAFETY LISTS

U.S. DOT - Appendix A Table 2 - Radionuclides
final RQ = 100 curies (3.7E 12 Bq)

ENVIRONMENTAL LISTS

CERCLA/SARA List of Radionuclides (Appendix B) and Their Reportable Quantities
final RQ = 100 curies (3.7E 12 Bq)

RHENIUM 188M   **RR-00374-7**

HEALTH AND SAFETY LISTS

U.S. DOT - Appendix A Table 2 - Radionuclides
12.7 hour half-life: Final RQ = 10 curies (3.7E 11 Bq); 64.0 hour half-life: Final RQ = 10 curies (3.7E 11 Bq)

ENVIRONMENTAL LISTS

CERCLA/SARA List of Radionuclides (Appendix B) and Their Reportable Quantities
12.7 hour half-life: Final RQ = 10 curies (3.7E 11 Bq); 64.0 hour half-life: Final RQ = 10 curies (3.7E 11 Bq)

RHENIUM 189   **15765-78-3**

HEALTH AND SAFETY LISTS

U.S. DOT - Appendix A Table 2 - Radionuclides
final RQ = 10 curies (3.7E 11 Bq)

ENVIRONMENTAL LISTS

CERCLA/SARA List of Radionuclides (Appendix B) and Their Reportable Quantities
final RQ = 10 curies (3.7E 11 Bq)

RHODAMINE WT   **6373-07-5**

HEALTH AND SAFETY LISTS

U.S. DOT - Appendix A Table 2 - Radionuclides
final RQ = 10 curies (3.7E 11 Bq)

ENVIRONMENTAL LISTS

CERCLA/SARA List of Radionuclides (Appendix B) and Their Reportable Quantities
final RQ = 10 curies (3.7E 11 Bq)

RHODAMINE WT   **37299-86-8**

HEALTH AND SAFETY LISTS

U.S. DOT - Appendix A Table 2 - Radionuclides
final RQ = 100 curies (3.7E 12 Bq)

ENVIRONMENTAL LISTS

CERCLA/SARA List of Radionuclides (Appendix B) and Their Reportable Quantities
final RQ = 100 curies (3.7E 12 Bq)

RHODANINE   **141-84-4**

HEALTH AND SAFETY LISTS

U.S. DOT - Appendix A Table 2 - Radionuclides
final RQ = 10 curies (3.7E 11 Bq)

ENVIRONMENTAL LISTS

CERCLA/SARA List of Radionuclides (Appendix B) and Their Reportable Quantities
final RQ = 10 curies (3.7E 11 Bq)

RHODINOL   **141-25-3**

HEALTH AND SAFETY LISTS

U.S. DOT - Appendix A Table 2 - Radionuclides
final RQ = 1000 curies (3.7E 13 Bq)

ENVIRONMENTAL LISTS

CERCLA/SARA List of Radionuclides (Appendix B) and Their Reportable Quantities
final RQ = 1000 curies (3.7E 13 Bq)

RHODIUM   **7440-16-6**

HEALTH AND SAFETY LISTS

U.S. DOT - Appendix A Table 2 - Radionuclides
final RQ = 1000 curies (3.7E 13 Bq)

ENVIRONMENTAL LISTS

CERCLA/SARA List of Radionuclides (Appendix B) and Their Reportable Quantities
final RQ = 1000 curies (3.7E 13 Bq)

RHODIUM 99   **15765-79-4**

HEALTH AND SAFETY LISTS

U.S. DOT - Appendix A Table 2 - Radionuclides
final RQ = 1000 curies (3.7E 13 Bq)

ENVIRONMENTAL LISTS

CERCLA/SARA List of Radionuclides (Appendix B) and Their Reportable Quantities
final RQ = 1000 curies (3.7E 13 Bq)

RHODIUM 99M   **RR-00373-6**

HEALTH AND SAFETY LISTS

U.S. DOT - Appendix A Table 2 - Radionuclides
final RQ = 1000 curies (3.7E 13 Bq)

ENVIRONMENTAL LISTS

CERCLA/SARA List of Radionuclides (Appendix B) and Their Reportable Quantities
final RQ = 1000 curies (3.7E 13 Bq)

RHODIUM 100   **15765-80-7**

ENVIRONMENTAL LISTS

List of Pesticide Product Inert Ingredients
[present]

RHODIUM 101   **14378-53-1**

ENVIRONMENTAL LISTS

List of Pesticide Product Inert Ingredients
[present]

RHODIUM 101M   **RR-00372-5**

INTERNATIONAL LISTS

Canada - WHMIS: Ingredient Disclosure
1% item 1382 (1467)

RHODIUM 102   **15765-82-9**

HEALTH AND SAFETY LISTS

NFPA - Flash Points
flash point > 212 degrees F (100 degrees C)

NFPA - Hazard Identification Ratings
health-0; flammability-1; reactivity-0

RHODIUM 102M   RR-00371-4

HEALTH AND SAFETY LISTS

ACGIH 1995 - Time Weighted Averages
metal: 1 mg/m3 TWA; insoluble compounds, as Rh: 1 mg/m3 TWA; soluble compounds, as Rh: 0.01 mg/m3

NIOSH 1990 - Pocket Guide - RELs
as Rh: 0.1 mg/m3 TWA

NIOSH 1990 - Pocket Guide - Target organs
none known

OSHA - Vacated PELs - Time Weighted Averages
metal fume and insoluble compounds, as Rh: 0.1 mg/m3 TWA; soluble compounds, as Rh: 0.001 mg/m3 TWA

OSHA - Final PELs - Time Weighted Averages
metal fume and insoluble compounds, as Rh: 0.1 mg/m3 TWA; soluble compounds, as Rh: 0.001 mg/m3 TWA

INTERNATIONAL LISTS

Australian Exposure Standards - Time Weighted Averages
metal, as Rh: 1 mg/m3 TWA; insoluble compounds, as Rh: 1 mg/m3 TWA; soluble compounds, as Rh: 0.01 mg/m3 TWA

Canada - WHMIS: Ingredient Disclosure
1% item 1384 (1469)

Canada - Alberta - 8 Hour Occupational Exposure Limit
as Rh: 1 mg/m3 TWA

Canada - Alberta - 15 Minute Occupational Exposure Limit
as Rh: 2 mg/m3 STEL

Canada - British Columbia - 8 Hour Exposure Limits
as Rh: 0.1 mg/m3 TWA

Canada - British Columbia - 15 Minute Exposure Limits
as Rh: 0.3 mg/m3 STEL

Canada - Ontario - OHSA - TWAEVs
1 mg/m3 TWAEV

Canada - Quebec - Time-Weighted Average Exposure Values
as Rh: 0.1 mg/m3 TWAEV

United Kingdom - Occupational Exposure Standards - TWAs
metal fume and dust, as Rh: 0.1 mg/m3 TWA; soluble salts: 0.001 mg/m3 TWA

United Kingdom - Occupational Exposure Standards - STELs
metal fume and dust, as Rh: 0.3 mg/m3 STEL; soluble salts, as Rh: 0.003 mg/m3 STEL

Israel - Time Weighted Averages
metal: 1 mg/m3 TWA; insoluble compounds, as Rh: 1 mg/m3 TWA; soluble compounds, as Rh: 0.01 mg/m3

Israel - Action Levels
metal: 0.5 mg/m3 AL; insoluble compounds, as Rh: 0.5 mg/m3 AL; soluble compounds, as Rh: 0.005 mg/m3 AL

Mexico - Instruction No. 10 - TWAs
1 mg/m3 TWA

STATE LISTS

California - Exposure Limits - PELs
metal: 0.1 mg/m3 PEL; insoluble compounds, as Rh: 0.1 mg/m3 PEL; soluble salts, as Rh: 0.001 mg/m3 PEL

California - Directors List of Hazardous Substances (8 CCR 339)
[present]

Florida Hazardous Substance List
[present]

Massachusetts Right To Know List
[present]

Minnesota Hazardous Substance List
[present]

Pennsylvania Right to Know List
[present]

PROPOSED REGULATIONS

Canada - Ontario - Proposed Occupational TWAEVs
as Rh: 0.1 mg/m3 TWAEV

Canada - Ontario - Proposed Occupational STEVs
as Rh: 0.3 mg/m3 STEV

RHODIUM 103M   RR-00366-7

HEALTH AND SAFETY LISTS

U.S. DOT - Appendix A Table 2 - Radionuclides
final RQ = 10 curies (3.7E 11 Bq)

ENVIRONMENTAL LISTS

CERCLA/SARA List of Radionuclides (Appendix B) and Their Reportable Quantities
final RQ = 10 curies (3.7E 11 Bq)

RHODIUM 105   14913-89-4

HEALTH AND SAFETY LISTS

U.S. DOT - Appendix A Table 2 - Radionuclides
final RQ = 100 curies (3.7E 12 Bq)

ENVIRONMENTAL LISTS

CERCLA/SARA List of Radionuclides (Appendix B) and Their Reportable Quantities
final RQ = 100 curies (3.7E 12 Bq)

RHODIUM 106M   RR-00365-6

HEALTH AND SAFETY LISTS

U.S. DOT - Appendix A Table 2 - Radionuclides
final RQ = 10 curies (3.7E 11 Bq)

ENVIRONMENTAL LISTS

CERCLA/SARA List of Radionuclides (Appendix B) and Their Reportable Quantities
final RQ = 10 curies (3.7E 11 Bq)

RHODIUM 107   15706-50-0

HEALTH AND SAFETY LISTS

U.S. DOT - Appendix A Table 2 - Radionuclides
final RQ = 10 curies (3.7E 11 Bq)

ENVIRONMENTAL LISTS

CERCLA/SARA List of Radionuclides (Appendix B) and Their Reportable Quantities
final RQ = 10 curies (3.7E 11 Bq)

RHODIUM COMPOUNDS, N.O.S.   RR-00611-1

HEALTH AND SAFETY LISTS

U.S. DOT - Appendix A Table 2 - Radionuclides
final RQ = 100 curies (3.7E 12 Bq)

ENVIRONMENTAL LISTS

CERCLA/SARA List of Radionuclides (Appendix B) and Their Reportable Quantities
final RQ = 100 curies (3.7E 12 Bq)

RHODIUM INSOLUBLE COMPOUNDS   RR-00047-5

HEALTH AND SAFETY LISTS

U.S. DOT - Appendix A Table 2 - Radionuclides
final RQ = 10 curies (3.7E 11 Bq)

ENVIRONMENTAL LISTS

CERCLA/SARA List of Radionuclides (Appendix B) and Their Reportable Quantities
final RQ = 10 curies (3.7E 11 Bq)

RHODIUM NITRATE   10139-58-9

HEALTH AND SAFETY LISTS

U.S. DOT - Appendix A Table 2 - Radionuclides
final RQ = 10 curies (3.7E 11 Bq)

ENVIRONMENTAL LISTS

CERCLA/SARA List of Radionuclides (Appendix B) and Their Reportable Quantities
final RQ = 10 curies (3.7E 11 Bq)

RHODIUM SOLUBLE COMPOUNDS   RR-00040-8

HEALTH AND SAFETY LISTS

U.S. DOT - Appendix A Table 2 - Radionuclides
final RQ = 1000 curies (3.7E 13 Bq)

ENVIRONMENTAL LISTS

CERCLA/SARA List of Radionuclides (Appendix B) and Their Reportable Quantities
final RQ = 1000 curies (3.7E 13 Bq)

RHODIUM SULFATE   10489-46-0

HEALTH AND SAFETY LISTS

U.S. DOT - Appendix A Table 2 - Radionuclides
final RQ = 100 curies (3.7E 12 Bq)

ENVIRONMENTAL LISTS

CERCLA/SARA List of Radionuclides (Appendix B) and Their Reportable Quantities
final RQ = 100 curies (3.7E 12 Bq)

RHODIUM TRICHLORIDE   10049-07-7

HEALTH AND SAFETY LISTS

U.S. DOT - Appendix A Table 2 - Radionuclides
final RQ = 10 curies (3.7E 11 Bq)

ENVIRONMENTAL LISTS

CERCLA/SARA List of Radionuclides (Appendix B) and Their Reportable Quantities
final RQ = 10 curies (3.7E 11 Bq)

RIBAVIRIN   36791-04-5

HEALTH AND SAFETY LISTS

U.S. DOT - Appendix A Table 2 - Radionuclides
final RQ = 1000 curies (3.7E 13 Bq)

ENVIRONMENTAL LISTS

CERCLA/SARA List of Radionuclides (Appendix B) and Their Reportable Quantities
final RQ = 1000 curies (3.7E 13 Bq)

RICE   RR-01135-8

INTERNATIONAL LISTS

Canada - WHMIS: Ingredient Disclosure
1% item 1383 (1468)

STATE LISTS

California - Directors List of Hazardous Substances (8 CCR 339)
[present]

RICE BRAN OIL   68553-81-1

INTERNATIONAL LISTS

Canada - Ontario - OHSA - TWAEVs
as Rh: 1 mg/m3 TWAEV

PROPOSED REGULATIONS

Canada - Ontario - Proposed Occupational TWAEVs
as Rh: 0.1 mg/m3 TWAEV

Canada - Ontario - Proposed Occupational STEVs
as Rh: 0.3 mg/m3 STEV

RICIN   9009-86-3

INTERNATIONAL LISTS

Canada - Ontario - OHSA - TWAEVs
as Rh: 0.01 mg/m3 TWAEV

PROPOSED REGULATIONS

Canada - Ontario - Proposed Occupational TWAEVs
as Rh: 0.001 mg/m3 TWAEV

Canada - Ontario - Proposed Occupational STEVs
as Rh: 0.003 mg/m3 STEV

RIDDELLINE   23246-96-0

HEALTH AND SAFETY LISTS

NIOSH 1990 - Pocket Guide - RELs
as Rh: 0.001 mg/m3 TWA

NIOSH 1990 - Pocket Guide - Target organs
eyes

INTERNATIONAL LISTS

Canada - Alberta - 8 Hour Occupational Exposure Limit
as Rh: 0.001 mg/m3 TWA

Canada - Alberta - 15 Minute Occupational Exposure Limit
0.003 mg/m3 STEL

Canada - British Columbia - 8 Hour Exposure Limits
as Rh: 0.001 mg/m3 TWA

Canada - British Columbia - 15 Minute Exposure Limits
as Rh: 0.003 mg/m3 STEL

Canada - Ontario - OHSA - TWAEVs
as Rh: 0.01 mg/m3 TWAEV

Canada - Quebec - Time-Weighted Average Exposure Values
as Rh: 0.001 mg/m3 TWAEV

Mexico - Instruction No. 10 - TWAs
0.01 mg/m3 TWA

PROPOSED REGULATIONS

Canada - Ontario - Proposed Occupational TWAEVs
as Rh: 0.001 mg/m3 TWAEV

Canada - Ontario - Proposed Occupational STEVs
as Rh: 0.003 mg/m3 STEV

RIFAMPICIN   13292-46-1

INTERNATIONAL LISTS

Canada - Ontario - OHSA - TWAEVs
as Rh: 0.01 mg/m3 TWAEV

PROPOSED REGULATIONS

Canada - Ontario - Proposed Occupational TWAEVs
as Rh: 0.001 mg/m3 TWAEV

Canada - Ontario - Proposed Occupational STEVs
as Rh: 0.003 mg/m3 STEV

ROCKWOOL   RR-01201-1

SEE ALSO:
RHODIUM

HEALTH AND SAFETY LISTS

NIOSH - Selected LD50s and LC50s
Oral, rat: LD50 = 1302 mg/kg

INTERNATIONAL LISTS

Canada - WHMIS: Ingredient Disclosure
1% item 1385 (1659)

Canada - Ontario - OHSA - TWAEVs
as Rh: 0.01 mg/m3 TWAEV

STATE LISTS

Massachusetts Right To Know List
[present]

PROPOSED REGULATIONS

Canada - Ontario - Proposed Occupational TWAEVs
as Rh: 0.001 mg/m3 TWAEV

Canada - Ontario - Proposed Occupational STEVs
as Rh: 0.003 mg/m3 STEV

RODENTICIDES, LIQUID OR SOLID, N.O.S.　　RR-00479-5

STATE LISTS

California - Air Bill 2588 Appendix A-II
9/90

California - Prop. 65 - Developmental Toxicity
developmental toxicity - initial date 4/1/90

California - Directors List of Hazardous Substances (8 CCR 339)
[present]

RONNEL　　299-84-3

ENVIRONMENTAL LISTS

List of Pesticide Product Inert Ingredients
[present] (includes rice hulls)

ROSIN　　8050-09-7

ENVIRONMENTAL LISTS

List of Pesticide Product Inert Ingredients
[present]

ROSIN CORE SOLDER PYROLYSIS PRODUCTS, AS FORMALDEHYDE　　RR-00074-8

INTERNATIONAL LISTS

Canada - WHMIS: Ingredient Disclosure
0.1% item 1386 (1470)

ROSIN, MALEATED, POLYMER WITH PENTAERYTHRITOL　　68333-69-7

HEALTH AND SAFETY LISTS

IARC - Group 3 (not classifiable)
[present]

NTP Chemical Status Reports - Testing Status and NTIS Number
Technical reports printed (no NTIS number given);　project leader assigned/study in design

ROSIN OIL　　8002-16-2

HEALTH AND SAFETY LISTS

IARC - Group 3 (not classifiable)
[present]

STATE LISTS

California - Directors List of Hazardous Substances (8 CCR 339)
[present]

ROSIN, PARTIALLY HYDROGENATED　　65997-06-0

HEALTH AND SAFETY LISTS

IARC - Group 2B (sufficient animal data)
[present]

OSHA - Possible Select Carcinogens
[present]

INTERNATIONAL LISTS

Canada - Quebec - Time-Weighted Average Exposure Values
1 fibre/cm3 TWAEV

Canada - Quebec - Carcinogens
C2 carcinogen: effect suspected in humans

German (DFG) - Carcinogens
as fibrous dust: animal evidence of carcinogenicity

STATE LISTS

California - Air Bill 2588 Appendix A-I
known or potential carcinogen: 9/89

ROSIN POLYMERS　　65997-05-9

STATE LISTS

NJ Right to Know List (Total)
sn 2752

ROSIN POLYMER WITH P-TERT-BUTYLPHENOL, FORMALDEHYDE AND GLYCEROL　　67700-49-6

HEALTH AND SAFETY LISTS

ACGIH 1995 - Time Weighted Averages
10 mg/m3 TWA

NIOSH - Selected LD50s and LC50s
Oral, rat: LD50 = 625 mg/kg Skin, rat: LD50 = 2000 mg/kg

NIOSH 1990 - Pocket Guide - RELs
10 mg/m3 TWA

NIOSH 1990 - Pocket Guide - IDLHs
5000 mg/m3 IDLH

NIOSH 1990 - Pocket Guide - Target organs
skin, liver, kidneys, blood plasma

OSHA - Vacated PELs - Time Weighted Averages
10 mg/m3 TWA

OSHA - Final PELs - Time Weighted Averages
15 mg/m3 TWA

INTERNATIONAL LISTS

Australian Exposure Standards - Time Weighted Averages
10 mg/m3 TWA

Canada - Alberta - 8 Hour Occupational Exposure Limit
10 mg/m3 TWA

Canada - Alberta - 15 Minute Occupational Exposure Limit
20 mg/m3 STEL

Canada - British Columbia - 8 Hour Exposure Limits
10 mg/m3 TWA

Canada - Ontario - OHSA - TWAEVs
10 mg/m3 TWAEV

Canada - Quebec - Time-Weighted Average Exposure Values
10 mg/m3 TWAEV

United Kingdom - Occupational Exposure Standards - TWAs
10 mg/m3 TWA

Israel - Time Weighted Averages
10 mg/m3 TWA

Israel - Action Levels
5 mg/m3 AL

Mexico - Instruction No. 10 - TWAs
10 mg/m3 TWA

STATE LISTS

California - Exposure Limits - PELs
10 mg/m3 PEL

California - Directors List of Hazardous Substances (8 CCR 339)
[present]

Florida Hazardous Substance List
[present]

Massachusetts Right To Know List
neurotoxin

Minnesota Hazardous Substance List
[present]

NJ Right to Know List (Total)
sn 1637

Pennsylvania Right to Know List
[present]

ROTENONE (COMMERCIAL)　　83-79-4

ENVIRONMENTAL LISTS

List of Pesticide Product Inert Ingredients
[present]

ROUGE　　RR-00006-6

HEALTH AND SAFETY LISTS

OSHA - Vacated PELs - Time Weighted Averages
as Formaldehyde: 0.1 mg/m3 TWA

INTERNATIONAL LISTS

Australian Exposure Standards - Time Weighted Averages
0.1 mg/m3 TWA

Canada - Alberta - 8 Hour Occupational Exposure Limit
0.1 mg/m3 TWA

Canada - Alberta - 15 Minute Occupational Exposure Limit
0.3 mg/m3 STEL

Canada - British Columbia - 8 Hour Exposure Limits
as formaldehyde: 0.1 mg/m3 TWA

Canada - British Columbia - 15 Minute Exposure Limits
0.3 mg/m3 STEL

Canada - Ontario - OHSA - TWAEVs
as formaldehyde: 0.1 mg/m3 TWAEV (listed as an agent of variable composition)

Canada - Quebec - Time-Weighted Average Exposure Values
0.1 mg/m3 TWAEV

United Kingdom - Occupational Exposure Standards - TWAs
as Formaldehyde: 0.1 mg/m3 TWA

United Kingdom - Occupational Exposure Standards - STELs
as Formaldehyde: 0.3 mg/m3 STEL

United Kingdom - Occupational Exposure Standards - Notes
sensitisation

Israel - Time Weighted Averages
as formaldehyde: 0.1 mg/m3 TWA

Israel - Action Levels
as formaldehyde: 0.05 mg/m3

Mexico - Instruction No. 10 - TWAs
0.1 mg/m3 TWA

Mexico - Instruction No. 10 - STELs
0.3 mg/m3 STEL

STATE LISTS

California - Exposure Limits - PELs
as formaldehyde: 0.1 mg/m3 PEL

California - Directors List of Hazardous Substances (8 CCR 339)
[present] (refers to smoke and fume product given off during smoldering)

Minnesota Hazardous Substance List
[present] as Formaldehyde

Pennsylvania Right to Know List
[present]

ROXARSONE   121-19-7

ENVIRONMENTAL LISTS

List of Pesticide Product Inert Ingredients
[present]

RUBBER INDUSTRY   RR-00548-1

HEALTH AND SAFETY LISTS

U.S. DOT - Substances From 49 CFR 172.101
regulated by DOT (UN1286)

U.S. DOT - Hazard Classes
DOT hazard class = 3

NFPA - Flash Points
flash point = 266 degrees F (130 degrees C)

NFPA - Hazard Identification Ratings
health-0; flammability-1; reactivity-0

STATE LISTS

NJ Right to Know List (Total)
sn 2753

RUBBER SOLVENT (NAPHTHA)   RR-00076-0

ENVIRONMENTAL LISTS

List of Pesticide Product Inert Ingredients
[present]

RUBIDIUM   7440-17-7

ENVIRONMENTAL LISTS

List of Pesticide Product Inert Ingredients
[present]

RUBIDIUM 79   14809-48-4

ENVIRONMENTAL LISTS

List of Pesticide Product Inert Ingredients
[present]

RUBIDIUM 81   18268-34-3

HEALTH AND SAFETY LISTS

ACGIH 1995 - Time Weighted Averages
5 mg/m3 TWA

U.S. DOT - Appendix B - Marine Pollutants
DOT regulated marine pollutant

NIOSH - Selected LD50s and LC50s
Oral, rat: LD50 = 60 mg/kg

NIOSH 1990 - Pocket Guide - RELs
5 mg/m3 TWA

NIOSH 1990 - Pocket Guide - Target organs
CNS, eyes, respiratory system

NTP Chemical Status Reports - Testing Status and NTIS Number
Technical reports printed (PB89139760/AS); Other chronic studies exist for which technical reports were not prepared

NTP Chemical Status Reports - Evidence of Carcinogenicity
male rat-equivocal evidence; female rat-no evidence; male mice-no evidence; female mice-no evidence

OSHA - Vacated PELs - Time Weighted Averages
5 mg/m3 TWA

OSHA - Final PELs - Time Weighted Averages
5 mg/m3 TWA

ENVIRONMENTAL LISTS

List of Pesticide Product Inert Ingredients
[present]

INTERNATIONAL LISTS

Australian Exposure Standards - Time Weighted Averages
5 mg/m3 TWA

Canada - Alberta - 8 Hour Occupational Exposure Limit
5 mg/m3 TWA

Canada - Alberta - 15 Minute Occupational Exposure Limit
10 mg/m3 STEL

Canada - British Columbia - 8 Hour Exposure Limits
5 mg/m3 TWA

Canada - British Columbia - 15 Minute Exposure Limits
10 mg/m3 STEL

Canada - Ontario - OHSA - TWAEVs
5 mg/m3 TWAEV

Canada - Quebec - Time-Weighted Average Exposure Values
5 mg/m3 TWAEV

United Kingdom - Occupational Exposure Standards - TWAs
5 mg/m3 TWA

United Kingdom - Occupational Exposure Standards - STELs
10 mg/m3 STEL

German (DFG) - MAK Values
total dust: 5 mg/m3 MAK

Israel - Time Weighted Averages
5 mg/m3 TWA

Israel - Action Levels
2.5 mg/m3 AL

Mexico - Instruction No. 10 - TWAs
5 mg/m3 TWA

Mexico - Instruction No. 10 - STELs
10 mg/m3 STEL

STATE LISTS

California - Exposure Limits - PELs
5 mg/m3 PEL

California - Directors List of Hazardous Substances (8 CCR 339)
[present]

Florida Hazardous Substance List
[present]

Massachusetts Right To Know List
[present]

Minnesota Hazardous Substance List
[present]

NJ Right to Know List (Total)
sn 1638

NJ Special Hazardous Substances
(corrosive)

Pennsylvania Right to Know List
[present]

RUBIDIUM 81M   RR-00364-5

HEALTH AND SAFETY LISTS

ACGIH 1995 - Time Weighted Averages
10 mg/m3 TWA (The value is for total dust containing no asbestos and < 1% crytaline silica)

OSHA - Vacated PELs - Time Weighted Averages
total dust: 10 mg/m3 TWA; respirable fraction: 5 mg/m3 TWA

OSHA - Final PELs - Time Weighted Averages
total dust: 15 mg/m3 TWA; respirable fraction: 5 mg/m3 TWA

INTERNATIONAL LISTS

Australian Exposure Standards - Time Weighted Averages
10 mg/m3 TWA

Canada - Alberta - 8 Hour Occupational Exposure Limit
respirable mass: 5 mg/m3 TWA; total mass: 10 mg/m3 TWA

Canada - British Columbia - 8 Hour Exposure Limits
nuisance dust: 10 mg/m3 TWA

Canada - British Columbia - 15 Minute Exposure Limits
20 mg/m3 STEL

Canada - Quebec - Time-Weighted Average Exposure Values
total dust: 10 mg/m3 TWAEV; respirable dust: 5 mg/m3 TWAEV

United Kingdom - Occupational Exposure Standards - TWAs
total inhalable dust: 10 mg/m3 TWA; respirable dust: 5 mg/m3 TWA

Israel - Time Weighted Averages
10 mg/m3 TWA (The value is for total dust containing no asbestos and < 1% crytaline silica)

Israel - Action Levels
5 mg/m3 AL

STATE LISTS

Minnesota Hazardous Substance List
[present] (includes inert or nuisance dust)

RUBIDIUM 82M   RR-00363-4

HEALTH AND SAFETY LISTS

NTP Chemical Status Reports - Testing Status and NTIS Number
Technical reports printed (PB89216543/AS)

NTP Chemical Status Reports - Evidence of Carcinogenicity
male rat-equivocal evidence; female rat-no evidence; male mice-no evidence; female mice-no evidence

RUBIDIUM 83   17056-36-9

HEALTH AND SAFETY LISTS

IARC - Group 1 (carcinogenic to humans)
[present]

OSHA - Select Carcinogens
[present]

INTERNATIONAL LISTS

United Kingdom - Maximum Exposure Limits - TWAs
dust: 8 mg/m3 TWA

STATE LISTS

Pennsylvania Right to Know List
special hazardous substance

Pennsylvania RTK - Special Hazardous Substances
[present]

RUBIDIUM 84   15765-86-3

HEALTH AND SAFETY LISTS

NIOSH - Health Standards - Exposure Limits
350 mg/m3 TWA; C (15 min) 1800 mg/m3 (Listed under 'Refined petroleum solvents')

NIOSH - Health Standards - Health Effects and Precautions
Eye, nose, and throat irritation; dermatitis; nervous system effects (Blood and urine monitoring required; prevent skin contact) (Listed under 'Refined petroleum solvents')

INTERNATIONAL LISTS

Canada - Alberta - 8 Hour Occupational Exposure Limit
400 ppm TWA; 1600 mg/m3 TWA

Canada - Alberta - 15 Minute Occupational Exposure Limit
500 ppm STEL; 2000 mg/m3 STEL

Canada - British Columbia - 8 Hour Exposure Limits
400 ppm TWA; 1600 mg/m3 TWA

Canada - Ontario - OHSA - TWAEVs
1600 mg/m3 TWAEV (listed as an agent of variable composition) As sum of components assayed by chromatographic procedure with reference to the bulk sample.

Canada - Quebec - Time-Weighted Average Exposure Values
400 ppm TWAEV; 1570 mg/m3 TWAEV

Israel - Time Weighted Averages
400 ppm TWA; 1590 mg/m3 TWA

Israel - Action Levels
200 ppm AL; 795 mg/m3 AL

Mexico - Instruction No. 10 - TWAs
400 ppm TWA; 1600 mg/m3 TWA

STATE LISTS

California - Exposure Limits - PELs
400 ppm PEL; 1600 mg/m3 PEL

California - Directors List of Hazardous Substances (8 CCR 339)
[present]

Florida Hazardous Substance List
[present]

Minnesota Hazardous Substance List
[present]

Pennsylvania Right to Know List
[present]

RUBIDIUM 86   14932-53-7

HEALTH AND SAFETY LISTS

U.S. DOT - Substances From 49 CFR 172.101
regulated by DOT (UN1423)

U.S. DOT - Hazard Classes
DOT hazard class = 8

STATE LISTS

NJ Right to Know List (Total)
sn 1639

RUBIDIUM 87   13982-13-3

HEALTH AND SAFETY LISTS

U.S. DOT - Appendix A Table 2 - Radionuclides
final RQ = 1000 curies (3.7E 13 Bq)

ENVIRONMENTAL LISTS

CERCLA/SARA List of Radionuclides (Appendix B) and Their Reportable Quantities
final RQ = 1000 curies (3.7E 13 Bq)

RUBIDIUM 88   14928-36-0

HEALTH AND SAFETY LISTS

U.S. DOT - Appendix A Table 2 - Radionuclides
final RQ = 100 curies (3.7E 12 Bq)

ENVIRONMENTAL LISTS

CERCLA/SARA List of Radionuclides (Appendix B) and Their Reportable Quantities
final RQ = 100 curies (3.7E 12 Bq)

RUBIDIUM 89   14191-65-2

HEALTH AND SAFETY LISTS

U.S. DOT - Appendix A Table 2 - Radionuclides
final RQ = 1000 curies (3.7E 13 Bq)

ENVIRONMENTAL LISTS

CERCLA/SARA List of Radionuclides (Appendix B) and Their Reportable Quantities
final RQ = 1000 curies (3.7E 13 Bq)

RUBIDIUM CHLORIDE   7791-11-9

HEALTH AND SAFETY LISTS

U.S. DOT - Appendix A Table 2 - Radionuclides
final RQ = 10 curies (3.7E 11 Bq)

ENVIRONMENTAL LISTS

CERCLA/SARA List of Radionuclides (Appendix B) and Their Reportable Quantities
final RQ = 10 curies (3.7E 11 Bq)

RUBIDIUM CHROMATE   13446-72-5

HEALTH AND SAFETY LISTS

U.S. DOT - Appendix A Table 2 - Radionuclides
final RQ = 10 curies (3.7E 11 Bq)

ENVIRONMENTAL LISTS

CERCLA/SARA List of Radionuclides (Appendix B) and Their Reportable Quantities
final RQ = 10 curies (3.7E 11 Bq)

RUBIDIUM DICHROMATE   13446-73-6

HEALTH AND SAFETY LISTS

U.S. DOT - Appendix A Table 2 - Radionuclides
final RQ = 10 curies (3.7E 11 Bq)

ENVIRONMENTAL LISTS

CERCLA/SARA List of Radionuclides (Appendix B) and Their Reportable Quantities
final RQ = 10 curies (3.7E 11 Bq)

RUBIDIUM HYDROXIDE   1310-82-3

HEALTH AND SAFETY LISTS

U.S. DOT - Appendix A Table 2 - Radionuclides
final RQ = 10 curies (3.7E 11 Bq)

ENVIRONMENTAL LISTS

CERCLA/SARA List of Radionuclides (Appendix B) and Their Reportable Quantities
final RQ = 10 curies (3.7E 11 Bq)

RUBIDIUM IODIDE   7790-29-6

HEALTH AND SAFETY LISTS

U.S. DOT - Appendix A Table 2 - Radionuclides
final RQ = 10 curies (3.7E 11 Bq)

ENVIRONMENTAL LISTS

CERCLA/SARA List of Radionuclides (Appendix B) and Their Reportable Quantities
final RQ = 10 curies (3.7E 11 Bq)

RUGULOSIN   23537-16-8

HEALTH AND SAFETY LISTS

U.S. DOT - Appendix A Table 2 - Radionuclides
final RQ = 1000 curies (3.7E 13 Bq)

ENVIRONMENTAL LISTS

CERCLA/SARA List of Radionuclides (Appendix B) and Their Reportable Quantities
final RQ = 1000 curies (3.7E 13 Bq)

RUTHENIUM 94   15125-02-7

HEALTH AND SAFETY LISTS

U.S. DOT - Appendix A Table 2 - Radionuclides
final RQ = 1000 curies (3.7E 13 Bq)

ENVIRONMENTAL LISTS

CERCLA/SARA List of Radionuclides (Appendix B) and Their Reportable Quantities
final RQ = 1000 curies (3.7E 13 Bq)

RUTHENIUM 97   15758-35-7

HEALTH AND SAFETY LISTS

NIOSH - Selected LD50s and LC50s
Oral, mouse: LD50 = 3800 mg/kg

INTERNATIONAL LISTS

Canada - WHMIS: Ingredient Disclosure
1% item 1387 (524)

RUTHENIUM 103   13968-53-1

STATE LISTS

Massachusetts Right To Know List
[present]

RUTHENIUM 105   14331-95-4

SEE ALSO:
RUBIDIUM
CHROMIUM

STATE LISTS

Massachusetts Right To Know List
[present]

RUTHENIUM 106   13967-48-1

HEALTH AND SAFETY LISTS

U.S. DOT - Substances From 49 CFR 172.101
regulated by DOT (UN2677)

U.S. DOT - Hazard Classes
DOT hazard class = 4.3

NIOSH - Selected LD50s and LC50s
Oral, rat: LD50 = 586 mg/kg

INTERNATIONAL LISTS

Canada - WHMIS: Ingredient Disclosure
1% item 1388 (997)

STATE LISTS

NJ Right to Know List (Total)
sn 1640

NJ Special Hazardous Substances
(corrosive)

RUTHENIUM CHLORIDE HYDROXIDE 16845-29-7

HEALTH AND SAFETY LISTS

NIOSH - Selected LD50s and LC50s
Oral, rat: LD50 = 4708 mg/kg

RUTHENIUM OXIDE 12036-10-1

HEALTH AND SAFETY LISTS

IARC - Group 3 (not classifiable)
[present]

RUTILE (TIO2) 1317-80-2

HEALTH AND SAFETY LISTS

U.S. DOT - Appendix A Table 2 - Radionuclides
final RQ = 1000 curies (3.7E 13 Bq)

ENVIRONMENTAL LISTS

CERCLA/SARA List of Radionuclides (Appendix B) and Their Reportable Quantities
final RQ = 1000 curies (3.7E 13 Bq)

RYANODINE 15662-33-6

HEALTH AND SAFETY LISTS

U.S. DOT - Appendix A Table 2 - Radionuclides
final RQ = 100 curies (3.7E 12 Bq)

ENVIRONMENTAL LISTS

CERCLA/SARA List of Radionuclides (Appendix B) and Their Reportable Quantities
final RQ = 100 curies (3.7E 12 Bq)

RYE FLOUR RR-01136-9

HEALTH AND SAFETY LISTS

U.S. DOT - Appendix A Table 2 - Radionuclides
final RQ = 10 curies (3.7E 11 Bq)

ENVIRONMENTAL LISTS

CERCLA/SARA List of Radionuclides (Appendix B) and Their Reportable Quantities
final RQ = 10 curies (3.7E 11 Bq)

SACCHARATED IRON OXIDE 8047-67-4

HEALTH AND SAFETY LISTS

U.S. DOT - Appendix A Table 2 - Radionuclides
final RQ = 100 curies (3.7E 12 Bq)

ENVIRONMENTAL LISTS

CERCLA/SARA List of Radionuclides (Appendix B) and Their Reportable Quantities
final RQ = 100 curies (3.7E 12 Bq)

SACCHARIN 81-07-2

HEALTH AND SAFETY LISTS

U.S. DOT - Appendix A Table 2 - Radionuclides
final RQ = 1 curie (3.7E 10 Bq)

ENVIRONMENTAL LISTS

CERCLA/SARA List of Radionuclides (Appendix B) and Their Reportable Quantities
final RQ = 1 curie (3.7E 10 Bq)

SAFFLOWER OIL 8001-23-8

INTERNATIONAL LISTS

Canada - WHMIS: Ingredient Disclosure
1% item 1389 (986)

SAFROLE 94-59-7

SEE ALSO:
RUTHENIUM

HEALTH AND SAFETY LISTS

NIOSH - Selected LD50s and LC50s
Oral, rat: LD50 = 4580 mg/kg

SALCOMINE 14167-18-1

STATE LISTS

Pennsylvania Right to Know List
[present]

SALICYLALDEHYDE 90-02-8

ENVIRONMENTAL LISTS

List of Pesticide Product Inert Ingredients
[present]

SALICYLAMIDE 65-45-2

ENVIRONMENTAL LISTS

List of Pesticide Product Inert Ingredients
[present]

SALICYLAZOSULFAPYRIDINE 599-79-1

HEALTH AND SAFETY LISTS

IARC - Group 3 (not classifiable)
[present]

STATE LISTS

California - Directors List of Hazardous Substances (8 CCR 339)
[present]

SALICYLIC ACID 69-72-7

HEALTH AND SAFETY LISTS

U.S. DOT - Appendix A Table 1 - Hazardous Substances
final RQ = 100 pounds (45.4 kg)

IARC - Group 2B (sufficient animal data)
[present]

NIOSH - Selected LD50s and LC50s
Oral, mouse: LD50 = 17 gm/kg

NTP Seventh Report - Suspect Carcinogens
suspect carcinogen

OSHA - Possible Select Carcinogens
[present]

ENVIRONMENTAL LISTS

CERCLA/SARA - Section 313 - Emission Reporting
form R reporting required for 0.1% de minimus concentration (only if it is being manufactured)

CERCLA/SARA - Hazardous Substances and their Reportable Quantities
final RQ = 100 pounds (45.4 kg)

EPA - Carcinogen Hazard Ranking for RQ Adjustment
Hazard ranking = Low

List of Pesticide Product Inert Ingredients
[present]

RCRA - U Series Wastes
waste number U202

RCRA - Hazardous Constituents-Appendix VIII
waste number U202

RCRA - Substances Banned From Land Disposal
[present]

STATE LISTS

California - Air Bill 2588 Appendix A-II
known or potential carcinogen

California - Prop. 65 - Cancer list
carcinogen - initial date 10/1/89

Florida Hazardous Substance List
[present]

Massachusetts Right To Know List
carcinogen; extraordinarily hazardous

Minnesota Hazardous Substance List
carcinogen

NJ Right to Know List (Total)
sn 1641

NJ Special Hazardous Substances
(carcinogen)

Pennsylvania Right to Know List
environmental hazard; special hazardous substance

Pennsylvania RTK - Special Hazardous Substances
[present]

SALITHION   3811-49-2

HEALTH AND SAFETY LISTS

NTP Chemical Status Reports - Testing Status and NTIS Number
Technical reports printed

ENVIRONMENTAL LISTS

List of Pesticide Product Inert Ingredients
[present]

SALTED FISH, CHINESE STYLE   RR-01707-2
SEE ALSO:
F039-HAZARDOUS WASTES

HEALTH AND SAFETY LISTS

U.S. DOT - Appendix A Table 1 - Hazardous Substances
final RQ = 100 pounds (45.4 kg)

FDA - Controlled Substances Act - Precursor chemicals
Threshold by base weight = 4 kilograms

IARC - Group 2B (sufficient animal data)
[present] (Overall evaluation based only on evidence of carcinogenicity in monograph (10, 1976) or in Supplement 4)

NFPA - Flash Points
flash point = 212 degrees F (100 degrees C)

NFPA - Hazard Identification Ratings
flammability-1; reactivity-0

NIOSH - Selected LD50s and LC50s
Oral, rat: LD50 = 1950 mg/kg

NTP Seventh Report - Suspect Carcinogens
suspect carcinogen

OSHA - Possible Select Carcinogens
[present]

ENVIRONMENTAL LISTS

CERCLA/SARA - Section 313 - Emission Reporting
form R reporting required for 0.1% de minimus concentration

CERCLA/SARA - Hazardous Substances and their Reportable Quantities
final RQ = 100 pounds (45.4 kg)

EPA - Carcinogen Hazard Ranking for RQ Adjustment
Hazard ranking = Low

List of Pesticide Product Inert Ingredients
[present]

RCRA - U Series Wastes
waste number U203

RCRA - Hazardous Constituents-Appendix VIII
waste number U203

RCRA - Basis for Listing - Appendix VII
Included in waste stream: F039

RCRA - Substances Banned From Land Disposal
[present]

RCRA - TSD Facilities Ground Water Monitoring
TM 8270 = 10 ug/L PQL

RCRA - Universal Treatment Standards (LDR)
WW: 0.081 mg/l; NWW: 22 mg/kg

INTERNATIONAL LISTS

Canada - WHMIS: Ingredient Disclosure
1% item 1390 (1472)

Canada - NPRI (National Pollutant Release Inventory)
[present]

STATE LISTS

California - Air Bill 2588 Appendix A-II
known or potential carcinogen

California - Prop. 65 - Cancer list
carcinogen - initial date 1/1/88

California - Prop. 65 - No Significant Risk Levels
no significant risk level = 3 ug/day

California - Directors List of Hazardous Substances (8 CCR 339)
[present]

Florida Hazardous Substance List
[present]

Massachusetts Right To Know List
carcinogen; extraordinarily hazardous

Minnesota Hazardous Substance List
carcinogen

NJ Right to Know List (Total)
sn 1642

NJ Special Hazardous Substances
(carcinogen; mutagen)

Pennsylvania Right to Know List
environmental hazard; special hazardous substance

Pennsylvania RTK - Special Hazardous Substances
[present]

SALT OF CYCLODIAMINE AND MINERAL ACID   RR-01220-4
SEE ALSO:
COBALT

ENVIRONMENTAL LISTS

CERCLA/SARA - Section 302 Extremely Hazardous Substances and TPQs
TPQ = 500/10,000 pounds

STATE LISTS

Florida Hazardous Substance List
effective March 13, 1992

Massachusetts Right To Know List
extraordinarily hazardous

Pennsylvania Right to Know List
environmental hazard

PROPOSED REGULATIONS

CERCLA/SARA - Proposed Hazardous Substance Additions
proposed RQ = 1 pound (.454 kg)

CERCLA/SARA - 1989 Proposed RQ Adjustments
proposed RQ = 100 pounds (45.4 kg)

SAMARIUM 141   14877-67-9

HEALTH AND SAFETY LISTS

NFPA - Flash Points
flash point = 172 degrees F (78 degrees C)

NFPA - Hazard Identification Ratings
health-0; flammability-2; reactivity-0

NIOSH - Selected LD50s and LC50s
Oral, rat: LD50 = 520 mg/kg Skin, rat: LD50 = 600 mg/kg

ENVIRONMENTAL LISTS

TSCA - Code of Federal Regulations Citations
40 CFR 712.30(x); 40 CFR 716.120(d)

TSCA - PAIR - Reporting List
Reporting Date: November 27, 1991

TSCA - Health and Safety Reporting List
Effective Date: September 30, 1991

INTERNATIONAL LISTS

Canada - WHMIS: Ingredient Disclosure
1% item 1391 (1473)

SAMARIUM 141M   RR-00362-3

HEALTH AND SAFETY LISTS

NIOSH - Selected LD50s and LC50s
Oral, rat: LD50 = 980 mg/kg

SAMARIUM 142   15701-12-9

HEALTH AND SAFETY LISTS

NTP Chemical Status Reports - Testing Status and NTIS Number
Two year studies: pathology quality assessment in progress; prechronic studies for which toxicity technical reports were not prepared

SAMARIUM 145   15065-02-8

HEALTH AND SAFETY LISTS

NFPA - Flash Points
flash point = 315 degrees F (157 degrees C)

NFPA - Hazard Identification Ratings
health-0; flammability-1; reactivity-0

NIOSH - Selected LD50s and LC50s
Oral, rat: LD50 = 891 mg/kg

ENVIRONMENTAL LISTS

CAA - HON Rule - SOCMI Chemicals
compliance by April 24, 1995

List of Pesticide Product Inert Ingredients
[present]

INTERNATIONAL LISTS

Canada - WHMIS: Ingredient Disclosure
0.1% item 1392 (130)

SAMARIUM 146   14280-31-0

HEALTH AND SAFETY LISTS

U.S. DOT - Appendix B - Marine Pollutants
DOT regulated marine pollutant

SAMARIUM 147   14392-33-7

HEALTH AND SAFETY LISTS

IARC - Group 1 (carcinogenic to humans)
[present]

OSHA - Select Carcinogens
[present]

SAMARIUM 151   15715-94-3

ENVIRONMENTAL LISTS

TSCA - Chemicals with Significant New Use Rules
PMN number: P-91-838

SAMARIUM 153   15766-00-4

HEALTH AND SAFETY LISTS

U.S. DOT - Appendix A Table 2 - Radionuclides
final RQ = 1000 curies (3.7E 13 Bq)

ENVIRONMENTAL LISTS

CERCLA/SARA List of Radionuclides (Appendix B) and Their Reportable Quantities
final RQ = 1000 curies (3.7E 13 Bq)

SAMARIUM 155   14391-31-2

HEALTH AND SAFETY LISTS

U.S. DOT - Appendix A Table 2 - Radionuclides
final RQ = 1000 curies (3.7E 13 Bq)

ENVIRONMENTAL LISTS

CERCLA/SARA List of Radionuclides (Appendix B) and Their Reportable Quantities
final RQ = 1000 curies (3.7E 13 Bq)

SAMARIUM 156   15759-70-3

HEALTH AND SAFETY LISTS

U.S. DOT - Appendix A Table 2 - Radionuclides
final RQ = 1000 curies (3.7E 13 Bq)

ENVIRONMENTAL LISTS

CERCLA/SARA List of Radionuclides (Appendix B) and Their Reportable Quantities
final RQ = 1000 curies (3.7E 13 Bq)

SAMARIUM NITRATE   10361-83-8

HEALTH AND SAFETY LISTS

U.S. DOT - Appendix A Table 2 - Radionuclides
final RQ = 100 curies (3.7E 12 Bq)

ENVIRONMENTAL LISTS

CERCLA/SARA List of Radionuclides (Appendix B) and Their Reportable Quantities
final RQ = 100 curies (3.7E 12 Bq)

SAMARIUM TRICHLORIDE   10361-82-7

HEALTH AND SAFETY LISTS

U.S. DOT - Appendix A Table 2 - Radionuclides
final RQ = 0.01 curies (3.7E 8 Bq)

ENVIRONMENTAL LISTS

CERCLA/SARA List of Radionuclides (Appendix B) and Their Reportable Quantities
final RQ = 0.01 curies (3.7E 8 Bq)

SANTALOL   115-71-9

HEALTH AND SAFETY LISTS

U.S. DOT - Appendix A Table 2 - Radionuclides
final RQ = 0.01 curies (3.7E 8 Bq)

ENVIRONMENTAL LISTS

CERCLA/SARA List of Radionuclides (Appendix B) and Their Reportable Quantities
final RQ = 0.01 curies (3.7E 8 Bq)

SANTICIZER 711   39393-37-8

HEALTH AND SAFETY LISTS

U.S. DOT - Appendix A Table 2 - Radionuclides
final RQ = 10 curies (3.7E 11 Bq)

ENVIRONMENTAL LISTS

CERCLA/SARA List of Radionuclides (Appendix B) and Their Reportable Quantities
final RQ = 10 curies (3.7E 11 Bq)

SARIN   107-44-8

HEALTH AND SAFETY LISTS

U.S. DOT - Appendix A Table 2 - Radionuclides
final RQ = 100 curies (3.7E 12 Bq)

ENVIRONMENTAL LISTS
CERCLA/SARA List of Radionuclides (Appendix B) and Their Reportable Quantities
final RQ = 100 curies (3.7E 12 Bq)

SAWDUST   RR-01137-0
HEALTH AND SAFETY LISTS
U.S. DOT - Appendix A Table 2 - Radionuclides
final RQ = 1000 curies (3.7E 13 Bq)
ENVIRONMENTAL LISTS
CERCLA/SARA List of Radionuclides (Appendix B) and Their Reportable Quantities
final RQ = 1000 curies (3.7E 13 Bq)

SCANDIUM 43   14276-61-0
HEALTH AND SAFETY LISTS
U.S. DOT - Appendix A Table 2 - Radionuclides
final RQ = 100 curies (3.7E 12 Bq)
ENVIRONMENTAL LISTS
CERCLA/SARA List of Radionuclides (Appendix B) and Their Reportable Quantities
final RQ = 100 curies (3.7E 12 Bq)

SCANDIUM 44   14391-94-7
HEALTH AND SAFETY LISTS
NIOSH - Selected LD50s and LC50s
Oral, rat: LD50 = 2160 mg/kg

SCANDIUM 44M   RR-00359-8
INTERNATIONAL LISTS
Canada - WHMIS: Ingredient Disclosure
1% item 1393 (1660)

SCANDIUM 46   13967-63-0
HEALTH AND SAFETY LISTS
NFPA - Flash Points
flash point > 212 degrees F (100 degrees C)
NFPA - Hazard Identification Ratings
flammability-1; reactivity-0

SCANDIUM 47   14391-96-9
ENVIRONMENTAL LISTS
List of Pesticide Product Inert Ingredients
[present]

SCANDIUM 48   14391-86-7
HEALTH AND SAFETY LISTS
NIOSH - Selected LD50s and LC50s
Inhalation, rat: LC50 = 150 mg/m3 10 mn Oral, rat: LD50 = 550 ug/kg Skin, mouse: LD50 = 1080 ug/kg
OSHA - List of Highly Hazardous Chemicals
threshhold quantity = 100 pounds
ENVIRONMENTAL LISTS
CERCLA/SARA - Section 302 Extremely Hazardous Substances and TPQs
TPQ = 10 pounds
STATE LISTS
Florida Hazardous Substance List
effective March 13, 1992
Massachusetts Right To Know List
extraordinarily hazardous
Pennsylvania Right to Know List
environmental hazard

PROPOSED REGULATIONS
CERCLA/SARA - Proposed Hazardous Substance Additions
proposed RQ = 1 pound (.454 kg)
CERCLA/SARA - 1989 Proposed RQ Adjustments
proposed RQ = 10 pounds (4.54 kg)

SCANDIUM 49   14391-97-0
ENVIRONMENTAL LISTS
List of Pesticide Product Inert Ingredients
[present]

SCOPOLAMINE HYDROBROMIDE   114-49-8
HEALTH AND SAFETY LISTS
U.S. DOT - Appendix A Table 2 - Radionuclides
final RQ = 1000 curies (3.7E 13 Bq)
ENVIRONMENTAL LISTS
CERCLA/SARA List of Radionuclides (Appendix B) and Their Reportable Quantities
final RQ = 1000 curies (3.7E 13 Bq)

SCOPOLAMINE HYDROBROMIDE TRIHYDRATE   6533-68-2
HEALTH AND SAFETY LISTS
U.S. DOT - Appendix A Table 2 - Radionuclides
final RQ = 100 curies (3.7E 12 Bq)
ENVIRONMENTAL LISTS
CERCLA/SARA List of Radionuclides (Appendix B) and Their Reportable Quantities
final RQ = 100 curies (3.7E 12 Bq)

SEAWEED   RR-01138-1
HEALTH AND SAFETY LISTS
U.S. DOT - Appendix A Table 2 - Radionuclides
final RQ = 10 curies (3.7E 11 Bq)
ENVIRONMENTAL LISTS
CERCLA/SARA List of Radionuclides (Appendix B) and Their Reportable Quantities
final RQ = 10 curies (3.7E 11 Bq)

2-SEC-BUTYL-4,6-DINITROPHENOL (DINOSEB)   88-85-7
HEALTH AND SAFETY LISTS
U.S. DOT - Appendix A Table 2 - Radionuclides
final RQ = 10 curies (3.7E 11 Bq)
ENVIRONMENTAL LISTS
CERCLA/SARA List of Radionuclides (Appendix B) and Their Reportable Quantities
final RQ = 10 curies (3.7E 11 Bq)

SECOBARBITAL SODIUM   309-43-3
HEALTH AND SAFETY LISTS
U.S. DOT - Appendix A Table 2 - Radionuclides
final RQ = 100 curies (3.7E 12 Bq)
ENVIRONMENTAL LISTS
CERCLA/SARA List of Radionuclides (Appendix B) and Their Reportable Quantities
final RQ = 100 curies (3.7E 12 Bq)

SELENATES, N.O.S.   RR-00485-3
HEALTH AND SAFETY LISTS
U.S. DOT - Appendix A Table 2 - Radionuclides
final RQ = 10 curies (3.7E 11 Bq)

ENVIRONMENTAL LISTS
CERCLA/SARA List of Radionuclides (Appendix B) and Their Reportable Quantities
final RQ - 10 curies (3.7E 11 Bq)

SELENIC ACID   7783-08-6

HEALTH AND SAFETY LISTS
U.S. DOT - Appendix A Table 2 - Radionuclides
final RQ - 1000 curies (3.7E 13 Bq)

ENVIRONMENTAL LISTS
CERCLA/SARA List of Radionuclides (Appendix B) and Their Reportable Quantities
final RQ - 1000 curies (3.7E 13 Bq)

SELENIUM   7782-49-2

ENVIRONMENTAL LISTS
TSCA - Code of Federal Regulations Citations
40 CFR 712.30(d)
TSCA - PAIR - Reporting List
Reporting Date: November 19, 1982

SELENIUM 70   19869-93-3

HEALTH AND SAFETY LISTS
NTP Chemical Status Reports - Testing Status and NTIS Number
Two year studies: pathology working group in progress; prechronic studies for which toxicity technical reports were not prepared

SELENIUM 73   15422-57-8

ENVIRONMENTAL LISTS
List of Pesticide Product Inert Ingredients
[present]

SELENIUM 73M   RR-00358-7
SEE ALSO:
F039-HAZARDOUS WASTES

HEALTH AND SAFETY LISTS
U.S. DOT - Appendix B - Marine Pollutants
DOT regulated marine pollutant
U.S. DOT - Appendix A Table 1 - Hazardous Substances
final RQ - 1000 pounds (454 kg)
NIOSH - Selected LD50s and LC50s
Oral, rat: LD50 - 25 mg/kg Skin, rat: LD50 - 80 mg/kg

ENVIRONMENTAL LISTS
CERCLA/SARA - Section 302 Extremely Hazardous Substances and TPQs
TPQ - 100/10,000 pounds
CERCLA/SARA - Section 313 - Emission Reporting
form R reporting required
CERCLA/SARA - Hazardous Substances and their Reportable Quantities
final RQ - 1000 pounds (454 kg)
Safe Drinking Water Act - MCLs
MCL - 0.007 mg/L
Safe Drinking Water Act - MCLGs
MCLG - 0.007 mg/L
RCRA - P Series Wastes
waste number P020
RCRA - Hazardous Constituents-Appendix VIII
waste number P020
RCRA - Basis for Listing - Appendix VII
Included in waste stream: F039
RCRA - Substances Banned From Land Disposal
[present]

RCRA - TSD Facilities Ground Water Monitoring
TM 8150 - 1 ug/L PQL; TM 8270 - 10 ug/L PQL
RCRA - Universal Treatment Standards (LDR)
WW: 0.066 mg/l; NWW: 2.5 mg/kg

INTERNATIONAL LISTS
Canada - Drinking Water Quality - MACs
0.01 mg/L MAC

STATE LISTS
California - Air Bill 2588 Appendix A-II
9/89
California - Prop. 65 - Developmental Toxicity
developmental toxicity - initial date 1/1/89
California - Prop. 65 - Reproductive - Male
male reproductive toxicity - initial date 1/1/89
California - Directors List of Hazardous Substances (8 CCR 339)
[present]
Florida Hazardous Substance List
effective March 13, 1992
Massachusetts Right To Know List
extraordinarily hazardous
NJ Right to Know List (Total)
sn 2354
Pennsylvania Right to Know List
environmental hazard

SELENIUM 75   14265-71-5

STATE LISTS
California - Prop. 65 - Developmental Toxicity
developmental toxicity - initial date 10/1/92

SELENIUM 79   15758-45-9

HEALTH AND SAFETY LISTS
U.S. DOT - Substances From 49 CFR 172.101
regulated by DOT (UN2630)
U.S. DOT - Hazard Classes
DOT hazard class - 6.1

STATE LISTS
NJ Right to Know List (Total)
sn 2761

SELENIUM 81   15422-58-9
SEE ALSO:
SELENIUM

HEALTH AND SAFETY LISTS
U.S. DOT - Substances From 49 CFR 172.101
regulated by DOT (UN1905)
U.S. DOT - Hazard Classes
DOT hazard class - 8

STATE LISTS
NJ Right to Know List (Total)
sn 1647
NJ Special Hazardous Substances
(corrosive)

SELENIUM 81M   RR-00357-6
SEE ALSO:
SELENIUM COMPOUNDS
F039-HAZARDOUS WASTES
K144-HAZARDOUS WASTES

HEALTH AND SAFETY LISTS
ACGIH 1995 - Time Weighted Averages
as Se: 0.2 mg/m3 TWA
U.S. DOT - Substances From 49 CFR 172.101
regulated by DOT (UN2658)

U.S. DOT - Hazard Classes
DOT hazard class = 6.1

U.S. DOT - Appendix A Table 1 - Hazardous Substances
final RQ = 100 pounds (45.4 kg) (no reporting or releases of this hazardous substance is required if the diameter of the solid metal released is equal to or exceeds 0.04 inches)

IARC - Group 3 (not classifiable)
[present]

NIOSH - Selected LD50s and LC50s
Oral, rat: LD50 = 6700 mg/kg

NIOSH 1990 - Pocket Guide - RELs
as Se: 0.2 mg/m3 TWA

NIOSH 1990 - Pocket Guide - Target organs
upper respiratory system, eyes, skin, liver, kidneys, blood

OSHA - Vacated PELs - Time Weighted Averages
as Se: 0.2 mg/m3 TWA

OSHA - Final PELs - Time Weighted Averages
as Se: 0.2 mg/m3 TWA

ENVIRONMENTAL LISTS

ATSDR Priority List
Rank (of 275): 113

CERCLA/SARA - Section 313 - Emission Reporting
form R reporting required for 1.0% de minimus concentration

CERCLA/SARA - Hazardous Substances and their Reportable Quantities
final RQ = 100 pounds (45.4 kg) (no reporting or releases of this hazardous substance is required if the diameter of the solid metal released is equal to or exceeds 0.04 inches)

Clean Water Act - Priority Pollutants
[present]

Clean Water Act - Toxic Pollutants
[present] (Listed under 'Selenium and compounds')

Safe Drinking Water Act - MCLs
MCL = 0.05 mg/L

Safe Drinking Water Act - MCLGs
MCLG = 0.05 mg/L

RCRA - D Series - Maximum Concentration of Contaminants
waste number D010; regulatory level = 1.0 mg/L

RCRA - D Series - Chronic Toxicity Reference Levels
chronic toxicity reference level = 0.01 mg/L

RCRA - Hazardous Constituents-Appendix VIII
hazardous constituent - no waste number

RCRA - Basis for Listing - Appendix VII
Included in waste stream: F039

RCRA - Substances Banned From Land Disposal
[present]

RCRA - TSD Facilities Ground Water Monitoring
TM 6010 = 750 ug/L PQL; TM 7740 = 20 ug/L PQL; TM 7741 = 20 ug/L PQL (all species in the ground water that contain this element are included)

RCRA - Universal Treatment Standards (LDR)
WW: 0.82 mg/l; NWW: 0.16 mg/l TCLP

TSCA - Code of Federal Regulations Citations
40 CFR 712.30(w)

TSCA - PAIR - Reporting List
Reporting Date: July 13, 1988

INTERNATIONAL LISTS

Australian Exposure Standards - Time Weighted Averages
as Se: 0.2 mg/m3 TWA (does not include hydrogen selenide)

Canada - WHMIS: Ingredient Disclosure
0.1% item 1398 (1479)

Canada - NPRI (National Pollutant Release Inventory)
[present]

Canada - Drinking Water Quality - MACs
0.01 mg/L MAC

Canada - Alberta - 8 Hour Occupational Exposure Limit
as Se: 0.2 mg/m3 TWA

Canada - Alberta - 15 Minute Occupational Exposure Limit
as Se: 0.6 mg/m3 STEL

Canada - British Columbia - 8 Hour Exposure Limits
as Se: 0.2 mg/m3 TWA

Canada - Ontario - OHSA - TWAEVs
as Se: 0.2 mg/m3 TWAEV

German (DFG) - MAK Values
total dust, as Se: 0.1 mg/m3 MAK

German (DFG) - Peak Limitations
10 x normal MAK (30 min average value); don't exceed during shift

Israel - Time Weighted Averages
as Se: 0.2 mg/m3 TWA

Israel - Action Levels
as Se: 0.1 mg/m3 AL

Mexico - Instruction No. 10 - TWAs
0.2 mg/m3 TWA

Mexico - Wastewater - Organic Toxic Pollutants and Heavy Metals
Listed under [Heavy Metals]

Mexico - Drinking Water - Ecological Criteria
0.01 mg/l

STATE LISTS

California - Air Bill 2588 Appendix A-I
[present]

California - Directors List of Hazardous Substances (8 CCR 339)
[present]

Florida Hazardous Substance List
[present]

Massachusetts Right To Know List
[present]

Minnesota Hazardous Substance List
[present] as Se

NJ Right to Know List (Total)
sn 1648

Pennsylvania Right to Know List
environmental hazard (for any compound of this substance) (any compound of this substance is also an environmental hazard)

PROPOSED REGULATIONS

Canada - Ontario - Proposed Occupational TWAEVs
as Se: 0.1 mg/m3 TWAEV

Canada - Ontario - Proposed Occupational STEVs
as Se: 1.0 mg/m3 STEV

SELENIUM 83   14687-60-6

HEALTH AND SAFETY LISTS

U.S. DOT - Appendix A Table 2 - Radionuclides
final RQ = 1000 curies (3.7E 13 Bq)

ENVIRONMENTAL LISTS

CERCLA/SARA List of Radionuclides (Appendix B) and Their Reportable Quantities
final RQ = 1000 curies (3.7E 13 Bq)

SELENIUM COMPOUNDS   RR-00612-2

HEALTH AND SAFETY LISTS

U.S. DOT - Appendix A Table 2 - Radionuclides
final RQ = 10 curies (3.7E 11 Bq)

ENVIRONMENTAL LISTS

CERCLA/SARA List of Radionuclides (Appendix B) and Their Reportable Quantities
final RQ = 10 curies (3.7E 11 Bq)

SELENIUM DIOXIDE 7446-08-4

HEALTH AND SAFETY LISTS
U.S. DOT - Appendix A Table 2 - Radionuclides
final RQ = 100 curies (3.7E 12 Bq)

ENVIRONMENTAL LISTS
CERCLA/SARA List of Radionuclides (Appendix B) and Their Reportable Quantities
final RQ = 100 curies (3.7E 12 Bq)

SELENIUM HEXAFLUORIDE 7783-79-1

HEALTH AND SAFETY LISTS
U.S. DOT - Appendix A Table 2 - Radionuclides
final RQ = 10 curies (3.7E 11 Bq)

ENVIRONMENTAL LISTS
CERCLA/SARA List of Radionuclides (Appendix B) and Their Reportable Quantities
final RQ = 10 curies (3.7E 11 Bq)

SELENIUM NITRIDE 12033-59-9

HEALTH AND SAFETY LISTS
U.S. DOT - Appendix A Table 2 - Radionuclides
final RQ = 10 curies (3.7E 11 Bq)

ENVIRONMENTAL LISTS
CERCLA/SARA List of Radionuclides (Appendix B) and Their Reportable Quantities
final RQ = 10 curies (3.7E 11 Bq)

SELENIUM OXIDE (VAN) 12640-89-0

HEALTH AND SAFETY LISTS
U.S. DOT - Appendix A Table 2 - Radionuclides
final RQ = 1000 curies (3.7E 13 Bq)

ENVIRONMENTAL LISTS
CERCLA/SARA List of Radionuclides (Appendix B) and Their Reportable Quantities
final RQ = 1000 curies (3.7E 13 Bq)

SELENIUM OXYCHLORIDE 7791-23-3

HEALTH AND SAFETY LISTS
U.S. DOT - Appendix A Table 2 - Radionuclides
final RQ = 1000 curies (3.7E 13 Bq)

ENVIRONMENTAL LISTS
CERCLA/SARA List of Radionuclides (Appendix B) and Their Reportable Quantities
final RQ = 1000 curies (3.7E 13 Bq)

SELENIUM SULFIDE 7446-34-6

HEALTH AND SAFETY LISTS
U.S. DOT - Appendix A Table 2 - Radionuclides
final RQ = 1000 curies (3.7E 13 Bq)

ENVIRONMENTAL LISTS
CERCLA/SARA List of Radionuclides (Appendix B) and Their Reportable Quantities
final RQ = 1000 curies (3.7E 13 Bq)

SELENIUM SULFIDE 7488-56-4
SEE ALSO:
SELENIUM COMPOUNDS

HEALTH AND SAFETY LISTS
IARC - Group 3 (not classifiable)
[present]

ENVIRONMENTAL LISTS
CERCLA/SARA - Section 313 - Emission Reporting
form R reporting required for 1.0% de minimus concentration
Clean Air Act (1990) - List of Hazardous Air Contaminants
[present] (includes any unique chemical substance that contains Selenium as part of that chemical's infrastructure)
Clean Water Act - Toxic Pollutants
[present] (Listed under 'Selenium and compounds')
RCRA - Hazardous Constituents-Appendix VIII
hazardous constituent - no waste number

INTERNATIONAL LISTS
Canada - WHMIS: Ingredient Disclosure
1% item 1409 (1494)
Canada - NPRI (National Pollutant Release Inventory)
[present]
Canada - Ontario - OHSA - TWAEVs
(except selenium hexafluoride and hydrogen selenide), as Se: 0.2 mg/m3 TWAEV
Canada - Quebec - Time-Weighted Average Exposure Values
as Se: 0.2 mg/m3 TWAEV
United Kingdom - Occupational Exposure Standards - TWAs
as Se: 0.1 mg/m3 TWA (does not include hydrogen selenide)

STATE LISTS
California - Air Bill 2588 Appendix A-I
[present]
California - Exposure Limits - PELs
as Se: 0.2 mg/m3 PEL
California - Directors List of Hazardous Substances (8 CCR 339)
[present]
NJ Right to Know List (Total)
sn 3148

PROPOSED REGULATIONS
Canada - Ontario - Proposed Occupational TWAEVs
as Se: 0.1 mg/m3 TWAEV (excluding selenium hexafluoride and hydrogen selenide)
Canada - Ontario - Proposed Occupational STEVs
as Se: 1.0 mg/m3 STEV

SELENIUM SULFIDE 56093-45-9
SEE ALSO:
SELENIUM

HEALTH AND SAFETY LISTS
U.S. DOT - Appendix A Table 1 - Hazardous Substances
final RQ = 10 pounds (4.54 kg)

ENVIRONMENTAL LISTS
CERCLA/SARA - Hazardous Substances and their Reportable Quantities
final RQ = 10 pounds (4.54 kg)
Clean Water Act - Hazardous Substances
[present]

INTERNATIONAL LISTS
Canada - WHMIS: Ingredient Disclosure
1% item 1396 (772)

STATE LISTS
Massachusetts Right To Know List
[present]
NJ Right to Know List (Total)
sn 1651
Pennsylvania Right to Know List
environmental hazard

1,1'-SELENOBISBENZENE   1132-39-4
SEE ALSO:
SELENIUM

HEALTH AND SAFETY LISTS
ACGIH 1995 - Time Weighted Averages
as Se: 0.05 ppm TWA; 0.16 mg/m3 TWA

U.S. DOT - Substances From 49 CFR 172.101
regulated by DOT (UN2194)

U.S. DOT - Hazard Classes
DOT hazard class = 2.3

U.S. DOT - Substances Which Are Poisonous by Inhalation
gaseous hazardous material poisonous by inhalation (UN2194)

NIOSH 1990 - Pocket Guide - RELs
as Se: 0.05 ppm TWA; 0.4 mg/m3 TWA

NIOSH 1990 - Pocket Guide - IDLHs
as Se: 5 ppm IDLH

NIOSH 1990 - Pocket Guide - Target organs
none known in humans

OSHA - Vacated PELs - Time Weighted Averages
as Se: 0.05 ppm TWA; 0.4 mg/m3 TWA

OSHA - Final PELs - Time Weighted Averages
as Se: 0.05 ppm TWA; 0.4 mg/m3 TWA

OSHA - List of Highly Hazardous Chemicals
threshhold quantity = 1000 pounds

INTERNATIONAL LISTS
Australian Exposure Standards - Time Weighted Averages
as Se: 0.05 ppm TWA; 0.16 mg/m3 TWA

Canada - Alberta - 8 Hour Occupational Exposure Limit
as Se: 0.05 ppm TWA; 0.16 mg/m3 TWA

Canada - Alberta - 15 Minute Occupational Exposure Limit
as Se: 0.15 ppm STEL; 0.48 mg/m3 STEL

Canada - British Columbia - 8 Hour Exposure Limits
as Se: 0.05 ppm TWA; 0.4 mg/m3 TWA

Canada - British Columbia - 15 Minute Exposure Limits
as Se: 0.05 ppm STEL; 0.4 mg/m3 STEL

Canada - Ontario - OHSA - TWAEVs
as Se: 0.025 ppm TWAEV; 0.01 mg/m3 TWAEV

Canada - Quebec - Time-Weighted Average Exposure Values
as Se: 0.05 ppm TWAEV; 0.16 mg/m3 TWAEV

Israel - Time Weighted Averages
as Se: 0.05 ppm TWA; 0.16 mg/m3 TWA

Israel - Action Levels
as Se: 0.025 ppm AL; 0.08 mg/m3 AL

STATE LISTS
California - Exposure Limits - PELs
0.05 ppm PEL; 0.4 mg/m3 PEL

Florida Hazardous Substance List
[present]

Massachusetts Right To Know List
[present]

Minnesota Hazardous Substance List
[present] as Se

NJ Right to Know List (Total)
sn 1650

Pennsylvania Right to Know List
[present]

SELENOMETHIONINE   1464-42-2
HEALTH AND SAFETY LISTS
U.S. DOT - Hazard Classes
Forbidden from transport by the DOT

SELENOUS ACID   7783-00-8
SEE ALSO:
SELENIUM

HEALTH AND SAFETY LISTS
U.S. DOT - Substances From 49 CFR 172.101
regulated by DOT (NA2811)

U.S. DOT - Hazard Classes
DOT hazard class = 6.1

SELSUN   RR-00504-9
SEE ALSO:
SELENIUM

HEALTH AND SAFETY LISTS
U.S. DOT - Substances From 49 CFR 172.101
regulated by DOT (UN2879)

U.S. DOT - Hazard Classes
DOT hazard class = 8

ENVIRONMENTAL LISTS
CERCLA/SARA - Section 302 Extremely Hazardous Substances and TPQs
TPQ = 500 pounds

STATE LISTS
Florida Hazardous Substance List
effective March 13, 1992

Massachusetts Right To Know List
extraordinarily hazardous

NJ Right to Know List (Total)
sn 1652

NJ Special Hazardous Substances
(corrosive)

Pennsylvania Right to Know List
environmental hazard

PROPOSED REGULATIONS
CERCLA/SARA - Proposed Hazardous Substance Additions
proposed RQ = 1 pound (.454 kg)

CERCLA/SARA - 1989 Proposed RQ Adjustments
proposed RQ = 100 pounds (45.4 kg)

SEMICARBAZIDE HYDROCHLORIDE   563-41-7
HEALTH AND SAFETY LISTS
NTP Chemical Status Reports - Testing Status and NTIS Number
Technical reports printed (PB82164955) (PB82165291)

NTP Chemical Status Reports - Evidence of Carcinogenicity
PB82165291:   male   mice-negative;   female   mice-negative; PB82164955: male rat-positive; female rat-positive; male mice-negative; female mice-positive

STATE LISTS
California - Air Bill 2588 Appendix A-I
known or potential carcinogen: 9/90

California - Prop. 65 - Cancer list
carcinogen - initial date 10/1/89

California - Directors List of Hazardous Substances (8 CCR 339)
[present]

Florida Hazardous Substance List
[present]

Massachusetts Right To Know List
carcinogen; extraordinarily hazardous

Minnesota Hazardous Substance List
carcinogen

NJ Right to Know List (Total)
sn 1653

NJ Special Hazardous Substances
(carcinogen)

Pennsylvania Right to Know List
environmental hazard; special hazardous substance
Pennsylvania RTK - Special Hazardous Substances
[present]

SENECIPHYLLINE   480-81-9
SEE ALSO:
SELENIUM

HEALTH AND SAFETY LISTS
U.S. DOT - Substances From 49 CFR 172.101
regulated by DOT (UN2657)
U.S. DOT - Hazard Classes
DOT hazard class = 6.1
U.S. DOT - Appendix A Table 1 - Hazardous Substances
final RQ = 10 pounds (4.54 kg)
NIOSH - Selected LD50s and LC50s
Oral, rat: LD50 = 138 mg/kg
NTP Seventh Report - Suspect Carcinogens
suspect carcinogen
OSHA - Possible Select Carcinogens
[present]

ENVIRONMENTAL LISTS
CERCLA/SARA - Hazardous Substances and their Reportable Quantities
final RQ = 10 pounds (4.54 kg)
EPA - Carcinogen Hazard Ranking for RQ Adjustment
Hazard ranking = Low
RCRA - U Series Wastes
waste number U205 (Reactive waste; Toxic waste)
RCRA - Hazardous Constituents-Appendix VIII
waste number U205
RCRA - Substances Banned From Land Disposal
[present]

INTERNATIONAL LISTS
Canada - WHMIS: Ingredient Disclosure
0.1% item 1397 (792)

STATE LISTS
Massachusetts Right To Know List
[present]
NJ Right to Know List (Total)
sn 1649
Pennsylvania Right to Know List
environmental hazard

SENKIRKINE   2318-18-5

STATE LISTS
Pennsylvania Right to Know List
environmental hazard; special hazardous substance
Pennsylvania RTK - Special Hazardous Substances
[present]

SEPIOLITE   15501-74-3
SEE ALSO:
SELENIUM

INTERNATIONAL LISTS
Canada - WHMIS: Ingredient Disclosure
1% item 1399 (1482)

SEPIOLITE   18307-23-8
SEE ALSO:
SELENIUM

INTERNATIONAL LISTS
Canada - WHMIS: Ingredient Disclosure
1% item 1400 (1483)

SESAME OIL   8008-74-0
SEE ALSO:
SELENIUM

HEALTH AND SAFETY LISTS
U.S. DOT - Appendix A Table 1 - Hazardous Substances
final RQ = 10 pounds (4.54 kg)

ENVIRONMENTAL LISTS
CERCLA/SARA - Section 302 Extremely Hazardous Substances and TPQs
TPQ = 1000/10,000 pounds
CERCLA/SARA - Hazardous Substances and their Reportable Quantities
final RQ = 10 pounds (4.54 kg)
RCRA - U Series Wastes
waste number U204
RCRA - Hazardous Constituents-Appendix VIII
waste number U204
RCRA - Substances Banned From Land Disposal
[present]

INTERNATIONAL LISTS
Canada - WHMIS: Ingredient Disclosure
1% item 1394 (131)

STATE LISTS
Florida Hazardous Substance List
effective March 13, 1992
Massachusetts Right To Know List
extraordinarily hazardous
NJ Right to Know List (Total)
sn 2762
Pennsylvania Right to Know List
environmental hazard

SESONE   136-78-7

HEALTH AND SAFETY LISTS
NTP Chemical Status Reports - Testing Status and NTIS Number
Technical reports printed (PB82164542)
NTP Chemical Status Reports - Evidence of Carcinogenicity
male mice-negative; female mice-negative

SETHOXYDIM [2-[1-(ETHOXYIMINO)BUTYL]-5-   74051-80-2
[2-(ETHYLTHIO)PROPYL]-3-HYDROXY-2-CYCLO-
HEXEN-1-ONE]

HEALTH AND SAFETY LISTS
IARC - Group 3 (not classifiable)
[present]
NIOSH - Selected LD50s and LC50s
Oral, mouse: LD50 = 225 mg/kg

ENVIRONMENTAL LISTS
CERCLA/SARA - Section 302 Extremely Hazardous Substances and TPQs
TPQ = 1000/10,000 pounds
TSCA - Code of Federal Regulations Citations
40 CFR 704.225(a)
TSCA - CAIR - Reporting List
reporting required by: manufacturer; distributor; importer; processor

STATE LISTS
California - Directors List of Hazardous Substances (8 CCR 339)
[present]
Florida Hazardous Substance List
effective March 13, 1992
Massachusetts Right To Know List
extraordinarily hazardous

NJ Right to Know List (Total)
sn 2765

Pennsylvania Right to Know List
environmental hazard

PROPOSED REGULATIONS

CERCLA/SARA - Proposed Hazardous Substance Additions
proposed RQ = 1 pound (.454 kg)

CERCLA/SARA - 1989 Proposed RQ Adjustments
proposed RQ = 100 pounds (45.4 kg)

SHALE   68476-95-9

HEALTH AND SAFETY LISTS

IARC - Group 3 (not classifiable)
[present]

SHALE OILS   68308-34-9

HEALTH AND SAFETY LISTS

IARC - Group 3 (not classifiable)
[present]

STATE LISTS

California - Directors List of Hazardous Substances (8 CCR 339)
[present]

SHELLAC   9000-59-3

HEALTH AND SAFETY LISTS

IARC - Group 3 (not classifiable)
[present]

SHIKIMIC ACID   138-59-0

INTERNATIONAL LISTS

German (DFG) - Carcinogens
as fibrous dust: suspected carcinogen

SILANE   7803-62-5

HEALTH AND SAFETY LISTS

NFPA - Flash Points
flash point = 491 degrees F (255 degrees C)

NFPA - Hazard Identification Ratings
health-0; flammability-1; reactivity-0

SILANE, (4-AMINOBUTYL)DIETHOXYMETHYL-   3037-72-7

HEALTH AND SAFETY LISTS

ACGIH 1995 - Time Weighted Averages
10 mg/m3 TWA

NIOSH - Selected LD50s and LC50s
Oral, rat: LD50 = 730 mg/kg

NIOSH 1990 - Pocket Guide - RELs
total: 10 mg/m3 TWA; respirable dust: 5 mg/m3 TWA

NIOSH 1990 - Pocket Guide - IDLHs
5000 mg/m3 IDLH

NIOSH 1990 - Pocket Guide - Target organs
none known

OSHA - Vacated PELs - Time Weighted Averages
total dust: 10 mg/m3 TWA; respirable fraction: 5 mg/m3 TWA

OSHA - Final PELs - Time Weighted Averages
total dust: 15 mg/m3 TWA; respirable fraction: 5 mg/m3 TWA

INTERNATIONAL LISTS

Australian Exposure Standards - Time Weighted Averages
10 mg/m3 TWA

Canada - Alberta - 8 Hour Occupational Exposure Limit
10 mg/m3 TWA

Canada - Alberta - 15 Minute Occupational Exposure Limit
20 mg/m3 STEL

Canada - British Columbia - 8 Hour Exposure Limits
10 mg/m3 TWA

Canada - British Columbia - 15 Minute Exposure Limits
20 mg/m3 STEL

Canada - Ontario - OHSA - TWAEVs
10 mg/m3 TWAEV

Canada - Quebec - Time-Weighted Average Exposure Values
10 mg/m3 TWAEV

United Kingdom - Occupational Exposure Standards - TWAs
10 mg/m3 TWA

United Kingdom - Occupational Exposure Standards - STELs
20 mg/m3 STEL

Israel - Time Weighted Averages
10 mg/m3 TWA

Israel - Action Levels
5 mg/m3 AL

Mexico - Instruction No. 10 - TWAs
15 mg/m3 TWA

STATE LISTS

California - Exposure Limits - PELs
total dust: 10 mg/m3 PEL; respirable fraction: 5 mg/m3 PEL

California - Directors List of Hazardous Substances (8 CCR 339)
[present]

Florida Hazardous Substance List
[present]

Massachusetts Right To Know List
[present]

Minnesota Hazardous Substance List
[present]

Pennsylvania Right to Know List
[present]

**SILANE, [(3-CHLOROPROPYL)DIMETHOXY[3-(OXI-   71808-64-5
RANYLMETHOXY)PROPYL]-**

ENVIRONMENTAL LISTS

CERCLA/SARA - Section 313 - Emission Reporting
form R reporting required

**SILANE, DIETHOXYMETHYL[3-(OXIRANYL-   2897-60-1
METHOXY)PROPYL]-**

ENVIRONMENTAL LISTS

List of Pesticide Product Inert Ingredients
[present]

**SILANE, (1,1-DIMETHYLETHOXY)DIMETHOXY (2-   RR-00950-7
METHYLPROPYL)-**

HEALTH AND SAFETY LISTS

U.S. DOT - Substances From 49 CFR 172.101
regulated by DOT (UN1288)

U.S. DOT - Hazard Classes
DOT hazard class = 3

IARC - Group 1 (carcinogenic to humans)
[present]

OSHA - Select Carcinogens
[present]

STATE LISTS

California - Air Bill 2588 Appendix A-II
known or potential carcinogen

California - Prop. 65 - Cancer list
carcinogen - initial date 4/1/90

California - Directors List of Hazardous Substances (8 CCR 339)
[present]

Minnesota Hazardous Substance List
carcinogen

NJ Right to Know List (Total)
sn 2766

Pennsylvania Right to Know List
special hazardous substance

Pennsylvania RTK - Special Hazardous Substances
[present]

SILANE, ETHENYLTRIETHOXY-   78-08-0

ENVIRONMENTAL LISTS

List of Pesticide Product Inert Ingredients
[present]

STATE LISTS

NJ Right to Know List (Total)
sn 2767

SILANE, ETHOXYDIMETHYL[3-(OXIRANYL-   17963-04-1
METHOXY)PROPYL]-

HEALTH AND SAFETY LISTS

IARC - Group 3 (not classifiable)
[present]

SILANE, TRIETHOXYPENTYL-   2761-24-2

HEALTH AND SAFETY LISTS

ACGIH 1995 - Time Weighted Averages
5 ppm TWA; 6.6 mg/m3 TWA

U.S. DOT - Substances From 49 CFR 172.101
regulated by DOT (UN2203)

U.S. DOT - Hazard Classes
DOT hazard class = 2.1

NFPA - Flash Points
gas (no flash point given)

NFPA - Hazard Identification Ratings
health-1; flammability-4; reactivity-3

NIOSH - Selected LD50s and LC50s
Inhalation, rat: LC50 = 9600 ppm 4 hr

OSHA - Vacated PELs - Time Weighted Averages
5 ppm TWA; 7 mg/m3 TWA

ENVIRONMENTAL LISTS

CAA - Flammable Substances for Accidental Release Prevention
threshold quantity = 10,000 lbs

INTERNATIONAL LISTS

Australian Exposure Standards - Time Weighted Averages
5 ppm TWA; 6.6 mg/m3 TWA

Canada - WHMIS: Ingredient Disclosure
1% item 1401 (1484)

Canada - Alberta - 8 Hour Occupational Exposure Limit
0.5 ppm TWA; 0.65 mg/m3 TWA

Canada - Alberta - 15 Minute Occupational Exposure Limit
1 ppm STEL; 1.3 mg/m3 STEL

Canada - British Columbia - 8 Hour Exposure Limits
0.5 ppm TWA; 7 mg/m3 TWA

Canada - British Columbia - 15 Minute Exposure Limits
1 ppm STEL; 2 mg/m3 STEL

Canada - Ontario - OHSA - TWAEVs
5 ppm TWAEV; 6.6 mg/m3 TWAEV

Canada - Quebec - Time-Weighted Average Exposure Values
5 ppm TWAEV; 6.6 mg/m3 TWAEV

United Kingdom - Occupational Exposure Standards - TWAs
0.5 ppm TWA; 0.7 mg/m3 TWA

United Kingdom - Occupational Exposure Standards - STELs
1 ppm STEL; 1.5 mg/m3 STEL

Israel - Time Weighted Averages
5 ppm TWA; 6.6 mg/m3 TWA

Israel - Action Levels
2.5 ppm AL; 3.3 mg/m3 AL

Mexico - Instruction No. 10 - TWAs
5 ppm TWA; 7 mg/m3 TWA

STATE LISTS

California - Exposure Limits - PELs
5 ppm PEL; 7 mg/m3 PEL

California - Directors List of Hazardous Substances (8 CCR 339)
[present]

Florida Hazardous Substance List
[present]

Massachusetts Right To Know List
[present]

Minnesota Hazardous Substance List
[present]

Pennsylvania Right to Know List
[present]

PROPOSED REGULATIONS

Canada - Ontario - Proposed Occupational TWAEVs
0.5 ppm TWAEV; 0.7 mg/m3 TWAEV

Canada - Ontario - Proposed Occupational STEVs
1 ppm STEV; 1.5 mg/m3 STEV

SILANE, TRIMETHOXY[2-(7-OXABICYCLO[4.1.0]   3388-04-3
HEPT-3-YL)ETHYL]

HEALTH AND SAFETY LISTS

NIOSH - Selected LD50s and LC50s
Oral, rat: LD50 = 6500 mg/kg Skin, rat: LD50 = 45 mg/kg

ENVIRONMENTAL LISTS

CERCLA/SARA - Section 302 Extremely Hazardous Substances and TPQs
TPQ = 1000 pounds

STATE LISTS

Florida Hazardous Substance List
effective March 13, 1992

Massachusetts Right To Know List
extraordinarily hazardous

Pennsylvania Right to Know List
environmental hazard

PROPOSED REGULATIONS

CERCLA/SARA - Proposed Hazardous Substance Additions
proposed RQ = 1 pound (.454 kg)

CERCLA/SARA - 1989 Proposed RQ Adjustments
proposed RQ = 1000 pounds (454 kg)

SILANETRIOL, METHYL-, TRIACETATE   4253-34-3
SEE ALSO:
GLYCIDOL (OXIRANEMETHANOL) AND ITS DERIVATIVES

ENVIRONMENTAL LISTS

EPA - Master Testing List
[present]

TSCA - Code of Federal Regulations Citations
40 CFR 716.120(c)

PROPOSED REGULATIONS

TSCA - Proposed Testing Rule for Glycidyl Ethers
member of Glycidyl subcategory III-B

SILICA, AMORPHOUS   7631-86-9
SEE ALSO:
GLYCIDOL (OXIRANEMETHANOL) AND ITS DERIVATIVES

ENVIRONMENTAL LISTS

EPA - Master Testing List
[present]

TSCA - Code of Federal Regulations Citations
40 CFR 716.120(c)

PROPOSED REGULATIONS

TSCA - Proposed Testing Rule for Glycidyl Ethers
member of Glycidyl subcategory III-A

SILICA, AMORPHOUS, DIATOMACEOUS EARTH   68855-54-9

ENVIRONMENTAL LISTS

TSCA - Chemicals with Significant New Use Rules
PMN number: P-89-906

SILICA, AMORPHOUS, PRECIPITATED AND GEL   112926-00-8

HEALTH AND SAFETY LISTS

NIOSH - Selected LD50s and LC50s
Oral, rat: LD50 = 23 gm/kg Skin, rabbit: LD50 = 10 gm/kg

SILICA, CRISTOBALITE   14464-46-1
SEE ALSO:
GLYCIDOL (OXIRANEMETHANOL) AND ITS DERIVATIVES

ENVIRONMENTAL LISTS

EPA - Master Testing List
[present]

TSCA - Code of Federal Regulations Citations
40 CFR 716.120(c)

PROPOSED REGULATIONS

TSCA - Proposed Testing Rule for Glycidyl Ethers
member of Glycidyl subcategory III-A

SILICA, CRYSTALLINE (GENERAL FORM)   RR-00087-3

ENVIRONMENTAL LISTS

List of Pesticide Product Inert Ingredients
[present]

SILICA FUME (AMORPHOUS)   69012-64-2

HEALTH AND SAFETY LISTS

NIOSH - Selected LD50s and LC50s
Oral, rat: LD50 = 12300 mg/kg Skin, rabbit: LD50 = 6300 mg/kg

SILICA, FUSED   60676-86-0

HEALTH AND SAFETY LISTS

NIOSH - Selected LD50s and LC50s
Oral, rat: LD50 = 2060 mg/kg

SILICA GEL   63231-67-4

HEALTH AND SAFETY LISTS

IARC - Group 3 (not classifiable)
[present]

NIOSH - Selected LD50s and LC50s
Oral, rat: LD50 = 3160 mg/kg

NIOSH 1990 - Pocket Guide - RELs
See Appendix D

NIOSH 1990 - Pocket Guide - Target organs
respiratory system

ENVIRONMENTAL LISTS

List of Pesticide Product Inert Ingredients
[present]

INTERNATIONAL LISTS

Australian Exposure Standards - Time Weighted Averages
2 mg/m3 TWA (respirable dust)

Canada - WHMIS: Ingredient Disclosure
1% item 1403 (1488)

Canada - Alberta - 8 Hour Occupational Exposure Limit
*respirable mass: 0.05 mg/m3 TWA; total mass: 0.15 mg/m3 TWA
(See additional requirements in Part 5)*

Canada - British Columbia - 8 Hour Exposure Limits
respirable mass: 1.2 mg/m3 TWA

Canada - Ontario - OHSA - TWAEVs
0.10 mg/m3 TWAEV (designated substance regulation)

Canada - Ontario - OHSA - CEVs
0.20 mg/m3 CEV (designated substance regulation)

Canada - Ontario - OHSA - Designated Substances
0.10 mg/m3 TWAEV; See Ontario Reg. 845 for full information

Canada - Quebec - Time-Weighted Average Exposure Values
6 mg/m3 TWAEV

United Kingdom - Occupational Exposure Standards - TWAs
total inhalable dust: 6 mg/m3 TWA; respirable dust: 3 mg/m3 TWA

German (DFG) - MAK Values
total dust: 4 mg/m3 MAK

German (DFG) - Pregnancy
no risk to embryo/fetus if exposure limits adhered to

STATE LISTS

California - Directors List of Hazardous Substances (8 CCR 339)
*[present] (exempt except when inhalable particulates are present or can
be generated) (applices to silica sand and flour, but not to naturally
occuring dirt and sand which have not gone through beneficiation)*

Florida Hazardous Substance List
[present]

Massachusetts Right To Know List
*[present] Exempt when encapsulated or if particulates are not present
and cannot be substantially generated through use of the product.*

Minnesota Hazardous Substance List
carcinogen

Pennsylvania Right to Know List
[present]

SILICA, TRIDYMITE   15468-32-3

INTERNATIONAL LISTS

Canada - WHMIS: Ingredient Disclosure
1% item 1402 (1489)

Canada - Alberta - 8 Hour Occupational Exposure Limit
respirable mass: 2 mg/m3 TWA; total mass: 5 mg/m3 TWA

Canada - Ontario - OHSA - TWAEVs
total dust: 4 mg/m3 TWAEV (listed as nuisance particulate)

German (DFG) - MAK Values
fine dust: 0.3 mg/m3 MAK

German (DFG) - Pregnancy
no risk to embryo/fetus if exposure limits adhered to

STATE LISTS

Minnesota Hazardous Substance List
[present]

Pennsylvania Right to Know List
[present]

SILICA-TRIPOLI   1317-95-9

HEALTH AND SAFETY LISTS

ACGIH 1995 - Time Weighted Averages
10 mg/m3 TWA

OSHA - Vacated PELs - Time Weighted Averages
6 mg/m3 TWA

OSHA - Final PELs - Time Weighted Averages
see Table Z-3

INTERNATIONAL LISTS

Australian Exposure Standards - Time Weighted Averages
10 mg/m3 TWA

Canada - Ontario - OHSA - TW-AEVs
total dust: 4 mg/m3 TWAEV (listed as nuisance particulates)

Israel - Time Weighted Averages
10 mg/m3 TWA (The value is for total dust containing no asbestos and < 1% crystalline silica)

Israel - Action Levels
2.5 mg/m3 AL

STATE LISTS

California - Exposure Limits - PELs
6 mg/m3 PEL

Minnesota Hazardous Substance List
[present] (includes inert or nuisance dust)

Pennsylvania Right to Know List
[present]

SILICIC ACID   1343-98-2
SEE ALSO:
SILICA, CRYSTALLINE (GENERAL FORM)

HEALTH AND SAFETY LISTS

ACGIH 1995 - Time Weighted Averages
0.05 mg/m3 TWA (this TLV is for the respirable fraction of dust)

NTP Seventh Report - Suspect Carcinogens
respirable dust: suspect carcinogen (Listed under 'Silica, crystalline')

OSHA - Vacated PELs - Time Weighted Averages
respirable dust: 0.05 mg/m3 TWA

OSHA - Final PELs - Time Weighted Averages
see Table Z-3

OSHA - Possible Select Carcinogens
[present]

INTERNATIONAL LISTS

Canada - WHMIS: Ingredient Disclosure
1% item 1405 (1490)

Canada - Alberta - 8 Hour Occupational Exposure Limit
respirable mass: 0.05 mg/m3 TWA; total mass: 0.15 mg/m3 TWA (See additional requirements in Part 5)

Canada - Quebec - Time-Weighted Average Exposure Values
respirable dust: 0.05 mg/m3

German (DFG) - MAK Values
fine dust: 0.15 mg/m3 MAK

German (DFG) - Pregnancy
no risk to embryo/fetus if exposure limits adhered to

Israel - Time Weighted Averages
total dust: 0.3 mg/m3 TWA; respirable dust: 0.1 mg/m3 TWA (Listed under 'Silica - Crystalline')

Israel - Action Levels
respirable dust: 0.025 mg/m3 AL

STATE LISTS

California - Exposure Limits - PELs
respirable dust: 0.05 mg/m3 PEL

Florida Hazardous Substance List
[present]

Massachusetts Right To Know List
carcinogen; extraordinarily hazardous

Minnesota Hazardous Substance List
[present]

Pennsylvania Right to Know List
[present]

SILICIC ACID (H4SIO4), BERYLLIUM SALT (1:2)   15191-85-2

HEALTH AND SAFETY LISTS

IARC - Group 2A (limited human data)
[present] (Listed under 'Silica')

NIOSH - Health Standards - Exposure Limits
respirable free silica: 50 ug/m3 TWA

NIOSH - Health Standards - Health Effects and Precautions
Chronic lung disease (Periodic chest X-ray and pulmonary function testing required)

OSHA - Possible Select Carcinogens
[present]

STATE LISTS

California - Air Bill 2588 Appendix A-I
known or potential carcinogen

California - Prop. 65 - Cancer list
carcinogen (airborne particles of respirable size) - initial date 10/1/88

SILICIC ACID, BERYLLIUM SALT   58500-38-2

HEALTH AND SAFETY LISTS

ACGIH 1995 - Time Weighted Averages
2 mg/m3 TWA (This value is for the respirable fraction of the silica dust)

SILICIC ACID (H6SI2O7), COBALT(2+) MAGNESIUM   75364-04-4
SALT (1:2:1)
SEE ALSO:
SILICA, CRYSTALLINE (GENERAL FORM)

HEALTH AND SAFETY LISTS

ACGIH 1995 - Time Weighted Averages
0.1 mg/m3 TWA (this TLV is for the respirable fraction of dust)

OSHA - Vacated PELs - Time Weighted Averages
respirable dust: 0.1 mg/m3 TWA

OSHA - Final PELs - Time Weighted Averages
see Table Z-3

ENVIRONMENTAL LISTS

List of Pesticide Product Inert Ingredients
[present]

INTERNATIONAL LISTS

Canada - WHMIS: Ingredient Disclosure
1% item 1404 (1487)

Canada - Alberta - 8 Hour Occupational Exposure Limit
respirable mass: 0.1 mg/m3 TWA; total mass: 0.3 mg/m3 TWA

Canada - Quebec - Time-Weighted Average Exposure Values
respirable dust: 0.1 mg/m3 TWAEV

United Kingdom - Occupational Exposure Standards - TWAs
respirable dust: 0.1 mg/m3 TWA

German (DFG) - MAK Values
fine dust: 0.3 mg/m3 MAK

German (DFG) - Pregnancy
no risk to embryo/fetus if exposure limits adhered to

Israel - Time Weighted Averages
total dust: 0.3 mg/m3 TWA; respirable dust: 0.1 mg/m3 TWA (Listed under 'Silica, Crystalline')

Israel - Action Levels
total dust: 0.075 mg/m3 AL; respirable dust: 0.025 mg/m3 AL

STATE LISTS

California - Exposure Limits - PELs
respirable dust: 0.1 mg/m3 PEL

Florida Hazardous Substance List
[present]

Massachusetts Right To Know List
[present]

Minnesota Hazardous Substance List
[present]

SILICIC ACID, ETHYL ESTER   11099-06-2

ENVIRONMENTAL LISTS

List of Pesticide Product Inert Ingredients
[present]

INTERNATIONAL LISTS

Australian Exposure Standards - Time Weighted Averages
10 mg/m3 TWA

Canada - Ontario - OHSA - TWAEVs
total dust: 4 mg/m3 TWAEV (listed as a nuisance particulate)

Canada - Quebec - Time-Weighted Average Exposure Values
total dust: 6 mg/m3 TWAEV

STATE LISTS

Minnesota Hazardous Substance List
[present] (includes inert or nuisance dust)

Pennsylvania Right to Know List
[present]

SILICIC ACID (H6SI2O7), HEXAETHYL ESTER   2157-42-8

SEE ALSO:
SILICA, CRYSTALLINE (GENERAL FORM)

HEALTH AND SAFETY LISTS

ACGIH 1995 - Time Weighted Averages
0.05 mg/m3 TWA (this TLV is for the respirable fraction of dust)

NTP Seventh Report - Suspect Carcinogens
respirable dust: suspect carcinogen (Listed under 'Silica, crystalline')

OSHA - Vacated PELs - Time Weighted Averages
respirable dust: 0.05 mg/m3 TWA

OSHA - Final PELs - Time Weighted Averages
see Table Z-3

OSHA - Possible Select Carcinogens
[present]

INTERNATIONAL LISTS

Canada - WHMIS: Ingredient Disclosure
1% item 1407 (1492)

Canada - Alberta - 8 Hour Occupational Exposure Limit
respirable mass: 0.05 mg/m3 TWA; total mass: 0.15 mg/m3 TWA (See additional requirements in Part 5)

Canada - Quebec - Time-Weighted Average Exposure Values
respirable dust: 0.05 mg/m3 TWAEV

German (DFG) - MAK Values
fine dust: 0.15 mg/m3 MAK

German (DFG) - Pregnancy
no risk to embryo/fetus if exposure limits adhered to

Israel - Time Weighted Averages
total dust: 0.3 mg/m3 TWA; respirable dust: 0.1 mg/m3 TWA (Listed under 'Silica - Crystalline')

Israel - Action Levels
total dust: 0.075 mg/m3 AL; respirable dust: 0.025 mg/m3 AL

STATE LISTS

California - Exposure Limits - PELs
respirable dust: 0.05 mg/m3 PEL

Florida Hazardous Substance List
[present]

Massachusetts Right To Know List
[present]

Minnesota Hazardous Substance List
[present]

Pennsylvania Right to Know List
[present]

SILICIC ACID, METHYL ESTER   12002-26-5

SEE ALSO:
SILICA, CRYSTALLINE (GENERAL FORM)

HEALTH AND SAFETY LISTS

ACGIH 1995 - Time Weighted Averages
0.1 mg/m3 TWA (this TLV is for the respirable fraction of dust)

OSHA - Vacated PELs - Time Weighted Averages
respirable dust (as quartz): 0.1 mg/m3 TWA

OSHA - Final PELs - Time Weighted Averages
see Table Z-3

INTERNATIONAL LISTS

Canada - WHMIS: Ingredient Disclosure
1% item 1408 (1493)

Canada - Alberta - 8 Hour Occupational Exposure Limit
respirable mass: 0.1 mg/m3 TWA; total mass: 0.3 mg/m3 TWA (See additional requirements in Part 5)

Canada - Quebec - Time-Weighted Average Exposure Values
respirable dust: 0.1 mg/m3 TWAEV

Israel - Time Weighted Averages
total dust: 0.3 mg/m3 TWA; respirable dust: 0.1 mg/m3 TWA (Listed under 'Silica-Crystalline')

Israel - Action Levels
total dust: 0.075 mg/m3 AL; respirable dust: 0.025 mg/m3 AL

STATE LISTS

California - Exposure Limits - PELs
respirable dust: 0.1 mg/m3 PEL

Florida Hazardous Substance List
[present]

Massachusetts Right To Know List
[present]

Minnesota Hazardous Substance List
[present]

NJ Right to Know List (Total)
sn 1664

Pennsylvania Right to Know List
[present]

SILICIC ACID (H8SI3O10), OCTAETHYL ESTER   4521-94-2

ENVIRONMENTAL LISTS

List of Pesticide Product Inert Ingredients
[present]

INTERNATIONAL LISTS

Canada - Quebec - Time-Weighted Average Exposure Values
6 mg/m3 TWAEV

SILICIC ACID (H8SI3O10), OCTAMETHYL ESTER   4421-95-8

STATE LISTS

Florida Hazardous Substance List
[present]

Massachusetts Right To Know List
carcinogen; extraordinarily hazardous

Pennsylvania Right to Know List
environmental hazard; special hazardous substance

Pennsylvania RTK - Special Hazardous Substances
[present]

SILICIC ACID (H4SIO4), TETRAKIS(2-ETHYLBUTYL) ESTER   78-13-7

STATE LISTS

Pennsylvania Right to Know List
environmental hazard; special hazardous substance

Pennsylvania RTK - Special Hazardous Substances
[present]

SILICOFLUORIDE COMPOUNDS, N.O.S.   RR-00613-3

STATE LISTS

Pennsylvania Right to Know List
special hazardous substance

Pennsylvania RTK - Special Hazardous Substances
[present]

SILICON   7440-21-3

STATE LISTS

Pennsylvania Right to Know List
[present]

SILICON 31   14276-49-4

STATE LISTS

Pennsylvania Right to Know List
[present]

SILICON 32   15092-72-5

STATE LISTS

Pennsylvania Right to Know List
[present]

SILICON CARBIDE   409-21-2

STATE LISTS

Pennsylvania Right to Know List
[present]

SILICON CARBIDE (SI2C3)   12327-32-1

STATE LISTS

Pennsylvania Right to Know List
[present]

SILICONE ESTER POLYACRYLATE   RR-01662-6

HEALTH AND SAFETY LISTS

NFPA - Flash Points
flash point = 335 degrees F (168 degrees C)
NFPA - Hazard Identification Ratings
health-1; flammability-1; reactivity-0

SILICON TETRACHLORIDE   10026-04-7

INTERNATIONAL LISTS

Canada - WHMIS: Ingredient Disclosure
1% item 1409 (1494)

STATE LISTS

NJ Right to Know List (Total)
sn 2438

SILICON TETRAFLUORIDE   7783-61-1

HEALTH AND SAFETY LISTS

ACGIH 1995 - Time Weighted Averages
10 mg/m3 TWA (The value is for total dust containing no asbestos and <1% crystalline silica)
U.S. DOT - Substances From 49 CFR 172.101
regulated by DOT (UN1346)
U.S. DOT - Hazard Classes
DOT hazard class = 4.1
OSHA - Vacated PELs - Time Weighted Averages
total dust: 10 mg/m3 TWA; respirable fraction: 5 mg/m3 TWA
OSHA - Final PELs - Time Weighted Averages
total dust: 15 mg/m3 TWA; respirable fraction: 5 mg/m3 TWA

INTERNATIONAL LISTS

Australian Exposure Standards - Time Weighted Averages
10 mg/m3 TWA
Canada - Alberta - 8 Hour Occupational Exposure Limit
respirable mass: 5 ppm TWA; total mass: 10 mg/m3 TWA
Canada - British Columbia - 8 Hour Exposure Limits
nuisance dust: 10 mg/m3 TWA
Canada - British Columbia - 15 Minute Exposure Limits
20 mg/m3 STEL

Canada - Ontario - OHSA - TWAEVs
total dust: 10 mg/m3 TWAEV (listed as nuisance particulate)
Canada - Quebec - Time-Weighted Average Exposure Values
total dust: 10 mg/m3 TWAEV; respirable dust: 5 mg/m3 TWAEV
United Kingdom - Occupational Exposure Standards - TWAs
total inhalable dust: 10 mg/m3 TWA; respirable dust: 5 mg/m3 TWA
Israel - Time Weighted Averages
10 mg/m3 TWA (The value is for total dust containing no asbestos and <1% crystalline silica)
Israel - Action Levels
5 mg/m3 AL
Mexico - Instruction No. 10 - TWAs
10 mg/m3 TWA; (nuisance particulate)
Mexico - Instruction No. 10 - STELs
20 mg/m3 STEL

STATE LISTS

Minnesota Hazardous Substance List
[present] (includes inert or nuisance dust)
NJ Right to Know List (Total)
sn 3125
Pennsylvania Right to Know List
[present]

SILKWORM PUPAE   RR-01139-2

HEALTH AND SAFETY LISTS

U.S. DOT - Appendix A Table 2 - Radionuclides
final RQ = 1000 curies (3.7E 13 Bq)

ENVIRONMENTAL LISTS

CERCLA/SARA List of Radionuclides (Appendix B) and Their Reportable Quantities
final RQ = 1000 curies (3.7E 13 Bq)

SILOXANES   RR-01713-0

HEALTH AND SAFETY LISTS

U.S. DOT - Appendix A Table 2 - Radionuclides
final RQ = 1 curie (3.7E 10 Bq)

ENVIRONMENTAL LISTS

CERCLA/SARA List of Radionuclides (Appendix B) and Their Reportable Quantities
final RQ = 1 curie (3.7E 10 Bq)

SILOXANES AND SILICONES, DI-ME, HYDROXY-TERMINATED   70131-67-8

HEALTH AND SAFETY LISTS

ACGIH 1995 - Time Weighted Averages
10 mg/m3 TWA (The value is for total dust containng no asbestos and <1% crystalline silica)
OSHA - Vacated PELs - Time Weighted Averages
total dust: 10 mg/m3 TWA; respirable fraction: 5 mg/m3 TWA
OSHA - Final PELs - Time Weighted Averages
total dust: 15 mg/m3 TWA; respirable fraction: 5 mg/m3 TWA

INTERNATIONAL LISTS

Australian Exposure Standards - Time Weighted Averages
10 mg/m3 TWA
Canada - Alberta - 8 Hour Occupational Exposure Limit
respirable mass: 5 mg/m3 TWA; total mass: 10 mg/m3 TWA
Canada - British Columbia - 8 Hour Exposure Limits
nuisance dust: 10 mg/m3 TWA
Canada - British Columbia - 15 Minute Exposure Limits
20 mg/m3 STEL
Canada - Ontario - OHSA - TWAEVs
total dust: 10 mg/m3 TWAEV (listed as a nuisance particulate)
Canada - Quebec - Time-Weighted Average Exposure Values
total dust: 10 mg/m3 TWAEV; respirable dust: 5 mg/m3 TWAEV

United Kingdom - Occupational Exposure Standards - TWAs
total inhalable dust: 10 mg/m3 TWA; respirable dust: 5 mg/m3 TWA

German (DFG) - MAK Values
fine dust (without fibers): 4 mg/m3 MAK

German (DFG) - Carcinogens
as fibrous dust: animal evidence of carcinogenicity

Israel - Time Weighted Averages
10 mg/m3 TWA (The value is for total dust containng no asbestos and < 1% crystalline silica)

Israel - Action Levels
5 mg/m3 AL

Mexico - Instruction No. 10 - TWAs
10 mg/m3 TWA; (nuisance particulate)

Mexico - Instruction No. 10 - STELs
20 mg/m3 STEL

STATE LISTS

Massachusetts Right To Know List
[present]

Minnesota Hazardous Substance List
[present] (includes inert or nuisance dust)

Pennsylvania Right to Know List
[present]

SILOXANES AND SILICONES, DIMETHYL, 3-HY-DROXYPROPYL METHYL, ETHOXYLATED   68937-54-2

STATE LISTS

Pennsylvania Right to Know List
[present]

SILOXANES AND SILICONES, DIMETHYL, METHYL HYDROGEN, REACTION PRODUCTS WITH POLYETHYLENE-POYPROPYLENE GLYCOL MONOACETATE ALLYL ETHER   68037-64-9

ENVIRONMENTAL LISTS

TSCA - Chemicals with Significant New Use Rules
PMN number: P-85-296

SILOXANES AND SILICONES, DIMETHYL, POLY-MERS WITH METHYL SILSESQUIOXANES AND POLYETHYLENE-POYPROPYLENE GLYCOL MONOBUTYL ETHER   68554-65-4

HEALTH AND SAFETY LISTS

U.S. DOT - Substances From 49 CFR 172.101
regulated by DOT (UN1818)

U.S. DOT - Hazard Classes
DOT hazard class = 8

NIOSH - Selected LD50s and LC50s
Inhalation, rat: LC50 = 8000 ppm 4 hr

INTERNATIONAL LISTS

Canada - WHMIS: Ingredient Disclosure
1% item 1410 (1582)

STATE LISTS

NJ Right to Know List (Total)
sn 1666

SILOXANES AND SILICONES, METHYL 3,3,3-TRIFLUOROPROPYL   63148-56-1

HEALTH AND SAFETY LISTS

U.S. DOT - Substances From 49 CFR 172.101
regulated by DOT (UN1859)

U.S. DOT - Hazard Classes
DOT hazard class = 2.3

U.S. DOT - Substances Which Are Poisonous by Inhalation
gaseous hazardous material poisonous by inhalation (UN1859)

INTERNATIONAL LISTS

Canada - WHMIS: Ingredient Disclosure
1% item 1411 (1590)

STATE LISTS

NJ Right to Know List (Total)
sn 1667

SILVER   7440-22-4

ENVIRONMENTAL LISTS

List of Pesticide Product Inert Ingredients
[present]

SILVER 102   14833-32-0

PROPOSED REGULATIONS

TSCA - ITC 33rd Report Priority Testing List
recommended for testing

TSCA - ITC 34th Report Priority Testing List
recommended for testing

SILVER 103   14967-69-2

ENVIRONMENTAL LISTS

TSCA - PAIR - Reporting List
Effective Date: October 12, 1993; Reporting Date: February 28, 1994

TSCA - Health and Safety Reporting List
Effective Date: October 12, 1993; Sunset Date: October 12, 2003

SILVER 104   15116-79-7

ENVIRONMENTAL LISTS

List of Pesticide Product Inert Ingredients
[present]

SILVER 104M   RR-00356-5

ENVIRONMENTAL LISTS

List of Pesticide Product Inert Ingredients
[present]

SILVER 105   14928-14-4

ENVIRONMENTAL LISTS

List of Pesticide Product Inert Ingredients
[present]

SILVER 106   14333-39-2

ENVIRONMENTAL LISTS

List of Pesticide Product Inert Ingredients
[present]

SILVER 106M   RR-00290-4

SEE ALSO:
DIETHYLENE GLYCOL BIS (PHENYLCARBONATE)
F039-HAZARDOUS WASTES

HEALTH AND SAFETY LISTS

ACGIH 1995 - Time Weighted Averages
metal: 0.1 mg/m3 TWA; soluble compounds, as Ag: 0.01 mg/m3 TWA

U.S. DOT - Appendix A Table 1 - Hazardous Substances
final RQ = 1000 pounds (454 kg) (no reporting of releases of this hazardous substance is required if the diameter of the solid metal released is equal to or exceeds 0.004 inches)

NIOSH - Selected LD50s and LC50s
Oral, mouse: LD50 = 100 mg/kg

NIOSH 1990 - Pocket Guide - RELs
as Ag: 0.01 mg/m3 TWA

NIOSH 1990 - Pocket Guide - Target organs
nasal septum, skin, eyes

OSHA - Vacated PELs - Time Weighted Averages
as Ag: 0.01 mg/m3 TWA
OSHA - Final PELs - Time Weighted Averages
0.01 mg/m3 TWA

ENVIRONMENTAL LISTS

ATSDR Priority List
Rank (of 275): 173
CERCLA/SARA - Section 313 - Emission Reporting
form R reporting required for 1.0% de minimus concentration
CERCLA/SARA - Hazardous Substances and their Reportable Quantities
final RQ = 1000 pounds (454 kg) (no reporting of releases of this hazardous substance is required if the diameter of the solid metal released is equal to or exceeds 0.004 inches)
Clean Water Act - Priority Pollutants
[present]
Clean Water Act - Toxic Pollutants
[present] (Listed under 'Silver and compounds')
Safe Drinking Water Act - MCLs
MCL = 0.05 mg/L
Safe Drinking Water Act - SMCLs
SMCL = 0.1 mg/L
RCRA - D Series - Maximum Concentration of Contaminants
waste number D011; regulatory level = 5.0 mg/L
RCRA - D Series - Chronic Toxicity Reference Levels
chronic toxicity reference level = 0.05 mg/L
RCRA - Hazardous Constituents-Appendix VIII
hazardous constituent - no waste number
RCRA - Basis for Listing - Appendix VII
Included in waste stream: F039
RCRA - Substances Banned From Land Disposal
[present]
RCRA - TSD Facilities Ground Water Monitoring
TM 6010 = 70 ug/L PQL; TM 7760 = 100 ug/L PQL (all species in the ground water that contain this element are included)
RCRA - Universal Treatment Standards (LDR)
WW: 0.43 mg/l; NWW: 0.30 mg/l TCLP
TSCA - Code of Federal Regulations Citations
40 CFR 712.30(w)
TSCA - PAIR - Reporting List
Reporting Date: July 13, 1988

INTERNATIONAL LISTS

Australian Exposure Standards - Time Weighted Averages
metal: 0.1 mg/m3 TWA; soluble compounds, as Ag: 0.01 mg/m3 TWA
Canada - WHMIS: Ingredient Disclosure
1% item 1414 (256)
Canada - NPRI (National Pollutant Release Inventory)
[present]
Canada - Alberta - 8 Hour Occupational Exposure Limit
0.1 mg/m3 TWA
Canada - Alberta - 15 Minute Occupational Exposure Limit
0.3 mg/m3 STEL
Canada - British Columbia - 8 Hour Exposure Limits
as Ag: 0.01 mg/m3 TWA
Canada - British Columbia - 15 Minute Exposure Limits
as Ag: 0.03 mg/m3 STEL
Canada - Ontario - OHSA - TWAEVs
0.1 mg/m3 TWAEV
Canada - Quebec - Time-Weighted Average Exposure Values
as Ag: 0.1 mg/m3 TWAEV
German (DFG) - MAK Values
total dust: 0.01 mg/m3 MAK

German (DFG) - Peak Limitations
10 x normal MAK (30 min average value); don't exceed during shift
Israel - Time Weighted Averages
metal: 0.1 mg/m3 TWA; Soluble compounds, as Ag: 0.01 mg/m3 TWA
Israel - Action Levels
metal: 0.05 mg/m3 AL; soluble compounds as Ag: 0.005 mg/m3 AL
Mexico - Instruction No. 10 - TWAs
0.1 mg/m3 TWA
Mexico - Wastewater - Organic Toxic Pollutants and Heavy Metals
Listed under [Heavy Metals]
Mexico - Drinking Water - Ecological Criteria
0.05 mg/l

STATE LISTS

California - Air Bill 2588 Appendix A-I
6/91
California - Exposure Limits - PELs
metal, as Ag: 0.01 mg/m3 PEL; soluble compounds, as Ag: 0.01 mg/m3 PEL
California - Directors List of Hazardous Substances (8 CCR 339)
[present]
Florida Hazardous Substance List
[present]
Massachusetts Right To Know List
[present]
Minnesota Hazardous Substance List
[present]
NJ Right to Know List (Total)
sn 1669
Pennsylvania Right to Know List
environmental hazard (any compound if this substance is also an environmental hazard)

SILVER 108M   RR-00355-4

HEALTH AND SAFETY LISTS

U.S. DOT - Appendix A Table 2 - Radionuclides
final RQ = 100 curies (3.7E 12 Bq)

ENVIRONMENTAL LISTS

CERCLA/SARA List of Radionuclides (Appendix B) and Their Reportable Quantities
final RQ = 100 curies (3.7E 12 Bq)

SILVER 110M   RR-00354-3

HEALTH AND SAFETY LISTS

U.S. DOT - Appendix A Table 2 - Radionuclides
final RQ = 1000 curies (3.7E 13 Bq)

ENVIRONMENTAL LISTS

CERCLA/SARA List of Radionuclides (Appendix B) and Their Reportable Quantities
final RQ = 1000 curies (3.7E 13 Bq)

SILVER 111   15760-04-0

HEALTH AND SAFETY LISTS

U.S. DOT - Appendix A Table 2 - Radionuclides
final RQ = 1000 curies (3.7E 13 Bq)

ENVIRONMENTAL LISTS

CERCLA/SARA List of Radionuclides (Appendix B) and Their Reportable Quantities
final RQ = 1000 curies (3.7E 13 Bq)

SILVER 112   14331-86-3

HEALTH AND SAFETY LISTS

U.S. DOT - Appendix A Table 2 - Radionuclides
final RQ = 1000 curies (3.7E 13 Bq)

ENVIRONMENTAL LISTS

CERCLA/SARA List of Radionuclides (Appendix B) and Their Reportable Quantities
final RQ = 1000 curies (3.7E 13 Bq)

SILVER 115   15760-07-3

HEALTH AND SAFETY LISTS

U.S. DOT - Appendix A Table 2 - Radionuclides
final RQ = 10 curies (3.7E 11 Bq)

ENVIRONMENTAL LISTS

CERCLA/SARA List of Radionuclides (Appendix B) and Their Reportable Quantities
final RQ = 10 curies (3.7E 11 Bq)

SILVER ACETYLIDE   13092-75-6

HEALTH AND SAFETY LISTS

U.S. DOT - Appendix A Table 2 - Radionuclides
final RQ = 1000 curies (3.7E 13 Bq)

ENVIRONMENTAL LISTS

CERCLA/SARA List of Radionuclides (Appendix B) and Their Reportable Quantities
final RQ = 1000 curies (3.7E 13 Bq)

SILVER AMMONIUM NITRATE   23606-32-8

HEALTH AND SAFETY LISTS

U.S. DOT - Appendix A Table 2 - Radionuclides
final RQ = 10 curies (3.7E 11 Bq)

ENVIRONMENTAL LISTS

CERCLA/SARA List of Radionuclides (Appendix B) and Their Reportable Quantities
final RQ = 10 curies (3.7E 11 Bq)

SILVER ARSENITE   7784-08-9

HEALTH AND SAFETY LISTS

U.S. DOT - Appendix A Table 2 - Radionuclides
final RQ = 10 curies (3.7E 11 Bq)

ENVIRONMENTAL LISTS

CERCLA/SARA List of Radionuclides (Appendix B) and Their Reportable Quantities
final RQ = 10 curies (3.7E 11 Bq)

SILVER AZIDE   13863-88-2

HEALTH AND SAFETY LISTS

U.S. DOT - Appendix A Table 2 - Radionuclides
final RQ = 10 curies (3.7E 11 Bq)

ENVIRONMENTAL LISTS

CERCLA/SARA List of Radionuclides (Appendix B) and Their Reportable Quantities
final RQ = 10 curies (3.7E 11 Bq)

SILVER CHLORITE   7783-91-7

HEALTH AND SAFETY LISTS

U.S. DOT - Appendix A Table 2 - Radionuclides
final RQ = 10 curies (3.7E 11 Bq)

ENVIRONMENTAL LISTS

CERCLA/SARA List of Radionuclides (Appendix B) and Their Reportable Quantities
final RQ = 10 curies (3.7E 11 Bq)

SILVER CHROMATE   7784-01-2

HEALTH AND SAFETY LISTS

U.S. DOT - Appendix A Table 2 - Radionuclides
final RQ = 100 curies (3.7E 12 Bq)

ENVIRONMENTAL LISTS

CERCLA/SARA List of Radionuclides (Appendix B) and Their Reportable Quantities
final RQ = 100 curies (3.7E 12 Bq)

SILVER COMPOUNDS   RR-00574-3

HEALTH AND SAFETY LISTS

U.S. DOT - Appendix A Table 2 - Radionuclides
final RQ = 1000 curies (3.7E 13 Bq)

ENVIRONMENTAL LISTS

CERCLA/SARA List of Radionuclides (Appendix B) and Their Reportable Quantities
final RQ = 1000 curies (3.7E 13 Bq)

SILVER CYANIDE   506-64-9

HEALTH AND SAFETY LISTS

U.S. DOT - Hazard Classes
Forbidden from transport by the DOT

SILVER HEPTAFLUOROBUTYRATE   3794-64-7
SEE ALSO:
SILVER

INTERNATIONAL LISTS

Canada - WHMIS: Ingredient Disclosure
1% item 1412 (1198)

SILVER NITRATE   7761-88-8
SEE ALSO:
SILVER
ARSENIC

HEALTH AND SAFETY LISTS

U.S. DOT - Substances From 49 CFR 172.101
regulated by DOT (UN1683)

U.S. DOT - Hazard Classes
DOT hazard class = 6.1

U.S. DOT - Appendix B - Marine Pollutants
DOT regulated marine pollutant

STATE LISTS

NJ Right to Know List (Total)
sn 1670

SILVER ORTHOARSENITE   RR-01312-7

HEALTH AND SAFETY LISTS

U.S. DOT - Hazard Classes
Forbidden from transport by the DOT

SILVER OXALATE   533-51-7

HEALTH AND SAFETY LISTS

U.S. DOT - Hazard Classes
Forbidden from transport by the DOT

SILVER OXIDE   20667-12-3

STATE LISTS

Massachusetts Right To Know List
[present]

SILVER OXIDE (AG2O2)   25455-73-6

ENVIRONMENTAL LISTS

CERCLA/SARA - Section 313 - Emission Reporting
form R reporting required for 1.0% de minimus concentration

Clean Water Act - Toxic Pollutants
[present] (Listed under 'Silver and compounds')

RCRA - Hazardous Constituents-Appendix VIII
hazardous constituent - no waste number

INTERNATIONAL LISTS

Canada - NPRI (National Pollutant Release Inventory)
[present]

United Kingdom - Occupational Exposure Standards - TWAs
as Ag: 0.01 mg/m3 TWA

STATE LISTS

California - Air Bill 2588 Appendix A-I
6/91

California - Directors List of Hazardous Substances (8 CCR 339)
[present] (compounds existing in stable emulsions or suspensions are exempt)

NJ Right to Know List (Total)
sn 3008

SILVER PENTAFLUOROPROPIONATE 509-09-1

SEE ALSO:
CYANIDE ANION
SILVER

HEALTH AND SAFETY LISTS

U.S. DOT - Substances From 49 CFR 172.101
regulated by DOT (UN1684)

U.S. DOT - Hazard Classes
DOT hazard class = 6.1

U.S. DOT - Appendix B - Marine Pollutants
DOT regulated marine pollutant

U.S. DOT - Appendix A Table 1 - Hazardous Substances
final RQ = 1 pound (0.454 kg)

NIOSH - Selected LD50s and LC50s
Oral, rat: LD50 = 123 mg/kg

ENVIRONMENTAL LISTS

CERCLA/SARA - Hazardous Substances and their Reportable Quantities
final RQ = 1 pound (0.454 kg)

RCRA - P Series Wastes
waste number P104

RCRA - Hazardous Constituents-Appendix VIII
waste number P104

RCRA - Substances Banned From Land Disposal
[present]

INTERNATIONAL LISTS

Canada - WHMIS: Ingredient Disclosure
1% item 1413 (587)

STATE LISTS

Massachusetts Right To Know List
[present]

NJ Right to Know List (Total)
sn 1671

Pennsylvania Right to Know List
environmental hazard

SILVER PICRATE 146-84-9

SEE ALSO:
SILVER

INTERNATIONAL LISTS

Canada - WHMIS: Ingredient Disclosure
1% item 1415 (939)

STATE LISTS

Minnesota Hazardous Substance List
[present]

SILVER SOLUBLE COMPOUNDS RR-00041-9

SEE ALSO:
SILVER
SILVER
SILVER COMPOUNDS

HEALTH AND SAFETY LISTS

U.S. DOT - Substances From 49 CFR 172.101
regulated by DOT (UN1493)

U.S. DOT - Hazard Classes
DOT hazard class = 5.1

U.S. DOT - Appendix A Table 1 - Hazardous Substances
final RQ = 1 pound (0.454 kg)

NIOSH - Selected LD50s and LC50s
Oral, mouse: LD50 = 50 mg/kg

ENVIRONMENTAL LISTS

CERCLA/SARA - Hazardous Substances and their Reportable Quantities
final RQ = 1 pound (0.454 kg)

Clean Water Act - Hazardous Substances
[present]

TSCA - Code of Federal Regulations Citations
40 CFR 712.30(w)

TSCA - PAIR - Reporting List
Reporting Date: July 13, 1988

INTERNATIONAL LISTS

Canada - WHMIS: Ingredient Disclosure
1% item 1416 (1197)

STATE LISTS

Florida Hazardous Substance List
[present]

Massachusetts Right To Know List
[present]

NJ Right to Know List (Total)
sn 1672

Pennsylvania Right to Know List
environmental hazard

SILVEX (2,4,5-TP) 93-72-1

HEALTH AND SAFETY LISTS

U.S. DOT - Appendix B - Marine Pollutants
DOT regulated severe marine pollutant

SIMAZINE 122-34-9

HEALTH AND SAFETY LISTS

U.S. DOT - Hazard Classes
Forbidden from transport by the DOT

SIMETHICONE 8050-81-5

SEE ALSO:
SILVER

HEALTH AND SAFETY LISTS

NIOSH - Selected LD50s and LC50s
Oral, rat: LD50 = 2820 mg/kg

SLACK WAX (PETROLEUM) 64742-61-6

ENVIRONMENTAL LISTS

List of Pesticide Product Inert Ingredients
[present]

SLAGWOOL   RR-01202-2
SEE ALSO:
SILVER

INTERNATIONAL LISTS
Canada - WHMIS: Ingredient Disclosure
1% item 1417 (1342)

SLUDGE ACID   64742-24-1
SEE ALSO:
SILVER

HEALTH AND SAFETY LISTS
U.S. DOT - Substances From 49 CFR 172.101
regulated by DOT (UN1347)
U.S. DOT - Hazard Classes
dry: Forbidden from transport by the DOT; wetted: DOT hazard class = 4.1

STATE LISTS
NJ Right to Know List (Total)
sn 1673

SMECTITE   12199-37-0

INTERNATIONAL LISTS
Canada - WHMIS: Ingredient Disclosure
1% item 1419 (255)
Canada - Alberta - 8 Hour Occupational Exposure Limit
as Ag: 0.01 mg/m3 TWA
Canada - Alberta - 15 Minute Occupational Exposure Limit
as Ag: 0.03 mg/m3 STEL
Canada - Ontario - OHSA - TWAEVs
0.01 mg/m3 TWAEV
Canada - Quebec - Time-Weighted Average Exposure Values
as Ag: 0.01 mg/m3 TWAEV
Mexico - Instruction No. 10 - TWAs
0.01 mg/m3 TWA

SOAPSTONE   14378-12-2
SEE ALSO:
F039-HAZARDOUS WASTES

HEALTH AND SAFETY LISTS
U.S. DOT - Appendix A Table 1 - Hazardous Substances
final RQ = 100 pounds (45.4 kg)
NIOSH - Selected LD50s and LC50s
Oral, rat: LD50 = 650 mg/kg

ENVIRONMENTAL LISTS
ATSDR Priority List
Rank (of 275): 253
CERCLA/SARA - Hazardous Substances and their Reportable Quantities
final RQ = 100 pounds (45.4 kg)
Clean Water Act - Hazardous Substances
[present]
Safe Drinking Water Act - MCLs
MCL = 0.05 mg/L
Safe Drinking Water Act - MCLGs
MCLG = 0.05 mg/L
RCRA - D Series - Maximum Concentration of Contaminants
waste number D017; regulatory level = 1.0 mg/L
RCRA - D Series - Chronic Toxicity Reference Levels
chronic toxicity reference level = 0.01 mg/L
RCRA - Hazardous Constituents-Appendix VIII
contained in F027 waste stream
RCRA - Basis for Listing - Appendix VII
Included in waste stream: F039

RCRA - TSD Facilities Ground Water Monitoring
TM 8150 = 2 ug/L PQL
RCRA - Universal Treatment Standards (LDR)
WW: 0.72 mg/l; NWW: 7.9 mg/kg

STATE LISTS
California - Directors List of Hazardous Substances (8 CCR 339)
[present]
Massachusetts Right To Know List
[present]
NJ Right to Know List (Total)
sn 1899
Pennsylvania Right to Know List
environmental hazard

SOAPSTONE   RR-00010-2

HEALTH AND SAFETY LISTS
IARC - Group 3 (not classifiable)
[present]
NIOSH - Selected LD50s and LC50s
Oral, rat: LD50 = 971 mg/kg

ENVIRONMENTAL LISTS
CERCLA/SARA - Section 313 - Emission Reporting
form R reporting required
Safe Drinking Water Act - MCLs
MCL = 0.004 mg/L
Safe Drinking Water Act - MCLGs
MCLG = 0.004 mg/L

INTERNATIONAL LISTS
Canada - Drinking Water Quality - IMACs
0.01 mg/L IMAC

STATE LISTS
Massachusetts Right To Know List
[present]

SODA LIME   8006-28-8

ENVIRONMENTAL LISTS
List of Pesticide Product Inert Ingredients
[present]

SODIUM   7440-23-5

ENVIRONMENTAL LISTS
List of Pesticide Product Inert Ingredients
[present]

SODIUM 22   13966-32-0

HEALTH AND SAFETY LISTS
IARC - Group 2B (sufficient animal data)
[present]
OSHA - Possible Select Carcinogens
[present]

INTERNATIONAL LISTS
Canada - Quebec - Time-Weighted Average Exposure Values
1 fibre/cm3 TWAEV
Canada - Quebec - Carcinogens
C2 carcinogen: effect suspected in humans
German (DFG) - Carcinogens
as fibrous dust: suspected carcinogen

STATE LISTS
California - Air Bill 2588 Appendix A-I
known or potential carcinogen: 9/89

SODIUM 24   13982-04-2

HEALTH AND SAFETY LISTS

U.S. DOT - Substances From 49 CFR 172.101
regulated by DOT (UN1906)

U.S. DOT - Hazard Classes
DOT hazard class = 8

STATE LISTS

NJ Right to Know List (Total)
sn 2770

SODIUM ACETATE   127-09-3

ENVIRONMENTAL LISTS

List of Pesticide Product Inert Ingredients
[present]

SODIUM ACID PHOSPHATE   7558-80-7

INTERNATIONAL LISTS

Canada - Quebec - Time-Weighted Average Exposure Values
total dust: 6 mg/m3 TWAEV; respirable dust: 3 mg/m3 TWAEV

SODIUM ACRYLATE   7446-81-3

HEALTH AND SAFETY LISTS

ACGIH 1995 - Time Weighted Averages
respirable dust: 3 mg/m3 TWA; inhalable dust: 6 mg/m3 TWA (The 'respirable dust' TLV is for the respirable fraction of dust for the substance listed) (The 'inhalable dust' TLV is for the dust containing no asbestos and < 1% crystalline silica)

NIOSH 1990 - Pocket Guide - RELs
total: 6 mg/m3 TWA; respirable dust: 3 mg/m3 TWA

NIOSH 1990 - Pocket Guide - Target organs
lungs, CVS

OSHA - Vacated PELs - Time Weighted Averages
total dust (less than 1% crystalline silica): 6 mg/m3 TWA; respirable dust (less than 1% crystalline silica): 3 mg/m3 TWA (Listed under 'Silicates')

OSHA - Final PELs - Time Weighted Averages
see Table Z-3

ENVIRONMENTAL LISTS

List of Pesticide Product Inert Ingredients
[present]

INTERNATIONAL LISTS

Australian Exposure Standards - Time Weighted Averages
6 mg/m3 TWA; respirable dust: 3 mg/m3 TWA

Canada - WHMIS: Ingredient Disclosure
1% item 1420 (1505)

Canada - Alberta - 8 Hour Occupational Exposure Limit
respirable mass: 3 mg/m3 TWA; total mass: 6 mg/m3 TWA

Canada - Ontario - OHSA - TWAEVs
total dust: 6 mg/m3 TWAEV; respirable dust: 3 mg/m3 TWAEV (listed as a mineral dust)

Israel - Time Weighted Averages
respirable dust: 3 mg/m3 TWA; total dust: 6 mg/m3 TWA (The 'respirable dust' TLV is for the respirable fraction of dust for the substance listed) (The 'total dust' TLV is for total dust containing no asbestos and < 1% crystalline silica)

Israel - Action Levels
respirable dust: 1.5 mg/m3 AL; total: 3 mg/m3 AL

STATE LISTS

California - Exposure Limits - PELs
total dust (<1% crystalline silica): 6 mg/m3 PEL; respirable dust (<1% crystalline silica): 3 mg/m3 PEL (Listed under 'Silicates')

California - Directors List of Hazardous Substances (8 CCR 339)
[present] (exempt except when inhalable dust is present or can be generated)

Florida Hazardous Substance List
[present]

Massachusetts Right To Know List
[present] exempt when encapsulated or if particulates are not present and cannot be substantially generated through use of the product

Minnesota Hazardous Substance List
[present]

Pennsylvania Right to Know List
[present]

SODIUM ALUMINATE (NAALO2)   1302-42-7

HEALTH AND SAFETY LISTS

U.S. DOT - Substances From 49 CFR 172.101
regulated by DOT (UN1907)

U.S. DOT - Hazard Classes
DOT hazard class = 8

STATE LISTS

NJ Right to Know List (Total)
sn 2772

SODIUM ALUMINUM FLUORIDE   15096-52-3

HEALTH AND SAFETY LISTS

U.S. DOT - Substances From 49 CFR 172.101
regulated by DOT (UN1428)

U.S. DOT - Hazard Classes
DOT hazard class = 4.3

U.S. DOT - Appendix A Table 1 - Hazardous Substances
final RQ = 10 pounds (4.54 kg)

ENVIRONMENTAL LISTS

CERCLA/SARA - Hazardous Substances and their Reportable Quantities
final RQ = 10 pounds (4.54 kg)

Clean Water Act - Hazardous Substances
[present]

INTERNATIONAL LISTS

Canada - Drinking Water Quality - AOs
<= 200 mg/L AO

STATE LISTS

California - Directors List of Hazardous Substances (8 CCR 339)
[present]

Florida Hazardous Substance List
[present]

Massachusetts Right To Know List
[present]

NJ Right to Know List (Total)
sn 1674

NJ Special Hazardous Substances
(reactive - second degree)

Pennsylvania Right to Know List
environmental hazard

SODIUM ALUMINUM HYDRIDE   13770-96-2

HEALTH AND SAFETY LISTS

U.S. DOT - Appendix A Table 2 - Radionuclides
final RQ = 10 curies (3.7E 11 Bq)

ENVIRONMENTAL LISTS

ATSDR Priority List
Rank (of 275): 198

CERCLA/SARA List of Radionuclides (Appendix B) and Their Reportable Quantities
final RQ = 10 curies (3.7E 11 Bq)

SODIUM ALUMINUM PHOSPHATE · 7785-88-8

HEALTH AND SAFETY LISTS

U.S. DOT - Appendix A Table 2 - Radionuclides
final RQ = 10 curies (3.7E 11 Bq)

ENVIRONMENTAL LISTS

CERCLA/SARA List of Radionuclides (Appendix B) and Their Reportable Quantities
final RQ = 10 curies (3.7E 11 Bq)

SODIUM ALUMINUM SILICATE · 1344-00-9

HEALTH AND SAFETY LISTS

NIOSH - Selected LD50s and LC50s
Oral, rat: LD50 = 3530 mg/kg

ENVIRONMENTAL LISTS

List of Pesticide Product Inert Ingredients
[present]

SODIUM AMALGAM · 11110-52-4

HEALTH AND SAFETY LISTS

NIOSH - Selected LD50s and LC50s
Oral, rat: LD50 = 8290 mg/kg

ENVIRONMENTAL LISTS

List of Pesticide Product Inert Ingredients
[present]

SODIUM AMIDE · 7782-92-5

ENVIRONMENTAL LISTS

TSCA - Code of Federal Regulations Citations
40 CFR 712.30(d)

TSCA - PAIR - Reporting List
Reporting Date: November 19, 1982

SODIUM AMMONIUM VANADATE · 12055-09-3

HEALTH AND SAFETY LISTS

U.S. DOT - Substances From 49 CFR 172.101
regulated by DOT (UN1819, UN2812)

U.S. DOT - Hazard Classes
DOT hazard class = 8

ENVIRONMENTAL LISTS

List of Pesticide Product Inert Ingredients
[present]

SODIUM ANTHRAQUINONE-1-SULFONATE · 128-56-3

HEALTH AND SAFETY LISTS

NIOSH - Selected LD50s and LC50s
Oral, rat: LD50 = 200 mg/kg

NIOSH 1990 - Pocket Guide - RELs
as F: 2.5 mg/m3 TWA (Listed under 'Flourides')

NIOSH 1990 - Pocket Guide - IDLHs
as F: 500 mg/m3 IDLH (Listed under 'Fluorides')

NIOSH 1990 - Pocket Guide - Target organs
eyes, respiratory system, CNS, skeleton, kidneys, skin (Listed under 'Fluorides')

INTERNATIONAL LISTS

Canada - WHMIS: Ingredient Disclosure
1% item 52 (914)

SODIUM ANTIMONATE · 11112-10-0

HEALTH AND SAFETY LISTS

U.S. DOT - Substances From 49 CFR 172.101
regulated by DOT (UN2835)

U.S. DOT - Hazard Classes
DOT hazard class = 4.3

INTERNATIONAL LISTS

Canada - WHMIS: Ingredient Disclosure
1% item 1421 (1011)

STATE LISTS

NJ Right to Know List (Total)
sn 1677

SODIUM ARSANILATE · 127-85-5

ENVIRONMENTAL LISTS

List of Pesticide Product Inert Ingredients
[present]

SODIUM ARSENATE · 7631-89-2

ENVIRONMENTAL LISTS

List of Pesticide Product Inert Ingredients
[present]

SODIUM ARSENITE · 7784-46-5

STATE LISTS

NJ Right to Know List (Total)
sn 1678

SODIUM ASCORBATE · 134-03-2

STATE LISTS

NJ Right to Know List (Total)
sn 1679

SODIUM AZIDE · 26628-22-8

HEALTH AND SAFETY LISTS

U.S. DOT - Substances From 49 CFR 172.101
regulated by DOT (UN2863)

U.S. DOT - Hazard Classes
DOT hazard class = 6.1

STATE LISTS

NJ Right to Know List (Total)
sn 1680

SODIUM BENZOATE · 532-32-1

HEALTH AND SAFETY LISTS

NIOSH - Selected LD50s and LC50s
Oral, rat: LD50 = 20 gm/kg

STATE LISTS

Massachusetts Right To Know List
[present]

SODIUM BICARBONATE · 144-55-8

INTERNATIONAL LISTS

Canada - WHMIS: Ingredient Disclosure
1% item 1422 (252)

SODIUM BIFLUORIDE · 1333-83-1

SEE ALSO:
ARSENIC

HEALTH AND SAFETY LISTS

U.S. DOT - Substances From 49 CFR 172.101
regulated by DOT (UN2473)

U.S. DOT - Hazard Classes
DOT hazard class = 6.1

STATE LISTS

NJ Right to Know List (Total)
sn 1681

SODIUM BISULFATE   7681-38-1

SEE ALSO:
  ARSENIC

HEALTH AND SAFETY LISTS

U.S. DOT - Substances From 49 CFR 172.101
  regulated by DOT (UN1685)

U.S. DOT - Hazard Classes
  DOT hazard class = 6.1

U.S. DOT - Appendix A Table 1 - Hazardous Substances
  final RQ = 1 pound (0.454 kg)

NTP Seventh Report - Known Carcinogens
  known carcinogen (Listed under 'Arsenic and certain arsenic compounds')

OSHA - Select Carcinogens
  [present]

ENVIRONMENTAL LISTS

CERCLA/SARA - Section 302 Extremely Hazardous Substances and TPQs
  TPQ = 1000/10,000 pounds

CERCLA/SARA - Hazardous Substances and their Reportable Quantities
  final RQ = 1 pound (0.454 kg)

Clean Water Act - Hazardous Substances
  [present]

EPA - Carcinogen Hazard Ranking for RQ Adjustment
  Hazard ranking = High

INTERNATIONAL LISTS

Canada - WHMIS: Ingredient Disclosure
  0.1% item 1423 (263)

STATE LISTS

Massachusetts Right To Know List
  extraordinarily hazardous

NJ Right to Know List (Total)
  sn 1682

NJ Special Hazardous Substances
  (carcinogen; teratogen)

Pennsylvania Right to Know List
  environmental hazard; special hazardous substance

Pennsylvania RTK - Special Hazardous Substances
  [present]

SODIUM BISULFITE   7631-90-5

SEE ALSO:
  ARSENIC
  INORGANIC ARSENIC

HEALTH AND SAFETY LISTS

U.S. DOT - Substances From 49 CFR 172.101
  regulated by DOT (UN2027, UN1686)

U.S. DOT - Hazard Classes
  DOT hazard class = 6.1

U.S. DOT - Appendix B - Marine Pollutants
  DOT regulated marine pollutant

U.S. DOT - Appendix A Table 1 - Hazardous Substances
  final RQ = 1 pound (0.454 kg)

NIOSH - Selected LD50s and LC50s
  Oral, rat: LD50 = 41 mg/kg Skin, rat: LD50 = 150 mg/kg

NTP Seventh Report - Known Carcinogens
  known carcinogen (Listed under 'Arsenic and certain arsenic compounds')

OSHA - Select Carcinogens
  [present]

ENVIRONMENTAL LISTS

ATSDR Priority List
  Rank (of 275): 174

CERCLA/SARA - Section 302 Extremely Hazardous Substances and TPQs
  TPQ = 500/10,000 pounds

CERCLA/SARA - Hazardous Substances and their Reportable Quantities
  final RQ = 1 pound (0.454 kg)

Clean Water Act - Hazardous Substances
  [present]

EPA - Carcinogen Hazard Ranking for RQ Adjustment
  Hazard ranking = High

INTERNATIONAL LISTS

Canada - WHMIS: Ingredient Disclosure
  0.1% item 1424 (267)

STATE LISTS

Massachusetts Right To Know List
  extraordinarily hazardous

NJ Right to Know List (Total)
  sn 1683

NJ Special Hazardous Substances
  (carcinogen; mutagen; teratogen)

Pennsylvania Right to Know List
  environmental hazard

SODIUM BOROHYDRIDE   16940-66-2

ENVIRONMENTAL LISTS

List of Pesticide Product Inert Ingredients
  [present]

SODIUM BROMATE   7789-38-0

HEALTH AND SAFETY LISTS

ACGIH 1995 - Ceiling Limits
  as hydrazoic acid vapor: C 0.11 ppm; as sodium azide: C 0.29 mg/m3

U.S. DOT - Substances From 49 CFR 172.101
  regulated by DOT (UN1687)

U.S. DOT - Hazard Classes
  DOT hazard class = 6.1

U.S. DOT - Appendix A Table 1 - Hazardous Substances
  final RQ = 1000 pounds (454 kg)

NIOSH - Selected LD50s and LC50s
  Oral, rat: LD50 = 27 mg/kg Skin, rabbit: LD50 = 20 mg/kg

NTP Chemical Status Reports - Testing Status and NTIS Number
  Technical reports printed (PB921335615)

NTP Chemical Status Reports - Evidence of Carcinogenicity
  male rat: no evidence; female rat: no evidence

OSHA - Vacated PELs - Ceiling Limits
  as HN3: C 0.1 ppm; as NaN3: C 0.3 mg/m3

OSHA - Vacated PELs - Skin Designation
  Prevent or reduce skin absorption

ENVIRONMENTAL LISTS

CERCLA/SARA - Section 302 Extremely Hazardous Substances and TPQs
  TPQ = 500 pounds (This material is a reactive solid. The TPQ does not default to 10,000 pounds for non-powder, non-molten, non-solvent form)

CERCLA/SARA - Section 313 - Emission Reporting
  form R reporting required

CERCLA/SARA - Hazardous Substances and their Reportable Quantities
  final RQ = 1000 pounds (454 kg)

RCRA - P Series Wastes
waste number P105
RCRA - Hazardous Constituents-Appendix VIII
waste number P105
RCRA - Substances Banned From Land Disposal
[present]

INTERNATIONAL LISTS

Australian Exposure Standards - Time Weighted Averages
Peak Limitation: 0.11 ppm; 0.3 mg/m3
Canada - WHMIS: Ingredient Disclosure
1% item 1425 (272)
Canada - Alberta - Ceiling Occupational Exposure Limit
C 0.1 ppm; C 0.27 mg/m3
Canada - British Columbia - Ceiling Exposure Limits
C 0.1 ppm; C 0.3 mg/m3
Canada - Ontario - OHSA - CEVs
0.1 ppm CEV; 0.26 mg/m3 CEV
Canada - Quebec - Ceiling Limits
P 0.11 ppm; P 0.3 mg/m3
United Kingdom - Occupational Exposure Standards - STELs
as NaN3: 0.3 mg/m3 STEL
German (DFG) - MAK Values
0.07 ppm MAK; 0.2 mg/m3 MAK
Israel - Ceiling Exposure Limits
C 0.11 ppm; C 0.29 mg/m3

STATE LISTS

California - Exposure Limits - Ceilings
C 0.1 ppm; C 0.3 mg/m3
California - Exposure Limits - Skin Notation
material may be absorbed through the skin, eyes or mucous membrane
California - Directors List of Hazardous Substances (8 CCR 339)
[present]
Florida Hazardous Substance List
[present]
Massachusetts Right To Know List
extraordinarily hazardous
Minnesota Hazardous Substance List
[present]
NJ Right to Know List (Total)
sn 1684
NJ Special Hazardous Substances
(mutagen)
Pennsylvania Right to Know List
environmental hazard

SODIUM BROMIDE   7647-15-6

HEALTH AND SAFETY LISTS

NIOSH - Selected LD50s and LC50s
Oral, rat: LD50 = 4070 mg/kg

ENVIRONMENTAL LISTS

List of Pesticide Product Inert Ingredients
[present]

SODIUM CACODYLATE   124-65-2

HEALTH AND SAFETY LISTS

NIOSH - Selected LD50s and LC50s
Oral, rat: LD50 = 4220 mg/kg

ENVIRONMENTAL LISTS

List of Pesticide Product Inert Ingredients
[present]

SODIUM CAPRYLATE   1984-06-1

HEALTH AND SAFETY LISTS

U.S. DOT - Substances From 49 CFR 172.101
regulated by DOT (UN2439)
U.S. DOT - Hazard Classes
DOT hazard class = 8
U.S. DOT - Appendix A Table 1 - Hazardous Substances
final RQ = 100 pounds (45.4 kg)

ENVIRONMENTAL LISTS

CERCLA/SARA - Hazardous Substances and their Reportable Quantities
final RQ = 100 pounds (45.4 kg)
Clean Water Act - Hazardous Substances
[present]
List of Pesticide Product Inert Ingredients
[present]

STATE LISTS

Massachusetts Right To Know List
[present]
NJ Right to Know List (Total)
sn 1703
NJ Special Hazardous Substances
(corrosive)
Pennsylvania Right to Know List
environmental hazard

SODIUM CARBONATE   497-19-8

HEALTH AND SAFETY LISTS

U.S. DOT - Substances From 49 CFR 172.101
regulated by DOT (UN2837)
U.S. DOT - Hazard Classes
DOT hazard class = 8

ENVIRONMENTAL LISTS

List of Pesticide Product Inert Ingredients
[present]

INTERNATIONAL LISTS

Canada - WHMIS: Ingredient Disclosure
1% item 1426 (305)

STATE LISTS

NJ Right to Know List (Total)
sn 1704
NJ Special Hazardous Substances
(corrosive)

SODIUM CARBOXYMETHYL CELLULOSE   9004-32-4

HEALTH AND SAFETY LISTS

ACGIH 1995 - Time Weighted Averages
5 mg/m3 TWA
U.S. DOT - Appendix A Table 1 - Hazardous Substances
final RQ = 5000 pounds (2270 kg)
NIOSH - Selected LD50s and LC50s
Oral, rat: LD50 = 2000 mg/kg
OSHA - Vacated PELs - Time Weighted Averages
5 mg/m3 TWA

ENVIRONMENTAL LISTS

CERCLA/SARA - Hazardous Substances and their Reportable Quantities
final RQ = 5000 pounds (2270 kg)
Clean Water Act - Hazardous Substances
[present]
List of Pesticide Product Inert Ingredients
[present]

TSCA - PAIR - Reporting List
Effective Date: January 26, 1994; Reporting Date: March 28, 1994
TSCA - Health and Safety Reporting List
Effective Date: January 26, 1994; Sunset Date: January 26, 2004

INTERNATIONAL LISTS

Australian Exposure Standards - Time Weighted Averages
5 mg/m3 TWA
Canada - WHMIS: Ingredient Disclosure
1% item 1427 (308)
Canada - Alberta - 8 Hour Occupational Exposure Limit
5 mg/m3 TWA
Canada - Alberta - 15 Minute Occupational Exposure Limit
10 mg/m3 STEL
Canada - Ontario - OHSA - TWAEVs
5 mg/m3 TWAEV
Canada - Quebec - Time-Weighted Average Exposure Values
5 mg/m3 TWAEV
United Kingdom - Occupational Exposure Standards - TWAs
5 mg/m3 TWA
Israel - Time Weighted Averages
5 mg/m3 TWA
Israel - Action Levels
2.5 mg/m3 AL

STATE LISTS

California - Exposure Limits - PELs
5 mg/m3 PEL
California - Directors List of Hazardous Substances (8 CCR 339)
[present]
Florida Hazardous Substance List
[present]
Massachusetts Right To Know List
[present]
Minnesota Hazardous Substance List
[present]
NJ Right to Know List (Total)
sn 1685
NJ Special Hazardous Substances
(corrosive)
Pennsylvania Right to Know List
environmental hazard

PROPOSED REGULATIONS

TSCA - ITC 31st Report Priority Testing List
designated to be tested

SODIUM CASEINATE   9005-46-3

HEALTH AND SAFETY LISTS

U.S. DOT - Substances From 49 CFR 172.101
regulated by DOT (UN1426)
U.S. DOT - Hazard Classes
DOT hazard class = 4.3

STATE LISTS

NJ Right to Know List (Total)
sn 2603

SODIUM CHLORATE   7775-09-9

HEALTH AND SAFETY LISTS

U.S. DOT - Substances From 49 CFR 172.101
regulated by DOT (UN1494)
U.S. DOT - Hazard Classes
DOT hazard class = 5.1

INTERNATIONAL LISTS

Canada - WHMIS: Ingredient Disclosure
1% item 1429 (320)

STATE LISTS

NJ Right to Know List (Total)
sn 1686

SODIUM CHLORIDE   7647-14-5

HEALTH AND SAFETY LISTS

NIOSH - Selected LD50s and LC50s
Oral, rat: LD50 = 3500 mg/kg

ENVIRONMENTAL LISTS

List of Pesticide Product Inert Ingredients
[present]

INTERNATIONAL LISTS

Canada - WHMIS: Ingredient Disclosure
0.1% item 1430 (341)

SODIUM CHLORITE   7758-19-2
SEE ALSO:
ARSENIC

HEALTH AND SAFETY LISTS

U.S. DOT - Substances From 49 CFR 172.101
regulated by DOT (UN1688)
U.S. DOT - Hazard Classes
DOT hazard class = 6.1
NIOSH - Selected LD50s and LC50s
Oral, rat: LD50 = 2600 mg/kg

ENVIRONMENTAL LISTS

CERCLA/SARA - Section 302 Extremely Hazardous Substances and TPQs
TPQ = 100/10,000 pounds

STATE LISTS

Florida Hazardous Substance List
effective March 13, 1992
Massachusetts Right To Know List
extraordinarily hazardous
NJ Right to Know List (Total)
sn 1687
Pennsylvania Right to Know List
environmental hazard

PROPOSED REGULATIONS

CERCLA/SARA - Proposed Hazardous Substance Additions
proposed RQ = 1 pound (.454 kg)
CERCLA/SARA - 1989 Proposed RQ Adjustments
proposed RQ = 100 pounds (45.4 kg)

SODIUM CHLOROACETATE   3926-62-3

ENVIRONMENTAL LISTS

List of Pesticide Product Inert Ingredients
[present]

SODIUM CHROMATE   7775-11-3

HEALTH AND SAFETY LISTS

NIOSH - Selected LD50s and LC50s
Inhalation, rat: LC50 = 2300 mg/m3 2 hr Oral, rat: LD50 = 4090 mg/kg

ENVIRONMENTAL LISTS

List of Pesticide Product Inert Ingredients
[present]

INTERNATIONAL LISTS
    Canada - WHMIS: Ingredient Disclosure
    1% item 1431 (392)

SODIUM CHROMATE DECAHYDRATE   13517-17-4
HEALTH AND SAFETY LISTS
    NIOSH - Selected LD50s and LC50s
    Oral, rat: LD50 = 27000 mg/kg
ENVIRONMENTAL LISTS
    List of Pesticide Product Inert Ingredients
    [present]

SODIUM CITRATE   68-04-2
ENVIRONMENTAL LISTS
    List of Pesticide Product Inert Ingredients
    [present]

SODIUM COPPER CYANIDE   RR-01313-8
HEALTH AND SAFETY LISTS
    U.S. DOT - Substances From 49 CFR 172.101
    regulated by DOT (UN2428, UN1495)
    U.S. DOT - Hazard Classes
    DOT hazard class = 5.1
    NIOSH - Selected LD50s and LC50s
    Oral, rat: LD50 = 1200 mg/kg
ENVIRONMENTAL LISTS
    List of Pesticide Product Inert Ingredients
    [present]
STATE LISTS
    Florida Hazardous Substance List
    [present]
    Massachusetts Right To Know List
    [present]
    NJ Right to Know List (Total)
    sn 1688
    NJ Special Hazardous Substances
    (reactive - second degree)
    Pennsylvania Right to Know List
    [present]

SODIUM CUMENESULFONATE   28348-53-0
HEALTH AND SAFETY LISTS
    NIOSH - Selected LD50s and LC50s
    Oral, rat: LD50 = 3000 mg/kg
ENVIRONMENTAL LISTS
    List of Pesticide Product Inert Ingredients
    [present]

SODIUM CYANIDE   143-33-9
HEALTH AND SAFETY LISTS
    U.S. DOT - Substances From 49 CFR 172.101
    regulated by DOT (UN1496, UN1908)
    U.S. DOT - Hazard Classes
    DOT hazard class = 5.1; solution: DOT hazard class = 8
    IARC - Group 3 (not classifiable)
    [present]
    NIOSH - Selected LD50s and LC50s
    Oral, rat: LD50 = 165 mg/kg
INTERNATIONAL LISTS
    Canada - WHMIS: Ingredient Disclosure
    1% item 1432 (403)

STATE LISTS
    Florida Hazardous Substance List
    [present]
    Massachusetts Right To Know List
    [present]
    NJ Right to Know List (Total)
    sn 1689
    NJ Special Hazardous Substances
    (corrosive; reactive - second degree)
    Pennsylvania Right to Know List
    [present]

SODIUM CYCLAMATE   139-05-9
HEALTH AND SAFETY LISTS
    U.S. DOT - Substances From 49 CFR 172.101
    regulated by DOT (UN2659)
    U.S. DOT - Hazard Classes
    DOT hazard class = 6.1
    NIOSH - Selected LD50s and LC50s
    Oral, rat: LD50 = 95 mg/kg
ENVIRONMENTAL LISTS
    EPA - Master Testing List
    [present]
    List of Pesticide Product Inert Ingredients
    [present]
STATE LISTS
    NJ Right to Know List (Total)
    sn 1690

SODIUM DECYLBENZENESULFONATE   1322-98-1
SEE ALSO:
    CHROMIUM
    CHROMIUM (VI) COMPOUNDS
    CHROMIUM COMPOUNDS
    CHROMIUM (VI) COMPOUNDS- WATER SOLUBLE
HEALTH AND SAFETY LISTS
    U.S. DOT - Appendix A Table 1 - Hazardous Substances
    final RQ = 10 pounds (4.54 kg)
    OSHA - Select Carcinogens
    [present]
ENVIRONMENTAL LISTS
    CERCLA/SARA - Hazardous Substances and their Reportable Quantities
    final RQ = 10 pounds (4.54 kg)
    Clean Water Act - Hazardous Substances
    [present]
    EPA - Carcinogen Hazard Ranking for RQ Adjustment
    Hazard ranking = High
    List of Pesticide Product Inert Ingredients
    [present]
INTERNATIONAL LISTS
    Canada - WHMIS: Ingredient Disclosure
    1% item 1433 (552)
STATE LISTS
    Massachusetts Right To Know List
    [present]
    NJ Right to Know List (Total)
    sn 1692
    NJ Special Hazardous Substances
    (carcinogen)
    Pennsylvania Right to Know List
    environmental hazard

SODIUM DEHYDROACETATE   4418-26-2

STATE LISTS

Massachusetts Right To Know List
[present]

SODIUM 2-DIAZO-1-NAPHTHOL-4-SULFONATE   RR-00760-3

ENVIRONMENTAL LISTS

List of Pesticide Product Inert Ingredients
[present]

SODIUM 2-DIAZO-1-NAPHTHOL-5-SULFONATE   RR-00761-4

HEALTH AND SAFETY LISTS

U.S. DOT - Appendix B - Marine Pollutants
DOT regulated severe marine pollutant

SODIUM DICAMBA [3,6-DICHLORO-2-METHOXY-   1982-69-0
BENZOIC ACID, SODIUM SALT]

ENVIRONMENTAL LISTS

List of Pesticide Product Inert Ingredients
[present]

SODIUM DICHLOROACETATE   2156-56-1

SEE ALSO:
CYANIDE ANION
CYANIDE ANION

HEALTH AND SAFETY LISTS

ACGIH 1995 - Ceiling Limits
C 5 mg/m3

ACGIH 1995 - Skin Designations
skin - potential for cutaneous absorption

U.S. DOT - Substances From 49 CFR 172.101
regulated by DOT (UN1689)

U.S. DOT - Hazard Classes
DOT hazard class = 6.1

U.S. DOT - Appendix B - Marine Pollutants
DOT regulated marine pollutant

U.S. DOT - Appendix A Table 1 - Hazardous Substances
final RQ = 10 pounds (4.54 kg)

NIOSH - Selected LD50s and LC50s
Oral, rat: LD50 = 6440 ug/kg

NIOSH 1990 - Pocket Guide - RELs
as CN: C 5 mg/m3 (10 min);   C 4.7 ppm (10 min) (Listed under 'Cyanides')

NIOSH 1990 - Pocket Guide - IDLHs
as CN: 50 mg/m3 IDLH (Listed under 'Cyanides')

NIOSH 1990 - Pocket Guide - Target organs
CVS, CNS, liver, kidneys, skin (Listed under "Cyanides")

NTP Chemical Status Reports - Testing Status and NTIS Number
Technical reports printed

ENVIRONMENTAL LISTS

CERCLA/SARA - Section 302 Extremely Hazardous Substances and TPQs
TPQ = 100 pounds (This material is a reactive solid. The TPQ does not default to 10,000 pounds for non-powder, non-molten, non-solution form)

CERCLA/SARA - Hazardous Substances and their Reportable Quantities
final RQ = 10 pounds (4.54 kg)

Clean Water Act - Hazardous Substances
[present]

EPA - Master Testing List
[present]

RCRA - P Series Wastes
waste number P106

RCRA - Hazardous Constituents-Appendix VIII
waste number P106

RCRA - Substances Banned From Land Disposal
[present]

TSCA - Code of Federal Regulations Citations
40 CFR 712.30(w);   40 CFR 716.120(a);   40 CFR 799.5000

TSCA - PAIR - Reporting List
Reporting Date: December 27, 1990

TSCA - Health and Safety Reporting List
Effective Date: October 29, 1990

TSCA - Substances Subject to Testing Consent Orders
Test for: Chemical Fate;   Terrestrial Effects;   Plant Uptake and Translocation

TSCA - Section 12(b) - Export Notification
export notification required - Section 4

INTERNATIONAL LISTS

Canada - Ontario - OHSA - TWAEVs
as CN: 5 mg/m3 TWAEV

Canada - Ontario - OHSA - Skin Notations
as CN: absorption through skin, mucous membranes, or eyes

Israel - Time Weighted Averages
as CN: 5 mg/m3 TWA

Israel - Action Levels
as CN: 2.5 mg/m3 AL

STATE LISTS

Florida Hazardous Substance List
[present]

Massachusetts Right To Know List
extraordinarily hazardous

Minnesota Hazardous Substance List
as Cn: skin

NJ Right to Know List (Total)
sn 1693

Pennsylvania Right to Know List
environmental hazard

SODIUM DICHLOROISOCYANURATE DIHYDRATE   51580-86-0

ENVIRONMENTAL LISTS

List of Pesticide Product Inert Ingredients
[present]

SODIUM DICHROMATE   10588-01-9

HEALTH AND SAFETY LISTS

NIOSH - Selected LD50s and LC50s
Oral, mouse: LD50 = 2000 mg/kg

ENVIRONMENTAL LISTS

List of Pesticide Product Inert Ingredients
[present]

INTERNATIONAL LISTS

Canada - WHMIS: Ingredient Disclosure
1% item 1434 (616)

SODIUM DIETHYLDITHIOCARBAMATE   148-18-5

INTERNATIONAL LISTS

Canada - WHMIS: Ingredient Disclosure
1% item 1435 (617)

SODIUM 1,4-DIHEXYL SULFOSUCCINATE   3006-15-3

HEALTH AND SAFETY LISTS

U.S. DOT - Substances From 49 CFR 172.101
regulated by DOT (UN3040)

U.S. DOT - Hazard Classes
DOT hazard class = 4.1

SODIUM DIISOPROPYLNAPHTHALENESULFONATE   1322-93-6

HEALTH AND SAFETY LISTS

U.S. DOT - Substances From 49 CFR 172.101
regulated by DOT (UN3041)

U.S. DOT - Hazard Classes
DOT hazard class = 4.1

SODIUM DIMETHYLDITHIOCARBAMATE   128-04-1

ENVIRONMENTAL LISTS

CERCLA/SARA - Section 313 - Emission Reporting
form R reporting required

SODIUM DINITRO-ORTHO-CRESOLATE   2312-76-7

ENVIRONMENTAL LISTS

List of Pesticide Product Inert Ingredients
[present]

SODIUM DINITRO-O-CRESOLATE   25641-53-6

STATE LISTS

Massachusetts Right To Know List
[present]

Pennsylvania Right to Know List
[present]

SODIUM DINONYL SULFOSUCCINATE   3246-20-6
SEE ALSO:
CHROMIUM
CHROMIUM COMPOUNDS
CHROMIUM (VI) COMPOUNDS

HEALTH AND SAFETY LISTS

U.S. DOT - Appendix A Table 1 - Hazardous Substances
final RQ = 10 pounds (4.54 kg)

NIOSH - Selected LD50s and LC50s
Oral, rat: LD50 = 50 mg/kg

OSHA - Select Carcinogens
[present]

ENVIRONMENTAL LISTS

CERCLA/SARA - Hazardous Substances and their Reportable Quantities
final RQ = 10 pounds (4.54 kg)

Clean Water Act - Hazardous Substances
[present]

EPA - Carcinogen Hazard Ranking for RQ Adjustment
Hazard ranking = High

INTERNATIONAL LISTS

Canada - WHMIS: Ingredient Disclosure
1% item 1436 (688)

STATE LISTS

California - Air Bill 2588 Appendix A-I
known or potential carcinogen: 6/91

Florida Hazardous Substance List
[present]

Massachusetts Right To Know List
[present]

NJ Right to Know List (Total)
sn 1695

NJ Special Hazardous Substances
(carcinogen)

Pennsylvania Right to Know List
environmental hazard

SODIUM DITHIONITE   7775-14-6

HEALTH AND SAFETY LISTS

IARC - Group 3 (not classifiable)
[present]

NIOSH - Selected LD50s and LC50s
Oral, rat: LD50 = 1500 mg/kg

NTP Chemical Status Reports - Testing Status and NTIS Number
Technical reports printed (PB293833/AS)

NTP Chemical Status Reports - Evidence of Carcinogenicity
male rat-negative; female rat-negative; male mice-negative; female mice-negative

SODIUM DODECYLBENZENESULFONATE   25155-30-0

ENVIRONMENTAL LISTS

List of Pesticide Product Inert Ingredients
[present]

SODIUM DODECYL DIPHENYL OXIDE SULFONATE   53467-00-8

ENVIRONMENTAL LISTS

List of Pesticide Product Inert Ingredients
[present]

SODIUM DODECYLPHENYL POLYOXYETHYLENE PHOSPHATES   RR-01140-5

HEALTH AND SAFETY LISTS

NIOSH - Selected LD50s and LC50s
Oral, rat: LD50 = 1000 mg/kg

ENVIRONMENTAL LISTS

CERCLA/SARA - Section 313 - Emission Reporting
form R reporting required

SODIUM EQUILIN SULFATE   16680-47-0

HEALTH AND SAFETY LISTS

U.S. DOT - Substances From 49 CFR 172.101
regulated by DOT (UN0234, UN1348)

U.S. DOT - Hazard Classes
DOT hazard class = 1.3C

U.S. DOT - Appendix B - Marine Pollutants
DOT regulated marine pollutant

NIOSH - Selected LD50s and LC50s
Oral, rat: LD50 = 26 mg/kg Skin, rat: LD50 = 200 mg/kg

STATE LISTS

NJ Right to Know List (Total)
sn 1696

SODIUM ERYTHORBATE   6381-77-7

INTERNATIONAL LISTS

Canada - WHMIS: Ingredient Disclosure
1% item 1437 (759)

SODIUM ESTRONE SULFATE   438-67-5

ENVIRONMENTAL LISTS

List of Pesticide Product Inert Ingredients
[present]

SODIUM ETHYLENEDIAMINETETRAACETATE   17421-79-3

HEALTH AND SAFETY LISTS

U.S. DOT - Substances From 49 CFR 172.101
regulated by DOT (UN1384)

U.S. DOT - Hazard Classes
DOT hazard class = 4.2

STATE LISTS

Florida Hazardous Substance List
[present]

Massachusetts Right To Know List
[present]

NJ Right to Know List (Total)
sn 1697

Pennsylvania Right to Know List
[present]

SODIUM O-(ETHYLMERCURITHIO)BENZOATE   54-64-8

HEALTH AND SAFETY LISTS

U.S. DOT - Appendix A Table 1 - Hazardous Substances
final RQ = 1000 pounds (454 kg)

NIOSH - Selected LD50s and LC50s
Oral, rat: LD50 = 1260 mg/kg

ENVIRONMENTAL LISTS

CERCLA/SARA - Hazardous Substances and their Reportable Quantities
final RQ = 1000 pounds (454 kg)

Clean Water Act - Hazardous Substances
[present]

List of Pesticide Product Inert Ingredients
[present]

INTERNATIONAL LISTS

Canada - WHMIS: Ingredient Disclosure
1% item 1438 (795)

STATE LISTS

California - Directors List of Hazardous Substances (8 CCR 339)
[present] (exempt when in solution)

Massachusetts Right To Know List
[present]

NJ Right to Know List (Total)
sn 1698

Pennsylvania Right to Know List
environmental hazard

SODIUM FLUORESCEIN   518-47-8

ENVIRONMENTAL LISTS

List of Pesticide Product Inert Ingredients
[present]

SODIUM FLUORIDE   7681-49-4

ENVIRONMENTAL LISTS

List of Pesticide Product Inert Ingredients
[present]

SODIUM FLUORIDES   RR-00490-0

HEALTH AND SAFETY LISTS

NTP Seventh Report - Known Carcinogens
known carcinogen (Listed under 'Conjugated estrogens')

OSHA - Select Carcinogens
[present]

STATE LISTS

Pennsylvania Right to Know List
special hazardous substance

Pennsylvania RTK - Special Hazardous Substances
[present]

SODIUM FLUOROACETATE   62-74-8

ENVIRONMENTAL LISTS

List of Pesticide Product Inert Ingredients
[present]

SODIUM FORMALDEHYDE SULFOXYLATE   149-44-0

HEALTH AND SAFETY LISTS

NTP Seventh Report - Known Carcinogens
known carcinogen (Listed under 'Conjugated estrogens')

OSHA - Select Carcinogens
[present]

STATE LISTS

Pennsylvania Right to Know List
special hazardous substance

Pennsylvania RTK - Special Hazardous Substances
[present]

SODIUM FORMATE   141-53-7

ENVIRONMENTAL LISTS

List of Pesticide Product Inert Ingredients
[present]

SODIUM GLUCONATE   527-07-1
SEE ALSO:
MERCURY

INTERNATIONAL LISTS

Canada - WHMIS: Ingredient Disclosure
0.1% item 1439 (871)

SODIUM HEXAMETAPHOSPHATE   10124-56-8

HEALTH AND SAFETY LISTS

NIOSH - Selected LD50s and LC50s
Oral, rat: LD50 = 6721 mg/kg

ENVIRONMENTAL LISTS

List of Pesticide Product Inert Ingredients
[present]

SODIUM HYDRIDE   7646-69-7

HEALTH AND SAFETY LISTS

U.S. DOT - Substances From 49 CFR 172.101
regulated by DOT (UN1690)

U.S. DOT - Hazard Classes
DOT hazard class = 6.1

U.S. DOT - Appendix A Table 1 - Hazardous Substances
final RQ = 1000 pounds (454 kg)

NIOSH - Selected LD50s and LC50s
Oral, rat: LD50 = 52 mg/kg

NIOSH 1990 - Pocket Guide - RELs
as F: 2.5 mg/m3 TWA (Listed under 'Fluorides')

NIOSH 1990 - Pocket Guide - IDLHs
as F: 500 mg/m3 IDLH (Listed under 'Flourides')

NIOSH 1990 - Pocket Guide - Target organs
eyes, respiratory system, CNS, skeleton, kidneys, skin (Listed under 'Flourides')

NTP Chemical Status Reports - Testing Status and NTIS Number
Technical reports printed (PB91178137)

NTP Chemical Status Reports - Evidence of Carcinogenicity
male rat-equivocal evidence; female rat-no evidence; male mice-no evidence; female mice-no evidence

ENVIRONMENTAL LISTS

CERCLA/SARA - Hazardous Substances and their Reportable Quantities
final RQ = 1000 pounds (454 kg)

Clean Water Act - Hazardous Substances
[present]

List of Pesticide Product Inert Ingredients
[present]

TSCA - Code of Federal Regulations Citations
40 CFR 712.30(w)

TSCA - PAIR - Reporting List
Reporting Date: July 13, 1988

INTERNATIONAL LISTS

Canada - WHMIS: Ingredient Disclosure
1% item 1440 (910)

STATE LISTS

Florida Hazardous Substance List
[present]

Massachusetts Right To Know List
[present]

NJ Right to Know List (Total)
sn 3038

Pennsylvania Right to Know List
environmental hazard

SODIUM HYDROSULFIDE   16721-80-5

STATE LISTS

NJ Right to Know List (Total)
sn 1699

SODIUM HYDROXIDE   1310-73-2

HEALTH AND SAFETY LISTS

ACGIH 1995 - Time Weighted Averages
0.05 mg/m3 TWA

ACGIH 1995 - Skin Designations
skin - potential for cutaneous absorption

U.S. DOT - Substances From 49 CFR 172.101
regulated by DOT (UN2629)

U.S. DOT - Hazard Classes
DOT hazard class = 6.1

U.S. DOT - Appendix A Table 1 - Hazardous Substances
final RQ = 10 pounds (4.54 kg)

NIOSH - Selected LD50s and LC50s
Oral, rat: LD50 = 100 ug/kg

NIOSH 1990 - Pocket Guide - RELs
0.05 mg/m3 TWA; 0.15 mg/m3 STEL

NIOSH 1990 - Pocket Guide - IDLHs
5 mg/m3 IDLH

NIOSH 1990 - Pocket Guide - Target organs
CVS, CNS, lungs, kidneys

NIOSH 1990 - Pocket Guide - Skin list
Potential for dermal absorption

OSHA - Vacated PELs - Time Weighted Averages
0.05 mg/m3 TWA

OSHA - Vacated PELs - Short Term Exposure Limits
0.15 mg/m3 STEL

OSHA - Vacated PELs - Skin Designation
Prevent or reduce skin absorption

OSHA - Final PELs - Time Weighted Averages
0.05 mg/m3 TWA

OSHA - Final PELs - Skin Notations
prevent or reduce skin absorption

ENVIRONMENTAL LISTS

CERCLA/SARA - Section 302 Extremely Hazardous Substances and TPQs
TPQ = 10/10,000 pounds

CERCLA/SARA - Section 313 - Emission Reporting
form R reporting required

CERCLA/SARA - Hazardous Substances and their Reportable Quantities
final RQ = 10 pounds (4.54 kg)

RCRA - P Series Wastes
waste number P058

RCRA - Hazardous Constituents-Appendix VIII
waste number P058

RCRA - Substances Banned From Land Disposal
[present]

TSCA - Code of Federal Regulations Citations
40 CFR 716.120(a)

TSCA - Health and Safety Reporting List
Effective Date: March 7, 1986

INTERNATIONAL LISTS

Australian Exposure Standards - Time Weighted Averages
0.05 mg/m3 TWA

Australian Exposure Standards - Short Term Exposure Limits
0.15 mg/m3 STEL

Australian Exposure Standards - Skin Effects
skin absorption

Canada - Alberta - 8 Hour Occupational Exposure Limit
0.05 mg/m3 TWA

Canada - Alberta - 15 Minute Occupational Exposure Limit
0.15 mg/m3 STEL

Canada - Alberta - Skin Designation
can be absorbed through the intact skin

Canada - British Columbia - 8 Hour Exposure Limits
0.05 mg/m3 TWA

Canada - British Columbia - 15 Minute Exposure Limits
0.15 mg/m3 STEL

Canada - British Columbia - Skin Notations
skin - potential for skin absorption

Canada - Ontario - OHSA - TWAEVs
0.05 mg/m3 TWAEV

Canada - Ontario - OHSA - STEVs
0.15 mg/m3 STEV

Canada - Ontario - OHSA - Skin Notations
absorption through skin, eyes, or mucous membranes

Canada - Quebec - Time-Weighted Average Exposure Values
0.05 mg/m3 TWAEV

Canada - Quebec - Short-term Exposure Values
0.15 mg/m3 STEV

Canada - Quebec - Skin Designations
absorbed through the skin

United Kingdom - Occupational Exposure Standards - TWAs
0.05 mg/m3 TWA

United Kingdom - Occupational Exposure Standards - STELs
0.15 mg/m3 STEL

United Kingdom - Occupational Exposure Standards - Notes
can be absorbed through skin

German (DFG) - MAK Values
total dust: 0.05 mg/m3 MAK

German (DFG) - Peak Limitations
2 x normal MAK (30 min. average value); don't exceed 4 times during shift

German (DFG) - Skin/Sensitizers
danger of cutaneous absorption

Israel - Time Weighted Averages
0.05 mg/m3 TWA

Israel - Short Term Exposure Limits
0.15 mg/m3 STEL

Israel - Action Levels
0.025 mg/m3 AL

Mexico - Instruction No. 10 - TWAs
0.05 mg/m3 TWA

Mexico - Instruction No. 10 - STELs
0.15 mg/m3 STEL

Mexico - Instruction No. 10 - Skin designation
skin - potential for cutaneous absorption

STATE LISTS

California - Exposure Limits - PELs
0.05 mg/m3 PEL

California - Exposure Limits - STELs
0.15 mg/m3 STEL

California - Exposure Limits - Skin Notation
material may be absorbed through the skin, eyes or mucous membrane

California - Directors List of Hazardous Substances (8 CCR 339)
[present]

Florida Hazardous Substance List
[present]

Massachusetts Right To Know List
extraordinarily hazardous

Minnesota Hazardous Substance List
skin

NJ Right to Know List (Total)
sn 1700

Pennsylvania Right to Know List
environmental hazard

SODIUM HYPOCHLORITE   7681-52-9

ENVIRONMENTAL LISTS

List of Pesticide Product Inert Ingredients
[present]

SODIUM HYPOCHLORITE (PENTAHYDRATE)   10022-70-5

HEALTH AND SAFETY LISTS

NIOSH - Selected LD50s and LC50s
Oral, mouse: LD50 = 11200 mg/kg

SODIUM HYPOPHOSPHATE   13721-43-2

ENVIRONMENTAL LISTS

List of Pesticide Product Inert Ingredients
[present]

SODIUM HYPOPHOSPHITE   7681-53-0

HEALTH AND SAFETY LISTS

U.S. DOT - Appendix A Table 1 - Hazardous Substances
final RQ = 5000 pounds (2270 kg) (Listed under 'Sodium phosphate, tribasic')

NIOSH - Selected LD50s and LC50s
Oral, mouse: LD50 = 7250 mg/kg

ENVIRONMENTAL LISTS

CERCLA/SARA - Hazardous Substances and their Reportable Quantities
final RQ = 5000 pounds (2270 kg) (Listed under 'Sodium phosphate, tribasic')

Clean Water Act - Hazardous Substances
[present] (Listed under 'Sodium phosphate, tribasic')

List of Pesticide Product Inert Ingredients
[present]

STATE LISTS

California - Directors List of Hazardous Substances (8 CCR 339)
[present] (Listed under 'Sodium phosphate, tribasic')

Massachusetts Right To Know List
[present]

NJ Right to Know List (Total)
sn 3045

Pennsylvania Right to Know List
environmental hazard

SODIUM IODIDE   7681-82-5

HEALTH AND SAFETY LISTS

U.S. DOT - Substances From 49 CFR 172.101
regulated by DOT (UN1427)

U.S. DOT - Hazard Classes
DOT hazard class = 4.3

STATE LISTS

Florida Hazardous Substance List
[present]

Massachusetts Right To Know List
[present]

NJ Right to Know List (Total)
sn 1702

NJ Special Hazardous Substances
(flammable - third degree; reactive - second degree)

Pennsylvania Right to Know List
[present]

SODIUM IRON(III) ETHYLENETETRAACETATE 15708-41-5

HEALTH AND SAFETY LISTS

U.S. DOT - Substances From 49 CFR 172.101
regulated by DOT (NA2922, UN2318, UN2949)

U.S. DOT - Hazard Classes
DOT hazard class = 4.2

U.S. DOT - Appendix A Table 1 - Hazardous Substances
final RQ = 5000 pounds (2270 kg)

ENVIRONMENTAL LISTS

CERCLA/SARA - Hazardous Substances and their Reportable Quantities
final RQ = 5000 pounds (2270 kg)

Clean Water Act - Hazardous Substances
[present]

INTERNATIONAL LISTS

Canada - WHMIS: Ingredient Disclosure
1% item 1441 (980)

STATE LISTS

California - Directors List of Hazardous Substances (8 CCR 339)
[present]

Florida Hazardous Substance List
[present]

Massachusetts Right To Know List
[present]

NJ Right to Know List (Total)
sn 1705

NJ Special Hazardous Substances
(corrosive)

Pennsylvania Right to Know List
environmental hazard

SODIUM ISOPROPYL ISOBUTYL   60874-90-0
NAPHTHALENESULFONATE

HEALTH AND SAFETY LISTS

ACGIH 1995 - Ceiling Limits
C 2 mg/m3

U.S. DOT - Substances From 49 CFR 172.101
regulated by DOT (UN1824, UN1823)

U.S. DOT - Hazard Classes
DOT hazard class = 8

U.S. DOT - Appendix A Table 1 - Hazardous Substances
final RQ = 1000 pounds (454 kg)

NIOSH 1990 - Pocket Guide - RELs
C 2 mg/m3

NIOSH 1990 - Pocket Guide - IDLHs
250 mg/m3 IDLH

NIOSH 1990 - Pocket Guide - Target organs
eyes, skin, respiratory system

NIOSH - Health Standards - Exposure Limits
C (15 min) 2 mg/m3

NIOSH - Health Standards - Health Effects and Precautions
Respiratory irritation (Prevent skin and eye contact)

OSHA - Vacated PELs - Ceiling Limits
C 2 mg/m3

OSHA - Final PELs - Time Weighted Averages
2 mg/m3 TWA

ENVIRONMENTAL LISTS

CERCLA/SARA - Hazardous Substances and their Reportable Quantities
final RQ = 1000 pounds (454 kg)

Clean Water Act - Hazardous Substances
[present]

List of Pesticide Product Inert Ingredients
[present]

INTERNATIONAL LISTS

Australian Exposure Standards - Time Weighted Averages
Peak Limitation: 2 mg/m3

Canada - WHMIS: Ingredient Disclosure
1% item 1442 (998)

Canada - Alberta - Ceiling Occupational Exposure Limit
C 2 mg/m3

Canada - British Columbia - Ceiling Exposure Limits
C 2 mg/m3

Canada - Ontario - OHSA - CEVs
2 mg/m3 CEV

Canada - Quebec - Ceiling Limits
P 2 mg/m3

United Kingdom - Occupational Exposure Standards - STELs
2 mg/m3 STEL

German (DFG) - MAK Values
total dust: 2 mg/m3 MAK

German (DFG) - Peak Limitations
2 x normal MAK (5 min momentary value);   don't exceed 8 times during shift

Israel - Ceiling Exposure Limits
C 2 mg/m3

Mexico - Instruction No. 10 - TWAs
2 mg/m3 TWA

STATE LISTS

California - Air Bill 2588 Appendix A-I
[present]

California - Exposure Limits - Ceilings
C 2 mg/m3

California - Directors List of Hazardous Substances (8 CCR 339)
[present]

Florida Hazardous Substance List
[present]

Massachusetts Right To Know List
[present]

Minnesota Hazardous Substance List
[present]

NJ Right to Know List (Total)
sn 1706

NJ Special Hazardous Substances
(corrosive)

Pennsylvania Right to Know List
environmental hazard

SODIUM LAURYL SULFATE   151-21-3

HEALTH AND SAFETY LISTS

AIHA - WEEL - Ceilings or Short Term Time Weighted Averages
2 mg/m3 STEL

U.S. DOT - Appendix A Table 1 - Hazardous Substances
final RQ = 100 pounds (45.4 kg)

ENVIRONMENTAL LISTS

CERCLA/SARA - Hazardous Substances and their Reportable Quantities
final RQ = 100 pounds (45.4 kg)

Clean Water Act - Hazardous Substances
[present]

INTERNATIONAL LISTS

Canada - WHMIS: Ingredient Disclosure
1% item 1443 (1013)

STATE LISTS

California - Directors List of Hazardous Substances (8 CCR 339)
[present]

Massachusetts Right To Know List
[present]

Minnesota Hazardous Substance List
[present]

NJ Right to Know List (Total)
sn 1707

Pennsylvania Right to Know List
environmental hazard

SODIUM LAURYL TRIETHOXY SULFATE   13150-00-0

HEALTH AND SAFETY LISTS

U.S. DOT - Appendix A Table 1 - Hazardous Substances
final RQ = 100 pounds (45.4 kg)

ENVIRONMENTAL LISTS

CERCLA/SARA - Hazardous Substances and their Reportable Quantities
final RQ = 100 pounds (45.4 kg)

Clean Water Act - Hazardous Substances
[present] (Listed under 'Sodium hypochlorite')

STATE LISTS

Massachusetts Right To Know List
[present]

NJ Right to Know List (Total)
sn 3039

Pennsylvania Right to Know List
environmental hazard

SODIUM LIGNO SULFONATE   8061-51-6

ENVIRONMENTAL LISTS

List of Pesticide Product Inert Ingredients
[present]

SODIUM MERCAPTOBENZOTHIAZOLE   2492-26-4

ENVIRONMENTAL LISTS

List of Pesticide Product Inert Ingredients
[present]

SODIUM METABISULFITE   7681-57-4

HEALTH AND SAFETY LISTS

NIOSH - Selected LD50s and LC50s
Oral, rat: LD50 = 4340 mg/kg

INTERNATIONAL LISTS

Canada - WHMIS: Ingredient Disclosure
1% item 1444 (1029)

SODIUM METABORATE   7775-19-1

ENVIRONMENTAL LISTS

List of Pesticide Product Inert Ingredients
[present]

SODIUM METASILICATE   6834-92-0

ENVIRONMENTAL LISTS

List of Pesticide Product Inert Ingredients
[present]

SODIUM METHYLATE   124-41-4

HEALTH AND SAFETY LISTS

NIOSH - Selected LD50s and LC50s
Oral, rat: LD50 = 1288 mg/kg

ENVIRONMENTAL LISTS

EPA - Master Testing List
[present]

List of Pesticide Product Inert Ingredients
[present]

INTERNATIONAL LISTS

Canada - WHMIS: Ingredient Disclosure
1% item 1446 (1539)

SODIUM N-METHYL-N-OLEOYLTAURINE   137-20-2

HEALTH AND SAFETY LISTS

NIOSH - Selected LD50s and LC50s
Oral, rat: LD50 = 1820 mg/kg

SODIUM N-METHYL-N-OLEYLTAURATE   7346-80-7

ENVIRONMENTAL LISTS

List of Pesticide Product Inert Ingredients
[present]

SODIUM MOLYBDATE   7631-95-0

HEALTH AND SAFETY LISTS

NIOSH - Selected LD50s and LC50s
Oral, rat: LD50 = 3120 mg/kg

ENVIRONMENTAL LISTS

List of Pesticide Product Inert Ingredients
[present]

SODIUM MONO, DI AND   RR-01142-7
TRIBUTYLNAPHTHALENESULFONATE

HEALTH AND SAFETY LISTS

ACGIH 1995 - Time Weighted Averages
5 mg/m3 TWA

OSHA - Vacated PELs - Time Weighted Averages
5 mg/m3 TWA

ENVIRONMENTAL LISTS

List of Pesticide Product Inert Ingredients
[present]

TSCA - PAIR - Reporting List
Effective Date: January 26, 1994;   Reporting Date: March 28, 1994

TSCA - Health and Safety Reporting List
Effective Date: January 26, 1994;   Sunset Date: January 26, 2004

INTERNATIONAL LISTS

Australian Exposure Standards - Time Weighted Averages
5 mg/m3 TWA

Canada - WHMIS: Ingredient Disclosure
1% item 1447 (1083)

Canada - Alberta - 8 Hour Occupational Exposure Limit
5 mg/m3 TWA

Canada - Alberta - 15 Minute Occupational Exposure Limit
10 mg/m3 STEL

Canada - Ontario - OHSA - TWAEVs
5 mg/m3 TWAEV

Canada - Quebec - Time-Weighted Average Exposure Values
5 ppm TWAEV

United Kingdom - Occupational Exposure Standards - TWAs
5 mg/m3 TWA

Israel - Time Weighted Averages
5 mg/m3 TWA

Israel - Action Levels
2.5 mg/m3 AL

STATE LISTS

California - Exposure Limits - PELs
5 mg/m3 PEL

California - Directors List of Hazardous Substances (8 CCR 339)
[present]

Florida Hazardous Substance List
[present]

Massachusetts Right To Know List
[present]

Minnesota Hazardous Substance List
[present]

NJ Right to Know List (Total)
sn 1708

Pennsylvania Right to Know List
[present]

PROPOSED REGULATIONS

TSCA - ITC 31st Report Priority Testing List
designated to be tested

SODIUM MONO AND DI-C8-13-ALKYL PHENOXY   RR-01141-6
BENZENE DISULFONATES

ENVIRONMENTAL LISTS

List of Pesticide Product Inert Ingredients
[present]

SODIUM MONOXIDE   12401-86-4

HEALTH AND SAFETY LISTS

NIOSH - Selected LD50s and LC50s
Oral, rat: LD50 = 1153 mg/kg

ENVIRONMENTAL LISTS

List of Pesticide Product Inert Ingredients
[present]

INTERNATIONAL LISTS

Canada - WHMIS: Ingredient Disclosure
1% item 1448 (1084)

SODIUM NITRATE   7631-99-4

HEALTH AND SAFETY LISTS

U.S. DOT - Substances From 49 CFR 172.101
regulated by DOT (UN1431, UN1289)

U.S. DOT - Hazard Classes
DOT hazard class = 4.2; solutions: DOT hazard class = 3

U.S. DOT - Appendix A Table 1 - Hazardous Substances
final RQ = 1000 pounds (454 kg)

ENVIRONMENTAL LISTS

CERCLA/SARA - Hazardous Substances and their Reportable Quantities
final RQ = 1000 pounds (454 kg)

CAA - HON Rule - SOCMI Chemicals
compliance by July 24, 1995

Clean Water Act - Hazardous Substances
[present]

STATE LISTS

California - Directors List of Hazardous Substances (8 CCR 339)
[present]

Massachusetts Right To Know List
[present]

NJ Right to Know List (Total)
sn 1709

Pennsylvania Right to Know List
environmental hazard

SODIUM NITRITE   7632-00-0

HEALTH AND SAFETY LISTS

NIOSH - Selected LD50s and LC50s
Oral, mammal: LD50 = 5190 mg/kg

ENVIRONMENTAL LISTS

List of Pesticide Product Inert Ingredients
[present]

TSCA - Code of Federal Regulations Citations
40 CFR 712.30(m);   40 CFR 716.120(a)

TSCA - PAIR - Reporting List
Reporting Date: February 26, 1985

TSCA - Health and Safety Reporting List
Effective Date: December 28, 1984

SODIUM NITROFERRICYANIDE   14402-89-2

ENVIRONMENTAL LISTS

List of Pesticide Product Inert Ingredients
[present]

SODIUM N-NONYLDIPHENYL ETHER SULFONATE   RR-01143-8
SEE ALSO:
MOLYBDENUM

ENVIRONMENTAL LISTS

List of Pesticide Product Inert Ingredients
[present]

INTERNATIONAL LISTS

Canada - WHMIS: Ingredient Disclosure
1% item 1450 (1165)

SODIUM   60883-84-3
NONYLMETHYLNAPHTHALENESULFONATE

ENVIRONMENTAL LISTS

List of Pesticide Product Inert Ingredients
[present]

SODIUM 1-OCTANESULFONATE   5324-84-5

ENVIRONMENTAL LISTS

List of Pesticide Product Inert Ingredients
[present]

SODIUM OCTYL PHENOXY DIETHOXYETHYL   RR-01144-9
SULFATE

HEALTH AND SAFETY LISTS

U.S. DOT - Substances From 49 CFR 172.101
regulated by DOT (UN1825)

U.S. DOT - Hazard Classes
DOT hazard class = 8

INTERNATIONAL LISTS

Canada - WHMIS: Ingredient Disclosure
1% item 1451 (1176)

STATE LISTS

NJ Right to Know List (Total)
sn 1710

NJ Special Hazardous Substances
(corrosive)

SODIUM OCTYL SULFATE   142-31-4

HEALTH AND SAFETY LISTS

U.S. DOT - Substances From 49 CFR 172.101
regulated by DOT (UN1498, UN1499)

U.S. DOT - Hazard Classes
DOT hazard class = 5.1

NIOSH - Selected LD50s and LC50s
Oral, rat: LD50 = 3236 mg/kg

ENVIRONMENTAL LISTS

List of Pesticide Product Inert Ingredients
[present]

INTERNATIONAL LISTS

Canada - WHMIS: Ingredient Disclosure
1% item 1452 (1211)

STATE LISTS

Florida Hazardous Substance List
[present]

Massachusetts Right To Know List
[present]

NJ Right to Know List (Total)
sn 1711

Pennsylvania Right to Know List
[present]

SODIUM OLEATE   143-19-1

HEALTH AND SAFETY LISTS

U.S. DOT - Substances From 49 CFR 172.101
regulated by DOT (UN1500)

U.S. DOT - Hazard Classes
DOT hazard class = 5.1

U.S. DOT - Appendix A Table 1 - Hazardous Substances
final RQ = 100 pounds (45.4 kg)

NIOSH - Selected LD50s and LC50s
Oral, rat: LD50 = 85 mg/kg

NTP Chemical Status Reports - Testing Status and NTIS Number
Approved for toxicology/carcinogenesis study;   Prechronic studies for which toxicity technical reports were not prepared

ENVIRONMENTAL LISTS

CERCLA/SARA - Section 313 - Emission Reporting
form R reporting required

CERCLA/SARA - Hazardous Substances and their Reportable Quantities
final RQ = 100 pounds (45.4 kg)

Clean Water Act - Hazardous Substances
[present]

List of Pesticide Product Inert Ingredients
[present]

INTERNATIONAL LISTS

Canada - WHMIS: Ingredient Disclosure
1% item 1453 (1218)

STATE LISTS

California - Directors List of Hazardous Substances (8 CCR 339)
[present]

Massachusetts Right To Know List
[present]

NJ Right to Know List (Total)
sn 2258

Pennsylvania Right to Know List
environmental hazard

SODIUM ALPHA-OLEFINSULFONATE   68439-56-5

HEALTH AND SAFETY LISTS

NIOSH - Selected LD50s and LC50s
Oral, rat: LD50 = 99 mg/kg

SODIUM N-OLEYL TAURINE   29169-69-5

ENVIRONMENTAL LISTS

List of Pesticide Product Inert Ingredients
[present]

SODIUM OMADINE   15922-78-8

ENVIRONMENTAL LISTS

List of Pesticide Product Inert Ingredients
[present]

SODIUM PENTABORATE   12007-92-0

ENVIRONMENTAL LISTS

List of Pesticide Product Inert Ingredients
[present]

SODIUM PENTACHLOROPHENATE   131-52-2

ENVIRONMENTAL LISTS

List of Pesticide Product Inert Ingredients
[present]

SODIUM PENTADECYLSULFATE   13393-71-0

HEALTH AND SAFETY LISTS

NIOSH - Selected LD50s and LC50s
Oral, rat: LD50 = 3200 mg/kg

ENVIRONMENTAL LISTS

List of Pesticide Product Inert Ingredients
[present]

SODIUM PERBORATE   7632-04-4

ENVIRONMENTAL LISTS

List of Pesticide Product Inert Ingredients
[present]

SODIUM PERCARBONATE   3313-92-6

ENVIRONMENTAL LISTS

List of Pesticide Product Inert Ingredients
[present]

SODIUM PERCARBONATE   4452-58-8

ENVIRONMENTAL LISTS

List of Pesticide Product Inert Ingredients
[present]

SODIUM PERCARBONATE   15630-89-4

HEALTH AND SAFETY LISTS

NIOSH - Selected LD50s and LC50s
Oral, mouse: LD50 = 870 mg/kg

INTERNATIONAL LISTS

German (DFG) - Peak Limitations
2 x normal MAK (30 min., average value) don't exceed 4 times during shift

German (DFG) - Skin/Sensitizers
danger of cutaneous absorption

German (DFG) - Pregnancy
no risk to embryo/fetus if exposure limits adhered to

SODIUM PERCHLORATE   7601-89-0

ENVIRONMENTAL LISTS

List of Pesticide Product Inert Ingredients
[present]

SODIUM PERMANGANATE   10101-50-5

HEALTH AND SAFETY LISTS

U.S. DOT - Substances From 49 CFR 172.101
regulated by DOT (UN2567)

U.S. DOT - Hazard Classes
DOT hazard class = 6.1

U.S. DOT - Appendix B - Marine Pollutants
DOT regulated severe marine pollutant

NIOSH - Selected LD50s and LC50s
Inhalation, rat: LD50 = 11700 ug/kg (8 hr) Oral, rat: LD50 = 126 mg/kg Skin, mouse: LD50 = 124 mg/kg

ENVIRONMENTAL LISTS

CERCLA/SARA - Section 313 - Emission Reporting
form R reporting required

RCRA - Hazardous Constituents-Appendix VIII
hazardous constituent - no waste number

STATE LISTS

Massachusetts Right To Know List
extraordinarily hazardous

NJ Right to Know List (Total)
sn 1712

Pennsylvania Right to Know List
environmental hazard

PROPOSED REGULATIONS

CERCLA/SARA - Proposed Hazardous Substance Additions
proposed RQ = 1 pound (.454 kg)

CERCLA/SARA - 1989 Proposed RQ Adjustments
proposed RQ = 100 pounds (45.4 kg)

SODIUM PEROXIDE   1313-60-6

ENVIRONMENTAL LISTS

List of Pesticide Product Inert Ingredients
[present]

SODIUM PERSULFATE   7775-27-1

ENVIRONMENTAL LISTS

List of Pesticide Product Inert Ingredients
[present]

SODIUM PHENOLATE   139-02-6

STATE LISTS

NJ Right to Know List (Total)
sn 1715

SODIUM O-PHENYLPHENOL   132-27-4

HEALTH AND SAFETY LISTS

U.S. DOT - Substances From 49 CFR 172.101
regulated by DOT (UN2467)

U.S. DOT - Hazard Classes
DOT hazard class = 5.1

STATE LISTS

NJ Right to Know List (Total)
sn 1713

SODIUM PHOSPHATE   7632-05-5
STATE LISTS
  NJ Right to Know List (Total)
  sn 1714

SODIUM PHOSPHATE DIBASIC   7558-79-4
HEALTH AND SAFETY LISTS
  U.S. DOT - Substances From 49 CFR 172.101
  regulated by DOT (UN1502)
  U.S. DOT - Hazard Classes
  DOT hazard class = 5.1
  NIOSH - Selected LD50s and LC50s
  Oral, rat: LD50 = 2100 mg/kg
STATE LISTS
  Florida Hazardous Substance List
  [present]
  Massachusetts Right To Know List
  [present]
  NJ Right to Know List (Total)
  sn 1716
  NJ Special Hazardous Substances
  (reactive - second degree)
  Pennsylvania Right to Know List
  [present]

SODIUM PHOSPHATE, TRIBASIC   7765-84-6
SEE ALSO:
  MANGANESE
HEALTH AND SAFETY LISTS
  U.S. DOT - Substances From 49 CFR 172.101
  regulated by DOT (UN1503)
  U.S. DOT - Hazard Classes
  DOT hazard class = 5.1
ENVIRONMENTAL LISTS
  List of Pesticide Product Inert Ingredients
  [present]
STATE LISTS
  NJ Right to Know List (Total)
  sn 1717

SODIUM PHOSPHATE TRIBISINE   14986-84-6
HEALTH AND SAFETY LISTS
  U.S. DOT - Substances From 49 CFR 172.101
  regulated by DOT (UN1504)
  U.S. DOT - Hazard Classes
  DOT hazard class = 5.1
STATE LISTS
  Florida Hazardous Substance List
  [present]
  Massachusetts Right To Know List
  [present]
  NJ Right to Know List (Total)
  sn 1718
  NJ Special Hazardous Substances
  (reactive - second degree)
  Pennsylvania Right to Know List
  [present]

SODIUM PHOSPHIDE   12058-85-4
HEALTH AND SAFETY LISTS
  U.S. DOT - Substances From 49 CFR 172.101
  regulated by DOT (UN1505)

U.S. DOT - Hazard Classes
DOT hazard class = 5.1
INTERNATIONAL LISTS
  Canada - Ontario - OHSA - TWAEVs
  as S2O8: 5 mg/m3 TWAEV (listed under 'Persulfates')
  United Kingdom - Occupational Exposure Standards - TWAs
  measured as S2O8: 1 mg/m3 TWA
STATE LISTS
  NJ Right to Know List (Total)
  sn 1721
PROPOSED REGULATIONS
  Canada - Ontario - Proposed Occupational TWAEVs
  2 mg/m3 TWAEV

SODIUM PICRAMATE   831-52-7
HEALTH AND SAFETY LISTS
  U.S. DOT - Substances From 49 CFR 172.101
  regulated by DOT (UN2497)
  U.S. DOT - Hazard Classes
  DOT hazard class = 8
ENVIRONMENTAL LISTS
  CAA - HON Rule - SOCMI Chemicals
  compliance by April 24, 1995
INTERNATIONAL LISTS
  Canada - WHMIS: Ingredient Disclosure
  1% item 1454 (1375)
STATE LISTS
  NJ Right to Know List (Total)
  sn 1722
  NJ Special Hazardous Substances
  (corrosive)

SODIUM PICRYL PEROXIDE   RR-01458-4
HEALTH AND SAFETY LISTS
  IARC - Group 2B (sufficient animal data)
  [present] (Degree of evidence in animals revised on the basis of data that appeared after the most recent monograph and/or on the basis of present criteria)
  NIOSH - Selected LD50s and LC50s
  Oral, rat: LD50 = 656 mg/kg
  OSHA - Possible Select Carcinogens
  [present]
ENVIRONMENTAL LISTS
  CERCLA/SARA - Section 313 - Emission Reporting
  form R reporting required
  List of Pesticide Product Inert Ingredients
  [present]
STATE LISTS
  California - Air Bill 2588 Appendix A-II
  known or potential carcinogen
  California - Prop. 65 - Cancer list
  carcinogen - initial date 1/1/90
  California - Prop. 65 - No Significant Risk Levels
  no significant risk level = 200 ug/day
  California - Directors List of Hazardous Substances (8 CCR 339)
  [present]
  Minnesota Hazardous Substance List
  carcinogen

SODIUM POLYACRYLATE   9003-04-7

ENVIRONMENTAL LISTS

   List of Pesticide Product Inert Ingredients
   [present]

SODIUM POLYMETHACRYLATE   67785-61-9

HEALTH AND SAFETY LISTS

   U.S. DOT - Appendix A Table 1 - Hazardous Substances
   final RQ = 5000 pounds (2270 kg)
   NIOSH - Selected LD50s and LC50s
   Oral, rat: LD50 = 17 gm/kg

ENVIRONMENTAL LISTS

   CERCLA/SARA - Hazardous Substances and their Reportable Quantities
   final RQ = 5000 pounds (2270 kg)
   Clean Water Act - Hazardous Substances
   [present]
   List of Pesticide Product Inert Ingredients
   [present]

STATE LISTS

   Massachusetts Right To Know List
   [present]
   NJ Right to Know List (Total)
   sn 1723
   Pennsylvania Right to Know List
   environmental hazard

SODIUM POLYSTYRENE SULFONATE   25704-18-1

STATE LISTS

   NJ Right to Know List (Total)
   sn 3043

SODIUM POTASSIUM ALUMINUM SILICATE   12736-96-8

HEALTH AND SAFETY LISTS

   NIOSH - Selected LD50s and LC50s
   Oral, mouse: LD50 = 3920 mg/kg

SODIUM PYRIDITHIONE   3811-73-2

HEALTH AND SAFETY LISTS

   U.S. DOT - Substances From 49 CFR 172.101
   regulated by DOT (UN1432)
   U.S. DOT - Hazard Classes
   DOT hazard class = 4.3

STATE LISTS

   NJ Right to Know List (Total)
   sn 1725

SODIUM RICINOLEATE   5323-95-5

HEALTH AND SAFETY LISTS

   U.S. DOT - Substances From 49 CFR 172.101
   regulated by DOT (UN0235, UN1349)
   U.S. DOT - Hazard Classes
   with less than 20% water: DOT hazard class = 1.3C; with 20% or more water: DOT hazard class = 4.1

STATE LISTS

   NJ Right to Know List (Total)
   sn 2780

SODIUM SACCHARIN   128-44-9

HEALTH AND SAFETY LISTS

   U.S. DOT - Hazard Classes
   Forbidden from transport by the DOT

SODIUM SALICYLATE   54-21-7

ENVIRONMENTAL LISTS

   List of Pesticide Product Inert Ingredients
   [present]

SODIUM SALT OF AN ALKYLATED, SULFONATED AROMATIC   RR-00213-1

ENVIRONMENTAL LISTS

   List of Pesticide Product Inert Ingredients
   [present]

SODIUM SALT OF N-COCO BETA AMINO BUTYRIC ACID   12788-84-0

HEALTH AND SAFETY LISTS

   NIOSH - Selected LD50s and LC50s
   Oral, rat; LD50 = 16 gm/kg

SODIUM SALT OF CRESOL SULFONIC ACID CONDENSED WITH UREA FORMALDEHYDE   68441-84-9

ENVIRONMENTAL LISTS

   List of Pesticide Product Inert Ingredients
   [present]

SODIUM SALT OF OLEIC ACID AMIDE OF PROTEIN HYDROLYSATE   RR-01145-0

HEALTH AND SAFETY LISTS

   NIOSH - Selected LD50s and LC50s
   Oral, mouse: LD50 = 870 mg/kg

INTERNATIONAL LISTS

   German (DFG) - MAK Values
   1 mg/m3 MAK
   German (DFG) - Peak Limitations
   2 x normal MAK (30 min., average value) don't exceed 4 times during shift
   German (DFG) - Skin/Sensitizers
   danger of cutaneous absorption
   German (DFG) - Pregnancy
   no risk to embryo/fetus if exposure limits adhered to

SODIUM SALTS OF NITRO-AROMATIC DERIVATIVES, N.O.S.   RR-00491-1

ENVIRONMENTAL LISTS

   List of Pesticide Product Inert Ingredients
   [present]

INTERNATIONAL LISTS

   Canada - WHMIS: Ingredient Disclosure
   1% item 1455 (1471)

SODIUM SALTS OF ALPHA-SULFO-OMEGA-HYDROXYPOLY(OXY-1,2-ETHANEDIYL) C11-14 ISOALKYL ETHERS, (C13-RICH)   78330-30-0

HEALTH AND SAFETY LISTS

   NIOSH - Selected LD50s and LC50s
   Oral, rat: LD50 = 14200 mg/kg

ENVIRONMENTAL LISTS

   List of Pesticide Product Inert Ingredients
   [present]

STATE LISTS

   California - Air Bill 2588 Appendix A-II
   known or potential carcinogen: 9/89
   California - Prop. 65 - Cancer list
   carcinogen - initial date 1/1/88
   Florida Hazardous Substance List
   [present]

STATE LISTS

Florida Hazardous Substance List
effective March 13, 1992

Massachusetts Right To Know List
extraordinarily hazardous

Pennsylvania Right to Know List
environmental hazard

PROPOSED REGULATIONS

CERCLA/SARA - Proposed Hazardous Substance Additions
proposed RQ = 1 pound (.454 kg)

CERCLA/SARA - 1989 Proposed RQ Adjustments
proposed RQ = 100 pounds (45.4 kg)

SODIUM-2,4,5-TRICHLOROPHENATE   136-32-3

ENVIRONMENTAL LISTS

List of Pesticide Product Inert Ingredients
[present]

SODIUM TRIDECYLPOLY(OXYETHYLENE)   54116-08-4
SULFATE

HEALTH AND SAFETY LISTS

U.S. DOT - Hazard Classes
Forbidden from transport by the DOT

SODIUM TRIDECYL SULFATE   3026-63-9

HEALTH AND SAFETY LISTS

NIOSH - Selected LD50s and LC50s
Oral, rat: LD50 = 764 mg/kg

SODIUM   1323-19-9
TRIISOPROPYLNAPHTHALENESULFONATE

ENVIRONMENTAL LISTS

List of Pesticide Product Inert Ingredients
[present]

SODIUM TUNGSTATE   13472-45-2

ENVIRONMENTAL LISTS

List of Pesticide Product Inert Ingredients
[present]

SODIUM VINYLBENZENESULFONATE, POLYMER   63182-08-1
WITH DIVINYLBENZENE

ENVIRONMENTAL LISTS

List of Pesticide Product Inert Ingredients
[present]

SODIUM XYLENE SULFONATE   1300-72-7

ENVIRONMENTAL LISTS

List of Pesticide Product Inert Ingredients
[present]

SOLASODINE   126-17-0

HEALTH AND SAFETY LISTS

NIOSH - Selected LD50s and LC50s
Oral, rat: LD50 = 1620 mg/kg

SOLVENT NAPHTHA (PETROLEUM), HEAVY   64742-96-7
ALIPHATIC

ENVIRONMENTAL LISTS

List of Pesticide Product Inert Ingredients
[present]

SOLVENT NAPHTHA (PETROLEUM), LIGHT   64742-89-8
ALIPHATIC

ENVIRONMENTAL LISTS

List of Pesticide Product Inert Ingredients
[present]

SOLVENT NAPHTHA (PETROLEUM), MEDIUM   64742-88-7
ALIPHATIC

ENVIRONMENTAL LISTS

List of Pesticide Product Inert Ingredients
[present]

SOLVENT YELLOW #72   4645-07-2

HEALTH AND SAFETY LISTS

NIOSH - Selected LD50s and LC50s
Oral, rat: LD50 = 1190 mg/kg

INTERNATIONAL LISTS

Canada - WHMIS: Ingredient Disclosure
1% item 1459 (1701)

SOOTS   RR-00028-2

ENVIRONMENTAL LISTS

List of Pesticide Product Inert Ingredients
[present]

SOOTS, TARS, AND MINERAL OILS   RR-00795-4

HEALTH AND SAFETY LISTS

NTP Chemical Status Reports - Testing Status and NTIS Number
*Two year studies: pathology quality assessment in progress;
prechronic studies for which toxicity technical reports were not prepared*

ENVIRONMENTAL LISTS

List of Pesticide Product Inert Ingredients
[present]

SORBALDEHYDE   142-83-6

INTERNATIONAL LISTS

Canada - WHMIS: Ingredient Disclosure
0.1% item 1460 (1497)

SORBIC ACID   110-44-1

ENVIRONMENTAL LISTS

List of Pesticide Product Inert Ingredients
[present]

SORBIC ACID, POTASSIUM SALT   590-00-1

ENVIRONMENTAL LISTS

List of Pesticide Product Inert Ingredients
[present]

STATE LISTS

NJ Right to Know List (Total)
sn 2575

SORBITAN DIOLEATE   29116-98-1

ENVIRONMENTAL LISTS

List of Pesticide Product Inert Ingredients
[present]

SORBITAN MONOHEXADECANOATE   26266-57-9

ENVIRONMENTAL LISTS

List of Pesticide Product Inert Ingredients
[present]

SORBITAN MONOLAURATE   1338-39-2

HEALTH AND SAFETY LISTS

IARC - Group 1 (carcinogenic to humans)
[present]

Massachusetts Right To Know List
carcinogen; extraordinarily hazardous

Pennsylvania Right to Know List
special hazardous substance

Pennsylvania RTK - Special Hazardous Substances
[present]

SODIUM SELENATE   13410-01-0

HEALTH AND SAFETY LISTS

NIOSH - Selected LD50s and LC50s
Oral, rat: LD50 = 1200 mg/kg

ENVIRONMENTAL LISTS

List of Pesticide Product Inert Ingredients
[present]

SODIUM SELENITE   7782-82-3

ENVIRONMENTAL LISTS

TSCA - Chemicals with Significant New Use Rules
PMN number: P-84-591

SODIUM SELENITE   10102-18-8

ENVIRONMENTAL LISTS

List of Pesticide Product Inert Ingredients
[present]

SODIUM SESQUICARBONATE   533-96-0

ENVIRONMENTAL LISTS

List of Pesticide Product Inert Ingredients
[present]

SODIUM SILICATE   1344-09-8

ENVIRONMENTAL LISTS

List of Pesticide Product Inert Ingredients
[present]

SODIUM SILICOFLUORIDE   16893-85-9

HEALTH AND SAFETY LISTS

U.S. DOT - Substances From 49 CFR 172.101
regulated by DOT (UN0203)

U.S. DOT - Hazard Classes
DOT hazard class = 1.3C

STATE LISTS

NJ Right to Know List (Total)
sn 2782

SODIUM STEARATE   822-16-2

ENVIRONMENTAL LISTS

List of Pesticide Product Inert Ingredients
[present]

SODIUM SULFAMATE   13845-18-6
SEE ALSO:
SELENIUM

HEALTH AND SAFETY LISTS

NIOSH - Selected LD50s and LC50s
Oral, rat: LD50 = 1600 ug/kg

NTP Chemical Status Reports - Testing Status and NTIS Number
Post peer review technical reports in progress

ENVIRONMENTAL LISTS

CERCLA/SARA - Section 302 Extremely Hazardous Substances and TPQs
TPQ = 100/10,000 pounds

INTERNATIONAL LISTS

Canada - WHMIS: Ingredient Disclosure
0.1% item 1456 (1477)

STATE LISTS

Florida Hazardous Substance List
effective March 13, 1992

Massachusetts Right To Know List
extraordinarily hazardous

NJ Right to Know List (Total)
sn 1726

Pennsylvania Right to Know List
environmental hazard

PROPOSED REGULATIONS

CERCLA/SARA - Proposed Hazardous Substance Additions
proposed RQ = 1 pound (.454 kg)

CERCLA/SARA - 1989 Proposed RQ Adjustments
proposed RQ = 100 pounds (45.4 kg)

SODIUM SULFATE (SOLUTION)   7757-82-6
SEE ALSO:
SELENIUM

HEALTH AND SAFETY LISTS

U.S. DOT - Appendix A Table 1 - Hazardous Substances
final RQ = 100 pounds (45.4 kg)

ENVIRONMENTAL LISTS

CERCLA/SARA - Hazardous Substances and their Reportable Quantities
final RQ = 100 pounds (45.4 kg)

Clean Water Act - Hazardous Substances
[present]

STATE LISTS

Massachusetts Right To Know List
[present]

NJ Right to Know List (Total)
sn 3047

Pennsylvania Right to Know List
environmental hazard

SODIUM SULFIDE   1313-82-2
SEE ALSO:
SELENIUM

HEALTH AND SAFETY LISTS

U.S. DOT - Substances From 49 CFR 172.101
regulated by DOT (NA2630)

U.S. DOT - Hazard Classes
DOT hazard class = 6.1

U.S. DOT - Appendix A Table 1 - Hazardous Substances
final RQ = 100 pounds (45.4 kg)

NIOSH - Selected LD50s and LC50s
Oral, rat: LD50 = 7 mg/kg

NTP Chemical Status Reports - Testing Status and NTIS Number
Technical reports printed

ENVIRONMENTAL LISTS

CERCLA/SARA - Section 302 Extremely Hazardous Substances and TPQs
TPQ = 100/10,000 pounds

CERCLA/SARA - Hazardous Substances and their Reportable Quantities
final RQ = 100 pounds (45.4 kg)

Clean Water Act - Hazardous Substances
[present] (Listed under 'Sodium selenite')

List of Pesticide Product Inert Ingredients
[present]

STATE LISTS

Florida Hazardous Substance List
effective March 13, 1992

Massachusetts Right To Know List
[present]

NJ Right to Know List (Total)
sn 1727

Pennsylvania Right to Know List
environmental hazard

SODIUM SULFIDE 1344-08-7

ENVIRONMENTAL LISTS

List of Pesticide Product Inert Ingredients
[present]

SODIUM SULFITE 7757-83-7

ENVIRONMENTAL LISTS

List of Pesticide Product Inert Ingredients
[present]

SODIUM SULFOSUCCINATE 20526-58-3

HEALTH AND SAFETY LISTS

U.S. DOT - Substances From 49 CFR 172.101
regulated by DOT (UN2674)

U.S. DOT - Hazard Classes
DOT hazard class = 6.1

NIOSH - Selected LD50s and LC50s
Oral, rat: LD50 = 125 mg/kg

ENVIRONMENTAL LISTS

List of Pesticide Product Inert Ingredients
[present]

STATE LISTS

Massachusetts Right To Know List
[present]

NJ Right to Know List (Total)
sn 1701

SODIUM SUPEROXIDE 12034-12-7

ENVIRONMENTAL LISTS

List of Pesticide Product Inert Ingredients
[present]

STATE LISTS

California - Exposure Limits - PELs
10 mg/m3 PEL

SODIUM N-(TALL-OIL ALKYL)-N-METHYLTAURINE 61791-41-1

ENVIRONMENTAL LISTS

List of Pesticide Product Inert Ingredients
[present]

SODIUM TARTRATE 868-18-8

HEALTH AND SAFETY LISTS

NIOSH - Selected LD50s and LC50s
Oral, mouse: LD50 = 5989 mg/kg

ENVIRONMENTAL LISTS

List of Pesticide Product Inert Ingredients
[present]

STATE LISTS

Massachusetts Right To Know List
[present]

Pennsylvania Right to Know List
environmental hazard

SODIUM TELLURITE 10102-20-2

HEALTH AND SAFETY LISTS

U.S. DOT - Substances From 49 CFR 172.101
regulated by DOT (UN1849, UN1385)

U.S. DOT - Hazard Classes
DOT hazard class = 8

STATE LISTS

Massachusetts Right To Know List
[present]

NJ Right to Know List (Total)
sn 1728

NJ Special Hazardous Substances
(corrosive)

SODIUM TETRADECYL SULFATE 1191-50-0

ENVIRONMENTAL LISTS

List of Pesticide Product Inert Ingredients
[present]

SODIUM TETRANITRATE RR-01459-5

ENVIRONMENTAL LISTS

List of Pesticide Product Inert Ingredients
[present]

SODIUM THIOCYANATE 540-72-7

ENVIRONMENTAL LISTS

List of Pesticide Product Inert Ingredients
[present]

SODIUM THIOSULFATE 7772-98-7

HEALTH AND SAFETY LISTS

U.S. DOT - Substances From 49 CFR 172.101
regulated by DOT (UN2547)

U.S. DOT - Hazard Classes
DOT hazard class = 5.1

STATE LISTS

NJ Right to Know List (Total)
sn 1729

SODIUM TOLUENESULFONATE 12068-03-0

ENVIRONMENTAL LISTS

List of Pesticide Product Inert Ingredients
[present]

SODIUM TRIBUTYLNAPHTHALENESULFONATE 64665-10-7

ENVIRONMENTAL LISTS

List of Pesticide Product Inert Ingredients
[present]

SODIUM TRICHLOROACETATE 650-51-1
SEE ALSO:
TELLURIUM

HEALTH AND SAFETY LISTS

NIOSH - Selected LD50s and LC50s
Oral, rat: LD50 = 83 mg/kg

ENVIRONMENTAL LISTS

CERCLA/SARA - Section 302 Extremely Hazardous Substances and TPQs
TPQ = 500/10,000 pounds

INTERNATIONAL LISTS

Canada - WHMIS: Ingredient Disclosure
1% item 1457 (1558)

OSHA - Select Carcinogens
[present]

STATE LISTS

California - Air Bill 2588 Appendix A-II
known or potential carcinogen

California - Prop. 65 - Cancer list
carcinogen - initial date 2/27/87

Minnesota Hazardous Substance List
carcinogen

Pennsylvania Right to Know List
special hazardous substance

Pennsylvania RTK - Special Hazardous Substances
[present]

SORBITAN MONOOCTADECANOATE 1338-41-6

STATE LISTS

California - Prop. 65 - Cancer list
carcinogen - initial date 2/27/87

California - Directors List of Hazardous Substances (8 CCR 339)
[present]

Pennsylvania Right to Know List
special hazardous substance

Pennsylvania RTK - Special Hazardous Substances
[present]

SORBITAN MONOOLEATE 1338-43-8

HEALTH AND SAFETY LISTS

NTP Chemical Status Reports - Testing Status and NTIS Number
Project leader assigned/study in design

INTERNATIONAL LISTS

Canada - WHMIS: Ingredient Disclosure
1% item 1461 (1499)

STATE LISTS

Florida Hazardous Substance List
[present]

Massachusetts Right To Know List
[present]

SORBITAN, 9-OCTADECENOATE 8007-43-0

HEALTH AND SAFETY LISTS

NIOSH - Selected LD50s and LC50s
Oral, rat: LD50 = 7360 mg/kg

ENVIRONMENTAL LISTS

List of Pesticide Product Inert Ingredients
[present]

INTERNATIONAL LISTS

Canada - WHMIS: Ingredient Disclosure
1% item 1462 (132)

SORBITAN, TALL-OIL FATTY ACID TRIESTERS, ETHOXYLATED 61790-88-3

HEALTH AND SAFETY LISTS

NIOSH - Selected LD50s and LC50s
Oral, rat: LD50 = 4920 mg/kg

SORBITAN, TRIOCTADECANOATE 26658-19-5

ENVIRONMENTAL LISTS

List of Pesticide Product Inert Ingredients
[present]

SORBITOL, TALL-OIL FATTY ACID SESQUIESTERS, ETHOXYLATED 68648-20-4

ENVIRONMENTAL LISTS

List of Pesticide Product Inert Ingredients
[present]

N-(SOYA ALKYL)-N-ETHYLMORPHOLINIUM ETHYLSULFATE 61791-34-2

ENVIRONMENTAL LISTS

List of Pesticide Product Inert Ingredients
[present]

N-(SOYA ALKYL)-N-METHYLMORPHOLINIUM SULFATE RR-01146-1

HEALTH AND SAFETY LISTS

NIOSH - Selected LD50s and LC50s
Oral, rat: LD50 = 31 gm/kg

ENVIRONMENTAL LISTS

List of Pesticide Product Inert Ingredients
[present]

SOYA ALKYL RESIN 63148-69-6

ENVIRONMENTAL LISTS

List of Pesticide Product Inert Ingredients
[present]

SOYA FATTY ACIDS 68308-53-2

ENVIRONMENTAL LISTS

List of Pesticide Product Inert Ingredients
[present]

SOYBEAN HULLS RR-01147-2

ENVIRONMENTAL LISTS

List of Pesticide Product Inert Ingredients
[present]

SOYBEAN LECITHIN 8002-43-5

ENVIRONMENTAL LISTS

List of Pesticide Product Inert Ingredients
[present]

SOYBEAN MEAL 68308-36-1

ENVIRONMENTAL LISTS

List of Pesticide Product Inert Ingredients
[present]

SOYBEAN OIL 8001-22-7

ENVIRONMENTAL LISTS

List of Pesticide Product Inert Ingredients
[present]

SOYBEAN OIL POLYMERIZED 68122-64-5

ENVIRONMENTAL LISTS

List of Pesticide Product Inert Ingredients
[present]

SOYBEAN OIL, POLYMER WITH PENTAERYTHRI-TOL, TOLUENEDIISOCYANATE AND TUNG OIL 67989-28-0

ENVIRONMENTAL LISTS

List of Pesticide Product Inert Ingredients
[present]

SOY FLOUR 68513-95-1

ENVIRONMENTAL LISTS

List of Pesticide Product Inert Ingredients
[present]

SPEARMINT OIL   8008-79-5

ENVIRONMENTAL LISTS

List of Pesticide Product Inert Ingredients
[present]

SPERM OIL   8002-24-2

ENVIRONMENTAL LISTS

List of Pesticide Product Inert Ingredients
[present]

SPIRAMYCIN   8025-81-8

ENVIRONMENTAL LISTS

List of Pesticide Product Inert Ingredients
[present]

SPIRONOLACTONE   52-01-7

HEALTH AND SAFETY LISTS

NFPA - Flash Points
flash point = 540 degrees F (282 degrees C)
NFPA - Hazard Identification Ratings
health-0; flammability-1; reactivity-0

ENVIRONMENTAL LISTS

List of Pesticide Product Inert Ingredients
[present]

STATE LISTS

Pennsylvania Right to Know List
[present]

STANNANE, ACETOXYTRIPHENYL-   900-95-8

ENVIRONMENTAL LISTS

List of Pesticide Product Inert Ingredients
[present]

STANNANE, FLUOROTRIBUTYL-   1983-10-4

ENVIRONMENTAL LISTS

List of Pesticide Product Inert Ingredients
[present]

STANNANE, TETRAMETHYL-   594-27-4

ENVIRONMENTAL LISTS

List of Pesticide Product Inert Ingredients
[present]

STANNANE, TETRAPHENYL-   595-90-4

HEALTH AND SAFETY LISTS

NIOSH - Selected LD50s and LC50s
Oral, rat: LD50 = 5 gm/kg

STANNANE, TRI-N-BUTYL-, HYDRIDE   688-73-3

HEALTH AND SAFETY LISTS

NFPA - Flash Points
*no. 1: flash point = 428 degrees F (220 degrees C); no. 2: flash
point = 460 degrees F (238 degrees C)*
NFPA - Hazard Identification Ratings
health-0; flammability-1; reactivity-0

STATE LISTS

Pennsylvania Right to Know List
[present]

STANNIC CHLORIDE   7646-78-8

HEALTH AND SAFETY LISTS

NIOSH - Selected LD50s and LC50s
Oral, rat: LD50 = 3550 mg/kg

STANNIC CHLORIDE, HYDRATED   10026-06-9

HEALTH AND SAFETY LISTS

IARC - Group 3 (not classifiable)
[present]

STATE LISTS

California - Directors List of Hazardous Substances (8 CCR 339)
[present]

STANNIC OXIDE   18282-10-5
SEE ALSO:
TIN

HEALTH AND SAFETY LISTS

U.S. DOT - Appendix B - Marine Pollutants
DOT regulated severe marine pollutant

ENVIRONMENTAL LISTS

CERCLA/SARA - Section 302 Extremely Hazardous Substances and
TPQs
TPQ = 500/10,000 pounds

INTERNATIONAL LISTS

Canada - WHMIS: Ingredient Disclosure
0.1% item 1655 (34)

STATE LISTS

Florida Hazardous Substance List
effective March 13, 1992
Massachusetts Right To Know List
extraordinarily hazardous
Pennsylvania Right to Know List
environmental hazard

PROPOSED REGULATIONS

CERCLA/SARA - Proposed Hazardous Substance Additions
proposed RQ = 1 pound (.454 kg)
CERCLA/SARA - 1989 Proposed RQ Adjustments
proposed RQ = 10 pounds (4.54 kg)

STANNIC PHOSPHIDE   12440-42-5
SEE ALSO:
TIN

ENVIRONMENTAL LISTS

CERCLA/SARA - Section 313 - Emission Reporting
form R reporting required

INTERNATIONAL LISTS

German (DFG) - MAK Values
*as TBTO: 0.002 ppm MAK; 0.05 mg/m3 MAK (Listed under 'Tri-n-
butyltin compounds')*
German (DFG) - Peak Limitations
*5 x normal MAK (30 min. average value); don't exceed 2 times during
shift (Listed under 'Tri-n-butyltin compounds')*
German (DFG) - Pregnancy
*no risk to embryo/fetus if exposure limits adhered to (Listed under 'Tri-
n-butyltin compounds')*

STANNOUS CHLORIDE   7772-99-8

HEALTH AND SAFETY LISTS

U.S. DOT - Substances Which Are Poisonous by Inhalation
liquid hazardous material poisonous by inhalation
NFPA - Flash Points
flash point < 70 degrees F (21 degrees C)
NFPA - Hazard Identification Ratings
health-2; reactivity-0

STATE LISTS

Florida Hazardous Substance List
[present]

Massachusetts Right To Know List
[present]

Pennsylvania Right to Know List
[present]

STANNOUS CHLORIDE DIHYDRATE   10025-69-1

HEALTH AND SAFETY LISTS

NFPA - Flash Points
flash point = 450 degrees F (232 degrees C)

NFPA - Hazard Identification Ratings
health-3; flammability-1; reactivity-0

STATE LISTS

Florida Hazardous Substance List
[present]

Massachusetts Right To Know List
[present]

Pennsylvania Right to Know List
[present]

STANNOUS FLUORIDE   7783-47-3
SEE ALSO:
TIN

STATE LISTS

California - Directors List of Hazardous Substances (8 CCR 339)
[present] (refers to coating containing tributyltin)

STANNOUS OXALATE   814-94-8
SEE ALSO:
TIN

HEALTH AND SAFETY LISTS

U.S. DOT - Substances From 49 CFR 172.101
regulated by DOT (UN2440, UN1827)

U.S. DOT - Hazard Classes
DOT hazard class = 8

NIOSH - Selected LD50s and LC50s
Inhalation, rat: LC50 = 2300 mg/m3 (10 mn)

STATE LISTS

Florida Hazardous Substance List
[present]

Massachusetts Right To Know List
[present]

NJ Right to Know List (Total)
sn 1859

NJ Special Hazardous Substances
(corrosive)

Pennsylvania Right to Know List
[present]

STARCH   9005-25-8
SEE ALSO:
TIN

STATE LISTS

NJ Right to Know List (Total)
sn 1731

NJ Special Hazardous Substances
(corrosive)

STARCH, HYDROGEN PHOSPHATE, 2-   53124-00-8
HYDROXYPROPYLETHER
SEE ALSO:
TIN

INTERNATIONAL LISTS

Canada - Alberta - 8 Hour Occupational Exposure Limit
2 mg/m3 TWA

Canada - Alberta - 15 Minute Occupational Exposure Limit
4 mg/m3 STEL

Mexico - Instruction No. 10 - TWAs
2 mg/m3 TWA

Mexico - Instruction No. 10 - STELs
4 mg/m3 STEL

STATE LISTS

Minnesota Hazardous Substance List
[present] as Sn (includes inorganic compounds except SnH4)

STARCH HYDROXYPROPYL ETHER   9049-76-7

HEALTH AND SAFETY LISTS

U.S. DOT - Substances From 49 CFR 172.101
regulated by DOT (UN1433)

U.S. DOT - Hazard Classes
DOT hazard class = 4.3

STARLICIDE   7745-89-3
SEE ALSO:
TIN

HEALTH AND SAFETY LISTS

NIOSH - Selected LD50s and LC50s
Oral, rat: LD50 = 700 mg/kg

NTP Chemical Status Reports - Testing Status and NTIS Number
Technical reports printed (PB82242553)

NTP Chemical Status Reports - Evidence of Carcinogenicity
male rat-equivocal; female rat-negative; male mice-negative; female mice-negative

INTERNATIONAL LISTS

Canada - WHMIS: Ingredient Disclosure
1% item 1569 (496)

STATE LISTS

NJ Right to Know List (Total)
sn 1733

STEARAMINOPROPYLDIMETHYL-.BETA.-HYDROX-   3758-54-1
YETHYLAMMONIUM PHOSPATE
SEE ALSO:
TIN

INTERNATIONAL LISTS

Canada - WHMIS: Ingredient Disclosure
1% item 1463 (545)

STEARATES   RR-00075-9
SEE ALSO:
TIN

HEALTH AND SAFETY LISTS

NIOSH - Selected LD50s and LC50s
Oral, rat: LD50 = 377 mg/kg

INTERNATIONAL LISTS

Canada - WHMIS: Ingredient Disclosure
1% item 1572 (905)

STEARIC ACID   57-11-4
SEE ALSO:
TIN

HEALTH AND SAFETY LISTS

NIOSH - Selected LD50s and LC50s
Oral, rat: LD50 = 3620 mg/kg

STEARIC ACID, LEAD (2+) SALT   1072-35-1

HEALTH AND SAFETY LISTS

ACGIH 1995 - Time Weighted Averages
10 mg/m3 TWA

© Van Nostrand Reinhold 1995

OSHA - Vacated PELs - Time Weighted Averages
total dust: 15 mg/m3 TWA; respirable fraction: 5 mg/m3 TWA
OSHA - Final PELs - Time Weighted Averages
total dust: 15 mg/m3 TWA; respirable fraction: 5 mg/m3 TWA
ENVIRONMENTAL LISTS
List of Pesticide Product Inert Ingredients
[present]
INTERNATIONAL LISTS
Australian Exposure Standards - Time Weighted Averages
10 mg/m3 TWA
Canada - Alberta - 8 Hour Occupational Exposure Limit
respirable mass: 5 mg/m3 TWA; total mass: 10 mg/m3 TWA
Canada - British Columbia - 8 Hour Exposure Limits
nuisance dust: 10 mg/m3 TWA
Canada - British Columbia - 15 Minute Exposure Limits
20 mg/m3 STEL
Canada - Ontario - OHSA - TWAEVs
total dust: 10 mg/m3 TWAEV (listed as a nuisance particulate)
Canada - Quebec - Time-Weighted Average Exposure Values
10 mg/m3 TWAEV
Israel - Time Weighted Averages
10 mg/m3 TWA
Israel - Action Levels
5 mg/m3 AL
STATE LISTS
Minnesota Hazardous Substance List
[present] (includes inert or nuisance dust)
Pennsylvania Right to Know List
[present]

STEARYL OCTANOATE   18312-31-7
ENVIRONMENTAL LISTS
List of Pesticide Product Inert Ingredients
[present]

STERIGMATOCYSTIN   10048-13-2
HEALTH AND SAFETY LISTS
NIOSH - Selected LD50s and LC50s
Oral, rat: LD50 = 218 mg/kg

STIBAMINE   6543-62-0
STATE LISTS
California - Directors List of Hazardous Substances (8 CCR 339)
[present]
Massachusetts Right To Know List
[present]

STIBANILIC ACID   554-76-7
HEALTH AND SAFETY LISTS
NIOSH - Selected LD50s and LC50s
Oral, rat: LD50 = 2835 mg/kg
ENVIRONMENTAL LISTS
List of Pesticide Product Inert Ingredients
[present]

STIBINE   7803-52-3
HEALTH AND SAFETY LISTS
ACGIH 1995 - Time Weighted Averages
10 mg/m3 TWA (does not include stearates of toxic metals)
INTERNATIONAL LISTS
Australian Exposure Standards - Time Weighted Averages
10 mg/m3 TWA

Canada - Ontario - OHSA - TWAEVs
total dust: 10 mg/m3 TWAEV (listed as a nuisance particulate)
Israel - Time Weighted Averages
10 mg/m3 TWA (does not include stearates of toxic metals)
Israel - Action Levels
5 mg/m3 AL (does not include stearates of toxic metals)
STATE LISTS
Minnesota Hazardous Substance List
[present] (includes inert or nuisance dust)

TRANS-STILBENE   103-30-0
HEALTH AND SAFETY LISTS
NFPA - Flash Points
flash point = 385 degrees F (196 degrees C)
NFPA - Hazard Identification Ratings
health-1; flammability-1; reactivity-0
ENVIRONMENTAL LISTS
List of Pesticide Product Inert Ingredients
[present]
INTERNATIONAL LISTS
Canada - WHMIS: Ingredient Disclosure
1% item 1464 (133)

STILBENE   588-59-0
SEE ALSO:
LEAD
LEAD STEARATE
HEALTH AND SAFETY LISTS
U.S. DOT - Appendix A Table 1 - Hazardous Substances
final RQ = 5000 pounds (2270 kg) (the RQ is subject to change when the assessment of potential carcinogenicity is completed) (Listed under 'Lead stearate')
ENVIRONMENTAL LISTS
CERCLA/SARA - Hazardous Substances and their Reportable Quantities
final RQ = 5000 pounds (2270 kg) (the RQ is subject to change when the assessment of potential carcinogenicity is completed) (Listed under 'Lead stearate')
Clean Water Act - Hazardous Substances
[present] (Listed under 'Lead stearate')
STATE LISTS
Massachusetts Right To Know List
[present]
Pennsylvania Right to Know List
environmental hazard

STODDARD SOLVENT   8052-41-3
ENVIRONMENTAL LISTS
List of Pesticide Product Inert Ingredients
[present]

STRAW OIL   RR-00883-3
HEALTH AND SAFETY LISTS
IARC - Group 2B (sufficient animal data)
[present] (Overall evaluation based only on evidence of carcinogenicity in monograph (10, 1976) or in Supplement 4)
NIOSH - Selected LD50s and LC50s
Oral, rat: LD50 = 120 mg/kg
OSHA - Possible Select Carcinogens
[present]
STATE LISTS
California - Air Bill 2588 Appendix A-II
known or potential carcinogen

California - Prop. 65 - Cancer list
carcinogen - initial date 4/1/88

California - Prop. 65 - No Significant Risk Levels
no significant risk level = 0.02 ug/day

California - Directors List of Hazardous Substances (8 CCR 339)
[present]

Florida Hazardous Substance List
[present]

Massachusetts Right To Know List
carcinogen; extraordinarily hazardous

Minnesota Hazardous Substance List
carcinogen

Pennsylvania Right to Know List
special hazardous substance

Pennsylvania RTK - Special Hazardous Substances
[present]

STREPTOMYCIN   57-92-1
SEE ALSO:
ANTIMONY

INTERNATIONAL LISTS

Canada - WHMIS: Ingredient Disclosure
1% item 1465 (1506)

STREPTOMYCIN SULFATE   3810-74-0
SEE ALSO:
ANTIMONY

INTERNATIONAL LISTS

Canada - WHMIS: Ingredient Disclosure
1% item 1466 (134)

STREPTOZOTOCIN   18883-66-4
SEE ALSO:
ANTIMONY

HEALTH AND SAFETY LISTS

ACGIH 1995 - Time Weighted Averages
0.1 ppm TWA; 0.51 mg/m3 TWA

U.S. DOT - Substances From 49 CFR 172.101
regulated by DOT (UN2676)

U.S. DOT - Hazard Classes
DOT hazard class = 2.3

U.S. DOT - Substances Which Are Poisonous by Inhalation
gaseous hazardous material poisonous by inhalation (UN2676)

NIOSH 1990 - Pocket Guide - RELs
0.1 ppm TWA; 0.5 mg/m3 TWA

NIOSH 1990 - Pocket Guide - IDLHs
40 ppm IDLH

NIOSH 1990 - Pocket Guide - Target organs
blood, liver, kidneys, lungs

OSHA - Vacated PELs - Time Weighted Averages
0.1 ppm TWA; 0.5 mg/m3 TWA

OSHA - Final PELs - Time Weighted Averages
0.1 ppm TWA; 0.5 mg/m3 TWA

OSHA - List of Highly Hazardous Chemicals
threshhold quantity = 500 pounds

INTERNATIONAL LISTS

Australian Exposure Standards - Time Weighted Averages
0.1 ppm TWA; 0.51 mg/m3 TWA

Canada - WHMIS: Ingredient Disclosure
1% item 1467 (1507)

Canada - Alberta - 8 Hour Occupational Exposure Limit
0.1 ppm TWA; 0.51 mg/m3 TWA

Canada - Alberta - 15 Minute Occupational Exposure Limit
0.3 ppm STEL; 1.5 mg/m3 STEL

Canada - British Columbia - 8 Hour Exposure Limits
0.1 ppm TWA; 0.5 mg/m3 TWA

Canada - British Columbia - 15 Minute Exposure Limits
0.3 ppm STEL; 1.5 mg/m3 STEL

Canada - Ontario - OHSA - TWAEVs
0.1 ppm TWAEV; 0.5 mg/m3 TWAEV

Canada - Quebec - Time-Weighted Average Exposure Values
0.1 ppm TWAEV; 0.51 mg/m3 TWAEV

United Kingdom - Occupational Exposure Standards - TWAs
0.1 ppm TWA; 0.5 mg/m3 TWA

United Kingdom - Occupational Exposure Standards - STELs
0.3 ppm STEL; 1.5 mg/m3 STEL

German (DFG) - MAK Values
0.1 ppm MAK; 0.5 mg/m3 MAK

German (DFG) - Peak Limitations
5 x normal MAK (30 min. average value); don't exceed 2 times during shift

Israel - Time Weighted Averages
0.1 ppm TWA; 0.51 mg/m3 TWA

Israel - Action Levels
0.05 ppm AL; 0.255 mg/m3 AL

Mexico - Instruction No. 10 - TWAs
0.1 ppm TWA; 0.5 mg/m3 TWA

Mexico - Instruction No. 10 - STELs
0.3 ppm STEL; 1.5 mg/m3 STEL

STATE LISTS

California - Exposure Limits - PELs
0.1 ppm PEL; 0.5 mg/m3 PEL

Florida Hazardous Substance List
[present]

Massachusetts Right To Know List
[present]

Minnesota Hazardous Substance List
[present]

NJ Right to Know List (Total)
sn 1735

Pennsylvania Right to Know List
[present]

STRONTIUM   7440-24-6

HEALTH AND SAFETY LISTS

NIOSH - Selected LD50s and LC50s
Oral, mouse: LD50 = 920 mg/kg

STRONTIUM 80   15701-15-2

ENVIRONMENTAL LISTS

CAA - HON Rule - SOCMI Chemicals
compliance by April 24, 1995

STRONTIUM 81   14809-49-5

HEALTH AND SAFETY LISTS

ACGIH 1995 - Time Weighted Averages
100 ppm TWA; 525 mg/m3 TWA

U.S. DOT - Appendix B - Marine Pollutants
DOT regulated marine pollutant

NFPA - Flash Points
flash point > 100 degrees F (38 degrees C)

NFPA - Hazard Identification Ratings
health-0; flammability-2; reactivity-0

NIOSH 1990 - Pocket Guide - RELs
350 mg/m3 TWA; C 1800 mg/m3 (15 min)

NIOSH 1990 - Pocket Guide - IDLHs
29,500 ppm IDLH

NIOSH 1990 - Pocket Guide - Target organs
skin, eyes, CNS, respiratory system
NIOSH - Health Standards - Exposure Limits
350 mg/m3 TWA; C (15 min) 1800 mg/m3 (Listed under 'Refined petroleum solvents')
NIOSH - Health Standards - Health Effects and Precautions
Eye, nose, and throat irritation; dermatitis; nervous system effects (Blood and urine monitoring required; prevent skin contact) (Listed under 'Refined petroleum solvents')
OSHA - Vacated PELs - Time Weighted Averages
100 ppm TWA; 525 mg/m3 TWA
OSHA - Final PELs - Time Weighted Averages
500 ppm TWA; 2900 mg/m3 TWA

ENVIRONMENTAL LISTS

List of Pesticide Product Inert Ingredients
[present]

INTERNATIONAL LISTS

Australian Exposure Standards - Time Weighted Averages
790 mg/m3 TWA
Australian Exposure Standards - Under Review
exposure limits under review
Canada - WHMIS: Ingredient Disclosure
1% item 1468 (1498)
Canada - Alberta - 8 Hour Occupational Exposure Limit
100 ppm TWA; 525 mg/m3 TWA
Canada - Alberta - 15 Minute Occupational Exposure Limit
200 ppm STEL; 1050 mg/m3 STEL
Canada - British Columbia - 8 Hour Exposure Limits
100 ppm TWA; 575 mg/m3 TWA
Canada - British Columbia - 15 Minute Exposure Limits
125 ppm STEL; 720 mg/m3 STEL
Canada - Ontario - OHSA - TWAEVs
525 mg/m3 TWAEV (listed as an agent of variable composition) As sum of components assayed by chromatographic procedure with reference to the bulk sample.
Canada - Quebec - Time-Weighted Average Exposure Values
100 ppm TWAEV; 525 mg/m3 TWAEV
United Kingdom - Occupational Exposure Standards - TWAs
100 ppm TWA; 575 mg/m3 TWA
United Kingdom - Occupational Exposure Standards - STELs
125 ppm STEL; 720 mg/m3 STEL
Israel - Time Weighted Averages
100 ppm TWA; 525 mg/m3 TWA
Israel - Action Levels
50 ppm AL; 262.5 mg/m3 AL
Mexico - Instruction No. 10 - TWAs
100 ppm TWA; 523 mg/m3 TWA
Mexico - Instruction No. 10 - STELs
200 ppm STEL; 1050 mg/m3 STEL

STATE LISTS

California - Exposure Limits - PELs
100 ppm PEL; 525 mg/m3 PEL
California - Directors List of Hazardous Substances (8 CCR 339)
[present]
Florida Hazardous Substance List
[present]
Massachusetts Right To Know List
[present]
Minnesota Hazardous Substance List
[present]
NJ Right to Know List (Total)
sn 1736
Pennsylvania Right to Know List
[present]

STRONTIUM 83   14809-51-9

HEALTH AND SAFETY LISTS

NFPA - Flash Points
flash point = 315 to 361 degrees F (157 to 183 degrees C)
NFPA - Hazard Identification Ratings
health-0; flammability-1; reactivity-0

STRONTIUM 85   13967-73-2

HEALTH AND SAFETY LISTS

NIOSH - Selected LD50s and LC50s
Oral, rat: LD50 = 9 gm/kg

STATE LISTS

Massachusetts Right To Know List
teratogen
NJ Right to Know List (Total)
sn 1737

STRONTIUM 85M   RR-00353-2

STATE LISTS

California - Air Bill 2588 Appendix A-II
6/91
California - Prop. 65 - Developmental Toxicity
developmental toxicity - initial date 1/1/91

STRONTIUM 87M   RR-00352-1

HEALTH AND SAFETY LISTS

U.S. DOT - Appendix A Table 1 - Hazardous Substances
final RQ = 1 pound (0.454 kg)
IARC - Group 2B (sufficient animal data)
[present] (Overall evaluation based only on evidence of carcinogenicity in monograph (17, 1978) or in Supplement 4)
NIOSH - Selected LD50s and LC50s
Oral, mouse: LD50 = 264 mg/kg
NTP Chemical Status Reports - Testing Status and NTIS Number
Chronic studies exist for which technical reports were not prepared
NTP Seventh Report - Suspect Carcinogens
suspect carcinogen
OSHA - Possible Select Carcinogens
[present]

ENVIRONMENTAL LISTS

CERCLA/SARA - Hazardous Substances and their Reportable Quantities
final RQ = 1 pound (0.454 kg)
EPA - Carcinogen Hazard Ranking for RQ Adjustment
Hazard ranking = High
RCRA - U Series Wastes
waste number U206
RCRA - Hazardous Constituents-Appendix VIII
waste number U206
RCRA - Substances Banned From Land Disposal
[present]

STATE LISTS

California - Air Bill 2588 Appendix A-II
known or potential carcinogen
California - Prop. 65 - Cancer list
carcinogen - initial date 1/1/88
California - Prop. 65 - No Significant Risk Levels
no significant risk level = 0.006 ug/day
Florida Hazardous Substance List
[present]
Massachusetts Right To Know List
carcinogen; extraordinarily hazardous

Minnesota Hazardous Substance List
carcinogen

NJ Right to Know List (Total)
sn 1738

NJ Special Hazardous Substances
(mutagen)

Pennsylvania Right to Know List
environmental hazard; special hazardous substance

Pennsylvania RTK - Special Hazardous Substances
[present]

STRONTIUM 89   14158-27-1

STATE LISTS

California - Directors List of Hazardous Substances (8 CCR 339)
[present]

NJ Right to Know List (Total)
sn 1739

PROPOSED REGULATIONS

Safe Drinking Water Act - Priority list
[present]

STRONTIUM 90   10098-97-2

HEALTH AND SAFETY LISTS

U.S. DOT - Appendix A Table 2 - Radionuclides
final RQ = 100 curies (3.7E 12 Bq)

ENVIRONMENTAL LISTS

CERCLA/SARA List of Radionuclides (Appendix B) and Their Reportable Quantities
final RQ = 100 curies (3.7E 12 Bq)

STRONTIUM 91   14331-91-0

HEALTH AND SAFETY LISTS

U.S. DOT - Appendix A Table 2 - Radionuclides
final RQ = 1000 curies (3.7E 13 Bq)

ENVIRONMENTAL LISTS

CERCLA/SARA List of Radionuclides (Appendix B) and Their Reportable Quantities
final RQ = 1000 curies (3.7E 13 Bq)

STRONTIUM 92   14928-29-1

HEALTH AND SAFETY LISTS

U.S. DOT - Appendix A Table 2 - Radionuclides
final RQ = 100 curies (3.7E 12 Bq)

ENVIRONMENTAL LISTS

CERCLA/SARA List of Radionuclides (Appendix B) and Their Reportable Quantities
final RQ = 100 curies (3.7E 12 Bq)

STRONTIUM ARSENITE   15195-06-9

HEALTH AND SAFETY LISTS

U.S. DOT - Appendix A Table 2 - Radionuclides
final RQ = 10 curies (3.7E 11 Bq)

ENVIRONMENTAL LISTS

CERCLA/SARA List of Radionuclides (Appendix B) and Their Reportable Quantities
final RQ = 10 curies (3.7E 11 Bq)

STRONTIUM CHLORATE   7791-10-8

HEALTH AND SAFETY LISTS

U.S. DOT - Appendix A Table 2 - Radionuclides
final RQ = 1000 curies (3.7E 13 Bq)

ENVIRONMENTAL LISTS

CERCLA/SARA List of Radionuclides (Appendix B) and Their Reportable Quantities
final RQ = 1000 curies (3.7E 13 Bq)

STRONTIUM CHLORIDE (SRCL2)   10476-85-4

HEALTH AND SAFETY LISTS

U.S. DOT - Appendix A Table 2 - Radionuclides
final RQ = 100 curies (3.7E 12 Bq)

ENVIRONMENTAL LISTS

CERCLA/SARA List of Radionuclides (Appendix B) and Their Reportable Quantities
final RQ = 100 curies (3.7E 12 Bq)

STRONTIUM CHROMATE   7789-06-2

HEALTH AND SAFETY LISTS

U.S. DOT - Appendix A Table 2 - Radionuclides
final RQ = 10 curies (3.7E 11 Bq)

ENVIRONMENTAL LISTS

CERCLA/SARA List of Radionuclides (Appendix B) and Their Reportable Quantities
final RQ = 10 curies (3.7E 11 Bq)

STRONTIUM DICHROMATE   RR-00556-1

HEALTH AND SAFETY LISTS

U.S. DOT - Appendix A Table 2 - Radionuclides
final RQ = 0.1 curies (3.7E 9 Bq)

ENVIRONMENTAL LISTS

ATSDR Priority List
Rank (of 275): 197

CERCLA/SARA List of Radionuclides (Appendix B) and Their Reportable Quantities
final RQ = 0.1 curies (3.7E 9 Bq)

STRONTIUM FLUOBORATE   13814-98-7

HEALTH AND SAFETY LISTS

U.S. DOT - Appendix A Table 2 - Radionuclides
final RQ = 10 curies (3.7E 11 Bq)

ENVIRONMENTAL LISTS

CERCLA/SARA List of Radionuclides (Appendix B) and Their Reportable Quantities
final RQ = 10 curies (3.7E 11 Bq)

STRONTIUM FLUORIDE   7783-48-4

HEALTH AND SAFETY LISTS

U.S. DOT - Appendix A Table 2 - Radionuclides
final RQ = 100 curies (3.7E 12 Bq)

ENVIRONMENTAL LISTS

CERCLA/SARA List of Radionuclides (Appendix B) and Their Reportable Quantities
final RQ = 100 curies (3.7E 12 Bq)

STRONTIUM NITRATE   10042-76-9

HEALTH AND SAFETY LISTS

U.S. DOT - Substances From 49 CFR 172.101
regulated by DOT (UN1691)

U.S. DOT - Hazard Classes
DOT hazard class = 6.1

U.S. DOT - Appendix B - Marine Pollutants
DOT regulated marine pollutant

STATE LISTS

NJ Right to Know List (Total)
sn 1740

STRONTIUM PERCHLORATE   13450-97-0

HEALTH AND SAFETY LISTS

U.S. DOT - Substances From 49 CFR 172.101
regulated by DOT (UN1506)

U.S. DOT - Hazard Classes
DOT hazard class = 5.1

STATE LISTS

NJ Right to Know List (Total)
sn 1741

STRONTIUM PEROXIDE   1314-18-7

SEE ALSO:
STRONTIUM

HEALTH AND SAFETY LISTS

NIOSH - Selected LD50s and LC50s
Oral, rat: LD50 = 2250 mg/kg

INTERNATIONAL LISTS

Canada - WHMIS: Ingredient Disclosure
1% item 1469 (526)

STRONTIUM PHOSPHIDE   12504-13-1

SEE ALSO:
CHROMIUM (VI) COMPOUNDS- WATER SOLUBLE
STRONTIUM
CHROMIUM (VI) COMPOUNDS
CHROMIUM COMPOUNDS

HEALTH AND SAFETY LISTS

ACGIH 1995 - Time Weighted Averages
as Cr: 0.0005 mg/m3 TWA

ACGIH 1995 - Carcinogens
as Cr: A2-suspected human carcinogen

U.S. DOT - Appendix A Table 1 - Hazardous Substances
final RQ = 10 pounds (4.54 kg)

NIOSH - Selected LD50s and LC50s
Oral, rat: LD50 = 3118 mg/kg

NTP Seventh Report - Known Carcinogens
known carcinogen (Listed under 'Chromium and certain chromium compounds')

OSHA - Select Carcinogens
[present]

ENVIRONMENTAL LISTS

CERCLA/SARA - Hazardous Substances and their Reportable Quantities
final RQ = 10 pounds (4.54 kg)

Clean Water Act - Hazardous Substances
[present]

EPA - Carcinogen Hazard Ranking for RQ Adjustment
Hazard ranking = High

INTERNATIONAL LISTS

Canada - WHMIS: Ingredient Disclosure
0.1% item 1470 (553)

STATE LISTS

California - Air Bill 2588 Appendix A-I
known or potential carcinogen: 6/91

Florida Hazardous Substance List
[present]

Massachusetts Right To Know List
carcinogen; extraordinarily hazardous

Minnesota Hazardous Substance List
carcinogen

NJ Right to Know List (Total)
sn 1742

NJ Special Hazardous Substances
(carcinogen)

Pennsylvania Right to Know List
environmental hazard; special hazardous substance

Pennsylvania RTK - Special Hazardous Substances
[present]

STRONTIUM SULFIDE   1314-96-1

STATE LISTS

Pennsylvania Right to Know List
[present]

STRYCHNINE   57-24-9

INTERNATIONAL LISTS

Canada - WHMIS: Ingredient Disclosure
1% item 1471 (890)

STRYCHNINE SALTS   RR-01709-4

HEALTH AND SAFETY LISTS

NIOSH - Selected LD50s and LC50s
Oral, rat: LD50 = 10600 mg/kg

STRYCHNINE, SULFATE   60-41-3

HEALTH AND SAFETY LISTS

U.S. DOT - Substances From 49 CFR 172.101
regulated by DOT (UN1507)

U.S. DOT - Hazard Classes
DOT hazard class = 5.1

NIOSH - Selected LD50s and LC50s
Oral, rat: LD50 = 2750 mg/kg

STATE LISTS

Florida Hazardous Substance List
[present]

Massachusetts Right To Know List
[present]

NJ Right to Know List (Total)
sn 1743

Pennsylvania Right to Know List
[present]

STYRENE   100-42-5

HEALTH AND SAFETY LISTS

U.S. DOT - Substances From 49 CFR 172.101
regulated by DOT (UN1508)

U.S. DOT - Hazard Classes
DOT hazard class = 5.1

STATE LISTS

NJ Right to Know List (Total)
sn 1744

STYRENE ACRYLATE COPOLYMER   25085-34-1

HEALTH AND SAFETY LISTS

U.S. DOT - Substances From 49 CFR 172.101
regulated by DOT (UN1509)

U.S. DOT - Hazard Classes
DOT hazard class = 5.1

STATE LISTS

Florida Hazardous Substance List
[present]

Massachusetts Right To Know List
[present]

NJ Right to Know List (Total)
sn 1745

Pennsylvania Right to Know List
[present]

STYRENE-BUTADIENE POLYMER   9003-55-8

HEALTH AND SAFETY LISTS

U.S. DOT - Substances From 49 CFR 172.101
regulated by DOT (UN2013)

U.S. DOT - Hazard Classes
DOT hazard class = 4.3

INTERNATIONAL LISTS

Canada - WHMIS: Ingredient Disclosure
1% item 1472 (1413)

STATE LISTS

NJ Right to Know List (Total)
sn 1746

STYRENE-DIVINYLBENZENE COPOLYMER   9003-70-7

STATE LISTS

Massachusetts Right To Know List
[present]

STYRENE OXIDE   96-09-3

HEALTH AND SAFETY LISTS

ACGIH 1995 - Time Weighted Averages
0.15 mg/m3 TWA

U.S. DOT - Substances From 49 CFR 172.101
regulated by DOT (UN1692)

U.S. DOT - Hazard Classes
DOT hazard class = 6.1

U.S. DOT - Appendix B - Marine Pollutants
DOT regulated marine pollutant

U.S. DOT - Appendix A Table 1 - Hazardous Substances
final RQ = 10 pounds (4.54 kg)

NIOSH - Selected LD50s and LC50s
Oral, rat: LD50 = 16 mg/kg

NIOSH 1990 - Pocket Guide - RELs
0.15 mg/m3 TWA

NIOSH 1990 - Pocket Guide - IDLHs
3 mg/m3 IDLH

NIOSH 1990 - Pocket Guide - Target organs
CNS

OSHA - Vacated PELs - Time Weighted Averages
0.15 mg/m3 TWA

OSHA - Final PELs - Time Weighted Averages
0.15 mg/m3 TWA

ENVIRONMENTAL LISTS

CERCLA/SARA - Section 302 Extremely Hazardous Substances and TPQs
TPQ = 100/10,000 pounds

CERCLA/SARA - Section 313 - Emission Reporting
form R reporting required

CERCLA/SARA - Hazardous Substances and their Reportable Quantities
final RQ = 10 pounds (4.54 kg)

Clean Water Act - Hazardous Substances
[present]

RCRA - P Series Wastes
waste number P108

RCRA - Hazardous Constituents-Appendix VIII
waste number P108

RCRA - Substances Banned From Land Disposal
[present]

INTERNATIONAL LISTS

Australian Exposure Standards - Time Weighted Averages
0.15 mg/m3 TWA

Canada - CEPA Schedule II Part II - Toxic Substances (Export)
[present]

Canada - Alberta - 8 Hour Occupational Exposure Limit
0.15 mg/m3 TWA

Canada - Alberta - 15 Minute Occupational Exposure Limit
0.45 mg/m3 STEL

Canada - British Columbia - 8 Hour Exposure Limits
0.15 mg/m3 TWA

Canada - British Columbia - 15 Minute Exposure Limits
0.45 mg/m3 STEL

Canada - Ontario - OHSA - TWAEVs
0.15 mg/m3 TWAEV

Canada - Quebec - Time-Weighted Average Exposure Values
0.15 mg/m3 TWAEV

United Kingdom - Occupational Exposure Standards - TWAs
0.15 mg/m3 TWA

United Kingdom - Occupational Exposure Standards - STELs
0.45 mg/m3 STEL

German (DFG) - MAK Values
total dust: 0.15 mg/m3 MAK

German (DFG) - Peak Limitations
2 x normal MAK (30 min. average value); don't exceed 4 times during shift

Israel - Time Weighted Averages
0.15 mg/m3 TWA

Israel - Action Levels
0.075 mg/m3 AL

Mexico - Instruction No. 10 - TWAs
0.15 mg/m3 TWA

Mexico - Instruction No. 10 - STELs
0.45 mg/m3 STEL

STATE LISTS

California - Exposure Limits - PELs
0.15 mg/m3 PEL

California - Directors List of Hazardous Substances (8 CCR 339)
[present]

Florida Hazardous Substance List
[present]

Massachusetts Right To Know List
extraordinarily hazardous

Minnesota Hazardous Substance List
[present]

NJ Right to Know List (Total)
sn 1747

Pennsylvania Right to Know List
environmental hazard

SUBSTITUTED 2-NITROBENZENESULFONAMIDE   RR-01734-5

ENVIRONMENTAL LISTS

CERCLA/SARA - Section 313 - Emission Reporting
form R reporting required

SUBSTITUTED 2-AMINOBENZENESULFONAMIDE   RR-01735-6

HEALTH AND SAFETY LISTS

NIOSH - Selected LD50s and LC50s
Oral, rat: LD50 = 5 mg/kg

ENVIRONMENTAL LISTS

CERCLA/SARA - Section 302 Extremely Hazardous Substances and TPQs
TPQ = 100/10,000 pounds

STATE LISTS

Florida Hazardous Substance List
effective March 13, 1992

Massachusetts Right To Know List
extraordinarily hazardous

NJ Right to Know List (Total)
sn 2789

Pennsylvania Right to Know List
environmental hazard

PROPOSED REGULATIONS

CERCLA/SARA - Proposed Hazardous Substance Additions
proposed RQ = 1 pound (.454 kg)

CERCLA/SARA - 1989 Proposed RQ Adjustments
proposed RQ = 100 pounds (45.4 kg)

SUBSTITUTED ACRYLAMIDE   RR-01645-5

HEALTH AND SAFETY LISTS

ACGIH 1995 - Time Weighted Averages
50 ppm TWA; 213 mg/m3 TWA

ACGIH 1995 - Short Term Exposure Limits
100 ppm STEL; 426 mg/m3 STEL

ACGIH 1995 - Skin Designations
skin - potential for cutaneous absorption

ACGIH 1995 - Biological Exposure Indices
Mandelic acid in urine: 800 mg/g creatinine, end of shift (Ns), 300 mg/g creatinine, prior to next shift (Ns); Phenylglyoxylic acid in urine: 240 mg/g creatinine, end/shift (Ns), 100 mg/g creatinine, prior/next shift; Styrene in venous blood: 0.55 mg/L, end/shift (Sq), 0.02 mg/L, prior/next shift (Sq)

AIHA - Odor Threshold Values
geometric mean air odor threshold = 0.14 ppm (detectable); 0.15 ppm (recognizable)

U.S. DOT - Substances From 49 CFR 172.101
regulated by DOT (UN2055)

U.S. DOT - Hazard Classes
DOT hazard class = 3

U.S. DOT - Appendix B - Marine Pollutants
inhibited form: DOT regulated marine pollutant

U.S. DOT - Appendix A Table 1 - Hazardous Substances
final RQ = 1000 pounds (454 kg)

IARC - Group 2B (sufficient animal data)
[present] (Other relevant data, as given in Supplement 7, influenced the making of the overall evaluation)

NFPA - Flash Points
flash point = 88 degrees F (31 degrees C)

NFPA - Hazard Identification Ratings
health-2; flammability-3; reactivity-2

NIOSH - Selected LD50s and LC50s
Inhalation, rat: LC50 = 24 gm/m3 Oral, rat: LD50 = 5000 mg/kg

NIOSH 1990 - Pocket Guide - RELs
50 ppm TWA; 215 mg/m3 TWA; 100 ppm STEL; 425 mg/m3 STEL

NIOSH 1990 - Pocket Guide - IDLHs
5000 ppm IDLH

NIOSH 1990 - Pocket Guide - Target organs
CNS, skin, eyes, respiratory system

NIOSH - Health Standards - Exposure Limits
50 ppm TWA; 213 mg/m3 TWA; C (15 min) 100 ppm; C (15 min) 426 mg/m3

NIOSH - Health Standards - Health Effects and Precautions
Nervous system effects; eye and respiratory system irritation; reproductive system effects (Prevent skin contact; warn workers of reproductive hazards)

NTP Chemical Status Reports - Testing Status and NTIS Number
Technical reports printed (PB300977/AS); prechronic studies for which toxicity technical reports were not prepared

NTP Chemical Status Reports - Evidence of Carcinogenicity
male rat-negative; female rat-negative; male mice-equivocal; female mice-negative

OSHA - Vacated PELs - Time Weighted Averages
50 ppm TWA; 215 mg/m3 TWA

OSHA - Vacated PELs - Short Term Exposure Limits
100 ppm STEL; 425 mg/m3 STEL

OSHA - Final PELs - Time Weighted Averages
100 ppm TWA; C 200 ppm

OSHA - Final PELs - Ceiling Limits
C 200 ppm

OSHA - Possible Select Carcinogens
[present]

ENVIRONMENTAL LISTS

ATSDR Priority List
Rank (of 275): 241

CERCLA/SARA - Section 313 - Emission Reporting
form R reporting required for 0.1% de minimus concentration

CERCLA/SARA - Hazardous Substances and their Reportable Quantities
final RQ = 1000 pounds (454 kg)

Clean Air Act (1990) - List of Hazardous Air Contaminants
[present]

CAA - HON Rule - SOCMI Chemicals
compliance by Oct. 24, 1994

CAA - HON Rule - Organic HAPs
[present]

Clean Water Act - Hazardous Substances
[present]

Safe Drinking Water Act - MCLs
MCL = 0.1 mg/L

Safe Drinking Water Act - MCLGs
MCLG = 0.1 mg/L

Safe Drinking Water Act - Monitoring
monitoring required

RCRA - TSD Facilities Ground Water Monitoring
TM 8020 = 1 ug/L PQL; TM 8240 = 5 ug/L PQL

TSCA - Code of Federal Regulations Citations
40 CFR 712.30(d)

TSCA - PAIR - Reporting List
Reporting Date: November 19, 1982

INTERNATIONAL LISTS

Australian Exposure Standards - Time Weighted Averages
50 ppm TWA; 213 mg/m3 TWA

Australian Exposure Standards - Short Term Exposure Limits
100 ppm STEL; 426 mg/m3 STEL

Canada - WHMIS: Ingredient Disclosure
0.1% item 1473 (1508)

Canada - NPRI (National Pollutant Release Inventory)
[present]

Canada - CEPA - Priority Substances List
estimated time for completion of assessment reports: 4 years

Canada - Alberta - 8 Hour Occupational Exposure Limit
50 ppm TWA; 213 mg/m3 TWA

Canada - Alberta - 15 Minute Occupational Exposure Limit
100 ppm STEL; 426 mg/m3 STEL

Canada - Alberta - Skin Designation
can be absorbed through the intact skin

Canada - British Columbia - 8 Hour Exposure Limits
50 ppm TWA

Canada - British Columbia - 15 Minute Exposure Limits
75 ppm STEL

Canada - British Columbia - Skin Notations
skin - potential for skin absorption

Canada - British Columbia - Carcinogens
carcinogen - 50 ppm TWA; 75 ppm STEL
Canada - Ontario - OHSA - TWAEVs
50 ppm TWAEV; 213 mg/m3 TWAEV (special exceptions may apply to this substance; see section 9 of the Regulation)
Canada - Ontario - OHSA - STEVs
200 ppm STEV; 852 mg/m3 STEV (Special exceptions may apply to this substance. See Section 9 of the Regulation.)
Canada - Quebec - Time-Weighted Average Exposure Values
50 ppm TWAEV; 213 mg/m3 TWAEV
Canada - Quebec - Short-term Exposure Values
100 ppm STEV; 426 mg/m3 STEV
Canada - Quebec - Skin Designations
absorbed through the skin
Canada - Quebec - Carcinogens
C3 carcinogen: effect detected in animals
United Kingdom - Maximum Exposure Limits - TWAs
100 ppm TWA; 420 mg/m3 TWA
United Kingdom - Maximum Exposure Limits - STELs
250 ppm STEL; 1050 mg/m3 STEL
German (DFG) - MAK Values
20 ppm MAK; 85 mg/m3 MAK
German (DFG) - Peak Limitations
2 x normal MAK (30 min. average value); don't exceed 4 times during shift
German (DFG) - Pregnancy
no risk to embryo/fetus if exposure limits adhered to
Israel - Time Weighted Averages
50 ppm TWA; 213 mg/m3 TWA
Israel - Short Term Exposure Limits
100 ppm STEL; 426 mg/m3 STEL
Israel - Action Levels
25 ppm AL; 106.5 mg/m3 AL
Mexico - Instruction No. 10 - TWAs
50 ppm TWA; 215 mg/m3 TWA
Mexico - Instruction No. 10 - STELs
100 ppm STEL; 425 mg/m3 STEL

STATE LISTS

California - Air Bill 2588 Appendix A-I
known or potential carcinogen
California - Exposure Limits - PELs
50 ppm PEL; 215 mg/m3 PEL
California - Exposure Limits - STELs
100 ppm STEL; 425 mg/m3 STEL
California - Exposure Limits - Ceilings
C 500 ppm
California - Directors List of Hazardous Substances (8 CCR 339)
[present]
Florida Hazardous Substance List
[present]
Massachusetts Right To Know List
[present]
Minnesota Hazardous Substance List
[present]
NJ Right to Know List (Total)
sn 1748
NJ Special Hazardous Substances
(flammable - third degree; mutagen; reactive - second degree)
Pennsylvania Right to Know List
environmental hazard

PROPOSED REGULATIONS

Canada - Ontario - Proposed Occupational TWAEVs
20 ppm TWAEV; 100 mg/m3 TWAEV

SUBSTITUTED ACRYLATED ALKOXYLATED ALIPHATIC POLYOL   RR-00237-9

ENVIRONMENTAL LISTS
List of Pesticide Product Inert Ingredients
[present]

SUBSTITUTED ALIPHATIC ACID HALIDE   RR-00920-1

HEALTH AND SAFETY LISTS
IARC - Group 3 (not classifiable)
[present]

ENVIRONMENTAL LISTS
List of Pesticide Product Inert Ingredients
[present]

SUBSTITUTED ALKYL HALIDE   RR-00942-7

ENVIRONMENTAL LISTS
List of Pesticide Product Inert Ingredients
[present]

SUBSTITUTED ALKYL PEROXYHEXANE CARBOXY-LATE (MIXED ISOMERS)   RR-00159-2

HEALTH AND SAFETY LISTS
IARC - Group 2A (limited human data)
[present] (Overall evaluation based only on evidence of carcinogenicity in monograph (36, 1985) or in Supplement 4) (Other relevant data, as given in Supplement 7 or in the monograph, influenced the making of the overall evaluation)
NFPA - Flash Points
flash point - 165 degrees F (74 degrees C)
NFPA - Hazard Identification Ratings
health-2; flammability-2; reactivity-0
NIOSH - Selected LD50s and LC50s
Oral, rat: LD50 = 2000 mg/kg Skin, rabbit: LD50 = 1060 mg/kg
NTP Chemical Status Reports - Testing Status and NTIS Number
Chronic studies exist for which technical reports were not prepared
OSHA - Possible Select Carcinogens
[present]

ENVIRONMENTAL LISTS
CERCLA/SARA - Section 313 - Emission Reporting
form R reporting required for 0.1% de minimus concentration
CERCLA/SARA - Hazardous Substances and their Reportable Quantities
final RQ = 1 pound (.454 kg)
Clean Air Act (1990) - List of Hazardous Air Contaminants
[present]
TSCA - Code of Federal Regulations Citations
40 CFR 712.30(d)
TSCA - PAIR - Reporting List
Reporting Date: November 19, 1982

INTERNATIONAL LISTS
Canada - WHMIS: Ingredient Disclosure
1% item 1474 (1320)
Canada - NPRI (National Pollutant Release Inventory)
[present]

STATE LISTS
California - Air Bill 2588 Appendix A-I
known or potential carcinogen
California - Prop. 65 - Cancer list
carcinogen - initial date 10/1/88
California - Prop. 65 - No Significant Risk Levels
no significant risk level = 4 ug/day
California - Directors List of Hazardous Substances (8 CCR 339)
[present]

Florida Hazardous Substance List
  [present]
Massachusetts Right To Know List
  carcinogen; extraordinarily hazardous
Minnesota Hazardous Substance List
  carcinogen
NJ Right to Know List (Total)
  sn 1749
NJ Special Hazardous Substances
  (carcinogen; mutagen; teratogen)
Pennsylvania Right to Know List
  environmental hazard

SUBSTITUTED AMINOBENZOIC ACID ESTER   RR-00922-3
  ENVIRONMENTAL LISTS
    TSCA - Chemicals with Significant New Use Rules
    PMN number: P-88-1937

SUBSTITUTED AROMATIC   RR-00210-8
  ENVIRONMENTAL LISTS
    TSCA - Chemicals with Significant New Use Rules
    PMN number: P-88-1938

SUBSTITUTED BENZENEDIAZONIUM   RR-01693-3
  ENVIRONMENTAL LISTS
    TSCA - Chemicals with Significant New Use Rules
    PMN number: P-90-1687

SUBSTITUTED BENZENEDICARBOXYLIC ACID ESTER   RR-01724-3
  ENVIRONMENTAL LISTS
    TSCA - Chemicals with Significant New Use Rules
    PMN number: P-86-346

SUBSTITUTED BENZENEDICARBOXYLIC ACID, POLY(ALKYL ACRYLATE) DERIVATIVE   RR-01012-8
  ENVIRONMENTAL LISTS
    TSCA - Chemicals with Significant New Use Rules
    PMN number: P-84-491

SUBSTITUTED BENZENESULFONIC ACID, ALKALI METAL SALT   RR-00968-7
  ENVIRONMENTAL LISTS
    TSCA - Chemicals with Significant New Use Rules
    PMN number: P-83-1222

2-SUBSTITUTED BENZOTRIAZOLE   RR-00990-5
  ENVIRONMENTAL LISTS
    TSCA - Chemicals with Significant New Use Rules
    PMN number: P-86-1493

SUBSTITUTED BIS(HYDROXYALKANE) POLYMER WITH EPICHLOROHYDRIN,ACRYLATE   RR-00977-8
  ENVIRONMENTAL LISTS
    TSCA - Chemicals with Significant New Use Rules
    PMN number: P-84-951

SUBSTITUTED CARBOHETEROCYCLIC BUTANE TETRACARBOXYLATE   RR-00991-6
  ENVIRONMENTAL LISTS
    TSCA - Chemicals with Significant New Use Rules
    PMN number: P-84-954

SUBSTITUTED CYCLOHEXYLDIAMINO ETHYL ESTER   RR-01222-6
  ENVIRONMENTAL LISTS

TSCA - Chemicals with Significant New Use Rules
  PMN number: P-93-533

SUBSTITUTED DIACRYLATE   RR-01569-0
  ENVIRONMENTAL LISTS
    TSCA - Chemicals with Significant New Use Rules
    PMN number: P-93-699

SUBSTITUTED DIALKYL OXAZOLONE   RR-00902-9
  ENVIRONMENTAL LISTS
    TSCA - Chemicals with Significant New Use Rules
    PMN number: P-86-1739

SUBSTITUTED DICHLOROBENZOTHIAZOLES   RR-01649-9
  ENVIRONMENTAL LISTS
    TSCA - Chemicals with Significant New Use Rules
    PMN number: P-89-776

SUBSTITUTED ETHANOLAMINE   RR-01216-8
  ENVIRONMENTAL LISTS
    TSCA - Chemicals with Significant New Use Rules
    PMN number: P-90-335

SUBSTITUTED ETHYL ALKENAMIDE   RR-01558-7
  ENVIRONMENTAL LISTS
    TSCA - Chemicals with Significant New Use Rules
    PMN number: P-84-1167

SUBSTITUTED HYDROXYALKYL ALKNENOATE, [[[[[(1-OXO-2-PROPENYL)OXY]ALKOXY] CARBONYLAMINO]SUBSTITUTED] AMINOCARBONYL]OXY-   RR-00979-0

  ENVIRONMENTAL LISTS
    TSCA - Chemicals with Significant New Use Rules
    PMN number: P-90-440

SUBSTITUTED HYDROXYLAMINE   RR-00944-9
  ENVIRONMENTAL LISTS
    TSCA - Chemicals with Significant New Use Rules
    PMN number: P-91-1243

SUBSTITUTED IMIDAZOLE   27136-73-8
  ENVIRONMENTAL LISTS
    TSCA - Chemicals with Significant New Use Rules
    PMN number: P-91-1464

SUBSTITUTED METHYLPYRIDINE   RR-00254-0
  ENVIRONMENTAL LISTS
    TSCA - Chemicals with Significant New Use Rules
    PMN number: P-86-1634

SUBSTITUTED NITRILE   RR-00898-0
  ENVIRONMENTAL LISTS
    TSCA - Chemicals with Significant New Use Rules
    PMN numbers: P-91-1190, P-91-1191

SUBSTITUTED NITROBENZENE   RR-01676-2
  ENVIRONMENTAL LISTS
    TSCA - Chemicals with Significant New Use Rules
    PMN number: P-91-490

SUBSTITUTED NITROPHENOL PESTICIDES, N O.S.   RR-00496-6
  ENVIRONMENTAL LISTS
    TSCA - Chemicals with Significant New Use Rules
    PMN number: P-86-1315

SUBSTITUTED OXIDE-ALKYLENE POLYMER, METHACRYLATE   RR-00229-9

ENVIRONMENTAL LISTS

  TSCA - Chemicals with Significant New Use Rules
    PMN number: P-86-1088

SUBSTITUTED OXIRANE   RR-00131-0

ENVIRONMENTAL LISTS

  TSCA - Chemicals with Significant New Use Rules
    PMN number: P-84-492

SUBSTITUTED 2-PHENOXYPYRIDINE   RR-00252-8

ENVIRONMENTAL LISTS

  List of Pesticide Product Inert Ingredients
    [present]

SUBSTITUTED PHENYL AZO SUBSTITUED BEN-ZENEDIAZONIUM SALT   RR-01671-7

ENVIRONMENTAL LISTS

  TSCA - Chemicals with Significant New Use Rules
    PMN numbers: P-83-24; P-83-49; P-83-272

SUBSTITUTED PHENYLIMINO CARBAMATE DERIVATIVE   RR-01215-7

ENVIRONMENTAL LISTS

  TSCA - Chemicals with Significant New Use Rules
    PMN number: P-83-603

SUBSTITUTED PHOSPHATE ESTER   RR-00197-8

ENVIRONMENTAL LISTS

  TSCA - Chemicals with Significant New Use Rules
    PMN number: P-92-1125

SUBSTITUTED POLYGLYCIDYL BENZENEAMINE   RR-00260-8

HEALTH AND SAFETY LISTS

  U.S. DOT - Substances From 49 CFR 172.101
    regulated by DOT (UN2779, UN2780, UN3013, UN3014)

  U.S. DOT - Hazard Classes
    toxic or toxic, flammable: DOT hazard class = 6.1; flammable, toxic: DOT hazard class = 3

STATE LISTS

  NJ Right to Know List (Total)
    sn 2791; sn 2792; sn 2793; sn 2794

SUBSTITUTED QUINOLINE   RR-01727-6

ENVIRONMENTAL LISTS

  TSCA - Chemicals with Significant New Use Rules
    PMN number: P-88-2566

SUBSTITUTED SPIRO OXAZINE   RR-01182-5

ENVIRONMENTAL LISTS

  TSCA - Chemicals with Significant New Use Rules
    PMN number: P-83-1157

SUBSTITUTED TRIAZINE ISOCYANURATE   RR-00926-7

ENVIRONMENTAL LISTS

  TSCA - Chemicals with Significant New Use Rules
    PMN numbers: P-83-23; P-83-75

SUBSTITUTED TRIAZOLE   RR-01243-1

ENVIRONMENTAL LISTS

  TSCA - Chemicals with Significant New Use Rules
    PMN number: P-92-652

SUBSTITUTED TRIPHENYLMETHANE   RR-00952-9

ENVIRONMENTAL LISTS

  TSCA - Chemicals with Significant New Use Rules
    PMN number: P-91-487

SUBTILISINS (PROTEOLYTIC ENZYMES)   9014-01-1

ENVIRONMENTAL LISTS

  TSCA - Chemicals with Significant New Use Rules
    PMN number: P-85-730

SUBTILISINS (PROTEOLYTIC ENZYMES AS 100% PURE ENZYME)   1395-21-7

ENVIRONMENTAL LISTS

  TSCA - Chemicals with Significant New Use Rules
    PMN number: P-83-394

SUCCINIC ACID   110-15-6

ENVIRONMENTAL LISTS

  TSCA - Chemicals with Significant New Use Rules
    PMN number: P-93-1183

SUCCINIC ACID PEROXIDE   123-23-9

ENVIRONMENTAL LISTS

  TSCA - Chemicals with Significant New Use Rules
    PMN number: P-92-283

SUCCINIC ANHYDRIDE   108-30-5

ENVIRONMENTAL LISTS

  TSCA - Chemicals with Significant New Use Rules
    PMN number: P-86-66

SUCCINIMIDE   123-56-8

ENVIRONMENTAL LISTS

  TSCA - Chemicals with Significant New Use Rules
    PMN number: P-90-1731

SUCCINONITRILE   110-61-2

ENVIRONMENTAL LISTS

  TSCA - Chemicals with Significant New Use Rules
    PMN number: P-87-1553

SUCROSE   57-50-1

HEALTH AND SAFETY LISTS

  NIOSH - Selected LD50s and LC50s
    Oral, rat: LD50 = 3700 mg/kg

  OSHA - Vacated PELs - Short Term Exposure Limits
    0.00006 mg/m3 STEL (60 min)

ENVIRONMENTAL LISTS

  List of Pesticide Product Inert Ingredients
    [present]

INTERNATIONAL LISTS

  Canada - Ontario - OHSA - CEVs
    0.00006 mg/m3 CEV

  United Kingdom - Occupational Exposure Standards - TWAs
    0.00006 mg/m3 TWA

  United Kingdom - Occupational Exposure Standards - STELs
    0.00006 mg/m3 STEL

  Mexico - Instruction No. 10 - TWAs
    0.00006 mg/m3 TWA

STATE LISTS

  California - Exposure Limits - STELs
    as pure crystalline proteolytic enzymes: 0.00006 mg/m3 STEL (sampled by a high volume sampler for at least 60 minutes)

California - Directors List of Hazardous Substances (8 CCR 339)
[present]

Florida Hazardous Substance List
effective March 13, 1992

SUCROSE OCTAACETATE   126-14-7

HEALTH AND SAFETY LISTS

ACGIH 1995 - Ceiling Limits
C 0.00006 mg/m3

INTERNATIONAL LISTS

Australian Exposure Standards - Time Weighted Averages
Peak Limitation: 0.00006 mg/m3 TWA

Australian Exposure Standards - Skin Effects
sensitiser

Canada - WHMIS: Ingredient Disclosure
0.1% item 1475 (1509)

Canada - Alberta - Ceiling Occupational Exposure Limit
C 0.00006 mg/m3

Canada - British Columbia - Ceiling Exposure Limits
C 0.00006 mg/m3

Canada - Quebec - Ceiling Limits
P 0.00006 mg/m3

Israel - Ceiling Exposure Limits
C 0.00006 mg/m3

STATE LISTS

California - Directors List of Hazardous Substances (8 CCR 339)
[present]

Minnesota Hazardous Substance List
[present] (proteolytic enzymes as 100% pure crystalline enzyme)

Pennsylvania Right to Know List
[present]

SUCROSE OCTANITRATE   30236-29-4

ENVIRONMENTAL LISTS

CAA - HON Rule - SOCMI Chemicals
compliance by Oct. 24, 1994

INTERNATIONAL LISTS

Canada - WHMIS: Ingredient Disclosure
1% item 1476 (135)

SUDAN G   85-86-9

HEALTH AND SAFETY LISTS

U.S. DOT - Organic Peroxides Table
Organic peroxide UN3102;   UN3116

STATE LISTS

NJ Right to Know List (Total)
sn 1751

SUDAN BROWN RR   6416-57-5

HEALTH AND SAFETY LISTS

IARC - Group 3 (not classifiable)
[present]

NIOSH - Selected LD50s and LC50s
Oral, rat: LD50 = 1510 mg/kg

NTP Chemical Status Reports - Testing Status and NTIS Number
Technical reports printed (PB90231135/AS)

NTP Chemical Status Reports - Evidence of Carcinogenicity
male rat-no evidence;   female rat-no evidence;   male mice-no evidence;   female mice-no evidence

INTERNATIONAL LISTS

Canada - WHMIS: Ingredient Disclosure
1% item 1477 (235)

STATE LISTS

California - Directors List of Hazardous Substances (8 CCR 339)
[present]

SUDAN RED 7B   6368-72-5

HEALTH AND SAFETY LISTS

NIOSH - Selected LD50s and LC50s
Oral, rat: LD50 = 14 gm/kg

SULFALLATE   95-06-7

HEALTH AND SAFETY LISTS

NFPA - Flash Points
flash point = 270 degrees F (132 degrees C)

NFPA - Hazard Identification Ratings
flammability-1;   reactivity-0

NIOSH - Selected LD50s and LC50s
Oral, mouse: LD50 = 129 mg/kg

NIOSH - Health Standards - Exposure Limits
6 ppm TWA;   20 mg/m3 TWA (Listed under 'Nitriles')

NIOSH - Health Standards - Health Effects and Precautions
Hepatic, renal, respiratory, cardiovascular, gastrointestinal, and nervous system effects (Periodic chest X-ray and pulmonary function testing required;   prevent skin and eye contact;   make first-aid kits and personnel available during use) (Listed under 'Nitriles')

ENVIRONMENTAL LISTS

CAA - HON Rule - SOCMI Chemicals
compliance by Oct. 24, 1994

STATE LISTS

Minnesota Hazardous Substance List
[present]

SULFAMETHAZINE   57-68-1

HEALTH AND SAFETY LISTS

ACGIH 1995 - Time Weighted Averages
10 mg/m3 TWA

NIOSH - Selected LD50s and LC50s
Oral, rat: LD50 = 29700 mg/kg

OSHA - Vacated PELs - Time Weighted Averages
total dust: 15 mg/m3 TWA;   respirable fraction: 5 mg/m3 TWA

OSHA - Final PELs - Time Weighted Averages
total dust: 15 mg/m3 TWA;   respirable fraction: 5 mg/m3 TWA

ENVIRONMENTAL LISTS

List of Pesticide Product Inert Ingredients
[present]

INTERNATIONAL LISTS

Australian Exposure Standards - Time Weighted Averages
10 mg/m3 TWA

Canada - Alberta - 8 Hour Occupational Exposure Limit
respirable mass: 5 mg/m3 TWA;   total mass: 10 mg/m3

Canada - British Columbia - 8 Hour Exposure Limits
nuisance dust: 10 mg/m3 TWA

Canada - British Columbia - 15 Minute Exposure Limits
20 mg/m3 STEL

Canada - Ontario - OHSA - TWAEVs
total dust: 10 mg/m3 TWAEV (listed as a nuisance particulate)

Canada - Quebec - Time-Weighted Average Exposure Values
10 mg/m3 TWAEV

United Kingdom - Occupational Exposure Standards - TWAs
10 mg/m3 TWA

United Kingdom - Occupational Exposure Standards - STELs
20 mg/m3 STEL

Israel - Time Weighted Averages
10 mg/m3 TWA

Israel - Action Levels
  5 mg/m3 AL
Mexico - Instruction No. 10 - TWAs
  10 mg/m3 TWA; (nuisance particulate)
Mexico - Instruction No. 10 - STELs
  20 mg/m3 STEL
STATE LISTS
  Minnesota Hazardous Substance List
    [present] (includes inert or nuisance dust)
  Pennsylvania Right to Know List
    [present]

SULFAMETHOXAZOLE   723-46-6
  ENVIRONMENTAL LISTS
    List of Pesticide Product Inert Ingredients
      [present]

SULFAMIC ACID   5329-14-6
  HEALTH AND SAFETY LISTS
    U.S. DOT - Hazard Classes
      Forbidden from transport by the DOT

SULFATE   RR-00518-5
  HEALTH AND SAFETY LISTS
    IARC - Group 3 (not classifiable)
      [present]
  INTERNATIONAL LISTS
    Canada - WHMIS: Ingredient Disclosure
      1% item 1478 (1500)

SULFATED BUTYL OLEATE   68422-69-5
  HEALTH AND SAFETY LISTS
    IARC - Group 3 (not classifiable)
      [present]

SULFATHIAZOLE   72-14-0
  HEALTH AND SAFETY LISTS
    IARC - Group 3 (not classifiable)
      [present]

SULFIDE   18496-25-8
  HEALTH AND SAFETY LISTS
    IARC - Group 2B (sufficient animal data)
      [present] (Overall evaluation based only on evidence of carcinogenicity in monograph (30, 1983) or in Supplement 4)
    NIOSH - Selected LD50s and LC50s
      Oral, rat: LD50 = 850 mg/kg Skin, rabbit: LD50 = 2200 mg/kg
    NTP Chemical Status Reports - Testing Status and NTIS Number
      Technical reports printed (PB286386/AS)
    NTP Chemical Status Reports - Evidence of Carcinogenicity
      male rat-positive; female rat-positive; male mice-positive; female mice-positive
    NTP Seventh Report - Suspect Carcinogens
      suspect carcinogen
    OSHA - Possible Select Carcinogens
      [present]
  STATE LISTS
    California - Air Bill 2588 Appendix A-II
      known or potential carcinogen
    California - Prop. 65 - Cancer list
      carcinogen - initial date 1/1/88
    California - Prop. 65 - No Significant Risk Levels
      no significant risk level = 4 ug/day

California - Directors List of Hazardous Substances (8 CCR 339)
      [present]
Florida Hazardous Substance List
      [present]
Massachusetts Right To Know List
      carcinogen; extraordinarily hazardous
Minnesota Hazardous Substance List
      carcinogen
NJ Right to Know List (Total)
      sn 1753
NJ Special Hazardous Substances
      (carcinogen)
Pennsylvania Right to Know List
      special hazardous substance
Pennsylvania RTK - Special Hazardous Substances
      [present]

SULFISOXAZOLE   127-69-5
  HEALTH AND SAFETY LISTS
    NTP Chemical Status Reports - Testing Status and NTIS Number
      Technical reports printed (call NCTR for documents)

SULFITES   RR-01552-1
  HEALTH AND SAFETY LISTS
    IARC - Group 3 (not classifiable)
      [present]
    NIOSH - Selected LD50s and LC50s
      Oral, rat: LD50 = 6370 mg/kg
  STATE LISTS
    California - Directors List of Hazardous Substances (8 CCR 339)
      [present]
    NJ Right to Know List (Total)
      sn 1754

SULFOACETIC ACID   123-43-3
  HEALTH AND SAFETY LISTS
    U.S. DOT - Substances From 49 CFR 172.101
      regulated by DOT (UN2967)
    U.S. DOT - Hazard Classes
      DOT hazard class = 8
    NIOSH - Selected LD50s and LC50s
      Oral, rat: LD50 = 3160 mg/kg
  ENVIRONMENTAL LISTS
    List of Pesticide Product Inert Ingredients
      [present]
  INTERNATIONAL LISTS
    Canada - WHMIS: Ingredient Disclosure
      1% item 1479 (136)
  STATE LISTS
    NJ Right to Know List (Total)
      sn 1770

SULFOLANE   126-33-0
  ENVIRONMENTAL LISTS
    Safe Drinking Water Act - SMCLs
      SMCL = 250 mg/L
  INTERNATIONAL LISTS
    Canada - Drinking Water Quality - AOs
      <= 500 mg/L AO
    Mexico - Drinking Water - Ecological Criteria
      500.0 mg/l

3-SULFOLENE   77-79-2

ENVIRONMENTAL LISTS

List of Pesticide Product Inert Ingredients
[present]

SULFFOMETURON METHYL   74222-97-2

HEALTH AND SAFETY LISTS

NIOSH - Selected LD50s and LC50s
Oral, mouse: LD50 = 4500 mg/kg

STATE LISTS

NJ Right to Know List (Total)
sn 1755

SULFONAMIDE   63-74-1

SEE ALSO:
F039-HAZARDOUS WASTES

ENVIRONMENTAL LISTS

RCRA - Basis for Listing - Appendix VII
Included in waste stream: F039

RCRA - TSD Facilities Ground Water Monitoring
TM 9030 = 10,000 ug/L PQL

RCRA - Universal Treatment Standards (LDR)
WW: 14 mg/l; NWW: Not applicable

INTERNATIONAL LISTS

Canada - Drinking Water Quality - AOs
<= 0.05 mg/L AO

Mexico - Drinking Water - Ecological Criteria
0.2 mg/l

SULFONAMIDE   RR-01244-2

HEALTH AND SAFETY LISTS

IARC - Group 3 (not classifiable)
[present]

NTP Chemical Status Reports - Testing Status and NTIS Number
Technical reports printed (PB288779/AS)

NTP Chemical Status Reports - Evidence of Carcinogenicity
male rat-negative; female rat-negative; male mice-negative; female mice-negative

SULFONATED CASTOR OIL, SODIUM SALT   68187-76-8

HEALTH AND SAFETY LISTS

IARC - Group 3 (not classifiable)
[present]

SULFONES   RR-01710-7

HEALTH AND SAFETY LISTS

NIOSH - Selected LD50s and LC50s
Oral, rat: LD50 = 3160 mg/kg Skin, rabbit: LD50 = 1570 mg/kg

SULFONIC ACIDS, PETROLEUM, SODIUM SALTS   68608-26-4

HEALTH AND SAFETY LISTS

NFPA - Flash Points
flash point = 350 degrees F (177 degrees C)

NFPA - Hazard Identification Ratings
health-2; flammability-1; reactivity-0

ENVIRONMENTAL LISTS

CAA - HON Rule - SOCMI Chemicals
compliance by Jan. 23, 1995

TSCA - Code of Federal Regulations Citations
40 CFR 712.30(x); 40 CFR 716.120(d)

TSCA - PAIR - Reporting List
Reporting Date: November 27, 1991

TSCA - Health and Safety Reporting List
Effective Date: September 30, 1991

STATE LISTS

Florida Hazardous Substance List
[present]

Massachusetts Right To Know List
[present]

Pennsylvania Right to Know List
[present]

SULFONYL BIS-(4-CHLOROBENZENE)   80-07-9

HEALTH AND SAFETY LISTS

NIOSH - Selected LD50s and LC50s
Oral, rat: LD50 = 2830 mg/kg

NTP Chemical Status Reports - Testing Status and NTIS Number
Technical reports printed (PB284656/AS)

NTP Chemical Status Reports - Evidence of Carcinogenicity
male rat-negative; female rat-negative; male mice-negative; female mice-negative

ENVIRONMENTAL LISTS

TSCA - Code of Federal Regulations Citations
40 CFR 712.30(x); 40 CFR 716.120(d)

TSCA - PAIR - Reporting List
Effective Date: September 30, 1991; Reporting Date: November 27, 1991

TSCA - Health and Safety Reporting List
Effective Date: September 30, 1991 Sunset Date: September 30, 2001

2,2'-SULFONYL BIS-ETHANOL   2580-77-0

HEALTH AND SAFETY LISTS

ACGIH 1995 - Time Weighted Averages
5 mg/m3 TWA

ACGIH 1995 - Carcinogens
A4-not classifiable as a human carcinogen

4,4'-SULFONYLDIPHENOL   80-09-1

HEALTH AND SAFETY LISTS

NIOSH - Selected LD50s and LC50s
Oral, rat: LD50 = 10500 mg/kg

5-SULFOSALICYLIC ACID   97-05-2

ENVIRONMENTAL LISTS

TSCA - Chemicals with Significant New Use Rules
PMN number: P-90-1732

SULFOTEP   3689-24-5

ENVIRONMENTAL LISTS

List of Pesticide Product Inert Ingredients
[present]

SULFOXIDE, 3-CHLOROPROPYL OCTYL   3569-57-1

PROPOSED REGULATIONS

TSCA - ITC 33rd Report Priority Testing List
recommended for testing

TSCA - ITC 34th Report Priority Testing List
recommended for testing

SULFUR   7704-34-9

ENVIRONMENTAL LISTS

List of Pesticide Product Inert Ingredients
[present]

SULFUR 35   15117-53-0

HEALTH AND SAFETY LISTS

NIOSH - Selected LD50s and LC50s
Oral, mouse: LD50 = 24 gm/kg

NTP Chemical Status Reports - Testing Status and NTIS Number
*Prechronic studies completed: chemials in review for further evaluation;
Approved for toxicology/carcinogenesis study*

ENVIRONMENTAL LISTS

TSCA - Code of Federal Regulations Citations
40 CFR 712.30(x); 40 CFR 716.120(d)

TSCA - PAIR - Reporting List
Reporting Date: November 27, 1991

TSCA - Health and Safety Reporting List
Effective Date: September 30, 1991

SULFUR CHLORIDE   12771-08-3

ENVIRONMENTAL LISTS

TSCA - Code of Federal Regulations Citations
40 CFR 712.30(x); 40 CFR 716.120(d)

TSCA - PAIR - Reporting List
Reporting Date: November 27, 1991

TSCA - Health and Safety Reporting List
Effective Date: September 30, 1991

SULFUR CHLORIDE AND CARBON TETRACHLO-
RIDE MIXTURES   RR-01166-5

ENVIRONMENTAL LISTS

TSCA - Code of Federal Regulations Citations
40 CFR 712.30(x); 40 CFR 716.120(d)

TSCA - PAIR - Reporting List
Reporting Date: November 27, 1991

TSCA - Health and Safety Reporting List
Effective Date: September 30, 1991

INTERNATIONAL LISTS

Canada - WHMIS: Ingredient Disclosure
1% item 1480 (1541)

SULFUR CHLORIDE PENTAFLUORIDE   RR-01170-1

HEALTH AND SAFETY LISTS

NIOSH - Selected LD50s and LC50s
Oral, rat: LD50 = 2450 mg/kg

SULFUR CHLORIDES   RR-01346-7

HEALTH AND SAFETY LISTS

ACGIH 1995 - Time Weighted Averages
0.2 mg/m3 TWA

ACGIH 1995 - Skin Designations
skin - potential for cutaneous absorption

U.S. DOT - Appendix B - Marine Pollutants
DOT regulated marine pollutant

U.S. DOT - Appendix A Table 1 - Hazardous Substances
final RQ = 100 pounds (45.4 kg)

NIOSH - Selected LD50s and LC50s
*Inhalation, rat: LC50 = 38 mg/m3 4 hr Oral, rat: LD50 = 5 mg/kg
Skin, rat: LD50 = 65 mg/kg*

NIOSH 1990 - Pocket Guide - RELs
0.2 mg/m3 TWA

NIOSH 1990 - Pocket Guide - IDLHs
35 mg/m3 IDLH

NIOSH 1990 - Pocket Guide - Target organs
CNS, CVS, respiratory system

NIOSH 1990 - Pocket Guide - Skin list
Potential for dermal absorption

OSHA - Vacated PELs - Time Weighted Averages
0.2 mg/m3 TWA

OSHA - Vacated PELs - Skin Designation
Prevent or reduce skin absorption

OSHA - Final PELs - Time Weighted Averages
0.2 mg/m3 TWA

OSHA - Final PELs - Skin Notations
prevent or reduce skin absorption

ENVIRONMENTAL LISTS

CERCLA/SARA - Section 302 Extremely Hazardous Substances and
TPQs
TPQ = 500 pounds

CERCLA/SARA - Hazardous Substances and their Reportable Quantities
final RQ = 100 pounds (45.4 kg)

RCRA - P Series Wastes
waste number P109

RCRA - Hazardous Constituents-Appendix VIII
waste number P109

RCRA - Substances Banned From Land Disposal
[present]

RCRA - TSD Facilities Ground Water Monitoring
TM 8270 = 10 ug/L PQL

INTERNATIONAL LISTS

Australian Exposure Standards - Time Weighted Averages
0.2 mg/m3 TWA

Australian Exposure Standards - Skin Effects
skin absorption

Canada - Alberta - 8 Hour Occupational Exposure Limit
0.2 mg/m3 TWA

Canada - Alberta - 15 Minute Occupational Exposure Limit
0.6 mg/m3 STEL

Canada - Alberta - Skin Designation
can be absorbed through the intact skin

Canada - British Columbia - 8 Hour Exposure Limits
0.2 mg/m3 TWA

Canada - British Columbia - 15 Minute Exposure Limits
0.6 mg/m3 STEL

Canada - British Columbia - Skin Notations
skin - potential for skin absorption

Canada - Ontario - OHSA - TWAEVs
0.2 mg/m3 TWAEV

Canada - Ontario - OHSA - Skin Notations
absorption through skin, eyes, or mucous membranes

Canada - Quebec - Time-Weighted Average Exposure Values
0.2 mg/m3 TWAEV

Canada - Quebec - Skin Designations
absorbed through the skin

United Kingdom - Occupational Exposure Standards - TWAs
0.2 mg/m3 TWA

United Kingdom - Occupational Exposure Standards - Notes
can be absorbed through skin

German (DFG) - MAK Values
0.015 ppm MAK; 0.2 mg/m3 MAK

German (DFG) - Peak Limitations
10 x normal MAK (30 min average value); don't exceed during shift

German (DFG) - Skin/Sensitizers
danger of cutaneous absorption

Israel - Time Weighted Averages
0.2 mg/m3 TWA

Israel - Action Levels
0.1 mg/m3 AL

Mexico - Instruction No. 10 - TWAs
0.2 mg/m3 TWA

Mexico - Instruction No. 10 - STELs
0.6 mg/m3 STEL

Mexico - Instruction No. 10 - Skin designation
skin - potential for cutaneous absorption

STATE LISTS

California - Exposure Limits - PELs
0.2 mg/m3 PEL

California - Exposure Limits - Skin Notation
material may be absorbed through the skin, eyes or mucous membrane

California - Directors List of Hazardous Substances (8 CCR 339)
[present]

Florida Hazardous Substance List
[present]

Massachusetts Right To Know List
extraordinarily hazardous

Minnesota Hazardous Substance List
skin

NJ Right to Know List (Total)
sn 1756

Pennsylvania Right to Know List
environmental hazard

SULFUR COATED UREA   RR-01148-3

HEALTH AND SAFETY LISTS

NIOSH - Selected LD50s and LC50s
Oral, rat: LD50 = 5660 mg/kg Skin, rabbit: LD50 = 8 mg/kg

ENVIRONMENTAL LISTS

CERCLA/SARA - Section 302 Extremely Hazardous Substances and TPQs
TPQ = 500 pounds

STATE LISTS

Florida Hazardous Substance List
effective March 13, 1992

Massachusetts Right To Know List
extraordinarily hazardous

Pennsylvania Right to Know List
environmental hazard

PROPOSED REGULATIONS

CERCLA/SARA - Proposed Hazardous Substance Additions
proposed RQ = 1 pound (.454 kg)

CERCLA/SARA - 1989 Proposed RQ Adjustments
proposed RQ = 100 pounds (45.4 kg)

SULFUR DICHLORIDE   10545-99-0

HEALTH AND SAFETY LISTS

U.S. DOT - Substances From 49 CFR 172.101
regulated by DOT (UN2448, UN1350, NA1350)

U.S. DOT - Hazard Classes
Forbidden from transport by the DOT

NFPA - Flash Points
flash point = 405 degrees F (207 degrees C)

NFPA - Hazard Identification Ratings
health-2; flammability-1; reactivity-0

ENVIRONMENTAL LISTS

List of Pesticide Product Inert Ingredients
[present]

INTERNATIONAL LISTS

Canada - Alberta - 8 Hour Occupational Exposure Limit
10 mg/m3 TWA

Canada - Alberta - 15 Minute Occupational Exposure Limit
20 mg/m3 STEL

STATE LISTS

California - Directors List of Hazardous Substances (8 CCR 339)
[present]

Florida Hazardous Substance List
[present]

Massachusetts Right To Know List
[present]

NJ Right to Know List (Total)
sn 1757

Pennsylvania Right to Know List
[present]

SULFUR DIOXIDE   7446-09-5

HEALTH AND SAFETY LISTS

U.S. DOT - Appendix A Table 2 - Radionuclides
final RQ = 1 curie (3.7E 10 Bq)

ENVIRONMENTAL LISTS

ATSDR Priority List
Rank (of 275): 198

CERCLA/SARA List of Radionuclides (Appendix B) and Their Reportable Quantities
final RQ = 1 curie (3.7E 10 Bq)

SULFUR FLUORIDE (SF5)   10546-01-7

HEALTH AND SAFETY LISTS

U.S. DOT - Appendix A Table 1 - Hazardous Substances
final RQ = 100 pounds (45.4 kg)

NFPA - Flash Points
flash point = 245 degrees F (118 degrees C)

NFPA - Hazard Identification Ratings
health-3; flammability-1; reactivity-1 (decomposes in water)

ENVIRONMENTAL LISTS

CERCLA/SARA - Hazardous Substances and their Reportable Quantities
final RQ = 100 pounds (45.4 kg)

Clean Water Act - Hazardous Substances
[present]

STATE LISTS

Massachusetts Right To Know List
[present]

NJ Right to Know List (Total)
sn 3048

Pennsylvania Right to Know List
environmental hazard

SULFUR HEXAFLUORIDE   2551-62-4

HEALTH AND SAFETY LISTS

U.S. DOT - Substances Which Are Poisonous by Inhalation
liquid hazardous material poisonous by inhalation

SULFURIC ACID   7664-93-9

HEALTH AND SAFETY LISTS

U.S. DOT - Substances Which Are Poisonous by Inhalation
gaseous hazardous material poisonous by inhalation

SULFURIC ACID   8014-95-7

HEALTH AND SAFETY LISTS

U.S. DOT - Substances From 49 CFR 172.101
regulated by DOT (UN1828)

U.S. DOT - Hazard Classes
DOT hazard class = 8

SULFURIC ACID, C8-10-ALKYL ESTERS, COM-   RR-01149-4
POUNDS WITH ISOPROPANOLAMINE
ENVIRONMENTAL LISTS
List of Pesticide Product Inert Ingredients
[present]

SULFURIC ACID DIPOTASSIUM SALT   7778-80-5
HEALTH AND SAFETY LISTS
U.S. DOT - Substances Which Are Poisonous by Inhalation
liquid hazardous material poisonous by inhalation (mono- or mixed with Carbon tetrachloride) (UN1828)
INTERNATIONAL LISTS
Canada - WHMIS: Ingredient Disclosure
1% item 1482 (684)
STATE LISTS
Pennsylvania Right to Know List
[present]

SULFURIC ACID, MONODECYL ESTER, SODIUM   142-87-0
SALT
HEALTH AND SAFETY LISTS
ACGIH 1995 - Time Weighted Averages
2 ppm TWA; 5.2 mg/m3 TWA
ACGIH 1995 - Short Term Exposure Limits
5 ppm STEL; 13 mg/m3 STEL
AIHA - Odor Threshold Values
geometric mean air odor threshold = 2.7 ppm (detectable); 4.4 ppm (recognizable)
U.S. DOT - Substances From 49 CFR 172.101
regulated by DOT (UN1079)
U.S. DOT - Hazard Classes
DOT hazard class = 2.3
U.S. DOT - Substances Which Are Poisonous by Inhalation
gaseous hazardous material poisonous by inhalation (UN1079)
IARC - Group 3 (not classifiable)
[present]
NIOSH - Selected LD50s and LC50s
Inhalation, rat: LC50 = 2520 ppm 1 hr
NIOSH 1990 - Pocket Guide - RELs
2 ppm TWA; 5 mg/m3 TWA; 5 ppm STEL; 10 mg/m3 STEL
NIOSH 1990 - Pocket Guide - IDLHs
100 ppm IDLH
NIOSH 1990 - Pocket Guide - Target organs
skin, eyes, respiratory system
NIOSH - Health Standards - Exposure Limits
0.5 ppm TWA; 1.3 mg/m3 TWA
NIOSH - Health Standards - Health Effects and Precautions
Respiratory effects (Pulmonary function testing required)
OSHA - Vacated PELs - Time Weighted Averages
2 ppm TWA; 5 mg/m3 TWA
OSHA - Vacated PELs - Short Term Exposure Limits
5 ppm STEL; 15 mg/m3 STEL
OSHA - Final PELs - Time Weighted Averages
5 ppm TWA; 13 mg/m3 TWA
OSHA - List of Highly Hazardous Chemicals
threshhold quantity = 1000 pounds
ENVIRONMENTAL LISTS
CERCLA/SARA - Section 302 Extremely Hazardous Substances and TPQs
TPQ = 500 pounds
CAA -Toxic Substances for Accidental Release Prevention
threshold quantity = 5,000 lbs
List of Pesticide Product Inert Ingredients
[present]

INTERNATIONAL LISTS
Australian Exposure Standards - Time Weighted Averages
2 ppm TWA; 5.2 mg/m3 TWA
Australian Exposure Standards - Short Term Exposure Limits
5 ppm STEL; 13 mg/m3 STEL
Canada - WHMIS: Ingredient Disclosure
1% item 1483 (773)
Canada - National Air Quality Objectives - Schedule I
desirable limits: 0-30 ug/m3 year; 0-150 ug/m3 day; 0-450 ug/m3 hour; acceptable limits: 30-60 ug/m3 year; 150-300 ug/m3 day; 450-900 ug/m3 hour; tolerable limits: 300-800 ug/m3 day
Canada - Alberta - 8 Hour Occupational Exposure Limit
2 ppm TWA; 5 mg/m3 TWA
Canada - Alberta - 15 Minute Occupational Exposure Limit
5 ppm STEL; 13 mg/m3 STEL
Canada - British Columbia - 8 Hour Exposure Limits
2 ppm TWA
Canada - British Columbia - 15 Minute Exposure Limits
5 ppm STEL
Canada - Ontario - OHSA - TWAEVs
2 ppm TWAEV; 5.2 mg/m3 TWAEV
Canada - Ontario - OHSA - STEVs
5 ppm STEV; 10.4 mg/m3 STEV
Canada - Quebec - Time-Weighted Average Exposure Values
2 ppm TWAEV; 5.2 mg/m3 TWAEV
Canada - Quebec - Short-term Exposure Values
5 ppm STEV; 13 mg/m3 STEV
United Kingdom - Occupational Exposure Standards - TWAs
2 ppm TWA; 5 mg/m3 TWA
United Kingdom - Occupational Exposure Standards - STELs
5 ppm STEL; 13 mg/m3 STEL
German (DFG) - MAK Values
2 ppm MAK; 5 mg/m3 MAK
German (DFG) - Peak Limitations
2 x normal MAK (5 min momentary value); don't exceed 8 times during shift
Israel - Time Weighted Averages
2 ppm TWA; 5.2 mg/m3 TWA
Israel - Short Term Exposure Limits
5 ppm STEL; 13 mg/m3 STEL
Israel - Action Levels
1 ppm AL; 2.6 mg/m3 AL
Mexico - Instruction No. 10 - TWAs
2 ppm TWA; 5 mg/m3 TWA
Mexico - Instruction No. 10 - STELs
5 ppm STEL; 10 mg/m3 STEL
STATE LISTS
California - Exposure Limits - PELs
2 ppm PEL; 5 mg/m3 PEL
California - Exposure Limits - STELs
5 ppm STEL; 10 mg/m3 STEL
California - Directors List of Hazardous Substances (8 CCR 339)
[present]
Florida Hazardous Substance List
[present]
Massachusetts Right To Know List
extraordinarily hazardous
Minnesota Hazardous Substance List
[present]
NJ Right to Know List (Total)
sn 1759
Pennsylvania Right to Know List
environmental hazard

PROPOSED REGULATIONS
CERCLA/SARA - Proposed Hazardous Substance Additions
proposed RQ = 1 pound (.454 kg)
CERCLA/SARA - 1989 Proposed RQ Adjustments
proposed RQ = 100 pounds (45.4 kg)

SULFURIZED ALKYLPHENOLS   RR-01016-2
STATE LISTS
Pennsylvania Right to Know List
[present]

SULFUR MONOCHLORIDE   10025-67-9
HEALTH AND SAFETY LISTS
ACGIH 1995 - Time Weighted Averages
1000 ppm TWA; 5970 mg/m3 TWA
U.S. DOT - Substances From 49 CFR 172.101
regulated by DOT (UN1080)
U.S. DOT - Hazard Classes
DOT hazard class = 2.2
OSHA - Vacated PELs - Time Weighted Averages
1000 ppm TWA; 6000 mg/m3 TWA
OSHA - Final PELs - Time Weighted Averages
1000 ppm TWA; 6000 mg/m3 TWA

INTERNATIONAL LISTS
Australian Exposure Standards - Time Weighted Averages
1000 ppm TWA; 5970 mg/m3 TWA
Canada - WHMIS: Ingredient Disclosure
1% item 1484 (956)
Canada - Alberta - 8 Hour Occupational Exposure Limit
1000 ppm TWA; 5973 mg/m3 TWA
Canada - Alberta - 15 Minute Occupational Exposure Limit
1250 ppm STEL; 7467 mg/m3 STEL
Canada - British Columbia - 8 Hour Exposure Limits
1000 ppm TWA; 6000 mg/m3 TWA
Canada - British Columbia - 15 Minute Exposure Limits
1250 ppm STEL; 7500 mg/m3 STEL
Canada - Ontario - OHSA - TWAEVs
1000 ppm TWAEV; 5970 mg/m3 TWAEV
Canada - Quebec - Time-Weighted Average Exposure Values
1000 ppm TWAEV; 5970 mg/m3 TWAEV
United Kingdom - Occupational Exposure Standards - TWAs
1000 ppm TWA; 6000 mg/m3 TWA
United Kingdom - Occupational Exposure Standards - STELs
1250 ppm STEL; 7500 mg/m3 STEL
German (DFG) - MAK Values
1000 ppm MAK; 6000 mg/m3 MAK
German (DFG) - Peak Limitations
2 x normal MAK (1 hour momentary value); don't exceed 3 times per shift
Israel - Time Weighted Averages
1000 ppm TWA; 5970 mg/m3 TWA
Israel - Action Levels
500 ppm AL; 2985 mg/m3 AL
Mexico - Instruction No. 10 - TWAs
1000 ppm TWA; 6000 mg/m3 TWA
Mexico - Instruction No. 10 - STELs
1250 ppm STEL; 7500 mg/m3 STEL

STATE LISTS
California - Exposure Limits - PELs
1000 ppm PEL; 6000 mg/m3 PEL
California - Directors List of Hazardous Substances (8 CCR 339)
[present]

Florida Hazardous Substance List
[present]
Massachusetts Right To Know List
[present]
Minnesota Hazardous Substance List
[present]
NJ Right to Know List (Total)
sn 1760
Pennsylvania Right to Know List
[present]

SULFUROUS ACID   7782-99-2
HEALTH AND SAFETY LISTS
ACGIH 1995 - Time Weighted Averages
1 mg/m3 TWA
ACGIH 1995 - Short Term Exposure Limits
3 mg/m3 STEL
AIHA - Odor Threshold Values
no geometric mean air odor threshold
U.S. DOT - Substances From 49 CFR 172.101
regulated by DOT (UN1832, UN1831, UN1830)
U.S. DOT - Hazard Classes
DOT hazard class = 8
U.S. DOT - Substances Which Are Poisonous by Inhalation
liquid hazardous material poisonous by inhalation (fuming, > 30% free SO3) (UN1831)
U.S. DOT - Appendix A Table 1 - Hazardous Substances
final RQ = 1000 pounds (454 kg)
IARC - Group 1 (carcinogenic to humans)
[present] (occupation exposures to strong inorganic acid mists containing sulfuric acid)
NIOSH - Selected LD50s and LC50s
Inhalation, rat: LC50 = 510 mg/m3 2 hr Oral, rat: LD50 = 2140 mg/kg
NIOSH 1990 - Pocket Guide - RELs
1 mg/m3 TWA
NIOSH 1990 - Pocket Guide - IDLHs
80 mg/m3 IDLH
NIOSH 1990 - Pocket Guide - Target organs
respiratory system, skin, eyes, teeth
NIOSH - Health Standards - Exposure Limits
1 mg/m3 TWA
NIOSH - Health Standards - Health Effects and Precautions
Pulmonary irritation (Prevent skin and eye contact)
OSHA - Vacated PELs - Time Weighted Averages
1 mg/m3 TWA
OSHA - Final PELs - Time Weighted Averages
1 mg/m3 TWA
OSHA - Select Carcinogens
[present]

ENVIRONMENTAL LISTS
CERCLA/SARA - Section 302 Extremely Hazardous Substances and TPQs
TPQ = 1000 pounds
CERCLA/SARA - Section 313 - Emission Reporting
form R reporting required for 1.0% de minimus concentration
CERCLA/SARA - Hazardous Substances and their Reportable Quantities
final RQ = 1000 pounds (454 kg)
Clean Water Act - Hazardous Substances
[present]
List of Pesticide Product Inert Ingredients
[present]

INTERNATIONAL LISTS

Australian Exposure Standards - Time Weighted Averages
1 mg/m3 TWA

Australian Exposure Standards - Short Term Exposure Limits
3 mg/m3 STEL

Canada - WHMIS: Ingredient Disclosure
1% item 1485 (138)

Canada - NPRI (National Pollutant Release Inventory)
[present]

Canada - Alberta - 8 Hour Occupational Exposure Limit
1 mg/m3 TWA

Canada - Alberta - 15 Minute Occupational Exposure Limit
3 mg/m3 STEL

Canada - British Columbia - 8 Hour Exposure Limits
1 mg/m3 TWA

Canada - Ontario - OHSA - TWAEVs
1 mg/m3 TWAEV

Canada - Quebec - Time-Weighted Average Exposure Values
1 mg/m3 TWAEV

Canada - Quebec - Short-term Exposure Values
3 mg/m3 STEV

United Kingdom - Occupational Exposure Standards - TWAs
1 mg/m3 TWA

German (DFG) - MAK Values
total dust: 1 mg/m3 MAK

German (DFG) - Peak Limitations
2 x normal MAK (5 min momentary value);   don't exceed 8 times during shift

Israel - Time Weighted Averages
1 mg/m3 TWA

Israel - Short Term Exposure Limits
3 mg/m3 STEL

Israel - Action Levels
0.5 mg/m3 AL

Mexico - Instruction No. 10 - TWAs
1 mg/m3 TWA

STATE LISTS

California - Air Bill 2588 Appendix A-I
6/91

California - Exposure Limits - PELs
1 mg/m3 PEL

California - Exposure Limits - STELs
3 mg/m3 STEL

California - Directors List of Hazardous Substances (8 CCR 339)
[present]

Florida Hazardous Substance List
[present]

Massachusetts Right To Know List
extraordinarily hazardous

Minnesota Hazardous Substance List
[present]

NJ Right to Know List (Total)
sn 1761

NJ Special Hazardous Substances
(corrosive;   reactive - second degree)

Pennsylvania Right to Know List
environmental hazard

SULFUR PENTAFLUORIDE   5714-22-7

HEALTH AND SAFETY LISTS

U.S. DOT - Appendix A Table 1 - Hazardous Substances
final RQ = 1000 pounds (454 kg)

NIOSH - Selected LD50s and LC50s
Inhalation, rat: LC50 = 347 ppm 1 hr

OSHA - List of Highly Hazardous Chemicals
65% to 80% by weight: threshhold quantity = 1000 pounds

ENVIRONMENTAL LISTS

CERCLA/SARA - Hazardous Substances and their Reportable Quantities
final RQ = 1000 pounds (454 kg)

CAA -Toxic Substances for Accidental Release Prevention
threshold quantity = 10,000 lbs (The mixture exemption in Section 68.115(b)(1) does not apply to this substance)

STATE LISTS

Massachusetts Right To Know List
[present]

NJ Right to Know List (Total)
sn 1762

Pennsylvania Right to Know List
environmental hazard

SULFUR TETRAFLUORIDE   7783-60-0

ENVIRONMENTAL LISTS

List of Pesticide Product Inert Ingredients
[present]

SULFUR TRIOXIDE   7446-11-9

HEALTH AND SAFETY LISTS

NIOSH - Selected LD50s and LC50s
Oral, rat: LD50 = 6600 mg/kg

ENVIRONMENTAL LISTS

List of Pesticide Product Inert Ingredients
[present]

SULFURYL CHLORIDE   7791-25-5

HEALTH AND SAFETY LISTS

NIOSH - Selected LD50s and LC50s
Oral, rat: LD50 = 1950 mg/kg

ENVIRONMENTAL LISTS

List of Pesticide Product Inert Ingredients
[present]

SULFURYL FLUORIDE   2699-79-8

ENVIRONMENTAL LISTS

TSCA - Chemicals with Significant New Use Rules
PMN number: P-89-708

SULPHUR-BRIDGED SUBSTITUTED PHENOLS   RR-00937-0

HEALTH AND SAFETY LISTS

ACGIH 1995 - Ceiling Limits
C 1 ppm; C 5.5 mg/m3

NIOSH 1990 - Pocket Guide - RELs
C 1 ppm; C 6 mg/m3

NIOSH 1990 - Pocket Guide - IDLHs
10 ppm IDLH

NIOSH 1990 - Pocket Guide - Target organs
skin, eyes, respiratory system

OSHA - Vacated PELs - Ceiling Limits
C 1 ppm; C 6 mg/m3

OSHA - Final PELs - Time Weighted Averages
1 ppm TWA; 6 mg/m3 TWA

ENVIRONMENTAL LISTS

List of Pesticide Product Inert Ingredients
[present]

INTERNATIONAL LISTS

Australian Exposure Standards - Time Weighted Averages
Peak Limitation: 1 ppm; 5.5 mg/m3

Canada - WHMIS: Ingredient Disclosure
1% item 1481 (525)
Canada - Alberta - Ceiling Occupational Exposure Limit
C 1 ppm; C 5.5 mg/m3
Canada - British Columbia - 8 Hour Exposure Limits
1 ppm TWA; 6 mg/m3 TWA
Canada - British Columbia - 15 Minute Exposure Limits
3 ppm STEL; 18 mg/m3 STEL
Canada - Ontario - OHSA - CEVs
1 ppm CEV; 6 mg/m3 CEV
Canada - Quebec - Ceiling Limits
P 1 ppm; P 5.5 mg/m3
United Kingdom - Occupational Exposure Standards - STELs
1 ppm STEL; 6 mg/m3 STEL
German (DFG) - MAK Values
1 ppm MAK; 6 mg/m3 MAK
German (DFG) - Peak Limitations
2 x normal MAK (5 min momentary value);   don't exceed 8 times during shift
Israel - Ceiling Exposure Limits
C 1 ppm; C 5.5 mg/m3
Mexico - Instruction No. 10 - TWAs
1 ppm TWA; 6 mg/m3 TWA
Mexico - Instruction No. 10 - STELs
3 ppm STEL; 18 mg/m3 STEL

STATE LISTS

California - Exposure Limits - Ceilings
C 1 ppm; C 6 mg/m3
California - Directors List of Hazardous Substances (8 CCR 339)
[present]
Florida Hazardous Substance List
[present]
Massachusetts Right To Know List
[present]
Minnesota Hazardous Substance List
[present]
NJ Right to Know List (Total)
sn 1758
NJ Special Hazardous Substances
(corrosive; reactive - second degree)
Pennsylvania Right to Know List
[present]

SULPROFOS   35400-43-2

HEALTH AND SAFETY LISTS

U.S. DOT - Substances From 49 CFR 172.101
regulated by DOT (UN1833)
U.S. DOT - Hazard Classes
DOT hazard class = 8

INTERNATIONAL LISTS

Canada - WHMIS: Ingredient Disclosure
1% item 1486 (137)

STATE LISTS

NJ Right to Know List (Total)
sn 1764
NJ Special Hazardous Substances
(corrosive)

SUNFLOWER SEEDS   68937-99-5

HEALTH AND SAFETY LISTS

ACGIH 1995 - Ceiling Limits
C 0.01 ppm; C 0.10 mg/m3

NIOSH - Selected LD50s and LC50s
Inhalation, rat: LC50 = 2 gm/m3 10 mn
NIOSH 1990 - Pocket Guide - RELs
C 0.01 ppm; C 0.1 mg/m3
NIOSH 1990 - Pocket Guide - IDLHs
1 ppm IDLH
NIOSH 1990 - Pocket Guide - Target organs
CNS, respiratory system
OSHA - Vacated PELs - Ceiling Limits
C 0.01 ppm; C 0.1 mg/m3 (Enforcement indefinitely stayed)
OSHA - Final PELs - Time Weighted Averages
0.025 ppm TWA; 0.25 mg/m3 TWA
OSHA - List of Highly Hazardous Chemicals
threshhold quantity = 250 pounds

INTERNATIONAL LISTS

Australian Exposure Standards - Time Weighted Averages
Peak Limitation: 0.01 ppm; 0.1 mg/m3
Canada - WHMIS: Ingredient Disclosure
1% item 1487 (1347)
Canada - Alberta - Ceiling Occupational Exposure Limit
C 0.01 ppm; C 0.1 mg/m3
Canada - British Columbia - 8 Hour Exposure Limits
0.025 ppm TWA; 0.25 mg/m3 TWA
Canada - British Columbia - 15 Minute Exposure Limits
0.075 ppm STEL; 0.75 mg/m3 STEL
Canada - Ontario - OHSA - CEVs
0.01 ppm CEV; 0.1 mg/m3 CEV
Canada - Quebec - Ceiling Limits
P 0.01 ppm; P 0.1 mg/m3
United Kingdom - Occupational Exposure Standards - TWAs
0.025 ppm TWA; 0.25 mg/m3 TWA
United Kingdom - Occupational Exposure Standards - STELs
0.075 ppm STEL; 0.75 mg/m3 STEL
German (DFG) - MAK Values
0.025 ppm MAK; 0.25 mg/m3 MAK
German (DFG) - Peak Limitations
2 x normal MAK (5 min momentary value);   don't exceed 8 times during shift
Israel - Ceiling Exposure Limits
C 0.01 ppm; C 0.10 mg/m3
Mexico - Instruction No. 10 - TWAs
0.025 ppm TWA; 0.25 mg/m3 TWA
Mexico - Instruction No. 10 - STELs
0.075 ppm STEL; 0.75 mg/m3 STEL

STATE LISTS

California - Exposure Limits - Ceilings
C 0.01 ppm; C 0.1 mg/m3
California - Directors List of Hazardous Substances (8 CCR 339)
[present]
Florida Hazardous Substance List
[present]
Massachusetts Right To Know List
[present]
Minnesota Hazardous Substance List
[present]

SUSPENDED PARTICULATE MATTER   RR-01557-6

HEALTH AND SAFETY LISTS

ACGIH 1995 - Ceiling Limits
C 0.1 ppm; C 0.44 mg/m3
U.S. DOT - Substances From 49 CFR 172.101
regulated by DOT (UN2418)

U.S. DOT - Hazard Classes
DOT hazard class = 2.3

U.S. DOT - Substances Which Are Poisonous by Inhalation
gaseous hazardous material poisonous by inhalation (UN2418)

OSHA - Vacated PELs - Ceiling Limits
C 0.1 ppm; C 0.4 mg/m3

OSHA - List of Highly Hazardous Chemicals
threshhold quantity = 250 pounds

ENVIRONMENTAL LISTS

CERCLA/SARA - Section 302 Extremely Hazardous Substances and TPQs
TPQ = 100 pounds

CAA -Toxic Substances for Accidental Release Prevention
threshold quantity = 2,500 lbs

INTERNATIONAL LISTS

Australian Exposure Standards - Time Weighted Averages
Peak Limitation: 0.1 ppm; 0.44 mg/m3

Canada - WHMIS: Ingredient Disclosure
1% item 1488 (1591)

Canada - Alberta - 8 Hour Occupational Exposure Limit
0.1 ppm TWA; 0.44 mg/m3 TWA

Canada - Alberta - 15 Minute Occupational Exposure Limit
0.3 ppm STEL; 1.3 mg/m3 STEL

Canada - British Columbia - 8 Hour Exposure Limits
0.1 ppm TWA; 0.4 mg/m3 TWA

Canada - British Columbia - 15 Minute Exposure Limits
0.3 ppm STEL; 1 mg/m3 STEL

Canada - Ontario - OHSA - CEVs
0.1 ppm CEV; 0.4 mg/m3 CEV

Canada - Quebec - Ceiling Limits
P 0.1 ppm; 0.44 mg/m3

United Kingdom - Occupational Exposure Standards - TWAs
0.1 ppm TWA; 0.4 mg/m3 TWA

United Kingdom - Occupational Exposure Standards - STELs
0.3 ppm STEL; 1 mg/m3 STEL

Israel - Ceiling Exposure Limits
C 0.1 ppm; C 0.44 mg/m3

Mexico - Instruction No. 10 - TWAs
0.1 ppm TWA; 0.4 mg/m3 TWA

Mexico - Instruction No. 10 - STELs
0.3 ppm STEL; 1 mg/m3 STEL

STATE LISTS

California - Exposure Limits - Ceilings
C 0.1 ppm; C 0.4 mg/m3

California - Directors List of Hazardous Substances (8 CCR 339)
[present]

Florida Hazardous Substance List
[present]

Massachusetts Right To Know List
extraordinarily hazardous

Minnesota Hazardous Substance List
[present]

NJ Right to Know List (Total)
sn 1766

Pennsylvania Right to Know List
environmental hazard

PROPOSED REGULATIONS

CERCLA/SARA - Proposed Hazardous Substance Additions
proposed RQ = 1 pound (.454 kg)

CERCLA/SARA - 1989 Proposed RQ Adjustments
proposed RQ = 100 pounds (45.4 kg)

SWAT 122-10-1

HEALTH AND SAFETY LISTS

U.S. DOT - Substances From 49 CFR 172.101
regulated by DOT (NA1829, UN 1829)

U.S. DOT - Hazard Classes
DOT hazard class = 8

U.S. DOT - Substances Which Are Poisonous by Inhalation
liquid hazardous material poisonous by inhalation (UN1829, NA1829)

OSHA - List of Highly Hazardous Chemicals
threshhold quantity = 1000 pounds

ENVIRONMENTAL LISTS

CERCLA/SARA - Section 302 Extremely Hazardous Substances and TPQs
TPQ = 100 pounds (This material is a reactive solid. The TPQ does not default to 10,000 pounds for non-powder, non-molten, non-solution form)

CAA -Toxic Substances for Accidental Release Prevention
threshold quantity = 10,000 lbs

INTERNATIONAL LISTS

Canada - WHMIS: Ingredient Disclosure
1% item 1489 (1696)

STATE LISTS

Massachusetts Right To Know List
extraordinarily hazardous

NJ Right to Know List (Total)
sn 1767

NJ Special Hazardous Substances
(corrosive)

Pennsylvania Right to Know List
environmental hazard

PROPOSED REGULATIONS

CERCLA/SARA - Proposed Hazardous Substance Additions
proposed RQ = 1 pound (.454 kg)

CERCLA/SARA - 1989 Proposed RQ Adjustments
proposed RQ = 100 pounds (45.4 kg)

SWEP 1918-18-9

HEALTH AND SAFETY LISTS

U.S. DOT - Substances From 49 CFR 172.101
regulated by DOT (UN1834)

U.S. DOT - Hazard Classes
DOT hazard class = 8

U.S. DOT - Substances Which Are Poisonous by Inhalation
liquid hazardous material poisonous by inhalation (UN1834)

INTERNATIONAL LISTS

Canada - WHMIS: Ingredient Disclosure
1% item 1490 (527)

STATE LISTS

Florida Hazardous Substance List
[present]

Massachusetts Right To Know List
[present]

NJ Right to Know List (Total)
sn 1768

NJ Special Hazardous Substances
(corrosive; reactive - second degree)

Pennsylvania Right to Know List
[present]

SYMPHYTINE   22571-95-5

HEALTH AND SAFETY LISTS

ACGIH 1995 - Time Weighted Averages
5 ppm TWA; 21 mg/m3 TWA

ACGIH 1995 - Short Term Exposure Limits
10 ppm STEL; 42 mg/m3 STEL

U.S. DOT - Substances From 49 CFR 172.101
regulated by DOT (UN2191)

U.S. DOT - Hazard Classes
DOT hazard class = 2.3

U.S. DOT - Substances Which Are Poisonous by Inhalation
gaseous hazardous material poisonous by inhalation (UN2191)

NIOSH - Selected LD50s and LC50s
Inhalation, rat: LC50 = 3020 ppm 1 hr Oral, rat: LD50 = 100 mg/kg

NIOSH 1990 - Pocket Guide - RELs
5 ppm TWA; 20 mg/m3 TWA; 10 ppm STEL; 40 mg/m3 STEL

NIOSH 1990 - Pocket Guide - IDLHs
1000 ppm IDLH

NIOSH 1990 - Pocket Guide - Target organs
CNS, respiratory system

OSHA - Vacated PELs - Time Weighted Averages
5 ppm TWA; 20 mg/m3 TWA

OSHA - Vacated PELs - Short Term Exposure Limits
10 ppm STEL; 40 mg/m3 STEL

OSHA - Final PELs - Time Weighted Averages
5 ppm TWA; 20 mg/m3 TWA

ENVIRONMENTAL LISTS

CERCLA/SARA - Section 313 - Emission Reporting
form R reporting required

INTERNATIONAL LISTS

Australian Exposure Standards - Time Weighted Averages
5 ppm TWA; 21 mg/m3 TWA

Australian Exposure Standards - Short Term Exposure Limits
10 ppm STEL; 42 mg/m3 STEL

Canada - Alberta - 8 Hour Occupational Exposure Limit
5 ppm TWA; 21 mg/m3 TWA

Canada - Alberta - 15 Minute Occupational Exposure Limit
10 ppm STEL; 42 mg/m3 STEL

Canada - British Columbia - 8 Hour Exposure Limits
5 ppm TWA; 20 mg/m3 TWA

Canada - British Columbia - 15 Minute Exposure Limits
10 ppm STEL; 40 mg/m3 STEL

Canada - Ontario - OHSA - TWAEVs
5 ppm TWAEV; 21 mg/m3 TWAEV

Canada - Ontario - OHSA - STEVs
10 ppm STEV; 42 mg/m3 STEV

Canada - Quebec - Time-Weighted Average Exposure Values
5 ppm TWAEV; 21 mg/m3 TWAEV

Canada - Quebec - Short-term Exposure Values
10 ppm STEV; 42 mg/m3 STEV

United Kingdom - Occupational Exposure Standards - TWAs
5 ppm TWA; 20 mg/m3 TWA

United Kingdom - Occupational Exposure Standards - STELs
10 ppm STEL; 40 mg/m3 STEL

Israel - Time Weighted Averages
5 ppm TWA; 21 mg/m3 TWA

Israel - Short Term Exposure Limits
10 ppm STEL; 42 mg/m3 STEL

Israel - Action Levels
2.5 ppm AL; 10.5 mg/m3 AL

Mexico - Instruction No. 10 - TWAs
5 ppm TWA; 20 mg/m3 TWA

Mexico - Instruction No. 10 - STELs
10 ppm STEL; 40 mg/m3 STEL

STATE LISTS

California - Exposure Limits - PELs
5 ppm PEL; 20 mg/m3 PEL

California - Exposure Limits - STELs
10 ppm STEL; 40 mg/m3 STEL

California - Directors List of Hazardous Substances (8 CCR 339)
[present]

Florida Hazardous Substance List
[present]

Massachusetts Right To Know List
[present]

Minnesota Hazardous Substance List
[present]

NJ Right to Know List (Total)
sn 1769

Pennsylvania Right to Know List
[present]

2,4,5-T   93-76-5

ENVIRONMENTAL LISTS

TSCA - Chemicals with Significant New Use Rules
PMN number: P-89-396

2,4,5-T AMINE   42589-07-1

HEALTH AND SAFETY LISTS

ACGIH 1995 - Time Weighted Averages
1 mg/m3 TWA

U.S. DOT - Appendix B - Marine Pollutants
DOT regulated severe marine pollutant

NIOSH - Selected LD50s and LC50s
Oral, rat: LD50 = 65 mg/kg Skin, rabbit: LD50 = 820 mg/kg

OSHA - Vacated PELs - Time Weighted Averages
1 mg/m3 TWA

ENVIRONMENTAL LISTS

CERCLA/SARA - Section 313 - Emission Reporting
form R reporting required

INTERNATIONAL LISTS

Australian Exposure Standards - Time Weighted Averages
1 mg/m3 TWA

Canada - Alberta - 8 Hour Occupational Exposure Limit
1 mg/m3 TWA

Canada - Alberta - 15 Minute Occupational Exposure Limit
3 mg/m3 STEL

Canada - Ontario - OHSA - TWAEVs
1 mg/m3 TWAEV

Canada - Quebec - Time-Weighted Average Exposure Values
1 mg/m3 TWAEV

Israel - Time Weighted Averages
1 mg/m3 TWA

Israel - Action Levels
0.5 mg/m3 AL

STATE LISTS

California - Exposure Limits - PELs
1 mg/m3 PEL

California - Directors List of Hazardous Substances (8 CCR 339)
[present]

Florida Hazardous Substance List
effective March 13, 1992

Massachusetts Right To Know List
[present]

Minnesota Hazardous Substance List
[present]

NJ Right to Know List (Total)
sn 1771

Pennsylvania Right to Know List
[present]

2,4,5-T SALTS   13560-99-1

ENVIRONMENTAL LISTS

List of Pesticide Product Inert Ingredients
[present]

TABUN   77-81-6

INTERNATIONAL LISTS

Canada - National Air Quality Objectives - Schedule I
*desirable limits: 0-60 ug/m3 year; acceptable limits: 60-70 ug/m3
year; 0-120 ug/m3 day; tolerable limits: 120-400 ug/m3 day*

TACKS   RR-01150-7

HEALTH AND SAFETY LISTS

NIOSH - Selected LD50s and LC50s
Oral, rat: LD50 = 31 mg/kg

STATE LISTS

NJ Right to Know List (Total)
sn 1772

TALC   14807-96-6

STATE LISTS

California - Directors List of Hazardous Substances (8 CCR 339)
[present]

TALC CONTAINING ASBESTOS FIBERS   RR-00029-3

HEALTH AND SAFETY LISTS

IARC - Group 3 (not classifiable)
[present]

TALL OIL   8002-26-4
SEE ALSO:
F039-HAZARDOUS WASTES

HEALTH AND SAFETY LISTS

ACGIH 1995 - Time Weighted Averages
10 mg/m3 TWA

U.S. DOT - Appendix A Table 1 - Hazardous Substances
final RQ = 1000 pounds (454 kg)

IARC - Group Unspecified
[present] (Listed under 'Chlorophenoxy herbicides')

NIOSH - Selected LD50s and LC50s
Oral, rat: LD50 = 300 mg/kg Skin, rat: LD50 = 1535 mg/kg

NIOSH 1990 - Pocket Guide - RELs
10 mg/m3 TWA

NIOSH 1990 - Pocket Guide - Target organs
skin, liver, GI tract

OSHA - Vacated PELs - Time Weighted Averages
10 mg/m3 TWA

OSHA - Final PELs - Time Weighted Averages
10 mg/m3 TWA

ENVIRONMENTAL LISTS

CERCLA/SARA - Hazardous Substances and their Reportable Quantities
final RQ = 1000 pounds (454 kg)

Clean Water Act - Hazardous Substances
[present]

RCRA - Hazardous Constituents-Appendix VIII
contained in F027 waste stream

RCRA - Basis for Listing - Appendix VII
Included in waste stream: F039

RCRA - TSD Facilities Ground Water Monitoring
TM 8150 = 2 ug/L PQL

RCRA - Universal Treatment Standards (LDR)
WW: 0.72 mg/l; NWW: 7.9 mg/kg

INTERNATIONAL LISTS

Australian Exposure Standards - Time Weighted Averages
10 mg/m3 TWA

Canada - Drinking Water Quality - MACs
0.28 mg/L MAC

Canada - Drinking Water Quality - AOs
<= 0.02 mg/L AO

Canada - Alberta - 8 Hour Occupational Exposure Limit
10 mg/m3 TWA

Canada - Alberta - 15 Minute Occupational Exposure Limit
20 mg/m3 STEL

Canada - British Columbia - 8 Hour Exposure Limits
10 mg/m3 TWA

Canada - British Columbia - 15 Minute Exposure Limits
20 mg/m3 STEL

Canada - Ontario - OHSA - TWAEVs
10 mg/m3 TWAEV

Canada - Quebec - Time-Weighted Average Exposure Values
10 mg/m3 TWAEV

Canada - Quebec - Carcinogens
C2 carcinogen: effect suspected in humans

United Kingdom - Occupational Exposure Standards - TWAs
10 mg/m3 TWA

United Kingdom - Occupational Exposure Standards - STELs
20 mg/m3 STEL

German (DFG) - MAK Values
total dust: 10 mg/m3 MAK

German (DFG) - Peak Limitations
5 x normal MAK (30 min. average value); don't exceed 2 times during shift

German (DFG) - Skin/Sensitizers
danger of cutaneous absorption

Israel - Time Weighted Averages
10 mg/m3 TWA

Israel - Action Levels
5 mg/m3 AL

Mexico - Instruction No. 10 - TWAs
10 mg/m3 TWA

STATE LISTS

California - Exposure Limits - PELs
10 mg/m3 PEL

California - Directors List of Hazardous Substances (8 CCR 339)
[present]

Florida Hazardous Substance List
[present]

Massachusetts Right To Know List
teratogen

Minnesota Hazardous Substance List
[present]

NJ Right to Know List (Total)
sn 1896

NJ Special Hazardous Substances
(mutagen)

Pennsylvania Right to Know List
environmental hazard

PROPOSED REGULATIONS
Safe Drinking Water Act - Priority list
[present]

TALL OIL, ETHOXYLATED   65071-95-6
HEALTH AND SAFETY LISTS
NIOSH - Selected LD50s and LC50s
Oral, rat: LD50 = 2 gm/kg
STATE LISTS
NJ Right to Know List (Total)
sn 1774

TALL OIL FATTY ACIDS   61790-12-3
HEALTH AND SAFETY LISTS
U.S. DOT - Appendix A Table 1 - Hazardous Substances
final RQ = 1000 pounds (454 kg)
ENVIRONMENTAL LISTS
CERCLA/SARA - Hazardous Substances and their Reportable Quantities
final RQ = 1000 pounds (454 kg)
Clean Water Act - Hazardous Substances
[present]
STATE LISTS
California - Directors List of Hazardous Substances (8 CCR 339)
[present]
Massachusetts Right To Know List
[present]
NJ Right to Know List (Total)
sn 2940
Pennsylvania Right to Know List
environmental hazard

TALL OIL FATTY ACIDS, ETHOXYLATED   61791-00-2
HEALTH AND SAFETY LISTS
NIOSH - Selected LD50s and LC50s
Inhalation, rat: LC50 = 304 mg/m3 10 mn Oral, rat: LD50 = 3700 ug/kg Skin, rat: LD50 = 18 mg/kg
ENVIRONMENTAL LISTS
CERCLA/SARA - Section 302 Extremely Hazardous Substances and TPQs
TPQ = 10 pounds
STATE LISTS
Florida Hazardous Substance List
effective March 13, 1992
Massachusetts Right To Know List
extraordinarily hazardous
NJ Right to Know List (Total)
sn 2796
Pennsylvania Right to Know List
environmental hazard
PROPOSED REGULATIONS
CERCLA/SARA - Proposed Hazardous Substance Additions
proposed RQ = 1 pound (.454 kg)
CERCLA/SARA - 1989 Proposed RQ Adjustments
proposed RQ = 10 pounds (4.54 kg)

TALL OIL FATTY ACID SOAP N-ALKYL(C16-C18)   RR-01151-8
TRIMETHYLENEDIAMINE
ENVIRONMENTAL LISTS
List of Pesticide Product Inert Ingredients
[present]

TALL OIL FATTY ACIDS, REACTION PRODUCT   68953-36-6
WITH TETRAETHYLENE PENTAMINE
HEALTH AND SAFETY LISTS
ACGIH 1995 - Time Weighted Averages
2 mg/m3 TWA (this TLV is for the respirable fraction of dust for Talc)
IARC - Group 3 (not classifiable)
[present]
NIOSH 1990 - Pocket Guide - RELs
respirable dust: 2 mg/m3 TWA
NIOSH 1990 - Pocket Guide - Target organs
lungs, CVS
NTP Chemical Status Reports - Testing Status and NTIS Number
Technical reports printed
OSHA - Vacated PELs - Time Weighted Averages
respirable dust (less than 1% crystalline silica): 2 mg/m3 TWA (Listed under 'Silicates')
OSHA - Final PELs - Time Weighted Averages
see Table Z-3
ENVIRONMENTAL LISTS
List of Pesticide Product Inert Ingredients
[present]
INTERNATIONAL LISTS
Australian Exposure Standards - Time Weighted Averages
containing no asbestos fibers: 2.5 mg/m3 TWA
Australian Exposure Standards - Under Review
exposure limits under review
Canada - Alberta - 8 Hour Occupational Exposure Limit
respirable mass: 2 mg/m3 TWA; total mass: 4 mg/m3 TWA
Canada - British Columbia - 8 Hour Exposure Limits
0.5 fibres/mL TWA
Canada - British Columbia - 15 Minute Exposure Limits
5 fibres/ml STEL
Canada - Ontario - OHSA - TWAEVs
respirable dust: 2 mg/m3 TWAEV (listed as mineral dust)
Canada - Quebec - Time-Weighted Average Exposure Values
1 fibers/m3 TWAEV (> 5 microns in length)
Canada - Quebec - Short-term Exposure Values
3 ppm STEV
United Kingdom - Occupational Exposure Standards - TWAs
total inhalable dust: 10 mg/m3 TWA; respirable dust: 1 mg/m3 TWA
German (DFG) - MAK Values
fine dust (without asbestos fibers): 2 mg/m3 MAK
German (DFG) - Pregnancy
no risk to embryo/fetus if exposure limits adhered to
Israel - Time Weighted Averages
total dust: 4 mg/m3 TWA; respirable dust: 2 mg/m3 TWA
Israel - Action Levels
respirable dust: 0.5 mg/m3 AL; total dust: 1.0 mg/m3 AL
STATE LISTS
California - Exposure Limits - PELs
respirable dust (containing no asbestos fibers and < 1% crystalline silica): 2 mg/m3 PEL (Listed under 'Silicates')
California - Directors List of Hazardous Substances (8 CCR 339)
[present] (exempt except when inhalable dust is present or can be generated)
Florida Hazardous Substance List
[present]
Massachusetts Right To Know List
[present] exempt when encapsulated or if particulates are not present and cannot be substantially generated through use of the product.
Minnesota Hazardous Substance List
[present]
Pennsylvania Right to Know List
[present]

TALL OIL FATTY ACIDS, REACTION PRODUCTS   **RR-01250-0**
WITH POLYAMINES, ALKYL SUBSTITUTED

HEALTH AND SAFETY LISTS

ACGIH 1995 - Time Weighted Averages
Use asbestos TLV-TWA (should not exceed 2 mg/m3 respirable dust)

IARC - Group 1 (carcinogenic to humans)
[present]

OSHA - Select Carcinogens
[present]

INTERNATIONAL LISTS

Canada - Alberta - 8 Hour Occupational Exposure Limit
2 f/cm3 TWA

Canada - Alberta - 15 Minute Occupational Exposure Limit
10 f/cm3 STEL

Canada - Quebec - Carcinogens
C1 carcinogen: effect detected in humans

Israel - Action Levels
0.1 f/sec AL

STATE LISTS

California - Air Bill 2588 Appendix A-I
known or potential carcinogen

California - Prop. 65 - Cancer list
carcinogen - initial date 4/1/90

Minnesota Hazardous Substance List
[present]

TALL OIL ROSIN   **8052-10-6**

ENVIRONMENTAL LISTS

List of Pesticide Product Inert Ingredients
[present]

TALLOW AMINE, ETHOXYLATED   **61791-26-2**

ENVIRONMENTAL LISTS

List of Pesticide Product Inert Ingredients
[present]

TALLOW BIS-HYDROXYETHYL GLYCINATE   **RR-01152-9**

ENVIRONMENTAL LISTS

List of Pesticide Product Inert Ingredients
[present]

TALLOW DIAMINE   **61791-53-5**

ENVIRONMENTAL LISTS

List of Pesticide Product Inert Ingredients
[present]

TALLOW OIL   **61789-97-7**

ENVIRONMENTAL LISTS

List of Pesticide Product Inert Ingredients
[present]

TAMOXIFEN CITRATE   **54965-24-1**

ENVIRONMENTAL LISTS

List of Pesticide Product Inert Ingredients
[present]

TANNIC ACID   **1401-55-4**

ENVIRONMENTAL LISTS

TSCA - Chemicals with Significant New Use Rules
PMN number: P-91-225

TANTALUM   **7440-25-7**

ENVIRONMENTAL LISTS

List of Pesticide Product Inert Ingredients
[present]

TANTALUM 172   **15759-26-9**

ENVIRONMENTAL LISTS

List of Pesticide Product Inert Ingredients
[present]

TANTALUM 173   **22095-77-8**

ENVIRONMENTAL LISTS

List of Pesticide Product Inert Ingredients
[present]

TANTALUM 174   **15758-54-0**

ENVIRONMENTAL LISTS

List of Pesticide Product Inert Ingredients
[present]

TANTALUM 175   **15759-28-1**

HEALTH AND SAFETY LISTS

NFPA - Flash Points
*flash point = 509 degrees F (265 degrees C); tallow oil: flash point
= 492 degrees F (256 degrees C)*

NFPA - Hazard Identification Ratings
health-0; flammability-1; reactivity-0

ENVIRONMENTAL LISTS

List of Pesticide Product Inert Ingredients
[present]

STATE LISTS

Pennsylvania Right to Know List
[present]

TANTALUM 176   **15758-55-1**

STATE LISTS

California - Air Bill 2588 Appendix A-II
9/90

California - Prop. 65 - Developmental Toxicity
developmental toxicity - initial date 7/1/90

California - Directors List of Hazardous Substances (8 CCR 339)
[present]

TANTALUM 177   **15759-27-0**

HEALTH AND SAFETY LISTS

IARC - Group 3 (not classifiable)
[present]

NFPA - Flash Points
flash point = 390 degrees F (199 degrees C)

NFPA - Hazard Identification Ratings
health-0; flammability-1; reactivity-0

NIOSH - Selected LD50s and LC50s
Oral, rat: LD50 = 2260 mg/kg

ENVIRONMENTAL LISTS

List of Pesticide Product Inert Ingredients
[present]

STATE LISTS

California - Directors List of Hazardous Substances (8 CCR 339)
[present]

TANTALUM 178   **RR-00289-1**

HEALTH AND SAFETY LISTS

ACGIH 1995 - Time Weighted Averages
As Ta: 5 mg/m3 TWA

NIOSH 1990 - Pocket Guide - RELs
as Ta: 5 mg/m3 TWA;   10 mg/m3 STEL
NIOSH 1990 - Pocket Guide - Target organs
none known in humans
OSHA - Vacated PELs - Time Weighted Averages
5 mg/m3 TWA
OSHA - Final PELs - Time Weighted Averages
5 mg/m3 TWA

INTERNATIONAL LISTS

Australian Exposure Standards - Time Weighted Averages
metal and oxide dusts: 5 mg/m3 TWA
Canada - WHMIS: Ingredient Disclosure
1% item 1491 (1554)
Canada - Alberta - 8 Hour Occupational Exposure Limit
5 mg/m3 TWA
Canada - Alberta - 15 Minute Occupational Exposure Limit
10 mg/m3 STEL
Canada - British Columbia - 8 Hour Exposure Limits
5 mg/m3 TWA
Canada - British Columbia - 15 Minute Exposure Limits
10 mg/m3 STEL
Canada - Ontario - OHSA - TWAEVs
total dust: 5 mg/m3 TWAEV (listed as nuisance particulates)
Canada - Quebec - Time-Weighted Average Exposure Values
metal and oxide dusts (as Ta): 5 mg/m3 TWAEV
United Kingdom - Occupational Exposure Standards - TWAs
5 mg/m3 TWA
United Kingdom - Occupational Exposure Standards - STELs
10 mg/m3 STEL
German (DFG) - MAK Values
total dust: 5 mg/m3 MAK
German (DFG) - Peak Limitations
10 x normal MAK (30 min average value); don't exceed during shift
Israel - Time Weighted Averages
5 mg/m3 TWA
Israel - Action Levels
2.5 mg/m3 AL
Mexico - Instruction No. 10 - TWAs
5 mg/m3 TWA
Mexico - Instruction No. 10 - STELs
10 mg/m3 STEL

STATE LISTS

California - Exposure Limits - PELs
metal dust, as Ta: 5 mg/m3 PEL
California - Directors List of Hazardous Substances (8 CCR 339)
[present]
Florida Hazardous Substance List
[present]
Massachusetts Right To Know List
[present]
Minnesota Hazardous Substance List
[present]
Pennsylvania Right to Know List
[present]

PROPOSED REGULATIONS

Canada - Ontario - Proposed Occupational STEVs
total dust: 10 mg/m3 STEV

TANTALUM 179   14391-27-6

HEALTH AND SAFETY LISTS

U.S. DOT - Appendix A Table 2 - Radionuclides
final RQ = 100 curies (3.7E 12 Bq)

ENVIRONMENTAL LISTS

CERCLA/SARA List of Radionuclides (Appendix B) and Their Reportable Quantities
final RQ = 100 curies (3.7E 12 Bq)

TANTALUM 180   15759-29-2

HEALTH AND SAFETY LISTS

U.S. DOT - Appendix A Table 2 - Radionuclides
final RQ = 100 curies (3.7E 12 Bq)

ENVIRONMENTAL LISTS

CERCLA/SARA List of Radionuclides (Appendix B) and Their Reportable Quantities
final RQ = 100 curies (3.7E 12 Bq)

TANTALUM 180M   RR-00351-0

HEALTH AND SAFETY LISTS

U.S. DOT - Appendix A Table 2 - Radionuclides
final RQ = 100 curies (3.7E 12 Bq)

ENVIRONMENTAL LISTS

CERCLA/SARA List of Radionuclides (Appendix B) and Their Reportable Quantities
final RQ = 100 curies (3.7E 12 Bq)

TANTALUM 182   13982-00-8

HEALTH AND SAFETY LISTS

U.S. DOT - Appendix A Table 2 - Radionuclides
final RQ = 100 curies (3.7E 12 Bq)

ENVIRONMENTAL LISTS

CERCLA/SARA List of Radionuclides (Appendix B) and Their Reportable Quantities
final RQ = 100 curies (3.7E 12 Bq)

TANTALUM 182M   RR-00348-5

HEALTH AND SAFETY LISTS

U.S. DOT - Appendix A Table 2 - Radionuclides
final RQ = 10 curies (3.7E 11 Bq)

ENVIRONMENTAL LISTS

CERCLA/SARA List of Radionuclides (Appendix B) and Their Reportable Quantities
final RQ = 10 curies (3.7E 11 Bq)

TANTALUM 183   14683-36-4

HEALTH AND SAFETY LISTS

U.S. DOT - Appendix A Table 2 - Radionuclides
final RQ = 1000 curies (3.7E 13 Bq)

ENVIRONMENTAL LISTS

CERCLA/SARA List of Radionuclides (Appendix B) and Their Reportable Quantities
final RQ = 1000 curies (3.7E 13 Bq)

TANTALUM 184   15701-21-0

HEALTH AND SAFETY LISTS

U.S. DOT - Appendix A Table 2 - Radionuclides
final RQ = 1000 curies (3.7E 13 Bq)

ENVIRONMENTAL LISTS

CERCLA/SARA List of Radionuclides (Appendix B) and Their Reportable Quantities
final RQ = 1000 curies (3.7E 13 Bq)

TANTALUM 185   15701-22-1

HEALTH AND SAFETY LISTS

U.S. DOT - Appendix A Table 2 - Radionuclides
final RQ = 1000 curies (3.7E 13 Bq)

ENVIRONMENTAL LISTS

CERCLA/SARA List of Radionuclides (Appendix B) and Their Reportable Quantities
final RQ = 1000 curies (3.7E 13 Bq)

TANTALUM 186   15701-16-3

HEALTH AND SAFETY LISTS

U.S. DOT - Appendix A Table 2 - Radionuclides
final RQ = 100 curies (3.7E 12 Bq)

ENVIRONMENTAL LISTS

CERCLA/SARA List of Radionuclides (Appendix B) and Their Reportable Quantities
final RQ = 100 curies (3.7E 12 Bq)

TANTALUM OXIDE   1314-61-0

HEALTH AND SAFETY LISTS

U.S. DOT - Appendix A Table 2 - Radionuclides
final RQ = 1000 curies (3.7E 13 Bq)

ENVIRONMENTAL LISTS

CERCLA/SARA List of Radionuclides (Appendix B) and Their Reportable Quantities
final RQ = 1000 curies (3.7E 13 Bq)

TARA GUM   39300-88-4

HEALTH AND SAFETY LISTS

U.S. DOT - Appendix A Table 2 - Radionuclides
final RQ = 10 curies (3.7E 11 Bq)

ENVIRONMENTAL LISTS

CERCLA/SARA List of Radionuclides (Appendix B) and Their Reportable Quantities
final RQ = 10 curies (3.7E 11 Bq)

TAR, COAL, HIGH-TEMP   65996-89-6

HEALTH AND SAFETY LISTS

U.S. DOT - Appendix A Table 2 - Radionuclides
final RQ = 1000 curies (3.7E 13 Bq)

ENVIRONMENTAL LISTS

CERCLA/SARA List of Radionuclides (Appendix B) and Their Reportable Quantities
final RQ = 1000 curies (3.7E 13 Bq)

TAR, COAL, LOW-TEMP   65996-90-9

HEALTH AND SAFETY LISTS

U.S. DOT - Appendix A Table 2 - Radionuclides
final RQ = 100 curies (3.7E 12 Bq)

ENVIRONMENTAL LISTS

CERCLA/SARA List of Radionuclides (Appendix B) and Their Reportable Quantities
final RQ = 100 curies (3.7E 12 Bq)

TAR OILS   8002-29-7

HEALTH AND SAFETY LISTS

U.S. DOT - Appendix A Table 2 - Radionuclides
final RQ = 10 curies (3.7E 11 Bq)

ENVIRONMENTAL LISTS

CERCLA/SARA List of Radionuclides (Appendix B) and Their Reportable Quantities
final RQ = 10 curies (3.7E 11 Bq)

TAR-OIL, WOOD   RR-00777-2

HEALTH AND SAFETY LISTS

U.S. DOT - Appendix A Table 2 - Radionuclides
final RQ = 1000 curies (3.7E 13 Bq)

ENVIRONMENTAL LISTS

CERCLA/SARA List of Radionuclides (Appendix B) and Their Reportable Quantities
final RQ = 1000 curies (3.7E 13 Bq)

TARS (POLYCYCLIC AROMATIC HYDROCARBONS)   RR-00516-3

HEALTH AND SAFETY LISTS

U.S. DOT - Appendix A Table 2 - Radionuclides
final RQ = 1000 curies (3.7E 13 Bq)

ENVIRONMENTAL LISTS

CERCLA/SARA List of Radionuclides (Appendix B) and Their Reportable Quantities
final RQ = 1000 curies (3.7E 13 Bq)

TARTARIC ACID   526-83-0

HEALTH AND SAFETY LISTS

ACGIH 1995 - Time Weighted Averages
as Ta: 5 mg/m3 TWA

NIOSH - Selected LD50s and LC50s
Oral, rat: LD50 = 8000 mg/kg

INTERNATIONAL LISTS

Canada - WHMIS: Ingredient Disclosure
1% item 1492 (1321)

Canada - Ontario - OHSA - TWAEVs
total dust: 5 mg/m3 TWAEV (listed as nuisance particulates)

Israel - Time Weighted Averages
5 mg/m3 TWA

Israel - Action Levels
2.5 mg/m3 AL

STATE LISTS

California - Exposure Limits - PELs
dust, as Ta: 5 mg/m3 PEL

California - Directors List of Hazardous Substances (8 CCR 339)
[present]

PROPOSED REGULATIONS

Canada - Ontario - Proposed Occupational STEVs
total dust: 10 mg/m3 STEV

TARTARIC ACID (D, I)   87-69-4

HEALTH AND SAFETY LISTS

NTP Chemical Status Reports - Testing Status and NTIS Number
Technical reports printed (PB82195546)

NTP Chemical Status Reports - Evidence of Carcinogenicity
male rat-negative; female rat-negative; male mice-negative; female mice-negative

TARTARIC ACID, LEAD(2+) SALT   815-84-9

HEALTH AND SAFETY LISTS

NTP Seventh Report - Known Carcinogens
known carcinogen (Listed under 'Soots, tars, and mineral oils')

OSHA - Select Carcinogens
[present]

STATE LISTS

Florida Hazardous Substance List
[present]

Massachusetts Right To Know List
carcinogen; extraordinarily hazardous

2,3,6-TBA   50-31-7

STATE LISTS

Florida Hazardous Substance List
[present]

Massachusetts Right To Know List
carcinogen; extraordinarily hazardous

TEAR GAS   RR-01521-4

HEALTH AND SAFETY LISTS

U.S. DOT - Substances From 49 CFR 172.101
regulated by DOT (UN1999)

U.S. DOT - Hazard Classes
DOT hazard class = 3

TEBUTHIURON   34014-18-1

HEALTH AND SAFETY LISTS

NFPA - Flash Points
flash point = 144 degrees F (62 degrees C)

NFPA - Hazard Identification Ratings
health-0; flammability-2; reactivity-0

STATE LISTS

Pennsylvania Right to Know List
[present]

TECHNETIUM 93   14119-14-3

SEE ALSO:
K022-HAZARDOUS WASTES

ENVIRONMENTAL LISTS

RCRA - Basis for Listing - Appendix VII
Included in waste stream: K022

TECHNETIUM 93M   RR-00347-4

ENVIRONMENTAL LISTS

CAA - HON Rule - SOCMI Chemicals
compliance by Oct. 24, 1994

TECHNETIUM 94   14809-55-3

HEALTH AND SAFETY LISTS

NFPA - Flash Points
flash point = 410 degrees F (210 degrees C)

NFPA - Hazard Identification Ratings
health-0; flammability-1; reactivity-0

ENVIRONMENTAL LISTS

List of Pesticide Product Inert Ingredients
[present]

TECHNETIUM 94M   RR-00346-3

STATE LISTS

Massachusetts Right To Know List
[present]

TECHNETIUM 96   14808-44-7

HEALTH AND SAFETY LISTS

NIOSH - Selected LD50s and LC50s
Oral, rat: LD50 = 650 mg/kg

STATE LISTS

NJ Right to Know List (Total)
sn 1776

TECHNETIUM 96M   RR-00344-1

STATE LISTS

NJ Right to Know List (Total)
sn 2800; candles, non-explosive: sn 2799

TECHNETIUM 97   15759-35-0

HEALTH AND SAFETY LISTS

NIOSH - Selected LD50s and LC50s
Oral, rat: LD50 = 644 mg/kg

ENVIRONMENTAL LISTS

CERCLA/SARA - Section 313 - Emission Reporting
form R reporting required

TECHNETIUM 97M   RR-00339-4

HEALTH AND SAFETY LISTS

U.S. DOT - Appendix A Table 2 - Radionuclides
final RQ = 100 curies (3.7E 12 Bq)

ENVIRONMENTAL LISTS

CERCLA/SARA List of Radionuclides (Appendix B) and Their Reportable Quantities
final RQ = 100 curies (3.7E 12 Bq)

TECHNETIUM 98   32025-58-4

HEALTH AND SAFETY LISTS

U.S. DOT - Appendix A Table 2 - Radionuclides
final RQ = 1000 curies (3.7E 13 Bq)

ENVIRONMENTAL LISTS

CERCLA/SARA List of Radionuclides (Appendix B) and Their Reportable Quantities
final RQ = 1000 curies (3.7E 13 Bq)

TECHNETIUM 99   14133-76-7

HEALTH AND SAFETY LISTS

U.S. DOT - Appendix A Table 2 - Radionuclides
final RQ = 10 curies (3.7E 11 Bq)

ENVIRONMENTAL LISTS

CERCLA/SARA List of Radionuclides (Appendix B) and Their Reportable Quantities
final RQ = 10 curies (3.7E 11 Bq)

TECHNETIUM 99M   RR-00338-3

HEALTH AND SAFETY LISTS

U.S. DOT - Appendix A Table 2 - Radionuclides
final RQ = 100 curies (3.7E 12 Bq)

ENVIRONMENTAL LISTS

CERCLA/SARA List of Radionuclides (Appendix B) and Their Reportable Quantities
final RQ = 100 curies (3.7E 12 Bq)

TECHNETIUM 101   14913-92-9

HEALTH AND SAFETY LISTS

U.S. DOT - Appendix A Table 2 - Radionuclides
final RQ = 10 curies (3.7E 11 Bq)

ENVIRONMENTAL LISTS

CERCLA/SARA List of Radionuclides (Appendix B) and Their Reportable Quantities
final RQ = 10 curies (3.7E 11 Bq)

TECHNETIUM 104   15701-17-4

HEALTH AND SAFETY LISTS

U.S. DOT - Appendix A Table 2 - Radionuclides
final RQ = 1000 curies (3.7E 13 Bq)

ENVIRONMENTAL LISTS

CERCLA/SARA List of Radionuclides (Appendix B) and Their Reportable Quantities
final RQ = 1000 curies (3.7E 13 Bq)

TEFLON   **25067-11-2**

HEALTH AND SAFETY LISTS

U.S. DOT - Appendix A Table 2 - Radionuclides
final RQ = 100 curies (3.7E 12 Bq)

ENVIRONMENTAL LISTS

CERCLA/SARA List of Radionuclides (Appendix B) and Their Reportable Quantities
final RQ = 100 curies (3.7E 12 Bq)

TEFLON DECOMPOSITION PRODUCTS   **RR-01640-0**

HEALTH AND SAFETY LISTS

U.S. DOT - Appendix A Table 2 - Radionuclides
final RQ = 100 curies (3.7E 12 Bq)

ENVIRONMENTAL LISTS

CERCLA/SARA List of Radionuclides (Appendix B) and Their Reportable Quantities
final RQ = 100 curies (3.7E 12 Bq)

TELLURIC ACID   **7803-68-1**

HEALTH AND SAFETY LISTS

U.S. DOT - Appendix A Table 2 - Radionuclides
final RQ = 10 curies (3.7E 11 Bq)

ENVIRONMENTAL LISTS

CERCLA/SARA List of Radionuclides (Appendix B) and Their Reportable Quantities
final RQ = 10 curies (3.7E 11 Bq)

TELLURIUM   **13494-80-9**

HEALTH AND SAFETY LISTS

U.S. DOT - Appendix A Table 2 - Radionuclides
final RQ = 10 curies (3.7E 11 Bq)

ENVIRONMENTAL LISTS

CERCLA/SARA List of Radionuclides (Appendix B) and Their Reportable Quantities
final RQ = 10 curies (3.7E 11 Bq)

TELLURIUM 116   **15125-45-8**

HEALTH AND SAFETY LISTS

U.S. DOT - Appendix A Table 2 - Radionuclides
final RQ = 100 curies (3.7E 12 Bq)

ENVIRONMENTAL LISTS

CERCLA/SARA List of Radionuclides (Appendix B) and Their Reportable Quantities
final RQ = 100 curies (3.7E 12 Bq)

TELLURIUM 121   **14304-79-1**

HEALTH AND SAFETY LISTS

U.S. DOT - Appendix A Table 2 - Radionuclides
final RQ = 1000 curies (3.7E 13 Bq)

ENVIRONMENTAL LISTS

CERCLA/SARA List of Radionuclides (Appendix B) and Their Reportable Quantities
final RQ = 1000 curies (3.7E 13 Bq)

TELLURIUM 121M   **RR-00335-0**

HEALTH AND SAFETY LISTS

U.S. DOT - Appendix A Table 2 - Radionuclides
final RQ = 1000 curies (3.7E 13 Bq)

ENVIRONMENTAL LISTS

CERCLA/SARA List of Radionuclides (Appendix B) and Their Reportable Quantities
final RQ = 1000 curies (3.7E 13 Bq)

TELLURIUM 123   **14304-80-4**

ENVIRONMENTAL LISTS

List of Pesticide Product Inert Ingredients
[present]

TELLURIUM 123M   **RR-00332-7**

INTERNATIONAL LISTS

Canada - British Columbia - 8 Hour Exposure Limits
2.5 mg/m3 TWA

TELLURIUM 125M   **RR-00330-5**

INTERNATIONAL LISTS

Canada - WHMIS: Ingredient Disclosure
1% item 1493 (139)

TELLURIUM 127   **13981-49-2**

HEALTH AND SAFETY LISTS

ACGIH 1995 - Time Weighted Averages
as Te: 0.1 mg/m3 TWA

NIOSH - Selected LD50s and LC50s
Oral, rat: LD50 = 83 mg/kg

NIOSH 1990 - Pocket Guide - RELs
as Te: 0.1 mg/m3 TWA

NIOSH 1990 - Pocket Guide - Target organs
skin, CNS

OSHA - Vacated PELs - Time Weighted Averages
as Te: 0.1 mg/m3 TWA

OSHA - Final PELs - Time Weighted Averages
as Te: 0.1 mg/m3 TWA

INTERNATIONAL LISTS

Australian Exposure Standards - Time Weighted Averages
as Te: 0.1 mg/m3 TWA

Canada - WHMIS: Ingredient Disclosure
1% item 1496 (1557)

Canada - Alberta - 8 Hour Occupational Exposure Limit
as Te: 0.1 mg/m3 TWA

Canada - Alberta - 15 Minute Occupational Exposure Limit
as Te: 0.3 mg/m3 STEL

Canada - Ontario - OHSA - TWAEVs
as Te: 0.1 mg/m3 TWAEV

Canada - Quebec - Time-Weighted Average Exposure Values
as Te: 0.1 mg/m3 TWAEV

German (DFG) - MAK Values
total dust, as Te: 0.1 mg/m3 MAK

German (DFG) - Peak Limitations
5 x normal MAK (30 min. average value); don't exceed 2 times during shift

Israel - Time Weighted Averages
as Te: 0.1 mg/m3 TWA

Israel - Action Levels
as Te: 0.05 mg/m3 AL

Mexico - Instruction No. 10 - TWAs
0.1 mg/m3 TWA

Mexico - Wastewater - Organic Toxic Pollutants and Heavy Metals
Listed under [Heavy Metals]

STATE LISTS

California - Directors List of Hazardous Substances (8 CCR 339)
[present]

Florida Hazardous Substance List
[present]

Massachusetts Right To Know List
extraordinarily hazardous

Minnesota Hazardous Substance List
[present] as Te (includes its compounds)
NJ Right to Know List (Total)
sn 1777
Pennsylvania Right to Know List
environmental hazard

PROPOSED REGULATIONS

CERCLA/SARA - Proposed Hazardous Substance Additions
proposed RQ = 1 pound (.454 kg)
CERCLA/SARA - 1989 Proposed RQ Adjustments
proposed RQ = 100 pounds (45.4 kg)

TELLURIUM 127M   RR-00329-2

HEALTH AND SAFETY LISTS

U.S. DOT - Appendix A Table 2 - Radionuclides
final RQ = 1000 curies (3.7E 13 Bq)

ENVIRONMENTAL LISTS

CERCLA/SARA List of Radionuclides (Appendix B) and Their Reportable Quantities
final RQ = 1000 curies (3.7E 13 Bq)

TELLURIUM 129   14269-71-7

HEALTH AND SAFETY LISTS

U.S. DOT - Appendix A Table 2 - Radionuclides
final RQ = 10 curies (3.7E 11 Bq)

ENVIRONMENTAL LISTS

CERCLA/SARA List of Radionuclides (Appendix B) and Their Reportable Quantities
final RQ = 10 curies (3.7E 11 Bq)

TELLURIUM 129M   RR-00326-9

HEALTH AND SAFETY LISTS

U.S. DOT - Appendix A Table 2 - Radionuclides
final RQ = 10 curies (3.7E 11 Bq)

ENVIRONMENTAL LISTS

CERCLA/SARA List of Radionuclides (Appendix B) and Their Reportable Quantities
final RQ = 10 curies (3.7E 11 Bq)

TELLURIUM 131   14683-12-6

HEALTH AND SAFETY LISTS

U.S. DOT - Appendix A Table 2 - Radionuclides
final RQ = 10 curies (3.7E 11 Bq)

ENVIRONMENTAL LISTS

CERCLA/SARA List of Radionuclides (Appendix B) and Their Reportable Quantities
final RQ = 10 curies (3.7E 11 Bq)

TELLURIUM 131M   RR-00324-7

HEALTH AND SAFETY LISTS

U.S. DOT - Appendix A Table 2 - Radionuclides
final RQ = 10 curies (3.7E 11 Bq)

ENVIRONMENTAL LISTS

CERCLA/SARA List of Radionuclides (Appendix B) and Their Reportable Quantities
final RQ = 10 curies (3.7E 11 Bq)

TELLURIUM 132   14234-28-7

HEALTH AND SAFETY LISTS

U.S. DOT - Appendix A Table 2 - Radionuclides
final RQ = 10 curies (3.7E 11 Bq)

ENVIRONMENTAL LISTS

CERCLA/SARA List of Radionuclides (Appendix B) and Their Reportable Quantities
final RQ = 10 curies (3.7E 11 Bq)

TELLURIUM 133   15759-52-1

HEALTH AND SAFETY LISTS

U.S. DOT - Appendix A Table 2 - Radionuclides
final RQ = 1000 curies (3.7E 13 Bq)

ENVIRONMENTAL LISTS

CERCLA/SARA List of Radionuclides (Appendix B) and Their Reportable Quantities
final RQ = 1000 curies (3.7E 13 Bq)

TELLURIUM 133M   RR-00323-6

HEALTH AND SAFETY LISTS

U.S. DOT - Appendix A Table 2 - Radionuclides
final RQ = 10 curies (3.7E 11 Bq)

ENVIRONMENTAL LISTS

CERCLA/SARA List of Radionuclides (Appendix B) and Their Reportable Quantities
final RQ = 10 curies (3.7E 11 Bq)

TELLURIUM 134   15701-09-4

HEALTH AND SAFETY LISTS

U.S. DOT - Appendix A Table 2 - Radionuclides
final RQ = 1000 curies (3.7E 13 Bq)

ENVIRONMENTAL LISTS

CERCLA/SARA List of Radionuclides (Appendix B) and Their Reportable Quantities
final RQ = 1000 curies (3.7E 13 Bq)

TELLURIUM COMPOUNDS, N.O.S.   RR-00614-4

HEALTH AND SAFETY LISTS

U.S. DOT - Appendix A Table 2 - Radionuclides
final RQ = 10 curies (3.7E 11 Bq)

ENVIRONMENTAL LISTS

CERCLA/SARA List of Radionuclides (Appendix B) and Their Reportable Quantities
final RQ = 10 curies (3.7E 11 Bq)

TELLURIUM DIOXIDE   7446-07-3

HEALTH AND SAFETY LISTS

U.S. DOT - Appendix A Table 2 - Radionuclides
final RQ = 1000 curies (3.7E 13 Bq)

ENVIRONMENTAL LISTS

CERCLA/SARA List of Radionuclides (Appendix B) and Their Reportable Quantities
final RQ = 1000 curies (3.7E 13 Bq)

TELLURIUM HEXAFLUORIDE   7783-80-4

HEALTH AND SAFETY LISTS

U.S. DOT - Appendix A Table 2 - Radionuclides
final RQ = 10 curies (3.7E 11 Bq)

ENVIRONMENTAL LISTS

CERCLA/SARA List of Radionuclides (Appendix B) and Their Reportable Quantities
final RQ = 10 curies (3.7E 11 Bq)

TELLURIUM TETRACHLORIDE   10026-07-0

HEALTH AND SAFETY LISTS

U.S. DOT - Appendix A Table 2 - Radionuclides
final RQ = 10 curies (3.7E 11 Bq)

ENVIRONMENTAL LISTS
  CERCLA/SARA List of Radionuclides (Appendix B) and Their Reportable Quantities
    final RQ = 10 curies (3.7E 11 Bq)

TEMAZEPAM   846-50-4
  HEALTH AND SAFETY LISTS
    U.S. DOT - Appendix A Table 2 - Radionuclides
      final RQ = 1000 curies (3.7E 13 Bq)
  ENVIRONMENTAL LISTS
    CERCLA/SARA List of Radionuclides (Appendix B) and Their Reportable Quantities
      final RQ = 1000 curies (3.7E 13 Bq)

TEMEPHOS   3383-96-8
  HEALTH AND SAFETY LISTS
    U.S. DOT - Appendix A Table 2 - Radionuclides
      final RQ = 1000 curies (3.7E 13 Bq)
  ENVIRONMENTAL LISTS
    CERCLA/SARA List of Radionuclides (Appendix B) and Their Reportable Quantities
      final RQ = 1000 curies (3.7E 13 Bq)

TEPP   107-49-3
  HEALTH AND SAFETY LISTS
    U.S. DOT - Appendix A Table 2 - Radionuclides
      final RQ = 1000 curies (3.7E 13 Bq)
  ENVIRONMENTAL LISTS
    CERCLA/SARA List of Radionuclides (Appendix B) and Their Reportable Quantities
      final RQ = 1000 curies (3.7E 13 Bq)

TERBACIL [5-CHLORO-3-(1,1-DIMETHYLETHYL)-6-   5902-51-2
METHYL-2,4(1H,3H)-PYRIMIDINEDIONE]
  INTERNATIONAL LISTS
    Canada - WHMIS: Ingredient Disclosure
      1% item 1494 (1556)
    Canada - British Columbia - 8 Hour Exposure Limits
      as Te: 0.1 mg/m3 TWA
    Canada - Ontario - OHSA - TWAEVs
      as Te: 0.1 mg/m3 TWAEV (except tellurium hexafluoride)
    Canada - Quebec - Time-Weighted Average Exposure Values
      as Te: 0.1 mg/m3 TWAEV
    United Kingdom - Occupational Exposure Standards - TWAs
      as Te: 0.1 mg/m3 TWA (does not include hydrogen telluride)
  STATE LISTS
    California - Exposure Limits - PELs
      as Te: 0.1 mg/m3 PEL
    California - Directors List of Hazardous Substances (8 CCR 339)
      [present]

TERBIUM 147   26209-85-8
  INTERNATIONAL LISTS
    Canada - WHMIS: Ingredient Disclosure
      1% item 1495 (774)

TERBIUM 149   15065-93-7
  SEE ALSO:
    TELLURIUM
  HEALTH AND SAFETY LISTS
    ACGIH 1995 - Time Weighted Averages
      as Te: 0.02 ppm TWA; 0.10 mg/m3 TWA
    U.S. DOT - Substances From 49 CFR 172.101
      regulated by DOT (UN2195)

U.S. DOT - Hazard Classes
  DOT hazard class = 2.3
U.S. DOT - Substances Which Are Poisonous by Inhalation
  gaseous hazardous material poisonous by inhalation (UN2195)
NIOSH 1990 - Pocket Guide - RELs
  as Te: 0.02 ppm TWA; 0.2 mg/m3 TWA
NIOSH 1990 - Pocket Guide - IDLHs
  as Te: 1 ppm IDLH
NIOSH 1990 - Pocket Guide - Target organs
  respiratory system
OSHA - Vacated PELs - Time Weighted Averages
  as Te: 0.02 ppm TWA; 0.2 mg/m3 TWA
OSHA - Final PELs - Time Weighted Averages
  as Te: 0.02 ppm TWA; 0.2 mg/m3 TWA
OSHA - List of Highly Hazardous Chemicals
  threshhold quantity = 250 pounds
ENVIRONMENTAL LISTS
  CERCLA/SARA - Section 302 Extremely Hazardous Substances and TPQs
    TPQ = 100 pounds
INTERNATIONAL LISTS
  Australian Exposure Standards - Time Weighted Averages
    as Te: 0.02 ppm TWA; 0.1 mg/m3 TWA
  Canada - WHMIS: Ingredient Disclosure
    1% item 1497 (957)
  Canada - Alberta - 8 Hour Occupational Exposure Limit
    as Te: 0.02 ppm TWA; 0.1 mg/m3 TWA
  Canada - British Columbia - 8 Hour Exposure Limits
    as Te: 0.02 ppm TWA; 0.2 mg/m3 TWA
  Canada - Ontario - OHSA - TWAEVs
    as Te: 0.01 ppm TWAEV; 0.10 mg/m3 TWAEV
  Canada - Quebec - Time-Weighted Average Exposure Values
    as Te: 0.02 ppm TWAEV; 0.1 mg/m3 TWAEV
  Israel - Time Weighted Averages
    as Te: 0.02 ppm TWA; 0.1 mg/m3 TWA
  Israel - Action Levels
    as Te: 0.01 ppm AL; 0.05 mg/m3 AL
  Mexico - Instruction No. 10 - TWAs
    0.02 ppm TWA; 0.2 mg/m3 TWA
STATE LISTS
  California - Exposure Limits - PELs
    0.02 ppm PEL; 0.2 mg/m3 PEL
  Florida Hazardous Substance List
    [present]
  Massachusetts Right To Know List
    extraordinarily hazardous
  Minnesota Hazardous Substance List
    [present] as Te
  NJ Right to Know List (Total)
    sn 1778
  Pennsylvania Right to Know List
    environmental hazard
PROPOSED REGULATIONS
  CERCLA/SARA - Proposed Hazardous Substance Additions
    proposed RQ = 1 pound (.454 kg)
  CERCLA/SARA - 1989 Proposed RQ Adjustments
    proposed RQ = 100 pounds (45.4 kg)

TERBIUM 150   15065-95-9
  INTERNATIONAL LISTS
    Canada - WHMIS: Ingredient Disclosure
      1% item 1498 (1583)

TERBIUM 151   14998-51-7

HEALTH AND SAFETY LISTS

NIOSH - Selected LD50s and LC50s
Oral, rat: LD50 = 2000 mg/kg

STATE LISTS

California - Air Bill 2588 Appendix A-II
9/90

California - Prop. 65 - Developmental Toxicity
developmental toxicity - initial date 4/1/90

NJ Right to Know List (Total)
sn 1779

NJ Special Hazardous Substances
(teratogen)

TERBIUM 153   14981-98-7

HEALTH AND SAFETY LISTS

ACGIH 1995 - Time Weighted Averages
10 mg/m3 TWA

U.S. DOT - Appendix B - Marine Pollutants
DOT regulated marine pollutant

NIOSH - Selected LD50s and LC50s
Oral, rat: LD50 = 1000 mg/kg Skin, rat: LD50 = 1370 mg/kg

OSHA - Vacated PELs - Time Weighted Averages
total dust: 10 mg/m3 TWA; respirable fraction: 5 mg/m3 TWA

OSHA - Final PELs - Time Weighted Averages
total dust: 15 mg/m3 TWA; respirable fraction: 5 mg/m3 TWA

ENVIRONMENTAL LISTS

CERCLA/SARA - Section 313 - Emission Reporting
form R reporting required

INTERNATIONAL LISTS

Australian Exposure Standards - Time Weighted Averages
10 mg/m3 TWA

Canada - Drinking Water Quality - IMACs
0.28 mg/L IMAC

Canada - Alberta - 8 Hour Occupational Exposure Limit
10 mg/m3 TWA

Canada - Alberta - 15 Minute Occupational Exposure Limit
20 mg/m3 STEL

Canada - British Columbia - 8 Hour Exposure Limits
10 mg/m3 TWA

Canada - British Columbia - 15 Minute Exposure Limits
20 mg/m3 STEL

Canada - Ontario - OHSA - TWAEVs
10 mg/m3 TWAEV

Canada - Quebec - Time-Weighted Average Exposure Values
10 mg/m3 TWAEV

Israel - Time Weighted Averages
10 mg/m3 TWA

Israel - Action Levels
5 mg/m3 AL

Mexico - Instruction No. 10 - TWAs
10 mg/m3 TWA

Mexico - Instruction No. 10 - STELs
20 mg/m3 STEL

STATE LISTS

California - Exposure Limits - PELs
total dust: 10 mg/m3 PEL; respirable fraction: 5 mg/m3 PEL

California - Directors List of Hazardous Substances (8 CCR 339)
[present]

Florida Hazardous Substance List
[present]

Massachusetts Right To Know List
neurotoxin

Minnesota Hazardous Substance List
[present]

NJ Right to Know List (Total)
sn 1780

Pennsylvania Right to Know List
[present]

TERBIUM 154   15758-64-2

HEALTH AND SAFETY LISTS

ACGIH 1995 - Time Weighted Averages
0.004 ppm TWA; 0.047 mg/m3 TWA

ACGIH 1995 - Skin Designations
skin - potential for cutaneous absorption

U.S. DOT - Substances From 49 CFR 172.101
regulated by DOT (NA3018)

U.S. DOT - Hazard Classes
DOT hazard class = 6.1

U.S. DOT - Substances Which Are Poisonous by Inhalation
gaseous hazardous material poisonous by inhalation (compressed gas mixtures) (UN1705)

U.S. DOT - Appendix B - Marine Pollutants
DOT regulated marine pollutant

U.S. DOT - Appendix A Table 1 - Hazardous Substances
final RQ = 10 pounds (4.54 kg)

NIOSH - Selected LD50s and LC50s
Oral, rat: LD50 = 500 ug/kg Skin, rat: LD50 = 2400 ug/kg

NIOSH 1990 - Pocket Guide - RELs
0.05 mg/m3 TWA

NIOSH 1990 - Pocket Guide - IDLHs
10 mg/m3 IDLH

NIOSH 1990 - Pocket Guide - Target organs
CNS, respiratory system, CVS, GI tract

NIOSH 1990 - Pocket Guide - Skin list
Potential for dermal absorption

OSHA - Vacated PELs - Time Weighted Averages
0.05 mg/m3 TWA

OSHA - Vacated PELs - Skin Designation
Prevent or reduce skin absorption

OSHA - Final PELs - Time Weighted Averages
0.05 mg/m3 TWA

OSHA - Final PELs - Skin Notations
prevent or reduce skin absorption

ENVIRONMENTAL LISTS

CERCLA/SARA - Section 302 Extremely Hazardous Substances and TPQs
TPQ = 100 pounds

CERCLA/SARA - Hazardous Substances and their Reportable Quantities
final RQ = 10 pounds (4.54 kg)

Clean Water Act - Hazardous Substances
[present]

RCRA - P Series Wastes
waste number P111

RCRA - Hazardous Constituents-Appendix VIII
waste number P111

RCRA - Substances Banned From Land Disposal
[present]

INTERNATIONAL LISTS

Australian Exposure Standards - Time Weighted Averages
0.004 ppm TWA; 0.047 mg/m3 TWA

Australian Exposure Standards - Skin Effects
skin absorption

Canada - Alberta - 8 Hour Occupational Exposure Limit
0.004 ppm TWA; 0.047 mg/m3 TWA

Canada - Alberta - Skin Designation
can be absorbed through the intact skin

Canada - British Columbia - 8 Hour Exposure Limits
0.004 ppm TWA; 0.05 mg/m3 TWA

Canada - British Columbia - 15 Minute Exposure Limits
0.01 ppm STEL; 0.2 mg/m3 STEL

Canada - British Columbia - Skin Notations
skin - potential for skin absorption

Canada - Ontario - OHSA - TWAEVs
0.004 ppm TWAEV; 0.047 mg/m3 TWAEV

Canada - Ontario - OHSA - Skin Notations
absorption through skin, eyes, or mucous membranes

Canada - Quebec - Time-Weighted Average Exposure Values
0.004 ppm TWAEV; 0.047 mg/m3 TWAEV

Canada - Quebec - Skin Designations
absorbed through the skin

United Kingdom - Occupational Exposure Standards - TWAs
0.004 ppm TWA; 0.05 mg/m3 TWA

United Kingdom - Occupational Exposure Standards - STELs
0.01 ppm STEL; 0.2 mg/m3 STEL

United Kingdom - Occupational Exposure Standards - Notes
can be absorbed through skin

German (DFG) - MAK Values
0.005 ppm MAK; 0.05 mg/m3 MAK

German (DFG) - Peak Limitations
10 x normal MAK (30 min average value); don't exceed during shift

German (DFG) - Skin/Sensitizers
danger of cutaneous absorption

Israel - Time Weighted Averages
0.004 ppm TWA; 0.047 mg/m3 TWA

Israel - Action Levels
0.002 ppm AL; 0.0235 mg/m3 AL

Mexico - Instruction No. 10 - TWAs
0.004 ppm TWA; 0.05 mg/m3 TWA

Mexico - Instruction No. 10 - Skin designation
skin - potential for cutaneous absorption

STATE LISTS

California - Exposure Limits - PELs
0.004 ppm PEL; 0.05 mg/m3 PEL

California - Exposure Limits - Skin Notation
material may be absorbed through the skin, eyes or mucous membrane

California - Directors List of Hazardous Substances (8 CCR 339)
[present]

Florida Hazardous Substance List
[present]

Massachusetts Right To Know List
extraordinarily hazardous; neurotoxin

Minnesota Hazardous Substance List
skin

NJ Right to Know List (Total)
sn 1781

Pennsylvania Right to Know List
environmental hazard

TERBIUM 155   1439-17-4

ENVIRONMENTAL LISTS

CERCLA/SARA - Section 313 - Emission Reporting
form R reporting required

TERBIUM 156   14391-10-7

HEALTH AND SAFETY LISTS

U.S. DOT - Appendix A Table 2 - Radionuclides
final RQ = 100 curies (3.7E 12 Bq)

ENVIRONMENTAL LISTS

CERCLA/SARA List of Radionuclides (Appendix B) and Their Reportable Quantities
final RQ = 100 curies (3.7E 12 Bq)

TERBIUM 156M   RR-00321-4

HEALTH AND SAFETY LISTS

U.S. DOT - Appendix A Table 2 - Radionuclides
final RQ = 100 curies (3.7E 12 Bq)

ENVIRONMENTAL LISTS

CERCLA/SARA List of Radionuclides (Appendix B) and Their Reportable Quantities
final RQ = 100 curies (3.7E 12 Bq)

TERBIUM 157   14391-18-5

HEALTH AND SAFETY LISTS

U.S. DOT - Appendix A Table 2 - Radionuclides
final RQ = 100 curies (3.7E 12 Bq)

ENVIRONMENTAL LISTS

CERCLA/SARA List of Radionuclides (Appendix B) and Their Reportable Quantities
final RQ = 100 curies (3.7E 12 Bq)

TERBIUM 158   15759-55-4

HEALTH AND SAFETY LISTS

U.S. DOT - Appendix A Table 2 - Radionuclides
final RQ = 10 curies (3.7E 11 Bq)

ENVIRONMENTAL LISTS

CERCLA/SARA List of Radionuclides (Appendix B) and Their Reportable Quantities
final RQ = 10 curies (3.7E 11 Bq)

TERBIUM 160   13981-29-8

HEALTH AND SAFETY LISTS

U.S. DOT - Appendix A Table 2 - Radionuclides
final RQ = 100 curies (3.7E 12 Bq)

ENVIRONMENTAL LISTS

CERCLA/SARA List of Radionuclides (Appendix B) and Their Reportable Quantities
final RQ = 100 curies (3.7E 12 Bq)

TERBIUM 161   14391-19-6

HEALTH AND SAFETY LISTS

U.S. DOT - Appendix A Table 2 - Radionuclides
final RQ = 10 curies (3.7E 11 Bq)

ENVIRONMENTAL LISTS

CERCLA/SARA List of Radionuclides (Appendix B) and Their Reportable Quantities
final RQ = 10 curies (3.7E 11 Bq)

TERBIUM CHLORIDE   10042-88-3

HEALTH AND SAFETY LISTS

U.S. DOT - Appendix A Table 2 - Radionuclides
final RQ = 100 curies (3.7E 12 Bq)

ENVIRONMENTAL LISTS

CERCLA/SARA List of Radionuclides (Appendix B) and Their Reportable Quantities
final RQ = 100 curies (3.7E 12 Bq)

TERBUFOS
13071-79-9

HEALTH AND SAFETY LISTS

U.S. DOT - Appendix A Table 2 - Radionuclides
final RQ = 10 curies (3.7E 11 Bq)

ENVIRONMENTAL LISTS

CERCLA/SARA List of Radionuclides (Appendix B) and Their Reportable Quantities
final RQ = 10 curies (3.7E 11 Bq)

TERBUTHYLAZINE
5915-41-3

HEALTH AND SAFETY LISTS

U.S. DOT - Appendix A Table 2 - Radionuclides
5.0 hour half-life: Final RQ = 1000 curies (3.7E 13 Bq); 24.4 hour half-life: Final RQ = 1000 curies (3.7E 13 Bq)

ENVIRONMENTAL LISTS

CERCLA/SARA List of Radionuclides (Appendix B) and Their Reportable Quantities
5.0 hour half-life: Final RQ = 1000 curies (3.7E 13 Bq); 24.4 hour half-life: Final RQ = 1000 curies (3.7E 13 Bq)

TERBUTRYN
886-50-0

HEALTH AND SAFETY LISTS

U.S. DOT - Appendix A Table 2 - Radionuclides
final RQ = 100 curies (3.7E 12 Bq)

ENVIRONMENTAL LISTS

CERCLA/SARA List of Radionuclides (Appendix B) and Their Reportable Quantities
final RQ = 100 curies (3.7E 12 Bq)

TEREPHTHALAMIDE, N,N'-DIMETHYL-N,N'-DINITROSO-
133-55-1

HEALTH AND SAFETY LISTS

U.S. DOT - Appendix A Table 2 - Radionuclides
final RQ = 10 curies (3.7E 11 Bq)

ENVIRONMENTAL LISTS

CERCLA/SARA List of Radionuclides (Appendix B) and Their Reportable Quantities
final RQ = 10 curies (3.7E 11 Bq)

TEREPHTHALIC ACID
100-21-0

HEALTH AND SAFETY LISTS

U.S. DOT - Appendix A Table 2 - Radionuclides
final RQ = 10 curies (3.7E 11 Bq)

ENVIRONMENTAL LISTS

CERCLA/SARA List of Radionuclides (Appendix B) and Their Reportable Quantities
final RQ = 10 curies (3.7E 11 Bq)

TEREPHTHALIC ACID, TETRACHLORO-, DIMETHYL ESTER
1861-32-1

HEALTH AND SAFETY LISTS

U.S. DOT - Appendix A Table 2 - Radionuclides
final RQ = 100 curies (3.7E 12 Bq)

ENVIRONMENTAL LISTS

CERCLA/SARA List of Radionuclides (Appendix B) and Their Reportable Quantities
final RQ = 100 curies (3.7E 12 Bq)

TEREPHTHALONITRILE
623-26-7

INTERNATIONAL LISTS

Canada - WHMIS: Ingredient Disclosure
1% item 1499 (528)

TEREPHTHALOYL CHLORIDE
100-20-9

HEALTH AND SAFETY LISTS

U.S. DOT - Appendix B - Marine Pollutants
DOT regulated severe marine pollutant

NIOSH - Selected LD50s and LC50s
Oral, rat: LD50 = 1600 ug/kg Skin, rabbit: LD50 = 1100 ug/kg

ENVIRONMENTAL LISTS

CERCLA/SARA - Section 302 Extremely Hazardous Substances and TPQs
TPQ = 100 pounds

INTERNATIONAL LISTS

Canada - Drinking Water Quality - IMACs
0.001 mg/L IMAC

STATE LISTS

Florida Hazardous Substance List
effective March 13, 1992

Massachusetts Right To Know List
extraordinarily hazardous; neurotoxin

NJ Right to Know List (Total)
sn 2801

Pennsylvania Right to Know List
environmental hazard

PROPOSED REGULATIONS

CERCLA/SARA - Proposed Hazardous Substance Additions
proposed RQ = 1 pound (.454 kg)

CERCLA/SARA - 1989 Proposed RQ Adjustments
proposed RQ = 100 pounds (45.4 kg)

TERGITOL 7
3282-85-7

HEALTH AND SAFETY LISTS

NIOSH - Selected LD50s and LC50s
Oral, rat: LD50 = 1845 mg/kg

TERGITOL NO. 4
139-88-8

STATE LISTS

Massachusetts Right To Know List
[present]

TERPENE HYDROCARBONS, N.O.S.
68956-56-9

HEALTH AND SAFETY LISTS

U.S. DOT - Substances From 49 CFR 172.101
regulated by DOT (UN2973)

U.S. DOT - Hazard Classes
DOT hazard class = 4.1

TERPENE POLYCHLORINATES (STROBANE6)
8001-50-1

HEALTH AND SAFETY LISTS

ACGIH 1995 - Time Weighted Averages
10 mg/m3 TWA

NFPA - Flash Points
flash point = 500 degrees F (260 degrees C)

NFPA - Hazard Identification Ratings
health-0; flammability-1; reactivity-0

NIOSH - Selected LD50s and LC50s
Oral, rat: LD50 = 18800 mg/kg

ENVIRONMENTAL LISTS

CAA - HON Rule - SOCMI Chemicals
compliance by Jan. 23, 1995

EPA - Master Testing List
[present]

RCRA - Universal Treatment Standards (LDR)
WW: 0.055 mg/l; NWW: 28 mg/kg

INTERNATIONAL LISTS

Canada - WHMIS: Ingredient Disclosure
1% item 1500 (140)

STATE LISTS

California - Air Bill 2588 Appendix A-I
6/91

Massachusetts Right To Know List
[present]

Pennsylvania Right to Know List
environmental hazard

PROPOSED REGULATIONS

TSCA - Proposed Substances for Developmental/ReproductiveTesting
proposed testing for: Reproductive Toxicity - oral

TERPENES AND TERPENOIDS, LIMONENE FRACTION, POLYMER WITH SUBSTITUTED CARBOPOLYCYCLES RR-00912-1

PROPOSED REGULATIONS

Safe Drinking Water Act - Priority list
[present]

O-TERPHENYL 84-15-1

INTERNATIONAL LISTS

Canada - WHMIS: Ingredient Disclosure
1% item 1501 (1562)

M-TERPHENYL 92-06-8

HEALTH AND SAFETY LISTS

NFPA - Flash Points
flash point = 356 degrees F (180 degrees C)

NFPA - Hazard Identification Ratings
health-3; flammability-1; reactivity-0

ENVIRONMENTAL LISTS

TSCA - Code of Federal Regulations Citations
40 CFR 712.30(x); 40 CFR 716.120(d)

TSCA - PAIR - Reporting List
Reporting Date: November 27, 1991

TSCA - Health and Safety Reporting List
Effective Date: September 30, 1991

STATE LISTS

Florida Hazardous Substance List
[present]

Massachusetts Right To Know List
[present]

Pennsylvania Right to Know List
[present]

P-TERPHENYL 92-94-4

HEALTH AND SAFETY LISTS

NIOSH - Selected LD50s and LC50s
Oral, rat: LD50 = 1430 mg/kg Skin, rabbit: LD50 = 3560 mg/kg

TERPHENYLS 26140-60-3

HEALTH AND SAFETY LISTS

NIOSH - Selected LD50s and LC50s
Oral, rat: LD50 = 1250 mg/kg Skin, guinea pig: LD50 = 650 mg/kg

INTERNATIONAL LISTS

Canada - WHMIS: Ingredient Disclosure
1% item 744 (1538)

ALPHA-TERPINEOL 98-55-5

HEALTH AND SAFETY LISTS

U.S. DOT - Substances From 49 CFR 172.101
regulated by DOT (UN2319)

U.S. DOT - Hazard Classes
DOT hazard class = 3

STATE LISTS

NJ Right to Know List (Total)
sn 2802

TERPINEOL 8006-39-1

HEALTH AND SAFETY LISTS

IARC - Group 3 (not classifiable)
[present]

STATE LISTS

California - Directors List of Hazardous Substances (8 CCR 339)
[present]

TERPINOLENE 586-62-9

ENVIRONMENTAL LISTS

TSCA - Chemicals with Significant New Use Rules
PMN number: P-88-1617

TERPINYL ACETATE 80-26-2

HEALTH AND SAFETY LISTS

NFPA - Flash Points
flash point = 325 degrees F (163 degrees C)

NFPA - Hazard Identification Ratings
health-0; flammability-1; reactivity-0

NIOSH - Selected LD50s and LC50s
Oral, rat: LD50 = 1900 mg/kg

INTERNATIONAL LISTS

Canada - WHMIS: Ingredient Disclosure
1% item 1503 (1564)

STATE LISTS

Massachusetts Right To Know List
[present]

TERRAZOLE 2593-15-9

HEALTH AND SAFETY LISTS

NFPA - Flash Points
flash point = 375 degrees F (191 degrees C)

NFPA - Hazard Identification Ratings
health-0; flammability-1; reactivity-0

NIOSH - Selected LD50s and LC50s
Oral, rat: LD50 = 2400 mg/kg

INTERNATIONAL LISTS

Canada - WHMIS: Ingredient Disclosure
1% item 1502 (1563)

STATE LISTS

Massachusetts Right To Know List
[present]

TESTOSTERONE 58-22-0

HEALTH AND SAFETY LISTS

NIOSH 1990 - Pocket Guide - Target organs
skin, eyes, respiratory system

INTERNATIONAL LISTS

Canada - WHMIS: Ingredient Disclosure
1% item 1504 (1565)

STATE LISTS
   Florida Hazardous Substance List
    [present]
   Massachusetts Right To Know List
    [present]

TESTOSTERONE CYPIONATE   58-20-8

HEALTH AND SAFETY LISTS
  ACGIH 1995 - Ceiling Limits
   C 0.53 ppm; C 5 mg/m3
  NIOSH 1990 - Pocket Guide - RELs
   C 5 mg/m3; C 0.5 ppm
  OSHA - Vacated PELs - Ceiling Limits
   C 0.5 ppm; C 5 mg/m3
  OSHA - Final PELs - Ceiling Limits
   C 1 ppm; C 9 mg/m3

INTERNATIONAL LISTS
  Australian Exposure Standards - Time Weighted Averages
   Peak Limitation: 0.5 ppm; 4.7 mg/m3
  Canada - Alberta - Ceiling Occupational Exposure Limit
   C 0.5 ppm; C 4.7 mg/m3
  Canada - British Columbia - Ceiling Exposure Limits
   C 1 ppm; C 9 mg/m3
  Canada - Ontario - OHSA - CEVs
   0.5 ppm CEV; 4.7 mg/m3 CEV
  Canada - Quebec - Ceiling Limits
   P 0.5 ppm; P 4.7 mg/m3
  United Kingdom - Occupational Exposure Standards - STELs
   0.5 ppm STEL; 5 mg/m3 STEL
  Israel - Ceiling Exposure Limits
   C 0.5 ppm; C 4.7 mg/m3
  Mexico - Instruction No. 10 - TWAs
   0.5 ppm TWA; 5 mg/m3 TWA

STATE LISTS
  California - Exposure Limits - Ceilings
   C 0.5 ppm; C 5 mg/m3
  California - Directors List of Hazardous Substances (8 CCR 339)
   [present]
  Massachusetts Right To Know List
   [present]
  Minnesota Hazardous Substance List
   [present]
  Pennsylvania Right to Know List
   [present]

TESTOSTERONE ENANTHATE   315-37-7

ENVIRONMENTAL LISTS
  List of Pesticide Product Inert Ingredients
   [present]

TESTOSTERONE ESTERS   RR-00796-5

HEALTH AND SAFETY LISTS
  NFPA - Flash Points
   flash point = 195 degrees F (91 degrees C)
  NFPA - Hazard Identification Ratings
   health-0; flammability-2; reactivity-0

TESTOSTERONE PROPIONATE   57-85-2

HEALTH AND SAFETY LISTS
  U.S. DOT - Substances From 49 CFR 172.101
   regulated by DOT (UN2541)
  U.S. DOT - Hazard Classes
   DOT hazard class = 3

NIOSH - Selected LD50s and LC50s
   Oral, rat: LD50 = 4390 mg/kg
STATE LISTS
  NJ Right to Know List (Total)
   sn 1785

TETRAAMMINEDICHLOROPLATINUM(II)   13933-32-9

HEALTH AND SAFETY LISTS
  NFPA - Flash Points
   flash point = 200 degrees F (93 degrees C)
  NFPA - Hazard Identification Ratings
   health-0; flammability-2; reactivity-0

TETRAAMYLBENZENE   2049-95-8

STATE LISTS
  California - Prop. 65 - Cancer list
   carcinogen - initial date 10/01/94

3,6,9,12-TETRAZAATETRADECANE-1,14-DIAMINE   4067-16-7

HEALTH AND SAFETY LISTS
  IARC - Group Unspecified
   [present] (Listed under 'Androgenic (anabolic) steroids')
STATE LISTS
  California - Air Bill 2588 Appendix A-II
   and esters: 9/89
  California - Prop. 65 - Cancer list
   carcinogen - initial date 4/1/88
  California - Directors List of Hazardous Substances (8 CCR 339)
   [present] (includes its esters)
  Florida Hazardous Substance List
   [present]
  Massachusetts Right To Know List
   carcinogen; extraordinarily hazardous
  NJ Right to Know List (Total)
   sn 1804
  NJ Special Hazardous Substances
   (teratogen)
  Pennsylvania Right to Know List
   special hazardous substance
  Pennsylvania RTK - Special Hazardous Substances
   [present]

TETRAAZIDO BENZENE QUINONE   22826-61-5

STATE LISTS
  California - Prop. 65 - Developmental Toxicity
   developmental toxicity - initial date 10/1/91

TETRABROMOBISHPENOL-B   92874-42-5

STATE LISTS
  California - Air Bill 2588 Appendix A-II
   9/90
  California - Prop. 65 - Developmental Toxicity
   developmental toxicity - initial date 4/1/90
  California - Directors List of Hazardous Substances (8 CCR 339)
   [present]

TETRABROMOBISPHENOL A   79-94-7

STATE LISTS
  Pennsylvania Right to Know List
   special hazardous substance
  Pennsylvania RTK - Special Hazardous Substances
   [present]

TETRABROMOBISPHENOL-A-BIS-2,3-DIBROMO-   21850-44-2
PROPYL ETHER

STATE LISTS

Massachusetts Right To Know List
carcinogen; extraordinarily hazardous

TETRABROMOBISPHENOL-A-BISETHOXYLATE   4162-45-2

INTERNATIONAL LISTS

Canada - WHMIS: Ingredient Disclosure
0.1% item 1505 (1600)

2,2',6,6'-TETRABROMOBISPHENOL A DIGLYCIDYL   3072-84-2
ETHER

HEALTH AND SAFETY LISTS

NFPA - Flash Points
flash point = 295 degrees F (146 degrees C)

NFPA - Hazard Identification Ratings
health-0; flammability-1; reactivity-0

TETRABROMOBISPHENOL-A DIACRYLATE   55205-38-4

HEALTH AND SAFETY LISTS

NIOSH - Selected LD50s and LC50s
Oral, rat: LD50 = 1600 mg/kg

TETRABROMOBISPHENOL-B   RR-00799-8

HEALTH AND SAFETY LISTS

U.S. DOT - Hazard Classes
Forbidden from transport by the DOT

TETRABROMOCATECHOL   488-47-1

ENVIRONMENTAL LISTS

TSCA - Section 12(b) - Export Notification
export notification required - Section 4

1,1,1,2-TETRABROMOETHANE   630-16-0

ENVIRONMENTAL LISTS

EPA - Master Testing List
[present]

TSCA - Code of Federal Regulations Citations
40 CFR 712.30(o); 40 CFR 716.120(a); 40 CFR 766.35(a)(2); 40 CFR 799.4000

TSCA - PAIR - Reporting List
Reporting Date: August 19, 1985

TSCA - Health and Safety Reporting List
Effective Date: June 20, 1985

TSCA - HDD/HDF - Chemicals Required for Testing
[present]

TSCA - Chemical Test Rules
Testing required by: manufacturers; importers; processors (40 CFR 799.4000)

TSCA - Section 12(b) - Export Notification
export notification required - Section 4

TETRABROMOETHANE   25167-20-8

ENVIRONMENTAL LISTS

TSCA - HDD/HDF - Chemicals Required for Testing
[present]

TSCA - Section 12(b) - Export Notification
export notification required - Section 4

TETRABUTYLAMMONIUM IODIDE   311-28-4

ENVIRONMENTAL LISTS

EPA - Master Testing List
[present]

TSCA - Code of Federal Regulations Citations
40 CFR 712.30(x); 40 CFR 716.120(d); 40 CFR 766.25; 40 CFR 766.35(a)(2)

TSCA - PAIR - Reporting List
Reporting Date: December 27, 1990

TSCA - Health and Safety Reporting List
Effective Date: October 29, 1990

TSCA - HDD/HDF - Chemicals Required for Testing
[present]

TSCA - Section 12(b) - Export Notification
export notification required - Section 4

1,4,5,8-TETRACARBOXYNAPHTHALENE   128-97-2

ENVIRONMENTAL LISTS

EPA - Master Testing List
[present]

PROPOSED REGULATIONS

TSCA - Proposed Testing Rule for Glycidyl Ethers
member of Glycidyl subcategory VI-B

TETRACHLORODIBENZOFURAN   30402-14-3

ENVIRONMENTAL LISTS

TSCA - Code of Federal Regulations Citations
40 CFR 712.30(x); 40 CFR 716.120

TSCA - PAIR - Reporting List
Effective Date: October 29, 1990 Reporting Date: December 27, 1990

TSCA - Health and Safety Reporting List
Effective Date: October 29, 1990; Sunset Date: October 29, 2000

TSCA - HDD/HDF - Chemicals Required for Testing
[present]

TSCA - Section 12(b) - Export Notification
export notification required - Section 4

TETRACHLOROAURIC ACID   16903-35-8

ENVIRONMENTAL LISTS

TSCA - HDD/HDF - Chemicals Required for Testing
[present]

3,3',4,4'-TETRACHLOROAZOBENZENE   14047-09-7

ENVIRONMENTAL LISTS

TSCA - HDD/HDF - Chemicals Required for Testing
[present]

TSCA - Section 12(b) - Export Notification
export notification required - Section 4

3,3',4,4'-TETRACHLOROAZOXYBENZENE   21232-47-3

HEALTH AND SAFETY LISTS

NTP Chemical Status Reports - Testing Status and NTIS Number
Prechronic studies completed: in review for further evaluation

1,2,4,5-TETRACHLOROBENZENE   95-94-3

HEALTH AND SAFETY LISTS

U.S. DOT - Substances From 49 CFR 172.101
regulated by DOT (UN2504)

U.S. DOT - Hazard Classes
DOT hazard class = 6.1

INTERNATIONAL LISTS

Canada - WHMIS: Ingredient Disclosure
1% item 1506 (1567)

STATE LISTS

NJ Right to Know List (Total)
sn 1805

1,2,3,4-TETRACHLOROBENZENE   634-66-2

HEALTH AND SAFETY LISTS

NIOSH - Selected LD50s and LC50s
Oral, rat: LD50 = 1990 mg/kg

1,2,4,5-TETRACHLOROBENZENE   959-43-3

HEALTH AND SAFETY LISTS

NIOSH - Selected LD50s and LC50s
Oral, rat: LD50 = 7500 mg/kg

TETRACHLOROBENZENES   RR-00143-4

ENVIRONMENTAL LISTS

ATSDR Priority List
Rank (of 275): 167

2,2',5,5'-TETRACHLOROBENZIDINE   15721-02-5
SEE ALSO:
GOLD

INTERNATIONAL LISTS

Canada - WHMIS: Ingredient Disclosure
1% item 1507 (141)

TETRACHLOROBIPHENYL   26914-33-0

HEALTH AND SAFETY LISTS

NTP Chemical Status Reports - Testing Status and NTIS Number
Prechronic studies completed: chemicals in review for further evaluation

INTERNATIONAL LISTS

Canada - WHMIS: Ingredient Disclosure
1% item 1508 (1570)

TETRACHLOROBISPHENOL-A   79-95-8

HEALTH AND SAFETY LISTS

NTP Chemical Status Reports - Testing Status and NTIS Number
Prechronic studies completed: chemicals in review for further evaluation

2,3,5,6-TETRACHLORO-2,5-CYCLOHEXADIENE-1,4-DIONE   118-75-2
SEE ALSO:
K149-HAZARDOUS WASTES
K150-HAZARDOUS WASTES
F022-HAZARDOUS WASTES
F026-HAZARDOUS WASTES
K151-HAZARDOUS WASTES
F024-HAZARDOUS WASTES
F039-HAZARDOUS WASTES
F025-HAZARDOUS WASTES

HEALTH AND SAFETY LISTS

U.S. DOT - Appendix A Table 1 - Hazardous Substances
final RQ = 5000 pounds (2270 kg)

NFPA - Flash Points
flash point = 311 degrees F (155 degrees C)

NFPA - Hazard Identification Ratings
health-1; flammability-1; reactivity-0

NIOSH - Selected LD50s and LC50s
Oral, rat: LD50 = 1500 mg/kg

NTP Chemical Status Reports - Testing Status and NTIS Number
Technical reports printed (PB91185330)

ENVIRONMENTAL LISTS

CERCLA/SARA - Hazardous Substances and their Reportable Quantities
final RQ = 5000 pounds (2270 kg)

CAA - HON Rule - SOCMI Chemicals
compliance by Oct. 24, 1994

RCRA - U Series Wastes
waste number U207

RCRA - Hazardous Constituents-Appendix VIII
waste number U207

RCRA - Basis for Listing - Appendix VII
Included in waste streams: F024, F025, F039, K085, K149, K150, K151

RCRA - Substances Banned From Land Disposal
[present]

RCRA - TSD Facilities Ground Water Monitoring
TM 8270 = 10 ug/L PQL

TSCA - Code of Federal Regulations Citations
40 CFR 712.30(d);   40 CFR 716.120(a);   40 CFR 799.1054

TSCA - PAIR - Reporting List
Reporting Date: November 19, 1982

TSCA - Chemicals with Significant New Use Rules
[present]

TSCA - HDD/HDF - Precursors Required for Reporting
[present]

TSCA - Multichemical Test Rules - Waste Constituents
hydrolysis testing for Chemical Fate

TSCA - Chemical Test Rules
Testing required by: manufacturers; importers; processors (40 CFR 799.1054)

TSCA - Section 12(b) - Export Notification
export notification required - Section 4 and Section 5

STATE LISTS

Massachusetts Right To Know List
[present]

Pennsylvania Right to Know List
environmental hazard

2,3,7,8-TETRACHLORODIBENZO-P-DIOXIN (TCDD)   1746-01-6
SEE ALSO:
K085-HAZARDOUS WASTES
F025-HAZARDOUS WASTES
F024-HAZARDOUS WASTES

HEALTH AND SAFETY LISTS

NIOSH - Selected LD50s and LC50s
Oral, rat: LD50 = 1167 mg/kg

ENVIRONMENTAL LISTS

RCRA - Basis for Listing - Appendix VII
Included in waste streams: F024, F025, K085

TSCA - Code of Federal Regulations Citations
40 CFR 712.30(d);   40 CFR 716.120(c)

TSCA - PAIR - Reporting List
Reporting Date: November 19, 1982

TETRACHLORODIBENZO-P-DIOXIN   41903-57-5

ENVIRONMENTAL LISTS

RCRA - Universal Treatment Standards (LDR)
WW: 0.055 mg/l;   NWW: 14 mg/kg

TETRACHLORODIBENZO-P-DIOXINS   RR-00626-8

HEALTH AND SAFETY LISTS

NFPA - Flash Points
flash point = 311 degrees F (155 degrees C)

NFPA - Hazard Identification Ratings
health-0; flammability-1; reactivity-0

INTERNATIONAL LISTS

Canada - CEPA - Priority Substances List
estimated time for completion of assessment reports: 4 years

2,3,7,8-TETRACHLORO DIBENZOFURANS   51207-31-9

HEALTH AND SAFETY LISTS
IARC - Group 3 (not classifiable)
[present]

TETRACHLORODIBENZOFURANS   RR-00508-3

ENVIRONMENTAL LISTS
ATSDR Priority List
Rank (of 275): 124

1,1,1,2-TETRACHLORO-2,2-DIFLUOROETHANE   76-11-9

HEALTH AND SAFETY LISTS
NIOSH - Selected LD50s and LC50s
Oral, rat: LD50 = 7432 mg/kg

ENVIRONMENTAL LISTS
TSCA - HDD/HDF - Chemicals Required for Testing
[present]

TSCA - Section 12(b) - Export Notification
export notification required - Section 4

1,1,2,2-TETRACHLORO-1,2-DIFLUOROETHANE   76-12-0

HEALTH AND SAFETY LISTS
NIOSH - Selected LD50s and LC50s
Oral, rat: LD50 = 4000 mg/kg

ENVIRONMENTAL LISTS
EPA - Master Testing List
[present]

TSCA - HDD/HDF - Chemicals Required for Testing
[present]

TSCA - Section 12(b) - Export Notification
export notification required - Section 4, Section 5

1,1,2,2-TETRACHLOROETHANE   79-34-5

HEALTH AND SAFETY LISTS
U.S. DOT - Appendix A Table 1 - Hazardous Substances
final RQ = 1 pound (0.454 kg)

IARC - Group 2B (sufficient animal data)
[present]

NIOSH - Selected LD50s and LC50s
Oral, rat: LD50 = 20 ug/kg Skin, rabbit: LD50 = 275 ug/kg

NIOSH - Health Standards - Exposure Limits
reduce exposure to lowest feasible concentration

NIOSH - Health Standards - Health Effects and Precautions
Chloracne; has produced tumors at many sites in animals

NIOSH - Health Standards - Carcinogenic Chemicals
potential human carcinogen

NTP Chemical Status Reports - Testing Status and NTIS Number
Technical reports printed (PB82163445) (PB82163684)

NTP Chemical Status Reports - Evidence of Carcinogenicity
PB82163684: male mice-equivocal; female mice-positive; PB82163445: male rat-positive; female rat-positive; male mice-positive; female mice-positive

NTP Seventh Report - Suspect Carcinogens
suspect carcinogen

OSHA - Possible Select Carcinogens
[present]

ENVIRONMENTAL LISTS
ATSDR Priority List
Rank (of 275): 052

CERCLA/SARA - Hazardous Substances and their Reportable Quantities
final RQ = 1 pound (0.454 kg)

Clean Air Act (1990) - List of Hazardous Air Contaminants
[present]

Clean Water Act - Priority Pollutants
[present]

Clean Water Act - Toxic Pollutants
[present]

Safe Drinking Water Act - MCLs
MCL = 0.00000003 mg/L

Safe Drinking Water Act - MCLGs
MCLG = Zero

EPA - Carcinogen Hazard Ranking for RQ Adjustment
Hazard ranking = High

RCRA - Hazardous Constituents-Appendix VIII
hazardous constituent - no waste number

RCRA - TSD Facilities Ground Water Monitoring
TM 8280 = 0.005 ug/L PQL

INTERNATIONAL LISTS
Canada - WHMIS: Ingredient Disclosure
0.1% item 1509 (1571)

German (DFG) - Carcinogens
animal evidence of carcinogenicity

Mexico - Wastewater - Organic Toxic Pollutants and Heavy Metals
Listed under [Polychlorinated Byphenyls]

Mexico - Drinking Water - Ecological Criteria
0.0000000001 mg/l Substance presents persistence, bioaccumulations or risk of cancer, reduce human exposure to a minimum; This level has been extrapolated by using a mathematic model

STATE LISTS
California - Air Bill 2588 Appendix A-I
known or potential carcinogen

California - Prop. 65 - Cancer list
carcinogen - initial date 1/1/88

California - Prop. 65 - Developmental Toxicity
developmental toxicity - initial date 4/1/91

California - Prop. 65 - No Significant Risk Levels
no significant risk level = 0.000005 ug/day

California - Directors List of Hazardous Substances (8 CCR 339)
[present]

Florida Hazardous Substance List
[present]

Massachusetts Right To Know List
carcinogen; extraordinarily hazardous; teratogen

Minnesota Hazardous Substance List
carcinogen

NJ Special Hazardous Substances
(carcinogen; teratogen)

Pennsylvania Right to Know List
environmental hazard: special hazardous substance

Pennsylvania RTK - Special Hazardous Substances
[present]

1,1,1,2-TETRACHLOROETHANE   630-20-6

ENVIRONMENTAL LISTS
ATSDR Priority List
Rank (of 275): 148

TETRACHLOROETHANES   25322-20-7

ENVIRONMENTAL LISTS
RCRA - Hazardous Constituents-Appendix VIII
hazardous constituent - no waste number

RCRA - Universal Treatment Standards (LDR)
WW: 0.000063 mg/l; NWW: 0.001 mg/kg

TETRACHLOROETHYLENE   127-18-4

ENVIRONMENTAL LISTS

ATSDR Priority List
Rank (of 275): 160

STATE LISTS

California - Air Bill 2588 Appendix A-I
known or potential carcinogen

TETRACHLOROFLUOROETHANE   134237-32-4

SEE ALSO:
F039-HAZARDOUS WASTES

ENVIRONMENTAL LISTS

RCRA - Hazardous Constituents-Appendix VIII
hazardous constituent - no waste number

RCRA - Basis for Listing - Appendix VII
Included in waste streams: F020, F022, F023, F026, F027, F028, F039

RCRA - Universal Treatment Standards (LDR)
WW: 0.000063 mg/l; NWW: 0.001 mg/kg

1,1,1,2-TETRACHLORO-2-FLUOROETHANE (HCFC-121A)   354-11-0

HEALTH AND SAFETY LISTS

ACGIH 1995 - Time Weighted Averages
500 ppm TWA; 4170 mg/m3 TWA

NIOSH 1990 - Pocket Guide - RELs
500 ppm TWA; 4170 mg/m3 TWA

NIOSH 1990 - Pocket Guide - IDLHs
15,000 ppm IDLH

NIOSH 1990 - Pocket Guide - Target organs
skin, respiratory system

OSHA - Vacated PELs - Time Weighted Averages
500 ppm TWA; 4170 mg/m3 TWA

OSHA - Final PELs - Time Weighted Averages
500 ppm TWA; 4170 mg/m3 TWA

INTERNATIONAL LISTS

Australian Exposure Standards - Time Weighted Averages
500 ppm TWA; 4170 mg/m3 TWA

Canada - WHMIS: Ingredient Disclosure
1% item 1510 (1572)

Canada - Alberta - 8 Hour Occupational Exposure Limit
500 ppm TWA; 4170 mg/m3 TWA

Canada - Alberta - 15 Minute Occupational Exposure Limit
625 ppm STEL; 5210 mg/m3 STEL

Canada - British Columbia - 8 Hour Exposure Limits
500 ppm TWA; 4170 mg/m3 TWA

Canada - British Columbia - 15 Minute Exposure Limits
625 ppm STEL; 5210 mg/m3 STEL

Canada - Ontario - OHSA - TWAEVs
500 ppm TWAEV; 4165 mg/m3 TWAEV

Canada - Quebec - Time-Weighted Average Exposure Values
500 ppm TWAEV; 4170 mg/m3 TWAEV

United Kingdom - Occupational Exposure Standards - TWAs
100 ppm TWA; 834 mg/m3 TWA

United Kingdom - Occupational Exposure Standards - STELs
100 ppm STEL; 834 mg/m3 STEL

German (DFG) - MAK Values
1000 ppm MAK; 8340 mg/m3 MAK

German (DFG) - Peak Limitations
2 x normal MAK (1 hour momentary value); don't exceed 3 times per shift

Israel - Time Weighted Averages
500 ppm TWA; 4170 mg/m3 TWA

Israel - Action Levels
250 ppm AL; 2085 mg/m3 AL

Mexico - Instruction No. 10 - TWAs
500 ppm TWA; 4170 mg/m3 TWA

Mexico - Instruction No. 10 - STELs
626 ppm STEL; 5210 mg/m3 STEL

STATE LISTS

California - Exposure Limits - PELs
500 ppm PEL; 4170 mg/m3 PEL

California - Directors List of Hazardous Substances (8 CCR 339)
[present]

Florida Hazardous Substance List
[present]

Massachusetts Right To Know List
[present]

Minnesota Hazardous Substance List
[present]

Pennsylvania Right to Know List
[present]

PROPOSED REGULATIONS

Canada - Ontario - Proposed Occupational TWAEVs
100 ppm TWAEV; 834 mg/m3 TWAEV

Canada - Ontario - Proposed Occupational STEVs
100 ppm STEV; 834 mg/m3 STEV

1,1,2,2-TETRACHLORO-1-FLUOROETHANE (HCFC-121)   354-14-3

HEALTH AND SAFETY LISTS

ACGIH 1995 - Time Weighted Averages
500 ppm TWA; 4170 mg/m3 TWA

NIOSH - Selected LD50s and LC50s
Inhalation, mouse: LC50 = 123 gm/m3 (2 hr) Oral, mouse: LD50 = 800 mg/kg

NIOSH 1990 - Pocket Guide - RELs
500 ppm TWA; 4170 mg/m3 TWA

NIOSH 1990 - Pocket Guide - IDLHs
15,000 ppm IDLH

NIOSH 1990 - Pocket Guide - Target organs
lungs, skin

NTP Chemical Status Reports - Testing Status and NTIS Number
Prechronic studies completed: in review for further evaluation

OSHA - Vacated PELs - Time Weighted Averages
500 ppm TWA; 4170 mg/m3 TWA

OSHA - Final PELs - Time Weighted Averages
500 ppm TWA; 4170 mg/m3 TWA

ENVIRONMENTAL LISTS

Class 1 Ozone Depletors
ozone depletion potential = 1.0

INTERNATIONAL LISTS

Australian Exposure Standards - Time Weighted Averages
500 ppm TWA; 4170 mg/m3 TWA

Canada - WHMIS: Ingredient Disclosure
1% item 1511 (1573)

Canada - Alberta - 8 Hour Occupational Exposure Limit
500 ppm TWA; 4170 mg/m3 TWA

Canada - Alberta - 15 Minute Occupational Exposure Limit
625 ppm STEL; 5210 mg/m3 STEL

Canada - British Columbia - 8 Hour Exposure Limits
500 ppm TWA; 4170 mg/m3 TWA

Canada - British Columbia - 15 Minute Exposure Limits
625 ppm STEL; 5210 mg/m3 STEL

Canada - Ontario - OHSA - TWAEVs
500 ppm TWAEV; 4165 mg/m3 TWAEV

Canada - Quebec - Time-Weighted Average Exposure Values
500 ppm TWAEV; 4170 mg/m3 TWAEV

United Kingdom - Occupational Exposure Standards - TWAs
100 ppm TWA; 834 mg/m3 TWA

United Kingdom - Occupational Exposure Standards - STELs
100 ppm STEL; 834 mg/m3 STEL

German (DFG) - MAK Values
200 ppm MAK; 1690 mg/m3 MAK

German (DFG) - Peak Limitations
5 x normal MAK (30 min. average value); don't exceed 2 times during shift

Israel - Time Weighted Averages
500 ppm TWA; 4170 mg/m3 TWA

Israel - Action Levels
250 ppm AL; 2085 mg/m3 AL

Mexico - Instruction No. 10 - TWAs
500 ppm TWA; 4170 mg/m3 TWA

Mexico - Instruction No. 10 - STELs
625 ppm STEL; 5210 mg/m3 STEL

STATE LISTS

California - Exposure Limits - PELs
500 ppm PEL; 4170 mg/m3 PEL

California - Directors List of Hazardous Substances (8 CCR 339)
[present]

Florida Hazardous Substance List
[present]

Massachusetts Right To Know List
[present]

Minnesota Hazardous Substance List
[present]

Pennsylvania Right to Know List
[present]

PROPOSED REGULATIONS

Canada - Ontario - Proposed Occupational TWAEVs
100 ppm TWAEV; 834 mg/m3 TWAEV

Canada - Ontario - Proposed Occupational STEVs
100 ppm STEV; 834 mg/m3 STEV

TETRACHLOROFLUOROPROPANE   134190-49-1

SEE ALSO:
K073-HAZARDOUS WASTES
K150-HAZARDOUS WASTES
K095-HAZARDOUS WASTES
BIS(2,4-DIMETHYLBUTYL) MALEATE
K030-HAZARDOUS WASTES
F039-HAZARDOUS WASTES
F024-HAZARDOUS WASTES
CHLORINATED ETHANES
K020-HAZARDOUS WASTES
K019-HAZARDOUS WASTES
F025-HAZARDOUS WASTES

HEALTH AND SAFETY LISTS

ACGIH 1995 - Time Weighted Averages
1 ppm TWA; 6.9 mg/m3 TWA

ACGIH 1995 - Skin Designations
skin - potential for cutaneous absorption

AIHA - Odor Threshold Values
geometric mean air odor threshold = 7.3 ppm (detectable)

U.S. DOT - Appendix B - Marine Pollutants
DOT regulated marine pollutant

U.S. DOT - Appendix A Table 1 - Hazardous Substances
final RQ = 100 pounds (45.4 kg)

IARC - Group 3 (not classifiable)
[present]

NIOSH - Selected LD50s and LC50s
Oral, rat: LD50 = 800 mg/kg

NIOSH 1990 - Pocket Guide - RELs
1 ppm TWA; 7 mg/m3 TWA

NIOSH 1990 - Pocket Guide - IDLHs
150 ppm IDLH (not considering carcinogenic effects)

NIOSH 1990 - Pocket Guide - Carcinogens
occupational carcinogen

NIOSH 1990 - Pocket Guide - Target organs
CNS, liver, kidneys

NIOSH 1990 - Pocket Guide - Skin list
Potential for dermal absorption

NIOSH - Health Standards - Exposure Limits
reduce exposure to lowest feasible concentration

NIOSH - Health Standards - Health Effects and Precautions
has produced tumors of the liver in animals; liver, gastrointestinal, and nervous system effects (Blood monitoring required; prevent skin contact)

NIOSH - Health Standards - Carcinogenic Chemicals
potential human carcinogen

NTP Chemical Status Reports - Testing Status and NTIS Number
Technical reports printed (PB277453/AS); prechronic studies completed: in review for further evaluation;

NTP Chemical Status Reports - Evidence of Carcinogenicity
male rat-equivocal; female rat-negative; male mice-positive; female mice-positive

OSHA - Vacated PELs - Time Weighted Averages
1 ppm TWA; 7 mg/m3 TWA

OSHA - Vacated PELs - Skin Designation
Prevent or reduce skin absorption

OSHA - Final PELs - Time Weighted Averages
5 ppm TWA; 35 mg/m3 TWA

OSHA - Final PELs - Skin Notations
prevent or reduce skin absorption

ENVIRONMENTAL LISTS

ATSDR Priority List
Rank (of 275): 105

CERCLA/SARA - Section 313 - Emission Reporting
form R reporting required for 0.1% de minimus concentration

CERCLA/SARA - Hazardous Substances and their Reportable Quantities
final RQ = 100 pounds (45.4 kg)

Clean Air Act (1990) - List of Hazardous Air Contaminants
[present]

CAA - HON Rule - SOCMI Chemicals
compliance by Jan. 23, 1995

CAA - HON Rule - Organic HAPs
[present]

Clean Water Act - Priority Pollutants
[present]

Safe Drinking Water Act - Monitoring
monitoring required

EPA - Carcinogen Hazard Ranking for RQ Adjustment
Hazard ranking = Low

EPA - Master Testing List
[present]

List of Pesticide Product Inert Ingredients
[present]

RCRA - U Series Wastes
waste number U209

RCRA - Hazardous Constituents-Appendix VIII
waste number U209

RCRA - Basis for Listing - Appendix VII
Included in waste streams: F024, F025, F039, K019, K020, K030, K073, K095, K150

RCRA - Substances Banned From Land Disposal
[present]

RCRA - TSD Facilities Ground Water Monitoring
TM 8010 = 0.5 ug/L PQL;   TM 8240 = 5 ug/L PQL

RCRA - Universal Treatment Standards (LDR)
WW: 0.57 mg/l;   NWW: 6.0 mg/kg

TSCA - Code of Federal Regulations Citations
40 CFR 716.120(a)

INTERNATIONAL LISTS

Australian Exposure Standards - Time Weighted Averages
1 ppm TWA;   6.9 mg/m3 TWA

Australian Exposure Standards - Skin Effects
skin absorption

Canada - WHMIS: Ingredient Disclosure
1% item 1513 (1575)

Canada - NPRI (National Pollutant Release Inventory)
[present]

Canada - CEPA - Priority Substances List
estimated time for completion of assessment reports: 4 years

Canada - Alberta - 8 Hour Occupational Exposure Limit
1 ppm TWA;   6.8 mg/m3 TWA

Canada - Alberta - 15 Minute Occupational Exposure Limit
5 ppm STEL;   34 mg/m3 STEL

Canada - British Columbia - 8 Hour Exposure Limits
5 ppm TWA;   35 mg/m3 TWA

Canada - British Columbia - 15 Minute Exposure Limits
10 ppm STEL;   70 mg/m3 STEL

Canada - British Columbia - Skin Notations
skin - potential for skin absorption

Canada - Ontario - OHSA - TWAEVs
1 ppm TWAEV;   7 mg/m3 TWAEV

Canada - Ontario - OHSA - Skin Notations
absorption through skin, eyes, or mucous membranes

Canada - Quebec - Time-Weighted Average Exposure Values
1 ppm TWAEV;   6.9 mg/m3 TWAEV

Canada - Quebec - Skin Designations
absorbed through the skin

German (DFG) - MAK Values
1 ppm MAK;   7 mg/m3 MAK

German (DFG) - Skin/Sensitizers
danger of cutaneous absorption

German (DFG) - Carcinogens
suspected carcinogen

Israel - Time Weighted Averages
1 ppm TWA;   6.9 mg/m3 TWA

Israel - Action Levels
0.5 ppm AL;   3.45 mg/m3 AL

Mexico - Instruction No. 10 - TWAs
5 ppm TWA;   35 mg/m3 TWA

Mexico - Instruction No. 10 - STELs
10 ppm STEL;   70 mg/m3 STEL

Mexico - Instruction No. 10 - Skin designation
skin - potential for cutaneous absorption

Mexico - Wastewater - Organic Toxic Pollutants and Heavy Metals
Listed under [Chlorinated Ethanes]

Mexico - Drinking Water - Ecological Criteria
0.002 mg/l Substance presents persistence, bioaccumulations or risk of cancer, reduce human exposure to a minimum;   This level has been extrapolated by using a mathematic model

STATE LISTS

California - Air Bill 2588 Appendix A-I
9/90

California - Prop. 65 - Cancer list
carcinogen - initial date 7/1/90

California - Prop. 65 - No Significant Risk Levels
no significant risk level = 3 ug/day

California - Exposure Limits - PELs
1 ppm PEL;   7 mg/m3 PEL

California - Exposure Limits - Skin Notation
material may be absorbed through the skin, eyes or mucous membrane

California - Directors List of Hazardous Substances (8 CCR 339)
[present]

Florida Hazardous Substance List
[present]

Massachusetts Right To Know List
carcinogen;   extraordinarily hazardous

Minnesota Hazardous Substance List
carcinogen;   skin

NJ Right to Know List (Total)
sn 1809

NJ Special Hazardous Substances
(carcinogen)

Pennsylvania Right to Know List
environmental hazard

PROPOSED REGULATIONS

Safe Drinking Water Act - Priority list
[present]

TETRACHLORONAPHTHALENE   1335-88-2
SEE ALSO:
F024-HAZARDOUS WASTES
F039-HAZARDOUS WASTES
F025-HAZARDOUS WASTES
K030-HAZARDOUS WASTES
K019-HAZARDOUS WASTES
K020-HAZARDOUS WASTES
K095-HAZARDOUS WASTES

HEALTH AND SAFETY LISTS

U.S. DOT - Appendix A Table 1 - Hazardous Substances
final RQ = 100 pounds (45.4 kg)

IARC - Group 3 (not classifiable)
[present]

NIOSH - Health Standards - Exposure Limits
handle with caution in the workplace

NIOSH - Health Standards - Health Effects and Precautions
Central nervous system effects;   possible liver and kidney effects

NTP Chemical Status Reports - Testing Status and NTIS Number
Technical reports printed (PB83218206);   Prechronic studies completed: in review for further evaluation

ENVIRONMENTAL LISTS

CERCLA/SARA - Section 313 - Emission Reporting
form R reporting required

CERCLA/SARA - Hazardous Substances and their Reportable Quantities
final RQ = 100 pounds (45.4 kg)

Safe Drinking Water Act - Monitoring
monitoring required

EPA - Carcinogen Hazard Ranking for RQ Adjustment
Hazard ranking = Low

RCRA - U Series Wastes
waste number U208

RCRA - Hazardous Constituents-Appendix VIII
waste number U208

RCRA - Basis for Listing - Appendix VII
Included in waste streams: F024, F025, F039, K019, K020, K030, K095

RCRA - Substances Banned From Land Disposal
[present]

RCRA - TSD Facilities Ground Water Monitoring
TM 8010 = 5 ug/L PQL; TM 8240 = 5 ug/L PQL

RCRA - Universal Treatment Standards (LDR)
WW: 0.057 mg/l; NWW: 6.0 mg/kg

TSCA - Code of Federal Regulations Citations
40 CFR 716.120(a)

TSCA - Health and Safety Reporting List
Effective Date: June 1, 1987

INTERNATIONAL LISTS

Canada - WHMIS: Ingredient Disclosure
1% item 1512 (1574)

STATE LISTS

California - Directors List of Hazardous Substances (8 CCR 339)
[present]

Massachusetts Right To Know List
[present]

Pennsylvania Right to Know List
environmental hazard

PROPOSED REGULATIONS

Safe Drinking Water Act - Priority list
[present]

2,3,5,6-TETRACHLORO-4-NITROANISOLE   2438-88-2

HEALTH AND SAFETY LISTS

U.S. DOT - Substances From 49 CFR 172.101
regulated by DOT (UN1702)

U.S. DOT - Hazard Classes
DOT hazard class = 6.1

ENVIRONMENTAL LISTS

ATSDR Priority List
Rank (of 275): 226

RCRA - Hazardous Constituents-Appendix VIII
hazardous constituent - no waste number

1,2,4,5-TETRACHLORO-3-NITROBENZENE   117-18-0
SEE ALSO:
F024-HAZARDOUS WASTES
F001-HAZARDOUS WASTES
K150-HAZARDOUS WASTES
K151-HAZARDOUS WASTES
F002-HAZARDOUS WASTES
K020-HAZARDOUS WASTES
BIS(2,4-DIMETHYLBUTYL) MALEATE
K147-HAZARDOUS WASTES
K073-HAZARDOUS WASTES
F025-HAZARDOUS WASTES
K116-HAZARDOUS WASTES
F039-HAZARDOUS WASTES
K016-HAZARDOUS WASTES
K019-HAZARDOUS WASTES

HEALTH AND SAFETY LISTS

ACGIH 1995 - Time Weighted Averages
25 ppm TWA; 170 mg/m3 TWA

ACGIH 1995 - Short Term Exposure Limits
100 ppm STEL; 685 mg/m3 STEL

ACGIH 1995 - Carcinogens
A3-animal carcinogen

ACGIH 1995 - Biological Exposure Indices
Perchloroethylene in end-exhaled air: (10 ppm), prior to the last shift of week; Perchloroethylene in blood: (1 mg/L), proir to the last shift of week; Trichloroacetic acid in urine: (7 mg/L), end of workweek (Ns, Sq)

AIHA - Odor Threshold Values
geometric mean air odor threshold = 47 ppm (detectable); 71 ppm (recognizable)

U.S. DOT - Substances From 49 CFR 172.101
regulated by DOT (UN1897)

U.S. DOT - Hazard Classes
DOT hazard class = 6.1

U.S. DOT - Appendix B - Marine Pollutants
DOT regulated marine pollutant

U.S. DOT - Appendix A Table 1 - Hazardous Substances
final RQ = 100 pounds (45.4 kg)

IARC - Group 2B (sufficient animal data)
[present]

NFPA - Flash Points
no flash point

NFPA - Hazard Identification Ratings
health-2; flammability-0; reactivity-0

NIOSH - Selected LD50s and LC50s
Inhalation, mouse: LC50 = 5200 ppm 4 hr Oral, rat: LD50 = 3005 mg/kg

NIOSH 1990 - Pocket Guide - RELs
Minimize workplace exposure of concentrations; limit number of workers exposed

NIOSH 1990 - Pocket Guide - IDLHs
500 ppm IDLH (not considering carcinogenic effects)

NIOSH 1990 - Pocket Guide - Carcinogens
occupational carcinogen

NIOSH 1990 - Pocket Guide - Target organs
liver, kidneys, eyes, CNS, upper respiratory system

NIOSH - Health Standards - Exposure Limits
minimize workplace exposure concentration

NIOSH - Health Standards - Health Effects and Precautions
has produced tumores of the liver in animals

NIOSH - Health Standards - Carcinogenic Chemicals
potential human carcinogen

NTP Chemical Status Reports - Testing Status and NTIS Number
Technical reports printed (PB272940/AS) (P87147054/AS)

NTP Chemical Status Reports - Evidence of Carcinogenicity
PB272940/AS: male rat-inadequate; female rat-inadequate; male mice-positive; female mice-positive; PB87147054/AS: male rat-clear evidence; female rat-some evidence; male mice-clear evidence; female mice-clear evidence

NTP Seventh Report - Suspect Carcinogens
suspect carcinogen

OSHA - Vacated PELs - Time Weighted Averages
25 ppm TWA; 170 mg/m3 TWA

OSHA - Final PELs - Time Weighted Averages
100 ppm TWA; C 200 ppm

OSHA - Final PELs - Ceiling Limits
C 200 ppm

OSHA - Possible Select Carcinogens
[present]

ENVIRONMENTAL LISTS

ATSDR Priority List
Rank (of 275): 023

CERCLA/SARA - Section 313 - Emission Reporting
form R reporting required for 0.1% de minimus concentration

CERCLA/SARA - Hazardous Substances and their Reportable Quantities
final RQ = 100 pounds (45.4 kg)

Clean Air Act (1990) - List of Hazardous Air Contaminants
[present]

CAA - HON Rule - SOCMI Chemicals
compliance by Oct. 24, 1994

CAA - HON Rule - Organic HAPs
[present]

Clean Water Act - Priority Pollutants
[present]

Clean Water Act - Toxic Pollutants
[present]

Safe Drinking Water Act - MCLs
MCL = 0.005 mg/L

Safe Drinking Water Act - MCLGs
MCLG = Zero

Safe Drinking Water Act - Monitoring
monitoring required

EPA - Carcinogen Hazard Ranking for RQ Adjustment
Hazard ranking = Low

RCRA - D Series - Maximum Concentration of Contaminants
waste number D039; regulatory level = 0.7 mg/L

RCRA - D Series - Chronic Toxicity Reference Levels
chronic toxicity reference level = 0.007 mg/L

RCRA - U Series Wastes
waste number U210

RCRA - Hazardous Constituents-Appendix VIII
waste number U210

RCRA - Basis for Listing - Appendix VII
Included in waste streams: F001, F002, F024, F025, F039, K016, K019, K020, K073, K116, K150, K151

RCRA - Substances Banned From Land Disposal
[present]

RCRA - TSD Facilities Ground Water Monitoring
TM 8010 = 0.5 ug/L PQL; TM 8240 = 5 ug/L PQL

RCRA - Universal Treatment Standards (LDR)
WW: 0.56 mg/l; NWW: 6.0 mg/kg

TSCA - Code of Federal Regulations Citations
40 CFR 716.120(a)

TSCA - Health and Safety Reporting List
Effective Date: June 1, 1987

INTERNATIONAL LISTS

Australian Exposure Standards - Time Weighted Averages
50 ppm TWA; 340 mg/m3 TWA

Australian Exposure Standards - Short Term Exposure Limits
150 ppm STEL; 1020 mg/m3 STEL

Australian Exposure Standards - Carcinogens
suspected carcinogen

Canada - WHMIS: Ingredient Disclosure
1% item 1249 (1356)

Canada - NPRI (National Pollutant Release Inventory)
[present]

Canada - CEPA - Priority Substances List
estimated time for completion of assessment reports: 4 years

Canada - Alberta - 8 Hour Occupational Exposure Limit
50 ppm TWA; 339 mg/m3 TWA

Canada - Alberta - 15 Minute Occupational Exposure Limit
100 ppm STEL; 678 mg/m3 STEL

Canada - Alberta - Skin Designation
can be absorbed through the intact skin

Canada - British Columbia - 8 Hour Exposure Limits
25 ppm TWA

Canada - British Columbia - 15 Minute Exposure Limits
75 ppm STEL

Canada - British Columbia - Skin Notations
skin - potential for skin absorption

Canada - British Columbia - Carcinogens
carcinogen - 25 ppm TWA; 75 mg/m3 STEL

Canada - Ontario - OHSA - TWAEVs
50 ppm TWAEV; 340 mg/m3 TWAEV

Canada - Ontario - OHSA - STEVs
200 ppm STEV; 1355 mg/m3 STEV

Canada - Quebec - Time-Weighted Average Exposure Values
50 ppm TWAEV; 339 mg/m3 TWAEV

Canada - Quebec - Short-term Exposure Values
200 ppm STEV; 1357 mg/m3 STEV

Canada - Quebec - Carcinogens
C3 carcinogen: effect detected in animals

United Kingdom - Occupational Exposure Standards - TWAs
50 ppm TWA; 335 mg/m3 TWA

United Kingdom - Occupational Exposure Standards - STELs
150 ppm STEL; 1000 mg/m3 STEL

German (DFG) - MAK Values
50 ppm MAK; 345 mg/m3 MAK

German (DFG) - Peak Limitations
2 x normal MAK (30 min. average value); don't exceed 4 times during shift

German (DFG) - Carcinogens
suspected carcinogen

German (DFG) - Pregnancy
no risk to embryo/fetus if exposure limits adhered to

Israel - Time Weighted Averages
50 ppm TWA; 339 mg/m3 TWA

Israel - Short Term Exposure Limits
200 ppm STEL; 1370 mg/m3 STEL

Israel - Action Levels
25 ppm AL; 169.5 mg/m3 AL

Mexico - Instruction No. 10 - TWAs
100 ppm TWA; 670 mg/m3 TWA

Mexico - Instruction No. 10 - STELs
200 ppm STEL; 1340 mg/m3 STEL

Mexico - Instruction No. 10 - Skin designation
skin - potential for cutaneous absorption

Mexico - Wastewater - Organic Toxic Pollutants and Heavy Metals
Listed under [Organic Toxic Pollutants]

Mexico - Drinking Water - Ecological Criteria
0.008 mg/l Substance presents persistence, bioaccumulations or risk of cancer, reduce human exposure to a minimum; This level has been extrapolated by using a mathematic model

STATE LISTS

California - Air Bill 2588 Appendix A-I
known or potential carcinogen

California - Prop. 65 - Cancer list
carcinogen - initial date 4/1/88

California - Prop. 65 - No Significant Risk Levels
no significant risk level = 14 ug/day

California - Exposure Limits - PELs
25 ppm PEL; 170 mg/m3 PEL

California - Exposure Limits - Ceilings
C 300 ppm

California - Directors List of Hazardous Substances (8 CCR 339)
[present]

Florida Hazardous Substance List
[present]

Massachusetts Right To Know List
carcinogen; extraordinarily hazardous

Minnesota Hazardous Substance List
skin

NJ Right to Know List (Total)
sn 1810

NJ Special Hazardous Substances
(carcinogen)

Pennsylvania Right to Know List
environmental hazard; special hazardous substance

Pennsylvania RTK - Special Hazardous Substances
[present]

PROPOSED REGULATIONS

ACGIH 1995 - Proposed Biological Exposure Indices
Perchloroethylene in end-exhaled air: 5 ppm, prior to the last shift of workweek Perchloroethylene in blood: 0.5 mg/L, prior to the last shift of workweek; Trichloroacetic acid in urine: 3.5 mg/L, end of workweek (Ns, Sq)

2,3,5,6-TETRACHLOROPHENOL   935-95-5

ENVIRONMENTAL LISTS

Class 2 Ozone Depletors
ozone depletion weight reserved

2,3,4,6-TETRACHLOROPHENOL   58-90-2

ENVIRONMENTAL LISTS

CERCLA/SARA - Section 313 - Emission Reporting
form R reporting required

TETRACHLOROPHENOL   25167-83-3

ENVIRONMENTAL LISTS

CERCLA/SARA - Section 313 - Emission Reporting
form R reporting required

2,3,4,6-TETRACHLOROPHENOL, SODIUM SALT   25567-55-9

ENVIRONMENTAL LISTS

Class 2 Ozone Depletors
ozone depletion weight reserved

TETRACHLOROPHTHALIC ANHYDRIDE   117-08-8

HEALTH AND SAFETY LISTS

ACGIH 1995 - Time Weighted Averages
2 mg/m3 TWA

NIOSH 1990 - Pocket Guide - RELs
2 mg/m3 TWA

NIOSH 1990 - Pocket Guide - Target organs
liver, skin

NIOSH 1990 - Pocket Guide - Skin list
Potential for dermal absorption

OSHA - Vacated PELs - Time Weighted Averages
2 mg/m3 TWA

OSHA - Vacated PELs - Skin Designation
Prevent or reduce skin absorption

OSHA - Final PELs - Time Weighted Averages
2 mg/m3 TWA

OSHA - Final PELs - Skin Notations
prevent or reduce skin absorption

ENVIRONMENTAL LISTS

TSCA - Code of Federal Regulations Citations
40 CFR 704.83; 40 CFR 712.30(d); 40 CFR 716.120(a)

TSCA - PAIR - Reporting List
Reporting Date: November 19, 1982

TSCA - Health and Safety Reporting List
Effective Date: October 4, 1982

INTERNATIONAL LISTS

Australian Exposure Standards - Time Weighted Averages
2 mg/m3 TWA

Canada - WHMIS: Ingredient Disclosure
1% item 1514 (1576)

Canada - Alberta - 8 Hour Occupational Exposure Limit
2 mg/m3 TWA

Canada - Alberta - 15 Minute Occupational Exposure Limit
4 mg/m3 STEL

Canada - British Columbia - 8 Hour Exposure Limits
2 mg/m3 TWA

Canada - British Columbia - 15 Minute Exposure Limits
4 mg/m3 STEL

Canada - Ontario - OHSA - TWAEVs
2 mg/m3 TWAEV

Canada - Quebec - Time-Weighted Average Exposure Values
2 mg/m3 TWAEV

United Kingdom - Occupational Exposure Standards - TWAs
2 mg/m3 TWA

United Kingdom - Occupational Exposure Standards - STELs
4 mg/m3 STEL

Israel - Time Weighted Averages
2 mg/m3 TWA

Israel - Action Levels
1 mg/m3 AL

Mexico - Instruction No. 10 - TWAs
2 mg/m3 TWA

Mexico - Instruction No. 10 - STELs
4 mg/m3 STEL

STATE LISTS

California - Exposure Limits - PELs
2 mg/m3 PEL

California - Exposure Limits - Skin Notation
material may be absorbed through the skin, eyes or mucous membrane

California - Directors List of Hazardous Substances (8 CCR 339)
[present]

Florida Hazardous Substance List
[present]

Massachusetts Right To Know List
[present]

Minnesota Hazardous Substance List
skin

Pennsylvania Right to Know List
[present]

1,1,2,3-TETRACHLOROPROPENE   10436-39-2

HEALTH AND SAFETY LISTS

NTP Chemical Status Reports - Testing Status and NTIS Number
Technical reports printed (PB287642/AS)

NTP Chemical Status Reports - Evidence of Carcinogenicity
male rat-negative; female rat-negative; male mice-negative; female mice-negative

TETRACHLOROTETRAFLUOROPROPANE   29255-31-0

ENVIRONMENTAL LISTS

TSCA - HDD/HDF - Precursors Required for Reporting
[present]

TETRACHLOROTRIFLUOROPROPANE   134237-37-9

ENVIRONMENTAL LISTS

ATSDR Priority List
Rank (of 275): 264

TETRACHLORODIFLUOROPROPANE   134237-39-1

SEE ALSO:
F020-HAZARDOUS WASTES
F027-HAZARDOUS WASTES
F023-HAZARDOUS WASTES
F039-HAZARDOUS WASTES

HEALTH AND SAFETY LISTS

U.S. DOT - Appendix A Table 1 - Hazardous Substances
final RQ = 10 pounds (4.54 kg)

NIOSH - Selected LD50s and LC50s
Oral, rat: LD50 = 140 mg/kg Skin, rabbit: LD50 = 250 mg/kg

ENVIRONMENTAL LISTS
  ATSDR Priority List
    Rank (of 275): 217
  CERCLA/SARA - Hazardous Substances and their Reportable Quantities
    final RQ = 10 pounds (4.54 kg)
  RCRA - Hazardous Constituents-Appendix VIII
    contained in F027 waste stream
  RCRA - Basis for Listing - Appendix VII
    Included in waste stream: F039
  RCRA - TSD Facilities Ground Water Monitoring
    TM 8270 = 10 ug/L PQL
  RCRA - Universal Treatment Standards (LDR)
    WW: 0.030 mg/l; NWW: 7.4 mg/kg
INTERNATIONAL LISTS
  Canada - Drinking Water Quality - MACs
    0.1 mg/L MAC
  Canada - Drinking Water Quality - AOs
    <= 0.001 mg/L AO
STATE LISTS
  Massachusetts Right To Know List
    [present]
  Pennsylvania Right to Know List
    environmental hazard

TETRACHLOROVINPHOS   961-11-5

HEALTH AND SAFETY LISTS
  U.S. DOT - Appendix B - Marine Pollutants
    DOT regulated marine pollutant
  NIOSH - Selected LD50s and LC50s
    Oral, rat: LD50 = 140 mg/kg
ENVIRONMENTAL LISTS
  ATSDR Priority List
    Rank (of 275): 161
  RCRA - Basis for Listing - Appendix VII
    Included in waste streams: F020, F023, F027, F028

TETRACOSAMETHYLCYCLODODECASILOXANE   18919-94-3

ENVIRONMENTAL LISTS
  RCRA - Hazardous Constituents-Appendix VIII
    hazardous constituent - no waste number

TETRACOSAMETHYLUNDECASILOXANE   107-53-9

HEALTH AND SAFETY LISTS
  NTP Chemical Status Reports - Testing Status and NTIS Number
    Technical reports printed (no NTIS number given)
ENVIRONMENTAL LISTS
  CAA - HON Rule - SOCMI Chemicals
    compliance by April 24, 1995
INTERNATIONAL LISTS
  Canada - WHMIS: Ingredient Disclosure
    0.1% item 1515 (236)

TETRACYANOMETHYLENE   670-54-2

ENVIRONMENTAL LISTS
  TSCA - Code of Federal Regulations Citations
    40 CFR 716.120(a)
  TSCA - Health and Safety Reporting List
    Effective Date: June 1, 1987
INTERNATIONAL LISTS
  Canada - WHMIS: Ingredient Disclosure
    1% item 1516 (1578)

TETRACYCLINE   60-54-8

ENVIRONMENTAL LISTS
  Class 1 Ozone Depletors
    ozone depletion potential = 1.0

TETRACYCLINE HYDROCHLORIDE   64-75-5

ENVIRONMENTAL LISTS
  Class 2 Ozone Depletors
    ozone depletion weight reserved

TETRACYCLINES   RR-01475-5

ENVIRONMENTAL LISTS
  Class 2 Ozone Depletors
    ozone depletion weight reserved

TETRADECAMETHYLCYCLOHEPTASILOXANE   107-50-6

HEALTH AND SAFETY LISTS
  IARC - Group 3 (not classifiable)
    [present]
  NIOSH - Selected LD50s and LC50s
    Oral, rat: LD50 = 4 gm/kg
  NTP Chemical Status Reports - Testing Status and NTIS Number
    Technical reports printed (PB278650/AS)
  NTP Chemical Status Reports - Evidence of Carcinogenicity
    male rat-negative; female rat-positive; male mice-positive; female mice-positive
ENVIRONMENTAL LISTS
  CERCLA/SARA - Section 313 - Emission Reporting
    form R reporting required for 1.0% de minimus concentration
STATE LISTS
  California - Air Bill 2588 Appendix A-II
    6/91
  California - Directors List of Hazardous Substances (8 CCR 339)
    [present]
  Massachusetts Right To Know List
    carcinogen; extraordinarily hazardous
  NJ Right to Know List (Total)
    sn 1813
  Pennsylvania Right to Know List
    environmental hazard

TETRADECAMETHYLHEXASILOXANE   107-52-8

ENVIRONMENTAL LISTS
  TSCA - PAIR - Reporting List
    Effective Date: October 12, 1993; Reporting Date: February 28, 1994
  TSCA - Health and Safety Reporting List
    Effective Date: October 12, 1993; Sunset Date: October 12, 2003

TETRADECANE   629-59-4

ENVIRONMENTAL LISTS
  TSCA - PAIR - Reporting List
    Effective Date: October 12, 1993; Reporting Date: February 28, 1994
  TSCA - Health and Safety Reporting List
    Effective Date: October 12, 1993; Sunset Date: October 12, 2003

TETRADECANE   64036-86-8

HEALTH AND SAFETY LISTS
  NIOSH - Selected LD50s and LC50s
    Oral, mouse: LD50 = 29 mg/kg

TERT-TETRADECANETHIOL   28983-37-1

HEALTH AND SAFETY LISTS

NIOSH - Selected LD50s and LC50s
Oral, rat: LD50 = 807 mg/kg

STATE LISTS

California - Prop. 65 - Developmental Toxicity
developmental toxicity (internal use) - initial date 10/1/91

Massachusetts Right To Know List
teratogen

NJ Right to Know List (Total)
sn 1814

NJ Special Hazardous Substances
(teratogen)

TETRADECANOL   27196-00-5

HEALTH AND SAFETY LISTS

NTP Chemical Status Reports - Testing Status and NTIS Number
Technical reports printed (PB90198540/AS)

NTP Chemical Status Reports - Evidence of Carcinogenicity
male rat-no evidence; female rat-no evidence; male mice-no evidence; female mice-no evidence

ENVIRONMENTAL LISTS

CERCLA/SARA - Section 313 - Emission Reporting
form R reporting required

STATE LISTS

California - Air Bill 2588 Appendix A-II
6/91

California - Prop. 65 - Developmental Toxicity
developmental toxicity (internal use) - initial date 1/1/91

1-TETRADECENE   1120-36-1

STATE LISTS

California - Prop. 65 - Developmental Toxicity
internal use: developmental toxicity - initial date 10/1/92

1-TETRADECYLPYRIDINIUM BROMIDE   1155-74-4

ENVIRONMENTAL LISTS

TSCA - PAIR - Reporting List
Effective Date: October 12, 1993; Reporting Date: February 28, 1994

TSCA - Health and Safety Reporting List
Effective Date: October 12, 1993; Sunset Date: October 12, 2003

TETRAETHANOL AMMONIUM HYDROXIDE   631-41-4

ENVIRONMENTAL LISTS

TSCA - PAIR - Reporting List
Effective Date: October 12, 1993; Reporting Date: February 28, 1994

TSCA - Health and Safety Reporting List
Effective Date: October 12, 1993; Sunset Date: October 12, 2003

TETRAETHOXYPROPANE   122-31-6

ENVIRONMENTAL LISTS

EPA - Master Testing List
[present]

TETRAETHYLAMMONIUM CHLORIDE   56-34-8

HEALTH AND SAFETY LISTS

NFPA - Flash Points
flash point = 212 degrees F (100 degrees C)

NFPA - Hazard Identification Ratings
health-0; flammability-1; reactivity-0

TETRAETHYLAMMONIUM PERCHLORATE   2567-83-1

HEALTH AND SAFETY LISTS

NFPA - Flash Points
flash point = 250 degrees F (121 degrees C)

NFPA - Hazard Identification Ratings
health-2; flammability-1; reactivity-0

STATE LISTS

Florida Hazardous Substance List
[present]

Massachusetts Right To Know List
[present]

Pennsylvania Right to Know List
[present]

TETRAETHYL DITHIOPYROPHOSPHATE   15108-81-3

HEALTH AND SAFETY LISTS

NFPA - Flash Points
flash point = 285 degrees F (141 degrees C)

NFPA - Hazard Identification Ratings
health-0; flammability-1; reactivity-0

TETRAETHYLENE GLYCOL   112-60-7

HEALTH AND SAFETY LISTS

NFPA - Flash Points
flash point = 230 degrees F (110 degrees C)

NFPA - Hazard Identification Ratings
health-0; flammability-1; reactivity-0

ENVIRONMENTAL LISTS

EPA - Master Testing List
[present]

TETRAETHYLENE GLYCOL DIACRYLATE   17831-71-9

ENVIRONMENTAL LISTS

List of Pesticide Product Inert Ingredients
[present]

TETRAETHYLENE GLYCOL MONOPHENYL ETHER   36366-93-5

INTERNATIONAL LISTS

Canada - WHMIS: Ingredient Disclosure
1% item 1517 (1001)

TETRAETHYLENEPENTAMINE   112-57-2

HEALTH AND SAFETY LISTS

NFPA - Flash Points
flash point = 190 degrees F (88 degrees C)

NFPA - Hazard Identification Ratings
health-0; flammability-2; reactivity-0

TETRA (2-ETHYLHEXYL) SILICATE   115-82-2

HEALTH AND SAFETY LISTS

NIOSH - Selected LD50s and LC50s
Oral, rat: LD50 = 2630 mg/kg

TETRAETHYL LEAD   78-00-2

HEALTH AND SAFETY LISTS

U.S. DOT - Hazard Classes
Forbidden from transport by the DOT

TETRAETHYL TIN   597-64-8

HEALTH AND SAFETY LISTS

U.S. DOT - Substances From 49 CFR 172.101
regulated by DOT (UN1703, UN1704)

U.S. DOT - Hazard Classes
DOT hazard class = 2.3

U.S. DOT - Substances Which Are Poisonous by Inhalation
gaseous hazardous material poisonous by inhalation (gases or mixtures) (UN1703)

1,1,1,2-TETRAFLUOROETHANE　　　　　　　　　811-97-2

HEALTH AND SAFETY LISTS

NFPA - Flash Points
flash point = 360 degrees F (182 degrees C)

NFPA - Hazard Identification Ratings
health-1; flammability-1; reactivity-0

NIOSH - Selected LD50s and LC50s
Oral, rat: LD50 = 29 gm/kg

ENVIRONMENTAL LISTS

CAA - HON Rule - SOCMI Chemicals
compliance by Oct. 24, 1994

TETRAFLUOROETHYLENE　　　　　　　　　116-14-3

HEALTH AND SAFETY LISTS

AIHA - WEEL - Time Weighted Averages
1 mg/m3 TWA

AIHA - WEEL - Skin Absorption Designations
skin absorber

STATE LISTS

Minnesota Hazardous Substance List
[present]

TETRAFLUOROHYDRAZINE　　　　　　　　　10036-47-2

ENVIRONMENTAL LISTS

List of Pesticide Product Inert Ingredients
[present]

TETRAFLUOROMETHANE　　　　　　　　　75-73-0

HEALTH AND SAFETY LISTS

U.S. DOT - Substances From 49 CFR 172.101
regulated by DOT (UN2320)

U.S. DOT - Hazard Classes
DOT hazard class = 8

NFPA - Flash Points
flash point = 325 degrees F (163 degrees C)

NFPA - Hazard Identification Ratings
health-2; flammability-1; reactivity-0

NIOSH - Selected LD50s and LC50s
Oral, rat: LD50 = 205 mg/kg Skin, rabbit: LD50 = 660 mg/kg

ENVIRONMENTAL LISTS

CAA - HON Rule - SOCMI Chemicals
compliance by Oct. 23, 1995

List of Pesticide Product Inert Ingredients
[present]

INTERNATIONAL LISTS

Canada - WHMIS: Ingredient Disclosure
1% item 1518 (1587)

STATE LISTS

Florida Hazardous Substance List
[present]

Massachusetts Right To Know List
[present]

NJ Right to Know List (Total)
sn 1816

NJ Special Hazardous Substances
(corrosive)

Pennsylvania Right to Know List
[present]

TETRAFLUOROPHTHALONITRILE　　　　　　　　　1835-49-0

HEALTH AND SAFETY LISTS

NFPA - Flash Points
flash point = 390 degrees F (199 degrees C)

NFPA - Hazard Identification Ratings
health-1; flammability-1; reactivity-0

TETRAGLYCIDYLAMINES　　　　　　　　　RR-00194-5
SEE ALSO:
LEAD

HEALTH AND SAFETY LISTS

ACGIH 1995 - Time Weighted Averages
as Pb: 0.1 mg/m3 TWA (For greater assurance of worker protection, biological monitoring is recommended.)

ACGIH 1995 - Skin Designations
as Pb: skin - potential for cutaneous absorption

U.S. DOT - Substances From 49 CFR 172.101
regulated by DOT (NA1649)

U.S. DOT - Hazard Classes
DOT hazard class = 6.1

U.S. DOT - Appendix B - Marine Pollutants
DOT regulated marine pollutant

U.S. DOT - Appendix A Table 1 - Hazardous Substances
final RQ = 10 pounds (4.54 kg)

NFPA - Flash Points
flash point = 200 degrees F (93 degrees C)

NFPA - Hazard Identification Ratings
health-3; flammability-2; reactivity-3

NIOSH - Selected LD50s and LC50s
Inhalation, rat: LC50 = 850 mg/m3 1 hr Oral, rat: LD50 = 12300 ug/kg

NIOSH 1990 - Pocket Guide - RELs
as Pb: 0.075 mg/m3 TWA

NIOSH 1990 - Pocket Guide - IDLHs
as Pb: 40 mg/m3 IDLH

NIOSH 1990 - Pocket Guide - Target organs
CNS, CVS, eyes, kidneys

NIOSH 1990 - Pocket Guide - Skin list
as Pb: Potential for dermal absorption

OSHA - Vacated PELs - Time Weighted Averages
as Pb: 0.075 mg/m3 TWA

OSHA - Vacated PELs - Skin Designation
Prevent or reduce skin absorption

OSHA - Final PELs - Time Weighted Averages
as Pb: 0.075 mg/m3 TWA

OSHA - Final PELs - Skin Notations
prevent or reduce skin absorption

ENVIRONMENTAL LISTS

CERCLA/SARA - Section 302 Extremely Hazardous Substances and TPQs
TPQ = 100 pounds

CERCLA/SARA - Hazardous Substances and their Reportable Quantities
final RQ = 10 pounds (4.54 kg)

CAA - HON Rule - SOCMI Chemicals
compliance by July 24, 1995

Clean Water Act - Hazardous Substances
[present]

RCRA - P Series Wastes
waste number P110

RCRA - Hazardous Constituents-Appendix VIII
waste number P110

RCRA - Substances Banned From Land Disposal
[present]

INTERNATIONAL LISTS

Australian Exposure Standards - Time Weighted Averages
as Pb: 0.1 mg/m3 TWA

Australian Exposure Standards - Skin Effects
as Pb: skin absorption

Canada - WHMIS: Ingredient Disclosure
1% item 1519 (1436)

Canada - CEPA Schedule II Part II - Toxic Substances (Export)
[present]

Canada - Alberta - 8 Hour Occupational Exposure Limit
as Pb: 0.1 mg/m3 TWA

Canada - Alberta - 15 Minute Occupational Exposure Limit
as Pb: 0.3 mg/m3 STEL

Canada - Alberta - Skin Designation
as Pb: can be absorbed through the intact skin

Canada - British Columbia - 8 Hour Exposure Limits
as Pb: 0.1 mg/m3 TWA

Canada - British Columbia - 15 Minute Exposure Limits
as Pb: 0.3 mg/m3 STEL

Canada - British Columbia - Skin Notations
skin - potential for skin absorption

Canada - Ontario - OHSA - TWAEVs
0.10 mg/m3 TWAEV (designated substance regulation)

Canada - Ontario - OHSA - STEVs
0.45 mg/m3 STEV (designated substance regulation)

Canada - Quebec - Time-Weighted Average Exposure Values
as Pb: 0.05 mg/m3 TWAEV

Canada - Quebec - Skin Designations
as Pb: absorbed through the skin

United Kingdom - Maximum Exposure Limits - TWAs
as Pb: 0.10 mg/m3 TWA (listed under 'Lead and compounds')

German (DFG) - MAK Values
as Pb: 0.05 mg/m3 MAK

German (DFG) - Peak Limitations
2 x normal MAK (30 min. average value); don't exceed 4 times during shift

German (DFG) - Skin/Sensitizers
danger of cutaneous absorption

German (DFG) - Pregnancy
classification not yet possible

Israel - Time Weighted Averages
as Pb: 0.1 mg/m3 TWA (For control of general room air, biologic monitoring is essential for personnel control)

Israel - Action Levels
as Pb: 0.05 mg/m3 AL

Mexico - Instruction No. 10 - TWAs
0.1 mg/m3 TWA

Mexico - Instruction No. 10 - STELs
0.5 mg/m3 STEL

Mexico - Instruction No. 10 - Skin designation
skin - potential for cutaneous absorption

STATE LISTS

California - Exposure Limits - PELs
as Pb: 0.075 mg/m3 PEL

California - Exposure Limits - Skin Notation
material may be absorbed through the skin, eyes or mucous membrane

Florida Hazardous Substance List
[present]

Massachusetts Right To Know List
extraordinarily hazardous

Minnesota Hazardous Substance List
as Pb: skin

NJ Right to Know List (Total)
sn 1817

NJ Special Hazardous Substances
(reactive - third degree)

Pennsylvania Right to Know List
environmental hazard

PROPOSED REGULATIONS

Canada - Ontario - Proposed Occupational TWAEVs
0.05 mg/m3 TWAEV

Canada - Ontario - Proposed Occupational STEVs
0.2 mg/m3 STEV

1,2,3,6-TETRAHYDROBENZALDEHYDE   100-50-5
SEE ALSO:
TIN

HEALTH AND SAFETY LISTS

NIOSH - Selected LD50s and LC50s
Oral, rat: LD50 = 16 mg/kg

ENVIRONMENTAL LISTS

CERCLA/SARA - Section 302 Extremely Hazardous Substances and TPQs
TPQ = 100 pounds

INTERNATIONAL LISTS

Canada - WHMIS: Ingredient Disclosure
1% item 1521 (1588)

STATE LISTS

Florida Hazardous Substance List
effective March 13, 1992

Massachusetts Right To Know List
extraordinarily hazardous

NJ Right to Know List (Total)
sn 2803

Pennsylvania Right to Know List
environmental hazard

PROPOSED REGULATIONS

CERCLA/SARA - Proposed Hazardous Substance Additions
proposed RQ = 1 pound (.454 kg)

CERCLA/SARA - 1989 Proposed RQ Adjustments
proposed RQ = 10 pounds (4.54 kg)

TETRAHYDROBENZALDEHYDE   1321-16-0

HEALTH AND SAFETY LISTS

AIHA - WEEL - Time Weighted Averages
1000 ppm TWA; 4240 mg/m3 TWA

ENVIRONMENTAL LISTS

TSCA - Code of Federal Regulations Citations
40 CFR 716.120(c)

TSCA - Health and Safety Reporting List
Effective Date: October 15, 1990; Sunset Date: November 9, 1993

1-TRANS-DELTA-9-TETRAHYDROCANNABINOL   1972-08-3
SEE ALSO:
FLUOROALKENES

HEALTH AND SAFETY LISTS

U.S. DOT - Substances From 49 CFR 172.101
regulated by DOT (UN1081)

U.S. DOT - Hazard Classes
DOT hazard class = 2.1

IARC - Group 3 (not classifiable)
[present]

NFPA - Flash Points
gas (no flash point given)

NFPA - Hazard Identification Ratings
health-2; flammability-4; reactivity-3

NIOSH - Selected LD50s and LC50s
Inhalation, rat: LC50 = 40000 ppm 4 hr
NTP Chemical Status Reports - Testing Status and NTIS Number
Two year studies: pathology working group in progress; prechronic studies for which toxicity technical reports were not prepared
OSHA - List of Highly Hazardous Chemicals
threshhold quantity = 5000 pounds

ENVIRONMENTAL LISTS

CAA - Flammable Substances for Accidental Release Prevention
threshold quantity = 10,000 lbs
TSCA - Code of Federal Regulations Citations
40 CFR 712.30(f); 40 CFR 716.120(c); 40 CFR 799.1700(a)(1)
TSCA - PAIR - Reporting List
Reporting Date: August 17, 1983
TSCA - Chemical Test Rules
Testing required by: manufacturers (40 CFR 799.1700) (Listed under 'Fluoroalkenes')
TSCA - Section 12(b) - Export Notification
export notification required - Section 4

STATE LISTS

Florida Hazardous Substance List
[present]
Massachusetts Right To Know List
[present]
NJ Right to Know List (Total)
sn 1819
NJ Special Hazardous Substances
(flammable - fourth degree; reactive - third degree)
Pennsylvania Right to Know List
[present]

1,2,3,4-TETRAHYDROCARBAZOLE   942-01-8

HEALTH AND SAFETY LISTS

U.S. DOT - Substances Which Are Poisonous by Inhalation
gaseous hazardous material poisonous by inhalation
NIOSH - Selected LD50s and LC50s
Inhalation, guinea pig: LC50 = 900 ppm (1 hr)
OSHA - List of Highly Hazardous Chemicals
threshhold quantity = 5000 pounds

INTERNATIONAL LISTS

Canada - WHMIS: Ingredient Disclosure
1% item 1522 (1589)

STATE LISTS

NJ Right to Know List (Total)
sn 1820

ENDO-TETRAHYDRODICYLCLOPENTADIENE   2825-83-4

HEALTH AND SAFETY LISTS

U.S. DOT - Substances From 49 CFR 172.101
regulated by DOT (UN1982)
U.S. DOT - Hazard Classes
DOT hazard class = 2.2

ENVIRONMENTAL LISTS

CAA - HON Rule - SOCMI Chemicals
compliance by Jan. 23, 1995

STATE LISTS

NJ Right to Know List (Total)
sn 1821

TETRAHYDRO-3,5-DIMETHYL-2H-1,3,5-THIADI-   533-74-4
AZINE-2-THIONE

HEALTH AND SAFETY LISTS

NIOSH - Selected LD50s and LC50s
Oral, mouse: LD50 = 56 mg/kg

TETRAHYDROFURAN   109-99-9

ENVIRONMENTAL LISTS

TSCA - Chemicals with Significant New Use Rules
PMN numbers: P-86-500; P-86-502

TETRAHYDROFURFURYL ALCOHOL   97-99-4

HEALTH AND SAFETY LISTS

NFPA - Flash Points
flash point = 135 degrees F (57 degrees C)
NFPA - Hazard Identification Ratings
health-2; flammability-2; reactivity-0

ENVIRONMENTAL LISTS

TSCA - Code of Federal Regulations Citations
40 CFR 712.30(x); 40 CFR 716.120(d)
TSCA - PAIR - Reporting List
Reporting Date: Novmeber 27, 1991
TSCA - Health and Safety Reporting List
Effective Date: September 30, 1991

INTERNATIONAL LISTS

Canada - WHMIS: Ingredient Disclosure
1% item 1523 (1592)

STATE LISTS

Pennsylvania Right to Know List
[present]

TETRAHYDROFURFURYLAMINE   4795-29-3

HEALTH AND SAFETY LISTS

U.S. DOT - Substances From 49 CFR 172.101
regulated by DOT (UN2498)
U.S. DOT - Hazard Classes
DOT hazard class = 3
NIOSH - Selected LD50s and LC50s
Inhalation, mouse: LC50 = 556 mg/m3 (4 hr) Oral, rat: LD50 = 2460 mg/kg Skin, rabbit: LD50 = 1770 mg/kg

STATE LISTS

Florida Hazardous Substance List
[present]
Massachusetts Right To Know List
[present]
NJ Right to Know List (Total)
sn 1822
Pennsylvania Right to Know List
[present]

TETRAHYDROFURFURYL METHACRYLATE   2455-24-5

HEALTH AND SAFETY LISTS

NTP Chemical Status Reports - Testing Status and NTIS Number
Post peer review technical reports in progress: prechronic studies for which toxicity technical reports were not prepared

TETRAHYDROFURFURYL OLEATE   RR-00885-5

HEALTH AND SAFETY LISTS

NIOSH - Selected LD50s and LC50s
Oral, rat: LD50 = 2650 mg/kg

TETRAHYDRONAPHTHALENE   119-64-2

HEALTH AND SAFETY LISTS

NFPA - Hazard Identification Ratings
reactivity-0

1,2,3,4-TETRAHYDRO-1-NAPHTHOL   529-33-9

ENVIRONMENTAL LISTS

CERCLA/SARA - Section 313 - Emission Reporting
form R reporting required

List of Pesticide Product Inert Ingredients
[present]

1,2,3,4-TETRAHYDRO-2-NAPHTHYLAMINE   2954-50-9

HEALTH AND SAFETY LISTS

ACGIH 1995 - Time Weighted Averages
200 ppm TWA; 590 mg/m3 TWA

ACGIH 1995 - Short Term Exposure Limits
250 ppm STEL; 737 mg/m3 STEL

AIHA - Odor Threshold Values
geometric mean air odor threshold = 31 ppm (detectable); 61 ppm (recognizable)

U.S. DOT - Substances From 49 CFR 172.101
regulated by DOT (UN2056)

U.S. DOT - Hazard Classes
DOT hazard class = 3

U.S. DOT - Appendix A Table 1 - Hazardous Substances
final RQ = 1000 pounds (454 kg)

NFPA - Flash Points
flash point = 6 degrees F (-14 degrees C)

NFPA - Hazard Identification Ratings
health-2; flammability-3; reactivity-1

NIOSH - Selected LD50s and LC50s
Inhalation, rat: LC50 = 21000 ppm 3 hr Oral, rat: LD50 = 2816 mg/kg

NIOSH 1990 - Pocket Guide - RELs
200 ppm TWA; 590 mg/m3 TWA; 250 ppm STEL; 735 mg/m3 STEL

NIOSH 1990 - Pocket Guide - IDLHs
20,000 ppm IDLH (lower explosive level)

NIOSH 1990 - Pocket Guide - Target organs
eyes, skin, CNS, respiratory system

NTP Chemical Status Reports - Testing Status and NTIS Number
Two year studies: pathology quality assessment in progress; prechronic studies for which toxicity technical reports were not prepared

OSHA - Vacated PELs - Time Weighted Averages
200 ppm TWA; 590 mg/m3 TWA

OSHA - Vacated PELs - Short Term Exposure Limits
250 ppm STEL; 735 mg/m3 STEL

OSHA - Final PELs - Time Weighted Averages
200 ppm TWA; 590 mg/m3 TWA

ENVIRONMENTAL LISTS

ATSDR Priority List
Rank (of 275): 268

CERCLA/SARA - Hazardous Substances and their Reportable Quantities
final RQ = 1000 pounds (454 kg)

CAA - HON Rule - SOCMI Chemicals
compliance by Oct. 24, 1994

EPA - Master Testing List
[present]

List of Pesticide Product Inert Ingredients
[present]

RCRA - U Series Wastes
waste number U213 (Ignitable waste)

RCRA - Hazardous Constituents-Appendix VIII
waste number U213 (Ignitable waste)

RCRA - Substances Banned From Land Disposal
[present]

TSCA - Health and Safety Reporting List
Effective Date: March 11, 1994; Sunset Date: March 11, 2004

TSCA - Multichemical Test Rules - Neurotoxicity
administrative stay for neurotoxicity tests effective June 27, 1994

INTERNATIONAL LISTS

Australian Exposure Standards - Time Weighted Averages
200 ppm TWA; 590 mg/m3 TWA

Australian Exposure Standards - Short Term Exposure Limits
250 ppm STEL; 737 mg/m3 STEL

Canada - WHMIS: Ingredient Disclosure
1% item 1524 (1593)

Canada - Alberta - 8 Hour Occupational Exposure Limit
200 ppm TWA; 590 mg/m3 TWA

Canada - Alberta - 15 Minute Occupational Exposure Limit
250 ppm STEL; 735 mg/m3 STEL

Canada - British Columbia - 8 Hour Exposure Limits
200 ppm TWA; 590 mg/m3 TWA

Canada - British Columbia - 15 Minute Exposure Limits
250 ppm STEL; 735 mg/m3 STEL

Canada - Ontario - OHSA - TWAEVs
200 ppm TWAEV; 590 mg/m3 TWAEV

Canada - Ontario - OHSA - STEVs
250 ppm STEV; 735 mg/m3 STEV

Canada - Quebec - Time-Weighted Average Exposure Values
100 ppm TWAEV; 300 mg/m3 TWAEV

United Kingdom - Occupational Exposure Standards - TWAs
200 ppm TWA; 590 mg/m3 TWA

United Kingdom - Occupational Exposure Standards - STELs
250 ppm STEL; 735 mg/m3 STEL

German (DFG) - MAK Values
200 ppm MAK; 590 mg/m3 MAK

German (DFG) - Peak Limitations
5 x normal MAK (30 min. average value); don't exceed 2 times during shift

German (DFG) - Pregnancy
no risk to embryo/fetus if exposure limits adhered to

Israel - Time Weighted Averages
200 ppm TWA; 590 mg/m3 TWA

Israel - Short Term Exposure Limits
250 ppm STEL; 737 mg/m3 STEL

Israel - Action Levels
100 ppm AL; 295 mg/m3 AL

Mexico - Instruction No. 10 - TWAs
200 ppm TWA; 590 mg/m3 TWA

Mexico - Instruction No. 10 - STELs
250 ppm STEL; 735 mg/m3 STEL

STATE LISTS

California - Exposure Limits - PELs
200 ppm PEL; 590 mg/m3 PEL

California - Exposure Limits - STELs
250 ppm STEL; 735 mg/m3 STEL

California - Directors List of Hazardous Substances (8 CCR 339)
[present]

Florida Hazardous Substance List
[present]

Massachusetts Right To Know List
[present]

Minnesota Hazardous Substance List
[present]

NJ Right to Know List (Total)
sn 1823

NJ Special Hazardous Substances
(flammable - third degree)

Pennsylvania Right to Know List
environmental hazard

PROPOSED REGULATIONS

Safe Drinking Water Act - Priority list
[present]

TSCA - ITC 32nd Report Priority Testing List
designated for dermal absorption testing

TETRAHYDRONAPHTHYL HYDROPEROXIDE   RR-01759-4

HEALTH AND SAFETY LISTS

AIHA - WEEL - Time Weighted Averages
2 ppm TWA; 8.4 mg/m3 TWA

NFPA - Flash Points
flash point = 167 degrees F (75 degrees C)

NFPA - Hazard Identification Ratings
health-2; flammability-2; reactivity-0

NIOSH - Selected LD50s and LC50s
Oral, rat: LD50 = 2500 mg/kg

ENVIRONMENTAL LISTS

List of Pesticide Product Inert Ingredients
[present]

INTERNATIONAL LISTS

Canada - WHMIS: Ingredient Disclosure
1% item 1525 (188)

STATE LISTS

Florida Hazardous Substance List
[present]

Massachusetts Right To Know List
[present]

Pennsylvania Right to Know List
[present]

TETRAHYDROPHTHALIC ACID ANHYDRIDE   85-43-8

HEALTH AND SAFETY LISTS

U.S. DOT - Substances From 49 CFR 172.101
regulated by DOT (UN2943)

U.S. DOT - Hazard Classes
DOT hazard class = 3

STATE LISTS

NJ Right to Know List (Total)
sn 1824

TETRAHYDROPHTHALIMIDE   85-40-5

ENVIRONMENTAL LISTS

TSCA - Code of Federal Regulations Citations
40 CFR 712.30(d)

TSCA - PAIR - Reporting List
Reporting Date: November 19, 1982

INTERNATIONAL LISTS

Canada - WHMIS: Ingredient Disclosure
1% item 1526 (1097)

TETRAHYDROPYRAN-2-METHANOL   100-72-1

HEALTH AND SAFETY LISTS

NFPA - Flash Points
flash point = 390 degrees F (199 degrees C)

NFPA - Hazard Identification Ratings
health-1; flammability-1; reactivity-0

1,2,3,6-TETRAHYDROPYRIDINE   694-05-3

HEALTH AND SAFETY LISTS

NFPA - Flash Points
flash point = 160 degrees F (71 degrees C)

NFPA - Hazard Identification Ratings
health-1; flammability-2; reactivity-0

NIOSH - Selected LD50s and LC50s
Oral, rat: LD50 = 2860 mg/kg Skin, rabbit: LD50 = 17 gm/kg

NTP Chemical Status Reports - Testing Status and NTIS Number
Approved for toxicology/carcinogenesis study

ENVIRONMENTAL LISTS

CAA - HON Rule - SOCMI Chemicals
compliance by July 24, 1995

List of Pesticide Product Inert Ingredients
[present]

INTERNATIONAL LISTS

Canada - WHMIS: Ingredient Disclosure
1% item 1527 (1594)

STATE LISTS

Pennsylvania Right to Know List
[present]

TETRAHYDROTHIOPHENE   110-01-0

HEALTH AND SAFETY LISTS

NIOSH - Selected LD50s and LC50s
Oral, rat: LD50 = 1620 mg/kg

2,2',4,4'-TETRAHYDROXYBENZOPHENONE   131-55-5

HEALTH AND SAFETY LISTS

NIOSH - Selected LD50s and LC50s
Oral, rat: LD50 = 4710 mg/kg

TETRAISOPROPYL TITANATE   546-68-9

HEALTH AND SAFETY LISTS

U.S. DOT - Organic Peroxides Table
Organic peroxide UN3106

TETRAKIS(2-CHLOROETHYL) ETHYLENE DIPHOSPHATE   33125-86-9

HEALTH AND SAFETY LISTS

U.S. DOT - Substances From 49 CFR 172.101
regulated by DOT (UN2698)

U.S. DOT - Hazard Classes
DOT hazard class = 8

NIOSH - Selected LD50s and LC50s
Oral, rat: LD50 = 5410 mg/kg

ENVIRONMENTAL LISTS

CAA - HON Rule - SOCMI Chemicals
compliance by Jan. 23, 1995

INTERNATIONAL LISTS

Canada - WHMIS: Ingredient Disclosure
1% item 1528 (237)

STATE LISTS

NJ Right to Know List (Total)
sn 1825

TETRAKIS(HYDROXYMETHYL)PHOSPHONIUM CHLORIDE   124-64-1

ENVIRONMENTAL LISTS

List of Pesticide Product Inert Ingredients
[present]

TETRAKIS(HYDROXYMETHYL)PHOSPHONIUM   55566-30-8
SULFATE

HEALTH AND SAFETY LISTS

NFPA - Flash Points
flash point = 200 degrees F (93 degrees C)

NFPA - Hazard Identification Ratings
health-1; flammability-2; reactivity-0

TETRAKIS(HYDROXYMETHYL) PHOSPHONIUM   RR-01548-5
SALTS

HEALTH AND SAFETY LISTS

U.S. DOT - Substances From 49 CFR 172.101
regulated by DOT (UN2410)

U.S. DOT - Hazard Classes
DOT hazard class = 3

STATE LISTS

NJ Right to Know List (Total)
sn 1826

N,N,N',N'-TETRAKIS(OXIRANYLMETHYL)-1,3-   65992-66-7
CYCLOHEXANEDIMETHANAMINE

HEALTH AND SAFETY LISTS

U.S. DOT - Substances From 49 CFR 172.101
regulated by DOT (UN2412)

U.S. DOT - Hazard Classes
DOT hazard class = 3

STATE LISTS

NJ Right to Know List (Total)
sn 1827

TETRALIN HYDROPEROXIDE   771-29-9

HEALTH AND SAFETY LISTS

NIOSH - Selected LD50s and LC50s
Oral, rat: LD50 = 1220 mg/kg

ENVIRONMENTAL LISTS

List of Pesticide Product Inert Ingredients
[present]

D-TETRALONE   529-34-0

HEALTH AND SAFETY LISTS

NIOSH - Selected LD50s and LC50s
Oral, rat: LD50 = 7460 mg/kg

ENVIRONMENTAL LISTS

List of Pesticide Product Inert Ingredients
[present]

1,1,3,3-TETRAMETHOXYPROPANE   102-52-3

ENVIRONMENTAL LISTS

TSCA - Code of Federal Regulations Citations
40 CFR 712.30(w); 40 CFR 716.120(a)

TSCA - PAIR - Reporting List
Reporting Date: February 14, 1989

TSCA - Health and Safety Reporting List
Effective Date: December 16, 1988; Sunset Date: November 9, 1993

PROPOSED REGULATIONS

TSCA - ITC 33rd Report Priority Testing List
recommended with intent-to-designate

TSCA - ITC 34th Report Priority Testing List
recommended with intent-to-designate

TETRAMETHYLAMMONIUM HYDROXIDE   75-59-2

HEALTH AND SAFETY LISTS

NTP Chemical Status Reports - Testing Status and NTIS Number
Technical reports printed (PB87204137/AS)

NTP Chemical Status Reports - Evidence of Carcinogenicity
male rat-no evidence; female rat-no evidence; male mice-no evidence; female mice-no evidence

TETRAMETHYLAMMONIUM SALTS OF ALKYLBEN-   RR-01682-0
ZENESULFONIC ACID

HEALTH AND SAFETY LISTS

NTP Chemical Status Reports - Testing Status and NTIS Number
Technical reports printed (PB87204137/AS)

NTP Chemical Status Reports - Evidence of Carcinogenicity
male rat-no evidence; female rat-no evidence; male mice-no evidence; female mice-no evidence

1,2,4,5-TETRAMETHYLBENZENE   95-93-2

HEALTH AND SAFETY LISTS

IARC - Group 3 (not classifiable)
[present]

1,2,3,4-TETRAMETHYLBENZENE   488-23-3

ENVIRONMENTAL LISTS

TSCA - Section 12(b) - Export Notification
P-84-7; export notification required - Section 5

1,2,3,5-TETRAMETHYLBENZENE   527-53-7

STATE LISTS

NJ Right to Know List (Total)
sn 1828

3,3',5,5'-TETRAMETHYLBIPHENYL-4,4'-DIOL   RR-00908-5

HEALTH AND SAFETY LISTS

NIOSH - Selected LD50s and LC50s
Oral, rat: LD50 = 810 mg/kg

N,N,N',N'-TETRAMETHYL-1,3-BUTANEDIAMINE   97-84-7

HEALTH AND SAFETY LISTS

NFPA - Flash Points
flash point = 170 degrees F (77 degrees C)

NFPA - Hazard Identification Ratings
health-0; flammability-2; reactivity-0

1,1,3,3-TETRAMETHYL BUTYL HYDROPEROXIDE   5809-08-5

HEALTH AND SAFETY LISTS

U.S. DOT - Substances From 49 CFR 172.101
regulated by DOT (UN1835)

U.S. DOT - Hazard Classes
DOT hazard class = 8

INTERNATIONAL LISTS

Canada - WHMIS: Ingredient Disclosure
1% item 1529 (1000)

STATE LISTS

NJ Right to Know List (Total)
sn 1829

NJ Special Hazardous Substances
(corrosive)

1,1,3,3-TETRAMETHYLBUTYLPEROXY-2-ETHYL   22288-43-3
HEXANOATE

ENVIRONMENTAL LISTS

TSCA - Chemicals with Significant New Use Rules
PMN number: P-92-1364

**ALPHA-(1,1,3,3-TETRAMETHYLBUTYL)PHENOXY-
OMEGA-POLYOXYPROPYLENE BLOCK POLYMER
WITH POLYOXYETHYLENE**   RR-01153-0

HEALTH AND SAFETY LISTS

NFPA - Flash Points
95%: flash point = 130 degrees F (54 degrees C)

NFPA - Hazard Identification Ratings
95%: health-0; flammability-2; reactivity-0

NIOSH - Selected LD50s and LC50s
Oral, rat: LD50 = 6989 mg/kg

TETRAMETHYLCYCLOTETRASILOXANE   2370-88-9

HEALTH AND SAFETY LISTS

NFPA - Flash Points
95%: flash point = 166 degrees F (74 degrees C)

NFPA - Hazard Identification Ratings
95%: health-0; flammability-2; reactivity-0

TETRAMETHYLDIVINYLDISILOXANE   2627-95-4

HEALTH AND SAFETY LISTS

NFPA - Flash Points
85.5%: flash point = 160 degrees F (71 degrees C)

NFPA - Hazard Identification Ratings
85.5%: health-0; flammability-2; reactivity-0

NIOSH - Selected LD50s and LC50s
Oral, rat: LD50 = 5157 mg/kg

TETRAMETHYLENEDIAMINE   110-60-1

ENVIRONMENTAL LISTS

TSCA - Chemicals with Significant New Use Rules
PMN number: P-88-972

TETRAMETHYLENE DIPEROXIDE DICARBAMIDE   RR-01460-8

INTERNATIONAL LISTS

Canada - WHMIS: Ingredient Disclosure
1% item 1530 (1596)

N,N,N',N'-TETRAMETHYL-1,6-HEXANEDIAMINE   111-18-2

HEALTH AND SAFETY LISTS

U.S. DOT - Organic Peroxides Table
Organic peroxide UN3105

STATE LISTS

NJ Right to Know List (Total)
sn 1830

TETRAMETHYL LEAD   75-74-1

HEALTH AND SAFETY LISTS

U.S. DOT - Organic Peroxides Table
Organic peroxide UN3115

STATE LISTS

NJ Right to Know List (Total)
technically pure: sn 2804

TETRAMETHYLMETHYLENEDIAMINE   51-80-9

ENVIRONMENTAL LISTS

List of Pesticide Product Inert Ingredients
[present]

TETRAMETHYL SILANE   75-76-3

ENVIRONMENTAL LISTS

TSCA - PAIR - Reporting List
*Effective Date: October 12, 1993;   Reporting Date: February 28,
1994*

TSCA - Health and Safety Reporting List
Effective Date: October 12, 1993;   Sunset Date: October 12, 2003

TETRAMETHYL SUCCINONITRILE   3333-52-6

ENVIRONMENTAL LISTS

TSCA - PAIR - Reporting List
*Effective Date: October 12, 1993;   Reporting Date: February 28,
1994*

TSCA - Health and Safety Reporting List
Effective Date: June 14, 1993

**(ANAD5) TETRAMETHYL THIURAM
MONOSULFIDE**   97-74-5

ENVIRONMENTAL LISTS

CAA - HON Rule - SOCMI Chemicals
compliance by Jan. 23, 1995

TETRAMETHYLUREA   632-22-4

HEALTH AND SAFETY LISTS

U.S. DOT - Hazard Classes
Forbidden from transport by the DOT

TETRANITRO-ANILINE   53014-37-2

HEALTH AND SAFETY LISTS

U.S. DOT - Substances Which Are Poisonous by Inhalation
liquid hazardous material poisonous by inhalation

TETRANITRO DIGLYCERIN   20600-96-8
SEE ALSO:
LEAD

HEALTH AND SAFETY LISTS

ACGIH 1995 - Time Weighted Averages
*as Pb: 0.15 mg/m3 TWA (For greater assurance of worker protection,
biological monitoring is recommended.)*

ACGIH 1995 - Skin Designations
as Pb: skin - potential for cutaneous absorption

U.S. DOT - Appendix B - Marine Pollutants
DOT regulated marine pollutant

NFPA - Flash Points
flash point = 100 degrees F (38 degrees C)

NFPA - Hazard Identification Ratings
health-3; flammability-3; reactivity-3

NIOSH - Selected LD50s and LC50s
Oral, rat: LD50 = 105 mg/kg

NIOSH 1990 - Pocket Guide - RELs
as Pb: 0.075 mg/m3 TWA

NIOSH 1990 - Pocket Guide - IDLHs
as Pb: 40 mg/m3 IDLH

NIOSH 1990 - Pocket Guide - Target organs
CNS, CVS, kidneys

NIOSH 1990 - Pocket Guide - Skin list
as Pb: Potential for dermal abosrption

OSHA - Vacated PELs - Time Weighted Averages
as Pb: 0.075 mg/m3 TWA

OSHA - Vacated PELs - Skin Designation
Prevent or reduce skin absorption

OSHA - Final PELs - Time Weighted Averages
as Pb: 0.075 mg/m3 TWA

OSHA - Final PELs - Skin Notations
prevent or reduce skin absorption

OSHA - List of Highly Hazardous Chemicals
threshhold quantity = 1000 pounds

ENVIRONMENTAL LISTS

CERCLA/SARA - Section 302 Extremely Hazardous Substances and TPQs
TPQ = 100 pounds

CERCLA/SARA - Hazardous Substances and their Reportable Quantities
final RQ = 10 pounds (4.54 kg)

CAA -Toxic Substances for Accidental Release Prevention
threshold quantity = 10,000 lbs

CAA - HON Rule - SOCMI Chemicals
compliance by Oct. 23, 1995

INTERNATIONAL LISTS

Australian Exposure Standards - Time Weighted Averages
as Pb: 0.15 mg/m3 TWA

Australian Exposure Standards - Skin Effects
as Pb: skin absorption

Canada - WHMIS: Ingredient Disclosure
1% item 1531 (1437)

Canada - CEPA Schedule II Part II - Toxic Substances (Export)
[present]

Canada - Alberta - 8 Hour Occupational Exposure Limit
as Pb: 0.15 mg/m3 TWA

Canada - Alberta - 15 Minute Occupational Exposure Limit
as Pb: 0.5 mg/m3 STEL

Canada - Alberta - Skin Designation
as Pb: can be absorbed through the intact skin

Canada - British Columbia - 8 Hour Exposure Limits
as Pb: 0.15 mg/m3 TWA

Canada - British Columbia - 15 Minute Exposure Limits
as Pb: 0.5 mg/m3 STEL

Canada - British Columbia - Skin Notations
skin - potential for skin absorption

Canada - Quebec - Time-Weighted Average Exposure Values
as Pb: 0.05 mg/m3 TWAEV

Canada - Quebec - Skin Designations
absorbed through the skin

German (DFG) - MAK Values
as Pb: 0.01 ppm MAK; 0.075 mg/m3 MAK

German (DFG) - Peak Limitations
2 x normal MAK (30 min. average value); don't exceed 4 times during shift

German (DFG) - Skin/Sensitizers
danger of cutaneous absorption

Israel - Time Weighted Averages
as Pb: 0.15 mg/m3 TWA (For control of general room air, biologic monotoring is essential for personnel control)

Israel - Action Levels
as Pb: 0.075 mg/m3 AL

Mexico - Instruction No. 10 - TWAs
0.15 mg/m3 TWA

Mexico - Instruction No. 10 - STELs
0.3 mg/m3 STEL

Mexico - Instruction No. 10 - Skin designation
skin - potential for cutaneous absorption

STATE LISTS

California - Exposure Limits - PELs
as Pb: 0.075 mg/m3 PEL

California - Exposure Limits - Skin Notation
material may be absorbed through the skin, eyes or mucous membrane

Florida Hazardous Substance List
[present]

Massachusetts Right To Know List
extraordinarily hazardous

Minnesota Hazardous Substance List
as Pb: skin

NJ Right to Know List (Total)
sn 1831

NJ Special Hazardous Substances
(flammable - third degree; reactive - third degree)

Pennsylvania Right to Know List
environmental hazard

PROPOSED REGULATIONS

CERCLA/SARA - Proposed Hazardous Substance Additions
proposed RQ = 1 pound (.454 kg)

CERCLA/SARA - 1989 Proposed RQ Adjustments
proposed RQ = 100 pounds (45.4 kg)

TETRANITROMETHANE   509-14-8

STATE LISTS

NJ Right to Know List (Total)
sn 1832

2,3,4,6-TETRANITROPHENOL   641-16-7

HEALTH AND SAFETY LISTS

U.S. DOT - Substances From 49 CFR 172.101
regulated by DOT (UN2749)

U.S. DOT - Hazard Classes
DOT hazard class = 3

ENVIRONMENTAL LISTS

CAA - Flammable Substances for Accidental Release Prevention
threshold quantity = 10,000 lbs

STATE LISTS

NJ Right to Know List (Total)
sn 1833

2,3,4,6-TETRANITROPHENYL METHYL NITRAMINE   RR-01375-2

HEALTH AND SAFETY LISTS

ACGIH 1995 - Time Weighted Averages
0.5 ppm TWA; 2.8 mg/m3 TWA

ACGIH 1995 - Skin Designations
skin - potential for cutaneous absorption

NIOSH - Selected LD50s and LC50s
Oral, rat: LD50 = 38900 ug/kg

NIOSH 1990 - Pocket Guide - RELs
3 mg/m3 TWA; 0.5 ppm TWA

NIOSH 1990 - Pocket Guide - IDLHs
5 ppm IDLH

NIOSH 1990 - Pocket Guide - Target organs
CNS

NIOSH 1990 - Pocket Guide - Skin list
Potential for dermal absorption

NIOSH - Health Standards - Exposure Limits
C (15 min) 1 ppm; C (15 min) 6 mg/m3 (Listed under 'Nitriles')

NIOSH - Health Standards - Health Effects and Precautions
Hepatic, renal, respiratory, cardiovascular, gastrointestinal, and nervous system effects (Periodic chest X-ray and pulmonary function testing required; prevent skin and eye contact; make first-aid kits and personnel available during use) (Listed under 'Nitriles')

OSHA - Vacated PELs - Time Weighted Averages
0.5 ppm TWA; 3 mg/m3 TWA

OSHA - Vacated PELs - Skin Designation
Prevent or reduce skin absorption

OSHA - Final PELs - Time Weighted Averages
0.5 ppm TWA; 3 mg/m3 TWA

OSHA - Final PELs - Skin Notations
prevent or reduce skin absorption

INTERNATIONAL LISTS

Australian Exposure Standards - Time Weighted Averages
0.5 ppm TWA; 2.8 mg/m3 TWA

Australian Exposure Standards - Skin Effects
skin absorption

Canada - WHMIS: Ingredient Disclosure
1% item 1532 (1597)

Canada - Alberta - 8 Hour Occupational Exposure Limit
0.5 ppm TWA; 2.8 mg/m3 TWA

Canada - Alberta - 15 Minute Occupational Exposure Limit
2 ppm STEL; 11 mg/m3 STEL

Canada - Alberta - Skin Designation
can be absorbed through the intact skin

Canada - British Columbia - 8 Hour Exposure Limits
0.5 ppm TWA; 3 mg/m3 TWA

Canada - British Columbia - 15 Minute Exposure Limits
2 ppm STEL; 9 mg/m3 STEL

Canada - British Columbia - Skin Notations
skin - potential for skin absorption

Canada - Ontario - OHSA - TWAEVs
0.5 ppm TWAEV; 2.8 mg/m3 TWAEV

Canada - Ontario - OHSA - Skin Notations
absorption through skin, eyes, or mucous membranes

Canada - Quebec - Time-Weighted Average Exposure Values
0.5 ppm TWAEV; 2.8 mg/m3 TWAEV

Canada - Quebec - Skin Designations
absorbed through the skin

United Kingdom - Occupational Exposure Standards - TWAs
0.5 ppm TWA; 3 mg/m3 TWA

United Kingdom - Occupational Exposure Standards - STELs
2 ppm STEL; 9 mg/m3 STEL

United Kingdom - Occupational Exposure Standards - Notes
can be absorbed through skin

German (DFG) - MAK Values
0.5 ppm MAK; 3 mg/m3 MAK

German (DFG) - Peak Limitations
2 x normal MAK (30 min. average value); don't exceed 4 times during shift

German (DFG) - Skin/Sensitizers
danger of cutaneous absorption

Israel - Time Weighted Averages
0.5 ppm TWA; 2.8 mg/m3 TWA

Israel - Action Levels
0.25 ppm AL; 1.4 mg/m3 AL

Mexico - Instruction No. 10 - TWAs
0.5 ppm TWA; 3 mg/m3 TWA

Mexico - Instruction No. 10 - STELs
2 ppm STEL; 9 mg/m3 STEL

Mexico - Instruction No. 10 - Skin designation
skin - potential for cutaneous absorption

STATE LISTS

California - Exposure Limits - PELs
0.5 ppm PEL; 3 mg/m3 PEL

California - Exposure Limits - Skin Notation
material may be absorbed through the skin, eyes or mucous membrane

California - Directors List of Hazardous Substances (8 CCR 339)
[present]

Florida Hazardous Substance List
[present]

Massachusetts Right To Know List
[present]

Minnesota Hazardous Substance List
skin

Pennsylvania Right to Know List
[present]

2,3,4,6-TETRANITROPHENYL METHYL NITRAMINE   RR-01376-3

HEALTH AND SAFETY LISTS

NIOSH - Selected LD50s and LC50s
Oral, mouse: LD50 = 818 mg/kg

2,3,4,6-TETRANITROPHENYLNITRAMINE   RR-01377-4

HEALTH AND SAFETY LISTS

NIOSH - Selected LD50s and LC50s
Oral, rat: LD50 = 1400 mg/kg

INTERNATIONAL LISTS

Canada - WHMIS: Ingredient Disclosure
1% item 1533 (1598)

TETRANITRORESORCINOL   RR-01461-9

HEALTH AND SAFETY LISTS

U.S. DOT - Substances From 49 CFR 172.101
regulated by DOT (UN0207)

U.S. DOT - Hazard Classes
DOT hazard class = 1.1D

STATE LISTS

NJ Right to Know List (Total)
sn 1835

2,3,5,6-TETRANITROSO-1,4-DINITROBENZENE   RR-01379-6

HEALTH AND SAFETY LISTS

U.S. DOT - Hazard Classes
Forbidden from transport by the DOT

2,3,5,6-TETRANITROSO NITROBENZENE   RR-01378-5

HEALTH AND SAFETY LISTS

ACGIH 1995 - Time Weighted Averages
0.005 ppm TWA; 0.04 mg/m3 TWA

ACGIH 1995 - Carcinogens
A2-suspected human carcinogen

U.S. DOT - Substances From 49 CFR 172.101
regulated by DOT (UN1510)

U.S. DOT - Hazard Classes
DOT hazard class = 5.1

U.S. DOT - Substances Which Are Poisonous by Inhalation
liquid hazardous material poisonous by inhalation (UN1510)

U.S. DOT - Appendix A Table 1 - Hazardous Substances
final RQ = 10 pounds (4.54 kg)

NIOSH - Selected LD50s and LC50s
Inhalation, rat: LC50 = 18 ppm 4 hr Oral, rat: LD50 = 130 mg/kg

NIOSH 1990 - Pocket Guide - RELs
1 ppm TWA; 8 mg/m3 TWA

NIOSH 1990 - Pocket Guide - IDLHs
5 ppm IDLH

NIOSH 1990 - Pocket Guide - Target organs
respiratory system, skin, eyes, blood, CNS

NTP Chemical Status Reports - Testing Status and NTIS Number
Technical reports printed (PB91113373)

NTP Chemical Status Reports - Evidence of Carcinogenicity
male rat-clear evidence; female rat-clear evidence; male mice-clear evidence; female mice-clear evidence

NTP Seventh Report - Suspect Carcinogens
suspect carcinogen

OSHA - Vacated PELs - Time Weighted Averages
1 ppm TWA; 8 mg/m3 TWA

OSHA - Final PELs - Time Weighted Averages
1 ppm TWA; 8 mg/m3 TWA

OSHA - Final PELs - Skin Notations
prevent or reduce skin absorption

ENVIRONMENTAL LISTS

CERCLA/SARA - Section 302 Extremely Hazardous Substances and TPQs
TPQ = 500 pounds

CERCLA/SARA - Hazardous Substances and their Reportable Quantities
final RQ = 10 pounds (4.54 kg)

CAA -Toxic Substances for Accidental Release Prevention
threshold quantity = 10,000 lbs

RCRA - P Series Wastes
waste number P112 (Reactive waste)

RCRA - Hazardous Constituents-Appendix VIII
waste number P112

RCRA - Substances Banned From Land Disposal
[present]

INTERNATIONAL LISTS

Australian Exposure Standards - Time Weighted Averages
1 ppm TWA; 8 mg/m3 TWA

Canada - WHMIS: Ingredient Disclosure
1% item 1534 (1601)

Canada - Alberta - 8 Hour Occupational Exposure Limit
1 ppm TWA; 8 mg/m3 TWA

Canada - Alberta - 15 Minute Occupational Exposure Limit
3 ppm STEL; 24 mg/m3 STEL

Canada - British Columbia - 8 Hour Exposure Limits
1 ppm TWA; 8 mg/m3 TWA

Canada - Ontario - OHSA - TWAEVs
1 ppm TWAEV; 8 mg/m3 TWAEV

Canada - Quebec - Time-Weighted Average Exposure Values
1 ppm TWAEV; 8 mg/m3 TWAEV

German (DFG) - MAK Values
1 ppm MAK; 8 mg/m3 MAK

German (DFG) - Carcinogens
animal evidence of carcinogenicity

Israel - Time Weighted Averages
1 ppm TWA; 8 mg/m3 TWA

Israel - Action Levels
0.5 ppm AL; 4 mg/m3 AL

Mexico - Instruction No. 10 - TWAs
1 ppm TWA; 8 mg/m3 TWA

STATE LISTS

California - Air Bill 2588 Appendix A-II
known or potential carcinogen: 9/90

California - Prop. 65 - Cancer list
carcinogen - initial date 7/1/90

California - Exposure Limits - PELs
1 ppm PEL; 8 mg/m3 PEL

California - Directors List of Hazardous Substances (8 CCR 339)
[present]

Florida Hazardous Substance List
[present]

Massachusetts Right To Know List
extraordinarily hazardous

Minnesota Hazardous Substance List
[present]

NJ Right to Know List (Total)
sn 1836

Pennsylvania Right to Know List
environmental hazard

2,4,8,10-TETRAOXA-3,9-DIPHOSPHASPIRO[5,5] UN-DECANE, 3,9-BIS[2,4,6-TRIS[1,1-DIMETHYLETHYL] PHENOXY]-   126505-35-9

HEALTH AND SAFETY LISTS

U.S. DOT - Hazard Classes
Forbidden from transport by the DOT

3,6,9,13-TETRAOXAHEXADEC-15-ENE-1,11-DIOL,BIS (4-METHYLBENZENESULFONATE)   RR-01728-7

HEALTH AND SAFETY LISTS

U.S. DOT - Hazard Classes
Forbidden from transport by the DOT

3,6,9,12-TETRAOXATETRADECANE-1,14-DIOL,BIS(4-METHYLBENZENESULFONATE)   41024-91-3

HEALTH AND SAFETY LISTS

U.S. DOT - Hazard Classes
Forbidden from transport by the DOT

3,6,9,12-TETRAOXATETRADECANE-1,14-DIOL,7-[(2-PROPENYLOXY)METHYL]-,BIS(4-METHYLBEN-ZENESULFONATE)   RR-01730-1

HEALTH AND SAFETY LISTS

U.S. DOT - Hazard Classes
Forbidden from transport by the DOT

2,5,8,11-TETRAOXATRIDECAN-13-OL   23783-42-8

HEALTH AND SAFETY LISTS

U.S. DOT - Hazard Classes
Forbidden from transport by the DOT

TETRAPROPYL LEAD   3440-75-3

HEALTH AND SAFETY LISTS

U.S. DOT - Hazard Classes
Forbidden from transport by the DOT

TETRAPROPYLORTHOTITANATE   3087-37-4

HEALTH AND SAFETY LISTS

U.S. DOT - Hazard Classes
Forbidden from transport by the DOT

TETRAPROPYL SUCCINIC ACID   RR-01154-1

ENVIRONMENTAL LISTS

TSCA - Chemicals with Significant New Use Rules
PMN number: P-91-65

TSCA - Section 12(b) - Export Notification
P-91-65; export notification required - Section 5

TETRASILOXANE, 1,1,1,3,5,7,7-OCTAMETHYL-3,5-BIS[3-(OXIRANYLMETHOXY)PROPYL]-   69155-42-6

ENVIRONMENTAL LISTS

TSCA - Chemicals with Significant New Use Rules
PMN number: P-93-1201

TETRASODIUM EDTA   64-02-8

ENVIRONMENTAL LISTS

TSCA - Chemicals with Significant New Use Rules
PMN number: P-93-1197

TSCA - Section 12(b) - Export Notification
P-93-1197; export notification required - Section 5

TETRASODIUM ETHYLENEDIAMINETETRAAC-ETATE TRIHYDRATE   67401-50-7

ENVIRONMENTAL LISTS

TSCA - Chemicals with Significant New Use Rules
PMN number: P-93-1205

TETRASODIUM PYROPHOSPHATE   7722-88-5

ENVIRONMENTAL LISTS

TSCA - PAIR - Reporting List
Effective Date: January 26, 1994;   Reporting Date: March 28, 1994

TSCA - Health and Safety Reporting List
Effective Date: January 26, 1994;   Sunset Date: January 26, 2004

PROPOSED REGULATIONS

TSCA - ITC 31st Report Priority Testing List
recommended for testing

TETRATRIACONTAMETHYLHEXADECASILOXANE   36938-50-8
SEE ALSO:
LEAD

INTERNATIONAL LISTS

Canada - WHMIS: Ingredient Disclosure
1% item 1535 (1438)

1-TETRAZENE-1-CARBOXIMIDIC ACID, 4-   109-27-3
(AMINOMETHYL)-, 2-NITROSOHYDRAZIDE

HEALTH AND SAFETY LISTS

U.S. DOT - Substances From 49 CFR 172.101
regulated by DOT (UN2413)

U.S. DOT - Hazard Classes
DOT hazard class = 3

STATE LISTS

NJ Right to Know List (Total)
sn 2805

TETRAZINE   70816-59-0

ENVIRONMENTAL LISTS

List of Pesticide Product Inert Ingredients
[present]

TETRAZOL-1-ACETIC ACID   21732-17-2
SEE ALSO:
GLYCIDOL (OXIRANEMETHANOL) AND ITS DERIVATIVES

ENVIRONMENTAL LISTS

EPA - Master Testing List
[present]

TSCA - Code of Federal Regulations Citations
40 CFR 716.120(c)

PROPOSED REGULATIONS

TSCA - Proposed Testing Rule for Glycidyl Ethers
member of Glycidyl subcategory III-C

TETRAZOLYL AZIDE   18432-28-5

ENVIRONMENTAL LISTS

List of Pesticide Product Inert Ingredients
[present]

TETRAZOLYL AZIDE   RR-01462-0

ENVIRONMENTAL LISTS

List of Pesticide Product Inert Ingredients
[present]

TETRYL   479-45-8

HEALTH AND SAFETY LISTS

ACGIH 1995 - Time Weighted Averages
5 mg/m3 TWA

NIOSH - Selected LD50s and LC50s
Oral, rat: LD50 = 4000 mg/kg

OSHA - Vacated PELs - Time Weighted Averages
5 mg/m3 TWA

ENVIRONMENTAL LISTS

List of Pesticide Product Inert Ingredients
[present]

INTERNATIONAL LISTS

Australian Exposure Standards - Time Weighted Averages
5 mg/m3 TWA

Canada - WHMIS: Ingredient Disclosure
1% item 1536 (1462)

Canada - Alberta - 8 Hour Occupational Exposure Limit
5 mg/m3 TWA

Canada - Alberta - 15 Minute Occupational Exposure Limit
10 mg/m3 STEL

Canada - Ontario - OHSA - TWAEVs
5 mg/m3 TWAEV

Canada - Quebec - Time-Weighted Average Exposure Values
5 mg/m3 TWAEV

United Kingdom - Occupational Exposure Standards - TWAs
5 mg/m3 TWA

Israel - Time Weighted Averages
5 mg/m3 TWA

Israel - Action Levels
2.5 mg/m3 AL

STATE LISTS

California - Exposure Limits - PELs
5 mg/m3 PEL

California - Directors List of Hazardous Substances (8 CCR 339)
[present]

Florida Hazardous Substance List
[present]

Massachusetts Right To Know List
[present]

Minnesota Hazardous Substance List
[present]

Pennsylvania Right to Know List
[present]

TEXTILE MANUFACTURING INDUSTRY   RR-01203-3

ENVIRONMENTAL LISTS

TSCA - PAIR - Reporting List
Effective Date: June 14, 1993

TSCA - Health and Safety Reporting List
Effective Date: October 12, 1993;   Sunset Date: October 12, 2003

THALIDOMIDE   50-35-1

HEALTH AND SAFETY LISTS

U.S. DOT - Hazard Classes
Forbidden from transport by the DOT

THALLIC OXIDE   1314-32-5

HEALTH AND SAFETY LISTS

U.S. DOT - Hazard Classes
Forbidden from transport by the DOT

THALLIUM   7440-28-0

HEALTH AND SAFETY LISTS

U.S. DOT - Substances From 49 CFR 172.101
regulated by DOT (UN0407)

U.S. DOT - Hazard Classes
DOT hazard class = 1.4C

STATE LISTS
  NJ Right to Know List (Total)
  sn 1838

THALLIUM 194   18235-46-6

HEALTH AND SAFETY LISTS
  U.S. DOT - Hazard Classes
  Forbidden from transport by the DOT

THALLIUM 194M   RR-00320-3

STATE LISTS
  NJ Right to Know List (Total)
  sn 2493

THALLIUM 195   26683-69-2

HEALTH AND SAFETY LISTS
  ACGIH 1995 - Time Weighted Averages
  1.5 mg/m3 TWA
  U.S. DOT - Substances From 49 CFR 172.101
  regulated by DOT (UN0208)
  U.S. DOT - Hazard Classes
  DOT hazard class = 1.1D
  NIOSH 1990 - Pocket Guide - RELs
  1.5 mg/m3 TWA
  NIOSH 1990 - Pocket Guide - Target organs
  eyes, CNS, skin, respiratory system; in animals: liver, kidneys
  NIOSH 1990 - Pocket Guide - Skin list
  Potential for dermal absorption
  OSHA - Vacated PELs - Time Weighted Averages
  1.5 mg/m3 TWA
  OSHA - Vacated PELs - Skin Designation
  Prevent or reduce skin absorption
  OSHA - Final PELs - Time Weighted Averages
  1.5 mg/m3 TWA
  OSHA - Final PELs - Skin Notations
  prevent or reduce skin absorption

ENVIRONMENTAL LISTS
  ATSDR Priority List
  Rank (of 275): 267

INTERNATIONAL LISTS
  Australian Exposure Standards - Time Weighted Averages
  1.5 mg/m3 TWA
  Australian Exposure Standards - Skin Effects
  sensitiser
  Canada - WHMIS: Ingredient Disclosure
  1% item 1537 (1604)
  Canada - Alberta - 8 Hour Occupational Exposure Limit
  1.5 mg/m3 TWA
  Canada - Alberta - 15 Minute Occupational Exposure Limit
  3.0 mg/m3 STEL
  Canada - Alberta - Skin Designation
  can be absorbed through the intact skin
  Canada - British Columbia - 8 Hour Exposure Limits
  1.5 mg/m3 TWA
  Canada - British Columbia - 15 Minute Exposure Limits
  3.0 mg/m3 STEL
  Canada - British Columbia - Skin Notations
  skin - potential for skin absorption
  Canada - Ontario - OHSA - TWAEVs
  1.5 mg/m3 TWAEV
  Canada - Ontario - OHSA - Skin Notations
  absorption through skin, eyes, or mucous membranes

Canada - Quebec - Time-Weighted Average Exposure Values
  1.5 mg/m3 TWAEV
United Kingdom - Occupational Exposure Standards - TWAs
  1.5 mg/m3 TWA
United Kingdom - Occupational Exposure Standards - STELs
  3 mg/m3 STEL
United Kingdom - Occupational Exposure Standards - Notes
  can be absorbed through skin
German (DFG) - MAK Values
  total dust: 1.5 mg/m3 MAK
German (DFG) - Skin/Sensitizers
  danger of cutaneous absorption; danger of sensitization (skin or respiratory)
Israel - Time Weighted Averages
  1.5 mg/m3 TWA
Israel - Action Levels
  0.75 mg/m3 AL
Mexico - Instruction No. 10 - TWAs
  1.5 mg/m3 TWA
Mexico - Instruction No. 10 - STELs
  3 mg/m3 STEL
Mexico - Instruction No. 10 - Skin designation
  skin - potential for cutaneous absorption

STATE LISTS
  California - Exposure Limits - PELs
  1.5 mg/m3 PEL
  California - Exposure Limits - Skin Notation
  material may be absorbed through the skin, eyes or mucous membrane
  California - Directors List of Hazardous Substances (8 CCR 339)
  [present]
  Florida Hazardous Substance List
  [present]
  Massachusetts Right To Know List
  [present]
  Minnesota Hazardous Substance List
  skin
  NJ Right to Know List (Total)
  sn 1839
  Pennsylvania Right to Know List
  [present]

THALLIUM 197   14107-52-9

HEALTH AND SAFETY LISTS
  IARC - Group 2B (sufficient animal data)
  [present]
  OSHA - Possible Select Carcinogens
  [present]

THALLIUM 198   15743-50-7

STATE LISTS
  California - Air Bill 2588 Appendix A-II
  [present]
  California - Prop. 65 - Developmental Toxicity
  developmental toxicity - initial date 7/1/87
  Massachusetts Right To Know List
  teratogen

THALLIUM 198M   RR-00319-0
SEE ALSO:
  THALLIUM

HEALTH AND SAFETY LISTS
  U.S. DOT - Appendix A Table 1 - Hazardous Substances
  final RQ = 100 pounds (45.4 kg)

NIOSH - Selected LD50s and LC50s
Oral, rat: LD50 = 44 mg/kg

ENVIRONMENTAL LISTS

CERCLA/SARA - Hazardous Substances and their Reportable Quantities
final RQ = 100 pounds (45.4 kg)

RCRA - P Series Wastes
waste number P113

RCRA - Hazardous Constituents-Appendix VIII
waste number P113

RCRA - Substances Banned From Land Disposal
[present]

INTERNATIONAL LISTS

Canada - WHMIS: Ingredient Disclosure
1% item 1547 (1322)

STATE LISTS

Massachusetts Right To Know List
[present]

NJ Right to Know List (Total)
sn 2807

Pennsylvania Right to Know List
environmental hazard

THALLIUM 199   **15064-66-1**

HEALTH AND SAFETY LISTS

ACGIH 1995 - Time Weighted Averages
elemental and soluble compounds, as Tl: 0.1 mg/m3 TWA

ACGIH 1995 - Skin Designations
elemental and soluble compounds, as Tl: skin - potential for cutaneous absorption

U.S. DOT - Appendix A Table 1 - Hazardous Substances
final RQ = 1000 pounds (454 kg) (no reporting of releases of this hazardous substance is required if the diameter of the pieces of the solid metal released is equal to or exceeds 0.004 inches)

NIOSH 1990 - Pocket Guide - RELs
as Tl: 0.1 mg/m3 TWA

NIOSH 1990 - Pocket Guide - IDLHs
as Tl: 20 mg/m3 IDLH

NIOSH 1990 - Pocket Guide - Target organs
eyes, CNS, lungs, liver, kidneys, GI tract, body hair

NIOSH 1990 - Pocket Guide - Skin list
as Tl: Potential for dermal absorption

OSHA - Vacated PELs - Time Weighted Averages
as Tl: 0.1 mg/m3 TWA

OSHA - Vacated PELs - Skin Designation
Prevent or reduce skin absorption

OSHA - Final PELs - Time Weighted Averages
as Tl: 0.1 mg/m3 TWA

OSHA - Final PELs - Skin Notations
prevent or reduce skin absorption

ENVIRONMENTAL LISTS

ATSDR Priority List
Rank (of 275): 252

CERCLA/SARA - Section 313 - Emission Reporting
form R reporting required for 1.0% de minimus concentration

CERCLA/SARA - Hazardous Substances and their Reportable Quantities
final RQ = 1000 pounds (454 kg) (no reporting of releases of this hazardous substance is required if the diameter of the pieces of the solid metal released is equal to or exceeds 0.004 inches)

Clean Water Act - Priority Pollutants
[present]

Clean Water Act - Toxic Pollutants
[present] (Listed under 'Thallium and compounds')

Safe Drinking Water Act - MCLs
MCL = 0.002 mg/L

Safe Drinking Water Act - MCLGs
MCLG = 0.0005 mg/L

RCRA - Hazardous Constituents-Appendix VIII
hazardous constituent - no waste number

RCRA - TSD Facilities Ground Water Monitoring
TM 6010 = 400 ug/L PQL; TM 7840 = 1000 ug/L PQL; TM 7841 = 10 ug/L PQL (all species in the ground water that contain this element are included)

RCRA - Universal Treatment Standards (LDR)
WW: 1.4 mg/l; NWW: 0.078 mg/l TCLP

TSCA - Code of Federal Regulations Citations
40 CFR 716.120(a)

TSCA - Health and Safety Reporting List
Effective Date: June 1, 1987

INTERNATIONAL LISTS

Australian Exposure Standards - Time Weighted Averages
soluble compounds, as Tl: 0.1 mg/m3 TWA

Australian Exposure Standards - Skin Effects
as Tl: skin absorption

Canada - WHMIS: Ingredient Disclosure
1% item 1542 (1606)

Canada - Alberta - 8 Hour Occupational Exposure Limit
as Tl: 0.1 mg/m3 TWA

Canada - Alberta - 15 Minute Occupational Exposure Limit
as Tl: 0.3 mg/m3 STEL

Canada - Alberta - Skin Designation
as Tl: can be absorbed through the intact skin

Canada - British Columbia - 8 Hour Exposure Limits
as Tl: 0.1 mg/m3 TWA

Canada - British Columbia - Skin Notations
skin - potential for skin absorption

Canada - Ontario - OHSA - TWAEVs
as Tl: 0.1 mg/m3 TWAEV

Canada - Ontario - OHSA - Skin Notations
absorption through skin, eyes, or mucous membranes

Canada - Quebec - Time-Weighted Average Exposure Values
as Tl: 0.1 mg/m3 TWAEV

Canada - Quebec - Skin Designations
as Tl: absorbed through the skin

German (DFG) - MAK Values
soluble compounds-total dust, as Tl: 0.1 mg/m3 MAK

German (DFG) - Peak Limitations
10 x normal MAK (30 min average value); don't exceed during shift

Israel - Time Weighted Averages
as Tl: 0.1 mg/m3 TWA

Israel - Action Levels
as Tl: 0.05 mg/m3 AL

Mexico - Instruction No. 10 - TWAs
0.1 mg/m3 TWA

Mexico - Drinking Water - Ecological Criteria
0.01 mg/l

STATE LISTS

California - Air Bill 2588 Appendix A-I
6/91

California - Directors List of Hazardous Substances (8 CCR 339)
[present]

Florida Hazardous Substance List
[present]

Massachusetts Right To Know List
[present]

Minnesota Hazardous Substance List
as Tl: skin (includes soluble compounds)

NJ Right to Know List (Total)
sn 1840
Pennsylvania Right to Know List
environmental hazard (any compound of this substance is also an environmental hazard)

THALLIUM 200   15720-55-5

HEALTH AND SAFETY LISTS
U.S. DOT - Appendix A Table 2 - Radionuclides
final RQ = 1000 curies (3.7E 13 Bq)

ENVIRONMENTAL LISTS
CERCLA/SARA List of Radionuclides (Appendix B) and Their Reportable Quantities
final RQ = 1000 curies (3.7E 13 Bq)

THALLIUM 201   15064-65-0

HEALTH AND SAFETY LISTS
U.S. DOT - Appendix A Table 2 - Radionuclides
final RQ = 100 curies (3.7E 12 Bq)

ENVIRONMENTAL LISTS
CERCLA/SARA List of Radionuclides (Appendix B) and Their Reportable Quantities
final RQ = 100 curies (3.7E 12 Bq)

THALLIUM 202   15720-57-7

HEALTH AND SAFETY LISTS
U.S. DOT - Appendix A Table 2 - Radionuclides
final RQ = 100 curies (3.7E 12 Bq)

ENVIRONMENTAL LISTS
CERCLA/SARA List of Radionuclides (Appendix B) and Their Reportable Quantities
final RQ = 100 curies (3.7E 12 Bq)

THALLIUM 204   13968-51-9

HEALTH AND SAFETY LISTS
U.S. DOT - Appendix A Table 2 - Radionuclides
final RQ = 100 curies (3.7E 12 Bq)

ENVIRONMENTAL LISTS
CERCLA/SARA List of Radionuclides (Appendix B) and Their Reportable Quantities
final RQ = 100 curies (3.7E 12 Bq)

THALLIUM(I) ACETATE   563-68-8

HEALTH AND SAFETY LISTS
U.S. DOT - Appendix A Table 2 - Radionuclides
final RQ = 10 curies (3.7E 11 Bq)

ENVIRONMENTAL LISTS
CERCLA/SARA List of Radionuclides (Appendix B) and Their Reportable Quantities
final RQ = 10 curies (3.7E 11 Bq)

THALLIUM BROMIDE   7789-40-4

HEALTH AND SAFETY LISTS
U.S. DOT - Appendix A Table 2 - Radionuclides
final RQ = 100 curies (3.7E 12 Bq)

ENVIRONMENTAL LISTS
CERCLA/SARA List of Radionuclides (Appendix B) and Their Reportable Quantities
final RQ = 100 curies (3.7E 12 Bq)

THALLIUM CHLORATE   13453-30-0

HEALTH AND SAFETY LISTS
U.S. DOT - Appendix A Table 2 - Radionuclides
final RQ = 100 curies (3.7E 12 Bq)

ENVIRONMENTAL LISTS
CERCLA/SARA List of Radionuclides (Appendix B) and Their Reportable Quantities
final RQ = 100 curies (3.7E 12 Bq)

THALLIUM (I) CHROMATE   13473-75-1

HEALTH AND SAFETY LISTS
U.S. DOT - Appendix A Table 2 - Radionuclides
final RQ = 10 curies (3.7E 11 Bq)

ENVIRONMENTAL LISTS
CERCLA/SARA List of Radionuclides (Appendix B) and Their Reportable Quantities
final RQ = 10 curies (3.7E 11 Bq)

THALLIUM COMPOUNDS   RR-00575-4

HEALTH AND SAFETY LISTS
U.S. DOT - Appendix A Table 2 - Radionuclides
final RQ = 1000 curies (3.7E 13 Bq)

ENVIRONMENTAL LISTS
CERCLA/SARA List of Radionuclides (Appendix B) and Their Reportable Quantities
final RQ = 1000 curies (3.7E 13 Bq)

THALLIUM(I) FLUORIDE   7789-27-7

HEALTH AND SAFETY LISTS
U.S. DOT - Appendix A Table 2 - Radionuclides
final RQ = 10 curies (3.7E 11 Bq)

ENVIRONMENTAL LISTS
CERCLA/SARA List of Radionuclides (Appendix B) and Their Reportable Quantities
final RQ = 10 curies (3.7E 11 Bq)

THALLIUM(I) IODIDE   7790-30-9

HEALTH AND SAFETY LISTS
U.S. DOT - Appendix A Table 2 - Radionuclides
final RQ = 10 curies (3.7E 11 Bq)

ENVIRONMENTAL LISTS
CERCLA/SARA List of Radionuclides (Appendix B) and Their Reportable Quantities
final RQ = 10 curies (3.7E 11 Bq)

THALLIUM(I) NITRATE   10102-45-1
SEE ALSO:
THALLIUM

HEALTH AND SAFETY LISTS
U.S. DOT - Appendix A Table 1 - Hazardous Substances
final RQ = 100 pounds (45.4 kg)
NIOSH - Selected LD50s and LC50s
Oral, mouse: LD50 = 35 mg/kg

ENVIRONMENTAL LISTS
CERCLA/SARA - Hazardous Substances and their Reportable Quantities
final RQ = 100 pounds (45.4 kg)
RCRA - U Series Wastes
waste number U214
RCRA - Hazardous Constituents-Appendix VIII
waste number U214

RCRA - Substances Banned From Land Disposal
[present]

INTERNATIONAL LISTS

Canada - WHMIS: Ingredient Disclosure
1% item 1538 (33)

STATE LISTS

Massachusetts Right To Know List
[present]

Pennsylvania Right to Know List
environmental hazard

THALLIUM NITRATE   16901-76-1
SEE ALSO:
    THALLIUM

INTERNATIONAL LISTS

Canada - WHMIS: Ingredient Disclosure
1% item 1539 (342)

THALLIUM SALT, N.O.S.   RR-00576-5

HEALTH AND SAFETY LISTS

U.S. DOT - Substances From 49 CFR 172.101
regulated by DOT (UN2573)

U.S. DOT - Hazard Classes
DOT hazard class = 5.1

U.S. DOT - Appendix B - Marine Pollutants
DOT regulated marine pollutant

STATE LISTS

NJ Right to Know List (Total)
sn 2808

THALLIUM(I) SELENIDE   12039-52-0

STATE LISTS

Massachusetts Right To Know List
[present]

THALLIUM, SOLUBLE COMPOUNDS   RR-01194-9

HEALTH AND SAFETY LISTS

U.S. DOT - Substances From 49 CFR 172.101
regulated by DOT (UN1707)

U.S. DOT - Hazard Classes
DOT hazard class = 6.1

U.S. DOT - Appendix B - Marine Pollutants
DOT regulated marine pollutant

ENVIRONMENTAL LISTS

CERCLA/SARA - Section 313 - Emission Reporting
form R reporting required for 1.0% de minimus concentration

Clean Water Act - Toxic Pollutants
[present] (Listed under 'Thallium and compounds')

RCRA - Hazardous Constituents-Appendix VIII
hazardous constituent - no waste number

STATE LISTS

California - Air Bill 2588 Appendix A-I
6/91

California - Directors List of Hazardous Substances (8 CCR 339)
[present]

NJ Right to Know List (Total)
sn 2809

THALLIUM SULFATE   10031-59-1
SEE ALSO:
    THALLIUM

INTERNATIONAL LISTS

Canada - WHMIS: Ingredient Disclosure
1% item 1543 (911)

THALLIUM(III) SULFATE   16222-66-5
SEE ALSO:
    THALLIUM

INTERNATIONAL LISTS

Canada - WHMIS: Ingredient Disclosure
1% item 1544 (1030)

THALLIUM, WATER-SOLUBLE COMPOUNDS, N.O.S.   RR-00048-6
SEE ALSO:
    THALLIUM

HEALTH AND SAFETY LISTS

U.S. DOT - Appendix B - Marine Pollutants
DOT regulated marine pollutant

U.S. DOT - Appendix A Table 1 - Hazardous Substances
final RQ = 100 pounds (45.4 kg)

NIOSH - Selected LD50s and LC50s
Oral, mouse: LD50 = 33 mg/kg

ENVIRONMENTAL LISTS

CERCLA/SARA - Hazardous Substances and their Reportable Quantities
final RQ = 100 pounds (45.4 kg)

RCRA - U Series Wastes
waste number U217

RCRA - Hazardous Constituents-Appendix VIII
waste number U217

RCRA - Substances Banned From Land Disposal
[present]

INTERNATIONAL LISTS

Canada - WHMIS: Ingredient Disclosure
1% item 1546 (1212)

STATE LISTS

Massachusetts Right To Know List
[present]

NJ Right to Know List (Total)
sn 1841

Pennsylvania Right to Know List
environmental hazard

THALLOUS CARBONATE   6533-73-9

HEALTH AND SAFETY LISTS

U.S. DOT - Substances From 49 CFR 172.101
regulated by DOT (UN2727)

U.S. DOT - Hazard Classes
DOT hazard class = 6.1

THALLOUS CHLORIDE   7791-12-0

STATE LISTS

NJ Right to Know List (Total)
sn 2810

THALLOUS MALONATE   2757-18-8
SEE ALSO:
    THALLIUM

HEALTH AND SAFETY LISTS

U.S. DOT - Appendix A Table 1 - Hazardous Substances
final RQ = 1000 pounds (454 kg)

ENVIRONMENTAL LISTS
CERCLA/SARA - Hazardous Substances and their Reportable Quantities
final RQ = 1000 pounds (454 kg)
RCRA - P Series Wastes
waste number P114
RCRA - Hazardous Constituents-Appendix VIII
waste number P114
RCRA - Substances Banned From Land Disposal
[present]

STATE LISTS
Massachusetts Right To Know List
[present]
Pennsylvania Right to Know List
environmental hazard

THALLOUS SULFATE   7446-18-6

INTERNATIONAL LISTS
United Kingdom - Occupational Exposure Standards - TWAs
as Tl: 0.1 mg/m3 TWA
United Kingdom - Occupational Exposure Standards - Notes
can be absorbed through skin

STATE LISTS
California - Exposure Limits - PELs
as Tl: 0.1 mg/m3 PEL
California - Exposure Limits - Skin Notation
material may be absorbed through the skin, eyes or mucous membrane

THEOBROMINE   83-67-0
SEE ALSO:
THALLIUM

HEALTH AND SAFETY LISTS
U.S. DOT - Substances From 49 CFR 172.101
regulated by DOT (NA1707)
U.S. DOT - Hazard Classes
DOT hazard class = 6.1
U.S. DOT - Appendix A Table 1 - Hazardous Substances
final RQ = 100 pounds (45.4 kg) (Listed under 'Thallium(I) sulfate')
NIOSH - Selected LD50s and LC50s
Oral, rat: LD50 = 16 mg/kg

ENVIRONMENTAL LISTS
CERCLA/SARA - Section 302 Extremely Hazardous Substances and TPQs
TPQ = 100/10,000 pounds
CERCLA/SARA - Hazardous Substances and their Reportable Quantities
final RQ = 100 pounds (45.4 kg) (Listed under 'Thallium(I) sulfate')
Clean Water Act - Hazardous Substances
[present]

INTERNATIONAL LISTS
Canada - WHMIS: Ingredient Disclosure
1% item 1549 (1529)

STATE LISTS
Florida Hazardous Substance List
effective March 13, 1992
Massachusetts Right To Know List
extraordinarily hazardous
NJ Right to Know List (Total)
sn 2913
Pennsylvania Right to Know List
environmental hazard

THEOPHYLLINE   58-55-9
SEE ALSO:
THALLIUM

INTERNATIONAL LISTS
Canada - WHMIS: Ingredient Disclosure
1% item 1548 (1531)

THIABENDAZOLE [2-(4-THIAZOLYL)-1H-BENIMIDA-   148-79-8
ZOLE, HYPOPHOSPHITE SALT]
INTERNATIONAL LISTS
Canada - WHMIS: Ingredient Disclosure
1% item 1551 (1605)

THIALDINE   638-17-5
SEE ALSO:
THALLIUM

HEALTH AND SAFETY LISTS
U.S. DOT - Appendix A Table 1 - Hazardous Substances
final RQ = 100 pounds (45.4 kg)
NIOSH - Selected LD50s and LC50s
Oral, mouse: LD50 = 21 mg/kg Skin, rat: LD50 = 117 mg/kg

ENVIRONMENTAL LISTS
CERCLA/SARA - Section 302 Extremely Hazardous Substances and TPQs
TPQ = 100/10,000 pounds
CERCLA/SARA - Hazardous Substances and their Reportable Quantities
final RQ = 100 pounds (45.4 kg)
RCRA - U Series Wastes
waste number U215
RCRA - Hazardous Constituents-Appendix VIII
waste number U215
RCRA - Substances Banned From Land Disposal
[present]

INTERNATIONAL LISTS
Canada - WHMIS: Ingredient Disclosure
1% item 1540 (393)

STATE LISTS
Florida Hazardous Substance List
effective March 13, 1992
Massachusetts Right To Know List
extraordinarily hazardous
NJ Right to Know List (Total)
sn 2811
Pennsylvania Right to Know List
environmental hazard

4-THIAPENTANAL   3268-49-3
SEE ALSO:
THALLIUM

HEALTH AND SAFETY LISTS
U.S. DOT - Appendix A Table 1 - Hazardous Substances
final RQ = 100 pounds (45.4 kg)
NIOSH - Selected LD50s and LC50s
Oral, mouse: LD50 = 24 mg/kg

ENVIRONMENTAL LISTS
CERCLA/SARA - Section 302 Extremely Hazardous Substances and TPQs
TPQ = 100/10,000 pounds
CERCLA/SARA - Hazardous Substances and their Reportable Quantities
final RQ = 100 pounds (45.4 kg)
RCRA - U Series Wastes
waste number U216

RCRA - Hazardous Constituents-Appendix VIII
waste number U216
RCRA - Substances Banned From Land Disposal
[present]

INTERNATIONAL LISTS
Canada - WHMIS: Ingredient Disclosure
1% item 1541 (529)

STATE LISTS
Florida Hazardous Substance List
effective March 13, 1992
Massachusetts Right To Know List
extraordinarily hazardous
NJ Right to Know List (Total)
sn 2812
Pennsylvania Right to Know List
environmental hazard

2-THIAZOLAMINE   96-50-4
SEE ALSO:
THALLIUM

HEALTH AND SAFETY LISTS
NIOSH - Selected LD50s and LC50s
Oral, rat: LD50 = 18800 ug/kg Skin, rat: LD50 = 57700 ug/kg

ENVIRONMENTAL LISTS
CERCLA/SARA - Section 302 Extremely Hazardous Substances and TPQs
TPQ = 100/10,000 pounds

INTERNATIONAL LISTS
Canada - WHMIS: Ingredient Disclosure
1% item 1545 (1072)

STATE LISTS
Florida Hazardous Substance List
effective March 13, 1992
Massachusetts Right To Know List
extraordinarily hazardous
NJ Right to Know List (Total)
sn 2813
Pennsylvania Right to Know List
environmental hazard

PROPOSED REGULATIONS
CERCLA/SARA - Proposed Hazardous Substance Additions
proposed RQ = 1 pound (.454 kg)
CERCLA/SARA - 1989 Proposed RQ Adjustments
proposed RQ = 100 pounds (45.4 kg)

THIAZOLE   288-47-1
SEE ALSO:
THALLIUM

HEALTH AND SAFETY LISTS
U.S. DOT - Appendix B - Marine Pollutants
DOT regulated marine pollutant
U.S. DOT - Appendix A Table 1 - Hazardous Substances
final RQ = 100 pounds (45.4 kg)
NIOSH - Selected LD50s and LC50s
Oral, rat: LD50 = 20300 ug/kg

ENVIRONMENTAL LISTS
CERCLA/SARA - Section 302 Extremely Hazardous Substances and TPQs
TPQ = 100/10,000 pounds
CERCLA/SARA - Hazardous Substances and their Reportable Quantities
final RQ = 100 pounds (45.4 kg)

Clean Water Act - Hazardous Substances
[present] (Listed under 'Thallium sulfate')
RCRA - P Series Wastes
waste number P115
RCRA - Hazardous Constituents-Appendix VIII
waste number P115
RCRA - Substances Banned From Land Disposal
[present]

INTERNATIONAL LISTS
Canada - WHMIS: Ingredient Disclosure
1% item 1550 (1530)

STATE LISTS
Massachusetts Right To Know List
extraordinarily hazardous
NJ Right to Know List (Total)
sn 2887
Pennsylvania Right to Know List
environmental hazard

2-THIAZOLIDINETHIONE   96-53-7
HEALTH AND SAFETY LISTS
IARC - Group 3 (not classifiable)
[present]

THIAZOLIUM, 3-[(4-AMINO-2-METHYL-5-PYRIMIDINYL)METHYL]-5-(2-HYDROXYETHYL)-4-METHYLCHLORIDE, MONOHYDROCHLORIDE   67-03-8
HEALTH AND SAFETY LISTS
IARC - Group 3 (not classifiable)
[present]
NTP Chemical Status Reports - Testing Status and NTIS Number
Two year studies: pathology quality assessment in progress; prechronic studies for which toxicity technical reports were not prepared

THIOACETAMIDE   62-55-5
ENVIRONMENTAL LISTS
CERCLA/SARA - Section 313 - Emission Reporting
form R reporting required

THIOACETIC ACID   507-09-5
HEALTH AND SAFETY LISTS
NFPA - Flash Points
flash point = 200 degrees F (93 degrees C)
NFPA - Hazard Identification Ratings
health-2; flammability-2; reactivity-1

STATE LISTS
Florida Hazardous Substance List
[present]
Massachusetts Right To Know List
[present]
Pennsylvania Right to Know List
[present]

4,4'-THIOBIS(6-TERT-BUTYL-M-CRESOL)   96-69-5
HEALTH AND SAFETY LISTS
U.S. DOT - Substances From 49 CFR 172.101
regulated by DOT (UN2785)
U.S. DOT - Hazard Classes
DOT hazard class = 6.1
NFPA - Flash Points
flash point = 142 degrees F (61 degrees C)
NFPA - Hazard Identification Ratings
flammability-2; reactivity-0

NIOSH - Selected LD50s and LC50s
Inhalation, rat: LC50 = 5820 mg/m3 4 hr Oral, rat: LD50 = 4400 mg/kg

ENVIRONMENTAL LISTS

TSCA - Code of Federal Regulations Citations
40 CFR 712.30(x); 40 CFR 716.120(d)

TSCA - PAIR - Reporting List
Reporting Date: November 27, 1991

TSCA - Health and Safety Reporting List
Effective Date: September 30, 1991

STATE LISTS

NJ Right to Know List (Total)
sn 1843

THIOCARBANILIDE   102-08-9

HEALTH AND SAFETY LISTS

NIOSH - Selected LD50s and LC50s
Oral, rat: LD50 = 480 mg/kg

THIOCARBAZIDE   2231-57-4

HEALTH AND SAFETY LISTS

NIOSH - Selected LD50s and LC50s
Oral, mouse: LD50 = 983 mg/kg

THIOCARBONYL TETRACHLORIDE   RR-01314-9

HEALTH AND SAFETY LISTS

NIOSH - Selected LD50s and LC50s
Oral, rat: LD50 = 300 mg/kg

THIOCYANIC ACID, (2-BENZOTHIAZOLYTHIO)   21564-17-0
METHYL ESTER

HEALTH AND SAFETY LISTS

NIOSH - Selected LD50s and LC50s
Oral, mouse: LD50 = 8224 mg/kg

ENVIRONMENTAL LISTS

List of Pesticide Product Inert Ingredients
[present]

4,4'-THIODIANILINE   139-65-1

HEALTH AND SAFETY LISTS

U.S. DOT - Appendix A Table 1 - Hazardous Substances
final RQ = 10 pounds (4.54 kg)

IARC - Group 2B (sufficient animal data)
[present] (Overall evaluation based only on evidence of carcinogenicity in monograph (7, 1974) or in Supplement 4)

NTP Seventh Report - Suspect Carcinogens
suspect carcinogen

OSHA - Possible Select Carcinogens
[present]

ENVIRONMENTAL LISTS

CERCLA/SARA - Section 313 - Emission Reporting
form R reporting required for 0.1% de minimus concentration

CERCLA/SARA - Hazardous Substances and their Reportable Quantities
final RQ = 10 pounds (4.54 kg)

EPA - Carcinogen Hazard Ranking for RQ Adjustment
Hazard ranking = Medium

RCRA - U Series Wastes
waste number U218

RCRA - Hazardous Constituents-Appendix VIII
waste number U218

RCRA - Substances Banned From Land Disposal
[present]

INTERNATIONAL LISTS

Canada - WHMIS: Ingredient Disclosure
0.1% item 1552 (1607)

STATE LISTS

California - Air Bill 2588 Appendix A-I
known or potential carcinogen

California - Prop. 65 - Cancer list
carcinogen - initial date 1/1/88

California - Prop. 65 - No Significant Risk Levels
no significant risk level = 0.1 ug/day

California - Directors List of Hazardous Substances (8 CCR 339)
[present]

Florida Hazardous Substance List
[present]

Massachusetts Right To Know List
carcinogen; extraordinarily hazardous

Minnesota Hazardous Substance List
carcinogen

NJ Right to Know List (Total)
sn 1844

NJ Special Hazardous Substances
(carcinogen)

Pennsylvania Right to Know List
environmental hazard; special hazardous substance

Pennsylvania RTK - Special Hazardous Substances
[present]

THIODICARB   59669-26-0

HEALTH AND SAFETY LISTS

U.S. DOT - Substances From 49 CFR 172.101
regulated by DOT (UN2436)

U.S. DOT - Hazard Classes
DOT hazard class = 3

STATE LISTS

NJ Right to Know List (Total)
sn 1845

4,4'-THIODIPHENOL   2664-63-3

HEALTH AND SAFETY LISTS

ACGIH 1995 - Time Weighted Averages
10 mg/m3 TWA

NTP Chemical Status Reports - Testing Status and NTIS Number
Galley or camera copy technical reports in progress; prechronic studies for which toxicity technical reports were not prepared

OSHA - Vacated PELs - Time Weighted Averages
total dust: 10 mg/m3 TWA; respirable fraction: 5 mg/m3 TWA

OSHA - Final PELs - Time Weighted Averages
total dust: 15 mg/m3 TWA; respirable fraction: 5 mg/m3 TWA

INTERNATIONAL LISTS

Australian Exposure Standards - Time Weighted Averages
10 mg/m3 TWA

Canada - WHMIS: Ingredient Disclosure
1% item 1553 (1608)

Canada - Alberta - 8 Hour Occupational Exposure Limit
10 mg/m3 TWA

Canada - Alberta - 15 Minute Occupational Exposure Limit
20 mg/m3 STEL

Canada - British Columbia - 8 Hour Exposure Limits
10 mg/m3 TWA

Canada - British Columbia - 15 Minute Exposure Limits
20 mg/m3 STEL

Canada - Ontario - OHSA - TWAEVs
10 mg/m3 TWAEV

Canada - Quebec - Time-Weighted Average Exposure Values
10 mg/m3 TWAEV

United Kingdom - Occupational Exposure Standards - TWAs
10 mg/m3 TWA

United Kingdom - Occupational Exposure Standards - STELs
20 mg/m3 STEL

Israel - Time Weighted Averages
10 mg/m3 TWA

Israel - Action Levels
5 mg/m3 AL

Mexico - Instruction No. 10 - TWAs
10 mg/m3 TWA;

Mexico - Instruction No. 10 - STELs
20 mg/m3 STEL

STATE LISTS

California - Exposure Limits - PELs
total dust: 10 mg/m3 PEL; respirable fraction: 5 mg/m3 PEL

California - Directors List of Hazardous Substances (8 CCR 339)
[present]

Florida Hazardous Substance List
[present]

Massachusetts Right To Know List
[present]

Minnesota Hazardous Substance List
[present]

Pennsylvania Right to Know List
[present]

BETA,BETA'-THIODIPROPIONITRILE   111-97-7

ENVIRONMENTAL LISTS

CAA - HON Rule - SOCMI Chemicals
compliance by April 24, 1995

THIODIPROPIONIC ACID   111-17-1

ENVIRONMENTAL LISTS

CERCLA/SARA - Section 302 Extremely Hazardous Substances and TPQs
TPQ = 1000/10,000 pounds

STATE LISTS

Florida Hazardous Substance List
effective March 13, 1992

Massachusetts Right To Know List
extraordinarily hazardous

NJ Right to Know List (Total)
sn 2818

Pennsylvania Right to Know List
environmental hazard

PROPOSED REGULATIONS

CERCLA/SARA - Proposed Hazardous Substance Additions
proposed RQ = 1 pound (.454 kg)

CERCLA/SARA - 1989 Proposed RQ Adjustments
proposed RQ = 100 pounds (45.4 kg)

THIOFANOX   39196-18-4

HEALTH AND SAFETY LISTS

U.S. DOT - Appendix B - Marine Pollutants
DOT regulated marine pollutant

THIOGLYCOLIC ACID   68-11-1

HEALTH AND SAFETY LISTS

NIOSH - Selected LD50s and LC50s
Oral, rat: LD50 = 1590 mg/kg Skin, rabbit: LD50 = 10 gm/kg

STATE LISTS

Massachusetts Right To Know List
[present]

NJ Right to Know List (Total)
sn 2819

BETA-THIOGUANIDINE DEOXYRIBOSIDE   789-61-7

HEALTH AND SAFETY LISTS

IARC - Group 2B (sufficient animal data)
[present] (Overall evaluation based only on evidence of carcinogenicity in monograph (27, 1982) or in Supplement 4)

NIOSH - Selected LD50s and LC50s
Oral, rat: LD50 = 1100 mg/kg

NTP Chemical Status Reports - Testing Status and NTIS Number
Technical reports printed (PB280360/AS)

NTP Chemical Status Reports - Evidence of Carcinogenicity
male rat-positive; female rat-positive; male mice-positive; female mice-positive

OSHA - Possible Select Carcinogens
[present]

ENVIRONMENTAL LISTS

CERCLA/SARA - Section 313 - Emission Reporting
form R reporting required for 0.1% de minimus concentration

INTERNATIONAL LISTS

Canada - WHMIS: Ingredient Disclosure
1% item 1554 (1611)

German (DFG) - Carcinogens
animal evidence of carcinogenicity

STATE LISTS

California - Air Bill 2588 Appendix A-II
known or potential carcinogen

California - Prop. 65 - Cancer list
carcinogen - initial date 4/1/88

California - Prop. 65 - No Significant Risk Levels
no significant risk level = 0.05 ug/day

California - Directors List of Hazardous Substances (8 CCR 339)
[present]

Florida Hazardous Substance List
[present]

Massachusetts Right To Know List
carcinogen; extraordinarily hazardous

Minnesota Hazardous Substance List
carcinogen

NJ Right to Know List (Total)
sn 1847

Pennsylvania Right to Know List
environmental hazard; special hazardous substance

Pennsylvania RTK - Special Hazardous Substances
[present]

THIOGUANINE   154-42-7

ENVIRONMENTAL LISTS

CERCLA/SARA - Section 313 - Emission Reporting
form R reporting required

PROPOSED REGULATIONS

Safe Drinking Water Act - Priority list
[present]

2-THIOLACTIC ACID   79-42-5

HEALTH AND SAFETY LISTS

NIOSH - Selected LD50s and LC50s
Oral, rat: LD50 = 3362 mg/kg

INTERNATIONAL LISTS
  Canada - WHMIS: Ingredient Disclosure
    1% item 1555 (1613)

THIOLS (N-ALKANE MONOTHIOLS)   RR-00553-8

INTERNATIONAL LISTS
  Canada - WHMIS: Ingredient Disclosure
    1% item 1557 (1615)

THIOMETON   640-15-3

INTERNATIONAL LISTS
  Canada - WHMIS: Ingredient Disclosure
    1% item 1556 (142)

THIONYL CHLORIDE   7719-09-7

HEALTH AND SAFETY LISTS
  U.S. DOT - Appendix A Table 1 - Hazardous Substances
    final RQ = 100 pounds (45.4 kg)
  NIOSH - Selected LD50s and LC50s
    Oral, rat: LD50 = 8500 ug/kg Skin, rabbit: LD50 = 39 mg/kg

ENVIRONMENTAL LISTS
  CERCLA/SARA - Section 302 Extremely Hazardous Substances and
  TPQs
    TPQ = 100/10,000 pounds
  CERCLA/SARA - Hazardous Substances and their Reportable Quantities
    final RQ = 100 pounds (45.4 kg)
  RCRA - P Series Wastes
    waste number P045
  RCRA - Hazardous Constituents-Appendix VIII
    waste number P045
  RCRA - Substances Banned From Land Disposal
    [present]

STATE LISTS
  Florida Hazardous Substance List
    effective March 13, 1992
  Massachusetts Right To Know List
    extraordinarily hazardous
  NJ Right to Know List (Total)
    sn 2820
  Pennsylvania Right to Know List
    environmental hazard

THIONYL DIFLUORIDE   7783-42-8

HEALTH AND SAFETY LISTS
  ACGIH 1995 - Time Weighted Averages
    1 ppm TWA; 3.8 mg/m3 TWA
  ACGIH 1995 - Skin Designations
    skin - potential for cutaneous absorption
  U.S. DOT - Substances From 49 CFR 172.101
    regulated by DOT (UN1940)
  U.S. DOT - Hazard Classes
    DOT hazard class = 8
  NIOSH - Selected LD50s and LC50s
    Oral, rat: LD50 = 114 mg/kg
  OSHA - Vacated PELs - Time Weighted Averages
    1 ppm TWA; 4 mg/m3 TWA
  OSHA - Vacated PELs - Skin Designation
    Prevent or reduce skin absorption

INTERNATIONAL LISTS
  Australian Exposure Standards - Time Weighted Averages
    1 ppm TWA; 3.8 mg/m3 TWA

Australian Exposure Standards - Skin Effects
    skin absorption
  Canada - WHMIS: Ingredient Disclosure
    1% item 1559 (143)
  Canada - Alberta - 8 Hour Occupational Exposure Limit
    1 ppm TWA; 3.8 mg/m3 TWA
  Canada - Alberta - 15 Minute Occupational Exposure Limit
    3 ppm STEL; 11 mg/m3 STEL
  Canada - British Columbia - 8 Hour Exposure Limits
    1 ppm TWA; 5 mg/m3 TWA
  Canada - Ontario - OHSA - TWAEVs
    1 ppm TWAEV; 3.8 mg/m3 TWAEV
  Canada - Ontario - OHSA - Skin Notations
    absorption through skin, eyes, or mucous membranes
  Canada - Quebec - Time-Weighted Average Exposure Values
    1 ppm TWAEV; 3.8 mg/m3 TWAEV
  Canada - Quebec - Skin Designations
    absorbed through the skin
  United Kingdom - Occupational Exposure Standards - TWAs
    1 ppm TWA; 5 mg/m3 TWA
  Israel - Time Weighted Averages
    1 ppm TWA; 3.8 mg/m3 TWA
  Israel - Action Levels
    0.5 ppm AL; 1.9 mg/m3 AL
  Mexico - Instruction No. 10 - TWAs
    1 ppm TWA; 5 mg/m3 TWA

STATE LISTS
  California - Exposure Limits - PELs
    1 ppm PEL; 3.8 mg/m3 PEL
  California - Exposure Limits - Skin Notation
    material may be absorbed through the skin, eyes or mucous membrane
  California - Directors List of Hazardous Substances (8 CCR 339)
    [present]
  Florida Hazardous Substance List
    [present]
  Massachusetts Right To Know List
    [present]
  Minnesota Hazardous Substance List
    [present]
  NJ Right to Know List (Total)
    sn 1848
  NJ Special Hazardous Substances
    (corrosive)
  Pennsylvania Right to Know List
    [present]

THIOPHANATE ETHYL [[1,2-PHENYLENEBIS   23564-06-9
(IMINOCARBONOTHIOYL)]BISCARBAMIC ACID DI-
ETHYL ESTER]

HEALTH AND SAFETY LISTS
  NTP Chemical Status Reports - Testing Status and NTIS Number
    Technical reports printed (PB281540/AS)
  NTP Chemical Status Reports - Evidence of Carcinogenicity
    male rat-equivocal; female rat-positive; male mice-inadequate; female mice-inadequate

STATE LISTS
  Massachusetts Right To Know List
    carcinogen; extraordinarily hazardous

THIOPHANATE-METHYL   23564-05-8

STATE LISTS
  California - Air Bill 2588 Appendix A-II
    9/90

California - Prop. 65 - Developmental Toxicity
developmental toxicity - initial date 7/1/90

THIOPHENE   110-02-1
HEALTH AND SAFETY LISTS
U.S. DOT - Substances From 49 CFR 172.101
regulated by DOT (UN2936)
U.S. DOT - Hazard Classes
DOT hazard class = 6.1
INTERNATIONAL LISTS
Canada - WHMIS: Ingredient Disclosure
1% item 1560 (144)
STATE LISTS
NJ Right to Know List (Total)
sn 1849

2-THIOPHENECARBOXALDEHYDE   98-03-3
STATE LISTS
Minnesota Hazardous Substance List
[present]

2,2'-(2,5-THIOPHENEDIYL)BIS(5-TERTIARYBUTYL-   7128-64-5
BENZOXAZOLE)
HEALTH AND SAFETY LISTS
NIOSH - Selected LD50s and LC50s
Oral, rat: LD50 = 40 mg/kg Skin, rat: LD50 = 179 mg/kg
STATE LISTS
Massachusetts Right To Know List
[present]
NJ Right to Know List (Total)
sn 2822

THIOPHOSGENE   463-71-8
HEALTH AND SAFETY LISTS
ACGIH 1995 - Ceiling Limits
C 1 ppm; C 4.9 mg/m3
U.S. DOT - Substances From 49 CFR 172.101
regulated by DOT (UN1836)
U.S. DOT - Hazard Classes
DOT hazard class = 8
U.S. DOT - Substances Which Are Poisonous by Inhalation
liquid hazardous material poisonous by inhalation (UN1836)
NIOSH - Selected LD50s and LC50s
Inhalation, rat: LC50 = 500 ppm 1 hr
OSHA - Vacated PELs - Ceiling Limits
C 1 ppm; C 5 mg/m3
OSHA - List of Highly Hazardous Chemicals
threshhold quantity = 250 pounds
INTERNATIONAL LISTS
Australian Exposure Standards - Time Weighted Averages
Peak Limitation: 1 ppm; 4.9 mg/m3
Canada - WHMIS: Ingredient Disclosure
1% item 1561 (530)
Canada - Alberta - Ceiling Occupational Exposure Limit
C 1 ppm; C 5 mg/m3
Canada - Ontario - OHSA - CEVs
1 ppm CEV; 5 mg/m3 CEV
Canada - Quebec - Ceiling Limits
P 1 ppm; P 5 mg/m3
United Kingdom - Occupational Exposure Standards - STELs
1 ppm STEL; 5 mg/m3 STEL
Israel - Ceiling Exposure Limits
C 1 ppm; C 4.9 mg/m3

STATE LISTS
California - Exposure Limits - Ceilings
C 1 ppm; C 5 mg/m3
California - Directors List of Hazardous Substances (8 CCR 339)
[present]
Florida Hazardous Substance List
[present]
Massachusetts Right To Know List
[present]
Minnesota Hazardous Substance List
[present]
NJ Right to Know List (Total)
sn 1850
NJ Special Hazardous Substances
(corrosive; reactive - second degree)
Pennsylvania Right to Know List
[present]

THIOPHOSPHORYL CHLORIDE   3982-91-0
INTERNATIONAL LISTS
Canada - WHMIS: Ingredient Disclosure
1% item 1562 (912)

THIOSEMICARBAZIDE   79-19-6
ENVIRONMENTAL LISTS
CERCLA/SARA - Section 313 - Emission Reporting
form R reporting required

THIOURACIL   141-90-2
ENVIRONMENTAL LISTS
CERCLA/SARA - Section 313 - Emission Reporting
form R reporting required

THIOUREA   62-56-6
HEALTH AND SAFETY LISTS
U.S. DOT - Substances From 49 CFR 172.101
regulated by DOT (UN2414)
U.S. DOT - Hazard Classes
DOT hazard class = 3
NFPA - Flash Points
flash point = 30 degrees F (-1 degrees C)
NFPA - Hazard Identification Ratings
health-2; flammability-3; reactivity-0
NTP Chemical Status Reports - Testing Status and NTIS Number
Prechronic studies for which toxicity technical reports were not pre-pared
STATE LISTS
Florida Hazardous Substance List
[present]
Massachusetts Right To Know List
[present]
NJ Right to Know List (Total)
sn 1851
NJ Special Hazardous Substances
(flammable - third degree)
Pennsylvania Right to Know List
[present]

THIOUREA, (2-CHLOROPHENYL)-   5344-82-1
ENVIRONMENTAL LISTS
TSCA - Code of Federal Regulations Citations
40 CFR 712.30(x); 40 CFR 716.120(d)
TSCA - PAIR - Reporting List
Reporting Date: November 27, 1991

TSCA - Health and Safety Reporting List
Effective Date: September 30, 1991

THIOUREA, (2-METHYLPHENYL)-   **614-78-8**

ENVIRONMENTAL LISTS

List of Pesticide Product Inert Ingredients
[present]

THIRAM   **137-26-8**

HEALTH AND SAFETY LISTS

U.S. DOT - Substances From 49 CFR 172.101
regulated by DOT (UN2474)

U.S. DOT - Hazard Classes
DOT hazard class = 6.1

U.S. DOT - Substances Which Are Poisonous by Inhalation
liquid hazardous material poisonous by inhalation (UN2474)

NIOSH - Selected LD50s and LC50s
Oral, rat: LD50 = 929 mg/kg

INTERNATIONAL LISTS

Canada - WHMIS: Ingredient Disclosure
1% item 1563 (1617)

STATE LISTS

NJ Right to Know List (Total)
sn 1852

THIRAM   **13718-26-8**

HEALTH AND SAFETY LISTS

U.S. DOT - Substances From 49 CFR 172.101
regulated by DOT (UN1837)

U.S. DOT - Hazard Classes
DOT hazard class = 8

NIOSH - Selected LD50s and LC50s
Oral, rat: LD50 = 750 mg/kg

INTERNATIONAL LISTS

Canada - WHMIS: Ingredient Disclosure
1% item 1564 (531)

THORIUM   **7440-29-1**

HEALTH AND SAFETY LISTS

U.S. DOT - Appendix A Table 1 - Hazardous Substances
final RQ = 100 pounds (45.4 kg)

NIOSH - Selected LD50s and LC50s
Oral, rat: LD50 = 9160 ug/kg

ENVIRONMENTAL LISTS

CERCLA/SARA - Section 302 Extremely Hazardous Substances and TPQs
TPQ = 100/10,000 pounds

CERCLA/SARA - Section 313 - Emission Reporting
form R reporting required

CERCLA/SARA - Hazardous Substances and their Reportable Quantities
final RQ = 100 pounds (45.4 kg)

RCRA - P Series Wastes
waste number P116

RCRA - Hazardous Constituents-Appendix VIII
waste number P116

RCRA - Substances Banned From Land Disposal
[present]

INTERNATIONAL LISTS

Canada - WHMIS: Ingredient Disclosure
1% item 85 (224)

STATE LISTS

Florida Hazardous Substance List
effective March 13, 1992

Massachusetts Right To Know List
extraordinarily hazardous

NJ Right to Know List (Total)
sn 2823

Pennsylvania Right to Know List
environmental hazard

THORIUM 226   **15571-75-2**

HEALTH AND SAFETY LISTS

IARC - Group 3 (not classifiable)
[present]

STATE LISTS

California - Directors List of Hazardous Substances (8 CCR 339)
[present]

THORIUM 227   **15623-47-9**

HEALTH AND SAFETY LISTS

U.S. DOT - Appendix A Table 1 - Hazardous Substances
final RQ = 10 pounds (4.54 kg)

IARC - Group 2B (sufficient animal data)
[present] (Overall evaluation based only on evidence of carcinogenicity in monograph (7, 1974) or in Supplement 4)

NIOSH - Selected LD50s and LC50s
Oral, rat: LD50 = 125 mg/kg

NTP Seventh Report - Suspect Carcinogens
suspect carcinogen

OSHA - Possible Select Carcinogens
[present]

ENVIRONMENTAL LISTS

CERCLA/SARA - Section 313 - Emission Reporting
form R reporting required for 0.1% de minimus concentration

CERCLA/SARA - Hazardous Substances and their Reportable Quantities
final RQ = 10 pounds (4.54 kg)

EPA - Carcinogen Hazard Ranking for RQ Adjustment
Hazard ranking = Medium

RCRA - U Series Wastes
waste number U219

RCRA - Hazardous Constituents-Appendix VIII
waste number U219

RCRA - Substances Banned From Land Disposal
[present]

INTERNATIONAL LISTS

Canada - WHMIS: Ingredient Disclosure
0.1% item 1565 (1619)

Canada - NPRI (National Pollutant Release Inventory)
[present]

German (DFG) - Carcinogens
suspected carcinogen

STATE LISTS

California - Air Bill 2588 Appendix A-I
known or potential carcinogen

California - Prop. 65 - Cancer list
carcinogen - initial date 1/1/88

California - Prop. 65 - No Significant Risk Levels
no significant risk level = 10 ug/day

California - Directors List of Hazardous Substances (8 CCR 339)
[present]

Florida Hazardous Substance List
[present]

Massachusetts Right To Know List
carcinogen; extraordinarily hazardous

Minnesota Hazardous Substance List
carcinogen

NJ Right to Know List (Total)
sn 1853

NJ Special Hazardous Substances
(carcinogen; mutagen)

Pennsylvania Right to Know List
environmental hazard; special hazardous substance

Pennsylvania RTK - Special Hazardous Substances
[present]

THORIUM 228   14274-82-9

HEALTH AND SAFETY LISTS

U.S. DOT - Appendix A Table 1 - Hazardous Substances
final RQ = 100 pounds (45.4 kg)

NIOSH - Selected LD50s and LC50s
Oral, rat: LD50 = 4600 ug/kg

ENVIRONMENTAL LISTS

CERCLA/SARA - Section 302 Extremely Hazardous Substances and TPQs
TPQ = 100/10,000 pounds

CERCLA/SARA - Hazardous Substances and their Reportable Quantities
final RQ = 100 pounds (45.4 kg)

RCRA - P Series Wastes
waste number P026

RCRA - Hazardous Constituents-Appendix VIII
waste number P026

TSCA - Code of Federal Regulations Citations
40 CFR 716.120(a)

TSCA - Health and Safety Reporting List
Effective Date: March 7, 1986

STATE LISTS

Florida Hazardous Substance List
effective March 13, 1992

Massachusetts Right To Know List
extraordinarily hazardous

NJ Right to Know List (Total)
sn 2824

Pennsylvania Right to Know List
environmental hazard

THORIUM 229   15594-54-4

ENVIRONMENTAL LISTS

CERCLA/SARA - Section 302 Extremely Hazardous Substances and TPQs
TPQ = 500/10,000 pounds

STATE LISTS

Florida Hazardous Substance List
effective March 13, 1992

Massachusetts Right To Know List
extraordinarily hazardous

NJ Right to Know List (Total)
sn 2825

Pennsylvania Right to Know List
environmental hazard

PROPOSED REGULATIONS

CERCLA/SARA - Proposed Hazardous Substance Additions
proposed RQ = 1 pound (.454 kg)

CERCLA/SARA - 1989 Proposed RQ Adjustments
proposed RQ = 100 pounds (45.4 kg)

THORIUM 230   14269-63-7

HEALTH AND SAFETY LISTS

ACGIH 1995 - Time Weighted Averages
1 mg/m3 TWA

U.S. DOT - Appendix A Table 1 - Hazardous Substances
final RQ = 10 pounds (4.54 kg)

IARC - Group 3 (not classifiable)
[present]

NIOSH - Selected LD50s and LC50s
Oral, rat: LD50 = 560 mg/kg

NIOSH 1990 - Pocket Guide - RELs
5 mg/m3 TWA

NIOSH 1990 - Pocket Guide - IDLHs
1500 mg/m3 IDLH

NIOSH 1990 - Pocket Guide - Target organs
skin, respiratory system

OSHA - Vacated PELs - Time Weighted Averages
5 mg/m3 TWA

OSHA - Final PELs - Time Weighted Averages
5 mg/m3 TWA

ENVIRONMENTAL LISTS

CERCLA/SARA - Section 313 - Emission Reporting
form R reporting required

CERCLA/SARA - Hazardous Substances and their Reportable Quantities
final RQ = 10 pounds (4.54 kg)

RCRA - U Series Wastes
waste number U244

RCRA - Hazardous Constituents-Appendix VIII
waste number U244

RCRA - Substances Banned From Land Disposal
[present]

TSCA - Code of Federal Regulations Citations
40 CFR 716.120(a)

TSCA - Health and Safety Reporting List
Effective Date: June 1, 1987

INTERNATIONAL LISTS

Australian Exposure Standards - Time Weighted Averages
1 mg/m3 TWA

Canada - WHMIS: Ingredient Disclosure
1% item 1566 (1620)

Canada - Alberta - 8 Hour Occupational Exposure Limit
5 mg/m3 TWA

Canada - Alberta - 15 Minute Occupational Exposure Limit
10 mg/m3 STEL

Canada - British Columbia - 8 Hour Exposure Limits
5 mg/m3 TWA

Canada - British Columbia - 15 Minute Exposure Limits
10 mg/m3 STEL

Canada - Ontario - OHSA - TWAEVs
5 mg/m3 TWAEV

Canada - Quebec - Time-Weighted Average Exposure Values
5 mg/m3 TWAEV

United Kingdom - Occupational Exposure Standards - TWAs
5 mg/m3 TWA

United Kingdom - Occupational Exposure Standards - STELs
10 mg/m3 STEL

German (DFG) - MAK Values
total dust: 5 mg/m3 MAK

German (DFG) - Peak Limitations
5 x normal MAK (30 min. average value); don't exceed 2 times during shift

German (DFG) - Pregnancy
classification not yet possible

Israel - Time Weighted Averages
1 mg/m3 TWA

Israel - Action Levels
0.5 mg/m3 AL

Mexico - Instruction No. 10 - TWAs
5 mg/m3 TWA

Mexico - Instruction No. 10 - STELs
10 mg/m3 STEL

STATE LISTS

California - Exposure Limits - PELs
5 mg/m3 PEL

California - Directors List of Hazardous Substances (8 CCR 339)
[present]

Florida Hazardous Substance List
[present]

Massachusetts Right To Know List
[present]

Minnesota Hazardous Substance List
[present]

NJ Right to Know List (Total)
sn 1854

Pennsylvania Right to Know List
environmental hazard

THORIUM 231   14932-40-2

HEALTH AND SAFETY LISTS

NIOSH - Selected LD50s and LC50s
Oral, rat: LD50 = 98 mg/kg

INTERNATIONAL LISTS

Canada - WHMIS: Ingredient Disclosure
1% item 1449 (1086)

THORIUM 234   15065-10-8

HEALTH AND SAFETY LISTS

U.S. DOT - Substances From 49 CFR 172.101
regulated by DOT (UN2975)

U.S. DOT - Hazard Classes
DOT hazard class = 7

U.S. DOT - Appendix A Table 2 - Radionuclides
final RQ = 0.001 curies (3.7E 7 Bq) (notification requirements for releases of mixtures or solutions can be found in 40 CFR 302.6(b))

ENVIRONMENTAL LISTS

ATSDR Priority List
Rank (of 275): 082

CERCLA/SARA List of Radionuclides (Appendix B) and Their Reportable Quantities
final RQ = 0.001 curies (3.7E 7 Bq) (notification requirements for releases of mixtures or solutions can be found in 40 CFR 302.6(b))

STATE LISTS

NJ Right to Know List (Total)
sn 1855

THORIUM DIOXIDE   1314-20-1

HEALTH AND SAFETY LISTS

U.S. DOT - Appendix A Table 2 - Radionuclides
final RQ = 100 curies (3.7E 12 Bq)

ENVIRONMENTAL LISTS

CERCLA/SARA List of Radionuclides (Appendix B) and Their Reportable Quantities
final RQ = 100 curies (3.7E 12 Bq)

THORIUM NITRATE   13823-29-5

HEALTH AND SAFETY LISTS

U.S. DOT - Appendix A Table 2 - Radionuclides
final RQ = 1 curie (3.7E 10 Bq)

ENVIRONMENTAL LISTS

ATSDR Priority List
Rank (of 275): 122

CERCLA/SARA List of Radionuclides (Appendix B) and Their Reportable Quantities
final RQ = 1 curie (3.7E 10 Bq)

THULIUM 162   15832-57-2

HEALTH AND SAFETY LISTS

U.S. DOT - Appendix A Table 2 - Radionuclides
final RQ = 0.01 curies (3.7E 8 Bq)

ENVIRONMENTAL LISTS

ATSDR Priority List
Rank (of 275): 091

CERCLA/SARA List of Radionuclides (Appendix B) and Their Reportable Quantities
final RQ = 0.01 curies (3.7E 8 Bq)

THULIUM 166   15690-75-2

HEALTH AND SAFETY LISTS

U.S. DOT - Appendix A Table 2 - Radionuclides
final RQ = 0.001 curies (3.7E 7 Bq)

ENVIRONMENTAL LISTS

CERCLA/SARA List of Radionuclides (Appendix B) and Their Reportable Quantities
final RQ = 0.001 curies (3.7E 7 Bq)

THULIUM 170   13981-30-1

HEALTH AND SAFETY LISTS

U.S. DOT - Appendix A Table 2 - Radionuclides
final RQ = 0.01 curies (3.7E 8 Bq)

ENVIRONMENTAL LISTS

ATSDR Priority List
Rank (of 275): 088

CERCLA/SARA List of Radionuclides (Appendix B) and Their Reportable Quantities
final RQ = 0.01 curies (3.7E 8 Bq)

THULIUM 171   14333-45-0

HEALTH AND SAFETY LISTS

U.S. DOT - Appendix A Table 2 - Radionuclides
final RQ = 100 curies (3.7E 12 Bq)

ENVIRONMENTAL LISTS

CERCLA/SARA List of Radionuclides (Appendix B) and Their Reportable Quantities
final RQ = 100 curies (3.7E 12 Bq)

THULIUM 172   15720-75-9

HEALTH AND SAFETY LISTS

U.S. DOT - Appendix A Table 2 - Radionuclides
final RQ = 100 curies (3.7E 12 Bq)

ENVIRONMENTAL LISTS

CERCLA/SARA List of Radionuclides (Appendix B) and Their Reportable Quantities
final RQ = 100 curies (3.7E 12 Bq)

THULIUM 173   14041-46-4
SEE ALSO:
   THORIUM

HEALTH AND SAFETY LISTS

   NTP Seventh Report - Known Carcinogens
      known carcinogen

   OSHA - Select Carcinogens
      [present]

ENVIRONMENTAL LISTS

   CERCLA/SARA - Section 313 - Emission Reporting
      form R reporting required for 1.0% de minimus concentration

INTERNATIONAL LISTS

   Canada - NPRI (National Pollutant Release Inventory)
      [present]

STATE LISTS

   California - Air Bill 2588 Appendix A-II
      known or potential carcinogen

   California - Prop. 65 - Cancer list
      carcinogen - initial date 2/27/87

   Florida Hazardous Substance List
      [present]

   Massachusetts Right To Know List
      carcinogen; extraordinarily hazardous

   Minnesota Hazardous Substance List
      carcinogen

   NJ Right to Know List (Total)
      sn 1856

   NJ Special Hazardous Substances
      (carcinogen)

   Pennsylvania Right to Know List
      environmental hazard; special hazardous substance

   Pennsylvania RTK - Special Hazardous Substances
      [present]

THULIUM 175   14041-47-5
SEE ALSO:
   THORIUM

HEALTH AND SAFETY LISTS

   U.S. DOT - Substances From 49 CFR 172.101
      regulated by DOT (UN2976)

   U.S. DOT - Hazard Classes
      DOT hazard class = 7

   NIOSH - Selected LD50s and LC50s
      Oral, mouse: LD50 = 1760 mg/kg

STATE LISTS

   Florida Hazardous Substance List
      [present]

   Massachusetts Right To Know List
      [present]

   NJ Right to Know List (Total)
      sn 1857

   Pennsylvania Right to Know List
      [present]

THULIUM CHLORIDE   13537-18-3

HEALTH AND SAFETY LISTS

   U.S. DOT - Appendix A Table 2 - Radionuclides
      final RQ = 1000 curies (3.7E 13 Bq)

ENVIRONMENTAL LISTS

   CERCLA/SARA List of Radionuclides (Appendix B) and Their Reportable Quantities
      final RQ = 1000 curies (3.7E 13 Bq)

THYMINE   65-71-4

HEALTH AND SAFETY LISTS

   U.S. DOT - Appendix A Table 2 - Radionuclides
      final RQ = 100 curies (3.7E 12 Bq)

ENVIRONMENTAL LISTS

   CERCLA/SARA List of Radionuclides (Appendix B) and Their Reportable Quantities
      final RQ = 100 curies (3.7E 12 Bq)

THYMOL   89-83-8

HEALTH AND SAFETY LISTS

   U.S. DOT - Appendix A Table 2 - Radionuclides
      final RQ = 10 curies (3.7E 11 Bq)

ENVIRONMENTAL LISTS

   CERCLA/SARA List of Radionuclides (Appendix B) and Their Reportable Quantities
      final RQ = 10 curies (3.7E 11 Bq)

TIGLIC ACID   80-59-1

HEALTH AND SAFETY LISTS

   U.S. DOT - Appendix A Table 2 - Radionuclides
      final RQ = 100 curies (3.7E 12 Bq)

ENVIRONMENTAL LISTS

   CERCLA/SARA List of Radionuclides (Appendix B) and Their Reportable Quantities
      final RQ = 100 curies (3.7E 12 Bq)

TIN   7440-31-5

HEALTH AND SAFETY LISTS

   U.S. DOT - Appendix A Table 2 - Radionuclides
      final RQ = 100 curies (3.7E 12 Bq)

ENVIRONMENTAL LISTS

   CERCLA/SARA List of Radionuclides (Appendix B) and Their Reportable Quantities
      final RQ = 100 curies (3.7E 12 Bq)

TIN 110   15700-33-1

HEALTH AND SAFETY LISTS

   U.S. DOT - Appendix A Table 2 - Radionuclides
      final RQ = 100 curies (3.7E 12 Bq)

ENVIRONMENTAL LISTS

   CERCLA/SARA List of Radionuclides (Appendix B) and Their Reportable Quantities
      final RQ = 100 curies (3.7E 12 Bq)

TIN 111   15720-78-2

HEALTH AND SAFETY LISTS

   U.S. DOT - Appendix A Table 2 - Radionuclides
      final RQ = 1000 curies (3.7E 13 Bq)

ENVIRONMENTAL LISTS

   CERCLA/SARA List of Radionuclides (Appendix B) and Their Reportable Quantities
      final RQ = 1000 curies (3.7E 13 Bq)

TIN 113   13966-06-8

HEALTH AND SAFETY LISTS

   NIOSH - Selected LD50s and LC50s
      Oral, mouse: LD50 = 4294 mg/kg

INTERNATIONAL LISTS

   Canada - WHMIS: Ingredient Disclosure
      1% item 1567 (532)

TIN 117M   RR-00313-4

HEALTH AND SAFETY LISTS
  NIOSH - Selected LD50s and LC50s
    Oral, mouse: LD50 = 3500 mg/kg

TIN 119M   RR-00311-2

HEALTH AND SAFETY LISTS
  NIOSH - Selected LD50s and LC50s
    Oral, rat: LD50 = 980 mg/kg

ENVIRONMENTAL LISTS
  List of Pesticide Product Inert Ingredients
    [present]

TIN 121   14683-06-8

INTERNATIONAL LISTS
  Canada - WHMIS: Ingredient Disclosure
    1% item 1568 (145)

TIN 121M   RR-00310-1

HEALTH AND SAFETY LISTS
  ACGIH 1995 - Time Weighted Averages
    *metal: 2 mg/m3 TWA; oxide and inorganic compounds, except SnH4,
    as Sn: 2 mg/m3 TWA; organic compounds, as Sn: 0.1 mg/m3 TWA*
  ACGIH 1995 - Short Term Exposure Limits
    organic compounds, as Sn: 0.2 mg/m3 STEL
  ACGIH 1995 - Skin Designations
    organic compounds, as Sn: skin - potential for cutaneous absorption
  NIOSH 1990 - Pocket Guide - RELs
    as Sn: 2 mg/m3 TWA
  NIOSH 1990 - Pocket Guide - IDLHs
    as Sn: 400 mg/m3 IDLH
  NIOSH 1990 - Pocket Guide - Target organs
    eyes, skin, respiratory system
  OSHA - Vacated PELs - Time Weighted Averages
    *inorganic compounds (except oxides), as Sn: 2 mg/m3 TWA; organic
    compounds, as Sn: 0.1 mg/m3 TWA*
  OSHA - Vacated PELs - Skin Designation
    organic compounds: Prevent or reduce skin absorption
  OSHA - Final PELs - Time Weighted Averages
    *inorganic compounds (except oxides), as Sn: 2 mg/m3 TWA; organic
    compounds, as Sn: 0.1 mg/m3 TWA*

ENVIRONMENTAL LISTS
  ATSDR Priority List
    Rank (of 275): 157
  RCRA - TSD Facilities Ground Water Monitoring
    *TM 7870 = 8000 ug/L PQL (all species in the ground water that
    contain this element are included)*

INTERNATIONAL LISTS
  Australian Exposure Standards - Time Weighted Averages
    *metal: 2 mg/m3 TWA; organic compounds, as Sn: 0.1 mg/m3 TWA;
    oxide and inorganic compounds (except SnH4), as Sn: 2 mg/m3 TWA*
  Australian Exposure Standards - Short Term Exposure Limits
    organic compounds, as Sn: 0.2 mg/m3 STEL
  Australian Exposure Standards - Skin Effects
    organic compounds as Sn: skin absorption
  Canada - WHMIS: Ingredient Disclosure
    1% item 1571 (804)
  Canada - Alberta - 8 Hour Occupational Exposure Limit
    2 mg/m3 TWA
  Canada - Alberta - 15 Minute Occupational Exposure Limit
    4 mg/m3 STEL
  Canada - Ontario - OHSA - TWAEVs
    as Sn: 2 mg/m3 TWAEV

Canada - Quebec - Time-Weighted Average Exposure Values
  as Sn: 2 mg/m3 TWAEV
German (DFG) - MAK Values
  *inorganic compounds-total dust, as Sn: 2 mg/m3 MAK; organic
  compounds-total dust, as Sn: 0.1 mg/m3 MAK*
German (DFG) - Peak Limitations
  *2 x normal MAK (30 min. average value); don't exceed 4 times during
  shift*
German (DFG) - Skin/Sensitizers
  organic compounds: danger of cutaneous absorption
Israel - Time Weighted Averages
  *metal: 2 mg/m3 TWA; Oxide and inorganic compounds, except SnH4,
  as Sn: 2 mg/m3 TWA; Organic compounds, as Sn: 0.1 mg/m3 TWA*
Israel - Action Levels
  *metal: 1 mg/m3 AL; oxide and inorganic compounds, except SnH4,
  as Sn: 1 mg/m3 AL; organic compounds, as Sn: 0.05 mg/m3 AL*
Mexico - Wastewater - Organic Toxic Pollutants and Heavy Metals
  Listed under [Heavy Metals]

STATE LISTS
  California - Exposure Limits - PELs
    2 mg/m3 PEL
  California - Directors List of Hazardous Substances (8 CCR 339)
    [present]
  Florida Hazardous Substance List
    [present]
  Massachusetts Right To Know List
    [present]
  Minnesota Hazardous Substance List
    [present]
  Pennsylvania Right to Know List
    [present]

TIN 123   14683-07-9

HEALTH AND SAFETY LISTS
  U.S. DOT - Appendix A Table 2 - Radionuclides
    final RQ = 100 curies (3.7E 12 Bq)

ENVIRONMENTAL LISTS
  CERCLA/SARA List of Radionuclides (Appendix B) and Their Reportable Quantities
    final RQ = 100 curies (3.7E 12 Bq)

TIN 123M   RR-00308-7

HEALTH AND SAFETY LISTS
  U.S. DOT - Appendix A Table 2 - Radionuclides
    final RQ = 1000 curies (3.7E 13 Bq)

ENVIRONMENTAL LISTS
  CERCLA/SARA List of Radionuclides (Appendix B) and Their Reportable Quantities
    final RQ = 1000 curies (3.7E 13 Bq)

TIN 125   14683-08-0

HEALTH AND SAFETY LISTS
  U.S. DOT - Appendix A Table 2 - Radionuclides
    final RQ = 10 curies (3.7E 11 Bq)

ENVIRONMENTAL LISTS
  CERCLA/SARA List of Radionuclides (Appendix B) and Their Reportable Quantities
    final RQ = 10 curies (3.7E 11 Bq)

TIN 126   15832-50-5

HEALTH AND SAFETY LISTS
  U.S. DOT - Appendix A Table 2 - Radionuclides
    final RQ = 100 curies (3.7E 12 Bq)

ENVIRONMENTAL LISTS

CERCLA/SARA List of Radionuclides (Appendix B) and Their Reportable Quantities
final RQ = 100 curies (3.7E 12 Bq)

TIN 127   15690-89-8

HEALTH AND SAFETY LISTS

U.S. DOT - Appendix A Table 2 - Radionuclides
final RQ = 10 curies (3.7E 11 Bq)

ENVIRONMENTAL LISTS

CERCLA/SARA List of Radionuclides (Appendix B) and Their Reportable Quantities
final RQ = 10 curies (3.7E 11 Bq)

TIN 128   16645-96-8

HEALTH AND SAFETY LISTS

U.S. DOT - Appendix A Table 2 - Radionuclides
final RQ = 1000 curies (3.7E 13 Bq)

ENVIRONMENTAL LISTS

CERCLA/SARA List of Radionuclides (Appendix B) and Their Reportable Quantities
final RQ = 1000 curies (3.7E 13 Bq)

TIN IV CHROMATE   10101-75-4

HEALTH AND SAFETY LISTS

U.S. DOT - Appendix A Table 2 - Radionuclides
final RQ = 10 curies (3.7E 11 Bq)

ENVIRONMENTAL LISTS

CERCLA/SARA List of Radionuclides (Appendix B) and Their Reportable Quantities
final RQ = 10 curies (3.7E 11 Bq)

TSCA - Chemicals with Significant New Use Rules
PMN number: P-87-304

TIN II CHROMATE   38455-77-5

HEALTH AND SAFETY LISTS

U.S. DOT - Appendix A Table 2 - Radionuclides
final RQ = 10 curies (3.7E 11 Bq)

ENVIRONMENTAL LISTS

CERCLA/SARA List of Radionuclides (Appendix B) and Their Reportable Quantities
final RQ = 10 curies (3.7E 11 Bq)

TIN COMPOUNDS, N.O.S.   RR-00615-5

HEALTH AND SAFETY LISTS

U.S. DOT - Appendix A Table 2 - Radionuclides
final RQ = 1000 curies (3.7E 13 Bq)

ENVIRONMENTAL LISTS

CERCLA/SARA List of Radionuclides (Appendix B) and Their Reportable Quantities
final RQ = 1000 curies (3.7E 13 Bq)

TSCA - Chemicals with Significant New Use Rules
PMN number: P-87-1265

TIN INORGANIC COMPOUNDS   RR-00043-1

HEALTH AND SAFETY LISTS

U.S. DOT - Appendix A Table 2 - Radionuclides
final RQ = 10 curies (3.7E 11 Bq)

ENVIRONMENTAL LISTS

CERCLA/SARA List of Radionuclides (Appendix B) and Their Reportable Quantities
final RQ = 10 curies (3.7E 11 Bq)

TINOPAL CBS   27344-41-8

HEALTH AND SAFETY LISTS

U.S. DOT - Appendix A Table 2 - Radionuclides
final RQ = 1 curie (3.7E 10 Bq)

ENVIRONMENTAL LISTS

CERCLA/SARA List of Radionuclides (Appendix B) and Their Reportable Quantities
final RQ = 1 curie (3.7E 10 Bq)

TIN ORGANIC COMPOUNDS   RR-00042-0

HEALTH AND SAFETY LISTS

U.S. DOT - Appendix A Table 2 - Radionuclides
final RQ = 100 curies (3.7E 12 Bq)

ENVIRONMENTAL LISTS

CERCLA/SARA List of Radionuclides (Appendix B) and Their Reportable Quantities
final RQ = 100 curies (3.7E 12 Bq)

TIN OXIDE   21651-19-4

HEALTH AND SAFETY LISTS

U.S. DOT - Appendix A Table 2 - Radionuclides
final RQ = 1000 curies (3.7E 13 Bq)

ENVIRONMENTAL LISTS

CERCLA/SARA List of Radionuclides (Appendix B) and Their Reportable Quantities
final RQ = 1000 curies (3.7E 13 Bq)

TIN PHOSPHIDE   25324-56-5

STATE LISTS

Massachusetts Right To Know List
[present]

TINUVIN 328 (BENZOTRIAZOLE UV ABSORBER)   25973-55-1

STATE LISTS

Massachusetts Right To Know List
[present]

TINUVIN 770 (AMINE LIGHT STABILIZER)   52829-07-9

INTERNATIONAL LISTS

Canada - WHMIS: Ingredient Disclosure
1% item 1570 (803)

STATE LISTS

California - Directors List of Hazardous Substances (8 CCR 339)
[present]

TITANATE [TI6O13(2-)], DIPOTASSIUM   12056-51-8

INTERNATIONAL LISTS

Canada - Alberta - 8 Hour Occupational Exposure Limit
except SnO4, as Sn: 2 mg/m3 TWA

Canada - Alberta - 15 Minute Occupational Exposure Limit
4 mg/m3 STEL

Canada - British Columbia - 8 Hour Exposure Limits
except SnH4 and SnO2, as Sn: 2 mg/m3 TWA

Canada - British Columbia - 15 Minute Exposure Limits
except SnH4 and SnO2: 4 mg/m3 STEL

Canada - British Columbia - Skin Notations
skin - potential for skin absorption

Canada - Ontario - OHSA - TWAEVs
as Sn: 2 mg/m3 TWAEV (except stannane)

Canada - Quebec - Time-Weighted Average Exposure Values
except SnH4, as Sn: 2 mg/m3 TWAEV

United Kingdom - Occupational Exposure Standards - TWAs
as Sn: 2 mg/m3 TWA (does not include SnH4)

United Kingdom - Occupational Exposure Standards - STELs
as Sn: 4 mg/m3 STEL (does not include SnH4)

Mexico - Instruction No. 10 - TWAs
2 mg/m3 TWA

Mexico - Instruction No. 10 - STELs
4 mg/m3 STEL

STATE LISTS

California - Exposure Limits - PELs
as Sn: 2 mg/m3 PEL

TITANIUM 7440-32-6

HEALTH AND SAFETY LISTS

NIOSH - Selected LD50s and LC50s
Oral, rat: LD50 = 5580 mg/kg

INTERNATIONAL LISTS

Canada - WHMIS: Ingredient Disclosure
0.1% item 675 (303)

TITANIUM 44 15749-33-4

HEALTH AND SAFETY LISTS

NIOSH 1990 - Pocket Guide - RELs
as Sn: 0.1 mg/m3 TWA

NIOSH 1990 - Pocket Guide - Target organs
CNS, eyes, liver, urinary tract, skin, blood

NIOSH 1990 - Pocket Guide - Skin list
as Sn: Potential for dermal absorption

INTERNATIONAL LISTS

Canada - Alberta - 8 Hour Occupational Exposure Limit
as Sn: 0.1 mg/m3 TWA

Canada - Alberta - 15 Minute Occupational Exposure Limit
as Sn: 0.2 mg/m3 STEL

Canada - Alberta - Skin Designation
as Sn: can be absorbed through the intact skin

Canada - British Columbia - 8 Hour Exposure Limits
as Sn: 0.1 mg/m3 TWA

Canada - British Columbia - 15 Minute Exposure Limits
as Sn: 0.2 mg/m3 STEL

Canada - Ontario - OHSA - TWAEVs
as Sn: 0.1 mg/m3 TWAEV

Canada - Ontario - OHSA - Skin Notations
absorption through skin, eyes, or mucous membranes

Canada - Quebec - Time-Weighted Average Exposure Values
as Sn: 0.1 mg/m3 TWAEV

Canada - Quebec - Skin Designations
absorbed through the skin

United Kingdom - Occupational Exposure Standards - TWAs
as Sn: 0.1 mg/m3 TWA (does not include Cyhexatin)

United Kingdom - Occupational Exposure Standards - STELs
as Sn: 0.2 mg/m3 STEL (does not include Cyhexatin)

United Kingdom - Occupational Exposure Standards - Notes
can be absorbed through skin

German (DFG) - MAK Values
total dust: 0.1 mg/m3 MAK

German (DFG) - Peak Limitations
2 x normal MAK (30 min. average value); don't exceed 4 times during shift

German (DFG) - Skin/Sensitizers
danger of cutaneous absorption

German (DFG) - Pregnancy
as Sn: classification not yet possible

Mexico - Instruction No. 10 - TWAs
0.1 mg/m3 TWA

Mexico - Instruction No. 10 - STELs
0.2 mg/m3 STEL

STATE LISTS

California - Exposure Limits - PELs
as Sn: 0.1 mg/m3 PEL

California - Exposure Limits - Skin Notation
material may be absorbed through the skin, eyes or mucous membrane

Minnesota Hazardous Substance List
as Sn: skin

TITANIUM 45 14392-00-8

INTERNATIONAL LISTS

German (DFG) - Pregnancy
no risk to embryo/fetus if exposure limits adhered to

TITANIUM DIOXIDE 13463-67-7
SEE ALSO:
TIN

HEALTH AND SAFETY LISTS

OSHA - Vacated PELs - Time Weighted Averages
as Sn: 2 mg/m3 TWA

INTERNATIONAL LISTS

Canada - Alberta - 8 Hour Occupational Exposure Limit
respirable mass: 5 mg/m3 TWA; total mass: 10 mg/m3 TWA

Canada - British Columbia - 8 Hour Exposure Limits
nuisance dust, mists, and fumes, as Sn: 10 mg/m3 TWA

Canada - British Columbia - 15 Minute Exposure Limits
20 mg/m3 STEL

Canada - Ontario - OHSA - TWAEVs
as Sn: 2 mg/m3 TWAEV

Israel - Time Weighted Averages
2 mg/m3 TWA

Israel - Action Levels
1 mg/m3 AL

STATE LISTS

California - Exposure Limits - PELs
as Sn: 2 mg/m3 PEL

TITANIUM HYDRIDE 11140-68-4

STATE LISTS

NJ Right to Know List (Total)
sn 1732

TITANIUM OXIDE (TIO) 12137-20-1

ENVIRONMENTAL LISTS

List of Pesticide Product Inert Ingredients
[present]

TITANIUM SULFATE 13825-74-6

ENVIRONMENTAL LISTS

List of Pesticide Product Inert Ingredients
[present]

TITANIUM SULFATE (TI(SIO4)2) 13693-11-3

ENVIRONMENTAL LISTS

TSCA - Code of Federal Regulations Citations
40 CFR 721.2184

TSCA - Chemicals with Significant New Use Rules
PMN number: P-90-0226

TSCA - Section 12(b) - Export Notification
P-90-226; export notification required - Section 5

INTERNATIONAL LISTS
German (DFG) - Carcinogens
as fibrous dust: animal evidence of carcinogenicity

TITANIUM TETRACHLORIDE   7550-45-0

HEALTH AND SAFETY LISTS
U.S. DOT - Substances From 49 CFR 172.101
regulated by DOT (UN2546, UN1352)
U.S. DOT - Hazard Classes
DOT hazard class = 4.1

ENVIRONMENTAL LISTS
ATSDR Priority List
Rank (of 275): 223

INTERNATIONAL LISTS
Mexico - Wastewater - Organic Toxic Pollutants and Heavy Metals
Listed under [Heavy Metals]

STATE LISTS
California - Directors List of Hazardous Substances (8 CCR 339)
[present]
NJ Right to Know List (Total)
sn 1860; granules or powder sponge: sn 2827

TITANIUM TRICHLORIDE   7705-07-9

HEALTH AND SAFETY LISTS
U.S. DOT - Appendix A Table 2 - Radionuclides
final RQ = 1 curie (3.7E 10 Bq)

ENVIRONMENTAL LISTS
CERCLA/SARA List of Radionuclides (Appendix B) and Their Reportable Quantities
final RQ = 1 curie (3.7E 10 Bq)

TITANOCENE DICHLORIDE   1271-19-8

HEALTH AND SAFETY LISTS
U.S. DOT - Appendix A Table 2 - Radionuclides
final RQ = 1000 curies (3.7E 13 Bq)

ENVIRONMENTAL LISTS
CERCLA/SARA List of Radionuclides (Appendix B) and Their Reportable Quantities
final RQ = 1000 curies (3.7E 13 Bq)

TOBACCO DUST   84961-66-0

HEALTH AND SAFETY LISTS
ACGIH 1995 - Time Weighted Averages
10 mg/m3 TWA
IARC - Group 3 (not classifiable)
[present]
NIOSH 1990 - Pocket Guide - Carcinogens
occupational carcinogen
NIOSH 1990 - Pocket Guide - Target organs
lungs
NTP Chemical Status Reports - Testing Status and NTIS Number
Technical reports printed (PB288780/AS)
NTP Chemical Status Reports - Evidence of Carcinogenicity
male rat-negative; female rat-negative; male mice-negative; female mice-negative
OSHA - Vacated PELs - Time Weighted Averages
total dust: 10 mg/m3 TWA
OSHA - Final PELs - Time Weighted Averages
total dust: 15 mg/m3 TWA

ENVIRONMENTAL LISTS
List of Pesticide Product Inert Ingredients
[present]

INTERNATIONAL LISTS
Australian Exposure Standards - Time Weighted Averages
10 mg/m3 TWA
Canada - British Columbia - 8 Hour Exposure Limits
nuisance dust, mists, and fumes, as Ti: 10 mg/m3 TWA
Canada - British Columbia - 15 Minute Exposure Limits
20 mg/m3 STEL
Canada - Ontario - OHSA - TWAEVs
total dust: 10 mg/m3 TWAEV (listed as nuisance particulates)
Canada - Quebec - Time-Weighted Average Exposure Values
as Ti, total dust: 10 mg/m3 TWAEV; respirable dust: 5 mg/m3 TWAEV
United Kingdom - Occupational Exposure Standards - TWAs
total inhalable dust: 10 mg/m3 TWA; respirable dust: 5 mg/m3 TWA
German (DFG) - MAK Values
fine dust: 6 mg/m3 MAK
German (DFG) - Pregnancy
no risk to embryo/fetus if exposure limits adhered to
Israel - Time Weighted Averages
10 mg/m3 TWA (The value is for total dust containing no asbestos and <1% crystalline silica)
Israel - Action Levels
5 mg/m3 AL
Mexico - Instruction No. 10 - TWAs
10 mg/m3 TWA; (nuisance particulate)
Mexico - Instruction No. 10 - STELs
20 mg/m3 STEL

STATE LISTS
Minnesota Hazardous Substance List
[present] (includes inert or nuisance dust)
Pennsylvania Right to Know List
[present]

PROPOSED REGULATIONS
Canada - Ontario - Proposed Occupational TWAEVs
total dust: 5 mg/m3 TWAEV

TOBACCO PRODUCTS, SMOKELESS   RR-00031-7

HEALTH AND SAFETY LISTS
U.S. DOT - Substances From 49 CFR 172.101
regulated by DOT (UN1871)
U.S. DOT - Hazard Classes
DOT hazard class = 4.1

STATE LISTS
NJ Right to Know List (Total)
sn 1862

TOBACCO SMOKE   RR-00030-6

INTERNATIONAL LISTS
Canada - Alberta - 8 Hour Occupational Exposure Limit
respirable mass: 5 mg/m3 TWA; total mass: 10 mg/m3 TWA

TOBRAMYCIN SULFATE   49842-07-1

INTERNATIONAL LISTS
Canada - WHMIS: Ingredient Disclosure
1% item 1573 (1532)

STATE LISTS
NJ Right to Know List (Total)
sn 1863
NJ Special Hazardous Substances
(corrosive)

TOE PUFFS, NITROCELLULOSE BASE   RR-01508-7

HEALTH AND SAFETY LISTS

U.S. DOT - Substances From 49 CFR 172.101
regulated by DOT (NA1760)

U.S. DOT - Hazard Classes
DOT hazard class = 8

ENVIRONMENTAL LISTS

List of Pesticide Product Inert Ingredients
[present]

TOLAZAMIDE   1156-19-0

HEALTH AND SAFETY LISTS

U.S. DOT - Substances From 49 CFR 172.101
regulated by DOT (UN1838)

U.S. DOT - Hazard Classes
DOT hazard class = 8

U.S. DOT - Substances Which Are Poisonous by Inhalation
liquid hazardous material poisonous by inhalation (UN1838)

NIOSH - Selected LD50s and LC50s
Inhalation, rat: LC50 = 460 mg/m3 4 hr

ENVIRONMENTAL LISTS

CERCLA/SARA - Section 302 Extremely Hazardous Substances and TPQs
TPQ = 100 pounds

CERCLA/SARA - Section 313 - Emission Reporting
form R reporting required for 1.0% de minimus concentration

CERCLA/SARA - Hazardous Substances and their Reportable Quantities
final RQ = 1 pound (.454 kg)

Clean Air Act (1990) - List of Hazardous Air Contaminants
[present]

CAA -Toxic Substances for Accidental Release Prevention
threshold quantity = 2,500 lbs

INTERNATIONAL LISTS

Canada - WHMIS: Ingredient Disclosure
1% item 1574 (1584)

Canada - NPRI (National Pollutant Release Inventory)
[present]

STATE LISTS

California - Air Bill 2588 Appendix A-I
6/91

Florida Hazardous Substance List
[present]

Massachusetts Right To Know List
extraordinarily hazardous

NJ Right to Know List (Total)
sn 1864

NJ Special Hazardous Substances
(corrosive)

Pennsylvania Right to Know List
environmental hazard

PROPOSED REGULATIONS

CERCLA/SARA - Proposed Hazardous Substance Additions
proposed RQ = 1 pound (.454 kg)

CERCLA/SARA - 1989 Proposed RQ Adjustments
proposed RQ = 100 pounds (45.4 kg)

TOLBUTAMIDE   64-77-7

HEALTH AND SAFETY LISTS

U.S. DOT - Substances From 49 CFR 172.101
regulated by DOT (UN2441)

U.S. DOT - Hazard Classes
DOT hazard class = 4.2

INTERNATIONAL LISTS

Canada - WHMIS: Ingredient Disclosure
1% item 1575 (1661)

STATE LISTS

NJ Right to Know List (Total)
sn 1865

NJ Special Hazardous Substances
(corrosive)

O-TOLIDINE-BASED DYES   RR-00520-9

HEALTH AND SAFETY LISTS

NTP Chemical Status Reports - Testing Status and NTIS Number
Technical reports printed (PB92129576/AS)

INTERNATIONAL LISTS

Canada - WHMIS: Ingredient Disclosure
1% item 1576 (685)

TOLUENE   108-88-3

ENVIRONMENTAL LISTS

List of Pesticide Product Inert Ingredients
[present]

TOLUENE 2,4-DIAMINE   95-80-7

HEALTH AND SAFETY LISTS

IARC - Group 1 (carcinogenic to humans)
[present]

OSHA - Select Carcinogens
[present]

STATE LISTS

California - Air Bill 2588 Appendix A-II
known or potential carcinogen

California - Prop. 65 - Cancer list
carcinogen (oral use of) - initial date 4/1/88

TOLUENE-3,5-DIAMINE   108-71-4

HEALTH AND SAFETY LISTS

IARC - Group 1 (carcinogenic to humans)
[present]

OSHA - Select Carcinogens
[present]

STATE LISTS

California - Air Bill 2588 Appendix A-I
known or potential carcinogen

California - Prop. 65 - Cancer list
carcinogen - initial date 4/1/88

California - Prop. 65 - Developmental Toxicity
developmental toxicity - initial date 4/1/88

California - Prop. 65 - Reproductive - Female
female reproductive toxicity - initial date 4/1/88

California - Prop. 65 - Reproductive - Male
male reproductive toxicity - initial date 4/1/88

2,6-TOLUENEDIAMINE DIHYDROCHLORIDE   15481-70-6

STATE LISTS

California - Air Bill 2588 Appendix A-II
9/90

California - Prop. 65 - Developmental Toxicity
developmental toxicity - initial date 7/1/90

2,5-TOLUENEDIAMINE SULFATE   6369-59-1

STATE LISTS

NJ Right to Know List (Total)
sn 2828

TOLUENE-2,6-DIISOCYANATE   91-08-7

HEALTH AND SAFETY LISTS

NTP Chemical Status Reports - Testing Status and NTIS Number
Technical reports printed (PB284610/AS)

NTP Chemical Status Reports - Evidence of Carcinogenicity
male rat-negative; female rat-negative; male mice-negative; female mice-negative

TOLUENE 2,4-DIISOCYANATE   584-84-9

HEALTH AND SAFETY LISTS

NTP Chemical Status Reports - Testing Status and NTIS Number
Technical reports printed (PB274483/AS)

NTP Chemical Status Reports - Evidence of Carcinogenicity
male rat-negative; female rat-negative; male mice-negative; female mice-negative

TOLUENE DIISOCYANATE   26471-62-5

HEALTH AND SAFETY LISTS

NIOSH - Health Standards - Exposure Limits
handle with caution in the workplace; minimize exposure

NIOSH - Health Standards - Health Effects and Precautions
Bladder cancer (Substitute less toxic dyes wherever possible)

NIOSH - Health Standards - Carcinogenic Chemicals
potential human carcinogen

STATE LISTS

Minnesota Hazardous Substance List
carcinogen

TOLUENE, 2,3-DINITRO-   602-01-7
SEE ALSO:
K151-HAZARDOUS WASTES
K149-HAZARDOUS WASTES
F039-HAZARDOUS WASTES
F024-HAZARDOUS WASTES
BIS(2,4-DIMETHYLBUTYL) MALEATE
F005-HAZARDOUS WASTES
F025-HAZARDOUS WASTES
K015-HAZARDOUS WASTES
K036-HAZARDOUS WASTES
K037-HAZARDOUS WASTES

HEALTH AND SAFETY LISTS

ACGIH 1995 - Time Weighted Averages
50 ppm TWA; 188 mg/m3 TWA

ACGIH 1995 - Skin Designations
skin - potential for cutaneous absorption

ACGIH 1995 - Biological Exposure Indices
(Hippuric acid in urine): (2.5 g/g creatinine), (end of shift) (B, Ns), (last four hours of shift); (Toluene in venous blood): (1 mg/L creatinine), (end of shift) (Sq); (Toluene in end-exhaled air): (Sq)

AIHA - Odor Threshold Values
geometric mean air odor threshold = 1.6 ppm (detectable); 11 ppm (recognizable)

U.S. DOT - Substances From 49 CFR 172.101
regulated by DOT (UN1294)

U.S. DOT - Hazard Classes
DOT hazard class = 3

U.S. DOT - Substances Which Are Poisonous by Inhalation
liquid hazardous material poisonous by inhalation (when mixed with 2-Amino-2,3-dimethylbutyronitrile)

U.S. DOT - Appendix A Table 1 - Hazardous Substances
final RQ = 1000 pounds (454 kg)

FDA - Controlled Substances Act - Essential chemicals
Import/Export threshold volume = 500 gallons; weight = 1,591 kilograms; Domestic threshold volume = 50 gallons; weight = 159 kilograms

IARC - Group 3 (not classifiable)
[present]

NFPA - Flash Points
flash point = 40 degrees F (4 degrees C)

NFPA - Hazard Identification Ratings
health-2; flammability-3; reactivity-0

NIOSH - Selected LD50s and LC50s
Inhalation, mouse: LC50 = 5320 ppm 8 hr Oral, rat: LD50 = 5000 mg/kg Skin, rabbit: LD50 = 12124 mg/kg

NIOSH 1990 - Pocket Guide - RELs
100 ppm TWA; 375 mg/m3 TWA; 150 ppm STEL; 560 mg/m3 STEL

NIOSH 1990 - Pocket Guide - IDLHs
2000 ppm IDLH

NIOSH 1990 - Pocket Guide - Target organs
CNS, liver, kidneys, skin

NIOSH - Health Standards - Exposure Limits
100 ppm TWA (8 hr); 375 mg/m3 TWA; C (10 min) 200 ppm; C (10 min) 750 mg/m3

NIOSH - Health Standards - Health Effects and Precautions
Central nervous system depressant

NTP Chemical Status Reports - Testing Status and NTIS Number
Technical reports printed (PB90256371); prechronic studies for which toxicity technical reports were not prepared

NTP Chemical Status Reports - Evidence of Carcinogenicity
male rat-no evidence; female rat-no evidence; male mice-no evidence; female mice-no evidence

OSHA - Vacated PELs - Time Weighted Averages
100 ppm TWA; 375 mg/m3 TWA

OSHA - Vacated PELs - Short Term Exposure Limits
150 ppm STEL; 560 mg/m3 STEL

OSHA - Final PELs - Time Weighted Averages
200 ppm TWA; C 300 ppm

OSHA - Final PELs - Ceiling Limits
C 300 ppm

ENVIRONMENTAL LISTS

ATSDR Priority List
Rank (of 275): 053

CERCLA/SARA - Section 313 - Emission Reporting
form R reporting required for 1.0% de minimus concentration

CERCLA/SARA - Hazardous Substances and their Reportable Quantities
final RQ = 1000 pounds (454 kg)

Clean Air Act (1990) - List of Hazardous Air Contaminants
[present]

CAA - HON Rule - SOCMI Chemicals
compliance by Oct. 24, 1994

CAA - HON Rule - Organic HAPs
[present]

Clean Water Act - Hazardous Substances
[present]

Clean Water Act - Priority Pollutants
[present]

Clean Water Act - Toxic Pollutants
[present]

Safe Drinking Water Act - MCLs
MCL = 1 mg/L

Safe Drinking Water Act - MCLGs
MCLG = 1 mg/L

Safe Drinking Water Act - Monitoring
monitoring required

List of Pesticide Product Inert Ingredients
[present]
RCRA - U Series Wastes
waste number U220
RCRA - Hazardous Constituents-Appendix VIII
waste number U220
RCRA - Basis for Listing - Appendix VII
Included in waste streams: F005, FO24, F025, F039, K015, K036, K037, K149, K151
RCRA - Substances Banned From Land Disposal
[present]
RCRA - TSD Facilities Ground Water Monitoring
TM 8020 = 2 ug/L PQL;   TM 8240 = 5 ug/L PQL
RCRA - Universal Treatment Standards (LDR)
WW: 0.080 mg/l; NWW: 10 mg/kg
TSCA - Code of Federal Regulations Citations
40 CFR 712.30(d);   40 CFR 716.120(a)
TSCA - PAIR - Reporting List
Reporting Date: November 19, 1982
TSCA - Health and Safety Reporting List
Effective Date: October 4, 1982

INTERNATIONAL LISTS

Australian Exposure Standards - Time Weighted Averages
100 ppm TWA; 377 mg/m3 TWA

Australian Exposure Standards - Short Term Exposure Limits
150 ppm STEL; 565 mg/m3 STEL

Australian Exposure Standards - Under Review
exposure limits under review

Canada - WHMIS: Ingredient Disclosure
1% item 1578 (1622)

Canada - NPRI (National Pollutant Release Inventory)
[present]

Canada - Drinking Water Quality - AOs
<= 0.024 mg/L AO

Canada - Alberta - 8 Hour Occupational Exposure Limit
100 ppm TWA; 375 mg/m3 TWA

Canada - Alberta - 15 Minute Occupational Exposure Limit
150 ppm STEL; 560 mg/m3 STEL

Canada - Alberta - Skin Designation
can be absorbed through the intact skin

Canada - British Columbia - 8 Hour Exposure Limits
100 ppm TWA; 375 mg/m3 TWA

Canada - British Columbia - 15 Minute Exposure Limits
150 ppm STEL; 560 mg/m3 STEL

Canada - British Columbia - Skin Notations
skin - potential for skin absorption

Canada - Ontario - OHSA - TWAEVs
100 ppm TWAEV; 376 mg/m3 TWAEV

Canada - Ontario - OHSA - STEVs
150 ppm STEV; 564 mg/m3 STEV

Canada - Quebec - Time-Weighted Average Exposure Values
100 ppm TWAEV; 377 mg/m3 TWAEV

Canada - Quebec - Short-term Exposure Values
150 ppm STEV; 565 mg/m3 STEV

United Kingdom - Occupational Exposure Standards - TWAs
50 ppm TWA; 188 mg/m3 TWA

United Kingdom - Occupational Exposure Standards - STELs
150 ppm STEL; 560 mg/m3 STEL

United Kingdom - Occupational Exposure Standards - Notes
can be absorbed through skin

German (DFG) - MAK Values
50 ppm MAK; 190 mg/m3 MAK

German (DFG) - Peak Limitations
5 x normal MAK (30 min. average value); don't exceed 2 times during shift

German (DFG) - Pregnancy
no risk to embryo/fetus if exposure limits adhered to

Israel - Time Weighted Averages
100 ppm TWA; 377 mg/m3 TWA

Israel - Short Term Exposure Limits
150 ppm STEL; 565 mg/m3 STEL

Israel - Action Levels
50 ppm AL; 188.5 mg/m3 AL

Mexico - Instruction No. 10 - TWAs
100 ppm TWA; 375 mg/m3 TWA

Mexico - Instruction No. 10 - STELs
150 ppm STEL; 560 mg/m3 STEL

Mexico - Instruction No. 10 - Skin designation
skin - potential for cutaneous absorption

Mexico - Wastewater - Organic Toxic Pollutants and Heavy Metals
Listed under [Aromatic Hydrocarbons]

Mexico - Drinking Water - Ecological Criteria
14.3 mg/l

STATE LISTS

California - Air Bill 2588 Appendix A-I
[present]

California - Prop. 65 - Developmental Toxicity
developmental toxicity - initial date 1/1/91

California - Exposure Limits - PELs
100 ppm PEL; 375 mg/m3 PEL

California - Exposure Limits - STELs
150 ppm STEL; 560 mg/m3 STEL

California - Exposure Limits - Ceilings
C 500 ppm

California - Exposure Limits - Skin Notation
material may be absorbed through the skin, eyes or mucous membrane

California - Directors List of Hazardous Substances (8 CCR 339)
[present]

Florida Hazardous Substance List
[present]

Massachusetts Right To Know List
[present]

Minnesota Hazardous Substance List
[present]

NJ Right to Know List (Total)
sn 1866

NJ Special Hazardous Substances
(flammable - third degree)

Pennsylvania Right to Know List
environmental hazard

PROPOSED REGULATIONS

ACGIH 1995 - Proposed Biological Exposure Indices
BEIs withdrawn based on reduction of Chemical Substances TLV from 100 ppm to 50 ppm. BEI revision under review.

TOLUENE, 2,5-DINITRO-   **619-15-8**
SEE ALSO:
K114-HAZARDOUS WASTES
K113-HAZARDOUS WASTES
K112-HAZARDOUS WASTES
K027-HAZARDOUS WASTES
TOLUENE 2,4-DIAMINE
K115-HAZARDOUS WASTES

HEALTH AND SAFETY LISTS
AIHA - WEEL - Time Weighted Averages
0.02 ppm TWA

AIHA - WEEL - Skin Absorption Designations
skin absorber

U.S. DOT - Substances From 49 CFR 172.101
regulated by DOT (UN1709)

U.S. DOT - Hazard Classes
DOT hazard class = 6.1

U.S. DOT - Appendix A Table 1 - Hazardous Substances
final RQ = 10 pounds (4.54 kg) (Listed under 'Toluenediamine')

IARC - Group 2B (sufficient animal data)
[present] (Overall evaluation based only on evidence of carcinogenicity in monograph (16, 1978) or in Supplement 4)

NIOSH - Selected LD50s and LC50s
Oral, rat: LD50 = 260 mg/kg

NTP Chemical Status Reports - Testing Status and NTIS Number
Technical reports printed (PB293593/AS)

NTP Chemical Status Reports - Evidence of Carcinogenicity
male rat-positive; female rat-positive; male mice-negative; female mice-positive

NTP Seventh Report - Suspect Carcinogens
suspect carcinogen

OSHA - Possible Select Carcinogens
[present]

ENVIRONMENTAL LISTS

CERCLA/SARA - Section 313 - Emission Reporting
form R reporting required for 0.1% de minimus concentration

CERCLA/SARA - Hazardous Substances and their Reportable Quantities
final RQ = 10 pounds (4.54 kg) (Listed under 'Toluenediamine')

Clean Air Act (1990) - List of Hazardous Air Contaminants
[present]

CAA - HON Rule - SOCMI Chemicals
compliance by Jan. 23, 1995

CAA - HON Rule - Organic HAPs
[present]

EPA - Carcinogen Hazard Ranking for RQ Adjustment
Hazard ranking = Medium

EPA - Master Testing List
[present]

RCRA - Hazardous Constituents-Appendix VIII
hazardous constituent - no waste number

RCRA - Basis for Listing - Appendix VII
Included in waste streams: K027, K112, K113, K114, K115

TSCA - Code of Federal Regulations Citations
40 CFR 712.30(f); 40 CFR 716.120(c)

TSCA - PAIR - Reporting List
Reporting Date: August 17, 1983

INTERNATIONAL LISTS

Canada - WHMIS: Ingredient Disclosure
0.1% item 1579 (1624)

Canada - NPRI (National Pollutant Release Inventory)
[present]

German (DFG) - Carcinogens
animal evidence of carcinogenicity

STATE LISTS

California - Air Bill 2588 Appendix A-I
known or potential carcinogen

California - Prop. 65 - Cancer list
carcinogen - initial date 1/1/88

California - Prop. 65 - No Significant Risk Levels
no significant risk level = 0.2 ug/day

California - Directors List of Hazardous Substances (8 CCR 339)
[present]

Florida Hazardous Substance List
[present]

Massachusetts Right To Know List
carcinogen; extraordinarily hazardous

Minnesota Hazardous Substance List
carcinogen

NJ Right to Know List (Total)
sn 0613

NJ Special Hazardous Substances
(carcinogen)

Pennsylvania Right to Know List
environmental hazard; special hazardous substance

Pennsylvania RTK - Special Hazardous Substances
[present]

PROPOSED REGULATIONS

TSCA - Proposed Substances for Developmental/Reproductive Testing
proposed testing for: Developmental Toxicity - oral; Reproductive Toxicity - oral

O-TOLUENESULFONAMIDE   88-19-7
SEE ALSO:
TOLUENE-3,5-DIAMINE

ENVIRONMENTAL LISTS

TSCA - Code of Federal Regulations Citations
40 CFR 712.30(f); 40 CFR 716.120(a)

TSCA - PAIR - Reporting List
Reporting Date: August 17, 1983

INTERNATIONAL LISTS

Canada - WHMIS: Ingredient Disclosure
0.1% item 1581 (1626)

TOLUENE SULFONAMIDE BISPHENOL A EPOXY   RR-00969-8
ADDUCT

HEALTH AND SAFETY LISTS

NTP Chemical Status Reports - Testing Status and NTIS Number
Technical reports printed (PB80217912)

NTP Chemical Status Reports - Evidence of Carcinogenicity
male rat-negative; female rat-negative; male mice-negative; female mice-negative

P-TOLUENESULFONIC ACID   104-15-4
SEE ALSO:
2,5-TOLUENEDIAMINE SULFATE

HEALTH AND SAFETY LISTS

NTP Chemical Status Reports - Testing Status and NTIS Number
Technical reports printed (PB287127/AS)

NTP Chemical Status Reports - Evidence of Carcinogenicity
male rat-negative; female rat-negative; male mice-negative; female mice-negative

ENVIRONMENTAL LISTS

TSCA - Code of Federal Regulations Citations
40 CFR 712.30(d); 40 CFR 716.120(c)

TSCA - PAIR - Reporting List
Reporting Date: November 19, 1982

TOLUENE SULFONIC ACID   25231-46-3

HEALTH AND SAFETY LISTS

U.S. DOT - Appendix A Table 1 - Hazardous Substances
final RQ = 100 pounds (45.4 kg) (Listed under 'Toluene diisocyanate')

NIOSH - Selected LD50s and LC50s
Oral, bird: LD50 = 100 mg/kg

ENVIRONMENTAL LISTS

CERCLA/SARA - Section 302 Extremely Hazardous Substances and TPQs
TPQ = 100 pounds

CERCLA/SARA - Section 313 - Emission Reporting
form R reporting required for 0.1% de minimus concentration

CERCLA/SARA - Hazardous Substances and their Reportable Quantities
final RQ = 100 pounds (45.4 kg) (Listed under 'Toluene diisocyanate')

CAA -Toxic Substances for Accidental Release Prevention
threshold quantity = 10,000 lbs (The mixture exemption in Section 68.115(b)(1) does not apply to the substance)

TSCA - Code of Federal Regulations Citations
40 CFR 716.120; 40 CFR 704.225(a)

TSCA - CAIR - Reporting List
reporting required by: manufacturer; distributer; importer; processor

TSCA - Health and Safety Reporting List
Effective Date: June 1, 1987

INTERNATIONAL LISTS

Canada - WHMIS: Ingredient Disclosure
0.1% item 1585 (1629)

Canada - NPRI (National Pollutant Release Inventory)
[present]

German (DFG) - MAK Values
0.01 ppm MAK; 0.07 mg.m3 MAK

German (DFG) - Peak Limitations
2 x normal MAK (5 min momentary value); don't exceed 8 times during shift

German (DFG) - Skin/Sensitizers
danger of sensitization (skin or respiratory)

STATE LISTS

California - Air Bill 2588 Appendix A-I
known or potential carcinogen

California - Directors List of Hazardous Substances (8 CCR 339)
[present] (Listed under 'Toluene diisocyanates')

Florida Hazardous Substance List
effective March 13, 1992

Massachusetts Right To Know List
extraordinarily hazardous

NJ Right to Know List (Total)
sn 1868

Pennsylvania Right to Know List
environmental hazard

P-TOLUENE SULFONYL CHLORIDE   **98-59-9**
SEE ALSO:
K027-HAZARDOUS WASTES
TOLUENE DIISOCYANATE

HEALTH AND SAFETY LISTS

ACGIH 1995 - Time Weighted Averages
0.005 ppm TWA; 0.036 mg/m3 TWA

ACGIH 1995 - Short Term Exposure Limits
0.02 ppm STEL; 0.14 mg/m3 STEL

U.S. DOT - Appendix A Table 1 - Hazardous Substances
final RQ = 100 pounds (45.4 kg) (Listed under 'Toluene diisocyanate')

NFPA - Flash Points
flash point = 260 degrees F (127 degrees C)

NFPA - Hazard Identification Ratings
health-3; flammability-1; reactivity-3 (reacts exothermically with water)

NIOSH - Selected LD50s and LC50s
Inhalation, rat: LC50 = 14 ppm 4 hr Oral, rat: LD50 = 5800 mg/kg

NIOSH 1990 - Pocket Guide - RELs
0.005 ppm TWA; 0.04 mg/m3 TWA; 0.02 ppm STEL; 0.15 mg/m3 STEL

NIOSH 1990 - Pocket Guide - IDLHs
10 ppm IDLH (not considering carcinogenic effects)

NIOSH 1990 - Pocket Guide - Carcinogens
occupational carcinogen

NIOSH 1990 - Pocket Guide - Target organs
skin, respiratory system

NTP Seventh Report - Suspect Carcinogens
suspect carcinogen

OSHA - Vacated PELs - Time Weighted Averages
0.005 ppm TWA; 0.04 mg/m3 TWA

OSHA - Vacated PELs - Short Term Exposure Limits
0.02 ppm STEL; 0.15 mg/m3 STEL

OSHA - Final PELs - Ceiling Limits
C 0.02 ppm; C 0.14 mg/m3

OSHA - Possible Select Carcinogens
[present]

ENVIRONMENTAL LISTS

CERCLA/SARA - Section 302 Extremely Hazardous Substances and TPQs
TPQ = 500 pounds

CERCLA/SARA - Section 313 - Emission Reporting
form R reporting required for 0.1% de minimus concentration

CERCLA/SARA - Hazardous Substances and their Reportable Quantities
final RQ = 100 pounds (45.4 kg) (Listed under 'Toluene diisocyanate')

Clean Air Act (1990) - List of Hazardous Air Contaminants
[present]

CAA -Toxic Substances for Accidental Release Prevention
threshold quantity = 10,000 lbs (The mixture exemption in Section 68.115(b)(1) does not apply to the substance)

CAA - HON Rule - SOCMI Chemicals
compliance by Jan. 23, 1995

CAA - HON Rule - Organic HAPs
[present]

TSCA - Code of Federal Regulations Citations
40 CFR 704.225(a); 40 CFR 716.120(a)

TSCA - CAIR - Reporting List
reporting required by: manufacturer; distributer; importer; processor

TSCA - Health and Safety Reporting List
Effective Date: June 1, 1987

INTERNATIONAL LISTS

Canada - WHMIS: Ingredient Disclosure
0.1% item 1584 (1628)

Canada - NPRI (National Pollutant Release Inventory)
[present]

Canada - Alberta - 8 Hour Occupational Exposure Limit
0.005 ppm TWA; 0.04 mg/m3 TWA

Canada - Alberta - Ceiling Occupational Exposure Limit
C 0.02 ppm; C 0.14 mg/m3

Canada - Alberta - Designated Substances
designated substance - requires code of practice

German (DFG) - MAK Values
0.01 ppm MAK; 0.07 mg/m3 MAK

German (DFG) - Peak Limitations
2 x normal MAK (5 min momentary value); don't exceed 8 times during shift

German (DFG) - Skin/Sensitizers
danger of sensitization (skin or respiratory)

Israel - Time Weighted Averages
0.005 ppm TWA; 0.036 mg/m3 TWA

Israel - Short Term Exposure Limits
0.02 ppm STEL; 0.14 mg/m3 STEL

Israel - Action Levels
0.0025 ppm AL; 0.018 mg/m3 AL

Mexico - Instruction No. 10 - TWAs
0.02 ppm TWA; 0.14 mg/m3 TWA

STATE LISTS

California - Air Bill 2588 Appendix A-I
known or potential carcinogen

California - Exposure Limits - PELs
0.005 ppm PEL; 0.04 mg/m3 PEL

California - Exposure Limits - STELs
0.02 ppm STEL; 0.15 mg/m3 STEL

California - Exposure Limits - Ceilings
C 0.02 ppm

California - Directors List of Hazardous Substances (8 CCR 339)
[present]

Florida Hazardous Substance List
[present]

Massachusetts Right To Know List
carcinogen; extraordinarily hazardous

Minnesota Hazardous Substance List
[present]

NJ Right to Know List (Total)
sn 1869

Pennsylvania Right to Know List
environmental hazard; special hazardous substance

Pennsylvania RTK - Special Hazardous Substances
[present]

TOLUHYDROQUINONE   95-71-6
SEE ALSO:
TOLUENE 2,4-DIISOCYANATE
K027-HAZARDOUS WASTES

HEALTH AND SAFETY LISTS

U.S. DOT - Substances From 49 CFR 172.101
regulated by DOT (UN2078)

U.S. DOT - Hazard Classes
DOT hazard class = 6.1

U.S. DOT - Appendix A Table 1 - Hazardous Substances
final RQ = 100 pounds (45.4 kg) (Listed under 'Toluene diisocyanate')

IARC - Group 2B (sufficient animal data)
[present] (Overall evaluation based only on evidence of carcinogenicity in monograph [39, 1986] or in supplement 4)

NIOSH - Selected LD50s and LC50s
Inhalation, mouse: LC50 = 9700 ppb 4 hr Oral, rat: LD50 = 4130 mg/kg

NIOSH - Health Standards - Exposure Limits
5 ppb TWA; 35 ug/m3 TWA; C (10 min) 20 ppb; C (10 min) 140 ug/m3 (Listed under 'Diisocyanates')

NIOSH - Health Standards - Health Effects and Precautions
Respiratory effects and sensitization, pulmonary irritation (Periodic chest X-ray and pulmonary function testing required) (Listed under 'Diisocyanates')

NTP Chemical Status Reports - Testing Status and NTIS Number
Technical reports printed (PB87115176)

NTP Chemical Status Reports - Evidence of Carcinogenicity
male rat-positive; female rat-positive; male mice-negative; female mice-positive

OSHA - Possible Select Carcinogens
[present]

ENVIRONMENTAL LISTS

CERCLA/SARA - Section 313 - Emission Reporting
form R reporting required for 0.1% de minimus concentration

CERCLA/SARA - Hazardous Substances and their Reportable Quantities
final RQ = 100 pounds (45.4 kg) (Listed under 'Toluene diisocyanate')

CAA -Toxic Substances for Accidental Release Prevention
threshold quantity = 10,000 lbs (The mixture exemption in Section 68.115(b)(1) doe not apply to this substance)

CAA - HON Rule - SOCMI Chemicals
compliance by Jan. 23, 1995

RCRA - U Series Wastes
waste number U223 (Reactive waste; Toxic waste)

RCRA - Hazardous Constituents-Appendix VIII
waste number U223

RCRA - Basis for Listing - Appendix VII
Included in waste stream: K027

RCRA - Substances Banned From Land Disposal
[present]

TSCA - Code of Federal Regulations Citations
40 CFR 716.120(a); 40 CFR 704.225(a)

TSCA - CAIR - Reporting List
reporting required by: manufacturer; importer; distributor; processor

TSCA - Health and Safety Reporting List
Effective Date: June 1, 1987

INTERNATIONAL LISTS

Canada - WHMIS: Ingredient Disclosure
0.1% item 1583 (1627)

Canada - NPRI (National Pollutant Release Inventory)
[present]

Canada - British Columbia - Ceiling Exposure Limits
C 0.02 ppm; C 0.14 mg/m3

Canada - Ontario - OHSA - TWAEVs
0.005 ppm TWAEV; 0.2 micromoles/m3 TWAEV (designated substance regulation)

Canada - Ontario - OHSA - CEVs
0.02 ppm CEV; 0.8 micromoles/m3 CEV

Canada - Ontario - OHSA - Designated Substances
0.005 ppm TWAEV; 0.2 micromole/m3 TWAEV; See Ontario Reg. 842 for full information.

Canada - Quebec - Time-Weighted Average Exposure Values
0.005 ppm TWAEV; 0.036 mg/m3 TWAEV

Canada - Quebec - Short-term Exposure Values
0.02 ppm STEV; 0.14 mg/m3 STEV

STATE LISTS

California - Air Bill 2588 Appendix A-I
known or potential carcinogen: 6/91

California - Prop. 65 - Cancer list
carcinogen - initial date 10/1/89

California - Prop. 65 - No Significant Risk Levels
no significant risk level = 20 ug/day

California - Directors List of Hazardous Substances (8 CCR 339)
[present]

Massachusetts Right To Know List
[present]

NJ Right to Know List (Total)
sn 3132

Pennsylvania Right to Know List
environmental hazard

P-TOLUIC ACID   99-94-5

HEALTH AND SAFETY LISTS

NIOSH - Selected LD50s and LC50s
Oral, rat: LD50 = 911 mg/kg

STATE LISTS

California - Directors List of Hazardous Substances (8 CCR 339)
[present] (Listed under 'Dinitrotoluene, all isomers')

O-TOLUIDINE   95-53-4

HEALTH AND SAFETY LISTS

NIOSH - Selected LD50s and LC50s
Oral, rat: LD50 = 517 mg/kg

STATE LISTS

California - Directors List of Hazardous Substances (8 CCR 339)
[present] (Listed under 'Dinitrotoluene, all isomers')

P-TOLUIDINE   106-49-0

HEALTH AND SAFETY LISTS

NIOSH - Selected LD50s and LC50s
Oral, rat: LD50 = 4870 mg/kg

INTERNATIONAL LISTS

Canada - WHMIS: Ingredient Disclosure
1% item 1586 (1630)

M-TOLUIDINE   108-44-1

ENVIRONMENTAL LISTS

TSCA - Chemicals with Significant New Use Rules
PMN number: P-90-113

O-TOLUIDINE HYDROCHLORIDE   636-21-5

HEALTH AND SAFETY LISTS

NFPA - Flash Points
flash point = 363 degrees F (184 degrees C)

NFPA - Hazard Identification Ratings
health-3; flammability-1; reactivity-1

NIOSH - Selected LD50s and LC50s
Oral, rat: LD50 = 2480 mg/kg

ENVIRONMENTAL LISTS

CAA - HON Rule - SOCMI Chemicals
compliance by April 24, 1995

List of Pesticide Product Inert Ingredients
[present]

STATE LISTS

Florida Hazardous Substance List
[present]

Massachusetts Right To Know List
[present]

Pennsylvania Right to Know List
[present]

TOLUIDINES   26915-12-8

INTERNATIONAL LISTS

Canada - WHMIS: Ingredient Disclosure
1% item 1587 (146)

STATE LISTS

NJ Right to Know List (Total)
sn 1870

NJ Special Hazardous Substances
(corrosive)

TOLUTRIAZOLE   136-85-6

HEALTH AND SAFETY LISTS

AIHA - WEEL - Ceilings or Short Term Time Weighted Averages
5 mg/m3 STEL

AIHA - WEEL - Skin Absorption Designations
skin absorber

ENVIRONMENTAL LISTS

CAA - HON Rule - SOCMI Chemicals
compliance by April 24, 1995

INTERNATIONAL LISTS

United Kingdom - Occupational Exposure Standards - STELs
5 mg/m3 STEL

STATE LISTS

Minnesota Hazardous Substance List
[present]

M-TOLYDIETHANOLAMINE   91-99-6

HEALTH AND SAFETY LISTS

NFPA - Flash Points
flash point = 342 degrees F (172 degrees C)

NFPA - Hazard Identification Ratings
flammability-1; reactivity-0

P-TOLYL ACETATE   140-39-6

HEALTH AND SAFETY LISTS

NIOSH - Selected LD50s and LC50s
Oral, rat: LD50 = 400 mg/kg

TOLYL ALDEHYDE   1334-78-7

SEE ALSO:
K112-HAZARDOUS WASTES
K113-HAZARDOUS WASTES
K114-HAZARDOUS WASTES

HEALTH AND SAFETY LISTS

ACGIH 1995 - Time Weighted Averages
2 ppm TWA; 8.8 mg/m3 TWA

ACGIH 1995 - Skin Designations
skin - potential for cutaneous absorption

ACGIH 1995 - Carcinogens
A2-suspected human carcinogen

AIHA - Odor Threshold Values
no geometric mean air odor threshold

U.S. DOT - Appendix A Table 1 - Hazardous Substances
final RQ = 100 pounds (45.4 kg)

IARC - Group 2B (sufficient animal data)
[present]

NFPA - Flash Points
flash point = 185 degrees F (85 degrees C)

NFPA - Hazard Identification Ratings
health-3; flammability-2; reactivity-0

NIOSH - Selected LD50s and LC50s
Oral, rat: LD50 = 670 mg/kg Skin, rabbit: LD50 = 3250 mg/kg

NIOSH 1990 - Pocket Guide - RELs
2 ppm TWA; 9 mg/m3 TWA

NIOSH 1990 - Pocket Guide - IDLHs
100 ppm IDLH (not considering carcinogenic effects)

NIOSH 1990 - Pocket Guide - Carcinogens
occupational carcinogen

NIOSH 1990 - Pocket Guide - Target organs
blood, kidneys, liver, CVS, skin, eyes

NIOSH 1990 - Pocket Guide - Skin list
Potential for dermal absorption

NTP Seventh Report - Suspect Carcinogens
suspect carcinogen

OSHA - Vacated PELs - Time Weighted Averages
5 ppm TWA; 22 mg/m3 TWA

OSHA - Vacated PELs - Skin Designation
Prevent or reduce skin absorption

OSHA - Final PELs - Time Weighted Averages
5 ppm TWA; 22 mg/m3 TWA

OSHA - Final PELs - Skin Notations
prevent or reduce skin absorption
OSHA - Possible Select Carcinogens
[present]

ENVIRONMENTAL LISTS
CERCLA/SARA - Section 313 - Emission Reporting
form R reporting required for 0.1% de minimus concentration
CERCLA/SARA - Hazardous Substances and their Reportable Quantities
final RQ = 100 pounds (45.4 kg)
Clean Air Act (1990) - List of Hazardous Air Contaminants
[present]
CAA - HON Rule - SOCMI Chemicals
compliance by Jan. 23, 1995
CAA - HON Rule - Organic HAPs
[present]
EPA - Carcinogen Hazard Ranking for RQ Adjustment
Hazard ranking = Low
RCRA - U Series Wastes
waste number U328
RCRA - Hazardous Constituents-Appendix VIII
waste number U328
RCRA - Basis for Listing - Appendix VII
Included in waste streams: K112, K113, K114
RCRA - Substances Banned From Land Disposal
[present]
RCRA - TSD Facilities Ground Water Monitoring
TM 8270 = 10 ug/L PQL
TSCA - Code of Federal Regulations Citations
40 CFR 716.120(a)
TSCA - Health and Safety Reporting List
Effective Date: March 7, 1986

INTERNATIONAL LISTS
Australian Exposure Standards - Time Weighted Averages
2 ppm TWA; 8.8 mg/m3 TWA
Australian Exposure Standards - Skin Effects
skin absorption
Australian Exposure Standards - Carcinogens
probable carcinogen
Canada - WHMIS: Ingredient Disclosure
0.1% item 1589 (1633)
Canada - British Columbia - 8 Hour Exposure Limits
5 ppm TWA; 22 mg/m3 TWA
Canada - British Columbia - 15 Minute Exposure Limits
10 ppm STEL; 44 mg/m3 STEL
Canada - Quebec - Time-Weighted Average Exposure Values
2 ppm TWAEV; 8.8 mg/m3 TWAEV
Canada - Quebec - Skin Designations
absorbed through the skin
Canada - Quebec - Carcinogens
C2 carcinogen: effect suspected in humans
United Kingdom - Occupational Exposure Standards - TWAs
2 ppm TWA; 9 mg/m3 TWA
United Kingdom - Occupational Exposure Standards - STELs
5 ppm STEL; 22 mg/m3 STEL
German (DFG) - Skin/Sensitizers
danger of cutaneous absorption
German (DFG) - Carcinogens
animal evidence of carcinogenicity
Israel - Time Weighted Averages
2 ppm TWA; 8.8 mg/m3 TWA
Israel - Action Levels
1 ppm AL; 4.4 mg/m3 AL

Mexico - Instruction No. 10 - Skin designation
skin - potential for cutaneous absorption
Mexico - Instruction No. 10 - Carcinogens
potential carcinogen in humans - limited epidemiological evidence

STATE LISTS
California - Air Bill 2588 Appendix A-I
known or potential carcinogen
California - Prop. 65 - Cancer list
carcinogen - initial date 1/1/88
California - Prop. 65 - No Significant Risk Levels
no significant risk level = 4 ug/day
California - Exposure Limits - PELs
2 ppm PEL; 9 mg/m3 PEL
California - Exposure Limits - Skin Notation
material may be absorbed through the skin, eyes or mucous membrane
California - Directors List of Hazardous Substances (8 CCR 339)
[present]
Florida Hazardous Substance List
[present]
Massachusetts Right To Know List
carcinogen; extraordinarily hazardous
Minnesota Hazardous Substance List
carcinogen; skin
NJ Right to Know List (Total)
sn 1442
NJ Special Hazardous Substances
(carcinogen; mutagen)
Pennsylvania Right to Know List
environmental hazard; special hazardous substance
Pennsylvania RTK - Special Hazardous Substances
[present]

O-TOLYL BIGUANIDE   93-69-6
SEE ALSO:
K112-HAZARDOUS WASTES
K114-HAZARDOUS WASTES
K113-HAZARDOUS WASTES

HEALTH AND SAFETY LISTS
ACGIH 1995 - Time Weighted Averages
2 ppm TWA; 8.8 mg/m3 TWA
ACGIH 1995 - Skin Designations
skin - potential for cutaneous absorption
ACGIH 1995 - Carcinogens
A2-suspected human carcinogen
AIHA - Odor Threshold Values
no geometric mean air odor threshold
U.S. DOT - Appendix A Table 1 - Hazardous Substances
final RQ = 100 pounds (45.4 kg)
NFPA - Flash Points
flash point = 188 degrees F (87 degrees C)
NFPA - Hazard Identification Ratings
health-3; flammability-2; reactivity-0
NIOSH - Selected LD50s and LC50s
Oral, rat: LD50 = 656 mg/kg
OSHA - Vacated PELs - Time Weighted Averages
2 ppm TWA; 9 mg/m3 TWA
OSHA - Vacated PELs - Skin Designation
Prevent or reduce skin absorption

ENVIRONMENTAL LISTS
CERCLA/SARA - Hazardous Substances and their Reportable Quantities
final RQ = 100 pounds (45.4 kg)
EPA - Carcinogen Hazard Ranking for RQ Adjustment
Hazard ranking = Low

RCRA - U Series Wastes
waste number U353

RCRA - Hazardous Constituents-Appendix VIII
waste number U353

RCRA - Basis for Listing - Appendix VII
Included in waste streams: K112, K113, K114

RCRA - Substances Banned From Land Disposal
[present]

TSCA - Code of Federal Regulations Citations
40 CFR 716.120(a)

TSCA - Health and Safety Reporting List
Effective Date: March 7, 1986

INTERNATIONAL LISTS

Australian Exposure Standards - Time Weighted Averages
2 ppm TWA; 8.8 mg/m3 TWA

Australian Exposure Standards - Skin Effects
skin absorption

Australian Exposure Standards - Carcinogens
probable carcinogen

Canada - WHMIS: Ingredient Disclosure
0.1% item 1590 (1634)

Canada - Quebec - Time-Weighted Average Exposure Values
2 ppm TWAEV; 8.8 mg/m3 TWAEV

Canada - Quebec - Skin Designations
absorbed through the skin

Canada - Quebec - Carcinogens
C2 carcinogen: effect suspected in humans

German (DFG) - Skin/Sensitizers
danger of cutaneous absorption

German (DFG) - Carcinogens
suspected carcinogen

Israel - Time Weighted Averages
2 ppm TWA; 8.8 mg/m3 TWA

Israel - Action Levels
1 ppm AL; 4.4 mg/m3 AL

STATE LISTS

California - Air Bill 2588 Appendix A-II
known or potential carcinogen: 9/90

California - Prop. 65 - Cancer list
carcinogen - initial date 1/1/90

California - Exposure Limits - PELs
2 ppm PEL; 9 mg/m3 PEL

California - Exposure Limits - Skin Notation
material may be absorbed through the skin, eyes or mucous membrane

California - Directors List of Hazardous Substances (8 CCR 339)
[present]

Florida Hazardous Substance List
[present]

Massachusetts Right To Know List
[present]

Minnesota Hazardous Substance List
carcinogen; skin

NJ Right to Know List (Total)
sn 1622

Pennsylvania Right to Know List
[present]

PROPOSED REGULATIONS

TSCA - ITC 32nd Report Priority Testing List
designated for dermal absorption testing

O-TOLYL P-TOLUENE SULFONATE
RR-00875-3

HEALTH AND SAFETY LISTS

ACGIH 1995 - Time Weighted Averages
2 ppm TWA; 8.8 mg/m3 TWA

ACGIH 1995 - Skin Designations
skin - potential for cutaneous absorption

AIHA - Odor Threshold Values
no geometric mean air odor threshold

NIOSH - Selected LD50s and LC50s
Oral, rat: LD50 = 450 mg/kg

OSHA - Vacated PELs - Time Weighted Averages
2 ppm TWA; 9 mg/m3 TWA

OSHA - Vacated PELs - Skin Designation
Prevent or reduce skin absorption

ENVIRONMENTAL LISTS

EPA - Master Testing List
[present]

TSCA - Health and Safety Reporting List
Effective Date: March 11, 1994; Sunset Date: March 11, 2004

INTERNATIONAL LISTS

Australian Exposure Standards - Time Weighted Averages
2 ppm TWA; 8.8 mg/m3 TWA

Australian Exposure Standards - Skin Effects
skin absorption

Canada - WHMIS: Ingredient Disclosure
1% item 1588 (1632)

Canada - Quebec - Time-Weighted Average Exposure Values
2 ppm TWAEV; 8.8 mg/m3 TWAEV

Canada - Quebec - Skin Designations
absorbed through the skin

Israel - Time Weighted Averages
2 ppm TWA; 8.8 mg/m3 TWA

Israel - Action Levels
1 ppm AL; 4.4 mg/m3 AL

STATE LISTS

California - Exposure Limits - PELs
2 ppm PEL; 9 mg/m3 PEL

California - Exposure Limits - Skin Notation
material may be absorbed through the skin, eyes or mucous membrane

California - Directors List of Hazardous Substances (8 CCR 339)
[present]

Massachusetts Right To Know List
[present]

Minnesota Hazardous Substance List
skin

NJ Right to Know List (Total)
sn 1318

Pennsylvania Right to Know List
[present]

PROPOSED REGULATIONS

TSCA - ITC 32nd Report Priority Testing List
designated for dermal absorption testing

TOLYL TRIAZOLE
29385-43-1

HEALTH AND SAFETY LISTS

U.S. DOT - Appendix A Table 1 - Hazardous Substances
final RQ = 100 pounds (45.4 kg)

NIOSH - Selected LD50s and LC50s
Oral, rat: LD50 = 2951 mg/kg

NTP Chemical Status Reports - Testing Status and NTIS Number
Technical reports printed (PB290908/AS); Prechronic studies completed: in review for further evaluation

NTP Chemical Status Reports - Evidence of Carcinogenicity
male rat-positive; female rat-positive; male mice-positive; female mice-positive

NTP Seventh Report - Suspect Carcinogens
suspect carcinogen

OSHA - Possible Select Carcinogens
[present]

ENVIRONMENTAL LISTS

CERCLA/SARA - Section 313 - Emission Reporting
form R reporting required for 0.1% de minimus concentration

CERCLA/SARA - Hazardous Substances and their Reportable Quantities
final RQ = 100 pounds (45.4 kg)

EPA - Carcinogen Hazard Ranking for RQ Adjustment
Hazard ranking = Low

RCRA - U Series Wastes
waste number U222

RCRA - Hazardous Constituents-Appendix VIII
waste number U222

RCRA - Substances Banned From Land Disposal
[present]

STATE LISTS

California - Air Bill 2588 Appendix A-II
known or potential carcinogen

California - Prop. 65 - Cancer list
carcinogen - initial date 1/1/88

California - Prop. 65 - No Significant Risk Levels
no significant risk level = 5 ug/day

California - Directors List of Hazardous Substances (8 CCR 339)
[present]

Florida Hazardous Substance List
[present]

Massachusetts Right To Know List
carcinogen; extraordinarily hazardous

Minnesota Hazardous Substance List
carcinogen

NJ Right to Know List (Total)
sn 1443

NJ Special Hazardous Substances
(carcinogen)

Pennsylvania Right to Know List
environmental hazard; special hazardous substance

Pennsylvania RTK - Special Hazardous Substances
[present]

TOTAL TRIHALOMETHANES   RR-00759-0

HEALTH AND SAFETY LISTS

U.S. DOT - Substances From 49 CFR 172.101
regulated by DOT (UN1708)

U.S. DOT - Hazard Classes
DOT hazard class = 6.1

INTERNATIONAL LISTS

Canada - Alberta - 8 Hour Occupational Exposure Limit
2 ppm TWA; 9 mg/m3 TWA

Canada - Alberta - 15 Minute Occupational Exposure Limit
4 ppm STEL; 18 mg/m3 STEL

Canada - Alberta - Skin Designation
can be absorbed through the intact skin

Canada - Alberta - Designated Substances
designated substance - requires code of practice

Canada - Ontario - OHSA - TWAEVs
2 ppm TWAEV; 9 mg/m3 TWAEV

Canada - Ontario - OHSA - Skin Notations
absorption through skin, eyes, or mucous membranes

PROPOSED REGULATIONS

Canada - Ontario - Proposed Occupational TWAEVs
1 ppm TWAEV; 4.5 mg/m3 TWAEV

TOXAPHENE   8001-35-2

HEALTH AND SAFETY LISTS

NIOSH - Selected LD50s and LC50s
Oral, rat: LD50 = 1600 mg/kg

TOXINS DERIVED FROM FUSARIUM GRAMIN-   RR-01703-8
EARUM, F. CULMORUM AND F. CROOKWELLENSE

HEALTH AND SAFETY LISTS

NFPA - Flash Points
flash point = 400 degrees F (204 degrees C)

NFPA - Hazard Identification Ratings
health-2; flammability-1; reactivity-0

NIOSH - Selected LD50s and LC50s
Oral, rat: LD50 = 3100 mg/kg

STATE LISTS

Florida Hazardous Substance List
[present]

Massachusetts Right To Know List
[present]

Pennsylvania Right to Know List
[present]

TOXINS DERIVED FROM FUSARIUM   RR-01704-9
SPOROTRICHOIDES

HEALTH AND SAFETY LISTS

NFPA - Flash Points
flash point = 195 degrees F (91 degrees C)

NFPA - Hazard Identification Ratings
health-1; flammability-2; reactivity-0

NIOSH - Selected LD50s and LC50s
Oral, rat: LD50 = 1900 mg/kg Skin, rabbit: LD50 = 2100 mg/kg

TOXINS DERIVED FROM FUSARIUM   RR-01706-1
MONILIFORME

ENVIRONMENTAL LISTS

TSCA - Code of Federal Regulations Citations
40 CFR 712.30(x); 40 CFR 716.120(d)

TSCA - PAIR - Reporting List
Reporting Date: November 27, 1991

TSCA - Health and Safety Reporting List
Effective Date: September 30, 1991

INTERNATIONAL LISTS

Canada - WHMIS: Ingredient Disclosure
1% item 1591 (191)

2,4,5-TP ACID ESTERS   32534-95-5

ENVIRONMENTAL LISTS

List of Pesticide Product Inert Ingredients
[present]

TRALOMETHRIN   66841-25-6

HEALTH AND SAFETY LISTS

NFPA - Flash Points
flash point = 363 degrees F (184 degrees C)

NFPA - Hazard Identification Ratings
health-1; flammability-1; reactivity-0

TRANSFORMER OIL   RR-00778-3

HEALTH AND SAFETY LISTS

NIOSH - Selected LD50s and LC50s
Oral, rat: LD50 = 675 mg/kg

ENVIRONMENTAL LISTS

List of Pesticide Product Inert Ingredients
[present]

TSCA - Code of Federal Regulations Citations
40 CFR 712.30(w);   40 CFR 716.120(a)

TSCA - PAIR - Reporting List
Reporting Date: June 13, 1989

TSCA - Health and Safety Reporting List
Effective Date: April 13, 1989

TREMOLITE   14567-73-8

ENVIRONMENTAL LISTS

Safe Drinking Water Act - MCLs
MCL = 0.10 mg/L (This is the sum of the concentrations of bromodichloromethane, dibromochloromethane, tribromomethane (bromoform), and trichloromethane (chloroform))

TREOSULPHAN   299-75-2

SEE ALSO:
K041-HAZARDOUS WASTES
F039-HAZARDOUS WASTES
K148-HAZARDOUS WASTES
K098-HAZARDOUS WASTES

HEALTH AND SAFETY LISTS

ACGIH 1995 - Time Weighted Averages
0.5 mg/m3 TWA

ACGIH 1995 - Short Term Exposure Limits
1 mg/m3 STEL

ACGIH 1995 - Skin Designations
skin - potential for cutaneous absorption

U.S. DOT - Appendix B - Marine Pollutants
DOT regulated severe marine pollutant

U.S. DOT - Appendix A Table 1 - Hazardous Substances
final RQ = 1 pound (0.454 kg)

IARC - Group 2B (sufficient animal data)
[present] (Overall evaluation based only on evidence of carcinogenicity in monograph (20, 1979) or in Supplement 4)

NIOSH - Selected LD50s and LC50s
Oral, rat: LD50 = 55 mg/kg Skin, rat: LD50 = 600 mg/kg

NIOSH 1990 - Pocket Guide - IDLHs
200 mg/m3 IDLH (not considering carcinogenic effects)

NIOSH 1990 - Pocket Guide - Carcinogens
occupational carcinogen

NIOSH 1990 - Pocket Guide - Target organs
CNS, skin

NIOSH 1990 - Pocket Guide - Skin list
Potential for dermal absorption

NTP Chemical Status Reports - Testing Status and NTIS Number
Technical reports printed (PB292290/AS)

NTP Chemical Status Reports - Evidence of Carcinogenicity
male rat-equivocal; female rat-equivocal; male mice-positive; female mice-positive

NTP Seventh Report - Suspect Carcinogens
suspect carcinogen

OSHA - Vacated PELs - Time Weighted Averages
0.5 mg/m3 TWA

OSHA - Vacated PELs - Short Term Exposure Limits
1 mg/m3 STEL

OSHA - Vacated PELs - Skin Designation
Prevent or reduce skin absorption

OSHA - Final PELs - Time Weighted Averages
0.5 mg/m3 TWA

OSHA - Final PELs - Skin Notations
prevent or reduce skin absorption

OSHA - Possible Select Carcinogens
[present]

ENVIRONMENTAL LISTS

ATSDR Priority List
Rank (of 275): 032

CERCLA/SARA - Section 302 Extremely Hazardous Substances and TPQs
TPQ = 500/10,000 pounds

CERCLA/SARA - Section 313 - Emission Reporting
form R reporting required for 0.1% de minimus concentration

CERCLA/SARA - Hazardous Substances and their Reportable Quantities
final RQ = 1 pound (0.454 kg)

Clean Air Act (1990) - List of Hazardous Air Contaminants
[present]

Clean Water Act - Hazardous Substances
[present]

Clean Water Act - Priority Pollutants
[present]

Clean Water Act - Toxic Pollutants
[present]

Safe Drinking Water Act - MCLs
MCL = 0.003 mg/L

Safe Drinking Water Act - MCLGs
MCLG = Zero

EPA - Carcinogen Hazard Ranking for RQ Adjustment
Hazard ranking = Medium

RCRA - D Series - Maximum Concentration of Contaminants
waste number D015; regulatory level = 0.5 mg/L

RCRA - D Series - Chronic Toxicity Reference Levels
chronic toxicity reference level = 0.005 mg/L

RCRA - P Series Wastes
waste number P123

RCRA - Hazardous Constituents-Appendix VIII
waste number P123

RCRA - Basis for Listing - Appendix VII
Included in waste streams: F039, K041, K098

RCRA - Substances Banned From Land Disposal
[present]

RCRA - TSD Facilities Ground Water Monitoring
TM 8080 = 2 ug/L PQL; TM 8250 = 10 ug/L PQL

RCRA - Universal Treatment Standards (LDR)
WW: 0.0095 mg/l; NWW: 2.6 mg/kg

INTERNATIONAL LISTS

Australian Exposure Standards - Time Weighted Averages
0.5 mg/m3 TWA

Australian Exposure Standards - Short Term Exposure Limits
1 mg/m3 STEL

Australian Exposure Standards - Skin Effects
skin absorption

Canada - Alberta - 8 Hour Occupational Exposure Limit
0.5 mg/m3 TWA

Canada - Alberta - 15 Minute Occupational Exposure Limit
1 mg/m3 STEL

Canada - Alberta - Skin Designation
can be absorbed through the intact skin

Canada - British Columbia - 8 Hour Exposure Limits
0.5 mg/m3 TWA

Canada - British Columbia - 15 Minute Exposure Limits
1 mg/m3 STEL
Canada - British Columbia - Skin Notations
skin - potential for skin absorption
Canada - Ontario - OHSA - TWAEVs
0.5 mg/m3 TWAEV (listed as an agent of variable composition) As sum of components assayed by chromatographic procedure with reference to the bulk sample.
Canada - Ontario - OHSA - STEVs
1 mg/m3 STEV (listed as 'Agent of variable composition')
Canada - Ontario - OHSA - Skin Notations
absorption through skin, eyes, or mucous membranes (listed under 'Agents of Variable Composition')
Canada - Quebec - Time-Weighted Average Exposure Values
0.5 mg/m3 TWAEV
Canada - Quebec - Short-term Exposure Values
1 mg/m3 STEV
Canada - Quebec - Skin Designations
absorbed through the skin
Canada - Quebec - Carcinogens
C3 carcinogen: effect detected in animals
German (DFG) - MAK Values
total dust: 0.5 mg/m3 MAK
German (DFG) - Peak Limitations
10 x normal MAK (30 min average value); don't exceed during shift
German (DFG) - Skin/Sensitizers
danger of cutaneous absorption
Israel - Time Weighted Averages
0.5 mg/m3 TWA
Israel - Short Term Exposure Limits
1 mg/m3 STEL
Israel - Action Levels
0.25 mg/m3 AL
Mexico - Instruction No. 10 - TWAs
0.5 mg/m3 TWA
Mexico - Instruction No. 10 - Skin designation
skin - potential for cutaneous absorption
Mexico - Wastewater - Organic Toxic Pollutants and Heavy Metals
Listed under [Polychlorinated Byphenyls]
Mexico - Drinking Water - Ecological Criteria
0.000007 mg/l

STATE LISTS

California - Air Bill 2588 Appendix A-I
known or potential carcinogen
California - Prop. 65 - Cancer list
carcinogen - initial date 1/1/88
California - Prop. 65 - No Significant Risk Levels
no significant risk level = 0.6 ug/day
California - Exposure Limits - PELs
0.5 mg/m3 PEL
California - Exposure Limits - STELs
1 mg/m3 STEL
California - Exposure Limits - Skin Notation
material may be absorbed through the skin, eyes or mucous membrane
California - Directors List of Hazardous Substances (8 CCR 339)
[present]
Florida Hazardous Substance List
[present]
Massachusetts Right To Know List
carcinogen; extraordinarily hazardous
Minnesota Hazardous Substance List
carcinogen; skin
NJ Right to Know List (Total)
sn 1871

NJ Special Hazardous Substances
(carcinogen; teratogen)
Pennsylvania Right to Know List
environmental hazard; special hazardous substance
Pennsylvania RTK - Special Hazardous Substances
[present]

TRIACONTAMETHYLCYCLOPENTADECASILOXANE   23523-14-0
HEALTH AND SAFETY LISTS
IARC - Group 3 (not classifiable)
[present]

TRIACONTAMETHYLTETRADECASILOXANE   2471-10-5
HEALTH AND SAFETY LISTS
IARC - Group 3 (not classifiable)
[present]

TRIADIMEFON [1-(4-CHLOROPHENOXY)-3,3-   43121-43-3
DIMETHYL-1-(1H-1,2,4-TRIAZOL-1-YL)-2-BU-
TANONE]
HEALTH AND SAFETY LISTS
IARC - Group 2B (sufficient animal data)
[present]
OSHA - Possible Select Carcinogens
[present]

TRIALKOXY SILANE   541-05-9
HEALTH AND SAFETY LISTS
U.S. DOT - Appendix A Table 1 - Hazardous Substances
final RQ = 100 pounds (45.4 kg)
ENVIRONMENTAL LISTS
CERCLA/SARA - Hazardous Substances and their Reportable Quantities
final RQ = 100 pounds (45.4 kg)
Clean Water Act - Hazardous Substances
[present]
STATE LISTS
California - Directors List of Hazardous Substances (8 CCR 339)
[present]
Massachusetts Right To Know List
[present]
NJ Right to Know List (Total)
sn 2939
Pennsylvania Right to Know List
environmental hazard

TRIALLATE   2303-17-5
STATE LISTS
Massachusetts Right To Know List
neurotoxin

TRIALLYLAMINE   102-70-5
HEALTH AND SAFETY LISTS
NFPA - Flash Points
flash point = 295 degrees F (146 degrees C)
NFPA - Hazard Identification Ratings
health-0; flammability-1; reactivity-0
STATE LISTS
Pennsylvania Right to Know List
[present]

TRIALLYL BORATE   1693-71-6

HEALTH AND SAFETY LISTS

IARC - Group 1 (carcinogenic to humans)
[present]

NTP Chemical Status Reports - Testing Status and NTIS Number
Technical reports printed (PB90226572)

NTP Chemical Status Reports - Evidence of Carcinogenicity
male rat-negative; female rat-negative

OSHA - Select Carcinogens
[present]

INTERNATIONAL LISTS

Canada - Alberta - 8 Hour Occupational Exposure Limit
0.2 f/cm3 TWA

Canada - Alberta - 15 Minute Occupational Exposure Limit
1 f/cm3 STEL

Canada - British Columbia - 8 Hour Exposure Limits
0.5 fibres/mL TWA

Canada - British Columbia - 15 Minute Exposure Limits
5 fibres/ml STEL

Canada - Quebec - Time-Weighted Average Exposure Values
1 fibre/cm3 TWAEV

Canada - Quebec - Short-term Exposure Values
35 fibres/cm3 STEV

Canada - Quebec - Carcinogens
C1 carcinogen: effect detected in humans

STATE LISTS

California - Exposure Limits - PELs
respirable dust (containing no Asbestos fibers): 2 mg/m3 PEL (Listed under 'Silicates')

California - Directors List of Hazardous Substances (8 CCR 339)
[present]

Florida Hazardous Substance List
[present]

TRIAMIPHOS   1031-47-6

HEALTH AND SAFETY LISTS

IARC - Group 1 (carcinogenic to humans)
[present]

OSHA - Select Carcinogens
[present]

STATE LISTS

California - Air Bill 2588 Appendix A-II
known or potential carcinogen

California - Prop. 65 - Cancer list
carcinogen - initial date 2/27/87

California - Directors List of Hazardous Substances (8 CCR 339)
[present]

Florida Hazardous Substance List
[present]

Massachusetts Right To Know List
carcinogen; extraordinarily hazardous

Minnesota Hazardous Substance List
carcinogen

NJ Special Hazardous Substances
(carcinogen)

Pennsylvania Right to Know List
special hazardous substance

Pennsylvania RTK - Special Hazardous Substances
[present]

TRIAMTERENE   396-01-0

ENVIRONMENTAL LISTS

TSCA - PAIR - Reporting List
Effective Date: October 12, 1993;   Reporting Date: February 28, 1994

TSCA - Health and Safety Reporting List
Effective Date: October 12, 1993;   Sunset Date: October 12, 2003

TRIAMYLBENZENE   RR-00886-6

ENVIRONMENTAL LISTS

TSCA - PAIR - Reporting List
Effective Date: October 12, 1993;   Reporting Date: February 28, 1994

TSCA - Health and Safety Reporting List
Effective Date: October 12, 1993;   Sunset Date: October 12, 2003

TRI-N-AMYL BORATE   621-78-3

ENVIRONMENTAL LISTS

CERCLA/SARA - Section 313 - Emission Reporting
form R reporting required

TRIARYL PHOSPHATES, ISOPROPYLATED   RR-01315-0

ENVIRONMENTAL LISTS

TSCA - PAIR - Reporting List
Effective Date: October 12, 1993;   Reporting Date: February 28, 1994

TSCA - Health and Safety Reporting List
Effective Date: October 12, 1993;   Sunset Date: October 12, 2003

TRIARYL PHOSPHATES, N.O.S.   RR-01316-1

ENVIRONMENTAL LISTS

CERCLA/SARA - Section 313 - Emission Reporting
form R reporting required

INTERNATIONAL LISTS

Canada - Drinking Water Quality - MACs
0.23 mg/L MAC

3,5,7-TRIAZA-1-AZONIATRICYCLODECANE-1-(3-   4080-31-3
CHLORO-2-PROPENYL)-, CHLORIDE

HEALTH AND SAFETY LISTS

U.S. DOT - Substances From 49 CFR 172.101
regulated by DOT (UN2610)

U.S. DOT - Hazard Classes
DOT hazard class = 3

NIOSH - Selected LD50s and LC50s
Inhalation, rat: LC50 = 554 ppm 8 hr Oral, rat: LD50 = 1030 mg/kg Skin, rabbit: LD50 = 400 mg/kg

INTERNATIONAL LISTS

Canada - WHMIS: Ingredient Disclosure
1% item 1592 (1635)

STATE LISTS

NJ Right to Know List (Total)
sn 1875

1,3,5-TRIAZIN-2-AMINE, 4-DIMETHYLAMINO-6-   RR-01564-5
SUBSTITUTED-

HEALTH AND SAFETY LISTS

U.S. DOT - Substances From 49 CFR 172.101
regulated by DOT (UN2609)

U.S. DOT - Hazard Classes
DOT hazard class = 6.1

INTERNATIONAL LISTS

Canada - WHMIS: Ingredient Disclosure
1% item 1593 (316)

STATE LISTS

NJ Right to Know List (Total)
sn 1876

S-TRIAZINE, 2-(ISOPROPYLAMINO)-4-(METHY-LAMINO)-6-(METHYLTHIO)   1014-69-3

HEALTH AND SAFETY LISTS

NIOSH - Selected LD50s and LC50s
Oral, rat: LD50 = 20 mg/kg Skin, rat: LD50 = 48 mg/kg

ENVIRONMENTAL LISTS

CERCLA/SARA - Section 302 Extremely Hazardous Substances and TPQs
TPQ = 500/10,000 pounds

STATE LISTS

Florida Hazardous Substance List
effective March 13, 1992

Massachusetts Right To Know List
extraordinarily hazardous

NJ Right to Know List (Total)
sn 2830

Pennsylvania Right to Know List
environmental hazard

PROPOSED REGULATIONS

CERCLA/SARA - Proposed Hazardous Substance Additions
proposed RQ = 1 pound (.454 kg)

CERCLA/SARA - 1989 Proposed RQ Adjustments
proposed RQ = 100 pounds (45.4 kg)

TRIAZINE PESTICIDES, N.O.S.   RR-00506-1

HEALTH AND SAFETY LISTS

NTP Chemical Status Reports - Testing Status and NTIS Number
Prechronic studies for which toxicity technical reports were not prepared; technical reports printed (no NTIS number given)

1,3,5-TRIAZINE-2,4,6-TRIAMINE, HYDROBROMIDE   29305-12-2

HEALTH AND SAFETY LISTS

NFPA - Flash Points
flash point = 270 degrees F (132 degrees C)

NFPA - Hazard Identification Ratings
health-0; flammability-1; reactivity-0

1,3,5-TRIAZINE-2,4,6(1H,3H,5H-TRIONE, 1,3,5-TRIS (ISOCYANATOMETHYLPHENYL)-   26603-40-7

HEALTH AND SAFETY LISTS

NFPA - Flash Points
flash point = 180 degrees F (82 degrees C)

NFPA - Hazard Identification Ratings
health-1; flammability-2; reactivity-0

INTERNATIONAL LISTS

Canada - WHMIS: Ingredient Disclosure
1% item 1594 (317)

1,3,5-TRIAZINE-1,3,5(2H,4H,6H)-TRIPROPANAMINE-N,N,N',N',N'',N''-HEXAMTHYL-   15875-13-5

HEALTH AND SAFETY LISTS

U.S. DOT - Appendix B - Marine Pollutants
DOT regulated marine pollutant

TRIAZOFOS   24017-47-8

HEALTH AND SAFETY LISTS

U.S. DOT - Appendix B - Marine Pollutants
DOT regulated severe marine pollutant

TRIAZOLAM   28911-01-5

HEALTH AND SAFETY LISTS

NIOSH - Selected LD50s and LC50s
Oral, rat: LD50 = 500 mg/kg

ENVIRONMENTAL LISTS

CERCLA/SARA - Section 313 - Emission Reporting
form R reporting required

List of Pesticide Product Inert Ingredients
[present]

1H-1,2,4-TRIAZOLE-3-CARBOXYLIC ACID, 1-(2, 4-DICHLORPHENYL)-5-(TRICHLOROMETHYL)-, ETHYL ESTER   103112-35-2

ENVIRONMENTAL LISTS

TSCA - Chemicals with Significant New Use Rules
PMN numbers: P-92-343; P-92-344

TRIBASIC SODIUM PHOSPHATE DODECAHYDRATE   10101-89-0

HEALTH AND SAFETY LISTS

NIOSH - Selected LD50s and LC50s
Oral, rat: LD50 = 1390 mg/kg

TRIBENURON METHYL [2-(((((4-METHOXY-6-METHYL-1,3,5-TRIAZIN-2-YL)-METHYLAMINO) CARBONYL)AMINO)SULFONYL)-, METHYL ESTER]   101200-48-0

HEALTH AND SAFETY LISTS

U.S. DOT - Substances From 49 CFR 172.101
regulated by DOT (UN2763, UN2764, UN2997, UN2998)

U.S. DOT - Hazard Classes
toxic or toxic, flammable: DOT hazard class = 6.1; flammable, toxic: DOT hazard class = 3

STATE LISTS

NJ Right to Know List (Total)
sn 2831; sn 2832; sn 2833; sn 2834

TRIBROMINATED POLYSTYRENE   57137-10-7

ENVIRONMENTAL LISTS

TSCA - Chemicals with Significant New Use Rules
PMN number: P-89-844

1,3,5-TRIBROMOBENZENE   626-39-1

ENVIRONMENTAL LISTS

TSCA - Code of Federal Regulations Citations
40 CFR 712.30(x); 40 CFR 716.120(d)

TSCA - PAIR - Reporting List
Reporting Date: December 27, 1990

TSCA - Health and Safety Reporting List
Effective Date: October 29, 1990; Sunset Date: November 9, 1993

3,4',5-TRIBROMOSALICYLANILIDE   87-10-5

HEALTH AND SAFETY LISTS

NIOSH - Selected LD50s and LC50s
Oral, rat: LD50 = 3250 mg/kg Skin, rabbit: LD50 = 2020 mg/kg

TRIBUTYL ALUMINUM   1116-70-7

HEALTH AND SAFETY LISTS

U.S. DOT - Appendix B - Marine Pollutants
DOT regulated marine pollutant

NIOSH - Selected LD50s and LC50s
Inhalation, rat: LC50 = 280 mg/m3 4 hr Oral, rat: LD50 = 64 mg/kg Skin, rat: LD50 = 1100 mg/kg

ENVIRONMENTAL LISTS
  CERCLA/SARA - Section 302 Extremely Hazardous Substances and
  TPQs
  TPQ = 500 pounds
STATE LISTS
  Florida Hazardous Substance List
  effective March 13, 1992
  Massachusetts Right To Know List
  extraordinarily hazardous
  NJ Right to Know List (Total)
  sn 2835
  Pennsylvania Right to Know List
  environmental hazard
PROPOSED REGULATIONS
  CERCLA/SARA - Proposed Hazardous Substance Additions
  proposed RQ = 1 pound (.454 kg)
  CERCLA/SARA - 1989 Proposed RQ Adjustments
  proposed RQ = 100 pounds (45.4 kg)

TRIBUTYLAMINE   102-82-9
STATE LISTS
  California - Air Bill 2588 Appendix A-II
  9/90
  California - Prop. 65 - Developmental Toxicity
  developmental toxicity - initial date 4/1/90
  NJ Right to Know List (Total)
  sn 1877
  NJ Special Hazardous Substances
  (teratogen)

TRIBUTYL BORATE   688-74-4
ENVIRONMENTAL LISTS
  List of Pesticide Product Inert Ingredients
  [present]

TRIBUTYL CITRATE   77-94-1
HEALTH AND SAFETY LISTS
  U.S. DOT - Appendix A Table 1 - Hazardous Substances
  final RQ = 5000 pounds (2270 kg) (Listed under 'Sodium phosphate,
  tribasic')
  NIOSH - Selected LD50s and LC50s
  Oral, rat: LD50 = 7400 mg/kg
ENVIRONMENTAL LISTS
  CERCLA/SARA - Hazardous Substances and their Reportable Quan-
  tities
  final RQ = 5000 pounds (2270 kg) (Listed under 'Sodium phosphate,
  tribasic')
  Clean Water Act - Hazardous Substances
  [present] (Listed under 'Sodium phosphate, tribasic')
STATE LISTS
  California - Directors List of Hazardous Substances (8 CCR 339)
  [present] (Listed under 'Sodium phosphate, tribasic')
  Massachusetts Right To Know List
  [present]
  NJ Right to Know List (Total)
  sn 3044
  Pennsylvania Right to Know List
  environmental hazard

2,4,6-TRI-SEC-BUTYLPHENOL, ETHOXYLATED,   109909-39-9
SULFATED, SODIUM SALT
ENVIRONMENTAL LISTS
  CERCLA/SARA - Section 313 - Emission Reporting
  form R reporting required

TRIBUTYL PHOSPHATE   126-73-8
ENVIRONMENTAL LISTS
  TSCA - Code of Federal Regulations Citations
  40 CFR 712.30(w); 40 CFR 716.120
  TSCA - PAIR - Reporting List
  Reporting Date: March 12, 1990
  TSCA - Health and Safety Reporting List
  Effective Date: January 11, 1990

TRIBUTYLPHOSPHINE   998-40-3
ENVIRONMENTAL LISTS
  TSCA - HDD/HDF - Precursors Required for Reporting
  [present]

S,S,S-TRIBUTYL PHOSPHOROTRITHIOITE   150-50-5
HEALTH AND SAFETY LISTS
  NIOSH - Selected LD50s and LC50s
  Oral, rat: LD50 = 410 mg/kg
ENVIRONMENTAL LISTS
  EPA - Master Testing List
  [present]
  TSCA - Code of Federal Regulations Citations
  40 CFR 712.30(x); 40 CFR 716.120(d); 40 CFR 766.25; 40 CFR
  766.35(a)(2)
  TSCA - PAIR - Reporting List
  Reporting Date: December 27, 1990
  TSCA - Health and Safety Reporting List
  Effective Date: October 29, 1990
  TSCA - HDD/HDF - Chemicals Required for Testing
  [present]
  TSCA - Section 12(b) - Export Notification
  export notification required - Section 4

TRIBUTYLTIN ACETATE   56-36-0
STATE LISTS
  Florida Hazardous Substance List
  [present]
  Massachusetts Right To Know List
  [present]
  NJ Right to Know List (Total)
  sn 1878
  Pennsylvania Right to Know List
  [present]

TRIBUTYLTIN BENZOATE   4342-36-3
HEALTH AND SAFETY LISTS
  U.S. DOT - Substances From 49 CFR 172.101
  regulated by DOT (UN2542)
  U.S. DOT - Hazard Classes
  DOT hazard class = 8
  NFPA - Flash Points
  flash point = 187 degrees F (86 degrees C)
  NFPA - Hazard Identification Ratings
  health-3; flammability-2; reactivity-0
  NIOSH - Selected LD50s and LC50s
  Oral, rat: LD50 = 540 mg/kg Skin, rabbit: LD50 = 250 mg/kg
INTERNATIONAL LISTS
  Canada - WHMIS: Ingredient Disclosure
  1% item 1595 (1639)
STATE LISTS
  Florida Hazardous Substance List
  [present]

Massachusetts Right To Know List
[present]
NJ Right to Know List (Total)
sn 1879
NJ Special Hazardous Substances
(corrosive)
Pennsylvania Right to Know List
[present]

TRIBUTYLTIN CHLORIDE   1461-22-9

HEALTH AND SAFETY LISTS

NFPA - Flash Points
flash point = 200 degrees F (93 degrees C)
NFPA - Hazard Identification Ratings
health-3; flammability-2; reactivity-1

INTERNATIONAL LISTS

Canada - WHMIS: Ingredient Disclosure
1% item 1596 (318)

STATE LISTS

Florida Hazardous Substance List
[present]
Massachusetts Right To Know List
[present]
Pennsylvania Right to Know List
[present]

TRIBUTYLTIN COMPOUNDS   RR-01198-3

HEALTH AND SAFETY LISTS

NFPA - Flash Points
flash point = 315 degrees F (157 degrees C)
NFPA - Hazard Identification Ratings
health-0; flammability-1; reactivity-0

TRIBUTYLTIN LINOLEATE   24124-25-2

ENVIRONMENTAL LISTS

List of Pesticide Product Inert Ingredients
[present]

TRIBUTYLTIN METHACRYLATE   2155-70-6

HEALTH AND SAFETY LISTS

ACGIH 1995 - Time Weighted Averages
0.2 ppm TWA; 2.2 mg/m3 TWA
NFPA - Flash Points
flash point = 295 degrees F (146 degrees C)
NFPA - Hazard Identification Ratings
health-3; flammability-1
NIOSH - Selected LD50s and LC50s
Inhalation, mouse: LC50 = 1300 mg/m3 (8 hr) Oral, rat: LD50 = 3000 mg/kg
NIOSH 1990 - Pocket Guide - RELs
0.2 ppm TWA; 2.5 mg/m3 TWA
NIOSH 1990 - Pocket Guide - IDLHs
125 ppm IDLH
NIOSH 1990 - Pocket Guide - Target organs
skin, eyes, respiratory system
OSHA - Vacated PELs - Time Weighted Averages
0.2 ppm TWA; 2.5 mg/m3 TWA
OSHA - Final PELs - Time Weighted Averages
5 mg/m3 TWA

ENVIRONMENTAL LISTS

EPA - Master Testing List
[present]

TSCA - Code of Federal Regulations Citations
40 CFR 712.30(q),(x); 40 CFR 716.120(a); 40 CFR 799.4360
TSCA - PAIR - Reporting List
Reporting Date: August 18, 1986, December 27, 1990
TSCA - Health and Safety Reporting List
Effective Date: June 18, 1986
TSCA - Chemical Test Rules
Testing required by: manufacturers; importers; processors (40 CFR 799.4360)
TSCA - Section 12(b) - Export Notification
export notification required - Section 4

INTERNATIONAL LISTS

Australian Exposure Standards - Time Weighted Averages
0.2 ppm TWA; 2.2 mg/m3 TWA
Canada - WHMIS: Ingredient Disclosure
1% item 1597 (1395)
Canada - Alberta - 8 Hour Occupational Exposure Limit
0.2 ppm TWA; 2.2 mg/m3 TWA
Canada - Alberta - 15 Minute Occupational Exposure Limit
0.4 ppm STEL; 4.4 mg/m3 STEL
Canada - British Columbia - 8 Hour Exposure Limits
5 mg/m3 TWA
Canada - British Columbia - 15 Minute Exposure Limits
5 mg/m3 STEL
Canada - Ontario - OHSA - TWAEVs
0.2 ppm TWAEV; 2.2 mg/m3 TWAEV
Canada - Quebec - Time-Weighted Average Exposure Values
0.2 ppm TWAEV; 0.22 mg/m3 TWAEV
United Kingdom - Occupational Exposure Standards - TWAs
5 mg/m3 TWA
United Kingdom - Occupational Exposure Standards - STELs
5 mg/m3 STEL
Israel - Time Weighted Averages
0.2 ppm TWA; 2.2 mg/m3 TWA
Israel - Action Levels
0.1 ppm AL; 1.1 mg/m3 AL
Mexico - Instruction No. 10 - TWAs
0.2 ppm TWA; 2.5 mg/m3 TWA
Mexico - Instruction No. 10 - STELs
0.4 ppm STEL; 5 mg/m3 STEL

STATE LISTS

California - Air Bill 2588 Appendix A-I
9/89
California - Exposure Limits - PELs
0.2 ppm PEL; 2.5 mg/m3 PEL
California - Directors List of Hazardous Substances (8 CCR 339)
[present]
Florida Hazardous Substance List
[present]
Massachusetts Right To Know List
[present]
Minnesota Hazardous Substance List
[present]
Pennsylvania Right to Know List
[present]

TRIBUTYLTIN NAPHTHENATE   85409-17-2

HEALTH AND SAFETY LISTS

NFPA - Hazard Identification Ratings
health-0; flammability-1; reactivity-0

TRI-N-BUTYLTIN SALICYLATE   4342-30-7

ENVIRONMENTAL LISTS

CERCLA/SARA - Section 313 - Emission Reporting
form R reporting required

STATE LISTS

California - Directors List of Hazardous Substances (8 CCR 339)
[present]

TRIBUTYOXYETHYL PHOSPHATE   78-51-3
SEE ALSO:
TIN

HEALTH AND SAFETY LISTS

NIOSH - Selected LD50s and LC50s
Oral, rat: LD50 = 99 mg/kg

TRI-C6-12-ALKYL METHYL AMMONIUM CHLORIDE   72749-59-8
SEE ALSO:
TIN

INTERNATIONAL LISTS

Canada - WHMIS: Ingredient Disclosure
1% item 1598 (283)

German (DFG) - MAK Values
as TBTO: 0.002 ppm MAK; 0.05 mg/m3 MAK (Listed under 'Tri-n-butyltin compounds')

German (DFG) - Peak Limitations
5 x normal MAK (30 min. average value); don't exceed 2 times during shift (Listed under 'Tri-n-butyltin compounds')

German (DFG) - Pregnancy
no risk to embryo/fetus if exposure limits adhered to (Listed under 'Tri-n-butyltin compounds')

TRICAPRYLIN   538-23-8
SEE ALSO:
TIN

HEALTH AND SAFETY LISTS

NIOSH - Selected LD50s and LC50s
Oral, rat: LD50 = 129 mg/kg

INTERNATIONAL LISTS

Canada - WHMIS: Ingredient Disclosure
1% item 1599 (533)

German (DFG) - MAK Values
as TBTO: 0.002 ppm MAK; 0.05 mg/m3 MAK (Listed under 'Tri-n-butyltin compounds')

German (DFG) - Peak Limitations
5 x normal MAK (30 min. average value); don't exceed 2 times during shift (Listed under 'Tri-n-butyltin compounds')

German (DFG) - Pregnancy
no risk to embryo/fetus if exposure limits adhered to (Listed under 'Tri-n-butyltin compounds')

TRICHLORFON   52-68-6

HEALTH AND SAFETY LISTS

U.S. DOT - Appendix B - Marine Pollutants
DOT regulated severe marine pollutant

TRICHLORMETHINE (TRIMUSTINE HCL)   817-09-4
SEE ALSO:
TIN

INTERNATIONAL LISTS

German (DFG) - MAK Values
as TBTO: 0.002 ppm MAK; 0.05 mg/m3 MAK (Listed under 'Tri-n-butyltin compounds')

German (DFG) - Peak Limitations
5 x normal MAK (30 min. average value); don't exceed 2 times during shift (Listed under 'Tri-n-butyltin compounds')

German (DFG) - Pregnancy
no risk to embryo/fetus if exposure limits adhered to (Listed under 'Tri-n-butyltin compounds')

TRICHLOROACETAMIDE   594-65-0
SEE ALSO:
TIN

ENVIRONMENTAL LISTS

CERCLA/SARA - Section 313 - Emission Reporting
form R reporting required

INTERNATIONAL LISTS

Canada - WHMIS: Ingredient Disclosure
1% item 1600 (1098)

German (DFG) - MAK Values
as TBTO: 0.002 ppm MAK; 0.05 mg/m3 MAK (Listed under 'Tri-n-butyltin compounds')

German (DFG) - Peak Limitations
5 x normal MAK (30 min. average value); don't exceed 2 times during shift (Listed under 'Tri-n-butyltin compounds')

German (DFG) - Pregnancy
no risk to embryo/fetus if exposure limits adhered to (Listed under 'Tri-n-butyltin compounds')

TRICHLOROACETIC ACID   76-03-9

INTERNATIONAL LISTS

German (DFG) - MAK Values
as TBTO: 0.002 ppm MAK; 0.05 mg/m3 MAK (Listed under 'Tri-n-butyltin compounds')

German (DFG) - Peak Limitations
5 x normal MAK (30 min. average value); don't exceed 2 times during shift (Listed under 'Tri-n-butyltin compounds')

German (DFG) - Pregnancy
no risk to embryo/fetus if exposure limits adhered to (Listed under 'Tri-n-butyltin compounds')

TRICHLOROACETONITRILE   545-06-2
SEE ALSO:
TIN

HEALTH AND SAFETY LISTS

NIOSH - Selected LD50s and LC50s
Oral, rat: LD50 = 137 mg/kg

INTERNATIONAL LISTS

Canada - WHMIS: Ingredient Disclosure
1% item 1601 (1475)

TRICHLOROACETYL CHLORIDE   76-02-8

HEALTH AND SAFETY LISTS

NIOSH - Selected LD50s and LC50s
Oral, rat: LD50 = 3000 mg/kg

ENVIRONMENTAL LISTS

List of Pesticide Product Inert Ingredients
[present]

TSCA - Code of Federal Regulations Citations
40 CFR 712.30(x); 40 CFR 716.120(d)

TSCA - PAIR - Reporting List
Reporting Date: December 27, 1990

TSCA - Health and Safety Reporting List
Effective Date: October 29, 1990

1,2,3-TRICHLOROBENZENE   87-61-6

ENVIRONMENTAL LISTS

List of Pesticide Product Inert Ingredients
[present]

1,3,5-TRICHLOROBENZENE   108-70-3
HEALTH AND SAFETY LISTS
NTP Chemical Status Reports - Testing Status and NTIS Number
Technical reports printed

1,2,4-TRICHLOROBENZENE   120-82-1
HEALTH AND SAFETY LISTS
U.S. DOT - Appendix B - Marine Pollutants
DOT regulated marine pollutant
U.S. DOT - Appendix A Table 1 - Hazardous Substances
final RQ = 100 pounds (45.4 kg)
IARC - Group 3 (not classifiable)
[present]
NIOSH - Selected LD50s and LC50s
Inhalation, rat: LC50 = 1300 mg/m3 8 hr Oral, rat: LD50 = 150 mg/kg Skin, rat: LD50 = 2000 mg/kg
NTP Chemical Status Reports - Testing Status and NTIS Number
Prechronic studies for which toxicity technical reports were not prepared

ENVIRONMENTAL LISTS
CERCLA/SARA - Section 313 - Emission Reporting
form R reporting required for 1.0% de minimus concentration
CERCLA/SARA - Hazardous Substances and their Reportable Quantities
final RQ = 100 pounds (45.4 kg)
Clean Water Act - Hazardous Substances
[present]

STATE LISTS
California - Air Bill 2588 Appendix A-II
6/91
California - Directors List of Hazardous Substances (8 CCR 339)
[present]
Massachusetts Right To Know List
[present]
NJ Right to Know List (Total)
sn 1882
Pennsylvania Right to Know List
environmental hazard

TRICHLOROBENZENE   12002-48-1
STATE LISTS
California - Prop. 65 - Cancer list
carcinogen - initial date 1/1/92

TRICHLOROBUTENE   51023-22-4
INTERNATIONAL LISTS
Canada - WHMIS: Ingredient Disclosure
1% item 1602 (1640)

TRICHLOROBUTYLENE   RR-01317-2
HEALTH AND SAFETY LISTS
ACGIH 1995 - Time Weighted Averages
1 ppm TWA; 6.7 mg/m3 TWA
U.S. DOT - Substances From 49 CFR 172.101
regulated by DOT (UN1839, UN2564)
U.S. DOT - Hazard Classes
DOT hazard class = 8
OSHA - Vacated PELs - Time Weighted Averages
1 ppm TWA; 7 mg/m3 TWA

ENVIRONMENTAL LISTS
EPA - Master Testing List
[present]
INTERNATIONAL LISTS
Australian Exposure Standards - Time Weighted Averages
1 ppm TWA; 6.7 mg/m3 TWA
Canada - WHMIS: Ingredient Disclosure
1% item 1603 (147)
Canada - Alberta - 8 Hour Occupational Exposure Limit
1 ppm TWA; 6.7 mg/m3 TWA
Canada - Alberta - 15 Minute Occupational Exposure Limit
2 ppm STEL; 13.4 mg/m3 STEL
Canada - Ontario - OHSA - TWAEVs
1 ppm TWAEV; 6.7 mg/m3 TWAEV
Canada - Quebec - Time-Weighted Average Exposure Values
1 ppm TWAEV; 6.7 mg/m3 TWAEV
Israel - Time Weighted Averages
1 ppm TWA; 6.7 mg/m3 TWA
Israel - Action Levels
0.5 ppm AL; 3.35 mg/m3 AL

STATE LISTS
California - Exposure Limits - PELs
1 ppm PEL; 5 mg/m3 PEL
California - Directors List of Hazardous Substances (8 CCR 339)
[present]
Florida Hazardous Substance List
[present]
Massachusetts Right To Know List
[present]
Minnesota Hazardous Substance List
[present]
NJ Right to Know List (Total)
sn 1883
NJ Special Hazardous Substances
(corrosive)
Pennsylvania Right to Know List
[present]

PROPOSED REGULATIONS
Canada - Ontario - Proposed Occupational TWAEVs
1 mg/m3 TWAEV

TRICHLOROBUTYLENE OXIDE   3083-25-8
HEALTH AND SAFETY LISTS
IARC - Group 3 (not classifiable)
[present]

PROPOSED REGULATIONS
Safe Drinking Water Act - Priority list
[present]

TRICHLORO(CHLOROMETHYL)SILANE   1558-25-4
HEALTH AND SAFETY LISTS
U.S. DOT - Substances From 49 CFR 172.101
regulated by DOT (UN2442)
U.S. DOT - Hazard Classes
DOT hazard class = 8
U.S. DOT - Substances Which Are Poisonous by Inhalation
liquid hazardous material poisonous by inhalation (UN2442)
NIOSH - Selected LD50s and LC50s
Inhalation, rat: LC50 = 475 mg/m3 4 hr Oral, rat: LD50 = 600 mg/kg

ENVIRONMENTAL LISTS

CERCLA/SARA - Section 302 Extremely Hazardous Substances and TPQs
TPQ = 500 pounds

CERCLA/SARA - Section 313 - Emission Reporting
form R reporting required

INTERNATIONAL LISTS

Canada - WHMIS: Ingredient Disclosure
1% item 1604 (534)

STATE LISTS

Florida Hazardous Substance List
effective March 13, 1992

Massachusetts Right To Know List
extraordinarily hazardous

NJ Right to Know List (Total)
sn 1884

Pennsylvania Right to Know List
environmental hazard

PROPOSED REGULATIONS

CERCLA/SARA - Proposed Hazardous Substance Additions
proposed RQ = 1 pound (.454 kg)

CERCLA/SARA - 1989 Proposed RQ Adjustments
proposed RQ = 100 pounds (45.4 kg)

TRICHLORO(DICHLOROPHENYL)SILANE   27137-85-5

ENVIRONMENTAL LISTS

CAA - HON Rule - SOCMI Chemicals
compliance by Oct. 23, 1995

Safe Drinking Water Act - Monitoring
monitoring required at discretion of the state

TSCA - Code of Federal Regulations Citations
40 CFR 712.30(d); 40 CFR 716.120(a); 40 CFR 799.1053

TSCA - PAIR - Reporting List
Reporting Date: November 19, 1982

TSCA - HDD/HDF - Precursors Required for Reporting
[present]

TSCA - Chemical Test Rules
Testing required by: manufacturers; processors (40 CFR 799.1053) (Listed under 'Trichlorobenzenes')

TSCA - Section 12(b) - Export Notification
export notification required - Section 4

STATE LISTS

Massachusetts Right To Know List
[present]

NJ Right to Know List (Total)
sn 1885

1,2,2-TRICHLORO-1,1-DIFLUOROETHANE   354-21-2

ENVIRONMENTAL LISTS

TSCA - Code of Federal Regulations Citations
40 CFR 712.30(d); 40 CFR 716.120(a)

TSCA - PAIR - Reporting List
Reporting Date: November 19, 1982

TSCA - HDD/HDF - Precursors Required for Reporting
[present]

INTERNATIONAL LISTS

Canada - WHMIS: Ingredient Disclosure
1% item 1606 (1642)

TRICHLORODIFLUOROPROPANE   134237-42-6

SEE ALSO:
F039-HAZARDOUS WASTES
K150-HAZARDOUS WASTES
F025-HAZARDOUS WASTES
F024-HAZARDOUS WASTES

HEALTH AND SAFETY LISTS

ACGIH 1995 - Ceiling Limits
C 5 ppm; C 37 mg/m3

AIHA - Odor Threshold Values
no geometric mean air odor threshold

U.S. DOT - Appendix A Table 1 - Hazardous Substances
final RQ = 100 pounds (45.4 kg)

NFPA - Flash Points
flash point = 222 degrees F (105 degrees C)

NFPA - Hazard Identification Ratings
health-2; flammability-1; reactivity-0

NIOSH - Selected LD50s and LC50s
Oral, rat: LD50 = 756 mg/kg

OSHA - Vacated PELs - Ceiling Limits
C 5 ppm; C 40 mg/m3

ENVIRONMENTAL LISTS

ATSDR Priority List
Rank (of 275): 176

CERCLA/SARA - Section 313 - Emission Reporting
form R reporting required for 1.0% de minimus concentration

CERCLA/SARA - Hazardous Substances and their Reportable Quantities
final RQ = 100 pounds (45.4 kg)

Clean Air Act (1990) - List of Hazardous Air Contaminants
[present]

CAA - HON Rule - SOCMI Chemicals
compliance by Oct. 24, 1994

CAA - HON Rule - Organic HAPs
[present]

Clean Water Act - Priority Pollutants
[present]

Safe Drinking Water Act - MCLs
MCL = 0.07 mg/L

Safe Drinking Water Act - MCLGs
MCLG = 0.07 mg/L

Safe Drinking Water Act - Monitoring
monitoring required at discretion of the state

EPA - Master Testing List
[present]

RCRA - Hazardous Constituents-Appendix VIII
hazardous constituent - no waste number

RCRA - Basis for Listing - Appendix VII
Included in waste streams: F024, F025, F039, K150

RCRA - TSD Facilities Ground Water Monitoring
TM 8270 = 10 ug/L PQL

RCRA - Universal Treatment Standards (LDR)
WW: 0.055 mg/l; NWW: 19 mg/kg

TSCA - Code of Federal Regulations Citations
40 CFR 712.30(d); 40 CFR 716.120(c); 40 CFR 799.1053

TSCA - PAIR - Reporting List
Reporting Date: November 19, 1982

TSCA - HDD/HDF - Precursors Required for Reporting
[present]

TSCA - Chemical Test Rules
Testing required by: manufacturers; processors (40 CFR 799.1053) (Listed under 'Trichlorobenzenes')

TSCA - Section 12(b) - Export Notification
export notification required - Section 4

INTERNATIONAL LISTS

Australian Exposure Standards - Time Weighted Averages
Peak Limitation: 5 ppm; 37 mg/m3

Canada - WHMIS: Ingredient Disclosure
1% item 1605 (1641)

Canada - NPRI (National Pollutant Release Inventory)
[present]

Canada - Alberta - Ceiling Occupational Exposure Limit
C 5 ppm; C 40 mg/m3

Canada - British Columbia - Ceiling Exposure Limits
C 5 ppm; C 40 mg/m3

Canada - Ontario - OHSA - CEVs
5 ppm CEV; 37 mg/m3 CEV

Canada - Quebec - Ceiling Limits
P 5 ppm; P 37 mg/m3

United Kingdom - Occupational Exposure Standards - TWAs
5 ppm TWA; 40 mg/m3 TWA

United Kingdom - Occupational Exposure Standards - STELs
5 ppm STEL; 40 mg/m3 STEL

Israel - Ceiling Exposure Limits
C 5 ppm; C 37 mg/m3

Mexico - Instruction No. 10 - TWAs
5 ppm TWA; 40 mg/m3 TWA

Mexico - Wastewater - Organic Toxic Pollutants and Heavy Metals
Listed under [Chlorinated Benzenes]

STATE LISTS

California - Air Bill 2588 Appendix A-I
6/91

California - Exposure Limits - Ceilings
C 5 ppm; C 40 mg/m3

Florida Hazardous Substance List
[present]

Massachusetts Right To Know List
[present]

Minnesota Hazardous Substance List
[present]

NJ Right to Know List (Total)
sn 1887

Pennsylvania Right to Know List
environmental hazard

TRICHLORO DIPHENYL OXIDE   57321-63-8

HEALTH AND SAFETY LISTS

U.S. DOT - Substances From 49 CFR 172.101
regulated by DOT (UN2321, UN2322)

U.S. DOT - Hazard Classes
DOT hazard class = 6.1

U.S. DOT - Appendix B - Marine Pollutants
liquid form: DOT regulated marine pollutant

ENVIRONMENTAL LISTS

ATSDR Priority List
Rank (of 275): 232

RCRA - Basis for Listing - Appendix VII
Included in waste stream: F085

INTERNATIONAL LISTS

Canada - CEPA - Priority Substances List
estimated time for completion of assessment reports: 4 years

German (DFG) - MAK Values
5 ppm MAK; 40 mg/m3 MAK

German (DFG) - Peak Limitations
10 x normal MAK (30 min average value); don't exceed during shift

German (DFG) - Pregnancy
classification not yet possible

STATE LISTS

NJ Right to Know List (Total)
sn 1886

1,1,1-TRICHLOROETHANE   71-55-6

HEALTH AND SAFETY LISTS

U.S. DOT - Substances From 49 CFR 172.101
regulated by DOT (UN2322)

U.S. DOT - Hazard Classes
DOT hazard class = 6.1

U.S. DOT - Appendix B - Marine Pollutants
DOT regulated marine pollutant

INTERNATIONAL LISTS

Canada - WHMIS: Ingredient Disclosure
1% item 1607 (1643)

STATE LISTS

NJ Right to Know List (Total)
sn 1888

1,1,2-TRICHLOROETHANE   79-00-5

HEALTH AND SAFETY LISTS

U.S. DOT - Appendix B - Marine Pollutants
DOT regulated marine pollutant

TRICHLOROETHANE   25323-89-1

ENVIRONMENTAL LISTS

TSCA - Code of Federal Regulations Citations
40 CFR 712.30(d); 40 CFR 716.120(a); 40 CFR 721.975

TSCA - PAIR - Reporting List
Reporting Date: November 19, 1982

TSCA - Health and Safety Reporting List
Effective Date: October 4, 1982

TSCA - Chemicals with Significant New Use Rules
[present]

TSCA - Section 12(b) - Export Notification
export notification required - Section 5

INTERNATIONAL LISTS

Canada - WHMIS: Ingredient Disclosure
1% item 1608 (1323)

TRICHLOROETHYLENE   79-01-6

HEALTH AND SAFETY LISTS

NIOSH - Selected LD50s and LC50s
Inhalation, mouse: LC50 = 30 mg/m3 2 hr

OSHA - List of Highly Hazardous Chemicals
threshhold quantity = 100 pounds

ENVIRONMENTAL LISTS

CERCLA/SARA - Section 302 Extremely Hazardous Substances and TPQs
TPQ = 100 pounds

STATE LISTS

Florida Hazardous Substance List
effective March 13, 1992

Massachusetts Right To Know List
extraordinarily hazardous

NJ Right to Know List (Total)
sn 2836

Pennsylvania Right to Know List
environmental hazard

PROPOSED REGULATIONS

CERCLA/SARA - Proposed Hazardous Substance Additions
proposed RQ = 1 pound (.454 kg)

CERCLA/SARA - 1989 Proposed RQ Adjustments
proposed RQ = 10 pounds (4.54 kg)

TRICHLOROETHYLSILANE 115-21-9

HEALTH AND SAFETY LISTS

U.S. DOT - Substances From 49 CFR 172.101
regulated by DOT (UN1766)

U.S. DOT - Hazard Classes
DOT hazard class = 8

U.S. DOT - Appendix B - Marine Pollutants
DOT regulated marine pollutant

OSHA - List of Highly Hazardous Chemicals
threshhold quantity = 2500 pounds

ENVIRONMENTAL LISTS

CERCLA/SARA - Section 302 Extremely Hazardous Substances and
TPQs
TPQ = 500 pounds

INTERNATIONAL LISTS

Canada - WHMIS: Ingredient Disclosure
1% item 559 (675)

STATE LISTS

Florida Hazardous Substance List
effective March 13, 1992

Massachusetts Right To Know List
extraordinarily hazardous

NJ Right to Know List (Total)
sn 0663

NJ Special Hazardous Substances
(corrosive)

Pennsylvania Right to Know List
environmental hazard

PROPOSED REGULATIONS

CERCLA/SARA - Proposed Hazardous Substance Additions
proposed RQ = 1 pound (.454 kg)

CERCLA/SARA - 1989 Proposed RQ Adjustments
proposed RQ = 100 pounds (45.4 kg)

TRICHLOROFLUOROETHANE 134237-34-6

ENVIRONMENTAL LISTS

Class 2 Ozone Depletors
ozone depletion weight reserved

TSCA - Chemicals with Significant New Use Rules
PMN number: P-92-595

TSCA - Section 12(b) - Export Notification
P-92-595; export notification required - Section 5

TRICHLOROFLUOROMETHANE 75-69-4

ENVIRONMENTAL LISTS

Class 2 Ozone Depletors
ozone depletion weight reserved

TRICHLOROFLUOROPROPANE 134190-51-5

INTERNATIONAL LISTS

Canada - WHMIS: Ingredient Disclosure
1% item 1609 (1324)

TRICHLOROMETHANE SULPHURYL CHLORIDE 594-42-3

SEE ALSO:
K020-HAZARDOUS WASTES
K019-HAZARDOUS WASTES
K029-HAZARDOUS WASTES
F039-HAZARDOUS WASTES
F001-HAZARDOUS WASTES
K028-HAZARDOUS WASTES
F024-HAZARDOUS WASTES
F002-HAZARDOUS WASTES
BIS(2,4-DIMETHYLBUTYL) MALEATE
K096-HAZARDOUS WASTES
F025-HAZARDOUS WASTES

HEALTH AND SAFETY LISTS

ACGIH 1995 - Time Weighted Averages
350 ppm TWA; 1910 mg/m3 TWA

ACGIH 1995 - Short Term Exposure Limits
450 ppm STEL; 2460 mg/m3 STEL

ACGIH 1995 - Biological Exposure Indices
Methyl chloroform in end-exhaled air: 40 ppm, prior to last shift of workweek; Trichloroacetic acid in urine: 10 mg/L, end of workweek (Ns, Sq); Total trichloroethanol in urine: 30 mg/L, end of shift at end of workweek (Ns, Sq); Total trichloroethanol in blood: 1 mg/L, end of shift at end of workweek (Ns)

AIHA - Odor Threshold Values
geometric mean air odor threshold = 390 ppm (detectable); 710 ppm (recognizable)

U.S. DOT - Substances From 49 CFR 172.101
regulated by DOT (UN2831)

U.S. DOT - Hazard Classes
DOT hazard class = 6.1

U.S. DOT - Appendix B - Marine Pollutants
DOT regulated marine pollutant

U.S. DOT - Appendix A Table 1 - Hazardous Substances
final RQ = 1000 pounds (454 kg)

IARC - Group 3 (not classifiable)
[present]

NFPA - Flash Points
no flash point

NFPA - Hazard Identification Ratings
health-2; flammability-1; reactivity-0

NIOSH - Selected LD50s and LC50s
Inhalation, rat: LC50 = 18000 ppm 4 hr Oral, rat: LD50 = 10300 mg/kg

NIOSH 1990 - Pocket Guide - RELs
C 350 ppm (15 min); C 1900 mg/m3 (15 min)

NIOSH 1990 - Pocket Guide - IDLHs
1000 ppm IDLH

NIOSH 1990 - Pocket Guide - Target organs
skin, CNS, CVS, eyes,

NIOSH - Health Standards - Exposure Limits
C (15 min) 350 ppm; C (15 min) 1910 mg/m3; action level = 200 ppm, 1910 mg/m3 TWA; handle with caution

NIOSH - Health Standards - Health Effects and Precautions
Central nervous system, liver, and cardiovascular effects (Medical warning of possible congenital abnormalities required)

NTP Chemical Status Reports - Testing Status and NTIS Number
Technical reports printed (PB265082/AS); Prechronic studies completed: In review for further evaluation; short term toxicity studies scheduled for peer review;

NTP Chemical Status Reports - Evidence of Carcinogenicity
male rat-inadequate; female rat-inadequate; male mice-inadequate; female mice-inadequate

OSHA - Vacated PELs - Time Weighted Averages
350 ppm TWA; 1900 mg/m3 TWA

OSHA - Vacated PELs - Short Term Exposure Limits
450 ppm STEL; 2450 mg/m3 STEL

OSHA - Final PELs - Time Weighted Averages
350 ppm TWA; 1900 mg/m3 TWA

ENVIRONMENTAL LISTS

ATSDR Priority List
Rank (of 275): 072

CERCLA/SARA - Section 313 - Emission Reporting
form R reporting required for 1.0% de minimus concentration

CERCLA/SARA - Hazardous Substances and their Reportable Quantities
final RQ = 1000 pounds (454 kg)

Clean Air Act (1990) - List of Hazardous Air Contaminants
[present]

Class 1 Ozone Depletors
ozone depletion potential = 0.1

CAA - HON Rule - SOCMI Chemicals
compliance by Jan. 23, 1995

CAA - HON Rule - Organic HAPs
[present]

Clean Water Act - Priority Pollutants
[present]

Safe Drinking Water Act - MCLs
MCL = 0.20 mg/L

Safe Drinking Water Act - MCLGs
MCLG = 0.20 mg/L

EPA - Master Testing List
[present]

List of Pesticide Product Inert Ingredients
[present]

RCRA - U Series Wastes
waste number U226

RCRA - Hazardous Constituents-Appendix VIII
waste number U226

RCRA - Basis for Listing - Appendix VII
Included in waste streams: F001, F002, F024, F025, F039, K019, K020, K028, K029, K073, K096

RCRA - Substances Banned From Land Disposal
[present]

RCRA - TSD Facilities Ground Water Monitoring
TM 8240 = 5 ug/L PQL

RCRA - Universal Treatment Standards (LDR)
WW: 0.054 mg/l; NWW: 6.0 mg/kg

TSCA - Code of Federal Regulations Citations
40 CFR 712.30(d); 40 CFR 716.120(a); 40 CFR 799.440; 40 CFR 799.5000

TSCA - PAIR - Reporting List
Reporting Date: November 19, 1982

TSCA - Health and Safety Reporting List
Effective Date: October 4, 1982

TSCA - Substances Subject to Testing Consent Orders
Test for: Health Effects

TSCA - Chemical Test Rules
Testing required by: manufacturers; processors (40 CFR 799.4400)

TSCA - Section 12(b) - Export Notification
export notification required - Section 4

INTERNATIONAL LISTS

Australian Exposure Standards - Time Weighted Averages
125 ppm TWA; 680 mg/m3 TWA

Canada - WHMIS: Ingredient Disclosure
0.1% item 1610 (1644)

Canada - CEPA - Priority Substances List
estimated time for completion of assessment reports: 4 years

Canada - CEPA Schedule I - Toxic Substances
that has the molecular formula CCl3-CH3: quantities that may be manufactured or imported

Canada - Alberta - 8 Hour Occupational Exposure Limit
350 ppm TWA; 1910 mg/m3 TWA

Canada - Alberta - 15 Minute Occupational Exposure Limit
450 ppm STEL; 2455 mg/m3 STEL

Canada - British Columbia - 8 Hour Exposure Limits
350 ppm TWA; 1900 mg/m3 TWA

Canada - British Columbia - 15 Minute Exposure Limits
450 ppm STEL; 2380 mg/m3 STEL

Canada - Ontario - OHSA - TWAEVs
350 ppm TWAEV; 1910 mg/m3 TWAEV

Canada - Ontario - OHSA - STEVs
450 ppm STEV; 2455 mg/m3 STEV

Canada - Quebec - Time-Weighted Average Exposure Values
350 ppm TWAEV; 1910 mg/m3 TWAEV

Canada - Quebec - Short-term Exposure Values
450 ppm STEV; 2460 mg/m3 STEV

United Kingdom - Maximum Exposure Limits - TWAs
350 ppm TWA; 1900 mg/m3 TWA

United Kingdom - Maximum Exposure Limits - STELs
450 ppm STEL; 2450 mg/m3 STEL

German (DFG) - MAK Values
200 ppm MAK; 1080 mg/m3 MAK

German (DFG) - Peak Limitations
5 x normal MAK (30 min. average value); don't exceed 2 times during shift

German (DFG) - Pregnancy
no risk to embryo/fetus if exposure limits adhered to

Israel - Time Weighted Averages
200 ppm TWA

Israel - Short Term Exposure Limits
450 ppm STEL; 2460 mg/m3 STEL

Israel - Action Levels
100 ppm AL; 955 mg/m3 AL

Mexico - Instruction No. 10 - TWAs
350 ppm TWA; 1900 mg/m3 TWA

Mexico - Instruction No. 10 - STELs
450 ppm STEL; 2450 mg/m3 STEL

Mexico - Wastewater - Organic Toxic Pollutants and Heavy Metals
Listed under [Chlorinated Ethanes]

Mexico - Drinking Water - Ecological Criteria
18.4 mg/l Substance presents persistence, bioaccumulations or risk of cancer, reduce human exposure to a minimum; This level has been extrapolated by using a mathematic model

STATE LISTS

California - Air Bill 2588 Appendix A-I
[present]

California - Exposure Limits - PELs
350 ppm PEL; 1900 mg/m3 PEL

California - Exposure Limits - STELs
450 ppm STEL; 2450 mg/m3 STEL

California - Exposure Limits - Ceilings
C 800 ppm

California - Directors List of Hazardous Substances (8 CCR 339)
[present]

Florida Hazardous Substance List
[present]

Massachusetts Right To Know List
[present]

Minnesota Hazardous Substance List
[present]

NJ Right to Know List (Total)
sn 1237

Pennsylvania Right to Know List
environmental hazard

PROPOSED REGULATIONS

Canada - Ontario - Proposed Occupational TWAEVs
50 ppm TWAEV; 300 mg/m3 TWAEV

Canada - Ontario - Proposed Occupational STEVs
90 ppm STEV; 500 mg/m3 STEV

TRICHLOROMETHANETHIOL   75-70-7

SEE ALSO:
K019-HAZARDOUS WASTES
F024-HAZARDOUS WASTES
F002-HAZARDOUS WASTES
F039-HAZARDOUS WASTES
CHLORINATED ETHANES
BIS(2,4-DIMETHYLBUTYL) MALEATE
K095-HAZARDOUS WASTES
K020-HAZARDOUS WASTES
F025-HAZARDOUS WASTES
K096-HAZARDOUS WASTES

HEALTH AND SAFETY LISTS

ACGIH 1995 - Time Weighted Averages
10 ppm TWA; 55 mg/m3 TWA

ACGIH 1995 - Skin Designations
skin - potential for cutaneous absorption

U.S. DOT - Appendix A Table 1 - Hazardous Substances
final RQ = 100 pounds (45.4 kg)

IARC - Group 3 (not classifiable)
[present]

NIOSH - Selected LD50s and LC50s
Oral, rat: LD50 = 580 mg/kg Skin, rabbit: LD50 = 3730 mg/kg

NIOSH 1990 - Pocket Guide - RELs
10 ppm TWA; 45 mg/m3 TWA

NIOSH 1990 - Pocket Guide - IDLHs
500 ppm IDLH (not considering carcinogenic effects)

NIOSH 1990 - Pocket Guide - Carcinogens
occupational carcinogen

NIOSH 1990 - Pocket Guide - Target organs
CNS, eyes, nose, liver, kidneys

NIOSH 1990 - Pocket Guide - Skin list
Potential for dermal absorption

NIOSH - Health Standards - Exposure Limits
reduce exposure to lowest feasible concentration

NIOSH - Health Standards - Health Effects and Precautions
Central nervous system effects; has produced liver tumors in animals

NIOSH - Health Standards - Carcinogenic Chemicals
potential human carcinogen

NTP Chemical Status Reports - Testing Status and NTIS Number
Technical reports printed (PB283337/AS)

NTP Chemical Status Reports - Evidence of Carcinogenicity
male rat-negative; female rat-negative; male mice-positive; female mice-positive

OSHA - Vacated PELs - Time Weighted Averages
10 ppm TWA; 45 mg/m3 TWA

OSHA - Vacated PELs - Skin Designation
Prevent or reduce skin absorption

OSHA - Final PELs - Time Weighted Averages
10 ppm TWA; 45 mg/m3 TWA

OSHA - Final PELs - Skin Notations
prevent or reduce skin absorption

ENVIRONMENTAL LISTS

ATSDR Priority List
Rank (of 275): 143

CERCLA/SARA - Section 313 - Emission Reporting
form R reporting required for 1.0% de minimus concentration

CERCLA/SARA - Hazardous Substances and their Reportable Quantities
final RQ = 100 pounds (45.4 kg)

Clean Air Act (1990) - List of Hazardous Air Contaminants
[present]

CAA - HON Rule - SOCMI Chemicals
compliance by Jan. 23, 1995

CAA - HON Rule - Organic HAPs
[present]

Clean Water Act - Priority Pollutants
[present]

Safe Drinking Water Act - MCLs
MCL = 0.005 mg/L

Safe Drinking Water Act - MCLGs
MCLG = 0.003 mg/L

Safe Drinking Water Act - Monitoring
monitoring required

EPA - Carcinogen Hazard Ranking for RQ Adjustment
Hazard ranking = Low

EPA - Master Testing List
[present]

RCRA - U Series Wastes
waste number U227

RCRA - Hazardous Constituents-Appendix VIII
waste number U227

RCRA - Basis for Listing - Appendix VII
Included in waste streams: F002, F024, F025, F039, K019, K020, K095, K073, K096

RCRA - Substances Banned From Land Disposal
[present]

RCRA - TSD Facilities Ground Water Monitoring
TM 8010 = 0.2 ug/L PQL; TM 8240 = 5 ug/L PQL

RCRA - Universal Treatment Standards (LDR)
WW: 0.054 mg/l; NWW: 6.0 mg/kg

TSCA - Code of Federal Regulations Citations
40 CFR 712.30(w); 40 CFR 716.120(a)

TSCA - PAIR - Reporting List
Reporting Date: June 13, 1989

TSCA - Health and Safety Reporting List
Effective Date: June 1, 1987

INTERNATIONAL LISTS

Australian Exposure Standards - Time Weighted Averages
10 ppm TWA; 55 mg/m3 TWA

Australian Exposure Standards - Skin Effects
skin absorption

Canada - WHMIS: Ingredient Disclosure
1% item 1611 (1645)

Canada - NPRI (National Pollutant Release Inventory)
[present]

Canada - Alberta - 8 Hour Occupational Exposure Limit
10 ppm TWA; 55 mg/m3 TWA

Canada - Alberta - 15 Minute Occupational Exposure Limit
20 ppm STEL; 109 mg/m3 STEL

Canada - Alberta - Skin Designation
can be absorbed through the intact skin

Canada - British Columbia - 8 Hour Exposure Limits
10 ppm TWA; 45 mg/m3 TWA

Canada - British Columbia - 15 Minute Exposure Limits
20 ppm STEL; 90 mg/m3 STEL

Canada - British Columbia - Skin Notations
skin - potential for skin absorption

Canada - Ontario - OHSA - TWAEVs
10 ppm TWAEV; 55 mg/m3 TWAEV

Canada - Ontario - OHSA - Skin Notations
absorption through skin, eyes, or mucous membranes
Canada - Quebec - Time-Weighted Average Exposure Values
10 ppm TWAEV;   55 mg/m3 TWAEV
Canada - Quebec - Skin Designations
absorbed through the skin
German (DFG) - MAK Values
10 ppm MAK;   55 mg/m3 MAK
German (DFG) - Peak Limitations
5 x normal MAK (30 min. average value);   don't exceed 2 times during shift
German (DFG) - Skin/Sensitizers
danger of cutaneous absorption
German (DFG) - Carcinogens
suspected carcinogen
Israel - Time Weighted Averages
10 ppm TWA;   55 mg/m3 TWA
Israel - Action Levels
5 ppm AL;   27.5 mg/m3 AL
Mexico - Instruction No. 10 - TWAs
10 ppm TWA;   45 mg/m3 TWA
Mexico - Instruction No. 10 - STELs
20 ppm STEL;   30 mg/m3 STEL
Mexico - Instruction No. 10 - Skin designation
skin - potential for cutaneous absorption
Mexico - Wastewater - Organic Toxic Pollutants and Heavy Metals
Listed under [Chlorinated Ethanes]
Mexico - Drinking Water - Ecological Criteria
0.006 mg/l Substance presents persistence, bioaccumulations or risk of cancer, reduce human exposure to a minimum;   This level has been extrapolated by using a mathematic model

STATE LISTS
California - Air Bill 2588 Appendix A-I
known or potential carcinogen: 6/91
California - Prop. 65 - Cancer list
carcinogen - initial date 10/1/90
California - Prop. 65 - No Significant Risk Levels
no significant risk level = 10 ug/day
California - Exposure Limits - PELs
10 ppm PEL;   45 mg/m3 PEL
California - Exposure Limits - Skin Notation
material may be absorbed through the skin, eyes or mucous membrane
California - Directors List of Hazardous Substances (8 CCR 339)
[present]
Florida Hazardous Substance List
[present]
Massachusetts Right To Know List
carcinogen;   extraordinarily hazardous
Minnesota Hazardous Substance List
carcinogen;   skin
NJ Right to Know List (Total)
sn 1889
NJ Special Hazardous Substances
(carcinogen)
Pennsylvania Right to Know List
environmental hazard

TRICHLOROMETHYL PERCHLORATE   **67632-66-0**
ENVIRONMENTAL LISTS
ATSDR Priority List
Rank (of 275): 137

TRICHLORONAPHTHALENE   **1321-65-9**
SEE ALSO:
K018-HAZARDOUS WASTES
F024-HAZARDOUS WASTES
F002-HAZARDOUS WASTES
K019-HAZARDOUS WASTES
BIS(2,4-DIMETHYLBUTYL) MALEATE
K149-HAZARDOUS WASTES
K020-HAZARDOUS WASTES
F025-HAZARDOUS WASTES
F001-HAZARDOUS WASTES
F039-HAZARDOUS WASTES

HEALTH AND SAFETY LISTS
ACGIH 1995 - Time Weighted Averages
50 ppm TWA;   269 mg/m3 TWA
ACGIH 1995 - Short Term Exposure Limits
100 ppm STEL;   537 mg/m3 STEL
ACGIH 1995 - Carcinogens
A5-not suspected as a human carcinogen
ACGIH 1995 - Biological Exposure Indices
Trichloroacetic acid in urine: 100 mg/g end of week (Ns);   Trichloroacetic acid and trichloroethanol in urine: 300 mg/g creatinine, end of shift at end of week (Ns);   Free trichloroethanol in blood: 4 mg/L, end of shift at end of week (Ns);   Trichloroethylene in blood: (Sq);   Trichloroethylene in end-exhaled air: (Sq)
AIHA - Odor Threshold Values
geometric mean air odor threshold = 82 ppm (detectable);   110 ppm (recognizable)
U.S. DOT - Substances From 49 CFR 172.101
regulated by DOT (UN1710)
U.S. DOT - Hazard Classes
DOT hazard class = 6.1
U.S. DOT - Appendix A Table 1 - Hazardous Substances
final RQ = 100 pounds (45.4 kg)
IARC - Group 3 (not classifiable)
[present]
NFPA - Flash Points
no flash point
NFPA - Hazard Identification Ratings
health-2;   flammability-1;   reactivity-0
NIOSH - Selected LD50s and LC50s
Inhalation, mouse: LC50 = 8450 ppm (4 hr)   Oral, rat: LD50 = 3670 mg/kg
NIOSH 1990 - Pocket Guide - RELs
25 ppm TWA
NIOSH 1990 - Pocket Guide - IDLHs
1000 ppm IDLH (not considering carcinogenic effects)
NIOSH 1990 - Pocket Guide - Carcinogens
occupational carcinogen
NIOSH 1990 - Pocket Guide - Target organs
respiratory system, heart, liver, kidneys, CNS, skin
NIOSH - Health Standards - Exposure Limits
25 ppm TWA
NIOSH - Health Standards - Health Effects and Precautions
Central nervous system effects;   has produced liver tumors in animals (Warn workers of hazards)
NIOSH - Health Standards - Carcinogenic Chemicals
potential human carcinogen
NTP Chemical Status Reports - Testing Status and NTIS Number
Technical reports printed (PB91111815) (PB264122/AS) (PB88218896/AS);   prechronic studies for which toxicity technical reports were not prepared

NTP Chemical Status Reports - Evidence of Carcinogenicity
PB91111815: male rat-inadequate; female rat-negative; male mice-positive; female mice-positive; PB264122/AS: male rat-negative; female rat-negative; male mice-positive; female mice-positive; PB88218896/AS: male rat-inadequate; female rat-inadequate

OSHA - Vacated PELs - Time Weighted Averages
50 ppm TWA; 270 mg/m3 TWA

OSHA - Vacated PELs - Short Term Exposure Limits
200 ppm STEL; 1080 mg/m3 STEL

OSHA - Final PELs - Time Weighted Averages
100 ppm TWA; C 200 ppm

OSHA - Final PELs - Ceiling Limits
C 200 ppm

ENVIRONMENTAL LISTS

ATSDR Priority List
Rank (of 275): 011

CERCLA/SARA - Section 313 - Emission Reporting
form R reporting required for 1.0% de minimus concentration

CERCLA/SARA - Hazardous Substances and their Reportable Quantities
final RQ = 100 pounds (45.4 kg)

Clean Air Act (1990) - List of Hazardous Air Contaminants
[present]

CAA - HON Rule - SOCMI Chemicals
compliance by Oct. 24, 1994

CAA - HON Rule - Organic HAPs
[present]

Clean Water Act - Hazardous Substances
[present]

Clean Water Act - Priority Pollutants
[present]

Clean Water Act - Toxic Pollutants
[present]

Safe Drinking Water Act - MCLs
MCL = 0.005 mg/L

Safe Drinking Water Act - MCLGs
MCLG = Zero

EPA - Carcinogen Hazard Ranking for RQ Adjustment
Hazard ranking = Low

RCRA - D Series - Maximum Concentration of Contaminants
waste number D040; regulatory level = 0.5 mg/L

RCRA - D Series - Chronic Toxicity Reference Levels
chronic toxicity reference level = 0.005 mg/L

RCRA - U Series Wastes
waste number U228

RCRA - Hazardous Constituents-Appendix VIII
waste number U228

RCRA - Basis for Listing - Appendix VII
Included in waste streams: F001, F002, F024, F025, F039, K018, K019, K020

RCRA - Substances Banned From Land Disposal
[present]

RCRA - TSD Facilities Ground Water Monitoring
TM 8010 = 1 ug/L PQL; TM 8240 = 5 ug/L PQL

RCRA - Universal Treatment Standards (LDR)
WW: 0.054 mg/l; NWW: 6.0 mg/kg

INTERNATIONAL LISTS

Australian Exposure Standards - Time Weighted Averages
50 ppm TWA; 270 mg/m3 TWA

Australian Exposure Standards - Short Term Exposure Limits
200 ppm STEL; 1080 mg/m3 STEL

Canada - WHMIS: Ingredient Disclosure
1% item 1612 (1646)

Canada - NPRI (National Pollutant Release Inventory)
[present]

Canada - CEPA - Priority Substances List
estimated time for completion of assessment reports: 4 years

Canada - Drinking Water Quality - MACs
0.05 mg/L MAC

Canada - Alberta - 8 Hour Occupational Exposure Limit
50 ppm TWA; 269 mg/m3 TWA

Canada - Alberta - 15 Minute Occupational Exposure Limit
200 ppm STEL; 1074 mg/m3 STEL

Canada - British Columbia - 8 Hour Exposure Limits
100 ppm TWA; 535 mg/m3 TWA

Canada - British Columbia - 15 Minute Exposure Limits
150 ppm STEL; 800 mg/m3 STEL

Canada - Ontario - OHSA - TWAEVs
50 ppm TWAEV; 268 mg/m3 TWAEV

Canada - Ontario - OHSA - STEVs
200 ppm STEV; 1075 mg/m3 STEV

Canada - Quebec - Time-Weighted Average Exposure Values
50 ppm TWAEV; 269 mg/m3 TWAEV

Canada - Quebec - Short-term Exposure Values
200 ppm STEV; 1070 mg/m3 STEV

United Kingdom - Maximum Exposure Limits - TWAs
100 ppm TWA; 535 mg/m3 TWA

United Kingdom - Maximum Exposure Limits - STELs
150 ppm STEL; 802 mg/m3 STEL

United Kingdom - Maximum Exposure Limits - Notes
can be absorbed through skin

German (DFG) - MAK Values
50 ppm MAK; 270 mg/m3 MAK

German (DFG) - Peak Limitations
5 x normal MAK (30 min. average value); don't exceed 2 times during shift

German (DFG) - Carcinogens
suspected carcinogen

German (DFG) - Pregnancy
no risk to embryo/fetus if exposure limits adhered to

Israel - Time Weighted Averages
50 ppm TWA; 269 mg/m3 TWA

Israel - Short Term Exposure Limits
200 ppm STEL; 1070 mg/m3 STEL

Israel - Action Levels
25 ppm AL; 134.5 mg/m3 AL

Mexico - Instruction No. 10 - TWAs
100 ppm TWA; 535 mg/m3 TWA

Mexico - Instruction No. 10 - STELs
200 ppm STEL; 1080 mg/m3 STEL

Mexico - Wastewater - Organic Toxic Pollutants and Heavy Metals
Listed under [Organic Toxic Pollutants]

Mexico - Drinking Water - Ecological Criteria
0.03 mg/l Substance presents persistence, bioaccumulations or risk of cancer, reduce human exposure to a minimum; This level has been extrapolated by using a mathematic model

STATE LISTS

California - Air Bill 2588 Appendix A-I
known or potential carcinogen

California - Prop. 65 - Cancer list
carcinogen - initial date 4/1/88

California - Prop. 65 - No Significant Risk Levels
ingestion: no significant risk level = 50 ug/day; inhalation: no significant risk level = 80 ug/day

California - Exposure Limits - PELs
25 ppm PEL; 135 mg/m3 PEL

California - Exposure Limits - STELs
200 ppm STEL;   1080 mg/m3 STEL

California - Exposure Limits - Ceilings
C 300 ppm

California - Directors List of Hazardous Substances (8 CCR 339)
[present]

Florida Hazardous Substance List
[present]

Massachusetts Right To Know List
carcinogen;   extraordinarily hazardous

Minnesota Hazardous Substance List
carcinogen

NJ Right to Know List (Total)
sn 1890

NJ Special Hazardous Substances
(carcinogen;   mutagen)

Pennsylvania Right to Know List
environmental hazard

TRICHLORONATE   327-98-0

HEALTH AND SAFETY LISTS

U.S. DOT - Substances From 49 CFR 172.101
regulated by DOT (UN1196)

U.S. DOT - Hazard Classes
DOT hazard class = 3

NFPA - Flash Points
flash point = 72 degrees F (22 degrees C)

NFPA - Hazard Identification Ratings
health-3;   flammability-3;   reactivity-2 (avoid water as an extinguishing agent)

NIOSH - Selected LD50s and LC50s
Inhalation, mouse: LC50 = 300 mg/m3 (2 hr) Oral, rat: LD50 = 1330 mg/kg

ENVIRONMENTAL LISTS

CERCLA/SARA - Section 302 Extremely Hazardous Substances and TPQs
TPQ = 500 pounds

STATE LISTS

Florida Hazardous Substance List
[present]

Massachusetts Right To Know List
extraordinarily hazardous

NJ Right to Know List (Total)
sn 0912

NJ Special Hazardous Substances
(corrosive;   flammable - third degree)

Pennsylvania Right to Know List
environmental hazard

PROPOSED REGULATIONS

CERCLA/SARA - Proposed Hazardous Substance Additions
proposed RQ = 1 pound (.454 kg)

CERCLA/SARA - 1989 Proposed RQ Adjustments
proposed RQ = 100 pounds (45.4 kg)

2,4,5-TRICHLORONITROBENZENE   89-69-0

ENVIRONMENTAL LISTS

Class 2 Ozone Depletors
ozone depletion weight reserved

1,1,1-TRICHLORO-2,2,3,3,3-   4259-43-2
PENTAFLUOROPROPANE
SEE ALSO:
   F039-HAZARDOUS WASTES
   F002-HAZARDOUS WASTES

HEALTH AND SAFETY LISTS

ACGIH 1995 - Ceiling Limits
C 1000 ppm;   C 5620 mg/m3

U.S. DOT - Appendix A Table 1 - Hazardous Substances
final RQ = 5000 pounds (2270 kg)

NIOSH - Selected LD50s and LC50s
Inhalation, mouse: LC50 = 10 pph 30 mn

NIOSH 1990 - Pocket Guide - RELs
C 1000 ppm;   C 5600 mg/m3

NIOSH 1990 - Pocket Guide - IDLHs
10,000 ppm IDLH

NIOSH 1990 - Pocket Guide - Target organs
CVS, skin

NTP Chemical Status Reports - Testing Status and NTIS Number
Technical reports printed (PB286187/AS)

NTP Chemical Status Reports - Evidence of Carcinogenicity
male rat-inadequate;   female rat-inadequate;   male mice-negative;   female mice-negative

OSHA - Vacated PELs - Ceiling Limits
C 1000 ppm;   C 5600 mg/m3

OSHA - Final PELs - Time Weighted Averages
1000 ppm TWA;   5600 mg/m3 TWA

ENVIRONMENTAL LISTS

ATSDR Priority List
Rank (of 275): 257

CERCLA/SARA - Section 313 - Emission Reporting
form R reporting required for 1.0% de minimus concentration

CERCLA/SARA - Hazardous Substances and their Reportable Quantities
final RQ = 5000 pounds (2270 kg)

Class 1 Ozone Depletors
ozone depletion potential = 1.0

CAA - HON Rule - SOCMI Chemicals
compliance by Oct. 24, 1994

Safe Drinking Water Act - Monitoring
monitoring required at discretion of the state

EPA - Master Testing List
[present]

List of Pesticide Product Inert Ingredients
[present]

RCRA - U Series Wastes
waste number U121

RCRA - Hazardous Constituents-Appendix VIII
waste number U121

RCRA - Basis for Listing - Appendix VII
Included in waste streams: F002, F039

RCRA - Substances Banned From Land Disposal
[present]

RCRA - TSD Facilities Ground Water Monitoring
TM 8010 = 10 ug/L PQL;   TM 8240 = 5 ug/L PQL

RCRA - Universal Treatment Standards (LDR)
WW: 0.020 mg/l;   NWW: 30 mg/kg

INTERNATIONAL LISTS

Australian Exposure Standards - Time Weighted Averages
Peak Limitation: 1000 ppm;   5620 mg/m3

Canada - WHMIS: Ingredient Disclosure
1% item 1613 (1647)

Canada - Alberta - 8 Hour Occupational Exposure Limit
1000 ppm TWA; 5619 mg/m3 TWA

Canada - Alberta - 15 Minute Occupational Exposure Limit
1250 ppm STEL; 7024 mg/m3 STEL

Canada - British Columbia - 8 Hour Exposure Limits
1000 ppm TWA; 5600 mg/m3 TWA

Canada - British Columbia - 15 Minute Exposure Limits
1250 ppm STEL; 7000 mg/m3 STEL

Canada - Ontario - OHSA - CEVs
1000 ppm CEV; 5600 mg/m3 CEV

Canada - Quebec - Ceiling Limits
P 1000 ppm; P 5620 mg/m3

United Kingdom - Occupational Exposure Standards - TWAs
1000 ppm TWA; 5600 mg/m3 TWA

United Kingdom - Occupational Exposure Standards - STELs
1250 ppm STEL; 7000 mg/m3 STEL

German (DFG) - MAK Values
1000 ppm MAK; 5600 mg/m3 MAK

German (DFG) - Peak Limitations
2 x normal MAK (1 hour momentary value); don't exceed 3 times per shift

German (DFG) - Pregnancy
no risk to embryo/fetus if exposure limits adhered to

Israel - Ceiling Exposure Limits
C 1000 ppm; C 5620 mg/m3

Mexico - Instruction No. 10 - TWAs
1000 ppm TWA; 5600 mg/m3 TWA

Mexico - Instruction No. 10 - STELs
1250 ppm STEL; 7000 mg/m3 STEL

Mexico - Wastewater - Organic Toxic Pollutants and Heavy Metals
Listed under [Halomethanes]

STATE LISTS

California - Exposure Limits - Ceilings
C 1000 ppm; C 5600 mg/m3

California - Directors List of Hazardous Substances (8 CCR 339)
[present]

Florida Hazardous Substance List
[present]

Massachusetts Right To Know List
[present]

Minnesota Hazardous Substance List
[present]

NJ Right to Know List (Total)
sn 1891

Pennsylvania Right to Know List
environmental hazard

PROPOSED REGULATIONS

Safe Drinking Water Act - Priority list
[present]

2,4,6-TRICHLOROPHENOL   88-06-2

ENVIRONMENTAL LISTS

Class 2 Ozone Depletors
ozone depletion weight reserved

2,4,5-TRICHLOROPHENOL   95-95-4

HEALTH AND SAFETY LISTS

ACGIH 1995 - Time Weighted Averages
0.1 ppm TWA; 0.76 mg/m3 TWA

U.S. DOT - Substances From 49 CFR 172.101
regulated by DOT (UN1670)

U.S. DOT - Hazard Classes
DOT hazard class = 6.1

U.S. DOT - Substances Which Are Poisonous by Inhalation
liquid hazardous material poisonous by inhalation (UN1670)

U.S. DOT - Appendix B - Marine Pollutants
DOT regulated marine pollutant

U.S. DOT - Appendix A Table 1 - Hazardous Substances
final RQ = 100 pounds (45.4 kg)

NIOSH - Selected LD50s and LC50s
Inhalation, mouse: LC50 = 296 mg/m3 (2 hr) Oral, rat: LD50 = 82600 ug/kg

NIOSH 1990 - Pocket Guide - RELs
0.1 ppm TWA; 0.8 mg/m3 TWA

NIOSH 1990 - Pocket Guide - IDLHs
10 ppm IDLH

NIOSH 1990 - Pocket Guide - Target organs
eyes, respiratory system, liver, kidneys, skin

OSHA - Vacated PELs - Time Weighted Averages
0.1 ppm TWA; 0.8 mg/m3 TWA

OSHA - Final PELs - Time Weighted Averages
0.1 ppm TWA; 0.8 mg/m3 TWA

OSHA - List of Highly Hazardous Chemicals
threshhold quantity = 150 pounds

ENVIRONMENTAL LISTS

CERCLA/SARA - Section 302 Extremely Hazardous Substances and TPQs
TPQ = 500 pounds

CERCLA/SARA - Section 313 - Emission Reporting
form R reporting required

CERCLA/SARA - Hazardous Substances and their Reportable Quantities
final RQ = 100 pounds (45.4 kg)

CAA -Toxic Substances for Accidental Release Prevention
threshold quantity = 10,000 lbs

CAA - HON Rule - SOCMI Chemicals
compliance by July 24, 1995

TSCA - Code of Federal Regulations Citations
40 CFR 799.5055(c),(d)(1),(2),(e)(1); 40 CFR 712.30(x)

TSCA - PAIR - Reporting List
Reporting Date: November 27, 1991

TSCA - Health and Safety Reporting List
Effective Date: September 30, 1991

INTERNATIONAL LISTS

Australian Exposure Standards - Time Weighted Averages
0.1 ppm TWA; 0.76 mg/m3 TWA

Canada - WHMIS: Ingredient Disclosure
1% item 1250 (1357)

Canada - Alberta - 8 Hour Occupational Exposure Limit
0.1 ppm TWA; 0.76 mg/m3 TWA

Canada - Alberta - 15 Minute Occupational Exposure Limit
0.3 ppm STEL; 2.28 mg/m3 STEL

Canada - British Columbia - 8 Hour Exposure Limits
0.1 ppm TWA; 0.8 mg/m3 TWA

Canada - Ontario - OHSA - TWAEVs
0.1 ppm TWAEV; 0.8 mg/m3 TWAEV

Canada - Quebec - Time-Weighted Average Exposure Values
0.1 ppm TWAEV; 0.76 mg/m3 TWAEV

Israel - Time Weighted Averages
0.1 ppm TWA; 0.76 mg/m3 TWA

Israel - Action Levels
0.05 ppm AL; 0.38 mg/m3 AL

Mexico - Instruction No. 10 - TWAs
0.1 ppm TWA; 0.8 mg/m3 TWA

STATE LISTS

California - Exposure Limits - PELs
0.1 ppm PEL; 0.8 mg/m3 PEL

California - Directors List of Hazardous Substances (8 CCR 339)
[present]

Florida Hazardous Substance List
[present]

Massachusetts Right To Know List
extraordinarily hazardous

Minnesota Hazardous Substance List
[present]

NJ Right to Know List (Total)
sn 1480

Pennsylvania Right to Know List
environmental hazard

3,4,5-TRICHLOROPHENOL   609-19-8

ENVIRONMENTAL LISTS

RCRA - P Series Wastes
waste number P118

RCRA - Hazardous Constituents-Appendix VIII
waste number P118

RCRA - Substances Banned From Land Disposal
[present]

TSCA - Code of Federal Regulations Citations
40 CFR 799.5055(c)

TSCA - Multichemical Test Rules - Waste Constituents
soil adsorption and hydrolysis testing for Chemical Fate; subchronic toxicity testing for Health Effects

TSCA - Section 12(b) - Export Notification
export notification required - Section 4

2,3,6-TRICHLOROPHENOL   933-75-5

HEALTH AND SAFETY LISTS

U.S. DOT - Hazard Classes
Forbidden from transport by the DOT

2,3,5-TRICHLOROPHENOL   933-78-8

HEALTH AND SAFETY LISTS

ACGIH 1995 - Time Weighted Averages
5 mg/m3 TWA

ACGIH 1995 - Skin Designations
skin - potential for cutaneous absorption

NIOSH 1990 - Pocket Guide - RELs
5 mg/m3 TWA

NIOSH 1990 - Pocket Guide - Target organs
skin, liver

NIOSH 1990 - Pocket Guide - Skin list
Potential for dermal absorption

OSHA - Vacated PELs - Time Weighted Averages
5 mg/m3 TWA

OSHA - Vacated PELs - Skin Designation
Prevent or reduce skin absorption

OSHA - Final PELs - Time Weighted Averages
5 mg/m3 TWA

OSHA - Final PELs - Skin Notations
prevent or reduce skin absorption

ENVIRONMENTAL LISTS

TSCA - Code of Federal Regulations Citations
40 CFR 704.83; 40 CFR 712.30(d); 40 CFR 716.120(a)

TSCA - PAIR - Reporting List
Reporting Date: November 19, 1982

TSCA - Health and Safety Reporting List
Effective Date: October 4, 1982

INTERNATIONAL LISTS

Australian Exposure Standards - Time Weighted Averages
5 mg/m3 TWA

Australian Exposure Standards - Skin Effects
skin absorption

Canada - WHMIS: Ingredient Disclosure
1% item 1615 (1648)

Canada - Alberta - 8 Hour Occupational Exposure Limit
5 mg/m3 TWA

Canada - Alberta - 15 Minute Occupational Exposure Limit
10 mg/m3 STEL

Canada - British Columbia - 8 Hour Exposure Limits
5 mg/m3 TWA

Canada - British Columbia - 15 Minute Exposure Limits
10 mg/m3 STEL

Canada - Ontario - OHSA - TWAEVs
5 mg/m3 TWAEV

Canada - Ontario - OHSA - Skin Notations
absorption through skin, eyes, or mucous membranes

Canada - Quebec - Time-Weighted Average Exposure Values
5 mg/m3 TWAEV

Canada - Quebec - Skin Designations
absorbed through the skin

German (DFG) - MAK Values
total dust: 5 mg/m3 MAK

German (DFG) - Skin/Sensitizers
danger of cutaneous absorption

Israel - Time Weighted Averages
5 mg/m3 TWA

Israel - Action Levels
2.5 mg/m3 AL

Mexico - Instruction No. 10 - TWAs
5 mg/m3 TWA

Mexico - Instruction No. 10 - STELs
10 mg/m3 STEL

STATE LISTS

California - Exposure Limits - PELs
5 mg/m3 PEL

California - Exposure Limits - Skin Notation
material may be absorbed through the skin, eyes or mucous membrane

California - Directors List of Hazardous Substances (8 CCR 339)
[present]

Florida Hazardous Substance List
[present]

Massachusetts Right To Know List
[present]

Minnesota Hazardous Substance List
[present]

Pennsylvania Right to Know List
[present]

2,3,4-TRICHLOROPHENOL   15950-66-0

HEALTH AND SAFETY LISTS

U.S. DOT - Appendix B - Marine Pollutants
DOT regulated marine pollutant

NIOSH - Selected LD50s and LC50s
Oral, rat: LD50 = 15 mg/kg Skin, rat: LD50 = 64 mg/kg

ENVIRONMENTAL LISTS

CERCLA/SARA - Section 302 Extremely Hazardous Substances and TPQs
TPQ = 500 pounds

STATE LISTS

Florida Hazardous Substance List
effective March 13, 1992

Massachusetts Right To Know List
extraordinarily hazardous

NJ Right to Know List (Total)
sn 2837

Pennsylvania Right to Know List
environmental hazard

PROPOSED REGULATIONS

CERCLA/SARA - Proposed Hazardous Substance Additions
proposed RQ = 1 pound (.454 kg)

CERCLA/SARA - 1989 Proposed RQ Adjustments
proposed RQ = 100 pounds (45.4 kg)

TRICHLOROPHENOL   25167-82-2

ENVIRONMENTAL LISTS

TSCA - HDD/HDF - Precursors Required for Reporting
[present]

2,4,5-(TRICHLOROPHENOXY) ACETIC ACID-ISO-   93-78-7
PROPYL ESTER

ENVIRONMENTAL LISTS

Class 1 Ozone Depletors
ozone depletion potential = 1.0

(2,4,5-TRICHLOROPHENOXY)-ACETIC ACID   RR-01622-8
ESTERS
SEE ALSO:
    K099-HAZARDOUS WASTES
    F039-HAZARDOUS WASTES
    K043-HAZARDOUS WASTES
    F023-HAZARDOUS WASTES
    F027-HAZARDOUS WASTES
    K105-HAZARDOUS WASTES
    F020-HAZARDOUS WASTES

HEALTH AND SAFETY LISTS

U.S. DOT - Appendix A Table 1 - Hazardous Substances
final RQ = 10 pounds (4.54 kg)

IARC - Group Unspecified
[present] (Listed under 'Chlorophenols')

NIOSH - Selected LD50s and LC50s
Oral, rat: LD50 = 820 mg/kg Skin, mammal: LD50 = 700 mg/kg

NTP Chemical Status Reports - Testing Status and NTIS Number
Technical reports printed (PB293770/AS)

NTP Chemical Status Reports - Evidence of Carcinogenicity
male rat-positive; female rat-negative; male mice-positive; female mice-positive

NTP Seventh Report - Suspect Carcinogens
suspect carcinogen

OSHA - Possible Select Carcinogens
[present]

ENVIRONMENTAL LISTS

ATSDR Priority List
Rank (of 275): 084

CERCLA/SARA - Section 313 - Emission Reporting
form R reporting required for 0.1% de minimus concentration

CERCLA/SARA - Hazardous Substances and their Reportable Quantities
final RQ = 10 pounds (4.54 kg)

Clean Air Act (1990) - List of Hazardous Air Contaminants
[present]

Clean Water Act - Hazardous Substances
[present] (Listed under 'Trichlorophenol')

Clean Water Act - Priority Pollutants
[present]

EPA - Carcinogen Hazard Ranking for RQ Adjustment
Hazard ranking = Low

RCRA - D Series - Maximum Concentration of Contaminants
waste number D042; regulatory level = 2.0 mg/L

RCRA - D Series - Chronic Toxicity Reference Levels
chronic toxicity reference level = 0.02 mg/L

RCRA - Hazardous Constituents-Appendix VIII
contained in F027 waste stream

RCRA - Basis for Listing - Appendix VII
Included in waste streams: F039, K043, K099, K105

RCRA - TSD Facilities Ground Water Monitoring
TM 8040 = 5 ug/L PQL; TM 8270 = 10 ug/L PQL

RCRA - Universal Treatment Standards (LDR)
WW: 0.035 mg/l; NWW: 7.4 mg/kg

INTERNATIONAL LISTS

Canada - Drinking Water Quality - MACs
0.005 mg/L MAC

Canada - Drinking Water Quality - AOs
<= 0.002 mg/L AO

Mexico - Wastewater - Organic Toxic Pollutants and Heavy Metals
Listed under [Chlorinated Phenols]

Mexico - Drinking Water - Ecological Criteria
0.01 mg/l Substance presents persistence, bioaccumulations or risk of cancer, reduce human exposure to a minimum; This level has been extrapolated by using a mathematic model

STATE LISTS

California - Air Bill 2588 Appendix A-I
known or potential carcinogen

California - Prop. 65 - Cancer list
carcinogen - initial date 1/1/88

California - Prop. 65 - No Significant Risk Levels
no significant risk level = 10 ug/day

California - Directors List of Hazardous Substances (8 CCR 339)
[present] (Listed under 'Trichlorophenols')

Florida Hazardous Substance List
[present]

Massachusetts Right To Know List
carcinogen; extraordinarily hazardous

Minnesota Hazardous Substance List
carcinogen

NJ Right to Know List (Total)
sn 1894

NJ Special Hazardous Substances
(carcinogen)

Pennsylvania Right to Know List
environmental hazard; special hazardous substance

Pennsylvania RTK - Special Hazardous Substances
[present]

2-(2,4,5-TRICHLOROPHENOXY) ETHYL 2,2-DICHLO-   136-25-4
PROPIONATE (ERBON)
SEE ALSO:
    TETRACHLORODIBENZO-P-DIOXINS
    F039-HAZARDOUS WASTES

HEALTH AND SAFETY LISTS

U.S. DOT - Appendix A Table 1 - Hazardous Substances
final RQ = 10 pounds (4.54 kg)

IARC - Group Unspecified
[present] (Listed under 'Chlorophenols')

NIOSH - Selected LD50s and LC50s
Oral, rat: LD50 = 820 mg/kg

ENVIRONMENTAL LISTS

ATSDR Priority List
Rank (of 275): 139

CERCLA/SARA - Section 313 - Emission Reporting
form R reporting required for 1.0% de minimus concentration

CERCLA/SARA - Hazardous Substances and their Reportable Quantities
final RQ = 10 pounds (4.54 kg)

Clean Air Act (1990) - List of Hazardous Air Contaminants
[present]

CAA - HON Rule - SOCMI Chemicals
compliance by Oct. 24, 1994

CAA - HON Rule - Organic HAPs
[present]

Clean Water Act - Hazardous Substances
[present] (Listed under 'Trichlorophenol')

RCRA - D Series - Maximum Concentration of Contaminants
waste number D041; regulatory level = 400.0 mg/L

RCRA - D Series - Chronic Toxicity Reference Levels
chronic toxicity reference level = 4 mg/L

RCRA - Hazardous Constituents-Appendix VIII
contained in F027 waste stream

RCRA - Basis for Listing - Appendix VII
Included in waste stream: F039

RCRA - TSD Facilities Ground Water Monitoring
TM 8270 = 10 ug/L PQL

RCRA - Universal Treatment Standards (LDR)
WW: 0.18 mg/l; NWW: 7.4 mg/kg

TSCA - HDD/HDF - Chemicals Required for Testing
[present]

TSCA - Section 12(b) - Export Notification
export notification required - Section 4

STATE LISTS

California - Air Bill 2588 Appendix A-I
known or potential carcinogen: 6/91

California - Directors List of Hazardous Substances (8 CCR 339)
[present] (Listed under 'Trichlorophenols')

Massachusetts Right To Know List
[present]

NJ Right to Know List (Total)
sn 1895

Pennsylvania Right to Know List
environmental hazard

**TRICHLOROPHENOXYPROPIONIC ACID ESTER (3-   6047-17-2
(2,4,5))**

HEALTH AND SAFETY LISTS

U.S. DOT - Appendix A Table 1 - Hazardous Substances
final RQ = 10 pounds (4.54 kg)

ENVIRONMENTAL LISTS

CERCLA/SARA - Hazardous Substances and their Reportable Quantities
final RQ = 10 pounds (4.54 kg)

Clean Water Act - Hazardous Substances
[present] (Listed under 'Trichlorophenol')

STATE LISTS

California - Directors List of Hazardous Substances (8 CCR 339)
[present] (Listed under 'Trichlorophenols')

Massachusetts Right To Know List
[present]

NJ Right to Know List (Total)
sn 2956

Pennsylvania Right to Know List
environmental hazard

1,2,3-TRICHLOROPROPANE   96-18-4

HEALTH AND SAFETY LISTS

U.S. DOT - Appendix A Table 1 - Hazardous Substances
final RQ = 10 pounds (4.54 kg) (Listed under 'Trichlorophenol')

ENVIRONMENTAL LISTS

CERCLA/SARA - Hazardous Substances and their Reportable Quantities
final RQ = 10 pounds (4.54 kg) (Listed under 'Trichlorophenol')

Clean Water Act - Hazardous Substances
[present] (Listed under 'Trichlorophenol')

TSCA - HDD/HDF - Chemicals Required for Testing
[present]

TSCA - Section 12(b) - Export Notification
export notification required - Section 4

STATE LISTS

California - Directors List of Hazardous Substances (8 CCR 339)
[present] (Listed under 'Trichlorophenols')

Massachusetts Right To Know List
[present]

NJ Right to Know List (Total)
sn 2928

Pennsylvania Right to Know List
environmental hazard

TRICHLOROPROPANE   25735-29-9

HEALTH AND SAFETY LISTS

U.S. DOT - Appendix A Table 1 - Hazardous Substances
final RQ = 10 pounds (4.54 kg) (Listed under 'Trichlorophenol')

ENVIRONMENTAL LISTS

CERCLA/SARA - Hazardous Substances and their Reportable Quantities
final RQ = 10 pounds (4.54 kg) (Listed under 'Trichlorophenol')

Clean Water Act - Hazardous Substances
[present] (Listed under 'Trichlorophenol')

STATE LISTS

California - Directors List of Hazardous Substances (8 CCR 339)
[present] (Listed under 'Trichlorophenols')

Massachusetts Right To Know List
[present]

NJ Right to Know List (Total)
sn 2927

Pennsylvania Right to Know List
environmental hazard

TRICHLOROSILANE   10025-78-2

HEALTH AND SAFETY LISTS

U.S. DOT - Appendix A Table 1 - Hazardous Substances
final RQ = 10 pounds (4.54 kg) (Listed under 'Trichlorophenol')

ENVIRONMENTAL LISTS

CERCLA/SARA - Hazardous Substances and their Reportable Quantities
final RQ = 10 pounds (4.54 kg) (Listed under 'Trichlorophenol')

Clean Water Act - Hazardous Substances
[present] (Listed under 'Trichlorophenol')

STATE LISTS

California - Directors List of Hazardous Substances (8 CCR 339)
[present] (Listed under 'Trichlorophenols')

Massachusetts Right To Know List
[present]

NJ Right to Know List (Total)
sn 2926

Pennsylvania Right to Know List
environmental hazard

TRICHLOROTETRAFLUOROPROPANE   134237-38-0

HEALTH AND SAFETY LISTS
U.S. DOT - Appendix A Table 1 - Hazardous Substances
final RQ = 10 pounds (4.54 kg)

ENVIRONMENTAL LISTS
ATSDR Priority List
Rank (of 275): 258
CERCLA/SARA - Hazardous Substances and their Reportable Quantities
final RQ = 10 pounds (4.54 kg)
Clean Water Act - Hazardous Substances
[present]
EPA - Carcinogen Hazard Ranking for RQ Adjustment
Hazard ranking = Low
RCRA - Basis for Listing - Appendix VII
Included in waste streams: F020, F023, F027, F028

STATE LISTS
California - Directors List of Hazardous Substances (8 CCR 339)
[present]
Massachusetts Right To Know List
[present]
NJ Right to Know List (Total)
sn 3011
Pennsylvania Right to Know List
environmental hazard

TRICHLORO-S-TRIAZINETRIONE   87-90-1

HEALTH AND SAFETY LISTS
NIOSH - Selected LD50s and LC50s
Oral, rat: LD50 = 495 mg/kg

STATE LISTS
NJ Right to Know List (Total)
sn 1803

TRICHLOROTRIETHYLAMINE HYDROCHLORIDE   RR-01541-8

INTERNATIONAL LISTS
Canada - Ontario - OHSA - TWAEVs
10 mg/m3 TWAEV

1,1,2-TRICHLORO-1,2,2-TRIFLUOROETHANE   76-13-1

STATE LISTS
California - Directors List of Hazardous Substances (8 CCR 339)
[present]

1,1,1-TRICHLORO-2,2,2-TRIFLUOROETHANE   354-58-5

HEALTH AND SAFETY LISTS
NIOSH - Selected LD50s and LC50s
Oral, rat: LD50 = 500 mg/kg

STATE LISTS
NJ Right to Know List (Total)
sn 1900

TRICHLOROTRIFLUOROPROPANE   134237-40-4
SEE ALSO:
F039-HAZARDOUS WASTES

HEALTH AND SAFETY LISTS
ACGIH 1995 - Time Weighted Averages
10 ppm TWA; 60 mg/m3 TWA

ACGIH 1995 - Skin Designations
skin - potential for cutaneous absorption
NFPA - Flash Points
flash point = 160 degrees F (71 degrees C)
NFPA - Hazard Identification Ratings
health-3; flammability-2; reactivity-0
NIOSH - Selected LD50s and LC50s
Inhalation, mouse: LC50 = 3400 mg/m3 (2 hr) Oral, rat: LD50 = 320 mg/kg Skin, rabbit: LD50 = 1770 mg/kg
NIOSH 1990 - Pocket Guide - RELs
10 ppm TWA; 60 mg/m3 TWA
NIOSH 1990 - Pocket Guide - IDLHs
1000 ppm IDLH (not considering carcinogenic effects)
NIOSH 1990 - Pocket Guide - Carcinogens
occupational carcinogen
NIOSH 1990 - Pocket Guide - Target organs
eyes, respiratory system, skin, CNS, liver
NIOSH 1990 - Pocket Guide - Skin list
Potential for dermal absorption
NTP Chemical Status Reports - Testing Status and NTIS Number
Prechronic studies for which toxicity technical reports were not prepared; technical reports printed (no NTIS number given)
NTP Chemical Status Reports - Evidence of Carcinogenicity
male rat: clear evidence; female rat: clear evidence; male mice: clear evidence; female mice: clear evidence
OSHA - Vacated PELs - Time Weighted Averages
10 ppm TWA; 60 mg/m3 TWA
OSHA - Final PELs - Time Weighted Averages
50 ppm TWA; 300 mg/m3 TWA

ENVIRONMENTAL LISTS
CERCLA/SARA - Section 313 - Emission Reporting
form R reporting required
Safe Drinking Water Act - Monitoring
monitoring required
RCRA - Hazardous Constituents-Appendix VIII
hazardous constituent - no waste number
RCRA - Basis for Listing - Appendix VII
Included in waste stream: F039
RCRA - TSD Facilities Ground Water Monitoring
TM 8010 = 10 ug/L PQL; TM 8240 = 5 ug/L PQL
RCRA - Universal Treatment Standards (LDR)
WW: 0.85 mg/l; NWW: 30 mg/kg
TSCA - Code of Federal Regulations Citations
40 CFR 716.120(a)
TSCA - Health and Safety Reporting List
Effective Date: June 1, 1987

INTERNATIONAL LISTS
Australian Exposure Standards - Time Weighted Averages
10 ppm TWA; 60 mg/m3 TWA
Australian Exposure Standards - Skin Effects
skin absorption
Canada - WHMIS: Ingredient Disclosure
1% item 1616 (1649)
Canada - Alberta - 8 Hour Occupational Exposure Limit
50 ppm TWA; 302 mg/m3 TWA
Canada - Alberta - 15 Minute Occupational Exposure Limit
75 ppm STEL; 452 mg/m3 STEL
Canada - British Columbia - 8 Hour Exposure Limits
50 ppm TWA; 300 mg/m3 TWA
Canada - British Columbia - 15 Minute Exposure Limits
75 ppm STEL; 450 mg/m3 STEL
Canada - Ontario - OHSA - TWAEVs
10 ppm TWAEV; 60 mg/m3 TWAEV

Canada - Ontario - OHSA - Skin Notations
absorption through skin, eyes, or mucous membranes
Canada - Quebec - Time-Weighted Average Exposure Values
10 ppm TWAEV;   60 mg/m3 TWAEV
Canada - Quebec - Skin Designations
absorbed through the skin
United Kingdom - Occupational Exposure Standards - TWAs
50 ppm TWA;   300 mg/m3 TWA
United Kingdom - Occupational Exposure Standards - STELs
75 ppm STEL;   450 mg/m3 STEL
German (DFG) - MAK Values
50 ppm MAK;   300 mg/m3 MAK
German (DFG) - Peak Limitations
5 x normal MAK (30 min. average value);   don't exceed 2 times during shift
German (DFG) - Carcinogens
animal evidence of carcinogenicity
Israel - Time Weighted Averages
10 ppm TWA;   60 mg/m3 TWA
Israel - Action Levels
5 ppm AL;   30 mg/m3 AL
Mexico - Instruction No. 10 - TWAs
50 ppm TWA;   300 mg/m3 TWA
Mexico - Instruction No. 10 - STELs
75 ppm STEL;   450 mg/m3 STEL

STATE LISTS

California - Prop. 65 - Cancer list
carcinogen - initial date 10/1/92
California - Exposure Limits - PELs
10 ppm PEL;   60 mg/m3 PEL
California - Directors List of Hazardous Substances (8 CCR 339)
[present]
Florida Hazardous Substance List
[present]
Massachusetts Right To Know List
[present]
Minnesota Hazardous Substance List
[present]
NJ Special Hazardous Substances
(mutagen)
Pennsylvania Right to Know List
environmental hazard

PROPOSED REGULATIONS

NTP - Proposed Additions to Annual Report on Carcinogens
proposed as a suspect carcinogen for the NTP 9th report
Safe Drinking Water Act - Priority list
[present]

T2-TRICHOTHECENE   **RR-01542-9**
SEE ALSO:
K017-HAZARDOUS WASTES

ENVIRONMENTAL LISTS

RCRA - Hazardous Constituents-Appendix VIII
hazardous constituent - no waste number
RCRA - Basis for Listing - Appendix VII
Included in waste stream: K017

TRICLOPYR, TRIETHYLAMMONIUM SALT   **57213-69-1**
HEALTH AND SAFETY LISTS

U.S. DOT - Substances From 49 CFR 172.101
regulated by DOT (UN1295)
U.S. DOT - Hazard Classes
DOT hazard class = 4.3

NFPA - Flash Points
flash point = 7 degrees F (-14 degrees C)
NFPA - Hazard Identification Ratings
health-3; flammability-4; reactivity-2 (avoid use of water)
NIOSH - Selected LD50s and LC50s
Inhalation, mouse: LC50 = 1500 mg/m3 (2 hr) Oral, rat: LD50 = 1030 mg/kg
OSHA - List of Highly Hazardous Chemicals
threshhold quantity = 5000 pounds

ENVIRONMENTAL LISTS

CAA - Flammable Substances for Accidental Release Prevention
threshold quantity = 10,000 lbs

INTERNATIONAL LISTS

Canada - WHMIS: Ingredient Disclosure
1% item 1617 (1650)

STATE LISTS

Florida Hazardous Substance List
[present]
Massachusetts Right To Know List
[present]
NJ Right to Know List (Total)
sn 1903
NJ Special Hazardous Substances
(corrosive; flammable - fourth degree; reactive - second degree)
Pennsylvania Right to Know List
[present]

(Z)-9-TRICOSENE   **27519-02-4**
ENVIRONMENTAL LISTS

Class 2 Ozone Depletors
ozone depletion weight reserved

TRICRESYL PHOSPHATE   **1330-78-5**
HEALTH AND SAFETY LISTS

U.S. DOT - Substances From 49 CFR 172.101
regulated by DOT (UN2468)
U.S. DOT - Hazard Classes
dry: DOT hazard class = 5.1
NIOSH - Selected LD50s and LC50s
Oral, rat: LD50 = 406 mg/kg Skin, rabbit: LD50 = 20 gm/kg

INTERNATIONAL LISTS

Canada - WHMIS: Ingredient Disclosure
1% item 1614 (148)

STATE LISTS

Florida Hazardous Substance List
[present]
Massachusetts Right To Know List
[present]
NJ Right to Know List (Total)
sn 1892
Pennsylvania Right to Know List
[present]

1-TRIDECANOL   **112-70-9**
HEALTH AND SAFETY LISTS

IARC - Group 3 (not classifiable)
[present]

TRIDECYL ACID PHOSPHATE   **5116-94-9**
SEE ALSO:
F002-HAZARDOUS WASTES
F039-HAZARDOUS WASTES

HEALTH AND SAFETY LISTS

ACGIH 1995 - Time Weighted Averages
1000 ppm TWA; 7670 mg/m3 TWA

ACGIH 1995 - Short Term Exposure Limits
1250 ppm STEL; 9590 mg/m3 STEL

NIOSH - Selected LD50s and LC50s
Oral, rat: LD50 = 43 gm/kg

NIOSH 1990 - Pocket Guide - RELs
1000 ppm TWA; 7600 mg/m3 TWA; 1250 ppm STEL; 9500 mg/m3 STEL

NIOSH 1990 - Pocket Guide - IDLHs
4500 ppm IDLH

NIOSH 1990 - Pocket Guide - Target organs
skin, heart

OSHA - Vacated PELs - Time Weighted Averages
1000 ppm TWA; 7600 mg/m3 TWA

OSHA - Vacated PELs - Short Term Exposure Limits
1250 ppm STEL; 9500 mg/m3 STEL

OSHA - Final PELs - Time Weighted Averages
1000 ppm TWA; 7600 mg/m3 TWA

ENVIRONMENTAL LISTS

CERCLA/SARA - Section 313 - Emission Reporting
form R reporting required for 1.0% de minimus concentration

Class 1 Ozone Depletors
ozone depletion potential = 0.8

CAA - HON Rule - SOCMI Chemicals
compliance by Oct. 24, 1994

List of Pesticide Product Inert Ingredients
[present]

RCRA - Basis for Listing - Appendix VII
Included in waste stream: F002, F039

RCRA - Universal Treatment Standards (LDR)
WW: 0.057; NWW: 30 mg/kg

TSCA - Code of Federal Regulations Citations
40 CFR 716.120(a)

TSCA - Health and Safety Reporting List
Effective Date: April 13, 1989

INTERNATIONAL LISTS

Australian Exposure Standards - Time Weighted Averages
1000 ppm TWA; 7670 mg/m3 TWA

Australian Exposure Standards - Short Term Exposure Limits
1250 ppm STEL; 9590 mg/m3 STEL

Canada - WHMIS: Ingredient Disclosure
1% item 1618 (1651)

Canada - Alberta - 8 Hour Occupational Exposure Limit
1000 ppm TWA; 7664 mg/m3 TWA

Canada - Alberta - 15 Minute Occupational Exposure Limit
1250 ppm STEL; 9580 mg/m3 STEL

Canada - British Columbia - 8 Hour Exposure Limits
1000 ppm TWA; 7600 mg/m3 TWA

Canada - British Columbia - 15 Minute Exposure Limits
1250 ppm STEL; 9500 mg/m3 STEL

Canada - Ontario - OHSA - TWAEVs
1000 ppm TWAEV; 7650 mg/m3 TWAEV

Canada - Ontario - OHSA - STEVs
1250 ppm STEV; 9560 mg/m3 STEV

Canada - Quebec - Time Weighted Average Exposure Values
1000 ppm TWAEV; 7670 mg/m3 TWAEV

Canada - Quebec - Short-term Exposure Values
1250 ppm STEV; 9590 mg/m3 STEV

United Kingdom - Occupational Exposure Standards - TWAs
1000 ppm TWA; 7600 mg/m3 TWA

United Kingdom - Occupational Exposure Standards - STELs
1250 ppm STEL; 9500 mg/m3 STEL

German (DFG) - MAK Values
500 ppm MAK; 3800 mg/m3 MAK

German (DFG) - Peak Limitations
2 x normal MAK (1 hour momentary value); don't exceed 3 times per shift

Israel - Time Weighted Averages
1000 ppm TWA; 7670 mg/m3 TWA

Israel - Short Term Exposure Limits
1250 ppm STEL; 9590 mg/m3 STEL

Israel - Action Levels
500 ppm AL; 3835 mg/m3 AL

Mexico - Instruction No. 10 - TWAs
1000 ppm TWA; 1600 mg/m3 TWA

Mexico - Instruction No. 10 - STELs
1250 ppm STEL; 9500 mg/m3 STEL

STATE LISTS

California - Air Bill 2588 Appendix A-I
[present]

California - Exposure Limits - PELs
1000 ppm PEL; 7600 mg/m3 PEL

California - Exposure Limits - STELs
1250 ppm STEL; 9500 mg/m3 STEL

California - Exposure Limits - Ceilings
C 2000 ppm

California - Directors List of Hazardous Substances (8 CCR 339)
[present]

Florida Hazardous Substance List
[present]

Massachusetts Right To Know List
[present]

Minnesota Hazardous Substance List
[present]

Pennsylvania Right to Know List
environmental hazard

PROPOSED REGULATIONS

Canada - Ontario - Proposed Occupational TWAEVs
500 ppm TWAEV; 4000 mg/m3 TWAEV

Canada - Ontario - Proposed Occupational STEVs
750 ppm STEV; 6000 mg/m3 STEV

TRIDECYL ALCOHOL, ETHOXYLATED   **24938-91-8**

HEALTH AND SAFETY LISTS

NTP Chemical Status Reports - Testing Status and NTIS Number
Prechronic studies completed: in review for further evaluation

TRIDECYL ALCOHOL, ETHOXYLATED,   **9046-01-9**
PHOSPHATED

ENVIRONMENTAL LISTS

Class 2 Ozone Depletors
ozone depletion weight reserved

TRIDECYLBENZENESULFONIC ACID, DIMETHY- **60883-89-8**
LAMINE SALT

HEALTH AND SAFETY LISTS

IARC - Group 3 (not classifiable)
[present]

TRIDECYLBENZENESULFONIC ACID, PROPY- **60883-90-1**
LAMINE SALT

ENVIRONMENTAL LISTS

CERCLA/SARA - Section 313 - Emission Reporting
form R reporting required

TRIDECYL PHOSPHITE 2929-86-4
  ENVIRONMENTAL LISTS
    List of Pesticide Product Inert Ingredients
    [present]

TRI-N-DECYL TRIMELLITATE 4130-35-2
  SEE ALSO:
  BISAZOBIPHENYL DYES
  HEALTH AND SAFETY LISTS
    U.S. DOT - Substances From 49 CFR 172.101
    regulated by DOT (UN2574)
    U.S. DOT - Hazard Classes
    DOT hazard class = 6.1
    U.S. DOT - Appendix B - Marine Pollutants
    less than 1% ortho isomer: DOT regulated marine pollutant; more than 1% ortho isomer: DOT regulated severe marine pollutant
    NIOSH - Selected LD50s and LC50s
    Oral, rat: LD50 = 5190 mg/kg Skin, cat: LD50 = 1500 mg/kg
    NTP Chemical Status Reports - Testing Status and NTIS Number
    Galley or camera copy technical reports in progress; prechronic studies for which toxicity technical reports were not prepared
  ENVIRONMENTAL LISTS
    EPA - Master Testing List
    [present]
    TSCA - Code of Federal Regulations Citations
    40 CFR 712.30(d); 40 CFR 716.120(c)
    TSCA - PAIR - Reporting List
    Reporting Date: November 19, 1982
  INTERNATIONAL LISTS
    Canada - WHMIS: Ingredient Disclosure
    1% item 1619 (1396)
  STATE LISTS
    NJ Right to Know List (Total)
    sn 3130

2,4,6-TRI(DIMETHYLAMINOMETHYL)PHENOL 90-72-2
  HEALTH AND SAFETY LISTS
    NFPA - Flash Points
    *flash point = 250 degrees F (121 degrees C); *See list description*
    NFPA - Hazard Identification Ratings
    health-0; flammability-1; reactivity-0
    NIOSH - Selected LD50s and LC50s
    Oral, rat: LD50 = 17200 mg/kg Skin, rabbit: LD50 = 5600 mg/kg
  ENVIRONMENTAL LISTS
    List of Pesticide Product Inert Ingredients
    [present]

TRIETHANOLAMINE 102-71-6
  ENVIRONMENTAL LISTS
    List of Pesticide Product Inert Ingredients
    [present]

TRIETHANOLAMINE DODECYLBENZOSULFONATE 27323-41-7
  ENVIRONMENTAL LISTS
    List of Pesticide Product Inert Ingredients
    [present]

TRIETHANOLAMINE LAURYL SULFATE 139-96-8
  ENVIRONMENTAL LISTS
    List of Pesticide Product Inert Ingredients
    [present]

TRIETHANOLAMINE OLEATE 2717-15-9
  ENVIRONMENTAL LISTS
    List of Pesticide Product Inert Ingredients
    [present]

TRIETHANOLAMINE PHOSPHATE 10017-56-8
  ENVIRONMENTAL LISTS
    List of Pesticide Product Inert Ingredients
    [present]

TRIETHANOLAMINE SALTS OF FATTY ACIDS RR-01562-3
  HEALTH AND SAFETY LISTS
    NFPA - Flash Points
    flash point = 455 degrees F (235 degrees C)
    NFPA - Hazard Identification Ratings
    health-0; flammability-1; reactivity-0

TRIETHANOLAMINE STEARATE 4568-28-9
  ENVIRONMENTAL LISTS
    List of Pesticide Product Inert Ingredients
    [present]

TRIETHANOLAMINE SULFATE 7376-31-0
  HEALTH AND SAFETY LISTS
    NIOSH - Selected LD50s and LC50s
    Oral, rat: LD50 = 1200 mg/kg Skin, rat: LD50 = 1280 mg/kg

TRIETHOXYETHYL SILANE 78-07-9
  HEALTH AND SAFETY LISTS
    ACGIH 1995 - Time Weighted Averages
    5 mg/m3 TWA
    IARC - Group 3 (not classifiable)
    [present]
    NFPA - Flash Points
    flash point = 354 degrees F (179 degrees C)
    NFPA - Hazard Identification Ratings
    health-2; flammability-1; reactivity-1
    NIOSH - Selected LD50s and LC50s
    Oral, rat: LD50 = 8 gm/kg
    NTP Chemical Status Reports - Testing Status and NTIS Number
    Two year studies: scheduled for peer review; prechronic studies for which toxicity technical reports were not prepared
  ENVIRONMENTAL LISTS
    CAA - HON Rule - SOCMI Chemicals
    compliance by Oct. 24, 1994
    EPA - Master Testing List
    [present]
    List of Pesticide Product Inert Ingredients
    [present]
    TSCA - Code of Federal Regulations Citations
    40 CFR 712.30(w); 40 CFR 716.120(a)
    TSCA - PAIR - Reporting List
    Reporting Date: June 13, 1989
    TSCA - Health and Safety Reporting List
    Effective Date: April 13, 1989
  INTERNATIONAL LISTS
    Canada - WHMIS: Ingredient Disclosure
    1% item 1621 (1663)
  STATE LISTS
    Florida Hazardous Substance List
    [present]
    Massachusetts Right To Know List
    [present]

Pennsylvania Right to Know List
[present]

TRIETHOXYSILANE   998-30-1

HEALTH AND SAFETY LISTS

U.S. DOT - Appendix A Table 1 - Hazardous Substances
final RQ = 1000 pounds (454 kg)

ENVIRONMENTAL LISTS

CERCLA/SARA - Hazardous Substances and their Reportable Quantities
final RQ = 1000 pounds (454 kg)

Clean Water Act - Hazardous Substances
[present]

List of Pesticide Product Inert Ingredients
[present]

STATE LISTS

California - Directors List of Hazardous Substances (8 CCR 339)
[present] (exempt when in solution)

Massachusetts Right To Know List
[present]

NJ Right to Know List (Total)
sn 1905

Pennsylvania Right to Know List
environmental hazard

3-(TRIETHOXYSILYL)PROPYLAMINE   919-30-2

ENVIRONMENTAL LISTS

List of Pesticide Product Inert Ingredients
[present]

TRIETHYLALLYLGERMANIUM   1793-90-4

ENVIRONMENTAL LISTS

List of Pesticide Product Inert Ingredients
[present]

TRIETHYLALUMINUM   97-93-8

ENVIRONMENTAL LISTS

List of Pesticide Product Inert Ingredients
[present]

TRIETHYLAMINE   121-44-8

ENVIRONMENTAL LISTS

TSCA - Chemicals with Significant New Use Rules
PMN numbers: P-92-156; P-92-157; P-92-159

TRIETHYLAMINE CITRATE   78871-22-4

ENVIRONMENTAL LISTS

List of Pesticide Product Inert Ingredients
[present]

TRIETHYLAMINE NITRILOTRIACETATE   RR-01155-2

ENVIRONMENTAL LISTS

List of Pesticide Product Inert Ingredients
[present]

1,2,4-TRIETHYLBENZENE   25340-18-5

HEALTH AND SAFETY LISTS

NIOSH - Selected LD50s and LC50s
Oral, rat: LD50 = 14 gm/kg Skin, rabbit: LD50 = 16 gm/kg

ENVIRONMENTAL LISTS

List of Pesticide Product Inert Ingredients
[present]

TRIETHYLBORANE   97-94-9

HEALTH AND SAFETY LISTS

NIOSH - Selected LD50s and LC50s
Inhalation, mouse: LC50 = 500 mg/m3 (2 hr)

ENVIRONMENTAL LISTS

CERCLA/SARA - Section 302 Extremely Hazardous Substances and TPQs
TPQ = 500 pounds

STATE LISTS

Florida Hazardous Substance List
effective March 13, 1992

Massachusetts Right To Know List
extraordinarily hazardous

NJ Right to Know List (Total)
sn 2838

Pennsylvania Right to Know List
environmental hazard

PROPOSED REGULATIONS

CERCLA/SARA - Proposed Hazardous Substance Additions
proposed RQ = 1 pound (.454 kg)

CERCLA/SARA - 1989 Proposed RQ Adjustments
proposed RQ = 100 pounds (45.4 kg)

TRIETHYL CITRATE   77-93-0

HEALTH AND SAFETY LISTS

NIOSH - Selected LD50s and LC50s
Oral, rat: LD50 = 1780 mg/kg Skin, rabbit: LD50 = 4000 mg/kg

TRIETHYLENE GLYCOL   112-27-6

INTERNATIONAL LISTS

Canada - WHMIS: Ingredient Disclosure
1% item 1622 (1664)

TRIETHYLENE GLYCOL BIS(2-ETHYLHEXANOATE)   94-28-0

HEALTH AND SAFETY LISTS

NFPA - Flash Points
ignites spontaneously in air

NFPA - Hazard Identification Ratings
health-3; flammability-4; reactivity-3 (do not use water, foam or halogenated extinguishing agents)

STATE LISTS

Florida Hazardous Substance List
[present]

Massachusetts Right To Know List
[present]

NJ Right to Know List (Total)
sn 1906

NJ Special Hazardous Substances
(flammable - third degree; reactive - third degree)

Pennsylvania Right to Know List
[present]

TRIETHYLENE GLYCOL BIS (2-ETHYLBUTYRATE)   95-08-9

HEALTH AND SAFETY LISTS

ACGIH 1995 - Time Weighted Averages
(1) ppm TWA; (4.1) mg/m3 TWA

ACGIH 1995 - Short Term Exposure Limits
(5) ppm STEL; (20.7) mg/m3 STEL

ACGIH 1995 - Skin Designations
skin - potential for cutaneous absorption

AIHA - Odor Threshold Values
geometric mean air odor threshold = 0.25 ppm (detectable); 2.8 ppm (recognizable)

U.S. DOT - Substances From 49 CFR 172.101
regulated by DOT (UN1296)

U.S. DOT - Hazard Classes
DOT hazard class = 3

U.S. DOT - Appendix A Table 1 - Hazardous Substances
final RQ = 5000 pounds (2270 kg)

NFPA - Flash Points
flash point = 16 degrees F (-7 degrees C)

NFPA - Hazard Identification Ratings
health-3; flammability-3; reactivity-0

NIOSH - Selected LD50s and LC50s
Inhalation, mammal: LC50 = 6 gm/m3 8 hr Oral, rat: LD50 = 460 mg/kg Skin, rabbit: LD50 = 570 mg/kg

NIOSH 1990 - Pocket Guide - RELs
See Appendix D

NIOSH 1990 - Pocket Guide - IDLHs
1000 ppm IDLH

NIOSH 1990 - Pocket Guide - Target organs
skin, eyes, respiratory system

OSHA - Vacated PELs - Time Weighted Averages
10 ppm TWA; 40 mg/m3 TWA

OSHA - Vacated PELs - Short Term Exposure Limits
15 ppm STEL; 60 mg/m3 STEL

OSHA - Final PELs - Time Weighted Averages
25 ppm TWA; 100 mg/m3 TWA

ENVIRONMENTAL LISTS

CERCLA/SARA - Section 313 - Emission Reporting
form R reporting required

CERCLA/SARA - Hazardous Substances and their Reportable Quantities
final RQ = 5000 pounds (2270 kg)

Clean Air Act (1990) - List of Hazardous Air Contaminants
[present]

CAA - HON Rule - SOCMI Chemicals
compliance by July 24, 1995

CAA - HON Rule - Organic HAPs
[present]

Clean Water Act - Hazardous Substances
[present]

List of Pesticide Product Inert Ingredients
[present]

TSCA - Code of Federal Regulations Citations
40 CFR 716.120(a)

TSCA - Health and Safety Reporting List
Effective Date: January 13, 1984

INTERNATIONAL LISTS

Australian Exposure Standards - Time Weighted Averages
3 ppm TWA; 12 mg/m3 TWA

Australian Exposure Standards - Short Term Exposure Limits
5 ppm STEL; 20 mg/m3 STEL

Canada - WHMIS: Ingredient Disclosure
1% item 1623 (1665)

Canada - Alberta - 8 Hour Occupational Exposure Limit
10 ppm TWA; 42 mg/m3 TWA

Canada - Alberta - 15 Minute Occupational Exposure Limit
15 ppm STEL; 63 mg/m3 STEL

Canada - British Columbia - 8 Hour Exposure Limits
25 ppm TWA; 100 mg/m3 TWA

Canada - British Columbia - 15 Minute Exposure Limits
40 ppm STEL; 160 mg/m3 STEL

Canada - Ontario - OHSA - TWAEVs
10 ppm TWAEV; 41 mg/m3 TWAEV

Canada - Ontario - OHSA - STEVs
15 ppm STEV; 62 mg/m3 STEV

Canada - Quebec - Time-Weighted Average Exposure Values
10 ppm TWAEV; 41 mg/m3 TWAEV

Canada - Quebec - Short-term Exposure Values
15 ppm STEV; 62 mg/m3 STEV

United Kingdom - Occupational Exposure Standards - TWAs
10 ppm TWA; 40 mg/m3 TWA

United Kingdom - Occupational Exposure Standards - STELs
15 ppm STEL; 60 mg/m3 STEL

German (DFG) - MAK Values
10 ppm MAK; 40 mg/m3 MAK

German (DFG) - Peak Limitations
2 x normal MAK (10 min momentary value);   don't exceed 4 times per shift

Israel - Time Weighted Averages
10 ppm TWA; 41 mg/m3 TWA

Israel - Short Term Exposure Limits
15 ppm STEL; 62 mg/m3 STEL

Israel - Action Levels
5 ppm AL; 20.5 mg/m3 AL

Mexico - Instruction No. 10 - TWAs
25 ppm TWA; 100 mg/m3 TWA

Mexico - Instruction No. 10 - STELs
40 ppm STEL; 160 mg/m3 STEL

STATE LISTS

California - Air Bill 2588 Appendix A-I
6/91

California - Exposure Limits - PELs
10 ppm PEL; 40 mg/m3 PEL

California - Exposure Limits - STELs
15 ppm STEL; 60 mg/m3 STEL

California - Directors List of Hazardous Substances (8 CCR 339)
[present]

Florida Hazardous Substance List
[present]

Massachusetts Right To Know List
[present]

Minnesota Hazardous Substance List
[present]

NJ Right to Know List (Total)
sn 1907

NJ Special Hazardous Substances
(flammable - third degree)

Pennsylvania Right to Know List
environmental hazard

PROPOSED REGULATIONS

ACGIH 1995 - Notice of Intended Changes
(skin) 1 ppm TWA; 4.1 mg/m3 TWA; 3 ppm STEL; 12.4 mg/m3 STEL; A4-not classifiable as a human carcinogen

Canada - Ontario - Proposed Occupational TWAEVs
2 ppm TWAEV; 8 mg/m3 TWAEV

Canada - Ontario - Proposed Occupational STEVs
10 ppm STEV; 40 mg/m3 STEV

TRIETHYLENE GLYCOL DIACRYLATE   1680-21-3

ENVIRONMENTAL LISTS

List of Pesticide Product Inert Ingredients
[present]

TRIETHYLENE GLYCOL DIGLYCIDYL ETHER   1954-28-5

ENVIRONMENTAL LISTS

List of Pesticide Product Inert Ingredients
[present]

TRIETHYLENE GLYCOL, DIMETHYL ETHER   112-49-2

HEALTH AND SAFETY LISTS

U.S. DOT - Appendix B - Marine Pollutants
DOT regulated marine pollutant

NFPA - Flash Points
flash point = 181 degrees F (83 degrees C)

NFPA - Hazard Identification Ratings
flammability-2; reactivity-0

TRIETHYLENE GLYCOL DIMETHACRYLATE   109-16-0

HEALTH AND SAFETY LISTS

NFPA - Flash Points
ignites spontaneously in air

NFPA - Hazard Identification Ratings
health-1; flammability-3; reactivity-3 (avoid use of water of halogenated extinguishing agents)

STATE LISTS

Florida Hazardous Substance List
[present]

Massachusetts Right To Know List
[present]

Pennsylvania Right to Know List
[present]

TRIETHYLENE GLYCOL MONOBUTYL ETHER   143-22-6

HEALTH AND SAFETY LISTS

NFPA - Flash Points
flash point = 303 degrees F (151 degrees C)

NFPA - Hazard Identification Ratings
health-0; flammability-1; reactivity-0

NIOSH - Selected LD50s and LC50s
Inhalation, rat: LC50 = 1300 ppm 6 hr Oral, rat: LD50 = 5900 mg/kg

ENVIRONMENTAL LISTS

List of Pesticide Product Inert Ingredients
[present]

TRIETHYLENE GLYCOL MONOETHYL ETHER   112-50-5

HEALTH AND SAFETY LISTS

NFPA - Flash Points
flash point = 350 degrees F (177 degrees C)

NFPA - Hazard Identification Ratings
health-1; flammability-1; reactivity-0

NIOSH - Selected LD50s and LC50s
Oral, rat: LD50 = 17 gm/kg

ENVIRONMENTAL LISTS

CAA - HON Rule - SOCMI Chemicals
compliance by Oct. 24, 1994

List of Pesticide Product Inert Ingredients
[present]

INTERNATIONAL LISTS

Canada - WHMIS: Ingredient Disclosure
1% item 1625 (1667)

STATE LISTS

Pennsylvania Right to Know List
[present]

TRIETHYLENE GLYCOL MONOMETHYL ETHER   112-35-6

ENVIRONMENTAL LISTS

TSCA - Code of Federal Regulations Citations
40 CFR 712.30(x); 40 CFR 716.120(d)

TSCA - PAIR - Reporting List
Reporting Date: November 27, 1991

TSCA - Health and Safety Reporting List
Effective Date: September 30, 1991

TRIETHYLENEMELAMINE   51-18-3

HEALTH AND SAFETY LISTS

NIOSH - Selected LD50s and LC50s
Oral, rat: LD50 = 6000 mg/kg

TRIETHYLENE TETRAMINE   112-24-3

HEALTH AND SAFETY LISTS

AIHA - WEEL - Time Weighted Averages
1 mg/m3 TWA

AIHA - WEEL - Skin Absorption Designations
skin absorber

NIOSH - Selected LD50s and LC50s
Oral, rat: LD50 = 500 mg/kg Skin, rabbit: LD50 = 1900 mg/kg

STATE LISTS

Minnesota Hazardous Substance List
[present]

TRIETHYL PHOSPHATE   78-40-0

HEALTH AND SAFETY LISTS

IARC - Group 3 (not classifiable)
[present]

STATE LISTS

California - Directors List of Hazardous Substances (8 CCR 339)
[present]

TRIETHYL PHOSPHITE   122-52-1

HEALTH AND SAFETY LISTS

NFPA - Flash Points
flash point = 232 degrees F (111 degrees C)

NFPA - Hazard Identification Ratings
health-1; flammability-1; reactivity-0

ENVIRONMENTAL LISTS

CAA - HON Rule - SOCMI Chemicals
compliance by Oct. 24, 1994

STATE LISTS

California - Air Bill 2588 Appendix A-I
9/90

O,O,O-TRIETHYL PHOSPHOROTHIOATE   126-68-1

HEALTH AND SAFETY LISTS

NIOSH - Selected LD50s and LC50s
Oral, rat: LD50 = 10837 mg/kg

ENVIRONMENTAL LISTS

TSCA - Code of Federal Regulations Citations
40 CFR 712.30(d)

TSCA - PAIR - Reporting List
Reporting Date: November 19, 1982

INTERNATIONAL LISTS

Canada - WHMIS: Ingredient Disclosure
1% item 1626 (726)

TRIFLUOROACETIC ACID   76-05-1

SEE ALSO:
GLYCOL ETHERS

HEALTH AND SAFETY LISTS

NFPA - Flash Points
flash point = 290 degrees F (143 degrees C)

NFPA - Hazard Identification Ratings
health-0; flammability-1; reactivity-0

NIOSH - Selected LD50s and LC50s
Oral, rat: LD50 = 6730 mg/kg Skin, rabbit: LD50 = 3540 mg/kg

ENVIRONMENTAL LISTS

TSCA - Code of Federal Regulations Citations
40 CFR 712.30(o); 40 CFR 716.120(a); 40 CFR 799.5000

TSCA - PAIR - Reporting List
Reporting Date: August 19, 1985

TSCA - Health and Safety Reporting List
Effective Date: June 20, 1985

TSCA - Substances Subject to Testing Consent Orders
Test for: Health Effects

TSCA - Section 12(b) - Export Notification
export notification required - Section 4

INTERNATIONAL LISTS

Canada - WHMIS: Ingredient Disclosure
1% item 1627 (835)

TRIFLUOROACETIC ACID ANHYDRIDE   407-25-0
SEE ALSO:
GLYCOL ETHERS

HEALTH AND SAFETY LISTS

NFPA - Flash Points
flash point = 275 degrees F (135 degrees C)

NFPA - Hazard Identification Ratings
health-0; flammability-1; reactivity-0

NIOSH - Selected LD50s and LC50s
Oral, rat: LD50 = 10610 mg/kg Skin, rabbit: LD50 = 8 gm/kg

ENVIRONMENTAL LISTS

CAA - HON Rule - SOCMI Chemicals
compliance by Oct. 23, 1995

EPA - Master Testing List
[present]

TSCA - Code of Federal Regulations Citations
40 CFR 712.30(o); 40 CFR 716.120(a); 40 CFR 799.5000; 40 CFR 799.4440

TSCA - PAIR - Reporting List
Reporting Date: August 19, 1985

TSCA - Health and Safety Reporting List
Effective Date: June 20, 1985

TSCA - Substances Subject to Testing Consent Orders
Test for: Health Effects

TSCA - Section 12(b) - Export Notification
export notification required - Section 4

TRIFLUOROACETYLCHLORIDE   354-32-5

HEALTH AND SAFETY LISTS

NFPA - Flash Points
flash point = 245 degrees F (118 degrees C)

NFPA - Hazard Identification Ratings
health-0; flammability-1; reactivity-0

NIOSH - Selected LD50s and LC50s
Oral, rat: LD50 = 11300 mg/kg Skin, rabbit: LD50 = 7100 mg/kg

ENVIRONMENTAL LISTS

CAA - HON Rule - SOCMI Chemicals
compliance by Oct. 24, 1994

EPA - Master Testing List
[present]

TSCA - Code of Federal Regulations Citations
40 CFR 712.30(o); 40 CFR 716.120(a); 40 CFR 799.4440; 40 CFR 799.5000

TSCA - PAIR - Reporting List
Reporting Date: August 19, 1985

TSCA - Health and Safety Reporting List
Effective Date: June 20, 1985

TSCA - Substances Subject to Testing Consent Orders
Test for: Health Effects

TSCA - Chemical Test Rules
Testing required by: manufacturers; processors (40 CFR 799.4440)

TSCA - Section 12(b) - Export Notification
export notification required - Section 4

TRIFLUOROBROMOMETHANE   75-63-8

HEALTH AND SAFETY LISTS

IARC - Group 3 (not classifiable)
[present]

INTERNATIONAL LISTS

Canada - WHMIS: Ingredient Disclosure
0.1% item 1628 (1668)

STATE LISTS

California - Directors List of Hazardous Substances (8 CCR 339)
[present]

TRIFLUOROETHANE   27987-06-0

HEALTH AND SAFETY LISTS

U.S. DOT - Substances From 49 CFR 172.101
regulated by DOT (UN2259)

U.S. DOT - Hazard Classes
DOT hazard class = 8

NFPA - Flash Points
flash point = 275 degrees F (135 degrees C)

NFPA - Hazard Identification Ratings
health-3; flammability-1; reactivity-0

NIOSH - Selected LD50s and LC50s
Oral, rat: LD50 = 2500 mg/kg Skin, rabbit: LD50 = 805 mg/kg

ENVIRONMENTAL LISTS

EPA - Master Testing List
[present]

List of Pesticide Product Inert Ingredients
[present]

INTERNATIONAL LISTS

Canada - WHMIS: Ingredient Disclosure
0.1% item 1629 (1669)

STATE LISTS

Florida Hazardous Substance List
[present]

Massachusetts Right To Know List
[present]

NJ Right to Know List (Total)
sn 1908

NJ Special Hazardous Substances
(corrosive)

Pennsylvania Right to Know List
[present]

2,2,2-TRIFLUOROETHANOL   75-89-8

HEALTH AND SAFETY LISTS

NFPA - Flash Points
flash point = 240 degrees F (115 degrees C)

NFPA - Hazard Identification Ratings
health-0; flammability-1; reactivity-1

ENVIRONMENTAL LISTS

EPA - Master Testing List
[present]

List of Pesticide Product Inert Ingredients
[present]

TSCA - Code of Federal Regulations Citations
40 CFR 712.30(x); 40 CFR 716.120(d)

TSCA - PAIR - Reporting List
Reporting Date: December 27, 1990

TSCA - Health and Safety Reporting List
Effective Date: October 29, 1990

STATE LISTS

California - Air Bill 2588 Appendix A-I
9/89

TRIFLUOROMETHANE   75-46-7

HEALTH AND SAFETY LISTS

U.S. DOT - Substances From 49 CFR 172.101
regulated by DOT (UN2523)

U.S. DOT - Hazard Classes
DOT hazard class = 3

NIOSH - Selected LD50s and LC50s
Oral, rat: LD50 = 3200 mg/kg

STATE LISTS

NJ Right to Know List (Total)
sn 1909

2-(TRIFLUOROMETHYL) ANILINE   88-17-5

ENVIRONMENTAL LISTS

RCRA - Hazardous Constituents-Appendix VIII
hazardous constituent - no waste number

RCRA - TSD Facilities Ground Water Monitoring
TM 8270 = 10 ug/L PQL

TRIFLOUROMETHYLPHENYLISOCYANATE   1548-13-6

HEALTH AND SAFETY LISTS

U.S. DOT - Substances From 49 CFR 172.101
regulated by DOT (UN2699)

U.S. DOT - Hazard Classes
DOT hazard class = 8

NIOSH - Selected LD50s and LC50s
Inhalation, rat: LC50 = 10 gm/m3 8 hr Oral, rat: LD50 = 200 mg/kg

INTERNATIONAL LISTS

Canada - WHMIS: Ingredient Disclosure
1% item 1630 (149)

STATE LISTS

NJ Right to Know List (Total)
sn 1911

NJ Special Hazardous Substances
(corrosive)

TRIFLUOROPROPYLMETHYLCYCLOTRISILOXANE   2374-14-3

INTERNATIONAL LISTS

Canada - WHMIS: Ingredient Disclosure
1% item 1631 (238)

TRIFLURALIN   1582-09-8

HEALTH AND SAFETY LISTS

U.S. DOT - Substances From 49 CFR 172.101
regulated by DOT (UN3057)

U.S. DOT - Hazard Classes
DOT hazard class = 2.2

U.S. DOT - Substances Which Are Poisonous by Inhalation
gaseous hazardous material poisonous by inhalation (UN3057)

TRIFORINE [N,N'-[1,4-PIPERAZINEDIYL-BIS(2,2,2-TRICHLOROETHYLIDENE)] BISFORMAMIDE]   26644-46-2

HEALTH AND SAFETY LISTS

ACGIH 1995 - Time Weighted Averages
1000 ppm TWA; 6090 mg/m3 TWA

U.S. DOT - Substances From 49 CFR 172.101
regulated by DOT (UN1009)

U.S. DOT - Hazard Classes
DOT hazard class = 2.2

NIOSH - Selected LD50s and LC50s
Inhalation, rat: LC50 = 416 gm/m3 8 hr

NIOSH 1990 - Pocket Guide - RELs
1000 ppm TWA; 6100 mg/m3 TWA

NIOSH 1990 - Pocket Guide - IDLHs
50,000 ppm IDLH

NIOSH 1990 - Pocket Guide - Target organs
heart, CNS

OSHA - Vacated PELs - Time Weighted Averages
1000 ppm TWA; 6100 mg/m3 TWA

OSHA - Final PELs - Time Weighted Averages
1000 ppm TWA; 6100 mg/m3 TWA

ENVIRONMENTAL LISTS

CERCLA/SARA - Section 313 - Emission Reporting
form R reporting required for 1.0% de minimus concentration

Class 1 Ozone Depletors
ozone depletion potential = 10.0

INTERNATIONAL LISTS

Australian Exposure Standards - Time Weighted Averages
1000 ppm TWA; 6090 mg/m3 TWA

Canada - WHMIS: Ingredient Disclosure
1% item 1632 (1670)

Canada - CEPA Schedule II Part II - Toxic Substances (Export)
[present] (with the molecular formula CF3Br)

Canada - CEPA Schedule I - Toxic Substances
that has the molecular formula CF3Br: (a) quantities that may be imported; (b) prohibited commercial, manufacturing or processing uses

Canada - Alberta - 8 Hour Occupational Exposure Limit
1000 ppm TWA; 6090 mg/m3 TWA

Canada - Alberta - 15 Minute Occupational Exposure Limit
1200 ppm STEL; 7308 mg/m3 STEL

Canada - British Columbia - 8 Hour Exposure Limits
1000 ppm TWA; 6100 mg/m3 TWA

Canada - British Columbia - 15 Minute Exposure Limits
1200 ppm STEL; 7300 mg/m3 STEL

Canada - Ontario - OHSA - TWAEVs
1000 ppm TWAEV; 6085 mg/m3 TWAEV

Canada - Quebec - Time-Weighted Average Exposure Values
1000 ppm TWAEV; 6090 mg/m3 TWAEV

United Kingdom - Occupational Exposure Standards - TWAs
1000 ppm TWA; 6100 mg/m3 TWA

United Kingdom - Occupational Exposure Standards - STELs
1200 ppm STEL; 7300 mg/m3 STEL

German (DFG) - MAK Values
1000 ppm MAK; 6100 mg/m3 MAK

German (DFG) - Peak Limitations
2 x normal MAK (1 hour momentary value); don't exceed 3 times per shift

Israel - Time Weighted Averages
1000 ppm TWA; 6090 mg/m3 TWA

Israel - Action Levels
500 ppm AL; 3045 mg/m3 AL

Mexico - Instruction No. 10 - TWAs
1000 ppm TWA; 6100 mg/m3 TWA

Mexico - Instruction No. 10 - STELs
1200 ppm STEL; 7200 mg/m3 STEL

STATE LISTS

California - Exposure Limits - PELs
1000 ppm PEL; 6100 mg/m3 PEL

California - Directors List of Hazardous Substances (8 CCR 339)
[present]

Florida Hazardous Substance List
[present]

Massachusetts Right To Know List
[present]

Minnesota Hazardous Substance List
[present]

NJ Right to Know List (Total)
sn 1912

Pennsylvania Right to Know List
[present]

TRIFORMOXIME TRINITRATE   RR-01464-2

HEALTH AND SAFETY LISTS

U.S. DOT - Substances From 49 CFR 172.101
regulated by DOT (UN2035)

U.S. DOT - Hazard Classes
DOT hazard class = 2.1

STATE LISTS

NJ Right to Know List (Total)
sn 1914

TRIHALOMETHANES   RR-01615-9

HEALTH AND SAFETY LISTS

NIOSH - Selected LD50s and LC50s
Inhalation, mouse: LC50 = 2900 mg/m3 (8 hr) Oral, rat: LD50 = 240 mg/kg Skin, rat: LD50 = 1680 mg/kg

TRIHEXYL PHOSPHITE   6095-42-7

HEALTH AND SAFETY LISTS

U.S. DOT - Substances From 49 CFR 172.101
regulated by DOT (UN1984, UN3136)

U.S. DOT - Hazard Classes
DOT hazard class = 2.2

STATE LISTS

NJ Right to Know List (Total)
sn 1915

TRIISOBUTYL ALUMINUM   100-99-2

HEALTH AND SAFETY LISTS

U.S. DOT - Substances From 49 CFR 172.101
regulated by DOT (UN2942)

U.S. DOT - Hazard Classes
DOT hazard class = 6.1

INTERNATIONAL LISTS

Canada - WHMIS: Ingredient Disclosure
1% item 1633 (1671)

STATE LISTS

NJ Right to Know List (Total)
sn 1917

TRIISOBUTYLENE   7756-94-7

HEALTH AND SAFETY LISTS

U.S. DOT - Substances Which Are Poisonous by Inhalation
liquid hazardous material poisonous by inhalation

TRIISOCYANATOISOCYANURATE OF   RR-00510-7
ISOPHORONEDIISOCYANATE, SOLUTION

ENVIRONMENTAL LISTS

TSCA - PAIR - Reporting List
Effective Date: October 12, 1993; Reporting Date: February 28, 1994

TSCA - Health and Safety Reporting List
Effective Date: October 12, 1993; Sunset Date: October 12, 2003

TRIISOPROPANOLAMINE   122-20-3

HEALTH AND SAFETY LISTS

IARC - Group 3 (not classifiable)
[present]

NIOSH - Selected LD50s and LC50s
Oral, mouse: LD50 = 5000 mg/kg

NTP Chemical Status Reports - Testing Status and NTIS Number
Technical reports printed (PB278610/AS)

NTP Chemical Status Reports - Evidence of Carcinogenicity
male rat-negative; female rat-negative; male mice-negative; female mice-positive

ENVIRONMENTAL LISTS

ATSDR Priority List
Rank (of 275): 146

CERCLA/SARA - Section 313 - Emission Reporting
form R reporting required for 1.0% de minimus concentration

CERCLA/SARA - Hazardous Substances and their Reportable Quantities
final RQ = 1 pound (.454 kg)

Clean Air Act (1990) - List of Hazardous Air Contaminants
[present]

INTERNATIONAL LISTS

Canada - Drinking Water Quality - IMACs
0.045 mg/L IMAC

STATE LISTS

California - Air Bill 2588 Appendix A-I
6/91

California - Directors List of Hazardous Substances (8 CCR 339)
[present]

Massachusetts Right To Know List
carcinogen; extraordinarily hazardous

NJ Right to Know List (Total)
sn 1918

Pennsylvania Right to Know List
environmental hazard

PROPOSED REGULATIONS

Safe Drinking Water Act - Priority list
[present]

TRIISOPROPYLBENZENE   717-74-8

ENVIRONMENTAL LISTS

CERCLA/SARA - Section 313 - Emission Reporting
form R reporting required

TRIISOPROPYL BORATE   5419-55-6

HEALTH AND SAFETY LISTS

U.S. DOT - Hazard Classes
Forbidden from transport by the DOT

2,4,6-TRIISOPROPYLPHENOL   732-26-3

INTERNATIONAL LISTS

Canada - Drinking Water Quality - MACs
0.35 mg/L MAC

TRILAURYL TRITHIOPHOSPHITE 1656-63-9

HEALTH AND SAFETY LISTS

NFPA - Flash Points
flash point = 320 degrees F (160 degrees C)

NFPA - Hazard Identification Ratings
flammability-1; reactivity-0 (decomposes in water)

TRILOSTANE 13647-35-3

HEALTH AND SAFETY LISTS

NFPA - Flash Points
may ignite spontaneously in air

NFPA - Hazard Identification Ratings
health-3; flammibility-4; reactivity-3 (do not use water, foam, or halogenated extinguishing agents)

STATE LISTS

Florida Hazardous Substance List
[present]

Massachusetts Right To Know List
[present]

NJ Right to Know List (Total)
sn 1919

NJ Special Hazardous Substances
(flammable - third degree; reactive - third degree)

Pennsylvania Right to Know List
[present]

TRIMELLITIC ACID 528-44-9

HEALTH AND SAFETY LISTS

U.S. DOT - Substances From 49 CFR 172.101
regulated by DOT (UN2324)

U.S. DOT - Hazard Classes
DOT hazard class = 3

STATE LISTS

NJ Right to Know List (Total)
sn 1920

TRIMELLITIC ANHYDRIDE 552-30-7

STATE LISTS

NJ Right to Know List (Total)
sn 2840

TRIMETHADIONE 127-48-0

HEALTH AND SAFETY LISTS

NFPA - Flash Points
flash point = 320 degrees F (160 degrees C)

NFPA - Hazard Identification Ratings
health-2; flammability-1; reactivity-0

NIOSH - Selected LD50s and LC50s
Oral, rat: LD50 = 6500 mg/kg

ENVIRONMENTAL LISTS

List of Pesticide Product Inert Ingredients
[present]

INTERNATIONAL LISTS

Canada - WHMIS: Ingredient Disclosure
1% item 1635 (1680)

STATE LISTS

Florida Hazardous Substance List
[present]

Massachusetts Right To Know List
[present]

Pennsylvania Right to Know List
[present]

TRIMETHOPRIM 738-70-5

HEALTH AND SAFETY LISTS

NFPA - Flash Points
flash point = 207 degrees F (97 degrees C)

NFPA - Hazard Identification Ratings
health-0; flammability-1; reactivity-0

TRIMETHOPRIM/SULFAMETHOXAZOLE 8064-90-2

HEALTH AND SAFETY LISTS

U.S. DOT - Substances From 49 CFR 172.101
regulated by DOT (UN2616)

U.S. DOT - Hazard Classes
DOT hazard class = 3

NFPA - Flash Points
flash point = 82 degrees F (28 degrees C)

NFPA - Hazard Identification Ratings
health-3; flammability-3; reactivity-1

NIOSH - Selected LD50s and LC50s
Oral, mouse: LD50 = 2500 mg/kg

STATE LISTS

Florida Hazardous Substance List
[present]

Massachusetts Right To Know List
[present]

NJ Right to Know List (Total)
sn 1921

Pennsylvania Right to Know List
[present]

TRIMETHOXYBOROXINE 102-24-9

ENVIRONMENTAL LISTS

List of Pesticide Product Inert Ingredients
[present]

TRIMETHOXYSILANE 2487-90-3

HEALTH AND SAFETY LISTS

NFPA - Flash Points
flash point = 398 degrees F (203 degrees C)

NFPA - Hazard Identification Ratings
health-0; flammability-1; reactivity-0

N-[3-(TRIMETHYOXYSILYL)PROPYL]-1,2-ETHANEDIAMINE 1760-24-3

STATE LISTS

California - Air Bill 2588 Appendix A-II
9/90

California - Prop. 65 - Developmental Toxicity
developmental toxicity - initial date 4/1/90

TRIMETHYLALUMINIUM 75-24-1

HEALTH AND SAFETY LISTS

NIOSH - Selected LD50s and LC50s
Oral, mouse: LD50 = 2500 mg/kg

TRIMETHYLAMINE 75-50-3

HEALTH AND SAFETY LISTS

ACGIH 1995 - Ceiling Limits
C 0.04 mg/m3

NIOSH - Selected LD50s and LC50s
Oral, rat: LD50 = 5600 mg/kg

NIOSH - Health Standards - Exposure Limits
Handle in the workplace as an extremely toxic substance

NIOSH - Health Standards - Health Effects and Precautions
Pulmonary edema; immunological sensitization; irritation of pulmonary tract, eyes, nose, and skin
NTP Chemical Status Reports - Testing Status and NTIS Number
Prechronic studies for which toxicity technical reports were not prepared
OSHA - Vacated PELs - Time Weighted Averages
0.005 ppm TWA; 0.04 mg/m3 TWA

ENVIRONMENTAL LISTS
List of Pesticide Product Inert Ingredients
[present]

INTERNATIONAL LISTS
Australian Exposure Standards - Time Weighted Averages
0.005 ppm TWA; 0.039 mg/m3 TWA

Australian Exposure Standards - Skin Effects
sensitiser

Canada - WHMIS: Ingredient Disclosure
0.1% item 1636 (239)

Canada - Alberta - 8 Hour Occupational Exposure Limit
0.05 mg/m3 TWA

Canada - Alberta - 15 Minute Occupational Exposure Limit
0.15 mg/m3 STEL

Canada - Ontario - OHSA - TWAEVs
0.005 ppm TWAEV; 0.039 mg/m3 TWAEV

Canada - Quebec - Time-Weighted Average Exposure Values
0.005 ppm TWAEV; 0.039 mg/m3 TWAEV

United Kingdom - Occupational Exposure Standards - TWAs
0.04 mg/m3 TWA

United Kingdom - Occupational Exposure Standards - Notes
sensitisation

German (DFG) - MAK Values
fume: 0.005 ppm MAK; 0.04 mg/m3 MAK

German (DFG) - Peak Limitations
2 x normal MAK (5 min momentary value); don't exceed 8 times during shift

German (DFG) - Skin/Sensitizers
danger of sensitization (skin or respiratory)

Israel - Time Weighted Averages
0.005 ppm TWA; 0.039 mg/m3 TWA

Israel - Action Levels
0.0025 ppm AL; 0.0195 mg/m3 AL

STATE LISTS
California - Exposure Limits - PELs
0.005 ppm PEL; 0.04 mg/m3 PEL

California - Directors List of Hazardous Substances (8 CCR 339)
[present]

Florida Hazardous Substance List
[present]

Massachusetts Right To Know List
[present]

Minnesota Hazardous Substance List
[present]

Pennsylvania Right to Know List
[present]

4,4',6-TRIMETHYLANGELICIN PLUS ULTRAVIOLET   90370-29-9
A RADIATION

HEALTH AND SAFETY LISTS
NIOSH - Selected LD50s and LC50s
Oral, rat: LD50 = 2140 mg/kg

STATE LISTS
California - Air Bill 2588 Appendix A-II
6/91

California - Prop. 65 - Developmental Toxicity
developmental toxicity - initial date 1/1/91
Massachusetts Right To Know List
teratogen
NJ Right to Know List (Total)
sn 1923

2,4,5-TRIMETHYLANILINE   137-17-7
HEALTH AND SAFETY LISTS
NIOSH - Selected LD50s and LC50s
Oral, rat: LD50 = 200 mg/kg

STATE LISTS
NJ Right to Know List (Total)
sn 1924

1,3,5-TRIMETHYLBENZENE   108-67-8
HEALTH AND SAFETY LISTS
NTP Chemical Status Reports - Testing Status and NTIS Number
Project leader assigned/study in design

1,2,3-TRIMETHYLBENZENE   526-73-8
HEALTH AND SAFETY LISTS
NIOSH - Selected LD50s and LC50s
Oral, rat: LD50 = 5160 mg/kg

TRIMETHYL BENZENE   25551-13-7
HEALTH AND SAFETY LISTS
U.S. DOT - Substances From 49 CFR 172.101
regulated by DOT (NA9269)
U.S. DOT - Hazard Classes
DOT hazard class = 6.1
U.S. DOT - Substances Which Are Poisonous by Inhalation
liquid hazardous material poisonous by inhalation (NA9269)
NIOSH - Selected LD50s and LC50s
Inhalation, rat: LC50 = 125 ppm 4 hr Oral, rat: LD50 = 9330 mg/kg Skin, rabbit: LD50 = 6300 mg/kg
OSHA - List of Highly Hazardous Chemicals
threshhold quantity = 1500 pounds

TRIMETHYL BORATE   121-43-7
HEALTH AND SAFETY LISTS
NIOSH - Selected LD50s and LC50s
Oral, rat: LD50 = 7460 mg/kg

TRIMETHYLCHLOROSILANE   75-77-4
HEALTH AND SAFETY LISTS
NFPA - Flash Points
ignites spontaneously in air
NFPA - Hazard Identification Ratings
flammability-3; reactivity-3 (do not use water, foam or halogenated extinguishing agents)
STATE LISTS
Florida Hazardous Substance List
[present]
Massachusetts Right To Know List
[present]
NJ Right to Know List (Total)
sn 1926
NJ Special Hazardous Substances
(flammable - third degree; reactive - third degree)
Pennsylvania Right to Know List
[present]

3,3,5-TRIMETHYLCYCLOHEXANOL 116-02-9

HEALTH AND SAFETY LISTS

ACGIH 1995 - Time Weighted Averages
5 ppm TWA; 12 mg/m3 TWA

ACGIH 1995 - Short Term Exposure Limits
15 ppm STEL; 36 mg/m3 STEL

AIHA - Odor Threshold Values
no geometric mean air odor threshold

AIHA - WEEL - Time Weighted Averages
1 ppm TWA; 2.4 mg/m3 TWA

U.S. DOT - Substances From 49 CFR 172.101
regulated by DOT (UN1083, UN1297)

U.S. DOT - Hazard Classes
anhydrous: DOT hazard class = 2.1; aqueous solutions: DOT hazard class = 3

U.S. DOT - Appendix A Table 1 - Hazardous Substances
final RQ = 100 pounds (45.4 kg)

NFPA - Flash Points
gas (no flash point given)

NFPA - Hazard Identification Ratings
health-3; flammability-4; reactivity-0

NIOSH - Selected LD50s and LC50s
Inhalation, mammal: LC50 = 19 gm/m3 (8 hr)

OSHA - Vacated PELs - Time Weighted Averages
10 ppm TWA; 24 mg/m3 TWA

OSHA - Vacated PELs - Short Term Exposure Limits
15 ppm STEL; 36 mg/m3 STEL

ENVIRONMENTAL LISTS

CERCLA/SARA - Hazardous Substances and their Reportable Quantities
final RQ = 100 pounds (45.4 kg)

CAA - Flammable Substances for Accidental Release Prevention
threshold quantity = 10,000 lbs

CAA - HON Rule - SOCMI Chemicals
compliance by July 24, 1995

Clean Water Act - Hazardous Substances
[present]

INTERNATIONAL LISTS

Australian Exposure Standards - Time Weighted Averages
10 ppm TWA; 24 mg/m3 TWA

Australian Exposure Standards - Short Term Exposure Limits
15 ppm STEL; 36 mg/m3 STEL

Canada - WHMIS: Ingredient Disclosure
1% item 1637 (1681)

Canada - Alberta - 8 Hour Occupational Exposure Limit
10 ppm TWA; 24 mg/m3 TWA

Canada - Alberta - 15 Minute Occupational Exposure Limit
20 ppm STEL; 48 mg/m3 STEL

Canada - Ontario - OHSA - TWAEVs
10 ppm TWAEV; 24 mg/m3 TWAEV

Canada - Ontario - OHSA - STEVs
15 ppm STEV; 36 mg/m3 STEV

Canada - Quebec - Time-Weighted Average Exposure Values
10 ppm TWAEV; 24 mg/m3 TWAEV

Canada - Quebec - Short-term Exposure Values
15 ppm STEV; 36 mg/m3 STEV

United Kingdom - Occupational Exposure Standards - TWAs
10 ppm TWA; 24 mg/m3 TWA

United Kingdom - Occupational Exposure Standards - STELs
15 ppm STEL; 36 mg/m3 STEL

Israel - Time Weighted Averages
10 ppm TWA; 24 mg/m3 TWA

Israel - Short Term Exposure Limits
15 ppm STEL; 36 mg/m3 STEL

Israel - Action Levels
5 ppm AL; 12 mg/m3 AL

STATE LISTS

California - Exposure Limits - PELs
10 ppm PEL; 24 mg/m3 PEL

California - Exposure Limits - STELs
15 ppm STEL; 36 mg/m3 STEL

California - Directors List of Hazardous Substances (8 CCR 339)
[present]

Florida Hazardous Substance List
[present]

Massachusetts Right To Know List
[present]

Minnesota Hazardous Substance List
[present]

NJ Right to Know List (Total)
sn 1927

NJ Special Hazardous Substances
(flammable - fourth degree)

Pennsylvania Right to Know List
environmental hazard

1,3,5-TRIMETHYLCYCLOHEXANE RR-00533-4

HEALTH AND SAFETY LISTS

IARC - Group 3 (not classifiable)
[present]

TRIMETHYLCYCLOHEXANOL 933-48-2

HEALTH AND SAFETY LISTS

IARC - Group 3 (not classifiable)
[present]

NIOSH - Selected LD50s and LC50s
Oral, rat: LD50 = 1250 mg/kg

NTP Chemical Status Reports - Testing Status and NTIS Number
Technical reports printed (PB293802/AS)

NTP Chemical Status Reports - Evidence of Carcinogenicity
male rat-positive; female rat-positive; male mice-equivocal; female mice-positive

INTERNATIONAL LISTS

German (DFG) - Carcinogens
animal evidence of carcinogenicity

STATE LISTS

California - Directors List of Hazardous Substances (8 CCR 339)
[present]

Massachusetts Right To Know List
carcinogen; extraordinarily hazardous

TRIMETHYLCYCLOHEXANOL 1321-60-4

SEE ALSO:
AROMATIC C9 FRACTION FROM PETROLEUM REFINING

HEALTH AND SAFETY LISTS

AIHA - Odor Threshold Values
geometric mean air odor threshold = 2.2 ppm (detectable)

U.S. DOT - Substances From 49 CFR 172.101
regulated by DOT (UN2325)

U.S. DOT - Hazard Classes
DOT hazard class = 3

NFPA - Flash Points
flash point = 122 degrees F (50 degrees C)

NFPA - Hazard Identification Ratings
health-0; flammability-2; reactivity-0

NIOSH - Selected LD50s and LC50s
Inhalation, rat: LC50 = 24 gm/m3 4 hr

ENVIRONMENTAL LISTS

Safe Drinking Water Act - Monitoring
monitoring required at discretion of the state

EPA - Master Testing List
[present]

TSCA - Code of Federal Regulations Citations
40 CFR 712.30(g); 40 CFR 716.120(a),(b); 40 CFR 799.2175

TSCA - PAIR - Reporting List
Reporting Date: October 8, 1984

TSCA - Health and Safety Reporting List
Effective Date: February 13, 1984

TSCA - Section 12(b) - Export Notification
export notification required - Section 4

INTERNATIONAL LISTS

Canada - WHMIS: Ingredient Disclosure
0.1% item 1641 (1685)

STATE LISTS

Massachusetts Right To Know List
[present]

NJ Right to Know List (Total)
sn 1928

TRIMETHYLCYCLOHEXANONE   2408-37-9
SEE ALSO:
AROMATIC C9 FRACTION FROM PETROLEUM REFINING

HEALTH AND SAFETY LISTS

NFPA - Flash Points
regular: flash point = 111 degrees F (44 degrees C); 90.5%: flash point = 128 degrees F (53 degrees C)

NFPA - Hazard Identification Ratings
regular & 90.5%: health-0; flammability-2; reactivity-0

ENVIRONMENTAL LISTS

TSCA - Code of Federal Regulations Citations
40 CFR 712.30(g); 40 CFR 716.120(a); 40 CFR 799.2175

TSCA - PAIR - Reporting List
Reporting Date: October 8, 1984

TSCA - Health and Safety Reporting List
Effective Date: February 13, 1984; Sunset Date: November 9, 1993

INTERNATIONAL LISTS

Canada - WHMIS: Ingredient Disclosure
1% item 1639 (1683)

TRIMETHYLCYCLOHEXYLAMINE   34216-34-7
SEE ALSO:
AROMATIC C9 FRACTION FROM PETROLEUM REFINING

HEALTH AND SAFETY LISTS

ACGIH 1995 - Time Weighted Averages
25 ppm TWA; 123 mg/m3 TWA

NIOSH - Selected LD50s and LC50s
Oral, rat: LD50 = 8970 mg/kg

OSHA - Vacated PELs - Time Weighted Averages
25 ppm TWA; 125 mg/m3 TWA

ENVIRONMENTAL LISTS

List of Pesticide Product Inert Ingredients
[present]

TSCA - Code of Federal Regulations Citations
40 CFR 712.30(g); 40 CFR 716.120(a),(b)

TSCA - PAIR - Reporting List
Reporting Date: October 8, 1984

TSCA - Health and Safety Reporting List
Effective Date: February 13, 1984

INTERNATIONAL LISTS

Australian Exposure Standards - Time Weighted Averages
25 ppm TWA; 123 mg/m3 TWA

Canada - WHMIS: Ingredient Disclosure
1% item 1638 (1682)

Canada - Alberta - 8 Hour Occupational Exposure Limit
25 ppm TWA; 123 mg/m3 TWA

Canada - Alberta - 15 Minute Occupational Exposure Limit
35 ppm STEL; 172 mg/m3 STEL

Canada - British Columbia - 8 Hour Exposure Limits
25 ppm TWA; 125 mg/m3 TWA

Canada - British Columbia - 15 Minute Exposure Limits
35 ppm STEL; 170 mg/m3 STEL

Canada - Ontario - OHSA - TWAEVs
25 ppm TWAEV; 123 mg/m3 TWAEV

Canada - Quebec - Time-Weighted Average Exposure Values
25 ppm TWAEV; 123 mg/m3 TWAEV

Israel - Time Weighted Averages
25 ppm TWA; 123 mg/m3 TWA

Israel - Action Levels
12.5 ppm AL; 61.5 mg/m3 AL

Mexico - Instruction No. 10 - TWAs
25 ppm TWA; 125 mg/m3 TWA

Mexico - Instruction No. 10 - STELs
35 ppm STEL; 170 mg/m3 STEL

STATE LISTS

California - Exposure Limits - PELs
25 ppm PEL; 125 mg/m3 PEL

California - Directors List of Hazardous Substances (8 CCR 339)
[present]

Florida Hazardous Substance List
[present]

Massachusetts Right To Know List
[present]

Minnesota Hazardous Substance List
[present]

NJ Right to Know List (Total)
sn 1929

Pennsylvania Right to Know List
[present]

2,2,5-TRIMETHYL-3-DICHLOROACETYL-1,3-OXAZOLIDINE   52836-31-4

HEALTH AND SAFETY LISTS

U.S. DOT - Substances From 49 CFR 172.101
regulated by DOT (UN2416)

U.S. DOT - Hazard Classes
DOT hazard class = 3

NFPA - Flash Points
flash point < 80 degrees F (27 degrees C)

NFPA - Hazard Identification Ratings
health-2; flammability-3; reactivity-1

NIOSH - Selected LD50s and LC50s
Oral, rat: LD50 = 6140 mg/kg Skin, rabbit: LD50 = 1980 mg/kg

INTERNATIONAL LISTS

Canada - WHMIS: Ingredient Disclosure
1% item 1642 (319)

STATE LISTS

Florida Hazardous Substance List
[present]

Massachusetts Right To Know List
[present]

NJ Right to Know List (Total)
sn 1930
NJ Special Hazardous Substances
(flammable - third degree)
Pennsylvania Right to Know List
[present]

TRIMETHYLENE GLYCOL DIPERCHLORATE   RR-01465-3

HEALTH AND SAFETY LISTS

U.S. DOT - Substances From 49 CFR 172.101
regulated by DOT (UN1298)
U.S. DOT - Hazard Classes
DOT hazard class = 3
NFPA - Flash Points
flash point = -18 degrees F (-28 degrees C)
NFPA - Hazard Identification Ratings
health-3; flammability-3; reactivity-2 (avoid use of water)

ENVIRONMENTAL LISTS

CERCLA/SARA - Section 302 Extremely Hazardous Substances and TPQs
TPQ = 1000 pounds
CERCLA/SARA - Section 313 - Emission Reporting
form R reporting required
CAA -Toxic Substances for Accidental Release Prevention
threshold quantity = 10,000 lbs
EPA - Master Testing List
[present]

STATE LISTS

Florida Hazardous Substance List
[present]
Massachusetts Right To Know List
extraordinarily hazardous
NJ Right to Know List (Total)
sn 1931
NJ Special Hazardous Substances
(corrosive; flammable - third degree; reactive - second degree)
Pennsylvania Right to Know List
environmental hazard

PROPOSED REGULATIONS

CERCLA/SARA - Proposed Hazardous Substance Additions
proposed RQ = 1 pound (.454 kg)
CERCLA/SARA - 1989 Proposed RQ Adjustments
proposed RQ = 1000 pounds (454 kg)

2,5,5-TRIMETHYLHEPTANE   1189-99-7

HEALTH AND SAFETY LISTS

NFPA - Flash Points
flash point = 190 degrees F (88 degrees C)
NFPA - Hazard Identification Ratings
health-2; flammability-2; reactivity-0
NIOSH - Selected LD50s and LC50s
Oral, rat: LD50 = 3250 mg/kg Skin, rabbit: LD50 = 2800 mg/kg

ENVIRONMENTAL LISTS

List of Pesticide Product Inert Ingredients
[present]

INTERNATIONAL LISTS

Canada - WHMIS: Ingredient Disclosure
1% item 1643 (1686)

STATE LISTS

Florida Hazardous Substance List
[present]

Massachusetts Right To Know List
[present]
Pennsylvania Right to Know List
[present]

N,N,N-TRIMETHYL-1-HEXADECANAMINIUM   112-02-7
CHLORIDE

HEALTH AND SAFETY LISTS

NFPA - Hazard Identification Ratings
health-0; reactivity-0

TRIMETHYLHEXAMETHYLENEDIAMINE   25620-58-0

ENVIRONMENTAL LISTS

CAA - HON Rule - SOCMI Chemicals
compliance by July 24, 1995

TRIMETHYLHEXAMETHYLENE DIISOCYANATE   28679-16-5

HEALTH AND SAFETY LISTS

NFPA - Flash Points
flash point = 165 degrees F (74 degrees C)
NFPA - Hazard Identification Ratings
health-2; flammability-2; reactivity-0

STATE LISTS

Florida Hazardous Substance List
[present]
Massachusetts Right To Know List
[present]
Pennsylvania Right to Know List
[present]

2,4,8-TRIMETHYL-6-NONANOL   RR-00591-4

ENVIRONMENTAL LISTS

CAA - HON Rule - SOCMI Chemicals
compliance by July 24, 1995

2,6,8-TRIMETHYL-4-NONYL POLYETHYLENE GLY-   60828-78-6
COL ETHER

HEALTH AND SAFETY LISTS

U.S. DOT - Substances From 49 CFR 172.101
regulated by DOT (UN2326)
U.S. DOT - Hazard Classes
DOT hazard class = 8

ENVIRONMENTAL LISTS

CAA - HON Rule - SOCMI Chemicals
compliance by Oct. 23, 1995

INTERNATIONAL LISTS

Canada - WHMIS: Ingredient Disclosure
1% item 1644 (1687)

STATE LISTS

NJ Right to Know List (Total)
sn 1932
NJ Special Hazardous Substances
(corrosive)

TRIMETHYLOLETHANE   77-85-0

ENVIRONMENTAL LISTS

List of Pesticide Product Inert Ingredients
[present]

TRIMETHYLOLPROPANE FATTY ACID   RR-00230-2
DIACRYLATE

HEALTH AND SAFETY LISTS

U.S. DOT - Hazard Classes
Forbidden from transport by the DOT

TRIMETHYLOLPROPANE PHOSPHITE   824-11-3

HEALTH AND SAFETY LISTS

NFPA - Flash Points
flash point < 131 degrees F (55 degrees C)

NFPA - Hazard Identification Ratings
health-0; flammability-2; reactivity-0

TRIMETHYLOLPROPANE, POLYMER WITH TOLUENE DIISOCYANATE AND POLYPROPYLENE GLYCOL   9040-80-6

ENVIRONMENTAL LISTS

List of Pesticide Product Inert Ingredients
[present]

TRIMETHYLOLPROPANE TRIACRYLATE   15625-89-5

HEALTH AND SAFETY LISTS

U.S. DOT - Substances From 49 CFR 172.101
regulated by DOT (UN2327)

U.S. DOT - Hazard Classes
DOT hazard class = 8

INTERNATIONAL LISTS

Canada - WHMIS: Ingredient Disclosure
1% item 1645 (1688)

STATE LISTS

NJ Right to Know List (Total)
sn 1933

NJ Special Hazardous Substances
(corrosive)

TRIMETHYLOPROPANE TRIMETHACRYLATE   3290-92-4

HEALTH AND SAFETY LISTS

U.S. DOT - Substances From 49 CFR 172.101
regulated by DOT (UN2328)

U.S. DOT - Hazard Classes
DOT hazard class = 6.1

INTERNATIONAL LISTS

Canada - WHMIS: Ingredient Disclosure
0.1% item 1646 (719)

STATE LISTS

NJ Right to Know List (Total)
sn 1934

2,2,4-TRIMETHYL-1,3-PENTANEDIOL   144-19-4

HEALTH AND SAFETY LISTS

NFPA - Flash Points
flash point = 199 degrees F (93 degrees C)

NFPA - Hazard Identification Ratings
health-0; flammability-2; reactivity-0

2,2,4-TRIMETHYL-1,3-PENTANEDIOL BENZOATE ISOBUTYRATE   35164-39-7

ENVIRONMENTAL LISTS

List of Pesticide Product Inert Ingredients
[present]

2,2,4-TRIMETHYL-1,3-PENTANEDIOL DIISOBUTYRATE   6846-50-0

ENVIRONMENTAL LISTS

List of Pesticide Product Inert Ingredients
[present]

2,2,4-TRIMETHYLPENTANE-1,3-DIOL MONOISOBUTYRATE   25265-77-4

ENVIRONMENTAL LISTS

TSCA - Chemicals with Significant New Use Rules
PMN number: P-88-2463

3,4,4-TRIMETHYL-2-PENTENE   598-96-9

HEALTH AND SAFETY LISTS

NIOSH - Selected LD50s and LC50s
Oral, rat: LD50 = 8390 ug/kg Skin, rat: LD50 = 929 mg/kg

ENVIRONMENTAL LISTS

CERCLA/SARA - Section 302 Extremely Hazardous Substances and TPQs
TPQ = 100/10,000 pounds

STATE LISTS

Florida Hazardous Substance List
effective March 13, 1992

Massachusetts Right To Know List
extraordinarily hazardous

Pennsylvania Right to Know List
environmental hazard

PROPOSED REGULATIONS

CERCLA/SARA - Proposed Hazardous Substance Additions
proposed RQ = 1 pound (.454 kg)

CERCLA/SARA - 1989 Proposed RQ Adjustments
proposed RQ = 100 pounds (45.4 kg)

2,4,4-TRIMETHYLPENTYL-2-PEROXYPHENOXY-ACETATE   RR-00757-8

ENVIRONMENTAL LISTS

List of Pesticide Product Inert Ingredients
[present]

2,3,5-TRIMETHYLPHENYL METHYLCARBAMATE   2655-15-4

HEALTH AND SAFETY LISTS

AIHA - WEEL - Time Weighted Averages
1 mg/m3 TWA

AIHA - WEEL - Skin Absorption Designations
skin absorber

NFPA - Flash Points
flash point = 300 degrees F (149 degrees C)

NFPA - Hazard Identification Ratings
health-0; flammability-1; reactivity-0

NIOSH - Selected LD50s and LC50s
Oral, rat: LD50 = 5190 mg/kg Skin, rabbit: LD50 = 5170 mg/kg

NTP Chemical Status Reports - Testing Status and NTIS Number
Approved for toxicology/carcinogenesis study

STATE LISTS

Minnesota Hazardous Substance List
[present]

TRIMETHYLPHOSPHATE   512-56-1

HEALTH AND SAFETY LISTS

AIHA - WEEL - Time Weighted Averages
1 mg/m3 TWA

STATE LISTS

Minnesota Hazardous Substance List
[present]

TRIMETHYL PHOSPHITE   121-45-9

HEALTH AND SAFETY LISTS

NFPA - Flash Points
flash point = 235 degrees F (113 degrees C)

NFPA - Hazard Identification Ratings
health-0; flammability-1; reactivity-0
NIOSH - Selected LD50s and LC50s
Oral, rat: LD50 = 3730 mg/kg Skin, rabbit: LD50 = 6300 mg/kg

4,5',8-TRIMETHYLPSORALEN   3902-71-4

HEALTH AND SAFETY LISTS

NFPA - Flash Points
flash point = 325 degrees F (163 degrees C)
NFPA - Hazard Identification Ratings
health-0; flammability-1; reactivity-0

TRIMETHYL SPIROPOLYHETEROCYCLIC NAPH-THALENE COMPOUND   RR-01181-4

HEALTH AND SAFETY LISTS

NFPA - Flash Points
flash point = 250 degrees F (121 degrees C)
NFPA - Hazard Identification Ratings
health-0; flammability-1; reactivity-0

ENVIRONMENTAL LISTS

EPA - Master Testing List
[present]
List of Pesticide Product Inert Ingredients
[present]

TRIMETHYLTHIOUREA   2489-77-2

HEALTH AND SAFETY LISTS

NFPA - Flash Points
flash point = 248 degrees F (120 degrees C)
NFPA - Hazard Identification Ratings
health-0; flammability-1; reactivity-0
NIOSH - Selected LD50s and LC50s
Oral, rat: LD50 = 3200 mg/kg

ENVIRONMENTAL LISTS

EPA - Master Testing List
[present]
List of Pesticide Product Inert Ingredients
[present]

TRIMETHYL THIOUREA   2782-91-4

HEALTH AND SAFETY LISTS

NFPA - Flash Points
flash point < 70 degrees F (21 degrees C)
NFPA - Hazard Identification Ratings
health-0; flammability-3; reactivity-0

STATE LISTS

Florida Hazardous Substance List
[present]
Massachusetts Right To Know List
[present]
Pennsylvania Right to Know List
[present]

TRIMETHYL TIN CHLORIDE   1066-45-1

HEALTH AND SAFETY LISTS

U.S. DOT - Organic Peroxides Table
Organic peroxide UN3115

STATE LISTS

NJ Right to Know List (Total)
sn 2844

1,3,5-TRIMETHYL-2,4,6-TRINITROBENZENE   602-96-0

ENVIRONMENTAL LISTS

CERCLA/SARA - Section 313 - Emission Reporting
form R reporting required

TRINITROACETIC ACID   RR-01466-4

HEALTH AND SAFETY LISTS

NIOSH - Selected LD50s and LC50s
Oral, rat: LD50 = 840 mg/kg Skin, rabbit: LD50 = 3388 mg/kg
NTP Chemical Status Reports - Testing Status and NTIS Number
Technical reports printed (PB285851/AS)
NTP Chemical Status Reports - Evidence of Carcinogenicity
male rat-positive; female rat-negative; male mice-negative; female mice-positive

ENVIRONMENTAL LISTS

EPA - Master Testing List
[present]

INTERNATIONAL LISTS

Canada - WHMIS: Ingredient Disclosure
0.1% item 1647 (1398)
German (DFG) - Skin/Sensitizers
danger of cutaneous absorption
German (DFG) - Carcinogens
suspected carcinogen

STATE LISTS

California - Air Bill 2588 Appendix A-I
9/89
Massachusetts Right To Know List
carcinogen; extraordinarily hazardous

TRINITROACETONITRILE   630-72-8

HEALTH AND SAFETY LISTS

ACGIH 1995 - Time Weighted Averages
2 ppm TWA; 10 mg/m3 TWA
AIHA - Odor Threshold Values
no geometric mean air odor threshold
U.S. DOT - Substances From 49 CFR 172.101
regulated by DOT (UN2329)
U.S. DOT - Hazard Classes
DOT hazard class = 3
NFPA - Flash Points
flash point = 130 degrees F (54 degrees C)
NFPA - Hazard Identification Ratings
health-0; flammability-2; reactivity-0
NIOSH - Selected LD50s and LC50s
Oral, rat: LD50 = 1600 mg/kg
OSHA - Vacated PELs - Time Weighted Averages
2 ppm TWA; 10 mg/m3 TWA

ENVIRONMENTAL LISTS

TSCA - Code of Federal Regulations Citations
40 CFR 712.30(d)
TSCA - PAIR - Reporting List
Reporting Date: November 19, 1982

INTERNATIONAL LISTS

Australian Exposure Standards - Time Weighted Averages
2 ppm TWA; 10 mg/m3 TWA
Canada - WHMIS: Ingredient Disclosure
1% item 1648 (1404)
Canada - Alberta - 8 Hour Occupational Exposure Limit
2 ppm TWA; 10 mg/m3 TWA
Canada - Alberta - 15 Minute Occupational Exposure Limit
5 ppm STEL; 25 mg/m3 STEL

Canada - Ontario - OHSA - TWAEVs
2 ppm TWAEV;   10 mg/m3 TWAEV
Canada - Quebec - Time-Weighted Average Exposure Values
2 ppm TWAEV;   10 mg/m3 TWAEV
United Kingdom - Occupational Exposure Standards - TWAs
2 ppm TWA;   10 mg/m3 TWA
Israel - Time Weighted Averages
2 ppm TWA;   10 mg/m3 TWA
Israel - Action Levels
1 ppm AL;   5 mg/m3 AL
Mexico - Instruction No. 10 - TWAs
2 ppm TWA;   10 mg/m3 TWA
Mexico - Instruction No. 10 - STELs
5 ppm STEL;   25 mg/m3 STEL

STATE LISTS
California - Exposure Limits - PELs
2 ppm PEL;   10 mg/m3 PEL
California - Directors List of Hazardous Substances (8 CCR 339)
[present]
Florida Hazardous Substance List
[present]
Massachusetts Right To Know List
[present]
Minnesota Hazardous Substance List
[present]
NJ Right to Know List (Total)
sn 1935
Pennsylvania Right to Know List
[present]

TRINITROAMINE COBALT   RR-01467-5
HEALTH AND SAFETY LISTS
IARC - Group 3 (not classifiable)
[present]

TRINITRO-ANILINE   26952-42-1
ENVIRONMENTAL LISTS
TSCA - Chemicals with Significant New Use Rules
PMN number: P-91-1456

TRINITROANISOLE   606-35-9
HEALTH AND SAFETY LISTS
NIOSH - Selected LD50s and LC50s
Oral, rat: LD50 = 316 mg/kg
NTP Chemical Status Reports - Testing Status and NTIS Number
Technical reports printed (PB288802/AS)
NTP Chemical Status Reports - Evidence of Carcinogenicity
male rat-negative;   female rat-positive;   male mice-negative;   female mice-negative
INTERNATIONAL LISTS
Canada - WHMIS: Ingredient Disclosure
1% item 1649 (1689)
STATE LISTS
Massachusetts Right To Know List
carcinogen;   extraordinarily hazardous

TRINITROBENZENE   99-35-4
HEALTH AND SAFETY LISTS
NIOSH - Selected LD50s and LC50s
Oral, rat: LD50 = 920 mg/kg

TRINITROBENZENESULFONIC ACID   2508-19-2
SEE ALSO:
TIN
HEALTH AND SAFETY LISTS
NIOSH - Selected LD50s and LC50s
Oral, rat: LD50 = 12600 ug/kg
ENVIRONMENTAL LISTS
CERCLA/SARA - Section 302 Extremely Hazardous Substances and TPQs
TPQ = 500/10,000 pounds
INTERNATIONAL LISTS
Canada - WHMIS: Ingredient Disclosure
0.1% item 1650 (535)
STATE LISTS
Florida Hazardous Substance List
effective March 13, 1992
Massachusetts Right To Know List
extraordinarily hazardous
Pennsylvania Right to Know List
environmental hazard
PROPOSED REGULATIONS
CERCLA/SARA - Proposed Hazardous Substance Additions
proposed RQ = 1 pound (.454 kg)
CERCLA/SARA - 1989 Proposed RQ Adjustments
proposed RQ = 100 pounds (45.4 kg)

TRINITROBENZOIC ACID   129-66-8
HEALTH AND SAFETY LISTS
U.S. DOT - Hazard Classes
Forbidden from transport by the DOT

TRINITROBENZOIC ACID (VAN)   35860-50-5
HEALTH AND SAFETY LISTS
U.S. DOT - Hazard Classes
Forbidden from transport by the DOT

TRINITROCHLOROBENZENE   28260-61-9
HEALTH AND SAFETY LISTS
U.S. DOT - Hazard Classes
Forbidden from transport by the DOT

TRINITRO-M-CRESOL   602-99-3
HEALTH AND SAFETY LISTS
U.S. DOT - Hazard Classes
Forbidden from transport by the DOT

2,4,6-TRINITRO-1,3-DIAZOBENZENE   29306-57-8
HEALTH AND SAFETY LISTS
U.S. DOT - Substances From 49 CFR 172.101
regulated by DOT (UN0153)
U.S. DOT - Hazard Classes
DOT hazard class = 1.1D
STATE LISTS
NJ Right to Know List (Total)
sn 1936

TRINITROETHANOL   918-54-7
HEALTH AND SAFETY LISTS
U.S. DOT - Substances From 49 CFR 172.101
regulated by DOT (UN0213)
U.S. DOT - Hazard Classes
DOT hazard class = 1.1D

STATE LISTS
NJ Right to Know List (Total)
sn 1937

TRINITROETHYLNITRATE   RR-01468-6

HEALTH AND SAFETY LISTS
U.S. DOT - Substances From 49 CFR 172.101
regulated by DOT (UN0214, UN1354)
U.S. DOT - Hazard Classes
*with less than 30% water: DOT hazard class = 1.1D; with 30% or
more water: DOT hazard class = 4.1*
U.S. DOT - Appendix A Table 1 - Hazardous Substances
final RQ = 10 pounds (4.54 kg)
NIOSH - Selected LD50s and LC50s
Oral, rat: LD50 = 450 mg/kg

ENVIRONMENTAL LISTS
ATSDR Priority List
Rank (of 275): 108
CERCLA/SARA - Hazardous Substances and their Reportable Quantities
final RQ = 10 pounds (4.54 kg)
RCRA - U Series Wastes
waste number U234 (Reactive waste; Toxic waste)
RCRA - Hazardous Constituents-Appendix VIII
waste number U234
RCRA - Substances Banned From Land Disposal
[present]
RCRA - TSD Facilities Ground Water Monitoring
TM 8270 = 10 ug/L PQL
TSCA - Chemicals with Significant New Use Rules
[present]
TSCA - Section 12(b) - Export Notification
export notification required - Section 5

STATE LISTS
Florida Hazardous Substance List
[present]
Massachusetts Right To Know List
[present]
NJ Right to Know List (Total)
sn 1938
NJ Special Hazardous Substances
(flammable - fourth degree; reactive - fourth degree)
Pennsylvania Right to Know List
environmental hazard

2,4,7-TRINITRO-FLUOREN-9-ONE   129-79-3

HEALTH AND SAFETY LISTS
U.S. DOT - Substances From 49 CFR 172.101
regulated by DOT (UN0386)
U.S. DOT - Hazard Classes
DOT hazard class = 1.1D

STATE LISTS
NJ Right to Know List (Total)
sn 1939

TRINITROFLUORENONE   25322-14-9

STATE LISTS
NJ Right to Know List (Total)
sn 1940

TRINITROMETHANE   517-25-9

HEALTH AND SAFETY LISTS
U.S. DOT - Substances From 49 CFR 172.101
regulated by DOT (UN0215, UN1355)

U.S. DOT - Hazard Classes
DOT hazard class = 1.1D

TRINITRONAPHTHALENE   55810-17-8

HEALTH AND SAFETY LISTS
U.S. DOT - Substances From 49 CFR 172.101
regulated by DOT (UN0155)
U.S. DOT - Hazard Classes
DOT hazard class = 1.1D

STATE LISTS
NJ Right to Know List (Total)
sn 1941

TRINITROPHENETOLE   4732-14-3

HEALTH AND SAFETY LISTS
U.S. DOT - Substances From 49 CFR 172.101
regulated by DOT (UN0216)
U.S. DOT - Hazard Classes
DOT hazard class = 1.1D; heavy metal salts: Forbidden from transport by the DOT
STATE LISTS
NJ Right to Know List (Total)
sn 1943

2,4,6-TRINITROPHENYL GUANIDINE   134282-42-1

HEALTH AND SAFETY LISTS
U.S. DOT - Hazard Classes
Forbidden from transport by the DOT

2,4,6-TRINITROPHENYL GUANIDINE   RR-01381-0

HEALTH AND SAFETY LISTS
U.S. DOT - Hazard Classes
Forbidden from transport by the DOT

2,4,6-TRINITROPHENYL NITRAMINE   RR-01382-1

HEALTH AND SAFETY LISTS
U.S. DOT - Hazard Classes
Forbidden from transport by the DOT

2,4,6-TRINITROPHENYL TRIMETHYLOL METHYL   RR-01383-2
NITRAMINE TRINITRATE

HEALTH AND SAFETY LISTS
NIOSH - Selected LD50s and LC50s
Oral, rat: LD50 = 9910 mg/kg
NTP Chemical Status Reports - Testing Status and NTIS Number
Technical reports printed (PB92-238864/AS)

INTERNATIONAL LISTS
German (DFG) - Carcinogens
suspected carcinogen

TRINITRORESORCINOL   82-71-3

HEALTH AND SAFETY LISTS
U.S. DOT - Substances From 49 CFR 172.101
regulated by DOT (UN0387)
U.S. DOT - Hazard Classes
DOT hazard class = 1.1D

STATE LISTS
NJ Right to Know List (Total)
sn 1942

2,4,6-TRINITROSO-3-METHYL NITRAMINOANISOLE   RR-01384-3
HEALTH AND SAFETY LISTS
U.S. DOT - Hazard Classes
Forbidden from transport by the DOT
NIOSH - Selected LD50s and LC50s
Inhalation, mouse: LC50 = 800 mg/m3 (2 hr)

TRINITROTETRAMINE COBALT NITRATE   RR-01469-7
HEALTH AND SAFETY LISTS
U.S. DOT - Substances From 49 CFR 172.101
regulated by DOT (UN0217)
U.S. DOT - Hazard Classes
DOT hazard class = 1.1D
STATE LISTS
NJ Right to Know List (Total)
sn 1944

2,4,6-TRINITROTOLUENE   118-96-7
HEALTH AND SAFETY LISTS
U.S. DOT - Substances From 49 CFR 172.101
regulated by DOT (UN0218)
U.S. DOT - Hazard Classes
DOT hazard class = 1.1D
STATE LISTS
NJ Right to Know List (Total)
sn 1945

2,4,6-TRINITRO-1,3,5-TRIAZIDO BENZENE   RR-01380-9
HEALTH AND SAFETY LISTS
U.S. DOT - Hazard Classes
Forbidden from transport by the DOT

TRI-(β-NITROXYETHYL) AMMONIUM NITRATE   RR-01463-1
STATE LISTS
NJ Right to Know List (Total)
sn 2178

TRI[NONYLPHENYL] PHOSPHITE   26523-78-4
HEALTH AND SAFETY LISTS
U.S. DOT - Hazard Classes
Forbidden from transport by the DOT

(Z,Z,Z)-TRI-9-OCTADECENOATE SORBITAN   26266-58-0
HEALTH AND SAFETY LISTS
U.S. DOT - Hazard Classes
Forbidden from transport by the DOT

TRIOCTYL PHOSPHITE   3028-88-4
HEALTH AND SAFETY LISTS
U.S. DOT - Substances From 49 CFR 172.101
regulated by DOT (UN0219, UN0394)
U.S. DOT - Hazard Classes
DOT hazard class = 1.1D
STATE LISTS
NJ Right to Know List (Total)
sn 1947

TRIORTHOCRESYL PHOSPHATE   78-30-8
HEALTH AND SAFETY LISTS
U.S. DOT - Hazard Classes
Forbidden from transport by the DOT

1,3,5-TRIOXANE   110-88-3
HEALTH AND SAFETY LISTS
U.S. DOT - Hazard Classes
Forbidden from transport by the DOT

TRIPELENNAMINE HYDROCHLORIDE   154-69-8
HEALTH AND SAFETY LISTS
ACGIH 1995 - Time Weighted Averages
0.5 mg/m3 TWA
ACGIH 1995 - Skin Designations
skin - potential for cutaneous absorption
U.S. DOT - Substances From 49 CFR 172.101
regulated by DOT (UN1356,UN0209)
U.S. DOT - Hazard Classes
DOT hazard class = 1.1D
NIOSH - Selected LD50s and LC50s
Oral, rat: LD50 = 795 mg/kg
NIOSH 1990 - Pocket Guide - RELs
0.5 mg/m3 TWA
NIOSH 1990 - Pocket Guide - Target organs
blood, liver, eyes, CVS, CNS, kidneys, skin
NIOSH 1990 - Pocket Guide - Skin list
Potential for dermal absorption
OSHA - Vacated PELs - Time Weighted Averages
0.5 mg/m3 TWA
OSHA - Vacated PELs - Skin Designation
Prevent or reduce skin absorption
OSHA - Final PELs - Time Weighted Averages
1.5 mg/m3 TWA
OSHA - Final PELs - Skin Notations
prevent or reduce skin absorption
ENVIRONMENTAL LISTS
ATSDR Priority List
Rank (of 275): 090
INTERNATIONAL LISTS
Australian Exposure Standards - Time Weighted Averages
0.5 mg/m3 TWA
Australian Exposure Standards - Skin Effects
skin absorption
Canada - WHMIS: Ingredient Disclosure
1% item 1651 (1690)
Canada - Alberta - 8 Hour Occupational Exposure Limit
0.5 mg/m3 TWA
Canada - Alberta - 15 Minute Occupational Exposure Limit
3 mg/m3 STEL
Canada - Alberta - Ceiling Occupational Exposure Limit
C 3 ppm
Canada - Alberta - Skin Designation
can be absorbed through the intact skin
Canada - British Columbia - Ceiling Exposure Limits
C 0.5 mg/m3
Canada - Ontario - OHSA - TWAEVs
0.01 ppm TWAEV; 0.1 mg/m3 TWAEV
Canada - Ontario - OHSA - Skin Notations
absorption through skin, eyes, or mucous membranes
Canada - Quebec - Time-Weighted Average Exposure Values
0.5 mg/m3 TWAEV
Canada - Quebec - Skin Designations
absorbed through the skin
United Kingdom - Occupational Exposure Standards - TWAs
0.5 mg/m3 TWA

German (DFG) - MAK Values
0.01 ppm MAK; 0.1 mg/m3 MAK (includes isomers in technical mixtures)

German (DFG) - Peak Limitations
2 x normal MAK (30 min. average value); don't exceed 4 times during shift

German (DFG) - Skin/Sensitizers
danger of cutaneous absorption

German (DFG) - Carcinogens
suspected carcinogen

Israel - Time Weighted Averages
0.5 mg/m3 TWA

Israel - Action Levels
0.25 mg/m3 AL

Mexico - Instruction No. 10 - TWAs
0.5 mg/m3 TWA

Mexico - Instruction No. 10 - STELs
3 mg/m3 STEL

STATE LISTS

California - Exposure Limits - PELs
0.5 mg/m3 PEL

California - Exposure Limits - Skin Notation
material may be absorbed through the skin, eyes or mucous membrane

California - Directors List of Hazardous Substances (8 CCR 339)
[present]

Florida Hazardous Substance List
[present]

Massachusetts Right To Know List
[present]

Minnesota Hazardous Substance List
skin

NJ Right to Know List (Total)
sn 1948

NJ Special Hazardous Substances
(flammable - fourth degree; reactive - fourth degree)

Pennsylvania Right to Know List
[present]

TRIPHENYL AMINE   603-34-9

HEALTH AND SAFETY LISTS

U.S. DOT - Hazard Classes
Forbidden from transport by the DOT

TRIPHENYL ANTIMONY   603-36-1

HEALTH AND SAFETY LISTS

U.S. DOT - Hazard Classes
Forbidden from transport by the DOT

TRIPHENYLENE   217-59-4

ENVIRONMENTAL LISTS

List of Pesticide Product Inert Ingredients
[present]

TRIPHENYLMETHANE   519-73-3

ENVIRONMENTAL LISTS

List of Pesticide Product Inert Ingredients
[present]

TRIPHENYL PHOSPHATE   115-86-6

HEALTH AND SAFETY LISTS

NFPA - Flash Points
flash point = 340 degrees F (171 degrees C)

NFPA - Hazard Identification Ratings
health-0; flammability-1; reactivity-0

TRIPHENYL PHOSPHITE   101-02-0

SEE ALSO:
BISAZOBIPHENYL DYES

HEALTH AND SAFETY LISTS

ACGIH 1995 - Time Weighted Averages
0.1 mg/m3 TWA

ACGIH 1995 - Skin Designations
skin - potential for cutaneous absorption

NFPA - Flash Points
flash point = 437 degrees F (225 degrees C)

NFPA - Hazard Identification Ratings
health-2; flammability-1; reactivity-0

NIOSH - Selected LD50s and LC50s
Oral, rat: LD50 = 1160 mg/kg

NIOSH 1990 - Pocket Guide - RELs
0.1 mg/m3 TWA

NIOSH 1990 - Pocket Guide - IDLHs
40 mg/m3 IDLH

NIOSH 1990 - Pocket Guide - Target organs
PNS, CNS

NIOSH 1990 - Pocket Guide - Skin list
Potential for dermal absorption

OSHA - Vacated PELs - Time Weighted Averages
0.1 mg/m3 TWA

OSHA - Vacated PELs - Skin Designation
Prevent or reduce skin absorption

OSHA - Final PELs - Time Weighted Averages
0.1 mg/m3 TWA

ENVIRONMENTAL LISTS

TSCA - Code of Federal Regulations Citations
40 CFR 712.30(d); 40 CFR 716.120(c)

TSCA - PAIR - Reporting List
Reporting Date: November 19, 1982

INTERNATIONAL LISTS

Australian Exposure Standards - Time Weighted Averages
0.1 mg/m3 TWA

Australian Exposure Standards - Skin Effects
skin absorption

Canada - WHMIS: Ingredient Disclosure
1% item 1620 (1397)

Canada - Alberta - 8 Hour Occupational Exposure Limit
0.1 mg/m3 TWA

Canada - Alberta - 15 Minute Occupational Exposure Limit
0.3 mg/m3 STEL

Canada - British Columbia - 8 Hour Exposure Limits
0.1 mg/m3 TWA

Canada - British Columbia - 15 Minute Exposure Limits
0.3 mg/m3 STEL

Canada - Ontario - OHSA - TWAEVs
0.1 mg/m3 TWAEV

Canada - Ontario - OHSA - Skin Notations
absorption through skin, eyes, or mucous membranes

Canada - Quebec - Time-Weighted Average Exposure Values
0.1 mg/m3 TWAEV

Canada - Quebec - Skin Designations
absorbed through the skin

United Kingdom - Occupational Exposure Standards - TWAs
0.1 mg/m3 TWA

United Kingdom - Occupational Exposure Standards - STELs
0.3 mg/m3 STEL

Israel - Time Weighted Averages
0.1 mg/m3 TWA

Israel - Action Levels
0.05 mg/m3 AL

Mexico - Instruction No. 10 - TWAs
0.1 mg/m3 TWA

Mexico - Instruction No. 10 - STELs
0.3 mg/m3 STEL

STATE LISTS

California - Air Bill 2588 Appendix A-I
9/89

California - Exposure Limits - PELs
0.1 mg/m3 PEL

California - Exposure Limits - Skin Notation
material may be absorbed through the skin, eyes or mucous membrane

California - Directors List of Hazardous Substances (8 CCR 339)
[present]

Florida Hazardous Substance List
[present]

Massachusetts Right To Know List
[present]

Minnesota Hazardous Substance List
skin

NJ Right to Know List (Total)
sn 1949

Pennsylvania Right to Know List
[present]

TRIPHENYL PHOSPHITE   101-33-7

HEALTH AND SAFETY LISTS

NFPA - Flash Points
flash point = 113 degrees F (45 degrees C)

NFPA - Hazard Identification Ratings
health-2; flammability-2; reactivity-0

STATE LISTS

Florida Hazardous Substance List
[present]

Massachusetts Right To Know List
[present]

Pennsylvania Right to Know List
[present]

TRIPHENYLSULFONIUM CHLORIDE   4270-70-6

HEALTH AND SAFETY LISTS

NTP Chemical Status Reports - Testing Status and NTIS Number
Prechronic studies for which toxicity technical reports were not prepared

TRIPHENYLTIN CHLORIDE   639-58-7

HEALTH AND SAFETY LISTS

ACGIH 1995 - Time Weighted Averages
5 mg/m3 TWA

NIOSH - Selected LD50s and LC50s
Oral, rat: LD50 = 3200 mg/kg

OSHA - Vacated PELs - Time Weighted Averages
5 mg/m3 TWA

INTERNATIONAL LISTS

Australian Exposure Standards - Time Weighted Averages
5 mg/m3 TWA

Canada - WHMIS: Ingredient Disclosure
1% item 1652 (1698)

Canada - Alberta - 8 Hour Occupational Exposure Limit
5 mg/m3 TWA

Canada - Alberta - 15 Minute Occupational Exposure Limit
10 mg/m3 STEL

Canada - Ontario - OHSA - TWAEVs
5 mg/m3 TWAEV

Canada - Quebec - Time-Weighted Average Exposure Values
5 mg/m3 TWAEV

Israel - Time Weighted Averages
5 mg/m3 TWA

Israel - Action Levels
2.5 mg/m3 AL

STATE LISTS

California - Exposure Limits - PELs
5 mg/m3 PEL

California - Directors List of Hazardous Substances (8 CCR 339)
[present]

Florida Hazardous Substance List
[present]

Massachusetts Right To Know List
[present]

Minnesota Hazardous Substance List
[present]

Pennsylvania Right to Know List
[present]

TRIPHENYLTIN COMPOUNDS   RR-01199-4
SEE ALSO:
ANTIMONY

HEALTH AND SAFETY LISTS

NIOSH - Selected LD50s and LC50s
Oral, rat: LD50 = 183 mg/kg

TRIPHENYLTIN HYDROXIDE   76-87-9

HEALTH AND SAFETY LISTS

IARC - Group 3 (not classifiable)
[present]

TRIPHENYLTIN HYDROXIDE   76-99-3

HEALTH AND SAFETY LISTS

NFPA - Flash Points
flash point > 212 degrees F (100 degrees C)

NFPA - Hazard Identification Ratings
health-0; flammability-1; reactivity-0

TRIPHOSPHORIC ACID, SODIUM SALT   13573-18-7
SEE ALSO:
BISAZOBIPHENYL DYES

HEALTH AND SAFETY LISTS

ACGIH 1995 - Time Weighted Averages
3 mg/m3 TWA

NFPA - Flash Points
flash point = 428 degrees F (220 degrees C)

NFPA - Hazard Identification Ratings
health-2; flammability-1; reactivity-0

NIOSH - Selected LD50s and LC50s
Oral, rat: LD50 = 3500 mg/kg

NIOSH 1990 - Pocket Guide - RELs
3 mg/m3 TWA

NIOSH 1990 - Pocket Guide - Target organs
blood

OSHA - Vacated PELs - Time Weighted Averages
3 mg/m3 TWA

OSHA - Final PELs - Time Weighted Averages
3 mg/m3 TWA

ENVIRONMENTAL LISTS

EPA - Master Testing List
[present]

TSCA - Code of Federal Regulations Citations
40 CFR 712.30(d); 40 CFR 716.120(a)
TSCA - PAIR - Reporting List
Reporting Date: November 19, 1982
INTERNATIONAL LISTS
Australian Exposure Standards - Time Weighted Averages
3 mg/m3 TWA
Canada - WHMIS: Ingredient Disclosure
1% item 1653 (1399)
Canada - Alberta - 8 Hour Occupational Exposure Limit
3 mg/m3 TWA
Canada - Alberta - 15 Minute Occupational Exposure Limit
6 mg/m3 STEL
Canada - British Columbia - 8 Hour Exposure Limits
3 mg/m3 TWA
Canada - British Columbia - 15 Minute Exposure Limits
6 mg/m3 STEL
Canada - Ontario - OHSA - TWAEVs
3 mg/m3 TWAEV
Canada - Quebec - Time-Weighted Average Exposure Values
3 mg/m3 TWAEV
United Kingdom - Occupational Exposure Standards - TWAs
3 mg/m3 TWA
United Kingdom - Occupational Exposure Standards - STELs
6 mg/m3 STEL
Israel - Time Weighted Averages
3 mg/m3 TWA
Israel - Action Levels
1.5 mg/m3 AL
Mexico - Instruction No. 10 - TWAs
3 mg/m3 TWA
Mexico - Instruction No. 10 - STELs
6 mg/m3 STEL
STATE LISTS
California - Air Bill 2588 Appendix A-I
9/89
California - Exposure Limits - PELs
3 mg/m3 PEL
California - Exposure Limits - Skin Notation
material may be absorbed through the skin, eyes or mucous membrane
California - Directors List of Hazardous Substances (8 CCR 339)
[present]
Florida Hazardous Substance List
[present]
Massachusetts Right To Know List
[present]
Minnesota Hazardous Substance List
[present]
Pennsylvania Right to Know List
[present]

TRIPOTASSIUM NITRILOTRIACETATE   2399-85-1
HEALTH AND SAFETY LISTS
NFPA - Flash Points
flash point = 425 degrees F (218 degrees C)
NFPA - Hazard Identification Ratings
health-0; flammability-1; reactivity-0
NIOSH - Selected LD50s and LC50s
Oral, rat: LD50 = 1600 mg/kg
ENVIRONMENTAL LISTS
List of Pesticide Product Inert Ingredients
[present]

INTERNATIONAL LISTS
Canada - WHMIS: Ingredient Disclosure
1% item 1654 (1405)
STATE LISTS
California - Air Bill 2588 Appendix A-I
9/89

TRIPROLIDINE   486-12-4
HEALTH AND SAFETY LISTS
NFPA - Flash Points
flash point = 210 degrees F (99 degrees C)
NFPA - Hazard Identification Ratings
health-1; flammability-1; reactivity-0

TRIPROPYL ALUMINUM   102-67-0
ENVIRONMENTAL LISTS
List of Pesticide Product Inert Ingredients
[present]

TRIPROPYLAMINE   102-69-2
SEE ALSO:
TIN
HEALTH AND SAFETY LISTS
NIOSH - Selected LD50s and LC50s
Oral, rat: LD50 = 135 mg/kg
ENVIRONMENTAL LISTS
CERCLA/SARA - Section 302 Extremely Hazardous Substances and TPQs
TPQ = 500/10,000 pounds
CERCLA/SARA - Section 313 - Emission Reporting
form R reporting required
INTERNATIONAL LISTS
Canada - WHMIS: Ingredient Disclosure
1% item 1656 (536)
STATE LISTS
Florida Hazardous Substance List
effective March 13, 1992
Massachusetts Right To Know List
extraordinarily hazardous
NJ Right to Know List (Total)
sn 1952
Pennsylvania Right to Know List
environmental hazard
PROPOSED REGULATIONS
CERCLA/SARA - Proposed Hazardous Substance Additions
proposed RQ = 1 pound (.454 kg)
CERCLA/SARA - 1989 Proposed RQ Adjustments
proposed RQ = 100 pounds (45.4 kg)

TRIPROPYLENE   13987-01-4
HEALTH AND SAFETY LISTS
U.S. DOT - Appendix B - Marine Pollutants
DOT regulated severe marine pollutant

TRIPROPYLENE GLYCOL   1638-16-0
SEE ALSO:
TIN
TIN
TIN ORGANIC COMPOUNDS
HEALTH AND SAFETY LISTS
U.S. DOT - Appendix B - Marine Pollutants
DOT regulated severe marine pollutant

NIOSH - Selected LD50s and LC50s
Oral, rat: LD50 = 46 mg/kg

NTP Chemical Status Reports - Testing Status and NTIS Number
Technical reports printed (PB287399/AS)

NTP Chemical Status Reports - Evidence of Carcinogenicity
male rat-negative; female rat-negative; male mice-negative; female mice-negative

ENVIRONMENTAL LISTS

CERCLA/SARA - Section 313 - Emission Reporting
form R reporting required

INTERNATIONAL LISTS

Canada - WHMIS: Ingredient Disclosure
0.1% item 1657 (999)

STATE LISTS

California - Prop. 65 - Cancer list
carcinogen - initial date 7/1/92

Massachusetts Right To Know List
[present]

NJ Right to Know List (Total)
sn 1953

TRIPROPYLENE GLYCOL DIACRYLATE   42978-66-5

STATE LISTS

Massachusetts Right To Know List
teratogen

TRIPROPYLENE GLYCOL METHYL ETHER   20324-33-8

HEALTH AND SAFETY LISTS

NIOSH - Selected LD50s and LC50s
Oral, rat: LD50 = 5190 mg/kg

TRIPROPYLENE GLYCOL MONOMETHYL ETHER   25498-49-1

HEALTH AND SAFETY LISTS

NIOSH - Selected LD50s and LC50s
Oral, rat: LD50 = 1220 mg/kg

TRIS(AZIRIDINYL)-P-BENZOQUINONE   68-76-8

HEALTH AND SAFETY LISTS

NTP Chemical Status Reports - Testing Status and NTIS Number
Technical reports printed (no NTIS number given)

TRIS(1-AZIRIDINYL)PHOSPHINE SULFIDE   52-24-4

HEALTH AND SAFETY LISTS

NFPA - Flash Points
ignites spontaneously in air

NFPA - Hazard Identification Ratings
flammability-3; reactivity-3 (do not use water, foam or halogenated extinguishing agents)

STATE LISTS

Florida Hazardous Substance List
[present]

Massachusetts Right To Know List
[present]

NJ Right to Know List (Total)
sn 1954

NJ Special Hazardous Substances
(flammable - third degree; reactive - third degree)

Pennsylvania Right to Know List
[present]

TRIS(1-AZIRIDINYL) PHOSPHINE OXIDE   545-55-1

HEALTH AND SAFETY LISTS

U.S. DOT - Substances From 49 CFR 172.101
regulated by DOT (UN2260)

U.S. DOT - Hazard Classes
DOT hazard class = 3

NFPA - Flash Points
flash point = 105 degrees F (41 degrees C)

NFPA - Hazard Identification Ratings
health-2; flammability-2; reactivity-0

NIOSH - Selected LD50s and LC50s
Inhalation, mouse: LC50 = 3800 mg/m3 (2 hr) Oral, rat: LD50 = 72 mg/kg Skin, rabbit: LD50 = 429 mg/kg

INTERNATIONAL LISTS

Canada - WHMIS: Ingredient Disclosure
1% item 1658 (1699)

STATE LISTS

Florida Hazardous Substance List
[present]

Massachusetts Right To Know List
[present]

NJ Right to Know List (Total)
sn 1955

NJ Special Hazardous Substances
(corrosive)

Pennsylvania Right to Know List
[present]

TRIS, BIS-BIFLUOROAMINO DIETHOXY PROPANE   39409-64-8
(TVOPA)

HEALTH AND SAFETY LISTS

U.S. DOT - Substances From 49 CFR 172.101
regulated by DOT (UN2057)

U.S. DOT - Hazard Classes
DOT hazard class = 3

NFPA - Flash Points
flash point = 75 degrees F (24 degrees C)

NFPA - Hazard Identification Ratings
health-0; flammability-3; reactivity-0

STATE LISTS

Florida Hazardous Substance List
[present]

Massachusetts Right To Know List
[present]

NJ Right to Know List (Total)
sn 1956

NJ Special Hazardous Substances
(flammable - third degree)

Pennsylvania Right to Know List
[present]

TRIS(2-BUTOXYETHYL) PHOSPHITE   2718-67-4

HEALTH AND SAFETY LISTS

NFPA - Flash Points
flash point = 285 degrees F (141 degrees C)

NFPA - Hazard Identification Ratings
health-0; flammability-1; reactivity-0

ENVIRONMENTAL LISTS

List of Pesticide Product Inert Ingredients
[present]

TRIS(2-CHLOROETHYL)AMINE   555-77-1

ENVIRONMENTAL LISTS

TSCA - PAIR - Reporting List
Effective Date: January 26, 1994; Reporting Date: March 28, 1994
TSCA - Health and Safety Reporting List
Effective Date: January 26, 1994; Sunset Date: January 26, 2004

PROPOSED REGULATIONS

TSCA - ITC 31st Report Priority Testing List
recommended for testing

1,2,3-TRIS(CHLOROMETHOXY)PROPANE   38571-73-2

ENVIRONMENTAL LISTS

TSCA - PAIR - Reporting List
Effective Date: January 26, 1994; Reporting Date: March 28, 1994
TSCA - Health and Safety Reporting List
Effective Date: January 26, 1994; Sunset Date: January 26, 2004

PROPOSED REGULATIONS

TSCA - ITC 31st Report Priority Testing List
recommended for testing

TRIS(2-CHLORO-1-PROPYL) PHOSPHATE   RR-01717-4

HEALTH AND SAFETY LISTS

NFPA - Flash Points
flash point = 250 degrees F (121 degrees C)
NFPA - Hazard Identification Ratings
health-0; flammability-1; reactivity-0
NIOSH - Selected LD50s and LC50s
Oral, rat: LD50 = 3300 mg/kg

ENVIRONMENTAL LISTS

List of Pesticide Product Inert Ingredients
[present]
TSCA - Code of Federal Regulations Citations
40 CFR 712.30(w); 40 CFR 716.120(a)
TSCA - PAIR - Reporting List
Reporting Date: June 13, 1989
TSCA - Health and Safety Reporting List
Effective Date: April 13, 1989

TRISDIBROMOPHENYL PHOSPHATE   49690-63-3

HEALTH AND SAFETY LISTS

IARC - Group 3 (not classifiable)
[present]

ENVIRONMENTAL LISTS

CERCLA/SARA - Section 313 - Emission Reporting
form R reporting required for 0.1% de minimus concentration

STATE LISTS

California - Air Bill 2588 Appendix A-II
known or potential carcinogen: 9/90
California - Prop. 65 - Cancer list
carcinogen - initial date 10/1/89
California - Directors List of Hazardous Substances (8 CCR 339)
[present]
Florida Hazardous Substance List
[present]
Massachusetts Right To Know List
carcinogen; extraordinarily hazardous
Minnesota Hazardous Substance List
carcinogen
NJ Special Hazardous Substances
(carcinogen; mutagen)
Pennsylvania Right to Know List
environmental hazard; special hazardous substance
Pennsylvania RTK - Special Hazardous Substances
[present]

TRIS(2,3-DIBROMOPROPYL)PHOSPHATE   126-72-7

HEALTH AND SAFETY LISTS

IARC - Group 1 (carcinogenic to humans)
[present]
NIOSH - Selected LD50s and LC50s
Oral, mouse: LD50 = 38 mg/kg
NTP Chemical Status Reports - Testing Status and NTIS Number
Technical reports printed (PB285702/AS)
NTP Chemical Status Reports - Evidence of Carcinogenicity
male rat-positive; female rat-positive; male mice-positive; female mice-positive
NTP Seventh Report - Suspect Carcinogens
suspect carcinogen
OSHA - Select Carcinogens
[present]
OSHA - Possible Select Carcinogens
[present]

ENVIRONMENTAL LISTS

RCRA - Hazardous Constituents-Appendix VIII
hazardous constituent - no waste number

STATE LISTS

California - Air Bill 2588 Appendix A-II
known or potential carcinogen
California - Prop. 65 - Cancer list
carcinogen - initial date 1/1/88
California - Prop. 65 - No Significant Risk Levels
no significant risk level = 0.06 ug/day
California - Directors List of Hazardous Substances (8 CCR 339)
[present]
Florida Hazardous Substance List
[present]
Massachusetts Right To Know List
carcinogen; extraordinarily hazardous
Minnesota Hazardous Substance List
carcinogen
NJ Special Hazardous Substances
(carcinogen; mutagen)
Pennsylvania Right to Know List
environmental hazard; special hazardous substance
Pennsylvania RTK - Special Hazardous Substances
[present]

PROPOSED REGULATIONS

NTP - Proposed Additions to Annual Report on Carcinogens
proposed as a known carcinogen for the NTP 9th report

TRIS(2,3-DICHLOROPROPYL) PHOSPHATE   78-43-3

HEALTH AND SAFETY LISTS

U.S. DOT - Substances From 49 CFR 172.101
regulated by DOT (UN2501)
U.S. DOT - Hazard Classes
DOT hazard class = 6.1
IARC - Group 3 (not classifiable)
[present]

INTERNATIONAL LISTS

Canada - WHMIS: Ingredient Disclosure
1% item 1659 (1325)

STATE LISTS

NJ Right to Know List (Total)
sn 0175

TRIS(DISUBSTITUTED ALKYL) HETEROCYCLE   RR-00954-1

HEALTH AND SAFETY LISTS

U.S. DOT - Hazard Classes
Forbidden from transport by the DOT

TRIS(2-ETHYLHEXYL)PHOSPHATE   78-42-2

ENVIRONMENTAL LISTS

List of Pesticide Product Inert Ingredients
[present]

TRIS (HYDROXYMETHYL) AMINOMETHANE   77-86-1

HEALTH AND SAFETY LISTS

IARC - Group 2B (sufficient animal data)
[present]
NIOSH - Selected LD50s and LC50s
Inhalation, rat: LC50 = 200 mg/m3 10 mn Oral, rat: LD50 = 5 mg/kg
Skin, rat: LD50 = 2 mg/kg
OSHA - Possible Select Carcinogens
[present]

ENVIRONMENTAL LISTS

CERCLA/SARA - Section 302 Extremely Hazardous Substances and TPQs
TPQ = 100 pounds

STATE LISTS

Florida Hazardous Substance List
effective March 13, 1992
Massachusetts Right To Know List
extraordinarily hazardous
NJ Right to Know List (Total)
sn 2847
Pennsylvania Right to Know List
environmental hazard

PROPOSED REGULATIONS

CERCLA/SARA - Proposed Hazardous Substance Additions
proposed RQ = 1 pound (.454 kg)
CERCLA/SARA - 1989 Proposed RQ Adjustments
proposed RQ = 10 pounds (4.54 kg)

TRISILOXANE, 1,1,1,3,5,5,5-HEPTAMETHYL-3-[3-   7422-52-8
(OXIRANYLMETHOXY)PROPYL]-

HEALTH AND SAFETY LISTS

IARC - Group 3 (not classifiable)
[present]

STATE LISTS

California - Directors List of Hazardous Substances (8 CCR 339)
[present]

TRIS ISOPROPYLATED PHENYL PHOSPHATE   68937-41-7

PROPOSED REGULATIONS

TSCA - ITC 33rd Report Priority Testing List
recommended with intent to designate

TRIS(2-METHYL-1-AZIRIDINYL)PHOSPHINE OXIDE   57-39-6

ENVIRONMENTAL LISTS

TSCA - Section 12(b) - Export Notification
P-86-1662; export notification required - Section 5

TRISODIUM EDTA   150-38-9

HEALTH AND SAFETY LISTS

U.S. DOT - Appendix A Table 1 - Hazardous Substances
final RQ = 10 pounds (4.54 kg)

IARC - Group 2A (limited human data)
[present] (Other relevant data, as given in Supplement 7, influenced the making of the overall evaluation)
NIOSH - Selected LD50s and LC50s
Oral, rat: LD50 = 1010 mg/kg
NTP Chemical Status Reports - Testing Status and NTIS Number
Technical reports printed (PB280271/AS)
NTP Chemical Status Reports - Evidence of Carcinogenicity
male rat-positive; female rat-positive; male mice-positive; female mice-positive
NTP Seventh Report - Suspect Carcinogens
suspect carcinogen
OSHA - Possible Select Carcinogens
[present]

ENVIRONMENTAL LISTS

CERCLA/SARA - Section 313 - Emission Reporting
form R reporting required for 0.1% de minimus concentration
CERCLA/SARA - Hazardous Substances and their Reportable Quantities
final RQ = 10 pounds (4.54 kg)
EPA - Carcinogen Hazard Ranking for RQ Adjustment
Hazard ranking = Medium
RCRA - U Series Wastes
waste number U235
RCRA - Hazardous Constituents-Appendix VIII
waste number U235
RCRA - Substances Banned From Land Disposal
[present]
RCRA - Universal Treatment Standards (LDR)
WW: 0.11 mg/l; NWW: 0.10 mg/kg
TSCA - Code of Federal Regulations Citations
40 CFR 704.205; 40 CFR 721.2200
TSCA - Chemicals with Significant New Use Rules
[present]
TSCA - Section 12(b) - Export Notification
export notification required - Section 5

INTERNATIONAL LISTS

Canada - WHMIS: Ingredient Disclosure
0.1% item 1660 (1400)

STATE LISTS

California - Air Bill 2588 Appendix A-II
known or potential carcinogen: 9/89
California - Prop. 65 - Cancer list
carcinogen - initial date 1/1/88
California - Prop. 65 - No Significant Risk Levels
no significant risk level = 0.3 ug/day
California - Directors List of Hazardous Substances (8 CCR 339)
[present]
Florida Hazardous Substance List
[present]
Massachusetts Right To Know List
carcinogen; extraordinarily hazardous
Minnesota Hazardous Substance List
carcinogen
NJ Right to Know List (Total)
sn 1957
NJ Special Hazardous Substances
(carcinogen)
Pennsylvania Right to Know List
environmental hazard; special hazardous substance
Pennsylvania RTK - Special Hazardous Substances
[present]

TRISODIUM NITRILOTRIACETATE MONOHYDRATE   18662-53-8

HEALTH AND SAFETY LISTS

NIOSH - Selected LD50s and LC50s
Oral, rat: LD50 = 2830 mg/kg

PROPOSED REGULATIONS

TSCA - ITC 33rd Report Priority Testing List
recommended with intent-to-designate

TSCA - ITC 34th Report Priority Testing List
recommended with intent-to-designate

TRISODIUM ORTHOVANADATE   13721-39-6

ENVIRONMENTAL LISTS

TSCA - Chemicals with Significant New Use Rules
PMN number: P-90-142

TRISODIUM PHOSPHATE   7601-54-9

HEALTH AND SAFETY LISTS

NIOSH - Selected LD50s and LC50s
Oral, rat: LD50 = 37 gm/kg

NTP Chemical Status Reports - Testing Status and NTIS Number
Technical reports printed (PB85171502/AS)

NTP Chemical Status Reports - Evidence of Carcinogenicity
male rat-equivocal evidence; female rat-no evidence; male mice-no evidence; female mice-some evidence

ENVIRONMENTAL LISTS

TSCA - Code of Federal Regulations Citations
40 CFR 712.30(x);   40 CFR 716.120(d)

TSCA - PAIR - Reporting List
Reporting Date: December 27, 1990

TSCA - Health and Safety Reporting List
Effective Date: October 29, 1990

TRISODIUM SULFOSUCCINATE   13419-59-5

HEALTH AND SAFETY LISTS

NIOSH - Selected LD50s and LC50s
Oral, rat: LD50 = 5900 mg/kg

ENVIRONMENTAL LISTS

List of Pesticide Product Inert Ingredients
[present]

TRISUBSTITUTED ANTHRACENE   RR-01720-9

SEE ALSO:
GLYCIDOL (OXIRANEMETHANOL) AND ITS DERIVATIVES

ENVIRONMENTAL LISTS

EPA - Master Testing List
[present]

TSCA - Code of Federal Regulations Citations
40 CFR 716.120(c)

PROPOSED REGULATIONS

TSCA - Proposed Testing Rule for Glycidyl Ethers
member of Glycidyl subcategory III-A

TRISUBSTITUTED HYDROQUINONE DIESTER   RR-01563-4

ENVIRONMENTAL LISTS

EPA - Master Testing List
[present]

TRISUBSTITUTED PHENOL   RR-00170-7

HEALTH AND SAFETY LISTS

IARC - Group 3 (not classifiable)
[present]

TRITHIOCYANURIC ACID   638-16-4

HEALTH AND SAFETY LISTS

NIOSH - Selected LD50s and LC50s
Oral, rat: LD50 = 2150 mg/kg

NTP Chemical Status Reports - Testing Status and NTIS Number
Technical reports printed (PB270938/AS)

NTP Chemical Status Reports - Evidence of Carcinogenicity
male rat-negative; female rat-negative; male mice-negative; female mice-negative

ENVIRONMENTAL LISTS

List of Pesticide Product Inert Ingredients
[present]

TRITIUM   10028-17-8

HEALTH AND SAFETY LISTS

NTP Chemical Status Reports - Testing Status and NTIS Number
Technical reports printed (PB266177/AS)

NTP Chemical Status Reports - Evidence of Carcinogenicity
PB266177/AS: male   rat-positive;   female   rat-positive;   PB266177/AS: male rat-equivocal; female rat-equivocal; male mice-negative; female mice-negative

STATE LISTS

California - Air Bill 2588 Appendix A-II
known or potential carcinogen: 6/91

California - Prop. 65 - Cancer list
carcinogen - initial date 4/1/89

California - Prop. 65 - No Significant Risk Levels
no significant risk level = 70 ug/day

Massachusetts Right To Know List
carcinogen; extraordinarily hazardous

TRITON CF 10   60864-33-7

INTERNATIONAL LISTS

Canada - WHMIS: Ingredient Disclosure
1% item 1661 (1291)

TRITONAL   RR-01331-0

HEALTH AND SAFETY LISTS

AIHA - WEEL - Ceilings or Short Term Time Weighted Averages
5 mg/m3 STEL

U.S. DOT - Appendix A Table 1 - Hazardous Substances
final RQ = 5000 pounds (2270 kg) (Listed under 'Sodium phosphate, tribasic')

ENVIRONMENTAL LISTS

CERCLA/SARA - Hazardous Substances and their Reportable Quantities
final RQ = 5000 pounds (2270 kg) (Listed under 'Sodium phosphate, tribasic')

Clean Water Act - Hazardous Substances
[present] (Listed under 'Sodium phosphate, tribasic')

List of Pesticide Product Inert Ingredients
[present]

STATE LISTS

California - Directors List of Hazardous Substances (8 CCR 339)
[present] (Listed under 'Sodium phosphate, tribasic')

Massachusetts Right To Know List
[present]

Minnesota Hazardous Substance List
[present]

NJ Right to Know List (Total)
sn 1724

Pennsylvania Right to Know List
environmental hazard

TRP-P-1   62450-06-0

ENVIRONMENTAL LISTS

List of Pesticide Product Inert Ingredients
[present]

TRP-P-2   62450-07-1

ENVIRONMENTAL LISTS

TSCA - Chemicals with Significant New Use Rules
PMN number: P-91-689

TRYPAN BLUE   72-57-1

ENVIRONMENTAL LISTS

TSCA - Chemicals with Significant New Use Rules
PMN number: P-92-329

L-TRYPTOPHAN   73-22-3

ENVIRONMENTAL LISTS

TSCA - Chemicals with Significant New Use Rules
PMN number: P-85-605

T-2 TOXIN   21259-20-1

HEALTH AND SAFETY LISTS

NIOSH - Selected LD50s and LC50s
Oral, rat: LD50 = 9500 mg/kg

TUNG OIL   8001-20-5

HEALTH AND SAFETY LISTS

U.S. DOT - Appendix A Table 2 - Radionuclides
final RQ = 100 curies (3.7E 12 Bq)

ENVIRONMENTAL LISTS

ATSDR Priority List
Rank (of 275): 172

CERCLA/SARA List of Radionuclides (Appendix B) and Their Reportable Quantities
final RQ = 100 curies (3.7E 12 Bq)

TUNGSTEN   7440-33-7

ENVIRONMENTAL LISTS

List of Pesticide Product Inert Ingredients
[present]

TUNGSTEN 176   15749-43-6

HEALTH AND SAFETY LISTS

U.S. DOT - Substances From 49 CFR 172.101
regulated by DOT (UN0390)

U.S. DOT - Hazard Classes
DOT hazard class = 1.1D

TUNGSTEN 177   15749-44-7

HEALTH AND SAFETY LISTS

IARC - Group 2B (sufficient animal data)
[present] (Overall evaluation based only on evidence of carcinogenicity in monograph (31, 1983) or in Supplement 4)

OSHA - Possible Select Carcinogens
[present]

STATE LISTS

California - Air Bill 2588 Appendix A-II
known or potential carcinogen

California - Prop. 65 - Cancer list
carcinogen - initial date 4/1/88

California - Prop. 65 - No Significant Risk Levels
no significant risk level = 0.03 ug/day

California - Directors List of Hazardous Substances (8 CCR 339)
[present]

Massachusetts Right To Know List
carcinogen; extraordinarily hazardous

Minnesota Hazardous Substance List
carcinogen (includes its acetate)

Pennsylvania Right to Know List
special hazardous substance

Pennsylvania RTK - Special Hazardous Substances
[present]

TUNGSTEN 178   15055-23-9

HEALTH AND SAFETY LISTS

IARC - Group 2B (sufficient animal data)
[present] (Overall evaluation based only on evidence of carcinogenicity in monograph (31, 1983) or in Supplement 4)

OSHA - Possible Select Carcinogens
[present]

STATE LISTS

California - Air Bill 2588 Appendix A-II
known or potential carcinogen

California - Prop. 65 - Cancer list
carcinogen - initial date 4/1/88

California - Prop. 65 - No Significant Risk Levels
no significant risk level = 0.2 ug/day

California - Directors List of Hazardous Substances (8 CCR 339)
[present]

Massachusetts Right To Know List
carcinogen; extraordinarily hazardous

Minnesota Hazardous Substance List
carcinogen (includes its acetate)

Pennsylvania Right to Know List
special hazardous substance

Pennsylvania RTK - Special Hazardous Substances
[present]

TUNGSTEN 179   14683-32-0

SEE ALSO:
BISAZOBIPHENYL DYES

HEALTH AND SAFETY LISTS

U.S. DOT - Appendix A Table 1 - Hazardous Substances
final RQ = 10 pounds (4.54 kg)

IARC - Group 2B (sufficient animal data)
[present] (Overall evaluation based only on evidence of carcinogenicity in monograph (8, 1975) or in Supplement 4)

OSHA - Possible Select Carcinogens
[present]

ENVIRONMENTAL LISTS

CERCLA/SARA - Section 313 - Emission Reporting
form R reporting required

CERCLA/SARA - Hazardous Substances and their Reportable Quantities
final RQ = 10 pounds (4.54 kg)

EPA - Carcinogen Hazard Ranking for RQ Adjustment
Hazard ranking = Medium

RCRA - U Series Wastes
waste number U236

RCRA - Hazardous Constituents-Appendix VIII
waste number U236

RCRA - Substances Banned From Land Disposal
[present]

TSCA - Code of Federal Regulations Citations
40 CFR 712.30(d);   40 CFR 716.120(c)

TSCA - PAIR - Reporting List
Reporting Date: November 19, 1982

STATE LISTS

California - Air Bill 2588 Appendix A-II
known or potential carcinogen

California - Prop. 65 - Cancer list
carcinogen - initial date 10/1/89

California - Directors List of Hazardous Substances (8 CCR 339)
[present]

Florida Hazardous Substance List
[present]

Massachusetts Right To Know List
carcinogen; extraordinarily hazardous

Minnesota Hazardous Substance List
carcinogen

NJ Right to Know List (Total)
sn 0465

NJ Special Hazardous Substances
(carcinogen)

Pennsylvania Right to Know List
environmental hazard; special hazardous substance

Pennsylvania RTK - Special Hazardous Substances
[present]

TUNGSTEN 181   15749-46-9

HEALTH AND SAFETY LISTS

NTP Chemical Status Reports - Testing Status and NTIS Number
Technical reports printed (PB285792/AS)

NTP Chemical Status Reports - Evidence of Carcinogenicity
male rat-negative; female rat-negative; male mice-negative; female mice-negative

TUNGSTEN 185   14932-41-3

HEALTH AND SAFETY LISTS

IARC - Group Unspecified
[present] (Listed under 'Toxins derived from Fusarium sporotri-chioides')

TUNGSTEN 187   14983-48-3

HEALTH AND SAFETY LISTS

NFPA - Flash Points
flash point = 552 degrees F (289 degrees C)

NFPA - Hazard Identification Ratings
health-0; flammability-1; reactivity-0

ENVIRONMENTAL LISTS

List of Pesticide Product Inert Ingredients
[present]

TUNGSTEN 188   24421-27-0

HEALTH AND SAFETY LISTS

ACGIH 1995 - Time Weighted Averages
insoluble compounds as W: 5 mg/m3 TWA; soluble compounds as W: 1 mg/m3 TWA

ACGIH 1995 - Short Term Exposure Limits
insoluble compounds as W: 10 mg/m3 STEL; soluble compounds as W: 3 mg/m3 STEL

OSHA - Vacated PELs - Time Weighted Averages
insoluble compounds as W: 5 mg/m3 TWA; soluble compounds as W: 1 mg/m3 TWA

OSHA - Vacated PELs - Short Term Exposure Limits
insoluble compounds as W: 10 mg/m3 STEL; soluble compounds as W: 3 mg/m3 STEL

INTERNATIONAL LISTS

Australian Exposure Standards - Time Weighted Averages
insoluble compounds as W: 5 mg/m3 TWA; soluble compounds as W: 1 mg/m3 TWA

Australian Exposure Standards - Short Term Exposure Limits
insoluble compounds as W: 10 mg/m3 STEL; soluble compounds as W: 3 mg/m3 STEL

Canada - WHMIS: Ingredient Disclosure
1% item 1664 (1703)

Israel - Time Weighted Averages
insoluble compounds as W: 5 mg/m3 TWA; soluble compounds as W: 1 mg/m3 TWA

Israel - Short Term Exposure Limits
insoluble compounds as W: 10 mg/m3 STEL; soluble compounds as W: 3 mg/m3 STEL

Israel - Action Levels
insoluble compounds as W: 2.5 mg/m3 AL; soluble compounds as W: .5 mg/m3 AL

Mexico - Wastewater - Organic Toxic Pollutants and Heavy Metals
Listed under [Heavy Metals]

STATE LISTS

California - Exposure Limits - PELs
metal, as W: 5 mg/m3 PEL; insoluble compounds as W: 5 mg/m3 PEL; soluble compounds as W: 1 mg/m3 PEL

California - Exposure Limits - STELs
insoluble compounds as W: 10 mg/m3 STEL; soluble compounds as W: 3 mg/m3 STEL

California - Directors List of Hazardous Substances (8 CCR 339)
[present]

Florida Hazardous Substance List
[present]

Massachusetts Right To Know List
[present]

Minnesota Hazardous Substance List
[present] as W (includes compounds)

Pennsylvania Right to Know List
[present]

TUNGSTEN ANTIMONATE   69600-03-9

HEALTH AND SAFETY LISTS

U.S. DOT - Appendix A Table 2 - Radionuclides
final RQ = 1000 curies (3.7E 13 Bq)

ENVIRONMENTAL LISTS

CERCLA/SARA List of Radionuclides (Appendix B) and Their Reportable Quantities
final RQ = 1000 curies (3.7E 13 Bq)

TUNGSTEN CARBIDE   12070-12-1

HEALTH AND SAFETY LISTS

U.S. DOT - Appendix A Table 2 - Radionuclides
final RQ = 100 curies (3.7E 12 Bq)

ENVIRONMENTAL LISTS

CERCLA/SARA List of Radionuclides (Appendix B) and Their Reportable Quantities
final RQ = 100 curies (3.7E 12 Bq)

TUNGSTEN COMPOUNDS, N.O.S.   RR-00616-6

HEALTH AND SAFETY LISTS

U.S. DOT - Appendix A Table 2 - Radionuclides
final RQ = 100 curies (3.7E 12 Bq)

ENVIRONMENTAL LISTS

CERCLA/SARA List of Radionuclides (Appendix B) and Their Reportable Quantities
final RQ = 100 curies (3.7E 12 Bq)

TUNGSTEN HEXAFLUORIDE   7783-82-6
HEALTH AND SAFETY LISTS
U.S. DOT - Appendix A Table 2 - Radionuclides
final RQ = 1000 curies (3.7E 13 Bq)
ENVIRONMENTAL LISTS
CERCLA/SARA List of Radionuclides (Appendix B) and Their Reportable Quantities
final RQ = 1000 curies (3.7E 13 Bq)

TUNGSTEN, INSOLUBLE COMPOUNDS   RR-00007-7
HEALTH AND SAFETY LISTS
U.S. DOT - Appendix A Table 2 - Radionuclides
final RQ = 100 curies (3.7E 12 Bq)
ENVIRONMENTAL LISTS
CERCLA/SARA List of Radionuclides (Appendix B) and Their Reportable Quantities
final RQ = 100 curies (3.7E 12 Bq)

TUNGSTEN, SOLUBLE COMPOUNDS   RR-00008-8
HEALTH AND SAFETY LISTS
U.S. DOT - Appendix A Table 2 - Radionuclides
final RQ = 10 curies (3.7E 11 Bq)
ENVIRONMENTAL LISTS
CERCLA/SARA List of Radionuclides (Appendix B) and Their Reportable Quantities
final RQ = 10 curies (3.7E 11 Bq)

TUNGSTIC ACID SALTS   RR-01623-9
HEALTH AND SAFETY LISTS
U.S. DOT - Appendix A Table 2 - Radionuclides
final RQ = 100 curies (3.7E 12 Bq)
ENVIRONMENTAL LISTS
CERCLA/SARA List of Radionuclides (Appendix B) and Their Reportable Quantities
final RQ = 100 curies (3.7E 12 Bq)

TUNGSTIC ACID (SOLUBLE TUNGSTEN)   7783-03-1
HEALTH AND SAFETY LISTS
U.S. DOT - Appendix A Table 2 - Radionuclides
final RQ = 10 curies (3.7E 11 Bq)
ENVIRONMENTAL LISTS
CERCLA/SARA List of Radionuclides (Appendix B) and Their Reportable Quantities
final RQ = 10 curies (3.7E 11 Bq)

TUNGSTIC OXIDE   1314-35-8
SEE ALSO:
ANTIMONY
TUNGSTEN
INTERNATIONAL LISTS
Canada - WHMIS: Ingredient Disclosure
1% item 1662 (253)

TURBO FUELS   RR-00779-4
ENVIRONMENTAL LISTS
TSCA - Code of Federal Regulations Citations
40 CFR 712.30(w)
TSCA - PAIR - Reporting List
Reporting Date: July 13, 1988

TURMERIC, OLEORESIN (CURCUMIN)   8024-37-1
INTERNATIONAL LISTS
Canada - WHMIS: Ingredient Disclosure
1% item 1663 (1702)
United Kingdom - Occupational Exposure Standards - TWAs
soluble, as W: 1 ppm TWA; insoluble, as W: 5 mg/m3 TWA
United Kingdom - Occupational Exposure Standards - STELs
soluble, as W: 3 mg/m3 STEL; insoluble, as W: 10 mg/m3 STEL
STATE LISTS
California - Directors List of Hazardous Substances (8 CCR 339)
[present]

TURPENTINE   8006-64-2
HEALTH AND SAFETY LISTS
U.S. DOT - Substances From 49 CFR 172.101
regulated by DOT (UN2196)
U.S. DOT - Hazard Classes
DOT hazard class = 2.3
U.S. DOT - Substances Which Are Poisonous by Inhalation
gaseous hazardous material poisonous by inhalation (UN2196)
STATE LISTS
NJ Right to Know List (Total)
sn 1961
NJ Special Hazardous Substances
(corrosive)

TURPENTINE (RESIN FROM PINUS SPECIES PARTICULARLY PALUSTRIS)   9005-90-7
HEALTH AND SAFETY LISTS
NIOSH - Health Standards - Exposure Limits
5 mg W/m3 TWA (Listed under 'Tungsten...')
NIOSH - Health Standards - Health Effects and Precautions
Lung and skin effects (Periodic chest X-ray and pulmonary function testing required) (Listed under 'Tungsten')
INTERNATIONAL LISTS
Canada - Alberta - 8 Hour Occupational Exposure Limit
as W: 5 mg/m3 TWA
Canada - Alberta - 15 Minute Occupational Exposure Limit
as W: 10 mg/m3 STEL
Canada - British Columbia - 8 Hour Exposure Limits
as W: 5 mg/m3 TWA
Canada - British Columbia - 15 Minute Exposure Limits
as W: 10 mg/m3 STEL
Canada - Ontario - OHSA - TWAEVs
as W: 5 mg/m3 TWAEV
Canada - Ontario - OHSA - STEVs
10 mg/m3 STEV
Canada - Quebec - Time-Weighted Average Exposure Values
as W: 5 mg/m3 TWAEV
Canada - Quebec - Short-term Exposure Values
10 mg/m3 STEV
Mexico - Instruction No. 10 - TWAs
5 mg/m3 TWA
Mexico - Instruction No. 10 - STELs
10 mg/m3 STEL

ULTRAMARINE BLUE   57455-37-5
HEALTH AND SAFETY LISTS
NIOSH - Health Standards - Exposure Limits
1 mg W/m3 TWA (Listed under 'Tungsten...')
NIOSH - Health Standards - Health Effects and Precautions
Lung and skin effects (Periodic chest X-ray and pulmonary function testing required) (Listed under 'Tungsten')

INTERNATIONAL LISTS
Canada - Alberta - 8 Hour Occupational Exposure Limit
  as W: 1 mg/m3 TWA
Canada - Alberta - 15 Minute Occupational Exposure Limit
  as W: 3 mg/m3 STEL
Canada - British Columbia - 8 Hour Exposure Limits
  as W: 1 mg/m3 TWA
Canada - British Columbia - 15 Minute Exposure Limits
  as W: 3 mg/m3 STEL
Canada - Ontario - OHSA - TWAEVs
  as W: 1 mg/m3 TWAEV
Canada - Ontario - OHSA - STEVs
  3 mg/m3 STEV
Canada - Quebec - Time-Weighted Average Exposure Values
  as W: 1 mg/m3 TWAEV
Canada - Quebec - Short-term Exposure Values
  as W: 3 mg/m3 STEV
Mexico - Instruction No. 10 - TWAs
  1 mg/m3 TWA
Mexico - Instruction No. 10 - STELs
  3 mg/m3 STEL

UNDECANAL   112-44-7

INTERNATIONAL LISTS
Canada - Ontario - OHSA - TWAEVs
  as W: 1 mg/m3 TWAEV

10-UNDECANAL   112-45-8

INTERNATIONAL LISTS
Canada - Ontario - OHSA - TWAEVs
  as W: 1 mg/m3 TWAEV
Canada - Ontario - OHSA - STEVs
  3 mg/m3 TWAEV

UNDECANAL, 2-METHYL-   110-41-8

HEALTH AND SAFETY LISTS
NIOSH - Selected LD50s and LC50s
  Oral, rat: LD50 = 840 mg/kg

UNDECANE   1120-21-4

STATE LISTS
Pennsylvania Right to Know List
  [present]

1-UNDECANETHIOL   5332-52-5

HEALTH AND SAFETY LISTS
NTP Chemical Status Reports - Testing Status and NTIS Number
  Technical reports printed

UNDECANOL   112-42-5

HEALTH AND SAFETY LISTS
ACGIH 1995 - Time Weighted Averages
  100 ppm TWA; 556 mg/m3 TWA
U.S. DOT - Substances From 49 CFR 172.101
  regulated by DOT (UN1299)
U.S. DOT - Hazard Classes
  DOT hazard class = 3
U.S. DOT - Appendix B - Marine Pollutants
  DOT regulated marine pollutant
NFPA - Flash Points
  flash point = 95 degrees F (35 degrees C)
NFPA - Hazard Identification Ratings
  health-1; flammability-3; reactivity-0

NIOSH - Selected LD50s and LC50s
  Inhalation, mouse: LC50 = 29 gm/m3 2 hr Oral, rat: LD50 = 5760
  mg/kg
NIOSH 1990 - Pocket Guide - RELs
  100 ppm TWA; 560 mg/m3 TWA
NIOSH 1990 - Pocket Guide - IDLHs
  1500 ppm IDLH
NIOSH 1990 - Pocket Guide - Target organs
  skin, eyes, kidneys, respiratory system
OSHA - Vacated PELs - Time Weighted Averages
  100 ppm TWA; 560 mg/m3 TWA
OSHA - Final PELs - Time Weighted Averages
  100 ppm TWA; 560 mg/m3 TWA
INTERNATIONAL LISTS
Australian Exposure Standards - Time Weighted Averages
  100 ppm TWA; 560 mg/m3 TWA
Australian Exposure Standards - Skin Effects
  sensitiser
Australian Exposure Standards - Under Review
  exposure limits under review
Canada - WHMIS: Ingredient Disclosure
  1% item 1665 (1561)
Canada - Alberta - 8 Hour Occupational Exposure Limit
  100 ppm TWA; 560 mg/m3 TWA
Canada - Alberta - 15 Minute Occupational Exposure Limit
  150 ppm STEL; 840 mg/m3 STEL
Canada - British Columbia - 8 Hour Exposure Limits
  100 ppm TWA; 560 mg/m3 TWA
Canada - British Columbia - 15 Minute Exposure Limits
  150 ppm STEL; 840 mg/m3 STEL
Canada - Ontario - OHSA - TWAEVs
  560 mg/m3 TWAEV (listed as an agent of variable composition)
Canada - Quebec - Time-Weighted Average Exposure Values
  100 ppm TWAEV; 556 mg/m3 TWAEV
United Kingdom - Occupational Exposure Standards - TWAs
  100 ppm TWA; 560 mg/m3 TWA
United Kingdom - Occupational Exposure Standards - STELs
  150 ppm STEL; 840 mg/m3 STEL
German (DFG) - MAK Values
  100 ppm MAK; 560 mg/m3 MAK
German (DFG) - Peak Limitations
  2 x normal MAK (5 min momentary value); don't exceed 8 times
  during shift
German (DFG) - Skin/Sensitizers
  danger of sensitization (skin or respiratory)
Israel - Time Weighted Averages
  100 ppm TWA; 556 mg/m3 TWA
Israel - Action Levels
  50 ppm AL; 278 mg/m3 AL
Mexico - Instruction No. 10 - TWAs
  100 ppm TWA; 560 mg/m3 TWA
STATE LISTS
California - Exposure Limits - PELs
  100 ppm PEL; 560 mg/m3 PEL
California - Directors List of Hazardous Substances (8 CCR 339)
  [present]
Florida Hazardous Substance List
  [present]
Massachusetts Right To Know List
  [present]
Minnesota Hazardous Substance List
  [present]
NJ Right to Know List (Total)
  sn 1962

NJ Special Hazardous Substances
(flammable - third degree)

2-UNDECANOL   1653-30-1

HEALTH AND SAFETY LISTS

U.S. DOT - Substances From 49 CFR 172.101
regulated by DOT (UN1300)

U.S. DOT - Hazard Classes
DOT hazard class = 3

U.S. DOT - Appendix B - Marine Pollutants
DOT regulated marine pollutant

ENVIRONMENTAL LISTS

List of Pesticide Product Inert Ingredients
[present]

INTERNATIONAL LISTS

Australian Exposure Standards - Time Weighted Averages
480 mg/m3 TWA

Australian Exposure Standards - Under Review
exposure limits under review

STATE LISTS

NJ Right to Know List (Total)
sn 2848

Pennsylvania Right to Know List
[present]

2-UNDECANONE   112-12-9

ENVIRONMENTAL LISTS

List of Pesticide Product Inert Ingredients
[present]

4-UNDECANONE, 2-METHYL-   19594-40-2

ENVIRONMENTAL LISTS

TSCA - Code of Federal Regulations Citations
40 CFR 712.30(x); 40 CFR 716.120(d)

TSCA - PAIR - Reporting List
Reporting Date: November 27, 1991

TSCA - Health and Safety Reporting List
Effective Date: September 30, 1991

9-UNDECENAL   143-14-6

ENVIRONMENTAL LISTS

TSCA - Code of Federal Regulations Citations
40 CFR 712.30(x); 40 CFR 716.120(d)

TSCA - PAIR - Reporting List
Reporting Date: November 27, 1991

TSCA - Health and Safety Reporting List
Effective Date: September 30, 1991

UNDERGROUND HAEMATITE MINING WITH EX-   RR-00526-5
POSURE TO RADON

ENVIRONMENTAL LISTS

TSCA - Code of Federal Regulations Citations
40 CFR 712.30(x); 40 CFR 716.120(d)

TSCA - PAIR - Reporting List
Reporting Date: November 27, 1991

TSCA - Health and Safety Reporting List
Effective Date: September 30, 1991

UNLEADED GASOLINE (WHOLLY VAPORIZED)   RR-00088-4

HEALTH AND SAFETY LISTS

U.S. DOT - Substances From 49 CFR 172.101
regulated by DOT (UN2330)

U.S. DOT - Hazard Classes
DOT hazard class = 3

NFPA - Flash Points
flash point = 149 degrees F (65 degrees C)

NFPA - Hazard Identification Ratings
health-0; flammability-2; reactivity-0

STATE LISTS

NJ Right to Know List (Total)
sn 1963

UNSATURATED AMINO ALKYL ESTER SALT   RR-00899-1

HEALTH AND SAFETY LISTS

NIOSH - Health Standards - Exposure Limits
C (15 min) 0.5 ppm; C (15 min) 3.9 mg/m3 (Listed under 'Thiols')

NIOSH - Health Standards - Health Effects and Precautions
Irritation; eye, skin, blood, and nervous system effects (Blood and urine monitoring required; prevent skin contact) (Listed under 'Thiols')

UNSATURATED AMINO ESTER SALT   RR-00900-7

HEALTH AND SAFETY LISTS

NIOSH - Selected LD50s and LC50s
Oral, rat: LD50 = 3000 mg/kg

INTERNATIONAL LISTS

Canada - WHMIS: Ingredient Disclosure
1% item 1666 (1704)

UNSATURATED ORGANIC COMPOUND   RR-00998-3

HEALTH AND SAFETY LISTS

NFPA - Flash Points
flash point = 235 degrees F (113 degrees C)

NFPA - Hazard Identification Ratings
health-1; flammability-1; reactivity-0

UNTREATED AND MILDLY-TREATED OILS   RR-00054-4

HEALTH AND SAFETY LISTS

NFPA - Flash Points
flash point = 192 degrees F (89 degrees C)

NFPA - Hazard Identification Ratings
health-0; flammability-2; reactivity-0

NIOSH - Selected LD50s and LC50s
Oral, rat: LD50 = 5000 mg/kg

URACIL MUSTARD   66-75-1

STATE LISTS

Pennsylvania Right to Know List
[present]

URANIUM   7440-61-1

ENVIRONMENTAL LISTS

TSCA - Code of Federal Regulations Citations
40 CFR 712.30(x); 40 CFR 716.120(d)

TSCA - PAIR - Reporting List
Reporting Date: November 27, 1991

TSCA - Health and Safety Reporting List
Effective Date: September 30, 1991

URANIUM 230   15743-51-8

HEALTH AND SAFETY LISTS

IARC - Group 1 (carcinogenic to humans)
[present] (Listed under 'Haematite and ferric oxide')

OSHA - Select Carcinogens
[present]

STATE LISTS

Pennsylvania Right to Know List
[present]

URANIUM 231   15700-08-0

STATE LISTS

California - Prop. 65 - Cancer list
carcinogen - initial date 4/1/88

URANIUM 232   14158-29-3

ENVIRONMENTAL LISTS

TSCA - Chemicals with Significant New Use Rules
PMN number: P-84-527

URANIUM 233   13968-55-3

ENVIRONMENTAL LISTS

TSCA - Chemicals with Significant New Use Rules
PMN number: P-84-537

URANIUM 234   13966-29-5

ENVIRONMENTAL LISTS

TSCA - Chemicals with Significant New Use Rules
PMN number: P-84-358

URANIUM 235   15117-96-1

HEALTH AND SAFETY LISTS

IARC - Group 1 (carcinogenic to humans)
[present] (Listed under 'Mineral oils')

NTP Seventh Report - Known Carcinogens
known carcinogen (Listed under 'Soots, tars, and mineral oils')

OSHA - Select Carcinogens
[present]

URANIUM 236   13982-70-2

HEALTH AND SAFETY LISTS

U.S. DOT - Appendix A Table 1 - Hazardous Substances
final RQ = 10 pounds (4.54 kg)

IARC - Group 2B (sufficient animal data)
[present]

NIOSH - Selected LD50s and LC50s
Oral, rat: LD50 = 7500 ug/kg

NTP Chemical Status Reports - Testing Status and NTIS Number
Chronic studies exist for which technical reports were not prepared

OSHA - Possible Select Carcinogens
[present]

ENVIRONMENTAL LISTS

CERCLA/SARA - Hazardous Substances and their Reportable Quantities
final RQ = 10 pounds (4.54 kg)

EPA - Carcinogen Hazard Ranking for RQ Adjustment
Hazard ranking = Medium

RCRA - U Series Wastes
waste number U237

RCRA - Hazardous Constituents-Appendix VIII
waste number U237

RCRA - Substances Banned From Land Disposal
[present]

STATE LISTS

California - Air Bill 2588 Appendix A-II
known or potential carcinogen

California - Prop. 65 - Cancer list
carcinogen - initial date 4/1/88

California - Prop. 65 - Developmental Toxicity
developmental toxicity - initial date 1/1/92

California - Prop. 65 - Reproductive - Female
female reproductive toxicity - initial date 1/1/92

California - Prop. 65 - Reproductive - Male
male reproductive toxicity - initial date 1/1/92

California - Directors List of Hazardous Substances (8 CCR 339)
[present]

Florida Hazardous Substance List
[present]

Massachusetts Right To Know List
carcinogen; extraordinarily hazardous

Minnesota Hazardous Substance List
carcinogen

NJ Special Hazardous Substances
(carcinogen; mutagen)

Pennsylvania Right to Know List
environmental hazard; special hazardous substance

Pennsylvania RTK - Special Hazardous Substances
[present]

URANIUM 237   14269-75-1

HEALTH AND SAFETY LISTS

ACGIH 1995 - Time Weighted Averages
soluble and insoluble compounds, as U: 0.2 mg/m3 TWA

ACGIH 1995 - Short Term Exposure Limits
soluble and insoluble compounds, as U: 0.6 mg/m3 STEL

U.S. DOT - Appendix A Table 2 - Radionuclides
final RQ = 0.1 curies (3.7E 9 Bq) (notification requirements for releases of mixtures or solutions can be found in 40 CFR 302.6(b))

NIOSH 1990 - Pocket Guide - RELs
as U: 0.2 mg/m3 TWA;   0.6 mg/m3 STEL

NIOSH 1990 - Pocket Guide - IDLHs
as U: 30 mg/m3 IDLH (not considering carcinogenic effects)

NIOSH 1990 - Pocket Guide - Carcinogens
occupational carcinogen

NIOSH 1990 - Pocket Guide - Target organs
skin, bone marrow, lymphatics

OSHA - Vacated PELs - Time Weighted Averages
soluble compounds, as U: 0.05 mg/m3 TWA; insoluble compounds, as U: 0.2 mg/m3 TWA

OSHA - Vacated PELs - Short Term Exposure Limits
insoluble compounds, as U: 0.6 mg/m3 STEL

OSHA - Final PELs - Time Weighted Averages
soluble compounds, as U: 0.05 mg/m3 TWA;   insoluble compounds, as U: 0.05 mg/m3 TWA

ENVIRONMENTAL LISTS

ATSDR Priority List
Rank (of 275): 076

CERCLA/SARA List of Radionuclides (Appendix B) and Their Reportable Quantities
final RQ = 0.1 curies (3.7E 9 Bq) (notification requirements for releases of mixtures or solutions can be found in 40 CFR 302.6(b))

INTERNATIONAL LISTS

Australian Exposure Standards - Time Weighted Averages
soluble and insoluble compounds, as U: 0.2 mg/m3 TWA

Australian Exposure Standards - Short Term Exposure Limits
soluble and insoluble compounds, as U: 0.6 mg/m3 STEL

Canada - WHMIS: Ingredient Disclosure
1% item 1668 (1707)

Canada - Drinking Water Quality - MACs
0.1 mg/L MAC

Canada - British Columbia - 8 Hour Exposure Limits
as U: 0.2 mg/m3 TWA

Canada - British Columbia - 15 Minute Exposure Limits
as U: 0.6 mg/m3 STEL
Canada - Ontario - OHSA - TWAEVs
as U: 0.2 mg/m3 TWAEV
Canada - Ontario - OHSA - STEVs
as U: 0.6 mg/m3 STEV
Canada - Quebec - Short-term Exposure Values
as U: 0.6 mg/m3 STEV
German (DFG) - MAK Values
total dust, as U: 0.25 mg/m3 MAK
German (DFG) - Peak Limitations
10 x normal MAK (30 min average value); don't exceed during shift
Israel - Time Weighted Averages
as U: 0.2 mg/m3 TWA
Israel - Short Term Exposure Limits
as U: 0.6 mg/m3 STEL
Israel - Action Levels
as U: 0.1 mg/m3 AL
Mexico - Instruction No. 10 - TWAs
0.2 mg/m3 TWA
Mexico - Instruction No. 10 - STELs
0.6 mg/m3 STEL
STATE LISTS
California - Directors List of Hazardous Substances (8 CCR 339)
[present] (exempt when in capsules as prescribed in 49 CFR 173.403 (z))
Florida Hazardous Substance List
[present]
Massachusetts Right To Know List
[present]
Minnesota Hazardous Substance List
[present] as U (includes natural compounds both soluble and insoluble)
NJ Right to Know List (Total)
sn 1969
Pennsylvania Right to Know List
[present]

URANIUM 239   13982-01-9
HEALTH AND SAFETY LISTS
U.S. DOT - Appendix A Table 2 - Radionuclides
final RQ = 1 curie (3.7E 10 Bq)
ENVIRONMENTAL LISTS
CERCLA/SARA List of Radionuclides (Appendix B) and Their Reportable Quantities
final RQ = 1 curie (3.7E 10 Bq)

URANIUM 240   15687-53-3
HEALTH AND SAFETY LISTS
U.S. DOT - Appendix A Table 2 - Radionuclides
final RQ = 1000 curies (3.7E 13 Bq)
ENVIRONMENTAL LISTS
CERCLA/SARA List of Radionuclides (Appendix B) and Their Reportable Quantities
final RQ = 1000 curies (3.7E 13 Bq)

URANIUM COMPOUNDS, N.O.S.   RR-00617-7
HEALTH AND SAFETY LISTS
U.S. DOT - Appendix A Table 2 - Radionuclides
final RQ = 0.01 curies (3.7E 8 Bq)
ENVIRONMENTAL LISTS
CERCLA/SARA List of Radionuclides (Appendix B) and Their Reportable Quantities
final RQ = 0.01 curies (3.7E 8 Bq)

URANIUM HEXAFLUORIDE   7783-81-5
HEALTH AND SAFETY LISTS
U.S. DOT - Appendix A Table 2 - Radionuclides
final RQ = 0.1 curies (3.7E 9 Bq)
ENVIRONMENTAL LISTS
ATSDR Priority List
Rank (of 275): 192
CERCLA/SARA List of Radionuclides (Appendix B) and Their Reportable Quantities
final RQ = 0.1 curies (3.7E 9 Bq)

URANIUM INSOLUBLE COMPOUNDS   RR-00045-3
HEALTH AND SAFETY LISTS
U.S. DOT - Appendix A Table 2 - Radionuclides
final RQ = 0.1 curies (3.7E 9 Bq) (notification requirements for releases of mixtures and solutions can be found in 40 CFR 302.6(b))
ENVIRONMENTAL LISTS
ATSDR Priority List
Rank (of 275): 117
CERCLA/SARA List of Radionuclides (Appendix B) and Their Reportable Quantities
final RQ = 0.1 curies (3.7E 9 Bq) (notification requirements for releases of mixtures and solutions can be found in 40 CFR 302.6(b))

URANIUM SOLUBLE COMPOUNDS   RR-00044-2
HEALTH AND SAFETY LISTS
U.S. DOT - Appendix A Table 2 - Radionuclides
final RQ = 0.1 curies (3.7E 9 Bq) (notification requirements for releases of mixtures or solutions can be found in 40 CFR 302.6(b))
ENVIRONMENTAL LISTS
ATSDR Priority List
Rank (of 275): 117
CERCLA/SARA List of Radionuclides (Appendix B) and Their Reportable Quantities
final RQ = 0.1 curies (3.7E 9 Bq) (notification requirements for releases of mixtures or solutions can be found in 40 CFR 302.6(b))

URANYL ACETATE   541-09-3
HEALTH AND SAFETY LISTS
U.S. DOT - Appendix A Table 2 - Radionuclides
final RQ = 0.1 curies (3.7E 9 Bq)
ENVIRONMENTAL LISTS
CERCLA/SARA List of Radionuclides (Appendix B) and Their Reportable Quantities
final RQ = 0.1 curies (3.7E 9 Bq)

URANYL NITRATE   10102-06-4
HEALTH AND SAFETY LISTS
U.S. DOT - Appendix A Table 2 - Radionuclides
final RQ = 100 curies (3.7E 12 Bq)
ENVIRONMENTAL LISTS
CERCLA/SARA List of Radionuclides (Appendix B) and Their Reportable Quantities
final RQ = 100 curies (3.7E 12 Bq)

URANYL NITRATE   36478-76-9
HEALTH AND SAFETY LISTS
U.S. DOT - Appendix A Table 2 - Radionuclides
final RQ = 1000 curies (3.7E 13 Bq)
ENVIRONMENTAL LISTS
CERCLA/SARA List of Radionuclides (Appendix B) and Their Reportable Quantities
final RQ = 1000 curies (3.7E 13 Bq)

URANYL NITRATE HEXAHYDRATE   13520-83-7

HEALTH AND SAFETY LISTS

U.S. DOT - Appendix A Table 2 - Radionuclides
final RQ = 1000 curies (3.7E 13 Bq)

ENVIRONMENTAL LISTS

CERCLA/SARA List of Radionuclides (Appendix B) and Their Reportable Quantities
final RQ = 1000 curies (3.7E 13 Bq)

UREA   57-13-6

INTERNATIONAL LISTS

Canada - WHMIS: Ingredient Disclosure
1% item 1667 (1706)

Canada - Ontario - OHSA - TWAEVs
as U: 0.2 mg/m3 TWAEV

Canada - Ontario - OHSA - STEVs
0.6 mg/m3 STEV

United Kingdom - Occupational Exposure Standards - TWAs
soluble, as U: 0.2 mg/m3 TWA

United Kingdom - Occupational Exposure Standards - STELs
as U: 0.6 mg/m3 STEL

STATE LISTS

California - Directors List of Hazardous Substances (8 CCR 339)
[present]

**UREA, CONDENSATE WITH POLY[OXY(METHYL-1,   RR-00214-2
2-ETHANEDIYL)]-ALPHA-[2-AMINOETHYLETHYL]-
X-(2-AMINOETHYLETHOXY)**

HEALTH AND SAFETY LISTS

U.S. DOT - Substances From 49 CFR 172.101
regulated by DOT (UN2977, UN2978)

U.S. DOT - Hazard Classes
DOT hazard class = 7

STATE LISTS

NJ Right to Know List (Total)
sn 1970

UREA-FORMALDEHYDE-RESINS   97380-66-0

INTERNATIONAL LISTS

Canada - Alberta - 8 Hour Occupational Exposure Limit
0.2 mg/m3 TWA

Canada - Alberta - 15 Minute Occupational Exposure Limit
as U: 0.6 mg/m3 STEL

Canada - Quebec - Time-Weighted Average Exposure Values
as U: 0.2 mg/m3 TWAEV

STATE LISTS

California - Exposure Limits - PELs
as U: 0.2 mg/m3 PEL

California - Exposure Limits - STELs
0.6 mg/m3 STEL

**UREA, (HEXAHYDRO-6-METHYL-2-OXO-4-PYRIM-   1129-42-6
IDINYL)-**

HEALTH AND SAFETY LISTS

NIOSH 1990 - Pocket Guide - RELs
as U: 0.05 mg/m3 TWA

NIOSH 1990 - Pocket Guide - IDLHs
as U: 20 mg/m3 IDLH (not considering carcinogenic effects)

NIOSH 1990 - Pocket Guide - Carcinogens
occupational carcinogen

NIOSH 1990 - Pocket Guide - Target organs
respiratory system, blood, liver, lymphatics, kidneys, skin, bone marrow

INTERNATIONAL LISTS

Canada - Alberta - 8 Hour Occupational Exposure Limit
as U: 0.2 mg/m3 TWA

Canada - Alberta - 15 Minute Occupational Exposure Limit
as U: 0.6 mg/m3 STEL

Canada - Quebec - Time-Weighted Average Exposure Values
as U: 0.05 mg/m3 TWAEV

STATE LISTS

California - Exposure Limits - PELs
as U: 0.05 mg/m3 PEL

UREA HYDROGEN PEROXIDE   RR-01338-7

HEALTH AND SAFETY LISTS

U.S. DOT - Appendix A Table 1 - Hazardous Substances
final RQ = 100 pounds (45.4 kg)

ENVIRONMENTAL LISTS

CERCLA/SARA - Hazardous Substances and their Reportable Quantities
final RQ = 100 pounds (45.4 kg)

Clean Water Act - Hazardous Substances
[present]

STATE LISTS

Massachusetts Right To Know List
[present]

NJ Right to Know List (Total)
sn 1975

Pennsylvania Right to Know List
environmental hazard

UREA, 1-(2-METHYLCYCLOHEXYL)-3-PHENYL-   1982-49-6

HEALTH AND SAFETY LISTS

U.S. DOT - Substances From 49 CFR 172.101
regulated by DOT (UN2981)

U.S. DOT - Hazard Classes
DOT hazard class = 7

U.S. DOT - Appendix A Table 1 - Hazardous Substances
final RQ = 100 pounds (45.4 kg)

ENVIRONMENTAL LISTS

CERCLA/SARA - Hazardous Substances and their Reportable Quantities
final RQ = 100 pounds (45.4 kg)

Clean Water Act - Hazardous Substances
[present]

INTERNATIONAL LISTS

Canada - WHMIS: Ingredient Disclosure
1% item 1669 (1213)

STATE LISTS

Florida Hazardous Substance List
[present]

Massachusetts Right To Know List
[present]

NJ Right to Know List (Total)
sn 1978

Pennsylvania Right to Know List
environmental hazard

UREA, N,N"-METHYLENEBIS-   13547-17-6

HEALTH AND SAFETY LISTS

U.S. DOT - Appendix A Table 1 - Hazardous Substances
final RQ = 100 pounds (45.4 kg)

ENVIRONMENTAL LISTS

CERCLA/SARA - Hazardous Substances and their Reportable Quantities
final RQ = 100 pounds (45.4 kg)

Clean Water Act - Hazardous Substances
[present]

STATE LISTS

Massachusetts Right To Know List
[present]

Pennsylvania Right to Know List
environmental hazard

UREA NITRATE   124-47-0

HEALTH AND SAFETY LISTS

U.S. DOT - Substances From 49 CFR 172.101
regulated by DOT (UN2980)

U.S. DOT - Hazard Classes
DOT hazard class = 7

STATE LISTS

NJ Right to Know List (Total)
sn 1980

UREA NITRATE (VAN)   17687-37-5

HEALTH AND SAFETY LISTS

AIHA - WEEL - Time Weighted Averages
10 mg/m3 TWA

NIOSH - Selected LD50s and LC50s
Oral, rat: LD50 = 14300 mg/kg

ENVIRONMENTAL LISTS

EPA - Master Testing List
[present]

List of Pesticide Product Inert Ingredients
[present]

STATE LISTS

Minnesota Hazardous Substance List
[present]

UREA PEROXIDE   124-43-6

ENVIRONMENTAL LISTS

TSCA - Chemicals with Significant New Use Rules
PMN number: P-84-482

UREA, POYLMER WITH FORMALDEHYDE   9011-05-6

ENVIRONMENTAL LISTS

EPA - Master Testing List
[present]

UREA, REACTION PRODUCTS WITH   68611-64-3
FORMALDEHYDE

ENVIRONMENTAL LISTS

TSCA - Code of Federal Regulations Citations
40 CFR 721.2490

TSCA - Chemicals with Significant New Use Rules
PMN number: P-89-303

TSCA - Section 12(b) - Export Notification
P-89-303; export notification required - Section 5

URETHANE   51-79-6

HEALTH AND SAFETY LISTS

U.S. DOT - Substances From 49 CFR 172.101
regulated by DOT (UN1511)

U.S. DOT - Hazard Classes
DOT hazard class = 5.1

URETHANE ACRYLATE   RR-00240-4

HEALTH AND SAFETY LISTS

NIOSH - Selected LD50s and LC50s
Oral, rat: LD50 = 5000 mg/kg

STATE LISTS

California - Directors List of Hazardous Substances (8 CCR 339)
[present]

URIC ACID   69-93-2

ENVIRONMENTAL LISTS

List of Pesticide Product Inert Ingredients
[present]

UROFOLLITROPIN   26995-91-5

STATE LISTS

NJ Right to Know List (Total)
sn 1984

USED ENGINE OILS   RR-00095-3

HEALTH AND SAFETY LISTS

U.S. DOT - Substances From 49 CFR 172.101
regulated by DOT (UN0220, UN1357)

U.S. DOT - Hazard Classes
DOT hazard class = 4.1

VACUUM TOWER CONDENSATE (PETROLEUM)   64741-49-7

STATE LISTS

NJ Right to Know List (Total)
sn 1985

VALERALDEHYDE   110-62-3

ENVIRONMENTAL LISTS

EPA - Master Testing List
[present]

List of Pesticide Product Inert Ingredients
[present]

TSCA - Code of Federal Regulations Citations
40 CFR 712.30(n); 40 CFR 716.20(b)(1); 40 CFR 716.120(a)

TSCA - PAIR - Reporting List
Reporting Date: August 1, 1985

TSCA - Health and Safety Reporting List
Effective Date: June 3, 1985 (Studies on agronomic plant growth or damage which demonstrate only that the resin stimulates plant growth or causes plant damage when applied as a fertilizer are exempt)

VALERIC ACID   109-52-4

ENVIRONMENTAL LISTS

EPA - Master Testing List
[present]

TSCA - Code of Federal Regulations Citations
40 CFR 712.30(n); 40 CFR 761.20(b)(1); 40 CFR 716.120(a)

TSCA - PAIR - Reporting List
Reporting Date: August 1, 1985

TSCA - Health and Safety Reporting List
Effective Date: June 3, 1985; Sunset Date: November 9, 1993 (Studies on agronomic plant growth or damage which demonstrate only that the resins stimulate plant growth or cause plant damage when applied as a fertilizer are exempt)

4-VALEROLACTONE   108-29-2

HEALTH AND SAFETY LISTS

U.S. DOT - Appendix A Table 1 - Hazardous Substances
final RQ = 100 pounds (45.4 kg)

IARC - Group 2B (sufficient animal data)
[present] (Overall evaluation based only on evidence of carcinogenicity in monograph (7, 1974) or in Supplement 4)
NIOSH - Selected LD50s and LC50s
Oral, mouse: LD50 = 2500 mg/kg
NTP Chemical Status Reports - Testing Status and NTIS Number
Post peer review technical reports in progress; project leader assigned/study in design
NTP Seventh Report - Suspect Carcinogens
suspect carcinogen
OSHA - Possible Select Carcinogens
[present]

ENVIRONMENTAL LISTS

CERCLA/SARA - Section 313 - Emission Reporting
form R reporting required for 0.1% de minimus concentration
CERCLA/SARA - Hazardous Substances and their Reportable Quantities
final RQ = 100 pounds (45.4 kg)
Clean Air Act (1990) - List of Hazardous Air Contaminants
[present]
EPA - Carcinogen Hazard Ranking for RQ Adjustment
Hazard ranking = Low
RCRA - U Series Wastes
waste number U238
RCRA - Hazardous Constituents-Appendix VIII
waste number U238
RCRA - Substances Banned From Land Disposal
[present]
TSCA - Code of Federal Regulations Citations
40 CFR 721.2550
TSCA - Chemicals with Significant New Use Rules
[present]
TSCA - Section 12(b) - Export Notification
export notification required - Section 5

INTERNATIONAL LISTS

Canada - WHMIS: Ingredient Disclosure
0.1% item 1670 (1708)
German (DFG) - Carcinogens
animal evidence of carcinogenicity

STATE LISTS

California - Air Bill 2588 Appendix A-I
known or potential carcinogen
California - Prop. 65 - Cancer list
carcinogen - initial date 1/1/88
California - Prop. 65 - Developmental Toxicity
developmental toxicity - initial date 10/01/94
California - Prop. 65 - No Significant Risk Levels
no significant risk level = 0.7 ug/day
California - Directors List of Hazardous Substances (8 CCR 339)
[present]
Florida Hazardous Substance List
[present]
Massachusetts Right To Know List
carcinogen; extraordinarily hazardous
Minnesota Hazardous Substance List
carcinogen
NJ Right to Know List (Total)
sn 1986
NJ Special Hazardous Substances
(carcinogen; mutagen; teratogen)
Pennsylvania Right to Know List
environmental hazard; special hazardous substance

Pennsylvania RTK - Special Hazardous Substances
[present]

VALERYL CHLORIDE   638-29-9

ENVIRONMENTAL LISTS

TSCA - Chemicals with Significant New Use Rules
PMN number: P-85-301

VALINOMYCIN   2001-95-8

ENVIRONMENTAL LISTS

List of Pesticide Product Inert Ingredients
[present]

VALPROATE   99-66-1

STATE LISTS

California - Air Bill 2588 Appendix A-II
9/90
California - Prop. 65 - Developmental Toxicity
developmental toxicity - initial date 4/1/90

VANADIC ACID, TRIISOBUTYL ESTER   19120-62-8
SEE ALSO:
PETROLEUM DISTILLATES (NAPHTHA)

STATE LISTS

California - Prop. 65 - Cancer list
carcinogen - initial dates 2/27/87

VANADIUM   7440-62-2

STATE LISTS

Massachusetts Right To Know List
carcinogen; extraordinarily hazardous

VANADIUM 47   14867-38-0

HEALTH AND SAFETY LISTS

ACGIH 1995 - Time Weighted Averages
50 ppm TWA; 176 mg/m3 TWA
AIHA - Odor Threshold Values
no geometric mean air odor threshold
U.S. DOT - Substances From 49 CFR 172.101
regulated by DOT (UN2058)
U.S. DOT - Hazard Classes
DOT hazard class = 3
NFPA - Flash Points
flash point = 54 degrees F (12 degrees C)
NFPA - Hazard Identification Ratings
health-1; flammability-3; reactivity-0
NIOSH - Selected LD50s and LC50s
Oral, rat: LD50 = 3200 mg/kg Skin, guinea pig: LD50 = 20 gm/kg
OSHA - Vacated PELs - Time Weighted Averages
50 ppm TWA; 175 mg/m3 TWA

ENVIRONMENTAL LISTS

TSCA - Code of Federal Regulations Citations
40 CFR 712.30(x); 40 CFR 716.120(d)
TSCA - PAIR - Reporting List
Reporting Date: November 27, 1991
TSCA - Health and Safety Reporting List
Effective Date: September 30, 1991

INTERNATIONAL LISTS

Australian Exposure Standards - Time Weighted Averages
50 ppm TWA; 176 mg/m3 TWA
Canada - WHMIS: Ingredient Disclosure
1% item 1671 (1709)
Canada - Alberta - 8 Hour Occupational Exposure Limit
50 ppm TWA; 175 mg/m3 TWA

Canada - Alberta - 15 Minute Occupational Exposure Limit
75 ppm STEL; 265 mg/m3 STEL

Canada - British Columbia - 8 Hour Exposure Limits
50 ppm TWA; 175 mg/m3 TWA

Canada - Ontario - OHSA - TWAEVs
50 ppm TWAEV; 175 mg/m3 TWAEV

Canada - Quebec - Time-Weighted Average Exposure Values
50 ppm TWAEV; 176 mg/m3 TWAEV

Israel - Time Weighted Averages
50 ppm TWA; 176 mg/m3 TWA

Israel - Action Levels
25 ppm AL; 88 mg/m3 AL

Mexico - Instruction No. 10 - TWAs
50 ppm TWA; 175 mg/m3 TWA

STATE LISTS

California - Exposure Limits - PELs
50 ppm PEL; 175 mg/m3 PEL

California - Directors List of Hazardous Substances (8 CCR 339)
[present]

Florida Hazardous Substance List
[present]

Massachusetts Right To Know List
[present]

Minnesota Hazardous Substance List
[present]

NJ Right to Know List (Total)
sn 1987

NJ Special Hazardous Substances
(flammable - third degree)

Pennsylvania Right to Know List
[present]

VANADIUM 48   14331-97-6

HEALTH AND SAFETY LISTS

NFPA - Flash Points
flash point = 205 degrees F (96 degrees C)

NFPA - Hazard Identification Ratings
health-2; flammability-1; reactivity-0

NIOSH - Selected LD50s and LC50s
Inhalation, mouse: LC50 = 4100 mg/kg (2 hr) Oral, mouse: LD50 = 600 mg/kg

ENVIRONMENTAL LISTS

List of Pesticide Product Inert Ingredients
[present]

INTERNATIONAL LISTS

Canada - WHMIS: Ingredient Disclosure
1% item 1672 (150)

STATE LISTS

Florida Hazardous Substance List
[present]

Massachusetts Right To Know List
[present]

NJ Right to Know List (Total)
sn 1988

NJ Special Hazardous Substances
(corrosive)

Pennsylvania Right to Know List
[present]

VANADIUM 49   14392-01-9

HEALTH AND SAFETY LISTS

NIOSH - Selected LD50s and LC50s
Oral, rat: LD50 = 8800 mg/kg

VANADIUM CARBIDE   11130-21-5

HEALTH AND SAFETY LISTS

U.S. DOT - Substances From 49 CFR 172.101
regulated by DOT (UN2502)

U.S. DOT - Hazard Classes
DOT hazard class = 8

INTERNATIONAL LISTS

Canada - WHMIS: Ingredient Disclosure
1% item 1673 (537)

STATE LISTS

NJ Right to Know List (Total)
sn 1989

NJ Special Hazardous Substances
(corrosive)

VANADIUM COMPOUNDS   RR-00093-1

HEALTH AND SAFETY LISTS

NIOSH - Selected LD50s and LC50s
Oral, rat: LD50 = 4 mg/kg Skin, rabbit: LD50 = 5 mg/kg

ENVIRONMENTAL LISTS

CERCLA/SARA - Section 302 Extremely Hazardous Substances and TPQs
TPQ = 1000/10,000 pounds

STATE LISTS

Florida Hazardous Substance List
effective March 13, 1992

Massachusetts Right To Know List
extraordinarily hazardous

NJ Right to Know List (Total)
sn 2850

Pennsylvania Right to Know List
environmental hazard

PROPOSED REGULATIONS

CERCLA/SARA - Proposed Hazardous Substance Additions
proposed RQ = 1 pound (.454 kg)

CERCLA/SARA - 1989 Proposed RQ Adjustments
proposed RQ = 100 pounds (45.4 kg)

VANADIUM OXYTRICHLORIDE   7727-18-6

STATE LISTS

California - Air Bill 2588 Appendix A-II
[present]

California - Prop. 65 - Developmental Toxicity
developmental toxicity - initial date 7/1/87

Massachusetts Right To Know List
teratogen

VANADIUM PENTOXIDE   1314-62-1

INTERNATIONAL LISTS

Canada - WHMIS: Ingredient Disclosure
1% item 1674 (1711)

VANADIUM TETRACHLORIDE   7632-51-1
SEE ALSO:
F039-HAZARDOUS WASTES

HEALTH AND SAFETY LISTS

NIOSH - Health Standards - Exposure Limits
1 mg V/m3 TWA (Listed under 'Vanadium')

NIOSH - Health Standards - Health Effects and Precautions
Eye, skin, and lung effects (Periodic chest X-ray and pulmonary function testing required)

OSHA - Final PELs - Time Weighted Averages
respirable dust, as V2O5: 0.5 mg/m3 TWA; fume, as V2O5: 0.1 mg/m3 TWA

ENVIRONMENTAL LISTS

ATSDR Priority List
Rank (of 275): 178

CERCLA/SARA - Section 313 - Emission Reporting
form R reporting required for 1.0% de minimus concentration (only fume or dust)

RCRA - Basis for Listing - Appendix VII
Included in waste stream: F039

RCRA - TSD Facilities Ground Water Monitoring
TM 6010 = 80 ug/L PQL; TM 7910 = 2000 ug/L PQL; TM 7911 = 40 ug/L PQL (all species in the ground water that contain this element are included)

RCRA - Universal Treatment Standards (LDR)
WW: 4.3 mg/l; NWW: 0.23 mg/l TCLP

INTERNATIONAL LISTS

Canada - WHMIS: Ingredient Disclosure
1% item 1675 (1712)

Canada - NPRI (National Pollutant Release Inventory)
[present] (as fume or dust)

Canada - Alberta - 8 Hour Occupational Exposure Limit
as V2O5, respirable dust and fume: 0.05 mg/m3 TWA

Canada - Alberta - 15 Minute Occupational Exposure Limit
as V2O5; respirable dust and fume: 0.15 mg/m3

Canada - British Columbia - 8 Hour Exposure Limits
dust, as V: 0.5 mg/m3 TWA

Canada - British Columbia - 15 Minute Exposure Limits
dust, as V: 1.5 mg/m3 STEL

Canada - British Columbia - Ceiling Exposure Limits
fume, as V: C 0.05 mg/m3

Mexico - Instruction No. 10 - TWAs
0.5 mg/m3 TWA

Mexico - Wastewater - Organic Toxic Pollutants and Heavy Metals
Listed under [Heavy Metals]

STATE LISTS

California - Air Bill 2588 Appendix A-I
fume or dust: 6/91

California - Directors List of Hazardous Substances (8 CCR 339)
[present]

Massachusetts Right To Know List
[present]

NJ Right to Know List (Total)
fume or dust: sn 1990

Pennsylvania Right to Know List
environmental hazard

PROPOSED REGULATIONS

Safe Drinking Water Act - Priority list
[present]

VANADIUM TRICHLORIDE   7718-98-1

HEALTH AND SAFETY LISTS

U.S. DOT - Appendix A Table 2 - Radionuclides
final RQ = 1000 curies (3.7E 13 Bq)

ENVIRONMENTAL LISTS

CERCLA/SARA List of Radionuclides (Appendix B) and Their Reportable Quantities
final RQ = 1000 curies (3.7E 13 Bq)

VANADIUM TRIOXIDE   1314-34-7

HEALTH AND SAFETY LISTS

U.S. DOT - Appendix A Table 2 - Radionuclides
final RQ = 10 curies (3.7E 11 Bq)

ENVIRONMENTAL LISTS

CERCLA/SARA List of Radionuclides (Appendix B) and Their Reportable Quantities
final RQ = 10 curies (3.7E 11 Bq)

VANADYL SULFATE   27774-13-6

HEALTH AND SAFETY LISTS

U.S. DOT - Appendix A Table 2 - Radionuclides
final RQ = 1000 curies (3.7E 13 Bq)

ENVIRONMENTAL LISTS

CERCLA/SARA List of Radionuclides (Appendix B) and Their Reportable Quantities
final RQ = 1000 curies (3.7E 13 Bq)

VANILLIN   121-33-5

HEALTH AND SAFETY LISTS

NIOSH - Health Standards - Exposure Limits
1 mg V/m3 TWA (Listed under 'Vanadium')

NIOSH - Health Standards - Health Effects and Precautions
Eye, skin, and lung effects (Periodic chest X-ray and pulmonary function testing required) (Listed under 'Vanadium')

VEGETABLE OIL   68956-68-3

HEALTH AND SAFETY LISTS

NIOSH - Health Standards - Exposure Limits
C (15 min) 0.05 mg V/m3 (Listed under 'Vanadium')

NIOSH - Health Standards - Health Effects and Precautions
Eye, skin, and lung effects (Periodic chest X-ray and pulmonary function testing required) (Listed under 'Vanadium')

VEGETABLE OIL MISTS   8008-89-7

HEALTH AND SAFETY LISTS

U.S. DOT - Substances From 49 CFR 172.101
regulated by DOT (UN2443)

U.S. DOT - Hazard Classes
DOT hazard class = 8

NIOSH - Selected LD50s and LC50s
Oral, rat: LD50 = 140 mg/kg

INTERNATIONAL LISTS

Canada - WHMIS: Ingredient Disclosure
1% item 1676 (1331)

STATE LISTS

NJ Right to Know List (Total)
sn 1992

NJ Special Hazardous Substances
(corrosive)

VERATRALDEHYDE   120-14-9

SEE ALSO:
VANADIUM

HEALTH AND SAFETY LISTS

ACGIH 1995 - Time Weighted Averages
respirable dust and fume, as V2O5: 0.05 mg/m3 TWA

U.S. DOT - Substances From 49 CFR 172.101
regulated by DOT (UN2862)

U.S. DOT - Hazard Classes
DOT hazard class = 6.1

U.S. DOT - Appendix A Table 1 - Hazardous Substances
final RQ = 1000 pounds (454 kg)

NIOSH - Selected LD50s and LC50s
Oral, rat: LD50 = 10 mg/kg

NIOSH 1990 - Pocket Guide - RELs
as V2O5: C 0.05 mg/m3 (15 min) (respirable dust); C 0.05 mg/m3 (15 min) (respirable fume)

NIOSH 1990 - Pocket Guide - IDLHs
respirable dust and fume, as V2O5: 70 mg/m3 IDLH

NIOSH 1990 - Pocket Guide - Target organs
skin, eyes, respiratory system

NTP Chemical Status Reports - Testing Status and NTIS Number
Approved for toxicology/carcinogenesis study; prechronic studies for which toxicity technical reports were not prepared

OSHA - Vacated PELs - Time Weighted Averages
respirable dust (as V2O5): 0.05 mg/m3 TWA; fume (as V2O5): 0.05 mg/m3 TWA

ENVIRONMENTAL LISTS

CERCLA/SARA - Section 302 Extremely Hazardous Substances and TPQs
TPQ = 100/10,000 pounds

CERCLA/SARA - Hazardous Substances and their Reportable Quantities
final RQ = 1000 pounds (454 kg)

Clean Water Act - Hazardous Substances
[present]

RCRA - P Series Wastes
waste number P120

RCRA - Hazardous Constituents-Appendix VIII
waste number P120

RCRA - Substances Banned From Land Disposal
[present]

TSCA - Code of Federal Regulations Citations
40 CFR 712.30(w)

TSCA - PAIR - Reporting List
Reporting Date: July 13, 1988

INTERNATIONAL LISTS

Australian Exposure Standards - Time Weighted Averages
respirable dust and fume, as V2O5: 0.05 mg/m3 TWA

Canada - WHMIS: Ingredient Disclosure
0.1% item 1677 (1352)

Canada - Ontario - OHSA - TWAEVs
as V2O5: 0.05 mg/m3 TWAEV

Canada - Quebec - Time-Weighted Average Exposure Values
fume and respirable dust: 0.05 mg/m3 TWAEV

United Kingdom - Occupational Exposure Standards - TWAs
total inhalable dust, as V: 0.5 mg/m3 TWA; fume and respirable dust, as V: 0.05 mg/m3 TWA

German (DFG) - MAK Values
fine dust: 0.05 mg/m3 MAK

German (DFG) - Peak Limitations
5 x normal MAK (30 min. average value); don't exceed 2 times during shift

Israel - Time Weighted Averages
as V2O5: 0.05 mg/m3 TWA

Israel - Action Levels
as V2O5: 0.025 mg/m3 AL

STATE LISTS

California - Exposure Limits - PELs
respirable dust and fume: 0.05 mg/m3 PEL

California - Directors List of Hazardous Substances (8 CCR 339)
[present]

Florida Hazardous Substance List
[present]

Massachusetts Right To Know List
extraordinarily hazardous

Minnesota Hazardous Substance List
[present] as V2O5

NJ Right to Know List (Total)
sn 1993

Pennsylvania Right to Know List
environmental hazard

PROPOSED REGULATIONS

ACGIH 1995 - Proposed Biological Exposure Indices
Vanadium in urine: 50 ug/g creatinine, end of shift

VERMICULITE   1318-00-9

HEALTH AND SAFETY LISTS

U.S. DOT - Substances From 49 CFR 172.101
regulated by DOT (UN2444)

U.S. DOT - Hazard Classes
DOT hazard class = 8

NIOSH - Selected LD50s and LC50s
Oral, rat: LD50 = 160 mg/kg

INTERNATIONAL LISTS

Canada - WHMIS: Ingredient Disclosure
1% item 1678 (1585)

STATE LISTS

Florida Hazardous Substance List
[present]

Massachusetts Right To Know List
[present]

NJ Right to Know List (Total)
sn 1994

NJ Special Hazardous Substances
(corrosive)

Pennsylvania Right to Know List
[present]

VINBLASTINE   865-21-4

HEALTH AND SAFETY LISTS

U.S. DOT - Substances From 49 CFR 172.101
regulated by DOT (UN2475)

U.S. DOT - Hazard Classes
DOT hazard class = 8

NIOSH - Selected LD50s and LC50s
Oral, rat: LD50 = 350 mg/kg

INTERNATIONAL LISTS

Canada - WHMIS: Ingredient Disclosure
1% item 1679 (1662)

STATE LISTS

NJ Right to Know List (Total)
sn 1995

NJ Special Hazardous Substances
(corrosive)

VINBLASTINE SULFATE   143-67-9

HEALTH AND SAFETY LISTS

U.S. DOT - Substances From 49 CFR 172.101
regulated by DOT (UN2860)

U.S. DOT - Hazard Classes
DOT hazard class = 6.1

NIOSH - Selected LD50s and LC50s
Oral, mouse: LD50 = 382 mg/kg

INTERNATIONAL LISTS

Canada - WHMIS: Ingredient Disclosure
0.1% item 1680 (1697)

STATE LISTS
NJ Right to Know List (Total)
sn 1996

VINCLOZOLIN [3-(3,5-DICHLOROPHENYL)-5-ETHENYL-5-METHYL-2,4-OXAZOLIDINEDIONE]   50471-44-8

HEALTH AND SAFETY LISTS

U.S. DOT - Substances From 49 CFR 172.101
regulated by DOT (UN2931)

U.S. DOT - Hazard Classes
DOT hazard class = 6.1

U.S. DOT - Appendix A Table 1 - Hazardous Substances
final RQ = 1000 pounds (454 kg)

ENVIRONMENTAL LISTS

CERCLA/SARA - Hazardous Substances and their Reportable Quantities
final RQ = 1000 pounds (454 kg)

Clean Water Act - Hazardous Substances
[present]

INTERNATIONAL LISTS

Canada - WHMIS: Ingredient Disclosure
1% item 1681 (1533)

STATE LISTS

California - Directors List of Hazardous Substances (8 CCR 339)
[present]

Massachusetts Right To Know List
[present]

NJ Right to Know List (Total)
sn 1997

Pennsylvania Right to Know List
environmental hazard

VINCRISTINE   57-22-7

HEALTH AND SAFETY LISTS

NIOSH - Selected LD50s and LC50s
Oral, rat: LD50 = 1580 mg/kg

ENVIRONMENTAL LISTS

EPA - Master Testing List
[present]

List of Pesticide Product Inert Ingredients
[present]

TSCA - Code of Federal Regulations Citations
40 CFR 712.30(x); 40 CFR 716.120(d)

TSCA - PAIR - Reporting List
Reporting Date: November 27, 1991

TSCA - Health and Safety Reporting List
Effective Date: September 30, 1991

VINCRISTINE SULFATE   2068-78-2

HEALTH AND SAFETY LISTS

ACGIH 1995 - Time Weighted Averages
10 mg/m3 TWA (except caster, cashew nut, or similar irritant oils)

OSHA - Vacated PELs - Time Weighted Averages
total dust: 15 mg/m3 TWA; respirable fraction: 5 mg/m3 TWA

OSHA - Final PELs - Time Weighted Averages
total dust: 15 mg/m3 TWA; respirable fraction: 5 mg/m3 TWA

INTERNATIONAL LISTS

Australian Exposure Standards - Time Weighted Averages
10 mg/m3 TWA (does not include castor oil, cashew nut or similar irritant oil)

Canada - Ontario - OHSA - TWAEVs
10 mg/m3 TWAEV (listed as nuisance particulates)

Canada - Quebec - Time-Weighted Average Exposure Values
except castor oil, peanut oil, and similar irritants, total dust: 10 mg/m3 TWAEV; respirable dust: 5 mg/m3 TWAEV

Israel - Time Weighted Averages
10 mg/m3 TWA (except caster, cashew nut, or similar irritant oils)

Israel - Action Levels
5 mg/m3 AL

Mexico - Instruction No. 10 - TWAs
8.1 mg/m3 TWA

STATE LISTS

Minnesota Hazardous Substance List
[present] (includes inert or nuisance dusts except cashew nut, castor or similar irritant oils)

Pennsylvania Right to Know List
[present]

VINYL ACETATE   108-05-4

INTERNATIONAL LISTS

Canada - British Columbia - 8 Hour Exposure Limits
10 mg/m3 TWA (except castor, cashew nut, or similar irritating oils)

VINYL ACETATE - ACRYLIC ESTER COPOLYMERS   63181-88-4

HEALTH AND SAFETY LISTS

NIOSH - Selected LD50s and LC50s
Oral, rat: LD50 = 2000 mg/kg

ENVIRONMENTAL LISTS

TSCA - Code of Federal Regulations Citations
40 CFR 712.30(x); 40 CFR 716.120(d)

TSCA - PAIR - Reporting List
Reporting Date: November 27, 1991

TSCA - Health and Safety Reporting List
Effective Date: September 30, 1991

INTERNATIONAL LISTS

Canada - WHMIS: Ingredient Disclosure
1% item 1682 (1713)

VINYL ACETATE - BUTYL ACRYLATE - ACRYLIC ACID TERPOLYMER   25085-41-0

ENVIRONMENTAL LISTS

List of Pesticide Product Inert Ingredients
[present]

VINYL ACETATE, CROTONIC ACID, VINYL NEODECANOATE, GLYCIDYL METHACRYLATE POLYMER   68928-85-8

HEALTH AND SAFETY LISTS

NTP Chemical Status Reports - Testing Status and NTIS Number
Chronic studies exist for which technical reports were not prepared

VINYL ACETATE-VINYL ALCOHOL CHLORIDE POLYMER   25086-48-0

HEALTH AND SAFETY LISTS

IARC - Group 3 (not classifiable)
[present]

STATE LISTS

California - Air Bill 2588 Appendix A-II
9/90

California - Prop. 65 - Developmental Toxicity
developmental toxicity - initial date 7/1/90

VINYLBENZENE-VEGETABLE OIL COPOLYMER   RR-01156-3

ENVIRONMENTAL LISTS

CERCLA/SARA - Section 313 - Emission Reporting
form R reporting required

VINYL BROMIDE   593-60-2
HEALTH AND SAFETY LISTS

NTP Chemical Status Reports - Testing Status and NTIS Number
Chronic studies exist for which technical reports were not prepared

VINYL BUTYRATE   123-20-6
HEALTH AND SAFETY LISTS

IARC - Group 3 (not classifiable)
[present]

STATE LISTS

California - Air Bill 2588 Appendix A-II
9/90

California - Prop. 65 - Developmental Toxicity
developmental toxicity - initial date 7/1/90

VINYL CHLORIDE   75-01-4
HEALTH AND SAFETY LISTS

ACGIH 1995 - Time Weighted Averages
10 ppm TWA; 35 mg/m3 TWA

ACGIH 1995 - Short Term Exposure Limits
15 ppm STEL; 53 mg/m3 STEL

ACGIH 1995 - Carcinogens
A3-animal carcinogen

AIHA - Odor Threshold Values
geometric mean air odor threshold = 0.12 ppm (detectable); 0.40 ppm (recognizable)

U.S. DOT - Substances From 49 CFR 172.101
regulated by DOT (UN1301)

U.S. DOT - Hazard Classes
DOT hazard class = 3

U.S. DOT - Appendix A Table 1 - Hazardous Substances
final RQ = 5000 pounds (2270 kg)

IARC - Group 3 (not classifiable)
[present]

NFPA - Flash Points
flash point = 18 degrees F (-8 degrees C)

NFPA - Hazard Identification Ratings
health-2; flammability-3; reactivity-2

NIOSH - Selected LD50s and LC50s
Inhalation, rat: LC50 = 4000 ppm 2 hr Oral, rat: LD50 = 2920 mg/kg Skin, rabbit: LD50 = 2335 mg/kg

NIOSH - Health Standards - Exposure Limits
C (15 min) 4 ppm; C (15 min) 15 mg/m3

NIOSH - Health Standards - Health Effects and Precautions
Irritation

OSHA - Vacated PELs - Time Weighted Averages
10 ppm TWA; 30 mg/m3 TWA

OSHA - Vacated PELs - Short Term Exposure Limits
20 ppm STEL; 60 mg/m3 STEL

ENVIRONMENTAL LISTS

CERCLA/SARA - Section 302 Extremely Hazardous Substances and TPQs
TPQ = 1000 pounds

CERCLA/SARA - Section 313 - Emission Reporting
form R reporting required for 1.0% de minimus concentration

CERCLA/SARA - Hazardous Substances and their Reportable Quantities
final RQ = 5000 pounds (2270 kg)

Clean Air Act (1990) - List of Hazardous Air Contaminants
[present]

CAA -Toxic Substances for Accidental Release Prevention
threshold quantity = 15,000 lbs

CAA - HON Rule - SOCMI Chemicals
compliance by Jan. 23, 1995

CAA - HON Rule - Organic HAPs
[present]

Clean Water Act - Hazardous Substances
[present]

RCRA - TSD Facilities Ground Water Monitoring
TM 8240 = 5 ug/L PQL

TSCA - Code of Federal Regulations Citations
40 CFR 716.120(c)

TSCA - Health and Safety Reporting List
Effective Date: February 10, 1986

INTERNATIONAL LISTS

Australian Exposure Standards - Time Weighted Averages
10 ppm TWA; 35 mg/m3 TWA

Australian Exposure Standards - Short Term Exposure Limits
20 ppm STEL; 70 mg/m3 STEL

Canada - WHMIS: Ingredient Disclosure
1% item 1683 (35)

Canada - NPRI (National Pollutant Release Inventory)
[present]

Canada - Alberta - 8 Hour Occupational Exposure Limit
10 ppm TWA; 35 mg/m3 TWA

Canada - Alberta - 15 Minute Occupational Exposure Limit
20 ppm STEL; 70 mg/m3 STEL

Canada - British Columbia - 8 Hour Exposure Limits
10 ppm TWA; 30 mg/m3 TWA

Canada - British Columbia - 15 Minute Exposure Limits
20 ppm STEL; 60 mg/m3 STEL

Canada - Ontario - OHSA - TWAEVs
10 ppm TWAEV; 35 mg/m3 TWAEV

Canada - Ontario - OHSA - STEVs
20 ppm STEV; 70 mg/m3 STEV

Canada - Quebec - Time-Weighted Average Exposure Values
10 ppm TWAEV; 35 mg/m3 TWAEV

Canada - Quebec - Short-term Exposure Values
20 ppm STEV; 70 mg/m3 STEV

United Kingdom - Occupational Exposure Standards - TWAs
10 ppm TWA; 30 mg/m3 TWA

United Kingdom - Occupational Exposure Standards - STELs
20 ppm STEL; 60 mg/m3 STEL

German (DFG) - MAK Values
10 ppm MAK; 35 mg/m3 MAK

German (DFG) - Peak Limitations
2 x normal MAK (5 min momentary value); don't exceed 8 times during shift

German (DFG) - Carcinogens
suspected carcinogen

German (DFG) - Pregnancy
classification not yet possible

Israel - Time Weighted Averages
10 ppm TWA; 35 mg/m3 TWA

Israel - Short Term Exposure Limits
20 ppm STEL 70 mg/m3 STEL

Israel - Action Levels
5 ppm AL; 17.5 mg/m3 AL

Mexico - Instruction No. 10 - TWAs
10 ppm TWA; 30 mg/m3 TWA

Mexico - Instruction No. 10 - STELs
20 ppm STEL; 60 mg/m3 STEL

STATE LISTS

California - Air Bill 2588 Appendix A-I
6/91

California - Exposure Limits - PELs
10 ppm PEL; 30 mg/m3 PEL

California - Exposure Limits - STELs
20 ppm STEL; 60 mg/m3 STEL

California - Directors List of Hazardous Substances (8 CCR 339)
[present]

Florida Hazardous Substance List
[present]

Massachusetts Right To Know List
extraordinarily hazardous

Minnesota Hazardous Substance List
[present]

NJ Right to Know List (Total)
sn 1998

NJ Special Hazardous Substances
(flammable - third degree; reactive - second degree)

Pennsylvania Right to Know List
environmental hazard

PROPOSED REGULATIONS

Canada - Ontario - Proposed Occupational TWAEVs
10 ppm TWAEV; 35 mg/m3 TWAEV

Canada - Ontario - Proposed Occupational STEVs
15 ppm STEV; 30 mg/m3 STEV

VINYL CHLORIDE-VINYL ACETATE COPOLYMER   9003-22-9

ENVIRONMENTAL LISTS

List of Pesticide Product Inert Ingredients
[present]

VINYL CHLOROACETATE   2549-51-1

ENVIRONMENTAL LISTS

List of Pesticide Product Inert Ingredients
[present]

4-VINYLCYCLOHEXENE   100-40-3

ENVIRONMENTAL LISTS

List of Pesticide Product Inert Ingredients
[present]

VINYL CYCLOHEXENE DIOXIDE   106-87-6

ENVIRONMENTAL LISTS

List of Pesticide Product Inert Ingredients
[present]

VINYL EPOXY ESTER   RR-01002-6

ENVIRONMENTAL LISTS

List of Pesticide Product Inert Ingredients
[present]

VINYL ETHYL ALCOHOL   627-27-0
SEE ALSO:
VINYL HALIDES

HEALTH AND SAFETY LISTS

ACGIH 1995 - Time Weighted Averages
5 ppm TWA; 22 mg/m3 TWA

ACGIH 1995 - Carcinogens
A2-suspected human carcinogen

U.S. DOT - Substances From 49 CFR 172.101
regulated by DOT (UN1085)

U.S. DOT - Hazard Classes
DOT hazard class = 2.1

IARC - Group 2A (limited human data)
[present] (Overall evaluation based only on evidence of carcinogenicity in monograph (39, 1986) or in Supplement 4) (Other relevant data, as given in Supplement 7 or in the monograph, influenced the making of the overall evaluation)

NFPA - Flash Points
no flash point

NFPA - Hazard Identification Ratings
health-2; flammability-0; reactivity-1

NIOSH - Selected LD50s and LC50s
Oral, rat: LD50 = 500 mg/kg

NIOSH - Health Standards - Exposure Limits
as cited for Vinyl chloride in 29 CFR 1910.1017 with goal of zero exposure (Listed under 'Vinyl halides')

NIOSH - Health Standards - Health Effects and Precautions
has produced liver and kidney tumors in animals (Listed under 'Vinyl halides')

NIOSH - Health Standards - Carcinogenic Chemicals
potential human carcinogen (Listed under 'Vinyl halides')

OSHA - Vacated PELs - Time Weighted Averages
5 ppm TWA; 20 mg/m3 TWA

OSHA - Possible Select Carcinogens
[present]

ENVIRONMENTAL LISTS

CERCLA/SARA - Section 313 - Emission Reporting
form R reporting reqired for 0.1% de minimus concentration

CERCLA/SARA - Hazardous Substances and their Reportable Quantities
final RQ = 1 pound (.454 kg)

Clean Air Act (1990) - List of Hazardous Air Contaminants
[present]

TSCA - Code of Federal Regulations Citations
40 CFR 712.30(d),(x)

TSCA - PAIR - Reporting List
Reporting Date: November 19, 1982, December 27, 1990

TSCA - Health and Safety Reporting List
Effective Date: October 29, 1990

INTERNATIONAL LISTS

Australian Exposure Standards - Time Weighted Averages
5 ppm TWA; 22 mg/m3 TWA

Australian Exposure Standards - Carcinogens
probable carcinogen

Canada - WHMIS: Ingredient Disclosure
0.1% item 1684 (343)

Canada - Alberta - 8 Hour Occupational Exposure Limit
5 ppm TWA; 22 mg/m3 TWA

Canada - Alberta - 15 Minute Occupational Exposure Limit
10 ppm STEL; 44 mg/m3 STEL

Canada - Alberta - Designated Substances
designated substance - requires code of practice

Canada - British Columbia - 8 Hour Exposure Limits
250 ppm TWA; 1100 mg/m3 TWA

Canada - Ontario - OHSA - TWAEVs
5 ppm TWAEV; 22 mg/m3 TWAEV

Canada - Quebec - Time-Weighted Average Exposure Values
5 ppm TWAEV; 22 mg/m3 TWAEV

Canada - Quebec - Carcinogens
C2 carcinogen: effect suspected in humans

Israel - Time Weighted Averages
5 ppm TWA; 22 mg/m3 TWA

Israel - Action Levels
2.5 ppm AL; 11 mg/m3 AL

STATE LISTS

California - Air Bill 2588 Appendix A-I
known or potential carcinogen

California - Prop. 65 - Cancer list
carcinogen - initial date 10/1/88

California - Exposure Limits - PELs
5 ppm PEL;   20 mg/m3 PEL

California - Directors List of Hazardous Substances (8 CCR 339)
[present]

Florida Hazardous Substance List
[present]

Massachusetts Right To Know List
carcinogen;   extraordinarily hazardous

Minnesota Hazardous Substance List
carcinogen

NJ Right to Know List (Total)
sn 1999

Pennsylvania Right to Know List
environmental hazard

PROPOSED REGULATIONS

Canada - Ontario - Proposed Occupational TWAEVs
1 ppm TWAEV;   4 mg/m3 TWAEV

VINYL ETHYL ETHER   109-92-2

HEALTH AND SAFETY LISTS

U.S. DOT - Substances From 49 CFR 172.101
regulated by DOT (UN2838)

U.S. DOT - Hazard Classes
DOT hazard class = 3

NFPA - Flash Points
flash point = 68 degrees F (20 degrees C)

NFPA - Hazard Identification Ratings
health-2;   flammability-3;   reactivity-2

NIOSH - Selected LD50s and LC50s
Oral, rat: LD50 = 8530 mg/kg

STATE LISTS

Florida Hazardous Substance List
[present]

Massachusetts Right To Know List
[present]

NJ Right to Know List (Total)
sn 2000

NJ Special Hazardous Substances
(flammable - third degree;   reactive - second degree)

Pennsylvania Right to Know List
[present]

VINYL 2-ETHYLHEXYL ETHER   103-44-6
SEE ALSO:
F025-HAZARDOUS WASTES
K029-HAZARDOUS WASTES
UNSATURATED AMINO ALKYL ESTER SALT
BIS(2,4-DIMETHYLBUTYL) MALEATE
F024-HAZARDOUS WASTES
F039-HAZARDOUS WASTES
K019-HAZARDOUS WASTES
K028-HAZARDOUS WASTES
K020-HAZARDOUS WASTES

HEALTH AND SAFETY LISTS

ACGIH 1995 - Time Weighted Averages
5 ppm TWA;   13 mg/m3 TWA

ACGIH 1995 - Carcinogens
A1-confirmed human carcinogen

AIHA - Odor Threshold Values
no geometric mean air odor threshold

U.S. DOT - Substances From 49 CFR 172.101
regulated by DOT (UN1086)

U.S. DOT - Hazard Classes
DOT hazard class = 2.1

U.S. DOT - Appendix A Table 1 - Hazardous Substances
final RQ = 1 pound (0.454 kg)

IARC - Group 1 (carcinogenic to humans)
[present]

NFPA - Flash Points
flash point = -108.4 degrees F (-78 degrees C)

NFPA - Hazard Identification Ratings
health-2;   flammability-4;   reactivity-2

NIOSH - Selected LD50s and LC50s
Oral, rat: LD50 = 500 mg/kg

NIOSH 1990 - Pocket Guide - RELs
Lowest reliably detectable concentration

NIOSH 1990 - Pocket Guide - Carcinogens
occupational carcinogen

NIOSH 1990 - Pocket Guide - Target organs
liver, CNS, blood, lymphatic system, respiratory system

NIOSH - Health Standards - Exposure Limits
lowest reliably detectable concentration

NIOSH - Health Standards - Health Effects and Precautions
Liver cancer (Liver function testing required)

NIOSH - Health Standards - Carcinogenic Chemicals
potential human carcinogen

NTP Seventh Report - Known Carcinogens
known carcinogen

OSHA - 29 CFR 1910 Specifically Regulated Chemicals
1 ppm TWA PEL;   5 ppm STEL;   0.5 ppm TWA action level;   Cancer suspect agent (see 29 CFR 1910.1017)

OSHA - Select Carcinogens
[present]

ENVIRONMENTAL LISTS

ATSDR Priority List
Rank (of 275): 005

CERCLA/SARA - Section 313 - Emission Reporting
form R reporting required for 0.1% de minimus concentration

CERCLA/SARA - Hazardous Substances and their Reportable Quantities
final RQ = 1 pound (0.454 kg)

Clean Air Act (1990) - List of Hazardous Air Contaminants
[present]

CAA - Flammable Substances for Accidental Release Prevention
threshold quantity = 10,000 lbs

CAA - HON Rule - SOCMI Chemicals
compliance by Oct. 24, 1994

CAA - HON Rule - Organic HAPs
[present]

Clean Water Act - Priority Pollutants
[present]

Clean Water Act - Toxic Pollutants
[present]

Safe Drinking Water Act - MCLs
MCL = 0.002 mg/L

Safe Drinking Water Act - MCLGs
MCLG = Zero

EPA - Carcinogen Hazard Ranking for RQ Adjustment
Hazard ranking = High

List of Pesticide Product Inert Ingredients
[present]

RCRA - D Series - Maximum Concentration of Contaminants
waste number D043; regulatory level = 0.2 mg/L

RCRA - D Series - Chronic Toxicity Reference Levels
chronic toxicity reference level = 0.002 mg/L

RCRA - U Series Wastes
waste number U043

RCRA - Hazardous Constituents-Appendix VIII
waste number U043

RCRA - Basis for Listing - Appendix VII
Included in waste streams: F024, F025, F039, K019, K020, K028, K029

RCRA - Substances Banned From Land Disposal
[present]

RCRA - TSD Facilities Ground Water Monitoring
TM 8010 = 2 ug/L PQL; TM 8240 = 10 ug/L PQL

RCRA - Universal Treatment Standards (LDR)
WW: 0.27 mg/l; NWW: 6.0 mg/kg

INTERNATIONAL LISTS

Australian Exposure Standards - Time Weighted Averages
5 ppm TWA; 13 mg/m3 TWA

Australian Exposure Standards - Carcinogens
confirmed carcinogen

Australian Exposure Standards - Under Review
exposure limits under review

Canada - WHMIS: Ingredient Disclosure
0.1% item 1685 (538)

Canada - NPRI (National Pollutant Release Inventory)
[present]

Canada - CEPA Schedule I - Toxic Substances
limited atmospheric releases from vinyl chloride and polyvinyl chloride plants

Canada - Alberta - 8 Hour Occupational Exposure Limit
2 ppm TWA; 5.2 mg/m3 TWA

Canada - Alberta - 15 Minute Occupational Exposure Limit
10 ppm STEL; 26 mg/m3 STEL

Canada - Alberta - Designated Substances
designated substance - requires code of practice

Canada - British Columbia - 8 Hour Exposure Limits
1 ppm TWA; 2.5 mg/m3 TWA

Canada - British Columbia - Carcinogens
carcinogen - 1 ppm TWA; 2.5 mg/m3 TWA

Canada - Ontario - OHSA - TWAEVs
2 ppm TWAEV; 5.2 mg/m3 TWAEV (designated substance regulation)

Canada - Ontario - OHSA - CEVs
10 ppm CEV; 26 mg/m3 CEV (designated substance regulation)

Canada - Ontario - OHSA - Designated Substances
2 ppm TWAEV; 5.2 mg/m3 TWAEV; See Ontario Reg. 846 for full information.

Canada - Quebec - Time-Weighted Average Exposure Values
1 ppm TWAEV; 2.5 mg/m3 TWAEV (substance of which the recirculation is prohibited)

Canada - Quebec - Short-term Exposure Values
5 ppm STEV; 13 mg/m3 STEV

Canada - Quebec - Carcinogens
C1 carcinogen: effect detected in humans

United Kingdom - Maximum Exposure Limits - TWAs
7 ppm TWA

German (DFG) - Carcinogens
proven carcinogen

Israel - Time Weighted Averages
1 ppm TWA

Israel - Action Levels
0.5 ppm AL; 6.5 mg/m3 AL

Mexico - Instruction No. 10 - TWAs
10 ppm TWA; 20 mg/m3 TWA

Mexico - Instruction No. 10 - Carcinogens
potentially carcinogenic contaminant

Mexico - Wastewater - Organic Toxic Pollutants and Heavy Metals
Listed under [Organic Toxic Pollutants]

Mexico - Drinking Water - Ecological Criteria
0.02 mg/l This level has been extrapolated by using a mathematic model

STATE LISTS

California - Air Bill 2588 Appendix A-I
known or potential carcinogen

California - Prop. 65 - Cancer list
carcinogen - initial date 2/27/87

California - Prop. 65 - No Significant Risk Levels
no significant risk level = 3 ug/day

California - Exposure Limits - PELs
1 ppm PEL; 0.5 ppm Action Level; see also section 5210 for chronic effects

California - Exposure Limits - STELs
5 ppm STEL

California - Exposure Limits - Skin Notation
material may be absorbed through the skin, eyes or mucous membrane

California - Directors List of Hazardous Substances (8 CCR 339)
[present]

Florida Hazardous Substance List
[present]

Massachusetts Right To Know List
carcinogen; extraordinarily hazardous; teratogen

Minnesota Hazardous Substance List
carcinogen

NJ Right to Know List (Total)
sn 2001

NJ Special Hazardous Substances
(carcinogen; flammable - fourth degree; mutagen)

Pennsylvania Right to Know List
environmental hazard; special hazardous substance

Pennsylvania RTK - Special Hazardous Substances
[present]

VINYL FLUORIDE   **75-02-5**

HEALTH AND SAFETY LISTS

IARC - Group 3 (not classifiable)
[present]

ENVIRONMENTAL LISTS

List of Pesticide Product Inert Ingredients
[present]

VINYL HALIDES   **RR-00527-6**

HEALTH AND SAFETY LISTS

U.S. DOT - Substances From 49 CFR 172.101
regulated by DOT (UN2589)

U.S. DOT - Hazard Classes
DOT hazard class = 6.1

INTERNATIONAL LISTS

Canada - WHMIS: Ingredient Disclosure
1% item 1686 (407)

STATE LISTS

NJ Right to Know List (Total)
sn 2002

VINYLIDENE CHLORIDE   75-35-4
HEALTH AND SAFETY LISTS

ACGIH 1995 - Time Weighted Averages
0.1 ppm TWA;   0.4 mg/m3 TWA

ACGIH 1995 - Carcinogens
A2-suspected human carcinogen

AIHA - WEEL - Time Weighted Averages
5 ppm TWA;   22 mg/m3 TWA

IARC - Group 3 (not classifiable)
[present]

NFPA - Flash Points
flash point = 61 degrees F (16 degrees C)

NFPA - Hazard Identification Ratings
health-0; flammability-3; reactivity-2

NIOSH - Selected LD50s and LC50s
Inhalation, mouse: LC50 = 27000 mg/m3 (8 hr) Oral, rat: LD50 = 2563 mg/kg Skin, rabbit: LD50 = 16640 mg/kg

NTP Chemical Status Reports - Testing Status and NTIS Number
Technical reports printed (PB87116182)

NTP Chemical Status Reports - Evidence of Carcinogenicity
male rat-inadequate; female rat-inadequate; male mice-inadequate; female mice-clear evidence

ENVIRONMENTAL LISTS

CAA - HON Rule - SOCMI Chemicals
compliance by Jan. 23, 1995

EPA - Master Testing List
[present]

TSCA - Code of Federal Regulations Citations
40 CFR 712.30(w);   40 CFR 716.120(a);   40 CFR 799.5000

TSCA - PAIR - Reporting List
Reporting Date: March 12, 1990

TSCA - Health and Safety Reporting List
Effective Date: January 11, 1990; Sunset Date: November 9, 1993

TSCA - Substances Subject to Testing Consent Orders
Test for: Health Effects and Chemical Fate

TSCA - Section 12(b) - Export Notification
export notification required - Section 4

STATE LISTS

California - Directors List of Hazardous Substances (8 CCR 339)
[present]

Florida Hazardous Substance List
[present]

Massachusetts Right To Know List
[present]

Minnesota Hazardous Substance List
[present]

Pennsylvania Right to Know List
[present]

VINYLIDENE CHLORIDE-VINYL CHLORIDE   9011-06-7
COPOLYMERS
HEALTH AND SAFETY LISTS

ACGIH 1995 - Time Weighted Averages
10 ppm TWA;   57 mg/m3 TWA

ACGIH 1995 - Skin Designations
skin - potential for cutaneous absorption

ACGIH 1995 - Carcinogens
A2-suspected human carcinogen

IARC - Group 3 (not classifiable)
[present]

NIOSH - Selected LD50s and LC50s
Inhalation, rat: LC50 = 800 ppm 4 hr Oral, rat: LD50 = 2130 mg/kg Skin, rabbit: LD50 = 620 mg/kg

NTP Chemical Status Reports - Testing Status and NTIS Number
Technical reports printed (PB90219957/AS)

NTP Chemical Status Reports - Evidence of Carcinogenicity
male rat-clear evidence; female rat-clear evidence; male mice-clear evidence; female mice-clear evidence

NTP Seventh Report - Suspect Carcinogens
suspect carcinogen

OSHA - Vacated PELs - Time Weighted Averages
10 ppm TWA; 60 mg/m3 TWA

OSHA - Vacated PELs - Skin Designation
Prevent or reduce skin absorption

ENVIRONMENTAL LISTS

TSCA - Code of Federal Regulations Citations
40 CFR 712.30(d)

TSCA - PAIR - Reporting List
Reporting Date: November 19, 1982

INTERNATIONAL LISTS

Australian Exposure Standards - Time Weighted Averages
10 ppm TWA; 57 mg/m3 TWA

Australian Exposure Standards - Skin Effects
skin absorption

Australian Exposure Standards - Carcinogens
suspected carcinogen

Canada - WHMIS: Ingredient Disclosure
0.1% item 1688 (775)

Canada - Alberta - 8 Hour Occupational Exposure Limit
10 ppm TWA; 57 mg/m3 TWA

Canada - Alberta - 15 Minute Occupational Exposure Limit
15 ppm STEL; 86 mg/m3 STEL

Canada - Alberta - Designated Substances
designated substance - requires code of practice

Canada - British Columbia - 8 Hour Exposure Limits
10 ppm TWA; 60 mg/m3 TWA

Canada - Ontario - OHSA - TWAEVs
10 ppm TWAEV; 57 mg/m3 TWAEV

Canada - Ontario - OHSA - Skin Notations
absorption through skin, eyes, or mucous membranes

Canada - Quebec - Time-Weighted Average Exposure Values
10 ppm TWAEV; 57 mg/m3 TWAEV (substance of which the recirculation is prohibited)

Canada - Quebec - Skin Designations
absorbed through the skin

Canada - Quebec - Carcinogens
C2 carcinogen: effect suspected in humans

German (DFG) - Carcinogens
animal evidence of carcinogenicity

Israel - Time Weighted Averages
10 ppm TWA; 57 mg/m3 TWA

Israel - Action Levels
5 ppm AL; 28.5 mg/m3 AL

Mexico - Instruction No. 10 - TWAs
10 ppm TWA; 60 mg/m3 TWA

Mexico - Instruction No. 10 - Carcinogens
potential carcinogen in humans - limited epidemiological evidence

STATE LISTS

California - Air Bill 2588 Appendix A-II
known or potential carcinogen: 9/90

California - Prop. 65 - Cancer list
carcinogen - initial date 7/1/90

California - Exposure Limits - PELs
10 ppm PEL; 60 mg/m3 PEL

California - Exposure Limits - Skin Notation
material may be absorbed through the skin, eyes or mucous membrane

California - Directors List of Hazardous Substances (8 CCR 339)
[present]

Florida Hazardous Substance List
[present]

Massachusetts Right To Know List
[present]

Minnesota Hazardous Substance List
carcinogen; skin

NJ Special Hazardous Substances
(mutagen)

Pennsylvania Right to Know List
[present]

VINYLIDENE FLUORIDE 75-38-7

ENVIRONMENTAL LISTS

TSCA - Chemicals with Significant New Use Rules
PMN number: P-85-527

VINYL ISOBUTYL ETHER 109-53-5

HEALTH AND SAFETY LISTS

NFPA - Flash Points
flash point = 100 degrees F (38 degrees C)

NFPA - Hazard Identification Ratings
health-0; flammability-2; reactivity-0

VINYL ISOOCTYL ETHER RR-00887-7

HEALTH AND SAFETY LISTS

U.S. DOT - Substances From 49 CFR 172.101
regulated by DOT (UN1302)

U.S. DOT - Hazard Classes
DOT hazard class = 3

NFPA - Flash Points
flash point < -50 degrees F (-46 degrees C)

NFPA - Hazard Identification Ratings
health-2; flammability-4; reactivity-2

NIOSH - Selected LD50s and LC50s
Oral, rat: LD50 = 6153 mg/kg

ENVIRONMENTAL LISTS

CAA - Flammable Substances for Accidental Release Prevention
threshold quantity = 10,000 lbs

STATE LISTS

Florida Hazardous Substance List
[present]

Massachusetts Right To Know List
[present]

NJ Right to Know List (Total)
sn 2004

NJ Special Hazardous Substances
(flammable - fourth degree; reactive - second degree)

Pennsylvania Right to Know List
[present]

VINYL METHYL ETHER 107-25-5

HEALTH AND SAFETY LISTS

NFPA - Flash Points
flash point = 135 degrees F (57 degrees C)

NFPA - Hazard Identification Ratings
health-2; flammability-2; reactivity-2

STATE LISTS

Florida Hazardous Substance List
[present]

Massachusetts Right To Know List
[present]

Pennsylvania Right to Know List
[present]

VINYL NITRATE POLYMER RR-01470-0

SEE ALSO:
VINYL HALIDES

HEALTH AND SAFETY LISTS

U.S. DOT - Substances From 49 CFR 172.101
regulated by DOT (UN1860)

U.S. DOT - Hazard Classes
DOT hazard class = 2.1

IARC - Group 3 (not classifiable)
[present]

NFPA - Flash Points
gas (no flash point given)

NFPA - Hazard Identification Ratings
health-1; flammability-4; reactivity-2

NIOSH - Health Standards - Exposure Limits
as cited for Vinyl chloride in 29 CFR 1910.1017 with goal of zero exposure (Listed under 'Vinyl halides')

NIOSH - Health Standards - Health Effects and Precautions
has produced liver and kidney tumors in animals (Listed under 'Vinyl halides')

NIOSH - Health Standards - Carcinogenic Chemicals
potential human carcinogen (Listed under 'Vinyl halides')

ENVIRONMENTAL LISTS

CAA - Flammable Substances for Accidental Release Prevention
threshold quantity = 10,000 lbs

EPA - Master Testing List
[present]

TSCA - Code of Federal Regulations Citations
40 CFR 712.30(f); 40 CFR 716.120(a); 40 CFR 799.1700(a)(1)

TSCA - PAIR - Reporting List
Reporting Date: August 17, 1983

TSCA - Health and Safety Reporting List
Effective Date: October 4, 1982

TSCA - Chemical Test Rules
Testing required by: manufacturers (40 CFR 799.1700) (Listed under 'Fluoroalkenes')

TSCA - Section 12(b) - Export Notification
export notification required - Section 4

STATE LISTS

Florida Hazardous Substance List
[present]

Massachusetts Right To Know List
[present]

NJ Right to Know List (Total)
sn 2005

NJ Special Hazardous Substances
(flammable - fourth degree; reactive - second degree)

Pennsylvania Right to Know List
[present]

VINYLNORBORNENE 3048-64-4

STATE LISTS

Minnesota Hazardous Substance List
carcinogen

VINYL OCTADECYL ETHER   RR-00835-5

SEE ALSO:
  F025-HAZARDOUS WASTES
  K029-HAZARDOUS WASTES
  P-TERT-AMYLPHENYL BUTYL ETHER
  BIS(2,4-DIMETHYLBUTYL) MALEATE
  K020-HAZARDOUS WASTES
  VINYL HALIDES
  F039-HAZARDOUS WASTES
  K019-HAZARDOUS WASTES
  F024-HAZARDOUS WASTES

HEALTH AND SAFETY LISTS

ACGIH 1995 - Time Weighted Averages
  5 ppm TWA;   20 mg/m3 TWA

ACGIH 1995 - Short Term Exposure Limits
  20 ppm STEL;   79 mg/m3 STEL

U.S. DOT - Substances From 49 CFR 172.101
  regulated by DOT (UN1303)

U.S. DOT - Hazard Classes
  DOT hazard class = 3

U.S. DOT - Appendix B - Marine Pollutants
  inhibited form: DOT regulated marine pollutant

U.S. DOT - Appendix A Table 1 - Hazardous Substances
  final RQ = 100 pounds (45.4 kg)

IARC - Group 3 (not classifiable)
  [present]

NFPA - Flash Points
  flash point = -19 degrees F (-28 degrees C)

NFPA - Hazard Identification Ratings
  health-2;   flammability-4;   reactivity-2

NIOSH - Selected LD50s and LC50s
  Inhalation, rat: LC50 = 6350 ppm 4 hr

NIOSH - Health Standards - Exposure Limits
  as cited for Vinyl chloride in 29 CFR 1910.1017 with goal of zero exposure (Listed under 'Vinyl halides')

NIOSH - Health Standards - Health Effects and Precautions
  has produced liver and kidney tumors in animals (Listed under 'Vinyl halides')

NIOSH - Health Standards - Carcinogenic Chemicals
  potential human carcinogen (Listed under 'Vinyl halides')

NTP Chemical Status Reports - Testing Status and NTIS Number
  Technical reports printed (PB82258393)

OSHA - Vacated PELs - Time Weighted Averages
  1 ppm TWA;   4 mg/m3 TWA

ENVIRONMENTAL LISTS

ATSDR Priority List
  Rank (of 275): 069

CERCLA/SARA - Section 313 - Emission Reporting
  form R reporting required for 1.0% de minimus concentration

CERCLA/SARA - Hazardous Substances and their Reportable Quantities
  final RQ = 100 pounds (45.4 kg)

Clean Air Act (1990) - List of Hazardous Air Contaminants
  [present]

CAA - Flammable Substances for Accidental Release Prevention
  threshold quantity = 10,000 lbs

CAA - HON Rule - SOCMI Chemicals
  compliance by Jan. 23, 1995

CAA - HON Rule - Organic HAPs
  [present]

Clean Water Act - Hazardous Substances
  [present]

Clean Water Act - Priority Pollutants
  [present]

Clean Water Act - Toxic Pollutants
  [present] (Listed under 'Dichloroethylenes')

Safe Drinking Water Act - MCLs
  MCL = 0.007 mg/L

Safe Drinking Water Act - MCLGs
  MCLG = 0.007 mg/L

EPA - Carcinogen Hazard Ranking for RQ Adjustment
  Hazard ranking = Low

EPA - Master Testing List
  [present]

RCRA - D Series - Maximum Concentration of Contaminants
  waste number D029;   regulatory level = 0.7 mg/L

RCRA - D Series - Chronic Toxicity Reference Levels
  chronic toxicity reference level = 0.007 mg/L

RCRA - U Series Wastes
  waste number U078

RCRA - Hazardous Constituents-Appendix VIII
  waste number U078

RCRA - Basis for Listing - Appendix VII
  Included in waste streams: F024, F025, F039, K019, K020, K029, K073

RCRA - Substances Banned From Land Disposal
  [present]

RCRA - TSD Facilities Ground Water Monitoring
  TM 8010 = 1 ug/L PQL;   TM 8240 = 5 ug/L PQL

RCRA - Universal Treatment Standards (LDR)
  WW: 0.025 mg/l;   NWW: 6.0 mg/kg

INTERNATIONAL LISTS

Australian Exposure Standards - Time Weighted Averages
  5 ppm TWA;   20 mg/m3 TWA

Australian Exposure Standards - Short Term Exposure Limits
  20 ppm STEL;   79 mg/m3 STEL

Australian Exposure Standards - Under Review
  exposure limits under review

Canada - WHMIS: Ingredient Disclosure
  1% item 1689 (539)

Canada - NPRI (National Pollutant Release Inventory)
  [present]

Canada - Alberta - 8 Hour Occupational Exposure Limit
  5 ppm TWA;   20 mg/m3 TWA

Canada - Alberta - 15 Minute Occupational Exposure Limit
  10 ppm STEL;   40 mg/m3 STEL

Canada - British Columbia - 8 Hour Exposure Limits
  10 ppm TWA;   40 mg/m3 TWA

Canada - British Columbia - 15 Minute Exposure Limits
  20 ppm STEL;   80 mg/m3 STEL

Canada - Ontario - OHSA - TWAEVs
  1 ppm TWAEV;   4 mg/m3 TWAEV

Canada - Ontario - OHSA - STEVs
  20 ppm STEV;   80 mg/m3 STEV

Canada - Quebec - Time-Weighted Average Exposure Values
  1 ppm TWAEV;   4 mg/m3 TWAEV

United Kingdom - Maximum Exposure Limits - TWAs
  10 ppm TWA;   40 mg/m3 TWA

German (DFG) - MAK Values
  2 ppm MAK;   8 mg/m3 MAK

German (DFG) - Peak Limitations
  2 x normal MAK (30 min. average value);   don't exceed 4 times during shift

German (DFG) - Carcinogens
  suspected carcinogen

German (DFG) - Pregnancy
  no risk to embryo/fetus if exposure limits adhered to

Israel - Time Weighted Averages
5 ppm TWA; 20 mg/m3 TWA

Israel - Short Term Exposure Limits
20 ppm STEL; 79 mg/m3 STEL

Israel - Action Levels
2.5 ppm AL; 10 mg/m3 AL

Mexico - Instruction No. 10 - TWAs
5 ppm TWA; 20 mg/m3 TWA

Mexico - Instruction No. 10 - STELs
20 ppm STEL; 80 mg/m3 STEL

Mexico - Wastewater - Organic Toxic Pollutants and Heavy Metals
Listed under [Dichloroethylenes]

Mexico - Drinking Water - Ecological Criteria
0.0003 mg/l Substance presents persistence, bioaccumulations or risk of cancer, reduce human exposure to a minimum

STATE LISTS

California - Air Bill 2588 Appendix A-I
[present]

California - Exposure Limits - PELs
1 ppm PEL; 4 mg/m3 PEL

California - Directors List of Hazardous Substances (8 CCR 339)
[present]

Florida Hazardous Substance List
[present]

Massachusetts Right To Know List
carcinogen; extraordinarily hazardous

Minnesota Hazardous Substance List
[present]

NJ Right to Know List (Total)
sn 2006

NJ Special Hazardous Substances
(carcinogen; flammable - fourth degree; mutagen; reactive - second degree)

Pennsylvania Right to Know List
environmental hazard

4-VINYLPYRIDINE   100-43-6

HEALTH AND SAFETY LISTS

IARC - Group 3 (not classifiable)
[present]

VINYLPYRROLIDINONE-STYRENE POLYMER   25086-29-7
SEE ALSO:
VINYL HALIDES

HEALTH AND SAFETY LISTS

U.S. DOT - Substances From 49 CFR 172.101
regulated by DOT (UN1959)

U.S. DOT - Hazard Classes
DOT hazard class = 2.1

IARC - Group 3 (not classifiable)
[present]

NFPA - Flash Points
gas (no flash point given)

NFPA - Hazard Identification Ratings
health-1; flammability-4; reactivity-2

NIOSH - Health Standards - Exposure Limits
as cited for Vinyl chloride in 29 CFR 1910.1017 with goal of zero exposure (Listed under 'Vinyl halides')

NIOSH - Health Standards - Health Effects and Precautions
has produced liver and kidney tumors in animals (Listed under 'Vinyl halides')

NIOSH - Health Standards - Carcinogenic Chemicals
potential human carcinogen (Listed under 'Vinyl halides')

NTP Chemical Status Reports - Testing Status and NTIS Number
Prechronic studies for which toxicity technical reports were not prepared

ENVIRONMENTAL LISTS

CAA - Flammable Substances for Accidental Release Prevention
threshold quantity = 10,000 lbs

EPA - Master Testing List
[present]

TSCA - Code of Federal Regulations Citations
40 CFR 712.30(d); 40 CFR 716.120(a); 40 CFR 799.1700(a)(1)

TSCA - PAIR - Reporting List
Reporting Date: November 19, 1982

TSCA - Health and Safety Reporting List
Effective Date: October 4, 1982

TSCA - Chemical Test Rules
Testing required by: manufacturers (40 CFR 799.1700) (Listed under 'Fluoroalkenes')

TSCA - Section 12(b) - Export Notification
export notification required - Section 4

INTERNATIONAL LISTS

German (DFG) - Carcinogens
suspected carcinogen

STATE LISTS

Florida Hazardous Substance List
[present]

Massachusetts Right To Know List
[present]

NJ Right to Know List (Total)
sn 2007

NJ Special Hazardous Substances
(flammable - fourth degree; reactive - second degree)

Pennsylvania Right to Know List
[present]

N-VINYL-2-PYRROLIDONE   88-12-0

HEALTH AND SAFETY LISTS

U.S. DOT - Substances From 49 CFR 172.101
regulated by DOT (UN1304)

U.S. DOT - Hazard Classes
DOT hazard class = 3

NFPA - Flash Points
flash point = 15 degrees F (-9 degrees C)

NFPA - Hazard Identification Ratings
health-2; flammability-3; reactivity-2

NIOSH - Selected LD50s and LC50s
Oral, rat: LD50 = 17 gm/kg Skin, rabbit: LD50 = 20 gm/kg

STATE LISTS

Florida Hazardous Substance List
[present]

Massachusetts Right To Know List
[present]

NJ Right to Know List (Total)
sn 2008

NJ Special Hazardous Substances
(flammable - third degree; reactive - second degree)

Pennsylvania Right to Know List
[present]

VINYL TOLUENE   25013-15-4

HEALTH AND SAFETY LISTS

NFPA - Flash Points
flash point = 140 degrees F (60 degrees C)

NFPA - Hazard Identification Ratings
health-1; flammability-2; reactivity-0

VINYLTRICHLOROSILANE   75-94-5

HEALTH AND SAFETY LISTS

U.S. DOT - Substances From 49 CFR 172.101
regulated by DOT (UN1087)

U.S. DOT - Hazard Classes
DOT hazard class = 2.1

NFPA - Flash Points
gas (no flash point given)

NFPA - Hazard Identification Ratings
health-2; flammability-4; reactivity-2

NIOSH - Selected LD50s and LC50s
Oral, rat: LD50 = 4900 mg/kg

ENVIRONMENTAL LISTS

CAA - Flammable Substances for Accidental Release Prevention
threshold quantity = 10,000 lbs

STATE LISTS

Florida Hazardous Substance List
[present]

Massachusetts Right To Know List
[present]

NJ Right to Know List (Total)
sn 2009

NJ Special Hazardous Substances
(flammable - fourth degree; reactive - second degree)

Pennsylvania Right to Know List
[present]

VINYLTRIS(METHOXYETHOXY)SILANE   1067-53-4

HEALTH AND SAFETY LISTS

U.S. DOT - Hazard Classes
Forbidden from transport by the DOT

VINYZENE   5579-85-1

HEALTH AND SAFETY LISTS

NIOSH - Selected LD50s and LC50s
*Inhalation, mouse: LC50 = 18 gm/m3 8 hr Oral, rat: LD50 = 4365
mg/kg Skin, rabbit: LD50 = 13372 mg/kg*

STATE LISTS

Massachusetts Right To Know List
[present]

VITAMIN A   68-26-8

HEALTH AND SAFETY LISTS

NFPA - Flash Points
flash point = 350 degrees F (177 degrees C)

NFPA - Hazard Identification Ratings
health-0; flammability-1; reactivity-0

VITAMIN E   1406-18-4

HEALTH AND SAFETY LISTS

NIOSH - Selected LD50s and LC50s
Oral, bird: LD50 = 100 mg/kg

VITAMIN E ACETATE   58-95-7

ENVIRONMENTAL LISTS

List of Pesticide Product Inert Ingredients
[present]

WALNUT SHELLS   RR-01157-4

HEALTH AND SAFETY LISTS

IARC - Group 3 (not classifiable)
[present]

NFPA - Flash Points
flash point = 209 degrees F (98 degrees C)

NFPA - Hazard Identification Ratings
health-0; flammability-1; reactivity-0

NIOSH - Selected LD50s and LC50s
*Inhalation, rat: LC50 = 3200 mg/m3 8 hr Oral, rat: LD50 = 1470
mg/kg Skin, rabbit: LD50 = 560 mg/kg*

ENVIRONMENTAL LISTS

CAA - HON Rule - SOCMI Chemicals
compliance by Oct. 23, 1995

List of Pesticide Product Inert Ingredients
[present]

INTERNATIONAL LISTS

Canada - WHMIS: Ingredient Disclosure
1% item 1690 (1716)

German (DFG) - Carcinogens
animal evidence of carcinogenicity

WARFARIN   81-81-2

HEALTH AND SAFETY LISTS

ACGIH 1995 - Time Weighted Averages
50 ppm TWA; 242 mg/m3 TWA

ACGIH 1995 - Short Term Exposure Limits
100 ppm STEL; 483 mg/m3 STEL

U.S. DOT - Substances From 49 CFR 172.101
regulated by DOT (UN2618)

U.S. DOT - Hazard Classes
DOT hazard class = 3

U.S. DOT - Appendix B - Marine Pollutants
DOT regulated marine pollutant

NFPA - Flash Points
flash point = 127 degrees F (53 degrees C)

NFPA - Hazard Identification Ratings
health-2; flammability-2; reactivity-2

NIOSH - Selected LD50s and LC50s
*Inhalation, mouse: LC50 = 3020 mg/m3 (8 hr) Oral, rat: LD50 = 4
gm/kg*

NIOSH 1990 - Pocket Guide - RELs
100 ppm TWA; 480 mg/m3 TWA

NIOSH 1990 - Pocket Guide - IDLHs
5000 ppm IDLH

NIOSH 1990 - Pocket Guide - Target organs
eyes, skin, respiratory system

NTP Chemical Status Reports - Testing Status and NTIS Number
Technical reports printed (PB90260035)

NTP Chemical Status Reports - Evidence of Carcinogenicity
*male rat-no evidence; female rat-no evidence; male mice-no evi-
dence; female mice-no evidence*

OSHA - Vacated PELs - Time Weighted Averages
100 ppm TWA; 480 mg/m3 TWA

OSHA - Final PELs - Time Weighted Averages
100 ppm TWA; 480 mg/m3 TWA

ENVIRONMENTAL LISTS

CAA - HON Rule - SOCMI Chemicals
compliance by April 24, 1995

TSCA - Health and Safety Reporting List
Effective Date: March 11, 1994; Sunset Date: March 11, 2004

INTERNATIONAL LISTS

Australian Exposure Standards - Time Weighted Averages
50 ppm TWA; 242 mg/m3 TWA

Australian Exposure Standards - Short Term Exposure Limits
100 ppm STEL; 483 mg/m3 STEL

Canada - WHMIS: Ingredient Disclosure
1% item 1691 (1717)

Canada - Alberta - 8 Hour Occupational Exposure Limit
50 ppm TWA; 242 mg/m3 TWA

Canada - Alberta - 15 Minute Occupational Exposure Limit
100 ppm STEL; 483 mg/m3 STEL

Canada - British Columbia - 8 Hour Exposure Limits
100 ppm TWA; 480 mg/m3 TWA

Canada - British Columbia - 15 Minute Exposure Limits
150 ppm STEL; 720 mg/m3 STEL

Canada - Ontario - OHSA - TWAEVs
50 ppm TWAEV; 241 mg/m3 TWAEV

Canada - Ontario - OHSA - STEVs
100 ppm STEV; 482 mg/m3 STEV

Canada - Quebec - Time-Weighted Average Exposure Values
50 ppm TWAEV; 242 mg/m3 TWAEV

Canada - Quebec - Short-term Exposure Values
100 ppm STEV; 483 mg/m3 STEV

United Kingdom - Occupational Exposure Standards - TWAs
100 ppm TWA; 480 mg/m3 TWA (does not include alpha-methyl-styrene)

United Kingdom - Occupational Exposure Standards - STELs
150 ppm STEL; 720 mg/m3 STEL (does not include alpha-methyl-styrene)

German (DFG) - MAK Values
100 ppm MAK; 480 mg/m3 MAK

German (DFG) - Peak Limitations
2 x normal MAK (10 min momentary value); don't exceed 4 times per shift

Israel - Time Weighted Averages
50 ppm TWA; 242 mg/m3 TWA

Israel - Short Term Exposure Limits
100 ppm STEL; 483 mg/m3 STEL

Israel - Action Levels
25 ppm AL; 121 mg/m3 AL

Mexico - Instruction No. 10 - TWAs
50 ppm TWA; 240 mg/m3 TWA

Mexico - Instruction No. 10 - STELs
100 ppm STEL; 485 mg/m3 STEL

STATE LISTS

California - Exposure Limits - PELs
50 ppm PEL; 240 mg/m3 PEL

California - Directors List of Hazardous Substances (8 CCR 339)
[present]

Florida Hazardous Substance List
[present]

Massachusetts Right To Know List
[present]

Minnesota Hazardous Substance List
[present]

NJ Right to Know List (Total)
sn 2010

Pennsylvania Right to Know List
[present]

PROPOSED REGULATIONS

TSCA - ITC 32nd Report Priority Testing List
designated for dermal absorption testing

WARFARIN SODIUM   129-06-6

HEALTH AND SAFETY LISTS

U.S. DOT - Substances From 49 CFR 172.101
regulated by DOT (UN1305)

U.S. DOT - Hazard Classes
DOT hazard class = 3

NFPA - Flash Points
flash point = 70 degrees F (21 degrees C)

NFPA - Hazard Identification Ratings
health-3; flammability-3; reactivity-2 (avoid use of water)

NIOSH - Selected LD50s and LC50s
Oral, rat: LD50 = 1280 mg/kg Skin, rabbit; LD50 = 680 mg/kg

INTERNATIONAL LISTS

Canada - WHMIS: Ingredient Disclosure
1% item 1692 (1718)

STATE LISTS

Florida Hazardous Substance List
[present]

Massachusetts Right To Know List
[present]

NJ Right to Know List (Total)
sn 2011

NJ Special Hazardous Substances
(corrosive; flammable - third degree; reactive - second degree)

Pennsylvania Right to Know List
[present]

WASTE ANESTHETIC GASES AND VAPORS   RR-00530-1

HEALTH AND SAFETY LISTS

NIOSH - Selected LD50s and LC50s
Oral, rat: LD50 = 2960 mg/kg Skin, rabbit: LD50 = 1500 mg/kg

WASTE CRANKCASE OILS   RR-01600-2

HEALTH AND SAFETY LISTS

NIOSH - Selected LD50s and LC50s
Oral, rat: LD50 = 500 mg/kg Skin, rabbit: LD50 = 3160 mg/kg

WASTE OIL   RR-01509-8

HEALTH AND SAFETY LISTS

NIOSH - Selected LD50s and LC50s
Oral, rat: 2000 mg/kg

STATE LISTS

California - Air Bill 2588 Appendix A-II
known or potential carcinogen: 9/89

California - Prop. 65 - Developmental Toxicity
developmental toxicity (in daily doses greater than 10,000 IU) - initial date 7/1/89

California - Directors List of Hazardous Substances (8 CCR 339)
[present]

Massachusetts Right To Know List
teratogen

WATER   7732-18-5

ENVIRONMENTAL LISTS

List of Pesticide Product Inert Ingredients
[present]

WATER GAS (CARBURETED)   RR-00803-7

HEALTH AND SAFETY LISTS

NTP Chemical Status Reports - Testing Status and NTIS Number
Prechronic studies for which toxicity technical reports were not prepared

ENVIRONMENTAL LISTS

List of Pesticide Product Inert Ingredients
[present]

WATER REACTIVE SOLID, N.O.S.   **RR-01510-1**

ENVIRONMENTAL LISTS

List of Pesticide Product Inert Ingredients
[present]

WAX, LIQUID   **RR-01511-2**

HEALTH AND SAFETY LISTS

ACGIH 1995 - Time Weighted Averages
0.1 mg/m3 TWA

U.S. DOT - Appendix B - Marine Pollutants
DOT regulated marine pollutant (includes its salts)

U.S. DOT - Appendix A Table 1 - Hazardous Substances
when present at concentrations > 0.3%: Final RQ = 100 pounds (45.4 kg)

NIOSH - Selected LD50s and LC50s
Inhalation, rat: LC50 = 320 mg/m3 8 hr Oral, rat: LD50 = 3 mg/kg Skin, rat: LD50 = 1400 mg/kg

NIOSH 1990 - Pocket Guide - RELs
0.1 mg/m3 TWA

NIOSH 1990 - Pocket Guide - IDLHs
350 mg/m3 IDLH

NIOSH 1990 - Pocket Guide - Target organs
CVS, blood

OSHA - Vacated PELs - Time Weighted Averages
0.1 mg/m3 TWA

OSHA - Final PELs - Time Weighted Averages
0.1 mg/m3 TWA

ENVIRONMENTAL LISTS

CERCLA/SARA - Section 302 Extremely Hazardous Substances and TPQs
TPQ = 500/10,000 pounds

CERCLA/SARA - Section 313 - Emission Reporting
form R reporting required

CERCLA/SARA - Hazardous Substances and their Reportable Quantities
when present at concentrations > 0.3%: Final RQ = 100 pounds (45.4 kg)

RCRA - P Series Wastes
when present at concentrations greater than 0.3%: waste number P001

RCRA - U Series Wastes
when present at concentrations of 0.3% or less: waste number U248

RCRA - Hazardous Constituents-Appendix VIII
when present at concentrations greater than 0.3%: waste number P001; when present at concentrations less than 0.3%: waste number U248

RCRA - Substances Banned From Land Disposal
[present]

INTERNATIONAL LISTS

Australian Exposure Standards - Time Weighted Averages
0.1 mg/m3 TWA

Canada - Alberta - 8 Hour Occupational Exposure Limit
0.1 mg/m3 TWA

Canada - Alberta - 15 Minute Occupational Exposure Limit
0.3 mg/m3 STEL

Canada - British Columbia - 8 Hour Exposure Limits
0.1 mg/m3 TWA

Canada - British Columbia - 15 Minute Exposure Limits
0.3 mg/m3 STEL

Canada - Ontario - OHSA - TWAEVs
0.1 mg/m3 TWAEV

Canada - Quebec - Time-Weighted Average Exposure Values
0.1 mg/m3 TWAEV

United Kingdom - Occupational Exposure Standards - TWAs
0.1 mg/m3 TWA

United Kingdom - Occupational Exposure Standards - STELs
0.3 mg/m3 STEL

German (DFG) - MAK Values
total dust: 0.5 mg/m3 MAK

German (DFG) - Peak Limitations
5 x normal MAK (30 min. average value); don't exceed 2 times during shift

Israel - Time Weighted Averages
0.1 mg/m3 TWA

Israel - Action Levels
0.05 mg/m3 AL

Mexico - Instruction No. 10 - TWAs
0.1 mg/m3 TWA

Mexico - Instruction No. 10 - STELs
0.3 mg/m3 STEL

STATE LISTS

California - Air Bill 2588 Appendix A-II
[present]

California - Prop. 65 - Developmental Toxicity
developmental toxicity - initial date 7/1/87

California - Exposure Limits - PELs
0.1 mg/m3 PEL

California - Directors List of Hazardous Substances (8 CCR 339)
[present]

Florida Hazardous Substance List
[present]

Massachusetts Right To Know List
extraordinarily hazardous; teratogen

Minnesota Hazardous Substance List
[present]

NJ Special Hazardous Substances
(teratogen)

Pennsylvania Right to Know List
environmental hazard

WAX, OXACERITE   **RR-00838-8**

HEALTH AND SAFETY LISTS

NIOSH - Selected LD50s and LC50s
Oral, rat: LD50 = 8700 ug/kg

ENVIRONMENTAL LISTS

CERCLA/SARA - Section 302 Extremely Hazardous Substances and TPQs
TPQ = 100/10,000 pounds

STATE LISTS

Florida Hazardous Substance List
effective March 13, 1992

Massachusetts Right To Know List
extraordinarily hazardous; teratogen

Pennsylvania Right to Know List
environmental hazard

PROPOSED REGULATIONS

CERCLA/SARA - Proposed Hazardous Substance Additions
proposed RQ = 1 pound (.454 kg)

CERCLA/SARA - 1989 Proposed RQ Adjustments
proposed RQ = 10 pounds (4.54 kg)

WELDING FUMES (NOC) RR-00009-9

STATE LISTS

Minnesota Hazardous Substance List
[present]

WHALE OIL 70983-99-2

INTERNATIONAL LISTS

Canada - CEPA - Priority Substances List
estimated time for completion of assessment reports: 3 years

WHEAT RR-01158-5

STATE LISTS

NJ Right to Know List (Total)
sn 2851

WHEAT GERM OIL 8006-95-9

ENVIRONMENTAL LISTS

List of Pesticide Product Inert Ingredients
[present]

WHEY 68608-58-2

HEALTH AND SAFETY LISTS

NFPA - Hazard Identification Ratings
health-2; flammability-4; reactivity-0

STATE LISTS

Pennsylvania Right to Know List
[present]

WHITE BEESWAX 8012-89-3

STATE LISTS

NJ Right to Know List (Total)
sn 2853

WHITE MINERAL OIL 8042-47-5

STATE LISTS

NJ Right to Know List (Total)
sn 2854

WHITE PHOSPHORUS 12185-10-3

HEALTH AND SAFETY LISTS

NFPA - Flash Points
flash point = 236 degrees F (113 degrees C)

NFPA - Hazard Identification Ratings
health-0; flammability-1; reactivity-0

WINTERGREEN OIL 68917-75-9

HEALTH AND SAFETY LISTS

ACGIH 1995 - Time Weighted Averages
5 mg/m3 TWA

IARC - Group 2B (sufficient animal data)
[present]

OSHA - Vacated PELs - Time Weighted Averages
5 mg/m3 TWA

OSHA - Possible Select Carcinogens
[present]

INTERNATIONAL LISTS

Australian Exposure Standards - Time Weighted Averages
5 mg/m3 TWA

Canada - Alberta - 8 Hour Occupational Exposure Limit
as total particulate: 5 mg/m3 TWA

Canada - Alberta - 15 Minute Occupational Exposure Limit
total particulate: 10 mg/m3 STEL

Canada - British Columbia - 8 Hour Exposure Limits
5.0 mg/m3 TWA

Canada - British Columbia - 15 Minute Exposure Limits
10 mg/m3 STEL

Canada - Ontario - OHSA - TWAEVs
total weight, oil free: 5 mg/m3 TWAEV (listed as an agent of variable composition)

Canada - Quebec - Time-Weighted Average Exposure Values
5 mg/m3 TWAEV

United Kingdom - Occupational Exposure Standards - TWAs
5 mg/m3 TWA

Israel - Time Weighted Averages
5 mg/m3 TWA (total particulate)

Israel - Action Levels
2.5 mg/m3 AL (total particulate)

STATE LISTS

California - Exposure Limits - PELs
total particulates: 5 mg/m3 PEL

Minnesota Hazardous Substance List
[present]

Pennsylvania Right to Know List
[present]

WOLLASTONITE (CA(SIO3)) 13983-17-0

HEALTH AND SAFETY LISTS

NFPA - Flash Points
flash point = 446 degrees F (230 degrees C)

NFPA - Hazard Identification Ratings
health-0; flammability-1; reactivity-0

WOOD DUST, ALLERGENIC - EXCLUDING WESTERN RED CEDAR RR-01736-7

ENVIRONMENTAL LISTS

List of Pesticide Product Inert Ingredients
[present]

WOOD DUST, ALL SOFT AND HARD WOODS RR-00514-1

ENVIRONMENTAL LISTS

List of Pesticide Product Inert Ingredients
[present]

WOOD DUSTS-HARD WOOD RR-00016-8

ENVIRONMENTAL LISTS

List of Pesticide Product Inert Ingredients
[present]

WOOD DUSTS-SOFT WOODS RR-00017-9

ENVIRONMENTAL LISTS

List of Pesticide Product Inert Ingredients
[present]

WOOD DUST, WESTERN RED CEDAR RR-00524-3

ENVIRONMENTAL LISTS

List of Pesticide Product Inert Ingredients
[present]

WOOD FILLER, LIQUID RR-01512-3

ENVIRONMENTAL LISTS

TSCA - PAIR - Reporting List
Effective Date: January 26, 1994; Reporting Date: June 28, 1994

TSCA - Health and Safety Reporting List
Effective Date: January 26, 1994; Sunset Date: January 26, 2004

WOOD INDUSTRIES - FURNITURE & CABINET   RR-00545-8
MAKING
ENVIRONMENTAL LISTS
List of Pesticide Product Inert Ingredients
[present]

WOOD PRESERVATIVE, LIQUID   RR-01513-4
HEALTH AND SAFETY LISTS
IARC - Group 3 (not classifiable)
[present]
NTP Chemical Status Reports - Testing Status and NTIS Number
Chronic studies exist for which technical reports were not prepared
ENVIRONMENTAL LISTS
List of Pesticide Product Inert Ingredients
[present]
INTERNATIONAL LISTS
Canada - Quebec - Time-Weighted Average Exposure Values
1 fibre/cm3 TWAEV
STATE LISTS
California - Directors List of Hazardous Substances (8 CCR 339)
[present]

XANTHAN GUM   11138-66-2
INTERNATIONAL LISTS
Canada - British Columbia - 8 Hour Exposure Limits
2.5 mg/m3 TWA
Canada - British Columbia - 15 Minute Exposure Limits
5 mg/m3 STEL

XENON   7440-63-3
HEALTH AND SAFETY LISTS
OSHA - Vacated PELs - Time Weighted Averages
5 mg/m3 TWA
OSHA - Vacated PELs - Short Term Exposure Limits
10 mg/m3 STEL
INTERNATIONAL LISTS
Canada - Alberta - 8 Hour Occupational Exposure Limit
5 mg/m3 TWA
Canada - Alberta - 15 Minute Occupational Exposure Limit
10 mg/m3 STEL
Canada - British Columbia - 8 Hour Exposure Limits
5 mg/m3 TWA
Canada - British Columbia - 15 Minute Exposure Limits
10 mg/m3 STEL
Canada - Quebec - Time-Weighted Average Exposure Values
hard and soft, except red cedar: 5 mg/m3 TWAEV
German (DFG) - Skin/Sensitizers
danger of sensitization (skin or respiratory) (except beech and oak dust)
German (DFG) - Carcinogens
suspect carcinogen (except beech and oak)
STATE LISTS
California - Exposure Limits - PELs
5 mg/m3 PEL (does not include western red cedar)
California - Exposure Limits - STELs
10 mg/m3 STEL (does not include western red cedar)

XENON 120   15151-08-3
HEALTH AND SAFETY LISTS
ACGIH 1995 - Time Weighted Averages
1 mg/m3 TWA

INTERNATIONAL LISTS
Australian Exposure Standards - Time Weighted Averages
1 mg/m3 TWA
Australian Exposure Standards - Skin Effects
sensitiser
Australian Exposure Standards - Under Review
exposure limits under review
Canada - Ontario - OHSA - TWAEVs
1 mg/m3 TWAEV
United Kingdom - Maximum Exposure Limits - TWAs
5 mg/m3 TWA
United Kingdom - Maximum Exposure Limits - Notes
capable of causing respiratory sensitisation
German (DFG) - Carcinogens
proven carcinogen
Israel - Time Weighted Averages
1 mg/m3 TWA
Israel - Action Levels
as beech and oak: 0.5 mg/m3 AL
Mexico - Instruction No. 10 - TWAs
1 mg/m3 TWA
STATE LISTS
Minnesota Hazardous Substance List
[present]
Pennsylvania Right to Know List
[present]

XENON 121   17913-80-3
HEALTH AND SAFETY LISTS
ACGIH 1995 - Time Weighted Averages
5 mg/m3 TWA
ACGIH 1995 - Short Term Exposure Limits
10 mg/m3 STEL
INTERNATIONAL LISTS
Australian Exposure Standards - Time Weighted Averages
5 mg/m3 TWA
Australian Exposure Standards - Short Term Exposure Limits
10 mg/m3 STEL
Australian Exposure Standards - Skin Effects
sensitiser
Australian Exposure Standards - Under Review
exposure limits under review
Canada - Ontario - OHSA - TWAEVs
5 mg/m3 TWAEV
Canada - Ontario - OHSA - STEVs
10 mg/m3 STEV
Israel - Time Weighted Averages
5 mg/m3 TWA
Israel - Short Term Exposure Limits
10 mg/m3 STEL
Israel - Action Levels
2.5 mg/m3 AL
Mexico - Instruction No. 10 - TWAs
5 mg/m3 TWA
Mexico - Instruction No. 10 - STELs
10 mg/m3 STEL
STATE LISTS
Minnesota Hazardous Substance List
[present]
Pennsylvania Right to Know List
[present]

XENON 122   15151-09-4

HEALTH AND SAFETY LISTS

OSHA - Vacated PELs - Time Weighted Averages
2.5 mg/m3 TWA

INTERNATIONAL LISTS

Canada - Alberta - 8 Hour Occupational Exposure Limit
2.5 mg/m3 TWA

Canada - Alberta - 15 Minute Occupational Exposure Limit
5 mg/m3 STEL

Canada - British Columbia - 8 Hour Exposure Limits
1 mg/m3 TWA

Canada - British Columbia - 15 Minute Exposure Limits
5 mg/m3 STEL

Canada - Quebec - Time-Weighted Average Exposure Values
2.5 mg/m3 TWAEV

STATE LISTS

California - Exposure Limits - PELs
2.5 mg/m3 PEL

XENON 123   15700-10-4

STATE LISTS

NJ Right to Know List (Total)
sn 2855

XENON 125   13994-18-8

HEALTH AND SAFETY LISTS

IARC - Group 1 (carcinogenic to humans)
[present]

OSHA - Select Carcinogens
[present]

STATE LISTS

Pennsylvania Right to Know List
special hazardous substance

Pennsylvania RTK - Special Hazardous Substances
[present]

XENON 127   13994-19-9

STATE LISTS

California - Air Bill 2588 Appendix A-I
9/89

NJ Right to Know List (Total)
sn 2879

XENON 129M   RR-00307-6

ENVIRONMENTAL LISTS

List of Pesticide Product Inert Ingredients
[present]

XENON 131M   RR-00288-0

HEALTH AND SAFETY LISTS

U.S. DOT - Substances From 49 CFR 172.101
regulated by DOT (UN2591, UN2036)

U.S. DOT - Hazard Classes
DOT hazard class = 2.2

XENON 133   14932-42-4

HEALTH AND SAFETY LISTS

U.S. DOT - Appendix A Table 2 - Radionuclides
final RQ = 100 curies (3.7E 12 Bq)

ENVIRONMENTAL LISTS

CERCLA/SARA List of Radionuclides (Appendix B) and Their Reportable Quantities
final RQ = 100 curies (3.7E 12 Bq)

XENON 133M   RR-00305-4

HEALTH AND SAFETY LISTS

U.S. DOT - Appendix A Table 2 - Radionuclides
final RQ = 10 curies (3.7E 11 Bq)

ENVIRONMENTAL LISTS

CERCLA/SARA List of Radionuclides (Appendix B) and Their Reportable Quantities
final RQ = 10 curies (3.7E 11 Bq)

XENON 135   14995-62-1

HEALTH AND SAFETY LISTS

U.S. DOT - Appendix A Table 2 - Radionuclides
final RQ = 100 curies (3.7E 12 Bq)

ENVIRONMENTAL LISTS

CERCLA/SARA List of Radionuclides (Appendix B) and Their Reportable Quantities
final RQ = 100 curies (3.7E 12 Bq)

XENON 135M   RR-00304-3

HEALTH AND SAFETY LISTS

U.S. DOT - Appendix A Table 2 - Radionuclides
final RQ = 10 curies (3.7E 11 Bq)

ENVIRONMENTAL LISTS

CERCLA/SARA List of Radionuclides (Appendix B) and Their Reportable Quantities
final RQ = 10 curies (3.7E 11 Bq)

XENON 138   15751-81-2

HEALTH AND SAFETY LISTS

U.S. DOT - Appendix A Table 2 - Radionuclides
final RQ = 100 curies (3.7E 12 Bq)

ENVIRONMENTAL LISTS

CERCLA/SARA List of Radionuclides (Appendix B) and Their Reportable Quantities
final RQ = 100 curies (3.7E 12 Bq)

O-XYLENE   95-47-6

HEALTH AND SAFETY LISTS

U.S. DOT - Appendix A Table 2 - Radionuclides
final RQ = 100 curies (3.7E 12 Bq)

ENVIRONMENTAL LISTS

CERCLA/SARA List of Radionuclides (Appendix B) and Their Reportable Quantities
final RQ = 100 curies (3.7E 12 Bq)

P-XYLENE   106-42-3

HEALTH AND SAFETY LISTS

U.S. DOT - Appendix A Table 2 - Radionuclides
final RQ = 1000 curies (3.7E 13 Bq)

ENVIRONMENTAL LISTS

CERCLA/SARA List of Radionuclides (Appendix B) and Their Reportable Quantities
final RQ = 1000 curies (3.7E 13 Bq)

TSCA - Chemicals with Significant New Use Rules
PMN number: P-87-1456

M-XYLENE   108-38-3

HEALTH AND SAFETY LISTS

U.S. DOT - Appendix A Table 2 - Radionuclides
final RQ = 1000 curies (3.7E 13 Bq)

ENVIRONMENTAL LISTS

CERCLA/SARA List of Radionuclides (Appendix B) and Their Reportable Quantities
final RQ = 1000 curies (3.7E 13 Bq)

M-XYLENE-ALPHA, ALPHA'-DIAMINE   1477-55-0

HEALTH AND SAFETY LISTS

U.S. DOT - Appendix A Table 2 - Radionuclides
final RQ = 1000 curies (3.7E 13 Bq)

ENVIRONMENTAL LISTS

CERCLA/SARA List of Radionuclides (Appendix B) and Their Reportable Quantities
final RQ = 1000 curies (3.7E 13 Bq)

XYLENE RANGE AROMATIC SOLVENT   68920-06-9

HEALTH AND SAFETY LISTS

U.S. DOT - Appendix A Table 2 - Radionuclides
final RQ = 1000 curies (3.7E 13 Bq)

ENVIRONMENTAL LISTS

CERCLA/SARA List of Radionuclides (Appendix B) and Their Reportable Quantities
final RQ = 1000 curies (3.7E 13 Bq)

TSCA - Chemicals with Significant New Use Rules
PMN number: P-88-63

XYLENES (O-, M-, P- ISOMERS)   1330-20-7

HEALTH AND SAFETY LISTS

U.S. DOT - Appendix A Table 2 - Radionuclides
final RQ = 100 curies (3.7E 12 Bq)

ENVIRONMENTAL LISTS

CERCLA/SARA List of Radionuclides (Appendix B) and Their Reportable Quantities
final RQ = 100 curies (3.7E 12 Bq)

XYLENESULFONIC ACID   25321-41-9

HEALTH AND SAFETY LISTS

U.S. DOT - Appendix A Table 2 - Radionuclides
final RQ = 10 curies (3.7E 11 Bq)

ENVIRONMENTAL LISTS

CERCLA/SARA List of Radionuclides (Appendix B) and Their Reportable Quantities
final RQ = 10 curies (3.7E 11 Bq)

TSCA - HDD/HDF - Chemicals Required for Testing
[present]

XYLENESULFONIC ACID, AMMONIUM SALT   26447-10-9

HEALTH AND SAFETY LISTS

U.S. DOT - Appendix A Table 2 - Radionuclides
final RQ = 10 curies (3.7E 11 Bq)

ENVIRONMENTAL LISTS

CERCLA/SARA List of Radionuclides (Appendix B) and Their Reportable Quantities
final RQ = 10 curies (3.7E 11 Bq)

XYLENESULFONIC ACID, CALCIUM SALT   28088-63-3
SEE ALSO:
XYLENES (O-, M-, P- ISOMERS)

HEALTH AND SAFETY LISTS

ACGIH 1995 - Time Weighted Averages
100 ppm TWA;   434 mg/m3 TWA

ACGIH 1995 - Short Term Exposure Limits
150 ppm STEL;   651 mg/m3 STEL

AIHA - Odor Threshold Values
geometric mean air odor threshold = 5.4 ppm (detectable)

U.S. DOT - Appendix A Table 1 - Hazardous Substances
final RQ = 1000 pounds (454 kg) (Listed under 'Xylene (mixed)')

NFPA - Flash Points
flash point = 90 degrees F (32 degrees C)

NFPA - Hazard Identification Ratings
health-2; flammability-3; reactivity-0

ENVIRONMENTAL LISTS

ATSDR Priority List
Rank (of 275): 242

CERCLA/SARA - Section 313 - Emission Reporting
form R reporting required for 1.0% de minimus concentration

CERCLA/SARA - Hazardous Substances and their Reportable Quantities
final RQ = 1000 pounds (454 kg) (Listed under 'Xylene (mixed)')

Clean Air Act (1990) - List of Hazardous Air Contaminants
[present]

CAA - HON Rule - SOCMI Chemicals
compliance by Oct. 24, 1994

CAA - HON Rule - Organic HAPs
[present]

Clean Water Act - Hazardous Substances
[present] (Listed under 'Xylene (mixed)')

Safe Drinking Water Act - Monitoring
monitoring required

TSCA - Code of Federal Regulations Citations
40 CFR 712.30(d);   40 CFR 716.120(a)

TSCA - PAIR - Reporting List
Reporting Date: November 19, 1982

TSCA - Health and Safety Reporting List
Effective Date: October 4, 1982

INTERNATIONAL LISTS

Australian Exposure Standards - Time Weighted Averages
80 ppm TWA;   350 mg/m3 TWA

Australian Exposure Standards - Short Term Exposure Limits
150 ppm STEL;   655 mg/m3 STEL

Canada - WHMIS: Ingredient Disclosure
1% item 1695 (1723)

Canada - NPRI (National Pollutant Release Inventory)
[present]

Canada - Quebec - Time-Weighted Average Exposure Values
100 ppm TWAEV;   434 mg/m3 TWAEV

Canada - Quebec - Short-term Exposure Values
150 ppm STEV;   651 mg/m3 STEV

Israel - Time Weighted Averages
100 ppm TWA;   434 mg/m3 TWA

Israel - Short Term Exposure Limits
150 ppm STEL;   651 mg/m3 STEL

Israel - Action Levels
50 ppm AL;   217 mg/m3 AL

STATE LISTS

California - Air Bill 2588 Appendix A-I
6/91

California - Directors List of Hazardous Substances (8 CCR 339)
[present] (Listed under 'Xylene, all isomers')

Florida Hazardous Substance List
[present]

Massachusetts Right To Know List
[present]

NJ Right to Know List (Total)
sn 2903

Pennsylvania Right to Know List
environmental hazard

XYLENESULFONIC ACID, MAGNESIUM SALT   36729-43-8
SEE ALSO:
    XYLENES (O-, M-, P- ISOMERS)

HEALTH AND SAFETY LISTS

ACGIH 1995 - Time Weighted Averages
100 ppm TWA;   434 mg/m3 TWA

ACGIH 1995 - Short Term Exposure Limits
150 ppm STEL;   651 mg/m3 STEL

AIHA - Odor Threshold Values
geometric mean air odor threshold = 2.1 ppm (detectable)

U.S. DOT - Appendix A Table 1 - Hazardous Substances
final RQ = 1000 pounds (454 kg) (Listed under 'Xylene (mixed)')

NFPA - Flash Points
flash point = 81 degrees F (27 degrees C)

NFPA - Hazard Identification Ratings
health-2;   flammability-3;   reactivity-0

NIOSH - Selected LD50s and LC50s
Inhalation, rat: LC50 = 4550 ppm 4 hr Oral, rat: LD50 = 5 gm/kg

ENVIRONMENTAL LISTS

CERCLA/SARA - Section 313 - Emission Reporting
form R reporting required for 1.0% de minimus concentration

CERCLA/SARA - Hazardous Substances and their Reportable Quantities
final RQ = 1000 pounds (454 kg) (Listed under 'Xylene (mixed)')

Clean Air Act (1990) - List of Hazardous Air Contaminants
[present]

CAA - HON Rule - SOCMI Chemicals
compliance by Oct. 24, 1994

CAA - HON Rule - Organic HAPs
[present]

Clean Water Act - Hazardous Substances
[present] (Listed under 'Xylene (mixed)')

Safe Drinking Water Act - Monitoring
monitoring required

EPA - Master Testing List
[present]

TSCA - Code of Federal Regulations Citations
40 CFR 712.30(d);   40 CFR 716.120(a)

TSCA - PAIR - Reporting List
Effective Date: January 26, 1994;   Reporting Date: March 28, 1994

TSCA - Health and Safety Reporting List
Effective Date: January 26, 1994;   Sunset Date: January 26, 2004

INTERNATIONAL LISTS

Australian Exposure Standards - Time Weighted Averages
80 ppm TWA;   350 mg/m3 TWA

Australian Exposure Standards - Short Term Exposure Limits
150 ppm STEL;   655 mg/m3 STEL

Canada - WHMIS: Ingredient Disclosure
0.1% item 1696 (1724)

Canada - NPRI (National Pollutant Release Inventory)
[present]

Canada - Quebec - Time-Weighted Average Exposure Values
100 ppm TWAEV;   434 mg/m3 TWAEV

Canada - Quebec - Short-term Exposure Values
150 ppm STEV;   651 mg/m3 STEV

Israel - Time Weighted Averages
100 ppm TWA;   434 mg/m3 TWA

Israel - Short Term Exposure Limits
150 ppm STEL;   651 mg/m3 STEL

Israel - Action Levels
50 ppm AL;   217 mg/m3 AL

STATE LISTS

California - Air Bill 2588 Appendix A-I
6/91

California - Directors List of Hazardous Substances (8 CCR 339)
[present] (Listed under 'Xylene, all isomers')

Florida Hazardous Substance List
[present]

Massachusetts Right To Know List
[present]

NJ Right to Know List (Total)
sn 2904

Pennsylvania Right to Know List
environmental hazard

PROPOSED REGULATIONS

TSCA - ITC 31st Report Priority Testing List
designated to be tested

XYLENESULFONIC ACID, POTASSIUM SALT   30346-73-7
SEE ALSO:
    XYLENES (O-, M-, P- ISOMERS)

HEALTH AND SAFETY LISTS

ACGIH 1995 - Time Weighted Averages
100 ppm TWA;   434 mg/m3 TWA

ACGIH 1995 - Short Term Exposure Limits
150 ppm STEL;   651 mg/m3 STEL

AIHA - Odor Threshold Values
geometric mean air odor threshold = 0.62 ppm (detectable)

U.S. DOT - Appendix A Table 1 - Hazardous Substances
final RQ = 1000 pounds (454 kg) (Listed under 'Xylene (mixed)')

NFPA - Flash Points
flash point = 81 degrees F (27 degrees C)

NFPA - Hazard Identification Ratings
health-2;   flammability-3;   reactivity-0

NIOSH - Selected LD50s and LC50s
Oral, rat: LD50 = 5 gm/kg Skin, rabbit: LD50 = 14100 mg/kg

ENVIRONMENTAL LISTS

ATSDR Priority List
Rank (of 275): 269

CERCLA/SARA - Section 313 - Emission Reporting
form R reporting required for 1.0% de minimus concentration

CERCLA/SARA - Hazardous Substances and their Reportable Quantities
final RQ = 1000 pounds (454 kg) (Listed under 'Xylene (mixed)')

Clean Air Act (1990) - List of Hazardous Air Contaminants
[present]

CAA - HON Rule - SOCMI Chemicals
compliance by Oct. 24, 1994

CAA - HON Rule - Organic HAPs
[present]

Clean Water Act - Hazardous Substances
[present] (Listed under 'Xylene (mixed)')

Safe Drinking Water Act - Monitoring
monitoring required

TSCA - Code of Federal Regulations Citations
40 CFR 712.30(d);   40 CFR 716.120(a)

TSCA - PAIR - Reporting List
Reporting Date: November 19, 1982

TSCA - Health and Safety Reporting List
Effective Date: October 4, 1982

INTERNATIONAL LISTS

Australian Exposure Standards - Time Weighted Averages
80 ppm TWA; 350 mg/m3 TWA

Australian Exposure Standards - Short Term Exposure Limits
150 ppm STEL; 655 mg/m3 STEL

Canada - WHMIS: Ingredient Disclosure
1% item 1694 (1722)

Canada - NPRI (National Pollutant Release Inventory)
[present]

Canada - Quebec - Time-Weighted Average Exposure Values
100 ppm TWAEV; 434 mg/m3 TWAEV

Canada - Quebec - Short-term Exposure Values
150 ppm STEV; 651 mg/m3 STEV

Israel - Time Weighted Averages
100 ppm TWA; 434 mg/m3 TWA

Israel - Short Term Exposure Limits
150 ppm STEL; 651 mg/m3 STEL

Israel - Action Levels
50 ppm AL; 217 mg/m3 AL

STATE LISTS

California - Air Bill 2588 Appendix A-I
6/91

California - Directors List of Hazardous Substances (8 CCR 339)
[present] (Listed under 'Xylene, all isomers')

Florida Hazardous Substance List
[present]

Massachusetts Right To Know List
[present]

NJ Right to Know List (Total)
sn 2902

Pennsylvania Right to Know List
environmental hazard

XYLENESULFONIC ACID, ZINC SALT   36729-46-1

HEALTH AND SAFETY LISTS

ACGIH 1995 - Ceiling Limits
C 0.1 mg/m3

ACGIH 1995 - Skin Designations
skin - potential for cutaneous absorption

NIOSH - Selected LD50s and LC50s
Inhalation, rat: LC50 = 700 ppm 1 hr Oral, rat: LD50 = 1600 mg/kg Skin, rabbit: LD50 = 2000 mg/kg

OSHA - Vacated PELs - Ceiling Limits
C 0.1 mg/m3

OSHA - Vacated PELs - Skin Designation
Prevent or reduce skin absorption

ENVIRONMENTAL LISTS

TSCA - Code of Federal Regulations Citations
40 CFR 712.30(f)

TSCA - PAIR - Reporting List
Reporting Date: August 17, 1983

INTERNATIONAL LISTS

Australian Exposure Standards - Time Weighted Averages
Peak Limitation: 0.1 mg/m3

Australian Exposure Standards - Skin Effects
skin absorption

Canada - WHMIS: Ingredient Disclosure
1% item 1697 (1725)

Canada - Alberta - Ceiling Occupational Exposure Limit
C 0.1 mg/m3

Canada - British Columbia - Ceiling Exposure Limits
C 0.1 mg/m3

Canada - Ontario - OHSA - CEVs
0.1 mg/m3 CEV

Canada - Ontario - OHSA - Skin Notations
absorption through skin, eyes, or mucous membranes

Canada - Quebec - Ceiling Limits
P 0.1 mg/m3

Canada - Quebec - Skin Designations
absorbed through the skin

Israel - Ceiling Exposure Limits
C 0.1 mg/m3

Mexico - Instruction No. 10 - TWAs
0.1 mg/m3 TWA

STATE LISTS

California - Exposure Limits - Ceilings
C 0.1 mg/m3

California - Exposure Limits - Skin Notation
material may be absorbed through the skin, eyes or mucous membrane

California - Directors List of Hazardous Substances (8 CCR 339)
[present]

Florida Hazardous Substance List
[present]

Massachusetts Right To Know List
[present]

Minnesota Hazardous Substance List
[present]

Pennsylvania Right to Know List
[present]

2,5-XYLENOL   95-87-4

ENVIRONMENTAL LISTS

List of Pesticide Product Inert Ingredients
[present]

3,5-XYLENOL   108-68-9

SEE ALSO:
F039-HAZARDOUS WASTES
F003-HAZARDOUS WASTES

HEALTH AND SAFETY LISTS

ACGIH 1995 - Time Weighted Averages
100 ppm TWA; 434 mg/m3 TWA

ACGIH 1995 - Short Term Exposure Limits
150 ppm STEL; 651 mg/m3 STEL

ACGIH 1995 - Biological Exposure Indices
Methylhippuric acids in urine: 1.5 g/g creatinine, end of shift

AIHA - Odor Threshold Values
geometric mean air odor threshold = 20 ppm (detectable); 40 ppm (recognizable)

U.S. DOT - Substances From 49 CFR 172.101
regulated by DOT (UN1307)

U.S. DOT - Hazard Classes
DOT hazard class = 3

U.S. DOT - Appendix A Table 1 - Hazardous Substances
final RQ = 1000 pounds (454 kg)

IARC - Group 3 (not classifiable)
[present]

NIOSH - Selected LD50s and LC50s
Inhalation, rat: LC50 = 5000 ppm 4 hr Oral, rat: LD50 = 4300 mg/kg

NIOSH 1990 - Pocket Guide - RELs
100 ppm TWA; 435 mg/m3 TWA; 150 ppm STEL; 655 mg/m3 STEL

NIOSH 1990 - Pocket Guide - IDLHs
1000 ppm IDLH

NIOSH 1990 - Pocket Guide - Target organs
CNS, eyes, blood, liver, kidneys, skin, GI tract

NIOSH - Health Standards - Exposure Limits
100 ppm TWA; 434 mg/m3 TWA; C (10 min) 200 ppm; C (10 min) 868 mg/m3

NIOSH - Health Standards - Health Effects and Precautions
Central nervous system depressant; respiratory irritation

NTP Chemical Status Reports - Testing Status and NTIS Number
Technical reports printed (PB87189684/AS)

NTP Chemical Status Reports - Evidence of Carcinogenicity
male rat-no evidence; female rat-no evidence; male mice-no evidence; female mice-no evidence

OSHA - Vacated PELs - Time Weighted Averages
100 ppm TWA; 435 mg/m3 TWA

OSHA - Vacated PELs - Short Term Exposure Limits
150 ppm STEL; 655 mg/m3 STEL

OSHA - Final PELs - Time Weighted Averages
100 ppm TWA; 435 mg/m3 TWA

ENVIRONMENTAL LISTS

ATSDR Priority List
Rank (of 275): 079

CERCLA/SARA - Section 313 - Emission Reporting
form R reporting required for 1.0% de minimus concentration

CERCLA/SARA - Hazardous Substances and their Reportable Quantities
final RQ = 1000 pounds (454 kg)

Clean Air Act (1990) - List of Hazardous Air Contaminants
[present]

CAA - HON Rule - SOCMI Chemicals
compliance by Oct. 24, 1994

CAA - HON Rule - Organic HAPs
[present]

Clean Water Act - Hazardous Substances
[present]

Safe Drinking Water Act - MCLs
MCL = 10 mg/L

Safe Drinking Water Act - MCLGs
MCLG = 10 mg/L

List of Pesticide Product Inert Ingredients
[present]

RCRA - U Series Wastes
waste number U239 (Ignitable waste; Toxic waste)

RCRA - Hazardous Constituents-Appendix VIII
waste number U239 (Ignitable waste; Toxic waste)

RCRA - Basis for Listing - Appendix VII
Included in waste stream: F039

RCRA - Substances Banned From Land Disposal
[present]

RCRA - TSD Facilities Ground Water Monitoring
TM 8020 = 5 ug/L PQL; TM 8240 = 5 ug/L PQL

RCRA - Universal Treatment Standards (LDR)
WW: 0.32 mg/l; NWW: 30 mg/kg

TSCA - Code of Federal Regulations Citations
40 CFR 712.30(d)

TSCA - PAIR - Reporting List
Reporting Date: November 19, 1982

INTERNATIONAL LISTS

Australian Exposure Standards - Time Weighted Averages
80 ppm TWA; 350 mg/m3 TWA

Australian Exposure Standards - Short Term Exposure Limits
150 ppm STEL; 655 mg/m3 STEL

Australian Exposure Standards - Under Review
exposure limits under review

Canada - NPRI (National Pollutant Release Inventory)
[present]

Canada - CEPA - Priority Substances List
estimated time for completion of assessment reports: 4 years

Canada - Drinking Water Quality - AOs
<= 0.3 mg/L AO

Canada - Alberta - 8 Hour Occupational Exposure Limit
100 ppm TWA; 434 mg/m3 TWA

Canada - Alberta - Skin Designation
can be absorbed through the intact skin

Canada - British Columbia - 8 Hour Exposure Limits
100 ppm TWA; 435 mg/m3 TWA

Canada - British Columbia - 15 Minute Exposure Limits
150 ppm STEL; 655 mg/m3 STEL

Canada - British Columbia - Skin Notations
skin - potential for skin absorption

Canada - Ontario - OHSA - TWAEVs
100 ppm TWAEV; 435 mg/m3 TWAEV

Canada - Ontario - OHSA - STEVs
150 ppm STEV; 650 mg/m3 STEV

Canada - Quebec - Time-Weighted Average Exposure Values
100 ppm TWAEV; 434 mg/m3 TWAEV

Canada - Quebec - Short-term Exposure Values
150 ppm STEV; 651 mg/m3 STEV

United Kingdom - Occupational Exposure Standards - TWAs
100 ppm TWA; 435 mg/m3 TWA

United Kingdom - Occupational Exposure Standards - STELs
150 ppm STEL; 650 mg/m3 STEL

United Kingdom - Occupational Exposure Standards - Notes
can be absorbed through skin

German (DFG) - MAK Values
100 ppm MAK; 440 mg/m3 MAK

German (DFG) - Peak Limitations
2 x normal MAK (30 min. average value); don't exceed 4 times during shift

German (DFG) - Pregnancy
classification not yet possible

Israel - Time Weighted Averages
100 ppm TWA; 434 mg/m3 TWA

Israel - Short Term Exposure Limits
150 ppm STEL; 651 mg/m3 STEL

Israel - Action Levels
50 ppm AL; 217 mg/m3 AL

Mexico - Instruction No. 10 - TWAs
100 ppm TWA; 435 mg/m3 TWA

Mexico - Instruction No. 10 - STELs
150 ppm STEL; 655 mg/m3 STEL

STATE LISTS

California - Air Bill 2588 Appendix A-I
[present]

California - Exposure Limits - PELs
100 ppm PEL; 435 mg/m3 PEL

California - Exposure Limits - STELs
150 ppm STEL; 655 mg/m3 STEL

California - Exposure Limits - Ceilings
C 300 ppm

California - Directors List of Hazardous Substances (8 CCR 339)
[present]

Florida Hazardous Substance List
[present]

Massachusetts Right To Know List
[present]

Minnesota Hazardous Substance List
[present] (includes all isomers)

NJ Right to Know List (Total)
sn 2014

NJ Special Hazardous Substances
(flammable - third degree)

Pennsylvania Right to Know List
environmental hazard

XYLENOL   1300-71-6

ENVIRONMENTAL LISTS

CAA - HON Rule - SOCMI Chemicals
compliance by April 24, 1995

List of Pesticide Product Inert Ingredients
[present]

2,6-XYLIDENE   87-62-7

ENVIRONMENTAL LISTS

List of Pesticide Product Inert Ingredients
[present]

2,4-XYLIDINE   95-68-1

ENVIRONMENTAL LISTS

List of Pesticide Product Inert Ingredients
[present]

2,5-XYLIDINE   95-78-3

ENVIRONMENTAL LISTS

List of Pesticide Product Inert Ingredients
[present]

XYLIDINE   1300-73-8

ENVIRONMENTAL LISTS

List of Pesticide Product Inert Ingredients
[present]

XYLYL BROMIDE   28258-59-5

ENVIRONMENTAL LISTS

List of Pesticide Product Inert Ingredients
[present]

P-XYLYL DIAZIDE   RR-01455-1

HEALTH AND SAFETY LISTS

NIOSH - Selected LD50s and LC50s
Oral, rat: LD50 = 444 mg/kg

XYLYLENE DICHLORIDE   28347-13-9

HEALTH AND SAFETY LISTS

NIOSH - Selected LD50s and LC50s
Oral, rat: LD50 = 608 mg/kg

YEAST   68876-77-7

HEALTH AND SAFETY LISTS

U.S. DOT - Substances From 49 CFR 172.101
regulated by DOT (UN2261)

U.S. DOT - Hazard Classes
DOT hazard class = 6.1

U.S. DOT - Appendix B - Marine Pollutants
DOT regulated marine pollutant

U.S. DOT - Appendix A Table 1 - Hazardous Substances
final RQ = 1000 pounds (454 kg)

ENVIRONMENTAL LISTS

CERCLA/SARA - Hazardous Substances and their Reportable Quantities
final RQ = 1000 pounds (454 kg)

CAA - HON Rule - SOCMI Chemicals
compliance by Oct. 23, 1995

Clean Water Act - Hazardous Substances
[present]

List of Pesticide Product Inert Ingredients
[present]

TSCA - Code of Federal Regulations Citations
40 CFR 716.120(a)

TSCA - Health and Safety Reporting List
Effective Date: June 1, 1987

INTERNATIONAL LISTS

Canada - WHMIS: Ingredient Disclosure
1% item 1698 (1729)

STATE LISTS

California - Directors List of Hazardous Substances (8 CCR 339)
[present]

Massachusetts Right To Know List
[present]

NJ Right to Know List (Total)
sn 2015

Pennsylvania Right to Know List
environmental hazard

YELLOW AB   85-84-7

HEALTH AND SAFETY LISTS

IARC - Group 2B (sufficient animal data)
[present]

NFPA - Flash Points
flash point = 206 degrees F (97 degrees C)

NFPA - Hazard Identification Ratings
health-3; flammability-1; reactivity-0

NIOSH - Selected LD50s and LC50s
Oral, rat: LD50 = 840 mg/kg

NTP Chemical Status Reports - Testing Status and NTIS Number
Technical reports printed (PB90256363)

NTP Chemical Status Reports - Evidence of Carcinogenicity
male rat-positive; female rat-positive

ENVIRONMENTAL LISTS

CERCLA/SARA - Section 313 - Emission Reporting
form R reporting required for 1.0% de minimus concentration

STATE LISTS

California - Air Bill 2588 Appendix A-II
6/91

California - Prop. 65 - Cancer list
carcinogen - initial date 1/1/91

Florida Hazardous Substance List
[present]

Massachusetts Right To Know List
[present]

NJ Right to Know List (Total)
sn 2016

Pennsylvania Right to Know List
environmental hazard

YELLOW OB   131-79-3

HEALTH AND SAFETY LISTS

IARC - Group 3 (not classifiable)
[present]

NIOSH - Selected LD50s and LC50s
Oral, rat: LD50 = 467 mg/kg

INTERNATIONAL LISTS

Canada - WHMIS: Ingredient Disclosure
1% item 1700 (1727)

German (DFG) - MAK Values
5 ppm MAK;   25 mg/m3 MAK

German (DFG) - Skin/Sensitizers
danger of cutaneous absorption

German (DFG) - Carcinogens
suspected carcinogen

YTTERBIUM 162   24347-38-4

HEALTH AND SAFETY LISTS

IARC - Group 3 (not classifiable)
[present]

INTERNATIONAL LISTS

Canada - WHMIS: Ingredient Disclosure
1% item 1701 (1728)

YTTERBIUM 166   14834-83-4

HEALTH AND SAFETY LISTS

ACGIH 1995 - Time Weighted Averages
0.5 ppm TWA;   2.5 mg/m3 TWA

ACGIH 1995 - Skin Designations
skin - potential for cutaneous absorption

ACGIH 1995 - Carcinogens
A2-suspected human carcinogen

U.S. DOT - Substances From 49 CFR 172.101
regulated by DOT (UN1711)

U.S. DOT - Hazard Classes
DOT hazard class = 6.1

NIOSH - Selected LD50s and LC50s
Oral, rabbit: LD50 = 600 mg/kg

NIOSH 1990 - Pocket Guide - RELs
2 ppm TWA;   10 mg/m3 TWA

NIOSH 1990 - Pocket Guide - IDLHs
150 ppm IDLH

NIOSH 1990 - Pocket Guide - Target organs
blood, lungs, liver, kidneys, CVS

NIOSH 1990 - Pocket Guide - Skin list
Potential for dermal absorption

OSHA - Vacated PELs - Time Weighted Averages
2 ppm TWA;   10 mg/m3 TWA

OSHA - Vacated PELs - Skin Designation
Prevent or reduce skin absorption

OSHA - Final PELs - Time Weighted Averages
5 ppm TWA;   25 mg/m3 TWA

OSHA - Final PELs - Skin Notations
prevent or reduce skin absorption

ENVIRONMENTAL LISTS

CAA - HON Rule - SOCMI Chemicals
compliance by April 24, 1995

TSCA - Health and Safety Reporting List
Effective Date: March 11, 1994;   Sunset Date: March 11, 2004

INTERNATIONAL LISTS

Australian Exposure Standards - Time Weighted Averages
0.5 ppm TWA;   2.5 mg/m3 TWA

Australian Exposure Standards - Skin Effects
skin absorption

Australian Exposure Standards - Carcinogens
probable carcinogen

Canada - WHMIS: Ingredient Disclosure
1% item 1699 (1726)

Canada - Alberta - 8 Hour Occupational Exposure Limit
5 ppm TWA;   25 mg/m3 TWA

Canada - Alberta - 15 Minute Occupational Exposure Limit
10 ppm STEL;   50 mg/m3 STEL

Canada - Alberta - Skin Designation
can be absorbed through the intact skin

Canada - British Columbia - 8 Hour Exposure Limits
5 ppm TWA;   25 mg/m3 TWA

Canada - British Columbia - 15 Minute Exposure Limits
10 ppm STEL;   50 mg/m3 STEL

Canada - British Columbia - Skin Notations
skin - potential for skin absorption

Canada - Ontario - OHSA - TWAEVs
2 ppm TWAEV;   10 mg/m3 TWAEV

Canada - Ontario - OHSA - Skin Notations
absorption through skin, eyes, or mucous membranes

Canada - Quebec - Time-Weighted Average Exposure Values
0.5 ppm TWAEV;   2.5 mg/m3 TWAEV

Canada - Quebec - Skin Designations
absorbed through the skin

Canada - Quebec - Carcinogens
C2 carcinogen: effect suspected in humans

United Kingdom - Occupational Exposure Standards - TWAs
2 ppm TWA;   10 mg/m3 TWA

United Kingdom - Occupational Exposure Standards - STELs
10 ppm STEL;   50 mg/m3 STEL

United Kingdom - Occupational Exposure Standards - Notes
can be absorbed through skin

German (DFG) - MAK Values
5 ppm MAK;   25 mg/m3 MAK (includes all isomers except 2,4-Xylidine)

German (DFG) - Skin/Sensitizers
danger of cutaneous absorption

Israel - Time Weighted Averages
0.5 ppm TWA;   2.5 mg/m3 TWA

Israel - Action Levels
0.25 ppm AL;   1.25 mg/m3 AL

Mexico - Instruction No. 10 - TWAs
5 ppm TWA;   25 mg/m3 TWA

STATE LISTS

California - Exposure Limits - PELs
2 ppm PEL;   10 mg/m3 PEL

California - Exposure Limits - Skin Notation
material may be absorbed through the skin, eyes or mucous membrane

California - Directors List of Hazardous Substances (8 CCR 339)
[present]

Florida Hazardous Substance List
[present]

Massachusetts Right To Know List
[present]

Minnesota Hazardous Substance List
skin

NJ Right to Know List (Total)
sn 2017

Pennsylvania Right to Know List
[present]

PROPOSED REGULATIONS

TSCA - ITC 32nd Report Priority Testing List
designated for dermal absorption testing

YTTERBIUM 167   14041-45-3

INTERNATIONAL LISTS

Canada - WHMIS: Ingredient Disclosure
1% item 1702 (344)

STATE LISTS

NJ Right to Know List (Total)
sn 2018

YTTERBIUM 169   14269-78-4

HEALTH AND SAFETY LISTS

U.S. DOT - Hazard Classes
Forbidden from transport by the DOT

YTTERBIUM 175   14041-44-2

HEALTH AND SAFETY LISTS

NIOSH - Selected LD50s and LC50s
Inhalation, rat: LC50 = 200 mg/m3 4 hr Oral, rat: LD50 = 1 gm/kg

ENVIRONMENTAL LISTS

CERCLA/SARA - Section 302 Extremely Hazardous Substances and
TPQs
TPQ = 100/10,000 pounds

STATE LISTS

Florida Hazardous Substance List
effective March 13, 1992

Massachusetts Right To Know List
extraordinarily hazardous

NJ Right to Know List (Total)
sn 2858

Pennsylvania Right to Know List
environmental hazard

PROPOSED REGULATIONS

CERCLA/SARA - Proposed Hazardous Substance Additions
proposed RQ = 1 pound (.454 kg)

CERCLA/SARA - 1989 Proposed RQ Adjustments
proposed RQ = 100 pounds (45.4 kg)

YTTERBIUM 177   14119-23-4

ENVIRONMENTAL LISTS

List of Pesticide Product Inert Ingredients
[present]

YTTERBIUM 178   29919-07-1

HEALTH AND SAFETY LISTS

IARC - Group 3 (not classifiable)
[present]

YTTERBIUM CHLORIDE   10361-91-8

HEALTH AND SAFETY LISTS

IARC - Group 3 (not classifiable)
[present]

STATE LISTS

California - Directors List of Hazardous Substances (8 CCR 339)
[present]

YTTERBIUM NITRATE   13768-67-7

HEALTH AND SAFETY LISTS

U.S. DOT - Appendix A Table 2 - Radionuclides
final RQ = 1000 curies (3.7E 13 Bq)

ENVIRONMENTAL LISTS

CERCLA/SARA List of Radionuclides (Appendix B) and Their Reportable Quantities
final RQ = 1000 curies (3.7E 13 Bq)

YTTRIUM   7440-65-5

HEALTH AND SAFETY LISTS

U.S. DOT - Appendix A Table 2 - Radionuclides
final RQ = 10 curies (3.7E 11 Bq)

ENVIRONMENTAL LISTS

CERCLA/SARA List of Radionuclides (Appendix B) and Their Reportable Quantities
final RQ = 10 curies (3.7E 11 Bq)

YTTRIUM 86   14809-53-1

HEALTH AND SAFETY LISTS

U.S. DOT - Appendix A Table 2 - Radionuclides
final RQ = 1000 curies (3.7E 13 Bq)

ENVIRONMENTAL LISTS

CERCLA/SARA List of Radionuclides (Appendix B) and Their Reportable Quantities
final RQ = 1000 curies (3.7E 13 Bq)

YTTRIUM 86M   RR-00303-2

HEALTH AND SAFETY LISTS

U.S. DOT - Appendix A Table 2 - Radionuclides
final RQ = 10 curies (3.7E 11 Bq)

ENVIRONMENTAL LISTS

CERCLA/SARA List of Radionuclides (Appendix B) and Their Reportable Quantities
final RQ = 10 curies (3.7E 11 Bq)

YTTRIUM 87   14274-68-1

HEALTH AND SAFETY LISTS

U.S. DOT - Appendix A Table 2 - Radionuclides
final RQ = 100 curies (3.7E 12 Bq)

ENVIRONMENTAL LISTS

CERCLA/SARA List of Radionuclides (Appendix B) and Their Reportable Quantities
final RQ = 100 curies (3.7E 12 Bq)

YTTRIUM 88   13982-36-0

HEALTH AND SAFETY LISTS

U.S. DOT - Appendix A Table 2 - Radionuclides
final RQ = 1000 curies (3.7E 13 Bq)

ENVIRONMENTAL LISTS

CERCLA/SARA List of Radionuclides (Appendix B) and Their Reportable Quantities
final RQ = 1000 curies (3.7E 13 Bq)

YTTRIUM 90   10098-91-6

HEALTH AND SAFETY LISTS

U.S. DOT - Appendix A Table 2 - Radionuclides
final RQ = 1000 curies (3.7E 13 Bq)

ENVIRONMENTAL LISTS

CERCLA/SARA List of Radionuclides (Appendix B) and Their Reportable Quantities
final RQ = 1000 curies (3.7E 13 Bq)

YTTRIUM 90M   RR-00302-1

INTERNATIONAL LISTS

Canada - WHMIS: Ingredient Disclosure
1% item 1703 (540)

YTTRIUM 91   14234-24-3

INTERNATIONAL LISTS

Canada - WHMIS: Ingredient Disclosure
1% item 1704 (1214)

YTTRIUM 91M   RR-00301-0

HEALTH AND SAFETY LISTS

ACGIH 1995 - Time Weighted Averages
metal and compounds, as Y: 1 mg/m3 TWA

NIOSH 1990 - Pocket Guide - RELs
as Y: 1 mg/m3 TWA

NIOSH 1990 - Pocket Guide - Target organs
eyes, lungs

OSHA - Vacated PELs - Time Weighted Averages
1 mg/m3 TWA

OSHA - Final PELs - Time Weighted Averages
1 mg/m3 TWA

INTERNATIONAL LISTS

Australian Exposure Standards - Time Weighted Averages
metal and compounds as Y: 1 mg/m3 TWA

Canada - WHMIS: Ingredient Disclosure
1% item 1706 (1732)

Canada - Alberta - 8 Hour Occupational Exposure Limit
1 mg/m3 TWA

Canada - Alberta - 15 Minute Occupational Exposure Limit
3 mg/m3 STEL

Canada - British Columbia - 8 Hour Exposure Limits
1 mg/m3 TWA

Canada - British Columbia - 15 Minute Exposure Limits
3 mg/m3 STEL

Canada - Ontario - OHSA - TWAEVs
as Y: 1 mg/m3 TWAEV

Canada - Quebec - Time-Weighted Average Exposure Values
1 mg/m3 TWAEV

United Kingdom - Occupational Exposure Standards - TWAs
1 mg/m3 TWA

United Kingdom - Occupational Exposure Standards - STELs
3 mg/m3 STEL

German (DFG) - MAK Values
total dust: 5 mg/m3 MAK

German (DFG) - Peak Limitations
10 x normal MAK (30 min average value); don't exceed during shift

Israel - Time Weighted Averages
as Y: 1 mg/m3 TWA

Israel - Action Levels
as Y: 0.5 mg/m3 AL

Mexico - Instruction No. 10 - TWAs
1 mg/m3 TWA

Mexico - Instruction No. 10 - STELs
3 mg/m3 STEL

STATE LISTS

Florida Hazardous Substance List
[present]

Massachusetts Right To Know List
[present]

Minnesota Hazardous Substance List
[present] as Y (includes metal and compounds)

Pennsylvania Right to Know List
[present]

YTTRIUM 92   15751-59-4

HEALTH AND SAFETY LISTS

U.S. DOT - Appendix A Table 2 - Radionuclides
final RQ = 10 curies (3.7E 11 Bq)

ENVIRONMENTAL LISTS

CERCLA/SARA List of Radionuclides (Appendix B) and Their Reportable Quantities
final RQ = 10 curies (3.7E 11 Bq)

YTTRIUM 93   14981-70-5

HEALTH AND SAFETY LISTS

U.S. DOT - Appendix A Table 2 - Radionuclides
final RQ = 1000 curies (3.7E 13 Bq)

ENVIRONMENTAL LISTS

CERCLA/SARA List of Radionuclides (Appendix B) and Their Reportable Quantities
final RQ = 1000 curies (3.7E 13 Bq)

YTTRIUM 94   15422-72-7

HEALTH AND SAFETY LISTS

U.S. DOT - Appendix A Table 2 - Radionuclides
final RQ = 10 curies (3.7E 11 Bq)

ENVIRONMENTAL LISTS

CERCLA/SARA List of Radionuclides (Appendix B) and Their Reportable Quantities
final RQ = 10 curies (3.7E 11 Bq)

YTTRIUM 95   15422-71-6

HEALTH AND SAFETY LISTS

U.S. DOT - Appendix A Table 2 - Radionuclides
final RQ = 10 curies (3.7E 11 Bq)

ENVIRONMENTAL LISTS

CERCLA/SARA List of Radionuclides (Appendix B) and Their Reportable Quantities
final RQ = 10 curies (3.7E 11 Bq)

YTTRIUM COMPOUNDS   RR-00618-8

HEALTH AND SAFETY LISTS

U.S. DOT - Appendix A Table 2 - Radionuclides
final RQ = 10 curies (3.7E 11 Bq)

ENVIRONMENTAL LISTS

CERCLA/SARA List of Radionuclides (Appendix B) and Their Reportable Quantities
final RQ = 10 curies (3.7E 11 Bq)

YUCCA EXTRACTIVES (FROM YUCCA,   90147-57-2
AGAVACEAE)

HEALTH AND SAFETY LISTS

U.S. DOT - Appendix A Table 2 - Radionuclides
final RQ = 100 curies (3.7E 12 Bq)

ENVIRONMENTAL LISTS

CERCLA/SARA List of Radionuclides (Appendix B) and Their Reportable Quantities
final RQ = 100 curies (3.7E 12 Bq)

ZEARALENONE   17924-92-4

HEALTH AND SAFETY LISTS

U.S. DOT - Appendix A Table 2 - Radionuclides
final RQ = 10 curies (3.7E 11 Bq)

ENVIRONMENTAL LISTS

CERCLA/SARA List of Radionuclides (Appendix B) and Their Reportable Quantities
final RQ = 10 curies (3.7E 11 Bq)

ZEIN 9010-66-6

HEALTH AND SAFETY LISTS

U.S. DOT - Appendix A Table 2 - Radionuclides
final RQ = 1000 curies (3.7E 13 Bq)

ENVIRONMENTAL LISTS

CERCLA/SARA List of Radionuclides (Appendix B) and Their Reportable Quantities
final RQ = 1000 curies (3.7E 13 Bq)

ZINC 7440-66-6

HEALTH AND SAFETY LISTS

U.S. DOT - Appendix A Table 2 - Radionuclides
final RQ = 100 curies (3.7E 12 Bq)

ENVIRONMENTAL LISTS

CERCLA/SARA List of Radionuclides (Appendix B) and Their Reportable Quantities
final RQ = 100 curies (3.7E 12 Bq)

ZINC 62 14833-23-9

HEALTH AND SAFETY LISTS

U.S. DOT - Appendix A Table 2 - Radionuclides
final RQ = 100 curies (3.7E 12 Bq)

ENVIRONMENTAL LISTS

CERCLA/SARA List of Radionuclides (Appendix B) and Their Reportable Quantities
final RQ = 100 curies (3.7E 12 Bq)

ZINC 63 14833-26-2

HEALTH AND SAFETY LISTS

U.S. DOT - Appendix A Table 2 - Radionuclides
final RQ = 1000 curies (3.7E 13 Bq)

ENVIRONMENTAL LISTS

CERCLA/SARA List of Radionuclides (Appendix B) and Their Reportable Quantities
final RQ = 1000 curies (3.7E 13 Bq)

ZINC 65 13982-39-3

HEALTH AND SAFETY LISTS

U.S. DOT - Appendix A Table 2 - Radionuclides
final RQ = 1000 curies (3.7E 13 Bq)

ENVIRONMENTAL LISTS

CERCLA/SARA List of Radionuclides (Appendix B) and Their Reportable Quantities
final RQ = 1000 curies (3.7E 13 Bq)

ZINC 69 13982-23-5

INTERNATIONAL LISTS

Canada - WHMIS: Ingredient Disclosure
1% item 1705 (1731)

Canada - Ontario - OHSA - TWAEVs
as Y: 1 mg/m3 TWAEV

STATE LISTS

California - Exposure Limits - PELs
as Y: 1 mg/m3 PEL

California - Directors List of Hazardous Substances (8 CCR 339)
[present]

ZINC 69M RR-00300-9

ENVIRONMENTAL LISTS

List of Pesticide Product Inert Ingredients
[present]

ZINC 71M RR-00299-3

HEALTH AND SAFETY LISTS

IARC - Group Unspecified
[present] (Listed under 'Toxins derived from Fusarium graminearum, F. culmorum and F. crookwellense')

NTP Chemical Status Reports - Testing Status and NTIS Number
Technical reports printed (PB83165753)

NTP Chemical Status Reports - Evidence of Carcinogenicity
male rat-negative; female rat-negative; male mice-positive; female mice-positive

STATE LISTS

California - Directors List of Hazardous Substances (8 CCR 339)
[present]

ZINC 72 15743-55-2

ENVIRONMENTAL LISTS

List of Pesticide Product Inert Ingredients
[present]

ZINC ACETATE 557-34-6
SEE ALSO:
ZINC COMPOUNDS

HEALTH AND SAFETY LISTS

U.S. DOT - Substances From 49 CFR 172.101
regulated by DOT (UN1436)

U.S. DOT - Hazard Classes
DOT hazard class = 4.3

U.S. DOT - Appendix A Table 1 - Hazardous Substances
final RQ = 1000 pounds (454 kg) (no reporting of releases of this hazardous substance is required if the diameter of the solid metal released is equal to or exceeds 0.004 inches)

ENVIRONMENTAL LISTS

ATSDR Priority List
Rank (of 275): 064

CERCLA/SARA - Section 313 - Emission Reporting
form R reporting required for 1.0% de minimus concentration (only fume or dust)

CERCLA/SARA - Hazardous Substances and their Reportable Quantities
final RQ = 1000 pounds (454 kg) (no reporting of releases of this hazardous substance is required if the diameter of the solid metal released is equal to or exceeds 0.004 inches)

Clean Water Act - Priority Pollutants
[present]

Clean Water Act - Toxic Pollutants
[present] (Listed under 'Zinc and compounds')

Safe Drinking Water Act - SMCLs
SMCL = 5 mg/L

List of Pesticide Product Inert Ingredients
[present]

RCRA - TSD Facilities Ground Water Monitoring
TM 6010 = 20 ug/L PQL; TM 7950 = 50 ug/L PQL (all species in the ground water that contain this element are included)

RCRA - Universal Treatment Standards (LDR)
WW: 2.61 mg/l; NWW: 5.3 mg/l TCLP

INTERNATIONAL LISTS

Canada - NPRI (National Pollutant Release Inventory)
[present] (as fume or dust)

Canada - CEPA Schedule III Part II - Restricted Substances (Ocean Dumping)
[present]

Canada - Drinking Water Quality - AOs
<= 5.0 mg/L AO

Mexico - Wastewater - Organic Toxic Pollutants and Heavy Metals
Listed under [Heavy Metals]

Mexico - Drinking Water - Ecological Criteria
5.0 mg/l

STATE LISTS

California - Air Bill 2588 Appendix A-I
[present]

California - Directors List of Hazardous Substances (8 CCR 339)
[present]

Florida Hazardous Substance List
[present]

Massachusetts Right To Know List
[present]

NJ Right to Know List (Total)
sn 2021

Pennsylvania Right to Know List
environmental hazard (any compound of this substance is also an environmental hazard)

PROPOSED REGULATIONS

Safe Drinking Water Act - Priority list
[present]

ZINC AMMONIUM CHLORIDE (ZN.CL4.2H4-N)   14639-97-5

HEALTH AND SAFETY LISTS

U.S. DOT - Appendix A Table 2 - Radionuclides
final RQ = 100 curies (3.7E 12 Bq)

ENVIRONMENTAL LISTS

CERCLA/SARA List of Radionuclides (Appendix B) and Their Reportable Quantities
final RQ = 100 curies (3.7E 12 Bq)

ZINC AMMONIUM CHLORIDE (ZN.CL5.3H4-N)   14639-98-6

HEALTH AND SAFETY LISTS

U.S. DOT - Appendix A Table 2 - Radionuclides
final RQ = 1000 curies (3.7E 13 Bq)

ENVIRONMENTAL LISTS

CERCLA/SARA List of Radionuclides (Appendix B) and Their Reportable Quantities
final RQ = 1000 curies (3.7E 13 Bq)

ZINC AMMONIUM CHLORIDE [VAN]   52628-25-8

HEALTH AND SAFETY LISTS

U.S. DOT - Appendix A Table 2 - Radionuclides
final RQ = 10 curies (3.7E 11 Bq)

ENVIRONMENTAL LISTS

CERCLA/SARA List of Radionuclides (Appendix B) and Their Reportable Quantities
final RQ = 10 curies (3.7E 11 Bq)

ZINC AMMONIUM NITRITE   63885-01-8

HEALTH AND SAFETY LISTS

U.S. DOT - Appendix A Table 2 - Radionuclides
final RQ = 100 curies (3.7E 12 Bq)

ENVIRONMENTAL LISTS

CERCLA/SARA List of Radionuclides (Appendix B) and Their Reportable Quantities
final RQ = 100 curies (3.7E 12 Bq)

ZINC AMMONIUM SULFATE   7783-24-6

HEALTH AND SAFETY LISTS

U.S. DOT - Appendix A Table 2 - Radionuclides
final RQ = 100 curies (3.7E 12 Bq)

ENVIRONMENTAL LISTS

CERCLA/SARA List of Radionuclides (Appendix B) and Their Reportable Quantities
final RQ = 100 curies (3.7E 12 Bq)

ZINC ARSENATE   1303-39-5

HEALTH AND SAFETY LISTS

U.S. DOT - Appendix A Table 2 - Radionuclides
final RQ = 100 curies (3.7E 12 Bq)

ENVIRONMENTAL LISTS

CERCLA/SARA List of Radionuclides (Appendix B) and Their Reportable Quantities
final RQ = 100 curies (3.7E 12 Bq)

ZINC ARSENITE   10326-24-6

HEALTH AND SAFETY LISTS

U.S. DOT - Appendix A Table 2 - Radionuclides
final RQ = 100 curies (3.7E 12 Bq)

ENVIRONMENTAL LISTS

CERCLA/SARA List of Radionuclides (Appendix B) and Their Reportable Quantities
final RQ = 100 curies (3.7E 12 Bq)

ZINC ASHES   RR-00517-4

SEE ALSO:
ZINC
ZINC COMPOUNDS
ZINC

HEALTH AND SAFETY LISTS

U.S. DOT - Appendix A Table 1 - Hazardous Substances
final RQ = 1000 pounds (454 kg)

NIOSH - Selected LD50s and LC50s
Oral, rat: LD50 = 2510 mg/kg

ENVIRONMENTAL LISTS

CERCLA/SARA - Hazardous Substances and their Reportable Quantities
final RQ = 1000 pounds (454 kg)

Clean Water Act - Hazardous Substances
[present]

STATE LISTS

Massachusetts Right To Know List
[present]

NJ Right to Know List (Total)
sn 2022

Pennsylvania Right to Know List
environmental hazard

ZINC BERYLLIUM SILICATE   39413-47-3

SEE ALSO:
ZINC AMMONIUM CHLORIDE [VAN]
ZINC

HEALTH AND SAFETY LISTS

U.S. DOT - Appendix A Table 1 - Hazardous Substances
final RQ = 1000 pounds (454 kg) (Listed under 'Zinc ammonium chloride')

ENVIRONMENTAL LISTS

CERCLA/SARA - Hazardous Substances and their Reportable Quantities
final RQ = 1000 pounds (454 kg) (Listed under 'Zinc ammonium chloride')

Clean Water Act - Hazardous Substances
[present] (Listed under 'Zinc ammonium chloride')

STATE LISTS

Massachusetts Right To Know List
[present]

Pennsylvania Right to Know List
environmental hazard

ZINC O,O-BIS(2-ETHYLHEXYL) PHOSPHORODITHIOATE   4259-15-8
SEE ALSO:
ZINC AMMONIUM CHLORIDE [VAN]
ZINC

HEALTH AND SAFETY LISTS

U.S. DOT - Appendix A Table 1 - Hazardous Substances
final RQ = 1000 pounds (454 kg) (Listed under 'Zinc ammonium chloride')

ENVIRONMENTAL LISTS

CERCLA/SARA - Hazardous Substances and their Reportable Quantities
final RQ = 1000 pounds (454 kg) (Listed under 'Zinc ammonium chloride')

Clean Water Act - Hazardous Substances
[present] (Listed under 'Zinc ammonium chloride')

STATE LISTS

Massachusetts Right To Know List
[present]

Pennsylvania Right to Know List
environmental hazard

ZINC BIS(PENTACHLOROPHENOL)   117-97-5

HEALTH AND SAFETY LISTS

U.S. DOT - Appendix A Table 1 - Hazardous Substances
final RQ = 1000 pounds (454 kg) (Listed under 'Zinc ammonium chloride')

ENVIRONMENTAL LISTS

CERCLA/SARA - Hazardous Substances and their Reportable Quantities
final RQ = 1000 pounds (454 kg) (Listed under 'Zinc ammonium chloride')

Clean Water Act - Hazardous Substances
[present] (Listed under 'Zinc ammonium chloride')

STATE LISTS

Massachusetts Right To Know List
[present]

NJ Right to Know List (Total)
sn 2859

Pennsylvania Right to Know List
environmental hazard

ZINC BORATE   1332-07-6

HEALTH AND SAFETY LISTS

U.S. DOT - Substances From 49 CFR 172.101
regulated by DOT (UN1512)

U.S. DOT - Hazard Classes
DOT hazard class = 5.1

STATE LISTS

NJ Right to Know List (Total)
sn 2023

ZINC BROMATE   14519-07-4

INTERNATIONAL LISTS

Canada - WHMIS: Ingredient Disclosure
1% item 1707 (1535)

ZINC BROMIDE   7699-45-8
SEE ALSO:
ZINC
ARSENIC

HEALTH AND SAFETY LISTS

U.S. DOT - Substances From 49 CFR 172.101
regulated by DOT (UN1712)

U.S. DOT - Hazard Classes
DOT hazard class = 6.1

STATE LISTS

NJ Right to Know List (Total)
sn 2024

ZINC CARBONATE   3486-35-9
SEE ALSO:
ARSENIC
ZINC

STATE LISTS

NJ Right to Know List (Total)
sn 2025

ZINC CHLORATE   10361-95-2

STATE LISTS

NJ Right to Know List (Total)
sn 2880

ZINC CHLORIDE   7646-85-7
SEE ALSO:
ZINC
BERYLLIUM

HEALTH AND SAFETY LISTS

NTP Seventh Report - Suspect Carcinogens
suspect carcinogen (Listed under 'Beryllium and certain beryllium compounds')

OSHA - Possible Select Carcinogens
[present]

STATE LISTS

Florida Hazardous Substance List
[present]

Massachusetts Right To Know List
carcinogen; extraordinarily hazardous

Pennsylvania Right to Know List
environmental hazard; special hazardous substance

Pennsylvania RTK - Special Hazardous Substances
[present]

ZINC CHROMATE   13530-65-9
SEE ALSO:
ZINC

ENVIRONMENTAL LISTS

EPA - Master Testing List
[present]

ZINC CHROMATE HYDROXIDE   15930-94-6
SEE ALSO:
ZINC

INTERNATIONAL LISTS

Canada - WHMIS: Ingredient Disclosure
1% item 1708 (776)

ZINC COMPOUNDS   RR-00578-7
SEE ALSO:
  ZINC COMPOUNDS
  ZINC

HEALTH AND SAFETY LISTS
  U.S. DOT - Appendix A Table 1 - Hazardous Substances
    final RQ = 1000 pounds (454 kg)

ENVIRONMENTAL LISTS
  CERCLA/SARA - Hazardous Substances and their Reportable Quantities
    final RQ = 1000 pounds (454 kg)
  Clean Water Act - Hazardous Substances
    [present]

STATE LISTS
  Massachusetts Right To Know List
    [present]
  NJ Right to Know List (Total)
    sn 2026
  Pennsylvania Right to Know List
    environmental hazard

ZINC CYANIDE   557-21-1

HEALTH AND SAFETY LISTS
  U.S. DOT - Substances From 49 CFR 172.101
    regulated by DOT (UN2469)
  U.S. DOT - Hazard Classes
    DOT hazard class = 5.1

STATE LISTS
  NJ Right to Know List (Total)
    sn 2882

ZINC DIBUTYL DITHIOCARBAMATE   136-23-2
SEE ALSO:
  ZINC

HEALTH AND SAFETY LISTS
  U.S. DOT - Appendix B - Marine Pollutants
    DOT regulated marine pollutant
  U.S. DOT - Appendix A Table 1 - Hazardous Substances
    final RQ = 1000 pounds (454 kg)

ENVIRONMENTAL LISTS
  CERCLA/SARA - Hazardous Substances and their Reportable Quantities
    final RQ = 1000 pounds (454 kg)
  Clean Water Act - Hazardous Substances
    [present]

STATE LISTS
  Massachusetts Right To Know List
    [present]
  NJ Right to Know List (Total)
    sn 2027
  Pennsylvania Right to Know List
    environmental hazard

ZINC, DICHLORO(4,4-DIMETHYL-5-((((METHYL AMINO)CARBONYL)OXY)IMINO)PENTANENITRILE) -, (T-4)-   58270-08-9
SEE ALSO:
  ZINC

HEALTH AND SAFETY LISTS
  U.S. DOT - Appendix A Table 1 - Hazardous Substances
    final RQ = 1000 pounds (454 kg)

ENVIRONMENTAL LISTS
  CERCLA/SARA - Hazardous Substances and their Reportable Quantities
    final RQ = 1000 pounds (454 kg)
  Clean Water Act - Hazardous Substances
    [present]
  List of Pesticide Product Inert Ingredients
    [present]

STATE LISTS
  Massachusetts Right To Know List
    [present]
  NJ Right to Know List (Total)
    sn 2028
  Pennsylvania Right to Know List
    environmental hazard

ZINC DICHROMATE   14018-95-2
SEE ALSO:
  ZINC

HEALTH AND SAFETY LISTS
  U.S. DOT - Substances From 49 CFR 172.101
    regulated by DOT (UN1513)
  U.S. DOT - Hazard Classes
    DOT hazard class = 5.1

STATE LISTS
  Florida Hazardous Substance List
    [present]
  Massachusetts Right To Know List
    [present]
  NJ Right to Know List (Total)
    sn 2029
  Pennsylvania Right to Know List
    environmental hazard

ZINC DIETHYLDITHIOCARBAMATE   14324-55-1
SEE ALSO:
  ZINC
  ZINC
  ZINC COMPOUNDS

HEALTH AND SAFETY LISTS
  ACGIH 1995 - Time Weighted Averages
    1 mg/m3 TWA
  ACGIH 1995 - Short Term Exposure Limits
    2 mg/m3 STEL
  U.S. DOT - Substances From 49 CFR 172.101
    regulated by DOT (UN2331, UN1840)
  U.S. DOT - Hazard Classes
    DOT hazard class = 8
  U.S. DOT - Appendix A Table 1 - Hazardous Substances
    final RQ = 1000 pounds (454 kg)
  NIOSH - Selected LD50s and LC50s
    Oral, rat: LD50 = 350 mg/kg
  NIOSH 1990 - Pocket Guide - RELs
    1 mg/m3 TWA; 2 mg/m3 STEL
  NIOSH 1990 - Pocket Guide - IDLHs
    4800 mg/m3 IDLH
  NIOSH 1990 - Pocket Guide - Target organs
    skin, eyes, respiratory system
  OSHA - Vacated PELs - Time Weighted Averages
    1 mg/m3 TWA
  OSHA - Vacated PELs - Short Term Exposure Limits
    2 mg/m3 STEL
  OSHA - Final PELs - Time Weighted Averages
    1 mg/m3 TWA

ENVIRONMENTAL LISTS

CERCLA/SARA - Hazardous Substances and their Reportable Quantities
final RQ = 1000 pounds (454 kg)

Clean Water Act - Hazardous Substances
[present]

INTERNATIONAL LISTS

Australian Exposure Standards - Time Weighted Averages
fume: 1 mg/m3 TWA

Australian Exposure Standards - Short Term Exposure Limits
fume: 2 mg/m3 STEL

Canada - WHMIS: Ingredient Disclosure
1% item 1709 (541)

Canada - Alberta - 8 Hour Occupational Exposure Limit
1 mg/m3 TWA

Canada - Alberta - 15 Minute Occupational Exposure Limit
fume: 2 mg/m3 STEL

Canada - British Columbia - 8 Hour Exposure Limits
1 mg/m3 TWA

Canada - British Columbia - 15 Minute Exposure Limits
2 mg/m3 STEL

Canada - Ontario - OHSA - TWAEVs
1 mg/m3 TWAEV

Canada - Ontario - OHSA - STEVs
2 mg/m3 STEV

Canada - Quebec - Time-Weighted Average Exposure Values
1 mg/m3 TWAEV

United Kingdom - Occupational Exposure Standards - TWAs
fume: 1 mg/m3 TWA

United Kingdom - Occupational Exposure Standards - STELs
fume: 2 mg/m3 STEL

Israel - Time Weighted Averages
1 mg/m3 TWA

Israel - Short Term Exposure Limits
2 mg/m3 STEL

Israel - Action Levels
0.5 mg/m3 AL

Mexico - Instruction No. 10 - TWAs
1 mg/m3 TWA

Mexico - Instruction No. 10 - STELs
2 mg/m3 STEL

STATE LISTS

California - Exposure Limits - PELs
fume: 1 mg/m3 PEL

California - Exposure Limits - STELs
fume: 2 mg/m3 STEL

Florida Hazardous Substance List
[present]

Massachusetts Right To Know List
[present]

Minnesota Hazardous Substance List
[present]

NJ Right to Know List (Total)
sn 2030

NJ Special Hazardous Substances
(corrosive)

Pennsylvania Right to Know List
environmental hazard

ZINC 2-ETHYLHEXOATE   136-53-8

SEE ALSO:
CHROMIUM
ZINC
CHROMIUM (VI) COMPOUNDS
ZINC COMPOUNDS
ZINC
CHROMIUM COMPOUNDS

HEALTH AND SAFETY LISTS

ACGIH 1995 - Time Weighted Averages
as Cr: 0.01 mg/m3 TWA

ACGIH 1995 - Carcinogens
as Cr: A1-confirmed human carcinogen

NTP Seventh Report - Known Carcinogens
known carcinogen (Listed under 'Chromium and certain chromium compounds')

OSHA - Vacated PELs - Ceiling Limits
as CrO3: C 0.1 mg/m3

OSHA - Select Carcinogens
[present]

INTERNATIONAL LISTS

Australian Exposure Standards - Time Weighted Averages
as Cr: 0.01 mg/m3 TWA

Australian Exposure Standards - Carcinogens
as Cr: confirmed carcinogen

Canada - WHMIS: Ingredient Disclosure
0.1% item 1710 (554)

Canada - Alberta - 8 Hour Occupational Exposure Limit
as Cr: 0.05 mg/m3 TWA

Canada - Alberta - 15 Minute Occupational Exposure Limit
as Cr: 0.15 mg/m3 STEL

Canada - Alberta - Designated Substances
designated substance - requires code of practice

Canada - British Columbia - 8 Hour Exposure Limits
as Cr: 0.05 mg/m3 TWA

Canada - British Columbia - Carcinogens
carcinogen - as Cr: 0.05 mg/m3 TWA

Canada - Quebec - Time-Weighted Average Exposure Values
as Cr: 0.01 mg/m3 TWAEV (substance of which the recirculation is prohibited)

Canada - Quebec - Carcinogens
as Cr: C1 carcinogen: effect detected in humans

German (DFG) - Carcinogens
proven carcinogen

Israel - Time Weighted Averages
as Cr: 0.01 mg/m3 TWA

Israel - Action Levels
as Cr: 0.005 mg/m3 AL

Mexico - Instruction No. 10 - TWAs
0.05 mg/m3 TWA

Mexico - Instruction No. 10 - Carcinogens
potential carcinogen in humans - limited epidemiological evidence

STATE LISTS

California - Exposure Limits - PELs
as Cr: 0.01 mg/m3 PEL

Florida Hazardous Substance List
[present]

Massachusetts Right To Know List
carcinogen; extraordinarily hazardous

Minnesota Hazardous Substance List
as Cr: carcinogen

NJ Right to Know List (Total)
sn 2031

NJ Special Hazardous Substances
(carcinogen)
Pennsylvania Right to Know List
environmental hazard; special hazardous substance
Pennsylvania RTK - Special Hazardous Substances
[present]

ZINC ETHYLPHENYLDITHIOCARBAMATE   14634-93-6

STATE LISTS

California - Exposure Limits - PELs
as Cr: 0.01 mg/m3 PEL

ZINC FLUORIDE   7783-49-5

ENVIRONMENTAL LISTS

CERCLA/SARA - Section 313 - Emission Reporting
form R reporting required for 1.0% de minimus concentration

Clean Water Act - Toxic Pollutants
[present] (Listed under 'Zinc and compounds')

INTERNATIONAL LISTS

Canada - NPRI (National Pollutant Release Inventory)
[present]

Canada - CEPA Schedule III Part II - Restricted Substances (Ocean Dumping)
[present]

STATE LISTS

California - Air Bill 2588 Appendix A-I
9/89

California - Directors List of Hazardous Substances (8 CCR 339)
[present] (exempt when present in motor oils at 2.5% or below. Zinc oxide is exempt when present as dust or when generated as a fume. Zinc stearate is exempt except when present as a dust)
NJ Right to Know List (Total)
sn 3012

ZINC FORMATE   557-41-5

SEE ALSO:
ZINC
CYANIDE ANION

HEALTH AND SAFETY LISTS

U.S. DOT - Substances From 49 CFR 172.101
regulated by DOT (UN1713)

U.S. DOT - Hazard Classes
DOT hazard class = 6.1

U.S. DOT - Appendix B - Marine Pollutants
DOT regulated marine pollutant

U.S. DOT - Appendix A Table 1 - Hazardous Substances
final RQ = 10 pounds (4.54 kg)

ENVIRONMENTAL LISTS

CERCLA/SARA - Hazardous Substances and their Reportable Quantities
final RQ = 10 pounds (4.54 kg)

Clean Water Act - Hazardous Substances
[present]

RCRA - P Series Wastes
waste number P121

RCRA - Hazardous Constituents-Appendix VIII
waste number P121

RCRA - Substances Banned From Land Disposal
[present]

INTERNATIONAL LISTS

Canada - WHMIS: Ingredient Disclosure
1% item 1711 (596)

STATE LISTS

Massachusetts Right To Know List
[present]

NJ Right to Know List (Total)
sn 2032

Pennsylvania Right to Know List
environmental hazard

ZINC HYDROSULFITE   7779-86-4

SEE ALSO:
ZINC COMPOUNDS
ZINC
ZINC

ENVIRONMENTAL LISTS

List of Pesticide Product Inert Ingredients
[present]

ZINC HYDROXIDE   20427-58-1

SEE ALSO:
ZINC

ENVIRONMENTAL LISTS

CERCLA/SARA - Section 302 Extremely Hazardous Substances and TPQs
TPQ = 100/10,000 pounds

STATE LISTS

Florida Hazardous Substance List
effective March 13, 1992

Massachusetts Right To Know List
extraordinarily hazardous

NJ Right to Know List (Total)
sn 2863

Pennsylvania Right to Know List
environmental hazard

PROPOSED REGULATIONS

CERCLA/SARA - Proposed Hazardous Substance Additions
proposed RQ = 1 pound (.454 kg)

CERCLA/SARA - 1989 Proposed RQ Adjustments
proposed RQ = 100 pounds (45.4 kg)

ZINC IRON OXIDE   12063-19-3

SEE ALSO:
CHROMIUM
ZINC

INTERNATIONAL LISTS

Canada - WHMIS: Ingredient Disclosure
0.1% item 1712 (689)

STATE LISTS

Florida Hazardous Substance List
[present]

Massachusetts Right To Know List
[present]

Minnesota Hazardous Substance List
carcinogen

Pennsylvania Right to Know List
environmental hazard

ZINC IRON YELLOW   68187-51-9

SEE ALSO:
ZINC
ZINC COMPOUNDS
ZINC

HEALTH AND SAFETY LISTS

NIOSH - Selected LD50s and LC50s
Oral, rat: LD50 = 3340 mg/kg

INTERNATIONAL LISTS

Canada - WHMIS: Ingredient Disclosure
1% item 1713 (703)

ZINC MERCAPTOBENZOTHIAZOLE   155-04-4
SEE ALSO:
ZINC COMPOUNDS
ZINC

ENVIRONMENTAL LISTS

List of Pesticide Product Inert Ingredients
[present]

ZINC NITRATE   7779-88-6
SEE ALSO:
ZINC

INTERNATIONAL LISTS

Canada - WHMIS: Ingredient Disclosure
1% item 1714 (874)

ZINC OCTOATE   557-09-5
SEE ALSO:
ZINC

HEALTH AND SAFETY LISTS

U.S. DOT - Appendix A Table 1 - Hazardous Substances
final RQ = 1000 pounds (454 kg)

ENVIRONMENTAL LISTS

CERCLA/SARA - Hazardous Substances and their Reportable Quantities
final RQ = 1000 pounds (454 kg)

Clean Water Act - Hazardous Substances
[present]

INTERNATIONAL LISTS

Canada - WHMIS: Ingredient Disclosure
1% item 1715 (913)

STATE LISTS

Massachusetts Right To Know List
[present]

NJ Right to Know List (Total)
sn 2034

Pennsylvania Right to Know List
environmental hazard

ZINC OXIDE   1314-13-2
SEE ALSO:
ZINC

HEALTH AND SAFETY LISTS

U.S. DOT - Appendix A Table 1 - Hazardous Substances
final RQ = 1000 pounds (454 kg)

ENVIRONMENTAL LISTS

CERCLA/SARA - Hazardous Substances and their Reportable Quantities
final RQ = 1000 pounds (454 kg)

Clean Water Act - Hazardous Substances
[present]

STATE LISTS

Massachusetts Right To Know List
[present]

NJ Right to Know List (Total)
sn 2035

Pennsylvania Right to Know List
environmental hazard

ZINC PERMANGANTE   23414-72-4
SEE ALSO:
ZINC

HEALTH AND SAFETY LISTS

U.S. DOT - Substances From 49 CFR 172.101
regulated by DOT (UN1931)

U.S. DOT - Hazard Classes
DOT hazard class = 9

U.S. DOT - Appendix A Table 1 - Hazardous Substances
final RQ = 1000 pounds (454 kg)

ENVIRONMENTAL LISTS

CERCLA/SARA - Hazardous Substances and their Reportable Quantities
final RQ = 1000 pounds (454 kg)

Clean Water Act - Hazardous Substances
[present]

STATE LISTS

Massachusetts Right To Know List
[present]

NJ Right to Know List (Total)
sn 2033

Pennsylvania Right to Know List
environmental hazard

ZINC PEROXIDE   1314-22-3
SEE ALSO:
ZINC COMPOUNDS
ZINC
ZINC

ENVIRONMENTAL LISTS

List of Pesticide Product Inert Ingredients
[present]

INTERNATIONAL LISTS

Canada - WHMIS: Ingredient Disclosure
1% item 1716 (1002)

ZINC PHENOLSULFONATE   127-82-2
SEE ALSO:
ZINC
ZINC COMPOUNDS

ENVIRONMENTAL LISTS

List of Pesticide Product Inert Ingredients
[present]

ZINC PHOSPHIDE   1314-84-7
SEE ALSO:
ZINC
IRON
ZINC
ZINC COMPOUNDS

ENVIRONMENTAL LISTS

List of Pesticide Product Inert Ingredients
[present]

ZINC PHOSPHIDE   51810-70-9
SEE ALSO:
ZINC

HEALTH AND SAFETY LISTS

NIOSH - Selected LD50s and LC50s
Oral, rat: LD50 = 540 mg/kg

ZINC PHOSPHITE   **14332-59-3**
SEE ALSO:
ZINC
ZINC
ZINC COMPOUNDS

HEALTH AND SAFETY LISTS
U.S. DOT - Substances From 49 CFR 172.101
regulated by DOT (UN1514)
U.S. DOT - Hazard Classes
DOT hazard class = 5.1
U.S. DOT - Appendix A Table 1 - Hazardous Substances
final RQ = 1000 pounds (454 kg)

ENVIRONMENTAL LISTS
CERCLA/SARA - Hazardous Substances and their Reportable Quantities
final RQ = 1000 pounds (454 kg)
Clean Water Act - Hazardous Substances
[present]

STATE LISTS
Massachusetts Right To Know List
[present]
NJ Right to Know List (Total)
sn 2036
Pennsylvania Right to Know List
environmental hazard

ZINC POTASSIUM CHROMATE   **11103-86-9**
SEE ALSO:
ZINC

HEALTH AND SAFETY LISTS
NIOSH - Selected LD50s and LC50s
Oral, mouse: LD50 = 2370 mg/kg

ENVIRONMENTAL LISTS
List of Pesticide Product Inert Ingredients
[present]

ZINC POTASSIUM CYANIDE   **14244-62-3**
SEE ALSO:
ZINC
ZINC COMPOUNDS
ZINC

HEALTH AND SAFETY LISTS
ACGIH 1995 - Time Weighted Averages
fume: 5 mg/m3 TWA; dust: 10 mg/m3 TWA (The value for Zinc oxide 'dust' is for total dust containing no asbestos and < 1% crystalline silica)
ACGIH 1995 - Short Term Exposure Limits
fume: 10 mg/m3 STEL
NIOSH - Selected LD50s and LC50s
Oral, mouse: LD50 = 7950 mg/kg
NIOSH 1990 - Pocket Guide - RELs
5 mg/m3 TWA; 10 mg/m3 STEL
NIOSH 1990 - Pocket Guide - Target organs
respiratory system
NIOSH - Health Standards - Exposure Limits
5 mg ZnO/m3 TWA; C (15 min) 15 mg ZnO/m3
NIOSH - Health Standards - Health Effects and Precautions
Metal fume fever
OSHA - Vacated PELs - Time Weighted Averages
fume: 5 mg/m3 TWA; total dust: 10 mg/m3 TWA; respirable fraction: 5 mg/m3 TWA
OSHA - Vacated PELs - Short Term Exposure Limits
fume: 10 mg/m3 STEL
OSHA - Final PELs - Time Weighted Averages
fume: 5 mg/m3 TWA; total dust: 15 mg/m3 TWA; respirable fraction: 5 mg/m3 TWA

ENVIRONMENTAL LISTS
List of Pesticide Product Inert Ingredients
[present]

INTERNATIONAL LISTS
Australian Exposure Standards - Time Weighted Averages
dust: 10 mg/m3 TWA; fume: 5 mg/m3 TWA
Australian Exposure Standards - Short Term Exposure Limits
fume: 10 mg/m3 STEL
Canada - WHMIS: Ingredient Disclosure
1% item 1717 (1326)
Canada - Alberta - 8 Hour Occupational Exposure Limit
as fume: 5 mg/m3 TWA; as dust: respirable mass: 5 mg/m3 TWA; total mass: 10 mg/m3 TWA
Canada - Alberta - 15 Minute Occupational Exposure Limit
10 mg/m3 STEL
Canada - British Columbia - 8 Hour Exposure Limits
fume: 5 mg/m3 TWA; dust: 10 mg/m3 TWA
Canada - British Columbia - 15 Minute Exposure Limits
fume: 10 mg/m3 STEL
Canada - Ontario - OHSA - TWAEVs
fume: 5 mg/m3 TWAEV; dust (listed as a nuisance particulate): 10 mg/m3 TWAEV
Canada - Ontario - OHSA - STEVs
fume: 10 mg/m3 STEV
Canada - Quebec - Time-Weighted Average Exposure Values
fume: 5 mg/m3 TWAEV; dust: 10 mg/m3 TWAEV
Canada - Quebec - Short-term Exposure Values
10 mg/m3 STEV
United Kingdom - Occupational Exposure Standards - TWAs
fume: 5 mg/m3 TWA
United Kingdom - Occupational Exposure Standards - STELs
fume: 10 mg/m3 STEL
German (DFG) - MAK Values
fine dust: 5 mg/m3 MAK
German (DFG) - Peak Limitations
10 x normal MAK (30 min average value); don't exceed during shift
Israel - Time Weighted Averages
fume: 5 mg/m3 TWA; dust: 10 mg/m3 TWA (The value for Zinc oxide 'dust' is for total dust containing no asbestos and < 1% crystalline silica)
Israel - Short Term Exposure Limits
fume: 10 mg/m3 STEL
Israel - Action Levels
fume: 2.5 mg/m3 AL; dust: 5 mg/m3 AL
Mexico - Instruction No. 10 - TWAs
as fume: 5 mg/m3 TWA as dust: 10 mg/m3 TWA
Mexico - Instruction No. 10 - STELs
10 mg/m3 STEL

STATE LISTS
California - Air Bill 2588 Appendix A-I
[present]
California - Exposure Limits - PELs
fume: 5 mg/m3 PEL
California - Exposure Limits - STELs
fume: 10 mg/m3 STEL
Florida Hazardous Substance List
[present]
Massachusetts Right To Know List
[present]
Minnesota Hazardous Substance List
[present] (includes inert or nuisance dust)

Pennsylvania Right to Know List
environmental hazard

PROPOSED REGULATIONS

Canada - Ontario - Proposed Occupational TWAEVs
total dust: 5 mg/m3 TWAEV

ZINC PROPIONATE   557-28-8
SEE ALSO:
ZINC
MANGANESE

HEALTH AND SAFETY LISTS

U.S. DOT - Substances From 49 CFR 172.101
regulated by DOT (UN1515)
U.S. DOT - Hazard Classes
DOT hazard class = 5.1

INTERNATIONAL LISTS

Canada - WHMIS: Ingredient Disclosure
1% item 1718 (1360)

STATE LISTS

NJ Right to Know List (Total)
sn 2038

ZINC PYRITHIONE   13463-41-7
SEE ALSO:
ZINC

HEALTH AND SAFETY LISTS

U.S. DOT - Substances From 49 CFR 172.101
regulated by DOT (UN1516)
U.S. DOT - Hazard Classes
DOT hazard class = 5.1

STATE LISTS

NJ Right to Know List (Total)
sn 2039

ZINC PYROPHOSPHATE   7446-26-6
SEE ALSO:
ZINC

HEALTH AND SAFETY LISTS

U.S. DOT - Appendix A Table 1 - Hazardous Substances
final RQ = 5000 pounds (2270 kg)

ENVIRONMENTAL LISTS

CERCLA/SARA - Hazardous Substances and their Reportable Quantities
final RQ = 5000 pounds (2270 kg)
Clean Water Act - Hazardous Substances
[present]

STATE LISTS

Massachusetts Right To Know List
[present]
NJ Right to Know List (Total)
sn 2040
Pennsylvania Right to Know List
environmental hazard

ZINC RESINATE   9010-69-9
SEE ALSO:
ZINC

HEALTH AND SAFETY LISTS

U.S. DOT - Substances From 49 CFR 172.101
regulated by DOT (UN1714)
U.S. DOT - Hazard Classes
DOT hazard class = 4.3
U.S. DOT - Appendix A Table 1 - Hazardous Substances
final RQ = 100 pounds (45.4 kg)

NIOSH - Selected LD50s and LC50s
Oral, rat: LD50 = 12 mg/kg

ENVIRONMENTAL LISTS

CERCLA/SARA - Section 302 Extremely Hazardous Substances and TPQs
TPQ = 500 pounds (This material is a reactive solid. The TPQ does not default to 10,000 pounds for non-powder, non-molten, non-solution form)
CERCLA/SARA - Hazardous Substances and their Reportable Quantities
final RQ = 100 pounds (45.4 kg)
Clean Water Act - Hazardous Substances
[present]
RCRA - P Series Wastes
when present at concentrations greater than 10%: waste number P122 (Reactive waste; Toxic waste)
RCRA - U Series Wastes
when present at concentrations of 10% or less: waste number U249
RCRA - Hazardous Constituents-Appendix VIII
when present at concentrations greater than 10%: waste number P122; when present at concentrations less than 10%: waste number U249
RCRA - Substances Banned From Land Disposal
[present]

INTERNATIONAL LISTS

Canada - WHMIS: Ingredient Disclosure
1% item 1719 (1414)

STATE LISTS

Florida Hazardous Substance List
effective March 13, 1992
Massachusetts Right To Know List
extraordinarily hazardous
NJ Right to Know List (Total)
sn 2041
Pennsylvania Right to Know List
environmental hazard

ZINC SELENATE   RR-00758-9
STATE LISTS

Pennsylvania Right to Know List
environmental hazard

ZINC SELENITE   13597-46-1
SEE ALSO:
ZINC

INTERNATIONAL LISTS

Canada - WHMIS: Ingredient Disclosure
1% item 1720 (1406)

ZINC SILICOFLUORIDE   16871-71-9
SEE ALSO:
ZINC
CHROMIUM
CHROMIUM COMPOUNDS
ZINC COMPOUNDS
CHROMIUM (VI) COMPOUNDS
ZINC

HEALTH AND SAFETY LISTS

ACGIH 1995 - Time Weighted Averages
as Cr: 0.01 mg/m3 TWA
ACGIH 1995 - Carcinogens
as Cr: A1-confirmed human carcinogen

INTERNATIONAL LISTS

Israel - Time Weighted Averages
as Cr: 0.01 mg/m3 TWA

Israel - Action Levels
as Cr: 0.005 mg/m3 AL

STATE LISTS

California - Exposure Limits - PELs
as Cr: 0.01 mg/m3 PEL

Minnesota Hazardous Substance List
as Cr: carcinogen

NJ Right to Know List (Total)
sn 2042

ZINC STEARATE   557-05-1

SEE ALSO:
ZINC

INTERNATIONAL LISTS

Canada - WHMIS: Ingredient Disclosure
1% item 1721 (597)

ZINC SULFATE   7733-02-0

SEE ALSO:
ZINC

INTERNATIONAL LISTS

Canada - WHMIS: Ingredient Disclosure
0.1% item 1722 (1451)

ZINC SULFATE, BASIC   59766-35-7

SEE ALSO:
ZINC

HEALTH AND SAFETY LISTS

NIOSH - Selected LD50s and LC50s
Oral, rat: LD50 = 177 mg/kg

INTERNATIONAL LISTS

Canada - WHMIS: Ingredient Disclosure
0.1% item 1723 (1460)

ZINC SULFIDE   1314-98-3

SEE ALSO:
ZINC

ENVIRONMENTAL LISTS

List of Pesticide Product Inert Ingredients
[present]

ZINC YELLOW (ZINC CHROMATE PIGMENT)   37300-23-5

SEE ALSO:
ZINC

HEALTH AND SAFETY LISTS

U.S. DOT - Substances From 49 CFR 172.101
regulated by DOT (UN2714)

U.S. DOT - Hazard Classes
DOT hazard class = 4.1

ENVIRONMENTAL LISTS

List of Pesticide Product Inert Ingredients
[present]

STATE LISTS

NJ Right to Know List (Total)
sn 2883

ZINEB   12122-67-7

STATE LISTS

NJ Right to Know List (Total)
sn 2864

ZIRAM   137-30-4

STATE LISTS

NJ Right to Know List (Total)
sn 2865

ZIRCON   14940-68-2

SEE ALSO:
ZINC

HEALTH AND SAFETY LISTS

U.S. DOT - Substances From 49 CFR 172.101
regulated by DOT (UN2855)

U.S. DOT - Hazard Classes
DOT hazard class = 6.1

U.S. DOT - Appendix A Table 1 - Hazardous Substances
final RQ = 5000 pounds (2270 kg)

ENVIRONMENTAL LISTS

CERCLA/SARA - Hazardous Substances and their Reportable Quantities
final RQ = 5000 pounds (2270 kg)

Clean Water Act - Hazardous Substances
[present]

INTERNATIONAL LISTS

Canada - WHMIS: Ingredient Disclosure
1% item 1724 (1496)

STATE LISTS

Massachusetts Right To Know List
[present]

NJ Right to Know List (Total)
sn 2043

Pennsylvania Right to Know List
environmental hazard

ZIRCONIUM   7440-67-7

SEE ALSO:
ZINC
ZINC
ZINC COMPOUNDS

HEALTH AND SAFETY LISTS

NFPA - Flash Points
flash point = 530 degrees F (277 degrees C)

NFPA - Hazard Identification Ratings
health-0; flammability-1; reactivity-0

OSHA - Vacated PELs - Time Weighted Averages
total dust: 10 mg/m3 TWA; respirable fraction: 5 mg/m3 TWA

OSHA - Final PELs - Time Weighted Averages
total dust: 15 mg/m3 TWA; respirable fraction: 5 mg/m3 TWA

ENVIRONMENTAL LISTS

List of Pesticide Product Inert Ingredients
[present]

INTERNATIONAL LISTS

Canada - WHMIS: Ingredient Disclosure
1% item 1725 (1504)

Canada - Alberta - 8 Hour Occupational Exposure Limit
respirable mass: 5 mg/m3 TWA; total mass: 10 mg/m3 TWA

Canada - British Columbia - 8 Hour Exposure Limits
10 mg/m3 TWA

Canada - British Columbia - 15 Minute Exposure Limits
20 mg/m3 STEL

Canada - Quebec - Time-Weighted Average Exposure Values
10 mg/m3 TWAEV

United Kingdom - Occupational Exposure Standards - TWAs
total inhalable dust: 10 mg/m3 TWA; respirable dust: 5 mg/m3 TWA

United Kingdom - Occupational Exposure Standards - STELs
total inhalable dust: 20 mg/m3 STEL

Mexico - Instruction No. 10 - TWAs
10 mg/m3 TWA; (nuisance particulate)
Mexico - Instruction No. 10 - STELs
20 mg/m3 STEL

STATE LISTS

California - Exposure Limits - PELs
10 mg/m3 PEL
Massachusetts Right To Know List
[present]
Minnesota Hazardous Substance List
[present] (includes inert or nuisance dust)
Pennsylvania Right to Know List
environmental hazard

ZIRCONIUM 86   15743-56-3
SEE ALSO:
ZINC COMPOUNDS
ZINC
ZINC

HEALTH AND SAFETY LISTS

U.S. DOT - Appendix A Table 1 - Hazardous Substances
final RQ = 1000 pounds (454 kg)
NIOSH - Selected LD50s and LC50s
Oral, rat: LD50 = 2949 mg/kg

ENVIRONMENTAL LISTS

CERCLA/SARA - Hazardous Substances and their Reportable Quantities
final RQ = 1000 pounds (454 kg)
Clean Water Act - Hazardous Substances
[present]

INTERNATIONAL LISTS

Canada - WHMIS: Ingredient Disclosure
1% item 1726 (1534)

STATE LISTS

Massachusetts Right To Know List
[present]
NJ Right to Know List (Total)
sn 2044
Pennsylvania Right to Know List
environmental hazard

ZIRCONIUM 88   14681-75-5

ENVIRONMENTAL LISTS

List of Pesticide Product Inert Ingredients
[present]

ZIRCONIUM 89   13981-27-6
SEE ALSO:
ZINC
ZINC COMPOUNDS
ZINC

ENVIRONMENTAL LISTS

List of Pesticide Product Inert Ingredients
[present]

ZIRCONIUM 93   15751-77-6
SEE ALSO:
CHROMIUM
CHROMIUM COMPOUNDS
ZINC COMPOUNDS
ZINC
CHROMIUM (VI) COMPOUNDS
ZINC
ZINC CHROMATE

HEALTH AND SAFETY LISTS

ACGIH 1995 - Time Weighted Averages
as Cr: 0.01 mg/m3 TWA
ACGIH 1995 - Carcinogens
as Cr: A1-confirmed human carcinogen

INTERNATIONAL LISTS

Canada - WHMIS: Ingredient Disclosure
1% item 1727 (1059)
Israel - Time Weighted Averages
as Cr: 0.01 mg/m3 TWA
Israel - Action Levels
as Cr: 0.005 mg/m3 AL

STATE LISTS

California - Exposure Limits - PELs
as Cr: 0.01 mg/m3 PEL
Minnesota Hazardous Substance List
as Cr: carcinogen

ZIRCONIUM 95   13967-71-0
SEE ALSO:
ZINC

HEALTH AND SAFETY LISTS

IARC - Group 3 (not classifiable)
[present]
NIOSH - Selected LD50s and LC50s
Oral, rat: LD50 = 5200 mg/kg

ENVIRONMENTAL LISTS

CERCLA/SARA - Section 313 - Emission Reporting
form R reporting required for 1.0% de minimus concentration

STATE LISTS

California - Air Bill 2588 Appendix A-II
known or potential carcinogen: 9/90
California - Prop. 65 - Cancer list
carcinogen - initial date 1/1/90
California - Directors List of Hazardous Substances (8 CCR 339)
[present]
Massachusetts Right To Know List
[present]
NJ Right to Know List (Total)
sn 2045
NJ Special Hazardous Substances
(carcinogen; mutagen)
Pennsylvania Right to Know List
environmental hazard

ZIRCONIUM 97   14928-30-4
SEE ALSO:
ZINC
ZINC COMPOUNDS
ZINC

HEALTH AND SAFETY LISTS

IARC - Group 3 (not classifiable)
[present]
NIOSH - Selected LD50s and LC50s
Oral, rat: LD50 = 1400 mg/kg

NTP Chemical Status Reports - Testing Status and NTIS Number
Technical reports printed (PB83202622)

NTP Chemical Status Reports - Evidence of Carcinogenicity
male rat-positive; female rat-negative; male mice-negative; female mice-equivocal

INTERNATIONAL LISTS

Canada - WHMIS: Ingredient Disclosure
1% item 1728 (1733)

STATE LISTS

Massachusetts Right To Know List
neurotoxin

ZIRCONIUM ACETYLACETONATE   17501-44-9
SEE ALSO:
ZIRCONIUM

INTERNATIONAL LISTS

Canada - WHMIS: Ingredient Disclosure
1% item 1729 (1734)

ZIRCONIUM[IV], [2,2-BIS[[2-PROPENY-LOXY]METHYL]-1-BUTANOLATO-1,2]TRIS(2-PROPENOATO-O-)   RR-01252-2

HEALTH AND SAFETY LISTS

ACGIH 1995 - Time Weighted Averages
as Zr: 5 mg/m3 TWA

ACGIH 1995 - Short Term Exposure Limits
as Zr: 10 mg/m3 STEL

U.S. DOT - Substances From 49 CFR 172.101
regulated by DOT (UN1358, UN1308, UN1932, UN2008, UN2009, UN2858)

U.S. DOT - Hazard Classes
finished metal sheets, strip or collected wire: DOT hazard class = 4.2; sheets, strip or wire between 18 and 254 microns thick: DOT hazard class = 4.2; dry powder: DOT hazard class = 4.2; wetted powder: DOT hazard class = 4.1; metal scrap: DOT hazard class = 4.2

NIOSH 1990 - Pocket Guide - RELs
as Zr: 5 mg/m3 TWA; 10 mg/m3 STEL

NIOSH 1990 - Pocket Guide - IDLHs
as Zr: 500 mg/m3 IDLH

NIOSH 1990 - Pocket Guide - Target organs
skin, respiratory system

OSHA - Vacated PELs - Time Weighted Averages
as Zr: 5 mg/m3 TWA

OSHA - Vacated PELs - Short Term Exposure Limits
as Zr: 10 mg/m3 STEL

OSHA - Final PELs - Time Weighted Averages
as Zr: 5 mg/m3 TWA

INTERNATIONAL LISTS

Australian Exposure Standards - Time Weighted Averages
as Zr: 5 mg/m3 TWA

Australian Exposure Standards - Short Term Exposure Limits
as Zr: 10 mg/m3 STEL

Canada - WHMIS: Ingredient Disclosure
1% item 1733 (1736)

Canada - Alberta - 8 Hour Occupational Exposure Limit
as Zr: 5 mg/m3 TWA

Canada - Alberta - 15 Minute Occupational Exposure Limit
as Zr: 10 mg/m3 STEL

Canada - British Columbia - 8 Hour Exposure Limits
as Zr: 5 mg/m3 TWA

Canada - British Columbia - 15 Minute Exposure Limits
as Zr: 10 mg/m3 STEL

Canada - Ontario - OHSA - TWAEVs
as Zr: 5 mg/m3 TWAEV

Canada - Ontario - OHSA - STEVs
as Zr: 10 mg/m3 STEV

Canada - Quebec - Time-Weighted Average Exposure Values
as Zr: 5 mg/m3 TWAEV

Canada - Quebec - Short-term Exposure Values
as Zr: 10 mg/m3 STEV

German (DFG) - MAK Values
total dust, as Zr: 5 mg/m3 MAK

German (DFG) - Peak Limitations
10 x normal MAK (30 min average value); don't exceed during shift

Israel - Time Weighted Averages
as Zr: 5 mg/m3 TWA

Israel - Short Term Exposure Limits
as Zr: 10 mg/m3 STEL

Israel - Action Levels
as Zr: 2.5 mg/m3 AL

Mexico - Instruction No. 10 - TWAs
5 mg/m3 TWA

Mexico - Instruction No. 10 - STELs
10 mg/m3 STEL

STATE LISTS

California - Directors List of Hazardous Substances (8 CCR 339)
[present]

Florida Hazardous Substance List
[present]

Massachusetts Right To Know List
[present]

Minnesota Hazardous Substance List
[present] as Zr

NJ Right to Know List (Total)
sn 2047

NJ Special Hazardous Substances
(flammable - fourth degree)

Pennsylvania Right to Know List
[present]

ZIRCONIUM CHLORIDE HYDROXIDE   10119-31-0

HEALTH AND SAFETY LISTS

U.S. DOT - Appendix A Table 2 - Radionuclides
final RQ = 100 curies (3.7E 12 Bq)

ENVIRONMENTAL LISTS

CERCLA/SARA List of Radionuclides (Appendix B) and Their Reportable Quantities
final RQ = 100 curies (3.7E 12 Bq)

ZIRCONIUM COMPOUNDS, N.O.S.   RR-00624-6

HEALTH AND SAFETY LISTS

U.S. DOT - Appendix A Table 2 - Radionuclides
final RQ = 10 curies (3.7E 11 Bq)

ENVIRONMENTAL LISTS

CERCLA/SARA List of Radionuclides (Appendix B) and Their Reportable Quantities
final RQ = 10 curies (3.7E 11 Bq)

ZIRCONIUM ETHYL HEXOATE   22464-99-9

HEALTH AND SAFETY LISTS

U.S. DOT - Appendix A Table 2 - Radionuclides
final RQ = 100 curies (3.7E 12 Bq)

ENVIRONMENTAL LISTS

CERCLA/SARA List of Radionuclides (Appendix B) and Their Reportable Quantities
final RQ = 100 curies (3.7E 12 Bq)

ZIRCONIUM HYDRIDE   7704-99-6

HEALTH AND SAFETY LISTS

U.S. DOT - Appendix A Table 2 - Radionuclides
final RQ = 1 curie (3.7E 10 Bq)

ENVIRONMENTAL LISTS

CERCLA/SARA List of Radionuclides (Appendix B) and Their Reportable Quantities
final RQ = 1 curie (3.7E 10 Bq)

ZIRCONIUM HYDRIDE (VAN)   11105-16-1

HEALTH AND SAFETY LISTS

U.S. DOT - Appendix A Table 2 - Radionuclides
final RQ = 10 curies (3.7E 11 Bq)

ENVIRONMENTAL LISTS

CERCLA/SARA List of Radionuclides (Appendix B) and Their Reportable Quantities
final RQ = 10 curies (3.7E 11 Bq)

ZIRCONIUM NAPHTHENATE   72854-21-8

HEALTH AND SAFETY LISTS

U.S. DOT - Appendix A Table 2 - Radionuclides
final RQ = 10 curies (3.7E 11 Bq)

ENVIRONMENTAL LISTS

CERCLA/SARA List of Radionuclides (Appendix B) and Their Reportable Quantities
final RQ = 10 curies (3.7E 11 Bq)

ZIRCONIUM NITRATE   13746-89-9
SEE ALSO:
ZIRCONIUM

INTERNATIONAL LISTS

Canada - WHMIS: Ingredient Disclosure
1% item 1730 (45)

ZIRCONIUM OXYCHLORIDE   7699-43-6

ENVIRONMENTAL LISTS

TSCA - Chemicals with Significant New Use Rules
PMN number: P-91-389

ZIRCONIUM PICRAMATE   63868-82-6
SEE ALSO:
ZIRCONIUM

INTERNATIONAL LISTS

Canada - WHMIS: Ingredient Disclosure
1% item 1731 (987)

ZIRCONIUM POTASSIUM FLUORIDE   16923-95-8

INTERNATIONAL LISTS

Canada - WHMIS: Ingredient Disclosure
1% item 1732 (1735)

United Kingdom - Occupational Exposure Standards - TWAs
as Zr: 5 mg/m3 TWA

United Kingdom - Occupational Exposure Standards - STELs
as Zr: 10 mg/m3 STEL

STATE LISTS

California - Exposure Limits - PELs
as Zr: 5 mg/m3 PEL

California - Exposure Limits - STELs
as Zr: 10 mg/m3 STEL

California - Directors List of Hazardous Substances (8 CCR 339)
[present]

ZIRCONIUM SODIUM LACTATE   10377-98-7
SEE ALSO:
ZIRCONIUM

ENVIRONMENTAL LISTS

List of Pesticide Product Inert Ingredients
[present]

ZIRCONIUM SULFATE   14475-73-1

STATE LISTS

NJ Right to Know List (Total)
sn 2048

ZIRCONIUM SULFATE   14644-61-2

HEALTH AND SAFETY LISTS

U.S. DOT - Substances From 49 CFR 172.101
regulated by DOT (UN1437)

U.S. DOT - Hazard Classes
DOT hazard class = 4.1

ZIRCONIUM TETRACHLORIDE   10026-11-6

ENVIRONMENTAL LISTS

List of Pesticide Product Inert Ingredients
[present]

ZIRCONYL NITRATE   13826-66-9
SEE ALSO:
ZIRCONIUM

HEALTH AND SAFETY LISTS

U.S. DOT - Substances From 49 CFR 172.101
regulated by DOT (UN2728)

U.S. DOT - Hazard Classes
DOT hazard class = 5.1

U.S. DOT - Appendix A Table 1 - Hazardous Substances
final RQ = 5000 pounds (2270 kg)

NIOSH - Selected LD50s and LC50s
Oral, rat: LD50 = 2290 mg/kg

ENVIRONMENTAL LISTS

CERCLA/SARA - Hazardous Substances and their Reportable Quantities
final RQ = 5000 pounds (2270 kg)

Clean Water Act - Hazardous Substances
[present]

STATE LISTS

Massachusetts Right To Know List
[present]

NJ Right to Know List (Total)
sn 2049

Pennsylvania Right to Know List
environmental hazard

ZIRCONIUM OXYCHLORIDE   7699-43-6
SEE ALSO:
ZIRCONIUM

HEALTH AND SAFETY LISTS

NIOSH - Selected LD50s and LC50s
Oral, rat: LD50 = 3500 mg/kg

ZIRCONIUM PICRAMATE   63868-82-6
SEE ALSO:
ZIRCONIUM

HEALTH AND SAFETY LISTS

U.S. DOT - Substances From 49 CFR 172.101
regulated by DOT (UN1517, UN0236)

U.S. DOT - Hazard Classes
DOT hazard class = 1.3C

STATE LISTS

Massachusetts Right To Know List
[present]

NJ Right to Know List (Total)
sn 2050

ZIRCONIUM POTASSIUM FLUORIDE   16923-95-8
SEE ALSO:
ZIRCONIUM

HEALTH AND SAFETY LISTS

U.S. DOT - Appendix A Table 1 - Hazardous Substances
final RQ = 1000 pounds (454 kg)

NIOSH - Selected LD50s and LC50s
Oral, mouse: LD50 = 98 mg/kg

ENVIRONMENTAL LISTS

CERCLA/SARA - Hazardous Substances and their Reportable Quantities
final RQ = 1000 pounds (454 kg)

Clean Water Act - Hazardous Substances
[present]

INTERNATIONAL LISTS

Canada - WHMIS: Ingredient Disclosure
1% item 1734 (900)

STATE LISTS

Massachusetts Right To Know List
[present]

NJ Right to Know List (Total)
sn 2051

Pennsylvania Right to Know List
environmental hazard

ZIRCONIUM SODIUM LACTATE   10377-98-7
SEE ALSO:
ZIRCONIUM

INTERNATIONAL LISTS

Canada - WHMIS: Ingredient Disclosure
1% item 1735 (1063)

ZIRCONIUM SULFATE   14475-73-1

HEALTH AND SAFETY LISTS

U.S. DOT - Substances From 49 CFR 172.101
regulated by DOT (NA9163)

U.S. DOT - Hazard Classes
DOT hazard class = 8

ZIRCONIUM SULFATE   14644-61-2
SEE ALSO:
ZIRCONIUM

HEALTH AND SAFETY LISTS

U.S. DOT - Appendix A Table 1 - Hazardous Substances
final RQ = 5000 pounds (2270 kg)

ENVIRONMENTAL LISTS

CERCLA/SARA - Hazardous Substances and their Reportable Quantities
final RQ = 5000 pounds (2270 kg)

Clean Water Act - Hazardous Substances
[present]

STATE LISTS

Massachusetts Right To Know List
[present]
NJ Right to Know List (Total)
sn 2052

Pennsylvania Right to Know List
environmental hazard

ZIRCONIUM TETRACHLORIDE   10026-11-6
SEE ALSO:
ZIRCONIUM

HEALTH AND SAFETY LISTS

U.S. DOT - Substances From 49 CFR 172.101
regulated by DOT (UN2503)

U.S. DOT - Hazard Classes
DOT hazard class = 8

U.S. DOT - Appendix A Table 1 - Hazardous Substances
final RQ = 5000 pounds (2270 kg)

NIOSH - Selected LD50s and LC50s
Oral, rat: LD50 = 1688 mg/kg

ENVIRONMENTAL LISTS

CERCLA/SARA - Hazardous Substances and their Reportable Quantities
final RQ = 5000 pounds (2270 kg)

Clean Water Act - Hazardous Substances
[present]

INTERNATIONAL LISTS

Canada - WHMIS: Ingredient Disclosure
1% item 1736 (1586)

STATE LISTS

California - Directors List of Hazardous Substances (8 CCR 339)
[present]

Florida Hazardous Substance List
[present]

Massachusetts Right To Know List
[present]

NJ Right to Know List (Total)
sn 2053

NJ Special Hazardous Substances
(corrosive)

Pennsylvania Right to Know List
environmental hazard

ZIRCONYL NITRATE   13826-66-9
SEE ALSO:
ZIRCONIUM

HEALTH AND SAFETY LISTS

NIOSH - Selected LD50s and LC50s
Oral, rat: LD50 = 2500 mg/kg

GLOSSARY

ACGIH - American Conference of Governmental Industrial Hygienists is an organization of professional personnel in governmental agencies or educational institutions engaged in occupational safety and health programs. ACGIH establishes recommended occupational exposure limits for chemical substances and physical agents. See TLV.

Action Levels - Levels of exposure at which OSHA regulations for protective programs must be put into effect.

Acute Effect - Adverse effect on a human or animal that has severe symptoms developing rapidly and coming quickly to a crisis.

Acute Exposure - A single exposure to a toxic substance that results in severe biological harm or death. Acute exposures are usually characterized as lasting no longer than a day.

Acute Toxicity - Acute effects resulting from a single dose of, or exposure to, a substance. Ordinarily used to denote effects in experimental animals.

Advisory - A nonregulatory document that communicates risk information to persons who may have to make risk management decisions.

Aerosol - A fine aerial suspension of particles sufficiently small in size to confer some degree of stability from sedimentation (e.g., smoke or fog).

Airborne Particulates - Total suspended particulate matter found in the atmosphere as solid particles or liquid droplets. Chemical composition of particulates varies widely, depending on location and time of year. Airborne particulates include: windblown dust, emissions from industrial processes, smoke from the burning of wood and coal, and the exhaust of motor vehicles.

Air Pollutant - Any substance in air that could, in high enough concentrations, harm man, animals, vegetation, or materials. Pollutants may include almost any natural or artificial composition of matter capable of being airborne. They may be in the form of solid particles, liquid droplets, gases, or in combinations of these forms. Generally, they fall into two main groups: (1) those emitted directly from identifiable sources and (2) those produced in the air by interaction between two or more primary pollutants, or by reaction with normal atmospheric constituents, with or without photoactivation. Exclusive of pollen, fog, and dust, which are of natural origin, about 100 contaminants have been identified and fall into the following categories: solids, sulfur compounds, volatile organic chemicals, nitrogen compounds, oxygen compounds, halogen compounds, radioactive compounds, and odors.

Air Quality Standards - The level of pollutants prescribed by regulations that may not be exceeded during a specified time in a defined area.

Allergic Reaction - An abnormal physiological response to chemical or physical stimuli.

Anhydrous - Free from water, especially water of crystallization.

Aqueous - Of, relating to, or resembling water.

Asphyxiant - A vapor or gas that can cause unconsciousness or death by suffocation (lack of oxygen). Most simple asphyxiants are harmful to the body only when they become so concentrated that they reduce oxygen in the air (normally about 21 percent) to dangerous levels (18 percent or lower). Asphyxiation is one of the principal potential hazards of working in confined and enclosed spaces.

ASTM - American Society for Testing and Materials is the world's largest source of voluntary consensus standards for materials, products, systems, and services. ASTM is a resource for sampling and testing methods, health and safety aspects of materials, safe performance guidelines, effects of physical and biological agents and chemicals.

ATSDR - Agency for Toxic Substances and Disease Registry (HHS)

Auto-ignition Temperature - The temperature to which a closed, or nearly closed container must be heated in order that the flammable liquid, when introduced into the container, will ignite spontaneously or burn.

Boiling Point (BP) - The temperature at which a liquid changes to a vapor state at a given pressure. The boiling point usually expressed in degrees Fahrenheit at sea level pressure (760 mmHg, or one atmosphere). For mixtures, the initial boiling point or the boiling range may be given. Flammable materials with low boiling points generally present special fire hazards.

C - Centigrade or Celsius; a thermometric scale, with 0 degrees as the melting point of ice and 100 degrees as the boiling point of water. See F.

Ceiling Limit - The maximum allowable human exposure limit for an airborne substance which is not to be exceeded even momentarily. Also see PEL and TLV.

CAA - The Clean Air Act was enacted to regulate/reduce air pollution. CAA is administered by the U.S. Environmental Protection Agency.

CAA Amendments - CAA Amendments 1990; expands EPA enforcement powers and adds restrictions on air toxics, ozone-depleting chemicals, stationary and mobile emissions sources, and emissions implicated in creating acid rain and global warming.

Carcinogen - A substance or agent capable of causing or producing cancer in mammals, including humans. A chemical is considered to be a carcinogen if:

(a) It has been evaluated by the International Agency for Research on Cancer (IARC) and found to be a carcinogen or potential carcinogen; or

(b) It is listed as a carcinogen or potential carcinogen in the "Annual Report on Carcinogens" published by the National Toxicology Program (NTP) (latest edition); or

(c) It is regulated by OSHA as a carcinogen.

Carcinogenicity - The ability to produce cancer.

CAS - Chemical Abstracts Service is an organization under the American Chemical Society. CAS abstracts and indexes chemical literature from all over the world in "Chemical Abstracts." CAS Registry Numbers are used to identify specific chemicals or mixtures.

CAS Registry Number (or CAS Number) - An identification number assigned by the Chemical Abstracts Service (CAS) of the American Chemical Society. The CAS Number is used in various databases, including Chemical Abstracts, for identification and retrieval.

cc - Cubic centimeter is a volume measurement in the metric system that is equal in capacity to one milliliter (ml). One quart is about 946 cubic centimeters.

CDC - Centers for Disease Control (DHHS)

Central Nervous System - The brain and spinal cord. These organs supervise and coordinate the activity of the entire nervous system. Sensory impulses are transmitted into the central nervous system, and motor impulses are transmitted out.

CERCLA - Comprehensive Environmental Response, Compensation, and Liability Act of 1980. The Act requires that the Coast Guard National Response Center be notified in the event of a hazardous substance release. The Act also provides for a fund (the Superfund) to be used for the cleanup of abandoned hazardous waste disposal sites.

CFC - see Chlorofluorocarbons

CFR - Code of Federal Regulations. A collection of the regulations that have been promulgated under United States Law.

Chemical Family - A group of single elements or compounds with a common general name. Example: acetone, methyl ethyl ketone (MEK), and methyl isobutyl ketone (MIBK) are of the "Ketone" family; acrolein, furfural, and acetaldehyde are of the "aldehyde" family.

Chemical Name - The name given to a chemical in the nomenclature system developed by the International Union of Pure and Applied Chemistry (IUPAC) or the Chemical Abstracts Service (CAS). The scientific designation of a chemical or a name that will clearly identify the chemical for hazard evaluation purposes.

Chemical Pneumonitis - Inflammation of the lungs caused by accumulation of fluids due to chemical irritation.

Chlorinated Hydrocarbons - Include a class of persistent, broad-spectrum insecticides that linger inthe environment and accumulate in the food chain. Among them are DDT, aldrin, dieldrin, heptachlor, chlordane, lindane, endrin, mirex, hexachloride, and toxaphene. Other examples include TCE, used as an industrial solvent.

Chlorinated Solvent - An organic solvent containing chlorine atoms, e.g., methylene chloride and 1,1,1-trichloromethane, which are used in aerosol spray containers and in traffic paint.

Chlorofluorocarbons - A family of inert, nontoxic, and easily liquified chemicals used in refrigeration, air conditioning, packaging, insulation, or as solvents and aerosol propellants. Because CFC's are not destroyed in the lower atmosphere, they drift into the upper atmosphere where their chlorine components destroy ozone.

Chronic Effect - An adverse effect on a human or animal body, with symptoms that develop slowly over a long period of time or that recur frequently. Also see Acute.

Chronic Exposure - Long-term contact with a substance.

Chronic Toxicity - Adverse (chronic) effects resulting from repeated doses of, or exposures to, a substance over a relatively prolonged period of time. Ordinarily used to denote effects in experimental animals.

Clean Water Act - Federal law enacted to regulate/reduce water pollution. CWA is administered by EPA.

Combustible - A term used by NFPA, DOT, and others to classify certain liquids that will burn, on the basis of flash points. Both NFPA and DOT generally define "combustible liquids" as having a flash point of100°F (37.8°C) or higher but below 200°F (93.3°C). Also see "flammable." Nonliquid substances such as wood and paper are classified as "ordinary combustibles" by NFPA.

Combustible Liquid - Any liquid having a flash point at or above 100°F (37.8°C), but below 200°F (93.3°C), except any mixture having components with flash points of 200°F (93.3°C) or higher, the total volume of which makes up ninety-nine percent (99total volume of the mixture.

Common Name - Any means used to identify a chemical other than its chemical name (e.g., code name, code number, trade name, brand name, or generic name). See Generic.

Contaminant - Any physical, chemical, biological, or radiological substance or matter that has an adverse effect on air, water or soil.

Corrosive - A chemical that causes visible destruction of, or irreversible alterations in, living tissue by chemical action at the site of contact. For example, a chemical is considered to be corrosive if, when tested on the intact skin of albino rabbits by the method described by the DOT in Appendix A to 49 CFR Part 173, it destroys or changes irreversibly the structure of the tissue at the site of contact following an exposure period of 4 hours. This term shall not refer to action on inanimate surfaces.

CWA - The Clean Water Act was enacted to regulate/reduce water pollution. It is administered by the U.S. Environmental Protection Agency.

Dermal - Relating to the skin.

Dermal Toxicity - Adverse effects resulting from skin exposure to a substance. Ordinarily used to denote effects in experimental animals.

Dermatitis - Inflammation of the skin.

DHHS - U.S. Department of Health and Human Services (formerly U.S. Department of Health, Education and Welfare). Includes NIOSH and the Public Health Service.

Effluent - Wastewater, treated or untreated, that flows out of a treatment plant, sewer, or industrial outfall. Generally refers to wastes discharged into surface waters.

Environmental Protection Agency (EPA) - Responsible for enforcing regulations related to the Resource Conservation and Recovery Act (RCRA), Toxic Substance Control Act (TSCA), Clean Air Act (CAA), Clean Water Act (CWA), Superfund, and others.

Environmental Toxicity - Information obtained as a result of tests designed to study the effects of a substance on aquatic and plant life.

EPA - U.S. Environmental Protection Agency.

Explosive - A chemical that causes a sudden, almost instantaneous release of pressure, gas, and heat when subjected to sudden shock, pressure, or high temperature.

Exposure (or Exposed) - State of being open and vulnerable to a hazardous chemical by inhalation, ingestion, skin contact, absorption, or any other course; includes potential (accidental or possible) exposure.

Extremely Hazardous Substance - Any of the 406 chemicals identified by EPA on the basis of toxicity, and listed under SARA Title III. The list is subject to revision.

F - Fahrenheit is a scale for measuring temperature. On the Fahrenheit scale, water boils at 212 degrees and freezes at 32 degrees.

f/cc - Fibers per cubic centimeter of air.

FDA - U.S. Food and Drug Administration.

Federal Register - This publication, issued every workday by the United States Printing Office, is the legal medium for recording and communicating the rules and regulations established by the executive branch of the federal government. Individuals or corporations cannot be held legally responsible for compliance with a regulation unless it has come into effect past the deadline published in the Register. In addition, executive agencies are required to publish in advance some types of proposed regulations.

Fetus - The developing human young in the uterus from the seventh week of gestation until birth.

Fibrosis - An abnormal thickening of fibrous connective tissue, usually in the lungs.

FIFRA - Federal Insecticide, Fungicide, and Rodenticide Act requires that certain useful poisons, such as chemical pesticides, sold to the public contain labels that carry health hazard warnings to protect users. It is administered by the EPA.

Flammable (or Inflammable) - Property of a chemical that includes one of the following categories:

(a) "Aerosol, flammable." An aerosol that, when tested by the method described in 16 CFR 1500.45, yields a flame projection exceeding 18 inches at full valve opening, or a flash-back (a flame extending back to the valve) at any degree of valve opening;

(b) "Gas, flammable." (1) A gas that, at ambient temperature and pressure, forms a flammable mixture with air at a concentration of 13 percent by volume or less; or (2) a gas that, at ambient temperature and pressure, forms a range of flammable mixtures with air wider than 12 percent by volume, regardless of the lower limit;

(c) "Liquid, flammable." Any liquid having a flash point below 100°F (37.8°C), except any mixture having components with flash points of 100°F (37.8°C), or higher, the total of which make up 99 percent or more of the total volume of mixture.

(d) "Solid, flammable." A solid, other than a blasting agent or explosive as defined in 1910.109(a), that is liable to cause fire through friction, absorption of moisture, spontaneous chemical change, or retained heat from manufacturing or processing, or which can be ignited readily and when ignited burns so vigorously and persistently as to create a serious hazard. A solid is a flammable solid if, when tested by the method described in 16 CFR1500.44, it ignites and burns with a self-sustained flame at a rate greater than one-tenth of an inch per second along its major axis.

Flashpoint - The minimum temperature at which a liquid gives off a vapor in sufficient concentration to ignite when tested by the following methods:

(a) Tagliabue Closed Tester (see American National Standard Method of Test for Flash Point by Tag Closed Tester, Z11.24 1979 [ASTM D56-79]).

(b) Pensky-Martens Closed Tester (see American National Standard Method of Test for Flash Point by Pensky-Martens Closed Tester, Z11.7-1979 [ASTM D93-79]).

(c) Setaflash Closed Tester (see American National Standard Method of Test for Flash Point by Setaflash Closed Tester [ASTM D 3278-78]).

Fluorides - Gaseous, solid, or dissolved compounds containing fluorine.

Fume - A solid condensation particle of extremely small

diameter, commonly generated from molten metal as metal fume.

g - Gram is a metric unit of weight. One ounce U.S. (avoirdupois) is about 28.4 grams.

Generic Name - A designation or identification used to identify a chemical by other than its chemical name (e.g., code name, code number, trade name, and brand name).

Genetic - Pertaining to or carried by genes. Hereditary.

Gestation - The development of the fetus in the uterus from conception to birth; pregnancy.

GI - Gastrointestinal

g/kg - Grams per kilogram is an expression of dose used in oral and dermal toxicology testing to denote grams of a substance dosed per kilogram of animal body weight. Also see "kg" (kilogram).

Halogen - Any of the group of five chemically related nonmetallic elements that includes bromine, fluorine, chlorine, iodine, and astatine.

Hazardous Chemical - Any chemical whose presence or use is a physical hazard or a health hazard.

Hazardous Substance - 1. Any material that poses a threat to human health and/or the environment. Typical hazardous substances are toxic, corrosive, ignitable, explosive, or chemically reactive. 2. Any substance designated by EPA to be reported if a designated quantity of the substance is spilled in the waters of the United States or if otherwise emitted to the environment.

Hazardous Waste - By-products of society that can pose a substantial or potential hazard to human health or the environment when improperly managed. Possesses at least one of four characteristics (ignitability, corrosivity, reactivity, or toxicity), or appears on special EPA lists.

HCS - Hazard Communication Standard is an OSHA regulation issued under 29 CFR Part 1910.1200.

Health Hazard - A chemical for which there is significant evidence, based on at least one study conducted in accordance withestablished scientific principles, that acute or chronic health effects may occur in exposed employees. The term "health hazard" includes chemicals that are carcinogens, toxic or highly toxic agents, reproductive toxins, irritants, corrosives, sensitizers, hepatotoxins, nephrotoxins, neurotoxins, agents that act on the hematopoietic system, and agents that damage the lungs, skin, eyes or mucous membranes.

Heavy Metals - Metallic elements with high atomic weights, e.g., mercury, chromium, cadmium, arsenic, and lead. They can damage living things at low concentrations and tend to accumulate in the food chain.

Hemoglobin - An iron-containing conjugated protein or respiratory pigment occurring in the red blood cells of vertebrates.

Hepatotoxin - A substance that causes injury to the liver.

HMTA - Hazardous Materials Transportation Act

HMTR - Hazardous Materials Transportation Regulations

HOC - Halogenated Organic Carbons

Hydrocarbons (HCs) - Chemical compounds that consist entirely of hydrogen and carbon.

IARC - International Agency for Research on Cancer.

Ignitable - Capable of being set afire.

Immediately Dangerous to Life and Health (IDLH) - The maximum level to which a healthy individual can be exposed to a chemical for 30 minutes and escape without suffering irreversible health effects or impairing symptoms. Used as a "level of concern."

Impervious - A material that does not allow another substance to pass through or penetrate it.

Ingestion - Taking in by the mouth.

Inhalation - Breathing in of a substance in the form of a gas, vapor, fume, mist, or dust.

Inhibitor - A chemical added to another substance to prevent an unwanted chemical change.

Inorganic Chemicals - Chemical substances of mineral origin, not of basically carbon structure.

Insoluble - Incapable of being dissolved in a liquid.

Irritant - A chemical, which is not corrosive, that causes a reversible inflammatory effect on living tissue by chemical action at the site of contact. A chemical is a skin irritant if, when tested on the intact skin of albino rabbits by the methods of 16 CFR 1500.41 for 4 hours exposure or by other appropriate techniques, it results in an empirical score of 5 or more. A chemical is an eye irritant if so determined under the procedure listed in 16 CFR 1500.42 or other appropriate techniques.

Irritating - As defined by DOT, a property of a liquid or solid substance which, upon contact with fire or when exposed to air, gives off dangerous or intensely irritating fumes (not including poisonous materials).

Isomer - One of two or more chemical substances that have the same molecular formula but different chemical and physical properties due to different arrangement of the atoms in the molecule.

Isotope - A variation of an element that has the same atomic number but a different weight because of its neutrons. Various isotopes of the same element may have different radioactive behaviors.

kg - Kilogram is a metric unit of weight, about 2.2 U.S. pounds. Also see "g/kg," "g," and "mg."

L - Liter is a metric unit of capacity. A U.S. quart is about 0.9 liters.

Land Bans - Prohibitions of specific toxic materials from disposal in landfills under RCRA.

LC - Lethal Concentration, the concentration of a substance being tested that will kill.

LC_{50} - The concentration of a material in air that will kill 50 percent of a group of test animals with a single exposure (usually 1 to 4 hours). The LC_{50} is expressed as parts of material per million parts of air, by volume (ppm) for gases and vapors, or as micrograms of material per liter of air (g/l) or milligrams of material per cubic meter of air (mg/m^3) for dusts and mists, as well as for gases and vapors.

LD - Lethal Dose is the quantity of a substance being tested that will kill.

LD_{50} - A single dose of a material expected to kill 50 percent of a group of test animals. The LD_{50} dose is usually expressed as milligrams or grams of material per kilogram of animal body weight (mg/kg or g/kg). The material may be administered by mouth or applied to the skin.

M - Meter is a unit of length in the metric system. One meter is about 39 inches.

m^3 - Cubic meter is a metric measure of volume, approximately 35.3 cubic feet or 1.3 cubic yards.

Material Safety Data Sheet (MSDS) - A compilation of information required under the OSHA Hazard Communication Standard on the identity of hazardous chemicals, health, and physical hazards, exposure limits, and precautions. Section 311 of SARA requires facilities to submit MSDSs under certain circumstances.

Maximum Contaminant Level (MCL) - The maximum permissible level of a contaminant in water delivered to any user of a public water system. MCLs are enforceable standards.

Metabolism - Physical and chemical processes of the body.

Meter - A unit of length; equivalent to 39.37 inches.

mg - Milligram is a metric unit of weight that is one-thousandth of a gram.

mg/kg - Milligrams of substance per kilogram of body weight, an expression of toxicological dose.

mg/m^3 - Milligrams per cubic meter, a unit for expressing concentrations of dusts, gases, or mists in air.

Micron - (Micrometer) A unit of length equal to one-millionth of a meter; approximately 0.000039 of an inch.

Microgram (ug) - A metric unit of weight that is one-millionth of a gram.

Mist - Suspended liquid droplets generated by condensation from the gaseous to the liquid state, or by breaking up a liquid into a dispersed state, such as splashing, foaming, or atomizing. Mist is formed when a finely divided liquid is suspended in air.

Mixture - Any combination of two or more chemicals if the combination is not, in whole or part, the result of a chemical reaction.

ml - Milliliter, a metric unit of capacity, equal in volume to 1 cubic centimeter (cc), or approximately one-sixteenth of a cubic inch. One-thousandth of a liter.

mmHg - Millimeters (mm) of mercury (Hg) is a unit of measurement for low pressures or partial vacuums.

MSDS - Material Safety Data Sheet.

MSHA - Mine Safety and Health Administration, U.S. Department of Labor.

Mutagen - A substance or agent capable of altering the genetic material in a living cell.

Narcosis - A state of stupor, unconsciousness, or arrested activity produced by the influence of narcotics or other chemicals.

Nausea - Tendency to vomit, feeling of sickness at the stomach.

NCI - National Cancer Institute, the part of the National Institutes of Health that studies cancer causes and prevention as well as diagnosis, treatment, and rehabilitation of cancer patients.

NFPA - National Fire Protection Association is an international membership organization which promotes/improves fire protection and prevention and establishes safeguards against loss of life and property by fire. Best known on the industrial scene for the National Fire Codes – 16 volumes of codes, standards, recommended practices and manuals developed (and periodically updated) by NFPA technical committees. Among these is NFPA 704M, the code for showing hazards of materials as they might be encountered under fire or related emergency conditions, using the familiar diamond-shaped label or placard with appropriate numbers or symbols.

Nephrotoxin - A substance that causes injury to the kidneys.

Neurotoxin - A material that causes injury to the nerve cells and may produce emotional or behavioral abnormalities.

ng - Nanogram, one-billionth of a gram.

NIOSH - National Institute for Occupational Safety and Health, U.S. Public Health Service, U.S. Department of Health and Human Services (DHHS). Among other activities, tests and certifies respiratory protective devices and air sampling detector tubes, recommends occupational exposure limits for various substances, and assists OSHA and MSHA in occupational safety and health investigations and research.

Nonflammable - Not easily ignited, or if ignited, not burning rapidly.

NRC - National Response Center, a notification center that must be called when significant oil or chemical spills or other environment-related accidents occur. The toll-free telephone number is 1-800-424-8802.

NTP - National Toxicology Program. The NTP publishes an Annual Report on Carcinogens.

Oral - Used in or taken into the body through the mouth.

Oral Toxicity - Adverse effects resulting from taking a substanceinto the body by mouth. Ordinarily used to denote effects in experimental animals.

Organic - 1. Refering to or derived from living organisms. 2. In chemistry, any compound containing carbon.

Organophosphates - Pesticide chemicals that contain phosphorus. Used to control insects, they are short-lived, but some can be toxic when first applied.

Organotins - Chemical compounds used in antifoulant paints to protect the hulls of boats and ships, buoys, and dock pilings from marine organisms such as barnacles.

OSHA - Occupational Safety and Health Administration, U.S. Department of Labor.

OSHA - Hazard Communication Standard - 29 CFR 1910.1200 in the Code of Federal Regulations.

Overexposure - Exposure to a hazardous material beyond the allowable exposure limits.

Oxidizer - A chemical other than a blasting agent or explosive that initiates or promotes combustion in other materials, causing fire either by itself of through the release of oxygen or other gases.

Particulates - Fine liquid or solid particles such as dust, smoke, mist, fumes, or smog, found in air or emissions.

PCBs - A group of toxic, persistent chemicals (polychlorinated biphenyls) used in transformers and capacitors for insulating purposes and in gas pipeline systems as a lubricant. Further sale of new use was banned by law in 1979.

PEL - Permissible Exposure Limit is an occupational exposure limit established by OSHA's regulatory authority. It may be a time-weighted average (TWA) limit or a maximum concentration exposure limit.

pH - The symbol relating the hydrogen ion (H+) concentration to that of a given standard solution. A pH of 7 is neutral. Numbers increasing from 7 to 14 indicate greater alkalinity. Numbers decreasing from 7 to 0 indicate greater acidity.

Phenols - Organic compounds that are by-products of petroleum refining, tanning, and textile, dye, and resin manfacturing. Low concentrations cause taste and odor problems in water; higher concentrations can kill aquatic life and humans.

Phosphates - Certain chemical compounds containing phosphorus

Physical Hazard - A chemical that is a combustible liquid, a compressed gas, explosive, flammable, an organic peroxide, an oxidizer, pyrophoric, unstable (reactive) or water-reactive, based on scientifically valid evidence.

Pneumoconiosis - A condition of the lung characterized by permanent deposition of particulate matter and tissue reaction to its presence. It may range from relatively harmless forms of iron oxide deposition to destructive forms of silicosis.

ppb - Parts per billion, a common way to express the concentration of a gas or vapor in air. Used to express extremely low concentrations of unusually toxic gases or vapors; also, the concentration of a particular substance in a liquid or solid.

ppm - Parts per million is the concentration of a gas or vapor in air parts by volume of the gas or vapor in a million parts of air; also, the concentration of a particulate in a liquid or solid.

Pulmonary - Relating to, or associated with, the lungs.

Pulmonary Edema - Fluid in the lungs.

Radiation - Any form of energy propagated as rays, waves, or streams of energetic particles. The term is frequently used in relation to the emission of rays from the nucleus of an atom.

Radioactive Substances - Substances that emit radiation

Registry of Toxic Effects of Chemical Substances (RTECS) - A NIOSH publication, one of the information sources OSHA recommends for hazard determination. RTECS provides data on toxicity for over 50,000 different chemicals. It has an extensive cross-reference, listing trade names and synonyms. It is available as hard copy, computer tape, microfiche, and on-line through the National Library of Medicine.

REL - The NIOSH Recommended Exposure Limit is the highest allowable airborne concentration which is not expected to injure workers. It may be expressed as a ceiling limit or as a time-weighted average (TWA).

Release - Any occurrence in which a regulated substance is emitted into air, water or soil.

Reproductive Toxin - Substance that affects either male or female reproductive systems.

Respiratory Protection - Devices that protect the wearer's respiratory system from overexposure by inhalation to airborne contaminants. Respiratory protection is used when a worker must work in an area where he/she may be exposed to concentration in excess of the allowable exposure limit.

Respiratory System - The breathing system that includes the lungs and the air passages (trachea or "windpipe," larynx, mouth, and nose), plus the associated nervous and circulatory supply.

Right to Know - A term applied to a variety of laws and regulations enacted by municipal, county, and state governments that provide for the availability of information on chemical hazards; also includes the OSHA Hazard Communication Standard. The different laws that have been enacted around the country vary greatly from the OSHA Standard. Some require that information be made available not only to employees, but to emergency personnel and the community as a whole. Many of the local and state laws require submission of work area surveys as well as annual activity reports. The basic intent of these laws is the same as the OSHA Standard.

Routes of Entry - The means by which material may gain access to the body, for example, inhalation, ingestion, and skin contact.

RCRA - Resource Conservation and Recovery Act, environmental legislation administered by the EPA and aimed at controlling the generation, treating, storage, transportation and disposal of hazardous wastes.

RTECS - See Registry of Toxic Effects of Chemical Substances.

SARA - Superfund Amendments and Reauthorization Act of 1986

SARA Title III - Section of SARA requiring public disclosure of chemical information and development of emergency response plans.

Sensitization - 1. A condition of being made sensitive to a specific substance (i.e., antigen) such as a protein or pollen; 2. The process of making a person susceptible to a substance by repeated injections, as a serum.

Sensitizer - A chemical that causes a substantial proportion of exposed people or animals to develop an allergic reaction in normal tissue after repeated exposure to the chemical.

Skin Absorption - Ability of some hazardous chemicals to pass directly through the skin and enter the bloodstream.

STEL - Short-Term Exposure Limit; a term used by ACGIH.

Subcutaneous - Beneath the layers of the skin.

Systemic Poison - A poison that spreads throughout the body, affecting all body systems and organs. Its adverse effect is not localized in one spot or area.

Systemic Toxicity - Adverse effects caused by a substance that affects the body in a general rather than local manner.

Synonym - Another name or names by which a material is known. Methyl alcohol, for example, is know as methanol or wood alcohol.

Target Organ Effects - Those effects that are recognized to be a result of exposure to a specific chemical.

TCL - Toxic Concentration Low; the lowest concentration of a gas or vapor capable of producing a defined toxic effect in a specified test species over a specified time.

TCLP - Toxicity Characteristic Leaching Procedure; required test under RCRA to determine toxicity and mobility characteristics of hazardous wastes.

TDL - Toxic Dose Low; lowest administered dose of material capable of producing a defined toxic effect in a specified test species.

Teratogen - A substance or agent, exposure to which by a pregnant female can result in malformations in the fetus.

Threshold Planning Quantity (TPQ) - The specified quantity for a chemical on the list of extremely hazardous substances of SARA Title III which triggers notification by facilities to the state emergency response commission.

TLV - Threshold Limit Value is a term used by ACGIH to express the airborne concentration of material to which nearly all persons can be exposed day after day without adverse effects. ACGIH expresses TLVs in three ways: (a) TLV-TWA: The allowable Time-Weighted Average concentration for a normal 8-hour workday or 40-hour workweek. (b) TLV-STEL: The Short-Term Exposure Limit, or maximum concentration for a continuous 15-minute exposure period (maximum of four such periods per day, with at least 60 minutes between exposure periods, and provided the daily TLV-TWA is not exceeded). (c) TLV-C: The ceiling exposure limit – the concentration that should not be exceeded even instantaneously.

Toxic - A chemical falling within any of the following categories: (a) A chemical that has a median lethal dose (LD_{50}) of more than 50 milligrams per kilogram but not more than 500 milligrams per kilogram of body weight when administered orally to albino rats weighing between 200 and 300 grams each. (b) A chemical that has a median lethal dose (LD_{50}) of more than 200 milligrams per kilogram but not more than 1,000 milligrams per kilogram of body weight when administered by continuous contact for 24 hours (or less if death occurs within 24 hours) with the bare skin of albino rabbits weighing between two and three kilograms each. (c) A chemical that has a median lethal concentration (LC_{50}) in air of more than 200 parts per million but not more than 2,000 parts per million by volume of gas or vapor, or more than two milligrams per liter of mist, fume, or dust, when administered by continuous inhalation for one hour (or less if death occurs within 1 hour) to albino rats weighing between 200 and 300 grams each.

Toxic Substance - Any substance that can cause acute or chronic injury to the human body, or which is suspected of being able to cause diseases or injury under some conditions.

Toxicity - The sum of adverse effects resulting from exposure to a material, generally, by the mouth, skin, or respiratory tract.

TPQ - See Threshhold Planning Quantity.

TSCA - Toxic Substances Control Act; Federal Environmental legislation administered by the EPA that regulates the manufacture, handling, and use of materials classified as "toxic substances."

TSDF - Treatment, Storage and/or Disposal Facility; facility regulated under RCRA.

TWA - Time-Weighted Average; the airborne concentration of a material to which a person is exposed, averaged over the total exposure time – generally the total workday (8 to 12 hours).

Vapor - The gaseous form of a solid or liquid substance as it evaporates.

Volatility - A measure of how quickly a substance forms a vapor at ordinary temperatures.

APPENDIX

Key to RCRA "F"- and "K"- Series Wastes

F001 The following spent halogenated solvents used in degreasing: Tetrachloroethylene, trichloroethylene, methylene chloride, 1,1,1-trichloroethane, carbon tetrachloride, and chlorinated fluorocarbons; all spent solvent mixtures/blends used in degreasing containing, before use, a total of ten percent or more (by volume) of one or more of the above halogenated solvents or those solvents listed in F002, F004, and F005; and still bottoms from the recovery of these spent solvents and spent solvent mixtures.

F002 The following spent halogenated solvents: Tetrachloroethylene, methylene chloride, trichloroethylene, 1,1,1-trichloroethane, chlorobenzene, 1,1,2-trichloro-1,2,2-trifluoroethane, ortho-dichlorobenzene, trichlorofluoromethane, and 1,1,2-trichloroethane; all spent solvent mixtures/blends containing, before use, a total of ten percent or more (by volume) of one or more of the above halogenated solvents or those listed in F001, F004, or F005; and still bottoms from the recovery of these spent solvent mixtures.

F003 The following spent non-halogenated solvents: Xylene, acetone, ethyl acetate, ethyl benzene, ethyl ether, methyl isobutyl ketone, n-butyl alcohol, cyclohexanone, and methanol; all spent solvent mixtures/blends containing, before use, only the above spent non-halogenated solvents; and all spent solvent mixtures/blends containing, before use, one or more of the above non-halogenated solvents, and, a total of ten percent or more (by volume) of one or more of those solvents listed in F001, F002, F004, and F005; and still bottoms from the recovery of these spent solvents and spent solvent mixtures.

F004 The following spent non-halogenated solvents: Toluene, methyl ethyl ketone, carbon disulfide, isobutanol, pyridine, benzene, 2-ethoxyethanol, and 2-nitropropane; all spent solvent mixtures/blends containing, before use, a total of ten percent or more (by volume) of one or more of the above non-halogenated solvents or those solvents listed in F001, F002, F004; and still bottoms from the recovery of these spent sovents and spent solvent mixtures.

F005 The following spent non-halogenated solvents: Toluene, methyl ethyl ketone, carbon disulfide, isobutanol, pyridine, benzene, 2-ethoxyethanol, and 2-nitropropane; all spent solvent mixtures/blends containing, before use, a total of ten percent or more (by volume) of one or more of the above non-halogenated solvents or those solvents listed in F001, F002, F004; and still bottoms from the recovery

of these spent sovents and spent solvent mixtures.

F006 Wastewater treatment sludges from electroplating operations except from the following processes: (1) Sulfuric acid anodizing of aluminum; (2) tin plating on carbon steel; (3) zinc plating (segregated basis) on carbon steel; (4) aluminum or zinc-aluminum plating on carbon steel; (5) cleaning/stripping associated with tin, zinc and aluminum plating on carbon steel; and (6) chemical etching and milling of aluminum.

F007 Spent cyanide plating bath solutions from electroplating operations.

F008 Plating bath residues from the bottom of plating baths from electroplating operations where cyanides are used in the process.

F009 Spent stripping and cleaning bath solutions from electroplating operations where cyanides are used in the process.

F010 Quenching bath residues from oil baths from metal heat treating operations where cyanides are used in the process.

F011 Spent cyanide solutions from salt bath pot cleaning from metal heat treating operations.

F012 Quenching waste water treatment sludges from metal heat treating operations where cyanides are used in the process.

F019 Wastewater treatment sludges from the chemical conversion coating of aluminum except from zirconium phosphating in aluminum can washing when such phosphating is an exclusive conversion coating process.

F020 Wastes (except wastewater and spent carbon from hydrogen chloride purification) from the production or manufacturing use (as a reactant, chemical intermediate, or component in a formulating process) or tri- or tetrachlorophenol, or of intermediates used to produce their pesticide derivatives. (This listing does not include wastes from the production of Hexachlorophene from highly purified 2,4, 5-trichlorophenol).

F021 Wastes (except wastewater and spent carbon from hydrogen chloride purification) from the production or manufacturing use (as a reactant, chemical intermediate, or component in a formulating process) of pentachlorophenol, or of intermediates used to produce its derivatives.

F022 Wastes (except wastewater and spent carbon from hydrogen chloride purification) from the production or manufacturing use (as a reactant, chemical intermediate, or component in a formulating process) of tetra-, penta-, or hexachlorobenzenes under alkaline conditions.

F023 Wastes (except wastewater and spent carbon from hydrogen chloride purification) from the production of materials on equipment previously used for the production or manufacturing use (as a reactant, chemical intermediate, or component in a formulating process) of tri- and tetrachlorophenols. (This listing does not include wastes from equipment used only for the production or use of hexachlorophene from highly purified 2,4,5-trichlorophenol).

F024 Process wastes, including but not limited to, distillation residues, heavy ends, tars, and reactor cleanout wastes, from the production of certain chlorinated aliphatic hydrocarbons by free radical catalyzed processes. These chlorinated aliphatic hydrocarbons are those having carbon chain lengths ranging from one to and including five, with varying amounts and positions of chlorine substitution. (This listing does not include wastewaters, waste water treatment sludges, spent catalysts, and wastes listed in 40 CFR 261.31 or 40 CFR 261.32)

F025 Condensed light ends, spent filters and filter aids, and spent desiccant wastes from the production of certain chlorinated aliphatic hydrocarbons, by free radical catalyzed processes. These chlorinated aliphatic hydrocarbons are those having carbon chain lengths ranging from one to and including five, with varying amounts and positions of chlorine substitution.

F026 Wastes (except wastewater and spent carbon from hydrogen chloride purification) from the production of materials on equipment previously used for the manufacturing use (as a reactant, chemical intermediate, or component in a formulating process) of tetra-, penta-, or hexachlorobenzene under alkaline conditions.

F027 Discarded unused formulations containing tri-, tetra-, or pentachlorophenol or discarded unused formulations containing compounds derived from these chlorophenols. (This listing does not include formulations containing hexachlorophene synthesized from prepurified 2,4,5-trichlorophenol as the sole component).

F028 Residues resulting from the incineration or thermal treatment of soil contaminated with EPA Hazardous Waste Numbers F020, F021, F022, F023, F026, and F027.

F032 Wastewaters, process residuals, preservative drippage, and spent formulations from wood preserving processes generated at plants that currently use or have previously used chlorophenolic formulations (except potentially cross-contaminated wastes that have had the F032 waste code deleted in accordance with 40 CFR 261.35 and where the generator does not resume or initiate use of chlorophenolic formulations). This listing does not include K001 bottom sediment sludge from the treatment of wastewater from wood preserving processes that use creosote and/or pentachlorophenol. (NOTE: The listing of wastewaters that have not come into contact with process contaminants is stayed administratively. The listing for plants that have previously used chlorophenolic formulations is administratively stayed whenever these wastes are covered by the F034 or F035 listings. These stays will remain in effect until further administrative action is taken).

F034 Wastewaters, process residuals, preservative drippage, and spent formulations from wood preserving processes generated at plants that use creosote formulations. This listing does not include K001 bottom sediment sludge from the treatment of wastewater from wood preserving processes that use creosote and/or pentachlorophenol. (NOTE: The listing of wastewaters that have not come into contact with process contaminants is stayed administratively. The stay will remain in effect until further administrative action is taken).

F035 Wastewaters, process residuals, preservative drippage, and spent formulations from wood preserving processes generated at plants that use inorganic preservatives containing arsenic or chromium. This listing does not include K001 bottom sediment sludge from the treatment of wastewater from wood preserving processes that use creosote and/or pentachlorophenol. (NOTE: The listing of wastewaters that have not come into contact with process contaminants is stayed administratively. The stay will remain in effect until further administrative action is taken).

F037 Petroleum refinery primary oil/water/solids separation sludge; Any sludge generated from the gravitational separation of oil/water/solids during the storage or treatment of process wastewaters and oily cooling wastewaters from petroleum refineries. Such sludges include, but are not limited to, those generated in: oil/water/solids separators; tanks and impoundments; ditches and other conveyances; sumps, and stormwater units receiving dry weather flow. Sludge generated in stormwater units that do not receive dry weather flow, sludges generated from non-contact once-through cooling waters segregated for treatment from other processes or oil cooling waters, sludges generated in aggressive biological treatment units as defined in 40 CFR 261.31 (b)(2) (including sludges generated in one or more additional units after wastewaters have been treated in aggressive biological treatment units) and K051 wastes are not included in this listing.

F038 Petroleum refinery secondary (emulsified) oil/water/solids separation sludge; Any sludge and/or float generated from the physical and/or chemical separation of oil/water/solids in process wastewaters and oil cooling wastewaters from petroleum refineries. Such wastes include, but are not limited to, all sludges and floats generated in: induced air flotation (IAF) units, tanks and impoundments, and all sludges generated in DAF units. Sludges generated in stormwater units that do not receive dry weather flow, sludges generated from non-contact once-through cooling waters segregated for treatment from other processes or oil cooling waters, sludges and floats generated in aggressive biological treatment units as defined in 40 CFR 261.31(b)(2) (including sludges and floats generated in one or more additional units after waste waters have been treated in aggressive biological treatment units) and F037, K048, and K051 wastes are not included in this listing.

F039 Leachate resulting from the treatment, storage, or disposal of wastes classified by more than one waste code under Subpart D, or from a mixture of wastes classified under Subparts C and D of this part. (Leachate resulting from the management of one or more of the following EPA Hazardous Wastes and non other hazardous wastes retains its hazardous waste code(s): F020, F021, F022, F023, F026, F027, and/or F028).

K001 Bottom sediment sludge from the treatment of wastewaters from wood preserving processes that use creosote and/or pentachlorophenol.

K002 Wastewater treatment sludge from the production of chrome yellow and orange pigments.

K003 Wastewater treatment sludge from the production of molybdate orange pigments.

K004 Wastewater treatment sludge from the production of zinc yellow pigments.

K005 Wastewater treatment sludge from the production of chrome green pigments.

K006 Wastewater treatment sludge from the production of chrome oxide green pigments (anhydrous and hydrated).

K007 Wastewater treatment sludge from the production of iron blue pigments.

K008 Oven residue from the production of chrome oxide green pigments.

K009 Distillation bottoms from the production of acetaldehyde from ethylene.

K010 Distillation side cuts from the production of acetaldehyde from ethylene.

K011 Bottom stream from wastewater stripper in the production of acrylonitrile.

K013 Bottom stream from acetonitrile column in the production of acrylonitrile.

K014 Bottoms from the acetonitrile purification column in the production of acrylonitrile.

K015 Still bottoms from the distillation of benzyl chloride.

K016 Heavy ends or distillation residues from the production of carbon tetrachloride.

K017 Heavy ends (still bottoms) from the purification column in the production of epichlorohydrin.

K018 Heavy ends from the fractionation column in ethyl chloride production.

K019 Heavy ends from the distillation of ethylene dichloride in ethylene dichloride production.

K020 Heavy ends from the distillation of vinyl chloride in vinyl chloride production.

K021 Aqueous spent antimony catalyst waste from fluoromethanes production.

K022 Distillation bottom tars from the production of phenol/acetone from cumene.

K023 Distillation light ends from the production of phthalic anhydride from naphthalene.

K024 Distillation bottoms from the production of phthalic anhydride from naphthalene.

K025 Distillation bottoms from the production of nitrobenzene by the nitration of benzene.

K026 Stripping still tails from the production of methyl ethyl pyridines.

K027 Centrifuge and distillation residues from toluene diisocyanate production.

K028 Spent catalyst from the hydrochlorinator reactor in the production of 1,1,1-trichloroethane.

K029 Waste from the product steam stripper in the production of 1,1,1-trichloroethane.

K030 Column bottoms or heavy ends from the combined production of trichloroethylene and perchloroethylene.

K031 By-product salts generated in the production of MSMA and cacodylic acid.

K032 Wastewater treatment sludge from the production of chlordane.

K033 Wastewater and scrub water from the chlorination of cyclopentadiene in the production of chlordane.

K034 Filter solids from the filtration of hexachloro-cyclopentadiene in the production of chlordane.

K035 Wastewater treatment sludges generated in the production of creosote.

K036 Still bottoms from toluene reclamation distillation in the production of disulfoton.

K037 Wastewater treatment sludges from the production of disulfoton.

K038 Wastewater from the washing and stripping of phorate production.

K039 Filter cake from the filtration of diethylphosphorodithioic acid in the production of phorate.

K040 Wastewater treatment sludge from the production of phorate.

K041 Wastewater treatment sludge from the production of toxaphene.

K042 Heavy ends or distillation residues from the distillation of tetrachlorobenzene in the production of 2,4,5-T.

K043 2,6-dichlorophenol waste from the production of 2,4-D.

K044 Wastewater treatment sludges from the manufacturing and processing of explosives.

K045 Spent carbon from treatment of wastewater containing explosives.

K046 Wastewater treatment sludges from the manufacturing, formulation and loading of lead-based initiating compounds.

K047 Pink/red water from TNT operations.

K048 Dissolved air flotation (DAF) float from the petroleum refining industry.

K049 Slop oil emulsion solids from the petroleum refining industry.

K050 Heat exchanger bundle cleaning sludge from the petroleum refining industry.

K051 API separator sludge from the petroleum refining industry.

K052 Tank bottoms (leaded) from the petroleum refining industry.

K060 Ammonia still lime sludge from coking operations.

K061 Emission control dust/sludge from the primary production of steel in electric furnaces.

K062 Spent pickle liquor generated by steel finishing operations of facilities within the iron and steel industry (SIC codes 331 and 332).

K064 Acid plant blowdown slurry/sludge resulting from the thickening of blowdown slurry from primary copper production.

K065 Surface impoundment solids contained in and dredged from suface impoundments at primary lead smelting facilities.

K066 Sludge from treatment of process wastewater and/or acid plant blowdown from primary zinc production.

K069 Emission control dust/sludge from secondary lead smelting.

K071 Brine purification muds from the mercury cell process in chlorine production, where separately prepurified brine is not used.

K073 Chlorinated hydrocarbon waste from the purification of step of the diaphragm cell process using graphite anodes in chlorine production.

K083 Distillation bottoms from aniline extraction.

K084 Wastewater treatment sludges generated during the production of veterinary pharmaceuticals from arsenic or organo-arsenic compounds.

K085 Distillation of fractionation column bottoms from the production of chlorobenzenes.

K086 Solvent washes and sludges, caustic washes and sludges, or water washes and sludges from cleaning tubs and equipment used in the formulation of ink from pigments, driers, soaps, and stabilizers containing chromium and lead.

K087 Decanter tank tar sludge from coking operations.

K088 Spent potliners from primary aluminum reduction.

K090 Emission control dust or sludge from ferro-chromiumsilicon production.

K091 Emission control dust or sludge from ferrochromium production.

K093 Distillation light ends from the production of phthalic anhydride from ortho-xylene.

K094 Distillation bottoms from the production of phthalic anhydride from ortho-xylene.

K095 Distillation bottoms from the production of 1,1,1-trichloroethane.

K096 Heavy ends from the heavy ends column from the production of 1,1,1-trichloroethane.

K097 Vacuum stripper discharge from the chlordane chlorinator in the production of chlordane.

K098 Untreated process wastewater from the production of toxaphene.

K099 Untreated wastewater from the production of 2,4-D.

K100 Waste leaching solution from acid leaching of emission control dust/sludge from secondary lead smelting.

K101 Distillation tar residues from the distillation of aniline-based compounds in the production of veterinary pharmaceuticals from arsenic or organo-arsenic compounds.

K102 Residue from the use of activated carbon for decolorization in the production of veterinary pharmaceuticals from arsenic or organo-arsenic compounds.

K103 Process residues from aniline extraction from the production of aniline.

K104 Combined wastewater streams generated from nitrobenzene/aniline production.

K105 Separated aqueous stream from the reactor product washing step in the production of chlorobenzenes.

K106 Wastewater treatment sludge from the mercury cell process in chlorine production.

K107 Column bottoms from product separation from the production of 1,1-dimethylhydrazine (UDMH) from carboxylic acid hydrazides.

K108 Condensed column overheads from product separation and condensed reactor vent gases from the production of 1,1-dimethylhydrazine from carboxylic acid hydrazides.

K109 Spent filter cartridges from product purification from the production of 1,1-dimethylhydrazine (UDMH) from carboxylic acid hydrazides.

K110 Condensed column overheads from intermediate separation from the production of 1,1-dimethylhydrazine (UDMH) from carboxylic acid hydrazides.

K111 Product washwaters from production of dinitrotoluene via nitration of toluene.

K112 Reaction by-product water from the drying column in the production of toluenediamine via hydrogenation of dinitrotoluene.

K113 Condensed liquid light ends from the purification of toluenediamine in the production of toluenediamine via hydrogenation of dinitrotoluene.

K114 Vicinals from the purification of toluenediamine in the production of toluenediamine via hydrogenation of dinitrotoluene.

K115 Heavy ends from the purification of toluenediamine in the production of toluenediamine via hydrogenation of dinitrotoluene.

K116 Organic condensate from the solvent recovery column in the production of toluene diisocyanate via phosgenation of toluenediamine.

K117 Wastewater from the reaction vent gas scrubber in the production of ethylene bromide via bromination of ethene.

K118 Spent absorbent solids from purification of ethylene dibromide in the production of ethylene dibromide via bromination of ether.

K123 Process wastewater (including supernates, filtrates, and washwaters) from the production of ethylenebisdithiocarbamic acid and its salts.

K124 Reactor vent scrubber water from the production of ethylenebisdithiocarbamic acid and its salts.

K125 Filtration, evaporation, and centrifugation solids from the production of ethylenebisdithiocarbamic acid and its salts.

K126 Baghouse dust and floor sweepings in milling and packaging operations from the production of ethylenebisdithiocarbamic acid and its salts.

K131 Wastewater from the reactor and spent sulfuric acid from the acid dryer from the production of methyl bromide.

K132 Spent absorbent and wastewater separator solids from the production of methyl bromide.

K136 Still bottoms from the purification of ethylene dibromide in the production of ethylene dibromide via bromination of ethene.

K141 Process residues form the recovery of coal tar, including, but not limited to, tar collecting sump residues from the production of coke from coal or the recovery of coke by-products produced from coal. This listing does not include K087 (decanter tank tar sludge from coking operations).

K142 Tar storage tank residues from the production of coke from coal or from the recovery of coke by-products produced from coal.

K143 Process residues from the recovery of light oil, including, but not limited to, those generated in stills, decanters, and wash oil recovery unites from the recovery of coke by-products produced from coal.

K144 Wastewater treatment sludges from light oil refining, including, but not limited to, intercepting or contamination sump sludges from the recovery of coke by-products produced from coal.

K145 Residues from naphthalene collection and recovery

operations from the recovery of coke by-products produced from coal.

K147 Tar storage tank residues from coal tar refining.

K148 Residues from coal tar distillation, including, but not limited to, still bottoms.

K149 Distillation bottoms from the production of alpha- (or methyl-) chlorinated toluenes, ring-chlorinated toluenes, benzoyl chlorides, and compounds with mixtures of these functional groups. (This waste does not include still bottoms from the distillation of benzyl chloride).

K150 Organic residuals, excluding spent carbon adsorbent, from the spent chlorine gas and hydrochloric acid recovery processes associated with the production of alpha- (or methyl-) chlorinated toluenes, benzoyl chlorides, and compounds with mixtures of these functional groups.

K151 Wastewater treatment sludges, excluding neutralization and biological sludges, generated during the treatment of wastewaters from the production of alpha- (or methyl-) chlorinated toluenes, benzoyl chlorides, and compounds with mixtures of these functional groups.